Thermal Barrier Coatings: Failure Theory and Evaluation Technology

Yichun Zhou · Li Yang · Wang Zhu

Thermal Barrier Coatings: Failure Theory and Evaluation Technology

Science Press
Beijing

Springer

Yichun Zhou
Xidian University
Xi'an, China

Li Yang
Xidian University
Xi'an, China

Wang Zhu
Xiangtan University
Xiangtan, China

ISBN 978-981-19-2722-5 ISBN 978-981-19-2723-2 (eBook)
https://doi.org/10.1007/978-981-19-2723-2

Jointly published with Science Press
The print edition is not for sale in China (Mainland). Customers from China (Mainland) please order the print book from: Science Press.

This Springer imprint is published by the registered company Springer Nature Singapore Pte Ltd.
The registered company address is: 152 Beach Road, #21-01/04 Gateway East, Singapore 189721, Singapore

Contents

Chapter 1
Introduction

The aeroengine is the "heart" of an aircraft, and these national treasures are an important indicators of national core competitiveness. The turbine inlet temperatures in both third- and fourth-generation aeroengines exceed the melting points of high-temperature metallic materials. A new generation of materials can yield a new generation of equipment. As the most feasible technology for improving the service temperature of gas turbine engines, thermal barrier coatings (TBCs) have become an indispensable thermal protection material for hot-end components (e.g., high-pressure turbine blades in aeroengines and gas turbines) and, to a remarkable extent, determine engine performance and development levels. All the world's aviation powers have listed TBCs as a key core technology in their major advancement plans. Likewise, China has categorized TBCs as a key technology urgently needed for aeroengines and gas turbines.

Operating in hot-end components such as engine turbine blades, TBCs are subjected to long-term extreme conditions such as impacts from 2000 K gas near the critical Mach number, centrifugal forces generated by rotations at 10,000–50,000 rpm, fatigue, creep, calcium-magnesium-aluminosilicate (CMAS) corrosion, solid particle erosion, and oxidation, accompanied by chemical reactions. As a result of extreme adverse conditions, coatings spall and fail by a variety of complex mechanisms, posing a multitude of new challenges to research on failure theories. For example, oxidation, thermal mismatch, growth stress, and high temperature, which are involved in the interfacial oxidation failure of TBCs, affect one another. The oxidation process and the resulting coating spallation are both a typical thermo-mechano-chemical coupling physical nonlinear problem and a geometric nonlinear problem with strain up to 10%. As another example, TBC melts (CMAS) corrode at high temperatures, and their relevant infiltration and performance evolution patterns and thermo-mechano-chemical coupling mechanisms are all new failure phenomena exhibited by TBCs, the mechanical nature of which has been poorly explained so far.

Turbine blades are geometrically complex, and TBCs with defects and multi-layer systems have a complex microstructure. During service, coatings and interfaces evolve in terms of composition, microstructure, and performance. Their nonlinear

Y. Zhou et al., *Thermal Barrier Coatings: Failure Theory and Evaluation Technology*,
https://doi.org/10.1007/978-981-19-2723-2_1

physical and geometric properties create new problems for mechanical performance characterization and numerical simulation techniques. In addition, high temperatures are inevitable service conditions for TBCs. The characterization of the coating performance at high temperatures and the accurate description of the loading conditions in complex high-temperature environments are issues to be considered in the failure analysis of TBCs.

Performance evaluation and service life prediction are the most urgent issues of the TBC application sector. The association of the service life with the material and service-environment parameters offers the most direct basis for the TBC production sector. Understanding this association is the ultimate goal of investigation into failure mechanisms and, more importantly, is the most difficult scientific problem at the forefront of the research in this field. Moreover, simulation and evaluation devices can offer the most reliable, direct experimental basis. They are the topical and key areas of research on TBC failure mechanism analysis and evaluation techniques, but technological embargos are often imposed by relevant countries.

Thus, spallation failure theories, numerical computational methods, mechanical performance characterizations, performance and service life evaluations, and pre-test-run simulation and evaluation devices all merit investigation in the study of TBC failure mechanisms and performance evaluation.

1.1 TBCs and the Corresponding Preparation Methods

1.1.1 TBC Materials and Structures

To meet the demand for increasing service temperatures in gas turbine engines, the National Aeronautics and Space Administration (NASA) of the United States put forward the concept of TBCs in 1953 [1–3]. TBCs offer thermal protection based on the following principle. Ceramic materials are characterized by high-temperature resistance, high thermal stability, low thermal conductivity, and high corrosion resistance. Because of these properties, ceramic materials are combined with metallic substrates in the form of coatings to insulate high-temperature metallic substrate materials from high-temperature gas, with the goal of reducing surface temperatures in hot-end metallic components and simultaneously enhancing their resistance to high-temperature oxidation and hot corrosion [4, 5]. TBCs have been satisfactorily applied in hot-end components such as turbine blades in aeroengines and gas turbines [1–7].

A typical TBC has a two-layer structure [8]. Figure 1.1 shows the geometry and structure of a TBC on a turbine blade. The blade substrate is generally made of a Ni-based high-temperature alloy (a directionally solidified alloy or a single-crystal alloy). The TBC contains a metallic bond coat (BC) layer and a thermally insulating ceramic layer. The ceramic layer, generally 90–300 μm in thickness, is currently commonly made of ZrO_2 ceramic stabilized with 7–8wt.% Y_2O_3. A coating with

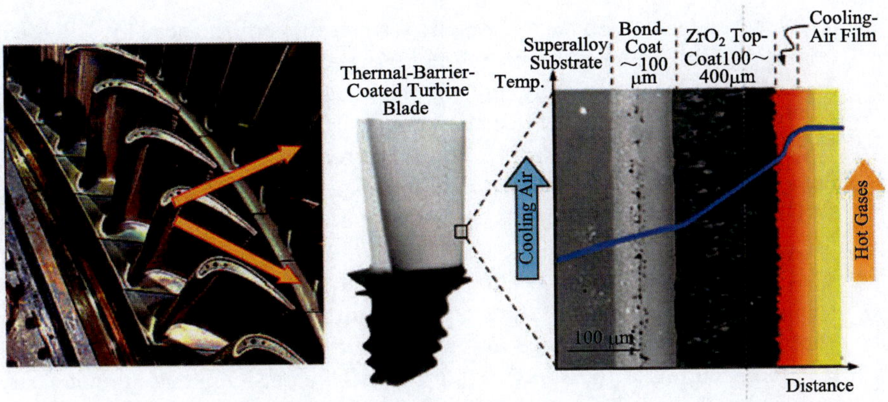

Fig. 1.1 Thermal-barrier-coated turbine blades and a cross-sectional view of a TBC system [3]

such ceramic layer is referred to as 7YSZ (7wt.% yttria-stabilized zirconia) coating. The metallic BC layer, generally 50–150 μm in thickness, is currently commonly made of MCrAlYX (M = Ni and/or Co; X = Hf, Ta, Si, etc.) or a diffused Pt/Ni–Al coating. The BC layer can effectively reduce the mismatch between the thermal, physical, and mechanical properties of the substrate and the ceramic layer while producing an oxidation film capable of improving the resistance of the substrate alloy to oxidation and corrosion. In addition, during preparation and service, the metallic elements in the BC layer and the oxygen in the coating and the external environment diffuse to the interface between the BC and ceramic layers and undergo oxidation reactions, forming thermal growth oxides (TGOs) that are composed primarily of α-Al_2O_3 and generally have a thickness of less than 10 μm [3, 9].

Yttria-stabilized zirconia (YSZ) coatings are prone to phase transformation and corrosion; hence, they are unable to operate safely at temperatures above 1200 °C [10, 11]. This has inspired the development of a series of new TBC material systems involving the following three main areas:

(1) Doping with rare-earth oxides. Rare-earth metallic oxides resistant to higher temperatures are added to YSZ coatings to reduce their thermal conductivity and enhance their temperature capabilities. The main rare-earth elements (REEs) used for this purpose include Hf, Ce, Sc, Gd, Nd, Yb, and La [8, 11, 12]. For example, ZrO_2-$(YNdGd)_2O_3$ and ZrO_2-$(YNdYb)_2O_3$, two TBCs doped with rare-earth metallic oxides, have 50–66% lower thermal conductivity and display exceptional corrosion resistance.
(2) New materials. $A_2B_2O_7$ (A = REE; B = Zr, Hf, or Ce) compounds, $LnMAl_{11}O_{19}$ or $LnTi_2Al_9O_{19}$ (Ln = La, Gd, Sm, or Yb; M = Mg, Mn, Zn, Cr, or Sm) magnetoplumbite compounds, rare-earth doped aluminates (RE_3AlO_{12}), and perovskites (e.g., $SrZrO_3$) are new materials with high melting points, low thermal conductivity, and high coefficients of thermal expansion (CTEs) [8, 13, 14]. For example, a $La_2Zr_2O_7$ TBC is characterized by a high melting point,

a low thermal conductivity, and excellent sintering performance [15, 16]. Shen et al. proposed Ta-based coatings (e.g., ZrO_2-$YTaO_4$, ZrO_2-$YbTaO_4$, and Y-Ta-Zr-O) that have extremely high phase-change temperatures and low CTEs, can meet the demand for service temperatures above 1400 °C, and display very high fracture toughness [16–19]. From a ferroelastic toughening perspective, the author of this book proposed new $A_6B_2O_{17}$ (A = Hf or Zr; B = Ta or Y) TBCs with high melting points, high thermal insulation capacities, and high fracture toughnesses [20].

However, none of these coating systems have been employed in real-world applications. Their most prominent shortcomings are low fracture toughness and thermal shock resistance lower than that of YSZ coatings. In addition, the spray-coating of films (e.g., Pd, Pt, $ZrSiO_4$, and SiOC) that are not easily wetted by CMAS [21] onto YSZ surfaces can reduce CMAS infiltration but fail to completely insulate YSZ surfaces from CMAS. Adding Al, Ti, or certain REEs to YSZ can allow YSZ to crystalize with CMAS and thus effectively inhibit erosion, [22] but the thermal insulation and mechanical properties of the YSZ are weakened.

(3) New structures. New TBC structures are designed with two- or multilayer, functionally graded, composite coatings [13, 14, 23, 24]. For example, the $La_2(Zr_{0.7}Ce_{0.3})_2O_7$/YSZ two-coating system designed in China exploits the resistance of $La_2(Zr_{0.7}Ce_{0.3})_2O_7$ to CMAS and the high thermal insulation capacity of YSZ [16, 25]. A functionally gradient TBC can be prepared by mixing a BC-layer material and a ceramic layer to achieve continuously gradient changes in composition and structure and hence eliminate the interface between the coating and BC-layer material. Functionally gradient TBCs are nonhomogeneous materials whose function gradually changes with composition and that are not prone to spallation, as shown in Fig. 1.2. To reduce preparation difficulties, functionally gradient TBCs are often made into gradient layers whose BC- and ceramic-layer composition gradually changes at a fixed ratio. Compound coatings are multilayer coating systems prepared based on functions to accommodate various complex environments (e.g., high temperatures, thermal stresses, corrosion, and oxidation) [26]. Figure 1.3 shows a representative compound TBC structure that contains a corrosion-resistant surface coating, a thermal insulation layer, a corrosion/oxidation-resistant layer, a thermal-stress control layer, and a diffusion barrier. Similar to functionally gradient coatings, compound coatings have been rarely employed in real-world applications due to preparation process difficulties and the unclear mechanism of regulating the interfacial properties of various coatings.

1.1.2 TBC Preparation Methods

Since NASA proposed the concept of TBCs in the 1950s, TBC preparation processes have advanced rapidly, from the earliest flame spraying technique to the multiple

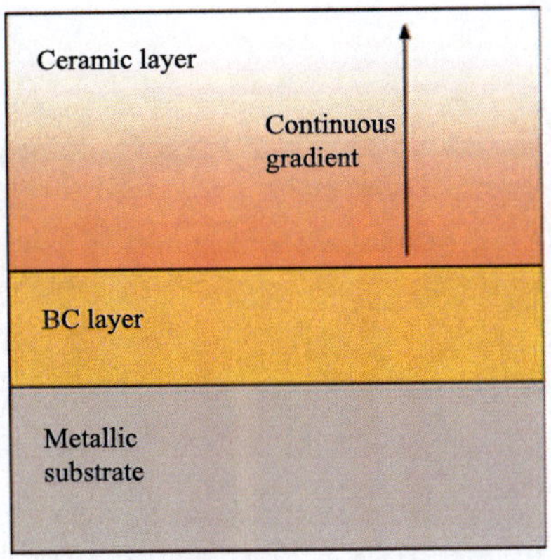

Fig. 1.2 Functionally gradient TBC

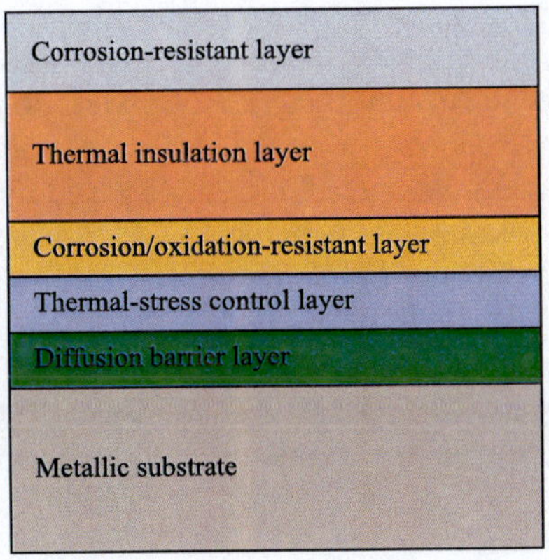

Fig. 1.3 Principle of compound coatings

preparation methods used today, including high-velocity oxygen fuel (HVOF) spraying, detonation spraying, ion plating, magnetron sputtering, laser cladding, arc evaporation, ion beam assisted deposition, chemical vapor deposition, plasma spraying (PS), and electron beam-physical vapor deposition (EB-PVD) [27–32]. These techniques can be employed to prepare TBCs of various types and for various purposes, enabling the coatings to be more widely applied in various types of thermal protection. PS and EB-PVD are currently the main, most widely used TBC preparation methods. The preparation processes directly affect the microstructure, various types of service performance, and reliability of coatings and have a direct impact on their quality.

(1) PS

PS is a process where a direct-current-driven plasma arc is used to heat metallic or ceramic powder to a molten or semi-molten state and to spray the molten material at a high speed and deposit it onto the preprocessed (e.g., sandblasted) surface of a workpiece, forming a firmly adhered surface layer [27]. Figure 1.4 shows the working

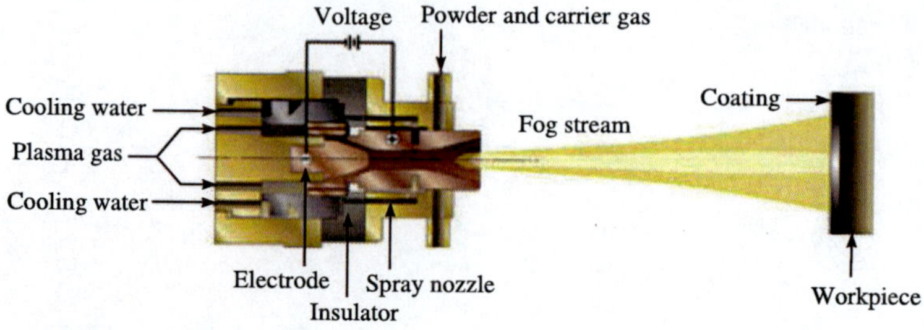

Fig. 1.4 Schematic diagram of the principle of APS

principle of PS. Plasma arcs are characterized by high temperatures (approximately 2×10^4 K at the center) and high speeds (spraying the material at a Mach number of approximately 1) and are stable and controllable. There are two main types of PS, namely, atmospheric PS (APS), which is primarily used to prepare the ceramic layer, and vacuum PS (VPS), which is employed to fabricate the BC layer.

As shown in Fig. 1.5a, upon impact on the substrate, molten or semi-molten particles spread out, solidify, and form splats on the surface of the substrate, leading to a layered structure. A PS TBC has a layered structure consisting of splats in a columnar or isometric crystal structure with grain sizes in the range of 50–200 nm, [28] resulting in a high porosity with a large number of interlayer defects (e.g., voids). Thus, these TBCs have a low thermal conductivity and good thermal insulation performance and are often used in stationary high-temperature components such as combustion chambers and guide vanes in gas turbines and aeroengines [7, 8].

However, PS coatings contain unmolten raw materials, impurities, and voids in large quantities. These defects lead to sulfuration and salt corrosion and, thereby, form crack sources in a coating during service, weakening its bond with the substrate and compromising its thermal shock resistance or even causing spallation failure [3]. In addition, PS coatings have high surface roughness and low thermal shock resistance and are, therefore, unable to meet the stringent requirements of rotating engine blades. As a result of ceramic-layer-related factors such as pores and impurities, PS coatings

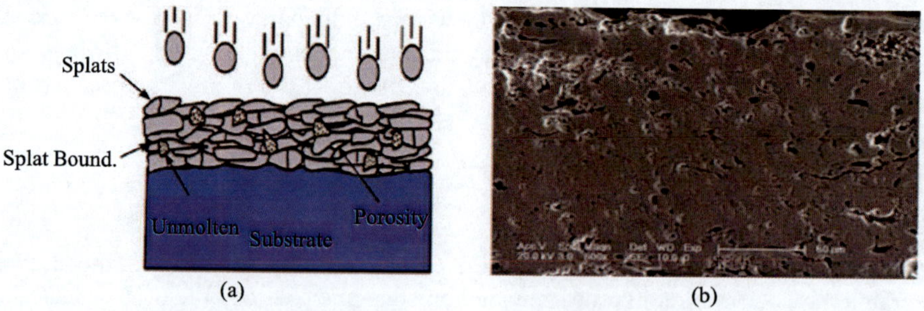

Fig. 1.5 **a** Schematic diagram of the deposition of a coating by APS. **b** Microstructure of a typical plasma spray fabricated coating

are inferior to EB-PVD coatings in terms of thermal cycling stability [3]. In the presence of low-grade fuels or in corrosive environments, corrosive gas and fused salts erode PS coatings through their pores and, consequently, accelerate coating failure.

(2) EB-PVD

EB-PVD is a new TBC preparation process developed by researchers from the United States, the United Kingdom, Germany, and the former Soviet Union in the 1980s. In this process, a focused high-power electron beam is used to heat a target material in a high vacuum chamber to rapidly melt, vaporize, and gasify it. The vaporized source material forms a cloud-like mass that moves to the surface of the workpiece and deposits to form a coating. Figure 1.6 shows a schematic diagram of the EB-PVD process. The workpiece is often heated to improve its bond with the coating.

EB-PVD TBCs have a columnar crystal structure, as shown in Fig. 1.7. Compared to the porous, layered PS coatings, this columnar crystal structure is able to withstand relatively large strains and imparts high thermal shock resistance to the coating. EB-PVD coatings have high surface smoothness and do not block the cooling gas channels in blades, thereby helping to maintain the aerodynamic performance of blades. These coatings are denser and more resistant to oxidation and hot corrosion. There is a smooth interface between the coating and the substrate, which are strongly bonded. However, EB-PVD coatings still have certain shortcomings. First, their deposition rates are low, and their columnar crystal structure leads to a high thermal conductivity. It is difficult to control the composition of relatively complex coating materials. Moreover, high-power electron beam devices and large vacuum chambers have relatively high operating costs. Thus, all TBCs for hot-end components operating in adverse conditions (e.g., the moving blades of an engine) are

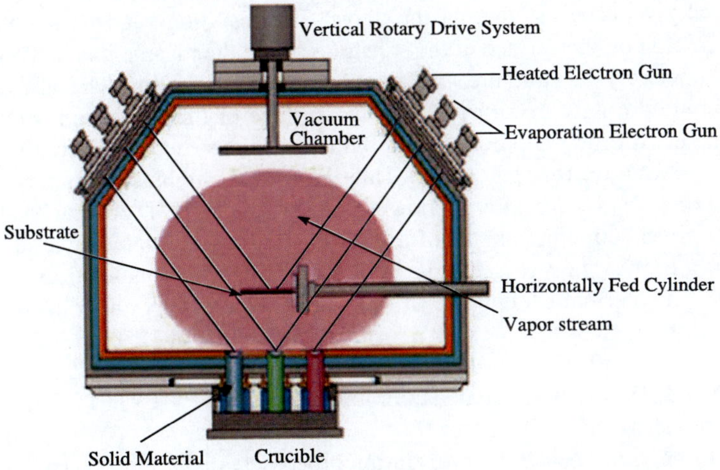

Fig. 1.6 Schematic diagram of a typical EB-PVD setup

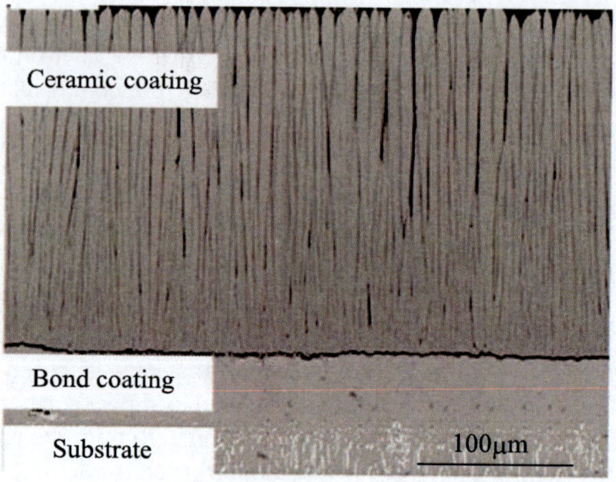

Fig. 1.7 Microstructure of an EB-PVD TBC

prepared by EB-PVD. This process also reflects the future development direction for higher-performance TBC preparation processes. All the world's powers in the engine field are racing to study EB-PVD but have imposed blockades on the relevant core technologies, particularly those in the equipment area, against other countries, especially China.

(3) PS-PVD

PS-PVD is a recently developed new process for coating preparation. In this process, a high-power (up to 200 kW) plasma spray gun is used in a low-pressure spray chamber to generate a plasma flame with a length of 2 m and a diameter of 0.2–0.4 m [33]. The vaporized state of the coating material and the formed coating structure vary with location along the direction of the plasma flame. The material powder melts rapidly and partially vaporizes in the plasma flame. The resulting vapor is deposited and forms a coating on the surface of the substrate above the plasma flame. This coating has a columnar crystal structure similar to that of a PVD coating. The non-vaporized molten liquid droplets are sprayed onto the surface of the substrate in front of the plasma flame, forming another coating, which displays the characteristic structure (i.e., layered structure) of a PS coating. Thus, PS-PVD combines the characteristics of PVD and PS [34]. Specifically, as shown in Fig. 1.8, a PS-PVD coating has the same columnar crystal structure as that of a PVD coating and the same layered structure as that of a PS coating. Accordingly, PS-PVD combines the advantages of PS and EB-PVD. The thermal cycling life of PS-PVD YSZ coatings under cyclic thermal conditions provided by a flame with a surface temperature of 1250 °C and a substrate temperature of 1050 °C can reach more than twice that of APS YSZ coatings [35]. Currently, PS-PVD remains at an exploratory stage, and PS-PVD TBCs have yet to be used in real-world applications.

In addition, other new TBC preparation processes, including suspension plasma spraying, [36] radio frequency plasma-assisted physical vapor deposition, [37] flame-assisted vapor deposition, [38] and HVOF spraying, [39] have been developed.

Fig. 1.8 The columnar crystal structure of a PVD coating and the layered structure of a PS coating

However, PS and EB-PVD remain the only TBC preparation processes currently used in real-world production. All the other processes are either at the theoretical or experimental research stage, and coatings prepared by these processes are inferior to those prepared by PS and EB-PVD in terms of comprehensive performance, preparation efficiency, and maturity.

1.2 TBC Spallation Failure and Its Main Influencing Factors

1.2.1 Service Conditions for TBCs

Hot-end components with TBCs operate in extremely harsh conditions. For example, typical service conditions for the TBCs of turbine blades in aeroengines include the following:

(1) Long-term high temperatures. High-temperature aeroengine components with TBCs, such as combustion chambers and turbine blades, are exposed to gas in combustion chambers for a long period of time. The heat release rate per unit volume (100,000 kW/m^{3}atm) of main combustion chambers in aeroengines is 1000 times higher than that (75 kW/m^{3}atm) of 1000-mW supercritical boiler units [40]. For an engine with a thrust-to-weight ratio (TWR) of 10, the gas inlet temperature of its turbine can reach 1700 °C; in comparison, this temperature increases to 2000 °C for an engine with a TWR of 15 [40]. More importantly, TBCs need to operate at this temperature for thousands or even tens of thousands of hours.

(2) High-pressure ratios. The gas pressure at the entrance of the gas inlet of an engine is approximately 1 bar. After the gas passes through and burns in an approximately 1-m-long gas generator (also referred to as core engine) composed of a high-pressure gas compressor, a combustion chamber, and a turbine, the gas pressure at the exit can reach 40–50 bar, more than twice the pressure of impounded water in the Three Gorges Dam [40]. Thus, TBCs are also under extremely high compressive loads.

(3) High rotational speeds. TBCs applied to the working blades of an engine rotate with the turbine at 20,000–50,000 rpm, which generates a tremendous centrifugal load. In addition, the supersonic high-temperature gas and the blades rotating at high speeds interact with one another, generating hot spot, turbulence, and wake effects on the TBC surfaces of the blades, which are also service loads on TBCs.

(4) Complex media. The high-temperature gas acting on TBC surfaces has high oxygen content and often contains corrosive gases, impurity particles, and corrosive solid media. These particles and media may originate from fuel impurities, impurity particles in the outside air, or debris produced within the engine [41]. These solid and gaseous media interact with coatings, resulting in complex failure modes such as oxidation, corrosion, and erosion, which are the main factors causing coating spallation.

(5) Complex stresses. The thermophysical and mechanical properties of the ceramic, BC, and TGO layers of a TBC and its substrate differ significantly. Thermal mismatch stresses are generated in these layers during service at high temperatures due to their CTE mismatch. The formation of a TGO layer leads to the formation of thermal growth stresses at the interface. When corrosion occurs, new chemical reactions lead to the formation of a stress field. In addition, effects such as phase transformation and sintering induce changes in the stress field of the coating. Thus, complex stresses constitute service load conditions for a TBC and, at the same time, are generated by the TBC under service conditions.

1.2.2 TBC Spallation Failure and Its Main Influencing Factors

The combined action of the abovementioned various loads (e.g., thermal, mechanical, and chemical loads) leads to various TBC failure modes (e.g., internal cracks, holes, and interface cracks) and ultimately causes TBCs to fail by spallation. The spallation of a coating exposes the substrate alloy to high-temperature gas. In addition, the spalled coating moves with the high-temperature gas within the components of the engine and even collides with the turbine blades, posing a fatal threat to the engine. Moreover, spallation is a very large bottleneck that restricts the application and development of coatings. Thus, researchers have conducted a series of investigations into the modes and mechanisms of coating spallation failure, and they have gradually realized that high-temperature oxidation, erosion, and CMAS corrosion are three key factors that cause coating spallation.

(1) **Interfacial oxidation**. For a TBC in service at high temperatures, oxygen molecules in air and oxygen atoms in its ceramic layer diffuse into the interface between the ceramic and BC layers and react with the metallic elements that have diffused from the BC layer, forming a TGO layer [3, 9]. As Al has the highest diffusivity, it is the first to undergo reactions, forming a dense TGO layer composed of Al_2O_3, as shown in Fig. 1.9. A dense TGO layer can effectively slow the diffusion of other metallic elements in the BC layer and impede oxidation in the BC layer. This is a favorable aspect of TGOs. On the other hand, the formation of TGOs also results in growth and thermal mismatch stresses and causes cracks to grow and propagate at the TGO/BC interface (I) and TGO/ceramic interface (II) as well as in the ceramic layer (III) and TGO (IV) layer, eventually causing the coating to spall. This is a highly unfavorable aspect of TGOs. More dangerously, when the Al content is insufficient or the TGO growth is inadequately dense, other metallic elements in the BC layer, such as Ni and Cr, also diffuse and oxidize, forming loose mixed oxides, [3, 9] which are extremely prone to producing crack initiation and propagation. As high temperatures are unavoidable service conditions for TBCs, high-temperature oxidation is considered to be the first key factor inducing coating spallation.

(2) **High-temperature erosion**. During service, TBCs are inevitably repeatedly impacted by hard-particle flow in the combustion chamber of an engine, leading to coating thickening, crack formation, or even spallation, as shown in Fig. 1.10, i.e., erosion failure [42–46]. These hard particles originate mainly from two sources. (1) Particles formed within the engine. These particles can be either

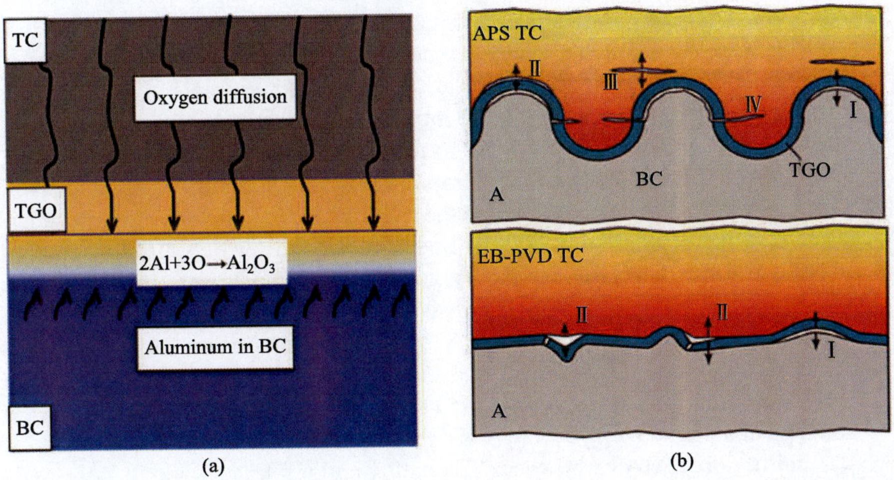

Fig. 1.9 **a** Schematic diagram of the interfacial oxidation of a TBC. **b** Two coating spallation failure modes induced by interfacial oxidation

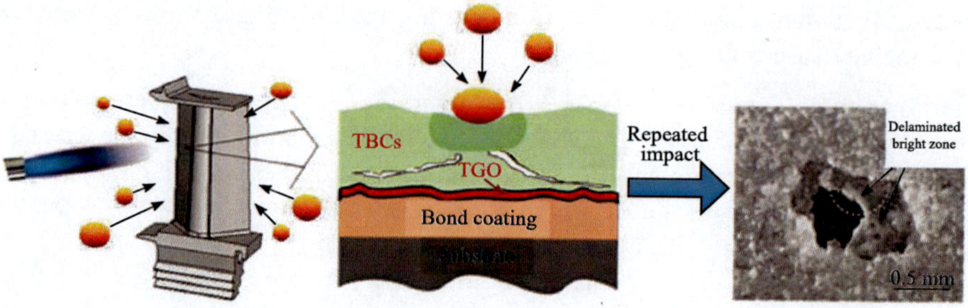

Fig. 1.10 Erosion damage and spallation of a TBC [14, 47]

carbon particles formed during the combustion process or particles formed from the erosion of the inner walls of the combustion chamber and the turbine blades of the engine. (2) External objects sucked into the gas turbine, such as sand particles, dust, and metallic (e.g., Al) particles. In some studies, TBC failure caused by particles of the first and second sources is referred to as erosion and impact damage by external objects, respectively. However, more studies refer to both types of damage as erosion and view the differences in damage caused by particles of different natures as different failure mechanisms. As flight conditions become increasingly complex due to haze and sand dust, TBC erosion failure phenomena and mechanisms become increasingly complicated. Thus, an increasing number of studies have regarded erosion as the second key influencing factor, after interfacial oxidation, for TBC spallation failure [47].

(3) **CMAS corrosion**. With respect to CMAS corrosion, as the service temperature of a coating further increases (above 1250 °C), CMAS (e.g., sand dust and volcanic ash) melts and infiltrates into the coating along its pores, inducing compositional, structural, and performance changes and eventually causing it to spall [10, 21, 48–50]. The service temperatures for engines currently in service and under development essentially exceed the melting points of CMAS. The time required for CMAS to infiltrate into the entire ceramic layer of a coating is far shorter than that required for the formation of a TGO layer [48]. In addition, the CMAS infiltration rate increases rapidly as the service temperature increases. Thus, CMAS corrosion has become the most dangerous and unavoidable factor causing coating spallation. YSZ, a material currently widely used to prepare TBCs, has low thermal conductivity, high thermal shock resistance, and high thermal stability but relatively low resistance to CMAS corrosion. At 1250 °C, CMAS can penetrate an EB-PVD YSZ coating within one minute [48, 49] and continues to erode the BC layer and the metallic substrate [50]. As shown in Fig. 1.11, according to Drexler et al. [51] from a thermochemical perspective, the erosion of a YSZ coating by CMAS is accompanied by the following processes. (I) Infiltration. CMAS wets the YSZ coating and infiltrates it via its pores. (II) Loss of Y. YSZ grains are dissolved in CMAS, resulting in a loss of Y. (III) YSZ delamination. CMAS penetrates the YSZ grain boundaries, and the delaminated

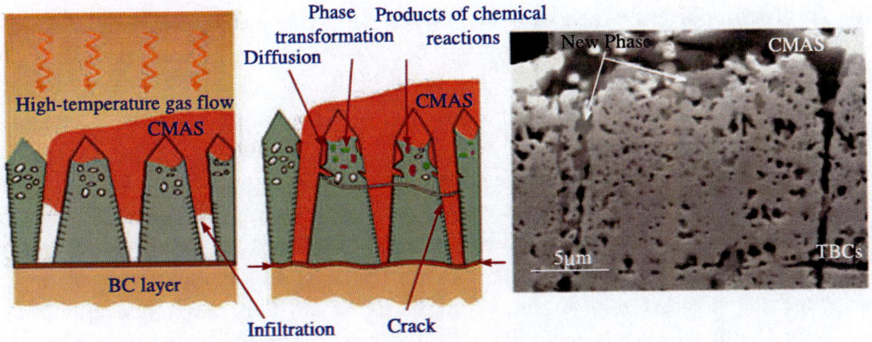

Fig. 1.11 Schematic diagram of the CMAS-induced spallation of a YSZ coating

YSZ grains are scattered within the CMAS. (IV) Diffusion and reactions. Y diffuses into CMAS while, at the same time, elements such as Si, Ca, and Mg diffuse from CMAS to YSZ grains and undergo chemical reactions. These processes consume the ceramic layer. Apart from this, the loss of the stabilizer (yttria) accelerates the transition of ZrO_2 from a tetragonal phase to a monoclinic phase, causing a volume expansion of approximately 5% [52, 53]. Moreover, the CMAS in the pores reduces the strain tolerance of the coating [54, 55]. Furthermore, the microstructure and the products of chemical reactions cause changes in the physical and mechanical properties [56] and eventually lead to coating spallation. Enhancing the resistance of YSZ TBCs to CMAS has become an urgent task in this field.

(4) **Sintering**. At high temperatures, the ceramic layer of TBC sinters. For an APS ceramic layer, high-temperature sintering causes grain growth, reduces the porosity and number of microcracks, and increases the CTE of the coating. For an EB-PVD coating, sintering reduces the distance between its columnar crystals and causes the feathery structure of its columnar crystals to weaken or even disappear, thereby reducing its thermal insulation performance and strain tolerance [57–59]. The sintering of the ceramic layer generally produces a planar compressive stress and thus results in the formation of cracks perpendicular to the interface in a coating system.

(5) **Phase transformation**. Ceramic-layer phase transformation is also one of the important causes of TBC failure [60, 61]. ZrO_2 ceramics have different crystal forms (e.g., tetragonal, cubic, and monoclinic) at different temperatures. The thermal cycling process of a TBC generally involves three stages, namely, warming, holding, and cooling, spanning a wide range of temperatures. For a TBC in service, a ZrO_2 ceramic undergoes phase transformation as temperature continuously changes. The temperature at which the monoclinic phase transitions to the tetragonal phase falls right in the operating temperature range of gas turbines. This phase transformation generally produces a volume change in the material and increases the stress within the coating, thereby promoting coating spallation and failure.

(6) **Interdiffusion between the BC layer and the substrate.** BC-layer and substrate alloys differ significantly in composition and concentration. Thus, for a TBC in service at temperatures above 1350 K, significant elemental interdiffusion generally occurs at the interface between the BC layer and substrate [62–65]. This interfacial elemental interdiffusion predominantly involves outward diffusion of Cr and Ni from the substrate to the BC layer and inward diffusion of Al, Cr, and Co from the BC layer to the substrate, accompanied by extensive formation of diffusion holes and harmful phases. As a consequence, the mechanical performance of the substrate deteriorates. Notably, interdiffusion engenders a substantial loss of Al in the BC layer. This inevitably reduces the purity of the Al_2O_3 among the TGOs at the interface between the ceramic and BC layers, thereby facilitating the formation of Ni-rich mixed oxides and reducing the oxidation resistance of the coating. With the monocrystallization of alloys at high temperatures and the increase in service temperatures for TBCs, interdiffusion between the BC layer and the substrate causes the monocrystal substrate to lose its unidirectional crystallization properties. A rapid decrease in the mechanical performance or even a complete loss of monocrystal properties resulting from changes in lattice orientation, microstructure, and composition is an issue related to TBCs on monocrystal substrates that warrants urgent attention.

1.3 Solid Mechanics Requirements and Challenges Generated by TBC Failure

1.3.1 Solid Mechanics Requirements Generated by TBC Failure

Through more than half a century of development since the proposal of the concept of TBCs in 1953, researchers have mastered material formulas for YSZ TBCs and, on this basis, developed a series of new TBC materials. In addition, researchers have mastered two mature preparation processes—PS and PVD—and, on this basis, developed a series of new preparation processes. The world's powers in the engine field, such as the United States, the United Kingdom, and France, have extensively used TBC technology in various types of military and civil aeroengines as well as gas turbines. China has also realized the application of TBC technology in aeroengines in mass production and verified the thermal insulation effects of TBCs. In addition, China has categorized TBCs as an indispensable thermal insulation and protection technology for the "two-engine" (i.e., aeroengines and gas turbines) special project. However, long-term exposure to high temperatures, high pressures, and high rotational speeds as well as the actions of various mechanical loads and media cause coatings to spall in modes such as oxidation, corrosion, and erosion, severely restricting the application and development of coatings. Studying theoretical and experimental methods and testing techniques for analyzing TBC failure mechanisms, analyzing, evaluating, and predicting the service performance, service life, and safe use of TBCs, and establishing the correlations between the performance and material composition and preparation process parameters are effective means and necessary requisites for addressing TBC spallation failure. Thus, to ensure safe use and sustainable development of TBCs, it is urgently necessary to develop failure theories, testing methods, and evaluation techniques to analyze coating spallation.

Solid mechanics is a discipline that studies the displacement, movement, stress, strain, and damage of deformable solids under external factors. The origin of solid mechanics can be traced back to the second millennium BC, when buildings, simple vehicles and vessels, and hunting implements began to be constructed with a mechanical philosophy in China and other ancient civilizations. The Zhaozhou Stone Arch Bridge built in China during the Sui Dynasty (591–599 AD) already embodied some basic concepts found in contemporary bar, plate, and shell design. The development of industrial technologies for constructing large machinery, long-span bridges, and large factory plants in the eighteenth century facilitated the rapid growth of solid mechanics. Thus far, solid mechanics has developed and spawned a number of subdisciplines, such as the mechanics of materials, elastic mechanics, plastic mechanics, fracture mechanics, structural mechanics, mechanics of composite materials, rock mechanics, computational mechanics of materials, and experimental solid mechanics. These subdisciplines differ regarding research ideas, fundamental assumptions, and research methods and each has its own focus in terms of research objects, yet they cannot be strictly separated from one another.

A TBC is composed of a deformable substrate and coating-system layers. Deformation, stress and strain fields, and microstructural and damage evolution patterns within the coating systems in various service environments (i.e., under various loads) are the principal factors that cause a TBC to spall. Thus, TBC spallation failure analysis mainly involves the following subdisciplines of solid mechanics:

(1) Mechanics of materials. As the earliest established subdiscipline of solid mechanics, the mechanics of materials is the study of the mechanical properties, deformation states, and failure patterns of materials under external forces and provides a basis for selecting materials and component dimensions in engineering design. However, it focuses primarily on one-dimensional (1D) bars and simple plates and shells, and it enlightens and lays a foundation for other subdisciplines of solid mechanics.

(2) Elastic mechanics. Also known as theory of elasticity, elastic mechanics focuses on stress and strain fields and relevant patterns in (three-dimensional, 3D) elastic objects under external forces. The first assumption of elastic mechanics is that the object of study is an ideal elastic body, which means that the object deforms under an external force, that there is a one-to-one correspondence between the stress and strain at each point within the object, and that after the external force is removed, the object returns to its original shape without any trace of deformation. The most basic idea of elastic mechanics is to hypothetically divide an object into an infinite number of microelements and consider their force balance and compatibility of deformation. In addition, elastic mechanics takes

into account the functional relationship between stress and strain during the deformation process of an object.

(3) Plastic mechanics. Also called theory of plasticity, plastic mechanics is the study of the relationships between plastic deformation and external forces as well as the stress and strain fields and relevant patterns in solid objects in a plastic deformation state under external forces. When an object is subjected to a sufficiently large external force, part or the entirety of its deformation exceeds the elastic range and the object then enters a plastic state. After the external force is removed, part or the entirety of the object's deformation does not vanish, and the object is unable to fully return to its original shape. Generally, plastic deformation tends to occur at locations where the shape of the object changes abruptly, near the point of action of concentrated forces, and near crack tips. Regarding research methods, similar to elastic mechanics, plastic mechanics begins with microelement analysis. Often, for an object subjected to forces, part of the object is in elastic state, and the other part is in plastic state. Thus, it is necessary to investigate the coexistence of elasticity and plasticity in objects.

(4) Fracture mechanics. Sometimes referred to as fracture theory, fracture mechanics is the examination of the stress and strain fields at crack tips in engineering structures and, on this basis, the analysis of crack propagation conditions and patterns. It is the latest established subdiscipline of solid mechanics. Many solid coatings (e.g., TBCs in this book) and their interfaces contain large numbers of defects (e.g., cracks, micropores, grain boundaries, dislocations, and impurities) that develop into macrocracks under thermal, mechanical, and alternating loads and in corrosive media. Thus, fracture theory can also be viewed as crack theory. The fracture toughness and crack propagation rate proposed in fracture theory are important metrics for predicting the critical crack size and estimating the life of components, with extensive applications in engineering structures. Studying crack propagation patterns, establishing fracture criteria, and controlling and preventing fracture failure are the objectives of fracture mechanics research.

(5) Experimental solid mechanics. Experimental solid mechanics is a subdiscipline that covers the testing of solid mechanics parameters and is referred to as experimental stress analysis. Its basic approach is to use advanced experimental methods (e.g., electrical measurements, photoelasticity, and modern optical, acoustic, and magnetic methods) and to establish or use mechanical, physical, and mathematical models to characterize the mechanical properties and damage parameters of materials, with the goal of providing parameters and a basis for analyzing failure mechanisms.

(6) Computational solid mechanics. This is a subdiscipline that solves various problems in solid mechanics by discrete numerical methods using electronic computers as a tool. Its basic approach is to employ certain discrete numerical methods to approximate the to-be-solved continuous function with a finite number of unknowns, based on the established physical and mathematical models. This approach effectively converts a differential equation problem to a linear algebraic problem that is then solved with a computer.

The theories of the abovementioned subdisciplines of solid mechanics provide ideas, basic theories, experimental methods, and computing tools for displacement, motion, stress, strain, and failure analysis of TBCs under service conditions. In addition, the development of the mechanics of composite materials offers a basis for addressing problems related to pores, cracks, and grain boundaries in TBCs, while the development of new subdisciplines of solid mechanics, such as structural mechanics, provides a reference for TBC stability and fracture analysis.

1.3.2 Solid Mechanics Challenges Presented by TBC Failure

While solid mechanics provides a theoretical system and research methods for studying TBC failure mechanisms, the extremely complex service conditions, geometric configuration, microstructure, and microstructural evolution patterns of TBCs also present tremendous challenges to solid mechanics.

(1) Thermo-mechano-chemical coupling theory. TBC spallation failure under service conditions is a typical thermo-mechano-chemical coupling process. For example, the following factors affect spallation under high-temperature oxidation and corrosion conditions. (I) Chemical field. Oxygen, CMAS, and various elements in a TBC system react with one another, forming new compounds (e.g., TGOs, $ZrSiO_4$, and spinel) that depend on the chemical-field parameters (e.g., chemical potential and concentration) of the system. (II) Temperature field. Both interfacial oxidation and CMAS corrosion occur at high temperatures and are thus directly related to the temperature field. (III) Stress field. As a result of the mismatch between the CTEs of the layers of a TBC as well as the mechanical loads to which the TBC is subjected during service, interfacial oxidation, erosion, and CMAS corrosion all occur in the presence of a stress field. (IV) Mutual facilitation between chemical, temperature, and stress fields. The new products from chemical reactions affect the temperature field of a system, while the thermal mismatch intensified by the constrained system and new materials alters the stress field of the system. The temperature field significantly affects chemical reactions in terms of, for example, element diffusion, infiltration depth, reaction speed, and thermal mismatch stress. The stress field significantly affects the diffusion of oxygen or elements involved in chemical reactions [66, 67]. (V) Crack propagation occurs under multifield (thermal, mechanical, and chemical fields) coupling. Oxygen diffusion, chemical reactions, and the stress field concentrate significantly at the locations of microdefects [68] and further facilitate crack propagation. Thus, thermo–mechano–chemical coupling is the intrinsic mechanism of microscopic damage and crack formation and propagation in TBCs. Figure 1.12 shows a schematic diagram depicting the thermo-mechano-chemical coupling problem, which corresponds to the chemical, mechanical, and thermal loads. Three commonly used variables, namely,

the extent of chemical reaction, strain, and temperature, are selected as generalized deformations. The work conjugates of the physical fields are the chemical potential, stress, and entropy, respectively, which affect one another under the action of multiple fields (i.e., thermal, mechanical, and chemical fields). Here, the thermal load is taken as an example. The work conjugate (i.e., entropy) of the generalized deformation (i.e., temperature) is obtained via the heat capacity, and temperature can also result in thermal expansion strain and affect the extent of chemical reaction. Apart from this, the chemical reaction absorbs or releases heat, whereas plastic deformation produces deformation heat, i.e., changes in entropy. The current theories of solid mechanics focus on problems relating to single mechanical or thermal loads or thermomechanical coupling. Establishing theories or models of solid mechanics that consider the thermal, mechanical, and chemical fields and their coupling to analyze the microstructural evolution patterns and crack propagation mechanisms during the oxidation and corrosion of TBCs is a tremendous challenge posed by TBCs to the theories of solid mechanics.

(2) Trans-scale theory. The TBC failure process is a typical trans-scale problem at both temporal and spatial scales. Here, we use oxidation as an example. At the temporal scale, a TBC oxidizes within several seconds at a high temperature or even during the preparation process. In contrast, it takes hundreds or even tens of thousands of hours for oxidation to cause a TBC to spall. In addition, the microstructure, performance, and damage of coatings evolve in real time. Thus, the TBC failure process is a typical trans-temporal-scale problem.

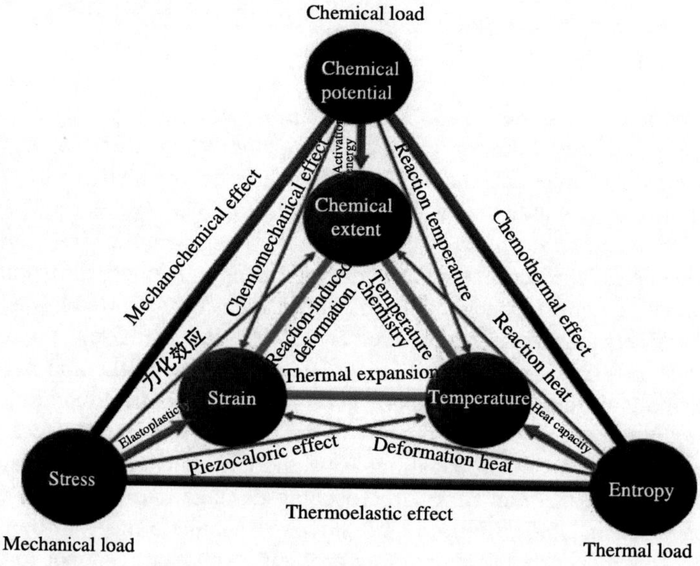

Fig. 1.12 Thermo-mechano-chemical coupling

At the spatial scale, in terms of micromechanisms, oxidation, phase transformation, and corrosion all occur on the atomic scale. In terms of macroscopic structures, the TGO layer is several micrometers in thickness, and the substrate material is several millimeters in thickness. In addition, there is a complex interfacial effect between the TGO layer and the substrate material. Thus, the TBC failure process is a typical trans-spatial-scale problem. Trans-scale physical and mechanical analysis models and experimental methods are required to analyze TBC failure mechanisms and their key influencing factors (e.g., stress field, interfacial fracture energy, microstructural evolution, and crystal structure). However, all the available theories of solid mechanics are established at the macroscopic continuum scale, and the physical models (e.g., those of quantum mechanics) are focused on micromechanical problems at the atomic and molecular levels. The establishment of trans-scale models between the atomic/molecular scale and the macroscopic continuum scale to analyze TBC spallation mechanisms across spatial and temporal scales is a great challenge posed by TBCs to solid mechanics.

(3) Numerical computational methods. Hot-end components (e.g., turbine blades) with TBCs, as shown in Fig. 1.1, have a complex warped structure. A multilayer TBC system spans from several micrometers to millimeters in scale and has a large number of pores, cracks, and grain boundaries. The interfaces between its layers are morphologically complex. The establishment of trans-scale numerical geometric models with complex geometry and microstructure is a challenge to numerically simulating the TBC failure process. For a TBC, its interfacial oxidation, corrosion, and erosion are accompanied by the growth of a TGO layer and products of corrosion or its erosion-induced spallation. Accounting for the growth or delamination process is another challenge to numerically simulating the TBC failure process. More importantly, both constitutive relations and failure theories considering thermo-mechano-chemical coupling involve physically nonlinear problems. Local strains can reach 10% when oxidation and corrosion occur, [69, 70] posing another geometrically nonlinear problem. Most of the available commercial finite element software programs are used to calculate and analyze linear problems, while very few are aimed at nonlinear coupling problems. Thus, numerically solving physically and geometrically nonlinear problems poses another challenge to the numerical simulation of the TBC failure process.

(4) Performance and failure characterization. Mechanical performance and failure (microstructural evolution) are two major mechanical problems in the thin-film coating field, and they provide a key basis for evaluating the spallation resistance of TBCs and understanding their spallation process. TBC systems with multilayer structures, complex interfacial morphologies, and microstructures that evolve with service conditions present a multitude of new challenges to mechanical performance and failure characterization. (I) Characterization of the performance of thin-film coating systems. Relatively mature mechanical testing methods have been developed to test the mechanical performance of bulk materials, including tensile methods for testing a material's Young's modulus, yield

strength, tensile strength, and stress–strain relation, a series of methods (e.g., the three-point bending method) for testing fracture toughness, and methods (e.g., X-ray and Raman methods) for testing residual stresses. However, the following difficult questions to be answered in the thin-film coating research field remain topical: whether and, if so, under what conditions the above methods are applicable to thin-film coating systems; which models, when applicable, should be used to characterize the intrinsic mechanical performance of coatings; how to overcome the scale and substrate effects; and how to develop new, effective methods for characterizing coatings, particularly their interfacial mechanical performance. (II) Signal detection under complex service conditions (e.g., high temperatures). TBCs operate under complex service conditions (e.g., high temperature, erosion, and CMAS corrosion) for long periods of time. However, characterization of mechanical performance at high temperatures by various detection methods (particularly contact detection methods) and signal detection using these methods at high temperatures are enormous challenges. In addition, testing at high temperatures produces multiple effects (e.g., temperature drift and sensor deformation). Overcoming these effects to obtain true detection test signals is another challenge. These challenges are formidable difficulties facing both the TBC field and experimental mechanics. (III) Identification and quantitative analysis of damage modes. For a TBC, various failure modes (e.g., interfacial oxidation, corrosion, sintering, surface cracks, and interface cracks) or microstructural evolution occur before spallation. Detecting and identifying these failure modes and quantitatively evaluating their extents can provide a direct basis for analyzing coating spallation mechanisms and are tremendous challenges facing experimental mechanics.

(5) Experimental equipment. It is infeasible to rely on conventional mechanical testing methods (e.g., tensile, bending, thermomechanical fatigue, and thermal shock methods) to comprehensively consider the complex geometric structure of TBCs and the complex couplings of various loads (e.g., thermal, mechanical, and chemical loads) to which TBCs are subjected or to explain the mechanisms by which TBCs spall prematurely and the relevant influencing factors. As early as the 1970s, NASA conducted a test run of TBCs in J-75 turbine engines with very high heat fluxes, verified their thermal insulation effects, and, on this basis, adjusted the formula for each component of the ceramic layer [71]. In fact, test runs in actual engines require tremendous labor and material resources. Thus, developing experimental simulation techniques for TBCs to simulate their complex service conditions and to nondestructively detect their failure process in real time or in situ can provide a powerful basis for correctly understanding their spallation mechanisms and for predicting their service life. Hence, experimental simulation equipment for TBCs has been a topical issue in this research field. However, it is very difficult to simultaneously simulate various complex conditions (e.g., high temperatures, erosion, and CMAS corrosion), particularly the centrifugal forces generated by high-speed rotation and the equivalent loading of their interaction with high-temperature gas. Research

institutes, such as NASA of the United States, the National Aerospace Laboratory (*Nationaal Lucht- en Ruimtevaartlaboratorium*) of the Netherlands, the National Energy Research Center of Germany, and the Aerospace Research Center of the National Research Council of Canada, have conducted extensive studies and developed experimental equipment for simulating high-temperature and erosion conditions for TBCs [72–76]. Nevertheless, currently there is a significant dearth of equipment for simulating working conditions involving high-speed rotation. In addition, there is a blockade on this type of equipment against China. Thus, for China, developing experimental simulation equipment for TBCs is an indispensable research element and a challenge for research on TBC failure mechanisms.

(6) Failure prediction. Material failure prediction has been a difficult problem facing engineering applications. Aeroengines and gas turbines require equipment with ultrahigh reliability. Against this application background, TBC failure time prediction is an even more difficult task and an inescapable hurdle. Thus, TBC failure prediction is a major topic of great interest to the engineering and academic communities. Accurate prediction of coatings' service life and establishment of its correlation with service conditions and material parameters provide a scientific basis for the safe application and optimum design of coatings. Regrettably, research on TBC failure prediction has thus far progressed slowly. Admittedly, failure problems are an important direction of solid mechanics research and fruitful achievements have been made in the past. For example, in mechanics of materials, strength, stiffness, and stability criteria have been established for 1D components under simple loads (e.g., tensile, compressive, and flexural loads) to guide engineering applications and designs. In elastic and plastic mechanics, strength theories have been established for 3D materials, particularly metallic materials. In fracture mechanics, crack and stress strength factors and their critical-value theories have been put forward to provide a theoretical basis for predicting material failure. However, TBCs operate under extremely complex loads, which are a result of multifield (thermal, mechanical, and chemical fields) coupling. The failure of a TBC with a complex microstructure is not simply a result of crack propagation but is instead accompanied by microstructural evolution and its interaction with the stress and strain fields. For a TBC, both the failure mode and the failure-inducing mechanism are intricate and complex. It is difficult to accurately predict the failure mode and time for a TBC based on conventional single strength criteria, crack propagation criteria, or stability criteria or based on failure criteria with a single damage parameter.

1.4 Content Overview

This book addresses the urgent needs and many challenges of solid mechanics posed by research on TBC spallation failure. Based on the author's nearly two decades of accumulated research findings and the research outcomes obtained by fellow researchers in this field, the author describes TBC failure theories and evaluation techniques from three aspects, namely, basic theories, performance characterization, and performance evaluation. To meet the different needs of researchers and application personnel, this book emphasizes the integration of "science" (basic theories) and "engineering" (engineering applications) and focuses on "theory-method-equipment" as a central theme of research. For example, Part I (failure theories) of this book introduces the basic theories and establishes frameworks for thermo-mechano-chemical coupling constitutive relations. In addition, Section I presents in detail the specific thermo-mechano-chemical coupling failure models for interfacial oxidation and CMAS corrosion in TBCs as well as their oxidation and corrosion patterns and mechanisms. This section aims to provide a basis for studying and developing thermo-mechano-chemical coupling theories in the TBC and relevant fields and to provide a scientific basis for relevant technical personnel responsible for TBC processes and applications to understand oxidation and corrosion mechanisms. In view of the challenges presented by TBCs to the available characterization methods and testing equipment of solid mechanics, Parts II (performance characterization) and III (evaluation techniques) of this book present introductions, analyses, and summaries with a focus on a system hierarchy of "theoretical model-test method-test equipment," which can inspire new ways of thinking in researchers and help facilitate their studies. Application personnel can directly use these techniques to characterize and evaluate TBCs. In short, this book is suitable for researchers engaged in the failure analysis of TBCs and related fields and, as a methodological book, for engineering personnel in the TBC research field, as well as for postgraduate and advanced undergraduate students in materials, mechanics, characterizations and other related fields.

This book is divided into three parts with a total of 15 chapters. Part I, which consists of Chaps. 2–7, introduces TBC failure theories. Specifically, Chaps. 2 and 3 give theoretical frameworks and nonlinear finite element theories, respectively, for thermo-mechano-chemical coupling. The thermo-mechano-chemical coupling failure theories and mechanisms of interfacial oxidation, CMAS corrosion, and erosion in TBCs are presented in Chaps. 4–7, respectively. Part II, which consists of Chaps. 8–13, focuses on the techniques for characterizing TBC performance and failure. Chapters 8–10 concisely describing various advanced methods for characterizing the basic mechanical properties, fracture toughness, and residual stresses of TBCs, respectively. Chapters 11–13 introduces nondestructive real-time characterization methods for key damage parameters including cracks, interfacial oxidation, and stress and strain fields. Part III, which consists of Chaps. 14–16, is devoted to the TBC performance evaluation techniques, including techniques for evaluating the comprehensive effects of thermal insulation and strength and for evaluating reliability and service life, and the progress in simulation assessment methods and testing platforms. By systematically introducing theories, methods, testing platforms, and relevant testing techniques, the author hopes to help readers to gain a comprehensive understanding of TBC spallation failure and relevant research methods and, on this basis, conduct relevant research to promote the safe application and sustainable development of TBCs in aeroengines and gas turbines in China.

This book is a culmination of over two decades of research on TBC failure theories and evaluation techniques, with contributions from more than 100 master's and doctoral students and achievements under long-term funding support, including two major projects from the National Natural Science Foundation of China, one project from the Outstanding Youth Foundation of China, one key project form National Natural Science Foundation of China, one major project from the Advanced Research Project of the 13th Five-year Plan of the Equipment Development Department of the Central Military Commission of China, two projects from the Basic Research Program for National Defence Technology, and multiple projects commissioned by relevant enterprises of the Aero Engine Corporation of China. Additionally, included in this book are the latest results and progress achieved by fellow researchers around the world in this research field. Detailed references are provided wherever applicable. The author hopes that the publication of this book will be helpful for the safe application and development of TBCs in China.

Special thanks go to Mr. LIU Daxiang, member of the Chinese Academy of Engineering, and member of the Standing Committee of the 10th National People's Congress of the People's Republic of China, for prefacing the book.

References

1. Miller RA. Thermal barrier coatings for aircraft engines: history and directions. J Therm Spray Technol. 1997;6(1):35.
2. Miller RA. Current status of thermal barrier coatings—an overview. Surf Coat Technol. 1987;30(1):1–11.
3. Padture NP, Gell M, Jordan EH. Thermal barrier coatings for gas-turbine engine applications. Science. 2002;296(5566):280–4.
4. Zhu DM, Miller RA. Thermal and environmental barrier coatings for advanced propulsion engine systems. NASA Technical Memorandum, 213129; 2004.
5. Cao X. Thermal barrier coating materials. Beijing: Science Press; 2007. (In Chinese).
6. Clarke DR, Oechsner M, Padture NP. Thermal-barrier coatings for more efficient gas-turbine engines. MRS Bull. 2012;37(10):891–8.
7. Wang T, Fang X. Thermal barrier coating strength theories and testing techniques. Xi'an: Xi'an Jiaotong University Press; 2007. (In Chinese).
8. Cao X. New materials and structures for thermal barrier coatings. Beijing: Science Press; 2015. (In Chinese).
9. Rabiei A, Evans AG. Failure mechanisms associated with the thermally grown oxide in plasma-sprayed thermal barrier coatings. Acta Mater. 2000;48(15):3963–76.
10. Levi CG, Hutchinson JW, Vidal-Sétif MH, et al. Environmental degradation of thermal-barrier coatings by molten deposits. MRS Bull. 2012;37(10):932–41.
11. Darolia R. Thermal barrier coatings technology: critical review, progress update, remaining challenges and prospects. Int Mater Rev. 2013;58(6):315–48.
12. Xue Z, Guo H, Gong S, Xu H. Novel ceramic materials for thermal barrier coatings. J Aeronaut Mater. 2018;38(02):10–20. (In Chinese with English abstract).
13. Vaßen R, Jarligo MO, Steinke T, et al. Overview on advanced thermal barrier coatings. Surf Coat Technol. 2010;205(4):938–42.
14. Ma W, Mack DE, Vaßen R, et al. Perovskite-type strontium zirconate as a new material for thermal barrier coatings. J Am Ceram Soc. 2008;91(8):2630–5.

15. Guo H, Gong S, Xu H. Research progress on new high/ultra-high temperature thermal barrier coatings and processing technologies. Acta Aeronautica et Astronautica Sinica. 2014;35(10):2722–32. (In Chinese with English abstract).
16. Mou R, Xu Z, He S, He L, Cao X. $La_2(Zr_{0.7}Ce_{0.3})_2O_7$-A new oxide ceramic material for thermal barrier coatings. J Mater Eng. 2009(07):67–71+78. (In Chinese with English abstract).
17. Shen Y, Leckie RM, Levi CG, et al. Low thermal conductivity without oxygen vacancies in equimolar $YO_{1.5}$+ $TaO_{2.5}$-and $YbO_{1.5}$+$TaO_{2.5}$-stabilized tetragonal zirconia ceramics. Acta Materialia, 2010; 58(13):4424–31.
18. Shian S, Sarin P, Gurak M, et al. The tetragonal-monoclinic, ferroelastic transformation in yttrium tantalate and effect of zirconia alloying. Acta Mater. 2014;69:196–202.
19. Macauley CA, Fernandez AN, Levi CG. Phase equilibria in the ZrO_2-YO_1. 5-$TaO_{2.5}$ system at 1500 °C. J Eur Ceramic Soc. 2017; 37(15):4888–901.
20. Tan ZY, Yang ZH, Zhu W, et al. Mechanical properties and calcium-magnesium-alumino-silicate(CMAS) corrosion behavior of a promising $Hf_6Ta_2O_{17}$ ceramic for thermal barrier coatings. Ceram Int. 2020;46(16):25242–8.
21. Rai AK, Bhattacharya RS, Wolfe DE, et al. CMAS-resistant thermal barrier coatings (TBC). Int J Appl Ceram Technol. 2010;7(5):662–74.
22. Aygun A, Vasiliev AL, Padture NP, et al. Novel thermal barrier coatings that are resistant to high-temperature attack by glassy deposits. Acta Mater. 2007;55(20):6734–45.
23. Cao XQ, Vassen R, Tietz F, et al. New double-ceramic-layer thermal barrier coatings based on zirconia-rare earth composite oxides. J Eur Ceram Soc. 2006;26(3):247–51.
24. Xu H, Guo H, Liu F, et al. Development of gradient thermal barrier coatings and their hot-fatigue behavior. Surf Coat Technol. 2000;130(1):133–9.
25. Zhou X, Zou B, He L, et al. Hot corrosion behaviour of La_2 $(Zr_{0.7}Ce_{0.3})_2O_7$ thermal barrier coating ceramics exposed to molten calcium magnesium aluminosilicate at different temperatures. Corrosion Sci. 2015; 100:566–78.
26. Takahashi M. Thermal Barrier Coatings Design for Gas Turbines. In: Proceedings of ITSC'95, Kobe; 1995. p. 83–8.
27. Wu Z. Thermal spraying techniques and their applications. Beijing: China Machine Press; 2005. (In Chinese).
28. Khan AN, Lu J. Manipulation of air plasma spraying parameters for the production of ceramic coatings. J Mater Process Technol. 2009;209(5):2508–14.
29. Lima RS, Kucuk A, Berndt CC. Bimodal distribution of mechanical properties on plasma sprayed nanostructured partially stabilized zirconia. Mater Sci Eng, A. 2002;327(2):224–32.
30. Sommer E, Terry SG, Sigle W, et al. Metallic precipitate formation during alumina growth in a FeCrAl-based thermal barrier coating model system. In: Materials science forum, vol. 369. Trans Tech Publications Ltd; 2001. p. 671–78.
31. Wada K, Yamaguchi N, Matsubara H. Effect of substrate rotation on texture evolution in ZrO_2–4 mol.% Y_2O_3 layers fabricated by EB-PVD. Surf Coat Technol. 2005; 191(2–3):367–74.
32. Garcia JRV, Goto T. Thermal barrier coatings produced by chemical vapor deposition. Sci Technol Adv Mater. 2003;4(4):397–402.
33. Refke A, Hawley D, Doesburg J, et al. LPPS thin film technology for the application of TBC systems. In: International thermal spray conference. Basel: DVS-Verlag, Düsseldorf; 2005. p. 438–43.
34. Li C, Guo H, Gao L, et al. Microstructures of yttria-stabilized zirconia coatings by plasma spray-physical vapor deposition. J Therm Spray Technol. 2015;24(3):534–41.
35. Rezanka S, Mauer G, Vaßen R. Improved thermal cycling durability of thermal barrier coatings manufactured by PS-PVD. J Therm Spray Technol. 2014;23(1–2):182–9.
36. Curry N, VanEvery K, Snyder T, et al. Thermal conductivity analysis and lifetime testing of suspension plasma-sprayed thermal barrier coatings. Coatings. 2014;4(3):630–50.
37. James AS, Matthews A. Developments in rf plasma-assisted physical vapour deposition partially yttria-stabilized zirconia thermal barrier coatings. Surf Coat Technol. 1990;43:436–45.
38. Choy KL, Vyas JD. Processing and Structural Characterisation of Y_2O_3 -ZrO_2 Films Deposited Using Flame Assisted Vapour Deposition Technique. Key Eng Mater. 1999;161–163:653–6.

39. Fuhui XIE DW. History and current status of thermal barrier coating research. Mater Rev. 2002; 3.
40. Liu D. The urgent need of thermal barrier coatings for aeroengines. In: The 124th Shuangqing forum of the national natural science foundation of China, 2014. (In Chinese).
41. Wu RT, Osawa M, Yokokawa T, et al. Degradation mechanisms of an advanced jet engine service-retired TBC component. J Solid Mech Mater Eng. 2010;4(2):119–30.
42. Wellman RG, Nicholls JR. A review of the erosion of thermal barrier coatings. J Phys D Appl Phys. 2007;40(16):R293.
43. Fleck NA, Zisis T. The erosion of EB-PVD thermal barrier coatings: the competition between mechanisms. Wear. 2010;268(11–12):1214–24.
44. Evans AG, Fleck NA, Faulhaber S, et al. Scaling laws governing the erosion and impact resistance of thermal barrier coatings. Wear. 2006;260(7–8):886–94.
45. Chen X, He MY, Spitsberg I, et al. Mechanisms governing the high temperature erosion of thermal barrier coatings. Wear. 2004;256(7–8):735–46.
46. Wellman RG, Deakin MJ, Nicholls JR. The effect of TBC morphology on the erosion rate of EB PVD TBCs. Wear. 2005;258(1–4):349–56.
47. Yang L, Zhou Y, Qi S. Research progress in erosion mechanisms of thermal barrier coatings. Adv Mech. 2012;42(6):704–21. (In Chinese with English abstract).
48. Zhao H, Levi CG, Wadley HNG. Molten silicate interactions with thermal barrier coatings. Surf Coat Technol. 2014;251:74–86.
49. Yin B, Liu Z, Yang L, et al. Factors influencing the penetration depth of molten volcanic ash in thermal barrier coatings: Theoretical calculation and experimental testing. Results Phys. 2019;13: 102169.
50. Mohan P, Yao B, Patterson T, et al. Electrophoretically deposited alumina as protective overlay for thermal barrier coatings against CMAS degradation. Surf Coat Technol. 2009;204(6–7):797–801.
51. Drexler JM, Shinoda K, Ortiz AL, et al. Air-plasma-sprayed thermal barrier coatings that are resistant to high-temperature attack by glassy deposits. Acta Mater. 2010;58(20):6835–44.
52. Li W, Zhao H, Zhong X, et al. Air plasma-sprayed yttria and yttria-stabilized zirconia thermal barrier coatings subjected to calcium-magnesium-alumino-silicate (CMAS). J Therm Spray Technol. 2014;23(6):975–83.
53. Xu GN, Yang L, Zhou YC, et al. A chemo-thermo-mechanically constitutive theory for thermal barrier coatings under CMAS infiltration and corrosion. J Mech Phys Solids. 2019;133: 103710.
54. Vidal-Setif MH, Chellah N, Rio C, et al. Calcium-magnesium-alumino-silicate (CMAS) degradation of EB-PVD thermal barrier coatings: characterization of CMAS damage on ex-service high pressure blade TBCs. Surf Coat Technol. 2012;208:39–45.
55. Xia J, Yang L, Wu RT, et al. On the resistance of rare earth oxide-doped YSZ to high temperature volcanic ash attack. Surf Coat Technol. 2016;307:534–41.
56. Xia J, Yang L, Wu RT, et al. Degradation mechanisms of air plasma sprayed free-standing yttria-stabilized zirconia thermal barrier coatings exposed to volcanic ash. Appl Surf Sci. 2019;481:860–71.
57. Wu Y, Luo H, Cai C, et al. Comparison of CMAS corrosion and sintering induced microstructural characteristics of APS thermal barrier coatings. J Mater Sci Technol. 2019;35(3):440–7.
58. Zhao X, Wang X, Xiao P. Sintering and failure behaviour of EB-PVD thermal barrier coating after isothermal treatment. Surf Coat Technol. 2006;200(20–21):5946–55.
59. Renteria AF, Saruhan B. Effect of ageing on microstructure changes in EB-PVD manufactured standard PYSZ top coat of thermal barrier coatings. J Eur Ceram Soc. 2006;26(12):2249–55.
60. Guo S, Kagawa Y. Isothermal and cycle properties of EB-PVD yttria- partially-stabilized zirconia thermal barrier coatings at 1150 and 1300°C. Ceram Int. 2007;33(3):373–8.
61. Cutler RA, Reynolds JR, Jones A. Sintering and characterization of polycrystalline monoclinic, tetragonal, and cubic zirconia. J Am Ceram Soc. 1992;75(8):2173–83.
62. Müller J, Neuschütz D. Efficiency of α-alumina as diffusion barrier between bond coat and bulk material of gas turbine blades. Vacuum. 2003;71(1–2):247–51.

63. Liang T, Guo H, Peng H, et al. Precipitation phases in the nickel-based superalloy DZ 125 with YSZ/CoCrAlY thermal barrier coating. J Alloy Compd. 2011;509(34):8542–8.
64. Renusch D, Schorr M, Schütze M. The role that bond coat depletion of aluminum has on the lifetime of APS-TBC under oxidizing conditions. Mater Corros. 2008;59(7):547–55.
65. Hesnawi A, Li H, Zhou Z, et al. Isothermal oxidation behaviour of EB-PVD MCrAlY bond coat. Vacuum. 2007;81(8):947–52.
66. Yang F, Fang DN, Liu B. A theoretical model and phase field simulation on the evolution of interface roughness in the oxidation process. Modell Simul Mater Sci Eng. 2011;20(1): 015001.
67. Shen Q, Li SZ, Yang L, et al. Coupled mechanical-oxidation modeling during oxidation of thermal barrier coatings. Comput Mater Sci. 2018;154:538–46.
68. Yang F, Liu B, Fang D. Interplay between fracture and diffusion behaviors: modeling and phase field computation. Comput Mater Sci. 2011;50(9):2554–60.
69. Shen Q, Yang L, Zhou YC, et al. Models for predicting TGO growth to rough interface in TBCs. Surf Coat Technol. 2017;325:219–28.
70. Shen Q, Yang L, Zhou YC, et al. Effects of growth stress in finite-deformation thermally grown oxide on failure mechanism of thermal barrier coatings. Mech Mater. 2017;114:228–42.
71. Zhou Y, Liu Q, Yang L, Wu D, Mao W. Failure mechanisms and life prediction of thermal barrier coatings. Acta Mech Solida Sin. 2010;31(05):504–31. (In Chinese with English abstract).
72. Robinson RC. NASA GRC's high pressure burner rig facility and materials test capabilities. 1999. Dynacs Engineering Company, Inc., Cleveland, Ohio, Report No. NASA/CR—1999-209411.
73. Zhu DM, Miller RA, Kuczmarski MA. Development and life prediction of erosion resistant turbine low conductivity thermal barrier coatings. 2010. NASA Glenn Research Center; Cleveland, OH, Report No. NASA/TM—2010-215669.
74. Vaßen R, Kagawa Y, Subramanian R, et al. Testing and evaluation of thermal-barrier coatings. MRS Bull. 2012;37(10):911.
75. Wanhill RJH, Mom AJA, Hersbach HJC, et al. NLR experience with high velocity burner rig testing 1979–1989. High Temp Technol. 1989;7(4):202–11.
76. Bruce RW. Development of 1232°C (2250 F) erosion and impact tests for thermal barrier coatings. Tribol Trans. 1998;41(4):399–410.

Chapter 2
Basic Theoretical Frameworks for Thermo–Mechano-Chemical Coupling in TBCs

Rigorous and harsh service conditions in aeroengines result in complex thermo–mechano-chemical coupling failure problems in coatings. To provide an in-depth understanding of the failure mechanisms of thermal barrier coatings (TBCs) under various failure modes, this chapter first introduces the relevant concepts and basic assumptions of continuum mechanics and then establishes theoretical frameworks for thermo–mechano-chemical coupling based on the small- and large-deformation theories, respectively. This chapter comprises three parts. Part 1 gives a brief introduction to continuum mechanics. Part 2 focuses on stress and strain measures, equilibrium equations, and constitutive equations under the small-deformation theory; defines field variables descriptive of thermal diffusion, material infiltration, and chemical reaction processes; and derives constitutive equations for thermomechanical and thermo–mechano-chemical coupling and governing equations for various field variables from the perspective of the laws of thermodynamics. Part 3 introduces several concepts such as deformation and velocity gradients and then presents stress and strain measures as well as mass conservation and force equilibrium equations for large deformation, followed by the establishment of energy functions for thermomechanical coupling and thermo–mechano-chemical coupling as well as the derivation of governing equations for various field variables based on the laws of thermodynamics.

2.1 Continuum Mechanics

Mechanics is one of the seven basic natural sciences and is closely related to the eight major application-oriented basic disciplines. Forces are ubiquitous from the vast universe down to tiny fundamental particles. The disciplines of both solid mechanics and materials science have played pivotal roles in modern industry. The crossover between the two disciplines becomes increasingly prominent with scientific and technological advancement. A series of important concepts and methods, such as continuum mechanics, stress, strain, bifurcation, fracture toughness, and

Y. Zhou et al., *Thermal Barrier Coatings: Failure Theory and Evaluation Technology*,
https://doi.org/10.1007/978-981-19-2723-2_2

finite element methods, have been created in solid mechanics. These outstanding achievements not only have shaped modern and contemporary civil construction, machinery manufacturing, and aerospace industries but also have provided examples or basic theoretical foundations for an expansive range of natural sciences such as nonlinear sciences, solid earth geophysics, and materials science and engineering.

The dimensions of various objects in nature, those as small as atoms, molecules, and electrons and as large as the universe, cover a vast range from nanometers to light years. The "materials" of interest to humans can be basically categorized into four ranges, namely (a) subangstroms to angstroms (i.e., the physical microscale), (b) several tenths of a nanometer to tens of nanometers (i.e., the nanoscale), (c) submicrons to decimillimeters (i.e., the mesoscale), and (d) millimeters and larger (i.e., the macroscale). Corresponding to the macroscale, the scale between angstroms and decimillimeters can be collectively referred to as microscale, which is further divided into three levels (in descending order of size): the mesoscale, nanoscale, and physical microscale. With respect to physical models, physical microscale models take into consideration complex electron quantum effects. Generally, quantum mechanics is used to studying the basic properties of materials at this scale. At the nanoscale, a material can be simulated as an array of particles bound by interatomic potentials. Generally, statistical mechanics and molecular dynamics are used to investigate the basic properties of materials of this scale. Continuum mechanics based on Newtonian mechanics is generally employed, using continuum elements with intrinsic microstructures, to examine the basic properties of mesoscale materials. See Table 2.1 for a summary [1].

Continuum mechanics [2–4] is a subdiscipline of mechanics that studies the macromechanical properties of the so-called "continua", including solids and fluids. The macromechanical properties of a material refer to those governed by Newtonian mechanics in a 3D Euclidean space and under the condition that time passes uniformly. Continuum mechanics makes no assumptions about material structure. Continuum mechanics is not contradictory to material structure theory; instead, the two complement each other. Material structure theory studies the properties of materials with special structures, whereas continuum mechanics investigates the common properties of many materials with different structures. Continuum

Table 2.1 Behavior of and research methods for materials of various scales

Scale	Research content	Research method
Physical microscale (angstroms and smaller)	Electron quantum effects	Quantum mechanics
Nanoscale (several tenths of a nanometer to tens of nanometers)	Arrays of particles bound by interatomic potentials	Statistical mechanics and molecular dynamics
Mesoscale (submicrons to decimillimeters)	Continuum elements	Continuum mechanics
Macroscale (millimeters and larger)	Continuum elements	Continuum mechanics

mechanics primarily aims to establish mechanical models for various materials, to determine their constitutive relations in the form of mathematical expressions, and to solve problems under given initial and boundary conditions. Generally, continuum mechanics includes the following basic areas. ① Deformation geometry. This discipline is the study of the geometric properties of the deformation of continua, as well as the determination of the changes in the spatial location and direction of each part of an object resulting from deformation and the changes in the distances between neighboring points. Here, deformation geometry includes important concepts such as configurations, deformation gradients, strain tensors, the basic deformation theorem, and the polar decomposition theorem. ② Kinematics. This discipline is mainly the study of the time rates of various quantities in continuum mechanics. Here, kinematics includes important concepts such as velocity gradients, deformation rates, and rotational speeds. ③ Basic equations. These equations are established based on the law of conservation, which is applicable to all materials. Examples include continuity, motion, and energy equations and entropy inequality. ④ Constitutive equations. Unique constitutive equations for the properties of materials are required to determine their responses to external factors. These equations allow us to mathematically obtain closed systems of equations and to solve problems under certain initial and boundary conditions. Thus, describing the constitutive relations of material properties from physical or from mathematical perspectives is essential. ⑤ Special theories. This field includes theories such as elasticity, viscous fluid, plasticity, viscoelasticity, thermoelastic solid, and thermoviscous fluid theories. ⑥ Problem solving.

The "continuum assumption" is the most fundamental assumption of continuum mechanics. This assumption states that the space occupied by a true fluid or solid can be considered to be approximately continuously, seamlessly replete with "mass points," whose macrophysical quantities (e.g., mass, velocity, pressure, and temperature) obey all the necessary laws of physics, such as the law of conservation of mass, Newton's laws of motion, the law of conservation of energy, the laws of thermodynamics, and the laws of thermal conduction and Fick's laws of diffusion. This assumption ignores the specific microstructure of a material (solid and fluid microstructures are under the scope of condensed matter physics) and instead represents its macrophysical quantities with a set of partial differential equations. The so-called "mass points" refer to molecular clusters (also known as microclusters) that are sufficiently large at the microscale and, at the same time, sufficiently small at the macroscale. On the one hand, molecular clusters should be large enough compared to the scale of molecular motion such that they contain large numbers of molecules and that definite values can be obtained by statistically averaging them. On the other hand, the molecular clusters must be sufficiently small compared to the characteristic scale of the problem under investigation so that the average physical quantities of one cluster can be viewed as uniform and invariant; thus, geometrically, one cluster can be approximated as a point. In addition, the period of time selected for statistical averaging must be sufficiently long at the microscale and sufficiently short at the macroscale. In other words, a sufficiently long period of time, during which microscale behaviors (e.g., collisions between molecules) occur many times, must be selected so that the definite values of the microscale properties can be obtained through statistical averaging over this period of time. On the other hand, for statistical averaging, a period of time much shorter than the characteristic time of the problem under study must be selected so that the period of time selected for averaging can be regarded as an instant.

Solids have definite shapes under no external forces. Solids include nondeformable rigid bodies and deformable solids. Rigid bodies are subjects of study under the scope of dynamics and control engineering. Solid mechanics in continuum mechanics examines the patterns of change in deformable solids under the action of external factors (e.g., stresses and strains) and primarily covers elastic and plastic problems. Elasticity refers to a material's ability to recover its original shape after deformation under a stress. Plasticity refers to a material's inability to return to its original shape after deformation under a stress and the existence of permanent deformation.

TBCs are generally more than several hundred micrometers or several millimeters in thickness. In other words, TBCs are mesoscale or macroscale materials. Thus, this book does not discuss the physical microscale or nanoscale material properties of TBCs but instead investigates their basic material properties at the mesoscale or macroscale with continuum mechanics based on Newtonian mechanics. The following basic assumptions are made for studying TBCs at the mesoscale: (a) coating materials are continuous, (b) objects are completely homogeneous, and (c) objects are isotropic. An additional assumption is needed when studying the deformation or stress state of a coating using small-deformation theory. This assumption states that the displacement and deformation of an object after it is stressed are very small; in other words, the displacement at each point of the whole object is much smaller than the original size of the object.

2.2 Theoretical Framework for Thermo–Mechano-Chemical Coupling Based on Small Deformation

2.2.1 *Strain and Stress Measures Based on Small Deformation [5, 6]*

(1) **Strain measures**

The location of each mass point within an object changes (i.e., undergoes a displacement) under loading. If the relative locations of all the points within an object remain the same as their initial relative locations after undergoing displacements, these displacements are referred to as rigid-body displacements. If there are changes in the relative locations of all the points within an object compared to their initial relative locations after undergoing displacements, then there is a change in the shape of the object. In this case, the object undergoes deformation, including a volumetric change and a shape distortion. A strain analysis examines the deformation of an object under a force (i.e., its strain state) and does not consider its rigid displacement.

The displacement at any point within an object can be represented by its components in the *x*-, *y*-, and *z*-directions, which are denoted by *u*, *v*, and *w*, respectively. Thus, the deformation state of an object can be determined as long as the displacement at each point within it is obtained. The displacements at different points within an object generally differ. Thus, *u*, *v*, and *w* should be functions of coordinates, i.e.,

$$u = u(x,\ y,\ z),\ v = v(x,\ y,\ z),\ \text{and}\ w = w(x,\ y,\ z) \tag{2.1}$$

To depict the material deformation at any point *P* within an object, three mutually perpendicular micro-line segments along the *x*-, *y*-, and *z*-directions, *PA*, *PB*, and *PC*, respectively, which pass through point *P* and are d*x*, d*y*, and d*z* in length, respectively, are taken, as shown in Fig. 2.1. When the object deforms under the action of an external force, the lengths of and the angles between these three micro-line segments change. The change in the relative length of each micro-line segment is referred to as a normal strain at point *P*, which is denoted by ε. The normal strains of micro-line segments *PA*, *PB*, and *PC*, which are along the *x*-, *y*-, and *z*-directions, respectively, are denoted by ε_x, ε_y, and ε_z, respectively. The normal strains resulting from elongation and contraction are defined to be positive and negative, respectively. The change in the angle between two micro-line segments is referred to as a shear strain at point *P* and is denoted by γ. The change in the angle between micro-line segments *PA* and *PB*, which are along the *x*- and *y*-directions, respectively, is denoted by γ_{xy}. Similarly, the change in the angle between micro-line segments *PB* and *PC*, which are along the *y*- and *z*-directions, respectively, is denoted by γ_{yz}, and the change in the angle between micro-line segments *PC* and *PA*, which are along the *z*- and *x*-directions, respectively, is denoted by γ_{zx}. The shear strains resulting from the decrease and increase in the angle between micro-line segments are defined to be positive and negative, respectively.

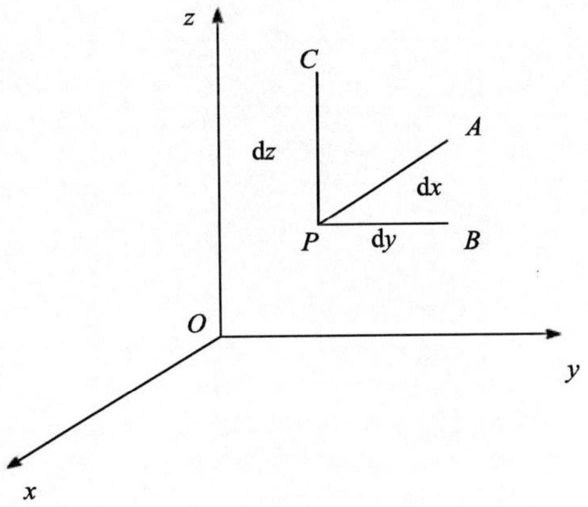

Fig. 2.1 Description of strain

We describe the extent of deformation of an object in a Cartesian coordinate system. For the convenience of description, let B and B' be the geometric states (i.e., configurations) before and after deformation, respectively, as shown in Fig. 2.2. Before the object deforms, the radius vectors of the end points $P(a_1, a_2, a_3)$ and $Q(a_1 + \mathrm{d}a_1, a_2 + \mathrm{d}a_2, a_3 + \mathrm{d}a_3)$ of any line element $\overrightarrow{PQ}$ are as follows:

$$\overrightarrow{OP} = \boldsymbol{a} = a_i \boldsymbol{e}_i \text{ and } \overrightarrow{OQ} = \boldsymbol{a} + \mathrm{d}\boldsymbol{a} = (a_i + \mathrm{d}a_i)\boldsymbol{e}_i \tag{2.2}$$

We use tensors to describe all physical quantities throughout the book. In Eq. (2.2), the subscript i indicates a summation from 1 to 3, i.e., $(a_i + \mathrm{d}a_i)\boldsymbol{e}_i = \sum_{i=1}^{3} (a_i + \mathrm{d}a_i)\boldsymbol{e}_i$. However, according to the Einstein summation convention, all summation symbols should be omitted. In addition, let us write both tensors and vectors in boldface. Thus, the line element $\overrightarrow{PQ}$ is expressed as follows:

$$\overrightarrow{PQ} = \overrightarrow{OQ} - \overrightarrow{OP} = \mathrm{d}\boldsymbol{a} = \mathrm{d}a_i \boldsymbol{e}_i \tag{2.3}$$

After the object deforms, points P and Q move to P' and Q', respectively, and their corresponding radius vectors are as follows:

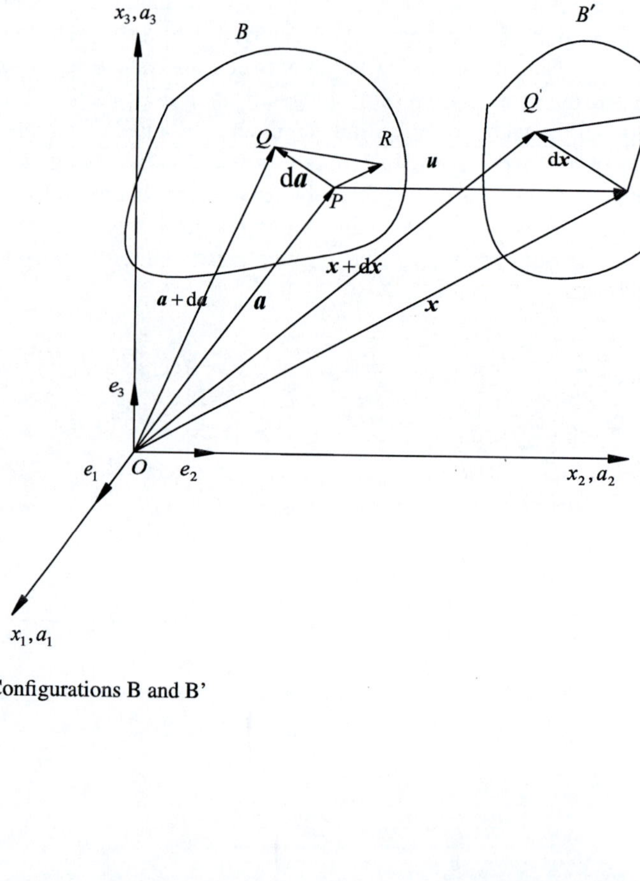

Fig. 2.2 Configurations B and B'

$$\overrightarrow{OP'} = \boldsymbol{x} = x_i \boldsymbol{e}_i \text{ and } \overrightarrow{OQ'} = \boldsymbol{x} + \mathrm{d}\boldsymbol{x} = (x_i + \mathrm{d}x_i)\boldsymbol{e}_i \tag{2.4}$$

where $\boldsymbol{x}$ should be a function of a_i, i.e., $x_m = x_m(a_i)$. Thus, the line segment after deformation is expressed as follows:

$$\overrightarrow{P'Q'} = \overrightarrow{OQ'} - \overrightarrow{OP'} = \mathrm{d}\boldsymbol{x} = \mathrm{d}x_i \boldsymbol{e}_i = \mathrm{d}x_1 \boldsymbol{e}_1 + \mathrm{d}x_2 \boldsymbol{e}_2 + \mathrm{d}x_3 \boldsymbol{e}_3 \tag{2.5}$$

The squares of the lengths of $\overrightarrow{PQ}$ and $\overrightarrow{P'Q'}$ are as follows:

$$\mathrm{d}s_0^2 = \mathrm{d}\boldsymbol{a} \cdot \mathrm{d}\boldsymbol{a} = \mathrm{d}a_i \mathrm{d}a_i = \delta_{ij} \mathrm{d}a_i \mathrm{d}a_j = \mathrm{d}a_1^2 + \mathrm{d}a_2^2 + \mathrm{d}a_3^2 \tag{2.6}$$

$$\mathrm{d}s^2 = \mathrm{d}\boldsymbol{x} \cdot \mathrm{d}\boldsymbol{x} = \mathrm{d}x_m \mathrm{d}x_m = \mathrm{d}x_1^2 + \mathrm{d}x_2^3 + \mathrm{d}x_3^2 \tag{2.7}$$

As $\boldsymbol{x}$ is a function of a_i, i.e., $x_m = x_m(a_i)$, we have

$$\mathrm{d}x_m = \frac{\partial x_m}{\partial a_i} \mathrm{d}a_i = \frac{\partial x_m}{\partial a_1} \mathrm{d}a_1 + \frac{\partial x_m}{\partial a_2} \mathrm{d}a_2 + \frac{\partial x_m}{\partial a_3} \mathrm{d}a_3 \tag{2.8}$$

By substituting Eq. (2.8) into Eq. (2.7), we have

$$\mathrm{d}s^2 = \frac{\partial x_m}{\partial a_i} \frac{\partial x_m}{\partial a_j} \mathrm{d}a_i \mathrm{d}a_j \tag{2.9}$$

By subtracting Eq. (2.6) from Eq. (2.9), we obtain the change in the square of the length of the line segment after deformation:

$$\mathrm{d}s^2 - \mathrm{d}s_0^2 = 2E_{ij} \mathrm{d}a_i \mathrm{d}a_j \tag{2.10}$$

where

$$E_{ij} = \frac{1}{2}\left(\frac{\partial x_m}{\partial a_i} \frac{\partial x_m}{\partial a_j} - \delta_{ij} \right) \tag{2.11}$$

Clearly, as shown in Fig. 2.2, $x_m(a_i) = a_m + u_m(a_i)$. By taking the derivative of this expression with respect to a_i, we have

$$\frac{\partial x_m}{\partial a_i} = \delta_{mi} + \frac{\partial u_m}{\partial a_i} \tag{2.12}$$

In the case of small deformation, the first derivatives of the displacement components are much smaller than 1, i.e.,

$$\left|\frac{\partial u_i}{\partial a_j}\right| << 1, \text{ and } \left|\frac{\partial u_j}{\partial x_j}\right| << 1, \tag{2.13}$$

By neglecting higher-order small quantities, we have

$$\frac{\partial u_i}{\partial a_j} = \frac{\partial u_i}{\partial x_k}\frac{\partial x_k}{\partial a_j} = \frac{\partial u_i}{\partial x_k}\left(\delta_{kj} + \frac{\partial u_k}{\partial a_j}\right) \approx \frac{\partial u_i}{\partial x_k}\delta_{kj} = \frac{\partial u_i}{\partial x_j} \tag{2.14}$$

Equation (2.12) is used to derive Eq. (2.14). Thus, when the small deformation of an object is described, there is no need to differentiate coordinates a_i and x_i. Hence, in the case of small deformation, Eq. (2.11) can be simplified as follows:

$$E_{ij} \approx \varepsilon_{ij} = \frac{1}{2}\left(\frac{\partial u_i}{\partial x_j} + \frac{\partial u_j}{\partial x_i}\right) \tag{2.15}$$

where ε_{ij} is referred to as the component of the Cauchy strain tensor or small-strain tensor. ε is a second-order symmetric tensor and has only six independent components, which, in a Cartesian coordinate system, are commonly written in the following forms:

$$\begin{gathered}
\varepsilon_{11} = \frac{\partial u_1}{\partial x_1}, \varepsilon_{12} = \varepsilon_{21} = \frac{1}{2}\left(\frac{\partial u_1}{\partial x_2} + \frac{\partial u_2}{\partial x_1}\right) \\
\varepsilon_{22} = \frac{\partial u_2}{\partial x_2}, \varepsilon_{23} = \varepsilon_{32} = \frac{1}{2}\left(\frac{\partial u_2}{\partial x_3} + \frac{\partial u_3}{\partial x_2}\right) \\
\varepsilon_{33} = \frac{\partial u_3}{\partial x_3} \text{ and } \varepsilon_{31} = \varepsilon_{13} = \frac{1}{2}\left(\frac{\partial u_3}{\partial x_1} + \frac{\partial u_1}{\partial x_3}\right)
\end{gathered} \tag{2.16}$$

The above is a set of linear differential equations referred to as strain–displacement equations or geometric equations. According to Eq. (2.16), the strain components can be determined by calculating the derivatives of the displacement equations, and the displacement components can be obtained by integrating the strain components.

Now, let us use strain tensor ε to determine the changes in the lengths of the line elements and the angles between the line elements after deformation. First, let us analyze the change in the length of a line element. The unit vector in the direction of line element $\overrightarrow{PQ}$ before deformation is given by

$$\boldsymbol{\nu} = \frac{\mathrm{d}\boldsymbol{a}}{\mathrm{d}s_0} = \frac{\mathrm{d}a_i}{\mathrm{d}s_0}\boldsymbol{e}_i = \nu_i\boldsymbol{e}_i \tag{2.17}$$

where $\nu_i = \frac{\mathrm{d}a_i}{\mathrm{d}s_0}$ is the direction cosine of line element $\overrightarrow{PQ}$. Here, we define $\lambda_\nu = \frac{\mathrm{d}s}{\mathrm{d}s_0}$ as the change in the length of a line segment after deformation and refer to it as the length ratio. Thus, based on Eqs. (2.10), (2.15), and (2.17), we have

$$\lambda_v = \frac{\mathrm{d}s}{\mathrm{d}s_0} = \left(1 + 2\varepsilon_{ij} v_i v_j\right)^{1/2} \approx 1 + \varepsilon_{ij} v_i v_j \tag{2.18}$$

The engineering normal strain ε_v of a line element in the $\boldsymbol{v}$ direction is often defined as the relative change in the length of the line element after deformation, i.e.,

$$\varepsilon_v = \frac{\mathrm{d}s - \mathrm{d}s_0}{\mathrm{d}s_0} = \lambda_v - 1 \tag{2.19}$$

Substitution of Eq. (2.18) into Eq. (2.19) yields

$$\varepsilon_v = \varepsilon_{ij} v_i v_j \tag{2.20}$$

which has the following expanded form:

$$\varepsilon_v = \varepsilon_{11} v_1 v_1 + \varepsilon_{22} v_2 v_2 + \varepsilon_{33} v_3 v_3 + 2\varepsilon_{12} v_1 v_2 + 2\varepsilon_{23} v_2 v_3 + 2\varepsilon_{31} v_3 v_1 \tag{2.21}$$

When $\boldsymbol{v} = \boldsymbol{e}_i$ ($i = 1, 2, 3$), based on Eq. (2.21), we have

$$\varepsilon_x = \varepsilon_{11} \quad \varepsilon_y = \varepsilon_{22} \quad \varepsilon_z = \varepsilon_{33} \tag{2.22}$$

Thus, the three diagonal components of strain tensor ε_{ij} are equal to the engineering normal strains of the three line elements along the coordinate directions, respectively. The engineering normal strains resulting from elongation and contraction are defined as positive and negative, respectively.

Now, let us discuss the directions of line elements. The unit vector of the line element $\overrightarrow{P'Q'}$ after deformation is given by

$$\boldsymbol{v}' = \frac{\mathrm{d}\boldsymbol{x}}{\mathrm{d}s} = \frac{\mathrm{d}x_i}{\mathrm{d}s}\boldsymbol{e}_i = v_i' \boldsymbol{e}_i \tag{2.23}$$

Its direction cosine is derived as follows:

$$v_i' = \frac{\mathrm{d}x_i}{\mathrm{d}s} = \frac{\partial x_i}{\partial a_j}\frac{\mathrm{d}a_j}{\mathrm{d}s_0}\frac{\mathrm{d}s_0}{\mathrm{d}s} = \frac{\partial x_i}{\partial a_j} v_j \frac{1}{\lambda_v} \tag{2.24}$$

Based on Eq. (2.12), the direction cosine of any line element after deformation can be expressed using displacement as follows:

$$v_i' = \left(\delta_{ji} + \frac{\partial u_i}{\partial a_j}\right) v_j \frac{1}{\lambda_v} \tag{2.25}$$

Based on Eqs. (2.18) and (2.20) and by neglecting the second-order small quantities, we have

$$\frac{1}{\lambda_\nu} = \frac{1}{1+\varepsilon_\nu} \approx 1 - \varepsilon_\nu \tag{2.26}$$

By substituting Eq. (2.26) into Eq. (2.24) and by neglecting the second-order small quantities, the direction cosine of the line segment after deformation is obtained:

$$\nu'_i = \nu_i + \frac{\partial u_i}{\partial a_j}\nu_j - \nu_i\varepsilon_\nu \tag{2.27}$$

According to the above equation, the direction cosine of a line element after deformation can be calculated based on the displacement gradient component $\frac{\partial u_i}{\partial a_j}$ and the normal strain ε_ν of the line element. For example, for a line element parallel to coordinate axis a_1 before deformation, its unit vector and direction cosines are as follows:

$$\boldsymbol{\nu} = \boldsymbol{e}_1 \text{ or } \nu_1 = 1, \qquad \nu_2 = \nu_3 = 0 \tag{2.28}$$

By substituting Eq. (2.28) into Eq. (2.20), we have

$$\varepsilon_\nu = \varepsilon_{ij}\nu_i\nu_j = \varepsilon_{11} \tag{2.29}$$

Based on Eq. (2.27), the direction cosines of the line element after deformation are as follows:

$$\nu'_1 = 1 + \frac{\partial u_1}{\partial a_1} - \varepsilon_{11} \approx 1, \ \ \nu'_2 = \frac{\partial u_2}{\partial a_1}, \quad \text{and} \quad \nu'_3 = \frac{\partial u_3}{\partial a_1} \tag{2.30}$$

The sum of squares of these three components is not strictly equal to 1, but the difference is only a second-order small quantity in the case of small deformation, which is allowable. Thus, the unit vector after deformation is given by

$$\boldsymbol{e'}_1 = \boldsymbol{e}_1 + \frac{\partial u_2}{\partial a_1}\boldsymbol{e}_2 + \frac{\partial u_3}{\partial a_1}\boldsymbol{e}_3 \tag{2.31}$$

Let $\frac{\pi}{2} - \theta_2$ be the angle between $\boldsymbol{e'}_1$ and $\boldsymbol{e}_2$. Then, we have

$$\cos\left(\frac{\pi}{2} - \theta_2\right) = \boldsymbol{e'}_1 \cdot \boldsymbol{e}_2 = \frac{\partial u_2}{\partial a_1} \tag{2.32}$$

When θ_2 is very small,

$$\cos\left(\frac{\pi}{2} - \theta_2\right) = \sin\theta_2 \approx \theta_2 \approx \frac{\partial u_2}{\partial a_1} \tag{2.33}$$

Similarly,

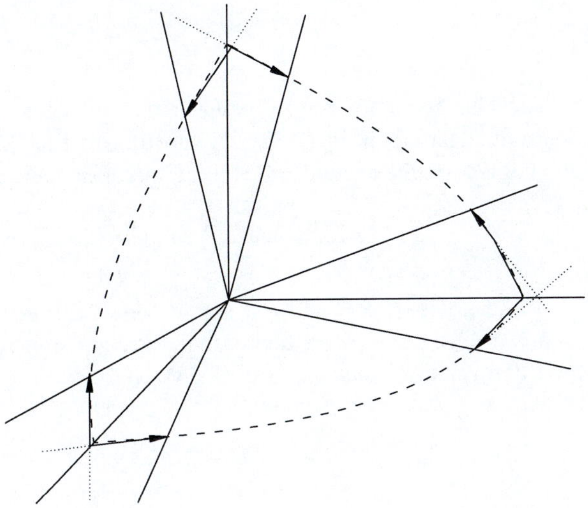

Fig. 2.3 Rotations of line elements after deformation

$$\cos\left(\frac{\pi}{2} - \theta_3\right) = e'_1 \cdot e_3 = \frac{\partial u_3}{\partial a_1} \approx \theta_3 \tag{2.34}$$

Equations (2.33) and (2.34) demonstrate that a line element perpendicular to axes a_2 and a_3 before deformation rotates toward axes a_2 and a_3 by $\frac{\partial u_2}{\partial a_1}$ and $\frac{\partial u_3}{\partial a_1}$, respectively, after deformation. Similarly, line elements along axes a_2 and a_3 rotate after deformation. See Fig. 2.3 for the angles and directions of rotations of these line elements. Figure 2.3 visualizes the rotations of the three line elements along different coordinate axes after deformation.

Finally, let us discuss the changes in the angles between line elements. We consider the two arbitrary line elements in Fig. 2.2, $\overrightarrow{PQ}$ and $\overrightarrow{PR}$. Their unit vectors are $\boldsymbol{v}$ and $\boldsymbol{t}$, respectively, and their direction cosines are v_i and t_i, respectively. The cosine of the angle between $\overrightarrow{PQ}$ and $\overrightarrow{PR}$ is given by

$$\cos(\boldsymbol{v}, \boldsymbol{t}) = \boldsymbol{v} \cdot \boldsymbol{t} = v_i t_i \tag{2.35}$$

After deformation, the two line elements change to $\overrightarrow{P'Q'}$ and $\overrightarrow{P'R'}$, respectively, which have unit vectors of $\boldsymbol{v}'$ and $\boldsymbol{t}'$, respectively, and direction cosines of v'_i and t'_i, respectively. Based on Eq. (2.30), the cosine of the angle between $\overrightarrow{P'Q'}$ and $\overrightarrow{P'R'}$ is given by

$$\cos(\boldsymbol{v}', \boldsymbol{t}') = \boldsymbol{v}' \cdot \boldsymbol{t}' = v'_i t'_i = \left(\delta_{mn} + \frac{\partial u_n}{\partial a_m} + \frac{\partial u_m}{\partial a_n} + \frac{\partial u_i}{\partial a_m}\frac{\partial u_i}{\partial a_n}\right) v_m t_n \frac{1}{\lambda_v}\frac{1}{\lambda_t} \tag{2.36}$$

Based on Eqs. (2.11) and (2.15), Eq. (2.36) is rewritten as follows:

$$\cos(\boldsymbol{\nu}', \boldsymbol{t}') = (\nu_m t_m + 2\varepsilon_{mn}\nu_m t_n)\frac{1}{\lambda_\nu \lambda_t} \tag{2.37}$$

The change in the angle between the line elements after deformation can be determined based on Eq. (2.37). By substituting Eqs. (2.26) and (2.15) into Eq. (2.37) and neglecting the second-order small quantities, we can simplify Eq. (2.37) into

$$\boldsymbol{\nu}' \cdot \boldsymbol{t}' = \cos(\boldsymbol{\nu}', \boldsymbol{t}') = (1 - \varepsilon_\nu - \varepsilon_t)\boldsymbol{\nu} \cdot \boldsymbol{t} + 2\nu_i \varepsilon_{ij} t_j \tag{2.38}$$

If line elements $\overrightarrow{PQ}$ and $\overrightarrow{PR}$ are perpendicular to each other before deformation, then $\boldsymbol{\nu} \cdot \boldsymbol{t} = 0$. Let θ be the decrease in the right angle between the two line elements after deformation. Based on Eq. (2.38), we have

$$\theta \approx \cos\left(\frac{\pi}{2} - \theta\right) = \cos(\boldsymbol{\nu}', \boldsymbol{t}') = 2\varepsilon_{ij}\nu_i t_j = 2\varepsilon_{\nu t} \tag{2.39}$$

The decrease in the right angle between two orthogonal line elements is generally defined as the engineering shear strain $\gamma_{\nu t}$, i.e.,

$$\gamma_{\nu t} = 2\varepsilon_{\nu t} = 2\varepsilon_{ij}\nu_i t_j \tag{2.40}$$

If $\boldsymbol{\nu}$ and $\boldsymbol{t}$ are unit vectors along the coordinate axes (e.g., $\nu_i = 1, t_j = 1 (i \neq j)$) and all the other direction cosines are zero, based on Eq. (2.40), we have

$$\gamma_{ij} = 2\varepsilon_{ij}, i \neq j \tag{2.41}$$

As the angle between $\boldsymbol{\nu}$ and $\boldsymbol{t}$ is the same as that between $\boldsymbol{t}$ and $\boldsymbol{\nu}$, we have

$$\varepsilon_{ij} = \varepsilon_{ji} \tag{2.42}$$

In other words, a strain tensor is symmetric and has only six independent components. According to the linear algebra theory, the matrix of these six symmetric strain components can be diagonalized. The three eigenvalues of the diagonalized matrix are referred to as principal strains that are known as, in descending order of magnitude, the first principal strain ε_1, the second principal strain ε_2, and the third principal strain ε_3.

Based on the above discussion, the six components ε_{ij} of a strain tensor have the following geometric meaning. When $i = j$, ε_{ij} is the normal strain of the line element along coordinate axis i. The normal strains resulting from elongation and contraction are positive and negative, respectively. When $i \neq j$, the shear strain between the orthogonal line elements along coordinate axes i and j is two times ε_{ij}. A shear strain resulting from sharpening (a decrease in the right angle) is positive, whereas a shear strain resulting from blunting (an increase in the right angle) is negative.

Based on Eqs. (2.20) and (2.23), Eqs. (2.37) and (2.39), and Eqs. (2.40) and (2.41), the strain tensor $\boldsymbol{\varepsilon}$ gives all the information regarding the deformation state of an object.

(2) **Stress measures**

External loads acting on objects can be categorized into volume and surface forces. Volume forces (e.g., gravitational and magnetic forces as well as, for moving objects, inertial forces) are distributed within the volumes of objects and are characterized by their direct proportion to the masses of objects. The force conditions generally vary from point to point within an object. To describe the volume force acting at a certain point *P* within an object, a small part of the object at point *P* is selected. This part includes point *P* and has a volume of ΔV, as shown in Fig. 1.4a.

Let $\Delta \boldsymbol{Q}$ be the volume force acting on ΔV. Thus, the average intensity of $\Delta \boldsymbol{Q}$ is $\Delta \boldsymbol{Q}/\Delta V$. Continuously reducing the selected small part of the object (i.e., continuously reducing ΔV) leads to continuous changes in both $\Delta \boldsymbol{Q}$ and $\Delta \boldsymbol{Q}/\Delta V$ (in terms of direction and magnitude) as well as the point of action. Now, let us infinitely reduce ΔV to allow it to approach point *P* and assume that $\Delta \boldsymbol{Q}$ is continuously distributed. Thus, $\Delta \boldsymbol{Q}/\Delta V$ approaches a certain limit $\boldsymbol{f}$, i.e.,

$$\lim_{\Delta V \to 0} \frac{\Delta \boldsymbol{Q}}{\Delta V} = \boldsymbol{f} \tag{2.43}$$

The limit vector $\boldsymbol{f} = f_i \boldsymbol{e}_i$ is the intensity of the volume force acting on the object at point *P*. As ΔV is a scalar, the direction of $\boldsymbol{f}$ is the limit direction of $\Delta \boldsymbol{Q}$. The projections of vector $\boldsymbol{f}$ onto the *x*-, *y*-, and *z*-coordinate axes, f_1, f_2, and f_3, respectively, are referred to as the volume-force components acting on the object at point *P*. The volume-force components along the positive and negative directions of a coordinate axis are positive and negative, respectively. The dimension of these components is [force][length]$^{-3}$.

Surface forces (e.g., fluid pressures and contact forces) are distributed on the surfaces of objects. Similarly, the surface force acting on the surface of an object generally varies from point to point. To depict the surface force acting on the surface of an object at a certain point *P*, a small part of the surface of the object at point *P* is selected. This small part contains point *P* and has an area of ΔS, as shown in Fig. 2.4b. Let $\Delta \boldsymbol{Q}$ be the surface force acting on ΔS. Thus, the average intensity of $\Delta \boldsymbol{Q}$ is $\Delta \boldsymbol{Q}/\Delta S$. Similar to the discussion on volume forces, let us infinitely reduce ΔS to allow it to approach point *P* and assume that $\Delta \boldsymbol{Q}$ is continuously distributed. Thus, $\Delta \boldsymbol{Q}/\Delta S$ approaches a certain limit $\boldsymbol{t}$, i.e.,

$$\lim_{\Delta S \to 0} \frac{\Delta \boldsymbol{Q}}{\Delta S} = \boldsymbol{t} \tag{2.44}$$

The limit vector $\boldsymbol{t} = t_i \boldsymbol{e}_i$ is the intensity of the surface force acting on the object at point *P*. As ΔS is a scalar, the direction of $\boldsymbol{t}$ is the limit direction of $\Delta \boldsymbol{Q}$. The projections of vector $\boldsymbol{t}$ onto the *x*-, *y*-, and *z*-coordinate axes, t_x, t_y, and t_z, respectively,

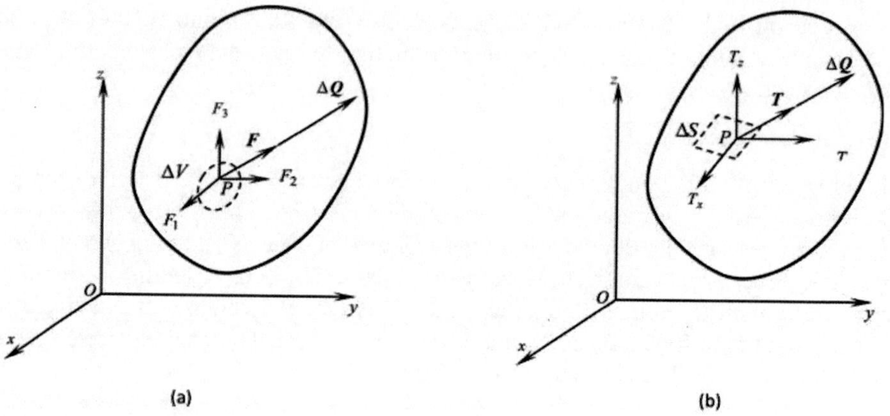

Fig. 2.4 External forces

are referred to as the surface-force components acting on the object at point P. The surface-force components along the positive and negative directions of a coordinate axis are positive and negative, respectively. The dimension of these components is [force][length]$^{-2}$.

An object undergoes deformation under the action of an external force. This deformation alters the intermolecular distances and forms an additional internal-force field within the object. If this internal-force field is sufficient to reach an equilibrium with the external force, deformation ceases, and the object reaches a stable equilibrium state. Now, let us discuss this additional internal-force field caused by external loading.

To accurately describe internal-force fields, Cauchy introduced an important concept—stress. Now, let us consider an object B, which is in an equilibrium state, in Fig. 2.5. An imaginary closed curved surface S is used to divide this object into an inner part and an outer part, which are referred to as the inner domain and the outer domain, respectively. P is an arbitrary point on the curved surface S. A surface element with an area of ΔS on S is selected with P as the centroid. Let $\boldsymbol{\nu}$ be the unit vector along the outer normal of the inner domain at point P (thus, $-\boldsymbol{\nu}$ is the unit vector along the outer normal of the outer domain at point P). Let $\Delta \boldsymbol{F}$ be the resultant of the forces from the outer domain acting on the inner domain via surface element ΔS. Generally, the direction of $\Delta \boldsymbol{F}$ differs from that of normal vector $\boldsymbol{\nu}$. Let us assume that as the surface element approaches point P and $\Delta S \to 0$, a limit value of $\Delta \boldsymbol{F}/\Delta S$ exists, and that the ratio of the resultant moment of the forces acting on the surface element to ΔS tends to zero. Thus, we define

$$\boldsymbol{\sigma}_\nu = \lim_{\Delta S \to 0} \frac{\Delta \boldsymbol{F}}{\Delta S} \tag{2.45}$$

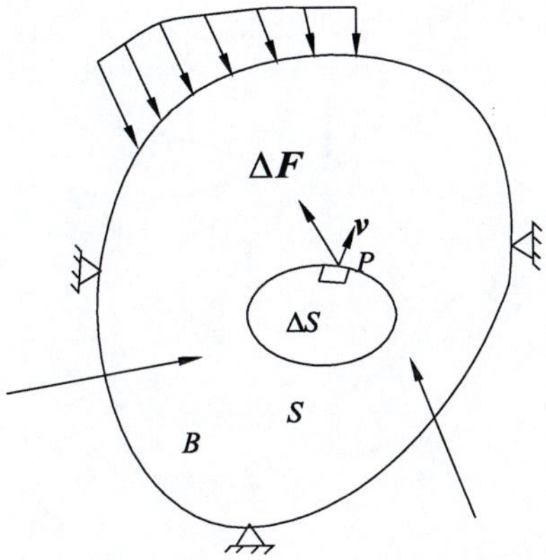

Fig. 2.5 Stress vector

as the stress vector acting from the outer domain on the surface element with a normal of $\boldsymbol{\nu}$ at point P in the inner domain or the stress vector from the outer domain to which the surface element with a normal of $\boldsymbol{\nu}$ at point P in the inner domain is subjected. If ΔS in Eq. (2.45) is set as the initial area of the surface element before deformation, then Eq. (2.45) gives the engineering stress (also known as nominal stress), which is often used in cases of small deformation. For large-deformation problems, ΔS should be set as the actual area of the surface element after deformation. In this case, Eq. (2.45) gives the true stress. This section discusses only cases of small deformation, i.e., cases in which objects sustain small changes in shape after deformation. See Sect. 2.3 or relevant textbooks [7–10] for cases of large deformation.

A comparison of Eqs. (2.44) and (2.45) shows that stress vector and surface-force vector are the same in terms of mathematical definition and physical dimension. The only difference is that stress is an unknown internal force acting on a cross-section of an object, whereas a surface force is a known external force acting on the outer surface of an object. As the internal cross-section infinitely approaches the outer surface, the value of the stress approaches that of the external surface force. The magnitude and direction of vector $\boldsymbol{\sigma}_\nu$ are related not only to the location of point P but also to the normal direction $\boldsymbol{\nu}$ of the surface element. Stress vectors acting on different normal surface elements at the same point differ, as shown in Fig. 2.6a. Conversely, as long as the surface elements on different curved surfaces pass through the same point and are in the same normal direction, the stress vectors acting on them are the same, as shown in Fig. 2.6b. Thus, stress vector $\boldsymbol{\sigma}_\nu$ is a function of location $\boldsymbol{r}$ and the orientation $\boldsymbol{\nu}$ of a certain surface passing through point P, i.e.,

$$\boldsymbol{\sigma}_\nu = \boldsymbol{\sigma}_\nu(\boldsymbol{r},\ \boldsymbol{\nu}) \tag{2.46}$$

Evidently, as long as the stress vector at the cross-section of any orientation that passes through point P is known, the stress state at point P can be determined. However, there are an infinite number of cross-sections of different orientations that

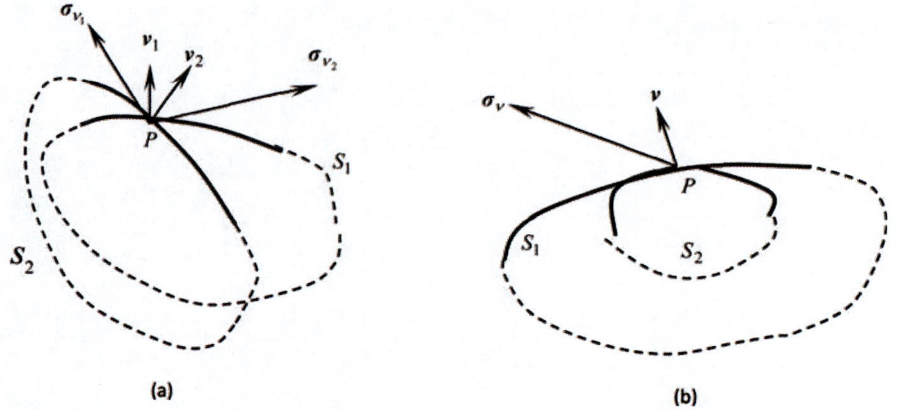

Fig. 2.6 Dependence of a stress vector on the normal direction ν

pass through point P, and it is impossible to consider them one by one. How can the stress state at a point be determined?

(3) Force equilibrium equations

Theoretically, the description of stress using Eq. (2.23) is sufficient. The use of Eq. (2.46) alone would clearly result in inconvenience since "there are an infinite number of cross-sections in different orientations that pass through point P, and it is impossible to consider them one by one." As mentioned in the previous section, the use of a coordinate system (e.g., a Cartesian coordinate system) is the best approach for discussing the stress state at any point P within an object. In a Cartesian coordinate system, let us select a regular hexahedral element within the neighborhood of point P with six cross-sections parallel to the coordinate planes (normal sections for short), as shown in Fig. 2.7. The three surface elements with outer normals in the same directions as the respective coordinate axes $x_i\,(i = 1, 2, 3)$ are referred to as positive surfaces, denoted by $\mathrm{d}S_i$. The unit normal vector of each of these surfaces is $\boldsymbol{\nu}_i = \boldsymbol{e}_i$, where $\boldsymbol{e}_i$ is the unit vector along the coordinate axis. The other three surface elements with outer normals opposite to the directions of the respective coordinate axes are referred to as negative surfaces. The unit normal vector of each of these surfaces is $-\boldsymbol{e}_i$. By decomposing the stress vector $\boldsymbol{\sigma}_{(i)}(i = 1, 2, 3)$ acting on each positive surface $\mathrm{d}S_i$ along the positive direction of the corresponding coordinate axis, we have.

$$
\begin{aligned}
\sigma_{(1)} &= \sigma_{11}\boldsymbol{e}_1 + \sigma_{12}\boldsymbol{e}_2 + \sigma_{13}\boldsymbol{e}_3 = \sigma_{1j}\boldsymbol{e}_j \\
\boldsymbol{\sigma}_{(2)} &= \sigma_{21}\boldsymbol{e}_1 + \sigma_{22}\boldsymbol{e}_2 + \sigma_{23}\boldsymbol{e}_3 = \sigma_{2j}\boldsymbol{e}_j \\
\boldsymbol{\sigma}_{(3)} &= \sigma_{31}\boldsymbol{e}_1 + \sigma_{32}\boldsymbol{e}_2 + \sigma_{33}\boldsymbol{e}_3 = \sigma_{3j}\boldsymbol{e}_j
\end{aligned} \tag{2.47}
$$

i.e.,

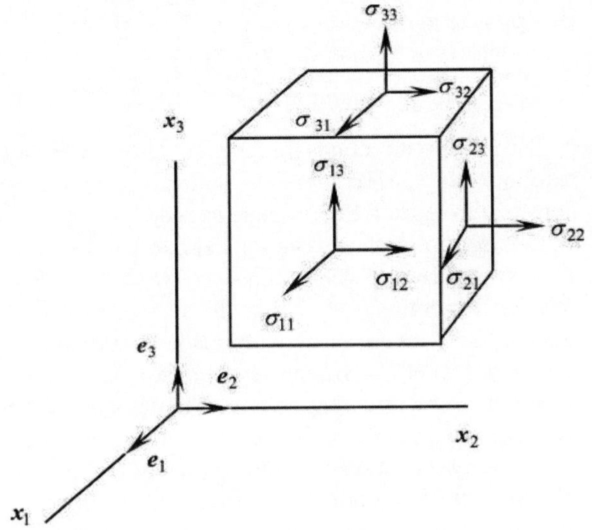

Fig. 2.7 Stress components in a rectangular coordinate system

$$\boldsymbol{\sigma}_{(i)} = \sigma_{ij}\boldsymbol{e}_j \tag{2.48}$$

The above equation contains a total of nine stress components, which can be given in the form of a matrix as follows:

$$(\sigma_{ij}) = \begin{pmatrix} \sigma_{11} & \sigma_{12} & \sigma_{13} \\ \sigma_{21} & \sigma_{22} & \sigma_{23} \\ \sigma_{31} & \sigma_{32} & \sigma_{33} \end{pmatrix} \tag{2.49}$$

The first index, i, indicates the normal direction of the surface element and is referred to as the surface-element index. The second index, j, indicates the decomposition direction of the stress and is referred to as the direction index. When $i = j$, the stress component is perpendicular to the surface element and is referred to as the normal stress. When $i \neq j$, the stress component acts within the surface-element plane and is referred to as the shear stress. The nine stress components in a Cartesian coordinate system are given by

$$(\sigma_{ij}) = \begin{pmatrix} \sigma_x & \tau_{xy} & \tau_{xz} \\ \tau_{yx} & \sigma_y & \tau_{yz} \\ \tau_{zx} & \tau_{zy} & \sigma_z \end{pmatrix} \tag{2.50}$$

The theory of elasticity stipulates that a stress vector $\boldsymbol{\sigma}_{(-i)}$ $(i = 1,\ 2,\ 3)$ acting on a negative surface should be decomposed along the negative direction of the coordinate axis and that when a microelement contracts toward its centroid to one point, the stresses on the negative and positive surfaces are the same in magnitude but opposite in direction, i.e.,

$$\boldsymbol{\sigma}_{(-i)} = -\boldsymbol{\sigma}_{(i)} = \sigma_{ij}(-\boldsymbol{e}_j) \tag{2.51}$$

The positive directions for the nine stress components σ_{ij} in Eq. (2.50) are defined as follows: the stress components acting on the positive surfaces and in the same directions as the respective coordinate axes are positive, and the stress components acting on the negative surfaces and opposite to the directions of the respective coordinate axes are positive. This definition correctly reflects the principle of action and reaction and the conventional view that “tension is positive, and compression is negative,” and facilitates a consistent mathematical treatment. However, note that the positive direction defined for shear stresses differs from that stipulated in the mechanics of materials. Stresses acting on any oblique plane that passes through point P can be expressed by σ_{ij}. Thus, the stress state at a point cannot be described with either one quantity (i.e., a scalar) or three quantities (i.e., a vector). Instead, nine quantities (i.e., nine stress components σ_{ij}) are required to comprehensively describe the stress state at a point. Based on the concept of tensors mentioned in the previous section, a second-order tensor $\boldsymbol{\sigma} = \sigma_{ij}\boldsymbol{e}_i\boldsymbol{e}_j$ can be used to describe the stress state at a point. Particular attention is warranted to preclude confusion between a stress vector on an oblique plane with a normal direction of $\boldsymbol{v}$ in Eq. (2.45) or (2.46) and a second-order stress tensor $\sigma = \sigma_{\mathrm{ij}}\boldsymbol{e}_{\mathrm{i}}\boldsymbol{e}_{\mathrm{j}}$.

This section is concerned only with the effect (i.e., deformation state) within materials under external loading in cases of small deformation. Thus, an object, in its entirety, should be in an equilibrium state under external loading. Now, let us discuss the static equilibrium problem of element bodies. Apart from the stress components acting on each of its surfaces, the selected element body is subjected to a volume force $\boldsymbol{F}$. Understanding the static equilibrium of an element body means determining the equilibrium conditions for the forces acting on it along the three coordinate axes as well as the equilibrium conditions for the moments of the forces about the three coordinate axes.

Here, Cartesian coordinates are selected as reference coordinates. An infinitesimal regular hexahedron with side lengths $\mathrm{d}x_1$, $\mathrm{d}x_2$, and $\mathrm{d}x_3$ in the neighborhood of an arbitrary point P is selected and referred to as a microelement for short (as shown in Fig. 2.8). A volume force $f_i(i = 1, 2, 3)$ acts on the microelement body at its centroid C. Let σ_{ij} be the stress components at the centroids of the three negative surfaces. There is an increment to each stress component at the centroid of a positive surface relative to that of the corresponding negative surface. By performing a Taylor series expansion and neglecting the higher-order small quantities, we can express a stress component at the centroid of a positive surface in terms of the corresponding negative-surface stress and its first derivative. For example, a negative-surface normal stress σ_{11} turns into $\sigma_{11} + \frac{\partial \sigma_{11}}{\partial x_1}\mathrm{d}x_1 + \cdots$ on a positive surface at a distance of $\mathrm{d}x_1$. Figure 1.8 shows the stress state of the microelement body. The force equilibrium condition for the microelement body along the x_1-direction is as follows:

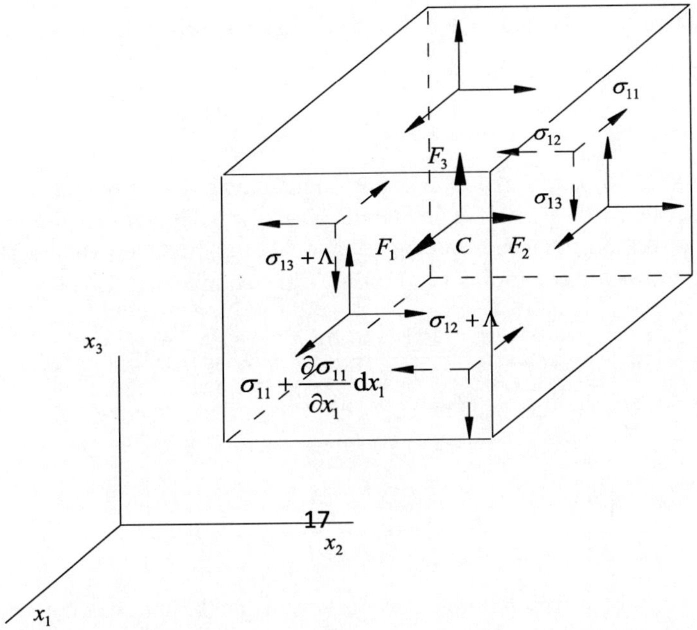

Fig. 2.8 Force equilibrium conditions

$$
\left(\sigma_{11} + \frac{\partial \sigma_{11}}{\partial x_1}\mathrm{d}x_1\right)\mathrm{d}x_2\mathrm{d}x_3 - \sigma_{11}\mathrm{d}x_2\mathrm{d}x_3 + \left(\sigma_{21} + \frac{\partial \sigma_{21}}{\partial x_2}\mathrm{d}x_2\right)\mathrm{d}x_3\mathrm{d}x_1 - \sigma_{21}\mathrm{d}x_3\mathrm{d}x_1
$$
$$
+\left(\sigma_{31} + \frac{\partial \sigma_{31}}{\partial x_3}\mathrm{d}x_3\right)\mathrm{d}x_1\mathrm{d}x_2 - \sigma_{31}\mathrm{d}x_1\mathrm{d}x_2 + f_1\mathrm{d}x_1\mathrm{d}x_2\mathrm{d}x_3 = 0 \tag{2.52}
$$

By combining the terms and then dividing the result by the area of the microelement and by taking the limit as the microelement approaches point (x_1,x_2,x_3), we have

$$
\frac{\partial \sigma_{11}}{\partial x_1} + \frac{\partial \sigma_{21}}{\partial x_2} + \frac{\partial \sigma_{31}}{\partial x_3} + f_1 = 0 \tag{2.53}
$$

Similarly, the force equilibrium conditions along the x_2- and x_3-directions are obtained as follows:

$$
\frac{\partial \sigma_{12}}{\partial x_1} + \frac{\partial \sigma_{22}}{\partial x_2} + \frac{\partial \sigma_{32}}{\partial x_3} + f_2 = 0 \tag{2.54}
$$

$$
\frac{\partial \sigma_{13}}{\partial x_1} + \frac{\partial \sigma_{23}}{\partial x_2} + \frac{\partial \sigma_{33}}{\partial x_3} + f_3 = 0 \tag{2.55}
$$

The above three equations can be rewritten concisely using index notation as follows:

$$\sigma_{ij,j} + f_i = 0 \quad \nabla \cdot \boldsymbol{\sigma} + \boldsymbol{f} = \boldsymbol{0} \tag{2.56}$$

Equation (2.56) is referred to as the differential equation of equilibrium, or equilibrium equation for short. This equation gives the relations that the first derivatives of the stress components and the volume-force components should satisfy. The following shows the commonly used forms of the equilibrium equation:

$$\begin{aligned} &\frac{\partial \sigma_x}{\partial x} + \frac{\partial \tau_{yx}}{\partial y} + \frac{\partial \tau_{zx}}{\partial z} + f_x = 0 \\ &\frac{\partial \tau_{xy}}{\partial x} + \frac{\partial \sigma_y}{\partial y} + \frac{\partial \tau_{zy}}{\partial z} + f_y = 0 \\ &\frac{\partial \tau_{xz}}{\partial x} + \frac{\partial \tau_{yz}}{\partial y} + \frac{\partial \sigma_z}{\partial z} + f_z = 0 \end{aligned} \tag{2.57}$$

Next, let us consider the moment equilibrium of the microelement body. The moments are taken about the axis passing through the centroid C and in the x_3-direction. The resultant moment of the stress and volume-force components whose lines of action pass through point C, or that are parallel to this axis, along this axis is zero. Thus, the moment equilibrium equation is given by

$$\begin{aligned} &\left(\sigma_{12} + \frac{\partial \sigma_{12}}{\partial x_1}\mathrm{d}x_1\right)\mathrm{d}x_2\mathrm{d}x_3 \cdot \frac{\mathrm{d}x_1}{2} + \sigma_{12}\mathrm{d}x_2\mathrm{d}x_3 \cdot \frac{\mathrm{d}x_1}{2} \\ &-\left(\sigma_{21} + \frac{\partial \sigma_{21}}{\partial x_2}\mathrm{d}x_2\right)\mathrm{d}x_1\mathrm{d}x_3 \cdot \frac{\mathrm{d}x_2}{2} - \sigma_{21}\mathrm{d}x_1\mathrm{d}x_3 \cdot \frac{\mathrm{d}x_2}{2} = 0 \end{aligned} \tag{2.58}$$

By neglecting the higher-order terms, we have only two terms left:

$$\sigma_{12}\mathrm{d}x_2\mathrm{d}x_3 \cdot \mathrm{d}x_1 - \sigma_{21}\mathrm{d}x_3\mathrm{d}x_1 \cdot \mathrm{d}x_2 = 0 \tag{2.59}$$

Therefore, we have

$$\sigma_{12} = \sigma_{21} \tag{2.60}$$

Similarly, by setting the resultant moment about each of the centroidal axes along the x_1- and x_2-directions, respectively, we have

$$\sigma_{23} = \sigma_{32} \text{ and } \sigma_{31} = \sigma_{13} \tag{2.61}$$

The above two equations can be collectively written as

$$\sigma_{ij} = \sigma_{ji} \tag{2.62}$$

The above equation shows the reciprocal theorem of shear stresses, which is also known as the symmetry of stress tensors.

Based on Eq. (2.49) or (2.50), nine stress components are required to completely describe the stress state at any point. However, according to the symmetry shown in Eq. (2.62), only six of the nine stress components are independent. Thus, only six stress components are required to describe the stress state at a point. Similar to a strain tensor, a stress tensor is symmetric and has only six independent components. According to linear algebra theory, the matrix comprising these six symmetric stress components can be diagonalized. The three eigenvalues of the diagonalized matrix are referred to as principal stresses, which are also known as, in descending order of magnitude, the first principal stress σ_1, the second principal stress σ_2, and the third principal stress σ_3. The planes of the three principal stresses are referred to as principal planes, whose normal directions are called principal directions. It is very easy to prove that the three principal directions are mutually perpendicular. Interested readers may refer to relevant textbooks and go through the proof themselves.

2.2.2 *Stress–Strain Constitutive Relations Based on Small Deformation [5, 6]*

We first assume that a material is isotropic, i.e., its properties are identical in any direction. Now, let us discuss the relation between the stress and strain components at an arbitrary point within an isotropic elastic body under linear elastic conditions. An infinitesimal regular hexahedron with all six faces parallel to the coordinate axes is taken, and normal stress σ_{xx} is applied in the x-axis direction. The normal strain of the hexahedron in the x-axis direction is then given by

$$\varepsilon_{xx} = \frac{\sigma_{xx}}{E} \tag{2.63}$$

where E is the elastic modulus of the material. Of particular note is that E_{ij} is used in Eqs. (2.11) and (2.15) to represent strain components.

For a specimen subjected to tension, as it elongates continuously along the tensile direction, its cross-sectional area continuously decreases. Conversely, for a specimen subjected to compression, as it continuously shortens along the axis of compression, its cross-sectional area continuously increases. The extent of the decrease or increase in the transverse area is evidently related to the extent of the unidirectional tension or compression. We assume that this relationship is linear at the stage where very small deformation takes place (i.e., the linear elastic stage). In other words, at an axial tensile (or compressive) strain of ε_{xx}, the extents of compression (or tension) in the y- and z-directions should be the same. Thus, the transverse strains ε_{yy} and ε_{zz} of the specimen are given by

$$\varepsilon_{yy} = \varepsilon_{zz} = -\nu\varepsilon_{xx} = -\nu\frac{\sigma_{xx}}{E} \tag{2.64}$$

where ν is the ratio of the transverse strain to axial strain in their linear relationship and is referred to as the transverse deformation coefficient or Poisson's ratio.

Similarly, if normal stress σ_{yy} is applied only along the y-axis direction, we have

$$\varepsilon_{yy} = \frac{\sigma_{yy}}{E} \quad \varepsilon_{zz} = \varepsilon_{xx} = -\nu\frac{\sigma_{yy}}{E} \tag{2.65}$$

If normal stress σ_{zz} is applied only along the z-axis direction, we have

$$\varepsilon_{zz} = \frac{\sigma_{zz}}{E} \quad \varepsilon_{xx} = \varepsilon_{yy} = -\nu\frac{\sigma_{zz}}{E} \tag{2.66}$$

The superposition principle is applicable under linear elastic conditions. If three normal stresses, σ_{xx}, σ_{yy}, and σ_{zz}, are applied simultaneously along the three respective coordinate axes, the resulting total strain is a linear superposition of the strains caused by the three stresses when applied individually, i.e.,

$$\begin{aligned} \varepsilon_{xx} &= \frac{1}{E}\left[\sigma_{xx} - \nu\left(\sigma_{yy} + \sigma_{zz}\right)\right] \\ \varepsilon_{yy} &= \frac{1}{E}\left[\sigma_{yy} - \nu\left(\sigma_{xx} + \sigma_{zz}\right)\right] \\ \varepsilon_{zz} &= \frac{1}{E}\left[\sigma_{zz} - \nu\left(\sigma_{xx} + \sigma_{yy}\right)\right] \end{aligned} \tag{2.67}$$

In a pure shear stress state, the shear stress τ_{xy} and the shear strain γ_{xy} follow a linear relationship, which can be derived based on Hooke's law:

$$\tau_{xy} = G\gamma_{xy} \tag{2.68}$$

where $G = \frac{E}{2(1+\nu)}$ is the shear modulus. In fact, in a pure shear stress state, the square $ABCD$ with a side length of $\sqrt{2}a$, as shown in Fig. 2.9, deforms into a parallelogram $A'B'C'D'$ with the principal stresses of $\sigma_1 = \tau$, $\sigma_2 = 0$, and $\sigma_3 = -\tau$. According to Eq. (2.67), the relative elongation of the horizontal diagonal in Fig. 2.9 is

$$\varepsilon_1 = \frac{1}{E}(\sigma_1 - \nu\sigma_3) = \frac{1+\nu}{E}\tau \tag{2.69}$$

The relative contraction of the vertical diagonal is

$$\varepsilon_3 = \frac{1}{E}(\sigma_3 - \nu\sigma_1) = -\frac{1+\nu}{E}\tau \tag{2.70}$$

Within the triangle $A'OB'$,

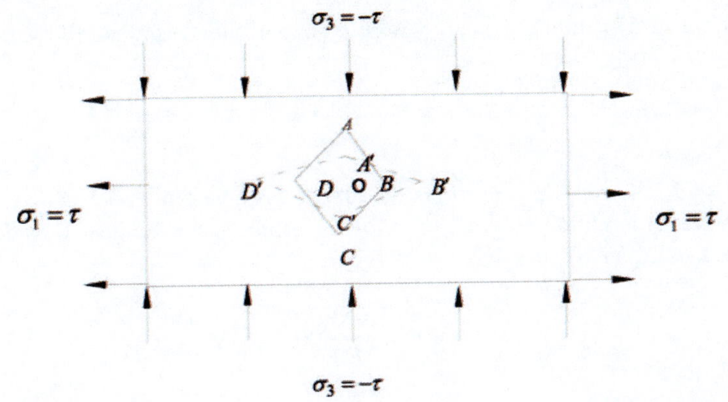

Fig. 2.9 Deformation caused by a pure shear stress state

$$OB' = a(1+\varepsilon_1) \text{ and } OA' = a(1+\varepsilon_3) \tag{2.71}$$

Thus,

$$A'B' = \sqrt{a^2(1+\varepsilon_1)^2 + (1+\varepsilon_3)^2 a^2} = a\sqrt{2(1+\varepsilon_1+\varepsilon_3+\cdots)} \tag{2.72}$$

As $\varepsilon_1 + \varepsilon_3 = 0$, we have

$$A'B' = \sqrt{2}a = AB \tag{2.73}$$

Let γ be the change in the right angle. Then, we have

$$\angle OA'B' = \frac{\pi}{4} + \frac{\gamma}{2}, \tan\left(\frac{\pi}{4} + \frac{\gamma}{2}\right) = \frac{OB'}{OA'} = \frac{a(1+\varepsilon_1)}{a(1+\varepsilon_3)} \tag{2.74}$$

As

$$\tan\left(\frac{\pi}{4} + \frac{\gamma}{2}\right) = \frac{1+\frac{\gamma}{2}}{1-\frac{\gamma}{2}} \approx 1+\gamma \text{ and } \frac{1+\varepsilon_1}{1+\varepsilon_3} \approx 1+\varepsilon_1-\varepsilon_3 \tag{2.75}$$

we have

$$\gamma = \frac{2(1+v)}{E}\tau \tag{2.76}$$

The above equation is the same as Eq. (2.68). Similarly, we have

$$\varepsilon_{xy} = \frac{1}{2G}\sigma_{xy} \quad \varepsilon_{yz} = \frac{1}{2G}\sigma_{yz} \quad \varepsilon_{zx} = \frac{1}{2G}\sigma_{zx} \tag{2.77}$$

Equations (2.67) and (2.77) can be written in a unified form as follows:

$$\varepsilon_{ij} = \frac{1+v}{E}\sigma_{ij} - \frac{v}{E}\delta_{ij}\Theta \text{ or } \boldsymbol{\varepsilon} = \frac{1+v}{E}\boldsymbol{\sigma} - \frac{v}{E}\Theta \boldsymbol{I} \tag{2.78}$$

where $\Theta = \sigma_{xx} + \sigma_{yy} + \sigma_{zz} = \sigma_{kk}$ is the first stress invariant and $\boldsymbol{I}$ is a unit tensor. Equation (2.78) is called Hooke's law for isotropic elastic bodies. By adding the three terms in Eq. (2.67), we have

$$e = \frac{1-2v}{E}\Theta \tag{2.79}$$

where $e = \varepsilon_{xx} + \varepsilon_{yy} + \varepsilon_{zz} = \varepsilon_{ii}$ is the volumetric strain. Equation (2.79) can be written in the following form:

$$\sigma_m = Ke \tag{2.80}$$

where $K = \frac{E}{3(1-2v)}$ is referred to as the bulk modulus of elasticity and $\sigma_m = \frac{1}{3}\sigma_{kk}$ is the mean stress. Equation (2.80) shows that the volumetric strain is directly proportional to the mean normal stress. This is the volumetric Hooke's law for isotropic elastic materials. The stress deviator tensor $\boldsymbol{S}$ and the strain deviator tensor $\boldsymbol{\varepsilon}'$ follow the relationship below

$$\boldsymbol{S} = 2G\boldsymbol{\varepsilon}' \text{ or } S_{ij} = 2G\varepsilon'_{ij} \tag{2.81}$$

Let $\lambda = \frac{Ev}{(1+v)(1-2v)}$ and $\mu = \frac{E}{2(1+v)} = G$. Then, Eq. (2.78) can be rewritten as follows:

$$\boldsymbol{\sigma} = \lambda e\boldsymbol{I} + 2\mu\boldsymbol{\varepsilon} \text{ or } \sigma_{ij} = \lambda e\delta_{ij} + 2\mu\varepsilon_{ij} \tag{2.82}$$

In the abovementioned elastic relations for isotropic bodies, there are five elastic constants (E, v, G, λ, and K), of which only two are independent (generally E and v, or λ and G, or K and G). For a given engineering material, E and v can be determined by a unidirectional tensile test, G by a thin-walled cylinder torsion test, and K by a hydrostatic pressure test. In an idealized extreme case, if $v = 1/2$, then $K = \infty$, which corresponds to the so-called incompressible material, with $G = E/3$. The incompressibility assumption is often used in the theory of plasticity. Materials with negative Poisson's ratios have yet to be discovered. All the measurements of the Poisson's ratio fall in the range of (0, 1/2). For metallic materials, the value of Poisson's ratio has a relatively insignificant impact on the calculated stresses and strains and, therefore, is often set to approximately 0.3.

Many engineering structures, such as dams, tunnels, thick-walled cylinders, rollers, and thin plates subjected to in-plane loads, can be simplified to 2D plane problems, i.e., plane stress and plane strain problems.

Plane stress problems focus on the study of thin plate-like elastic bodies. Here, let us consider a thin plate with a uniform thickness h much smaller than the other two dimensions. The plate is laterally subjected to a surface force that is applied on the side surface and is parallel to the plate plane and does not vary along the thickness direction. In addition, the volume force acts parallel to the plate plane and does not vary along the thickness direction. The middle plane of the plate at $z = 0$ is taken as plane xoy. A straight line perpendicular to the middle plane is set as the z-axis. The two surfaces of the plate at $z = \pm\frac{h}{2}$ are free and subjected to no external forces. Thus, $\sigma_z = \sigma_{zy} = \sigma_{zx} = 0$. Then, the stress–strain relations of the thin plate in a plane stress state are given by.

$$\varepsilon_x = \frac{1}{E}\left(\sigma_x - \nu\sigma_y\right), \quad \varepsilon_y = \frac{1}{E}\left(\sigma_y - \nu\sigma_x\right), \quad \varepsilon_z = -\frac{\nu}{E}\left(\sigma_x + \sigma_y\right)$$
$$\varepsilon_{xy} = \frac{1}{2G}, \quad \sigma_{xy} = \frac{1+\nu}{E}\sigma_{xy}, \quad \text{and } \varepsilon_{yz} = \varepsilon_{xz} = 0 \tag{2.83}$$

For plane strain problems, let us examine a long cylinder with a uniform cross-section. The axis of the cylinder is parallel to the z-axis, and the axial dimension of the cylinder is much greater than the other two dimensions. This cylinder is laterally subjected to a surface force that is perpendicular to the z-axis and does not vary in the z-direction. If the cylinder is infinitely long or is finitely long but is rigidly constrained at both ends (i.e., its axial displacement is constrained), then any cross-section of the cylinder can be viewed as a plane of symmetry. Thus, each point within the cylinder can move only along the x- and y-directions but not in the z-direction. In other words, the displacement components u and v are independent of the z-coordinate and are functions only of x and y, while the displacement component w is zero. Thus, the strain is given by

$$\varepsilon_x = \frac{\partial u}{\partial x}, \quad \varepsilon_y = \frac{\partial v}{\partial y}, \quad \varepsilon_{xy} = \frac{1}{2}\left(\frac{\partial v}{\partial x} + \frac{\partial u}{\partial y}\right), \quad \varepsilon_z = \varepsilon_{yz} = \varepsilon_{zx} = 0 \tag{2.84}$$

As $\varepsilon_z = 0$, $\sigma_z = \nu\left(\sigma_x + \sigma_y\right)$. Thus, the stress–strain relations of an elastic body in a plane strain state are obtained from Eq. (2.78) as follows:

$$\varepsilon_x = \frac{1}{E}\left[\left(1-\nu^2\right)\sigma_x - \nu(1+\nu)\sigma_y\right], \varepsilon_y = \frac{1}{E}\left[\left(1-\nu^2\right)\sigma_y - \nu(1+\nu)\sigma_x\right], \varepsilon_{xy} = \frac{1}{2G}\sigma_{xy} \tag{2.85}$$

We introduce the following symbols:

$$E' = \frac{E}{1-\nu^2}, \quad \nu' = \frac{\nu}{1-\nu} \tag{2.85}$$

to rewrite the constitutive equations for an elastic body in a plane strain state as follows:

$$\varepsilon_x = \frac{1}{E'}\left(\sigma_x - \nu'\sigma_y\right),\ \varepsilon_y = \frac{1}{E'}\left(\sigma_y - \nu'\sigma_x\right),\ \varepsilon_{xy} = \frac{1}{2G}\sigma_{xy} \tag{2.86}$$

Thus, the constitutive relations for plane problems can be written in a unified form as shown in Eq. (2.86), where

$$E' = \begin{cases} E & \text{for plane stress} \\ \dfrac{E}{1-\nu^2} & \text{for plane strain} \end{cases}$$

$$\nu' = \begin{cases} \nu & \text{for plane stress} \\ \dfrac{\nu}{1-\nu} & \text{for plane strain} \end{cases}$$

Thus, the two types of plane problems are mathematically treated in the same manner.

2.2.3 Constitutive Theory for Thermomechanical Coupling Based on Small Deformation [11]

When temperature is not considered, the state (i.e., deformation state) of an elastic solid can be completely determined by the strain tensor $\boldsymbol{\varepsilon}$. However, in thermoelastic problems, an additional physical quantity—temperature—is needed to determine the state of an elastic solid. In other words, the strain tensor $\boldsymbol{\varepsilon}$ and temperature T constitute a complete set of state quantities of an elastic solid. Any of the other quantities that depend on the state can be expressed as a function of state quantities $\boldsymbol{\varepsilon}$ and T. In addition, there is always a reversible process between any pair of starting and ending states of an elastic solid, regardless of whether the actual change in its state is irreversible or not, thereby avoiding the involvement of irreversible thermodynamics [12, 13].

Let us take from an object an arbitrary part with a volume of Ω, a surface area of Γ, and a surface outer normal direction of $\boldsymbol{\nu}$. The object is subjected to a surface-force vector of $\boldsymbol{t} = \boldsymbol{\sigma} \cdot \boldsymbol{\nu}$. Let $\boldsymbol{f}$ be the volume-force vector per unit volume and $\dot{\boldsymbol{u}}$ be the velocity. Then, the power of the external forces is given by

$$\dot{A} = \int_{\Gamma} \boldsymbol{t} \cdot \dot{\boldsymbol{u}} d\Gamma + \int_{\Omega} \boldsymbol{f} \cdot \dot{\boldsymbol{u}} d\Omega = \int_{\Gamma} \sigma_{ij}\nu_j\dot{u}_i d\Gamma + \int_{\Omega} f_i\dot{u}_i d\Omega \tag{2.87}$$

By transforming the surface integral in Eq. (2.87) to a volume integral using Gauss' theorem, we have

$$\dot{A} = \int_{\Omega} (\sigma_{ij,j}\dot{u}_i + \sigma_{ij}\dot{u}_{i,j})d\Omega + \int_{\Omega} f_i\dot{u}_i d\Omega \tag{2.88}$$

Let r_m and $\boldsymbol{q}$ be the heat generation rate per unit mass and the heat flux density vector, respectively. Then, the rate of increase in the thermal energy of the object is given by

$$\dot{Q} = -\int_{\Gamma} \boldsymbol{q} \cdot \boldsymbol{n} d\Gamma + \int_{\Omega} \rho r_m d\Omega = -\int_{\Omega} q_{i,i} d\Omega + \int_{\Omega} \rho r_m d\Omega \tag{2.89}$$

The first and second terms on the rightmost side of the above equation are the heat transferred into the object from the boundary and the heat generated by the heat source within the object, respectively. ρ is the density of the object. The kinetic energy of the system is given by

$$K = \int_{\Omega} \frac{1}{2}\rho\dot{\boldsymbol{u}} \cdot \dot{\boldsymbol{u}} d\Omega \tag{2.90}$$

Let e be the internal energy per unit mass. Then, the rate of change in the total energy is as follows:

$$\dot{K} + \dot{E} = \frac{d}{dt}\int_{\Omega} \frac{1}{2}\rho\dot{\boldsymbol{u}} \cdot \dot{\boldsymbol{u}} d\Omega + \int_{\Omega} \rho\dot{e} d\Omega = \int_{\Omega} (\dot{u}_i\ddot{u}_i + \dot{e})\rho d\Omega \tag{2.91}$$

Of particular note is that K is used here to represent kinetic energy and differs from the bulk modulus in Eq. (2.80) in Sect. 2.2.2 and that E is used here to represent internal energy and differs from the strain component in Eqs. (2.11) and (2.15) in Sect. 2.2.1 and Young's modulus in Sect. 2.2.2.

According to the first law of thermodynamics, the rate of change in the total energy is equal to the sum of the power of the external forces and the rate of increase in heat. Therefore, based on Eqs. (2.88), (2.89), and (2.91), we have

$$\int_{\Omega} \rho(\dot{u}_i\ddot{u}_i + \dot{e})d\Omega = \int_{\Omega} (\sigma_{ij,j}\dot{u}_i + \sigma_{ij}\dot{u}_{i,j} + f_i\dot{u}_i)d\Omega - \int_{\Omega} q_{i,i}d\Omega + \int_{\Omega} \rho r_m d\Omega \tag{2.92}$$

By using the equation of motion,

$$\sigma_{ij\cdot j} + f_i = \rho\ddot{u}_i \tag{2.93}$$

Equation (2.92) can be simplified to

$$\int_{\Omega} \rho(\dot{e} - r_m) d\Omega = \int_{\Omega} \sigma_{ij} \dot{u}_{i,j} d\Omega - \int_{\Omega} q_{i,i} d\Omega = \int_{\Omega} \sigma_{ij} \dot{\varepsilon}_{ij} d\Omega - \int_{\Omega} q_{i,i} d\Omega \tag{2.94}$$

$\dot{u}_{i,j} = \dot{\varepsilon}_{ij}$ is applied in the above equation and holds for any part Ω of the object. Thus, we have

$$\rho(\dot{e} - r_m) = -\boldsymbol{q} \cdot \nabla + \boldsymbol{\sigma} : \dot{\boldsymbol{\varepsilon}} \ \rho(\dot{e} - r_m) = \sigma_{ij} \dot{\varepsilon}_{ij} - q_{i,i} \tag{2.95}$$

The above equation is the differential form of the first law of thermodynamics for deformable continua.

Let T and s_m be the absolute temperature and the entropy density per unit mass, respectively, i.e.,

$$S = \int_{\Omega} \rho s_m d\Omega \tag{2.96}$$

According to the second law of thermodynamics, we have

$$T\dot{S} = T\dot{S}_{\mathrm{e}} + T\dot{S}_{\mathrm{i}} \geqslant \dot{Q} \tag{2.97}$$

where $\dot{S}_{\mathrm{e}}$ is the rate of entropy supply from external sources to the system (also known as entropy flow), $\dot{S}[0, 1]$ is the entropy production within the system (also known as the entropy production rate), and $\dot{Q}$ is the rate at which the system absorbs heat from external sources. Thus, we have

$$\dot{S} \geqslant -\int_{\Gamma} \frac{1}{T} \boldsymbol{q} \cdot d\boldsymbol{\Gamma} + \int_{\Omega} \frac{1}{T} \rho r_m d\Omega \tag{2.98}$$

where $\boldsymbol{q}/T$ and γ/T are referred to as the entropy flow and entropy source, respectively. The above expression is called entropy inequality or Clausius–Duhem inequality and is known as the integral form of the second law of thermodynamics. The ">" and "=" signs in Eq. (2.98) correspond to irreversible processes (e.g., plastic deformation) and reversible processes (e.g., elastic deformation), respectively. The right-hand side of Eq. (2.98) shows that the absorption of heat by the domain Ω from the neighborhood or the external domain leads to an increase in its total entropy. The excess of the actual increase in entropy over the increase in entropy reflected by the right-hand side of Eq. (2.98), that is, the entropy production rate $\dot{S}_i$ in Eq. (2.97), is irreversible. Thus, from Eqs. (2.97) and (2.98), we have

$$\dot{S} = \int_{\Omega} \dot{s}_m \rho d\Omega = -\int_{\Gamma} \frac{1}{T} \boldsymbol{q} \cdot d\boldsymbol{\Gamma} + \int_{\Omega} \frac{1}{T} \rho r_m d\Omega + \dot{S}_i \tag{2.99}$$

In addition, $\dot{S} \geqslant 0$. Let κ_{m} be the entropy production rate per unit mass. Then,

$$\dot{S}_i = \int_{\Omega} \rho \kappa_m d\Omega, \quad \kappa_{\mathrm{m}} \geq 0 \tag{2.100}$$

Equation (2.99) is the integral form of the entropy balance equation (or the rate of entropy balance). By transforming the first integral in Eq. (2.99) into a volume integral, the differential form of the second law of thermodynamics can be obtained as follows:

$$\rho T \dot{s}_m = \rho r_m - q_{i,i} + \frac{q_i}{T} T_{,i} + \rho T \kappa_{\mathrm{m}} \tag{2.101}$$

i.e.,

$$T \kappa_m = T \kappa_{th} + T \kappa_{\mathrm{int}} \geq 0 \tag{2.102}$$

where

$$T \kappa_{\mathrm{th}} = -\frac{q_i}{\rho T} T_{,i}, \; T \kappa_{\mathrm{int}} = T \dot{s}_m - (r_m - \frac{1}{\rho} q_{i,i}) \tag{2.103}$$

Equation (2.102) indicates that the entropy production rate κ_{m} can be divided into two parts, namely κ_{th} and κ_{int}. κ_{th} is the rate of entropy production resulting from thermal conduction. κ_{int}, referred to as the intrinsic entropy production rate, results from the excess of the entropy increase rate $\dot{s}_m$ over the rate of absorption of heat from the neighborhood or external sources per unit mass. The second law of thermodynamics requires that $\kappa_{\mathrm{m}} = \kappa_{\mathrm{th}} + \kappa_{\mathrm{int}} \geq 0$. However, a stronger assumption that $\kappa_{\mathrm{th}} \geq 0$ and $\kappa_{\mathrm{int}} \geq 0$ is often used, i.e.,

$$T \kappa_{\mathrm{th}} \geqslant 0 \quad T \kappa_{\mathrm{int}} \geqslant 0 \tag{2.104}$$

$T\kappa$, $T\kappa_{\mathrm{th}}$, and $T\kappa_{\mathrm{int}}$ are the total dissipation rate, the heat dissipation rate, and the intrinsic dissipation rate, respectively, per unit mass. Based on Eq. (2.103), $T \kappa_{th} \geq 0$ means that the heat flow vector $\boldsymbol{q}$ and the temperature gradient $T_{,i}$ must form an obtuse angle. For materials with isotropic thermal properties, Fourier's law of heat conduction states that

$$\boldsymbol{q} = -k \nabla T \tag{2.105}$$

where k is the thermal conductivity. The above equation shows that the heat flow vector q_i is directly proportional to the temperature gradient $T_{,i}$ but in the opposite direction.

Based on the differential form of the first law of thermodynamics in Eq. (1.95), the second equation in Eq. (2.103) can be rewritten as

$$T\kappa_{\text{int}} = T\dot{s}_m - \left(\dot{e} - \frac{1}{\rho}\boldsymbol{\sigma} : \dot{\boldsymbol{\varepsilon}}\right) \tag{2.106}$$

The expression in the parentheses on the right-hand side of Eq. (2.106) represents the excess of the rate of increase in the internal energy per unit mass over the deformation power per unit mass and is exactly the rate of absorption of heat per unit mass reflected by the expression in the parentheses on the right-hand side of the second equation in Eq. (2.103).

If we, for the sake of convenience, choose temperature T instead of entropy s_m as the independent variable, we can use the Helmholtz free energy ψ per unit mass instead of the internal energy e as the thermodynamic function, which is defined as follows:

$$\psi = e - Ts_m \tag{2.107}$$

Thus,

$$\dot{\psi} + s_m\dot{T} = \dot{e} - T\dot{s}_m \tag{2.108}$$

Equation (2.106) can be rewritten as

$$T\kappa_{\text{int}} = -s_m\dot{T} - \left(\dot{\psi} - \frac{1}{\rho}\boldsymbol{\sigma} : \dot{\boldsymbol{\varepsilon}}\right) \tag{2.109}$$

The state of a deformable body changes with temperature T and strain $\boldsymbol{\varepsilon}$, and a heat flow occurs only in the presence of a temperature gradient ∇T. Thus, generally speaking, the Helmholtz free energy ψ should be a function of T, $\boldsymbol{\varepsilon}$, and ∇T. On the other hand, when a deformable body deforms, it may undergo microstructural changes (e.g., plastic strain $\boldsymbol{\varepsilon}^{\text{p}}$ and internal damage). Thus, the relevant state quantities, including the stress $\boldsymbol{\sigma}$, internal energy e, entropy s_m, and Helmholtz free energy ψ, should also be functions of the internal variable ξ. It has been experimentally found that in an isothermal process, the influence of inelastic strains on free energy is negligible [14, 15]. Therefore, in classical plasticity theories, the dependence of state quantities (e.g., $\boldsymbol{\sigma}$, e, s_m, and ψ) on $\boldsymbol{\varepsilon}$ and $\boldsymbol{\varepsilon}^{\text{p}}$ is determined based on their dependence on the difference between $\boldsymbol{\varepsilon}$ and $\boldsymbol{\varepsilon}^{\text{p}}$, i.e., the elastic strain $\boldsymbol{\varepsilon}^{\text{e}} = \boldsymbol{\varepsilon} - \boldsymbol{\varepsilon}^{\text{p}}$. Hence,

$$\boldsymbol{\sigma} = \boldsymbol{\sigma}(\boldsymbol{\varepsilon}^{\text{e}}, \xi, T, \nabla T) \quad e = e(\boldsymbol{\varepsilon}^{\text{e}}, \xi, T, \nabla T)$$

$$s_m = s_m(\boldsymbol{\varepsilon}^{\text{e}}, \xi, T, \nabla T), \ \psi = \psi(\boldsymbol{\varepsilon}^{\text{e}}, \xi, T, \nabla T) \tag{2.110}$$

Thus, we have

$$\dot{\psi} = \frac{\partial \psi}{\partial \boldsymbol{\varepsilon}^{\mathrm{e}}} : \dot{\boldsymbol{\varepsilon}}^{\mathrm{e}} + \frac{\partial \psi}{\partial T}\dot{T} + \frac{\partial \psi}{\partial \nabla T} \cdot \nabla \dot{T} + \frac{\partial \psi}{\partial \xi}\dot{\xi} \tag{2.111}$$

By substituting Eq. (2.108) into the first law of thermodynamics in Eq. (2.95) to eliminate $\dot{e}$ and based on Eq. (2.111), we have

$$\begin{aligned} &\left(\rho\frac{\partial \psi}{\partial \boldsymbol{\varepsilon}^{\mathrm{e}}} - \boldsymbol{\sigma}\right) : \dot{\boldsymbol{\varepsilon}}^{\mathrm{e}} + \rho\left(\frac{\partial \psi}{\partial T} + s_m\right)\dot{T} + \rho\frac{\partial \psi}{\partial \nabla T} \cdot \nabla \dot{T} + \rho(T\dot{s}_m - r_m) + \\ &\nabla \cdot \boldsymbol{q} - \rho\frac{\partial \psi}{\partial \dot{\boldsymbol{\varepsilon}}^{\mathrm{e}}} : \dot{\boldsymbol{\varepsilon}}^{\mathrm{p}} + \rho\frac{\partial \psi}{\partial \xi}\dot{\xi} = 0 \end{aligned} \tag{2.112}$$

The above equation holds for any $\dot{\varepsilon}^{\,e}$, $\dot{T}$, $\nabla\dot{T}$, so we have

$$\boldsymbol{\sigma} = \rho\frac{\partial \psi}{\partial \boldsymbol{\varepsilon}^{\mathrm{e}}} \tag{2.113}$$

$$s_m = -\frac{\partial \psi}{\partial T} \tag{2.114}$$

$$\frac{\partial \psi}{\partial \nabla T} = \mathbf{0} \tag{2.115}$$

$$\rho(T\dot{s}_m - r_m) + \nabla \cdot \boldsymbol{q} + \rho\frac{\partial \psi}{\partial \xi}\dot{\xi} - \rho\frac{\partial \psi}{\partial \boldsymbol{\varepsilon}_e} : \dot{\boldsymbol{\varepsilon}}^{p} = 0 \tag{2.116}$$

Equation (2.113) is the constitutive relation of deformable bodies. Equation (2.114) is the expression of the entropy density. Equation (2.115) shows that the Helmholtz free energy ψ is independent of ∇T and is a function of only $\boldsymbol{\varepsilon}$ and T. Equation (2.116) is the thermal conduction equation. By substituting Eqs. (2.113) and (2.114) into Eq. (2.116) and eliminating s_m, we have

$$\nabla \cdot \boldsymbol{q} = \rho r_m + \rho T\left(\frac{\partial^2 \psi}{\partial \boldsymbol{\varepsilon}_e \partial T} : \dot{\boldsymbol{\varepsilon}}_e + \frac{\partial^2 T}{\partial T^2}\dot{T}\right) + \rho\left(T\frac{\partial^2 \psi}{\partial \xi \partial T} - \frac{\partial \psi}{\partial \xi}\right)\dot{\xi} + \boldsymbol{\sigma} : \dot{\boldsymbol{\varepsilon}}^{p} \tag{2.117}$$

The terms containing $\dot{\boldsymbol{\varepsilon}}$ reflect the influence of object deformation on the temperature field. Thus, Eq. (2.117) is the thermal conduction equation that accounts for the thermomechanical coupling effect.

For small deformation and small temperature change, Eq. (2.113) can be simplified to a linear thermoelastic constitutive equation. Let an object have an initial temperature of T_0 and zero initial strain and stress in the reference state and a temperature of T and a strain of $\boldsymbol{\varepsilon}$ in the actual state (for the sake of convenience, the superscript "e" in $\boldsymbol{\varepsilon}^{\mathrm{e}}$ of an object in the elastic state is omitted, and $\boldsymbol{\varepsilon}$ alone is used to represent elastic strain). The temperature change is given by

$$\theta = T - T_0 \tag{2.118}$$

By nondimensionalizing the above equation, we have

$$\theta' = \frac{\theta}{T_0} = \frac{T - T_0}{T_0} \tag{2.119}$$

For a small temperature change, $|\theta'| << 1$.

For a small strain ($|\boldsymbol{\varepsilon}| << 1$), the density ρ is independent of $\boldsymbol{\varepsilon}$. Equation (2.113) can be rewritten as

$$\boldsymbol{\sigma} = \frac{\partial(\rho\psi)}{\partial\boldsymbol{\varepsilon}} \tag{2.120}$$

By expanding the per-unit-volume free energy $\rho\psi$ into a series with respect to $\boldsymbol{\varepsilon}$ and θ' near the reference state and by neglecting the higher-order terms, we have

$$\rho\psi = \rho\psi_0 + D_{ij}\varepsilon_{ij} + b'\theta' + \frac{1}{2}C_{ijkl}\varepsilon_{ij}\varepsilon_{kl} + b'_{ij}\varepsilon_{ij}\theta' + \frac{1}{2}d'\theta'^2 \tag{2.121}$$

where ψ_0 is the free energy in the reference state and can be set to 0. For homogeneous materials, the coefficients D_{ij}, b', C_{ijkl}, b'_{ij}, and d' are all independent of the coordinate x_i. By substituting Eq. (2.121) into Eq. (2.120) and letting $b'_{ij} = T_o b_{ij}$, we have

$$\sigma_{ij} = D_{ij} + C_{ijkl}\varepsilon_{kl} + b_{ij}\theta \tag{2.122}$$

Assuming that the initial stress in the reference state ($\varepsilon_{kl} = 0, \theta = 0$) is 0, $D_{ij} = 0$. Then, the constitutive relation for linear thermoelastic materials is obtained as follows:

$$\sigma_{ij} = C_{ijkl}\varepsilon_{kl} + b_{ij}\theta \quad \boldsymbol{\sigma} = \boldsymbol{C} : \boldsymbol{\varepsilon} + \boldsymbol{b}\theta \tag{2.123}$$

If there is an initial stress (i.e., residual stress), $D_{ij} \neq 0$, which is exactly the residual stress. For isothermal elastic problems ($\theta = 0$), Eq. (2.123) is simplified to the generalized Hooke's law. For a general anisotropic case, the fourth-order elasticity tensor C_{ijkl} has 21 independent components. For isotropic materials, the constitutive relation is simplified to

$$\sigma_{ij} = C_{ijkl}\varepsilon_{kl} = 2G\varepsilon_{ij} + \lambda(tr\boldsymbol{\varepsilon})\delta_{ij} \quad \boldsymbol{\sigma} = 2G\boldsymbol{\varepsilon} + \lambda(tr\boldsymbol{\varepsilon})\boldsymbol{I} \tag{2.124}$$

Equation (2.124) contains only two independent elastic constants, G and λ.

For fully constrained ($\varepsilon_{kl} = 0$) thermoelastic problems, Eq. (2.123) is simplified to

$$\sigma_{ij} = b_{ij}\theta \tag{2.125}$$

As temperature θ is a scalar, the above equation requires that b_{ij}, similar to σ_{ij}, be a second-order symmetric tensor, and this term is referred to as the thermal modulus. There are six independent thermal moduli in a general anisotropic case. We know that a second-order symmetric tensor has three principal directions and three corresponding principal components. For isotropic materials, the values of thermal moduli are independent of the direction. Thus, b_{ij} should be a spherical tensor with three principal components equal in value and can be denoted by

$$b_{ij} = \beta\delta_{ij}, \qquad \boldsymbol{b} = \beta\boldsymbol{I} \tag{2.126}$$

The substitution of Eqs. (2.124) and (2.126) into Eq. (2.123) yields a constitutive relation for isotropic thermoelastic materials with stresses expressed in terms of strains:

$$\sigma_{ij} = 2G\varepsilon_{ij} + \lambda(tr\boldsymbol{\varepsilon})\delta_{ij} + \beta\theta\delta_{ij} \quad \boldsymbol{\sigma} = 2G\boldsymbol{\varepsilon} + \lambda(tr\boldsymbol{\varepsilon})\boldsymbol{I} + \beta\theta\boldsymbol{I} \tag{2.127}$$

A constitutive relation for isotropic thermoelastic materials with strains expressed in terms of stresses can also be obtained:

$$\varepsilon_{ij} = \frac{1}{2G}\left(\sigma_{ij} - \frac{\lambda}{2G + 3\lambda}\sigma_{kk}\delta_{ij}\right) + \alpha\theta\delta_{ij} \tag{2.128}$$

$$\boldsymbol{\varepsilon} = \frac{1}{2G}\left[\boldsymbol{\sigma} - \frac{\lambda}{2G + 3\lambda}(tr\boldsymbol{\sigma})\boldsymbol{I}\right] + \alpha\theta\boldsymbol{I} \tag{2.129}$$

where

$$\alpha = -\frac{1 - 2v}{\mathrm{E}}\beta \text{ or } \beta = -\frac{\mathrm{E}}{1 - 2v}\alpha \tag{2.130}$$

As seen in Eq. (2.129), the physical meaning of α is the normal strain resulting from an increase in temperature by one degree in the absence of stresses ($\sigma_{ij} = 0$), i.e., the coefficient of linear expansion. Equation (2.128) or (2.129), which expresses strains in terms of stresses, is referred to as the constitutive relation, whereas Eq. (2.127), which expresses stresses in terms of strains, is called the inverse constitutive relation. This is because strains can be superposed but stresses cannot in the case of small deformation.

Equation (2.127) is expanded in a Cartesian coordinate system as follows:

$$\begin{aligned} \sigma_x &= 2G\varepsilon_x + \lambda\varepsilon_v - \tfrac{\alpha \mathrm{E}\theta}{1-2v}, \quad \tau_{yz} = G\gamma_{yz} \\ \sigma_y &= 2G\varepsilon_y + \lambda\varepsilon_v - \tfrac{\alpha \mathrm{E}\theta}{1-2v}, \quad \tau_{xz} = G\gamma_{xz} \\ \sigma_z &= 2G\varepsilon_z + \lambda\varepsilon_v - \tfrac{\alpha \mathrm{E}\theta}{1-2v}, \quad \tau_{xy} = G\gamma_{xy} \end{aligned} \tag{2.131}$$

where $\varepsilon_v = \varepsilon_x + \varepsilon_y + \varepsilon_z$ is the volumetric strain. Equation (2.129) has the following expanded form in a Cartesian coordinate system:

$$\begin{aligned}
\varepsilon_x &= \frac{1}{\mathrm{E}}\left[\sigma_x - \nu\left(\sigma_y + \sigma_z\right)\right] + \alpha\theta, \quad \gamma_{yz} = \frac{1}{G}\tau_{yz} \\
\varepsilon_y &= \frac{1}{\mathrm{E}}\left[\sigma_y - \nu(\sigma_z + \sigma_x)\right] + \alpha\theta, \quad \gamma_{xz} = \frac{1}{G}\tau_{xz} \\
\varepsilon_z &= \frac{1}{\mathrm{E}}\left[\sigma_z - \nu\left(\sigma_x + \sigma_y\right)\right] + \alpha\theta, \quad \gamma_{xy} = \frac{1}{G}\tau_{xy}
\end{aligned} \tag{2.132}$$

By substituting Eq. (2.121) into Eq. (2.117) (noting that $\dot{\xi} = 0$, $\dot{\boldsymbol{\varepsilon}}^{\mathrm{p}} = \mathbf{0}$) and letting $d' = T_0^2 d$, we can obtain a thermal conduction equation for linear thermoelastic bodies as follows:

$$\nabla \cdot \boldsymbol{q} = T_0\left(\boldsymbol{b} : \dot{\boldsymbol{\varepsilon}} + d\dot{T}\right) + \rho r_m \tag{2.133}$$

Since $|\theta'| << 1$, T is approximated by T_0 in the above equation. By substituting Eqs. (2.105) and (2.126) into Eq. (2.133) and using Eq. (2.130), we have

$$(-T_0 d)\dot{T} - (kT_{,i})_{,i} = \rho r_m - \frac{\mathrm{E}\alpha}{1-2\nu} T_0 \dot{\varepsilon}_{kk} \tag{2.134}$$

The above equation can also be rewritten as follows:

$$\rho C_v \frac{\partial T}{\partial t} = \left(kT_{,i}\right)_{,i} + \rho r_m - \frac{\mathrm{E}\alpha}{1-2\nu} T_0 \dot{\varepsilon}_{kk} \tag{2.135}$$

where C_{v} is the specific heat in the absence of strains, or the constant-volume specific heat. For a solid, its constant-pressure specific heat C_{p} differs insignificantly from its C_{v}. Hence, C_{p} and C_{v} are not distinguished in subsequent analyses. For isotropic materials, pure shear deformation does not generate a thermal effect, and only volumetric changes ($\dot{\varepsilon}_{kk} \neq 0$) lead to a coupling between strain and temperature fields. The coupling term $-\frac{\mathrm{E}\alpha}{1-2\nu} T_0 \dot{\varepsilon}_{kk}$ can be viewed as an additional quantity of the heat source ρr_m, which releases heat during compression ($\dot{\varepsilon}_{kk} < 0$) and absorbs heat during expansion ($\dot{\varepsilon}_{kk} > 0$).

In general, thermoelastic theories that account for the thermo-solid coupling effect, the temperature, displacement, and stress within an elastic body can be determined only by simultaneously solving a thermal conduction equation (which contains an additional strain-rate term, as shown in Eq. 2.134) and a thermal stress problem (which involves equilibrium and geometric equations as well as a thermoelastic constitutive relation that contains an additional temperature term, as shown in Eq. 2.129), which is a difficult task. Many real-world engineering problems can be simplified to static problems in elasticity involving slow loading (i.e., $\dot{\varepsilon}_{kk} = 0$). Therefore, the additional strain term in the thermal conduction equation can be neglected, and thus, these problems reduce to uncoupled thermoelastic problems that consider only the influence of temperature on elastic deformation, which are also known as thermal stress problems.

2.2.4 *Constitutive Theory for Thermo–Mechano-Chemical Coupling Based on Small Deformation [16]*

(1) **Governing equations for infiltration and chemical reaction fields**

Fick's laws are often used to describe the diffusion and migration of materials. Fick's first law states that the flow of a diffusate per unit time through a unit cross-sectional area perpendicular to the diffusion direction (referred to as the diffusion flux j) is inversely proportional to the gradient of the chemical potential μ at the cross-section. It is mathematically expressed as follows:

$$j = -m\nabla\mu \tag{2.136}$$

where m is the migration rate of the diffusate and μ is the chemical potential, which is related to the material concentration. The driving force for material diffusion or infiltration originates from the decrease in the internal energy of the system. The negative sign in Eq. (2.136) means that a material always migrates from high to low chemical potential, i.e., in the opposite direction of the chemical potential gradient. Let us assume that the chemical potential μ is directly proportional to the material concentration c, i.e., the higher the material concentration is, the higher the stored potential energy is:

$$\mu = k_\mu c \tag{2.137}$$

where k_μ is a constant of proportionality. Substituting Eq. (2.137) into Eq. (2.136) gives

$$j = -m\nabla k_\mu c \tag{2.138}$$

If k_μ is independent of the spatial coordinates (i.e., k_μ has the same value at each point within the material), k_μ can be taken out of the gradient operator ∇, i.e.,

$$j = -mk_\mu\nabla c \tag{2.139}$$

Equation (2.138) indicates that a material always diffuses or migrates from high to low concentration. A comparison of Eqs. (2.139) and (2.105) shows that the governing equations for thermal conduction and material diffusion have similar forms and that in each case, the flow direction goes from a place where the field variable has a large value to a place where the field variable has a low value.

For an object with a fixed volume, the change in its material concentration c per unit time is equal to the difference between the amounts of material flowing into and out of it through its boundary per unit time minus the amount of material consumed by chemical reactions within it per unit time, [17] i.e.,

$$\int_{\Omega} \dot{c} \mathrm{d}\Omega = -\int_{\Gamma} \boldsymbol{n} \cdot \boldsymbol{j} \mathrm{d}\Gamma - \int_{\Omega} N \mathrm{d}\Omega \tag{2.140}$$

This equation is referred to as the concentration conservation equation, where $\boldsymbol{n}$ is the outer normal direction of the boundary surface of the object, with the negative sign in front of it signifying that the material diffuses from the outside of the object into the inside of the object in the direction opposite to the outer normal direction of the surface of the object. The last term in Eq. (2.140) represents the consumption due to chemical reactions, where N is the material concentration consumed by chemical reactions per unit time. By transforming the surface integral into a volume integral using Green's theorem and by removing the integral signs considering that the concentration conservation equation holds for regions of any volumes within the object, we have

$$\dot{c} = -\nabla \cdot \boldsymbol{j} - N \tag{2.141}$$

Substitution of Eq. (2.139) into Eq. (2.141) gives the following governing equation for the concentration field:

$$\dot{c} = k_{\mu} \nabla \cdot (m \nabla c) - N \tag{2.142}$$

If the migration rate m is independent of spatial coordinates, it can be taken out of the gradient operator. Thus, we have

$$\dot{c} = k_{\mu} m \nabla^2 c - N \tag{2.143}$$

$k_{\mu} m$ is denoted as the material-diffusion coefficient D, and Eq. (2.143) becomes the common diffusion equation.

Two basic problems need to be considered for a chemical reaction—one is the direction and extent of the reaction, which falls under the scope of chemical thermodynamics, and the other is the rate of the reaction, which is a problem in the area of chemical kinetics. Let us take the combination of hydrogen and oxygen in water as an example. The molar Gibbs free energy of this reaction is –237.2 kJ/mol, suggesting a strong reaction trend. However, no traces of water formed from this reaction can be detected even for several years at normal temperature simply because the reaction rate is too low under this condition. At normal temperatures, the bond coat (BC) layer of a TBC hardly reacts with oxygen in air. However, at high temperatures, the BC layer oxidizes to form a layer of thermal growth oxides (TGOs). A reaction proceeding at a rate directly proportional to the square of the concentration of the reactant (or the product of the concentrations of the two reactants) is called a second-order reaction. For example, the oxidation reaction of the BC layer of a TBC is given by [18]

$$\dot{n} = k_n(1 - n) \cdot c \tag{2.144}$$

where *n* is the volume fraction of the product and represents the extent of the chemical reaction (in this case, *n* is the volume fraction of TGOs). Of particular note is that *n* is used here to represent the volume fraction of the product, whereas boldfaced $\boldsymbol{n}$ is used to represent the outer normal direction of an arbitrary domain Ω. In Eq. (2.144), $\dot{n}$ denotes derivation with respect to time, (1-*n*) is the volume fraction of the unoxidized part of the BC layer, k_n is the reaction rate constant, and *c* is the oxygen concentration. Temperature has a more significant effect on the rate of a chemical reaction than concentration. Generally, the reaction rate constant increases rapidly as temperature rises. By summarizing extensive experimental data, Arrhenius et al. established the following empirical equation to describe the relationship between the reaction rate constant and the reaction temperature:

$$k_n = A_{\mathrm{P}}\exp[-E_{\mathrm{a}}/(RT)] \tag{2.145}$$

where E_{a} is the activation energy, which can generally be viewed as a constant independent of temperature (unit: $J \cdot \mathrm{mol}^{-1}$), and *R* is the ideal gas constant, which is generally set to 8.314 $J \cdot \mathrm{mol}^{-1} \cdot K^{-1}$. The consumption term in Eq. (2.140) is related to the amount of the product of the reaction, *n*, through the following equation:

$$N = M\frac{\partial n}{\partial t} \tag{2.146}$$

For interfacial oxidation, *M* in Eq. (2.146) is the molar concentration of oxygen in Al_2O_3. Substituting Eq. (2.146) into Eq. (2.143) yields

$$\dot{c} = k_\mu m \nabla^2 c - M\frac{\partial n}{\partial t} \tag{2.147}$$

(2) **Energy conservation and constitutive equations**

We ignore the impact of a system's internal kinetic energy on its total energy and consider that changes in its internal energy are derived from the work done by external forces, the heat absorbed from external sources, and the amount of material that infiltrates its boundary. In this case, the rate of change in the internal energy within an arbitrary domain Ω is given by [19]

$$\int_{\Omega} \dot{e}_{\mathrm{v}} \mathrm{d}\Omega = \int_{\Omega} (\boldsymbol{\sigma} \cdot \boldsymbol{n}) \cdot \boldsymbol{v} \mathrm{d}\Omega + \int_{\Omega} \boldsymbol{f} \cdot \boldsymbol{v} \mathrm{d}\Omega - \int_{\Gamma} \boldsymbol{q} \cdot \boldsymbol{n} \mathrm{d}\Gamma + \int_{\Omega} r_{\mathrm{v}} \mathrm{d}\Omega - \int_{\Gamma} \mu \boldsymbol{j} \cdot \boldsymbol{n} \mathrm{d}\Gamma \tag{2.148}$$

where e_{v} is the internal energy of the material per unit volume and r_{v} is the heat received by the material per unit volume from the heat source. On the right-hand side of Eq. (2.148), the first term represents the power of the surface force acting on the surface of domain Ω, the second term represents the power of the volume force, the third and fourth terms represent the rates of change in the system's internal energy caused by the heat flow and heat source, respectively, and the final term represents the rate of change in the system's internal energy due to the diffusion or infiltration of external material into domain Ω via its boundary. As Ω is an arbitrary domain, Eq. (2.148) can be written in the following local form:

$$\dot{e}_{\mathrm{v}} = (\nabla \cdot \boldsymbol{\sigma}) \cdot \boldsymbol{v} + \boldsymbol{\sigma} : \nabla \boldsymbol{v} + \boldsymbol{f}_{\mathrm{v}} \cdot \boldsymbol{v} - \nabla \cdot \boldsymbol{q} + r_{\mathrm{v}} - \nabla \mu \cdot \boldsymbol{j} - \mu \nabla \cdot \boldsymbol{j} \tag{2.149}$$

By substituting the stress equilibrium equation in Eq. (2.56) and the concentration conservation equation in Eq. (2.141) into Eq. (2.149), we have

$$\dot{e}_{\mathrm{v}} = \boldsymbol{\sigma} : \dot{\boldsymbol{\varepsilon}} - \nabla \cdot \boldsymbol{q} + r_{\mathrm{v}} - \nabla \mu \cdot \boldsymbol{j} + \mu \dot{c} \tag{2.150}$$

Here, we introduce the second law of thermodynamics, which is expressed as follows:

$$\int_{\Omega} \dot{s}_{v} \mathrm{d}\Omega \geq - \int_{\Gamma} (\boldsymbol{q} \cdot \boldsymbol{n}/T) \mathrm{d}\Gamma + \int_{\Omega} (r_{v}/T) \mathrm{d}\Omega \tag{2.151}$$

where s_{v} is entropy per unit volume. The two terms on the right-hand side of Eq. (2.151) represent the rates of change in the entropy caused by the heat flow and heat source, respectively. By transforming the surface integral in Eq. (2.151) into a volume integral using Green's theorem and by removing the integral signs, we have

$$\dot{s}_{\mathrm{v}} \geq -\nabla \cdot (\boldsymbol{q}/T) + r_{v}/T \tag{2.152}$$

By introducing an entropy production rate per unit volume, we can rewrite Eq. (2.152) as follows:

$$\dot{s}_{v} = -\nabla \cdot (\boldsymbol{q}/T) + r_{v}/T + \kappa_{v} = (\boldsymbol{q} \cdot \nabla T)/T^{2} + (-\nabla \cdot \boldsymbol{q} + r_{v})/T + \kappa_{v} \tag{2.153}$$

where κ_{v} is the entropy production rate of the material per unit volume. Substituting Eq. (2.150) into Eq. (2.153) gives

$$\dot{s}_v = (\boldsymbol{q} \cdot \nabla T)/T^2 + (\dot{e}_v - \boldsymbol{\sigma} : \dot{\boldsymbol{\varepsilon}} - \mu \dot{c} + \nabla \mu \cdot \boldsymbol{j})/T + \kappa_v \tag{2.154}$$

By replacing the internal energy with the Helmholtz free energy ψ, as shown in Eq. (2.107), we have

$$T\kappa_v = -\dot{\psi} - \dot{T} s_v + \boldsymbol{\sigma} : \dot{\boldsymbol{\varepsilon}} + \mu \dot{c} - (\boldsymbol{q} \cdot \nabla T)/T - \nabla \mu \cdot \boldsymbol{j} \geq 0 \tag{2.155}$$

The Helmholtz free energy ψ is defined as a function of temperature T, concentration c, elastic strain $\boldsymbol{\varepsilon}^e$, internal variable ξ, and the extent of the reaction n:

$$\psi = \psi(T, \boldsymbol{\varepsilon}^e, c, \xi, n) \tag{2.156}$$

Taking the derivative of the above equation with respect to time t yields

$$\dot{\psi} = (\partial \psi / \partial T)\dot{T} + (\partial \psi / \partial \boldsymbol{\varepsilon}^e) : \dot{\boldsymbol{\varepsilon}}^e + (\partial \psi / \partial c)\dot{c} + (\partial \psi / \partial n)\dot{n} + (\partial \psi / \partial \xi)\dot{\xi} \tag{2.157}$$

The strain can be decomposed as follows:

$$\boldsymbol{\varepsilon} = \boldsymbol{\varepsilon}^e + \boldsymbol{\varepsilon}_T(T) + \boldsymbol{\varepsilon}_{diff}(c) + \boldsymbol{\varepsilon}_{chem}(n) + \boldsymbol{\varepsilon}_\xi(\xi) \tag{2.158}$$

where $\boldsymbol{\varepsilon}_T(T)$ is the thermal strain, $\boldsymbol{\varepsilon}_{diff}(c)$ is the strain caused by material diffusion, $\boldsymbol{\varepsilon}_{chem}(n)$ is the strain caused by chemical reactions, and $\boldsymbol{\varepsilon}_\xi(\xi)$ is the strain caused by the remaining internal variables (e.g., phase transformation). By substituting Eqs. (2.157) and (2.158) into Eq. (2.155) and organizing the terms, we have

$$\begin{aligned}
&(\boldsymbol{\sigma} - \partial \psi / \partial \boldsymbol{\varepsilon}_e) : \dot{\boldsymbol{\varepsilon}}_e + (\boldsymbol{\sigma} : (\partial \boldsymbol{\varepsilon}_\xi / \partial \xi) - \partial \psi / \partial \xi)\dot{\xi} + (\boldsymbol{\sigma} : (\partial \boldsymbol{\varepsilon}_T / \partial T) - \partial \psi / \partial T - s_v)\dot{T} + \\
&(\boldsymbol{\sigma} : (\partial \boldsymbol{\varepsilon}_{diff} / \partial c) - \partial \psi / \partial c + \mu)\dot{c} + (\boldsymbol{\sigma} : (\partial \boldsymbol{\varepsilon}_{chem} / \partial n) - \partial \psi / \partial n)\dot{n} - \\
&(\boldsymbol{q} \cdot \nabla T)/T - \nabla \mu \cdot \boldsymbol{j} = \kappa_v T \geq 0
\end{aligned} \tag{2.159}$$

In the above equation, $\boldsymbol{\varepsilon}^e$, T, and c can take any values within their respective value ranges. Thus, the coefficients of $\dot{\boldsymbol{\varepsilon}}^e$, $\dot{T}$, and $\dot{c}$ must be zero. Thus, three generalized constitutive relations, which correspond to stress and strain, entropy and temperature, and chemical potential and concentration, respectively, are obtained as follows:

$$\boldsymbol{\sigma} = \partial \psi / \partial \boldsymbol{\varepsilon}^e \tag{2.160}$$

$$s_v = \boldsymbol{\sigma} : (\partial \boldsymbol{\varepsilon}_T / \partial T) - \partial \psi / \partial T \tag{2.161}$$

$$\mu = -\boldsymbol{\sigma} : (\partial \boldsymbol{\varepsilon}_{diff} / \partial c) + \partial \psi / \partial c \tag{2.162}$$

Substitution of the above three equations into Eq. (2.159) gives

$$\begin{aligned}\kappa_v T = -(\boldsymbol{q}\cdot\nabla T)/T - \nabla\mu\cdot\boldsymbol{j} + (\boldsymbol{\sigma}:(\partial\boldsymbol{\varepsilon}_{\text{chem}}/\partial n) - \partial\psi/\partial n)\dot{n} + \\ (\boldsymbol{\sigma}:(\partial\boldsymbol{\varepsilon}_{\xi}/\partial\xi) - \partial\psi/\partial\xi)\dot{\xi}\end{aligned} \tag{2.163}$$

Equation (2.163) is an expression of the total dissipation rate of the system. The four terms on the right-hand side of Eq. (2.163) are defined, from left to right, as the rate of thermal dissipation, the rate of dissipation due to infiltration or diffusion, the rate of dissipation due to chemical reactions, and the rate of dissipation due to the remaining irreversible physical processes (e.g., plastic deformation). The four rates of dissipation are given by

$$T\kappa_{\text{chem}} = (\boldsymbol{\sigma}:(\partial\boldsymbol{\varepsilon}_{\text{chem}}/\partial n) - \partial\psi/\partial n)\dot{n} \tag{2.164}$$

$$T\kappa_T = -(\boldsymbol{q}\cdot\nabla T)/T \tag{2.165}$$

$$T\kappa_{diff} = -\nabla\mu\cdot\boldsymbol{j} \tag{2.166}$$

$$T\kappa_{\xi} = (\boldsymbol{\sigma}:(\partial\boldsymbol{\varepsilon}_{\xi}/\partial\xi) - \partial\psi/\partial\xi)\dot{\xi} \tag{2.167}$$

If the chemical reaction, heat-flow, and material-diffusion processes are also irreversible, then $T\kappa_{\text{chem}} \geq 0$, $T\kappa_T \geq 0$, and $T\kappa_{diff} \geq 0$. We neglect the coupling between the temperature field and other field variables and substitute Eq. (2.105) into Eq. (2.165) to obtain

$$T\kappa_T = (k_q\nabla T\cdot\nabla T)/T \geq 0 \tag{2.168}$$

Neglecting the coupling between the concentration field and other field variables, we substitute Eqs. (2.137) and (2.39) into Eq. (2.166) to obtain

$$T\kappa_{diff} = mk_{\mu}^{2}\nabla c\cdot\nabla c \geq 0 \tag{2.169}$$

As demonstrated in Eqs. (2.168) and (2.169), the rates of dissipation due to heat flow and material diffusion are nonnegative; that is, these two processes are irreversible.

If the chemical reaction processes are also considered irreversible, it is commonly assumed that $\dot{n}$ and its coefficient in Eq. (2.164) have a directly proportional relationship:

$$\dot{n} = k_{chem}(\boldsymbol{\sigma}:(\partial\boldsymbol{\varepsilon}_{\text{chem}}/\partial n) - \partial\psi/\partial n) \tag{2.170}$$

where k_{chem} is the coefficient of proportionality and has a positive value. By substituting Eq. (2.170) into Eq. (2.164), we have

$$T\kappa_{\text{chem}} = k_{chem}(\boldsymbol{\sigma}:(\partial\boldsymbol{\varepsilon}_{\text{chem}}/\partial n) - \partial\psi/\partial n)^2 \geq 0 \tag{2.171}$$

We obtain an expression of the free energy function ψ by performing a second-order series expansion of ψ with respect to independent variables, including temperature T, concentration c, elastic strain $\boldsymbol{\varepsilon}^e$, internal variable ξ, and the extent of the reaction n: [20].

$$\begin{aligned}
\psi = {} & \psi_0 + \sigma_{\text{m}}^0\varepsilon + s_{ij}^0 e_{ij} + \mu^0 c - s^0 T - A^0 n - \Gamma^0\xi + \frac{1}{2}K\varepsilon^2 + \varepsilon R(c - c_0) - \\
& 3\alpha K\varepsilon(T - T_0) - 3K\varepsilon\beta(n - n_0) - \varepsilon\zeta(\xi - \xi_0) + Ge_{ij}e_{ij} + \frac{1}{2}O(c - c_0)^2 \\
& + (T - T_0)\nu_0(c - c_0) - \\
& \zeta^n(c - c_0)(n - n_0) + \zeta^\mu\left(\xi - \xi^0\right)(c - c_0) - \frac{1}{2}C\frac{(T - T_0)^2}{T_0} - \frac{T - T_0}{T_0}L_n(n - n_0) \\
& - \frac{T - T_0}{T_0}\zeta^T\left(\xi - \xi^0\right) - \\
& \frac{1}{2}\alpha_n(n - n_0)^2 - \zeta^A\left(\xi - \xi^0\right)(n - n_0) - \frac{1}{2}\pi^0(\xi - \xi_0)^2
\end{aligned} \tag{2.172}$$

where

$$\sigma_{\text{m}} = \frac{1}{3}\sigma_{kk}, s_{ij} = \sigma_{ij} - \sigma_m\delta_{ij}, \varepsilon = \varepsilon_{kk}, e_{ij} = \varepsilon_{ij} - \frac{1}{3}\varepsilon\delta_{ij} \tag{2.173}$$

In Eq. (2.172), K and G are the bulk modulus and the shear modulus, respectively; σ_{m}^0 and s_{ij}^0 are the residual spherical stress tensor and the residual stress deviator tensor, respectively; μ^0, s^0, A^0, Γ^0, R, ζ, O, ν_0, ζ^{n}, ζ^μ, C, L_n, ζ^T, α_n, ζ^A, and π^0 are constants of proportionality; and 3α and 3β are the coefficient of thermal expansion and the coefficient of chemical expansion, respectively.

By substituting Eq. (2.172) into Eq. (2.160), we obtain the constitutive relation as follows:

$$\begin{aligned}
\boldsymbol{\sigma} = {} & \sigma_{\text{m}}^0\boldsymbol{I} + \boldsymbol{s}^0 + K\varepsilon\boldsymbol{I} + 2G\boldsymbol{e} + R(c - c_0)\boldsymbol{I} - 3\alpha K(T - T_0)\boldsymbol{I} \\
& - 3\beta K(n - n_0)\boldsymbol{I} - \zeta(\xi - \xi_0)\boldsymbol{I}
\end{aligned} \tag{2.174}$$

Substitution of Eq. (2.172) into Eq. (2.162) yields

$$\mu = \mu_0 + \varepsilon R + \text{O}(c - c_0) + \nu_0(T - T_0) - \zeta^n(n - n_0) + \zeta^\mu(\xi - \xi_0) \tag{2.175}$$

By substituting Eq. (2.175) into Eq. (2.136) and then substituting the resulting equation into Eq. (2.141), we obtain the governing equation for the concentration as follows:

$$\dot{c} = \mathrm{m}\nabla^2\big(\varepsilon R + \mathrm{O}(c - c_0) + \nu_0(T - T_0) - \zeta^n(n - n_0) + \zeta^\mu(\xi - \xi_0)\big) - M\frac{\partial n}{\partial t} \tag{2.176}$$

If we neglect the influence of other field variables on the evolution of the concentration c and retain only the term $\mathrm{O}(c - c_0)$ (noting that here O is a coefficient, not a usual higher-order term), Eq. (2.143) is obtained.

Practice problem: Derive the expression of the governing equation for the reaction field n.

By substituting Eq. (2.107) into Eq. (2.150) and considering Eqs. (2.161) and (2.162), we can obtain the following governing equation for the temperature field T:

$$\begin{aligned}
&\frac{\partial\psi}{\partial n}\dot{n} + \frac{\partial\psi}{\partial\xi}\dot{\xi} + \dot{T}\boldsymbol{\sigma} : \frac{\partial\boldsymbol{\varepsilon}_T}{\partial T} + T\begin{bmatrix}\frac{\mathrm{d}\left(\boldsymbol{\sigma} : \frac{\partial\boldsymbol{\varepsilon}_T}{\partial T}\right)}{\mathrm{d}t} - \frac{\partial^2\psi}{\partial T^2}\dot{T} - \frac{\partial^2\psi}{\partial T\partial c}\dot{c} - \\ \frac{\partial^2\psi}{\partial T\partial n}\dot{n} - \frac{\partial^2\psi}{\partial T\partial\xi}\dot{\xi} - \frac{\partial^2\psi}{\partial T\partial\boldsymbol{\varepsilon}_e} : \boldsymbol{\varepsilon}_e\end{bmatrix} \\
&= \boldsymbol{\sigma} : (\dot{\boldsymbol{\varepsilon}} - \dot{\boldsymbol{\varepsilon}}_e) + k\nabla^2 T + r_v + \\
&m\left(\nabla\left(\begin{matrix}-\boldsymbol{\sigma} : \frac{\partial\boldsymbol{\varepsilon}_{diff}}{\partial c} + \\ \frac{\partial\psi}{\partial c}\end{matrix}\right)\right)^2 - \dot{c}\boldsymbol{\sigma} : \frac{\partial\boldsymbol{\varepsilon}_{diff}}{\partial c}
\end{aligned} \tag{2.177}$$

Practice problem: Derive Eq. (2.177) in detail.

2.3 Theoretical Framework for Thermo–Mechano-Chemical Coupling Based on Large Deformation

2.3.1 Kinematic Description [9]

Strains up to 8–10% can occur in TBCs during interfacial oxidation or calcium–magnesium aluminosilicate (CMAS) corrosion, for which the small-deformation theory is no longer applicable. In this case, the large-deformation theory, also known as the geometrically nonlinear theory, is required. When the motion and deformation of an object are described, a configuration is first needed as a reference for various physical quantities. This configuration is often referred to as the reference configuration, in which $\boldsymbol{X}$ and variable X_A are used to represent the position vector of a material point and material coordinates, respectively. Then, it is also necessary to describe the position of each material point at any time point t, referred to as the current configuration, in which $\boldsymbol{x}$ and variable x_i are used to represent the position vector and spatial coordinates, respectively, of material point $\boldsymbol{X}$ at time point t.

The motion of an object can be described as $\boldsymbol{x} = \boldsymbol{\chi}(\boldsymbol{X},\mathrm{t})$. The function $\boldsymbol{\chi}(\boldsymbol{X},\mathrm{t})$ maps the reference configuration to the current configuration at time point t. Thus, the displacement of a material point is given by

$$\boldsymbol{u} = \boldsymbol{\chi}(\boldsymbol{X}, t) - \boldsymbol{\chi}(\boldsymbol{X}, 0) = \boldsymbol{x} - \boldsymbol{X} \tag{2.178}$$

In large-deformation theory, the deformation gradient is an important physical quantity regarding deformation characteristics. It is defined as follows:

$$\boldsymbol{F} = \frac{\partial \boldsymbol{x}}{\partial \boldsymbol{X}} \tag{2.179}$$

The deformation gradient is usually decomposed in the analysis of large elasto-plastic deformation. Similarly, under multifield (thermal, mechanical, and chemical fields) coupling, deformation is described using three different configurations, namely a reference configuration, an intermediate configuration, and a current configuration. Let $\boldsymbol{X}$, $\overline{\boldsymbol{X}}$, and $\boldsymbol{x}$ be the position coordinates in the three configurations, respectively. Thus, the deformation gradient $\boldsymbol{F}$, the elastic deformation gradient $\boldsymbol{F}^{\mathrm{e}}$, and the inelastic deformation gradient $\boldsymbol{F}^{\mathrm{i}}$ are given by

$$\boldsymbol{F} = \frac{\partial \boldsymbol{x}}{\partial \boldsymbol{X}}, \boldsymbol{F}^{e} = \frac{\partial \boldsymbol{x}}{\partial \overline{\boldsymbol{X}}}, \boldsymbol{F}^{i} = \frac{\partial \overline{\boldsymbol{X}}}{\partial \boldsymbol{X}} \tag{2.180}$$

The deformation gradient $\boldsymbol{F}$ is decomposed by using the rule for finding the derivative of a composite function:

$$\boldsymbol{F} = \boldsymbol{F}^{e} \cdot \boldsymbol{F}^{i} \tag{2.181}$$

This decomposition of the deformation gradient has already been widely used in the analysis of large elastoplastic deformation and has recently been extensively employed to analyze large deformation under multifield coupling. For example, Nguyen et al. [21] decomposed the deformation gradient into an elastic deformation gradient and an inelastic deformation gradient and, on this basis, studied large deformation in soft materials under thermo–mechano-chemical coupling. Yadegari et al. [22] decomposed the deformation gradient into an elastic deformation gradient, a thermal deformation gradient, a plastic deformation gradient, and a phase-transformation deformation gradient.

The ratio J of the volume of a volume element before deformation ($d\Omega_0$) to its volume after deformation ($d\Omega$) can be expressed as follows using the determinant of the deformation gradient:

$$
\begin{aligned}
J &= \frac{\mathrm{d}\Omega}{\mathrm{d}\Omega_0} = \det \boldsymbol{F} = \det(\boldsymbol{F}^{\mathrm{e}} \cdot \boldsymbol{F}^{\mathrm{i}}) \\
&= \det \boldsymbol{F}^{\mathrm{e}} \cdot \det \boldsymbol{F}^{\mathrm{i}} = J^e \cdot J^i
\end{aligned}
\tag{2.182}
$$

where J^e and J^i are the Jacobian determinants of the elastic and inelastic deformation gradients, respectively, reflecting the changes in the volume of the volume element due to elastic and inelastic deformations, respectively. The material derivative of the volume ratio is often used in equilibrium equations in continuum mechanics. It is expressed as follows:

$$
\begin{aligned}
\dot{J} &= \frac{\partial J}{\partial F_{iA}} \frac{dF_{iA}}{dt} = \frac{\partial J}{\partial F_{iA}} \frac{\partial}{\partial t}\left(\frac{\partial x_i}{\partial X_A}\right) \\
&= \frac{\partial J}{\partial F_{iA}} \frac{\partial v_i}{\partial X_A} = J \frac{\partial X_A}{\partial x_i} \frac{\partial v_i}{\partial X_A} \\
&= J \frac{\partial v_i}{\partial x_i} = J div \boldsymbol{v}
\end{aligned}
\tag{2.183}
$$

Here, the relation $\frac{\partial J}{\partial F_{iA}} = J \frac{\partial X_A}{\partial x_i}$ is used. Interested readers can prove this relation as a practice problem using the theory of determinants and referring to pages 165–166 in Ref. [9]. Based on the equation of motion, the velocity $\boldsymbol{v}$ and velocity gradient $\boldsymbol{l}$ of an object are calculated as follows:

$$
\boldsymbol{v} = \dot{\boldsymbol{\chi}}(\boldsymbol{X}, t) \tag{2.184}
$$

$$
\boldsymbol{l} = grad \boldsymbol{v} = \dot{\boldsymbol{F}} \cdot \boldsymbol{F}^{-1} \tag{2.185}
$$

By substituting the decomposition expression of the deformation gradient in Eq. (2.181) into Eq. (2.185), we obtain the decomposition expression of the velocity gradient:

$$
\begin{aligned}
\boldsymbol{l} &= \dot{\boldsymbol{F}} \cdot \boldsymbol{F}^{-1} \\
&= \left(\dot{\boldsymbol{F}}^{\mathrm{e}} \cdot \boldsymbol{F}^{\mathrm{i}} + \boldsymbol{F}^{\mathrm{e}} \cdot \dot{\boldsymbol{F}}^{\mathrm{i}}\right) \cdot \left(\boldsymbol{F}^{\mathrm{i}^{-1}} \cdot \boldsymbol{F}^{e^{-1}}\right) \\
&= \dot{\boldsymbol{F}}^{\mathrm{e}} \cdot \boldsymbol{F}^{e^{-1}} + \boldsymbol{F}^{\mathrm{e}} \cdot \dot{\boldsymbol{F}}^{\mathrm{i}} \cdot \boldsymbol{F}^{\mathrm{i}-1} \cdot \boldsymbol{F}^{e^{-1}} \\
&= \boldsymbol{l}^{\mathrm{e}} + \boldsymbol{F}^{\mathrm{e}} \cdot \overline{\boldsymbol{L}}^{\mathrm{i}} \cdot \boldsymbol{F}^{e^{-1}}
\end{aligned}
\tag{2.186}
$$

where $\overline{\boldsymbol{L}}^{\mathrm{i}}$ is the inelastic velocity gradient in the intermediate configuration. Equation (2.186) can be understood by the fact that the velocity gradient can be decomposed into an elastic velocity gradient $\boldsymbol{l}^{\mathrm{e}}$ and an inelastic velocity gradient $\boldsymbol{l}^{\mathrm{i}}$, expressed as follows:

$$l^{\mathrm{e}} = \dot{F}^{\mathrm{e}} \cdot F^{e^{-1}} \text{ and } l^{\mathrm{i}} = F^{\mathrm{e}} \cdot \overline{L}^{\mathrm{i}} \cdot F^{e^{-1}} \tag{2.187}$$

Thus, Eq. (2.186) can be understood by the fact that the velocity gradient can be decomposed into an elastic part and an inelastic part:

$$l = l^{\mathrm{e}} + l^{\mathrm{i}} \tag{2.188}$$

The velocity gradient can be additively decomposed into a symmetric part, which is referred to as the deformation rate or strain rate and is often denoted by $\boldsymbol{d}$, and an antisymmetric part, which is called the spin and is often denoted by $\boldsymbol{w}$. Thus, based on Eq. (2.188), the corresponding elastic deformation rate, inelastic deformation rate, elastic spin, and inelastic spin are given by

$$\begin{aligned} d^e &= \frac{1}{2}(l^e + l^{eT}) \qquad & d^i &= \frac{1}{2}(l^i + l^{iT}) \\ w^e &= \frac{1}{2}(l^e - l^{eT}) \qquad & w^i &= \frac{1}{2}(l^i - l^{iT}) \end{aligned} \tag{2.189}$$

Hence, the deformation and spin can also each be decomposed into an elastic part and an inelastic part:

$$d = d^{\mathrm{e}} + d^{\mathrm{i}} \text{ and } w = w^{\mathrm{e}} + w^{\mathrm{i}} \tag{2.190}$$

This section focuses on the description of deformation in the multifield (thermal, mechanical, and chemical fields) coupling theory. By decomposing the deformation gradient into elastic and inelastic deformation gradients, we can define the elastic and inelastic velocity gradients accordingly and finally decompose the deformation rate into elastic and inelastic deformation rates. Readers interested in the theoretical knowledge in this area are advised to refer to monographs on continuum mechanics such as Advanced Solid Mechanics (Volume I) by Huang and Huang [9] The Mechanics and Thermodynamics of Continua by Gurtin et al. [23], and Introduction to Continuum Mechanics by Reddy [24].

2.3.2 *Stress and Strain Measures*

Appropriate state quantities must be selected as needed to measure the deformation at each point within an object. There are a variety of quantities suitable for measuring deformation, including deformation tensors and strain tensors.

Deformation tensors mainly include the right Cauchy–Green deformation tensor $\boldsymbol{C}$ (also known as the Green deformation tensor), the Cauchy deformation tensor $\boldsymbol{c}$, the left Cauchy–Green deformation tensor $\boldsymbol{B}$ (also known as the Finger deformation tensor), and the Piola deformation tensor $\boldsymbol{C}^{-1}$, which are given by

$$\boldsymbol{C} = \boldsymbol{F}^T \cdot \boldsymbol{F} \tag{2.191}$$

$$\boldsymbol{c} = \boldsymbol{F}^{-T} \cdot \boldsymbol{F}^{-1} \tag{2.192}$$

$$\boldsymbol{B} = \boldsymbol{F} \cdot \boldsymbol{F}^T \tag{2.193}$$

$$\boldsymbol{C}^{-1} = \boldsymbol{F}^{-1} \cdot \boldsymbol{F}^{-T} \tag{2.194}$$

Note that the Cauchy deformation tensor $\boldsymbol{c}$ differs from the material concentration c in Sect. 2.2.4.

Strain tensors mainly include the Green strain tensor $\boldsymbol{E}$, the Almansi strain tensor $\boldsymbol{e}$, the Hill strain tensor $\tilde{\boldsymbol{E}}$, and the Seth strain tensor $\boldsymbol{E}^{(n)}$, which are given by

$$\boldsymbol{E} = \frac{1}{2}(\boldsymbol{C} - \boldsymbol{I}) \tag{2.195}$$

$$\boldsymbol{e} = \frac{1}{2}(\boldsymbol{i} - \boldsymbol{c}) \tag{2.196}$$

$$\tilde{\boldsymbol{E}} = f(\lambda_{(i)})\boldsymbol{N}_i\boldsymbol{N}_i \tag{2.197}$$

$$\boldsymbol{E}^{(n)} = \frac{1}{2n}(\boldsymbol{C}^n - \boldsymbol{I}) \ \ (n \neq 0) \tag{2.198}$$

where $\boldsymbol{I}$ is the unit tensor in the reference configuration, $\boldsymbol{i}$ is the unit tensor in the current configuration, $\lambda_{(i)} = \sqrt{C_{(i)}}$ is the principal length ratio, and $f(\lambda_{(i)})$ is a monotonic, single-valued, continuous, and differentiable function of $\lambda_{(i)}$ that can be used to define new strain tensors. Regarding the Seth strain measure, different Seth strains are formed with different values of n. Note that the Green strain tensor $\boldsymbol{E}$ here differs from Young's modulus E and internal energy E discussed earlier and that the Almansi strain tensor $\boldsymbol{e}$ here differs from the internal energy per unit mass $\boldsymbol{e}$ discussed earlier.

The abovementioned deformation tensors and strain tensors are expressed using the deformation gradient tensor and its inverse tensor. In practical applications, these deformation and strain tensors are often expressed using the displacement gradient. The displacement gradient and the deformation gradient are related through the following equation:

$$\boldsymbol{F} = \frac{\partial \boldsymbol{x}}{\partial \boldsymbol{X}} = \frac{\partial \boldsymbol{u}}{\partial \boldsymbol{X}} + \boldsymbol{I} \tag{2.199}$$

The Green strain tensor $\boldsymbol{E}$ and the Almansi strain tensor $\boldsymbol{e}$ are expressed in terms of displacement as follows:

$$\boldsymbol{E} = \frac{1}{2}(\boldsymbol{u}\nabla_X + \nabla_X\boldsymbol{u} + \nabla_X\boldsymbol{u} \cdot \boldsymbol{u}\nabla_X) \tag{2.200}$$

$$\boldsymbol{e} = \frac{1}{2}(\boldsymbol{u}\nabla_x + \nabla_x\boldsymbol{u} + \nabla_x\boldsymbol{u} \cdot \boldsymbol{u}\nabla_x) \tag{2.201}$$

where ∇_X and ∇_x are the gradient operators in the reference and current configurations, respectively, also known as Hamilton operators.

Stress is a force per unit area and is a measure of a material's load-bearing capacity. All materials must be designed to operate within their respective load-bearing capacities. Thus, accurate prediction of the stress state of a material is vitally important for its design. A variety of stress measures can be defined with large-deformation theory, with some defined in the reference configuration and some in the current configuration. Commonly used stress measures include the Cauchy stress tensor $\boldsymbol{\sigma}$, the first Piola–Kirchhoff stress tensor $\boldsymbol{P}$, and the second Piola–Kirchhoff stress tensor $\boldsymbol{T}$ (noting that the $\boldsymbol{T}$ here differs from temperature T). Familiarity with the definitions of various stress tensors and their transformation relationships helps us to improve our proficiency in selecting suitable stress tensors for specific problems of mechanics. For example, in many cases, it is more convenient to use the equilibrium equations in the reference configuration than those in the current configuration, and hence, the first Piola–Kirchhoff stress tensor must be used to establish the equilibrium equations.

Under the small-deformation assumption, the Cauchy stress is often adopted to measure the stress state of an object. In cases of large deformation, the Cauchy stress is used to represent the stress state in the current configuration and to measure the force per unit area of deformation. Here, let us examine a continuum that changes from an undeformed state to a deformed state after deformation $\boldsymbol{\chi}(\boldsymbol{X}, t)$. The force in the current configuration is given by

$$d\boldsymbol{f} = \boldsymbol{\sigma} \cdot d\boldsymbol{a} \tag{2.202}$$

where $\boldsymbol{\sigma}$ is the Cauchy stress tensor in the current configuration and $d\boldsymbol{a}$ is a surface element in the current configuration. Let $d\boldsymbol{A}$ be the corresponding surface element in the reference configuration. Thus, the force is expressed as

$$d\boldsymbol{f} = \boldsymbol{P} \cdot d\boldsymbol{A} \tag{2.203}$$

where $\boldsymbol{P}$ is the first Piola–Kirchhoff stress tensor. The relationship between the surface elements in the current and reference configurations, respectively, is requisite for establishing the relationship between the first Piola–Kirchhoff stress tensor and the Cauchy stress tensor. According to Nanson's equation, the surface elements in the current and reference configurations are related through the following equation:

$$d\boldsymbol{a} = J\boldsymbol{F}^{-T} \cdot d\boldsymbol{A} \tag{2.204}$$

By substituting Eq. (2.204) into Eq. (2.202) and simultaneously solving the resulting equation and Eq. (2.203), we have

$$\boldsymbol{P} \cdot d\boldsymbol{A} = \boldsymbol{\sigma} \cdot d\boldsymbol{a} = \boldsymbol{\sigma} \cdot J\boldsymbol{F}^{-T} \cdot d\boldsymbol{A} \tag{2.205}$$

Thus, the relation between the first Piola–Kirchhoff stress tensor and the Cauchy stress tensor is given by

$$\boldsymbol{P} = J\boldsymbol{\sigma} \cdot \boldsymbol{F}^{-T} \tag{2.206}$$

As seen in the above equation, similar to the deformation gradient, the first Piola–Kirchhoff stress tensor is in fact a two-point tensor.

Similar to the relationship between line elements, the force $d\boldsymbol{F}$ on the surface element $d\boldsymbol{A}$ in the reference configuration and the force $d\boldsymbol{f}$ on the surface element $d\boldsymbol{a}$ in the current configuration follow the relation below

$$d\boldsymbol{F} = \boldsymbol{F}^{-1} \cdot d\boldsymbol{f} \tag{2.207}$$

The force in the reference configuration can be expressed using the second Piola–Kirchhoff stress tensor:

$$d\boldsymbol{F} = \boldsymbol{T} \cdot d\boldsymbol{A} \tag{2.208}$$

Based on Eqs. (2.203), (2.206), and (2.207), the relation between the first and second Piola–Kirchhoff stress tensors is given by

$$\boldsymbol{T} = \boldsymbol{F}^{-1} \cdot \boldsymbol{P} \tag{2.209}$$

By combining Eqs. (2.206) and (2.209), we obtain the relation between the second Piola–Kirchhoff stress tensor and the Cauchy stress tensor as follows:

$$\boldsymbol{T} = J\boldsymbol{F}^{-1} \cdot \boldsymbol{\sigma} \cdot \boldsymbol{F}^{-T} \tag{2.210}$$

In large-deformation theory, stress measures can be described with different stresses and transformed from one another. In practical applications, convenient stress measures can be flexibly selected according to actual problems.

2.3.3 *Mass Conservation and Force Equilibrium Equations*

(1) **Mass conservation equation**

The mass $m(\Omega)$ of a material domain Ω is given by

$$m(\Omega) = \int_{\Omega} \rho(\boldsymbol{X}, t) d\Omega \tag{2.211}$$

where $\rho(\boldsymbol{X}, t)$ is the density. Mass conservation requires that the mass of an arbitrary material domain be a constant. Thus, no material passes through the boundary of a material domain, nor is the conversion from mass to energy considered. According to the principle of conservation of mass, the material time derivative of $m(\Omega)$ is zero, i.e.,

$$\frac{\mathrm{D}m(\Omega)}{\mathrm{D}t} = \frac{\mathrm{D}}{\mathrm{D}t} \int_{\Omega} \rho(\boldsymbol{X}, t) d\Omega = 0 \tag{2.212}$$

By applying the Reynolds transport theorem to the above equation, we have

$$\int_{\Omega} (\frac{\mathrm{D}\rho}{\mathrm{D}t} + \rho \mathrm{div}(\boldsymbol{v})) d\Omega = 0 \tag{2.213}$$

Since the above equation holds for any subdomain, we have

$$\frac{\mathrm{D}\rho}{\mathrm{D}t} + \rho \mathrm{div}(\boldsymbol{v}) = 0 \tag{2.214}$$

Equation (2.214) is the mass conservation equation in the current configuration. Now, let us establish the mass conservation equation in the reference configuration. By integrating Eq. (2.212) with respect to time, we obtain an algebraic equation for density:

$$\int_{\Omega} \rho d\Omega = \int_{\Omega_0} \rho_0 d\Omega_0 = \text{Constant} \tag{2.215}$$

By transforming the integral on the left-hand side of the above equation to that over the reference domain, we have

$$\int_{\Omega_0} (\rho J - \rho_0) d\Omega_0 = 0 \tag{2.216}$$

As the above equation holds for anysubdomain, we have

$$\rho(\boldsymbol{X}, t) J(\boldsymbol{X}, t) = \rho_0(\boldsymbol{X}) \tag{2.217}$$

Equation (2.217) is the mass conservation equation in the reference configuration.

(2) **Force equilibrium equations**

Force equilibrium equations can be established based on the conservation of linear momentum, and they differ in different configurations. Here, let us establish force equilibrium equations in the current and reference configurations, respectively. Conservation of linear momentum can be described as follows: the time derivative of the linear momentum of an object is equal to the external force applied to the object. Thus, the linear momentum conservation equation in the current configuration is given by

$$\frac{D\boldsymbol{p}}{\mathrm{Dt}} = \boldsymbol{F} \tag{2.218}$$

In Eq. (2.218), $\boldsymbol{p}$ is the momentum of an object and is equal to the product of its mass and velocity, as expressed below

$$\boldsymbol{p}(t) = \int\limits_{\Omega} \rho \boldsymbol{v}(\boldsymbol{x}, t) d\Omega \tag{2.219}$$

and $\boldsymbol{F}$ is the external force, which is expressed in the current configuration as follows:

$$\boldsymbol{F}(t) = \int\limits_{\Omega} \rho \boldsymbol{f}(\boldsymbol{x}, t) d\Omega + \int\limits_{\Gamma} \boldsymbol{t}(\boldsymbol{x}, t) d\Gamma \tag{2.220}$$

The first and second terms on the right-hand side of the above equation represent the volume force and surface force, respectively. To preclude confusion with the second Piola–Kirchhoff stress tensor $\boldsymbol{T}$, $\boldsymbol{t}(\boldsymbol{x}, t)$ is used here to represent the surface force. By substituting Eqs. (2.219) and (2.220) into Eq. (2.218), we have

$$\frac{D}{Dt} \int\limits_{\Omega} \rho \boldsymbol{v}(\boldsymbol{x}, t) d\Omega = \int\limits_{\Omega} \rho \boldsymbol{f}(\boldsymbol{x}, t) d\Omega + \int\limits_{\Gamma} \boldsymbol{t}(\boldsymbol{x}, t) d\Gamma \tag{2.221}$$

Using the Reynolds transport theorem, the left-hand side of the above equation can be rewritten as follows:

$$\int\limits_{\Omega} \left(\frac{D}{Dt}(\rho \boldsymbol{v}) + div(\boldsymbol{v}) \rho \boldsymbol{v} \right) d\Omega = \int\limits_{\Omega} [\rho \frac{D}{Dt} \boldsymbol{v} + \boldsymbol{v} \left(\frac{D\rho}{Dt} + \rho div(\boldsymbol{v}) \right)] d\Omega \tag{2.222}$$

Based on the Cauchy relation and Gauss' theorem, the second term on the right-hand side of Eq. (2.221) is expressed as

$$\int\limits_{\Gamma} \boldsymbol{t}(\boldsymbol{x}, t) d\Gamma = \int\limits_{\Gamma} \boldsymbol{n} \cdot \boldsymbol{\sigma} d\Gamma = \int\limits_{\Omega} div \boldsymbol{\sigma} d\Omega \tag{2.223}$$

where $\boldsymbol{\sigma}$ is the Cauchy stress. According to the mass conservation equation, we have

$$\int_{\Omega} (div\boldsymbol{\sigma} + \rho \boldsymbol{f} - \rho \dot{\boldsymbol{v}}) d\Omega = \boldsymbol{0} \tag{2.224}$$

The above equation holds for any region. Thus, the force equilibrium equation in the current configuration is obtained as follows:

$$div\boldsymbol{\sigma} + \rho \boldsymbol{f} = \rho \dot{\boldsymbol{v}} \tag{2.225}$$

where $\dot{\boldsymbol{v}}$ is the material derivative of the velocity. For the TBCs studied in this book, loading is generally slow if the shock wave effect is not considered. Thus, the inertial force is negligible, and the force equilibrium equation becomes

$$div\boldsymbol{\sigma} + \rho \boldsymbol{f} = \boldsymbol{0} \tag{2.226}$$

Physical quantities in the reference configuration are often used in large-deformation analysis. In addition, description in the reference configuration is more convenient to use when solving equilibrium equations. Therefore, we derive in the following the force equilibrium equation in the reference configuration. Similarly, according to the conservation of linear momentum, we have

$$\frac{D\boldsymbol{p}}{Dt} = \boldsymbol{F} \tag{2.227}$$

where $\boldsymbol{p}$ is the momentum of an object and is the product of its mass and velocity, which is given by

$$\boldsymbol{p}(t) = \int_{\Omega_0} \rho_0 \boldsymbol{v}(\boldsymbol{X}, t) d\Omega_0 \tag{2.228}$$

Let $\boldsymbol{F}$ be the external force. Its expression in the reference configuration is given by

$$\boldsymbol{F}(t) = \int_{\Omega_0} \rho_0 \boldsymbol{f}(\boldsymbol{X}, t) d\Omega_0 + \int_{\Gamma_0} \boldsymbol{t}_0(\boldsymbol{X}, t) d\Gamma_0 \tag{2.229}$$

The two terms on the right-hand side of the above equation represent the volume force and surface force, respectively. As thermo–mechano-chemical coupling behaviors (e.g., interfacial oxidation or CMAS corrosion) are absent in the reference configuration, the external force does not include the force due to changes in mass. Substituting Eqs. (2.228) and (2.229) into Eq. (2.227) yields

$$\frac{d}{dt}\int_{\Omega_0} \rho_0 \boldsymbol{v}(\boldsymbol{X}, t) d\Omega_0 = \int_{\Omega_0} \rho_0 \boldsymbol{f}(\boldsymbol{X}, t) d\Omega_0 + \int_{\Gamma_0} \boldsymbol{t}_0(\boldsymbol{X}, t) d\Gamma_0 \tag{2.230}$$

By moving the material derivative to the inside of the integral on the left-hand side of the above equation, we have

$$\frac{d}{dt}\int_{\Omega_0} \rho_0 \boldsymbol{v}(\boldsymbol{X}, t) d\Omega_0 = \int_{\Omega_0} \rho_0 \frac{\partial \boldsymbol{v}(\boldsymbol{X}, t)}{\partial t} d\Omega_0 \tag{2.231}$$

Based on the Cauchy relation and Gauss' theorem, the second term on the right-hand side of Eq. (2.230) is expressed as

$$\int_{\Gamma_0} \boldsymbol{t}_0(\boldsymbol{X}, t) d\Gamma_0 = \int_{\Gamma_0} \boldsymbol{P} \cdot \boldsymbol{N} d\Gamma_0 = \int_{\Omega_0} \mathrm{Div} \boldsymbol{P}^T d\Omega_0 \tag{2.232}$$

where $\boldsymbol{P}$ is the first Piola–Kirchhoff stress and $\mathbf{N}$ is the outer normal direction of domain Ω_0 in the reference configuration. By substituting Eqs. (2.231) and (2.232) into Eq. (2.230), we obtain the force equilibrium equation in the reference configuration:

$$\int_{\Omega_0} (\boldsymbol{Div} \boldsymbol{P}^T + \rho_0 \boldsymbol{f} - \rho_0 \frac{\partial \boldsymbol{v}(\boldsymbol{X}, t)}{\partial t}) d\Omega_0 = \mathbf{0} \tag{2.233}$$

The above equation holds for any region. Thus, the force equilibrium equation in the reference configuration becomes:

$$\mathrm{Div} \boldsymbol{P}^T + \rho_0 \boldsymbol{f} = \rho_0 \frac{\partial \boldsymbol{v}(\boldsymbol{X}, t)}{\partial t} \tag{2.234}$$

By ignoring the inertial force, we obtain the force equilibrium equation in the reference configuration as follows:

$$\mathrm{Div} \boldsymbol{P}^T + \rho_0 \boldsymbol{f} = \mathbf{0} \tag{2.235}$$

Since the first Piola–Kirchhoff stress is not a symmetric tensor, the second Piola–Kirchhoff stress is often used to establish the force equilibrium equation in the reference configuration. Considering the transformation relation between the first and second Piola–Kirchhoff stresses in Eq. (2.209) (i.e., $\boldsymbol{P} = \boldsymbol{F} \cdot \boldsymbol{T}$), we can write the force equilibrium equation as follows:

$$\mathrm{Div}(\boldsymbol{T} \cdot \boldsymbol{F}^T) + \rho_0 \boldsymbol{f} = \mathbf{0} \tag{2.236}$$

By cross multiplying the position vector $\boldsymbol{x}$ with each term of the linear momentum conservation equation, we obtain an integral form of the angular momentum conservation equation:

$$\frac{D}{Dt}\int_{\Omega} \rho \boldsymbol{x} \times \boldsymbol{v} d\Omega = \int_{\Omega} \rho \boldsymbol{x} \times \boldsymbol{f} d\Omega + \int_{\Gamma} \boldsymbol{x} \times \boldsymbol{t} d\Gamma \tag{2.237}$$

To facilitate derivation, the above equation is written in the following component form in the Cartesian coordinate system:

$$\frac{D}{Dt}\int_{\Omega} \rho e_{ijk} x_i v_j d\Omega = \int_{\Gamma} e_{ijk} x_i t_j d\Gamma + \int_{\Omega} \rho e_{ijk} x_i f_j d\Omega \tag{2.238}$$

where e_{ijk} is a permutation symbol. Based on the Reynolds transport theorem, the material derivative on the left-hand side of the above equation can be rewritten as follows:

$$\begin{aligned} &\int_{\Omega} [\rho e_{ijk} \frac{D}{Dt}(x_i v_j) + e_{ijk} x_i v_j (\frac{D\rho}{Dt} + \rho v_{i,i})] d\Omega \\ &= \int_{\Omega} e_{ijk}(x_i \sigma_{pj})_{,p} d\Omega + \int_{\Omega} \rho e_{ijk} x_i f_j d\Omega + \int_{\Omega} \rho \gamma e_{ijk} x_i v_j d\Omega \end{aligned} \tag{2.239}$$

According to the conservation of mass in the reference configuration, the above equation can be simplified to

$$\begin{aligned} \int_{\Omega} \rho e_{ijk}(v_i v_j + x_i \frac{Dv_j}{Dt}) d\Omega &= \int_{\Omega} e_{ijk}(x_i \sigma_{pj})_{,p} d\Omega + \int_{\Omega} \rho e_{ijk} x_i f_j d\Omega \\ &= \int_{\Omega} e_{ijk}(x_i \sigma_{pj,p} + \delta_{ip}\sigma_{pj} + \rho x_i f_j) d\Omega \end{aligned} \tag{2.240}$$

As $e_{ijk} v_i v_j = 0$, the above equation can be further simplified to

$$\int_{\Omega} e_{ijk}[x_i(\sigma_{pj,p} + \rho f_j - \frac{Dv_j}{Dt}) + \sigma_{ij}] d\Omega = 0 \tag{2.241}$$

Equation (2.241) contains a force equilibrium equation that holds for any region. Thus,

$$e_{ijk}\sigma_{ij} = 0 \tag{2.242}$$

To satisfy the above relation, the Cauchy stress must be a symmetric tensor, i.e.,

$$\boldsymbol{\sigma} = \boldsymbol{\sigma}^T \tag{2.243}$$

Here, we do not use the method for derivation in the current configuration to derive angular momentum conservation in the reference configuration. Instead, based on the large-deformation stress measures and their transformation relations, the angular momentum conservation in the current configuration is directly transformed to a description in the reference configuration. We know that the second Piola–Kirchhoff stress is also a symmetric tensor, i.e., $\boldsymbol{T} = \boldsymbol{T}^T$.

According to the relation between the first Piola–Kirchhoff stress $\boldsymbol{P}$ and the second Piola–Kirchhoff stress $\boldsymbol{T}$ (i.e., $\boldsymbol{T} = \boldsymbol{F}^{-1} \cdot \boldsymbol{P}$), we have

$$\boldsymbol{P} \cdot \boldsymbol{F}^T = \boldsymbol{F} \cdot \boldsymbol{P}^T \tag{2.244}$$

2.3.4 *Constitutive Theory for Thermomechanical Coupling Based on Large Deformation [18, 25, 26]*

(1) **Thermomechanical coupling constitutive equation**

The first law of thermodynamics for an arbitrary domain Ω in the current configuration is given by

$$\dot{E} = -\int_{\Gamma} \boldsymbol{h} \cdot d\Gamma + \int_{\Omega} r_m \rho d\Omega + \int_{\Omega} \boldsymbol{\sigma} : \boldsymbol{d} d\Omega \tag{2.245}$$

where $\boldsymbol{h}$ is the heat flow passing through unit area per unit time, r_m is the heat absorbed per unit mass of material from external sources per unit time, and $\boldsymbol{d}$ is the rate of deformation in the current configuration. The relation between $\boldsymbol{d}$ and the Green strain tensor is given by

$$\dot{\boldsymbol{E}} = \boldsymbol{F}^T \cdot \boldsymbol{d} \cdot \boldsymbol{F} \tag{2.246}$$

In Eq. (2.245), the first term represents the heat that flows into domain Ω via surface Γ per unit time, the second term is the heat applied by external sources to domain Ω per unit time, and the third term represents the deformation power in domain Ω. We use Green's theorem to transform the surface integral along Γ in the first term on the right-hand side of Eq. (2.245) into a volume integral in domain Ω and, considering that Ω is an arbitrary domain, we have

$$\rho \dot{e} = -\boldsymbol{h} \cdot \nabla + \rho r_m + \boldsymbol{\sigma} : \boldsymbol{d} \tag{2.247}$$

The above equation is referred to as the differential form of the first law of thermodynamics.

Let s_m be the entropy per unit mass. Thus, the total entropy in domain Ω is given by

$$S = \int_{\Omega} \rho s_m dv \tag{2.248}$$

Let T be the temperature (absolute temperature $T > 0$). According to the second law of thermodynamics, the following must hold:

$$\dot{S} \geq -\int_{\Gamma} \frac{1}{T}\boldsymbol{h} \cdot d\boldsymbol{\Gamma} + \int_{\Omega} \frac{1}{T} r_m \rho d\Omega \tag{2.249}$$

The above expression is referred to as the entropy inequality, where $\boldsymbol{h}/T$ and $\frac{1}{T}r_m$ are called the entropy flow and entropy source, respectively. The "$>$" and "$=$" signs in Eq. (2.249) correspond to irreversible and reversible processes, respectively. The right-hand side of Eq. (2.249) indicates that the absorption of heat by the domain Ω from its neighborhood or external sources leads to an increase in its total entropy. The excess of the increase in entropy over the increase reflected by the right-hand side of Eq. (2.249) is irreversible and is called the total entropy production rate. The entropy production rate $\dot{S}_i$ in domain Ω is defined as the difference between the two sides of Eq. (2.249):

$$\dot{S} = \int_{\Omega} \rho \dot{s}_m d\Omega = -\int_{\Gamma} \frac{1}{T}\boldsymbol{h} \cdot d\boldsymbol{\Gamma} + \int_{\Omega} \frac{1}{T} r_m \rho d\Omega + \dot{S}_i,\ \dot{S}_i \geq 0 \tag{2.250}$$

Let κ_m be the entropy production rate per unit mass. Then, the above equation is rewritten as follows:

$$\dot{S}_i = \int_{\Omega} \rho \kappa_m d\Omega,\ \kappa_m \geq 0 \tag{2.1.251}$$

Equation (2.250) is referred to as the integral form of the entropy balance equation. We use Green's theorem to transform the surface integral along surface Γ in domain Ω in the first term on the right-hand side of Eq. (2.250) into a volume integral in domain Ω and obtain the following:

$$\rho \dot{s}_m = -\frac{1}{T}\boldsymbol{h} \cdot \nabla + \rho\left(\frac{1}{T} r_m + \kappa_m\right) \tag{2.252}$$

The above equation is referred to as the differential form of the entropy balance equation. Now, let us simplify the first term on the right-hand side of Eq. (2.252) by using the following tensor analysis equation:

$$(c\boldsymbol{h}) \cdot \nabla = c(\boldsymbol{h} \cdot \nabla) + (c\nabla) \cdot \boldsymbol{h} \tag{2.253}$$

Let $c = 1/T$. Then, Eq. (2.252) is transformed to

$$T\dot{s}_m = -\frac{1}{\rho}\boldsymbol{h} \cdot \nabla + \frac{1}{\rho T}(T\nabla) \cdot \boldsymbol{h} + r_m + T\kappa_m \tag{2.254}$$

$$T\kappa_m = T\kappa_{th} + T\kappa_{\text{int}} \geq 0 \tag{2.255}$$

where

$$T\kappa_{th} = -\frac{1}{\rho T}(T\nabla) \cdot \boldsymbol{h} \tag{2.256}$$

$$T\kappa_{\text{int}} = T\dot{s}_m - \left[-\frac{1}{\rho}\boldsymbol{h} \cdot \nabla + r_m\right] \tag{2.257}$$

Equation (2.255) shows that the entropy production rate κ_m can be divided into two parts, namely κ_{th}, which is the rate of entropy production caused by thermal production, and κ_{int}, which is the rate of entropy production caused by the excess of the entropy production rate $\dot{s}_m$ over the rate at which heat is absorbed from the neighborhood and external sources per unit mass. κ_{int} is called the intrinsic entropy production rate. The second law of thermodynamics requires that $\kappa_m = \kappa_{th} + \kappa_{\text{int}} \geq 0$. However, the following stronger assumption is often made:

$$\kappa_{\text{th}} \geq 0 \tag{2.258}$$

$$\kappa_{\text{int}} \geq 0 \tag{2.259}$$

$T\kappa_m, T\kappa_{th}$, and $T\kappa_{\text{int}}$ are the total dissipation rate per unit mass, the heat dissipation rate per unit mass, and the intrinsic dissipation rate per unit mass, respectively. Based on the differential form of the first law of thermodynamics in Eqs. (2.247) and (2.257) is rewritten as follows:

$$T\kappa_{\text{int}} = T\dot{s}_m - \left[-\frac{1}{\rho}(\boldsymbol{\sigma} : \boldsymbol{d}) + \dot{e}\right] \tag{2.260}$$

The part within the square brackets on the right-hand side of Eq. (2.260) represents the excess of the rate of increase in internal energy per unit mass over the deformation power per unit mass. For the sake of convenience, we use temperate T instead of entropy s_m as the independent variable. Then, we can use the Helmholtz free energy ψ per unit mass shown in Eq. (2.107) instead of the internal energy e as the thermodynamic function. Thus, Eq. (2.260) can be rewritten as follows:

$$T\kappa_{\text{int}} = -s_m\dot{T} - \left[\dot{\psi} - \frac{1}{\rho}\boldsymbol{\sigma} : \boldsymbol{d}\right] \tag{2.261}$$

(2) **Construction of thermomechanical coupling energy function and derivation of its governing equations**

The deformation power w is defined as

$$w = J\boldsymbol{\sigma} : \boldsymbol{d} \tag{2.262}$$

Combining Eqs. (2.210) and (2.246) with the above equation, we have

$$w = \left(\boldsymbol{F} \cdot \boldsymbol{T} \cdot \boldsymbol{F}^T\right) : \boldsymbol{d} = \boldsymbol{T} : \left(\boldsymbol{F}^T \cdot \boldsymbol{d} \cdot \boldsymbol{F}\right) = \boldsymbol{T} : \dot{\boldsymbol{E}} \tag{2.263}$$

The so-called elasticity means that of the two parts of a material's dissipation rate per unit mass $T\kappa_m$, only the heat dissipation rate $T\kappa_{th}$ is positive or zero, whereas the intrinsic dissipation rate $T\kappa_{\text{int}}$, as shown in Eq. (2.261), is constant at zero:

$$\rho T\kappa_{\text{int}} = -s_m\rho\dot{T} - \left[\rho\dot{\psi} - \frac{1}{J}\boldsymbol{T} : \dot{\boldsymbol{E}}\right] = 0 \tag{2.264}$$

By combining Eq. (2.217) and multiplying the above equation with ρ_0/ρ, we have

$$\rho_0\dot{\psi} = -\rho_0 s_m\dot{T} + \boldsymbol{T} : \dot{\boldsymbol{E}} \tag{2.265}$$

Equatiom (2.265) represents the rate of change in the Helmholtz free energy ψ in the reference configuration. ψ can also be viewed as a function of two state quantities, namely Green strain tensor $\boldsymbol{E}$ and temperature T, i.e., $\psi(\boldsymbol{E}, T)$. The rate of change in ψ is given by

$$\dot{\psi} = \frac{\partial\psi}{\partial\boldsymbol{E}} : \dot{\boldsymbol{E}} + \frac{\partial\psi}{\partial T}\dot{T} \tag{2.266}$$

Comparison of Eqs. (2.265) and (2.266) gives

$$\boldsymbol{T} = \rho_0\frac{\partial\psi}{\partial\boldsymbol{E}} \tag{2.267}$$

$$s_m = -\frac{\partial\psi}{\partial T} \tag{2.268}$$

For an isotropic material, its Helmholtz free energy can depend only on $\boldsymbol{E}$ through three invariants of $\boldsymbol{E}$. The following three invariants are selected for $\boldsymbol{E}$:

$$I_1 = I_1(\boldsymbol{E}) = \text{tr}\boldsymbol{E}$$

$$I_2 = I_2(\boldsymbol{E}') = \frac{1}{2}\mathrm{tr}(\boldsymbol{E}' \cdot \boldsymbol{E}') = \frac{1}{2}\mathrm{tr}\left(\boldsymbol{E}'^2\right) \tag{2.269}$$

$$I_3 = I_3(\boldsymbol{E}') = \frac{1}{3}\mathrm{tr}(\boldsymbol{E}' \cdot \boldsymbol{E}' \cdot \boldsymbol{E}') = \frac{1}{3}\mathrm{tr}\left(\boldsymbol{E}'^3\right)$$

where $\boldsymbol{E}' = \boldsymbol{E} - (\mathrm{tr}\boldsymbol{E})/3$ is the deviator of $\boldsymbol{E}$. It is easy to prove that the derivatives of I_1, I_2, and I_3 with respect to independent variable $\boldsymbol{E}$ are as follows (with a prime denoting a deviator):

$$\frac{\partial I_1}{\partial \boldsymbol{E}} = \boldsymbol{I}, \frac{\partial I_1}{\partial \boldsymbol{E}} = \boldsymbol{I} \text{ and } \frac{\partial I_3}{\partial \boldsymbol{E}} = \overline{\boldsymbol{I}} : (\boldsymbol{E}' \cdot \boldsymbol{E}') \tag{2.270}$$

where $\boldsymbol{I}$ is the unit tensor and $\overline{\boldsymbol{I}}$ is the special identity tensor. In the reference configuration, *A*, *B*, *C*, and *D* are used as subscripts to represent the components of $\boldsymbol{I}$ and $\overline{\boldsymbol{I}}$. Thus, the component of $\boldsymbol{I}$ is $I_{AB} = \delta_{AB}$, and the component of $\overline{\boldsymbol{I}}$ is $\overline{I}_{ABCD} = \frac{1}{2}(\delta_{AC}\delta_{BD} + \delta_{BC}\delta_{AD})$. Let temperature $T = T_0$ in the reference configuration. By performing a Taylor series expansion of the Helmholtz free energy $\psi(\boldsymbol{E}, T)$ near the reference configuration and retaining only up to the quadratic term, we have

$$\begin{aligned}\rho_0\psi(\boldsymbol{E}, T) = -\rho_0 s_0(T - T_0) - \frac{1}{2T_0}\rho_0 c_{v0}(T - T_0)^2 - p_0 I_1 - \\ \alpha K I_1(T - T_0) + \frac{1}{2}K I_1^2 + 2G I_2\end{aligned} \tag{2.271}$$

where ρ_0 is the mass density per unit volume in the reference configuration. There are six terms on the right-hand side of Eq. (2.271), each of which contains an unknown constant (i.e., s_0, c_{v0}, p_0, α, K, and G, respectively). p_0 is the same pressure in all three directions in the reference state. In the absence of initial stresses in the reference state, $p_0 = 0$. $\alpha/3$ is the coefficient of linear expansion for temperature. K is the bulk modulus, G is the shear modulus, and c_{v0} is the specific heat at constant volume in the reference state.

By substituting $\psi(\boldsymbol{E}, \theta)$ in Eq. (2.271) into the constitutive relations in Eqs. (2.267) and (2.268) and considering Eq. (2.270), we have

$$\boldsymbol{T} = -p_0\boldsymbol{I} + K[I_1(\boldsymbol{E}) - \alpha(T - T_0)]\boldsymbol{I} + 2G\boldsymbol{E}' \tag{2.272}$$

$$s_m = s_0 + c_{v0}\frac{1}{T}(T - T_0) + \frac{1}{\rho_0}K\alpha I_1(\boldsymbol{E}) \tag{2.273}$$

Of particular note is that here, the boldfaced **T** represents the second Piola–Kirchhoff stress tensor; in comparison, temperature is denoted by a nonboldfaced T. The inverse of Eq. (2.272) can be easily calculated. By taking the trace and deviator of each side of Eq. (2.272), we have

$$I_1(\boldsymbol{T}) = -3p_0 + 3K[I_1(\boldsymbol{E}) - \alpha(T - T_0)] \tag{2.274}$$

$$\boldsymbol{T}' = 2G\boldsymbol{E}' \tag{2.275}$$

Using the above two equations, the expressions of Eq. (2.273) with $\boldsymbol{T}$ and T as independent variables are obtained as follows:

$$\boldsymbol{E} = \frac{1}{3}\alpha(T - T_0)\boldsymbol{I} + \frac{1}{9K}(I_1(\boldsymbol{E}) + 3p_0)\boldsymbol{I} + \frac{1}{2G}\boldsymbol{T}' \tag{2.276}$$

$$s_m = s_0 + \left(c_{v0}\frac{1}{T_0} + \frac{1}{\rho_0}K\alpha^2\right)(T - T_0) + \frac{1}{3\rho_0}\alpha[I_1(\boldsymbol{T}) + 3p_0] \tag{2.277}$$

By substituting Eq. (2.107) into Eq. (2.247) and considering (2.267) and (2.268), the following governing equation for the temperature field T is obtained:

$$\rho\left(\frac{\partial\psi}{\partial\boldsymbol{E}} : \dot{\boldsymbol{E}} - T\left(\frac{\partial^2\psi}{\partial T^2}\dot{T} + \frac{\partial^2\psi}{\partial T\partial\boldsymbol{E}} : \dot{E}\right)\right) = k(\nabla T)^2 + \rho r_m + \sigma : \boldsymbol{d} \tag{2.278}$$

2.3.5 *Constitutive Theory for Thermo–Mechano-Chemical Coupling Based on Large Deformation*

(1) **Thermo–mechano-chemical coupling constitutive equations**

The rate of change in total energy is equal to the sum of the power of external forces, the power of thermal energy, and the power of chemical energy. According to the first law of thermodynamics, the principle of conservation of energy is expressed as

$$\frac{D}{Dt}(K + E) = W + Q + C \tag{2.279}$$

where K is the kinetic energy, E is the internal energy, W is the power of the external forces, Q is the power supplied by the heat source and heat flow, and C is the power supplied by the chemical reactions.

Each energy component can be described using the reference or current configuration. Here, we focus on the description of conservation equations in the reference configuration. Kinetic energy is primarily the energy generated from the macroscopic motion of an object. The kinetic energy K in the reference configuration is given by

$$K = \frac{1}{2}\int_{\Omega_0} \rho_0 \boldsymbol{v} \cdot \boldsymbol{v} d\Omega_0 \tag{2.280}$$

where $\boldsymbol{v}$ is the velocity in the reference configuration. The internal energy in the reference configuration is expressed as

$$E = \int_{\Omega_0} \rho_0 e d\Omega_0 \tag{2.281}$$

where e is the energy per unit mass.

The power of the volume force per unit volume, $\boldsymbol{f}$, and the surface force per unit surface area, $\boldsymbol{t}_0$, within a domain is given by

$$W = \int_{\Gamma_0} \boldsymbol{t}_0 \cdot \boldsymbol{v} d\Gamma_0 + \int_{\Omega_0} \rho_0 \boldsymbol{f} \cdot \boldsymbol{v} d\Omega_0 \tag{2.282}$$

By transforming the integral of surface forces along the boundary in Eq. (2.282) to a volume integral using the Cauchy relation and Gauss' theorem, we have

$$\begin{aligned} W &= \int_{\Gamma_0} \boldsymbol{t}_0 \cdot \boldsymbol{v} d\Gamma_0 + \int_{\Omega_0} \rho_0 \boldsymbol{f} \cdot \boldsymbol{v} d\Omega_0 \\ &= \int_{\Gamma_0} (\boldsymbol{N} \cdot \boldsymbol{P}^T) \cdot \boldsymbol{v} d\Gamma_0 + \int_{\Omega_0} \rho_0 \boldsymbol{f} \cdot \boldsymbol{v} d\Omega_0 \\ &= \int_{\Omega_0} [\nabla \cdot (\boldsymbol{P}^T \cdot \boldsymbol{v}) + \rho \boldsymbol{f} \cdot \boldsymbol{v}] d\Omega_0 \\ &= \int_{\Omega_0} [(\nabla \cdot \boldsymbol{P}^T + \rho \boldsymbol{f}) \cdot \boldsymbol{v} + \boldsymbol{P}^T : \nabla \boldsymbol{v}] d\Omega_0 \\ &= \int_{\Omega_0} [\rho_0 \frac{D\boldsymbol{v}}{Dt} \cdot \boldsymbol{v} + \boldsymbol{P}^T : \nabla \boldsymbol{v}] d\Omega_0 \end{aligned} \tag{2.283}$$

Of particular note is that here, ∇ is the operator in the reference configuration. In Eq. (2.283), $\boldsymbol{P}$ is the first Piola–Kirchhoff stress.

The power supplied by the heat source and heat flow is expressed as

$$Q = \int_{\Omega_0} \rho_0 r_m d\Omega_0 - \int_{\Gamma_0} \boldsymbol{q} \cdot \boldsymbol{N} d\Gamma_0 = \int_{\Omega_0} (\rho_0 r_m - \nabla \cdot \boldsymbol{q}) d\Omega_0 \tag{2.284}$$

where r_m is the heat source per unit mass and $\boldsymbol{q}$ is the heat flow per unit area.

By referring to the power of the heat flow, we obtain the power generated by chemical diffusion by

$$C = -\int_{\Gamma_0} \mu \boldsymbol{j} \cdot \boldsymbol{N} d\Gamma_0 \tag{2.285}$$

where $\boldsymbol{j}$ is the diffusion flow of the infiltrating material concentration.

By substituting Eqs. (2.280)–(2.285) into Eq. (2.279), we obtain an integral expression of energy conservation:

$$\begin{aligned} &\frac{D}{Dt} \int_{\Omega_0} \rho_0 (\frac{1}{2} \boldsymbol{v} \cdot \boldsymbol{v} + e) d\Omega_0 \\ &= \frac{1}{2} \frac{D}{Dt} \int_{\Omega_0} \rho_0 \boldsymbol{v} \cdot \boldsymbol{v} d\Omega_0 + \int_{\Omega_0} (\boldsymbol{P}^T : \dot{\boldsymbol{F}}) d\Omega_0 \\ &+ \int_{\Omega_0} (-\nabla \cdot \boldsymbol{q} + \rho_0 \kappa) d\Omega_0 - \int_{\Omega_0} (\mu \nabla \cdot \boldsymbol{j} + \boldsymbol{j} \cdot \nabla \mu) d\Omega_0 \end{aligned} \tag{2.286}$$

By simplifying the above equation using the Reynolds transport theorem and Gauss' theorem and then by reorganizing the terms in the resulting equation, we obtain a differential equation of energy conservation as follows:

$$\rho_0 \dot{e} = \boldsymbol{P}^T : \dot{\boldsymbol{F}} - \mu Div \boldsymbol{j} - \boldsymbol{j} \cdot \nabla \mu - Div \boldsymbol{q} + r_m \rho_0 \tag{2.287}$$

According to the second law of thermodynamics, we obtain the total entropy as

$$S = \int_{\Omega_0} s_m \rho_0 d\Omega_0 \tag{2.288}$$

Thus, the second law of thermodynamics can be written as follows:

$$\frac{D}{Dt} \int_{\Omega_0} \rho_0 s_m d\Omega_0 \geqslant -\int_{\Gamma_0} \frac{\boldsymbol{q} \cdot \boldsymbol{N}}{T} d\Gamma_0 + \int_{\Omega_0} \frac{r_m}{T} \rho_0 d\Omega_0 \tag{2.289}$$

The first and second terms on the right-hand side of the above equation represent the entropy flow and entropy source, respectively. Equation (2.289) is the integral form of the second law of thermodynamics and is also referred to as the entropy inequality or the Clausius–Duhem inequality. Through an integral transformation, a local inequality for the second law of thermodynamics is obtained:

$$\rho_0 \dot{s}_m \geq \frac{r_m \rho}{T} - \nabla \cdot (\frac{\boldsymbol{q}}{T}) = \frac{1}{T}(r_m \rho_0 - \nabla \cdot \boldsymbol{q}) + \frac{1}{T^2}\boldsymbol{q} \cdot \nabla T \tag{2.290}$$

According to the energy conservation equation in Eq. (2.285), the right-hand side of the inequality (2.290) can be rewritten in the following form:

$$\begin{aligned} &\frac{r_m \rho_0}{T} - \nabla \cdot (\frac{\boldsymbol{q}}{T}) = \frac{1}{T}(r_m \rho_0 - \nabla \cdot \boldsymbol{q}) + \frac{1}{T^2}\boldsymbol{q} \cdot \nabla T \\ &= \frac{1}{T}(\rho_0 \dot{e} - \boldsymbol{P}^T : \dot{\boldsymbol{F}} + \mu \nabla \cdot \boldsymbol{j} + \boldsymbol{j} \cdot \nabla \mu + \frac{1}{T}\boldsymbol{q} \cdot \nabla T) \end{aligned} \tag{2.291}$$

Combining Eqs. (2.290) and (2.291), we can obtain the entropy inequality:

$$\rho_0 \dot{s}_m \geq \frac{1}{T}(\rho_0 \dot{e} - \boldsymbol{P}^T : \dot{\boldsymbol{F}} + \mu \nabla \cdot \boldsymbol{j} + \boldsymbol{j} \cdot \nabla \mu + \frac{1}{T}\boldsymbol{q} \cdot \nabla T) \tag{2.292}$$

Ultimately, the entropy inequality is written as follows:

$$\rho_o(\dot{e} - T\dot{s}_m) - \boldsymbol{P}^T : \dot{\boldsymbol{F}} + \mu \nabla \cdot \boldsymbol{j} + \boldsymbol{j} \cdot \nabla \mu + \frac{1}{T}\boldsymbol{q} \cdot \nabla T \leq 0 \tag{2.293}$$

Based on Eqs. (2.107) and (2.108), the above entropy inequality is rewritten as

$$\rho_0(\dot{\psi} - s_m \dot{T}) - \boldsymbol{P}^T : \dot{\boldsymbol{F}} + \mu \nabla \cdot \boldsymbol{j} + \boldsymbol{j} \cdot \nabla \mu + \frac{1}{T}\boldsymbol{q} \cdot \nabla T \leq 0 \tag{2.294}$$

The second term on the left-hand side of Eq. (2.294) represents the deformation power. We know that the stress is conjugate, or work conjugate, to the deformation rate or strain rate with respect to the deformation power. Each stress description should correspond to a relevant strain description. For example, the first Piola–Kirchhoff stress is conjugate to the deformation gradient rate, the second Piola–Kirchhoff stress is conjugate to the Green strain rate, and the Cauchy stress is conjugate to the deformation rate. Thus, the deformation power can be written as the double dot product of the stress and its corresponding conjugate strain (deformation) rate. The deformation power per unit volume in the reference configuration, w^{int}, can be expressed as

$$w^{\text{int}} = J\boldsymbol{\sigma} : \boldsymbol{d} = \boldsymbol{P} : \dot{\boldsymbol{F}} = \boldsymbol{T} : \dot{\boldsymbol{E}} \tag{2.295}$$

By substituting the decomposition expression of the deformation gradient in Eq. (2.180) into the deformation power in the above equation, we have

$$
\begin{aligned}
&\boldsymbol{P} : \dot{\boldsymbol{F}} \\
&= \boldsymbol{P} : \left(\dot{\boldsymbol{F}}^{\mathrm{e}} \cdot \boldsymbol{F}^{\mathrm{i}} + \boldsymbol{F}^{\mathrm{e}} \cdot \dot{\boldsymbol{F}}^{\mathrm{i}}\right) \\
&= \left(\boldsymbol{P} \cdot \boldsymbol{F}^{\mathrm{i}^{\mathrm{T}}}\right) : \dot{\boldsymbol{F}}^{\mathrm{e}} + \left(\boldsymbol{F}^{\mathrm{eT}} \cdot \boldsymbol{P}\right) : \dot{\boldsymbol{F}}^{\mathrm{i}} \\
&= \left(J\boldsymbol{\sigma} \cdot \boldsymbol{F}^{\mathrm{e}^{-\mathrm{T}}}\right) : \dot{\boldsymbol{F}}^{\mathrm{e}} + \left(\boldsymbol{F}^{\mathrm{eT}} \cdot J\boldsymbol{\sigma} \cdot \boldsymbol{F}^{\mathrm{e}^{-\mathrm{T}}} \cdot \boldsymbol{F}^{\mathrm{i}^{-\mathrm{T}}}\right) : \dot{\boldsymbol{F}}^{\mathrm{i}} \\
&= \left(J\boldsymbol{F}^{\mathrm{e}^{-1}} \cdot \boldsymbol{\sigma} \cdot \boldsymbol{F}^{\mathrm{e}^{-\mathrm{T}}}\right) : \boldsymbol{F}^{\mathrm{e}^{\mathrm{T}}} \cdot \boldsymbol{F}^{\mathrm{e}} + \left(\boldsymbol{F}^{\mathrm{eT}} \cdot J\boldsymbol{\sigma} \cdot \boldsymbol{F}^{\mathrm{e}^{-\mathrm{T}}} \cdot \boldsymbol{F}^{\mathrm{i}^{-\mathrm{T}}}\right) : \boldsymbol{F}^{\mathrm{i}} \\
&= \boldsymbol{T} : \dot{\boldsymbol{E}}^{e} + \left(\boldsymbol{C}^{\mathrm{e}} \cdot \boldsymbol{T}\right) : \left(\dot{\boldsymbol{F}}^{\mathrm{i}} \cdot \boldsymbol{F}^{\mathrm{i}^{-1}}\right) \\
&= \boldsymbol{T} : \dot{E}^{e} + \boldsymbol{M}^{e} : \boldsymbol{L}^{i}
\end{aligned}
\tag{2.296}
$$

where $\boldsymbol{T}$ is the second Piola–Kirchhoff stress and $\boldsymbol{M}^e$ is the Mandel stress, which is often used to analyze large plastic deformation. $\boldsymbol{T}$ and $\boldsymbol{M}^e$ are given by

$$
\boldsymbol{T} = J\boldsymbol{F}^{\mathrm{e}^{-1}} \cdot \boldsymbol{\sigma} \cdot \boldsymbol{F}^{\mathrm{e}^{-\mathrm{T}}} \text{ and } \boldsymbol{M}^{e} = \boldsymbol{C}^{\mathrm{e}} \cdot \boldsymbol{S} = J\boldsymbol{F}^{\mathrm{eT}} \cdot \boldsymbol{\sigma} \cdot \boldsymbol{F}^{\mathrm{e}^{-\mathrm{T}}} \tag{2.297}
$$

Equation (2.296) shows that as a result of the decomposition of the deformation gradient, the deformation power can be decomposed into two parts, namely elastic deformation power and inelastic deformation power.

Let us assume that the Helmholtz free energy is a thermodynamic function of Green strain tensor $\boldsymbol{E}^e$, temperature T, and infiltrating material concentration c, i.e.,

$$
\psi = \hat{\psi}(\boldsymbol{E}^e, T, c) \tag{2.298}
$$

Based on Eq. (2.298), the relation of the Helmholtz free energy with $\boldsymbol{E}^e$, T, and c can be written as

$$
\dot{\psi} = \frac{\partial \hat{\psi}}{\partial \boldsymbol{E}^{\mathrm{e}}} : \dot{\boldsymbol{E}}^{\mathrm{e}} + \frac{\partial \hat{\psi}}{\partial T}\dot{T} + \frac{\partial \hat{\psi}}{\partial c}\dot{c} \tag{2.299}
$$

By substituting Eqs. (2.299) and (2.296) into the entropy inequality in Eq. (2.294) and then reorganizing the terms in the resulting equation, we have

$$
\rho_0(\frac{\partial \hat{\psi}}{\partial \boldsymbol{E}^{\mathrm{e}}} : \dot{\boldsymbol{E}}^{\mathrm{e}} + \frac{\partial \hat{\psi}}{\partial \theta}\dot{\theta} + \frac{\partial \hat{\psi}}{\partial c}\dot{c} + s_m\dot{\theta}) - \boldsymbol{T} : \dot{\boldsymbol{E}}^{e} - \boldsymbol{M}_e : \boldsymbol{L}^g + \mu\dot{c} + \boldsymbol{j} \cdot \nabla\mu + \frac{1}{T}\boldsymbol{q} \cdot \nabla T \leq 0 \tag{2.300}
$$

By combining like terms, the above equation is transformed to

$$
\begin{aligned}
&(\rho_0\frac{\partial \hat{\psi}}{\partial \boldsymbol{E}^{\mathrm{e}}} - \boldsymbol{T}) : \dot{\boldsymbol{E}}^{\mathrm{e}} + \rho_0(\frac{\partial \hat{\psi}}{\partial \theta} + s_m)\dot{\theta} + (\rho_0\frac{\partial \hat{\psi}}{\partial c} + \mu)\dot{c} \\
&\qquad - \boldsymbol{M}_e : \boldsymbol{L}^g + \boldsymbol{j} \cdot \nabla\mu + \frac{1}{T}\boldsymbol{q} \cdot \nabla T \leq 0
\end{aligned}
\tag{2.301}
$$

The entropy inequality holds for any values of all the variables (e.g., $\boldsymbol{E}^e$, T, and c). Thus, the coefficient of each variable must be zero. On this basis, the thermo–mechano-chemical coupling constitutive relations are obtained:

$$\begin{cases} \boldsymbol{T} = \rho_0 \dfrac{\partial \hat{\psi}}{\partial \boldsymbol{E}^{\mathrm{e}}} \\ s_m = -\dfrac{\partial \hat{\psi}}{\partial T} \\ \mu = \rho_0 \dfrac{\partial \hat{\psi}}{\partial c} \end{cases} \tag{2.302}$$

All the above discussion is based on the transformation of the Helmholtz free energy. In fact, multiple transforms are available to facilitate the solving of specific problems. These transforms are called Legendre transforms and are summarized in .Table 2.2

Practice problem: List the Legendre transforms for thermo–mechano-chemical coupling constitutive relations in the current configuration.

Table 2.2 Legendre transforms for thermo–mechano-chemical coupling constitutive relations in the reference configuration

Thermodynamic function	Name	Independent variable	Relation	Constitutive relation (response function)
e	Internal energy per unit mass	$\boldsymbol{E}, s_m, c$	$\rho_0 \dot{e}(\boldsymbol{E}, s_m, c) = \rho_0 T \dot{s}_m + \boldsymbol{T} : \dot{\boldsymbol{E}} + \mu \dot{c}$	$\boldsymbol{T} = \rho_0 \left(\frac{\partial e}{\partial \boldsymbol{E}}\right)_{s_m,c}$, $T = \left(\frac{\partial e}{\partial s_m}\right)_{\boldsymbol{E},c}$, $\mu = \rho_0 \left(\frac{\partial e}{\partial c}\right)_{\boldsymbol{E},s_m}$
ψ	Helmholtz free energy per unit mass	$\boldsymbol{E}, T, c$	$\psi(\boldsymbol{E}, T, c) = e(\boldsymbol{E}, s_m, c) - T s_m$ $\rho_0 \dot{\psi}(\boldsymbol{E}, T, c) = -\rho_0 s_m \dot{T} + \boldsymbol{T} : \dot{\boldsymbol{E}} + \mu \dot{c}$	$\boldsymbol{T} = \rho_0 \left(\frac{\partial \psi}{\partial \boldsymbol{E}}\right)_{T,c}$, $s_m = -\left(\frac{\partial \psi}{\partial T}\right)_{\boldsymbol{E},c}$, $\mu = \rho_0 \left(\frac{\partial \psi}{\partial c}\right)_{\boldsymbol{E},T}$
ψ_G	Gibbs free energy per unit mass	$\boldsymbol{T}, T, c$	$\psi_G(\boldsymbol{T}, T, c) = \psi(\boldsymbol{E}, T, c) - \frac{1}{\rho_0} \boldsymbol{T} : \boldsymbol{E}$ $\rho_0 \dot{\psi}_G(\boldsymbol{T}, T, c) = -\rho_0 s_m \dot{T} - \boldsymbol{E} : \dot{\boldsymbol{T}} + \mu \dot{c}$	$\boldsymbol{E} = -\rho_0 \left(\frac{\partial \psi_G}{\partial \boldsymbol{T}}\right)_{T,c}$, $s_m = -\left(\frac{\partial \psi_G}{\partial T}\right)_{\boldsymbol{T},c}$, $\mu = \rho_0 \left(\frac{\partial \psi_G}{\partial c}\right)_{\boldsymbol{T},T}$
ψ_H	Entropy per unit mass	$\boldsymbol{T}, s_m, c$	$\psi_H(\boldsymbol{T}, s_m) = e(\boldsymbol{E}, s_m) - \frac{1}{\rho_0} \boldsymbol{T} : \boldsymbol{E}$ $\rho_0 \dot{\psi}_H(\boldsymbol{T}, s_m) = \rho_0 T \dot{s}_m - \dot{\boldsymbol{T}} : \boldsymbol{E} + \mu \dot{c}$	$\boldsymbol{E} = -\rho_0 \left(\frac{\partial \psi_H}{\partial \boldsymbol{T}}\right)_{s_m,c}$, $T = \left(\frac{\partial \psi_H}{\partial s_m}\right)_{\boldsymbol{T},c}$, $\mu = \rho_0 \left(\frac{\partial \psi_H}{\partial c}\right)_{\boldsymbol{T},s_m}$

(2) Construction of thermo–mechano-chemical coupling free energy and derivation of its governing equations

Here, the influence of the consumption by chemical reactions on the concentration of the infiltrating material is considered. In a chemical reaction process, the concentration of the infiltrating material follows the law of conservation. That is, the rate of change in the concentration of the infiltrating material in a certain region within an object is equal to the amount of the infiltrating material entering this region through the boundary minus the amount of the infiltrating material consumed by the reaction. Thus, the equation of conservation of the concentration of the infiltrating material in the reference configuration is given by

$$\int_{\Omega_0} \frac{\partial c(\boldsymbol{X}, t)}{\partial t} d\Omega_0 = -\int_{\Gamma_0} \boldsymbol{j} \cdot d\boldsymbol{\Gamma}_0 - \int_{\Omega_0} N d\Omega_0 \tag{2.303}$$

where c is the concentration of the infiltrating material, $\boldsymbol{j}$ is the oxygen flux per unit reference area, and N is the oxygen concentration consumed by the oxidation reaction. By using the Gauss integral transformation, we transform Eq. (2.303) to

$$\int_{\Omega_0} \frac{\partial c}{\partial t} d\Omega_0 = -\int_{\Omega_0} (Div\boldsymbol{j} + N) d\Omega_0 \tag{2.304}$$

The above equation holds for any region in the reference configuration. Thus, we have

$$\frac{\partial c}{\partial t} + Div\boldsymbol{j} = -N \tag{2.305}$$

The above equation appears to be exactly the same as Eq. (2.141). In fact, Eq. (2.141) is for cases of small deformation, and Eq. (2.14) shows that the material coordinates are the same as Euler coordinates, so no distinction is made between the two. However, Eq. (2.305) is in the reference configuration, i.e., the material coordinates in the initial state. Similar to Eq. (2.136), the diffusion flux of the concentration of the infiltrating material satisfies Fick's first law:

$$\boldsymbol{j} = -m\nabla\mu \tag{2.306}$$

Equation (2.306) differs from Eq. (2.136) in that ∇ in Eq. (2.306) is the operator in the reference configuration. A constitutive relation is required to establish the relationship between the chemical potential and concentration.

By considering three processes, namely heat flow, deformation, and material infiltration, the free energy is decomposed as follows [15]:

$$\hat{\psi} = \psi^{\mathrm{e}}(\boldsymbol{E}_e) + \psi^{\mathrm{e}\theta}(\boldsymbol{E}_e, \theta) + \psi^{\mathrm{diff}}(\theta, c) \tag{2.307}$$

where $\psi^{\mathrm{e}}(\boldsymbol{E}_e)$ is the elastic strain energy, $\psi^{\mathrm{e}\theta}(\boldsymbol{E}_e, \theta)$ is the contribution of the coupling between the temperature θ and elastic strain $\boldsymbol{E}_e$ to the free energy, $\psi^{\theta\mathrm{c}}(\theta, c)$ is the contribution of the coupling between the temperature θ and concentration c to the free energy, and $\psi^{\mathrm{diff}}(\theta, c)$ is the diffusion energy affected by the temperature θ.

The isotropic elastic strain energy is constructed as

$$\psi^{\mathrm{e}}(\boldsymbol{E}_e) = G|\boldsymbol{E}_{\mathrm{e}}|^2 + \frac{1}{2}\left(K - \frac{2}{3}G\right)(\mathrm{tr}\boldsymbol{E}_{\mathrm{e}})^2 \tag{2.308}$$

where G is the shear modulus and K is the bulk modulus.

$\psi^{\mathrm{e}\theta}(\boldsymbol{E}_e, T)$ is given by

$$\psi^{\mathrm{e}\theta}(\boldsymbol{E}_e, T) = -(T - T_0)3K\alpha\mathrm{tr}\boldsymbol{E}_{\mathrm{e}} + c_R(T - T_0) - c_R T\ln\left(\frac{T}{T_0}\right) \tag{2.309}$$

where α is the coefficient of thermal expansion, T_0 is the reference temperature, and c_R is the specific heat of the material.

$\psi^{\mathrm{diff}}(T, c)$ is constructed as

$$\psi^{\mathrm{diff}}(T, c) = \mu_0 c + RTc\left(\ln\frac{c}{c_0} - 1\right) \tag{2.310}$$

where μ_0 is the reference chemical potential, R is the universal gas constant, and c_0 is introduced to ensure dimensional consistency and can be set to any reference concentration.

By substituting Eqs. (2.308), (2.309), and (2.310) into Eq. (2.307), we obtain the following free energy function:

$$\hat{\psi} = G|\boldsymbol{E}_{\mathrm{e}}|^2 + \frac{1}{2}\left(K - \frac{2}{3}G\right)(\mathrm{tr}\boldsymbol{E}_{\mathrm{e}})^2 - (T - T_0)3K\alpha\mathrm{tr}\boldsymbol{E}_{\mathrm{e}} + c_R(T - T_0) - c_R T\ln\left(\frac{T}{T_0}\right) + \mu_0 c + RTc\left(\ln\frac{c}{c_0} - 1\right) \tag{2.311}$$

Substitution of the above equation into Eq. (2.302) yields the following stress–strain constitutive relation:

$$\boldsymbol{T} = 2G\boldsymbol{E}^{\mathrm{e}} + K(\mathrm{tr}\boldsymbol{E}^{\mathrm{e}})\boldsymbol{I} - 3K\alpha(T - T_0)\boldsymbol{I} \tag{2.312}$$

The governing equation for the entropy s_m is expressed as

$$s_m = c_R \ln\left(\frac{T}{T_0}\right) + 3K\alpha\mathrm{tr}\boldsymbol{E}^{\mathrm{e}} - Rc_R(\ln c_R - 1) \tag{2.313}$$

The governing equation for the infiltration potential μ is given by

$$\mu = \mu_0 + RT \ln \frac{c}{c_0} \tag{2.314}$$

By substituting the above equation into Eq. (2.306) and then substituting the resulting equation into Eq. (2.305), we obtain the following governing equation for the concentration c:

$$\dot{c} = RDiv\left[m \ln \frac{c}{c_0} \nabla T + \frac{T}{c} \nabla c\right] - N \tag{2.315}$$

We substitute Eq. (2.107) into Eq. (2.287) and combine the resulting equation with Eq. (2.302) to obtain the governing equation for the temperature field T as follows:

$$\begin{aligned} &\rho_0\left(\frac{\partial \hat{\psi}}{\partial \boldsymbol{E}^{\mathrm{e}}} : \dot{\boldsymbol{E}}^{\mathrm{e}} + \frac{\partial \hat{\psi}}{\partial c}\dot{c} - T\left(\frac{\partial^2 \hat{\psi}}{\partial T^2}\dot{T} + \frac{\partial^2 \hat{\psi}}{\partial T \partial \boldsymbol{E}^e} : \dot{\boldsymbol{E}}^e + \frac{\partial^2 \hat{\psi}}{\partial T \partial c}\dot{c}\right)\right) \\ &= \boldsymbol{P}^T : \boldsymbol{F} + m\rho_0^2 \frac{\partial \hat{\psi}}{\partial c} \nabla^2 \left(\frac{\partial \hat{\psi}}{\partial c}\right) + \\ &m\rho_0 \left(\nabla \frac{\partial \hat{\psi}}{\partial c}\right)^2 + k\nabla^2 T + r_m \rho_0 \end{aligned} \tag{2.316}$$

Practice problem: Derive Eq. (2.316) in detail.

2.4 Summary and Outlook

Failure problems, such as corrosion, oxidation, and erosion, are inevitable in TBCs operating at high temperatures. These failure problems are often accompanied by the concurrent progression of heat transfer, chemical reactions, and/or stress and strain evolution, which affect one another and are therefore a thermo–mechano-chemical coupling physical process. In this chapter, to theoretically explain and analyze multifield coupling failure problems, we introduce concepts such as strain and stress measures from the perspectives of small- and large-deformation theories. Subsequently, we present the corresponding stress–strain constitutive relations as well as mass conservation and force equilibrium equations. Based on the laws of thermodynamics, we establish energy functions for thermo–mechano-chemical coupling and derive governing equations for various field variables. See Table 2.3 for a summary. These functions and equations, when combined with definite conditions (i.e., initial and boundary conditions), can be used to solve specific problems. They provide theoretical support for investigating various failure problems in TBCs in the later chapters of the book. For example, in Chap. 4, by considering geometric nonlinearity, we study in detail the interfacial oxidation problems in TBCs at high temperatures using the thermo–mechano-chemical coupling constitutive relations under large-deformation elasticity theory. In Chap. 5, we investigate stress and strain fields and interfacial growth related to the interfacial oxidation problems in TBCs operating at high temperatures by considering physical nonlinearity (i.e., nonlinear constitutive relations) and geometric linearity (i.e., small-deformation theory). Moreover, we thoroughly study crack propagation patterns under the assumption that crack propagation during oxidation at high temperatures is a process involving the evolution of internal variables. In Chap. 6, we examine in detail CMAS infiltration as well as the extents of the resulting changes in stress and strain fields and in TBC phase structure during CMAS corrosion of TBCs at high temperatures using the thermo–mechano-chemical coupling constitutive relations under the assumption of geometric linearity (i.e., the small-deformation theory).

While the contents of this chapter are based on TBCs, they are also applicable to other materials. The basic ideas introduced in this chapter can be used to study the corrosion behavior of various metals and ceramics. However, this chapter also has certain limitations: all the thermodynamic functions are expanded up to the quadratic term, assuming that the independent variables change insignificantly. This is because the thermo–mechano-chemical coupling theory has been developed only in recent years and is a brand-new field with multitudes of novel phenomena, theoretical foundations, and experimental methods to be explored. It is hoped that an increasing number of talented researchers will engage in tackling scientific problems of thermo–mechano-chemical coupling encountered in their journey to meet major national needs.

Table. 2.3 Governing equations for thermo–mechano-chemical coupling

Small-deformation theory	
Geometric equation	$E_{ij} \approx \varepsilon_{ij} = \frac{1}{2}\left(\frac{\partial u_i}{\partial x_j} + \frac{\partial u_j}{\partial x_i}\right)$
Equilibrium equations	$\sigma_{ji,j} + f_i = 0 \quad \nabla \cdot \boldsymbol{\sigma} + \boldsymbol{f} = 0$
Constitutive equation	$\varepsilon_{ij} = \frac{1+\nu}{E}\sigma_{ij} - \frac{\nu}{E}\delta_{ij}\Theta$
Thermoelastic constitutive equations	$\boldsymbol{\varepsilon} = \frac{1}{2G}\left[\boldsymbol{\sigma} - \frac{\lambda}{2G+3\lambda}(tr\boldsymbol{\sigma})\boldsymbol{I}\right] + \alpha\theta\boldsymbol{I}$ $\rho C_v \frac{\partial T}{\partial t} = \left(kT_{,i}\right)_{,i} + \rho r_m - \frac{E\alpha}{1-2\nu}T_0\dot{\varepsilon}_{kk}$
Thermo–mechano-chemical coupling constitutive equations	$\sigma = \sigma_{\mathrm{m}}^{0}\mathbf{1} + \boldsymbol{s}^0 + K\varepsilon\mathbf{1} + 2G\boldsymbol{e} + R(c - c_0)\mathbf{1} - 3\alpha K(T - T_0)\mathbf{1} -$ $3\beta K(n - n_0)\mathbf{1} - \zeta(\xi - \xi_0)\mathbf{1}$ $\frac{\partial \psi}{\partial n}\dot{n} + \frac{\partial \psi}{\partial \xi}\dot{\xi} + \dot{T}\boldsymbol{\sigma} : \frac{\partial \boldsymbol{\varepsilon}_T}{\partial T} + T\left[\begin{array}{c}\frac{\mathrm{d}\left(\boldsymbol{\sigma} : \frac{\partial \boldsymbol{\varepsilon}_T}{\partial T}\right)}{\mathrm{d}t} - \frac{\partial^2\psi}{\partial T^2}\dot{T} - \frac{\partial^2\psi}{\partial T\partial c}\dot{c} - \\ \frac{\partial^2\psi}{\partial T\partial n}\dot{n} - \frac{\partial^2\psi}{\partial T\partial \xi}\dot{\xi} - \frac{\partial^2\psi}{\partial T\partial \boldsymbol{\varepsilon}_e} : \boldsymbol{\varepsilon}_e\end{array}\right] =$ $\dot{c} =$ $\boldsymbol{\sigma} : (\dot{\boldsymbol{\varepsilon}} - \dot{\boldsymbol{\varepsilon}}_e) + k\nabla^2 T + r_v + m\left(\nabla\left(\begin{array}{c}-\boldsymbol{\sigma} : \frac{\partial \boldsymbol{\varepsilon}_{diff}}{\partial c} + \\ \frac{\partial \psi}{\partial c}\end{array}\right)\right)^2 - \dot{c}\boldsymbol{\sigma} : \frac{\partial \boldsymbol{\varepsilon}_{diff}}{\partial c}$ $\mathrm{m}\nabla^2\left(\begin{array}{c}\varepsilon R + \mathrm{O}(c - c_0) + \nu_0(T - T_0) - \zeta^n(n - n_0) + \\ \zeta^\mu(\xi - \xi_0)\end{array}\right) - M\frac{\partial n}{\partial t}$ $\dot{n} = k_{chem}(\boldsymbol{\sigma} : (\partial\boldsymbol{\varepsilon}_{\mathrm{chem}}/\partial n) - \partial\psi/\partial n)$

(continued)

Table. 2.3 (continued)

Small-deformation theory	
Large-deformation theory	
Geometric equation	$\boldsymbol{E} = \frac{1}{2}(\boldsymbol{u}\nabla_X + \nabla_X \boldsymbol{u} + \nabla_X \boldsymbol{u} \cdot \boldsymbol{u}\nabla_X)$
Equilibrium equation	$div\boldsymbol{\sigma} + \rho \boldsymbol{f} = \rho \dot{\boldsymbol{v}}$
Mass conservation equation	$\frac{D\rho}{Dt} + \rho \mathrm{div}(\boldsymbol{v}) = 0$
Thermomechanical coupling constitutive equations	$\boldsymbol{T} = -p_0\boldsymbol{I} + K[I_1(\boldsymbol{E}) - \alpha(T - T_0)]\boldsymbol{I} + 2G\boldsymbol{E}'$ $\rho\left(\frac{\partial\psi}{\partial \boldsymbol{E}} : \dot{\boldsymbol{E}} - T\left(\frac{\partial^2\psi}{\partial T^2}\dot{T} + \frac{\partial^2\psi}{\partial T \partial \boldsymbol{E}} : \dot{E}\right)\right) = k(\nabla T)^2 + \rho r_m + \boldsymbol{\sigma} : \boldsymbol{d}$
Thermo–mechano-chemical coupling constitutive equations	$\boldsymbol{T} = 2G\boldsymbol{E}^e + K(\mathrm{tr}\boldsymbol{E}^e)\boldsymbol{I} - 3K\alpha(T - T_0)\boldsymbol{I}$ $\rho_0\left(\begin{array}{c}\frac{\partial\hat{\psi}}{\partial \boldsymbol{E}^e} : \dot{\boldsymbol{E}}^e + \frac{\partial\hat{\psi}}{\partial c}\dot{c} - \\ T\left(\begin{array}{c}\frac{\partial^2\hat{\psi}}{\partial T^2}\dot{T} + \frac{\partial^2\hat{\psi}}{\partial T \partial \boldsymbol{E}^e} : \dot{\boldsymbol{E}}^e + \\ \frac{\partial^2\hat{\psi}}{\partial T \partial c}\dot{c}\end{array}\right)\end{array}\right) = \boldsymbol{P}^T : \boldsymbol{F} + m\rho_0^2\frac{\partial\hat{\psi}}{\partial c}\nabla^2\left(\frac{\partial\hat{\psi}}{\partial c}\right) +$ $m\rho_0\left(\nabla\frac{\partial\hat{\psi}}{\partial c}\right)^2 + k\nabla^2 T + r_m\rho_0$ $\dot{c} = RDiv\left[m \ln\frac{c}{c_0}\nabla T + \frac{T}{c}\nabla c\right] - N$ $\dot{n} = k_n(1 - n) \cdot c$

References

1. Yang, W. Macroscopic and microscopic fracture mechanics. Beijing: National Defense Industry Press;1995. (In Chinese).
2. Truesdell C. The elements of continuum mechanics. New York: Springer; 1966.
3. Oden JT. Finite elements of nonlinear continua. New York: McGraw-Hill; 1972.
4. Tokuoka T. Introduction to theoretical continuum mechanics. Beijing: Sciences Press;1982. (In Chinese), (Translated by Z. Zhao, T. Miao, and C. Cheng).
5. Zhou Y. Solid mechanics of materials, vol. I. Beijing: Sciences Press;2005. (In Chinese).
6. Zhou YC, Yang L, Huang YL. Micro- and macromechanical properties of materials. CRC Press, Taylor & Francis Group;2014
7. Huang K. Nonlinear continuum mechanics. Beijing: Sciences Press;1989. (In Chinese).
8. Guo Z. Theory of nonlinear elasticity. Beijing: Sciences Press;1980. (In Chinese).
9. Huang K, Huang Y. Advanced solid mechanics, vol. I. Beijing: Tsinghua University Press;2013. (In Chinese).
10. Huang K, Xue M, Lu W. Tensor analysis. Beijing: Tsinghua University Press;1986. (In Chinese).
11. Zhou Y. Solid mechanics in materials, vol. II. Beijing: Sciences Press;2005. (In Chinese).
12. Lu M, Luo X. Foundations of elasticity. Beijing: Tsinghua University Press;1990. (In Chinese).
13. Du Q, Yu S, Yao Z. Theory of elasticity. Beijing: Sciences Press;1986. (In Chinese).
14. Allen DH. Thermomechanical coupling in inelastic solids. Appl Mech Rev. 1991;44(8):373–81.
15. Rosakis P, Rosakis AJ, Ravichandran G, Hodowany J. A thermodynamic internal variable model for the partition of plastic work into heat and stored energy in metals. J Mech Phys Solids. 2000;48:581–607.
16. Xu GN, Yang L, Zhou YC, Pi ZP, Zhu W. A chemo-thermo-mechanically constitutive theory for thermal barrier coatings under CMAS infiltration and corrosion. J Mech Phys Solids. 2019;133: 103710.
17. Hille TS, Turteltaub S, Suiker ASJ. Oxide growth and damage evolution in thermal barrier coatings. Eng Fract Mech. 2011;78(10):2139–52.
18. Shen Q, Li SZ, Yang L, et al. Coupled mechanical-oxidation modeling during oxidation of thermal barrier coatings. Comput Mater Sci. 2018;538–46.
19. Loeffel K, Anand L. A chemo-thermo-mechanically coupled theory for elastic–viscoplastic deformation, diffusion, and volumetric swelling due to a chemical reaction. Int J Plast. 2011;27(9):1409–31.
20. Zhang X, Zhong Z. A coupled theory for chemically active and deformable solids with mass diffusion and heat conduction. J Mech Phys Solids. 2017;49–75.
21. Nguyen TA, Lejeunes S, Eyheramendy D, et al. A thermodynamical framework for the thermo-chemo-mechanical couplings in soft materials at finite strain. Mech Mater. 2016;158–71.
22. Yadegari S, Turteltaub S, Suiker ASJ. Coupled thermomechanical analysis of transformation-induced plasticity in multiphase steels. Mech Mater. 2012;53:1–14.
23. Gurtin ME, Fried E, Anand L. The mechanics and thermodynamics of continua. Cambridge University Press;2010.
24. Reddy JN. An introduction to continuum mechanics. Cambridge University Press;2013.
25. Shen Q, Yang L, Zhou YC, Wei YG, Zhu W. Effects of growth stress in finite-deformation thermally grown oxide on failure mechanism of thermal barrier coatings. Mech Mater. 2017;114:228–42.
26. Shen Q, Yang L, Zhou YC, Wei YG, Wang NG. Models for predicting TGO growth to rough interface in TBCs. Surf Coat Technol. 2017;325:219–28.

Chapter 3
Nonlinear FEA of TBCs on Turbine Blades

Chapter 2 introduces the basic theoretical thermo–mechano-chemical coupling frameworks for thermal barrier coatings (TBCs). Based on these theories, we can examine the failure mechanisms of TBCs. However, it is impossible to directly obtain analytical solutions for the temperature, displacement, and stress fields of TBCs on complex turbine blade structures during operation. This is mainly because TBCs on turbine blades have highly complex structures with nonlinear erosion, corrosion, and phase-transformation problems. In addition, it is often costly and time-consuming to study TBC failure problems through experimentation. Thus, numerical simulations have become an important technical means for solving these complex problems.

With the development of computer technology, the finite element (FE) method (FEM) has become an indispensable tool in the scientific research and industrial fields, greatly facilitating the research and development of modern industrial products. In recent years, as numerical simulation techniques continuously mature, nonlinear FE analysis (FEA) has been extensively used as an effective means for solving myriad complex problems. The nonlinear FEM is increasingly employed to supersede prototype tests. For example, powerful nonlinear dynamics software programs, including LS-DYNA and ABAQUS, are used to simulate test processes such as target shooting with bullets, blasting, and jet flow in the military field. These programs have provided strong support for the research and development of new weapons. In the fields of machinery and electronics, drop simulation analysis can be used to examine the mechanical performance of electronic products and to predict their failures and thus provide guidance on product design and optimization. In automotive design, complete vehicle tests have been replaced by collision simulations to evaluate both initial design concepts and final design details (e.g., the placement of accelerometers for the determination of when airbags are deployed, the positioning of internal shock absorbers, the selection of materials, and the construction of collision criteria).

Studying TBC problems by nonlinear FEA requires an understanding of its basic concepts. FE programs in many commercial software applications are basically "black boxes." Without understanding the concepts of FEA and the problems,

Y. Zhou et al., *Thermal Barrier Coatings: Failure Theory and Evaluation Technology*,
https://doi.org/10.1007/978-981-19-2723-2_3

analyzers often find themselves at a loss when faced with complex problems. Thus, in this chapter, we introduce nonlinear FE numerical simulation methods from three perspectives, namely FEA principles, FE modeling, and mesh generation, with a view toward helping readers to better understand the FEM and to solve their own problems.

3.1 FEA Principles

The basic idea of the FEM is to discretize a continuous solution domain into a set of a finite number of elements connected to one another in a certain way. As elements can be combined via different modes of connection and can have different shapes, the FEM can model solution domains with complex geometries. The FEM assumes an approximate function within each element to represent, in a piecewise manner, the unknown function to be solved in the whole solution domain. The approximate function within each element is expressed in terms of the values of the unknown function or its derivative at different nodes of the element and the corresponding interpolation functions. This way, the value of the unknown field function or its derivative at each node becomes a new unknown quantity. Thus, a continuous problem with infinite degrees of freedom is transformed into a discrete problem with finite degrees of freedom. Once these unknown quantities are solved, the approximate values of the field functions within each element can be calculated using the interpolation functions, thus obtaining an approximate solution for the whole solution domain.

The use of the FEM to solve problems in continuum mechanics is conceptually described as follows. Based on the principle of virtual work, the differential equations of equilibrium, geometric equations, and physical equations for the whole structure are established using the variational method on each element obtained from discretizing the structure, so the stresses, strains, and displacements of each element are obtained, and then the internal stresses and strains of the structure are calculated. The functional variational principle is the theoretical basis of the FEM [1].

3.1.1 Functional Variational Principle

Functional integral equations can be generally established for some physical problems in a domain Ω. Let

$$J(u) = \int_{\Omega} F(u)\mathrm{d}\Omega \tag{3.1}$$

where u is a family of functions, $F(u)$ is a known function, and $J(u)$ is referred to as the functional of u (i.e., the function of the functions). The subsequent analysis is conducted in a Cartesian coordinate system as an example. Thus, $u = u(x, y, z)$. When u changes, we have

$$\tilde{u} = u + \delta u \tag{3.2}$$

Thus, its functional is given by

$$J(\tilde{u}) = J(u + \delta u) \tag{3.3}$$

If $J(u)$ satisfies high-order differentiability conditions, the above equation can be expanded to

$$\begin{aligned} J(\tilde{u}) &= J(u) + \delta J(u) + \frac{1}{2}\delta^2 J(u) + \cdots \\ &= J(u) + \frac{\mathrm{d}J}{\mathrm{d}u}\delta u + \frac{1}{2}\frac{\mathrm{d}^2 J}{\mathrm{d}u^2}\delta^2 u + \cdots \end{aligned} \tag{3.4}$$

δJ is often called the variation of $J(u)$, and we have

$$\delta J = \frac{\mathrm{d}J}{\mathrm{d}u}\delta u \tag{3.5}$$

Now, let us establish a certain functional:

$$\begin{aligned} J(u) = &\int\limits_{\Omega} \left\{ \frac{1}{2}\left[p(x,\text{y},\text{z})\left[(\frac{\partial u}{\partial x})^2 + (\frac{\partial u}{\partial y})^2 + (\frac{\partial u}{\partial z})^2 \right]\right] - f(x,\text{y},\text{z})u \right\} \mathrm{d}V \\ &+ \int\limits_{\Gamma} \left\{ \frac{1}{2}\gamma(x,\text{y},\text{z})u^2 - q(x,\text{y},\text{z})u \right\} \mathrm{d}S \end{aligned} \tag{3.6}$$

Let Γ be the boundary of domain Ω. The functional $J(u)$ exists under two conditions: $u(x, y, z)$ is piecewise smooth, and $\frac{\partial u}{\partial x}, \frac{\partial u}{\partial y}, \frac{\partial u}{\partial z}$ exist. Thus, the variation of $J(u)$ is given by

$$\begin{aligned} \delta J(u) = &\int\limits_{\Omega} \left\{ \left[p(x,\text{y},\text{z})\left(\frac{\partial u}{\partial x}\frac{\partial(\delta \mathrm{u})}{\partial x} + \frac{\partial u}{\partial y}\frac{\partial(\delta u)}{\partial y} + \frac{\partial u}{\partial z}\frac{\partial(\delta u)}{\partial z} \right)\right] - f(x,\text{y},\text{z})\delta u \right\} \mathrm{d}V \\ &+ \int\limits_{\Gamma} [\gamma(x,\text{y},\text{z})u - q(x,\text{y},\text{z})]\delta u \mathrm{dS} \end{aligned} \tag{3.7}$$

By using the Gauss–Green integral formula, we have

$$\delta J(u) = \int_{\Gamma} \left[p(x,y,z)\frac{\partial u}{\partial n} + \gamma(x,y,z)u - q(x,y,z) \right] \delta u \mathrm{d}S$$
$$- \int_{\Omega} \left\{ \left[\frac{\partial}{\partial x}\left(p(x,y,z)\frac{\partial u}{\partial x} \right) + \frac{\partial}{\partial y}\left(p(x,y,z)\frac{\partial u}{\partial y} \right) + \frac{\partial}{\partial z}\left(p(x,y,z)\frac{\partial u}{\partial z} \right) \right] + f(x,y,z) \right\} \delta u \mathrm{d}V \tag{3.8}$$

If $J(u)$ has an extreme value in $u^*(x, y, z)$, then

$$\delta J(u^*) = 0 \tag{3.9}$$

Due to the arbitrariness of δu, based on Eq. (3.8), the conditions for the above equation to hold are as follows:

$$\begin{cases} \frac{\partial}{\partial x}\left(p(x, y, z)\frac{\partial u}{\partial x}\right) + \frac{\partial}{\partial y}\left(p(x, y, z)\frac{\partial u}{\partial y}\right) + \frac{\partial}{\partial z}\left(p(x, y, z)\frac{\partial u}{\partial z}\right) + f(x, y, z) = 0 \quad (x, y, z) \in \Omega \\ p(x, y, z)\frac{\partial u(x,y,z)}{\partial n} + \gamma(x, y, z)u(x, y, z) - q(x, y, z) = 0 \quad (x, y, z) \in \Gamma \end{cases} \tag{3.10}$$

Here, $\boldsymbol{n}$ is the outer normal direction of the boundary Γ. In Eq. (3.10), the first equation is the differential governing equation, and the second equation is the boundary conditions. Thus, the extreme solution of the functional $J(u)$ is the solution of the physical problem in Eq. (3.10). Hence, the differential boundary-value problem is transformed into finding the extreme value of an integral functional. Note that the functional integral equation already contains the second- and third-type boundary conditions, i.e., the second equation in Eq. (3.10), also known as natural boundary conditions. The first-type boundary conditions (i.e., coercive boundary conditions) also need to be considered when solving the equation. Here, the readers are invited to reflect on why the value of $J(u)$ in $u^*(x, y, z)$ should be found. The fundamental physical idea is that a physical state in nature is based on the principle of least action and the principle of minimum energy.

The Ritz method is an approximation method commonly used to find the minimum of a functional. Let us assume that the function u on which the functional $J(u)$ depends has the following form:

$$u(x,y,z) = \sum_{i=1}^{n} a_i \phi_i(x, y, z) \tag{3.11}$$

where φ_i is a set of linearly independent functions that meet the boundary conditions (referred to as basis functions) and a_i is the coefficient to be determined. Equation (3.11) is a trial function. With the substitution of Eq. (3.11) into the functional expression in Eq. (3.6), the functional is transformed into a function of parameter a_i:

$$J(u) = J(a_1, a_2, \ldots, a_n) \tag{3.12}$$

Thus, based on Eq. (3.9), parameter a_i satisfies the following system of equations:

$$. \tag{3.13}$$

Equation (3.13) is a system of n linear algebraic equations, from which n values of a_i can be solved. With the substitution of a_i into Eq. (3.11), an approximate solution to the boundary-value problem of the differential equation described in Eq. (3.10) is obtained. The Ritz method gives only one optimal solution for the trial function, and the accuracy of the solution depends on the selected trial function. In general, the use of a large range of trial functions with a large number of coefficients to be determined results in a good solution accuracy. Polynomial functions are often used as trial functions to facilitate differential and integral operations.

The use of the Ritz method to solve a differential equation requires that a functional corresponding to the differential equation be found. It is challenging to establish a functional, similar to the one in Eq. (3.6), for a specific physical problem. TBC failure problems often entail the coupling of multiple factors (e.g., thermal, mechanical, and chemical factors), and their governing equations are usually nonlinear (e.g., geometric and physical nonlinearities, which are introduced in detail in Chaps. 4 and 5, respectively). It is exceedingly difficult to find functionals for complex nonlinear problems of thermo–mechano-chemical coupling in thermal barriers. In other words, it is a daunting task to obtain an approximate solution by the Ritz method.

The method of weighted residuals (MWR) is often used to solve the boundary-value problem of a differential equation for which a functional either is difficult to find or does not exist. First, let the approximate solution of the differential equation in Eq. (3.10) take the following form:

$$u(x, y, z) = \sum_{j=1}^{n} a_j \phi_j(x, y, z) \tag{3.14}$$

where a_j is the coefficient to be determined and $\varphi_j(x,y,z)$ is the basis function of a complete sequence of functions satisfying the boundary conditions. The approximate solution in Eq. (3.14) satisfies the boundary conditions but does not necessarily satisfy the differential equation. Substitution of this approximate solution into the differential equation results in the following residual:

$$\begin{aligned} R(x, y, z) = {} & \frac{\partial}{\partial x}\left(p(x, y, z)\frac{\partial u}{\partial x}\right) + \frac{\partial}{\partial y}\left(p(x, y, z)\frac{\partial u}{\partial y}\right) \\ & + \frac{\partial}{\partial z}\left(p(x, y, z)\frac{\partial u}{\partial z}\right) + f(x, y, z) \neq 0 \end{aligned} \tag{3.15}$$

If $u(x,y,z)$ is an exact solution, the residual should be constant at zero. In fact, it is very difficult to find exact solutions for most problems. Thus, we relax the condition that the weighted integral of the residual in the solution domain is equal to zero. Then, the following approximate solution can be obtained:

$$\int_{\Omega}\left\{\left[\frac{\partial}{\partial x}\left(p(x,y,z)\frac{\partial u}{\partial x}\right)+\frac{\partial}{\partial y}\left(p(x,y,z)\frac{\partial u}{\partial y}\right)+\frac{\partial}{\partial z}\left((p(x,y,z)\frac{\partial u}{\partial z}\right)\right]+f(x,y,z)\right\} w_i(x,y,z)\mathrm{d}V=0 \tag{3.16}$$

where $w_i(x,y,z)$ is a weight function. Through the selection of n weight functions and the substitution of Eq. (3.14) into Eq. (3.16), a system of n algebraic equations with a_j as an unknown quantity can be generated and is solved for n values of a_j. With the substitution of the obtained values of a_j into Eq. (3.14), an approximate solution for the differential equation is obtained. The readers are invited to understand the difference between the Ritz method and the Galerkin method. It can be found that when it is impossible to obtain an exact solution for a real physical state, the solution that best approximates the real situation should be sought instead. Based on the above analysis of the calculation of the extreme values of functionals and the features of the Galerkin method in Eq. (3.9), the following two aspects of the FEM can be summarized: (1) The theoretical basis of the FEM is the functional variational principle, the essence of which is based on the principle of least action and the principle of minimum energy. Thus, the essence of the FEM is derived from the laws of nature. (2) The FEM gives an approximate solution. In other words, the solution obtained by the FEM does not satisfy the differential equation describing a specific physical problem, but instead, the integral of the weighted function and differential equation over a certain domain is zero; that is to say, we ought not to pursue a perfect solution when solving a specific problem.

The accuracy of an approximate solution obtained by the MWR depends heavily on the selected weight function. Based on the selected weight function, the MWR can be classified into the collocation method, the method of least squares (MLS), the method of moments, and the Galerkin method. The Galerkin method, which selects a weight function as the basis function, i.e., $w_i(x,y,z) = \varphi_i(x,y,z)$, has been widely used due to its high accuracy.

While the Ritz method and the Galerkin method emerged early, it is not easy to select basis functions when solving complex real-world problems. In particular, it is exceptionally difficult to find a basis function that satisfies both natural boundary conditions and sufficient smoothness conditions and that can be incorporated into the Galerkin weighted integral form for solution domains with complex boundary shapes (e.g., TBCs on turbine blades). Due to the development of computer technology, it is now possible to divide a solution domain into a finite number of subdomains, to further apply the Galerkin method to the subdomains, and use a simple basis function to approximate the solution function of the differential equation to determine an approximate distribution of the solutions for each element. This is the so-called FEM, which has been extensively used in solid and structural mechanics.

3.1.2 Weak Form of the Eulerian Formulation

Selecting a suitable mesh description is important in FEA. Lagrangian meshes are an appealing choice and are most widely used in solid mechanics due to their ability to easily handle complex boundary conditions and track material points, thus enabling an accurate description of history-dependent materials. Two approaches are generally used in the Lagrangian FEA development.

(1) Lagrangian measures are used to formulate stress and strain equations, and derivative and integral operations are performed using the corresponding Lagrangian coordinates $\boldsymbol{X}$. This is referred to as the complete Lagrangian formulation. This approach is often used to solve large-deformation problems of solids but is certainly also appli cable to small-deformation problems.
(2) Eulerian measures are employed to formulate stress and strain equations, and derivative and integral operations are carried out with the corresponding Eulerian coordinates $\boldsymbol{x}$. This is called the complete Eulerian formulation. This approach is often used to solve small-deformation problems of solids and is naturally also applicable to large-deformation analysis.

The Eulerian and Lagrangian formulations are completely equivalent and fully interconvertible through a transformation of the deformation gradient.

As mentioned in the previous section, to solve momentum equations using the FEM, it is necessary to first derive their discrete forms. To this end, these equations first need to bc transformed into their variational form, which is often referred to as the weak form. Correspondingly, momentum equations and force boundary conditions are referred to as the strong form [2].

First, let us define the variational function and trial function spaces. The variational function space is defined as

$$\delta v_j(X) \in u_0,\ \ u_0 = \left\{ \delta v_j \middle| \, \delta v_j \in C^0(X), \delta v_j = 0 \text{ on } \Gamma_{v_i} \right\} \tag{3.17}$$

Here, the function C^0 means that the derivative is piecewise differentiable and that discontinuity occurs at some points, on line segments, and on surfaces for 1D, 2D, and 3D functions, respectively. In Eq. (3.17), $\boldsymbol{X}$ is the material coordinates in the Lagrangian system, and v_j is the velocity (i.e., the time derivative of the displacement u_j). The strong form of the momentum equations, including the momentum equation, the surface-force boundary condition, and the internal continuity condition, are given by

$$\frac{\partial \sigma_{ji}}{\partial x_j} + \rho b_i = \rho \dot{v}_i \quad \text{within} \quad \Omega \tag{3.18}$$

$$n_j \sigma_{ji} = \bar{t}_i \ \text{within} \quad \Gamma_{t_i} \tag{3.19}$$

$$n_j \sigma_{ji} = \bar{t}_i \text{ within } \Gamma_{\text{int}} \tag{3.20}$$

Here, Γ_{int} is the set of all the stress-discontinuity surfaces within the object, i.e., the internal continuity condition for a static problem at the material interface $[[\boldsymbol{n} \cdot \boldsymbol{\sigma}]] = \boldsymbol{0}$ or, in the form of components, $[[n_i \sigma_{ij}]] = n_i^A \sigma_{ij}^B + n_i^B \sigma_{ij}^B = 0$; $\bar{t}_i$ is the surface force; b_i is the volume force per unit density; and ρ is the density. By integrating the product of the variational function δv_i and the momentum equation in the current configuration, we have

$$\int_{\Omega} \delta v_i \left(\frac{\partial \sigma_{ji}}{\partial x_j} + \rho b_i - \rho \dot{v}_i \right) \mathrm{d}\Omega = 0 \tag{3.21}$$

Under the small-deformation theory, the current configuration is the same as the reference configuration. Thus, all the integral domains are denoted by Ω, which has a boundary Γ. The first term on the left-hand side of Eq. (3.21) can be expanded as follows by applying the basic principle of calculus:

$$\int_{\Omega} \delta v_i \frac{\partial \sigma_{ji}}{\partial x_j} \mathrm{d}\Omega = \int_{\Omega} \left[\frac{\partial}{\partial x_j} (\delta v_i \sigma_{ji}) - \frac{\partial (\delta v_i)}{\partial x_j} \sigma_{ji} \right] \mathrm{d}\Omega \tag{3.22}$$

Let us assume that discontinuity occurs on a finite set of surface Γ_{int}. Then, according to Gauss's theorem, we have

$$\int_{\Omega} \left[\frac{\partial}{\partial x_j} (\delta v_i \sigma_{ji}) \right] \mathrm{d}\Omega = \int_{\Gamma_{\text{int}}} \delta v_i [[n_j \sigma_{ji}]] \mathrm{d}\Gamma + \int_{\Gamma} \delta v_i n_j \sigma_{ji} \mathrm{d}\Gamma \tag{3.23}$$

Based on the surface-force continuity condition in Eq. (3.20), the first integral on the right-hand side of Eq. (3.23) is zero. Let us apply Eq. (3.19) to the second integral on the right-hand side of Eq. (3.23). Since the integral of the variational function over the whole surface-force boundary is zero, Eq. (3.23) becomes

$$\int_{\Omega} \left[\frac{\partial}{\partial x_j} (\delta v_i \sigma_{ji}) \right] \mathrm{d}\Omega = \sum_{i=1}^{n_{SD}} \int_{\Gamma_{t_i}} \delta v_i \bar{t}_i \mathrm{d}\Gamma \tag{3.24}$$

where n_{SD} is the number of boundaries. By substituting Eq. (3.24) into Eq. (3.22) and applying integration by parts, we have

$$\int_{\Omega} \delta v_i \frac{\partial \sigma_{ji}}{\partial x_j} \mathrm{d}\Omega = \sum_{i=1}^{n_{SD}} \int_{\Gamma_{t_i}} \delta v_i \bar{t}_i \mathrm{d}\Gamma - \int_{\Omega} \frac{\partial (\delta v_i)}{\partial x_j} \sigma_{ji} \mathrm{d}\Omega \tag{3.25}$$

Further substitution of the above equation into Eq. (3.21) yields

$$\int_{\Omega} \frac{\partial(\delta v_i)}{\partial x_j}\sigma_{ji}\mathrm{d}\Omega - \int_{\Omega} \delta v_i \rho b_i \mathrm{d}\Omega - \sum_{i=1}^{n_{SD}} \int_{\Gamma_{t_i}} \delta v_i \bar{t}_i \mathrm{d}\Gamma + \int_{\Omega} \delta v_i \rho \dot{v}_i \mathrm{d}\Omega = 0 \quad (3.26)$$

Equation (3.26) is the weak form of the momentum equation, the surface-force boundary condition, and the internal continuity condition. It can be found that each term in the weak form is a virtual power. Within the first integral is the rate of change in internal virtual work per unit volume or the internal virtual power. Thus, the total internal virtual power δp^{int} is given by

$$\delta p^{\mathrm{int}} = \int_{\Omega} \delta D_{ij}\sigma_{ji}\mathrm{d}var\Omega = \int_{\Omega} \frac{\partial(\delta v_i)}{\partial x_j}\sigma_{ji}\mathrm{d}\Omega = \int_{\Omega} \delta D : \sigma \mathrm{d}\Omega \quad (3.27)$$

where $\boldsymbol{D}$ is the deformation rate. The second and third terms are the power generated by the external force acting on the object and the power generated by a specific surface force, respectively. Thus, external virtual power δp^{exp} is defined as follows:

$$\delta p^{\mathrm{ext}} = \int_{\Omega} \delta v_i \rho b_i \mathrm{d}\Omega + \sum_{i=1}^{n_{SD}} \int_{\Gamma_{t_i}} \delta v_i \bar{t}_i \mathrm{d}\Gamma = \int_{\Omega} \delta v \cdot \rho b \mathrm{d}\Omega + \sum_{i=1}^{n_{SD}} \int_{\Gamma_{t_i}} \delta v_i e_j \cdot \bar{t} \mathrm{d}\Gamma \quad (3.28)$$

The last term is the power generated by the inertial force, which is thus defined as inertial virtual power or dynamic virtual power:

$$\delta p^{\mathrm{kin}} = \int_{\Omega} \delta v_i \rho \dot{v}_i \mathrm{d}\Omega \quad (3.29)$$

By substituting Eqs. (3.27)–(3.29) into Eq. (3.26), we have

$$\delta p = \delta p^{\mathrm{int}} - \delta p^{\mathrm{ext}} + \delta p^{\mathrm{kin}} = 0 \ \forall \delta v_i \in u_0 \quad (3.30)$$

Equation (3.30) is the weak form of the momentum equation. It can be seen that when an object that is in an equilibrium state under external forces is subjected to an infinitesimal virtual velocity δv_i, the external virtual power of the external force is equal to the sum of the physical internal power resulting from δv_i and the inertial virtual power. Thus, Eq. (3.30) is also referred to as the principle of virtual power.

3.1.3 FE Discretization of the Eulerian Formulation

The discrete equation of the FE model is derived from the weak form by applying FE interpolation to the variational term and trial function. In this section, the weak form of the momentum equation is transformed into a discrete FE equation. The current domain Ω is divided into element domains Ω_e. All the element domains collectively form the whole domain, i.e., $\Omega = \bigcup_e \Omega_e$. Let x_{iI} ($I = n_N$, where the subscript denotes the node number) be the node coordinates in the current configuration. In the FEM, motion $\boldsymbol{x}(\boldsymbol{X},t)$ is approximately expressed as

$$x_i(\mathrm{X}, t) = N_I(\mathrm{X}) x_{iI}(t) \text{ or } \mathrm{x}(\mathrm{X}, t) = N_I(\mathrm{X}) \mathrm{x}_I(t) \tag{3.31}$$

Here, $N_I(\boldsymbol{X})$ is an interpolation function or shape function. Once the element type and nodes are determined, so is the shape function. $\boldsymbol{x}_I$ is the position vector of node I. Equation (3.31) is written for a node with an initial position $\boldsymbol{X}_J$ as follows:

$$\mathrm{x}(\mathrm{X}_J, t) = \mathrm{x}_I(t) N_I(\mathrm{X}_J) = \mathrm{x}_I(t) \delta_{IJ} = \mathrm{x}_J(t) \tag{3.32}$$

The displacement is the difference between the current and original positions of the node. Thus, a displacement field is given by

$$u_i(\boldsymbol{X}, t) = x_i(\boldsymbol{X}, t) - X_i = u_{iI}(t) N_I(\boldsymbol{X}) \text{ or } \boldsymbol{u}(\boldsymbol{X}, t) = \boldsymbol{u}_I(t) N_I(\boldsymbol{X}) \tag{3.33}$$

The velocity is the material time derivative of the displacement of the node:

$$v_i(\boldsymbol{X}, t) = \frac{\partial u_i(\boldsymbol{X}, t)}{\partial t} = \dot{u}_{iI}(t) N_I(\boldsymbol{X}) = v_{iI}(t) N_I(\boldsymbol{X}) \text{ or } \boldsymbol{v}(\boldsymbol{X}, t) = \dot{\boldsymbol{u}}_I(t) N_I(\boldsymbol{X}) \tag{3.34}$$

The variational function or variable is not a function of time. Therefore, the variational function is approximated as follows:

$$\delta v_i(\boldsymbol{X}) = \delta v_{iI}(t) N_I(\boldsymbol{X}) \text{ or } \delta \boldsymbol{v}(\boldsymbol{X}) = \delta \boldsymbol{u}_I(t) N_I(\boldsymbol{X}) \tag{3.35}$$

where δv_{iI} is the velocity of a virtual node. At the first step of establishing a discrete FE equation, the variational function is substituted into the principle of virtual power:

$$\delta v_{iI} \int_{\Omega} \frac{\partial N_I(\boldsymbol{X})}{\partial x_j} \sigma_{ji} \mathrm{d}\Omega - \delta v_{iI} \int_{\Omega} N_I(\boldsymbol{X}) \rho b_i \mathrm{d}\Omega$$
$$- \sum_{i=1}^{n_{SD}} \delta v_{iI} \int_{\Gamma_{t_i}} N_I(\boldsymbol{X}) \bar{t}_i \mathrm{d}\Gamma + \delta v_{iI} \int_{\Omega} N_I(\boldsymbol{X}) \rho \dot{v}_i \mathrm{d}\Omega = 0 \tag{3.36}$$

In the above equation, the stress is a function of the trial velocity and trial displacement. According to the definition of variation, the virtual velocity at any location with a specific velocity must be zero, i.e., $\delta v_i = 0$ on Γ_{v_i}. Thus, only the nodes that are not on Γ_{v_i} have arbitrary virtual velocities. On this basis, the weak form of the momentum equation is given by

$$\begin{aligned} &\int_{\Omega} \frac{\partial N_I(\boldsymbol{X})}{\partial x_j} \sigma_{ji} \mathrm{d}\Omega - \int_{\Omega} N_I(\boldsymbol{X}) \rho b_i \mathrm{d}\Omega - \sum_{i=1}^{n_{SD}} \int_{\Gamma_{t_i}} N_I(\boldsymbol{X}) \bar{t}_i \mathrm{d}\Gamma \\ &- \int_{\Omega} N_I(\boldsymbol{X}) \rho \dot{v}_i \mathrm{d}\Omega = 0 \forall (I, i) \notin \Gamma_{v_i} \end{aligned} \tag{3.37}$$

Similarly, nodal forces are defined corresponding to each term in the virtual power equation. These nodal forces can be found in most FE software (e.g., ANSYS and ABAQUS). The internal virtual power is given by

$$\delta p^{\mathrm{int}} = \delta v_{iI} f_{iI}^{\mathrm{int}} = \int_{\Omega} \frac{\partial (\delta v_i)}{\partial x_j} \sigma_{ji} \mathrm{d}\Omega = \delta v_{iI} \int_{\Omega} \frac{\partial N_i}{\partial x_j} \sigma_{ji} \mathrm{d}\Omega \tag{3.38}$$

Based on the above equation, internal nodal forces are given by

$$f_{iI}^{\mathrm{int}} - \int_{\Omega} \frac{\partial N_i}{\partial x_j} \sigma_{ji} \mathrm{d}\Omega \tag{3.39}$$

These nodal forces represent the stresses of the object. These expressions can be applied to both the whole mesh and an arbitrary element or element set. Similarly, external nodal forces are defined in the form of external virtual power:

$$\begin{aligned} \delta p^{\mathrm{ext}} = \delta v_{iI} f_{iI}^{\mathrm{ext}} &= \int_{\Omega} \delta v_i \rho b_i \mathrm{d}\Omega + \sum_{i=1}^{n_{SD}} \int_{\Gamma_{t_i}} \delta v_i \bar{t}_i \mathrm{d}\Gamma \\ &= \delta v_{iI} \int_{\Omega} N_i \rho b_i \mathrm{d}\Omega + \sum_{i=1}^{n_{SD}} \delta v_{iI} \int_{\Gamma_{t_i}} N_i \bar{t}_i \mathrm{d}\Gamma \end{aligned} \tag{3.40}$$

Thus, external nodal forces are given by

$$f_{iI}^{\mathrm{ext}} = \int_{\Omega} N_i \rho b_i \mathrm{d}\Omega + \sum_{i=1}^{n_{SD}} \int_{\Gamma_{t_i}} N_i \bar{t}_i \mathrm{d}\Gamma \tag{3.41}$$

External nodal forces correspond to externally applied loads. Similarly, inertial forces are defined as follows:

$$\delta p^{\mathrm{kin}} = \delta v_{iI} f_{iI}^{\mathrm{kin}} = \int_{\Omega} \delta v_i \rho \dot{v}_i d\Omega = \delta v_{iI} \int_{\Omega} N_I \rho \dot{v}_i d\Omega = \delta v_{iI} \int_{\Omega} N_I N_J \rho \dot{v}_{iJ} d\Omega \tag{3.42}$$

Here, $\dot{v}_i(\boldsymbol{X}, t) = \dot{v}_{iJ}(t)\mathrm{N}_J(\boldsymbol{X})$ is applied. Hence, inertial forces are defined as follows:

$$f_{iI}^{\mathrm{kin}} = \int_{\Omega} \rho N_I N_J d\Omega \dot{v}_{jJ}(t) \tag{3.43}$$

A mass matrix is defined as follows:

$$M_{ijIJ} = \delta_{ij} \int_{\Omega} \rho N_I N_J d\Omega \tag{3.44}$$

According to Eqs. (3.40) and (3.41), inertial forces are expressed as

$$f_{iI}^{\mathrm{kin}} = M_{ijIJ} \dot{v}_{jJ}(t) \tag{3.45}$$

With the expressions of internal, external, and inertial nodal forces, the discrete expression of the weak form in Eq. (3.36) can be written concisely as follows:

$$\delta v_{iI} (f_{iI}^{\mathrm{int}} - f_{iI}^{\mathrm{ext}} + M_{ijIJ} \dot{v}_{jJ}) = 0 \; \forall \delta v_{iI} \notin \Gamma_{v_i} \tag{3.46}$$

Since δv is arbitrary, the following equation, expressed in a matrix form, is derived:

$$\boldsymbol{Ma} + \boldsymbol{f}^{\mathrm{int}} = \boldsymbol{f}^{\mathrm{ext}} \tag{3.47}$$

Equation (3.47) is the discrete momentum equation, which is quadratic with respect to time. As it is not discrete in time, this equation is also called the semidiscrete momentum equation. It can be found that the discrete momentum equation in Eq. (3.47) is a system of ordinary differential equations with respect to nodal velocities. If the acceleration is zero, Eq. (3.47) is referred to as the discrete equilibrium equation, which is a system of nonlinear algebraic equations with respect to stresses and nodal displacements.

Generally, when an FE model is established, the shape function is expressed in terms of parent element coordinate $\boldsymbol{\xi}$, which is often referred to as the element coordinate:

$$x_i(\boldsymbol{\xi}, t) = x_{iI}(t) N_I(\boldsymbol{\xi}) \quad \text{or} \quad \boldsymbol{x}(\boldsymbol{\xi}, t) = \boldsymbol{x}_I(t) N_I(\boldsymbol{\xi}) \tag{3.48}$$

The FE approximation of motion $\boldsymbol{x}(\boldsymbol{\xi}, t)$ involves the mapping of the parent domain of an element to its current domain. Apart from that, discontinuity is inadmissible; the conditions for such mapping include the following: (1) $\boldsymbol{x}(\boldsymbol{\xi}, t)$ must be in a one-to-one correspondence, (2) $\boldsymbol{x}(\boldsymbol{\xi}, t)$ must be at least C^0 in space, and (3) the Jacobian determinant of the element must be positive, i.e.,

$$J_\xi = \det(x_{,\xi}) > 0 \tag{3.49}$$

These conditions ensure that $\boldsymbol{x}(\boldsymbol{\xi}, t)$ is reversible.

The calculation of nodal forces is most important in FE calculations. Here, the calculation procedure is described with internal nodal forces as an example.

(1) $\boldsymbol{f}^{\text{int}} = \boldsymbol{0}$.
(2) For each integration point $\boldsymbol{\xi}_Q$ (on the parent element):

- (i) For each node I, calculate matrix $[B_{Ij}] = \left[\frac{\partial N_I(\boldsymbol{\xi}_Q)}{\partial x_j}\right]$.
- (ii) Calculate matrix $\boldsymbol{L} = [L_{ij}] = [v_{iI} B_{Ij}] = \boldsymbol{v}_I \boldsymbol{B}_I^T$.
- (iii) Calculate matrix $\boldsymbol{D} = \frac{1}{2}(\boldsymbol{L}^T + \boldsymbol{L})$.
- (iv) Calculate the deformation gradient $\boldsymbol{F}$ and the strain tensor $\boldsymbol{E}$.
- (v) Based on the constitutive equations, calculate the Cauchy stress $\boldsymbol{\sigma}$ or the second Piola–Kirchhoff stress $\boldsymbol{S} = J\boldsymbol{F}^{-1}\boldsymbol{\sigma}\boldsymbol{F}^{-T}$, $J = \det(\boldsymbol{F})$.
- (vi) If $\boldsymbol{S}$ is obtained, calculate $\boldsymbol{\sigma}$ using $\boldsymbol{\sigma} = J^{-1}\boldsymbol{F}\boldsymbol{S}\boldsymbol{F}^T$.
- (vii) For each node I, calculate $\boldsymbol{f}_I^{\text{int}} + \boldsymbol{B}_I^T \boldsymbol{\sigma} J_\xi \overline{w}_Q \rightarrow \boldsymbol{f}_I^{\text{int}}$, where $\overline{w}_Q$ is the weight of an integration point in the parent element, i.e., $\int_{-1}^{1} f(\xi)d\xi = \sum_{Q=1}^{n_Q} \overline{w}_Q f(\xi_Q)$ for the parent element with n_Q integration points at coordinates ξ_Q. Interested readers are referred to the numerical integration methods.

The cycle is terminated. Here, boldface indicates a matrix, and the superscript T denotes transpose.

3.1.4 Weak Form of the Lagrangian Formulation

The Lagrangian formulation is highly suitable for geometrically nonlinear or large-deformation problems in the reference configuration. Large deformation occurs within a TBC during failure processes (e.g., oxidation and erosion). In this case, the small-deformation assumption may not apply to analyzing the stress field of the TBC. In recent years, the author of the book has proposed theoretical TBC models based on the large-deformation assumption (see Chap. 4 for details) and verified that results closer to the actual situation can be obtained under the large-deformation assumption for some problems of TBCs. To carry out FEA in the framework of the large-deformation theory, it is necessary to establish an FE discrete momentum equation for large deformation. First, let us define the variational function and trial function spaces. The variational function space is defined as follows:

$$\delta\boldsymbol{u}(\boldsymbol{X}) \in u_0, \boldsymbol{u}(\boldsymbol{X}, t) \in u \tag{3.50}$$

where u is the space of the kinematically admissible displacement, u_0 is the same space with an additional zero displacement boundary condition, $\boldsymbol{X}$ is the material coordinates in the Lagrangian system, and $\boldsymbol{u}$ is the displacement. The force equilibrium equations in the reference configuration are presented in Chap. 2. Here, we further integrate the product of the variational function and the momentum equation in the initial configuration to obtain

$$\int_{\Omega_0} \delta u_i \left(\frac{\partial P_{ji}}{\partial X_j} + \rho_0 b_i - \rho_0 \ddot{u}_i \right) \mathrm{d}\Omega_0 = 0 \tag{3.51}$$

where Ω_0 is the integration domain in the reference configuration and has a boundary Γ_0 and $\boldsymbol{P}$ is the first Piola–Kirchhoff stress, which is introduced in Chap. 2. Since Eq. (3.51) contains a derivative of the Piola–Kirchhoff stress, this weak form requires that the trial displacement have C^1 continuity. Here, a C^1 function is continuous and differentiable and has a first-order derivative that is continuous everywhere. By applying the derivative product rule to eliminate the derivative of the Piola–Kirchhoff nominal stress, we have

$$\int_{\Omega_0} \delta u_i \frac{\partial P_{ji}}{\partial X_j} \mathrm{d}\Omega_0 = \int_{\Omega_0} \left[\frac{\partial}{\partial X_j} (\delta u_i P_{ji}) - \frac{\partial (\delta u_i)}{\partial X_j} P_{ji} \right] \mathrm{d}\Omega_0 \tag{3.52}$$

Thus, based on Gauss's theorem, we have

$$\int_{\Omega_0} \frac{\partial}{\partial X_j} (\delta u_i P_{ji}) \mathrm{d}\Omega_0 = \int_{\Gamma_{\mathrm{int}}^0} \delta u_i [[n_j^0 P_{ji}]] \mathrm{d}\Gamma_0 + \int_{\Gamma_0} \delta u_i n_j^0 P_{ji} \mathrm{d}\Gamma_0 \tag{3.53}$$

According to the surface-force continuity condition $[[\boldsymbol{n}^0 \cdot \boldsymbol{P}]] = \boldsymbol{0}$ or, in the form of components, $[[n_j^0 P_{ji}]] = n_j^{0A} P_{ji}^B + n_j^{0B} P_{ji}^B = 0$ ($\boldsymbol{n}^0$ is the outer normal of the internal interface Γ_{int}^0), the first integral on the right-hand side of Eq. (3.53) is zero. Since $\delta u_i = 0$ on $\Gamma_{\mathrm{u_i}}^0$ and $\Gamma_{\mathrm{t_i}}^0 = \Gamma^0 - \Gamma_{\mathrm{u_i}}^0$, the second term on the right-hand side of Eq. (3.53) can be simplified to the surface-force boundary. Thus, we have

$$\int_{\Omega_0} \frac{\partial}{\partial X_j} (\delta u_i P_{ji}) \mathrm{d}\Omega_0 = \int_{\Gamma_0} \delta u_i n_j^0 P_{ji} \mathrm{d}\Gamma_0 = \sum_{i=1}^{n_{SD}} \int_{\Gamma_{ti}^0} \delta u_i \bar{t}_i^0 \mathrm{d}\Gamma_0 \tag{3.54}$$

For the deformation gradient F_{ij}, we have

$$\delta F_{ij} = \delta\left(\frac{\partial u_i}{\partial X_j}\right) = \frac{\partial (\delta u_i)}{\partial X_j} \tag{3.55}$$

Further, we have

$$\int_{\Omega_0} (\delta F_{ij} P_{ji} - \delta u_i \rho_0 b_i + \delta u_i \rho_0 \ddot{u}_i) \mathrm{d}\Omega_0 - \sum_{i=1}^{n_{SD}} \int_{\Gamma_{\mathrm{int}}^0} \delta u_i \bar{t}_i^0 \mathrm{d}\Gamma_0 = 0 \tag{3.56}$$

or

$$\int_{\Omega_0} (\delta F{:}P - \rho_0 \delta u \cdot b + \rho_0 \delta u \cdot \ddot{u}) \mathrm{d}\Omega_0 - \sum_{i=1}^{n_{SD}} \int_{\Gamma_{\mathrm{int}}^0} \delta(u \cdot e_i)(e_i \cdot \bar{t}_i^0) \mathrm{d}\Gamma_0 = 0 \tag{3.57}$$

The above equation is the weak form of the momentum equation, the surface-force boundary condition, and the internal continuity condition for large deformation. It can be found that each term in the weak form represents virtual work. The internal, external, and inertial virtual work parameters per unit volume are respectively given by

$$\delta w^{\mathrm{int}} = \int_{\Omega_0} \delta F^T{:}P \mathrm{d}\Omega_0 \tag{3.58}$$

$$\delta w^{\mathrm{ext}} = \int_{\Omega_0} (\rho_0 \delta \boldsymbol{u} \cdot \boldsymbol{b}) \mathrm{d}\Omega_0 + \sum_{i=1}^{n_{SD}} \int_{\Gamma_{\mathrm{int}}^0} \delta(\boldsymbol{u} \cdot \boldsymbol{e}_i)(\boldsymbol{e}_i \cdot \bar{\boldsymbol{t}}_i^0) \mathrm{d}\Gamma_0 \tag{3.59}$$

$$\delta w^{\mathrm{kin}} = \int_{\Omega_0} (\rho_0 \delta u \cdot \ddot{u}) \mathrm{d}\Omega_0 \tag{3.60}$$

Equation (3.57) is the principle of virtual work:

$$\delta w^{\mathrm{int}} + \delta w^{\mathrm{kin}} - \delta w^{\mathrm{ext}} = 0 \tag{3.61}$$

3.1.5 FE Discretization of the Lagrangian Formulation

Similar to the weak form for small deformation, the shape function is a function of initial coordinates. The trial displacement field is given by

$$u_i(\boldsymbol{X}, t) = u_{iI}(t) N_I(\boldsymbol{X}) \text{ or } \boldsymbol{u}(X, t) = \boldsymbol{u}_I(t) N_I(\boldsymbol{X}) \tag{3.62}$$

Since the variational function or variable is not a function of time, we have

$$\delta u_i(\boldsymbol{X}, t) = \delta u_{iI}(t) N_I(\boldsymbol{X}) \text{ or } \delta \boldsymbol{u}(X, t) = \delta \boldsymbol{u}_I(t) N_I(\boldsymbol{X}) \tag{3.63}$$

The velocity is the material time derivative of the displacement:

$$\dot{u}_i(X,t) = \dot{u}_{iI}(t)N_I(X) \tag{3.64}$$

The corresponding acceleration is given by

$$\ddot{u}_i(X,t) = \ddot{u}_{iI}(t)N_I(X) \tag{3.65}$$

Thus, the deformation gradient is expressed as

$$F_{ij} = \frac{\partial x_i}{\partial X_j} = \frac{\partial N_I}{\partial X_j}x_{iI} \tag{3.66}$$

$$\delta F_{ij} = \frac{\partial N_I}{\partial X_j}\delta x_{iI} = \frac{\partial N_I}{\partial X_j}\delta u_{iI} \tag{3.67}$$

Here, $\delta x_{iI} = \delta(X_{iI} + u_{iI}) = \delta u_{iI}$ is applied. For the sake of convenience, Eqs. (3.66) and (3.67) are sometimes rewritten as follows:

$$F_{ij} = B^0_{jI}x_{iI} \quad B^0_{jI} = \frac{\partial N_I}{\partial X_j} \quad \boldsymbol{F} = \boldsymbol{x}\boldsymbol{B}_0^T \tag{3.68}$$

$$\delta F_{ij} = \frac{\partial N_I}{\partial X_j}\delta x_{iI} = \frac{\partial N_I}{\partial X_j}\delta u_{iI} \quad \delta\boldsymbol{F} = \delta\boldsymbol{u}\boldsymbol{B}_0^T \tag{3.69}$$

Here, boldface indicates a matrix, and the superscript T denotes the transpose. Based on the above equation, internal nodal forces can be defined in the form of internal virtual work:

$$\delta w^{\text{int}} = \delta u_{iI} f_{iI}^{\text{int}} = \int_{\Omega_0} \delta F_{ij} P_{ji} \mathrm{d}\Omega_0 = \delta u_{iI} \int_{\Omega_0} \frac{\partial N_I}{\partial X_j} P_{ji} \mathrm{d}\Omega_0 \tag{3.70}$$

Equation (3.69) is used in the last step of Eq. (3.70). Then, based on the arbitrariness of δu_{iI}, the internal nodal forces are obtained as

$$f_{iI}^{\text{int}} = \int_{\Omega_0} \frac{\partial N_I}{\partial X_j} P_{ji} \mathrm{d}\Omega_0 = \int_{\Omega_0} B^0_{jI} P_{ji} \mathrm{d}\Omega_0 \text{or} \boldsymbol{f}^{\text{int,T}} = \int_{\Omega_0} \boldsymbol{B}_0^T \boldsymbol{P} \mathrm{d}\Omega_0 \tag{3.71}$$

The above equation can be transformed into Eq. (3.39), which is described earlier in the Eulerian formulation. Interested readers can verify the transformation themselves. The external nodal forces are defined in the form of internal virtual work:

$$\delta w^{\text{ext}} = \delta u_{iI} f_{iI}^{\text{ext}} = \int\limits_{\Omega_0} (\delta u_i \rho_0 b_i) \text{d}\Omega_0 + \int\limits_{\Gamma_{t_i}^0} \delta u_i \bar{t}_i^0 \text{d}\Gamma_0$$

$$= \delta u_{iI} \left[\int\limits_{\Omega_0} (N_I \rho_0 b_i) \text{d}\Omega_0 + \int\limits_{\Gamma_{t_i}^0} N_I \bar{t}_i^0 \text{d}\Gamma_0 \right] \quad (3.72)$$

Thus, we have

$$f_{iI}^{\text{ext}} = \int\limits_{\Omega_0} (N_I \rho_0 b_i) \text{d}\Omega_0 + \int\limits_{\Gamma_{t_i}^0} N_I \bar{t}_i^0 \text{d}\Gamma_0 \quad (3.73)$$

Similarly, Eq. (3.73) can be transformed into Eq. (3.41), which is described earlier in the Eulerian formulation. Based on Eq. (3.60), it is defined that nodal forces are equivalent to inertial forces. Thus, we have

$$\delta w^{\text{kin}} = \delta u_{iI} f_{iI}^{\text{kin}} = \int\limits_{\Omega_0} (\delta u_i \rho_0 \ddot{u}_i) \text{d}\Omega_0 = \delta u_{iI} \int\limits_{\Omega_0} (\rho_0 N_I N_J) \text{d}\Omega_0 \ddot{u}_{iJ} = \delta u_{iI} M_{ijIJ} \ddot{u}_{jJ} \quad (3.74)$$

Since δu_{iI} in the above equation is arbitrary, we have

$$M_{ijIJ} = \delta_{ij} \int\limits_{\Omega_0} (\rho_0 N_I N_J) \text{d}\Omega_0 \quad (3.75)$$

Equation (3.75) can be transformed into Eq. (3.44), which is described earlier in the Eulerian formulation. Substitution of the above expression into the weak form in Eq. (3.61) yields

$$\delta u_{iI} (f_{iI}^{\text{int}} - f_{iI}^{\text{ext}} + M_{ijIJ} \ddot{u}_{jJ}) = 0 \ \forall I, i \notin \Gamma_{u_i} \quad (3.76)$$

Since Eq. (3.76) is applicable to arbitrary values of nodal displacement components unconstrained by the displacement boundary conditions, we have

$$M_{ijIJ} \ddot{u}_{jJ} + f_{iI}^{\text{int}} = f_{iI}^{\text{ext}} \ \forall I, i \notin \Gamma_{u_i} \quad (3.77)$$

Similar to that of the Eulerian formulation, the calculation procedure is :

(1) $\boldsymbol{f}^{\text{int}} = \boldsymbol{0}$.
(2) For each integration point $\boldsymbol{\xi}_Q$ (on the parent element):

 (i) For each node I, calculate matrix, $[B_{Ij}^0] = \left[\frac{\partial N_I(\boldsymbol{\xi}_Q)}{\partial X_j} \right]$.

(ii) Calculate matrix $\boldsymbol{H} = \boldsymbol{B}_I^0 \boldsymbol{u}_I$ $[H_{ij}] = \left[\frac{\partial N_I}{\partial X_j} u_{iI}\right]$.
(iii) Calculate matrix $\boldsymbol{F} = \boldsymbol{I} + \boldsymbol{H}$ $J = \det(\boldsymbol{F})$, where $\boldsymbol{I}$ is the unit matrix.
(iv) Calculate the strain tensor $\boldsymbol{E} = \frac{1}{2}\left(\boldsymbol{H} + \boldsymbol{H}^T + \boldsymbol{H}^T\boldsymbol{H}\right)$.
(v) Calculate $\dot{E} = \frac{\Delta E}{\Delta t}$, $\dot{F} = \frac{\Delta F}{\Delta t}$, $D = \mathrm{sym}\left(\dot{F}F^{-1}\right)$, if needed.
(vi) Based on the constitutive equations, calculate the Cauchy stress $\boldsymbol{\sigma}$ or the second Piola–Kirchhoff stress $\boldsymbol{S} = J\boldsymbol{F}^{-1} \cdot \boldsymbol{\sigma} \cdot \boldsymbol{F}^{-T}$, $J = \det(\boldsymbol{F})$.
(vii) Calculate $\boldsymbol{P} = \boldsymbol{S}\boldsymbol{F}^T$ or $\boldsymbol{P} = J\boldsymbol{F}^{-1}\boldsymbol{\sigma}$.
(viii) For each node I, calculate $\boldsymbol{f}_I^{\mathrm{int}} + \boldsymbol{B}_I^{0T}\boldsymbol{P} J_\xi^0 \overline{w}_Q \rightarrow \boldsymbol{f}_I^{\mathrm{int}}$, where $\overline{w}_Q$ is, similarly, the weight of the integral on the parent element and $J_\xi^0 = \det\left(\boldsymbol{X}_{,\xi}\right) > 0$.

The cycle is terminated. Here, boldface indicates a matrix, and the superscript T denotes the transpose.

3.1.6 *Weak Form of the Arbitrary Lagrangian–Eulerian Formulation*

Lagrangian meshes are often used in FEA. However, Lagrangian mesh cells deform with the material. In an FE model that simulates a real-world morphology (e.g., thermal growth oxide (TGO) evolution and calcium–magnesium aluminosilicate (CMAS) infiltration), the material in the mesh region deforms and twists significantly, reducing the approximation accuracy of the involved cells and thereby preventing the model calculation from converging. To effectively eliminate nonconvergent calculations resulting from excessive mesh deformation, a combination of Lagrangian and Eulerian meshes, also known as the arbitrary Lagrangian–Eulerian (ALE) formulation, is used in deformed mesh regions, taking advantage of the Eulerian mesh for large deformation.

Both the mesh and material move in the ALE formulation. The motion of the material is described as follows:

$$x = \varphi(X, t) \tag{3.78}$$

where $\boldsymbol{X}$ is the material coordinates and function φ maps the initial configuration of the object to the current configuration. Let us consider another domain $\hat{\Omega}$, which is referred to as the reference domain or the ALE domain. Let χ be the initial value of the particle position. Thus, we have

$$\chi = \varphi(\boldsymbol{X}, 0) \tag{3.79}$$

$\boldsymbol{\chi}$ is the reference or ALE coordinates. Then, the motion of the mesh is described as follows:

$$x = \widehat{\varphi}(\chi, t) \tag{3.80}$$

Through function composition, the ALE and material coordinates are linked as follows:

$$\chi = \widehat{\varphi}^{-1}(x, t) = \hat{\varphi}^{-1}(\varphi(X, t), t) = \psi(X, t) \tag{3.81}$$

The above-established relationship between the material and ALE coordinates is a function of time. By considering a function $f(\chi, t)$, we have

$$\frac{\mathrm{D}f(\chi, t)}{\mathrm{D}t} \equiv \dot{f}(\chi, t) = \frac{\partial f(\chi, t)}{\partial t} + \frac{\partial f(\chi, t)}{\partial \chi_i}\frac{\partial \psi_i(X, t)}{\partial t} = f_{t[\chi]} + \frac{\partial f}{\partial \chi_i}\frac{\partial \chi_i}{\partial t} \tag{3.82}$$

Now, let us define the reference particle velocity w_i as

$$w_i = \frac{\partial \psi_i(X, t)}{\partial t} = \left.\frac{\partial \chi_i}{\partial t}\right|_{[X]} \tag{3.83}$$

Then, the material time derivative or total derivative is expressed as

$$\frac{\mathrm{D}f(\chi, t)}{\mathrm{D}t} \equiv f_{,t[\chi]} + \frac{\partial f}{\partial \chi_i} w_i \tag{3.84}$$

Thus, the material velocity is given by

$$v_j = \frac{\partial \varphi_j(X, t)}{\partial t} = \frac{\partial \widehat{\varphi}_j(\chi, t)}{\partial t} + \frac{\partial \widehat{\varphi}_j(\chi, t)}{\partial \chi_i}\frac{\partial \psi_i(X, t)}{\partial t} = \hat{v}_j + \frac{\partial x_j}{\partial \chi_i} w_i \tag{3.85}$$

Here, the mesh velocity $\hat{v}_j$ is as follows:

$$\hat{v}_j = \left.\frac{\partial \widehat{\varphi}_j(\chi, t)}{\partial t}\right|_{[\chi]} \tag{3.86}$$

Now, let us define convection velocity $\boldsymbol{c}$ as the difference between the material and mesh velocities:

$$c_i = v_i - \hat{v}_i = \frac{\partial x_i}{\partial \chi_j}\frac{\partial \chi_j}{\partial t} = \frac{\partial x_i}{\partial \chi_j} w_j \tag{3.87}$$

Thus,

$$\left.\frac{\partial f}{\partial \chi_i}\right|_t = \left.\frac{\partial f}{\partial x_j}\right|_t \left.\frac{\partial x_j}{\partial \chi_i}\right|_t \frac{Df}{Dt} = f_{,t[X]} + \left.\frac{\partial f}{\partial x_j}\frac{\partial x_j}{\partial \chi_i}\frac{\partial \chi_i}{\partial t}\right|_{[X]}$$
$$= f_{,t[X]} + f_{,j}\frac{\partial x_j}{\partial \chi_i}w_i = f_{,t[X]} + f_{,j}c_j \tag{3.88}$$

Equation (3.88) can also be written as follows:

$$\frac{Df}{Dt} = f_{[t]} + \boldsymbol{c}\cdot \mathrm{grad} f = f_{,t[X]} + \boldsymbol{c}\cdot\nabla f \tag{3.89}$$

All the nonindependent variables are defined as functions of cell coordinates. The ALE domain is divided into elements. The ALE coordinates of the element e are given by

$$\boldsymbol{\chi}(\boldsymbol{\xi}^e) = \boldsymbol{\varphi}^{\mathrm{e}}(\boldsymbol{\xi}^e) = \boldsymbol{\chi}_I N_I(\boldsymbol{\xi}^e) \tag{3.90}$$

where $\boldsymbol{\xi}^e$ is the coordinates of the element e. The motion of the mesh is given by

$$\boldsymbol{x}(\boldsymbol{\xi}^e) = \widehat{\boldsymbol{\varphi}}^h(\boldsymbol{\chi}(\boldsymbol{\xi}^e),t) = \boldsymbol{x}_I(t)N_I(\boldsymbol{\xi}^e) \tag{3.91}$$

The mesh velocity is obtained as follows:

$$\hat{\boldsymbol{v}} = \frac{\partial\widehat{\boldsymbol{\varphi}}^h(\boldsymbol{\chi}(\boldsymbol{\xi}^e),t)}{\partial t} = \dot{\boldsymbol{x}}_I(t)N_I(\boldsymbol{\xi}^e) = \hat{\boldsymbol{v}}_I(t)N_I(\boldsymbol{\xi}^e) \tag{3.92}$$

In the ALE formulation, density is also a nonindependent variable and is given by

$$\rho(\boldsymbol{\xi}^e,\mathrm{t}) = \rho_I(t)N_I^{\rho}(\boldsymbol{\xi}^e) \tag{3.93}$$

where $N_I^{\rho}(\boldsymbol{\xi}^e)$ is the shape function of density and may differ from the shape function of the motion of the mesh.

The material time derivative of the velocity is derived from Eq. (3.89) as follows:

$$\dot{v} = \frac{D\boldsymbol{v}}{Dt} = v_{t[X]} + \boldsymbol{c}\cdot\mathrm{grad}\boldsymbol{v} = v_{t[\gamma]} + \boldsymbol{c}\cdot\nabla\boldsymbol{v} \tag{3.94}$$

The material velocity is obtained through interpolation:

$$\boldsymbol{v}(\boldsymbol{\chi}(\boldsymbol{\xi}),t) = \boldsymbol{v}_I(t)N_I(\boldsymbol{\xi}) \tag{3.95}$$

Through similar interpolation, the transfer velocity is obtained as follows:

$$\boldsymbol{c}(\boldsymbol{\chi}(\boldsymbol{\xi}),t)=\boldsymbol{c}_I(t)N_I(\boldsymbol{\xi})=\left(\boldsymbol{v}_I(t)-\hat{\boldsymbol{v}}_I(t)\right)N_I(\boldsymbol{\xi}) \tag{3.96}$$

Thus, the interpolation of the material time derivative is obtained:

$$\dot{\boldsymbol{v}}=\frac{d\boldsymbol{v}_I(t)}{dt}N_I(\xi)+\boldsymbol{c}(\xi,t)\cdot\nabla N_I(\xi)\boldsymbol{v}_I(t) \tag{3.97}$$

The time derivative of density is similarly obtained as follows:

$$\dot{\rho}=\frac{d\rho_I(t)}{dt}N_I^\rho(\xi)+\boldsymbol{c}(\xi,t)\cdot\nabla N_I^\rho(\xi)\rho_I(t) \tag{3.98}$$

Now, let us analyze the weak form and the FE matrix equation. First, we look at the weak form of the continuity equation. Let $\rho\in C^0$ be a solution. The continuity equation in the Eulerian description (i.e., the current configuration) is given by

$$\dot{\rho}+\rho v_{j,j}=0 \tag{3.99}$$

By applying the ALE formulation in Eq. (3.89) to Eq. (3.99), we have

$$\rho_{t[z]}+\rho_{,j}c_j+\rho v_{j,j}=0 \quad \rho_{f[\chi]}+\boldsymbol{c}\cdot\nabla\rho+\rho\nabla\cdot\boldsymbol{v}=0 \tag{3.100}$$

By multiplying the above equation with the variational function $\delta\tilde{\rho}\in C^0$, we obtain the following weak form of the continuity equation:

$$\int_\Omega \delta\tilde{\rho}\rho_{,t[\chi]}d\Omega+\int_\Omega \delta\tilde{\rho}\rho_{,j}c_j d\Omega+\int_\Omega \delta\tilde{\rho}\rho v_{j,j}d\Omega=0 \tag{3.101}$$

The momentum equation in the Eulerian description is transformed into

$$\rho\dot{v}_i=\sigma_{ji,j}+\rho b_i \tag{3.102}$$

By applying Eq. (3.89) to Eq. (3.201), we transform the momentum equation into

$$\rho\left\{v_{i,t[\chi]}+c_j v_{i,j}\right\}=\sigma_{ji,j}+\rho b_i \text{ or } \rho\left[v_{,t[\chi]}+c\cdot\text{grad}\boldsymbol{v}\right]=\text{div}(\boldsymbol{\sigma})+\rho\boldsymbol{b} \tag{3.103}$$

Further, by using the variational function $\delta\tilde{v}_i\in u_0$, where u_0 is defined the same as that in Eq. (3.17), the following weak form is obtained:

$$\begin{aligned}\int_\Omega \delta\tilde{v}_i\rho\dot{v}_i\mathrm{d}\Omega&=\int_\Omega \delta\tilde{v}_i\rho v_{i,t[\chi]}\mathrm{d}\Omega+\int_\Omega \delta\tilde{v}_i\rho c_j v_{i,j}d\Omega=\int_\Omega \delta\tilde{v}_i\sigma_{ij,j}\mathrm{d}\Omega+\int_\Omega \delta\tilde{v}_i\rho b_i\mathrm{d}\Omega\\&=-\int_\Omega \delta\tilde{v}_{i,j}\sigma_{ij}\mathrm{d}\Omega+\int_\Omega \delta\tilde{v}_i\rho b_i\mathrm{d}\Omega+\int_{\Gamma_t} \delta\tilde{v}_i\bar{t}_i\mathrm{d}\Gamma\end{aligned} \tag{3.104}$$

Here, we directly give the FE matrix equation. Interested readers may derive this equation themselves. The continuity equation is given by

$$\boldsymbol{M}^{\rho}\frac{\mathrm{d}\boldsymbol{\rho}}{\mathrm{d}t}+\boldsymbol{L}^{\rho}\boldsymbol{\rho}+\boldsymbol{K}^{\rho}\boldsymbol{\rho}=\boldsymbol{0} \tag{3.105}$$

where $\boldsymbol{M}^{\rho}$, $\boldsymbol{L}^{\rho}$, and $\boldsymbol{K}^{\rho}$ are the capacity matrix, the transformation matrix, and the scatter matrix, respectively.

$$\begin{aligned}\boldsymbol{M}^{\rho}=[M^{\rho}_{IJ}]=\int_{\Omega}\overline{N}^{\rho}_{I}N^{\rho}_{J}d\Omega\ \ \boldsymbol{L}^{\rho}=[L^{\rho}_{IJ}]\\ =\int_{\Omega}\overline{N}^{\rho}_{I}c_{i}N^{\rho}_{J,j}d\Omega\ \ \boldsymbol{K}^{\rho}=[K^{\rho}_{IJ}]=\int_{\Omega}\overline{N}^{\rho}_{I}v_{i,i}N^{\rho}_{J}d\Omega\end{aligned} \tag{3.106}$$

$$\boldsymbol{\rho}=[\rho_{J}]\ \ \frac{d\boldsymbol{\rho}}{dt}=\left[\frac{d\rho_{J}}{dt}\right] \tag{3.107}$$

Here, $\overline{N}_{I}=N_{I}+\tau c_{j}N_{I,j}$, where τ is a time parameter introduced by numerical instability.

The discrete form of the momentum equation is given by

$$\boldsymbol{M}\frac{\mathrm{d}\boldsymbol{v}}{\mathrm{d}t}+\boldsymbol{L}\boldsymbol{v}+\boldsymbol{f}^{\mathrm{int}}=\boldsymbol{f}^{\mathrm{ext}} \tag{3.108}$$

where $\boldsymbol{M}$ and $\boldsymbol{L}$ are the generalized mass matrix and the transfer matrix, respectively, corresponding to the velocity described in the reference configuration; $\boldsymbol{f}^{\mathrm{int}}$ and $\boldsymbol{f}^{\mathrm{ext}}$ are the vectors of the internal and external forces, respectively.

$$\boldsymbol{M}=\mathbf{I}[M_{IJ}]=(\int_{\Omega}\rho\overline{N}_{I}N_{J}\mathrm{d}\Omega)\mathbf{I} \tag{3.109}$$

$$\boldsymbol{L}=\mathbf{I}[L_{IJ}]=(\int_{\Omega}\rho\overline{N}_{I}c_{i}N_{J,i}\mathrm{d}\Omega)\mathbf{I} \tag{3.110}$$

$$\boldsymbol{f}^{\mathrm{int}}=[f^{\mathrm{int}}_{iI}]=\int_{\Omega}\overline{N}_{I,j}\sigma_{ij}\mathrm{d}\Omega \tag{3.111}$$

$$\boldsymbol{f}^{\mathrm{ext}}=[f^{ext}_{iI}]=\int_{\Omega}(\overline{N}_{I}\rho b_{i})\mathrm{d}\Omega+\int_{\Gamma_{t_i}}\overline{N}_{I}\bar{t}_{i}\mathrm{d}\Gamma \tag{3.112}$$

$$\boldsymbol{v}=[v_{J}]\ \ \frac{d\boldsymbol{v}}{dt}=\left[\frac{dv_{J}}{dt}\right] \tag{3.113}$$

3.1.7 Initial and Boundary Conditions

Initial displacement and velocity conditions are also required to obtain an FE solution. Initial conditions can be represented by initial nodal values. Here, the 1D case is illustrated as an example. Its initial conditions are given by

$$u_i(0) = u_0(X_I) \tag{3.114}$$

$$\dot{u}_i(0) = \dot{u}_0(X_I) \tag{3.115}$$

Thus, the initial conditions of an undeformed object at rest in the initial state are given by

$$u_I(0) = 0 \text{ and } \dot{u}_I(0) = 0 \ \forall I \tag{3.116}$$

For more complex initial conditions, nodal displacements and velocities can be obtained by least-squares fitting of initial data. The square of the difference between the FE interpolation function and the initial data is minimized. Let

$$M = \frac{1}{2}\int_{X_a}^{X_b}\left[\sum_I u_I(0)N_I(X) - u_0(X)\right]^2 \rho_0 A_0 \mathrm{d}X \tag{3.117}$$

where A_0 is the cross-sectional area of a 1D bar. To find the minimum of Eq. (3.117), let us set its derivative with respect to the initial nodal displacement to zero:

$$0 = \frac{\partial M}{\partial u_K(0)} = \int_{X_a}^{X_b} N_K(X)\left[\sum_I u_I(0)N_I(X) - u_0(X)\right]^2 \rho_0 A_0 \mathrm{d}X \tag{3.118}$$

By applying the mass matrix, we can rewrite the above equation as follows:

$$\boldsymbol{Mu}(0) = \boldsymbol{g} \text{ where } g_K = \int_{X_a}^{X_b} N_K(X)u_0(X)\rho_0 A_0 \mathrm{d}X \tag{3.119}$$

Similarly, least-squares fits can be obtained for other initial conditions. Boundary conditions often include natural and coercive boundary conditions. Natural boundary conditions are already included in the weak form, and coercive boundary conditions are displacement boundaries. Boundary nodal displacement can be defined as follows:

$$u_i(X_I,t) = \overline{U}_i(X_I,t) \text{ on } \Gamma_{u_i} \tag{3.120}$$

where $\overline{U}_i$ is the boundary displacement. In FE software (e.g., ABAQUS, ANSYS, and COMSOL), these discrete governing equations and boundary conditions are programmed and, combined with discrete meshes and input material parameters, are used to perform FEA to produce numerical results for the analyzed problem.

3.2 FE Modeling of TBCs on Turbine Blades

3.2.1 *Geometric Characteristics of Turbine Blades*

The pursuit of a high thrust-to-weight ratio for modern aeroengines has resulted in a continuous increase in turbine inlet temperatures. Amid this background, complex cooling structures and TBC technology are increasingly used to ensure the safe and reliable operation of turbine blades in high-temperature fuel gas environments, representing the development trend of turbine blades. Structurally, turbine blades have gradually evolved from solid blades to air-cooled blades with complex structures (e.g., complex internal cavities, film cooling holes, and pin fins). Both guide and moving blades have spatial surfaces as a result of the high-thrust blade design and performance demands (e.g., improved energy conversion efficiency) of turbine blades. The blade body is the main part of a turbine blade and is located within the cascade flow channel. The geometric properties of a blade body include its external and internal profiles. The external profile of a blade body is a spatial freeform surface fitted by several sets of cross-sectional curves parallel to each other at different flow-channel heights following aerodynamic calculations. The internal profile of a blade body is also a spatial surface formed by several blade-shaped cross-sectional curves stacked in a certain pattern. Internally, a blade body contains cooling structures such as longitudinal ribs (baffles), pin fins, and impingement sleeves, forming a serpentine internal flow cooling cavity. In addition, film holes are distributed on a blade body and connected to the internal cavity of the blade. Cooling airflow moves from within the cavity to the exterior of the blade, forming a cooling air layer, which prevents direct heat transfer between the high-temperature fuel gas and the turbine blade. Moreover, to enhance heat transfer to facilitate the absorption of heat from the internal walls of a blade, the internal structures of the blade body are cooled by means of impingement cooling and heat transfer enhancement by disturbing the flow. Figure 3.1 shows the typical structure of a turbine blade. The geometric characteristics of the turbine blade structure [3] are summarized as follows:

a. Blade body. The blade body includes an external surface and an internal surface. The external surface of the blade body consists of a convex-side surface (the blade body's suction side, which is convex), a concave-side surface (the blade body's pressure side, which is concave), a leading-edge surface, and a trailing-edge surface. The internal surface of the blade body consists of an internal convex-side surface (corresponding to the external convex side), an internal concave-side surface (corresponding to the external concave side), an internal leading-edge

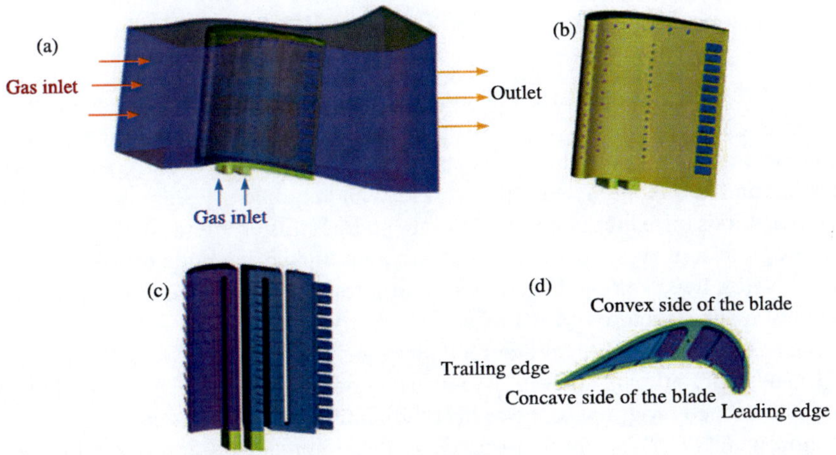

Fig. 3.1 Geometric structure of a TBC on a turbine blade. **a** Whole structure. **b** TBC on the blade. **c** Cooling channels. **d** Middle cross-section

surface (corresponding to the external leading edge), and an internal trailing-edge surface (corresponding to the external trailing edge).

b. Platform and tenon. The tenon connects the blade and the disk. A tree-shaped tenon with two teeth is often used. The platform, consisting of surfaces, is placed between the tenon and the blade body.
c. Film holes. The film holes are the channels through which the cooling air flows from inside the blade body to the mainstream region outside the blade body. The film holes, often circular and prepared at certain angles, collectively form an array.
d. Baffles (longitudinal ribs). The baffles, often rectangular drawn structures, are placed inside the blade body. They separate the internal cooling channels into multiple cavities connected in series or parallel.
e. Pin fins. The pin fins often comprise a group of cylindrical structures and are placed inside the blade body to disturb the flow and facilitate heat transfer.
f. Trailing edge: The trailing edge discharges the air after heat transfer and simultaneously cools the tail of the blade. The trailing edge often consists of a long-strip solid array along the height of the blade and is smoothly connected to the interior of the blade body.
g. Impingement sleeves. The impingement sleeves are solid structures with a uniform cross-sectional area and provide impingement cooling inside the blade body. There is often a fixed distance between the impingement sleeves and the internal surface of the blade body. Impingement sleeves are constructed in a way similar to that in which the interior of a blade body is constructed.
h. TBC. The TBC is a ceramic coating with low thermal conductivity and a thickness of several hundred microns. It is deposited on the surface of the blade to provide thermal insulation.

3.2.2 *Parametric Modeling of Turbine Blades*

Before FEA, it is necessary to establish a 3D model of a blade based on its data. In FEA and structural optimization of a TBC on a turbine blade, the shape or characteristic parameters of the blade need to be constantly adjusted. However, repeated manual revisions and modeling increase the workload and complexity. Parametric modeling design methods have thus been developed to address this problem. With the continual increase in design accuracy and complexity, automated methods for designing air-cooled blades have emerged, significantly improving the accuracy and speed of the design of blade structures with TBCs.

The currently available commercial software such as SolidWorks, Unigraphics (UG), and ProE, all have relevant parametric modeling modules. Based on independently developed programs alone or in combination with the secondary development function of UG/CATIA, many researchers have conducted extensive work on the parametric design of air-cooled turbine blades. Zhou [4] and Liu et al. [5] each designed blade profiles using independently developed parametric design software programs that allow the geometric parameters of blade cascades to be further modified to meet the blade requirements. Yu et al. [6] realized automated modeling of a multicavity reverse-flow-type air-cooled turbine blade using a combination of mathematical analysis and feature modeling techniques with the Application Programming Interface (API) system provided by SolidWorks as a secondary development interface. By analyzing the factors affecting the geometric characteristics of turbine blades as well as their aerodynamic performance and structural strength, Liu and Ma [7] determined the control parameters for the cross-sectional shape of a blade and formulated a CATIA-based modeling method for aeroturbine blades. Han [8] combined the API of the secondary development module of UG and an independently developed Visual C program into a tool capable of automatically reading computational models, changing parameters, and generating and outputting new models. Shi [9] built a blade-body cooling structure through secondary development using UG based on data processing in MATLAB, thereby realizing rapid modeling of turbine blades. By analyzing geometric solid models established in UG and their topological relations, Li [10] and Song [11] each studied the modeling processes of turbine cascades, blade-body profiles, and turbine channels and developed a UG-based geometric modeling system for turbines that was then used to parametrically model turbine blades in UG. As MATLAB and UG are common tools used by researchers, here, we use a study conducted at the Harbin Institute of Technology (HIT) [12] as an example to briefly introduce the parametric design process of a blade equipped with a cooling structure.

(1) Parametric design of a blade's external shape and cooling channels

The external shape of the turbine blade body is a spatial surface with a complex morphology formed by stacking the blade profile lines of multiple cross-sections obtained from aerodynamic calculations along the height of the blade in a certain pattern. The main function of the blade body is to convert some of the internal energy of air flowing by the blade to kinetic energy and to alter the direction of the airflow.

To ensure the working temperature of a blade, film holes connected to the interior of the blade are prepared on its surface. Cooling air flows out of the film holes and forms a cooling air film on the surface of the blade. A first-stage guide vane of a certain gas turbine is used as the reference blade in the parametric design introduced in this section.

The parametric modeling process of the external shape of a blade body can be divided into two steps, namely the generation of the cross-sectional lines of the blade body and the generation of a solid blade model. Currently, many software programs (e.g., the HIT Blade Editor and NUMECA) can be employed to generate blade profile lines based on relevant aerodynamic parameters. Blades can be modeled based on blade-shape data in UG. Generally, blade-shape design software programs can export the coordinates of the pressure and suction sides at different cross-sections. In traditional practice, a designer first imports blade-shape coordinates of different cross-sections into UG to form blade profile lines and generates a pressure side and a suction side by skinning, followed by stitching the surfaces to produce the external shape of the blade. This process requires multiple steps that involve repetitive operations. A MATLAB program has been developed to reduce the workload. This program reconstructs blade-shape data in a counterclockwise or clockwise direction based on a combination of coordinate data for both the pressure and suction sides at the same cross-section. Another program has been developed using the UG/OPEN API. This program reads processed discrete points, generates blade profile lines using the nonuniform rational B-spline curve method, and then directly generates a solid blade model using the function UF_MODL_create_thru_curves1. In practice, a user needs only to input the number of cross-sections and call the MATLAB program to complete data processing. Subsequently, he or she can call the program developed in UG to generate a blade model. In this way, all the manual operations are automated by programs, improving the design speed. Figure 3.2 shows the processed data format and an example of a generated blade model.

Hollow turbine blades are often fabricated to facilitate cooling. A blade's internal-cavity profile lines can be obtained by offsetting its external profile lines along the spatial normal direction based on the wall thickness in the MATLAB program. Considering factors such as strength, not all the blades have uniform wall thicknesses. In the program, users can set different blade thicknesses along the height of the blade as needed. Since a blade has a thin trailing edge, after the blade thickness reaches a certain value, the blade's internal profile lines cross each other, as shown in the top image in Fig. 3.3. The trailing edge is chamfered during data processing. Based on the input diameter of the trailing edge of the internal cavity, an arc can be directly formed at the trailing edge, as shown in the bottom image in Fig. 3.3. With the input of the wall thicknesses at different blade heights and the diameter of the trailing edge based on the external-shape data for the blade, coordinate data for the internal profile line of each cross-section of the blade can be obtained. The internal shape of the blade is constructed in a way similar to that in which its external shape is constructed. Here, to facilitate subsequent trimming, the internal shape is constructed in such a way that it exceeds the upper and lower end walls of the blade. Figure 3.4 shows the completed structure.

Number of cross-sections of the blade
Number of points on the first profile line
Coordinates of the discrete point(*x*,*y*,*z*)
Number of points on the second profile line
Coordinates of the discrete point(*x*,*y*,*z*)
Number of points on the third profile line
Coordinates of the discrete point(*x*,*y*,*z*)
……

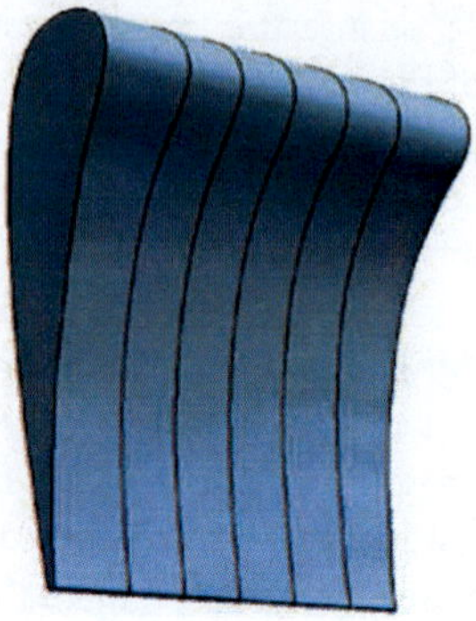

Fig. 3.2 Output data format and an example of a generated blade model

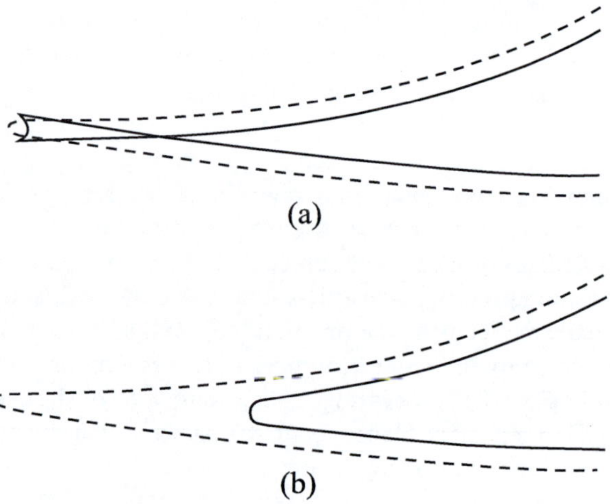

Fig. 3.3 Trailing edge of the internal shape of the blade

The interior of a blade is often divided into multiple cavities by baffles. There are many ways to position baffles. In some relevant studies, baffles are positioned based on the chord length of the blade. Relevant parameters include the chordwise location L of the baffle, the angle α between the baffle and the chordwise direction, and the thickness d of the baffle, as shown in the top image in Fig. 3.5. Here, considering that some baffles are not straight, the position of a baffle is determined using the dimensionless ratios (L_1/Lp and L_2/Ls) of its centerline to the blade's cross-sectional profile lines, as shown in the bottom image in Fig. 3.6. Data for the central cross-section of the baffle can be derived from the positional parameters at each cross-section of the blade. Then, data points for the surface of the baffle can be obtained by interpolation on the two sides based on the thickness. Boolean operations are

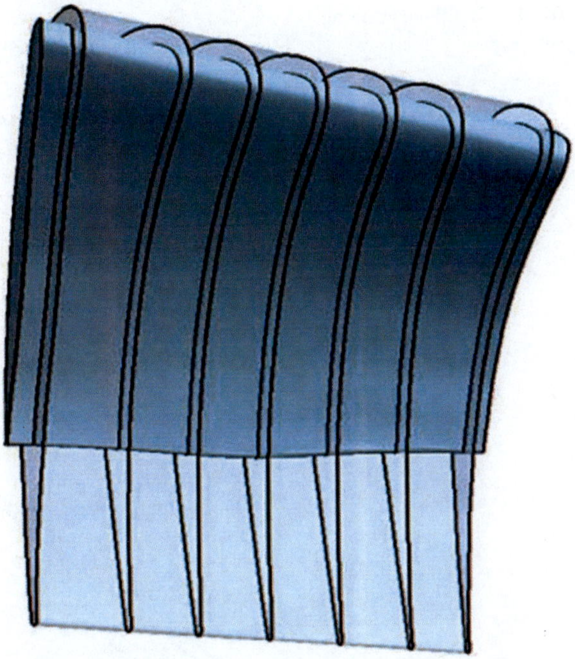

Fig. 3.4 Internal shape of the blade

performed on the baffle drawn using the baffle program in UG with the internal shape and solid model of the blade, followed by the chamfering of some local positions, to obtain complete internal cavities of the guide vane, as shown in Fig. 3.6.

There is a large centrifugal force on a moving turbine blade, for which impingement cooling is not often used considering its strength. Air-cooling channels are often constructed into a serpentine shape with turns, which can also be viewed as a result of partitioning by multiple baffles with different heights only at those turns. Compared to a stationary blade, a moving blade requires two additional baffle-positioning parameters, $h1$ and $h2$, which are the lengths by which the baffles are shortened at the upper and lower ends, respectively. The other operations remain the same as those of a stationary blade. Figure 3.7 shows the final serpentine channels and the added positioning parameters. The above parametric design of both the internal cavities and baffles of the blade is based on its external profile lines. Thus, it is possible to integrate the data processing programs and to call the MATLAB program to simultaneously generate data for the blade's external shape, internal shape, and baffles.

(2) Parametric design of a flow-disturbing cooling structure

Flow-disturbing structures are often added to a blade's internal cavities to increase the heat transfer area and to enhance the flow disturbance to strengthen the cooling effect. Ribs are the most commonly used flow-disturbing structures. Of various ribs, those with rectangular cross-sections are often used, though those with circular cross-sections are also available. The insides of blades, particularly the two sides of the

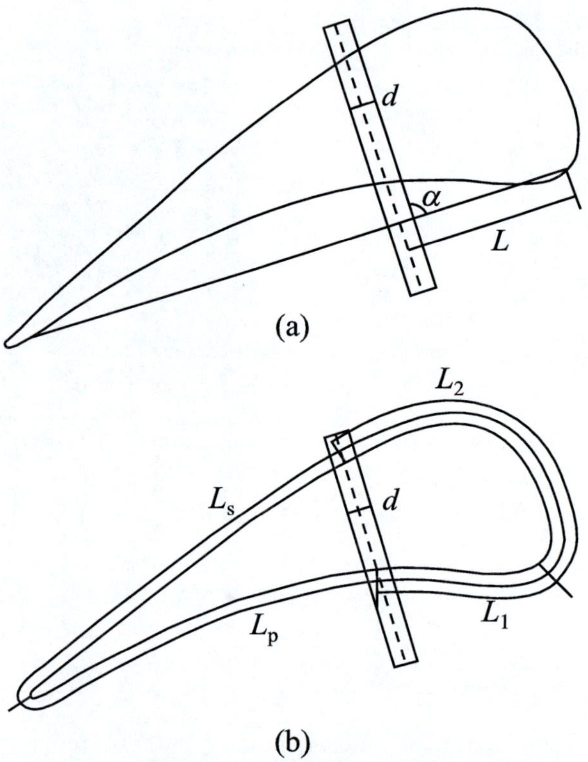

Fig. 3.5 Baffle-positioning parameters

Fig. 3.6 Internal cavities of a guide vane

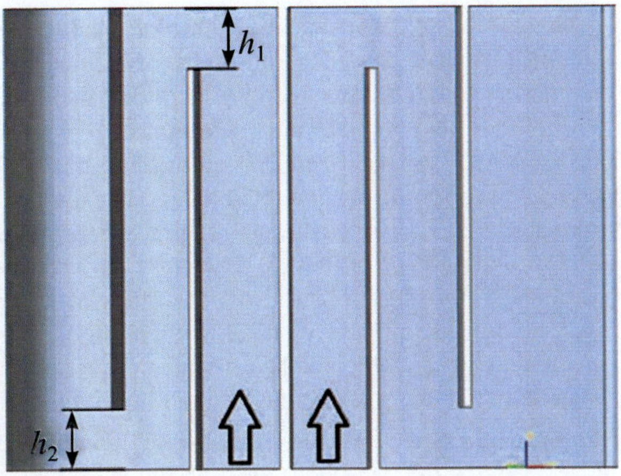

Fig. 3.7 Serpentine cooling channels

serpentine channels in moving blades, are covered with ribs. The design of these ribs using conventional geometric drawing methods involves a tedious procedure with a very large workload, so it is necessary to carry out a parametric design. The basic parameters of rectangular ribs include the rib height *h*, rib width *b*, rib spacing *S*, and rib angle α, as shown in Fig. 3.8a. It is not possible to completely position ribs by relying on these parameters alone, as it is also necessary to know the initial rib height S0 and the rib position *x* in the direction of the blade's chord. Some ribs do not run through the entire width of the cooling channel but are shorter than it. In this case, an additional rib length parameter *l* is needed. Figure 3.8b shows the parameters used to express ribs.

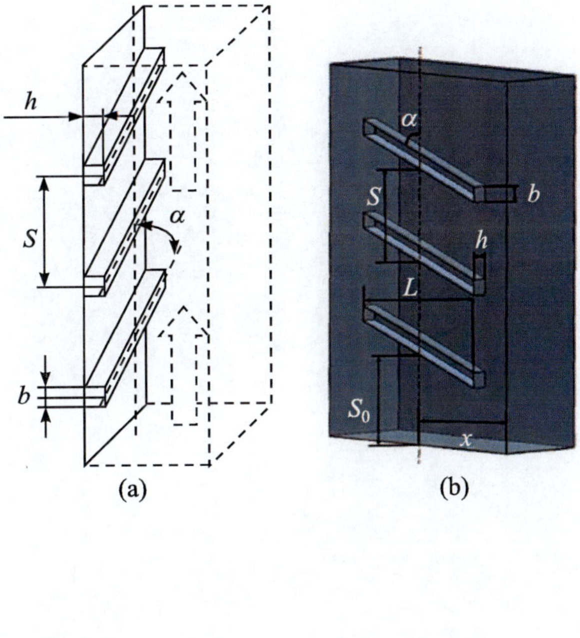

Fig. 3.8 Parametric expression of ribs

The cooling channels inside a turbine blade have complex shapes. It is more difficult to model ribs in cooling channels than on a plane. The difficulty in the parametric design of ribs lies in positioning them on a spatial surface. Rib design requires the data points for the blade's internal cross-sectional profile lines that have been obtained from the parameterization of the blade's internal cavities. Data for other surfaces can be similarly extracted to facilitate the addition of ribs to these surfaces if needed. The chordwise position of a rib, x, is determined using the method previously employed to position baffles or, alternatively, the coordinates can be manually input. Specifically, x is determined using the ratios of its centerline to the blade's profile lines. Ribs inside a blade are often straight, in other words, the sides of the ribs are on the same plane, which is the reference for various parameters of the ribs. A blade is spatially twisted and has no defined bottom plane. Thus, the plane determined by the leading-edge point, trailing-edge point, and central point of the blade root's cross-sectional line is used as the bottom plane. The coordinates of the central point on the side of the first rib can be obtained by moving upward a distance of $S0$ along the normal direction of this plane. Subsequently, through successive translation of the rib spacing S and the rib width b, the reference points for the side profile lines of all the ribs can be obtained. The profile-line directions of the ribs are calculated using the angle of inclination of the blade. On this basis, the data for the side profile lines of the ribs can be obtained. Figure 3.9 shows a schematic diagram. All the above operations have been implemented through MATLAB programming.

It is necessary to maintain consistency in the heights of the ribs when drawing them on the surface after determining their basic positions. To this end, the commonly used conventional approach is to first make the ribs into slightly larger rectangular plates, then to generate a new surface by inwardly offsetting the blade's internal surface by the rib height, and subsequently trim the rib plates. This approach has

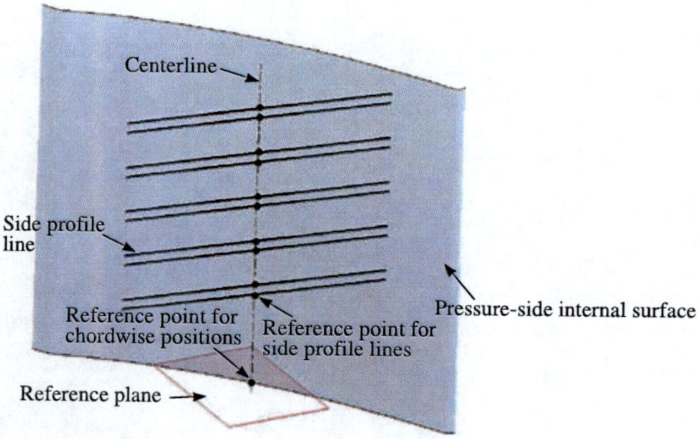

Fig. 3.9 Side profile lines of ribs

been programmed by means of software secondary development to complete the parametric modeling of the pin ribs. First, the surface where the top surfaces of the ribs are located is obtained by offsetting the blade's internal surface toward the two sides. The side profile lines of each rib are then drawn using the function UF_CURVE_create_line and projected to the top and bottom surfaces of the rib to obtain its four boundaries. The vertices of the boundaries are extracted to form the rib's cross-sectional profile line. Subsequently, the geometric structure of the rib is determined by sweeping along its boundaries using the function UF_MODL_create_sweep. Figure 3.10 shows a schematic diagram of rib modeling.

Apart from parallel ribs, other forms of ribs (e.g., V-shaped and discontinuous ribs) are used in many cases. The parametric expression of V-shaped ribs is similar to that of straight ribs except that a V-shaped angle is used instead of an angle of inclination. Discontinuous ribs can be considered to consist of multiple rows of parallel ribs that differ in position and initial height. Figure 3.11 shows the parametric expressions of V-shaped ribs and discontinuous ribs.

In parametric modeling, the same design concept is used for various ribs. Data points for the surface where ribs are located are required. Based on their design parameters, the spatial locations of the profile lines of ribs are first determined by processing data in MATLAB. Then, a UG program is called to perform sweep shaping. A whole row of ribs can be modeled at one time. The parameters (e.g., rib spacing, rib height, and the number of ribs) can be altered at will, thus significantly improving the design and modification efficiencies. Figure 3.12 schematically shows the ribs modeled using this design method for a stationary blade's internal cavities and a moving blade's serpentine channels.

In addition to rectangular ribs, flow-disturbing structures include cylindrical pin fins. These pin fins are often used in the trailing-edge slots to enhance flow disturbance and to increase the heat transfer area, thereby effectively cooling the thin trailing

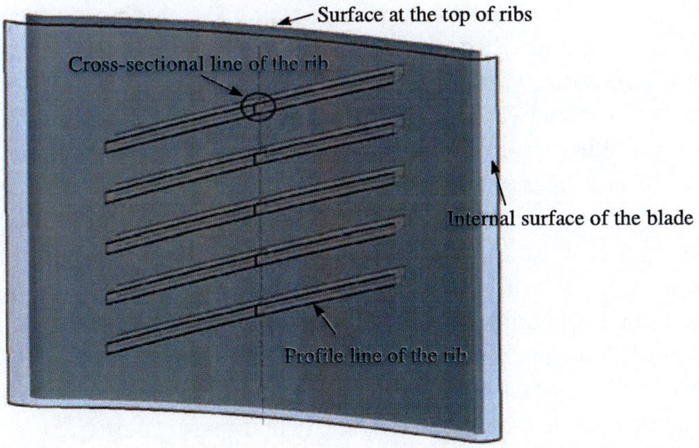

Fig. 3.10 Schematic diagram of ribs

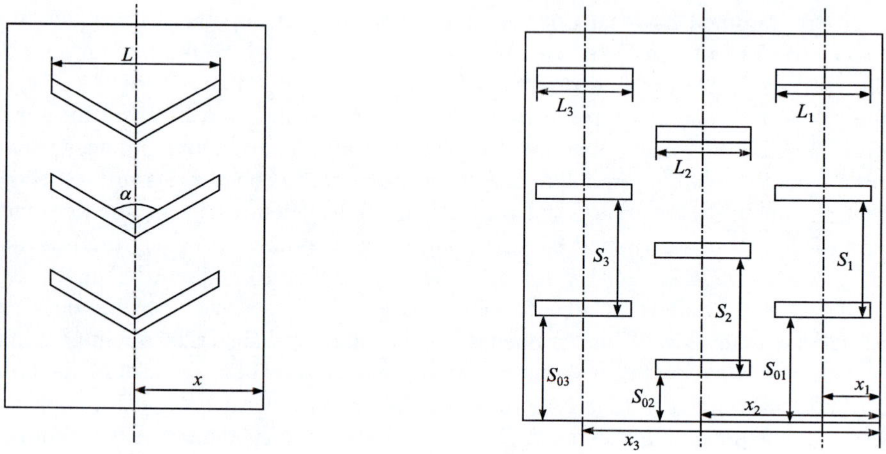

Fig. 3.11 Parametric expressions of V-shaped ribs and discontinuous ribs

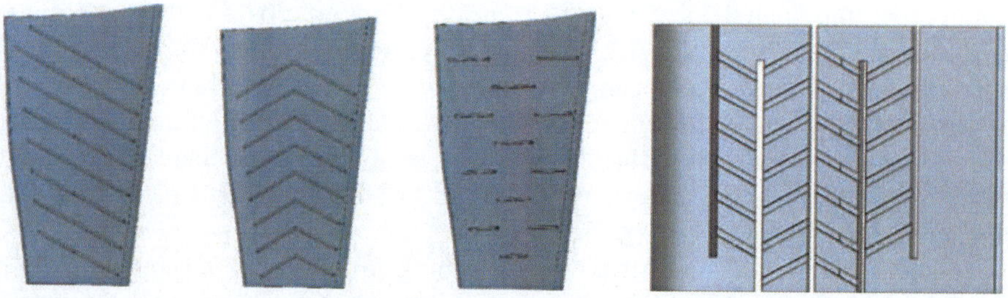

Fig. 3.12 An application example of ribs in stationary and moving blades

edge. Cylindrical pin fins have the advantage of high strength as they are connected to both the suction and pressure sides of a blade. Pin fins can be arranged in inline and cross configurations, with the cross configuration generally used to enhance the heat transfer performance. Figure 3.13 shows the basic parameters of cylindrical pin fins, including the radial fin spacing *Sy*, the chordwise fin spacing *Sx*, the initial chordwise position *x* of the first row of pin fins, the distance *S0* between the first pin fin and the bottom plane, the cylinder diameter *d*, and the pin height *h*. The parametric design of cylindrical pin fins somewhat differs from that of ribs but is similar to that of circular film holes. Please see the parametric design method for circular film holes for details. Figure 3.14 shows a rib array often used at a blade's trailing edge and its design parameters. The parallel ribs on the two sides cross one another, forming a net-like channel with turns, which enhances the disturbance of the cooling airflow inside it and increases the heat transfer coefficient. An analysis of the structure of the rib array shows that it can be considered to consist of two rows of parallel ribs stacked together and can thus be modeled using the parametric design method previously used to model ribs.

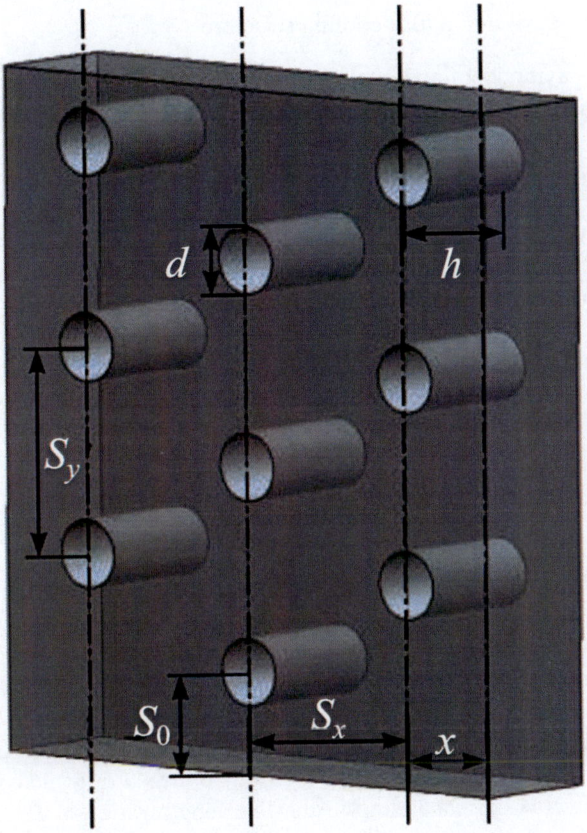

Fig. 3.13 Design parameters for pin fins

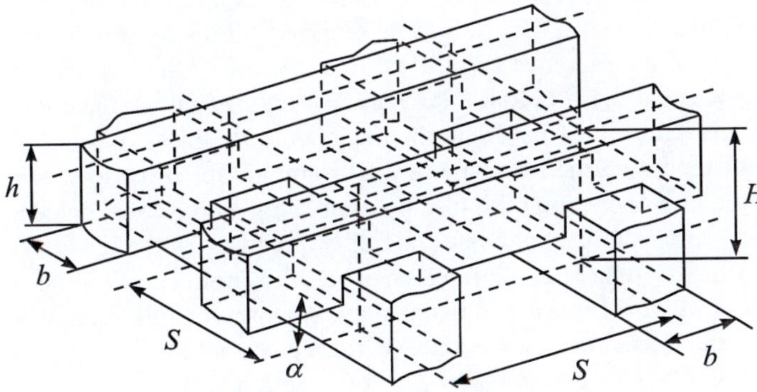

Fig. 3.14 Design parameters for a rib array

(3) Parametric design of a film-cooling structure

Film cooling, a mode of external cooling, is widely applied in modern high-temperature turbine blades. In this cooling mode, cooling air flows out of a blade's film holes and forms a cooling air film on its surface and thus prevents high-temperature fuel gas from heating its surface. The surface of a full-film-cooling turbine blade is covered with a large number of film holes. To increase the coverage effect of the air film, film holes are usually drilled at certain angles, with some inclined chordwise and some inclined radially, and are drilled at compound angles in some special locations. In addition, film holes come in various shapes. In addition to circular film holes, there are trapezoidal, fan-shaped, and expanded film holes. Modeling film holes is an exceedingly complex process. In parametric design, it is necessary to first determine the locations of film holes, then to determine their axes based on their compound angles, and subsequently to model them based on their cross-sectional shapes.

The point of intersection between the axis of a film hole and the surface of the blade is used as the reference point for the film hole. The coordinates of the reference point are determined based on its line length ratio-defined position on the surface of the blade, as shown in Fig. 3.15. First, it must be determined whether a film hole is on the suction side or the pressure side. The position of the line connecting the film holes in a row is determined based on the points at which it intersects the blade root and tip. The positions of the points of intersection are determined according to the ratios of their corresponding line lengths to the profile lines, i.e., S_{12}/S_2 (chordwise location 1) and S_{11}/S_1 (chordwise location 2), respectively. Then, the positions of the first and last film holes are determined according to the starting and ending positions of the film-hole row corresponding to the line length ratios of L_1/L (longitudinal location 1) and L_2/L (longitudinal location 2), respectively, on the line connecting the film holes. Finally, the connecting line is segmented based on the number of film holes to determine the coordinates of the centers of all other film holes.

The axial direction of a film hole is determined following the determination of the coordinates of its reference point. First, it is necessary to establish a body-fitted coordinate system at the film hole. The normal vector of the surface of the blade at the film hole is determined based on the blade's profile lines. A coordinate system is then established with the film hole as the origin and the directions of the normal vector and the line connecting the film holes as the y- and z-axes, respectively. The x- and z-axes represent the chordwise and radial directions of the blade, respectively, as shown in Fig. 3.16a. The axial vector of the film hole can then be determined based on its chordwise angle α and radial angle β. However, it is inconvenient to rely directly on α and β to determine the axial vector. Thus, another angle, γ, is introduced here. Figure 3.16b shows the relationships of γ with α and β. The axial vector can be easily determined based on α and γ. The three angles are related through the following conversion equation:

$$\gamma = \arctan\sqrt{(\cos^2\alpha - 1)\cos^2\beta/(\cos^2\beta - 1)} \tag{3.121}$$

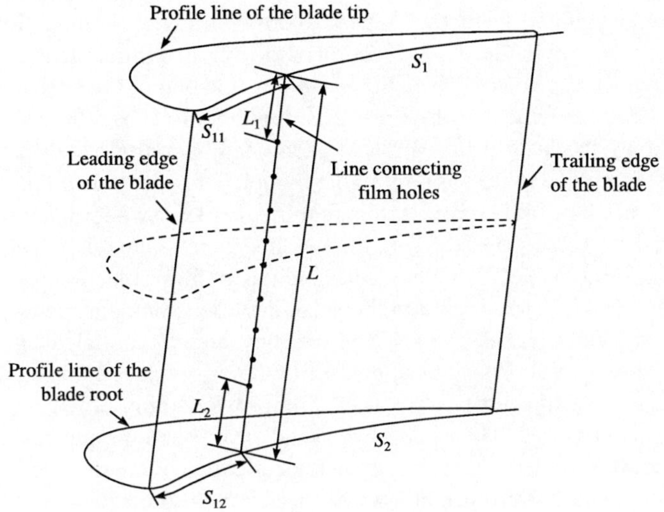

Fig. 3.15 Method for determining the coordinates of film holes

Both the coordinates and axial direction of a film hole discussed above are determined using programs written in MATLAB. The center coordinates and axial directions of film holes in a row can be accurately determined based solely on the cross-sectional profile-line data of the blade and input parameters such as the number and angles of film holes. Next, a solid model is established for a film hole based on its shape parameters. For circular film holes, in addition to diameter d, length L is added as a control parameter to prevent the drilling of film holes into other walls. In

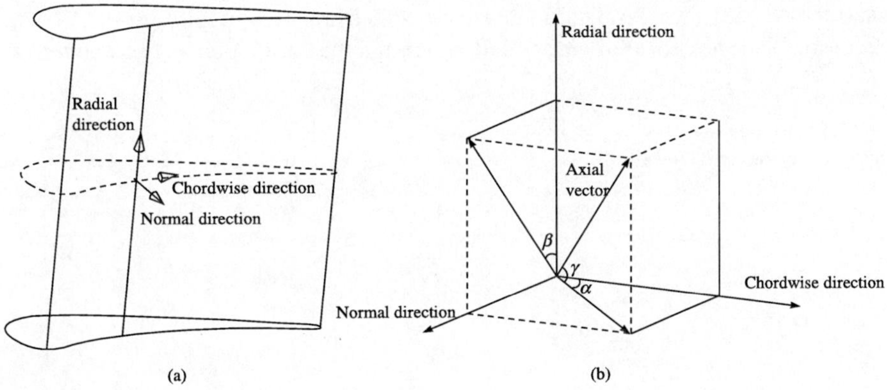

Fig. 3.16 Determination of the axis of a film hole

some studies, the guide-line method is used to draw a film hole in modeling software. Specifically, as shown in Fig. 3.17, a section of the axis of a film hole is created based on its center and axial vector. Then, the beginning and end of the axis are connected to form a guide line, which is subsequently imported into UG, followed by drawing a cylinder along the broken line. Finally, a Boolean operation is performed on the cylinder and the blade to obtain a film hole model. This method can produce a row of film holes at one time. However, this requires manual sweeping and sometimes may not produce results. In addition, this method is inapplicable to trapezoidal or variable-cross-section (VCS) holes. Here, a circular film hole is shaped using the function UF_MODL_create_cyl1 in the secondary development module of UG. A whole row of cylindrical film holes can be conveniently modeled by calling the program. Figure 3.18 shows an example of film holes prepared at an angle.

Film holes of various shapes are used to improve cooling efficiency. Expanded holes are frequently used. Their inlets have small cross-sections and can thus control the flow of cooling air, whereas their expanded outlets can increase the film coverage. Other types of holes that are currently widely used include trapezoidal and VCS holes. Parametric design methods vary for holes of different shapes. First, let us look at the direction of a film hole. A cylindrical film hole is circumferentially symmetric. As a result, its direction can be determined as long as its axial direction is known. In contrast, for an expanded or trapezoidal hole, in addition to its axial direction, its expansion direction is required to position it. The specific orientation of an expanded film hole can be determined based on its axial direction and the radial direction of the blade. The vectors in these two directions can be obtained by processing data in MATLAB. A trapezoidal film hole has a uniform cross-sectional area and is thus relatively easy to model. Figure 3.19 shows the cross-section control parameters for a trapezoidal film hole, which include, with the centerline of the film hole as a reference, the length L_1 of the longer base, the length L_2 of the shorter base, the distance d_1 between the longer base and the center, and the distance d_2 between the shorter base and the center. Positioning an expanded hole is a complex process. This is because a blade has a spatial surface and, as a result, the line of intersection between an

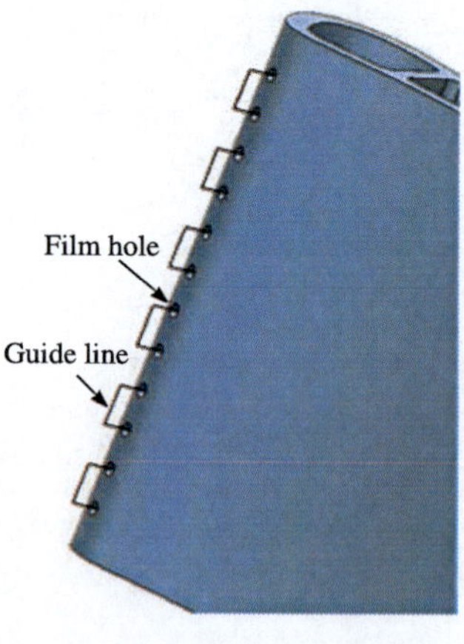

Fig. 3.17 Film holes generated using guide lines

Fig. 3.18 An example of circular film holes

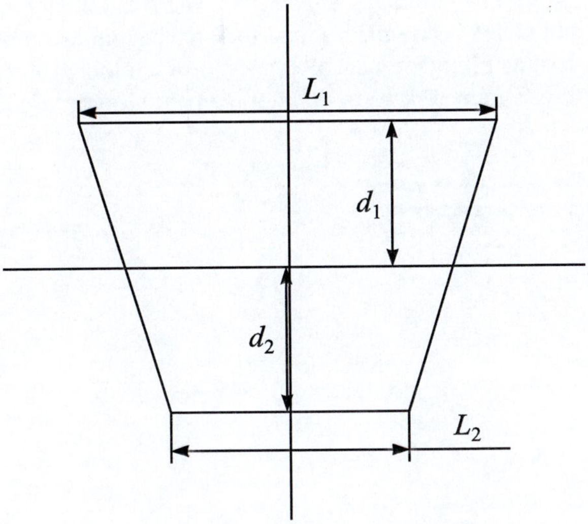

Fig. 3.19 Parameters of a trapezoidal hole

expanded hole and the blade's surface has an irregular shape and is difficult to directly control. Here, an expanded hole is positioned by controlling the shapes of three cross-sections perpendicular to its axis. Figure 3.20 shows the positional parameters of the cross-sections. The location where the expanded hole transitions and the locations of its front and back planes are determined by moving distances of D_1, D_2, and D_3, respectively, from its center (i.e., the point of intersection between its axis and the external surface of the blade) along the axis. The shape-control parameters for each cross-section are the same as those for a trapezoidal hole. Finally, a solid model of the film hole can be established using the function UF_MODL_create_thru_curves based on its cross-sectional line data. The design method for VCS film holes is similar to that for expanded holes, except that VCS film holes have variable cross-sectional shapes.

Full-coverage turbine blades are often covered with multiple rows of film holes. The program has thus been improved accordingly to simultaneously model multiple rows of film holes of the same type. This program requires the cross-sectional profile lines of the shape of the blade. Figure 3.21 shows the input format for the positional parameters of film holes. The coordinates and axial vector of each film hole are calculated by calling the MATLAB program. A solid model can be rapidly established by calling programs for film holes of different shapes in UG. The whole process requires only several minutes to complete and thus has a high-design efficiency. Figure 3.22 shows an example of film holes of various types generated on a stationary turbine blade.

Impingement cooling is primarily used inside the internal cavities of stationary blades. In this cooling mode, cooling air forms a jet flow through small holes, and the jet flow then impinges on the internal wall of the blade. The locations and sizes of impingement sleeves inside blades depend primarily upon two parameters, namely the impingement distance d and impingement sleeve thickness e, as shown in Fig. 3.23a. In normal cases, both the impingement distance and thickness are uniform for impingement sleeves. There are small numbers of impingement sleeves and plates inside blades, and they have simple control parameters. The parametric design of impingement sleeves by programming is time- and effort-intensive. Therefore, parameterization of impingement sleeves is directly carried out in modeling

Fig. 3.20 Cross-section control parameters

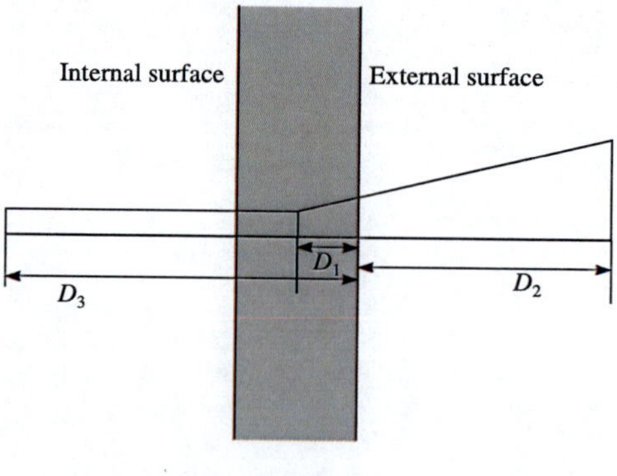

Fig. 3.21 Input parameter format for film holes

Number of cross-sectional profile lines of the blade	Total number of columns of film holes
Pressure side/suction side	Number of film holes in the first column
Chordwise location 1	Chordwise location 2
Radial location 1	Radial location 2
Chordwise angle	Chordwise angle
Pressure side/suction side	Number of film holes in the second column
⋮	⋮
Pressure side/suction side	Number of film holes in the third column
⋮	⋮

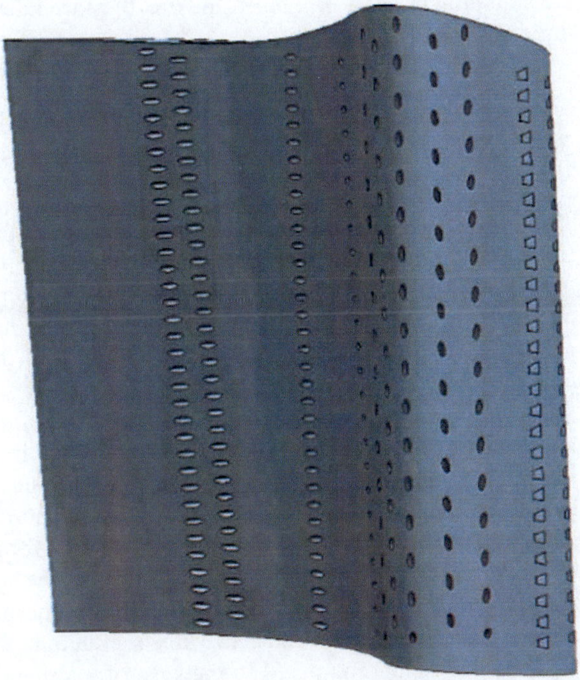

Fig. 3.22 An example of generated film holes

software, and the shape of an impingement sleeve can be obtained by simply offsetting the internal surface of the blade. Most impingement holes on an impingement sleeve are circular and are arranged in either an inline configuration or a cross configuration. When impingement holes are placed, it is also necessary to consider their impact on film cooling. Figure 3.23b shows the basic control parameters for impingement holes, including diameter d, transverse spacing Cx, and longitudinal spacing Cy. The axial direction of each impingement hole is perpendicular to the internal

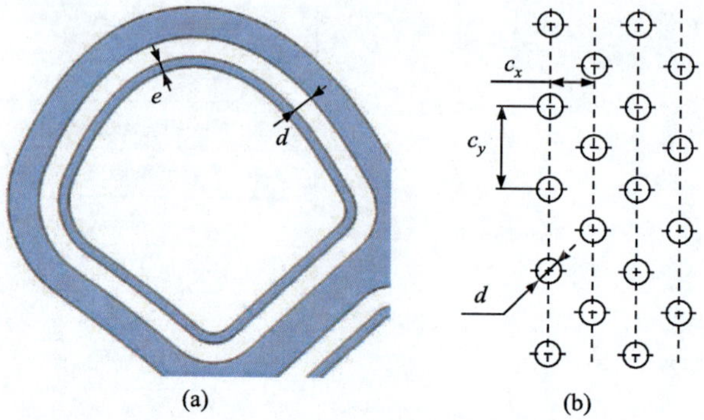

Fig. 3.23 Design parameters for impingement holes

wall of the blade. An analysis of the structure of impingement holes shows that they are similar to circular film holes and that they are more easily controlled in the axial direction.

3.3 Mesh Generation for Turbine Blades

To perform FEA, it is necessary to discretize the geometric model established for a turbine blade into a finite number of elements, i.e., an FE mesh. Based on whether the topology of the elements follows a regular pattern, FE meshes can be categorized into structured and unstructured meshes. The nodes in a structured mesh are numbered according to a certain rule. Topological information (e.g., neighboring relationships between nodes) can be determined for a structured mesh based on the node numbers. In contrast, the number of elements for each internal node contained in an unstructured mesh is indeterminate. Nodes in unstructured meshes are not structurally limited, unlike those in structured meshes. As long as the mesh distributions on the boundaries are specified, an unstructured mesh can be automatically generated between the boundaries without the need for partitioning it and transferring information between its subpartitions. Therefore, the boundaries can be satisfactorily treated. The generation methods for unstructured meshes have high automaticity and computational adaptivity, and hence, unstructured meshes are favored by researchers [13, 14]. However, compared to the case of a structured mesh, the same cuboidal space needs to be divided into five tetrahedrons in an unstructured mesh to meet the same computational requirements. Thus, an unstructured mesh generates significantly more cells than a structured mesh. For example, when a small-scale structure (e.g., a high-Reynolds-number boundary layer or a film hole) is simulated, sufficient mesh resolution is required for the direction perpendicular to the boundary layer. This requirement results in a sharp increase in the number of cells in an unstructured mesh and an increase in the time required for the iterative calculation of the flow field, thereby demanding extensive computational memory. Compared to unstructured meshes, structured meshes have low storage and access costs as well as high orthogonality, boundary closeness, and numerical computational accuracy. However, the available techniques fail to support the fully automated generation of structured meshes for air-cooled blades with a complex geometry due to the characteristics of structured meshes. This section introduces the principles and methods of generating unstructured meshes as well as the partitioning strategies and methods for structured meshes.

3.3.1 Generation of Unstructured Meshes

Currently, point-insertion and node-connection algorithms are used to generate unstructured meshes. The general approach is to gradually add new mesh nodes or create new elements within a domain with its boundaries as input until the domain is completely meshed. Relevant algorithms include the quadtree/octree algorithm, the advancing front algorithm, and the Delaunay triangulation algorithm [15].

(1) Quadtree/octree algorithms

In 1980, the research team led by Shephard became the first to introduce the quadtree/octree algorithm into the mesh generation field and employed this algorithm to generate triangular and tetrahedral elements [16, 17]. Here, a 2D case is used as an example. A rectangular bounding box covering the problem domain is used as a starting point. The bounding box is recursively quartered along the axial direction, with the squares obtained after each decomposition referred to as quadrants. Finally, the quadrants that fall outside the problem domain are removed, and the quadrants that intersect with the boundaries of the problem domain are specifically treated to obtain an initial quadtree that covers the problem domain. Bachmann et al. [18] improved the early modified quadtree by expressing topological entities within objects in a more definitive manner in the quadtree data structure and, thereby, eliminated the limitation of the topological complexity of geometries. Yiu et al. [19] and Greaves and Borthwick [20] described in detail the generation methods, unit coding systems, and data processing methods for quadtrees. Correspondingly, octrees can be obtained by extending quadtrees to 3D cases.

Some processing is required to apply decomposed quadtrees/octrees to numerical simulations. To ensure a smooth transition between elements of different sizes, the depths of neighboring quadrants/octants should not exceed 1. In addition, the levels of neighboring quadrants/octants within a domain should not exceed 1. To accurately discretize the boundaries of a domain, the size of the elements near the boundaries often needs to be refined based on the geometric characteristics of the boundaries to reduce errors in geometric characterization. The main advantage of the quadtree/octree algorithm is that set operations, which have a wide range of applications (e.g., union, intersection, and difference operations on two objects), can be very easily implemented, whereas the processing of these operations using other mesh methods is difficult or computationally expensive. Furthermore, due to its orderliness and hierarchy, the quadtree/octree algorithm can considerably facilitate and be particularly useful for attaining a balance between display accuracy and speed as well as eliminating hidden lines and hidden surfaces.

(2) Advancing front algorithm

The advancing front algorithm begins to generate a mesh at the boundary of a model and continuously advances toward the interior of the region. The fronts refer to the mesh edges (in 2D cases) or mesh surfaces (in 3D cases) that separate the meshed and unmeshed regions. In this algorithm, a set of front edges is established, with the initial set of front edges consisting of the mesh edges or surfaces at the boundary. This set of front edges is continuously updated as new elements are generated one by one. The advancing front algorithm is sometimes referred to as the wavefront algorithm. The mesh generation process for the problem domain is completed when the set of front edges is empty. The classical advancing front algorithm is recursively implemented, with each recursive process involving three steps: (a) seek a suitable front from the current boundary or restraint, (b) insert new nodes or directly link available nodes within the region to form a new element with the front, and (c) update the front. When a new element is determined, it is usually necessary to consider the geometric qualities of the new element and the region excluding the new element.

Meshes generated using the advancing front algorithm often are of high quality and are highly adaptive to geometric boundaries. However, there is a reliability issue with these meshes regarding the closure of the final region. The conventional advancing front algorithm generally fails to generate a mesh for the final remaining unmeshed region when it is small in size. In a 2D case, a reliable strategy for polygon triangulation can be adopted to close the remaining region. In a 3D case, some additional points may need to be added to successfully close the remaining region. While the classical advancing front algorithm is capable of generating planar or surface triangular meshes and solid tetrahedral meshes, the advancing front approach is also suitable for generating planar or surface quadrilateral meshes, [21, 22] solid hexahedral meshes, and viscous boundary-layer meshes [23–25].

(3) Delaunay mesh generation method

The Delaunay method is a theory proposed by Dirichlet in 1850 to divide a plane into convex polygons based on a set of known points, i.e., a Voronoi diagram [26]. Following this method, Delaunay triangulation was proposed in 1934 [27] but not introduced to the mesh generation field until the late 1970s and early 1980s. The basic principle of Delaunay triangulation is described below.

Let $P = \{p_1, p_2, \ldots, p_n)$ be a point set that consists of n points that do not coincide with one another in a given Euclidean space E^d. There always exists a continuous space $V(p_i)$ in the vicinity of each $p_i \in P$ satisfying the following condition:

$$V(p_i) = \{x{:}|p_i - x| \leqslant |p_i - x|, \forall j \neq i\} \tag{3.122}$$

where $V(p_i)$ is referred to as the Voronoi cell of point p_i. The Voronoi cells of all the points collectively constitute the Voronoi diagram of point set P. A dual diagram (i.e., a Delaunay triangulation, denoted by T) of the Voronoi diagram can be obtained by connecting the neighboring points in the Voronoi diagram, as shown in Fig. 3.24. In the Voronoi diagram, a $(d + 1)$ neighboring points form a simplex t. The sphere circumscribing t refers to a hypersphere that passes through all the vertices of t. In a 3D case, as t is a tetrahedron, T can also be called a Delaunay tetrahedralization. For the sake of simplicity, unless specified otherwise, a 3D or higher-dimensional T is still called a Delaunay triangulation. In addition, the Delaunay triangulation of a 2D point set has the "maximum–minimum angle" property; that is, among the results of all triangulations, the triangles from the Delaunay triangulation have the maximum–minimum internal angle. Although the "maximum–minimum angle" property is not applicable to 3D problems, the tetrahedral elements in the Delaunay triangulation of a reasonably distributed 3D point set usually have similarly good geometric properties and can meet the needs of numerical simulation. The Delaunay method has the advantages of rigorous mathematical principles, high generation efficiency, insusceptibility to causing spatial mesh penetration, and relatively simple data structure. However, the Delaunay method has the disadvantage of overly high complexity. To ensure the consistency of the boundary and the integrity of the object surface, it is necessary to control the placement of points on the object surface to prevent it from being penetrated.

A geometric model of a certain guide vane in a flow channel, as shown in Fig. 3.25, has been established to facilitate the FEA of the thermal insulation performance of TBCs on turbine blades and their thermal stresses under service conditions. The interior of the guide vane consists of a forward cooling cavity and an aft cooling cavity. The exterior of the guide vane is covered with a 300-μm-thick TBC. There are 14 rows of film holes on the blade, consisting of one row (H_{14}) of trailing-edge grooves and 13 rows (H_1–H_{13}) of film holes.

An unstructured tetrahedral mesh is generated for the guide vane using the octree mesh generation algorithm in the mesh generation software ICEM. This procedure involves the following basic steps:

(1) Import the geometric model of the guide vane; revise the geometric model to connect and close the points, lines, and planes of the geometry.
(2) Name the geometric model of the blade with various surface mesh parts based on their geometric properties; set and name a volume mesh.
(3) Set mesh sizes and densities for different parts based on the FE calculation accuracy requirements and the geometric features of the model; refine the mesh for locations such as film holes and the trailing edge.
(4) Generate a mesh automatically with ICEM using the octree algorithm based on the preset mesh parameters. The quality of the mesh is checked. If the quality of the mesh does not meet the requirements, the parameters are adjusted to regenerate a mesh. Once the quality of the mesh meets the solution requirements, the mesh is smoothed and exported in a format accepted by the solver.

Figure 3.26 shows the generated mesh, which has a total of 9.21 million cells. A prismatic boundary-layer mesh is generated for the interface between the blade surface and the fluid domain to help more accurately capture the flow and heat

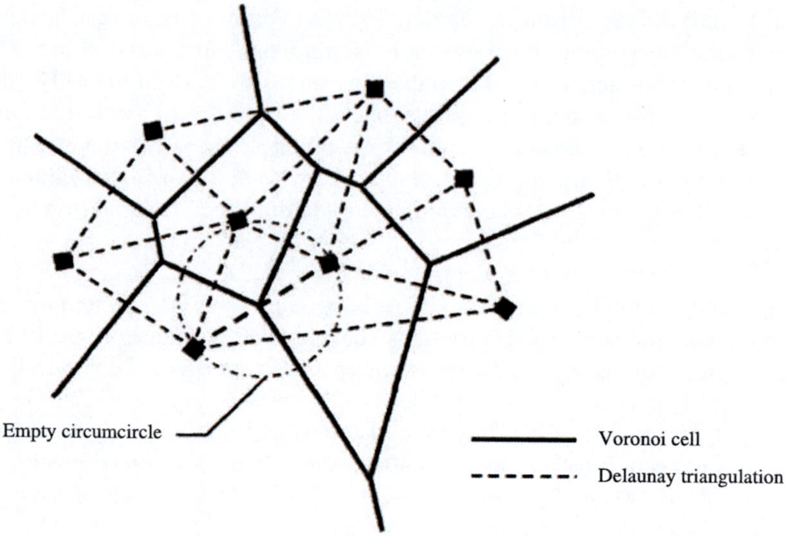

Fig. 3.24 A 2D Voronoi diagram and its dual Delaunay triangulation diagram

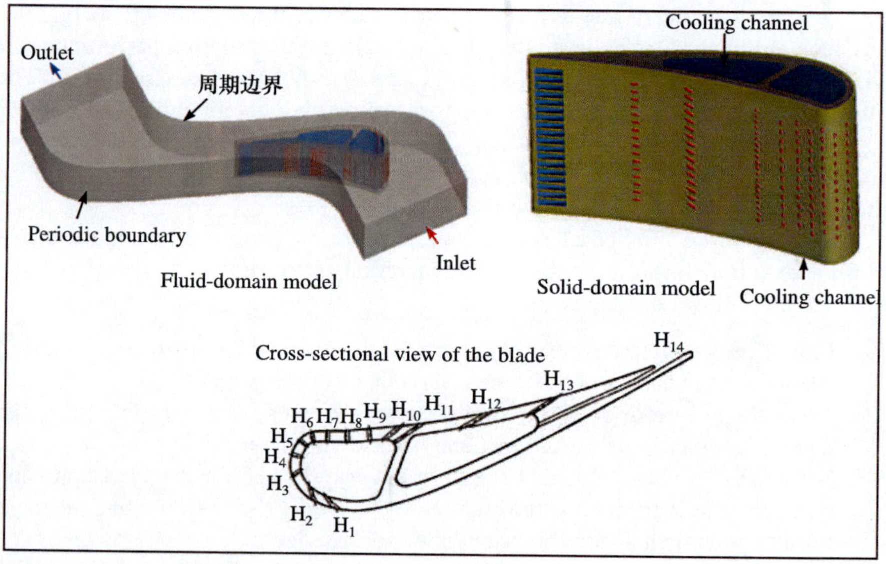

Fig. 3.25 Turbine blade model

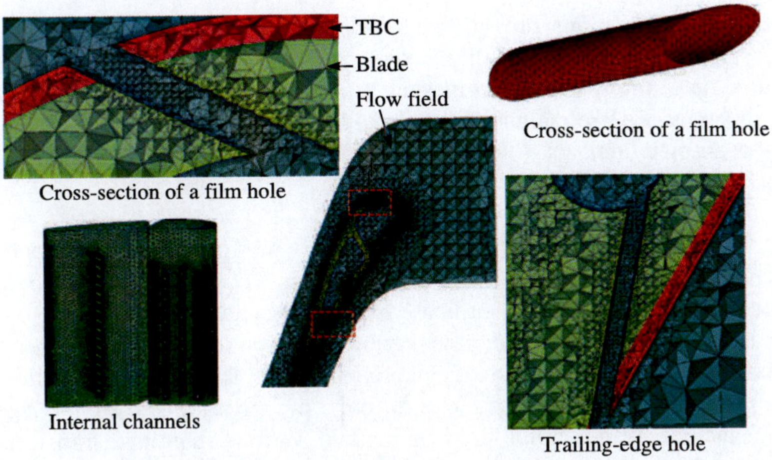

Fig. 3.26 Unstructured mesh generated for a turbine blade

transfer conditions of the boundary layer and, thereby, to improve the computational accuracy. The mesh generated for the mainstream region, which is distant from the blade, is sparse. The mesh is refined for the local regions near the blade, particularly those where there are significant changes in the airflow, such as the film holes and the boundary layer on the blade surface. The mesh is further refined for each row of film holes. As a result, the mesh completely represents the complex geometric features of the blade, including details such as the cooling channels, the pin fins, the trailing-edge grooves, the film holes, and the TBC, thus meeting the FEA requirements.

3.3.2 Structured Meshes for Turbine Blades

Structured mesh generation algorithms are based on parameterization and mapping techniques. These algorithms transform the problem of generating meshes for irregular regions in physical spaces through mapping functions into a problem of generating meshes for regular regions in parameter spaces. There are two main ways to determine mapping functions, namely algebraic interpolation and by solving partial differential equations (PDEs) [28]. The basic procedure for generating a structured network involves the following steps: (1) map the parametric equation of the shape boundary to a computational domain to form a regular computational boundary, (2) generate a structured mesh within the regular computational boundary, and (3) map the mesh within the computational domain backward to the physical domain. Thus, the ultimate goal of mesh generation is to use a certain mathematical method to transform coordinates from a physical domain to a computational domain, i.e.,

$$\xi = \xi(x, y) \quad \eta = \eta(x, y) \tag{3.123}$$

These transformations are all mathematically realized in simply connected domains. Relatively complex multiple connected meshes can be viewed as combinations of simple simply connected meshes. Currently, algebraic methods and differential equations are two main types of mature methods used to generate meshes for simply connected domains.

(1) Algebraic methods

An algebraic mesh generation method transforms a mesh for an irregular region in a physical space to a mesh for a rectangular region in a computational space through intermediate interpolation or coordinate transformation using some algebraic relations based on the known boundary values of the region in the physical space. Interpolation calculations are the core of algebraic methods. Different interpolation algorithms produce algebraic meshes with different properties. Some simple algebraic methods include the direct line-drawing method, various coordinate transformation methods, and the double-boundary method for regular boundaries. However, these methods have poor adaptability. Comparatively, algebraic mesh generation methods that construct interpolation formulas based on certain interpolation basis functions are highly versatile and capable of generating meshes with good properties. The transfinite interpolation method is a representative, mature, effective algebraic mesh generation method with strong adaptability and high speed.

(2) Differential equation methods

Differential equation methods are another type of classic mesh generation method. By taking advantage of the analytic properties of differential equations (e.g., the smoothness of harmonic functions and orthogonal invariance in transformations), these methods transform coordinates from physical spaces to computational spaces. Meshes generated by differential equation methods are superior to algebraic meshes in terms of smoothness, reasonableness, and versatility. Based on the equations used, differential equation methods for mesh generation are classified into elliptic, hyperbolic, and parabolic equation methods, of which elliptic equation methods are the most widely used in practice.

The generation of a structured mesh for a complex geometric model often involves three steps: (1) create partitions to represent the geometric model, (2) establish mapping relations to bridge the geometric model and the partitions, and (3) generate a structured mesh in the partitions. Thus, creating a reasonably partitioned topology based on the analysis of a geometric model is the most fundamental and important step in generating its structured mesh. Based on the arrangement of partitions, the topology of the computational domain can be categorized into parallel and nested patterns.

(1) Parallel topology

A parallel topology is formed by the stacking of multiple hexahedral topologies. Each hexahedron shares at most two faces with the neighboring bodies in each of the three topological directions. There can be an arbitrary number of hexahedrons arranged in each direction. As demonstrated in Fig. 3.27a, multiple hexahedral topologies

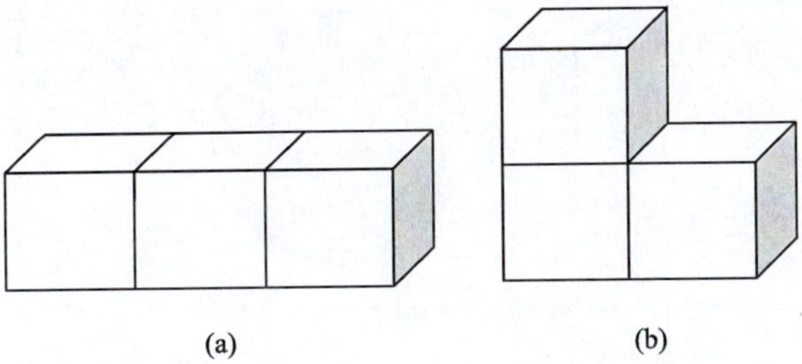

Fig. 3.27 Parallel topologies generated for structured meshes

can be arranged along one direction, with two neighboring hexahedrons sharing one connecting face. Alternatively, as shown in Fig. 3.27b, multiple hexahedrons can be arranged along three directions, with each hexahedron sharing at most two faces with its neighboring hexahedrons. A parallel hexahedral topology conforming to a geometric model of interest is constructed from scratch, step by step, and through addition.

(2) Nested topology

A nested topology is formed by placing a smaller hexahedron in a larger hexahedral topology and connecting the eight vertices of one hexahedron to the corresponding vertices of the other hexahedron, as shown in Fig. 3.28. In this structure, an even smaller hexahedron can be further placed in the smaller hexahedron, and this process can be repeated. Thus, a nested topology contains at least five hexahedral topologies. The nested topology is widely used in practice and has many variations. From an overall perspective, a nested topology can be generated for a geometric model through multiple nesting processes.

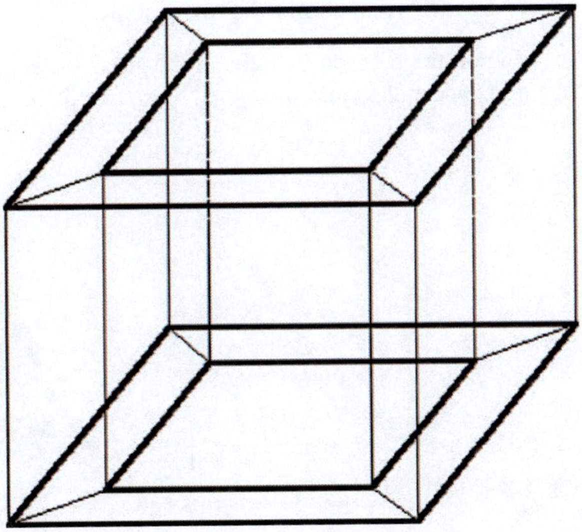

Fig. 3.28 A nested topology

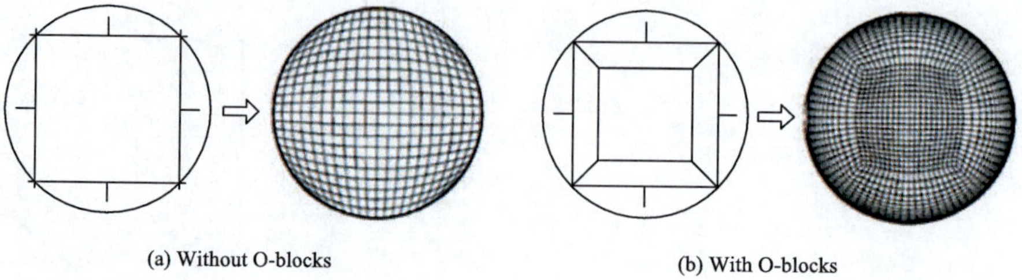

Fig. 3.29 O-block-based partitioning

O-blocks are the most commonly used to generate structured meshes. For geometric models of cylinders or circular holes, O-blocks can be used to prevent sparsely clustered and skewed mesh cells on curves or surfaces within partitions, thereby improving the mesh quality; they can also be employed to generate ideal boundary-layer mesh cells near surfaces, as shown in Fig. 3.29.

Y-partitioning is another common partitioning mode and is primarily used for geometric models with triangular cross-sections. Figure 3.30 shows the Y-partitioning strategy. Y-partitioning can also be performed in a nested pattern, as shown in Fig. 3.31. As demonstrated in Fig. 3.31, nested Y-partitioning moves tilted mesh cells from the boundary of the model to its center and significantly improves the mesh cells at the boundary of the model.

The flexible use of O- and Y-partitioning in combination with parallel and nested topologies can provide ideas for partitioning geometries when generating structured meshes. In practice, the focus is placed on establishing the topological relations between geometry and its partitions, instead of the actual geometry.

Based on the above concept of structured meshes, ICEM can be used to generate a structured mesh for the geometric model of a guide vane shown in Fig. 3.32. This procedure involves the following main steps:

(1) Import the geometric model of the guide vane, and revise the geometric model to connect and close its points, lines, and planes.

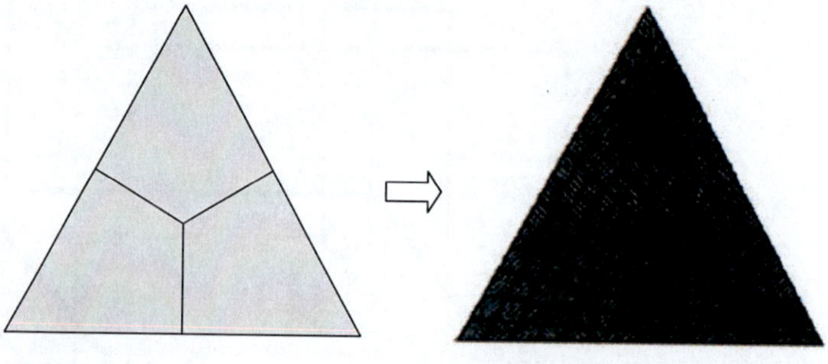

Fig. 3.30 Y-partitioning

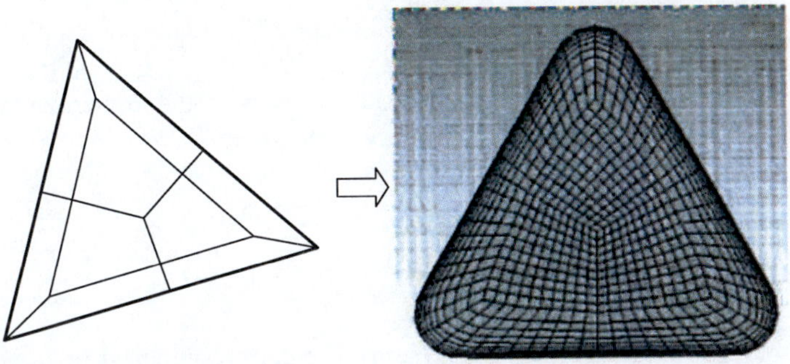

Fig. 3.31 Nested Y-partitioning

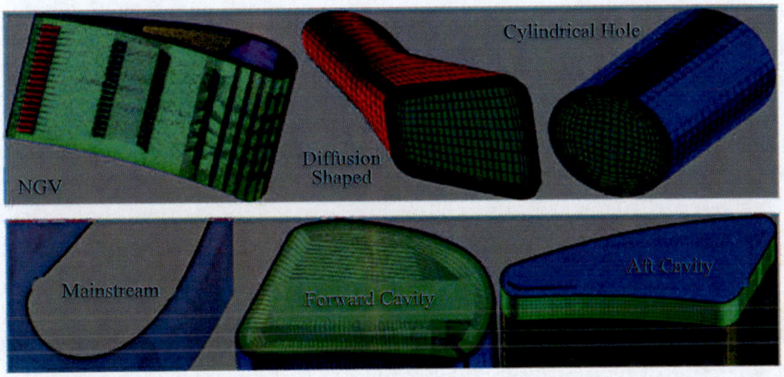

Fig. 3.32 Structured meshes [29]

(2) Name the geometric model of the blade with surface mesh parts based on its geometric properties, and set and name a volume mesh.
(3) Generate blocks, partition the blocks using O-blocks and parallel and nested topologies based on the geometry of the guide vane, and remove redundant blocks.
(4) Correlate the points, lines, and planes of each part of the geometry of the blade with those of the blocks.
(5) Set the number of nodes and the distribution pattern of nodes for the boundary lines of each block, observe the quality of the mesh using Premesh, and adjust the mesh parameters until a mesh with the required quality is obtained.
(6) Generate a true mesh, and export it in a file format accepted by the solver.

Figure 3.32 shows a guide turbine blade with cooling channels and multiple rows of film holes. Blocks in ICEM are used to generate a computational mesh. O-blocks are used for the flow field and the blade. Meshes are first separately generated for the film holes and the internal cooling cavities and then nested together. It can be found that the generated meshes are uniform and orderly and that they completely reflect the geometric features of the external shape of the blade and its film holes and internal cooling cavities. In addition, a boundary layer is generated at the fluid–solid interface, which improves the computational accuracy of the numerical simulation. Therefore, the generated meshes are of very high quality.

3.4 Image-Based FE Modeling

A TBC is a multilayered structural system, typically composed of a substrate, a bond coat (BC) layer, a TGO layer, and a ceramic top coat (TC) layer. In FEA, the multilayered and microscopic structure of a TBC plays an important role in reflecting its true stresses and failure during operation. Atmospheric plasma spraying (APS) and electron beam-physical vapor deposition (EB-PVD) are two processes commonly used to prepare TBCs. An APS coating is characterized by a layered microstructure, as shown in Fig. 3.33a, a high bonding strength, and a high interlayer porosity, which lowers its thermal expansion coefficient and thermal conductivity and, thereby, improves its insulation effect. An EB-PVD coating [30] displays a typical columnar crystal structure, as shown in Fig. 3.33b. The gaps existing between columnar crystals increase the strain tolerance of the coating and give it good thermal shock resistance. In addition, TBCs constantly operate at high temperatures. Oxygen from the external environment diffuses toward the interior of a TBC through the gaps or pores in its TC layer and reacts with alloy elements between its BC and TC layers to form high-temperature oxides, i.e., TGOs, that thicken continuously with service time. Figure 3.34 shows the morphological evolution of the TGOs in a TBC. Since the material parameters of the TGOs differ significantly from those of the TC and BC layers, high thermal stresses are generated during cooling. As they grow and thicken, the TGOs undergo morphological changes and become folded or wavy under the action of reaction rates and creep. These irregular TGO morphologies lead to stress concentration and further produce crack initiation and propagation.

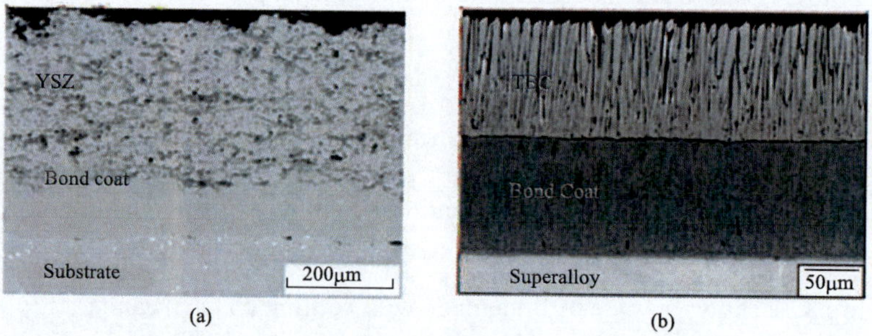

Fig. 3.33 Microstructural images of TBCs: **a** an APS coating [31]; **b** an EB-PVD coating [32]

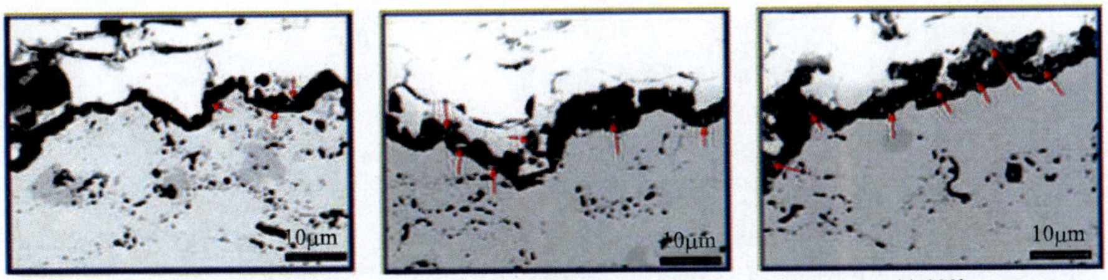

(a) 1h (b) 50h (c) 100h

Fig. 3.34 Pattern of change in the TGOs in an APS TBC with oxidation time [33]

3.4.1 Image-Based FEM

The microstructure and interfacial morphology of a TBC are vitally important for describing the mechanisms and evolution of its interfacial failure. We propose the use of image processing tools (Photoshop and CorelDraw) in combination with modeling software (UG and ABAQUS/COMSOL) to establish geometric models that reflect the true morphologies of TBCs. This method relies on scanning electron microscopy (SEM) images of the cross-sections of TBCs. By taking advantage of the file-format compatibility between multiple software programs, we can effectively and rapidly convert SEM images into geometric model images that can be used in FE software. Roughly speaking, this procedure involves the selection of an SEM image that clearly shows the TBC of interest, the extraction and vectorization of the profile of the TBC using such software as Photoshop, CorelDraw, and UG, and, finally, the exportation of the vectorized data into the FE software ABAQUS/COMSOL to generate a geometric model. The detailed steps are described below.

(1) Obtain a binary profile image in Photoshop through the following steps:

   ① Select an SEM image that clearly shows the profile of the TBC of interest, save it in.jpg format, and export it into Photoshop.
   ② Convert the SEM image into a grayscale image, increase its black/white contrast, segment it based on a threshold, and produce a threshold-segmented image by preserving the black regions.
   ③ Produce a morphological image by corroding the threshold-segmented image to remove impurities and by preserving the regions.
   ④ Extract a binary profile image from the morphological image of the TGO and export it in.jpg or.png format.

The principle of processing an SEM image in Photoshop is described as follows. As a digital image, an SEM image is composed of 2D pixels in software. Let us assume that in a 2D coordinate system, there are n pixels in the x-direction and m pixels in the y-direction. Thus, a 2D pixel matrix $\boldsymbol{F}$ with elements $f(i, j)$ can be formed:

$$F = \begin{bmatrix} f(0,0) & f(0,1) & \cdots & f(0,n-1) \\ f(1,0) & f(1,1) & \cdots & f(1,n-1) \\ \cdots & \cdots & \cdots & \cdots \\ f(\mathrm{m}-1,0) & f(\mathrm{m}-1,1) & \cdots & f(\mathrm{m}-1,n-1) \end{bmatrix} \tag{3.124}$$

This 2D matrix contains the grayscale information for the whole SEM image. Threshold segmentation is subsequently performed. That is, a specific gray value H is selected, and the gray values higher than H in the pixel matrix $\boldsymbol{F}$ are set to 0 (representing white), while those lower than H are set to 1 (representing black). This process can be described by the following equation:

$$f(i,j) = \begin{cases} 0 \ (f(i,j) \leqslant \mathrm{H}) \\ 1 \ (f(i,j) \geqslant \mathrm{H}) \end{cases} \quad (0 \leqslant i \leqslant m-1, 0 \leqslant j \leqslant n-1) \tag{3.125}$$

Based on Eqs. (3.125) and (3.126), a morphological image following threshold segmentation is obtained. Then, this image is corroded to remove impurities, and a profile image is extracted.

(2) Vectorize the binary profile image in CorelDraw to convert it into a vector image, and export it in.dwg or.dxf format.

Converting the binary image into a vector image in CorelDraw is a key step in geometric modeling. A bitmap differs from a vector image. A bitmap is an image composed of 2D or 3D pixels and becomes distorted when enlarged or reduced. In contrast, a vector image is recorded by a series of computer commands and characters and stores a geometric shape. A vector image does not become distorted when enlarged or reduced. Vector images are the basis of geometric modeling [34]. A digital image must be vectorized before it can be used to establish a geometric model for computer-aided design (CAD) or computer-aided engineering (CAE) software. Figure 3.35 shows the basic principle of vectorization. Let us select several points of interest from the image. First, we find the two points that are the furthest apart, which are A and H, and connect them. Thus, the whole image is divided into two parts by line segment AH. Now, in each part, we find the point with the longest perpendicular distance from line segment AH. Here, the upper part in Fig. 3.35 is used as an example for illustration. Point L is the furthest from line segment AH. Then, it is judged whether the distance between L and AH is shorter than a specific value *t*. If this distance is greater than *t*, the upper part is further divided into two smaller parts. Points A and L and points H and L are connected. In each part, the point that is the furthest away from the corresponding line segment (AL or LH) is determined. M and K are found to be the points that are the furthest away from AL and LH, respectively. Subsequently, it is judged whether the distance between M and AL and the distance between K and LH are shorter than *t*. If so, the other points are connected, the relative positions are recorded by the computer, and a line segment is generated. This process is repeated until line segments in the other parts are successively generated, and the whole image becomes a connected, closed whole HIJ...EFG. [35]

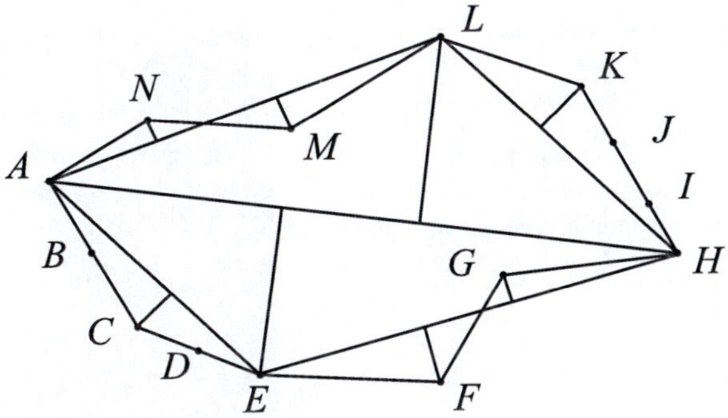

Fig. 3.35 Schematic diagram of the conversion of a bitmap into a vector image

(3) Generate a vector profile image of the original shape by scaling the vectorized profile image to the same size as the original SEM image based on its scale and by subsequently performing a geometry cleanup, and save and export the image in.stp or.CATPart format.
(4) Establish a true geometric model of the TBC in ABAQUS/COMSOL.

3.4.2 2D TGO Interface Modeling

Fourier series can be used to extract approximate interfacial profiles from images. Let us consider a 28-μm-long region of the overall profile with an average roughness $R_l = 1.052$. Here, the discussed region is assumed to be periodic and representative of the whole TGO layer. Next, let us briefly introduce the yttria-stabilized zirconia (YSZ)/BC interface. Let $P(x)$ and $P^*(x)$ be the discrete and Fourier-fitted functions of the interface, respectively, where x is the position along the oxide interface. To simplify the Fourier series that define the average profile of the oxides, only the first N spatial frequencies are considered. For cases where $k = 1 - N$, k/w is set to a fixed value, where w is the width of the analyzed profile. Thus, the average profile of the TGOs can be determined by the amplitude parameters (a_k and b_k) of the $(2 \times N)$ Fourier series as follows:

$$P_N^*(x) = \frac{a_0}{2} + \sum_{k=1}^{N}\left[a_k \cos\left(k\frac{2\pi x}{w}\right) + b_k \sin\left(k\frac{2\pi x}{w}\right)\right] \approx P^*(x) \tag{3.126}$$

The quality of the rough interface model is tested by comparing the measured average distribution of the oxides and the Fourier transforms with different numbers of terms N. As demonstrated in Fig. 3.36, the average profile of the oxides obtained by image processing and analysis at $N = 16$ agrees well with the measurement.

The geometric shape of the oxides in Fig. 3.37a is the 28-μm-long interfacial region of interest here. To establish an FE mesh for the interfacial region, it is necessary to first establish an FE mesh for the initial, undeformed interfacial region, as

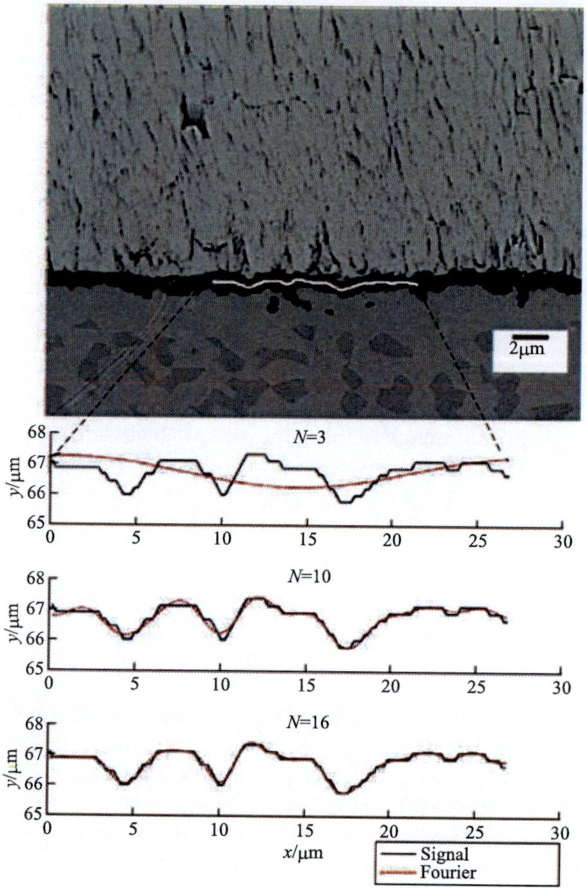

Fig. 3.36 Comparison of the average profiles (solid lines) of the oxides obtained by image processing and analysis and their Fourier approximations with different numbers of terms ($N = 3$, 10, and 16) (dotted lines)

shown in Fig. 3.37b. Then, the whole oxide layer is subjected to Fourier decomposition. Based on Eq. (3.128), each node is replaced along the out-of-plane z-direction. To ensure the smoothness of the mesh and the upper and lower surfaces, all the nodes outside the oxide layer are also displaced along the z-direction. As a result of linear function modulation, the upper and lower displacements of the TBC region of the model become zero. The following system of equations is used to satisfy these criteria:

$$\begin{cases} Z(x) = P_N^*(x) \cdot f(z), \quad \text{where } f(z) = \frac{z-z_3}{z_2-z_3}, \text{ for } z_2 < z \leqslant z_3 \\ Z(x) = P_N^*(x), \text{ for } z_1 < z \leqslant z_2 \\ Z(x) = P_N^*(x) \cdot g(z), \quad \text{where } g(z) = \frac{z}{z_1}, \text{ for } 0 < z \leqslant z_1 \end{cases} \tag{3.127}$$

where Z is the out-of-plane offset displacement of the new node relative to the old node z and z_i ($i = 1, 2, 3$) is the position of the interface. For different layers,

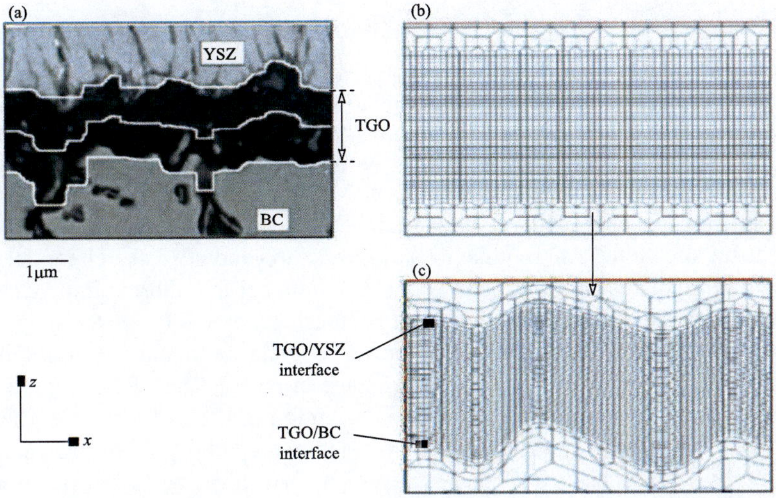

Fig. 3.37 **a** Interfacial details of a typical EB-PVD TBC. **b** FE mesh generated for the initial undeformed interfacial region. **c** Corresponding deformed FE mesh

$Z_1 = h_{BC}$, $Z_2 = h_{BC} + h_{TGO}$, and $z_3 = h_{BC} + h_{TGO} + h_{YSZ}$, where h_i ($i = BC, TGO, YSZ$) is the average thickness of each TBC layer. The lower left corner of the image in Fig. 3.37a is selected as the origin of the coordinates. The eventually obtained deformed mesh is consistent with the expected morphology of the TGOs, as shown in Figs. 3.37 and 3.38.

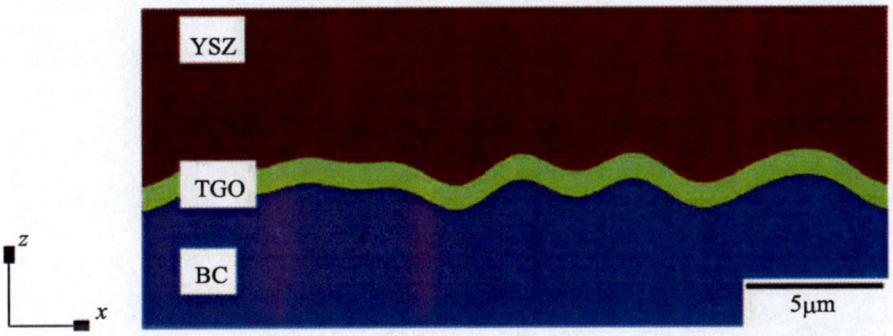

Fig. 3.38 A 2D model of the TBC system obtained by Fourier representations (the ceramic, TGO, and BC layers are red, yellow, and blue, respectively)

3.4.3 Porous Ceramic Layer Modeling

The image-based method can also be employed to generate porous FE structures for APS TBCs. Figure 3.39a shows an SEM image of the true cross-sectional microstructure of a certain TBC (8 mol%-YSZ). Figure 3.39b shows a solid model generated by the following procedure: (1) the SEM image is converted into a denoized grayscale image in Photoshop, (2) the interface is tracked in the Photoshop-processed image using the AutoCAD module in SolidWorks to generate a sketch model of the microstructure of the coating, and (3) the sketch model is imported into COMSOL for solid modeling. Figure 3.39c gives an FE mesh generated for the coating.

The image-based method can be used to generate columnar FE structures for EB-PVD TBCs. Figure 3.40a is an SEM image of an EB-PVD TBC. Figure 3.40b gives the corresponding grayscale value distribution of the SEM image, where the x-axis represents the gray value (0–255) of each pixel and the y-axis represents the frequencies or probabilities of the gray values corresponding to the pixels in the image. The gray values of the pixels corresponding to different compositions differ. The gray values of the pixels corresponding to the columnar crystal material are primarily concentrated within a small area, whereas the gray values of the pixels corresponding to the columnar crystal gaps are mainly concentrated within a large area. Figure 3.40c shows a sketch model generated for the microstructure of the EB-PVD coating.

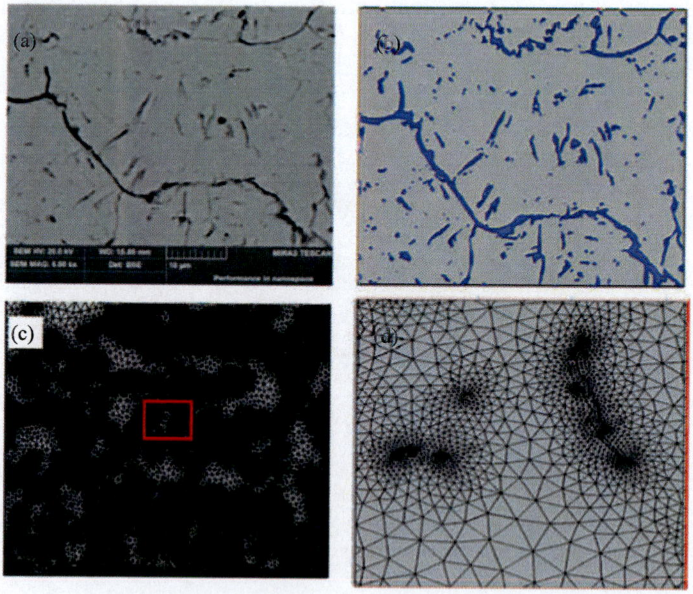

Fig. 3.39 **a** SEM image of a PS-TBC at 5000x. **b** Solid model. **c** Mesh model of the microstructure

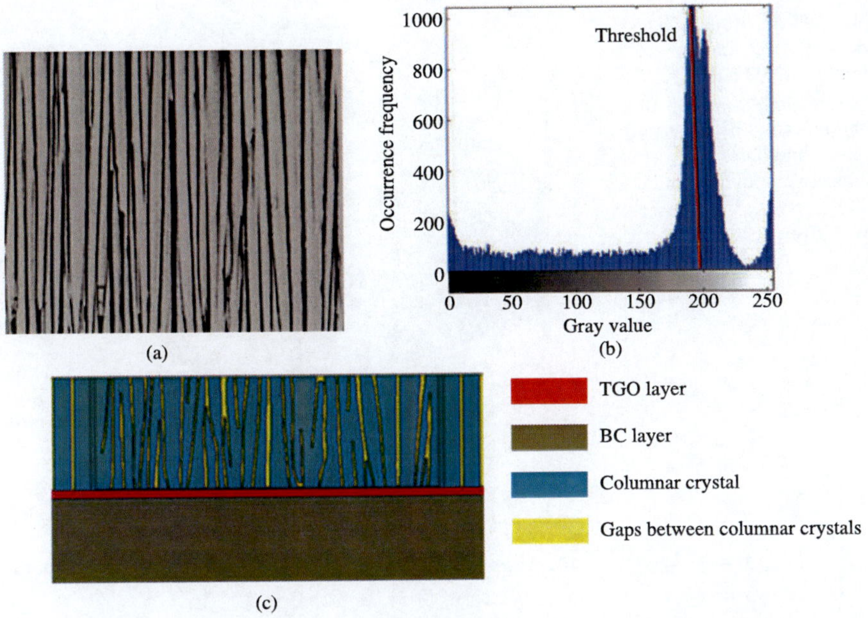

Fig. 3.40 **a** Original SEM image. **b** Grayscale histogram. **c** Geometric model of the true morphology [7]

3.4.4 D3 TGO Interface Modeling Method

It is impossible to directly measure the 3D morphology of the actual YSZ/BC interface in a TBC system of interest. Thus, let us assume that 2D Fourier analysis is also suitable for reproducing 3D roughness profiles. A new Fourier series $R^*(y)$ is established in the y-direction within a plane orthogonal with the surface of the coating shown in Fig. 3.41. $R^*(y)$ is consistent with $P^*(x)$ in terms of $\left(k\frac{2\pi}{w}\right)$ and $k = 1 - N$. The amplitude parameters (c_k and d_k) of each term are set to arbitrary values to obtain a complex 3D surface morphology consistent with that usually observed in an actual TBC system.

$$Q(x,y) = P_N^*(x)R_N^*(y) = P_N^*(x)\left\{\frac{c_0}{2} + \sum_N^k\left[c_k\cos\left(k\frac{2\pi y}{w}\right) + d_k\sin\left(k\frac{2\pi y}{w}\right)\right]\right\} \tag{3.128}$$

where $P_N^*(x)$ and $R_N^*(y)$ are the first N terms in the Fourier series $P^*(x)$ and $R^*(y)$, respectively. Thus, the Z-coordinate of each point within the 3D mesh is governed by the following system of equations:

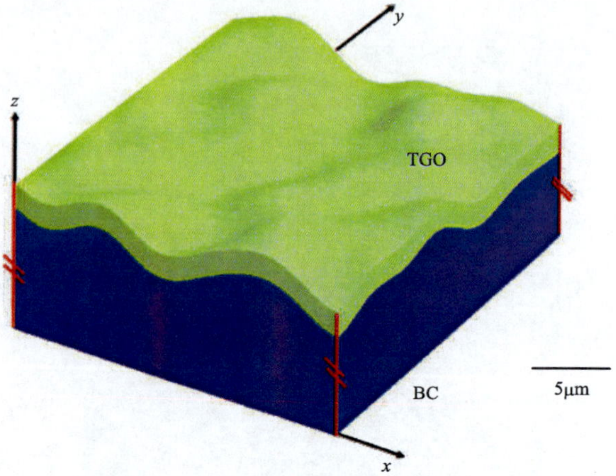

Fig. 3.41 A detailed 3D view of a TBC model generated by Fourier representations at $N = 16$ (the TGO and BC layers are yellow and blue, respectively; the ceramic TC layer is omitted for the sake of clarity)

$$\begin{cases} Z(x, y) = Q(x, y) \cdot f(z), \quad \text{where } f(z) = \frac{Z - z_3}{z_2 - z_3}, \text{ for } z_2 < z \leqslant Z_3 \\ Z(x, y) = Q(x, y), \quad \text{for } z_1 < z \leqslant z_2 \\ Z(x, y) = Q(x, y) \cdot g(z), \text{ where } g(z) = \frac{z}{z_1}, \text{ for } 0 < z \leqslant z_1 \end{cases} \tag{3.129}$$

Figure 5.5 shows the 3D model generated for the TGO layer. The final 3D model is highly realistic compared with an actual coating without a ceramic TC layer. However, note that as the TGO boundary conditions differ, the roughness of the known oxidized TC/TGO interface differs completely from that of the BC/TGO interface.

3.5 Summary and Outlook

The FEM is one of the greatest contributions of the twentieth century and has played a tremendous role in solving design and scientific problems in numerous engineering projects. Aeroengines, known as the heart of the modern industry, have a very complex structure. The development of aeroengines would be unarguably impossible without the FEM. TBCs on turbine blades are one of the most crucial core components of aeroengines, and their failure problems are characterized by geometric complexity and the physical nonlinearity of thermo–mechano-chemical coupling. The nonlinear FEM is an important technique for investigating TBC failure problems.

This chapter introduces the functional variational principle from the perspective of the theoretical basis of the FEM. The essence of the functional variational principle is based on the principle of least action and the principle of minimum energy. Thus, the essence of the FEM originates from the laws of nature. Approximate solutions are another core idea of the FEM. In other words, solutions obtained by the

FEM do not satisfy the differential equations that describe specific physical problems; instead, the integrals of the weighted functions and differential equations over a certain domain are zero. After establishing the theoretical basis of the FEM, this chapter introduces the weak form and matrix equations from three aspects, namely the Eulerian formulation, the Lagrangian formulation, and the adaptive mesh formulation (the arbitrary Lagrangian–Eulerian formulation). Despite short presentations, both the logic of the physical concepts and the mathematical derivation at each step are highly rigorous. The FEM is eventually represented by linear algebraic equations and, as long as these equations are solved, all physical quantities can be determined. In addition, this chapter gives a brief introduction to the geometric characteristics of turbine blades, the parametric modeling methods for complex structures (e.g., serpentine channels, ribs, and film holes) within turbine blades, the generation of FE meshes for turbine blades, and image-based FE modeling. A TBC has a multilayered structure of layers with significantly different thicknesses. The metallic material in the substrate is several millimeters in thickness, whereas the thickness of the TGO layer is only several microns. Moreover, the geometric, physical, and mechanical properties of the layers vary. Further, TBCs operate in very harsh conditions. Consequently, these problems with a very high degree of coupling have presented immense difficulties to the generation of meshes for TBCs on turbine blades. We have carried out research on all these problems. Interested readers are encouraged to refer to the relevant dissertations [36–39] and patents [34]. Furthermore, constitutive relations and computational methods are not covered in this chapter in the introduction of the FEM, and the interested readers may refer to relevant monographs. While its contents are organized for TBCs, this chapter presents basic research work and is equally applicable to other materials and fields.

References

1. Yang X, Li X. Modern finite element method: theories, techniques, and engineering applications. Beijing: Beihang University Press; 2007. (In Chinese).
2. Belytschko T, Liu WK, Moran B. Nonlinear finite elements for continua and structures. Chinese edition translated by Z. Zhuang: Tsinghua University Press; 2002. (In Chinese).
3. Yue M. A study on structured-mesh parameterization methods for composite air-cooled turbine blades. Nanjing: Nanjing University of Aeronautics and Astronautics; 2018. (In Chinese).
4. Zhou Z. Optimization of high subsonic axial compressor blades. J Propuls Technol. 2004;01:58–61. (In Chinese with English abstract).
5. Liu L, Zhou Z, Qiu M. Studies of shock structure in supersonic cascade and profile optimization design. J Propuls Technol. 2013;34(8):1050–5. (In Chinese with English abstract).
6. Yu K, Li L, Yue Z. Parametric design for cooling turbine blades based on analytic and feature modeling. J Propuls Technol. 2007;28(6):637–40. (In Chinese with English abstract).
7. Liu S, Ma H. Parameterization of modeling the aero-turbine blade based on CATIA. Ordnance Ind Autom. 2015;4:56–9. (In Chinese with English abstract).
8. Han X. Parameterization of cooling structures in turbine blades and optimum design of ribbed channels. Harbin: Harbin Institute of Technology; 2011. (In Chinese).
9. Shi Z. Parametric design and numerical investigation of composite cooling structures in gas turbine blades. Harbin: Harbin Institute of Technology; 2015. (In Chinese).

10. Li J. A study on parametric feature modeling methods for guide turbine blades. Chengdu: University of Electronic Science and Technology of China; 2009. (In Chinese).
11. Song H. A study on CFD mesh preprocessing for turbines based on UG. Nanjing: Nanjing University of Aeronautics and Astronautics; 2007. (In Chinese).
12. Dyedov V, Einstein DR, Jiao X, et al. Variational generation of prismatic boundary-layer meshes for biomedical computing. Int J Numer Meth Eng. 2009;79(8):907.
13. Sharov D, Luo H, Baum JD, et al. Unstructured Navier-Stokes grid generation at corners and ridges. Int J Numer Meth Fluids. 2013;43(6–7):717.
14. Zhao D. Parallel unstructured mesh generation approaches for large-scale numerical simulations. Zhejiang: Zhejiang University; 2016. (In Chinese).
15. Yerry MA, Shephard MS. Automatic three-dimensional mesh generation by the modified-octree technique. Int J Numer Meth Eng. 1984;20(11):1965–90.
16. Shephard MS, Georges MK. Automatic three-dimensional mesh generation by the modified-octree technique. Int J Numer Meth Eng. 1991;32(4):709–49.
17. Bachmann PL, et al. Robust geometrically based automatic two-dimension mesh generation. Int J Numer Methods Eng. 1987;24:1043–78.
18. Yiu KF, Greaves AM, Cruz S et al. Quadtree grid generation: information handling, boundary fitting and CFD applications. Comput Fluids. 1996;25(8):759–69.
19. Greaves AM, Borthwick AG. Hierarchical tree-based finite element mesh generation. Int J Numer Meth Eng. 1999;45:447–71.
20. Blacker TD, Stephenson MB. Paving: a new approach to automated quadrilateral mesh generation. Int J Numer Meth Eng. 1991;32:811–47.
21. Cass IL, Benzley SE, Meyers RJ, et al. Generalized 3-D paving: an automated quadrilateral surface mesh generation algorithm. Int J Numer Meth Eng. 1996;39:1475–89.
22. Owen SJ, Saigal S, Morph H. An indirect approach to advancing front hex meshing. Int J Numer Meth Eng. 2000;49(1–2):289–312.
23. Kallinderis Y, Ward S. Prismatic grid generation for three-dimensional complex geometries. AIAA J. 1993;3l(10):1850–56.
24. Sharov D, Nakahashi K. Hybrid prismatic/tetrahedral grid generation for viscous flow applications. AIAA J. 1998;36(2):157–62.
25. Voronoi G. Nouvelles applications des paramètres continus à la théorie des formes quadratiques. Deuxième mémoire. recherches sur les parallélloèdres primitifs. J Fur Die Rne Und Angewandte Mathematik, 1908;134(134):198–287.
26. Delaunay B. Sur la sphre vide. lzvestia Akademii Nauk SSSR. 1934;7:793–800.
27. Prapamonthon P, Xu H, Yang W, et al. Numerical study of the effects of thermal barrier coating and turbulence intensity on cooling performances of a nozzle guide vane. Energies. 2017;10(3):1–16.
28. Hass DD, Parrish PA, Wadley HNG. Electron beam directed vapor deposition of thermal barrier coatings. J Vac Sci Technol. 1998;16(6):3396–401.
29. Dong H, Yang GJ, Li CX, et al. Effect of TGO thickness on thermal cyclic lifetime and failure mode of plasma-sprayed TBCs. J Am Ceram Soc. 2014;97(4):1226–32.
30. Evans AG, Clarke DR, Levi CG. The influence of oxides on the performance of advanced gas turbines. J Eur Ceram Soc. 2008;28(7):1405–19.
31. Ma K, Schoenung JM. Isothermal oxidation behavior of cryomilled NiCrAlY bond coat: homogeneity and growth rate of TGO. Surf Coat Technol. 2011;205(21–22):5178–85.
32. Xiangtan University. A finite element modeling method based on the true microstructure of materials. China, CN201410279062.0. 2014-9-24. (In Chinese).
33. Yang L, Li HL, Zhou YC, et al. Erosion failure mechanism of EB-PVD thermal barrier coatings with real morphology. Wear. 2017;392:99–108.
34. Zhu W, Zhang ZB, Yang L, et al. Spallation of thermal barrier coatings with real thermally grown oxide morphology under thermal stress. Mater Des. 2018;146:180–93.
35. Dong J. Finite element simulations of erosion failures in APS-TBCs considering true microstructures. Xiangtan: Xiangtan University; 2015. (In Chinese).

36. Li H. Finite element simulations of erosion failures in TBCs prepared by EB-PVD based on true microstructures. Xiangtan: Xiangtan University; 2016. (In Chinese).
37. Zhang Z. Finite element analysis of thermal stresses and interfacial failures in thermal barrier coatings based on the true morphology of TGO interfaces. Xiangtan: Xiangtan University; 2016. (In Chinese).
38. Shen Q. A theoretical study on the growth and damage of thermal barrier coatings with interfacial oxidation considering large deformation and thermo–mechano–chemical coupling. Xiangtan: Xiangtan University; 2018. (In Chinese).
39. Jin Y. Finite element analysis of the failure mechanisms of thermal barrier coatings prepared by EB-PVD based on single-particle erosion. Xiangtan: Xiangtan University; 2018. (In Chinese).

Chapter 4
Geometric Nonlinearity Theory for the Interfacial Oxidation of TBCs

Spallation is a major bottleneck limiting the safe application of thermal barrier coatings (TBCs) in high-performance aeroengines, and interfacial oxidation is the main factor causing TBC spallation. In high-temperature service conditions, thermal growth oxide (TGO) growth and stress evolution resulting from interfacial oxidation are the primary causes of TBC spallation. Thus, prediction of the evolutionary patterns of the TGO growth and stress distribution is vitally important for accurately understanding TBC failure induced by interfacial oxidation. Interfacial oxidation of TBCs is a complex thermo–mechano-chemical multifield coupling phenomenon. In addition, TGO growth is accompanied by large deformation, i.e., this process is geometrically nonlinear. Consequently, interfacial oxidation of TBCs is, in essence, a large-deformation or geometrically nonlinear thermo–mechano-chemical coupling problem. To deepen the understanding of the TGO-growth-induced failure mechanism of TBCs, this chapter establishes a theoretical thermo–mechano-chemical coupling growth model for interfacial oxidation of TBCs based on the large-deformation theory and simulates the evolutionary patterns of the morphology of the TGO layer and stress during interfacial oxidation of TBCs using the finite element method (FEM).

This chapter consists of four sections. Section 4.1 introduces interfacial oxidation, stress evolution, and the subsequent spallation patterns. Section 4.2 decouples the oxidation problem of multifield coupling; provides the governing equations for oxygen infiltration, TGO growth, and stress and strain evolution during large deformation; and establishes FE models to predict the infiltration, chemical, stress, and strain fields. Section 4.3 derives the governing equations for the mutually coupled thermal, mechanical, and chemical fields based on large-deformation theory and from the perspective of the laws of thermodynamics, as well as predicts TGO growth and stress distribution based on the results produced by the analytical and FE models.

Y. Zhou et al., *Thermal Barrier Coatings: Failure Theory and Evaluation Technology*,
https://doi.org/10.1007/978-981-19-2723-2_4

4.1 Interfacial Oxidation Phenomenon and Failure

4.1.1 Characteristics and Patterns of Interfacial Oxidation

For a TBC operating at high temperatures, oxygen molecules in air and oxygen atoms in the TC layer inevitably diffuse to the TC/BC interface and react with the Al in the BC layer to form Al_2O_3. The formation of a dense TGO layer can effectively slow the oxidation of the BC layer and improve the oxidation resistance of the coating. However, the continuous thickening of the TGO layer increases the stress in the coating and makes it prone to cracking. Thus, TGOs are a "double-edged sword" in TBCs. When used properly, TGOs can effectively extend a coating's service life; otherwise, TGOs may shorten a coating's service life to well below the expected goal. Hence, it is particularly important to correctly understand the mechanism of interfacial oxidation of TBCs. Extensive research has been conducted in this area.

The main approach used in experimental research on interfacial oxidation is described as follows. First, a TBC specimen is placed in a high-temperature furnace or thermal cycling furnace. Then, an oxidation temperature is set. Subsequently, the specimen is oxidized for various periods of time. Finally, the oxidation pattern of the coating is determined by scanning electron microscopy (SEM) characterization or weighing. Figures 4.1 [1] and 4.2 [2] show SEM microstructures of the interfacial oxidation of TBCs prepared by electron beam-physical vapor deposition (EB-PVD) and atmospheric plasma spraying (APS), respectively. As demonstrated in the two figures, there is a significant increase in the thickness of the TGO layer with the number of cycles or isothermal oxidation time. It is generally agreed that TGO growth follows a parabolic pattern. However, Chen et al. [3] found three stages of TGO growth in their study on high-temperature oxidation of TBCs with a MCrAlY (M = Ni or Ni–Co) BC layer. In the first stage, the TGOs grew rapidly with an approximately parabolic growth curve. In the second stage, the TGOs grew linearly. In the third stage, the TGOs grew at an accelerating rate, and the TGO layer thickened significantly. Research finds that in TBCs with a MCrAlY BC layer, TGO growth always involves these three stages instead of following a parabolic pattern. This is mainly because as the oxidation time increases, the amount of Al continuously decreases, while other metals in the BC layer undergo oxidation to form layers or clusters of Cr and Ni oxides and spinel. These oxides offer no protection against oxidation. Thus, there is a sharp increase in the thickness of the TGO layer once the Al is exhausted. In contrast, TGO growth is relatively stable in a NiAl or PtAl BC layer and follows a parabolic pattern. To deepen the understanding of the mechanism of the interfacial oxidation of TBCs, Poza et al. analyzed the microstructure of TGOs in TBCs subjected to oxidation at 950 °C and 1050 °C for different periods of time with transmission electron microscopy (TEM). They found that the TGOs formed at 950 °C were composed primarily of α-Al_2O_3, whereas those formed at 1050 °C contained other spinel oxides. Experimental research provides a clear understanding of the phenomenon of interfacial oxidation of TBCs and lays a foundation for further mechanistic analysis.

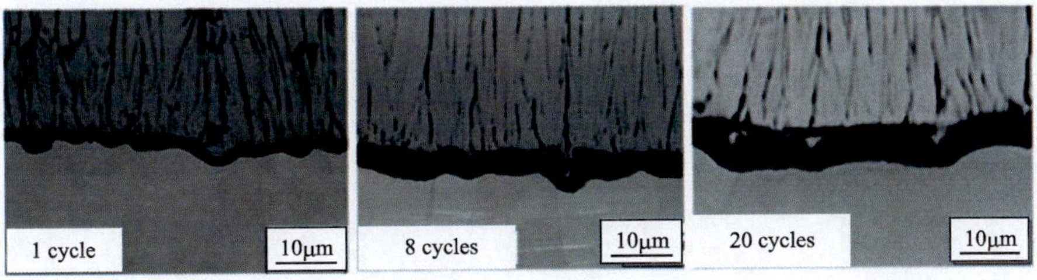

Fig. 4.1 Variation in TGOs in an EB-PVD TBC with oxidation time (each thermal cycle lasted for 60 min; the coating was heated to 1121 °C in 5 min and was then held at this temperature for 45 min, after which the coating was allowed to cool in air for 10 min) [1]

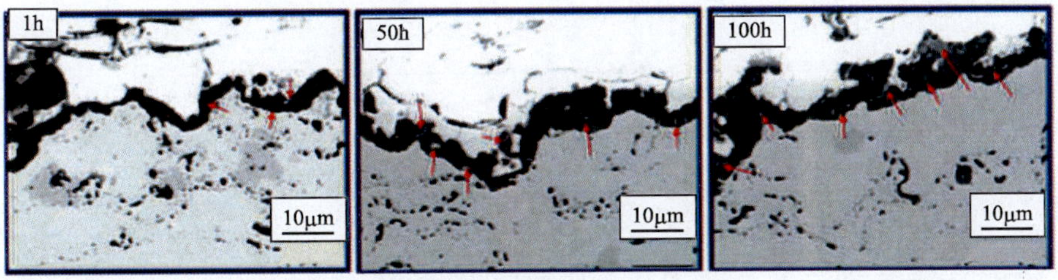

Fig. 4.2 Variation in TGOs in an APS TBC with oxidation time [2]

The main objective of the theoretical analysis of interfacial oxidation is to mathematically describe this process from the perspective of the nature of high-temperature oxidation, to establish geometric models, to solve mathematical equations to generate analytical expressions, and to predict TGO growth patterns. In essence, interfacial oxidation is the diffusion reaction of O and Al, and this phenomenon can be described using diffusion–reaction equations. On this basis, and together with boundary and initial conditions, a theoretical interfacial oxidation model can be mathematically formulated. Based on Wagner's high-temperature oxidation theory, Martena et al. established theoretical isothermal and thermal cycling oxidation models for TGOs and found good agreement between the results produced by these models and experimental measurements [4]. Yuan et al. proposed an oxidation–diffusion model based on interfacial oxidation experiments [5]. This model accounts for both the oxidation on the surface of the BC layer and the mutual diffusion between the BC layer and the substrate and is capable of effectively predicting the diffusion and distribution of each element in a coating. Osorio et al. established a 1D diffusion–reaction model for TGO growth and solved it using a nonsymmetrical radial basis function method in conjunction with the finite difference method; the obtained TGO growth pattern was found to be consistent with the experimental results [6]. In fact, interfacial oxidation of TBCs not only depends on time but also is closely related to temperature. As shown in Fig. 4.3, as the temperature rises, the thickness of the TGO layer increases significantly. In addition, the thickness varies considerably with the oxidation time [7].

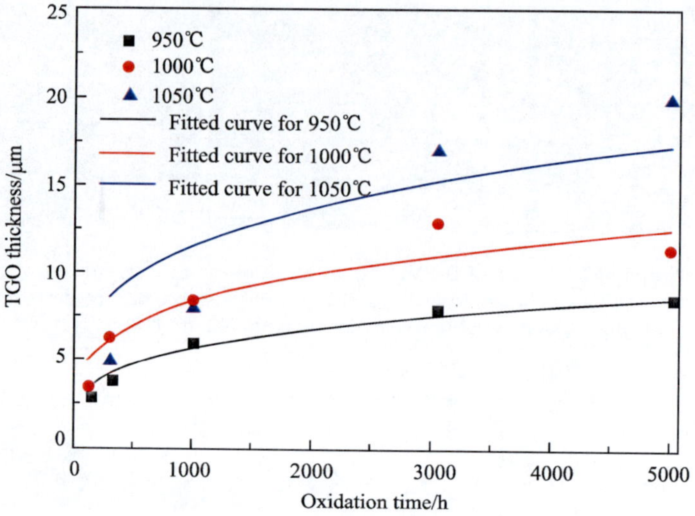

Fig. 4.3 TGO growth patterns at different temperatures [7]

By comprehensively considering temperature and time, Beck et al. [7] obtained the following theoretical equation:

$$h = k_p t^n \tag{4.1}$$

where h is the thickness of the TGO layer, t is the oxidation time, n is the oxidation exponent, and k_p is an oxidation coefficient related to temperature given by

$$k_p = k_0 \exp\left(-\frac{Q}{RT}\right) \tag{4.2}$$

The development of the FEM to simulate the interfacial oxidation of TBCs based on theoretical research can not only substantially reduce financial and material resource demands but also effectively predict the thickness and morphology of the TGO layer. Freborg et al. simulated TGO growth using an element replacement method; this method gradually transforms the elements of the BC layer into TGO elements during the oxidation process [8]. According to the classical Wagner's high-temperature oxidation theory, Freborg et al. included the relationship between TGO thickness and time in an FE program through a subroutine to simulate TGO growth [8]. This method has been widely used by researchers. Well-known researchers in the TBC field, including Evans, [9] Bäker, [10] Busso, [11], and Ranjbar-Far, [12] have all employed this method to simulate TGO growth. From the perspective of the nature of the interfacial oxidation of TBCs, Hill et al. [13] successfully reconstructed the TGO growth process by introducing a TGO volume fraction variable in combination with the diffusion–reaction equation [13]. Following Hill et al.'s success, Gupta et al. simulated TGO growth patterns with true interfacial morphologies [14].

4.1.2 *Stress Field Induced by Interfacial Oxidation*

Stress is the principal factor causing crack initiation and propagation in coatings. Thus, determining the stress distribution in a TBC is the basis for analyzing its failure due to interfacial oxidation. During the interfacial oxidation process, there are two main stress sources in a TBC. (1) The mismatch between the material parameters (e.g., thermal expansion coefficient and elastic modulus) of the TBC layers leads to thermal mismatch stress during cooling. (2) The TGO growth process is accompanied by significant volumetric deformation, which causes growth stress under the constraint of the coating at the interfaces. The combined action of these two types of stress causes coating failure. Researchers have dedicated considerable time to analyzing the stresses in coatings through experimentation, theoretical modeling, and FE calculations, with the goal of obtaining accurate coating stress information and providing a reliable basis for predicting the failure mechanisms and service lives of coatings. In the following, we analyze the achievements made in recent years in stress evolution in TBCs from three aspects, namely experimental characterization, theoretical modeling, and FE simulation.

Currently, there are two main experimental characterization techniques for stress evolution, X-ray diffraction (XRD) and photoluminescence piezospectroscopy (PLPS). XRD primarily measures stress based on the shift of XRD peaks. The presence of compressive stress reduces the interplanar spacing and causes the position of the associated diffraction peak to shift toward larger angles. Conversely, the presence of tensile stress increases the interplanar spacing, thereby shifting the position of the diffraction peak toward smaller angles. The stress in coating specimens can be determined by measuring diffraction peak shifts. Tanaka et al. [15] studied the distribution of the thermal stress in the ceramic layer of a TBC by XRD. Compressive stress was found in the ceramic layer, and the stress increased with the thickness of the ceramic layer. When the distance from the TBC top surface exceeded 40 μm, the stress tended to stabilize at approximately –220 MPa. As X-rays have a limited capacity to penetrate the ceramic layer, this technique is unreliable in the investigation of the stress state in the TGO layer. PLPS is predominantly used to characterize the stress evolution in TGO layers. This technique measures the residual stress in a TGO layer primarily based on the shift of the characteristic spectral peak of the fluorescence emitted by Cr^{3+} ions under the illumination of a laser source that penetrates the ceramic layer. Christensen et al. successfully characterized the stress evolution in the TGO layer of a TBC during high-temperature oxidation by PLPS [16]. Following Christensen et al.'s work, Schlichting et al. analyzed the effects of the thickness of the ceramic layer on the stress in the TGO layer by PLPS [17]. Wen et al. employed PLPS to analyze the evolutionary patterns of the stress in the TGO layer at different temperatures [18]. Wang et al. applied PLPS to stress detection in TBCs on real turbine blades [19]. They determined the evolutionary patterns of the stress in different regions of the TGO layer and found that the stress changed considerably with the curvature of the blade [19]. Today, PLPS is a highly mature technique and offers an indispensable tool for characterizing the stresses in TBCs.

Despite this, further research is required to investigate its applications in blades with both large areas and high curvatures. See Chap. 12 for details.

Theoretical stress evolution models provide important guidance for engineering applications. A simple theoretical model can be used to predict the stress state in a coating and to analyze the potential locations prone to failure. The results can then be fed back to the fabrication process to improve the coating's service life. Based on a simple planar model, Martena et al. analyzed the distribution of the thermal mismatch stress in a coating as a function of the TGO thickness [4]. They found a continuous increase in the tensile stress in the BC layer and a continuous decrease in the compressive stress in the TGO layer with increasing TGO thickness. Evans et al. calculated analytical solutions for the thermal mismatch stress and growth stress in a coating using a circular model. They found that the circumferential thermal mismatch stress in the TGO layer is compressive, that the radial thermal mismatch stress at the TC/TGO interface is tensile, and that the radial thermal mismatch stress in the TC layer transitions from compressive stress to tensile stress as the TGO thickness increases. Figure 4.4 shows the distribution of the thermal mismatch stress at a rough interface. As demonstrated in Fig. 4.4, the stress is tensile at the peak and compressive at the trough. The radial growth stress in the TC layer is compressive, whereas the circumferential growth stress in the TC layer is tensile. The radial growth stress in the TGO layer is compressive, whereas the circumferential growth stress in the TGO layer is tensile and increases as the TGO thickness increases.

Hsueh and Fuller [21] analyzed the stress states at the concave and convex interfaces of a coating by establishing simple concave and convex concentric circular models. They found that as the TGO thickness increases, tensile stress appears in both the trough region of the TC layer (the concave model) and the peak region of the BC layer (the convex model). Based on a spherical model, Sun et al. [22] examined the effects of interfacial roughness, TGO creep, and TGO growth on the growth stress in a coating. They found that the growth stress depends on TGO growth during the initial stage of formation, whereas creep plays a dominant role in the stress evolution during the later stage. Subsequently, they investigated the effects of anisotropic TGO growth on the stress distribution in a coating [23]. They found that the radial stress at both the TGO/BC and TGO/TC interfaces is tensile in the presence of a large circumferential growth strain in the TGO layer and that the radial stress at both the TGO/BC and TGO/TC interfaces is compressive when the radial growth strain plays a dominant role in the TGO layer. Theoretical stress evolution models all determine stress distributions in coatings based on simple geometric models in conjunction with the constitutive and equilibrium equations and boundary conditions. However, the complex geometric structure of TBCs necessitates the use of the FEM.

A multitude of factors, including TGO growth, interfacial morphology, TGO thickness, high-temperature creep, and material parameters, can affect the stress evolution in a TBC during interfacial oxidation. As a result, it is impossible to account for all the factors by relying solely on theoretical modeling. Thus, researchers across the world often employ the FEM to analyze the stress state in a coating. When simulating the stress distribution in a coating, sinusoidal functions or semicircles are generally used to represent the rough interfaces. Ranjbar-Far et al. [12] established FE

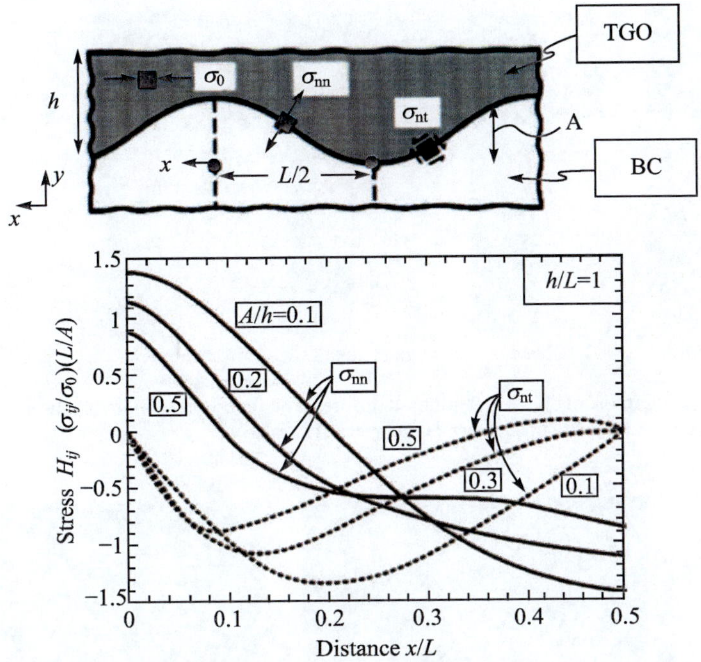

Fig. 4.4 Distribution of the thermal mismatch stress at the TGO/TC interface (H_{ij} is a dimensionless stress component; σ_0 is the planar thermal mismatch stress; σ_{nn} is the tensile stress perpendicular to the interfacial direction; σ_{nt} is the shear stress parallel to the interface; A is the interfacial amplitude; L is the wavelength; h is the TGO thickness)

models to describe semicircular and sinusoidal interfacial morphologies and, on this basis, examined the effects of several factors, including TGO growth, material creep in each layer at high temperatures, and the plastic deformation of the BC layer, on the stress distribution in a coating. They determined the σ_{22} distributions at different roughnesses, as shown in Fig. 4.5. As the amplitude increases from 5 to 30 μm, the maximum tensile stress increases from 87 to 290 MPa, and the maximum compressive stress increases from –93 to –350 MPa. Yang et al. [24] established a 3D FE model for a TBC on a turbine blade, with which they analyzed the stress distribution in the blade under thermal cycling and predicted spallation locations. Following Yang et al.'s study, [24] Zhu et al. [25] further examined the effects of the morphology of the TGO layer on the stress distribution in an actual blade; they concluded that the stress at the TGO/BC interface was tensile at peaks and compressive at troughs and that the stress increased significantly as the roughness increased. Owing to the development of image processing technology, it is now possible to convert SEM images into geometric models in the FE modeling of coatings. This technique is crucially important for studying the true stress state in a coating. Using this technique, Gupta et al. [14] successfully imported the true interfacial morphology of a TBC into an FE model and analyzed the corresponding TGO growth and growth stress

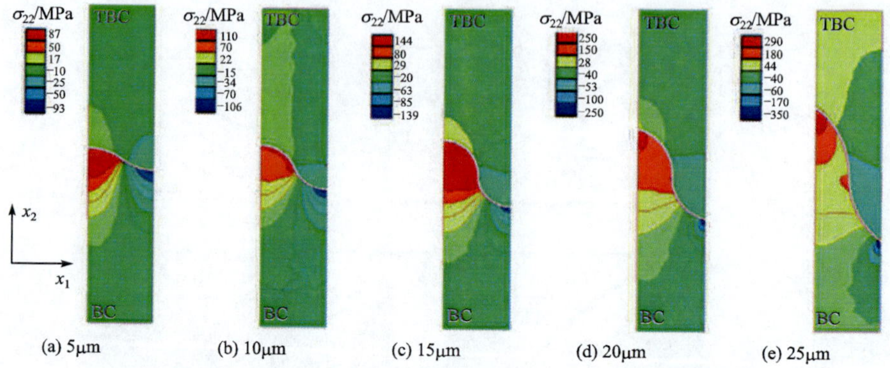

Fig. 4.5 Contour plots of σ_{22} distributions at different roughnesses (semicircular amplitudes) after one thermal cycle without considering TGO growth [12]

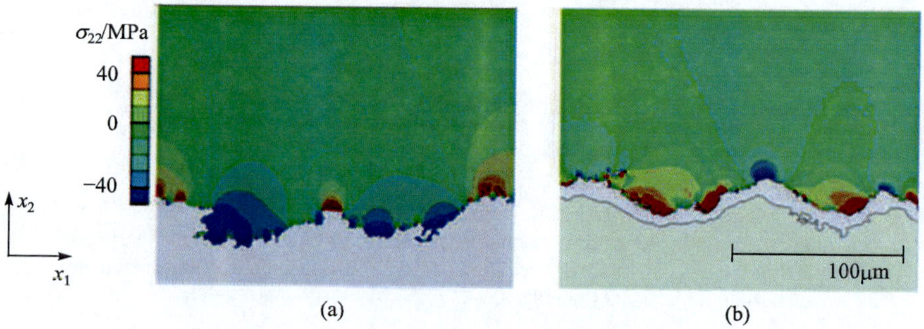

Fig. 4.6 Contour plots of σ_{22} distributions considering true interfacial morphology [14]

distribution; they found large stresses at the TGO/TC interface, as shown in Fig. 4.6. Nayebpashaee et al. [26] used this technique to investigate the effects of different oxides on the stress in coatings. Wang et al. reviewed the applications of the FEM in the TBC field. Bäker and Seiler [27] presented problems and development directions in the application of the FEM in the TBC field. The above research works offer important guidance for the FE simulation of the stress evolution in TBCs.

4.1.3 Coating Spallation Induced by Interfacial Oxidation

As the main cause of TBC failure, interfacial oxidation has long garnered the attention of researchers across the world, many of whom have extensively studied the failure mechanisms caused by interfacial oxidation. The difference between APS and EB-PVD, which are two TBC preparation processes, results in differences in both the morphology of the TGO layer and the failure mechanism of the TBCs prepared by these two processes. Researchers have examined the interfacial-oxidation-induced failure of these two types of TBCs and determined the corresponding failure mechanisms.

As shown in Fig. 4.7a, there are four main failure modes for APS TBCs [28]. Failure mode I: During cooling, the BC layer is in a tensile stress state at peaks. As the TGO thickens, the tensile stress further increases, resulting in crack initiation and propagation at peaks of the TGO/BC interface. Failure mode II: The mismatch in the thermal expansion coefficients and elastic moduli of the TC and TGO layers produces tensile stress at peaks of the TC layer, which, in turn, leads to crack formation at the TGO/TC interface. Failure mode III: Cracks are formed in the ceramic layer near peaks when the tensile stress exceeds the cracking strength of the ceramic layer. Failure mode IV: When the stress at the crack tips at peaks of the TGO/BC interface exceeds the critical stress of the TGO layer, the cracks propagate in and penetrate the TGO layer and extend to the TC layer.

Compared to APS TBCs, EB-PVD TBCs have a columnar crystal structure and a high strain tolerance. Cracks do not easily propagate in the ceramic layer of the TBC but generally appear at the TC/TGO or BC/TGO interface, as shown in Fig. 4.7b. There are three failure modes for EB-PVD TBCs. Failure mode I: Similar to failure model I of APS TBCs, after cooling, the peaks of the BC layer are subjected to tensile stress, which increases as the TGO thickness increases; after the tensile stress increases to a certain level, cracks initiate and propagate along the TGO/BC interface. Failure mode II: Irregular TGO growth leads to significant interfacial oxidation of local areas and the formation of a very large stress, causing cracks to propagate. Failure mode III: The interfaces in an EB-PVD TBC are relatively smooth, and the compressive stress in the TGO layer causes the TBC to buckle toward the BC layer.

As shown in Fig. 4.7, the failure locations of the APS TBC and the EB-PVD TBC are both near the TGO layer. Thus, the TGO growth and stress evolution during this process are the key factors causing interfacial oxidation failure in TBCs. Hence, elucidation of the TGO growth pattern and stress evolution in TBCs is the basis for accurately understanding TBC failure induced by interfacial oxidation.

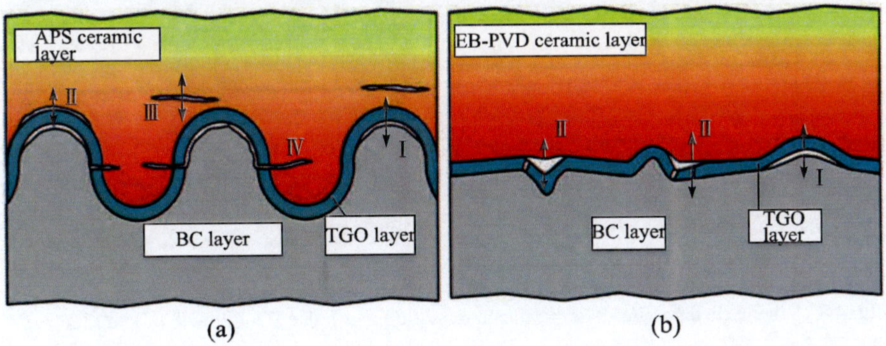

Fig. 4.7 Schematic diagrams of the failure modes of **a** an APS TBC and **b** an EB-PVD TBC

4.2 TGO Growth Model Based on Diffusion Reaction

4.2.1 Governing Equations

(1) Governing equations for TGO growth

Figure 4.8a shows the interfacial oxidation process in a TBC. At high temperatures, oxygen from the air diffuses to the TC/BC interface via the TC layer and reacts with the Al in the BC layer, forming Al_2O_3. As oxygen diffuses much faster than Al in the TGO layer, the reaction is considered to occur at the TGO/BC interface and only at high temperatures, i.e., the high-temperature oxidation reaction occurs only at the temperature-holding stage during thermal cycling. Hille et al. [13] established a TGO growth model based on the diffusion–reaction equation. This model assumes that TGO growth is a process in which the BC layer is continuously oxidized, as shown in Fig. 4.8b. A variable n is introduced to represent the volume fraction and hence the growth process of TGOs. $n = 1$ and $n = 0$ indicate that this region is composed completely of TGO and the BC layer, respectively, and $0 < n < 1$ indicates that this region consists of a mixture of TGO and the BC layer. Thus, the volume fraction of the BC layer is given by $(1 - n)$. In other words, the total volume fraction of the TGO and BC layers must be 1 in any region. This approach is conceptually similar to the use of order parameters in the phase-field method to describe the microstructural evolution of a material. The phase-field method has been used to study the evolution reaction of materials.

The formation of TGOs is controlled by the effective reaction between oxygen and Al. The distribution of the oxygen concentration c is governed by the following diffusion–reaction equation [13].

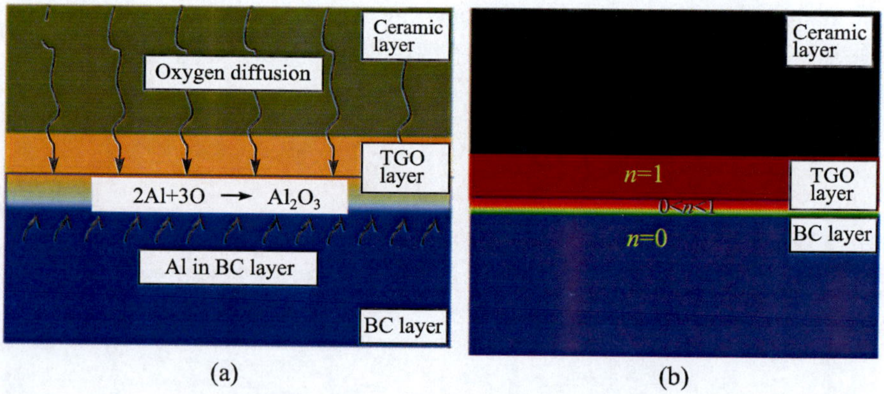

Fig. 4.8 Schematic diagrams of **a** interfacial oxidation and **b** the TGO growth model

$$\frac{\partial c}{\partial t} - \nabla \cdot (D \nabla c) = -S \tag{4.3}$$

where D is the diffusion coefficient of oxygen and S is the amount of oxygen consumed by the reaction (often referred to as the reaction term). Similarly, there should be a diffusion–reaction equation for Al. However, considering that TGO growth depends mainly on oxygen, Al diffusion is neglected here. The reaction term can be represented by a reaction model as follows. Let us assume that the formation of TGOs is directly proportional to the oxygen and Al concentrations and, in particular, that the Al concentration is equal to the volume fraction of the BC layer, i.e., $(1 - n)$. Thus, the TGO reaction is given by

$$\frac{\partial n}{\partial t} = \zeta (1 - n) c \tag{4.4}$$

where ζ is the TGO reaction rate constant. The reaction term and the formation of TGOs follow the relationship:

$$S = M \frac{\partial n}{\partial t} \tag{4.5}$$

where M is the molar concentration of oxygen in Al_2O_3. Thus, the following governing equations for the oxygen concentration c and the TGO volume fraction n during the TGO growth process are derived from Eqs. (4.3)–(4.5):

$$\begin{cases} \dfrac{\partial c}{\partial t} - \nabla \cdot (D \nabla c) = -M \zeta (1 - n) c \\ \dfrac{\partial n}{\partial t} = \zeta (1 - n) c \end{cases} \tag{4.6}$$

The concentration or flux at the boundary needs to be defined as the boundary condition for the diffusion–reaction equation. As oxygen diffuses much faster in the TC layer than in the TGO layer, the boundary conditions, excluding c, for TGO formation are given directly at the surface of the TGO layer. As a TBC is oxidized during the preparation process, it is necessary to define a very thin TGO layer for the initial stage, i.e., $n = 1$. Table 4.1 summarizes the performance parameters involved in the TGO growth model.

(2) **Governing equations for stress evolution during large deformation**

Interfacial oxidation of TBCs is accompanied by large deformation. In large-deformation analysis, it is necessary to first define a deformation gradient to describe the deformation state of the material. Let us set the pre- and post-oxidation states

Table 4.1 Material parameters of the TGO growth model [13]

	BC	TGO
Diffusion coefficient of oxygen ($m^2\ s^{-1}$)	2.0×10^{-14}	2.0×10^{-14}
Reaction rate of oxygen ($m^3\ mol^{-1}\ s^{-1}$)		1.0×10^{-5}
Molar concentration of oxygen ($mol\ m^{-3}$)		1.11×10^{5}

of a TBC as its reference and current configurations, respectively. In addition, let $\boldsymbol{X}$ and $\boldsymbol{x}$ be the coordinates of an arbitrary material point in the reference and current configurations, respectively. Thus, the deformation gradient $\boldsymbol{F}$ is defined as follows:

$$\boldsymbol{F} = \frac{\partial \boldsymbol{x}}{\partial \boldsymbol{X}} = \boldsymbol{F}^{\mathrm{e}} \cdot \boldsymbol{G} \tag{4.7}$$

where $\boldsymbol{F}^{\mathrm{e}}$ is the elastic deformation gradient and $\boldsymbol{G}$ is the growth deformation gradient. The deformation gradient is decomposed under the assumption that there exists an intermediary configuration that allows unrestrained TGO growth, i.e., a configuration in a zero-stress state. This approach draws on the large-deformation analysis method used by researchers to investigate the growth of biological tissues and cells [29]. In fact, this decomposition of the deformation gradient has, for a long time, been applied in the analysis of large elastoplastic deformation and has recently been widely used in the large-deformation analysis of multifield coupling. For example, N'Guyen et al. [30] decomposed the deformation gradient into an elastic deformation gradient and an inelastic deformation gradient and, on this basis, studied thermo–mechano chemical coupling in soft materials under large deformation. Yadegari et al. [31] decomposed the total deformation gradient into elastic, thermal, plastic, and transformation deformation gradients and, on this basis, examined the multifield coupling problem of multiphase steels.

The deformation of a material can be represented by the Green–Lagrange strain tensor $\boldsymbol{E}$, which is related to the deformation gradient $\boldsymbol{F}$ by

$$\begin{aligned} \boldsymbol{E} &= \frac{1}{2}(\boldsymbol{F}^{\mathrm{T}} \cdot \boldsymbol{F} - \boldsymbol{I}) = \frac{1}{2}[(\boldsymbol{I} + \nabla_0 \boldsymbol{u}) \cdot (\boldsymbol{I} + \nabla_0 \boldsymbol{u})^{\mathrm{T}} - \boldsymbol{I}] \\ &= \frac{1}{2}[(\nabla_0 \boldsymbol{u}) + (\nabla_0 \boldsymbol{u})^{\mathrm{T}} + (\nabla_0 \boldsymbol{u}) \cdot (\nabla_0 \boldsymbol{u})^{\mathrm{T}}] \end{aligned} \tag{4.8}$$

where $\boldsymbol{F}^{\mathrm{T}}$ is the transpose of the deformation gradient $\boldsymbol{F}$, $\boldsymbol{I}$ is the second-order unit tensor, $\boldsymbol{u}$ is the displacement of the material, and ∇_0 is the gradient operator in the reference configuration.

In a 2D case, the components of the Green–Lagrange strain tensor $\boldsymbol{E}$ take the following forms:

$$
\begin{aligned}
E_{11} &= \frac{\partial u_1}{\partial X_1} + \frac{1}{2}[(\frac{\partial u_1}{\partial X_1})^2 + (\frac{\partial u_2}{\partial X_1})^2] \\
E_{22} &= \frac{\partial u_2}{\partial X_2} + \frac{1}{2}[(\frac{\partial u_1}{\partial X_2})^2 + (\frac{\partial u_2}{\partial X_2})^2] \\
E_{12} = E_{21} &= \frac{1}{2}[\frac{\partial u_1}{\partial X_2} + \frac{\partial u_2}{\partial X_1} + \frac{\partial u_1}{\partial X_1}\frac{\partial u_1}{\partial X_2} + \frac{\partial u_2}{\partial X_1}\frac{\partial u_2}{\partial X_2}]
\end{aligned} \tag{4.9}
$$

With the volume and inertial forces neglected, the force equilibrium equation for large deformation is given by

$$Div\, \boldsymbol{P} = 0 \tag{4.10}$$

where $\boldsymbol{P}$ is the first Piola–Kirchhoff stress, and Div is the divergence operator in the reference configuration. The first Piola–Kirchhoff stress can be transformed through the deformation gradient into the second Piola–Kirchhoff stress. The relevant relationship is expressed as

$$\boldsymbol{P} = \boldsymbol{T} \cdot \boldsymbol{F}^{\mathrm{T}} \tag{4.11}$$

Thus, the force equilibrium equation in Eq. (4.10) can be expressed with the second Piola–Kirchhoff stress and the deformation gradient as follows:

$$Div\, \boldsymbol{T} \cdot \boldsymbol{F}^{\mathrm{T}} = 0 \tag{4.12}$$

During interfacial oxidation of a TBC, the relationship between the second Piola–Kirchhoff stress and the elastic Green–Lagrange strain tensor $\boldsymbol{E}^{\mathrm{e}}$ in the TGO layer is determined by the constitutive relation of the TGO layer. Based on the first and second laws of thermodynamics, the following constitutive relation of the TGO layer is obtained:

$$\boldsymbol{T} = \frac{\partial \psi}{\partial \boldsymbol{E}^{\mathrm{e}}} \tag{4.13}$$

where ψ is the total Helmholtz free energy of the system.

The high-temperature oxidation reaction equation for a metal is given by

$$M + O_2 \rightarrow M_a O_b \tag{4.14}$$

During the oxidation process, the crystal structure of the metal changes, causing the object to undergo internal deformation. The degree of deformation is represented by the ratio of the volume after deformation to that before deformation, $J = \det(\boldsymbol{F})$, which is referred to as the Pilling–Bedworth ratio (PBR). The PBR is expressed as

$$J_{PB} = \frac{V_{M_aO_b}}{aV_M} = \frac{M_{M_aO_b}\rho_M}{a\rho_{M_aO_b}M_M} \tag{4.15}$$

Based on the mechano-chemical coupling constitutive relation established by Cui et al. [32] for large deformation in lithium batteries, the free energy is rewritten in the form shown in Eq. (4.16), where the growth strain induced by oxidation is defined using the PBR in Eq. (4.15) and the elastic strain energy is defined using a hyperelastic model, in contrast to the hypoelastic model used by Cui et al.

$$\psi = J^G\left[\frac{1}{2}\mu(I^e - 3) + \lambda(\ln J^e)^2 - \mu \ln J^e\right] \tag{4.16}$$

where μ and λ are Lame's constants, I^e is the first invariant of the elastic Green–Lagrange strain tensor $\boldsymbol{E}^e$, $J^e = \det(\boldsymbol{F}^e)$ is the Jacobian determinant of the elastic deformation gradient tensor, and $J^G = \det(\boldsymbol{G})$ is the Jacobian determinant of the growth deformation gradient tensor. During the oxidation reaction process, the BC layer gradually undergoes oxidation and transforms into TGO. Accordingly, in the established TGO growth model, a value of n between 0 and 1 means that there is a mixed layer between the BC and TGO layers. Thus, Lame's constants μ and λ are given by

$$\begin{aligned} \mu &= n \cdot \mu_{\text{TGO}} + (1-n) \cdot \mu_{\text{BC}} \\ \lambda &= n \cdot \lambda_{\text{TGO}} + (1-n) \cdot \lambda_{\text{BC}} \end{aligned} \tag{4.17}$$

where μ_{TGO} and λ_{TGO} are Lame's constants of the TGO layer and μ_{BC} and λ_{BC} are Lame's constants of the BC layer.

The Jacobian determinant J^G of the growth deformation gradient tensor is derived from the PBR. Here, to facilitate calculation, J^G is expressed as follows:

$$J^G = [1 + \beta(c - c_0)]^3 \tag{4.18}$$

where c is the oxygen concentration, c_0 is the reference concentration, and β is a constant. This chapter mainly considers the expansion deformation of TBCs caused by oxygen diffusion. Later in this book, research results and comparisons with experimental data show that compared to that caused by TGO growth deformation, the expansion deformation of TBCs due to oxygen diffusion is small or not the main cause of the oxidation failure of TBCs. The effects of TGO growth deformation on TBCs are investigated in detail in this chapter.

Based on Eqs. (4.13) and (4.16), the following expression of the second Piola–Kirchhoff stress in a TBC during its interfacial oxidation process is obtained:

$$T = J^G \left[\mu I + \left(\lambda \ln J^e - \mu \right) \overset{-1}{C^e} \right] \tag{4.19}$$

where I is the unit tensor and $\overset{-1}{C^e}$ is the elastic deformation tensor, which is given by

$$\overset{-1}{C^e} = \overset{-1}{F^e} \cdot \overset{-1}{F^{eT}} \tag{4.20}$$

where $\overset{-1}{F^e}$ is the inverse of the elastic deformation gradient tensor and $\overset{-1}{F^{eT}}$ is the inverse of the transpose of the elastic deformation gradient tensor.

The above focuses on an introduction to the derivation of the governing equations for the stress field and constitutive relations with respect to the interfacial oxidation of TBCs based on large-deformation theory. In the following, we briefly analyze the governing equations for the stress field and constitutive relations in the case of small deformation. Let ε^{ox} be the growth strain resulting from TGO growth. Thus, the total strain ε can be represented by a superposition of the elastic strain ε^{el} and the growth strain ε^{ox} as follows:

$$\varepsilon = \varepsilon^{el} + \varepsilon^{ox} \tag{4.21}$$

For comparison, the governing equations for the stress field and the constitutive relations under small deformation are also analyzed. Extensive research has been conducted in this area. Only a brief introduction is presented here, and the detailed derivation can be found in the work of Zhou et al. [33, 34]. Interfacial oxidation of a TBC is a long-term, slow process that can be viewed as quasi-static, so inertial forces can be ignored. The force equilibrium equation under small deformation and without considering the volume force is expressed as

$$div\sigma = 0 \tag{4.22}$$

Under small deformation, the constitutive relation of a coating is given by

$$\sigma = C : (\varepsilon - \varepsilon^{ox}) = C : [\varepsilon - \beta(c - c_0) I] \tag{4.23}$$

where C is the elastic stiffness tensor.

Equations (4.12) and (4.19) constitute the force equilibrium and constitutive equations, respectively, under large deformation, while Eqs. (4.22) and (4.23) are the force equilibrium and constitutive equations, respectively, for the case of small deformation. The stress distribution under small or large deformation can be determined once the boundary conditions are given.

4.2.2 FE Simulation

(1) **FE simulation of large deformation induced by TGO growth**

It is nearly impossible to solve the above equations for real-world problems associated with interfaces in TBCs. Here, we employ the FEM to analyze the TGO growth and stress evolution in coatings. Currently, owing to its openness to users, the FE software COMSOL allows researchers to compile their own theories in the corresponding weak forms and include them in the software to solve complex theoretical equations. Please refer to Chap. 2 for the detailed theoretical basis. Here, the results are directly provided.

Trial functions are introduced to transform the TGO growth model and the strong forms of the governing equations for the stress field under large deformation into the corresponding weak forms. The main variables in the TGO growth model include the oxygen concentration c and the TGO volume fraction n. Let δc and δn be their respective trial functions. The displacement $\boldsymbol{u}$ is the main variable of the stress field and can be used to determine other variables of the stress field. Let δu_i be the trial function of displacement. In the case of large deformation, Eq. (4.6) is transformed by integration into the corresponding weak form as follows:

$$\int_{\Omega} (\delta c \frac{\partial c}{\partial t} + D\frac{\partial c}{\partial x}\frac{\partial \delta c}{\partial x} + D\frac{\partial c}{\partial y}\frac{\partial \delta c}{\partial y} + \delta c M\zeta(1-n)c)d\Omega = 0 \tag{4.24}$$

$$\int_{\Omega} (\delta n \frac{\partial n}{\partial t} - \delta n \zeta(1-n)c)d\Omega = 0 \tag{4.25}$$

Equation (4.12) is transformed by integration into the following weak form:

$$\begin{aligned} \int_{\Omega} \nabla \cdot (\boldsymbol{F}^T \cdot \boldsymbol{T}) \cdot \delta \boldsymbol{u} d\Omega &= \int_{\Omega} [-\delta \nabla u \cdot (\boldsymbol{F}^T \cdot \boldsymbol{T}) + \nabla \cdot (\boldsymbol{F}^T \cdot \boldsymbol{T} \cdot \delta \boldsymbol{u})] d\Omega \\ &= \int_{\Omega} [-(\delta \nabla \boldsymbol{u} + \nabla \boldsymbol{u} \cdot \delta \nabla \boldsymbol{u}) \cdot \boldsymbol{T}] d\Omega + \int_{\Gamma} \boldsymbol{n} \cdot (\boldsymbol{F}^T \cdot \boldsymbol{T} \cdot \delta \boldsymbol{u}) d\Gamma = 0 \end{aligned} \tag{4.26a}$$

Considering that there are no external forces acting on the boundaries, the second term in Eq. (4.26a) can be omitted to yield

$$\int_{\Omega} [-(\delta \nabla \boldsymbol{u} + \nabla \boldsymbol{u} \cdot \delta \nabla \boldsymbol{u}) \cdot \boldsymbol{T}] d\Omega = 0 \tag{4.26b}$$

For small deformation, Eq. (4.22) can be transformed into the following weak form:

$$\int_{\Omega} \delta\varepsilon_{ij}\sigma_{ij}d\Omega = 0 \tag{4.27}$$

Thus, Eqs. (4.24), (4.25), and (4.26) are the weak forms for the case of large deformation, while Eqs. (4.24), (4.25), and (4.27) are the weak forms under small-deformation conditions. The weak forms are then implemented in the weak-form module of the mathematical module in COMSOL, and the variables to be solved (i.e., the oxygen concentration c and the TGO volume fraction n) and the 2D displacement components u and v are defined. Subsequently, a geometric model for the interfacial oxidation of the TBC is generated and solved in COMSOL under the given boundary conditions for the oxidation process to determine the TGO growth and stress evolution in the coating.

The oxidation interfaces in TBCs are generally uneven. Figure 4.9a shows a schematic diagram of a geometric model established based on common sinusoidal interfaces. To simplify the calculation, this geometric model is generated without considering the substrate because the substrate has a very insignificant impact on the oxidation reaction that occurs mainly at the TGO/BC interface. This geometric model consists of a 200-μm-thick TC layer, a 100-μm-thick BC layer, and a 1-μm-thick TGO layer. The sinusoidal TGO interfaces (TGO/BC and TGO/TC) each have an amplitude of 15 μm and a wavelength of 60 μm. Figure 4.9b shows an FE mesh composed of triangular plane strain elements generated for the geometric model. The mesh for the key TGO growth regions, particularly the sinusoidal interfaces (e.g., peaks and troughs), is refined to ensure an accurate solution. In addition, an adaptive mesh is used in the simulation of the oxidation process to effectively solve the problem of nonconvergence caused by mesh distortion resulting from the large strain accompanying the TGO growth process. See Chap. 2 for details on meshing.

The boundary conditions of the FE model consist of those of the TGO growth model and those of the stress field. The established geometric model reflects one cycle. Periodic boundary conditions are imposed on the two boundaries of the model,

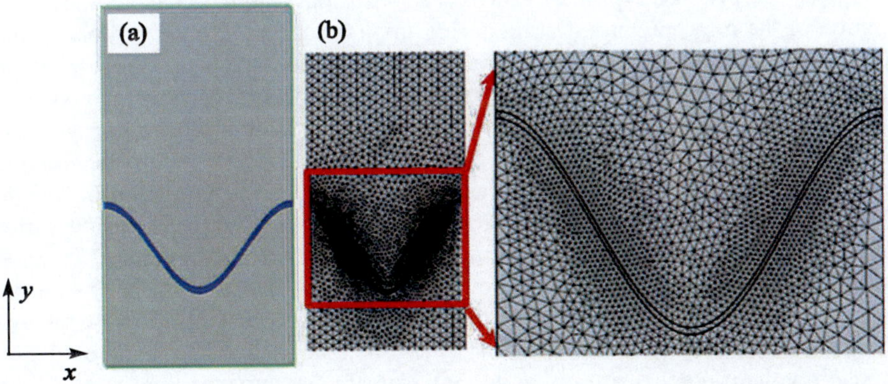

Fig. 4.9 **a** A geometric model of TGO growth. **b** An FE mesh generated for the geometric model

Table 4.2 Mechanical property parameters of the TBC [25]

	BC	TGO	TC
Young's modulus (GPa)	110	320	22
Poisson's ratio	0.33	0.25	0.12

i.e., $u(x=0, y)=u(x=60, y)$; the bottom of the model is fixed along the y-direction, i.e., $v=0$; the left boundary is symmetric; and the model has a free surface. n is set to 1 and 0 for TGO and BC layers, respectively. The oxygen concentration is set to 0 and 1.55 mol/m^3 at the bottom of the model (i.e., $c(x, y=0)=0$) and the surface of the TGO layer (i.e., $c(x, y=300)=1.55$ mol/m^3), respectively, and the initial oxygen concentration is set to 0. The oxidation temperature and time are set to 1000 °C and 1000 h, respectively. Table 4.2 summarizes the mechanical properties of the material of each layer of the TBC at 1000 °C.

(2) **Evolutionary pattern of the morphology of the TGO layer**

Figure 4.10 shows the evolutionary pattern of the morphology of the TGO layer with time (the TGO layer is shown in red with $n=1$, the BC layer is shown in blue with $n=0$, and there is a mixed layer between the red and blue layers with n between 0 and 1). As the oxidation time extends (10, 100, 500, and 1000 h of oxidation), the TGO thickness increases continuously but to different extents at the peak (1.3, 2.3, 7.0, and 11.2 μm) and at the trough (1.2, 1.9, 4.8, and 7.0 μm). Clearly, TGOs grow much faster and hence the TGO layer is thicker at the peak than at the trough. The nonuniform TGO growth increases the roughness of the coating, thereby increasing the stress in the coating. In their experimental studies on interfacial oxidation of TBCs, Zhou and Cherkaoui [33] and Hsueh and Ruller [35] found that coatings were thicker at peaks than at troughs, a finding that is consistent with the results of the above FE simulation. Thus, establishing a TGO growth model can help to more accurately determine the interfacial oxidation pattern in a TBC, which is not possible with a parabolic model.

TGO growth depends primarily on oxygen diffusion. Figure 4.11 shows the distribution of the oxygen concentration during the interfacial evolution process in the TBC. The maximum oxygen concentration is 1.55 mol/m^3 and mainly appears at the upper surface of the TGO layer (i.e., the red region). The minimum oxygen concentration is 0 and appears in the blue region. As the oxidation reaction progresses, oxygen continuously diffuses toward the BC layer while reacting with the Al in the BC layer. It can be found that interfacial oxidation (i.e., the formation of TGOs) can occur only in the regions to which oxygen diffuses. Oxygen diffusion varies significantly between locations. Specifically, oxygen diffuses faster at peaks than at troughs, thereby causing TGOs to grow faster at peaks than at troughs. The curvature of the interface is the main cause of the difference between the diffusion rates of the peaks and troughs. Zhou et al. [34] established theoretical interfacial oxidation models with different curvatures and found that a higher curvature leads to a higher oxidation rate. This is because the oxygen concentration is the highest at sharp peaks,

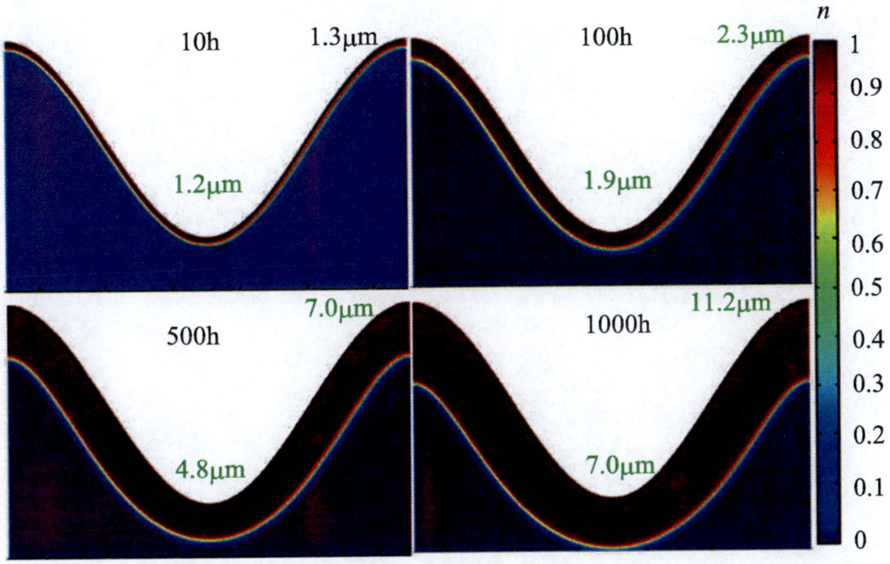

Fig. 4.10 TGO growth pattern

followed by that on planes, and that at troughs is the lowest. Consequently, oxygen diffuses faster, and thus, the TGO layer is thicker at peaks than at troughs.

Figure 4.12 shows the variation in the TGO thickness of the $n = 1$ region with time at different locations. The black and red curves represent the variation in the TGO thicknesses at the peak and the trough, respectively, with oxidation time, the blue curve shows the variation in the average of the TGO thicknesses at the peak and trough with oxidation time, and the scatter points represent experimental results from the literature [7, 36, 37]. As demonstrated in Fig. 4.12, regardless of whether at the peak or the trough, TGO growth always follows a nearly parabolic pattern, mainly because interfacial oxidation conforms to the diffusion equation. It can be directly seen that at 1000 h of oxidation, the TGO layer at the peak is approximately 4 μm thicker than that at the trough, and such a significant difference in thickness is highly likely to cause spallation of the coating.

(3) **Evolutionary pattern of the stress field in a coating**

Interfacial oxidation induces high growth stress in a TBC. It is generally believed that the stress in the y-direction is the primary cause of crack initiation and propagation. Thus, in the following, we mainly analyze the distribution and evolutionary patterns of the stress in the y-direction and focus on the stress distribution under large deformation. To facilitate comparison, the stress state in the current configuration, i.e., the Cauchy stress, is analyzed for the case of large deformation. Figure 4.13 shows the contour plots of the Cauchy stress in the y-direction, σ_{yy}, in the BC and TGO layers determined based on the large-deformation theory (the tensile and compressive stresses are shown in red and blue, respectively). There is tensile stress at the

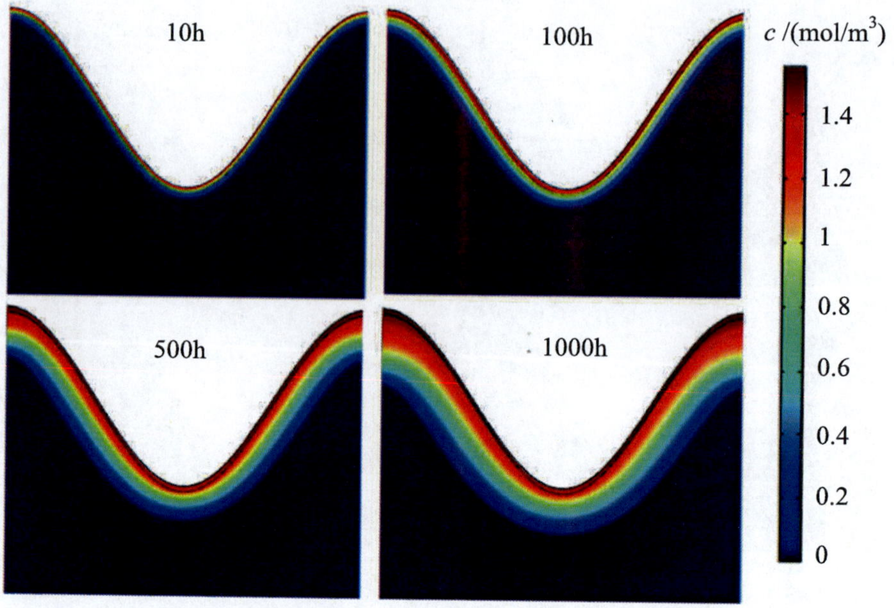

Fig. 4.11 Distribution of oxygen concentration

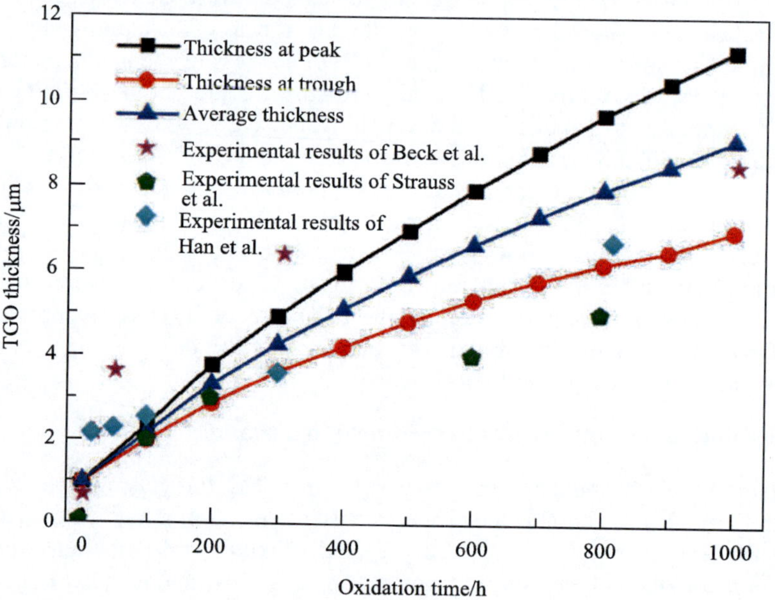

Fig. 4.12 Variation in TGO thickness with time

peak and compressive stress in the middle region between the peak and the trough. As a result of the change in the curvature of the interface, σ_{yy} in the BC and TGO layers transitions from tensile stress at the peak to compressive stress at the trough. As the TGO thickness increases, the maximum tensile stress in the BC and TGO layers increases from 409 to 1178 MPa, whereas the maximum compressive stress decreases from –4.0 to –2.9 GPa. A similar stress transition pattern was previously discovered by Sun et al. [22] In addition, the maximum compressive stress always occurs in the middle region, whereas the maximum tensile stress appears at the peak during the initial stage of oxidation and moves to the middle region as the oxidation process progresses, which may be a result of nonuniform TGO growth. The distribution of σ_{yy} in the TC layer is shown in Fig. 4.14, where red and blue regions are located at the trough and the peak, respectively, suggesting that σ_{yy} in the TC layer is tensile at the trough and compressive at the peak. As the oxidation time extends, the maximum tensile stress increases from 50 to 174 MPa, while the maximum compressive stress increases from –139 to –383 MPa. Based on the large-deformation theory, Al-Athel et al. [38] studied the stress distribution in coatings during the anisotropic TGO growth process and found that the maximum tensile and compressive stresses in the TC layer were 58 and –235 MPa, respectively.

The stress distributions under small deformation are similar to those under large deformation; only the magnitudes of the stress differ. Thus, the stress distributions under small deformation are not described here. The variation in the in-plane stress

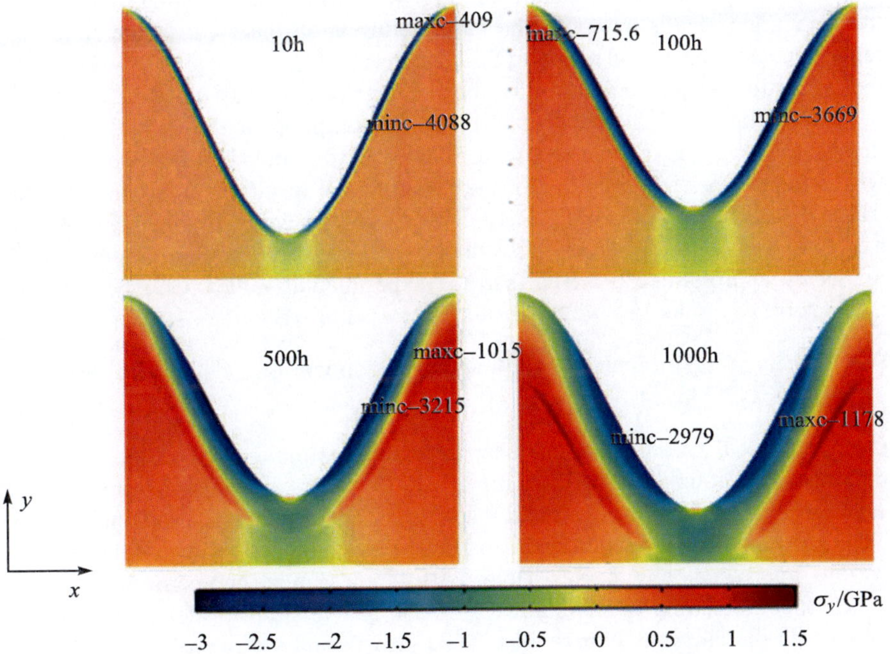

Fig. 4.13 Distribution of the *y*-direction stress in the TGO and BC layers

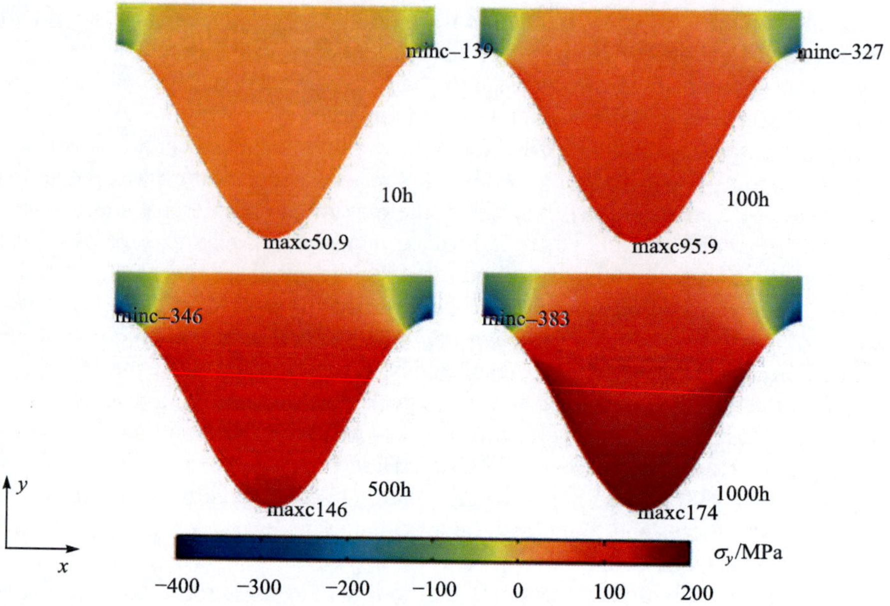

Fig. 4.14 Distribution of the *y*-direction stress in the TC layer

in the TGO layer with time is shown in Fig. 4.15, where the red solid squares show the simulation results based on the large-deformation theory, the blue solid stars represent the experimental results of Sridharan et al. [39], and the pink solid circles are the simulation results obtained under the small-deformation assumption. Regardless of whether based on the small-deformation assumption or the large-deformation theory, the stress first increases with oxidation time and then tends to stabilize, with a trend consistent with that of the experimental results. In addition, the stress values based on the large-deformation theory are in general agreement with the experimental values, whereas those under the small-deformation assumption are higher by approximately 1 GPa than the experimental values. Clearly, the stress analysis based on the large-deformation theory is more accurate.

(4) **Effects of the interfacial roughness on the morphology and stress evolution of the TGO layer**

Over the past decades, there has been a continuous recognition in the TBC field that interfacial roughness significantly affects the stress evolution and the extent of TGO growth in a coating. To investigate the effects of interfacial roughness on the TGO growth and stress evolution in coatings, we established geometric models with various interfacial roughnesses. Here, let us look at nine geometric models with different combinations of amplitude (5, 10, and 15 μm) and wavelength (40, 60, and 80 μm) in the sinusoidal curve. The TGO growth and stress evolution in each of these nine models are calculated.

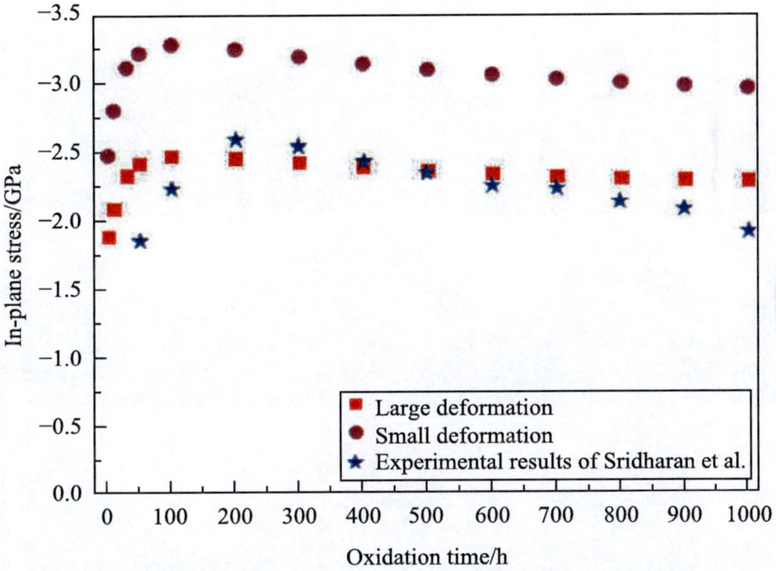

Fig. 4.15 Comparison of in-plane stresses in the TGO layer obtained based on the large-deformation theory and under the small-deformation assumption, respectively, with the corresponding experimental results

Figure 4.16 shows thc variation in the morphology of the TGO layer of the TBC with the interfacial roughness at 1000 h of interfacial oxidation (the lower-left and upper-right images correspond to the highest and lowest interfacial roughnesses, respectively). The nonuniformity in TGO growth becomes increasingly prominent as the interfacial roughness increases. With an amplitude of 5 μm, as the wavelength increases from 40 to 80 μm, the TGO thickness decreases from 10 to 8.5 μm at the peak but changes only slightly from 7.2 to 7.8 μm at the trough. Similarly, with a wavelength of 40 μm, as the amplitude increases, the TGO thickness decreases from 7.2 to 6.5 μm at the trough but increases from 10.0 to 17.8 μm at the peak. Clearly, the interfacial roughness affects the TGO growth significantly more at the peak than at the trough. In other words, high interfacial roughness results in significantly nonuniform TGO growth in a coating, making the coating prone to stress concentrations and damage that lead to crack initiation and propagation and ultimately cause coating spallation. However, the interfacial roughness is a key factor ensuring the interfacial bonding performance of a coating. Thus, finding the optimum interfacial roughness is the key to improving the service life of a coating.

Figure 4.17 shows the variation in TGO thickness with oxidation time. Figure 4.17a, b gives the TGO thickness at the peak and trough, respectively. The TGO growth curves demonstrate that as the interfacial roughness increases, the TGO thickness increases at the peak but decreases at the trough. In addition, the TGO growth curves corresponding to the trough are approximately parabolic. In comparison, there is a significant difference between the TGO growth curves corresponding

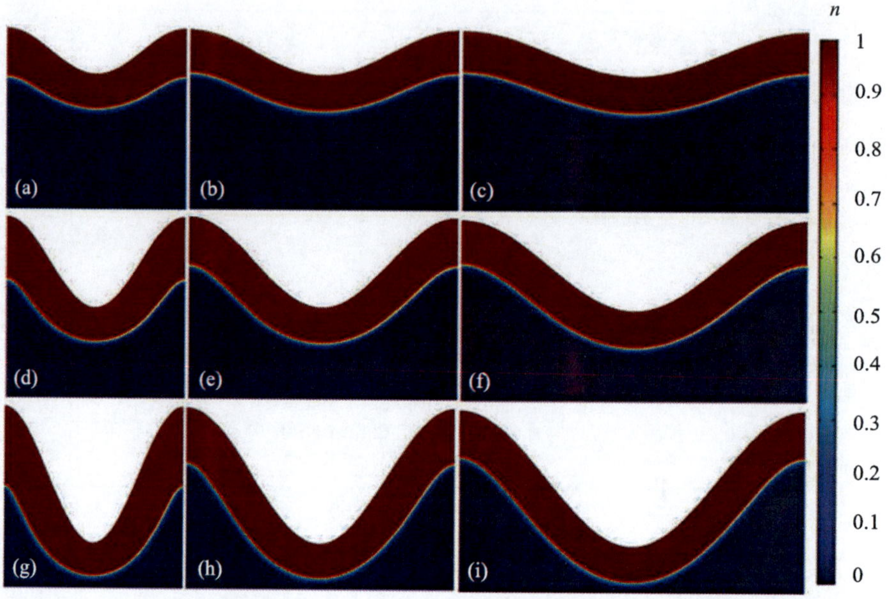

Fig. 4.16 Variation in TGO thickness with the interfacial roughness: **a** amplitude $A = 5$ μm, wavelength $L = 40$ μm; **b** $A = 5$ μm, $L = 60$ μm; **c** $A = 5$ μm, $L = 80$ μm; **d** $A = 10$ μm, $L = 40$ μm; **e** $A = 10$ μm, $L = 60$ μm; **f** $A = 10$ μm, $L = 80$ μm; **g** $A = 15$ μm, $L = 40$ μm; **h** $A = 15$ μm, $L = 60$ μm; **i** $A = 15$ μm, $L = 80$ μm

to the peak at different interfacial roughnesses. Specifically, the TGO growth curve corresponding to the peak can still be approximated as parabolic when the interfacial roughness is low, whereas there is an approximately linear relationship between the TGO thickness and the interfacial roughness when the interfacial roughness is high. This type of TGO growth is extremely detrimental to a coating because the coating readily spalls.

The variation in the distribution of the stress in the *y*-direction, σ_{yy}, in the TGO and BC layers of the TBC with interfacial roughness at 1000 h of oxidation is determined based on the large-deformation theory, as shown in Fig. 4.18, where the tensile and compressive stresses are shown in red and blue, respectively. The region between the peak and the trough is subjected to compressive stress, whereas tensile stress is present at the peak of the TGO/BC interface. Both the tensile and compressive stresses change significantly with the interfacial roughness. With an amplitude of 5 μm, as the wavelength increases from 40 to 80 μm, the maximum tensile stress increases from 480 to 753 MPa, and the maximum compressive stress increases from –713 to –1673 MPa. Similarly, with a wavelength of 40 μm, as the amplitude increases from 5 to 15 μm, the maximum tensile stress in the TGO and BC layers increases from 753 to 1475 MPa, and the maximum compressive stress increases from –1673 to –3111 MPa. As the interfacial roughness increases, the location of the maximum tensile stress gradually shifts from the peak to the vicinity of the peak of the TGO/BC interface.

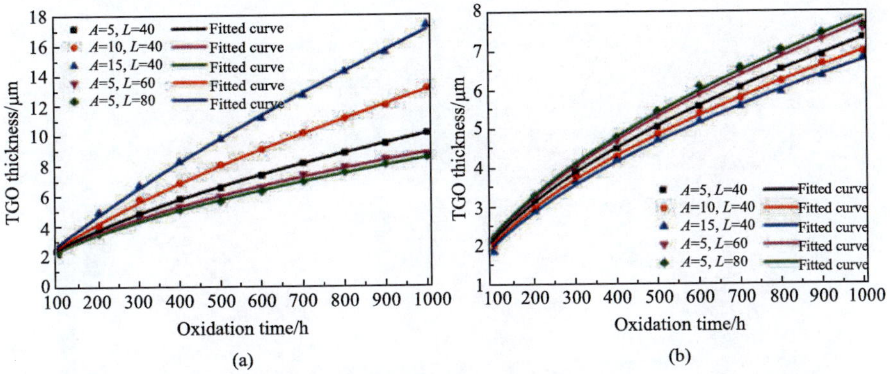

Fig. 4.17 Variation in TGO thickness with interfacial roughness: **a** at peak; **b** at trough (A = amplitude, L = wavelength)

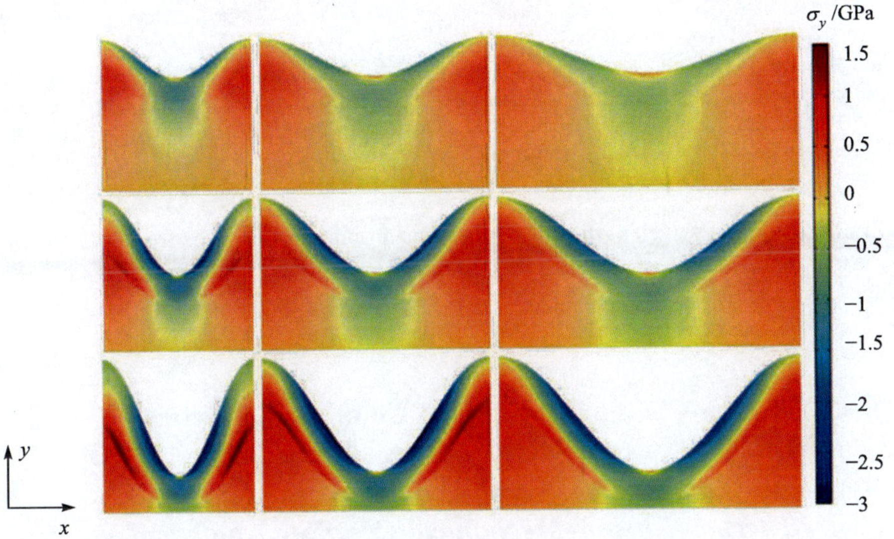

Fig. 4.18 Variation in the distribution of the *y*-direction stress in TGO and BC layers with the interfacil roughness

The variation in the distribution of σ_{yy} in the TC layer of the TBC at 1000 h of oxidation is obtained using the large-deformation theory, as presented in Fig. 4.19, with the tensile and compressive stresses shown in red and blue, respectively. The peak is subjected to compressive stress, whereas tensile stress is present at the trough of the TGO/BC interface. Both the compressive and tensile stresses change significantly with the interfacial roughness. With an amplitude of 5 μm, as the wavelength increases from 40 to 80 μm, the maximum tensile stress increases from 105 to 225 MPa, and the maximum compressive stress increases from –138 to –286 MPa.

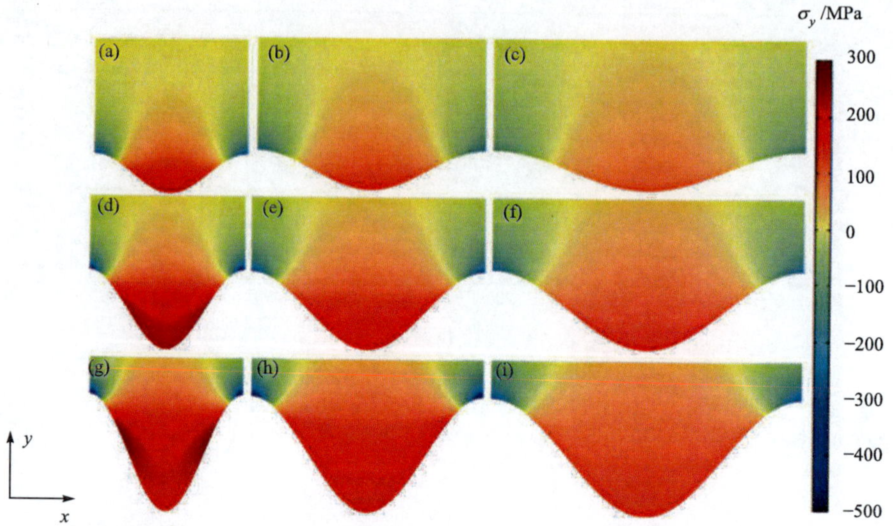

Fig. 4.19 Variation in the distribution of the *y*-direction stress in the TC layer with the interfacial roughness: **a** amplitude $A = 5$ μm, wavelength $L = 40$ μm; **b** $A = 5$ μm, $L = 60$ μm; **c** $A = 5$ μm, $L = 80$ μm; **d** $A = 10$ μm, $L = 40$ μm; **e** $A = 10$ μm, $L = 60$ μm; **f** $A = 10$ μm, $L = 80$ μm; **g** $A = 15$ μm, $L = 40$ μm; **h** $A = 15$ μm, $L = 60$ μm; **i** $A = 15$ μm, $L = 80$ μm

Similarly, with a wavelength of 40 μm, as the amplitude increases from 5 to 15 μm, the maximum tensile stress in the TGO and BC layers increases from 225 to 281 MPa, and the maximum compressive stress increases from –286 to –467 MPa.

4.3 Thermo–Chemo–Mechanical Coupling Analytical Model for Interfacial Oxidation of TBCs

4.3.1 Thermo–Chemo–Mechanical Coupling Analytical Growth Model for Interfacial Oxidation

(1) Stress analysis based on the large-deformation theory

In this section, the stress field in a coating is calculated based on the large-deformation theory using a simple cylindrical model. The oxidation interfaces in TBCs are often uneven. To describe such unevenness, the rough interfaces in a coating are simplified into regular curves, based on which the stress distribution in the coating is solved analytically. In FE simulations, sinusoidal curves or semicircles are used to represent interfacial roughness. However, in theoretical models, concentric circles or spheres are used instead to analyze the stress state in a coating.

In this section, simple cylindrical models, as shown in Fig. 4.20, are used to analyze the distribution of the stress field in a TBC undergoing large deformation due to interfacial oxidation. The convex model in Fig. 4.20a mainly shows the peak locations of the interfaces and consists of, from the inside out, a BC layer, a TGO layer, and a TC layer. The concave model in Fig. 4.20b mainly shows the trough locations of the interfaces and consists of, from the inside out, a TC layer, a TGO layer, and a BC layer. In both the convex and concave models, it is assumed that TGOs grow toward the BC layer.

The unoxidized state of the TBC is defined as the reference configuration, while its state following TGO growth is defined as the current configuration. Let A, B, and C be the respective radii of the layers before oxidation and a, b, and c be their respective radii after oxidation. In a cylindrical coordinate system, it is assumed that no deformation occurs along the length (i.e., axial) direction, i.e., the component

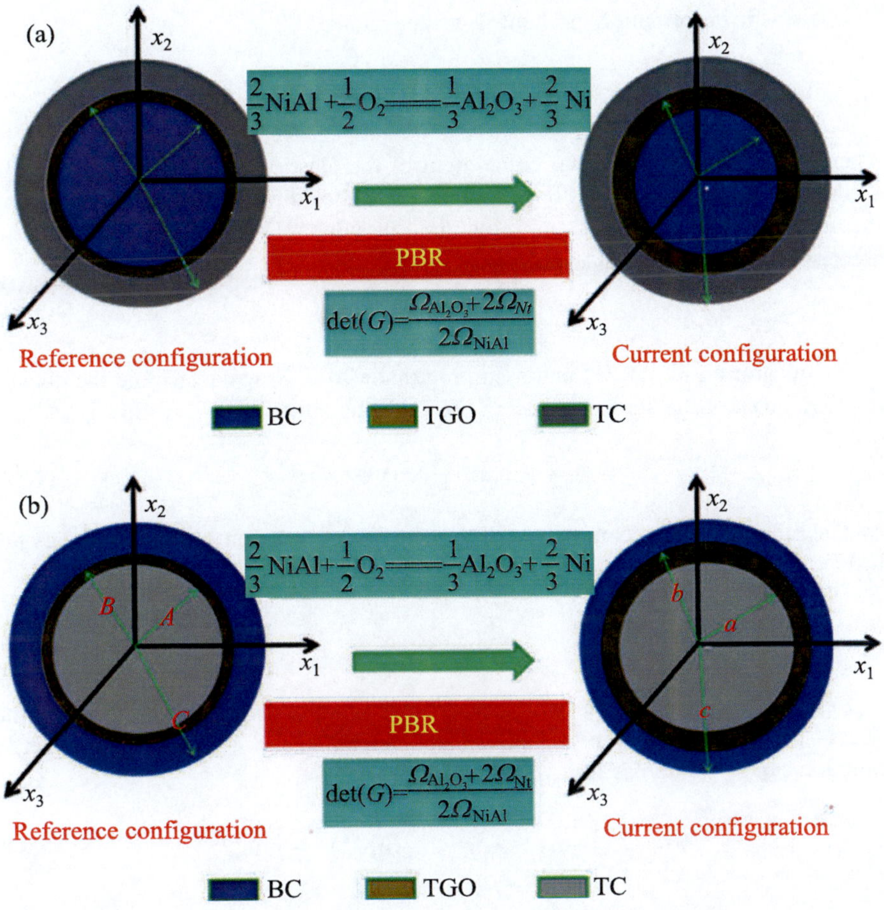

Fig. 4.20 Cylindrical models for growth stress: **a** convex model; **b** concave model

of the deformation gradient $\boldsymbol{F}$ in the length direction is 1. Thus, the deformation gradient $\boldsymbol{F}$ is expressed as

$$\boldsymbol{F} = diag\left(\frac{\partial r}{\partial R}, \frac{r}{R}, 1\right) \tag{4.28}$$

where r and R are the radii in the current and reference configurations, respectively.

The growth deformation is primarily a result of the volumetric deformation induced by TGO growth. Hence, the growth deformation gradient is given by

$$\boldsymbol{F}^g = diag(g_1, g_2, 1) \tag{4.29}$$

where g_1 and g_2 are the radial and circumferential components of the growth deformation gradient, respectively.

The elastic deformation gradient is written as

$$\boldsymbol{F}^e = diag(\alpha^{-1}, \alpha, 1) \tag{4.30}$$

where α is the circumferential component of the elastic deformation gradient. To facilitate subsequent calculations, it is assumed that the TGO layer is elastic and incompressible. Under this assumption, the following relationship is obtained:

$$\frac{\partial r}{\partial R} \cdot \frac{r}{R} = g_1 \cdot g_2 \tag{4.31}$$

By integrating Eq. (4.31) and assuming that $r(b) = B$, we determine the change in the radius a of the convex model in the current configuration as follows:

$$\alpha^2 = g_1 g_2 A^2 - (g_1 g_2 - 1) B^2 \tag{4.32}$$

The circumferential component of the elastic deformation gradient satisfies the following relationship:

$$\frac{r}{R} = g_2 \cdot \alpha \tag{4.33}$$

Here, an isotropic neo-Hookean constitutive model is employed to describe the stress–strain relationship in the TGO layer. Based on Eq. (4.16), the free-energy function can be written as follows:

$$\psi = J^G \left[\frac{1}{2} \mu (I^e - 3) + \frac{\lambda}{2} (\ln J^e)^2 - \mu \ln J^e \right] \tag{4.34}$$

where J^{G} is the Jacobian determinant of the growth gradient, J^{e} is the Jacobian determinant of the elastic deformation gradient, I^{e} is the first invariant of the Green strain tensor, and μ and λ are Lame's constants.

The radial and circumferential components of the Cauchy stress are given by

$$\sigma_{\theta\theta} = \mu(\alpha^2 - 1) \quad \sigma_{rr} = \mu(\alpha^{-2} - 1) \tag{4.35}$$

The force equilibrium equation in a cylindrical coordinate system is expressed as

$$\frac{\partial \sigma_{rr}}{\partial r} + \frac{\sigma_{rr} - \sigma_{\theta\theta}}{r} = 0 \tag{4.36}$$

Substituting Eq. (4.35) into Eq. (4.36) and transforming the derivative with respect to the variable r into the derivative with respect to α, we can rewrite Eq. (4.36) as

$$\frac{d\sigma_{rr}}{d\alpha} = -\frac{\mu(1 + \alpha^2)}{\alpha^3} \tag{4.37}$$

By integrating the above equation, we obtain the following expression of the radial stress in the convex model:

$$\sigma_{rr} = \frac{\mu}{2}\left[\left(\frac{1}{\alpha^2} - \frac{1}{\alpha_A^2}\right) + \ln\frac{\alpha^2}{\alpha_A^2}\right] + \sigma_{rr}|_{\alpha(A)} \tag{4.38}$$

Based on Eq. (4.35), the circumferential stress is expressed as

$$\sigma_{\theta\theta} = \sigma_{rr} + \mu(\alpha^2 - \alpha^{-2}) \tag{4.39}$$

where α is the elastic elongation, α_A is the elongation of radius A, and $\sigma_{rr}|_{\alpha(A)}$ depends on the boundary conditions. Based on the work of Hsueh and Fuller, [21] the radial thermal mismatch stress varies with the TGO thickness. Hence, when the boundary conditions are determined based on the thermal mismatch stress, the value of $\sigma_{rr}|_{\alpha(A)}$ varies with the TGO thickness h_{TGO}, as shown below

$$\begin{aligned} \sigma_{rr}|_{\alpha=\alpha_A} &= 983\ \mathrm{MPa} \quad h_{TGO} = 10\,\mu\mathrm{m} \\ \sigma_{rr}|_{\alpha=\alpha_A} &= 700\ \mathrm{MPa} \quad h_{TGO} = 5\,\mu\mathrm{m} \\ \sigma_{rr}|_{\alpha=\alpha_A} &= 370\ \mathrm{MPa} \quad h_{TGO} = 1\,\mu\mathrm{m} \end{aligned} \tag{4.40}$$

Therefore, the stresses in the TC layer in the convex model are given by

$$\sigma_{rr} = \frac{(C^2 - R^2)B^2}{(C^2 - B^2)R^2} \cdot \sigma_{rr}^B \quad \sigma_{\theta\theta} = -\frac{(C^2 + R^2)B^2}{(C^2 - B^2)R^2} \cdot \sigma_{rr}^B \tag{4.41}$$

where σ_B^{rr} is the radial stress at the TGO/TC interface and B and C are the radii of the TGO and TC layers, respectively.

The derivation steps for the concave model are the same as those for the convex model and thus are not repeated here. The expressions of the stress components in the TGO layer are derived as follows:

$$\sigma_{rr} = -\frac{\mu}{2}\left[\left(\frac{1}{\alpha^2} - \frac{1}{\alpha_B^2}\right) + \ln\frac{\alpha^2}{\alpha_B^2}\right] + \sigma_{rr}|_{\alpha(B)} \tag{4.42}$$

$$\sigma_{\theta\theta} = \sigma_{rr} + \mu(\alpha^2 - \alpha^{-2}) \tag{4.43}$$

where $\sigma_{rr}|_{\alpha(B)}$ is determined by the following boundary conditions:

$$\sigma_{rr}|_{\alpha=\alpha_B} = -1007\ \text{MPa} \quad h_{\text{TGO}} = 10\ \mu\text{m}$$
$$\sigma_{rr}|_{\alpha=\alpha_B} = -896\ \text{MPa} \quad h_{\text{TGO}} = 5\ \mu\text{m}$$
$$\sigma_{rr}|_{\alpha=\alpha_B} = -506\ \text{MPa} \quad h_{\text{TGO}} = 1\ \mu\text{m}$$

Similarly, the stress distribution in the BC layer of the concave model is determined based on the theory of elasticity as follows:

$$\sigma_{rr} = \frac{(C^2 - R^2)B^2}{(C^2 - B^2)R^2} \cdot \sigma_{rr}^B \quad \sigma_{\theta\theta} = -\frac{(C^2 + R^2)B^2}{(C^2 - B^2)R^2} \cdot \sigma_{rr}^B \tag{4.44}$$

where σ_{rr}^B is the stress at the TGO/BC interface and B and C are the radii of the TGO and BC layers of the concave model, respectively.

Practice problem: Derive Eqs. (4.42) and (4.43).

(2) **Stress analysis under the small-deformation assumption**

To facilitate the comparison of the results considering large and small deformations, respectively, the stress distribution in the coating is determined under the small-deformation assumption. The geometric models in Fig. 4.20 used for stress analysis based on the large-deformation theory are employed again to analyze the stress state induced by TGO growth under the small-deformation assumption. Thus, the analysis is further simplified into a plane strain problem. In the cylindrical coordinate system, the radial and circumferential strain components are expressed as

$$\varepsilon_r^e = \frac{\partial u_r}{\partial R} - \varepsilon_r^g \quad \varepsilon_\theta^e = \frac{u_r}{R} - \varepsilon_\theta^g \tag{4.45}$$

where u_r is the radial displacement. The radial and circumferential growth strain components are given by

$$\varepsilon_r^g = g_1 - 1 \quad \varepsilon_\theta^g = g_2 - 1 \tag{4.46}$$

where g_1 and g_2 are the radial and circumferential components of the growth deformation gradient tensor, respectively.

Based on the plane strain assumption, the stress–strain relationship is written as

$$\sigma_{rr} = \frac{2\mu}{1-v}(\varepsilon_r^e + v\varepsilon_\theta^e) \quad \sigma_{\theta\theta} = \frac{2\mu}{1-v}(\varepsilon_\theta^e + v\varepsilon_r^e) \tag{4.47}$$

where μ is the shear modulus and v is Poisson's ratio.

The force equilibrium equation in the cylindrical coordinate system is given by

$$\frac{\partial\sigma_{rr}}{\partial r} + \frac{\sigma_{rr} - \sigma_{\theta\theta}}{R} = 0 \tag{4.48}$$

By combining Eqs. (4.45), (4.46), and (4.47) and by substituting the resulting stress expression into Eq. (4.48), we can write the force equilibrium equation in terms of displacement as follows:

$$\frac{d^2u_r}{dR^2} + \frac{1}{R}\frac{du_r}{dR} - \frac{u_r}{R^2} = (1+v)\frac{d(g_1-1)}{dR} \tag{4.49}$$

By integrating the above equation, we obtain the following expression of the displacement in the coating:

$$u_r = C_1R + \frac{C_2}{R} + (1+v)\frac{1}{R}\int_a^R (g_1-1)RdR \tag{4.50}$$

where the lower bound of integration a is the radius of the BC layer of the convex model. In this book, the stress state under large deformation is the main area of interest. Therefore, only the convex model is used here in the derivation for small-deformation analysis. Note that the derivation for the concave model is the same as that for the convex model and thus is not repeated here. The constants in Eq. (4.50) are determined by the boundary conditions. Not only TGO growth produces stress but also the thermal mismatch between the TGO layer and the material of each layer induces residual thermal stress. Based on the theoretical model established by Hsueh and Fuller [21] for the residual stress in TBCs, residual thermal stresses are introduced based on the preset stress boundary conditions for different TGO thicknesses, as shown below

$$\begin{aligned}
&\sigma_{rr}|_{R=a} = 983\ \text{MPa} \quad \sigma_{rr}|_{R=b} = -400\ \text{MPa} \quad h_{TGO} = 10\ \mu\text{m} \\
&\sigma_{rr}|_{R=a} = 700\ \text{MPa} \quad \sigma_{rr}|_{R=b} = -200\ \text{MPa} \quad h_{TGO} = 5\ \mu\text{m} \\
&\sigma_{rr}|_{R=a} = 370\ \text{MPa} \quad \sigma_{rr}|_{R=b} = -50\ \text{MPa} \quad h_{TGO} = 1\ \mu\text{m}
\end{aligned} \tag{4.51}$$

The current displacement can be determined by the boundary conditions. However, large deformation does occur. Thus, after the displacement is obtained using the small-deformation theory, the strain is updated, i.e., second-order terms (or higher-order terms) are added, and then the stress is determined based on the constitutive relations. The specific expressions are shown below

$$\begin{cases} \varepsilon_r^e = \dfrac{\partial u_r}{\partial R} - \varepsilon_r^g + \dfrac{1}{2}(\dfrac{\partial u_r}{\partial R})^2 - \dfrac{1}{2}(\varepsilon_r^g)^2 \\ \varepsilon_\theta^e = \dfrac{u_r}{R} - \varepsilon_\theta^g + \dfrac{1}{2}(\dfrac{u_r}{R})^2 - \dfrac{1}{2}(\varepsilon_\theta^g)^2 \end{cases} \tag{4.52}$$

Based on the stress constitutive relations, the following expressions of the stress components are obtained:

$$\begin{cases} \sigma_{rr} = \dfrac{2(\mu - \lambda \ln(J^e))}{1-\nu}(\varepsilon_r^e + \nu\varepsilon_\theta^e) \\ \sigma_{\theta\theta} = \dfrac{2(\mu - \lambda \ln(J^e))}{1-\nu}(\varepsilon_\theta^e + \nu\varepsilon_r^e) \end{cases} \tag{4.53}$$

For the case of small deformation, the Jacobian determinant of the elastic deformation gradient is equal to $(\varepsilon_r^e + 1)(\varepsilon_\theta^e + 1)$. Thus, the following expression is obtained based on the small-deformation assumption:

$$\ln J^e \approx \varepsilon_r^e + \varepsilon_\theta^e + \varepsilon_r^e \cdot \varepsilon_\theta^e \tag{4.54}$$

Substituting Eq. (4.54) into Eq. (4.53) yields the expressions of the stress components in the TGO layer as follows:

$$\sigma_{rr} = \frac{2(\mu - \lambda(\varepsilon_r^e + \varepsilon_\theta^e + \varepsilon_r^e \cdot \varepsilon_\theta^e))}{1-\nu}(\varepsilon_r^e + \nu\varepsilon_\theta^e) \tag{4.55}$$

$$\sigma_{\theta\theta} = \frac{2(\mu - \lambda(\varepsilon_r^e + \varepsilon_\theta^e + \varepsilon_r^e \cdot \varepsilon_\theta^e))}{1-\nu}(\varepsilon_\theta^e + \nu\varepsilon_r^e) \tag{4.56}$$

The stresses in the TC and BC layers are determined with the same method as that for obtaining the stresses under large deformation.

(3) **Stress distribution in the coating**

Table 4.3 summarizes the material parameters of each layer at 1000 °C. In each model, radii A and C are set to 10 and 60 μm, respectively, while radius B is set to different values to represent different TGO thicknesses. It is generally believed that the stresses perpendicular to the interfaces are the main cause of coating failure. Hence, only the radial stress distribution is analyzed here.

Figure 4.21a, b shows the stress distributions in the convex and concave models, respectively, determined based on the large-deformation theory. As demonstrated in Fig. 4.21a, in the convex model, the radial stress transitions from tensile stress at the

Table 4.3 Material parameters of each layer at 1000 °C [25]

Property	BC	TGO	TC
Young's modulus (GPa)	110	320	22
Poisson's ratio	0.33	0.25	0.12

TGO/BC interface to compressive stress at the TGO/TC interface. In comparison, in the concave model, the radial stress in the coating transitions from compressive stress at the TGO/BC interface to tensile stress at the TGO/TC interface, as shown in Fig. 4.21b. According to the stress distributions in the models, the maximum tensile stress in the coating appears at the troughs of the TGO/TC interface and the peaks of the TGO/BC interface, meaning that these locations are where cracks may initiate and where the coating is prone to failure. These findings are highly consistent with the FE simulation results obtained in the previous section. In addition, as shown in Fig. 4.21, both the maximum tensile and compressive stresses increase as the TGO thickness increases. Specifically, when the TGO thickness is 1.0, 5.0, and 10.0 μm, the maximum tensile stress is 370, 700, and 983 MPa, respectively, and the maximum compressive stress is 506, 896, and 1007 MPa, respectively.

Figure 4.22a, b shows the stress distributions in the convex and concave models, respectively, under small deformation. As demonstrated in Fig. 4.22a, in the convex model, the stresses under large and small deformations have similar distribution patterns and differ only in magnitude. As shown in Fig. 4.22b, in the concave model, the stresses under large and small deformations are similar in distribution but change differently as the TGO thickness increases; specifically, an increase in the TGO thickness reduces the stress at the TGO/TC interface and may transform the tensile stress into compressive stress. Similar to the stress distribution under large deformation, the maximum tensile stress in the coating appears at the troughs of the TGO/TC interface and the peaks of the TGO/BC interface, where the coating is at risk of spallation. Through experimentation, Busso and Qian found that cracks often initiate at the troughs of the TC/TGO interface and the peaks of the TGO/BC interface [40].

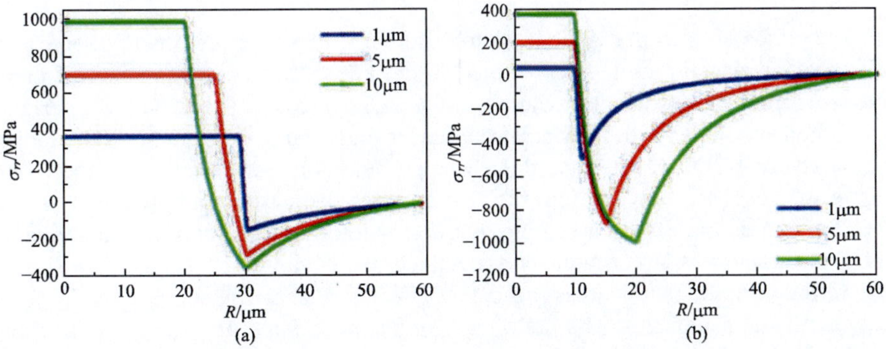

Fig. 4.21 Distributions of radial growth stress under large deformation: **a** convex model; **b** concave model

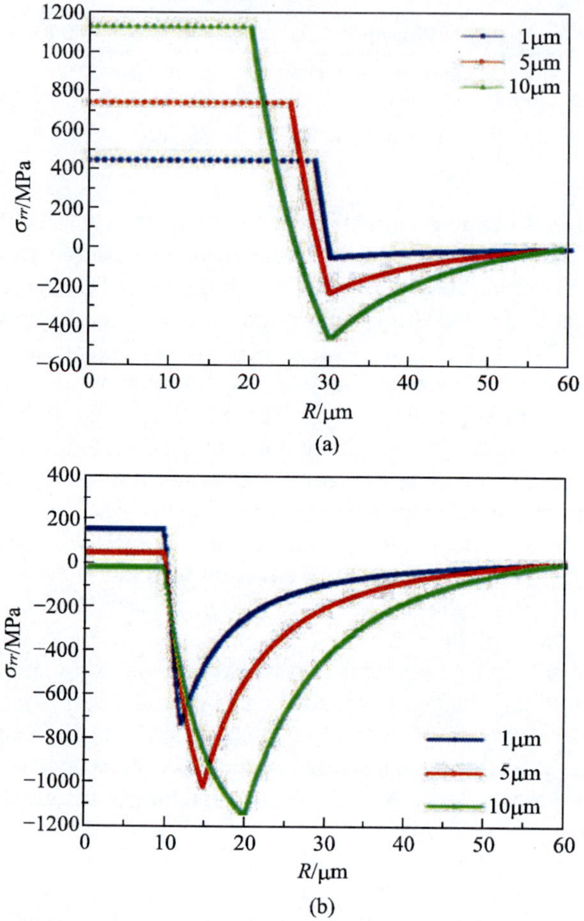

Fig. 4.22 Distributions of radial growth stress under small deformation: **a** convex model; **b** concave model

The radius of the inner circle is set to different values to approximately represent different interfacial roughnesses, with a larger radius representing a smoother interface. Figure 4.23 shows the variation in the radial stress at the TGO/BC interface in the convex model with interfacial roughness under large deformation. The radial stress at the TGO/BC interface decreases considerably as the radius of the inner circle increases (i.e., as the interfacial roughness decreases) and increases as the TGO thickness increases, indicating that higher interfacial roughness and a thicker TGO layer lead to a larger radial stress at the TGO/BC interface. Figure 4.24 shows the variation in the radial stress at the TGO/TC interface in the concave model with the interfacial roughness. The interfacial roughness has an insignificant impact on the radial stress at the TGO/TC interface; the TGO thickness is the primary factor affecting the stress distribution.

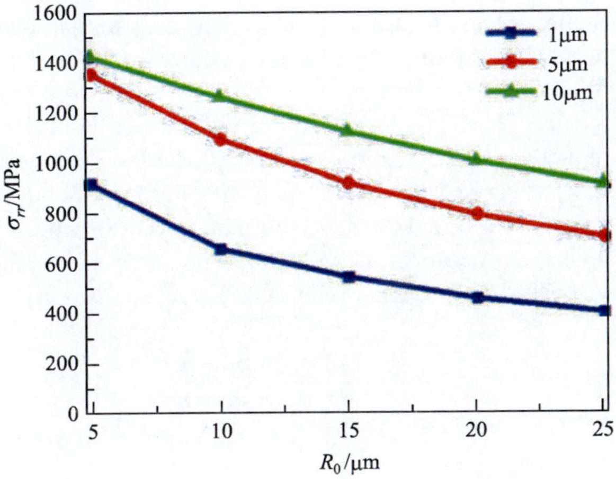

Fig. 4.23 Variations in the radial growth stress at the TGO/BC interface in the convex model with interfacial curvature under large deformation

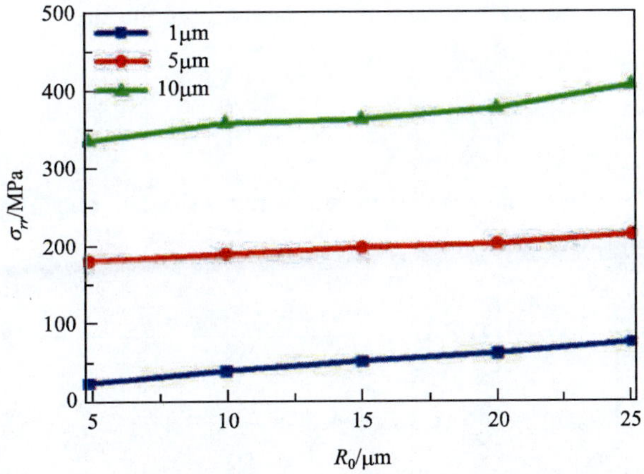

Fig. 4.24 Variations in the radial growth stress at the TGO/TC interface in the concave model with interfacial curvature under large deformation

(4) **Analytical solution for TGO growth**

Interfacial roughness has been a focal area of research on the interfacial failure of TBCs. This is because rough interfaces can cause stress concentrations, thereby resulting in coating spallation; in addition, a high stress level may accelerate TGO growth. Tang and Schoenung used a spherical model to analyze the effects of thermal mismatch stress on accelerated TGO growth, [41] but their model fails to account for the effect of growth stress. Tolpygo and Clarke [42] experimentally studied the relationship between the morphological evolution of the TGO layer and the residual stress in a coating. They found that the morphology of the TGO layer does not change with oxidation time when there is no significant change in the stress in the coating and that a sharp decline in the stress in the coating suggests TGO wrinkling or crack formation in the coating. Moreover, by investigating the pattern of interfacial oxidation in a two-layer structure composed of an oxidation layer and a metallic substrate,

Saillard et al. [43] found that stress is the principal factor inducing interfacial roughness. These studies demonstrate that stress may lead to unstable growth in local areas of the oxidation layer.

Here, based on the relationship between the TGO growth and the growth stress in a coating, a creep model is established to analyze the evolutionary pattern of the morphology of the TGO layer with the interfacial roughness. In the above cylindrical models, a TGO layer with an initial thickness of h_0 is assumed to consist of a viscoelastic material, and a plane strain problem is assumed. Without considering viscosity, the stresses in the TGO layer are given by

$$\begin{cases} \sigma_r = m - n/r^2 \\ \sigma_\theta = m + n/r^2 \\ u_r = 1/2G[(1-2\nu)mr + n/r] \end{cases} \tag{4.57}$$

where G is the shear modulus, ν is Poisson's ratio, and constants m and n are determined by the following boundary conditions:

$$\begin{cases} \sigma_r|_{r=a} = p \\ \sigma_r|_{r=a+h} = q \end{cases} \tag{4.58}$$

Then, the displacements at the TGO/BC and TGO/TC interfaces are given as follows:

$$\begin{cases} u_a = \dfrac{1}{2G}\left[(1-2\nu)a\dfrac{p(2ah_0+h_0^2)+(q-p)(a+h_0)^2}{2ah_0+h_0^2} + \dfrac{a(q-p)(a+h_0)^2}{2ah_0+h_0^2}\right] \\ u_{a+h_0} = \dfrac{1}{2G}\left[(1-2\nu)(a+h_0)\dfrac{p(2ah_0+h_0^2)+(q-p)(a+h_0)^2}{2ah_0+h_0^2} + \dfrac{a^2(q-p)(a+h_0)}{2ah_0+h_0^2}\right] \end{cases} \tag{4.59}$$

Thus, the change in TGO thickness is expressed as

$$\Delta u = u_{a+h_0} - u_a \tag{4.60}$$

Substituting Eq. (4.59) into Eq. (4.60), we obtain the final expression of the change in TGO thickness as follows:

$$\Delta u = \frac{h_0}{2G}\left[(1-2\nu)\frac{p(2ah_0+h_0^2)+(q-p)(a+h_0)^2}{2ah_0+h_0^2} - \frac{a(q-p)(a+h_0)}{2ah_0+h_0^2}\right] \tag{4.61}$$

Because the TGO layer is viewed as a viscoelastic material, the TGO thickness varies with time under stress. Based on the similarity theory, [44] the following expressions of the displacement in the phase space are obtained through a Laplace transform of the phase-space variable s:

$$\begin{aligned}\overline{u}_a(t) = \frac{1}{2\overline{G}}[1 - 2\overline{v}(s)a\frac{\overline{p}(s)(2ah_0 + h_0^2) + (\overline{q}(s) - \overline{p}(s))(a + h_0)^2}{2ah_0 + h_0^2}\\ + \frac{a(\overline{q}(s) - \overline{p}(s))(a + h_0)^2}{2ah_0 + h_0^2}]\end{aligned} \tag{4.62}$$

$$\begin{aligned}\overline{u}_{a+h_0}(t) = \frac{1}{2\overline{G}}[1 - 2\overline{v}(s)(a + h_0)\frac{\overline{p}(s)(2ah_0 + h_0^2) + (\overline{q}(s) - \overline{p}(s))(a + h_0)^2}{2ah_0 + h_0^2}\\ + \frac{a(\overline{q}(s) - \overline{p}(s))(a + h_0)^2}{2ah_0 + h_0^2}]\end{aligned} \tag{4.63}$$

The phase-space variable in the above equations is expressed as [45]

$$\begin{cases}\overline{p}(s) = p/s, \quad \overline{q}(s) = q/s\\ 1 - 2\overline{v} = \dfrac{3\overline{P}''\overline{Q}'}{2\overline{P}'\overline{Q}'' + \overline{P}''\overline{Q}'}\\ \dfrac{1 - 2\overline{v}}{2\overline{G}(s)} = \dfrac{3\overline{P}''\overline{P}'}{2\overline{P}'\overline{Q}'' + \overline{P}''\overline{Q}'}\\ \overline{P}' = 1 + \dfrac{\eta_1}{\eta_2}s, \overline{Q}' = \eta_1 s, \overline{P}'' = 1, \overline{Q}'' = 2K = \dfrac{2E}{3(1 - 2\nu)}\end{cases} \tag{4.64}$$

where η_1 and η_2 are the viscoelastic constants of the TGO layer.

The displacement of the TGO layer can be determined through an inverse Laplace transform. Thus, the TGO thickness is expressed as

$$h(t) = h_0 + \Delta u(t) \tag{4.65}$$

Finally, a dimensionless expression of the TGO thickness is obtained as follows:

$$\begin{aligned}\frac{h}{h_0} &= 1 + \frac{\Delta u}{h_0}\\ &= 1 - \frac{1}{1 - k_a}\frac{1}{1 + k_n}\left\{\sqrt{k_a}(k_q - k_p)(1 + k_n)^2\frac{t}{t_f} - 3(k_q - k_a k_p)(1 - e^{-t/t_f})\right\}\end{aligned} \tag{4.66}$$

In the above equation, the first term is the ratio of the initial TGO thickness to the radius of the model; the second term is the bulk modulus; the fourth term is the ratio of the bulk modulus to the viscoelastic constants; the fifth and sixth terms are the ratio of the boundary stress to the bulk modulus; and the final term is the ratio of the viscoelastic parameters. α, K, k_α, k_n, k_q, k_p, and t_f are given by

$$\alpha = \frac{h_0}{a}, K = \frac{E}{3(1-2\nu)}, k_a = \frac{\alpha^2}{(1+\alpha)^2}, k_n = \frac{4K}{\eta_2}, k_q = \frac{q}{4K},$$
$$k_p = \frac{p}{4K}, t_f = \frac{\eta_1}{\eta_2}(1+\frac{1}{k_n}) \tag{4.67}$$

(5) **TGO oxidation pattern**

In Sect. 3.2.2, the evolutionary patterns of the morphology of the TGO layer at different interfacial roughnesses are numerically analyzed using the FEM. As shown in Fig. 4.16, TGOs grow fast at the peak, and a higher interfacial roughness leads to faster TGO growth, while there is an insignificant change in the TGO growth rate at the trough as the interfacial roughness changes. The following equation is used to fit the TGO growth curves corresponding to different interfacial roughnesses:

$$h = a_1 \cdot t^{b_1} \tag{4.68}$$

where h is the TGO thickness, t is the oxidation time, and a_1 and b_1 are fitting parameters (see Table 4.4 for their specific values). The TGO growth curves show that as the interfacial roughness increases, the TGO thickness increases at the peak but decreases at the trough. In addition, as the interfacial roughness increases, the TGO growth curve corresponding to the peak transitions from a parabolic curve to a straight line, whereas the TGO growth curve corresponding to the trough remains parabolic. A sinusoidal curve is used to represent interfacial roughness in a coating. Thus, the interfacial curvature can be calculated using the following equation:

$$\kappa = \frac{|y''|}{(1+y'^2)^{\frac{3}{2}}}. \tag{4.69}$$

where y′ and y″ are the first-and second-order derivatives of the sinusoidal curve, respectively.

Thus, the radius of curvature of the TGO growth creep model can be calculated using Eq. (4.69). In addition, the peak and trough of the FE model correspond to the convex and concave models, respectively. The convex and concave models can

Table 4.4 Initial TGO thicknesses for different interfacial roughnesses (A = amplitude, L = wavelength)

Location	$A = 5$ μm, $L = 60$ μm	$A = 5$ μm, $L = 40$ μm	$A = 15$ μm, $L = 40$ μm
Peak	2.31 μm	2.39 μm	2.56 μm
Trough	2.12 μm	2.07 μm	1.81 μm

be differentiated based on the stresses at the TGO/BC and TGO/TC interfaces. The initial TGO thickness varies with the interfacial roughness (Table 4.4). According to the FE results obtained in the previous section, with the change in the interfacial roughness, the q and p at the peak change from –57.7 to –383 MPa and from 205 to 904 MPa, respectively, whereas the q and p at the trough change from 42.3 to 130 MPa and from –213 to –950 MPa, respectively. Thus, the parameters of the TGO creep model (i.e., Eqs. (4.58) and (4.67)) can be approximately expressed with the stresses in the coating. For the convex model, $k_p = -0.0001$ and $k_q = 0.0003$, and for the concave model, $k_p = -0.0003$ and $k_q = 0.0001$.

The TGO thickness can be calculated using Eq. (4.66). Figures 4.25 and 4.26 show the TGO growth curves corresponding to different interfacial roughnesses for the convex and concave models, respectively. The TGO thickness increases linearly with oxidation time. This is mainly because a dimensionless quantity (i.e., h/h_0 in Eq. (4.66)) is used to describe the change in the TGO thickness. See Table 4.4 for the values of the initial TGO thickness h_0. Evidently, the FE results obtained in the previous section generally agree with the results predicted by the TGO growth creep model. In addition, the TGO growth curves corresponding to the peak and trough show that the radius of curvature significantly affects TGO growth, i.e., the interfacial roughness is a major factor affecting the morphological evolution of the TGO layer. A small interfacial radius (h_0/a) (i.e., an interface with high roughness) leads to a large TGO thickness. Sandblasting and polishing can effectively reduce the interfacial roughness in local areas, thereby extending the coating's service life. Ni et al. [46] found that these processes could effectively improve the resistance of coatings to oxidation and that compared to the case of unprocessed original specimens, TGOs grew significantly more slowly in specimens treated with the above processes. However, the majority of the TGO growth models fail to account for the effect of interfacial roughness. The TGO growth creep model quantifies the effect of interfacial roughness on TGO growth. A comparison of the TGO growth curves corresponding to the peak (Fig. 4.25) and trough (Fig. 4.26) shows that at the same radius of curvature, the TGO thickness is significantly larger at the peak than at the trough, thus resulting in nonuniform TGO growth or unstable TGO growth in local areas. The accelerated TGO growth in local areas may lead to stress concentration, which, in turn, causes crack formation in the coating and accelerates the coating spallation.

4.3.2 *Thermo–Chemo–Mechanical Coupling Growth Constitutive Relations for Interfacial Oxidation*

(1) **Kinematic description of thermo–mechano-chemical coupling during interfacial oxidation**

Here, the assumption that the total deformation gradient of a coating can be decomposed into elastic and growth deformation gradients is applied again. Thus, the total

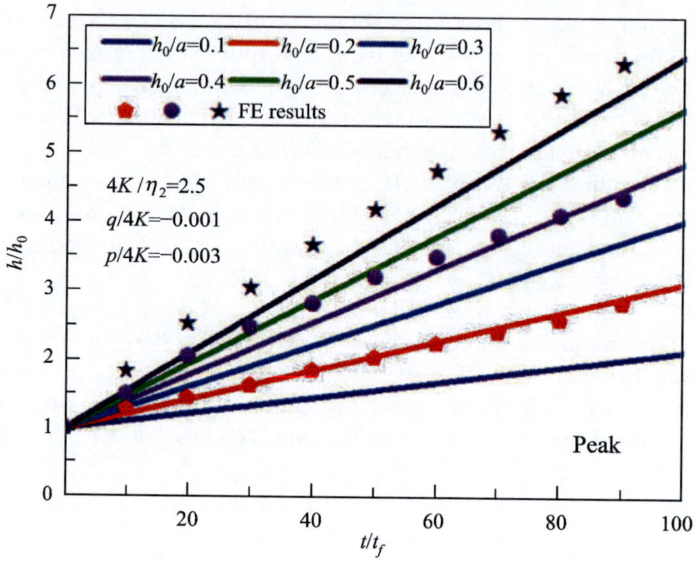

Fig. 4.25 Variation in TGO thickness at the peak with interfacial roughness

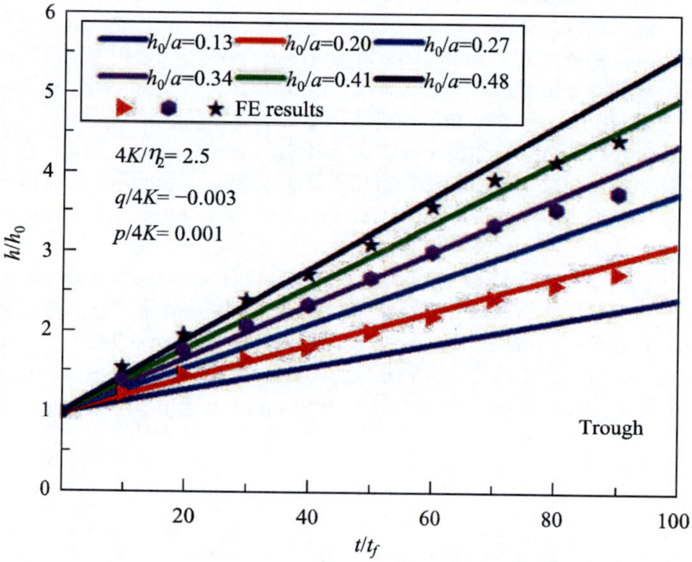

Fig. 4.26 Variation in TGO thickness at the trough with the interfacial roughness

deformation can be described based on three different configurations, namely a reference configuration, an intermediate configuration, and a current configuration. Let $\boldsymbol{X}$, $\overline{\boldsymbol{X}}$, and $\boldsymbol{x}$ be the position coordinates in the three configurations, respectively. Then, the deformation gradient $\boldsymbol{F}$, the elastic deformation gradient $\boldsymbol{F}^{\mathrm{e}}$, and the growth deformation gradient $\boldsymbol{F}^{\mathrm{g}}$ are given by

$$\boldsymbol{F} = \frac{\partial \boldsymbol{x}}{\partial \boldsymbol{X}}, \boldsymbol{F}^{e} = \frac{\partial \boldsymbol{x}}{\partial \overline{\boldsymbol{X}}}, \boldsymbol{F}^{g} = \frac{\partial \overline{\boldsymbol{X}}}{\partial \boldsymbol{X}} \tag{4.70}$$

The decomposed expression of the deformation gradient $\boldsymbol{F}$ assumed in the previous two chapters can be obtained using the derivative rules for composite functions:

$$\boldsymbol{F} = \boldsymbol{F}^{e} \cdot \boldsymbol{F}^{g} \tag{4.71}$$

The deformation gradient is decomposed under the assumption that there exists an intermediary configuration that allows unrestrained TGO growth, i.e., a configuration in a zero-stress state. This approach draws on the large-deformation analysis method used by researchers to investigate the growth of biological tissues and cells [29]. In fact, this decomposition of the deformation gradient has, for a long time, been applied in the analysis of large elastoplastic deformation and has recently been widely used in the large-deformation analysis of multifield coupling. For example, N'Guyen et al. [30] decomposed the deformation gradient into an elastic deformation gradient and an inelastic deformation gradient and, on this basis, studied thermo–mechano-chemical coupling in soft materials under large deformation. Yadegari et al. [31] decomposed the total deformation gradient into elastic, thermal, plastic, and transformation deformation gradients and, on this basis, examined the multifield coupling problem of multiphase steels.

The ratio J of the volume of a volume element before deformation, $d\Omega_0$, to that after deformation, $d\Omega$, can be expressed using the determinant of the deformation gradient as follows:

$$\begin{aligned} J = \frac{\mathrm{d}\Omega}{\mathrm{d}\Omega_0} &= \det \boldsymbol{F} = \det(\boldsymbol{F}^{\mathrm{e}} \cdot \boldsymbol{F}^{\mathrm{g}}) \\ &= \det \boldsymbol{F}^{\mathrm{e}} \cdot \det \boldsymbol{F}^{\mathrm{g}} = J^{e} \cdot J^{g} \end{aligned} \tag{4.72}$$

where J^e is the Jacobian determinant of the elastic deformation gradient and reflects the volume change of the volume element caused by elastic deformation and J^g is the Jacobian determinant of the growth deformation gradient and reflects the volume expansion caused by TGO growth. The material derivative of the volume ratio, shown below, is often used in equilibrium equations in continuum mechanics.

$$
\begin{aligned}
\dot{J} &= \frac{\partial J}{\partial F_{iA}}\frac{dF_{iA}}{dt} = \frac{\partial J}{\partial F_{iA}}\frac{\partial}{\partial t}\left(\frac{\partial x_i}{\partial X_A}\right) \\
&= \frac{\partial J}{\partial F_{iA}}\frac{\partial v_i}{\partial X_A} = J\frac{\partial X_A}{\partial x_i}\frac{\partial v_i}{\partial X_A} \\
&= J\frac{\partial v_i}{\partial x_i} = Jdiv\boldsymbol{v}
\end{aligned}
\tag{4.73}
$$

A mass point at $\boldsymbol{X}$ occupies different locations $\boldsymbol{x}$ at different time points t. The equation of motion of an object is given by

$$
\boldsymbol{x} = \chi(\boldsymbol{X}, t) \tag{4.74}
$$

Then, the velocity $\boldsymbol{v}$ and velocity gradient $\boldsymbol{l}$ are defined as follows:

$$
\boldsymbol{v} = \dot{\chi}(\boldsymbol{X}, t) \tag{4.75}
$$

$$
\boldsymbol{l} = grad\ \boldsymbol{v} = \dot{\boldsymbol{F}} \cdot \boldsymbol{F}^{-1} \tag{4.76}
$$

By substituting the decomposed expression of the deformation gradient in Eq. (4.71) into Eq. (4.76), the following decomposed expression of the velocity gradient is obtained:

$$
\begin{aligned}
\boldsymbol{l} &= \dot{\boldsymbol{F}} \cdot \boldsymbol{F}^{-1} \\
&= (\dot{\boldsymbol{F}}^{\mathrm{e}} \cdot \boldsymbol{F}^{\mathrm{g}} + \boldsymbol{F}^{\mathrm{e}} \cdot \dot{\boldsymbol{F}}^{\mathrm{g}}) \cdot (\boldsymbol{F}^{\mathrm{g}-1} \cdot \boldsymbol{F}^{e-1}) \\
&= \dot{\boldsymbol{F}}^{\mathrm{e}} \cdot \boldsymbol{F}^{e-1} + \boldsymbol{F}^{\mathrm{e}} \cdot \dot{\boldsymbol{F}}^{\mathrm{g}} \cdot \boldsymbol{F}^{\mathrm{g}-1} \cdot \boldsymbol{F}^{e-1} \\
&= \boldsymbol{l}^{\mathrm{e}} + \boldsymbol{F}^{\mathrm{e}} \cdot \overline{\boldsymbol{L}}^{\mathrm{g}} \cdot \boldsymbol{F}^{e-1}
\end{aligned}
\tag{4.77}
$$

where $\overline{\boldsymbol{L}}^{\mathrm{g}}$ is the growth velocity gradient in the intermediate configuration. Equation (4.77) can be understood as that the velocity gradient can be decomposed into an elastic velocity gradient $\boldsymbol{l}^{\mathrm{e}}$ and an oxidation growth velocity gradient $\boldsymbol{l}^{\mathrm{g}}$, which are expressed as

$$
\boldsymbol{l}^{\mathrm{e}} = \dot{\boldsymbol{F}}^{\mathrm{e}} \cdot \boldsymbol{F}^{e-1} \quad \boldsymbol{l}^{\mathrm{g}} = \boldsymbol{F}^{\mathrm{e}} \cdot \overline{\boldsymbol{L}}^{\mathrm{g}} \cdot \boldsymbol{F}^{e-1} \tag{4.78}
$$

Thus, Eq. (4.77) can be understood as that the velocity gradient can be decomposed into an elastic part and a growth part:

$$
\boldsymbol{l} = \boldsymbol{l}^{\mathrm{e}} + \boldsymbol{l}^{\mathrm{g}} \tag{4.79}
$$

The velocity gradient can be additively decomposed into a symmetric part, which is referred to as the deformation rate or strain rate (often denoted by $\boldsymbol{d}$), and an antisymmetric part, which is referred to as the spin rate (often denoted by $\boldsymbol{w}$). Then, based on Eq. (4.78), the expressions of the elastic deformation rate, the growth deformation rate, the elastic spin rate, and the growth spin rate are obtained as follows:

$$\begin{aligned} \boldsymbol{d}^e &= \frac{1}{2}(\boldsymbol{l}^e + \boldsymbol{l}^{eT}) \qquad & \boldsymbol{d}^g &= \frac{1}{2}(\boldsymbol{l}^g + \boldsymbol{l}^{gT}) \\ \boldsymbol{w}^e &= \frac{1}{2}(\boldsymbol{l}^e - \boldsymbol{l}^{eT}) \qquad & \boldsymbol{w}^g &= \frac{1}{2}(\boldsymbol{l}^g - \boldsymbol{l}^{gT}) \end{aligned} \tag{4.80}$$

Thus, the deformation rate and the spin rate each can similarly be decomposed into an elastic part and a growth part:

$$\boldsymbol{d} = \boldsymbol{d}^{\mathrm{e}} + \boldsymbol{d}^{\mathrm{g}} \quad \boldsymbol{w} = \boldsymbol{w}^{\mathrm{e}} + \boldsymbol{w}^{\mathrm{g}} \tag{4.81}$$

The right polar decomposition of the elastic deformation gradient yields

$$\boldsymbol{F}^e = \boldsymbol{R}^e \cdot \boldsymbol{U}^e \tag{4.82}$$

where $\boldsymbol{U}^e$ is the right stretch tensor and $\boldsymbol{R}^e$ is the orthogonal tensor. Thus, the elastic Green deformation tensor $\boldsymbol{C}^e$ is given by

$$\boldsymbol{C}^e = \boldsymbol{F}^{eT} \cdot \boldsymbol{F}^e = \boldsymbol{U}^2 \tag{4.83}$$

Based on the elastic Green deformation tensor, the elastic Green strain tensor $\boldsymbol{E}^e$ is defined as follows:

$$\boldsymbol{E}^e = \frac{1}{2}(\boldsymbol{C}^e - \boldsymbol{I}) = \frac{1}{2}(\boldsymbol{F}^{eT} \cdot \boldsymbol{F}^e - \boldsymbol{I}) \tag{4.84}$$

By taking the material derivative of Eq. (4.84), the material derivative $\dot{\boldsymbol{E}}^e$ of the Green strain tensor $\boldsymbol{E}^e$ is obtained as follows:

$$\begin{aligned} \dot{\boldsymbol{E}}^e &= \frac{1}{2}\dot{\boldsymbol{C}}^e \\ &= \frac{1}{2}(\dot{\boldsymbol{F}}^{eT} \cdot \boldsymbol{F}^e + \boldsymbol{F}^{eT} \cdot \dot{\boldsymbol{F}}^e) \\ &= \frac{1}{2}(\boldsymbol{F}^{eT} \cdot \boldsymbol{l}^{eT} \cdot \boldsymbol{F}^e + \boldsymbol{F}^{eT} \cdot \boldsymbol{l}^e \cdot \boldsymbol{F}^e) \\ &= \boldsymbol{F}^{eT} \cdot \boldsymbol{d}^e \cdot \boldsymbol{F}^e \end{aligned} \tag{4.85}$$

Equation (4.85) establishes the relationship between the material derivative of the Green strain tensor and the elastic strain rate.

This section mainly provides a description of deformation in the thermo–chemo–mechanical multifield coupling theory for the interfacial oxidation of TBCs. By decomposing the deformation gradient into an elastic deformation gradient and a growth deformation gradient, elastic and growth velocity gradients are defined. Finally, the deformation rate is decomposed into an elastic deformation rate and a growth deformation rate. Readers who are interested in theories in this area can refer to monographs on continua, such as the Advanced Solid Mechanics by Huang, [47] the Mechanics and Thermodynamics of Continua by Gurtin, 48] and the Introduction to Continuum Mechanics by Reddy [49].

(2) **Law of conservation of mass for multifield coupling interfacial oxidation**

The mass conservation equation is one of the most fundamental equations in continuum mechanics. Interfacial oxidation leads to volume expansion in a TBC and a change in its mass. Let us assume that a volume element $d\Omega_0$ before oxidation has a mass as follows:

$$dM_0 = \rho_0(\boldsymbol{X}, t)d\Omega_0 \tag{4.86}$$

where $\rho_0(\boldsymbol{X}, t)$ is the density in the reference configuration. Then, the mass M_0 of the whole volume Ω_0 is given by

$$M_0 = \int\limits_{\Omega_0} \rho_0(\boldsymbol{X}, t)d\Omega_0 \tag{4.87}$$

Let us assume that the mass of a volume element $d\Omega$ after oxidation is as follows:

$$dM = \rho(\boldsymbol{x}, t)d\Omega \tag{4.88}$$

where $\rho(\boldsymbol{x}, t)$ is the density in the reference configuration. Then, the mass M of the whole volume Ω is given by

$$M = \int\limits_{\Omega} \rho(\boldsymbol{x}, t)d\Omega \tag{4.89}$$

It is assumed that interfacial oxidation of a TBC occurs only during the stage where the reference configuration transitions to the intermediate configuration, oxidation leads only to a volume change, and the oxide density remains constant throughout the oxidation process. Thus, the mass in the intermediate configuration must be equal to that in the current configuration. Hence, the mass M of a volume element $\overline{\Omega}$ in the intermediate configuration can be expressed as

$$dM = \rho_0(\boldsymbol{X}, t)d\overline{\Omega} \tag{4.90}$$

The ratio of the volume after oxidation to that before oxidation, J_G, can be expressed using the Jacobian determinant of the growth deformation gradient. Based on Eqs. (4.86) and (4.90), the following relationship between J_G and the mass of the microelement is obtained:

$$J_G = \det \boldsymbol{F}^g = \frac{d\overline{\Omega}}{d\Omega_0} = \frac{dM}{dM_0} \tag{4.91}$$

The mass change after oxidation can be written in the form of the difference between the mass in the current configuration and that in the reference configuration. Thus, the integral form of the mass conservation equation is given by

$$\int\limits_{\Omega} \rho(\boldsymbol{x}, t) d\Omega - \int\limits_{\Omega_0} \rho_0(\boldsymbol{X}, t) d\Omega_0 = M - M_0 \tag{4.92}$$

Since $d\Omega = Jd\Omega_0$, the differential form of the mass conservation equation in the reference configuration is as follows:

$$\dot{\rho} + \rho div\boldsymbol{v} = \gamma\rho \tag{4.93}$$

where γ is the rate of mass change caused by interfacial oxidation. The mass conservation equation in the reference configuration is given by

$$\frac{D(\rho J)}{Dt} = \gamma J\rho \tag{4.94}$$

The mass in the intermediate configuration is equal to that in the current configuration, i.e.,

$$dM = \rho_0(\boldsymbol{X}) d\overline{\Omega} = \rho(\boldsymbol{x}, t) d\Omega \tag{4.95}$$

Based on the volume ratios between the reference, intermediate, and current configurations, Eq. (4.95) can be transformed to

$$\rho_0(\boldsymbol{X}) J_G d\Omega_0 = \rho(\boldsymbol{x}, t) J d\Omega_0 \tag{4.96}$$

where J_G is the Jacobian determinant of the growth deformation gradient. Equation (4.96) can be simplified to

$$\rho_0(\boldsymbol{X}) = \rho(\boldsymbol{x}, t) J_{\boldsymbol{F}^e} \tag{4.97}$$

where $J_{\boldsymbol{F}^e}$ is the Jacobian determinant of the elastic deformation gradient. By taking the material derivative of Eq. (4.97), we have

$$\dot{\rho} = -\rho \frac{\dot{J}_{F^e}}{J_{F^e}} \tag{4.98}$$

By substituting Eq. (4.98) into Eq. (4.97) and based on the relationship between the Jacobian determinant and its material derivative in Eq. (4.73), we have

$$\frac{\dot{J}}{J} - \frac{\dot{J}_{\mathrm{F}^e}}{J_{\mathrm{F}^e}} = \gamma \tag{4.99}$$

Equation (4.99) can be further simplified to the following relation between the determinant of the growth deformation gradient and the growth rate:

$$\frac{\dot{J}_{\mathrm{G}}}{J_{\mathrm{G}}} = \gamma \tag{4.100}$$

For isotropic growth, it is assumed that the growth deformation gradient has the following form:

$$\boldsymbol{F}^g = \begin{pmatrix} g & 0 & 0 \\ 0 & g & 0 \\ 0 & 0 & g \end{pmatrix} \tag{4.101}$$

where g is the elongation ratio along the three directions. Thus, the relationship between the elongation ratio and the growth rate is given by

$$\frac{3\dot{g}}{g} = \gamma \tag{4.102}$$

Thereby, the mass conservation equation for the interfacial oxidation of TBCs is established. Notably, different from mass conservation in conventional continuum mechanics, interfacial oxidation leads to a continuous increase in the total mass, and therefore, it is necessary to consider the TGO growth, which results in the forms of the mass conservation equation in Eqs. (4.93) and (4.94). In addition, the correlation between the growth deformation gradient and the growth rate is established, laying a foundation for describing the deformation due to interfacial oxidation.

(3) **Governing equations for multifield coupling interfacial oxidation**

Force equilibrium equations can be derived from the conservation of linear momentum. In addition, the equilibrium equation varies between configurations. Here, force equilibrium equations in the current and reference configurations are separately established. The conservation of linear momentum can be described as the time derivative of the linear momentum of an object that is equal to the external force applied to the object. Thus, the linear momentum conservation equation in the current configuration is given by

$$\frac{D\boldsymbol{p}}{\mathrm{Dt}} = \boldsymbol{b} \tag{4.103}$$

where $\boldsymbol{p}$ is the momentum of the object and is calculated as the product of the object's mass and velocity, as shown below

$$\boldsymbol{p}(t) = \int_{\Omega} \rho \boldsymbol{v}(\boldsymbol{x}, t) d\Omega \tag{4.104}$$

The external force $\boldsymbol{b}$ during interfacial oxidation is expressed as

$$\boldsymbol{b}(t) = \int_{\Omega} \rho \boldsymbol{f}(\boldsymbol{x}, t) d\Omega + \int_{\Gamma} \boldsymbol{t}(\boldsymbol{x}, t) d\Gamma + \int_{\Omega} \gamma \rho \boldsymbol{v} d\Omega \tag{4.105}$$

The first, second, and third terms on the right-hand side of Eq. (4.105) represent the volume force, the surface force, and the external force caused by the mass change resulting from interfacial oxidation, respectively. By substituting Eqs. (4.104) and (4.105) into Eq. (4.103), we have

$$\frac{D}{Dt} \int_{\Omega} \rho \boldsymbol{v}(\boldsymbol{x}, t) d\Omega = \int_{\Omega} \rho \boldsymbol{f}(\boldsymbol{x}, t) d\Omega + \int_{\Gamma} \boldsymbol{t}(\boldsymbol{x}, t) d\Gamma + \int_{\Omega} \gamma \rho \boldsymbol{v} d\Omega \tag{4.106}$$

Based on the Reynolds transport theorem, the left-hand side of Eq. (4.106) can be rewritten as follows:

$$\int_{\Omega} \left[\frac{D}{Dt} (\rho \boldsymbol{v}) + div(\boldsymbol{v}) \rho \boldsymbol{v} \right] d\Omega = \int_{\Omega} \left\{ \rho \frac{D}{Dt} \boldsymbol{v} + \boldsymbol{v} \left[\frac{D\rho}{Dt} + \rho div(\boldsymbol{v}) \right] \right\}]d\Omega \tag{4.107}$$

Based on the Cauchy relation and Gauss's theorem, the second term on the right-hand side of Eq. (4.105) can be rewritten as follows:

$$\int_{\Gamma} \boldsymbol{t}(\boldsymbol{x}, t) d\Gamma = \int_{\Gamma} \boldsymbol{n} \cdot \boldsymbol{\sigma} d\Gamma = \int_{\Omega} div \boldsymbol{\sigma} d\Omega \tag{4.108}$$

where $\boldsymbol{\sigma}$ is the Cauchy stress. Based on the mass conservation equation in Eq. (4.93), we have

$$\int_{\Omega} (div \boldsymbol{\sigma} + \rho \boldsymbol{f} - \rho \dot{\boldsymbol{v}}) d\Omega = 0 \tag{4.109}$$

The above equation holds for any region. Thus, the following force equilibrium equation in the current configuration is obtained:

$$div\boldsymbol{\sigma} + \rho \boldsymbol{f} = \rho \dot{\boldsymbol{v}} \tag{4.110}$$

where $\dot{\boldsymbol{v}}$ is the material derivative of the velocity. Normally, the load is applied slowly, so the inertial force is negligible. Thus, the force equilibrium equation is given by

$$div\boldsymbol{\sigma} + \rho \boldsymbol{f} = \boldsymbol{0} \tag{4.111}$$

Physical quantities in the reference configuration are often used in large-deformation analysis, and it is more convenient to use the reference configuration description when solving equilibrium equations. Therefore, the derivation of the force equilibrium equation in the reference configuration is given below. Similarly, based on the conservation of linear momentum, we have

$$\frac{D\boldsymbol{p}}{\mathrm{Dt}} = \boldsymbol{b} \tag{4.112}$$

where $\boldsymbol{p}$ is the momentum of the object and is calculated as the product of the object's mass and velocity, as shown below

$$\boldsymbol{p}(t) = \int\limits_{\Omega_0} \rho_0 \boldsymbol{v}(\boldsymbol{X}, t) d\Omega_0 \tag{4.113}$$

Let $\boldsymbol{b}$ be the external force during interfacial oxidation:

$$\boldsymbol{b}(t) = \int\limits_{\Omega_0} \rho_0 \boldsymbol{f}(\boldsymbol{X}, t) d\Omega_0 + \int\limits_{\Gamma_0} \boldsymbol{t}_0(\boldsymbol{X}, t) d\Gamma_0 \tag{4.114}$$

The terms on the right-hand side of Eq. (4.114) represent the volume and surface forces, respectively. Since interfacial oxidation does not occur in the reference configuration, the external forces do not include the force caused by the mass change. Substituting Eqs. (4.113) and (4.114) into Eq. (4.112) gives

$$\frac{d}{dt} \int\limits_{\Omega_0} \rho_0 \boldsymbol{v}(\boldsymbol{X}, t) d\Omega_0 = \int\limits_{\Omega_0} \rho_0 \boldsymbol{f}(\boldsymbol{X}, t) d\Omega_0 + \int\limits_{\Gamma_0} \boldsymbol{t}_0(\boldsymbol{X}, t) d\Gamma_0 \tag{4.115}$$

By moving the material derivative on the left-hand side of Eq. (4.115) inside the integral, we have

$$\frac{d}{dt} \int\limits_{\Omega_0} \rho_0 \boldsymbol{v}(\boldsymbol{X}, t) d\Omega_0 = \int\limits_{\Omega_0} \rho_0 \frac{\partial \boldsymbol{v}(\boldsymbol{X}, t)}{\partial t} d\Omega_0 \tag{4.116}$$

Based on the Cauchy relation and Gauss's theorem, the second term on the right-hand side of Eq. (4.114) can be rewritten as follows:

$$\int_{\Gamma_0} \boldsymbol{t}_0(\boldsymbol{X}, t) d\Gamma_0 = \int_{\Gamma_0} \boldsymbol{P} \cdot \boldsymbol{N} d\Gamma_0 = \int_{\Omega_0} \mathrm{Div} \boldsymbol{P}^T d\Omega_0 \tag{4.117}$$

where $\boldsymbol{P}$ is the first Piola–Kirchhoff stress. By substituting Eqs. (4.116) and (4.117) into Eq. (4.115), we obtain the following force equilibrium equation in the reference configuration:

$$\int_{\Omega_0} \left[DivP^T + \rho_0 f - \rho_0 \frac{\partial v(X, t)}{\partial t} \right] d\Omega_0 = 0 \tag{4.118}$$

The above equation holds for any region. Thus, the force equilibrium equation in the reference configuration can be obtained:

$$\boldsymbol{Div} \boldsymbol{P}^T + \rho_0 \boldsymbol{f} = \rho_0 \frac{\partial \boldsymbol{v}(\boldsymbol{X}, t)}{\partial t} \tag{4.119}$$

By ignoring the inertial force, the force equilibrium equation in the reference configuration can be written as follows:

$$\mathbf{Div} \boldsymbol{P}^T + \rho_0 \boldsymbol{f} = \mathbf{0} \tag{4.120}$$

Because the first Piola–Kirchhoff stress is not a symmetric tensor, the second Piola–Kirchhoff stress is generally used in the expression of the force equilibrium equation in the reference configuration. Based on the transformation relationship between the first and second Piola–Kirchhoff stresses (i.e., $\boldsymbol{P} = \boldsymbol{F} \cdot \boldsymbol{T}$), the following force equilibrium equation is obtained:

$$\mathbf{Div}(\boldsymbol{T} \cdot \boldsymbol{F}^T) + \rho_0 \boldsymbol{f} = \mathbf{0} \tag{4.121}$$

By cross multiplying each term in the linear momentum conservation equation with position vector $\boldsymbol{x}$, we obtain the integral form of the angular momentum conservation equation as follows:

$$\frac{D}{Dt} \int_{\Omega} \rho \boldsymbol{x} \times \boldsymbol{v} d\Omega = \int_{\Omega} \rho \boldsymbol{x} \times \boldsymbol{f} d\Omega + \int_{\Gamma} \boldsymbol{x} \times \boldsymbol{t} d\Gamma + \int_{\Omega} \rho \gamma \boldsymbol{x} \times \boldsymbol{v} d\Omega \tag{4.122}$$

To facilitate the derivation, the above equation is rewritten in the following form of components:

$$\frac{D}{Dt}\int\limits_{\Omega} \rho e_{ijk} x_i v_j d\Omega = \int\limits_{\Gamma} e_{ijk} x_i t_j d\Gamma + \int\limits_{\Omega} \rho e_{ijk} x_i f_j d\Omega + \int\limits_{\Omega} \rho \gamma e_{ijk} x_i v_j d\Omega \tag{4.123}$$

where e_{ijk} is the permutation symbol. Based on the Reynolds transport theorem, the material derivative on the left-hand side of Eq. (4.123) can be rewritten as

$$\begin{aligned} &\int\limits_{\Omega} \left[\rho e_{ijk} \frac{D}{Dt}(x_i v_j) + e_{ijk} x_i v_j (\frac{D\rho}{Dt} + \rho v_{i,i}) \right] d\Omega \\ &= \int\limits_{\Omega} e_{ijk}(x_i \sigma_{pj})_{,p} d\Omega + \int\limits_{\Omega} \rho e_{ijk} x_i f_j d\Omega + \int\limits_{\Omega} \rho \gamma e_{ijk} x_i v_j d\Omega \end{aligned} \tag{4.124}$$

Based on the mass conservation equation in the reference configuration, Eq. (4.124) can be simplified to

$$\begin{aligned} \int\limits_{\Omega} \rho e_{ijk}(v_i v_j + x_i \frac{Dv_j}{Dt}) d\Omega &= \int\limits_{\Omega} e_{ijk}(x_i \sigma_{pj})_{,p} d\Omega + \int\limits_{\Omega} \rho e_{ijk} x_i f_j d\Omega \\ &= \int\limits_{\Omega} e_{ijk}(x_i \sigma_{pj,p} + \delta_{ip}\sigma_{pj} + \rho x_i f_j) d\Omega \end{aligned} \tag{4.125}$$

Since $e_{ijk} v_i v_j = 0$, Eq. (4.125) can bc further simplified to

$$\int\limits_{\Omega} e_{ijk} \left[x_i (\sigma_{pj,p} + \rho f_j - \frac{Dv_j}{Dt}) + \sigma_{ij} \right] d\Omega = 0 \tag{4.126}$$

Equatiom (4.126) contains a force equilibrium equation that holds for any region. Thus, we have

$$e_{ijk}\sigma_{ij} = 0 \tag{4.127}$$

To satisfy the above relationship, the Cauchy stress must be a symmetric tensor, i.e.,

$$\boldsymbol{\sigma} = \boldsymbol{\sigma}^T \tag{4.128}$$

Based on the relationship between the first Piola–Kirchhoff stress $\boldsymbol{P}$ and the second Piola–Kirchhoff stress $\boldsymbol{T}$ (i.e., $\boldsymbol{T} = \boldsymbol{F}^{-1} \cdot \boldsymbol{P}$), we have

$$\boldsymbol{P} \cdot \boldsymbol{F}^T = \boldsymbol{F} \cdot \boldsymbol{P}^T \tag{4.129}$$

Interfacial oxidation of a TBC mainly involves the diffusion of oxygen in air to the TC/BC interface via the TC layer and the subsequent reaction between the oxygen and the Al in the BC layer to form Al_2O_3. During the oxidation process, the oxygen concentration follows the law of conservation, i.e., the rate of change in the oxygen concentration is equal to the oxygen flux into the coating through the interface minus the amount of oxygen consumed by the reaction. Thus, the oxygen concentration conservation equation is given by

$$\int_{\Omega_0} \frac{\partial c(\boldsymbol{X}, t)}{\partial t} d\Omega_0 = -\int_{\Gamma_0} \boldsymbol{j} d\Gamma_0 - \int_{\Omega_0} S d\Omega_0 \tag{4.130}$$

where c is the oxygen concentration, $\boldsymbol{j}$ is the oxygen flux per unit reference area, and S is the amount of oxygen consumed by the oxidation reaction. Through a Gaussian integral transform, Eq. (4.130) is transformed to

$$\int_{\Omega_0} \frac{\partial c}{\partial t} d\Omega_0 = -\int_{\Omega_0} (Div\,\boldsymbol{j} + S) d\Omega_0 \tag{4.131}$$

The above equation holds for any region in the reference configuration. Thus, we have

$$\frac{\partial c}{\partial t} + Div\,\boldsymbol{j} = -S \tag{4.132}$$

The oxygen diffusion flux follows Fick's first law of diffusion:

$$\boldsymbol{j} = -mc\nabla\mu \tag{4.133}$$

where m is the migration rate of oxygen and μ is the chemical potential of oxygen, which is conjugate with the oxygen concentration. The constitutive relations are required for establishing the relationship between the chemical potential and the concentration. Next, based on the conservation of energy, the relationship between the concentration and the chemical potential in the chemical field is derived in detail.

A variable n is introduced to represent the volume fraction and hence the growth process of TGOs. $n = 1$ means that this region is composed completely of TGO, $n = 0$ indicates that this region is composed completely of the BC layer, and $0 < n < 1$ means that this region is composed of a mixture of TGO and the BC layer. Then, the volume fraction of the BC layer is given by $(1 - n)$. In other words, in any region of the TGO and BC layers, their total volume fraction must be 1. Based on Eqs. (4.4) and (4.5), the governing equations for the thermo–mechano-chemical multifield coupling chemical field in a TBC during interfacial oxidation (i.e., a thermo–mechano-chemical coupling model for the oxygen concentration and TGO growth) are obtained as follows:

$$\begin{cases} \dfrac{\partial c}{\partial t} - \text{Div}(mc\nabla\mu) = -M\zeta(1-n)c \\ \dfrac{\partial n}{\partial t} = \zeta(1-n)c \end{cases} \tag{4.134}$$

(4) **Thermo–chemo–mechanical multifield coupling constitutive relations for interfacial oxidation**

All physical processes must conform to the law of the conservation of energy, and the total energy of a TBC during interfacial oxidation is no exception. According to the first law of thermodynamics, the rate of change in the total energy of a TBC is equal to the sum of the rates of change in the work done by the external force and in other types of energy such as thermal energy and chemical energy. Thus, the principle of the conservation of energy for the interfacial oxidation of TBCs is expressed as

$$\frac{D}{Dt}(K+U) = W + Q + C \tag{4.135}$$

where K is the kinetic energy, U is the internal energy, W is the power of the external force, Q is the power supplied by the heat source and heat flow, and C is the power supplied by the chemical reaction.

Each type of energy can be described in the reference configuration or the current configuration. Here, we focus on the description of the conservation equations in the reference configuration. The kinetic energy K is primarily the energy generated by the macroscopic motion of an object and, in the reference configuration, can be expressed as

$$K = \frac{1}{2}\int_{\Omega_0} \rho_0 \boldsymbol{v} \cdot \boldsymbol{v} d\Omega_0 \tag{4.136}$$

where $\boldsymbol{v}$ is the velocity in the reference configuration.

The internal energy mainly includes other forms of energy (e.g., energy involved in microscopic molecular motion and strain energy) and, in the reference configuration, is given by

$$U = \int_{\Omega_0} \rho_0 e d\Omega_0 \tag{4.137}$$

where e is the internal energy per unit mass.

Within the domain, the power of the volume force $\boldsymbol{f}$ per unit volume and the surface force $\boldsymbol{t}_0$ per unit surface area is expressed as

$$W = \int_{\Gamma_0} \boldsymbol{t}_0 \cdot \boldsymbol{v} d\Gamma_0 + \int_{\Omega_0} \rho_0 \boldsymbol{f} \cdot \boldsymbol{v} d\Omega_0 + \int_{\Omega_0} \rho_0 \gamma (e + \frac{1}{2} \boldsymbol{v} \cdot \boldsymbol{v}) d\Omega_0 \tag{4.138}$$

The third term on the right-hand side of Eq. (4.138) represents the increase in the energy, consisting of internal energy and kinetic energy, caused by interfacial oxidation. By transforming the boundary integral of the surface force in Eq. (4.138) into a volume integral based on the Cauchy relation and Gauss's theorem, we have

$$\begin{aligned} W &= \int_{\Gamma_0} \boldsymbol{t}_0 \cdot \boldsymbol{v} d\Gamma_0 + \int_{\Omega_0} \rho_0 \boldsymbol{f} \cdot \boldsymbol{v} d\Omega_0 + \int_{\Omega_0} \rho_0 \gamma (e + \frac{1}{2} \boldsymbol{v} \cdot \boldsymbol{v}) d\Omega_0 \\ &= \int_{\Gamma_0} (\boldsymbol{N} \cdot \boldsymbol{P}^T) \cdot \boldsymbol{v} d\Gamma_0 + \int_{\Omega_0} \rho_0 \boldsymbol{f} \cdot \boldsymbol{v} d\Omega_0 + \int_{\Omega_0} \rho_0 \gamma (e + \frac{1}{2} \boldsymbol{v} \cdot \boldsymbol{v}) d\Omega_0 \\ &= \int_{\Omega_0} [\nabla \cdot (\boldsymbol{P}^T \cdot \boldsymbol{v}) + \rho \boldsymbol{f} \cdot \boldsymbol{v}] d\Omega_0 + \int_{\Omega_0} \rho_0 \gamma (e + \frac{1}{2} \boldsymbol{v} \cdot \boldsymbol{v}) d\Omega_0 \\ &= \int_{\Omega_0} [(\nabla \cdot \boldsymbol{P}^T + \rho \boldsymbol{f}) \cdot \boldsymbol{v} + \boldsymbol{P}^T : \nabla \boldsymbol{v} + \rho_0 \gamma (e + \frac{1}{2} \boldsymbol{v} \cdot \boldsymbol{v})] d\Omega_0 \\ &= \int_{\Omega_0} [\rho_0 \frac{D\boldsymbol{v}}{Dt} \cdot \boldsymbol{v} + \boldsymbol{P}^T : \nabla \boldsymbol{v} + \rho_0 \gamma (e + \frac{1}{2} \boldsymbol{v} \cdot \boldsymbol{v})] d\Omega_0 \end{aligned} \tag{4.139}$$

where $\boldsymbol{P}$ is the first Piola–Kirchhoff stress.

The power supplied by the heat source and heat flow is given by

$$Q = \int_{\Omega_0} \rho_0 r_m d\Omega_0 - \int_{\Gamma_0} \boldsymbol{q} \cdot \boldsymbol{N} d\Gamma_0 = \int_{\Omega_0} (\rho_0 r_m - \nabla \cdot \boldsymbol{q}) d\Omega_0 \tag{4.140}$$

where r_m is the heat source per unit mass and $\boldsymbol{q}$ is the heat flow per unit area.

Based on the heat-flow power, the power generated by chemical diffusion is expressed as

$$C = -\int_{\Gamma_0} \mu \boldsymbol{j} \cdot \boldsymbol{N} d\Gamma_0 \tag{4.141}$$

where $\boldsymbol{j}$ is the diffusion flow of oxygen.

Substituting Eqs. (4.136)–(4.141) into Eq. (4.135), we obtain the integral expression of the energy conservation equation for interfacial oxidation of TBCs as follows:

$$\frac{D}{Dt}\int_{\Omega_0} \rho_0(\frac{1}{2}\boldsymbol{v}\cdot\boldsymbol{v}+e)d\Omega_0$$
$$=\frac{1}{2}\frac{D}{Dt}\int_{\Omega_0} \rho_0\boldsymbol{v}\cdot\boldsymbol{v}d\Omega_0+\int_{\Omega_0}(\boldsymbol{P}^T:\dot{\boldsymbol{F}})d\Omega_0+\int_{\Omega_0}\rho_0\gamma(e+\frac{1}{2}\boldsymbol{v}\cdot\boldsymbol{v})d\Omega_0 \tag{4.142}$$
$$+\int_{\Omega_0}(-\nabla\cdot\boldsymbol{q}+\rho_0 r_m)d\Omega_0-\int_{\Omega_0}(\mu\nabla\cdot\boldsymbol{j}+\boldsymbol{j}\cdot\nabla\mu)d\Omega_0$$

By simplifying Eq. (4.142) based on the Reynolds transport theorem and Gauss's theorem and reorganizing the terms in the resulting equation, we obtain the following differential energy conservation equation:

$$\rho_0\dot{e}=\boldsymbol{P}^T:\dot{\boldsymbol{F}}-\mu Div\boldsymbol{j}-\boldsymbol{j}\cdot\nabla\mu-Div\boldsymbol{q}+r_m\rho_0 \tag{4.143}$$

The interfacial oxidation of a TBC also needs to conform to the principle of entropy increase, i.e., the second law of thermodynamics. Let η be the entropy per unit volume. Then, the total entropy is given by

$$H=\int_{\Omega_0} s_m\rho_0 d\Omega_0 \tag{4.144}$$

Thus, the second law of thermodynamics can be written in the following form:

$$\frac{D}{Dt}\int_{\Omega_0}\rho_0 s_m d\Omega_0\geq-\int_{\Gamma_0}\frac{\boldsymbol{q}\cdot\boldsymbol{N}}{\theta}d\Gamma_0+\int_{\Omega_0}\frac{r_m}{\theta}\rho_0 d\Omega_0 \tag{4.145}$$

where θ is the temperature and the first and second terms on the right-hand side represent the entropy flow and entropy source, respectively. Equation (4.145) is the integral form of the second law of thermodynamics and is also known as the entropy inequality or the Clausius–Duhem inequality. Through an integral transform of Eq. (4.145), the following local inequality of the second law of thermodynamics is obtained:

$$\rho_0\dot{s}_m\geq\frac{r_m\rho}{\theta}-\nabla\cdot(\frac{\boldsymbol{q}}{\theta})=\frac{1}{\theta}(r_m\rho_0-\nabla\cdot\boldsymbol{q})+\frac{1}{\theta^2}\boldsymbol{q}\cdot\nabla\theta \tag{4.146}$$

According to the energy conservation equation in Eq. (4.147), the right-hand side of Eq. (4.146) can be rewritten in the following form:

$$\frac{r_m\rho_0}{\theta}-\nabla\cdot(\frac{\boldsymbol{q}}{\theta})=\frac{1}{\theta}(r_m\rho_0-\nabla\cdot\boldsymbol{q})+\frac{1}{\theta^2}\boldsymbol{q}\cdot\nabla\theta$$
$$=\frac{1}{\theta}(\rho_0\dot{e}-\boldsymbol{P}^T:\dot{\boldsymbol{F}}+\mu\nabla\cdot\boldsymbol{j}+\boldsymbol{j}\cdot\nabla\mu+\frac{1}{\theta}\boldsymbol{q}\cdot\nabla\theta) \tag{4.147}$$

Based on Eqs. (4.146) and (4.147), the entropy inequality can be rewritten as follows:

$$\rho_0 \dot{s}_m \geq \frac{1}{\theta}(\rho_0 \dot{e} - \boldsymbol{P}^T : \dot{\boldsymbol{F}} + \mu \nabla \cdot \boldsymbol{j} + \boldsymbol{j} \cdot \nabla \mu + \frac{1}{\theta} \boldsymbol{q} \cdot \nabla \theta) \tag{4.148}$$

Finally, the entropy inequality for the interfacial evolution of TBCs is given by

$$\rho_o(\dot{e} - \theta \dot{s}_m) - \boldsymbol{P}^T : \dot{\boldsymbol{F}} + \mu \nabla \cdot \boldsymbol{j} + \boldsymbol{j} \cdot \nabla \mu + \frac{1}{\theta} \boldsymbol{q} \cdot \nabla \theta \leq 0 \tag{4.149}$$

For convenience, the Helmholtz free energy is introduced as a thermodynamic function and is related to the internal energy, entropy, and temperature through the following equation:

$$\psi = e - \theta s_m \tag{4.150}$$

where ψ is the Helmholtz free energy per unit mass.

Taking the derivative of Eq. (4.150), we obtain the following relationship between the Helmholtz free energy and the internal energy in the form of rates:

$$\dot{\psi} = \dot{e} - \dot{\theta} s_m - \theta \dot{s}_m \tag{4.151}$$

By substituting Eq. (4.151) into the entropy inequality for interfacial oxidation of TBCs in Eq. (4.149), we have

$$\rho_0(\dot{\psi} - s_m \dot{\theta}) - \boldsymbol{P}^T : \dot{\boldsymbol{F}} + \mu \nabla \cdot \boldsymbol{j} + \boldsymbol{j} \cdot \nabla \mu + \frac{1}{\theta} \boldsymbol{q} \cdot \nabla \theta \leq 0 \tag{4.152}$$

The second term on the left-hand side of Eq. (4.152) represents the deformation power. We know that for the deformation power, the stress is conjugate, or work conjugate, with the deformation rate or the strain rate. The stress descriptions should match the corresponding strain descriptions. For example, the first Piola–Kirchhoff stress is conjugate with the deformation gradient rate, the second Piola–Kirchhoff stress is conjugate with the Green strain rate, and the Cauchy stress is conjugate with the deformation rate. Thus, the deformation power can be written as the double dot product of the stress and the corresponding conjugate strain (deformation) rate. Let w^{int} be the deformation power per unit volume in the reference configuration. Then, we have

$$w = J\boldsymbol{\sigma} : \boldsymbol{d} = \boldsymbol{P} : \dot{\boldsymbol{F}} = \boldsymbol{T} : \dot{\boldsymbol{E}} \tag{4.153}$$

For thermo–mechano-chemical multifield coupling interfacial oxidation of TBCs, this conjugate relationship similarly exists between various stress measures and strain-rate measures. The second term on the left-hand side of Eq. (4.152) represents the deformation power. However, when previously describing the deformation induced by interfacial oxidation, we decomposed the total deformation gradient into an elastic deformation gradient and a growth deformation gradient. Further analysis is required to determine the form and meaning of the deformation power discussed here because the deformation power plays a crucial role in the subsequent constitutive relations. Substituting Eq. (4.71) into the deformation power gives

$$\begin{aligned} &\boldsymbol{P} : \dot{\boldsymbol{F}} \\ &= \boldsymbol{P} : (\dot{\boldsymbol{F}}^{\mathrm{e}} \cdot \boldsymbol{F}^{\mathrm{g}} + \boldsymbol{F}^{\mathrm{e}} \cdot \dot{\boldsymbol{F}}^{\mathrm{g}}) \\ &= (\boldsymbol{P} \cdot \boldsymbol{F}^{\mathrm{gT}}):\dot{\boldsymbol{F}}^{\mathrm{e}} + (\boldsymbol{F}^{\mathrm{eT}} \cdot \boldsymbol{P}):\dot{\boldsymbol{F}}^{\mathrm{g}} \\ &= (J\boldsymbol{\sigma} \cdot \boldsymbol{F}^{\mathrm{e-T}}):\dot{\boldsymbol{F}}^{\mathrm{e}} + (\boldsymbol{F}^{\mathrm{eT}} \cdot J\boldsymbol{\sigma} \cdot \boldsymbol{F}^{\mathrm{e-T}} \cdot \boldsymbol{F}^{\mathrm{g-T}}):\dot{\boldsymbol{F}}^{\mathrm{g}} \\ &= (J\boldsymbol{F}^{\mathrm{e-1}} \cdot \boldsymbol{\sigma} \cdot \boldsymbol{F}^{\mathrm{e-T}}):\boldsymbol{F}^{\mathrm{eT}} \cdot \dot{\boldsymbol{F}}^{\mathrm{e}} + (\boldsymbol{F}^{\mathrm{eT}} \cdot J\boldsymbol{\sigma} \cdot \boldsymbol{F}^{\mathrm{e-T}} \cdot \boldsymbol{F}^{\mathrm{g-T}}):\dot{\boldsymbol{F}}^{\mathrm{g}} \\ &= \boldsymbol{T}:\dot{\boldsymbol{E}}^{e} + (\boldsymbol{C}^{\mathrm{e}} \cdot \boldsymbol{T}) : (\dot{\boldsymbol{F}}^{\mathrm{g}} \cdot \boldsymbol{F}^{\mathrm{g-1}}) \\ &= \boldsymbol{T}:\dot{\boldsymbol{E}}^{e} + \boldsymbol{M}^{e} : \boldsymbol{L}^{g} \end{aligned} \tag{4.154}$$

where $\boldsymbol{T}$ is the second Piola–Kirchhoff stress and $\boldsymbol{M}^e$ is the Mandel stress, which are often used in large plastic deformation analysis. $\boldsymbol{T}$ and $\boldsymbol{M}^e$ are expressed as

$$\boldsymbol{T} = J\boldsymbol{F}^{\mathrm{e}^{-1}} \cdot \boldsymbol{\sigma} \cdot \boldsymbol{F}^{\mathrm{e}^{-\mathrm{T}}} \quad \boldsymbol{M}^{e} = \boldsymbol{C}^{\mathrm{e}} \cdot \boldsymbol{S} = J\boldsymbol{F}^{\mathrm{eT}} \cdot \boldsymbol{\sigma} \cdot \boldsymbol{F}^{\mathrm{e}^{-\mathrm{T}}} \tag{4.155}$$

Equation (4.154) shows that as a result of the decomposition of the deformation gradient, the deformation power can be decomposed into two parts, namely an elastic deformation power and a deformation power caused by interfacial oxidation.

During the interfacial oxidation of a TBC, the Helmholtz free energy is a thermodynamic function of the Green strain tensor, temperature, and oxygen concentration. Thus, the Helmholtz free energy can be expressed as a function of these independent variables:

$$\psi = \hat{\psi}(\boldsymbol{E}^{e}, \theta, c) \tag{4.156}$$

where $\boldsymbol{E}^e$ is the elastic Green strain tensor, θ is the temperature, and c is the oxygen concentration. Based on Eq. (4.156), the relationship of the Helmholtz free energy with the Green strain tensor, temperature, and oxygen concentration is given by

$$\dot{\psi} = \frac{\partial \hat{\psi}}{\partial \boldsymbol{E}^{\mathrm{e}}} : \dot{\boldsymbol{E}}^{\mathrm{e}} + \frac{\partial \hat{\psi}}{\partial \theta}\dot{\theta} + \frac{\partial \hat{\psi}}{\partial c}\dot{c} \tag{4.157}$$

By substituting Eqs. (4.157) and (4.154) into the entropy inequality in Eq. (4.152) and reorganizing the terms in the resulting equation, we have

$$\rho_0(\frac{\partial \hat{\psi}}{\partial \boldsymbol{E}^{\mathrm{e}}} : \dot{\boldsymbol{E}}^{\mathrm{e}} + \frac{\partial \hat{\psi}}{\partial \theta}\dot{\theta} + \frac{\partial \hat{\psi}}{\partial c}\dot{c} + s_m\dot{\theta}) - \boldsymbol{T} : \dot{\boldsymbol{E}}^{e} - \boldsymbol{M}_e : \boldsymbol{L}^g + \mu\dot{c} + \boldsymbol{j}\cdot\nabla\mu + \frac{1}{\theta}\boldsymbol{q}\cdot\nabla\theta \leq 0 \tag{4.158}$$

By combining like terms, we transform Eq. (4.158) to

$$\begin{aligned}(\rho_0\frac{\partial \hat{\psi}}{\partial \boldsymbol{E}^{\mathrm{e}}} - \boldsymbol{T}) : \dot{\boldsymbol{E}}^{\mathrm{e}} + \rho_0(\frac{\partial \hat{\psi}}{\partial \theta} + s_m)\dot{\theta} + (\rho_0\frac{\partial \hat{\psi}}{\partial c} + \mu)\dot{c} \\ -\boldsymbol{M}_e : \boldsymbol{L}^g + \boldsymbol{j}\cdot\nabla\mu + \frac{1}{\theta}\boldsymbol{q}\cdot\nabla\theta \leq 0\end{aligned} \tag{4.159}$$

The entropy inequality holds for any values of the independent variables (i.e., the Green strain tensor, temperature, and oxygen concentration), and, consequently, their coefficients must be zero. Thus, the following thermo–mechano-chemical coupling constitutive relations for interfacial oxidation of TBCs are obtained:

$$\begin{cases} \boldsymbol{T} = \rho_0\dfrac{\partial \hat{\psi}}{\partial \boldsymbol{E}^{\mathrm{e}}} \\ s_m = -\dfrac{\partial \hat{\psi}}{\partial \theta} \\ \mu = \rho_0\dfrac{\partial \hat{\psi}}{\partial c} \end{cases} \tag{4.160}$$

Hence, only the last three terms on the left-hand side of Eq. (4.159) remain to represent the entropy increase inequality. In fact, these three terms represent the energy dissipation caused by forces, the energy dissipation caused by chemical reactions, and the entropy increase caused by heat transfer, respectively, during the interfacial oxidation process. Thus, the entropy inequality for interfacial oxidation of TBCs can be simplified as follows:

$$\boldsymbol{M}_e : \boldsymbol{L}^g - \boldsymbol{j}\cdot\nabla\mu - \frac{1}{\theta}\boldsymbol{q}\cdot\nabla\theta \geq 0 \tag{4.161}$$

Based on the thermo–mechano-chemical coupling constitutive relations for interfacial oxidation, the Helmholtz free energy rate is given by

$$\dot{\psi} = \frac{1}{\rho_0}\boldsymbol{T} : \dot{\boldsymbol{E}}^{\mathrm{e}} - s_m\dot{\theta} + \frac{1}{\rho_0}\mu\dot{c} \tag{4.162}$$

Based on the relationship between the internal energy and the Helmholtz free energy, the energy conservation equation can be written as follows:

$$\begin{aligned}\rho_0\dot{e} &= \rho_0\theta\dot{s}_m + \boldsymbol{T} : \dot{\boldsymbol{E}}^{\mathrm{e}} + \mu\dot{c} \\ &= \boldsymbol{P} : \dot{\boldsymbol{F}} - \mu Div\boldsymbol{j} - \boldsymbol{j}\cdot\nabla\mu - Div\boldsymbol{q} + r_m\rho_0\dot{c}\end{aligned} \tag{4.163}$$

By simplifying the deformation power and based on the constitutive relations, we can rewrite Eq. (4.163) as follows:

$$\rho_0 \theta \dot{s}_m = \boldsymbol{M}_e : \boldsymbol{L}^g - \mu Div \boldsymbol{j} - \boldsymbol{j} \cdot \nabla \mu - Div \boldsymbol{q} + r_m \rho_0 \dot{c} \tag{4.164}$$

The constitutive relations for the temperature field can be expressed in terms of internal energy and specific heat or in terms of entropy and Helmholtz free energy. Then, the specific heat C can be given by

$$C = \frac{\partial e}{\partial \theta} = \frac{\partial \hat{\psi}}{\partial \theta} + s_m + \theta \frac{\partial s_m}{\partial \theta} = -\theta \frac{\partial^2 \hat{\psi}}{\partial \theta^2} \tag{4.165}$$

Based on the constitutive relations in Eqs. (4.160), (4.165) can be rewritten as follows:

$$\theta \dot{s}_m = -\theta \frac{\partial \boldsymbol{T}}{\partial \theta} : \dot{\boldsymbol{C}}^e + C\dot{\theta} - \theta \frac{\partial \mu}{\partial \theta} \dot{c} \tag{4.166}$$

By moving the terms containing the specific heat in the above equation to one side of the equation, we obtain the governing equation for the temperature field as follows:

$$C\dot{\theta} = \theta \dot{s}_m + \theta \frac{\partial \boldsymbol{T}}{\partial \theta} : \dot{\boldsymbol{C}}^e + \theta \frac{\partial \mu}{\partial \theta} \dot{c} \tag{4.167}$$

where $\boldsymbol{T}$ is the previously mentioned second Piola–Kirchhoff stress and $\boldsymbol{C}^e$ is the Green deformation tensor. In the above equation, the first term represents the temperature change caused by the heat transfer and heat source, the second term represents the temperature change caused by the stress, and the third term represents the effect of the chemical potential on the temperature change.

By substituting Eq. (4.164) into Eq. (4.167), we obtain a governing equation for the temperature field. This equation clearly shows the coupling effects of various fields on the temperature field, including the effects of the stress and chemical fields on the temperature. The ultimate governing equation for the temperature field is given by

$$C\dot{\theta} = \boldsymbol{M}_e : \boldsymbol{L}^g - \boldsymbol{j} \cdot \nabla \mu - Div \boldsymbol{q} + r_m \rho_0 + \theta \frac{\partial \boldsymbol{T}}{\partial \theta} : \dot{\boldsymbol{C}}^e + \theta \frac{\partial \mu}{\partial \theta} \dot{c} \tag{4.168}$$

This section focuses mainly on establishing the thermo–mechano-chemical coupling energy conservation equation and entropy inequality for the interfacial oxidation of TBCs based on the principle of conservation of energy and the principle of entropy increase. On this basis, the thermo–mechano-chemical multifield coupling constitutive relations for the interfacial oxidation of TBCs are obtained, and the governing equation for the temperature field is provided.

The thermo–mechano-chemical coupling Helmholtz free energy during interfacial oxidation of a TBC mainly includes the thermoelastic strain energy and the chemical energy caused by material diffusion. The specific form of this Helmholtz free energy can be referred to as the free-energy form of the thermo–mechano-chemical multifield coupling large-deformation theory for interfacial oxidation of TBCs established by Loeffel and Anand [50]. However, note that a hyperelastic model, shown below, is used to represent the thermoelastic strain energy:

$$\hat{\psi} = J^G\left[\frac{1}{2}G(I^e - 3) + \lambda(\ln J^e)^2 - G\ln J^e - 3K\alpha(\theta - \theta_0)tr(\boldsymbol{E}^e) + C(\theta - \theta_0)\right] \\ +\mu_0 c + R\theta c(\ln c - 1) \tag{4.169}$$

where G and λ are Lame's constants, I^e is the first invariant of the elastic Green strain tensor $\boldsymbol{E}^e$, K is the bulk modulus, α is the thermal expansion coefficient, C is the specific heat, R is the universal gas constant, J^e is the Jacobian determinant of the elastic deformation gradient tensor, and J^G is the Jacobian determinant of the growth deformation gradient tensor.

By substituting the specific form of the Helmholtz free energy in Eq. (4.169) into the thermo–mechano-chemical coupling constitutive relations for interfacial oxidation of Eq. (4.16), we obtain the following specific forms of the constitutive relations for interfacial oxidation:

$$\begin{cases} \boldsymbol{T} = J^G\left[G\boldsymbol{1} + (\lambda\ln J^e - G)\overset{-1}{\boldsymbol{C}^e} - 3K\alpha(\theta - \theta_0)\boldsymbol{1}\right] \\ s_m = J^G[3K\alpha(tr\boldsymbol{E}^e) - C] - Rc(\ln c - 1) \\ \mu = \mu_0 + R\theta\ln c - \frac{1}{3}\beta tr\boldsymbol{T} + \frac{1}{3}\beta C(\theta - \theta_0) \end{cases} \tag{4.170}$$

The above equation shows that the variables of all the fields are coupled with one another during the interfacial oxidation of a TBC. The chemical and temperature fields affect the stress field, and at the same time, the stress field affects the chemical and temperature fields. In short, the field variables are all coupled with one another.

If the contribution of the temperature field to the chemical potential is not taken into consideration, the constitutive relation for the chemical field can be written as follows:

$$\mu = \mu_0 + R\theta\ln c - \frac{1}{3}\beta \mathrm{tr}\boldsymbol{T} \tag{4.171}$$

By substituting Eq. (4.171) into the previously introduced TGO growth model in Eq. (4.134), we obtain the following theoretical thermo–mechano-chemical coupling growth model for the interfacial oxidation of TBCs:

$$\begin{cases} \frac{\partial c}{\partial t} + \nabla \cdot \left(-D_0 \nabla c + \frac{\beta c}{R\theta} \nabla T_m \right) = -M\zeta(1-n)c \\ \frac{\partial n}{\partial t} = \zeta(1-n)c \end{cases} \tag{4.172}$$

The initial and boundary conditions for each field are required to solve the thermo–mechano-chemical multifield coupling problems of the interfacial oxidation of TBCs. The conditions for the stress field include the displacement and surface-force boundary conditions. The conditions for the chemical field include the concentration boundary conditions as well as the boundary and initial conditions for the diffusion flux. The conditions for the temperature field include the temperature and heat-flow boundary conditions. These boundary conditions can be described as follows:

$$\begin{cases} \chi = \tilde{\chi} & \text{At boundary } \Gamma_1 \\ \boldsymbol{T} \cdot \boldsymbol{N} = \boldsymbol{t}_0 & \text{At boundary } \Gamma_2 \\ |\mathrm{c} = \breve{\mathrm{c}} & \text{At boundary } \Gamma_c \\ | - \mathrm{D}(\nabla \mu) \cdot \boldsymbol{N} = \breve{j} & \text{At boundary } \Gamma_j \\ |\theta = \breve{\theta} & \text{At boundary } \Gamma_\theta \\ | - \mathrm{r_m}(\nabla \theta) \cdot \boldsymbol{N} = \breve{\mathrm{q}} & \text{At boundary } \Gamma_q \end{cases} \tag{4.173}$$

The initial conditions for thermo–mechano-chemical multifield coupling interfacial oxidation are given by

$$\chi(\boldsymbol{X}, 0) = \chi_0(\boldsymbol{X}) c(\boldsymbol{X}, 0) = c_0(\boldsymbol{X}) \theta(\boldsymbol{X}, 0) = \theta_0(\boldsymbol{X}) n(\boldsymbol{X}, 0) = n_0(\boldsymbol{X}) \tag{4.174}$$

4.3.3 Analysis of the Thermo–Mechano-Chemical Coupling Growth Patterns and Mechanisms During Interfacial Oxidation

The theoretical thermo–mechano-chemical coupling framework for interfacial oxidation of TBCs includes the governing and constitutive equations and boundary conditions for each physical field. Complex nonlinear coupling equations can be solved only with the FEM. In fact, the FEM transforms the governing equation for a physical field from a strong form to a weak form, subsequently discretizes the weak form of the governing equation with finite elements (FEs), and ultimately transforms the governing equation into a system of linear equations and solves it. In the FE software COMSOL, a coupling problem can be solved as long as the weak forms of the physical fields are provided. We introduce trial functions to transform the strong forms of thermo–mechano-chemical coupling interfacial oxidation of a TBC in Eqs. (4.172) and (4.12) into weak forms. The coupling growth of TGOs involves two main variables, namely the oxygen concentration c and the TGO volume fraction n. Let δc

and δn be their respective trial functions. The displacement $\boldsymbol{u}$ is the main variable of the stress field, other variables of which can be determined based on displacement. Let $\delta \mu_i$ be the trial function of the displacement.

$$\begin{cases} \int_\Omega (\delta c \frac{\partial c}{\partial t} + D \frac{\partial c}{\partial x} \frac{\partial \delta c}{\partial x} + D \frac{\partial c}{\partial y} \frac{\partial \delta c}{\partial y} - \delta c \frac{D\beta c}{R\theta} \nabla T_m + \delta c M \zeta (1-n) c) d\Omega = 0 \\ \int_\Omega (\delta n \frac{\partial n}{\partial t} - \delta n \zeta (1-n) c) d\Omega = 0 \\ \int_\Omega [-(\delta \nabla \boldsymbol{u} + \nabla \boldsymbol{u} \cdot \delta \nabla \boldsymbol{u}) \cdot \boldsymbol{T}] d\Omega = 0 \end{cases} \tag{4.175}$$

The above weak forms are then written into the mathematical module of COMSOL. The defined main variables of the physical fields include the oxygen concentration c, the TGO volume fraction n, and displacement components u and v for 2D cases. A thermo–mechano-chemical multifield coupling geometric model is then established for interfacial oxidation of the TBC and subsequently solved in COMSOL based on the boundary conditions provided for the oxidation process to determine the coupling growth of TGOs and the stress evolution in the coating. The interfacial oxidation considered here is an isothermal process, i.e., the oxidation temperature remains constant. Accordingly, the effects of temperature on stress and oxygen diffusion can be ignored.

(1) **Morphological evolution of TGOs based on multifield coupling**

To facilitate analysis, the coefficient of the stress gradient in the thermo–mechano-chemical coupling growth model for interfacial oxidation of TBCs is referred to as the coupling factor and denoted by κ; this term is set to different values to study the effect of stress on TGO growth. Figure 4.27 shows a TGO growth pattern that accounts for thermo–mechano-chemical coupling with $\kappa = 2 \times 10^{-24}$. The TGO layer is shown in red ($n = 1$), the BC layer is shown in blue ($n = 0$), and there is a mixed layer between the red and blue layers ($0 < n < 1$). As the oxidation time extends, the TGO thickness increases continuously but to different extents at the peak and trough. Specifically, at 10, 100, 500, and 1000 h of oxidation, the TGO thickness is 1.2, 2.8, 7.5, and 10.5 μm at the peak and 1.1, 1.7, 4.5, and 6.8 μm at the trough, respectively. Clearly, TGOs grow significantly faster at the peak than at the trough, resulting in a larger TGO thickness at the peak than at the trough. Nonuniform TGO growth increases the roughness and thus the stress of a coating. In their experimental study on the interfacial oxidation of TBCs, Tang and Schoenung [41] found that accelerated local growth could occur in a coating and that the coating was thicker at the peak than at the trough. These findings are consistent with those obtained from the above FE simulations. Thus, establishing a coupling TGO growth theory helps to more accurately determine the interfacial oxidation patterns of TBCs, while the parabolic model leads to uniform TGO growth.

To systematically examine the effect of stress on TGO growth, TGO growth patterns with different κ values are simulated. Figure 4.28 shows the morphologies of the TGO layer corresponding to different κ values (0, 2×10^{-24}, 5×10^{-23}, and 8×10^{-23}) at 1000 h of oxidation. A κ value of 0 means that coupling is not taken

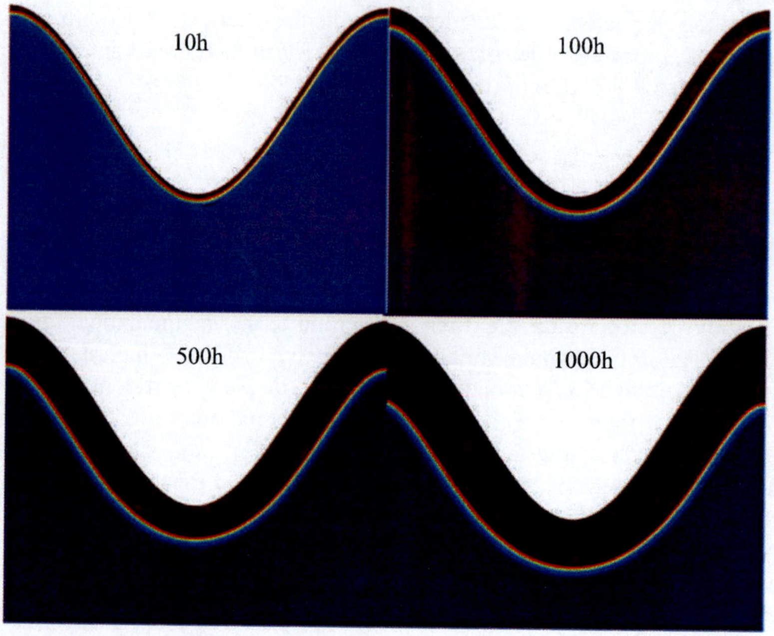

Fig. 4.27 TGO growth pattern

into consideration, i.e., the stress does not affect TGO growth. A higher κ value means that the stress more significantly affects TGO growth. As demonstrated in Fig. 4.28, the stress significantly inhibits TGO growth, i.e., as the κ increases, there is a continuous decrease in the TGO thickness. At $\kappa = 8 \times e^{-23}$, the TGO thickness is 5.9 μm at the peak and 4.2 μm at the trough, almost half of those when coupling is not considered, clearly indicating that the stress in the coating has a significant inhibiting effect on TGO growth. In addition, as κ increases, there is a decrease, to some extent, in the nonuniformity of TGO growth. Figure 4.29a, b shows the variation in the TGO thickness at the trough and peak, respectively, with κ and oxidation time. As demonstrated by the TGO growth curves, the TGO thickness at both the trough and peak decreases as κ increases. Moreover, the TGO growth curves corresponding to the trough are approximately parabolic, whereas the TGO thickness at the peak increases nearly linearly with κ but at a slow growth rate. Therefore, the stress in a coating plays a significantly inhibiting role in TGO growth and can reduce the nonuniformity of TGO growth and the stress level in the coating.

TGO growth depends primarily on the diffusion–reaction equation, whereas the effect of stress on TGO growth is mainly reflected by its inhibition of oxygen diffusion. Figure 4.30 shows the variation in the distribution of oxygen concentration with κ during the interfacial oxidation process of a TBC. The maximum oxygen concentration is 1.55 mol/m^3 and mainly appears at the upper surface of the TGO layer. The minimum oxygen concentration is 0 and occurs in the blue region. As the oxidation

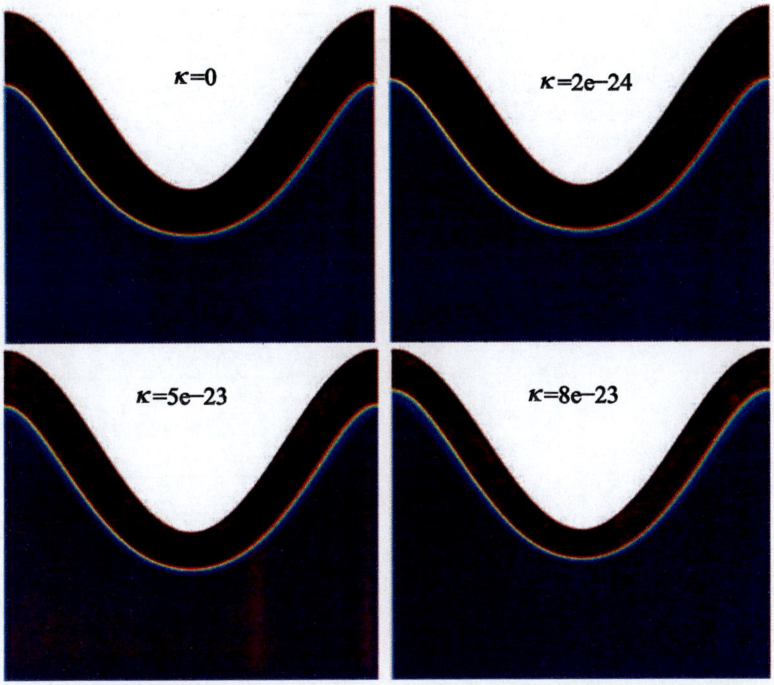

Fig. 4.28 Morphologies of the TGO layer at 1000 h of oxidation with different coupling factor κ values

reaction progresses, oxygen continuously diffuses toward the BC layer while reacting with Al in the BC layer. A comparison with Fig. 4.28 shows that interfacial oxidation occurs possibly only in the regions to which oxygen diffuses, ensuring TGO growth. The oxygen diffusion at different locations varies significantly. Specifically, oxygen diffuses faster at the peak than at the trough, resulting in a higher TGO growth rate at the peak than at the trough. The curvature of the interface is the main cause of the difference between the diffusion rates of the peak and trough. Yang et al. [51] established theoretical interfacial oxidation models with different curvatures, from which they found that a higher curvature leads to a higher oxidation rate. This is because the oxygen concentration is the highest at sharp peaks, followed by that on planes, and that at troughs is the lowest, resulting in a higher oxygen diffusion rate at peaks than at troughs, which in turn leads to a thicker TGO layer at peaks than at troughs. As κ increases, the oxygen diffusion rate decreases. Clearly, the stress in the coating affects TGO growth by affecting the distribution of oxygen content. A larger stress gradient has a more significant inhibiting effect on oxygen diffusion. In fact, researchers have analyzed the effect of stress on diffusion [32] and found that stress both inhibits and, sometimes, promotes diffusion, depending primarily on the stress state of the coating—tensile stress promotes oxygen diffusion while compressive stress inhibits oxygen diffusion.

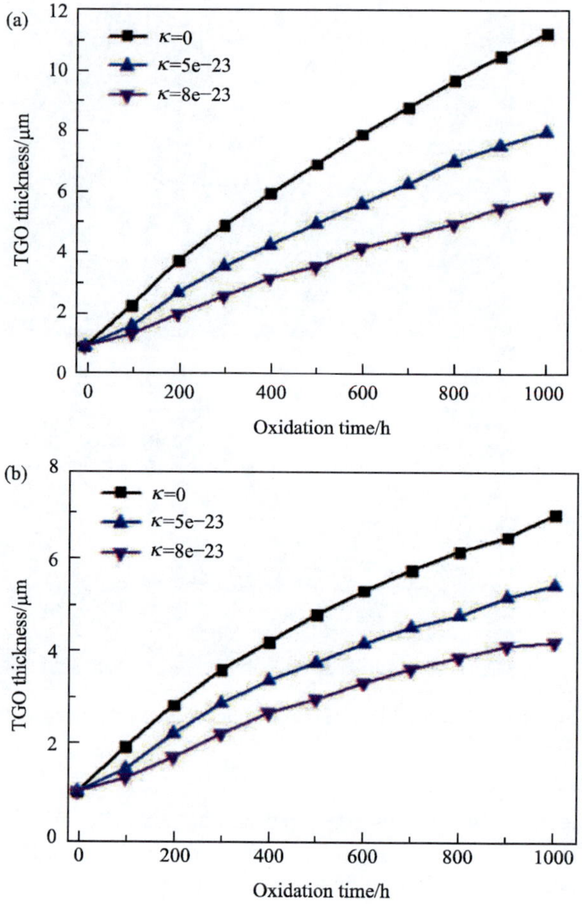

Fig. 4.29 TGO growth curves with different κ values: **a** peak and **b** trough

(2) **Stress distribution based on multifield coupling**

Figure 4.31 shows the variation in von Mises stress distribution in the TGO and BC layers with κ. The von Mises stress gradually decreases from the TC/TGO interface to the TGO/BC interface and is higher at the trough than at the peak. In addition, the von Mises stress is mainly formed in the TGO growth region and the BC layer close to the TGO layer. The regions distant from the TGOs are in a zero-stress state. This is mainly because the stress is formed as a result of TGO growth, and no oxidation reaction occurs in the regions distant from the TGOs. Moreover, κ significantly affects the stress distribution. As κ increases, the overall stress distribution in the TGO and BC layers remains the same, but there is a significant decrease in the stress distribution area. This is mainly because the stress affects TGO growth, resulting in a decrease in the area where TGO growth induces the formation of stress. A comparison of the stress patterns corresponding to different κ values reveals that the stress decreases as κ increases. Specifically, the maximum von Mises stress in the

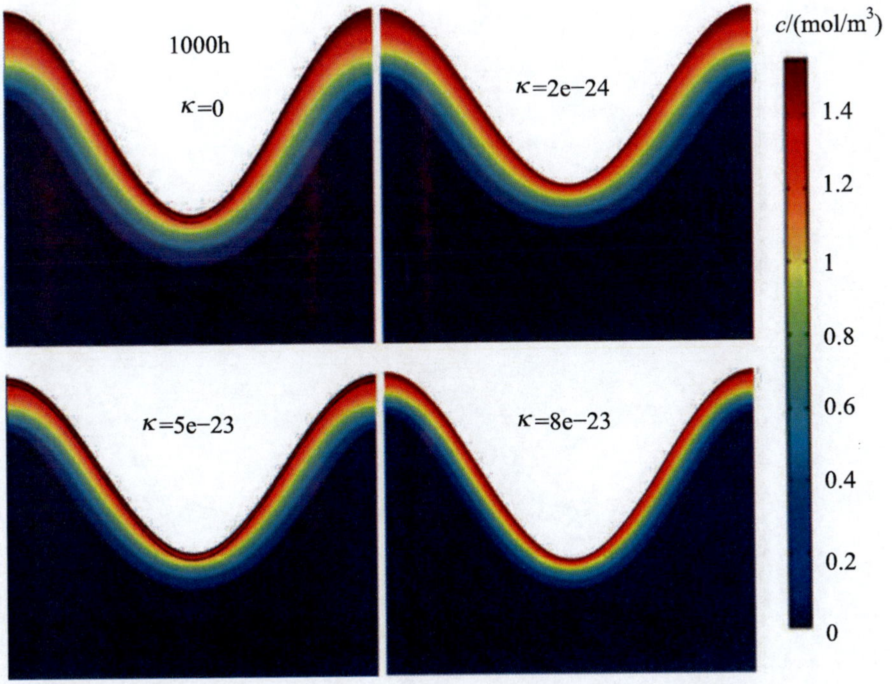

Fig. 4.30 Distributions of the oxygen concentration at 1000 h of oxidation with different κ values

TGO and BC layers is 6, 5.3, 5.0, and 4.8 GPa when $\kappa = 0, 2 \times 10^{-24}, 5 \times 10^{-23}$, and 8×10^{-23}, respectively. Evidently, the consideration of thermo–mechano-chemical coupling leads to a relaxation of the stress distribution and a reduction in the stress distribution area in the coating. The stress state predicted with the consideration of coupling is closer to the true stress state than that predicted without the consideration of coupling.

The thermo–mechano-chemical multifield coupling growth model for interfacial oxidation shows that the average stress in a coating is the main factor affecting the TGO growth and the distribution of the oxygen content. Figure 4.32 shows the variation in the average stress distribution in TGO and BC layers with κ at 1000 h of oxidation (the tensile and compressive stresses are shown in red and dark blue, respectively). The TGO layer is mainly in a tensile stress state, and as κ increases, there is a continuous decrease in the stress distribution area in the TGO layer, with the maximum stress appearing on the surface of the TGO layer at the trough. From the surface of the TGO layer to the BC layer, the average stress gradually transitions from tensile stress to compressive stress and reaches its maximum value at the TGO/BC layer. TGO growth mainly occurs at the TGO/BC interface, and the oxidation interface moves continuously with TGO growth. Figure 4.32 shows that the interfacial oxidation region of the TBC is primarily subjected to compressive stress, which inhibits oxygen diffusion and thus TGO growth. Therefore, the stress

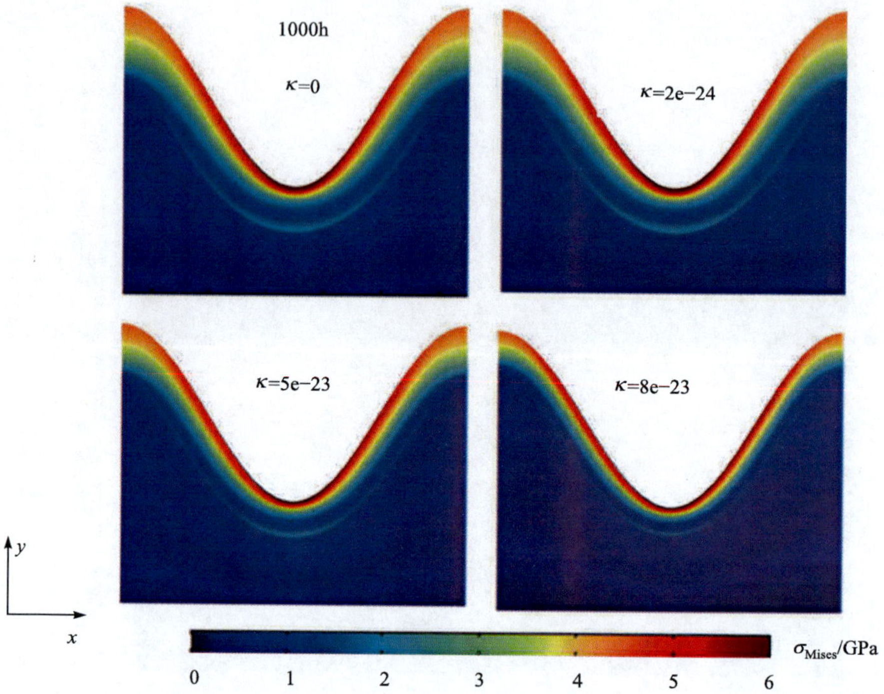

Fig. 4.31 Von Mises stress distributions in TGO and BC layers at 1000 h of oxidation with different κ values

distribution area decreases as κ increases. Because TGO growth is inhibited, there is a corresponding decrease in the stress in the coating. The maximum average stress in the TGO and BC layers is 5, 4.5, 4.2, and 3.8 GPa when $\kappa = 0, 2 \times 10^{-24}, 5 \times 10^{-23}$, and 8×10^{-23}, respectively.

Figure 4.33 shows the displacement of the TGO surface in the x-direction. Clearly, the displacement in the x-direction is antisymmetric about the centerline, i.e., the displacement in the x-direction is positive on the left side of the centerline and negative on the right side of the centerline. This means that the peak and trough continuously approach each other as interfacial oxidation progresses, which may explain the buckling instability of the coating. During the interfacial oxidation process of a TBC, folds appear on the surface of the TGO layer. As the oxidation time extends, cracks are formed in the folds, resulting in coating failure. The displacement distribution patterns show that as the peak and trough continuously approach each other, folds are highly likely to be induced in the TGO layer. At 1000 h of oxidation, the displacement in the x-direction decreases continuously as coupling increases. The maximum displacement in the x-direction is 0.11, 0.85, and 0.65 μm when $\kappa = 0$, 5×10^{-23}, and 8×10^{-23}, respectively. In addition, the location of the maximum

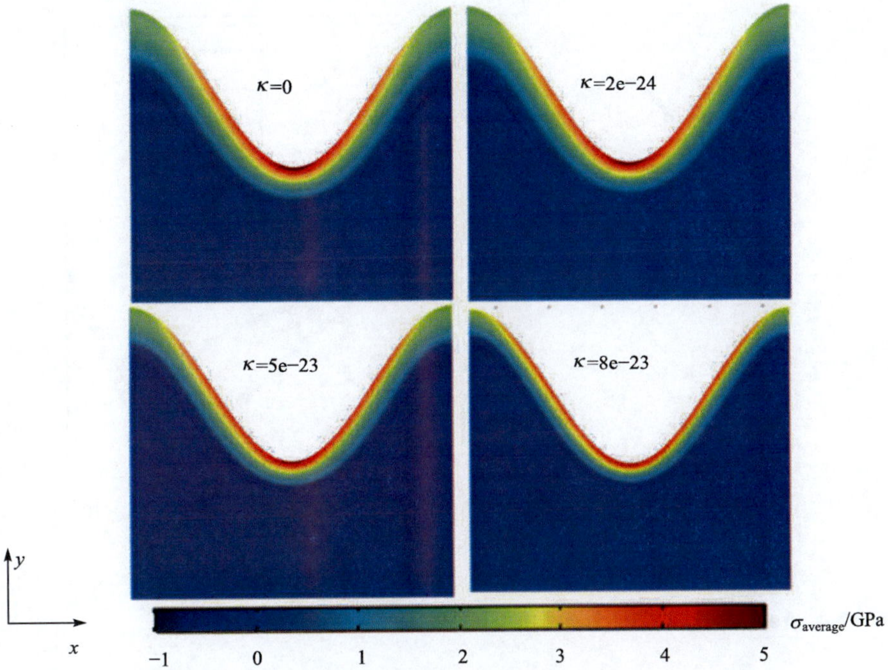

Fig. 4.32 Average stress distributions in the TGO and BC layers at 1000 h of oxidation with different κ values

displacement continuously moves away from the center of symmetry as κ increases, reducing the possibility of inducing folding or buckling. Clarke observed through SEM that significant folding occurred on the surfaces of metallic alloys after oxidation and that the severity of folding increased as the oxidation time extended [52]. This book focuses on theoretical analyses and FE simulations, and the corresponding experimental characterization of coupling is continually improved in future research.

Interfacial oxidation induces the formation of a large growth stress in a TBC. It is generally agreed that the stress in the *y*-direction is the main cause of crack initiation and propagation. Thus, in the following, we focus on analyzing the distribution and evolutionary patterns of the stress in the *y*-direction. Figure 4.34 shows the variation in the Cauchy stress distribution in the *y*-direction in TGO and BC layers with κ at 1000 h of oxidation (the tensile and compressive stresses are shown in red and blue, respectively; a darker color indicates a higher stress). Despite the different κ values (i.e., the different degrees of influence of TGO growth on stress), the corresponding stress distributions share some similarities. The four images in Fig. 4.34 demonstrate that tensile stress appears at the peak, whereas compressive stress occurs in the middle region between the peak and trough. As a result of the change in the curvature of the interface, the Cauchy stress in the *y*-direction in the BC and TGO layers transitions

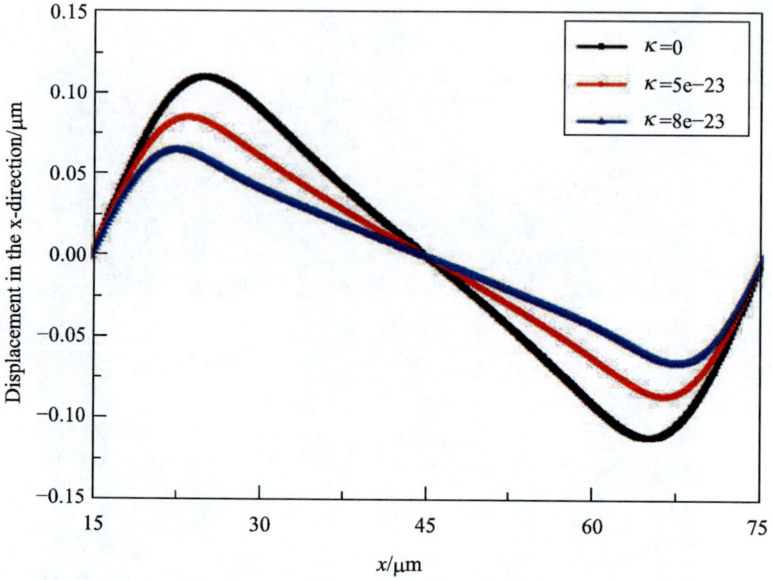

Fig. 4.33 Displacement of the surface of the TGO layer in the *x*-direction

from tensile stress at the peak to compressive stress at the trough. The majority of the regions in the coating are in a tensile stress state. Despite similar distributions, the stress, especially the tensile stress, decreases as κ increases. The maximum Cauchy tensile stress in the *y*-direction in TGO and BC layers is 1178, 1050, 936, and 850 MPa when $\kappa = 0$, 2×10^{-24}, 5×10^{-23}, and 8×10^{-23}, respectively. In addition, as κ increases, there is a decrease in both tensile and compressive stress areas, which is mainly a result of the decrease in the TGO thickness. The results reveal that the stress when considering coupling is lower than that when not considering coupling. This has an enormous impact on failure mechanisms and service life predictions but requires further experimental validation.

Figure 4.35 shows the variation in the Cauchy stress distribution in the *y*-direction in the TC layer with κ at 1000 h of oxidation (similarly, the tensile and compressive stresses are shown in red and blue, and a darker color indicates a higher stress). The four images in Fig. 4.35 show that the stress distributions in the TC layer differ significantly from those in TGO and BC layers. In the TC layer, tensile stress appears at the trough, whereas compressive stress occurs at the peak. In addition, the tensile stress area in the TC layer is significantly larger than the compressive stress area. Both the tensile and compressive Cauchy stresses in the TC layer decrease as κ increases. Specifically, when $\kappa = 0$, 2×10^{-24}, 5×10^{-23}, and 8×10^{-23}, the maximum tensile Cauchy stress in the *y*-direction in the TGO and BC layers is 200, 185, 120, and 93 MPa, respectively, whereas the maximum compressive Cauchy stress in the *y*-direction in the TGO and BC layers is –400, –360, –296, and –226 MPa, respectively. Al-Athel et al. [38] studied the stress distribution in a coating during an anisotropic

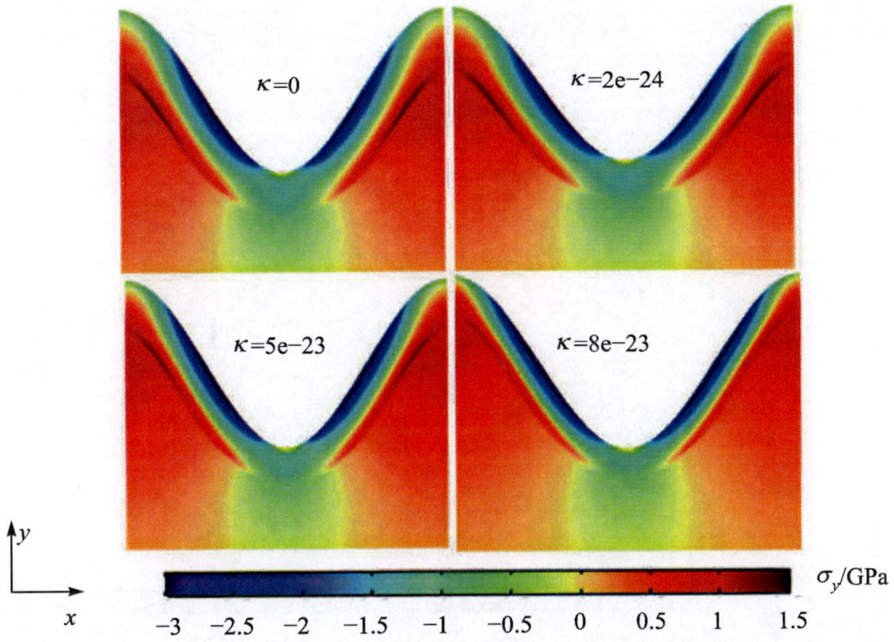

Fig. 4.34 Distributions of the *y*-direction stress in the TGO and BC layers at 1000 h of oxidation with different κ values

TGO growth process based on the large-deformation theory, and they found that the maximum tensile and compressive stresses in the TC layer were 58 and –235 MPa, respectively, which are in general agreement with the simulation results obtained with a κ value of 8×10^{-23}.

This chapter mainly examines the volume expansion resulting from oxygen diffusion, as well as studies the effect of stress on the diffusion process and realizes mechano-chemical coupling by introducing an oxygen–stress coupling term into the Helmholtz free energy and introducing a stress coupling term into the diffusion equation. However, it is subsequently found that the relevant research findings are not completely consistent with the experimental results. This is mainly because the expansion stress resulting from TGO growth caused by the interfacial oxidation of a TBC is the main part of the interfacial stress and is the main cause of interfacial failure. Therefore, it is inaccurate to consider the volume expansion due to oxygen diffusion and the effect of stress on the diffusion process by introducing an oxygen–stress coupling term into the Helmholtz free energy, leading to results inconsistent with experimental findings. In this chapter, we realize mechano-chemical coupling by introducing an experimentally determined coupling coefficient for the extent of the chemical reaction and the stress into the Helmholtz free energy to consider the effects of TGO growth stress and stress on the chemical reaction.

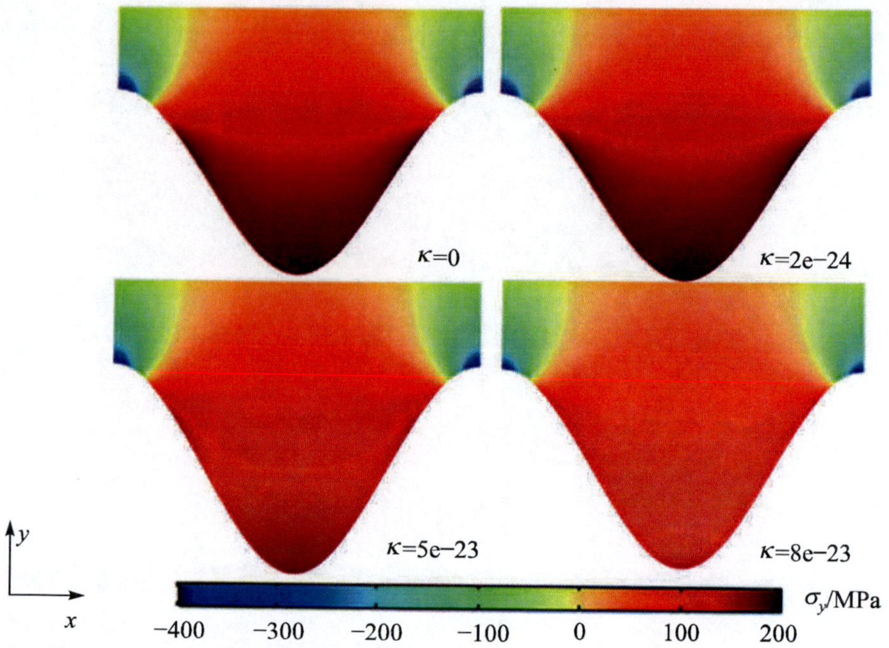

Fig. 4.35 Distributions of the *y*-direction stress in the TC layer at 1000 h of oxidation with different κ values

References

1. Sridharan S, et al. Damage evolution in an electron beam physical vapor deposited thermal barrier coating as a function of cycle temperature and time. Mater Sci Eng A. 2005;393(1–2):51–62.
2. Ma K, Schoenung JM. Isothermal oxidation behavior of cryomilled NiCrAlY bond coat: homogeneity and growth rate of TGO. Surf Coat Technol. 2011;205(21–22):5178–85.
3. Chen WR, et al. The growth and influence of thermally grown oxide in a thermal barrier coating. Surf Coat Technol. 2006;201(3–4):1074–9.
4. Martena M, et al. Modelling of TBC system failure: Stress distribution as a function of TGO thickness and thermal expansion mismatch. Eng Fail Anal. 2006;13(3):409–26.
5. Yuan K, et al. Modeling of microstructural evolution and lifetime prediction of MCrAlY coatings on nickel based superalloys during high temperature oxidation. Surf Coat Technol. 2013;232:204–15.
6. Osorio JD, et al. Diffusion–reaction of aluminum and oxygen in thermally grown Al_2O_3 oxide layers. Heat Mass Transf. 2013;50(4):483–92.
7. Beck T, et al. Damage mechanisms and lifetime behavior of plasma-sprayed thermal barrier coating systems for gas turbines—Part II: modeling. Surf Coat Technol. 2008;202(24):5901–8.
8. Freborg A, et al. Modeling oxidation induced stresses in thermal barrier coatings. Mater Sci Eng A. 1998;245(2):182–90.
9. Karlsson AM, Evans AG. A numerical model for the cyclic instability of thermally grown oxides in thermal barrier systems. Acta Mater. 2001;49(10):1793–804.
10. Rösler J, Bäker M, Volgmann M. Stress state and failure mechanisms of thermal barrier coatings: role of creep in thermally grown oxide. Acta Mater. 2001;49(18):3659–70.

11. Busso EP, et al. The influence of bondcoat and topcoat mechanical properties on stress development in thermal barrier coating systems. Acta Mater. 2009;57(8):2349–61.
12. Ranjbar-Far M, et al. Simulation of the effect of material properties and interface roughness on the stress distribution in thermal barrier coatings using finite element method. Mater Des. 2010;31(2):772–81.
13. Hille TS, Turteltaub S, Suiker ASJ. Oxide growth and damage evolution in thermal barrier coatings. Eng Fract Mech. 2011;78(10):2139–52.
14. Gupta M, et al. A diffusion-based oxide layer growth model using real interface roughness in thermal barrier coatings for lifetime assessment. Surf Coat Technol. 2015;271:181–91.
15. Tanaka M, et al. Residual stress measurement of an EB-PVD Y2O3-ZrO2 thermal barrier coating by micro-Raman spectroscopy. Surf Coat Technol. 2009;204(5):657–60.
16. Christensen RJ, et al. Nondestructive evaluation of the oxidation stresses through thermal barrier coatings using Cr3+ piezospectroscopy. Appl Phys Lett. 1996;69(24):3754–6.
17. Schlichting KW, et al. Application of Cr3+ photoluminescence piezo-spectroscopy to plasma-sprayed thermal barrier coatings for residual stress measurement. Mater Sci Eng, A. 2000;291(1–2):68–77.
18. Wen M, Jordan EH, Gell M. Evolution of photo-stimulated luminescence of EB-PVD/(Ni, Pt)Al thermal barrier coatings. Mater Sci Eng A. 2005;398(1–2):99–107.
19. Wang X, Lee G, Atkinson A. Investigation of TBCs on turbine blades by photoluminescence piezospectroscopy. Acta Mater. 2009;57(1):182–95.
20. Evans AG, He MY, Hutchinson JW. Mechanics-based scaling laws for the durability of thermal barrier coatings. Prog Mater Sci. 2001;46(3–4):249–71.
21. Hsueh CH, Fuller ER. Analytical modeling of oxide thickness effects on residual stresses in thermal barrier coatings. Scripta Mater. 2000;42(8):781–7.
22. Sun Y, et al. Local stress evolution in thermal barrier coating system during isothermal growth of irregular oxide layer. Surf Coat Technol. 2013;216:237–50.
23. Sun Y, et al. Local stress around cap-like portions of anisotropically and nonuniformly grown oxide layer in thermal barrier coating system. J Mater Sci. 2013;48(17):5962–82.
24. Yang L, et al. Finite Element Simulation on Thermal Fatigue of a Turbine Blade with Thermal Barrier Coatings. J Mater Sci Technol. 2014;30(4):371–80.
25. Zhu W, et al. The effect of morphology of thermally grown oxide on the stress field in a turbine blade with thermal barrier coatings. Surf Coat Technol. 2015;276:160–7.
26. Nayebpashaee N, et al. Finite element simulation of residual stress and failure mechanism in plasma sprayed thermal barrier coatings using actual microstructure as the representative volume. Surf Coat Technol. 2016;291:103–14.
27. Bäker M, Seiler P. A guide to finite element simulations of thermal barrier coatings. J Therm Spray Technol. 2017;26(6):1146–60.
28. Padture NP, Gell M, Jordan EH. Thermal barrier coatings for gas-turbine engine applications. Science. 2002;296(5566):280–4.
29. Ben Amar M, Goriely A. Growth and Instability in Elastic Tissues. J Mech Phys Solids. 2005;53(10):2284–319.
30. Guyen NTA, et al. A thermodynamical framework for the thermo-mechano-chemical couplings in soft materials at finite strain. Mech Mater 2016;95:158–171.
31. Yadegari S, Turteltaub S, Suiker ASJ. Coupled thermomechanical analysis of transformation-induced plasticity in multiphase steels. Mech Mater. 2012;53:1–14.
32. Cui Z, Gao F, Qu J. A finite deformation stress-dependent chemical potential and its applications to lithium ion batteries. J Mech Phys Solids. 2012;60(7):1280–95.
33. Zhou H, Cherkaoui M. A finite element analysis of the reactive element effect in oxidation of chromia-forming alloys. Phil Mag. 2010;90(25):3401–20.
34. Zhou H, Qu J, Cherkaoui M. Finite element analysis of oxidation induced metal depletion at oxide–metal interface. Comput Mater Sci. 2010;48(4):842–7.
35. Hsueh CH, Fuller ER. Residual stresses in thermal barrier coatings: effects of interface asperity curvature/height and oxide thickness. Mater Sci Eng, A. 2000;283(1–2):46–55.

36. Li C, Wang T, Liu XJ, Zheng ZH, Li Q. Evolution of mechanical properties of thermal barrier coatings subjected to thermal exposure by instrumented indentation testing. Ceram Int. 2016;42(8):10242–50.
37. Strauss D, Müller G, Schumacher G, Engelko V, Stamm W, Clemens D, Quaddakers WJ. Oxide scale growth on MCrAlY bond coatings after pulsed electron beam treatment and deposition of EB-PVDTBC. Surf Coat Technol. 2001;135(2–3):196–201.
38. Al-Athel K, et al. Modeling decohesion of a top-coat from a thermally-growing oxide in a thermal barrier coating. Surf Coat Technol. 2013;222:68–78.
39. Sridharan S, et al. Stress variation with thermal cycling in the thermally grown oxide of an EB-PVD thermal barrier coating. Surf Coat Technol. 2004;179(2–3):286–96.
40. Busso EP, Qian ZQ. A mechanistic study of microcracking in transversely isotropic ceramic–metal systems. Acta Mater. 2006;54(2):325–38.
41. Tang F, Schoenung JM. Local accumulation of thermally grown oxide in plasma-sprayed thermal barrier coatings with rough top-coat/bond-coat interfaces. Scripta Mater. 2005;52(9):905–9.
42. Tolpygo VK, Clarke DR. Morphological evolution of thermal barrier coatings induced by cyclic oxidation. Surf Coat Technol. 2003;163–164:81–6.
43. Saillard A, Cherkaoui M, Kadiri HE. Stress-induced roughness development during oxide scale growth on a metallic alloy for SOFC interconnects. Modell Simul Mater Sci Eng. 2011;19(1): 015009.
44. Sorbjan Z. Similarity theory. Wiley StatsRef: Statistics Reference Online. 2014.
45. Lockett FJ. Nonlinear viscoelastic solids. 1972: Academic Press.
46. Ni LY, Wu ZL, Zhou CG. Effects of surface modification on isothermal oxidation behavior of HVOF-sprayed NiCrAlY coatings. Progress Nat Sci Mater Int. 2011;21(2):173–9.
47. Huang K, Huang Y. Advanced Solid Mechanics (I). Tsinghua University Press, 2013.
48. Gurtin ME, Fried E, Anand L. The mechanics and thermodynamics of Continua. 2010: Cambridge University Press.
49. Reddy JN. An introduction to continuum mechanics. 2013: Cambridge University Press.
50. Loeffel K, Anand L. A chemo-thermo-mechanically coupled theory for elastic–viscoplastic deformation, diffusion, and volumetric swelling due to a chemical reaction. Int J Plast. 2011;27(9):1409–31.
51. Yang F, Fang DN, Liu B. A theoretical model and phase field simulation on the evolution of interface roughness in the oxidation process. Modell Simul Mater Sci Eng. 2012;20(1): 015001.
52. Clarke DR. Stress generation during high-temperature oxidation of metallic alloys. Curr Opin Solid State Mater Sci. 2002;6(3):237–44.

Chapter 5
Physically Nonlinear Coupling Growth and Damage Caused by Interfacial Oxidation in TBCs

A thin thermal growth oxide (TGO) layer composed primarily of α-Al_2O_3 [1, 2] is formed between the top coat (TC) and bond coat (BC) layers of a thermal barrier coating (TBC) during preparation and use. On the one hand, owing to its dense structure, the TGO layer is capable of protecting the BC layer from further oxidation. On the other hand, excessive and nonuniform TGO growth during thermal exposure generates enormous stresses near the TGO layer [3–6]. Experiments have shown that the oxidation kinetics of TGOs generally follow parabolic trends [2, 7]. Some researchers [8] have found that TGO growth involves three stages: TGO growth conforms to a parabolic trend during the first stage, follows an approximately linear pattern during the second stage, and accelerates during the third stage. Reportedly, external tensile loading can increase the oxidation reaction rate in a TBC [9]. Therefore, the oxidation process of a TBC not only depends on oxygen diffusion but also is affected by other factors such as stress (strain) and temperature, making the problem physically nonlinear. TGO growth results in a volume change and thermal expansion mismatch, which, in turn, produce extensive residual stress near the TGO layer [10–12]. Many experimental studies have shown that the rough geometry of the TGO layer causes cracks to form at the TC/TGO and TGO/BC interfaces and that the cracks connect the undulating peaks, eventually resulting in TBC failure [2, 11, 13–16].

This chapter focuses on discussing the modeling of complex interfaces caused by physically nonlinear interfacial oxidation, establishing thermodynamically consistent coupling theories, and describing coupling failure mechanisms.

Y. Zhou et al., *Thermal Barrier Coatings: Failure Theory and Evaluation Technology*,
https://doi.org/10.1007/978-981-19-2723-2_5

5.1 Physically Nonlinear Model for Thermo–Mechano–Chemical Coupling Growth Caused by Interfacial Oxidation in TBCs

5.1.1 Model Framework

A physically nonlinear model for thermo–mechano–chemical coupling growth caused by interfacial oxidation in TBCs is established based on the laws of thermodynamics. As the chemical reaction process is much slower than the mechanical deformation process, the effects of inertia can be ignored. Thus, the force equilibrium equation is given by

$$\nabla \cdot \boldsymbol{\sigma} + \boldsymbol{f} = \boldsymbol{0}, \tag{5.1}$$

Let $\widehat{c}$, be the absolute oxygen concentration (i.e., the number of oxygen molecules per unit volume) and $\boldsymbol{j}$ be the vector of the oxygen diffusion flux (i.e., the number of oxygen molecules passing per unit time and per unit area). Then, the oxygen mass conservation equation can be written as follows:

$$\dot{\widehat{c}} = -\nabla \cdot \boldsymbol{j} - N \tag{5.2}$$

where $\dot{\widehat{c}} = dc/dt$ is the rate of change in the absolute oxygen concentration.

Let Ω be an arbitrary open domain in a TBC and $\boldsymbol{n}$ be the outward unit vector perpendicular to the boundary $\partial\Omega$ of Ω. We obtain the following form for the internal energy equilibrium equation in Ω based on the first law of thermodynamics: [17, 18]

$$\frac{\mathrm{d}}{\mathrm{d}t}\int_{\Omega} e\mathrm{dV} = \int_{\Omega} \boldsymbol{f} \cdot \boldsymbol{v}\mathrm{dV} + \int_{\partial\Omega} \boldsymbol{t} \cdot \boldsymbol{v}\mathrm{dA} + \int_{\Omega} r\mathrm{dV} - \int_{\partial\Omega} \boldsymbol{q} \cdot \boldsymbol{n}\mathrm{dA} - \int_{\partial\Omega} \mu \boldsymbol{j} \cdot \boldsymbol{n}\mathrm{dA}, \tag{5.3}$$

By combining Eqs. (5.1), (5.2), and (5.3) and considering the principle of divergence, we have

$$\dot{e} = \boldsymbol{\sigma} : \dot{\boldsymbol{\varepsilon}} + r - \nabla \cdot \boldsymbol{q} + \mu\dot{\widehat{c}} + \mu N - \boldsymbol{j} \cdot \nabla\mu, \tag{5.4}$$

By combining the above equation with Eqs. (2.107) and (2.153) in Chap. 2, we obtain the following:

$$\theta\gamma = \boldsymbol{\sigma} : \dot{\boldsymbol{\varepsilon}} - \dot{\psi} - \eta\dot{\theta} + \mu\dot{\widehat{c}} + \mu N - \boldsymbol{j} \cdot \nabla\mu - \frac{\boldsymbol{q}}{\theta} \cdot \nabla\theta \geq 0 \tag{5.5}$$

By combining the above equation with Eq. (2.146) in Chap. 2, we have

$$\theta\gamma = \boldsymbol{\sigma} : \dot{\boldsymbol{\varepsilon}} - \dot{\psi} - \eta\dot{\theta} + \mu\dot{\widehat{c}} + \mu \mathrm{M}\dot{n} - \boldsymbol{j}\cdot\nabla\mu - \frac{\boldsymbol{q}}{\theta}\cdot\nabla\theta \geq 0. \tag{5.6}$$

The strain is decomposed as follows:

$$\boldsymbol{\varepsilon} = \underbrace{\boldsymbol{\varepsilon}^{e} + \boldsymbol{\varepsilon}^{\theta} + \boldsymbol{\varepsilon}^{n}}_{\boldsymbol{\varepsilon}^{E}} + \underbrace{\boldsymbol{\varepsilon}^{vp}}_{\boldsymbol{\varepsilon}^{IE}}, \tag{5.7}$$

where $\boldsymbol{\varepsilon}^{e}$ is the elastic strain, $\boldsymbol{\varepsilon}^{n}$ is the TGO growth strain, $\boldsymbol{\varepsilon}^{\theta}$ is the thermal expansion strain, $\boldsymbol{\varepsilon}^{vp}$ is the strain induced by viscoelasticity, $\boldsymbol{\varepsilon}^{E} = \boldsymbol{\varepsilon}^{e} + \boldsymbol{\varepsilon}^{\theta} + \boldsymbol{\varepsilon}^{n}$is the generalized elastic strain, and $\boldsymbol{\varepsilon}^{IE} = \boldsymbol{\varepsilon}^{vp}$ is the inelastic strain.

The Helmholtz free energy density can be viewed as a function of the state quantities:

$$\psi = \psi\left(\varepsilon_{ij}^{E}, \widehat{c}, n, \theta, \kappa_{\beta}\right), \tag{5.8}$$

where κ_{β} is the internal variable of viscoelasticity. Substitution of Eq. (5.8) into the dissipation inequality in Eq. (5.6) yields

$$\begin{aligned} \theta\gamma = &\left(\sigma_{ij} - \frac{\partial\psi}{\partial\varepsilon_{ij}^{E}}\right)\dot{\varepsilon}_{ij} - \left(\eta + \frac{\partial\psi}{\partial\theta}\right)\dot{\theta} + \left(\mu - \frac{\partial\psi}{\partial\widehat{c}}\right)\dot{\widehat{c}} + \left(\mu\mathrm{M} - \frac{\partial\psi}{\partial n}\right)\dot{n} \\ &- \boldsymbol{j}\cdot\nabla\mu - \tfrac{\boldsymbol{q}}{\theta}\cdot\nabla\theta + \tfrac{\partial\psi}{\partial\varepsilon_{ij}^{E}}\dot{\varepsilon}_{ij}^{vp} - \tfrac{\partial\psi}{\partial\kappa_{\beta}}\dot{\kappa}_{\beta} \geq 0. \end{aligned} \tag{5.9}$$

For the strain tensor, the temperature and absolute oxygen concentration can take any values, and, as a result, the corresponding coefficients must be zero. Then, we obtain the following constitutive relations:

$$\sigma_{ij} = \tfrac{\partial\psi}{\partial\varepsilon_{ij}^{E}},\ \eta = -\tfrac{\partial\psi}{\partial\theta},\ \mu = \frac{\partial\psi}{\partial\widehat{c}}. \tag{5.10}$$

By introducing the thermo–mechano–chemical coupling Helmholtz free energy [18, 19], we have

$$\begin{aligned} \psi\left(\boldsymbol{\varepsilon}^{E}, \widehat{c}, n, \theta, \kappa_{\beta}\right) = &\ \psi^{0} + \sigma^{0} : \varepsilon + 1/2\Lambda\left(\varepsilon_{kk}^{E}\right)^{2} + \Upsilon\boldsymbol{\varepsilon}^{E} : \boldsymbol{\varepsilon}^{E} + \mu^{0}\widehat{c} + 1/2\mathrm{U}\left(\widehat{c} - \widehat{c}^{0}\right)^{2} \\ &- \mathrm{A}^{0}n - 1/2a\left(n - n^{0}\right)^{2} + \rho\mathrm{C}\left(\theta - \theta\ln\frac{\theta}{\theta*}\right) \\ &- 3\mathrm{K}\alpha^{\theta}\left(\theta - \theta^{0}\right)\varepsilon_{kk}^{E} - 3\mathrm{K}\alpha^{\mathrm{n}}\left(n - n^{0}\right)\varepsilon_{kk}^{E} \\ &- \zeta\left(\widehat{c} - \widehat{c}^{0}\right)\left(n - n^{0}\right) + \psi^{vp}(\kappa_{\beta}), \end{aligned} \tag{5.11}$$

where the superscript "0" indicates the initial state, Λ and Υ are Lamé's constants, K is the bulk modulus of the material, μ^0 is the initial chemical potential, U is the quadratic coefficient of the diffusion energy, A^0 is the coefficient of the chemical reaction energy term, a is the quadratic coefficient of the chemical reaction energy, ρ is the density, C is the specific heat, θ^* is any constant temperature, α^θ is the coefficient of thermal expansion, α^n is the coupling coefficient between the chemical reaction and the strain, ζ is the coupling coefficient between the chemical reaction and the diffusion, and $\psi^{vp}(\kappa_\beta)$ is the part of the Helmholtz free energy dependent on κ_β.

Substitution of the Helmholtz free energy function into the constitutive relations yields the following isotropic constitutive relations:

$$\begin{cases} \sigma - \sigma^0 = \Lambda\varepsilon_{kk}^E \boldsymbol{I} + 2\Upsilon\varepsilon^E - 3K\alpha^n(n - n^0)\boldsymbol{I} - 3K\alpha^\theta(n - n^\theta)\boldsymbol{I} \\ \mu - \mu^0 = U\left(\hat{c} - \hat{c}^0\right) - \zeta(n - n^0) \\ \eta = \rho C \ln\frac{\theta}{\theta *} + 3K\alpha^\theta\varepsilon_{kk}^E \end{cases} \tag{5.12}$$

Assuming that both the initial volume fraction n^0 and initial stress $\boldsymbol{\sigma}^0$ of the TGO layer are zero, we have

$$\begin{cases} \boldsymbol{\sigma} = \Lambda\varepsilon_{kk}^E \boldsymbol{I} + 2\Upsilon\boldsymbol{\varepsilon}^E - 3K\alpha^n n\,\boldsymbol{I} - 3K\alpha^\theta(\theta - \theta^0)\boldsymbol{I} \\ \mu - \mu^0 = U\left(\hat{c} - \hat{c}^0\right) - \zeta n \\ \eta = \rho C \ln\dfrac{\theta}{\theta *} + 3K\alpha^\theta\boldsymbol{\varepsilon}_{kk}^E \end{cases} \tag{5.13}$$

In addition, we have

$$\boldsymbol{\varepsilon}^E = \frac{1}{2\Upsilon}\left(\boldsymbol{\sigma} - \frac{\Lambda}{2\Upsilon + 3\Lambda}\sigma_{kk}\boldsymbol{I}\right) + \alpha^n n\boldsymbol{I} + \alpha^\theta(\theta - \theta^0)\boldsymbol{I} = \boldsymbol{\varepsilon}^e + \boldsymbol{\varepsilon}^n + \boldsymbol{\varepsilon}^\theta, \tag{5.14}$$

where

$$\boldsymbol{\varepsilon}^e = \tfrac{1}{2\Upsilon}\left(\boldsymbol{\sigma} - \tfrac{\Lambda}{2\Upsilon + 3\Lambda}\sigma_{kk}\boldsymbol{I}\right),\ \boldsymbol{\varepsilon}^n = \alpha^n n\boldsymbol{I},\ \boldsymbol{\varepsilon}^\theta = \alpha^\theta(\theta - \theta^0)\boldsymbol{I}. \tag{5.15}$$

The dissipation inequality can be written in the following form:

$$\phi^{vp} + \phi^c + \phi^n + \phi^\theta \geq 0. \tag{5.16}$$

Assuming that these four dissipation mechanisms are independent of one another, each type of dissipation is positive definite, i.e.,

$$\phi^{vp} \geq 0,\ \phi^c \geq 0,\ \phi^n \geq 0,\ \phi^\theta \geq 0. \tag{5.17}$$

The viscoelastic dissipation potential energy is given by

$$\phi^{vp} = \frac{\partial \psi}{\partial \varepsilon_{ij}^{E}} \dot{\varepsilon}_{ij}^{vp} - \frac{\partial \psi}{\partial \kappa_\beta} \dot{\kappa}_\beta = \sigma_{ij} \dot{\varepsilon}_{ij}^{vp} - K_\beta \dot{\kappa}_\beta \geq 0, \tag{5.18}$$

where $K_\beta = \partial \psi / \partial \kappa_\beta$ is the conjugate force. According to the assumption of the maximum dissipation, the viscoelastic dissipation potential is maximized under the constraint of the yield criterion $f\left(\sigma_{ij}, K_\beta, \dot{\varepsilon}_{ij}^{vp}, \dot{\kappa}_\beta\right)$. By limiting the viscoelastic dissipation potential energy with the Lagrange multiplier λ^{vp}, we obtain the following:

$$\Pi^{vp} = \sigma_{ij} \dot{\varepsilon}_{ij}^{vp} - K_\beta \dot{\kappa}_\beta - \lambda^{vp} f\left(\sigma_{ij}, K_\beta, \dot{\varepsilon}_{ij}^{vp}, \dot{\kappa}_\beta\right) \geq 0. \tag{5.19}$$

The conditions for the maximum value of the above equation are as follows:

$$\frac{\partial \Pi^{vp}}{\partial \sigma_{ij}} = 0, \ \frac{\partial \Pi^{vp}}{\partial K_\beta} = 0. \tag{5.20}$$

i.e.,

$$\dot{\varepsilon}_{ij}^{vp} = \lambda^{vp} \frac{\partial f}{\partial \sigma_{ij}}, \ \dot{\kappa}_\beta = -\lambda^{vp} \frac{\partial f}{\partial K_\beta}, \tag{5.21}$$

The relevant loading/unloading criteria are as follows:

$$\begin{cases} \lambda^{vp} \geq 0, f = 0 & \text{For viscoplastic}, \ \text{和} \ \lambda^{vp} f = 0. \\ \lambda^{vp} = 0, f < 0 & \text{For elastic} \end{cases} \tag{5.22}$$

In the theory of rate-independent plasticity, the Lagrange multipliers are independent of the material. However, in the theory of rate-dependent viscoplasticity, the Lagrange multiplier λ^{vp} is defined by the constitutive equations in the theory of viscoplasticity. In Perzyna's theory of viscoplasticity, λ^{vp} is in the form of $\lambda^{vp} = \langle F(f) \rangle / \tau$, where $F(f)$ is a function of the yield criterion, the symbols $\langle \cdot \rangle$ represent Macaulay brackets (i.e., $\langle x \rangle = (x + |x|)/2$), and τ is the relaxation time. Then, we have

$$\dot{\varepsilon}_{ij}^{vp} = \frac{\langle F(f) \rangle}{\tau} \frac{\partial f}{\partial \sigma_{ij}}, \ \dot{\kappa}_\beta = -\frac{\langle F(f) \rangle}{\tau} \frac{\partial f}{\partial K_\beta}. \tag{5.23}$$

As the viscoplastic deformation of a TBC follows Norton's law of creep, we have the following: [20]

$$f = \sigma_{eff}, \ \kappa_\beta = 0, \ F(f) = \Gamma \sigma_{eff}^{\lambda}, \ K_\beta = 0, \tag{5.24}$$

where $\sigma_{eff} = \sqrt{3/2 s_{ij} s_{ij}}$ is the effective plastic strain, $s_{ij} = \sigma_{ij} - \delta_{ij} \sigma_{kk}/3$ is the deviatoric stress, and Γ and λ are positive coefficients ($\lambda \geq 1$). Then, we can obtain the following:

$$\dot{\varepsilon}_{ij}^{vp} = B(\sigma_{eff})^{\lambda} n_{ij}^{D}, \text{ where } B = \frac{\Gamma}{\tau},\ n_{ij}^{D} = \frac{3s_{ij}}{2\sigma_{eff}}. \tag{5.25}$$

Then, the diffusion dissipation potential energy can be written as follows:

$$\phi^{c} = -\boldsymbol{j} \cdot \nabla \mu \geq 0. \tag{5.26}$$

To ensure that the dot product of two vectors is greater than zero, the angle between them needs to be acute. In addition, according to Fick's law of isotropic diffusion (i.e., when the chemical potential gradient is not very large), the two vectors are linearly related to each other [21]. Thus, we obtain the following relationship:

$$\boldsymbol{j} = -\mathrm{m}\nabla \mu, \tag{5.27}$$

where m is a positive constant. By combining Eqs. (5.2), (5.13), and (5.27) and Eq. (2.146) in Chap. 2, we obtain the following oxygen diffusion equation:

$$\widehat{c} = \mathrm{mU}\nabla^{2}\widehat{c} - \mathrm{m}\zeta\nabla^{2} n - \mathrm{M}\dot{n}. \tag{5.28}$$

Then, the oxidation-reaction dissipation potential energy can be written as follows:

$$\phi^{n} = \left(\mu \mathrm{M} - \frac{\partial \psi}{\partial n}\right)\dot{n} \geq 0. \tag{5.29}$$

To satisfy the inequality in Eq. (5.29) at all times, we assume that $\dot{n}$ is linearly related to $\left(\mu \mathrm{M} - \frac{\partial \psi}{\partial n}\right)$, which can be understood as the driving force for the oxidation reaction. Thus, we have

$$\dot{n} = \mathrm{k}_{0}\left(\mu \mathrm{M} - \frac{\partial \psi}{\partial n}\right), \tag{5.30}$$

where k_0 is a positive parameter. By combining Eqs. (5.11), (5.13), and (5.30), we obtain the following oxidation-reaction equation:

$$\dot{n} = \mathrm{k}_{0}\left[(\mathrm{MU} + \zeta)\left(\widehat{c} - \widehat{c}^{\,0}\right) + (a - \mathrm{M}\zeta)n + 3\mathrm{K}\alpha^{\mathrm{n}}\boldsymbol{\varepsilon}_{kk}^{E} + \mathrm{A}^{0} + \mathrm{M}\mu^{0}\right]. \tag{5.31}$$

$0 \leq n \leq 1$, and $\dot{n} = 0$ when $n = 1$. We introduce $\mathrm{k}_0 = (1 - n)\mathrm{k}_1$, where k_1 is a positive constant, to restrict the increase in n. Then, we obtain the following:

$$\dot{n} = (1-n)\left[k_1(MU+\zeta)\left(\widehat{c}-\widehat{c}^0\right) + k_1(a-M\zeta)n + 3k_1K\alpha^n\boldsymbol{\varepsilon}_{kk}^E + k_1\left(A^0 + M\mu^0\right)\right]. \tag{5.32}$$

We assume that in the regions far from the interface, the oxygen concentration is equal to the reference oxygen concentration (i.e., $\widehat{c} = \widehat{c}_R$) and that the volumetric strain is equal to the reference strain in the BC layer (i.e., $\boldsymbol{\varepsilon}_{kk}^E = \varepsilon_R$). Because the oxidation rate is zero and no oxidation occurs (i.e., $\dot{n} = 0, n = 0$) in the regions of the BC layer far from the interface, Eq. (5.32) needs to satisfy the following condition:

$$\left[k_1(MU+\zeta)\left(\widehat{c}_R-\widehat{c}^0\right) + 3k_1K\alpha^n\varepsilon_R + k_1\left(A^0 + M\mu^0\right)\right] = 0. \tag{5.33}$$

Equation (5.32) is rewritten as follows:

$$\dot{n} = (1-n)\left[k_1(MU+\zeta)\left(\widehat{c}-\widehat{c}_R\right) + k_1(a-M\zeta)n + 3k_1K\alpha^n\left(\boldsymbol{\varepsilon}_{kk}^E - \varepsilon_R\right)\right]. \tag{5.34}$$

The relative oxygen concentration is defined as $c = \widehat{c} - \widehat{c}_R$, where $\widehat{c}_R$ is the critical oxygen concentration for the oxidation reaction. The coupled oxygen diffusion and TGO growth equations can be written as follows:

$$\begin{cases} \dot{c} = D\nabla^2 c - D^*\nabla^2 n - M\dot{n} \\ \dot{n} = (1-n)\left(Gc + G^*n + \varsigma\left(\boldsymbol{\varepsilon}_{kk}^E - \varepsilon_R\right)\right) \end{cases} \tag{5.35}$$

where $D = mU > 0$ and is the oxygen diffusion coefficient; $D^* = m\zeta > 0$ and is the chemical coefficient of diffusion, reflecting the influence of the oxidation gradient on the oxygen concentration $c = \widehat{c} - \widehat{c}_R$, i.e., D^* is the coefficient influencing the uniformity of TGO growth or interfacial oxidation on the oxygen concentration (if the TGO growth is highly uniform or linear, i.e., $\nabla^2 n = 0$, this component has no coupling effect); M is the oxygen-absorption term; $G = k_1(MU + \zeta) > 0$ and is the coefficient of the chemical reaction, reflecting the influence of the oxygen concentration on the TGO growth; $G^* = k_1(a - M\zeta)$ and is the coefficient influencing the effect of the chemical reaction itself on the extent of the chemical reaction and can also be understood as the self-interaction coefficient of the chemical reaction; and $\varsigma = 3k_1K\alpha^n > 0$, is the coefficient of influence of the volumetric strain on the chemical reaction and is referred to as the mechanochemical coupling coefficient, reflecting the commonly referenced mechanochemical coupling. Here, because only isotropic materials are considered, only the contribution of the volumetric strain exists. Readers can analyze anisotropic materials by following the above approach.

When $D^* = 0\ \text{mol}/(\text{m}^2 \cdot \text{s})$, $G^* = 0\ \text{s}^{-1}$, and $\varsigma = 0\text{s}^{-1}$ (i.e., $\zeta = 0\text{N} \cdot \text{m/mol}$, $a = \text{M}\zeta$, $k_1 \to 0\ \text{m}^2/(\text{N} \cdot \text{s})$, $U \to \infty\ \text{N} \cdot \text{m}^3/\text{mol}^{-2}$, $m \to 0\ \text{mol}/(\text{N} \cdot \text{s} \cdot \text{m})$, $D = mN$, and $G = k_1\text{MN}$), the coupled oxygen diffusion and TGO growth equations degenerate to a decoupled form consistent with Eq. (4.4) in Chap. 4, as shown below

$$\begin{cases} \dot{c} = D\nabla^2 c - M\dot{n} \\ \dot{n} = G(1 - n)c \end{cases} \tag{5.36}$$

Further, the heat dissipation potential energy is given by

$$\phi^\theta = -\frac{\boldsymbol{q}}{\theta} \cdot \nabla\theta \geq 0. \tag{5.37}$$

Similar to the analysis of Eq. (5.26), we can obtain Fourier's law of isotropic heat conduction as follows:

$$\boldsymbol{q} = -k_q \nabla\theta, \tag{5.38}$$

Assuming that the thermal conductivity k_q is a constant, by combining Eq. (2.107) from Chap. 2 and Eq. (5.4), we obtain the following:

$$\dot{\psi} + \dot{\theta}\eta + \theta\dot{\eta} = \boldsymbol{\sigma} : \dot{\boldsymbol{\varepsilon}} + r - \nabla \cdot \boldsymbol{q} + \mu\dot{\overline{c}} + \mu N - \boldsymbol{j} \cdot \nabla\mu. \tag{5.39}$$

Based on Eqs. (5.8), (5.10), and (5.13), we have

$$\dot{\psi} = \boldsymbol{\sigma} : \dot{\boldsymbol{\varepsilon}} - \boldsymbol{\sigma} : \dot{\boldsymbol{\varepsilon}}^{vp} + \mu\dot{\overline{c}} - \eta\dot{\theta} + \frac{\partial\psi}{\partial n}\dot{n},\ \eta = \rho C\frac{\dot{\theta}}{\theta} + 3K\alpha^\theta \dot{\boldsymbol{\varepsilon}}_{kk}^E. \tag{5.40}$$

Combining Eqs. (5.38), (5.39), and (5.40), we obtain the following heat conduction equation:

$$\rho C\dot{\theta} = k_q \nabla^2\theta + r + Q \tag{5.41}$$

where $Q = \sigma{:}\dot{\epsilon}^{vp} + \mu N - j \cdot \nabla\mu - 3K\alpha^\theta\theta\dot{\epsilon}_{kk}^E - \frac{\partial\psi}{\partial n}\dot{n}$. The interfacial oxidation layer (i.e., the TGO layer) is very thin, and TGO grows at temperatures as high as 900–1000 °C. Therefore, the chemical reaction has little effect on temperature (i.e., $r = 0$), and the TGO growth process lasts for a long period of time (i.e., several tens to thousands of hours). Thus, both the rate of the temperature increase and the temperature gradient are negligible. Therefore, the temperature is constant here. On this basis, we further obtain from Eq. (5.41) that $Q = 0$; i.e., the effect of Q is negligible. Then, Eq. (5.41) can be simplified to

$$\rho C\dot{\theta} = k_q \nabla^2\theta + r. \tag{5.42}$$

The above equation is retained here only as a reference for readers in other fields. The displacement and stress boundary conditions are given by

$$\boldsymbol{\sigma} \cdot \boldsymbol{n} = \boldsymbol{t},\ \boldsymbol{u} = \overline{\boldsymbol{u}}. \tag{5.43}$$

where $\overline{\boldsymbol{u}}$ is the displacement at the boundary. The boundary conditions for the diffusion field are as follows:

$$c = \overline{c},\ \boldsymbol{j} \cdot \boldsymbol{n} = \overline{j}. \tag{5.44}$$

where $\overline{c}$ is the relative oxygen concentration at the boundary and $\overline{j}$ is the oxygen diffusion flux at the boundary. The boundary conditions for the temperature field are as follows:

$$\theta = \overline{\theta},\ \boldsymbol{q} \cdot \boldsymbol{n} = \overline{q}. \tag{5.45}$$

where $\overline{\theta}$ is the thermodynamic temperature at the boundary and $\overline{q}$ is the heat flux at the boundary. The initial conditions for each field are set as follows:

$$\boldsymbol{u}(\boldsymbol{x}, 0) = \boldsymbol{u}^0(\boldsymbol{x}),\ c(\boldsymbol{x}, 0) = c^0(\boldsymbol{x}),\ n(\boldsymbol{x}, 0) = n^0(\boldsymbol{x}),\ \theta(\boldsymbol{x}, 0) = \theta^0(\boldsymbol{x}). \tag{5.46}$$

where $\boldsymbol{u}^0(\boldsymbol{x})$ is the initial displacement, $c^0(\boldsymbol{x})$ is the initial oxygen concentration, $n^0(\boldsymbol{x})$ is the initial volume fraction of TGOs, and $\theta^0(\boldsymbol{x})$ is the initial thermodynamic temperature.

5.1.2 Numerical Implementation

Figure 5.1 shows a 2D geometric model established for a TBC with a cosine interface shape. This geometric model is composed of a 200-μm-thick TC layer, a BC layer with $A_0 = 100\,\mu\text{m}$, and a h-μm-thick TGO layer (h evolves and is set to an initial value of 0). The width x_0 of the TBC is 40 μm. During the preparation of the ceramic TC layer, the BC transition layer is often roughened. In other words, a TC layer with an overly smooth surface results in a weak interfacial bond. On the other hand, a TC layer with an excessively rough surface results in TGO localization. Chapter 4 discusses the effects of the roughness of the BC layer on TGO growth in detail. Here, analyzing the effects of physical nonlinearity on TGO growth and its stress field, thermo–mechano–chemical coupling, and crack propagation is the focus. Therefore, the geometric shape of the upper boundary of the TGO layer is assumed to follow the cosine function $y = A\cos(2\pi x / x_0) + A_0$, where the amplitude A of the cosine function represents the roughness of the BC layer. Figure 5.1 shows the coordinate system. To focus on the TGO growth and its stress field, thermo–mechano–chemical coupling is not considered in the examination of the TGO growth and its stress and strain fields. In other words, analysis is performed under decoupled conditions. Thus, A is set to a constant of 10 μm. To focus on thermo–mechano–chemical coupling, the effects of roughness are not considered (i.e., A is set to 0) in the analysis of the problem of thermo–mechano–chemical coupling. Thus, the effects of geometry on TGO growth can be distinguished from the effects of thermo–mechano–chemical coupling, which can help us understand and analyze the effects of thermo–mechano–chemical coupling.

Table 5.1 summarizes the mechanical property parameters of the TBC. Here, $\dot{\varepsilon}_{ij}^{vp} = \mathrm{B}(\sigma_{eff})^{\lambda} n_{ij}^{D}$ in Eq. (5.25) is used to represent creep, where λ and B are creep parameters. In subsequent discussions, the subscript indicates the layer corresponding to the variable (i.e., TC, TGO, and BC indicate the ceramic, oxidation, and transition layers, respectively). The strain tensor in each layer can be written as follows:

$$\begin{aligned} \boldsymbol{\varepsilon}_{\mathrm{TC}} &= \boldsymbol{\varepsilon}_{\mathrm{TC}}^{e} + \boldsymbol{\varepsilon}_{\mathrm{TC}}^{\theta} + \boldsymbol{\varepsilon}_{\mathrm{TC}}^{vp}, \\ \boldsymbol{\varepsilon}_{\mathrm{TGO}} &= \boldsymbol{\varepsilon}_{\mathrm{TGO}}^{e} + \boldsymbol{\varepsilon}_{\mathrm{TGO}}^{\theta} + \boldsymbol{\varepsilon}_{\mathrm{TGO}}^{n} + \boldsymbol{\varepsilon}_{\mathrm{TGO}}^{vp}, \\ \boldsymbol{\varepsilon}_{\mathrm{BC}} &= \boldsymbol{\varepsilon}_{\mathrm{BC}}^{e} + \boldsymbol{\varepsilon}_{\mathrm{BC}}^{\theta} + \boldsymbol{\varepsilon}_{\mathrm{BC}}^{vp} \end{aligned} \tag{5.47}$$

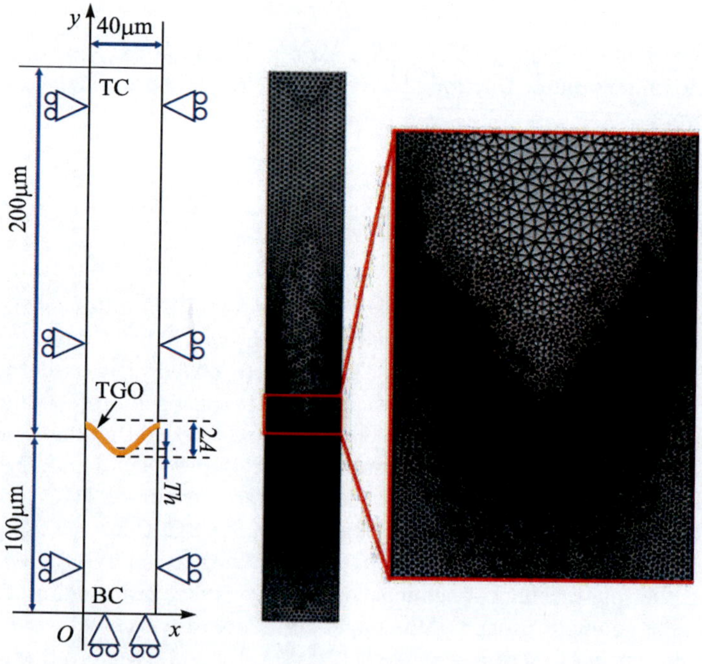

Fig. 5.1 Schematic diagram of a geometric model of a TBC and its mesh

Table 5.1 Mechanical properties of the TBC [23–27]

	TC	TGO	BC	Substrate
Young's modulus E (GPa)	22	320	110	–
Poisson's ratio ν	0.12	0.25	0.33	–
Coefficient of thermal expansion $\alpha^n(\mathrm{K}^{-1})$	12.2×10^{-6}	8.0×10^{-6}	17.6×10^{-6}	16.0×10^{-6}
Creep parameter $\mathrm{B}(\mathrm{s}^{-1} \cdot \mathrm{MPa}^{-\lambda})$	1.8×10^{-12}	3.5×10^{-12}	2.2×10^{-16}	–
Creep parameter λ	1	1	3	–

The TC/TGO interface satisfies the following boundary-traction continuity condition:

$$\boldsymbol{n}_{\mathrm{TC/TGO}} \cdot \boldsymbol{\sigma}_{\mathrm{TC}} = \boldsymbol{n}_{\mathrm{TC/TGO}} \cdot \boldsymbol{\sigma}_{\mathrm{TGO}}, \tag{5.48}$$

where $\boldsymbol{n}_{\mathrm{TC/TGO}}$ is the outward unit vector perpendicular to the TC/TGO interface. According to the Voigt assumption [22], for the TGO/BC mixed region model, we have

$$\boldsymbol{\varepsilon} = \boldsymbol{\varepsilon}_{\mathrm{TGO}} = \boldsymbol{\varepsilon}_{\mathrm{BC}}, \ \boldsymbol{\sigma} = n\boldsymbol{\sigma}_{\mathrm{TGO}} + (1 - n)\boldsymbol{\sigma}_{\mathrm{BC}}. \tag{5.49}$$

Because the TBC is much thinner than the substrate, we assume that the TBC does not affect the deformation of the substrate and that the deformation of the TBC is consistent with that of the substrate [23]. Here, the generalized plane strain assumption is used. The displacement of the TBC in the y-direction at the bottom and the displacement of the TBC in the x-direction on the left side are fixed; i.e., $v(x, y)|_{y=0\,\mu\mathrm{m}} = 0\mu\mathrm{m}$, and $u(x, y)|_{x=0\,\mu\mathrm{m}} = 0\,\mu\mathrm{m}$. The displacement of the TBC in the x-direction on the right side is fixed as follows: $u(x, y)|_{x=40\,\mu\mathrm{m}} = \alpha_{\mathrm{s}}^{\theta}(\theta - \theta^0)40\,\mu\mathrm{m}$, where $\alpha_{\mathrm{s}}^{\theta}$ is the coefficient of thermal expansion of the substrate. The deformation of the substrate in the z-direction acts on the TBC system through generalized plane strain conditions; i.e., $\varepsilon_{zz} = \alpha_{\mathrm{s}}^{\theta}(\theta - \theta^0)$. The top of the substrate is free; i.e., $\boldsymbol{\sigma} \cdot \boldsymbol{n} = 0$. Considering the isothermal oxidation process of the TBC, the TBC oxidation temperature θ is set to 1000 °C, and the reference temperature to the room temperature $\theta^{\,0}$ is set to 25 °C.

Since the TC is a porous material, the diffusion rate in the TC layer is several orders of magnitude higher than those in the TGO and BC layers [28], and oxidation does not occur in the TC layer. Therefore, the TC layer is transparent to oxygen [29, 30]. The relative oxygen concentration at the TC/TGO interface is set to 1.55 mol/m^3 [23] and the initial relative oxygen concentration in the BC layer to 0 mol/m^3. Moreover, to simplify the model, we assume that the TGO and BC layers have approximately the same oxygen diffusion coefficient (i.e., $\mathrm{D}^{\mathrm{BC}} \approx \mathrm{D}^{\mathrm{TGO}}$) [31]. Table 5.2 summarizes the specific values of the TGO growth parameters. Further, the initial volume fractions of

Table 5.2 TGO growth parameters [23]

Parameter	Value
Oxygen diffusion coefficient in the BC layer	$D_0^{BC} = 7.0 \times 10^{-14}\ m^2s^{-1}$
Oxygen diffusion coefficient in the TGO layer	$D_0^{TGO} = 7.0 \times 10^{-14}\ m^2s^{-1}$
Oxygen diffusion coefficient in the TC layer	$D_0^{TC} \to \infty\ m^2s^{-1}$
	$G = 210^{-5}\ m^3mol^{-1}\ s^{-1}$
Ratio of the number of moles of absorbed oxygen to the volume of the formed TGOs	$M = 1.1110^5\ molm^{-3}$
Stress coupling coefficient of the chemical reaction	$\alpha_n = 910^{-3}$

TGO in the TGO and BC layers are set to 1 and 0, respectively. Finally, the coupled governing equations are imported into the PDE module of COMSOL and solved using the COMSOL Multiphysics5.4 package.

5.1.3 *Results and Discussion*

First, the TGO growth and the resulting stress and strain fields are discussed. Here, the coupling is not considered; i.e., the equations are decoupled. The roughness of the BC layer is considered by setting A to 10 μm, and the results are shown in Figs. 5.2, 5.3, 5.4 and 5.5. Figure 5.2 shows the evolution of the relative oxygen concentration and TGO volume fraction n based on the TBC oxidation model. As the oxidation time increases, the oxygen diffusion zone increases, and the oxygen diffusion zone is larger at the peak than at the trough. As demonstrated in Fig. 5.2, in the decoupled TBC oxidation model, the TGO growth depends only on oxygen. The interface at $n = 0.5$ is defined as the TGO/BC interface. Figure 5.3 shows the variations in TGO thickness at the peak and trough with oxidation time. At 500 h of oxidation, the TGO thickness is 2.04 μm at the peak and 1.72 μm at the trough.

The difference in the TGO thicknesses at the peak and trough corresponds to nonuniform TGO growth in the experiment. A variety of factors, including nonuniform oxygen diffusion and nonuniform stress distribution, may lead to nonuniform growth. Figure 5.4 shows the mean stress $\sigma_m = (\sigma_{xx} + \sigma_{yy} + \sigma_{zz})/3$ and the volumetric strain $\varepsilon_{kk} = \varepsilon_{xx} + \varepsilon_{yy} + \varepsilon_{zz}$. The mean stress is positive in the TC layer but negative in both the TGO and BC layers (i.e., the stress is compressive in both of these layers). Due to creep, i.e., $\dot{\varepsilon}_{ij}^{vp} = B(\sigma_{eff})^{\lambda} n_{ij}^{D}$ in Eq. (5.25), the absolute value of the mean stress decreases with time. The volumetric strain is positive (i.e., expansion occurs) in each domain. Due to the creep, the volumetric strain in the TC layer decreases with time, whereas the volumetric strains in both the TGO and BC layers increase with increasing time.

Second, the effects of thermo–mechano–chemical coupling are investigated. To focus on the "coupling," the roughness of the BC layer is not taken into consideration. Thus, the amplitude A of the function corresponding to the geometric shape of

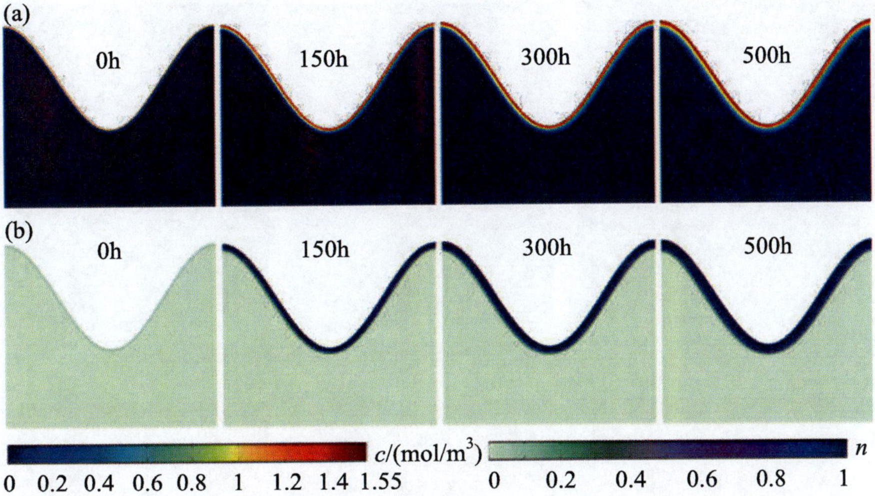

Fig. 5.2 Evolution of the relative oxygen concentration (**a**) and TGO volume fraction **b**

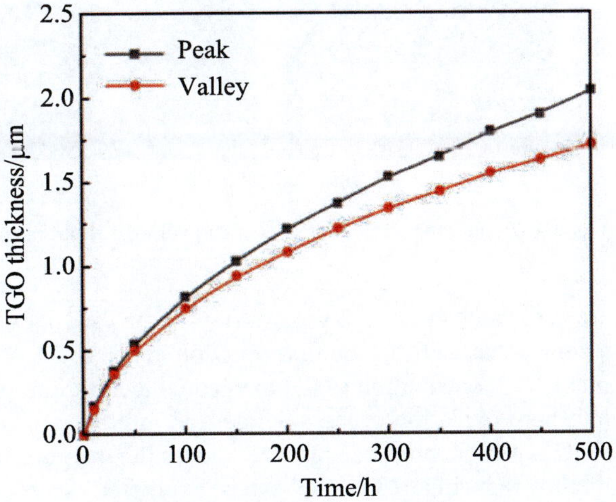

Fig. 5.3 Variations in the TGO thickness at the peak and trough with time

the BC layer is set to 0 (i.e., $y = A\cos(2\pi x / x_0) + A_0$, where $x_0 = 40\,\mu m$ and $A_0 = 100\,\mu m$), and the results are shown in Figs. 5.5, 5.6, 5.7, 5.8, 5.9 and 5.10. Now, the effects of the coefficient of the influence of oxidation on diffusion in Eq. set (5.35) (i.e., coefficient D^* of the term $\nabla^2 n$), that is, the positive correlation between the oxidation and diffusion, are discussed. As shown in Fig. 5.5a, the TGO grows faster as D^* increases. The interface at $n = 0.5$ is defined as the TGO/BC interface. As seen in Fig. 5.5b, c, under the influence of D^*, there is a slow change in the relative oxygen concentration and an even slower change in the volume fraction of TGO near the TGO/BC interface. In the first equation in in Eq. set (5.35), $\nabla^2 n$ and $\dot{n}$

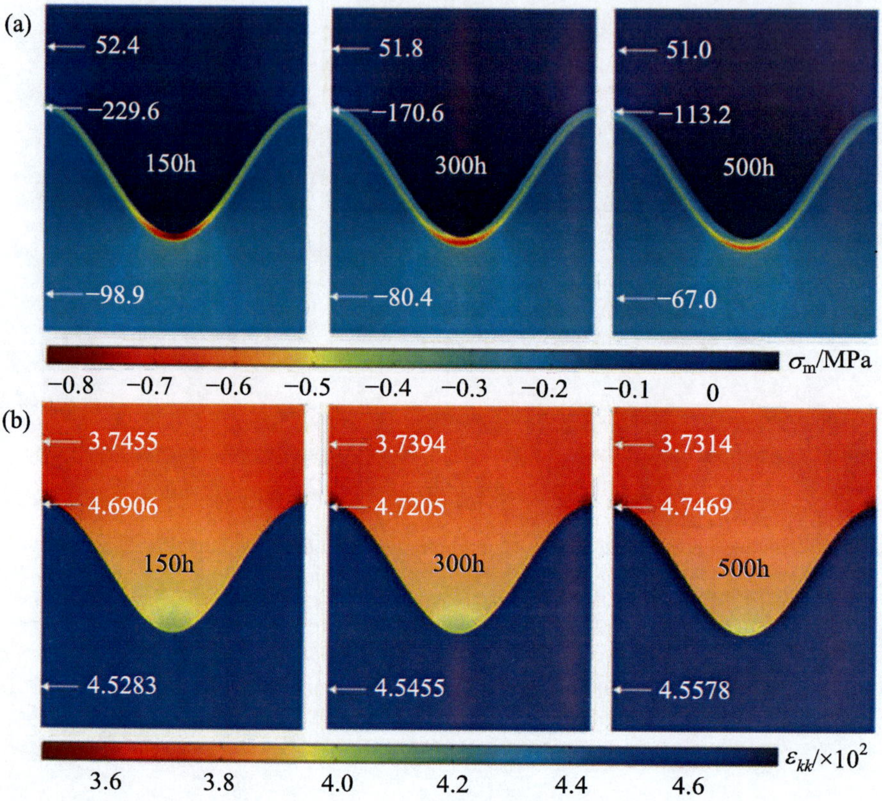

Fig. 5.4 Evolutions of mean stress (**a**) and volumetric strain (**b**)

are zero in the pure TGO regions (i.e., in the case of complete oxidation) because n is a constant at 1. Thus, the first equation in Eq. set (5.46) indicates that $D^*\nabla^2 n - M\dot{n}$ makes no contribution to the oxygen concentration inside the TGO layer, resulting in a pure oxidation process. Under the influence of the diffusion coefficient of the TGO layer, the presence of $\nabla^2 n$ reduces the oxygen concentration c at the interface, thereby reducing $\dot{n}$. Thus, D^* can be expressed as a coefficient of the hindering effect of the interface on the oxygen-absorption process of the oxidation reaction. Because $D^*\nabla^2 n$ hinders the absorption of oxygen, oxygen can diffuse further, resulting in a contraction of the completely oxidized regions and an expansion of the incompletely oxidized regions. The expansion of the incompletely oxidized regions reduces the maximum mean stress in the TGO layer, as shown in Fig. 5.5d.

Next, the effects of G^*, which is the self-interaction coefficient of the chemical reaction, are discussed. As shown in Fig. 5.6, the TGO growth rate increases quickly at a positive G^* and decreases at a negative G^*. As seen in the second equation in Eq. set (5.35), $G^* > 0$ accelerates the TGO growth at the interface and reduces the

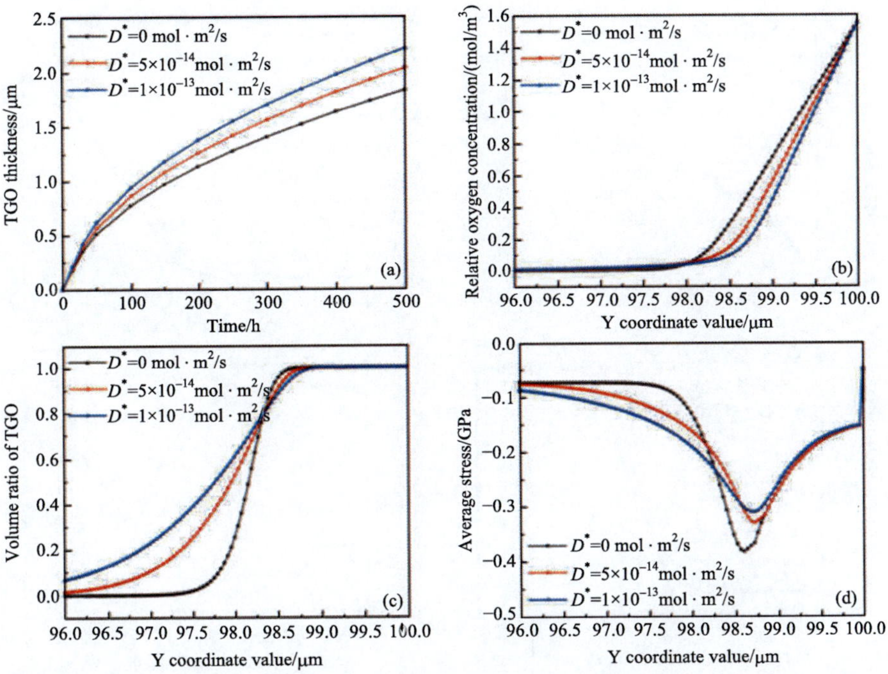

Fig. 5.5 Effects of the chemical coefficient of diffusion (the TC/TGO interface is located at $y = 100$). **a** Relation between the TGO thickness and oxidation time. **b** Relation between the relative oxygen concentration and *y*-coordinate at 500 h of oxidation. **c** Relation between the TGO volume fraction and *y*-coordinate at 500 h of oxidation. **d** Relation between the mean stress and *y*-coordinate at 500 h of oxidation

number of incompletely oxidized regions, while $G^* < 0$ does the opposite. A change in $\dot{n}$ further leads to a change in the oxygen concentration.

Here, the effects of the mechanochemical coupling coefficient $\varsigma > 0s^{-1}$ on the extent of oxidation are investigated. Creep causes the stress to change with time, which complicates the analysis of the coefficient. To more clearly examine the contribution of each term in Eq. set (5.35), first, the case involving only pure elastic deformation but no creep is discussed. The volumetric strain in a region far from the interface is chosen as the reference strain; i.e., $\varepsilon_R = 0.044779$. As shown in Fig. 5.7, the oxidation rate at the interface increases as the strain and ς increase. This is mainly because the volumetric expansion of the newly formed TGOs accelerates the oxidation reaction in the incompletely oxidized regions. Without the consideration of creep, the maximum mean stress is independent of the width of an incompletely oxidized region.

The earlier analysis shows that $G^* < 0$ reduces the TGO growth rate at the interface. If the effects of the incompletely oxidized regions on the interface could be eliminated by adjusting G^*, ς could be used to reflect the effects of external forces on TGO growth, thereby allowing us to determine the effects of stress on TGO growth

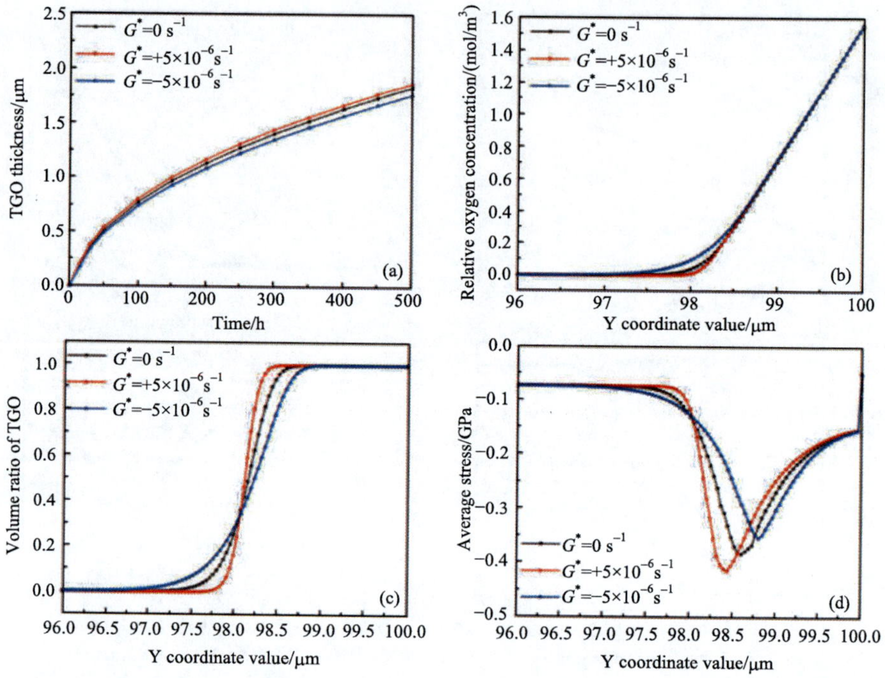

Fig. 5.6 Effects of G^*. **a** Relation between the TGO thickness and oxidation time. **b** Relation between the relative oxygen concentration and *y*-coordinate at 500 h of oxidation. **c** Relation between the TGO volume fraction and *y*-coordinate at 500 h of oxidation. **d** Relation between the mean stress and *y*-coordinate at 500 h of oxidation

by only experimentally measuring this single coefficient. Here, $G^* = -0.000266$ is chosen. As seen in Fig. 5.8, after the interfacial effects are eliminated, the variation in TGO thickness with time under coupled strain conditions is consistent with that under decoupled conditions.

Now, the effects of external loading on TGO growth are discussed. Here, external loading is described using the displacement boundary condition on the right side of the TBC in Fig. 5.9. Therefore, the right-side boundary condition of the TBC is changed to $u(x, y)\big|_{x=40\mu m} = \alpha_s^\theta(\theta - \theta^0)40\,\mu m + u^*$, where u^* is the displacement due to external loading. $\varsigma = 1 \times 10^{-2} s^{-1}$ and $G^* = -0.000266$ are chosen. As shown in Fig. 5.9, a tensile load ($u^* > 0$) promotes TGO growth, reduces the relative oxygen concentration required for the oxidation reaction, and contracts the incompletely oxidized regions. Conversely, a compressive load ($u^* < 0$) inhibits TGO growth, increases the relative oxygen concentration required for the oxidation reaction, and significantly widens the TGO/BC mixed layer. There are more regions where an oxidation reaction begins to occur but fewer completely oxidized regions under compression than under tension or no load. A high external load has a significant effect on the oxidation reaction. Some first-principles calculations [32, 33] have shown that tensile stress/strain can increase the extent of oxidation reactions.

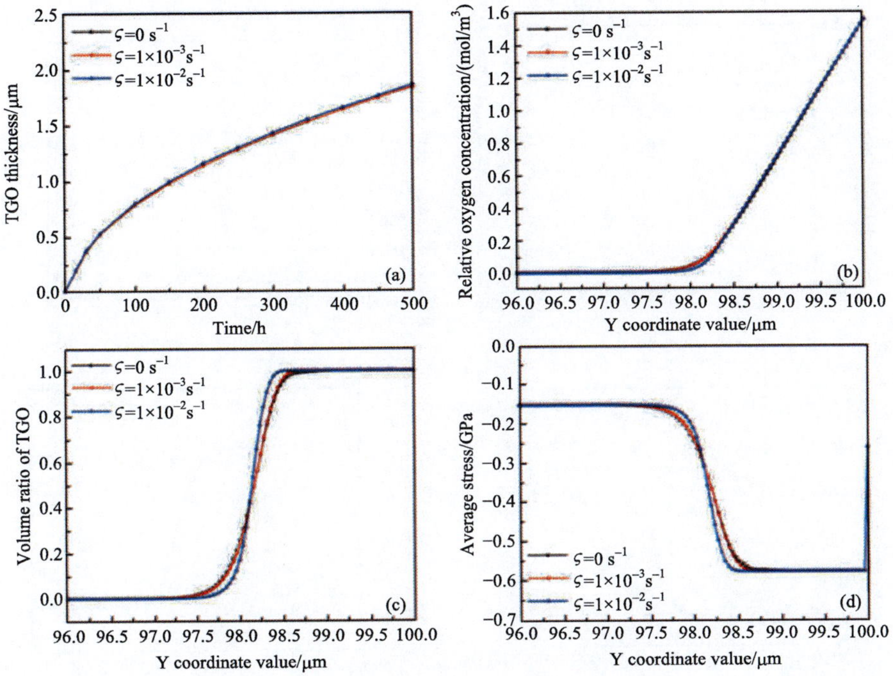

Fig. 5.7 Effects of ς. **a** Relation between the TGO thickness and oxidation time. **b** Relation between the relative oxygen concentration and *y*-coordinate at 500 h of oxidation. **c** Relation between the TGO volume fraction and *y*-coordinate at 500 h of oxidation. **d** Relation between the mean stress and *y*-coordinate at 500 h of oxidation

The tensile volumetric strain of a material can be understood as an increase in the interatomic spacing, which provides a higher probability for other atoms to enter the material and to bond with atoms in the material in chemical reactions. Conversely, a material under compressive volumetric strain has a higher density of atoms, which makes it difficult for the atoms in the material to bond with other atoms in chemical reactions.

Here, the effects of the creep on the oxidation reaction are examined. $\varsigma = 4 \times 10^{-2}\ s^{-1}$ and $G^* = -0.000266$ are chosen. As shown in Fig. 5.10, creep accelerates the oxidation reaction in the TBC. In addition, creep increases the generalized elastic strain in the TBC system, resulting in a decrease in the relative oxygen concentration required for the oxidation reaction, which, in turn, increases the oxidation reaction rate. As time increases, both the decrease in the relative oxygen concentration due to creep and its promoting effect on the oxidation reaction become increasingly prominent. In addition, the oxidation rate decreases as the TGO thickness increases. As time increases, diffusion slows TGO growth, whereas creep accelerates TGO growth. Under the combined action of these two factors, TGO growth follows an approximately linear pattern within a certain period of time.

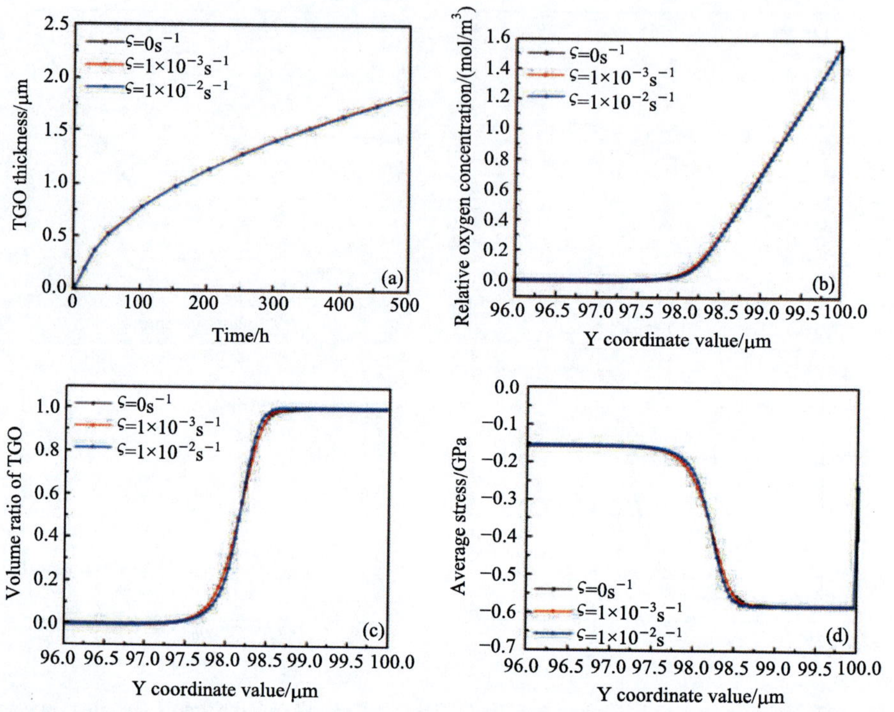

Fig. 5.8 Effects of strain and ς after the elimination of the effects of incompletely oxidized regions. **a** Relation between the TGO thickness and oxidation time. **b** Relation between the relative oxygen concentration and *y*-coordinate at 500 h of oxidation. **c** Relation between the TGO volume fraction and *y*-coordinate at 500 h of oxidation. **d** Relation between the mean stress and *y*-coordinate at 500 h of oxidation

5.1.4 Analytical Coupling Model for Interfacial Oxidation

To obtain an analytical model for oxide growth in TBCs, three additional assumptions are made:

(1) The oxidation reaction rate is much higher than the diffusion rate; i.e., there is no TGO/BC mixed layer. Therefore, the continuous transformation in the previous model is replaced by an abrupt change from the TGO layer to the BC layer.
(2) The oxygen concentration is linearly distributed in the TGO layer.
(3) To simplify the model, the effects of creep are ignored.

Let $\widehat{c}_{\mathrm{I}}$ be the oxygen concentration at the TC/TGO interface and $\widehat{c}_{\mathrm{R}}$ be the oxygen concentration (i.e., the critical oxygen concentration for the oxidation reaction) at the TGO/BC interface. Figure 5.11 shows a simplified 1D distribution of the oxygen concentration.

Regarding the uncoupled model for oxidation in TBCs, the second equation in Eq. (5.36) can be written as

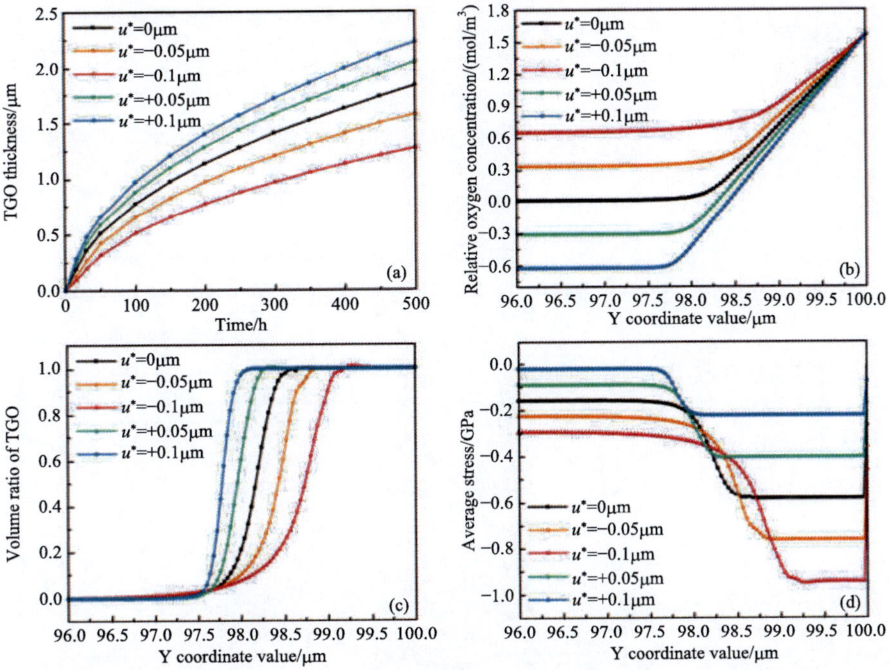

Fig. 5.9 Effects of external loading. **a** Relation between the TGO thickness and oxidation time. **b** Relation between the relative oxygen concentration and *y*-coordinate at 500 h of oxidation. **c** Relation between the TGO volume fraction and *y*-coordinate at 500 h of oxidation. **d** Relation between the mean stress and *y*-coordinate at 500 h of oxidation

$$\dot{n} = \mathrm{G}\left(\widehat{c} - \widehat{c}_R\right). \tag{5.50}$$

and the first equation in Eq. (5.36) can be written as

$$\frac{\mathrm{d}}{\mathrm{dt}}\int_{\Omega} \widehat{c}\mathrm{dV} = \int_{\Omega} \mathrm{D}\nabla^2\widehat{c}\mathrm{dV} - \int_{\Omega} \mathrm{M}\dot{n}\mathrm{dV}. \tag{5.51}$$

Using the divergence theorem, we have

$$\mathrm{d}\int_{\Omega} \widehat{c}\mathrm{dV} = \int_{\partial\Omega} \mathrm{D}\nabla\widehat{c}\cdot d\boldsymbol{A}\mathrm{dt} - \int_{\Omega} \mathrm{Md}n\mathrm{dV}. \tag{5.52}$$

For a TGO-growth microelement d(*Th*), we have

$$\frac{\widehat{c}_I - \widehat{c}_R}{2(Th + \mathrm{d}(Th))}\mathrm{d}(Th)\mathrm{dSd}(Th) = \mathrm{D}\frac{\widehat{c}_I - \widehat{c}_R}{Th}\mathrm{dSdt} - \mathrm{M}(1 - 0)\mathrm{dSd}(Th), \tag{5.53}$$

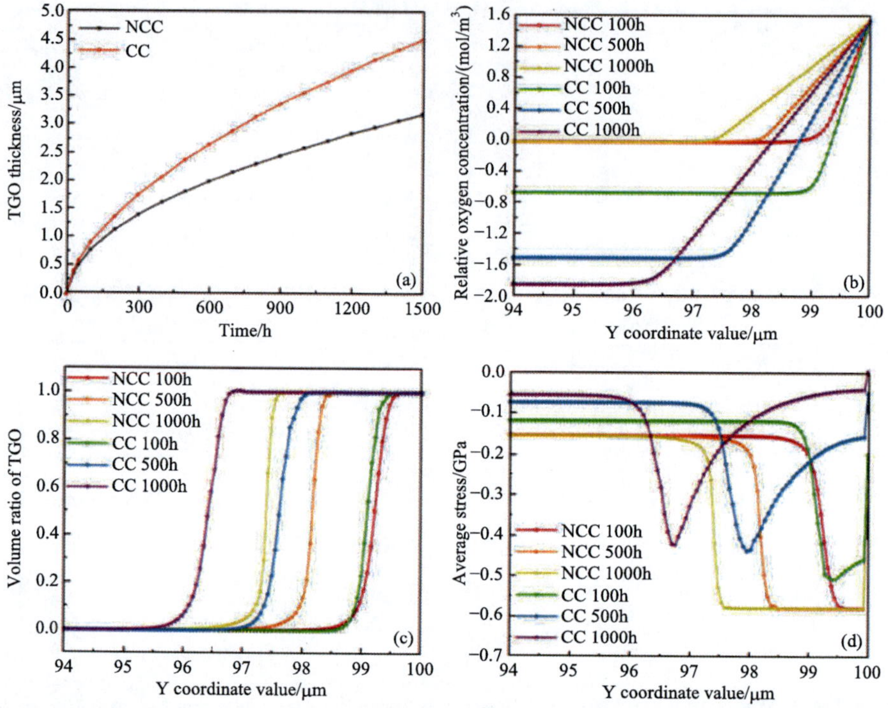

Fig. 5.10 Effects of creep (cc and ncc denote the cases with and without the consideration of the creep effect, respectively). **a** Relation between the TGO thickness and oxidation time. **b** Relation between the relative oxygen concentration and *y*-coordinate at 500 h of oxidation. **c** Relation between the TGO volume fraction and *y*-coordinate at 500 h of oxidation. **d** Relation between the mean stress and *y*-coordinate at 500 h of oxidation

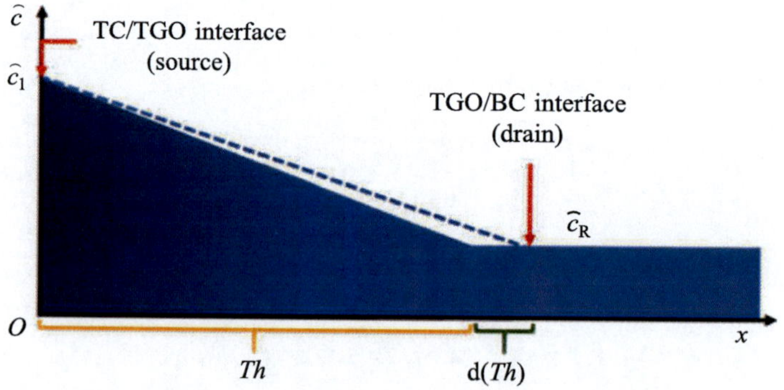

Fig. 5.11 Schematic diagram of oxygen diffusion

where dS is a region at the TGO/BC interface. The first term on the right-hand side of the above equation represents the increase in the amount of oxygen in the microelement, the second term represents the amount of oxygen diffusing from the left boundary of the microelement to the microelement, and the third term represents the amount of oxygen absorbed by the oxidation reaction. Then, we obtain the following equation:

$$\mathrm{M}Th\mathrm{d}(Th) = \mathrm{D}\left(\widehat{c}_I - \widehat{c}_R\right)\mathrm{dt}. \tag{5.54}$$

By integrating the above equation with the initial condition $Th(t = 0) = 0$, we have

$$Th = \sqrt{\frac{\mathrm{D}\Delta\widehat{c}}{\mathrm{M}}t}. \tag{5.55}$$

where $\Delta\widehat{c} = \widehat{c}_I - \widehat{c}_R$ is the change in the amount of oxygen. As seen in the above equation, during the diffusion process, the amount of oxygen absorbed by the oxidation reaction is much greater than that diffusing in the TGO layer. Therefore, the diffusion process can be simplified to a fixed oxygen-release source and a mobile oxygen-absorption source. Almost all the oxygen diffuses from the release source to the absorption source, between which the chemical potential remains constant and the oxygen concentration is linearly distributed, also proving the correctness of assumption (2). This simple model explains why the oxidation kinetics of TGOs follow a parabolic trend (i.e., the common parabolic equation).

Regarding the coupled model for oxidation in TBCs, Eq. set (5.35) can be written as follows:

$$\begin{cases} \dot{\widehat{c}} = \mathrm{D}\nabla^2\widehat{c} - \mathrm{M}\dot{n} \\ \dot{n} = (1 - n)\left(\mathrm{G}\left(\widehat{c} - \widehat{c}_R\right) + \varsigma\left(\boldsymbol{\varepsilon}_{kk}^E - \varepsilon_R\right)\right) \end{cases} \tag{5.56}$$

We introduce $\widehat{c}_* = \widehat{c}_R - \frac{\varsigma}{\mathrm{G}}\left(\boldsymbol{\varepsilon}_{kk}^E - \varepsilon_R\right)$ and write the second equation in Eq. set (5.56) as

$$\dot{n} = \mathrm{G}\left(\widehat{c} - \widehat{c}_*\right). \tag{5.57}$$

By comparing Eqs. (5.50) and (5.57), we change the condition required for the oxidation reaction to

$$\widehat{c} = \widehat{c}_* = \widehat{c}_R - \frac{\varsigma}{\mathrm{G}}\left(\boldsymbol{\varepsilon}_{kk}^E - \varepsilon_R\right) \tag{5.58}$$

where $\frac{\varsigma}{G}\left(\boldsymbol{\varepsilon}_{kk}^{E} - \varepsilon_{R}\right)$ reflects the effect of external loading on the oxidation reaction. Similarly, we can obtain the following:

$$Th = \sqrt{\frac{D\left(\Delta\hat{c} + \frac{\varsigma}{G}\left(\varepsilon_{kk}^{E} - \varepsilon_{R}\right)\right)}{M}t} \tag{5.59}$$

Note that this model is not applicable for high compressive loads because the requirement that the oxidation reaction rate is much higher than the diffusion reaction rate cannot be met.

5.1.5 Comparison with Experimental Results

To analyze the correctness of the above model, interfacial oxidation experiments under external loading are designed. Dumbbell-shaped (gauge length, 20 mm; shoulder radius, 26 mm; gauge diameter, 8 mm; cross-sectional diameter of the thread, 15 mm) and cylindrical (radius, 3 mm; length, 2 mm) Ni-based high-temperature alloy K465 substrates were used in high-temperature tensile and compressive creep tests, respectively, as shown in Fig. 5.12. A yttria-stabilized zirconia (YSZ) coating was deposited on the NiCrAlY transition layer by plasma spraying.

Figure 5.13 shows scanning electron microscopy (SEM) images of the TBC at 100 h of oxidation ((a): under no external load; (c): under a tensile load of 100 MPa). Figure 5.14a–d shows SEM images of the TBC under no external load at 300 h of oxidation, under a tensile load of 100 MPa at 300 h of oxidation, under no external load at 400 h of oxidation, and under a tensile load of 100 MPa at 400 h of oxidation, respectively. As clearly seen in these SEM images, thicker TGO layers were formed from different durations of oxidation under an external load of 100 MPa than under no external load.

In the comparison of the experimental results, the finite element (FE) numerical solutions, and the results produced by the simple theoretical model, the effects

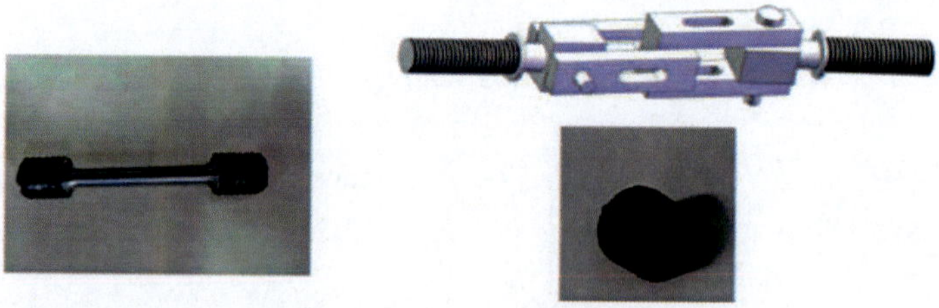

Tensile test Compressive test

Fig. 5.12 Photographs and schematic diagram of specimens

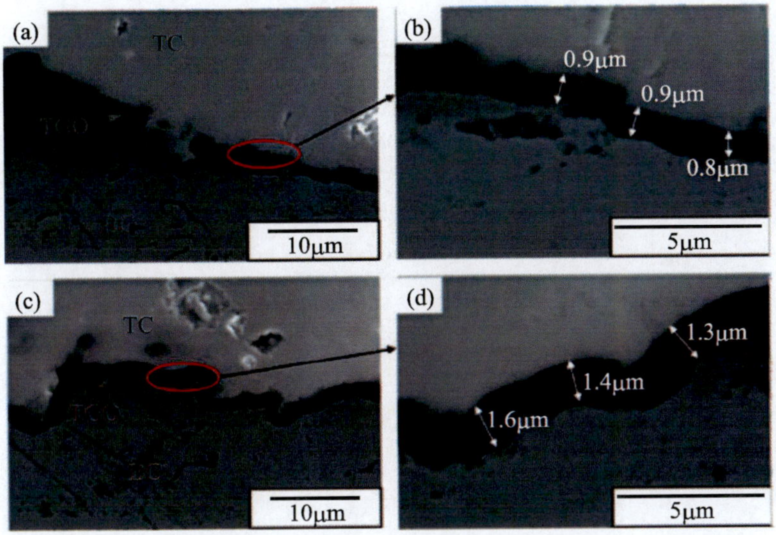

Fig. 5.13 SEM images of the TBC at 100 h of oxidation. **a** Under no external load. **b** Close-up of the image in **a**. **c** Under an external tensile load of 100 MPa. **d** Close-up of the image in **c**

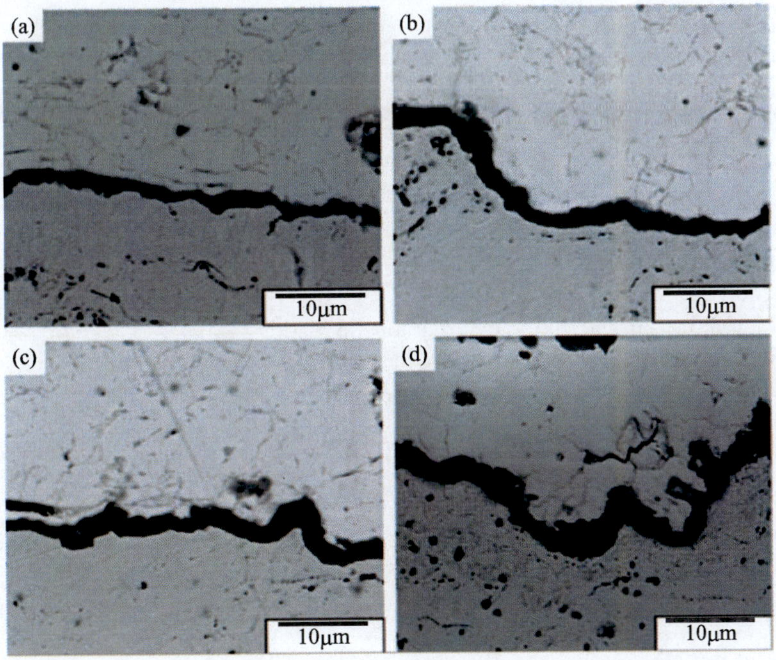

Fig. 5.14 SEM images of the TBC. **a** Under no external load at 300 h of oxidation. **b** Under an external tensile load of 100 MPa at 300 of oxidation. **c** Under no external load at 300 h of oxidation. **d** Under an external tensile load of 100 MPa at 400 of oxidation

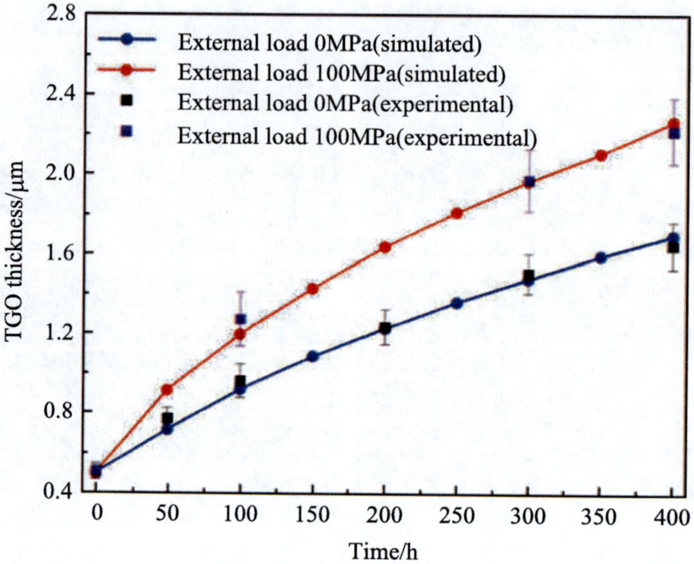

Fig. 5.15 Comparison of the variations in the simulated and experimental TGO thicknesses with oxidation time

of creep are ignored. The following parameters are selected: $D^{TGO} = D^{BC} = 7 \times 10^{-14}\,m^2s^{-1}$, $\varsigma = 0.036\,s^{-1}$, $G^* = -0.000266$, and $\varepsilon_R = 0.044779$. Figure 5.15 compares the numerical simulation and experimental results, showing a good agreement between the two.

Now, the analytical solutions produced by the simple model and the experimental results are compared. Tension is assumed to produce only an additional strain in the tensile direction (i.e., the *x*-direction), while the additional strain formed in the *z*-direction due to interlayer constraints is very small and can be ignored. Therefore, the additional strain due to the external tensile load is $\varepsilon_{ext} = 100\,\text{MPa}/E_{\text{substrate}} = 1.25 \times 10^{-3}$. According to the principle of the superposition of strains, the additional strains due to an applied load can be expressed as

$$\varepsilon_{xx}^* = \varepsilon_{ext},\ \varepsilon_{yy}^* \neq 0,\ \varepsilon_{zz}^* = 0, \tag{5.60}$$

where ε_{xx}^* is the additional strain of ε_{xx} in the BC layer, $\varepsilon_{zz}{}^*$ is the additional strain in the BC layer, and ε_{yy}^* is the additional strain of ε_{yy} in the BC layer. As the TBC is free in the *y*-direction, the stress in the *y*-direction is zero; i.e.,

$$\sigma_{yy}^* = \Lambda_{BC}\left(\varepsilon_{xx}^* + \varepsilon_{yy}^* + \varepsilon_{zz}^*\right) + 2\Upsilon_{BC}\varepsilon_{yy}^* = 0 \rightarrow \varepsilon_{yy}^* = -\frac{\Lambda_{BC}}{\Lambda_{BC}+2\Upsilon_{BC}}\varepsilon_{ext} \tag{5.61}$$

Thus, we have

$$\varepsilon_{kk}^{E} - \varepsilon_{R} = \varepsilon_{xx}^{*} + \varepsilon_{yy}^{*} + \varepsilon_{zz}^{*} = \frac{2\Upsilon_{BC}}{\Lambda_{BC}+2\Upsilon_{BC}}\varepsilon_{ext} \quad (5.62)$$

Then, considering the initial TGO thickness, we need to revise Eq. (5.59) to

$$Th^2 - Th_0^2 = \frac{D\left(\Delta\hat{c} + \frac{\varsigma}{G}\left(\varepsilon_{kk}^{E} - \varepsilon_{R}\right)\right)}{M}t \quad (5.63)$$

where Th_0 is the initial thickness of the TGO layer (here, $Th_0 = 0.5\,\mu m$) and $\Delta\hat{c} = 1.55$ mol/m^3. In addition, because the TGO layer is completely oxidized, a slightly higher diffusion coefficient ($D = 1.3 \times 10^{-13}$ m^2 s^{-1}) is needed. Figure 5.16 shows the theoretical modeling and experimental results. The theoretical and experimental values of the TGO thickness differ slightly at $\varsigma = 0.036$ s^{-1} but are highly consistent with each other at $\varsigma = 0.042$ s^{-1}. The theoretical model does not take the oxidation time into account and, therefore, neglects the decrease in the time required for the oxidation reaction due to tensile loading and ignores the effect of incompletely oxidized regions. Thus, a value of ς higher than that in Eq. set (5.35) when using Eq. (5.59) is needed to analyze the effect of stress on the oxidation reaction in a TBC.

Figure 5.17a and c shows SEM images of the TBC under no external load and under an external compressive load of 50 MPa at 100 h of oxidation, respectively. Figure 5.18 shows SEM images of the TBC under no external load and under an external compressive load of 100 MPa at 100 h of oxidation, respectively. Under

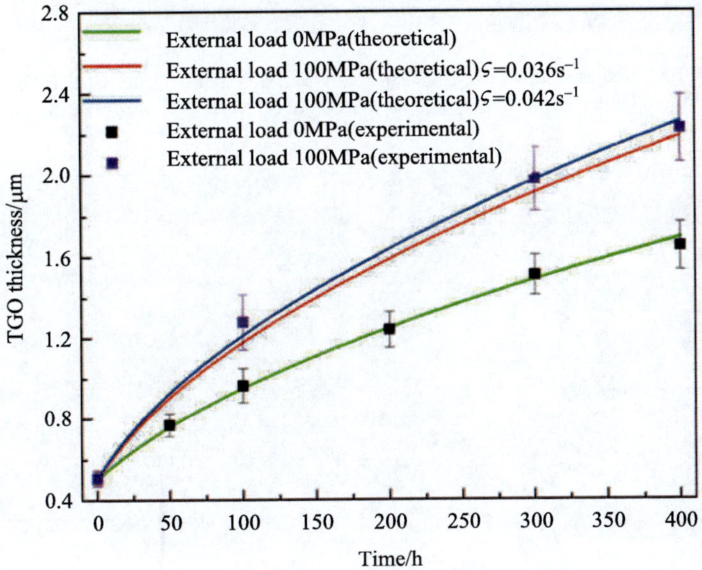

Fig. 5.16 Comparison of the variations in the theoretical and experimental TGO thicknesses with the oxidation time

an external compressive load of 50 or 100 MPa, a large number of incompletely oxidized regions appeared at the location of interfacial oxidation in the TBC, and oxidized regions appeared across a large area in the BC layer. The oxidized regions in the BC layer were more prominent under an external compressive load of 100 MPa than under that of 50 MPa.

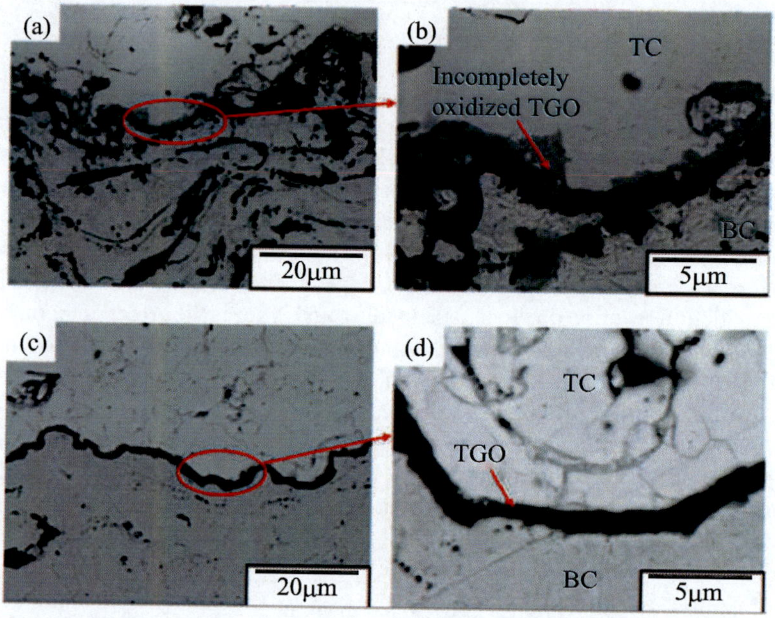

Fig. 5.17 SEM images of the TBC at 100 of oxidation. **a** Under no external load. **b** Close-up of the image in **a**. **c** Under an external compressive load of 50 MPa. **d** Close-up of the image in **c**

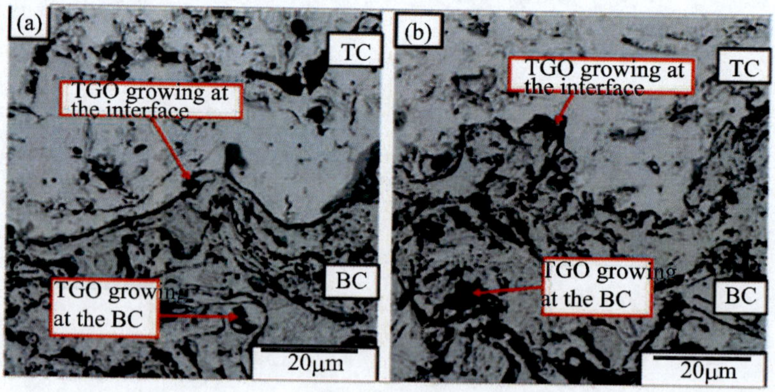

Fig. 5.18 Two SEM images of the TBC at 100 h of oxidation under an external compressive load of 100 MPa

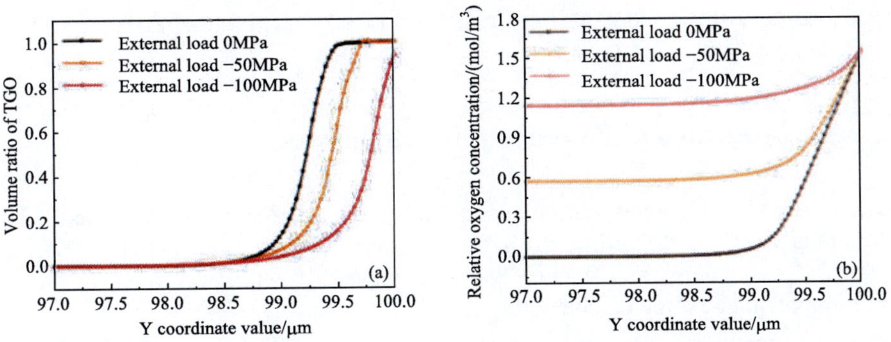

Fig. 5.19 **a** Relation between the TGO volume fraction and *y*-coordinate at 100 h of oxidation. **b** Relation between the relative oxygen concentration and *y*-coordinate at 100 h of oxidation

In addition, no phenomena similar to the oxidation of the entire BC layer occurs under no external load or under an external tensile load, which can be explained based on Fig. 5.19. An external compressive load slows the oxidation reaction, resulting in incomplete oxidation across a large area and incomplete depletion of the oxygen at the TGO/BC interface. As a result, the oxygen diffuses to the BC layer. Under a large external compressive load, the oxygen concentration in the BC layer is very high. The BC layer is unable to withstand the compressive stress near the defects and cracks. Consequently, the generalized elastic strains are larger near the defects and cracks than in other regions, leading to oxidation reactions near the defects and cracks. As a result, a large number of oxidized regions appear in the BC layer, and the entire BC layer can even be oxidized.

The above results obtained from the numerical simulations, the simple theoretical model, and the experiments all indicate that significant mechanochemical coupling exists and that there is a significant difference in the extent of oxidation and a complete difference in the mode of oxidation. We know that the oxidation of the whole BC layer causes rapid TBC spallation, that is, an extremely short TBC service life. As soon as a TBC spalls, its metallic substrate becomes exposed to high temperatures. In any advanced engine, its turbine inlet temperature exceeds the melting point of the metallic substrate material. As a result, TBC spallation causes rapid fracture in the turbine blades, thereby causing disastrous accidents such as engine explosion. Very clearly, the mechanochemical coupling results further suggest that the high-temperature oxidation experiments conducted by some researchers or engineering technicians with simple specimens in high-temperature furnaces are completely unable to reflect the problems of interfacial oxidation in actual turbine blades in highly complex stress states. Hence, extensive experimental simulations as close as possible to the actual working conditions are needed to reflect real-world interfacial oxidation conditions. Chapter 16 presents experimental simulation devices for TBCs on turbine blades in aeroengines.

5.2 Interfacial Oxidation Failure Theory that Integrates the CZM and PFM

5.2.1 *Integrated CZM and PFM Framework*

Here, a model integrates the cohesion zone model (CZM) and phase-field fracture model (PFM). According to Griffith's theory of fracture, the sum of the brittle fracture strain energy and fracture energy is regarded as a fixed value. The "total" pseudopotential energy can be defined as

$$W = \int_{\Omega\backslash\Gamma} w_{bulk} \mathrm{d}V + \int_{\Gamma} g_c \mathrm{d}\Gamma, \quad (5.64)$$

where W is the "total" pseudopotential energy, w_{bulk} is the degraded elastic energy per unit volume, g_c is the crack energy release rate, Ω is the domain, and Γ is the crack surface.

As shown in Fig. 5.20a, two types of failure often occur as a result of the interfacial oxidation in a TBC. (1) Interfacial failure, e.g., cracks at the TC/TGO interface (the TGO/BC interface is a TGO/BC mixed layer and lacks definite boundaries; the cracks at the TGO/BC interface are therefore considered bulk cracks). (2) Brittle failure in the bulk material (interfacial failure in a TBC occurs during cooling instead of at high temperatures and is thus considered brittle failure), e.g., cracks in the TC, BC, and TGO layers. Therefore, two failure mechanism models are needed to correctly understand the oxidation failure process in TBCs. A phase-field model (PFM) is often employed to describe brittle failure in bulk materials, while a cohesive zone model (CZM) is used to describe interfacial failure. To couple the PFM for brittle fracture and a CZM, the critical energy release rate is decomposed into two parts (as shown in Fig. 5.20b), namely, the critical energy release rate in the bulk material, which describes the failure of the bulk material using the phase field, and the critical energy release rate at the interface, which describes the interfacial failure using a CZM. We rewrite Eq. (5.64) as

$$W = \int_{\Omega\backslash\Gamma} w_{bulk} \mathrm{d}V + \int_{\Gamma^b} g_c^b \mathrm{d}\Gamma + \int_{\Gamma^i} g_c^i \mathrm{d}\Gamma, \quad (5.65)$$

where Γ^b is the bulk crack surface and Γ^i is the interface crack surface. In this way, a PFM and a CZM are used to describe the failure of the bulk material (i.e., Γ^b) and the failure of the interface crack (i.e., Γ^i), respectively. Using Eq. (5.65), an integrated CZM and PFM theoretical framework is established for interfacial oxidation failure.

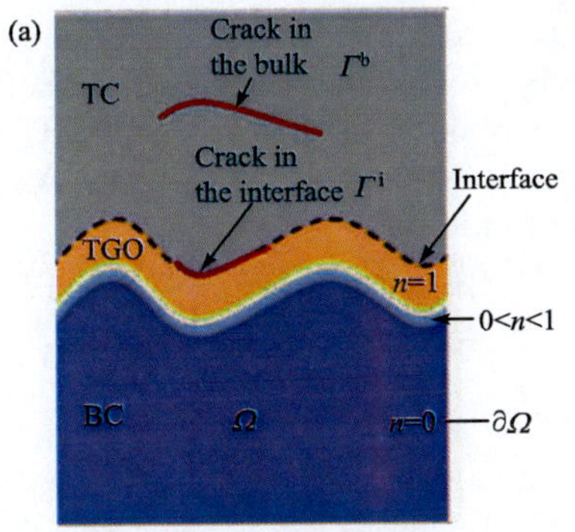

Fig. 5.20 Schematic diagram of a TBC containing cracks in the bulk and at the interface: **a** discrete cracks in the domain; **b** cracks in the domain continuously described by the phase field

5.2.2 *Introduction to PFM*

Now, the PFM for the bulk crack surface Γ^b is discussed. Miehe et al. [34] introduced a field variable d, which can be understood as a damage parameter, to describe the failure state of a material. d describes discrete cracks through continuous cracks so that a stress-field singularity near the crack tip can be prevented when using the PFM to simulate the cracking process. $d = 0$ means that the material is undamaged, and $d = 1$ indicates that the material completely fails. Figure 5.21a shows an infinitely long bar with a sharp crack at $x = 0$. As shown in Fig. 5.21b, this type of discrete crack can be described as follows:

$$d(x) = \begin{cases} 1 & x = 0 \\ 0 & x \neq 0. \end{cases} \tag{5.66}$$

Fig. 5.21 Schematic diagrams of cracks: **a** 1D bar truncated by a crack; **b** a discrete crack; **c** a continuous crack

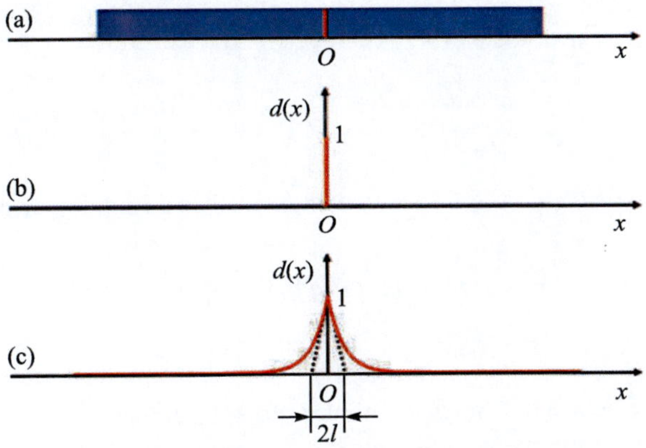

This discrete crack can also be approximated by the exponential function $\exp[-|x|/l]$, which is described by a global continuous crack-field variable (l is a parameter controlling the width of the continuous crack), in Fig. 5.21c. The crack propagation area increases as the field length scale parameter l increases. As $l \to 0$, the continuous crack transforms to a discrete crack.

Under the Dirichlet boundary condition of $d(0) = 1$ and $d(\pm\infty) = 1$, $\exp[-|x|/l]$ is the solution to the following equation:

$$d(x) - l^2 d''(x) = 0. \tag{5.67}$$

Equation (5.67) can be derived from the variational principle of the Euler equation expressed as

$$\mathrm{d} = \mathrm{Arg}\left\{ \inf_{\mathrm{d} \in W} I(d) \right\} \qquad I(d) = \frac{1}{2}\int_{\Omega} \left\{ d^2 + l^2 d'^2 \right\} \mathrm{dV}. \tag{5.68}$$

To eliminate the effect of the crack length parameter, we nondimensionalize the crack surface area as follows:

$$\Gamma^b(d) = \frac{1}{l} I(d) = \frac{1}{2l}\int_{\Omega} \{d^2 + l^2 d'^2\} \mathrm{dV} = \int_{\Omega} \gamma(d, d') \mathrm{dV}, \tag{5.69}$$

where $\gamma(d, d')$ is the crack surface energy density. If we expand the case of a crack in a 1D bar to that of a crack in a 2D plate material or a 3D bulk material, we can rewrite Eq. (5.67) as

$$\begin{cases} d(\boldsymbol{x},\mathrm{t}) - l^2 \Delta d(\boldsymbol{x},\mathrm{t}) = 0 \ \text{ in } \Omega \\ d(\boldsymbol{x},\mathrm{t}) = 1 \ \text{ on } \Gamma^b \\ \nabla d(\boldsymbol{x},\mathrm{t}) \cdot \boldsymbol{n} = 0 \ \text{ on } \partial\Omega \end{cases} \tag{5.70}$$

Equation set (5.70) can be written as follows:

$$\gamma(d, \nabla d) = \frac{1}{2l}(d^2 + l^2 |\nabla d|^2). \tag{5.71}$$

Equation (5.65) can be expressed as a function of these variables as follows:

$$W = \int_{\Omega\backslash\Gamma} w_{bulk}(\boldsymbol{\varepsilon}^e, d) \mathrm{dV} + \int_{\Gamma^b} g_c^b(\boldsymbol{u}, d) \mathrm{d}\Gamma + \int_{\Gamma^i} g_c^i(\boldsymbol{\delta}, \boldsymbol{\alpha}, d) \mathrm{d}\Gamma, \tag{5.72}$$

where $\boldsymbol{\delta}$ is the discontinuous displacement vector at the interface and $\boldsymbol{\alpha}$ is the maximum discontinuous displacement vector at the interface. We first analyze only crack propagation in a bulk material. Then, we can simplify Eq. (5.72) to

$$W_b = \int_{\Omega \backslash \Gamma} w_{bulk}\left(\boldsymbol{\varepsilon}^e, d\right) \mathrm{dV} + \int_{\Gamma^b} g_c^b(\boldsymbol{u}, d) \mathrm{d}\Gamma, \tag{5.73}$$

where $w_{bulk}(\boldsymbol{\varepsilon}^e, d)$ can be written as

$$w_{bulk}\left(\boldsymbol{\varepsilon}^e, d\right) = [g(d) + \kappa]\tilde{\psi}\left(\boldsymbol{\varepsilon}^e\right), \tag{5.74}$$

$$\tilde{\psi}\left(\boldsymbol{\varepsilon}^e\right) = \frac{1}{2}\lambda\left(\varepsilon_{kk}^e\right)^2 + \mu\boldsymbol{\varepsilon}^e : \boldsymbol{\varepsilon}^e, \tag{5.75}$$

where g(d) is the extent of the degradation of the elastic energy due to crack evolution, $\tilde{\psi}(\boldsymbol{\varepsilon}^e)$ is the effective elastic energy, $\boldsymbol{\varepsilon}^e$ is the elastic strain tensor, and λ and μ are Lamé's constants. μ is customarily used in the academic literature to represent a Lamé's constant and should not be confused with the chemical potential μ that is also conventionally used in the academic literature. A small positive parameter $\kappa \approx 0$ is used to prevent a singularity in the equation. The monotonically decreasing function g(d) should satisfy the following conditions:

$$g(0) = 1,\ g(1) = 0,\ g'(1) = 0. \tag{5.76}$$

The first two conditions correspond to the cases where the bulk material is intact and completely fractured, respectively. The last condition ensures that the crack driving force is zero when the bulk material is completely fractured. The following shows a simple function with the above properties:

$$g(d) = (1 - d)^2. \tag{5.77}$$

Thus, we have the constitutive relations in the following forms:

$$\boldsymbol{\sigma} = \left[(1 - d)^2 + \kappa\right]\tilde{\boldsymbol{\sigma}},\ \ \tilde{\boldsymbol{\sigma}} = \lambda\varepsilon_{kk}^e \boldsymbol{I} + 2\mu\boldsymbol{\varepsilon}^e. \tag{5.78}$$

where $\boldsymbol{\sigma}$ is the true Cauchy stress that accounts for damage and $\tilde{\boldsymbol{\sigma}}$ is the effective stress. The last term in Eq. (5.73) can be written as

$$\int_{\Gamma^b} g_c^b(\boldsymbol{u}, d)\mathrm{d}\Gamma = \int_{\Omega} \overline{g}_c^b \gamma(d, \nabla d)\mathrm{dV} = \int_{\Omega} \frac{\overline{g}_c^b}{2l}\left(d^2 + l^2|\nabla d|^2\right)\mathrm{dV}, \tag{5.79}$$

where $\overline{g}_c^b$ needs to meet the scale-dependent energy criterion that $\overline{g}_c^b = 256\sigma_c^2 l / 27E$[35] and l is governed by the critical fracture stress and Young's modulus E. Substituting Eqs. (5.74), (5.75), and (5.79) into Eq. (5.73) and then applying variational analysis to the resulting equation with respect to d, by the variational principle, we have

$$\delta W_b = \left\{ \frac{\overline{g}_c^b}{l}[d - l^2 \Delta d] - 2(1-d)\tilde{\psi} \right\} \delta d$$
$$\Rightarrow \frac{\overline{g}_c^b}{l}[d - l^2 \Delta d] - 2(1-d)\tilde{\psi} = 0 \tag{5.80}$$

To prevent crack closure, the crack driving force must be kept at the historical maximum value; i.e.,

$$[d - l^2 \Delta d] = (1-d)\tilde{H}(\boldsymbol{x}, t), \ \tilde{H} = \max_{s \in [0.t]} \frac{2\tilde{\psi}(\boldsymbol{\varepsilon}^e(\boldsymbol{x}, s))}{\overline{g}_c^b / l}, \tag{5.81}$$

where $\tilde{H}$ is the crack driving force and $\boldsymbol{x}$ is the position vector. This driving force is indistinguishable for tensile and compressive strains. However, under real-world physical conditions, compressive strains cannot produce cracks; only tensile and shear strains can. Therefore, we need to distinguish between tensile and compressive strains. Miehe et al. [36] proposed the use of the following equation to address the above problem:

$$\tilde{H} = \max_{s \in [0.t]} \zeta \left\langle \sum_{a=1}^{3} \left(\frac{\langle \tilde{\sigma}_a(\boldsymbol{x}, s) \rangle}{\sigma_c} \right)^2 - 1 \right\rangle, \tag{5.82}$$

where $\zeta > 0$ is the material parameter for the slope of the applied quadratic driving force function, $\tilde{\sigma}_a$ is the material parameter of the effective Cauchy principal stress, and σ_c is the critical fracture stress. As shown in Fig. 5.22, the bulk material gradually fails as the strain increases. In addition, compared to Eqs. (5.81), and (5.82) retains the linearly elastic properties of the unfractured region and prevents cracking under pure compression.

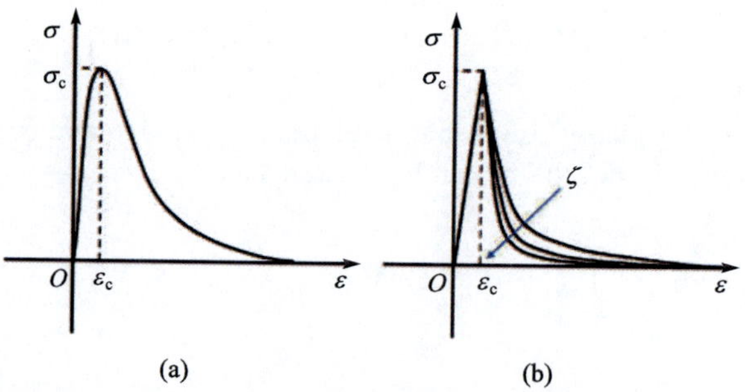

Fig. 5.22 Schematic diagrams of different qualitative characteristics of the driving force determined by uniformity tests: **a** using the energy criterion in Eq. (5.81); **b** using the energy criterion for the maximum principal stress in Eq. (5.82)

5.2.3 Introduction to CZM for Phase-Field Crack Interactions

Now, the CZM for the interface crack surface Γ^i is discussed. In 2D cases, interfacial vectors can be categorized into two types, namely, normal and tangential vectors, which correspond to Type I and II cracks, respectively. For a CZM that obeys the bilinear separation law, each mode of fracture has three independent parameters, namely, the interfacial stiffness k, the critical crack opening displacement δ_c, and the critical energy release rate for the interface $\overline{g}_c^i$. Here, the "softening" effect of phase-field cracks on the interface is considered. According to the formula established by Paggi and Reinoso [37], there is a linear relation between δ_c and the damage (d); that is, $\delta_c(d) = (1-d)\delta_c^0 + d\delta_c^1$. In the subsequent discussion, superscripts "0" and "1" indicate cases where $d = 0$ and $d = 1$, respectively; i.e., $\delta_c^0 = \delta_c(d = 0)$, and $\delta_c^1 = \delta_c(d = 1)$. In addition, we assume that the critical interfacial energy release rates I in different models (g_{Ic}^i and g_{IIc}^i) remain unchanged under the influence of bulk cracks (d) and that the ratio of the critical separation displacement δ_0 to δ_c remains unchanged under the influence of bulk cracks; i.e., $\delta_0(d)/\delta_c(d) = \delta_0^0/\delta_c^0$. Based on models I and II, the interfacial displacement δ_0 can be divided into a normal component and a tangential component; i.e., $\boldsymbol{\delta} = [\delta_n, \delta_t]$. As shown in Fig. 5.23, the bilinear traction–separation law for models I and II can be expressed as follows:

$$\sigma = \begin{cases} \mathrm{k}_n\delta_n,\ \delta_n < \delta_{n0} \\ \dfrac{\delta_{nc}-\delta_n}{\delta_{nc}-\delta_{n0}}\sigma_0,\ \delta_{n0} \le \delta_n \le \delta_{nc}, \\ 0,\ \delta_n > \delta_{nc} \end{cases} \quad \tau = \begin{cases} \mathrm{k}_t\delta_t,\ \delta_t < \delta_{t0} \\ \dfrac{\delta_{tc}-\delta_t}{\delta_{tc}-\delta_{t0}}\tau_0,\ \delta_{t0} \le \delta_t \le \delta_{tc}, \\ 0,\ \delta_t > \delta_{tc} \end{cases} \tag{5.83}$$

where σ and τ are the tractions in models I and II, respectively, the first subscripts n and t indicate the normal component and the tangential component, respectively, and the second subscripts 0 and c denote the starting point of the damage and the

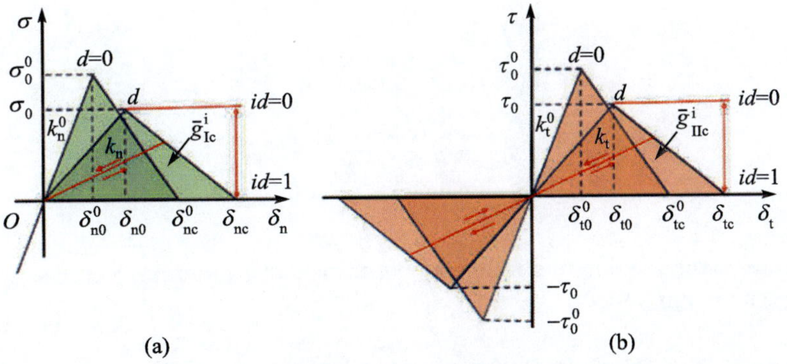

Fig. 5.23 Schematic diagrams of the phase-field coupling bilinear separation–traction CZM: **a** Type I separation–traction law; **b** Type II separation–traction law

fracture point, respectively. The critical separation strengths in models I and II (σ_0 and τ_0, respectively) correspond to their respective critical separation displacements (δ_{no} and δ_{to}, respectively). In addition, the interfacial failure (*id*) can be defined as follows: $id = 0$ for points δ_{n0} and δ_{t0}, and $id = 1$ for points δ_{nc} and δ_{tc}.

In cases where $\delta_n < 0$, there are no interface cracks [38], and bulk damage does not affect the compression performance; i.e.,

$$\sigma = \mathrm{k}_n^0 \delta_n. \tag{5.84}$$

As the critical energy release rate for the interface is assumed to remain unchanged under the influence of bulk damage (*d*), the interfacial stiffness $\mathbf{k} = [\mathrm{k}_n, \mathrm{k}_t]$ is governed by the following conditions:

$$\begin{cases} g_{\mathrm{Ic}}^i = \frac{1}{2}\mathrm{k}_n \delta_{n0}(d)\delta_{nc}(d) = \frac{1}{2}\mathrm{k}_n^0 \delta_{n0}^0 \delta_{nc}^0 \\ g_{\mathrm{IIc}}^i = \frac{1}{2}\mathrm{k}_t \delta_{t0}(d)\delta_{tc}(d) = \frac{1}{2}\mathrm{k}_t^0 \delta_{t0}^0 \delta_{tc}^0 \end{cases}. \tag{5.85}$$

$\delta_0(d)$ and $\delta_c(d)$ need to satisfy the assumption that $\delta_0(d)/\delta_c(d) = \delta_0^0/\delta_c^0$. k_n and k_t can be expressed as

$$\mathrm{k}_n(d) = \mathrm{k}_n^0\left(\frac{\delta_{nc}^0}{(1-d)\delta_{nc}^0 + d\delta_{nc}^1}\right)^2, \quad \mathrm{k}_t(d) = \mathrm{k}_t^0\left(\frac{\delta_{tc}^0}{(1-d)\delta_{tc}^0 + d\delta_{tc}^1}\right)^2. \tag{5.86}$$

In addition, the energy release rates g_{I}^i and g_{II}^i in models I and II, respectively, can be expressed as follows:

$$\begin{gathered} g_{\mathrm{I}}^i(\delta_n, d) = \begin{cases} \frac{1}{2}\mathrm{k}_n \delta_n^2, \ \delta_n < \delta_{n0} \\ g_{\mathrm{Ic}}^i - \frac{1}{2}\frac{(\delta_{nc}-\delta_n)^2}{\delta_{nc}-\delta_{n0}}\sigma_0, \ \delta_{n0} \le \delta_n \le \delta_{nc} \\ g_{\mathrm{Ic}}^i, \ \delta_n > \delta_{nc} \end{cases}, \\ g_{\mathrm{II}}^i(\delta_t, d) = \begin{cases} \frac{1}{2}\mathrm{k}_t \delta_t^2, \ \delta_t < \delta_{t0} \\ g_{\mathrm{IIc}}^i - \frac{1}{2}\frac{(\delta_{tc}-\delta_t)^2}{\delta_{tc}-\delta_{t0}}\tau_0, \ \delta_{t0} \le \delta_t \le \delta_{tc} \\ g_{\mathrm{IIc}}^i, \ \delta_t > \delta_{tc} \end{cases}. \end{gathered} \tag{5.87}$$

Under mixed loading, the following quadratic stress criterion is used to govern damage initiation [39]:

$$\left(\frac{\langle\sigma\rangle}{\sigma_0}\right)^2 + \left(\frac{\tau}{\tau_0}\right)^2 = 1. \tag{5.88}$$

Finally, the Benzeggagh–Kenane criterion is used for crack propagation: [40]

$$g^i_{\mathrm{Ic}}+\left(g^i_{\mathrm{IIc}}-g^i_{\mathrm{Ic}}\right)\left(\frac{g^i_{\mathrm{II}}}{g^i_{\mathrm{I}}+g^i_{\mathrm{II}}}\right)^{\eta}=g^i_{\mathrm{Ic}}+g^i_{\mathrm{IIc}}. \tag{5.89}$$

where η is the Benzeggagh–Kenane coefficient.

Interfacial oxidation results in TBC failure. The stress formed from TGO growth due to oxidation is the main cause of this type of failure, whereas the stress formed due to the thermal mismatch between the coating and substrate plays only a minor role [10]. Thus, the temperature stress, thermal diffusion process, and viscoplastic deformation are ignored at high temperatures. Then, the following is derived from Eq. (5.6):

$$\theta\gamma=\boldsymbol{\sigma}:\dot{\boldsymbol{\varepsilon}}-\dot{\psi}+\mu\dot{c}+\mu\mathrm{M}\dot{n}-\boldsymbol{j}\cdot\nabla\mu\geq 0. \tag{5.90}$$

Here, the Helmholtz free energy density can be regarded as a function of these state variables as follows:

$$\psi=\psi(\boldsymbol{\varepsilon},d,c,n). \tag{5.91}$$

Substitution of the Helmholtz free energy density in Eq. (5.91) into the dissipation inequality in Eq. (5.90) gives

$$\theta\gamma=\left(\sigma_{ij}-\frac{\partial\psi}{\partial\varepsilon_{ij}}\right)\dot{\varepsilon}_{ij}+\left(\mu-\frac{\partial\psi}{\partial c}\right)\dot{c}-\frac{\partial\psi}{\partial d}\dot{d}+\left(\mu\mathrm{M}-\frac{\partial\psi}{\partial n}\right)\dot{n}-\boldsymbol{j}\cdot\nabla\mu\geq 0. \tag{5.92}$$

Under the local dissipation assumption, we obtain the following constitutive equations:

$$\sigma_{ij}=\tfrac{\partial\psi}{\partial\varepsilon_{ij}},\ \mu=\tfrac{\partial\psi}{\partial c}. \tag{5.93}$$

The Helmholtz free energy density can be written as follows:

$$\psi=\psi(\varepsilon,d,c,n)=\psi\left(\varepsilon^e(\varepsilon,n),d,c\right)=\left[(1-d)^2+\kappa\right]\tilde{\psi}\left(\varepsilon^e\right)+\psi_c(c) \tag{5.94}$$

Here, the total strain is decomposed as follows: $\boldsymbol{\varepsilon}=\boldsymbol{\varepsilon}^e+\boldsymbol{\varepsilon}^n$, $\boldsymbol{\varepsilon}^n=\alpha_n n\boldsymbol{I}$, where $\boldsymbol{\varepsilon}^n$ is the expansion strain due to TGO growth, α_n is the expansion coefficient, $\psi_c(c)=\frac{1}{2}Uc^2$ is the contribution of the chemical part to the Helmholtz free energy density, and U is the proportionality constant for c. Substituting Eq. (5.94) into Eq. (5.93), we have

$$\boldsymbol{\sigma}=\left[(1-d)^2+\kappa\right](\lambda\varepsilon_{kk}\boldsymbol{I}+2\mu\boldsymbol{\varepsilon}-3\mathrm{K}\alpha_n n\boldsymbol{I}),\ \mu=\mathrm{N}c, \tag{5.95}$$

where K is the bulk modulus. Here, the increase in the surface energy during the crack propagation process is considered a complete dissipation process and can be expressed as follows:

$$-\frac{\partial \psi}{\partial d}\dot{d} = (1-d)\tilde{\psi}(\boldsymbol{\varepsilon}^e)\dot{d} \geq 0. \tag{5.96}$$

To allow the dissipation inequality in Eq. (5.92) to always be greater than zero, we make the following assumptions:

$$\begin{cases} \boldsymbol{j}_d = -m\nabla\mu \\ \dot{n} = k_0\left(\mu \mathrm{M} - \frac{\partial \psi}{\partial n}\right), \end{cases} \tag{5.97}$$

where m and k_0 are both positive parameters. Combining Eqs. (5.2), (5.94), (5.95), and (5.97), we obtain the following oxygen diffusion equation and governing equation for TGO growth:

$$\begin{cases} \dot{c} = m\mathrm{U}\nabla^2 c - \mathrm{M}\dot{n} \\ \dot{n} = k_0\left\{\mathrm{MU}c + 3\left[(1-d)^2 + \kappa\right]\mathrm{K}\alpha_n \varepsilon_{kk}^e\right\} \end{cases}. \tag{5.98}$$

Similar to the way in which we treat Eq. (5.31), we introduce $\mathrm{k}_0 = (1-n)\mathrm{k}_1$ (k_1 is a positive constant) to limit the increase in n. Then, we can obtain the following:

$$\begin{cases} \dot{c} = m\mathrm{U}\nabla^2 c - \mathrm{M}\dot{n} \\ \dot{n} = k_1(1-n)\left\{\mathrm{MN}c + 3\left[(1-d)^2 + \kappa\right]\mathrm{K}\alpha_n \boldsymbol{\varepsilon}_{kk}^e\right\} \end{cases}. \tag{5.99}$$

By neglecting the strain term and substituting the variables, we have

$$\begin{cases} \dot{c} = \mathrm{D}\nabla^2 c - \mathrm{M}\dot{n} \\ \dot{n} = \mathrm{G}(1-n)c \end{cases}, \tag{5.100}$$

where $\mathrm{D} = m\mathrm{U}$ is the oxygen diffusion coefficient and $\mathrm{G} = k_1\mathrm{MU}$ is the chemical reaction coefficient.

Finally, boundary and initial conditions need to be provided to improve the theory. The boundary conditions for the displacement and phase fields can be written as follows:

$$\boldsymbol{\sigma} \cdot \boldsymbol{n} = \boldsymbol{t},\ \boldsymbol{u} = \overline{\boldsymbol{u}},\ \nabla \cdot \boldsymbol{d} = 0, \tag{5.101}$$

The boundary conditions for the diffusion field are as follows:

$$c = \overline{c},\ \boldsymbol{j}_d \cdot \boldsymbol{n} = \overline{j}_d. \tag{5.102}$$

where $\overline{c}$ is the relative oxygen concentration at the interface and $\overline{j}_d$ is the vector of the oxygen diffusion flux at the interface. The initial condition for each field is set as follows:

$$\boldsymbol{u}(\boldsymbol{x}, 0) = \boldsymbol{u}^0(\boldsymbol{x}),\ c(\boldsymbol{x}, 0) = c^0(\boldsymbol{x}),\ n(\boldsymbol{x}, 0) = n^0(\boldsymbol{x}),\ d(\boldsymbol{x}, 0) = d^0(\boldsymbol{x}), \tag{5.103}$$

where $d^0(\boldsymbol{x})$ is the initial phase-field crack field.

5.2.4 Numerical Implementation

Figure 5.24 shows a 2D geometric model established for a TBC with a cosine interfacial shape. To simplify the calculations, the substrate is not considered. As the oxidation reaction occurs at the TGO/BC interface, the effect of the substrate on the TBC can be ignored. The geometric model contains a 200-μm-thick TC layer and a 100-μm-thick BC layer. The system has a width of 40 μm.

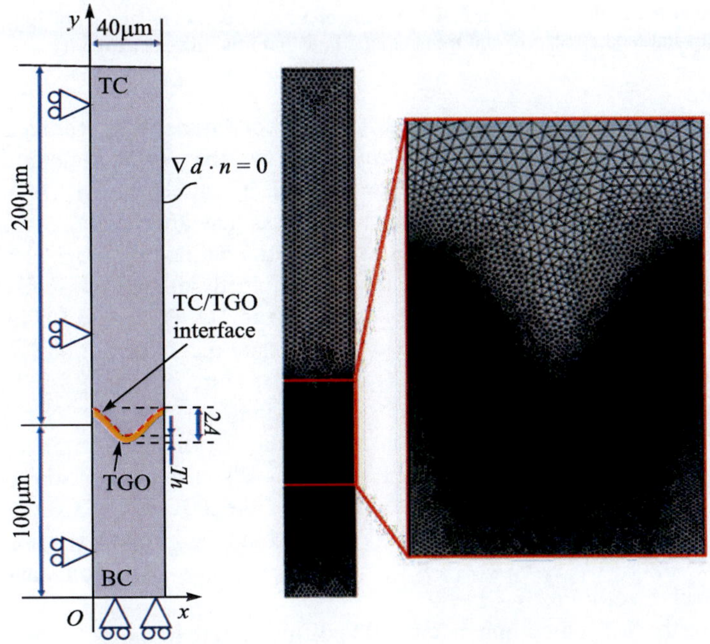

Fig. 5.24 Schematic diagram of the boundary conditions and mesh of the model for crack propagation in the TBC

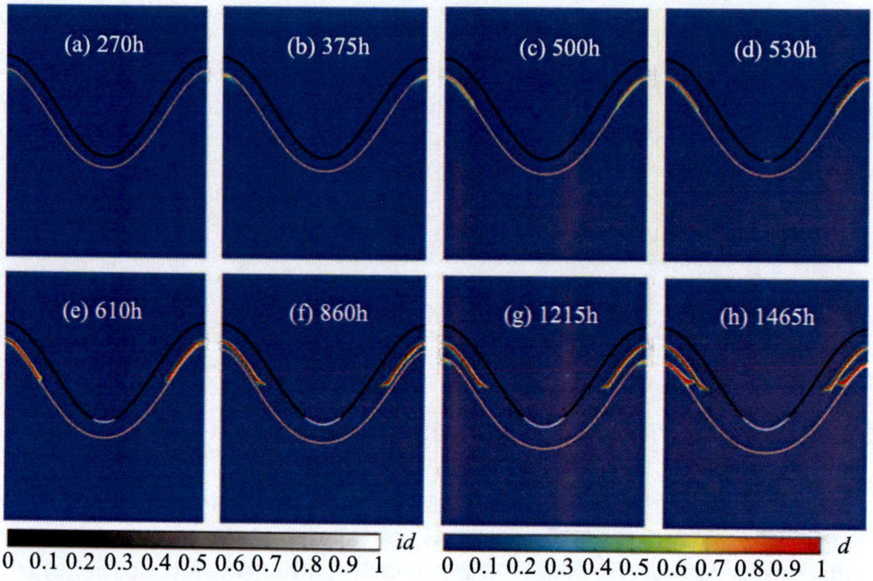

Fig. 5.25 Simulated crack initiation and propagation in the TBC: **a** visible damage appears at the peak of the BC layer at 270 h; **b** a bulk crack appears for the first time in the TGO/BC mixed region near the BC layer at 375 h; **c** visible damage appears at the trough of the TC/TGO interface at 500 h; **d** a crack appears at the trough of the TC/TGO interface at 530 h; **e** the first bulk crack propagates towards the inside of the TGO layer at 610 h; **f** the first bulk crack stabilizes at 860 h; **g** the second bulk crack is initiated at 1215 h; and **h** the second bulk crack stabilizes at 1465 h

The TGO thickness is 0 under the initial conditions. The geometric shape of the upper boundary of the TGO layer follows the cosine function of $y = \mathrm{A}\cos\left(2\pi x \big/ x_0\right) + A_0$ ($x_0 = 40\,\mu\text{m}$, $A_0 = 100\,\mu\text{m}$, and $\mathrm{A} = 10\,\mu\text{m}$). The maximum size of the mesh boundaries in regions where cracks may appear (i.e., regions where there are neighboring TGOs) is $l/2$. The crack width defined in Fig. 5.21 is $2l$. This model contains 99,912 cells. A CZM model combined with the phase field is used for the TC/TGO interface, and a PFM is used for the TC, TGO, and BC layers.

Table 5.1 summarizes the mechanical properties of the TBC. Similarly, the Voigt assumption is used for the TGO/BC mixed layer; i.e., $\boldsymbol{\varepsilon} = \boldsymbol{\varepsilon}_{\mathrm{TGO}} = \boldsymbol{\varepsilon}_{\mathrm{BC}}$, and $\boldsymbol{\sigma} = n\boldsymbol{\sigma}_{\mathrm{TGO}} + (1-n)\boldsymbol{\sigma}_{\mathrm{BC}}$, where the subscript indicates the layer corresponding to the variable. As thermal expansion is not considered, the plane strain assumption is adopted for the system. Figure 5.24 shows the displacements at the bottom of the TBC in the y-direction and on the left side of the TBC in the x-direction; i.e., $v(x, y)\big|_{y=0} = 0$, and $u(x, y)|_{x=0} = 0$. Periodic boundary conditions are used for the left and right sides; i.e., $\boldsymbol{u}(x, y)\big|_{x=0} = \boldsymbol{u}(x, y)\big|_{x=60\,\mu\text{m}}$. The boundary stress at the top is set to zero; i.e., $\boldsymbol{\sigma} \cdot \boldsymbol{n} = \boldsymbol{0}$.

Because the TC is a porous material, the diffusion rate in the TC layer is several orders of magnitude higher than those in the TGO and BC layers. In addition, no oxidation reactions occur in the TC layer. Thus, the TC layer is transparent to oxygen.

Table 5.3 TGO growth parameters [23]

Parameter	Value
Oxygen diffusion coefficient in the BC layer	$D_0^{BC} = 2.0 \times 10^{-13}\ m^2s^{-1}$
Oxygen diffusion coefficient in the TGO layer	$D_0^{TGO} = 2.0 \times 10^{-13}\ m^2s^{-1}$
Oxygen diffusion coefficient in the TC layer	$D_0^{TC} \to \infty\ m^2s^{-1}$
Oxidation reaction constant	$G = 110^{-4}\ m^3mol^{-1}\ s^{-1}$
Ratio of the number of moles of absorbed oxygen to the volume of the formed TGOs	$M = 1.1110^5\ molm^{-3}$
Stress coupling coefficient of the chemical reaction	$\alpha_n = 810^{-3}$

The oxygen concentration at the TC/TGO interface is set to 1.55 mol/m^3, and the initial oxygen concentration in the BC layer is set to 0 mol/m^3. Moreover, to simplify the model, the oxygen diffusion coefficients in the TGO and BC layers are assumed to be approximately equal ($D^{BC} \approx D^{TGO}$). Table 5.3 summarizes the specific values of the TGO growth parameters. Further, the initial volume fractions of TGOs in the TGO and BC layers are set to 1 and 0, respectively.

Table 5.4 summarizes the material parameters of the TBC failure model. As ceramics are brittle materials, the ζ in the TC layer, ζ^{TC}, is set to a large brittle fracture value of 0.5. In addition, the ζ in the plastic BC layer, ζ^{BC}, is set to a small value of 0.1, and the ζ in the TGO layer, ζ^{TGO}, is set to a value of 0.2, which is between the values of ζ^{TC} and ζ^{BC}. All the parameters of the TGO/BC mixed layer are linearly distributed; i.e., $P = nP_{TGO} + (1 - n)P_{BC}$, where P, P_{TGO}, and P_{BC} are parameters of the TGO/BC mixed layer, the TGO layer, and the BC layer, respectively. Finally, the governing equations are solved using the solid-mechanics and PDE modules in COMSOL Multiphysics 5.5.

5.2.5 *Results and Discussion*

As shown in Fig. 5.25, the TGO/BC interface (n = 0.5) and TC/TGO interface are shown in pink and black, respectively. If the color of the TC/TGO interface turns white, the TC/TGO interface is completed disconnected. During phase-field crack propagation, the material turns from blue to red. First, the tensile stress at the peak of the BC layer damages the material. When the TGO thickness at the peak increases to 3.2 μm at 375 h, a bulk crack appears for the first time in the TGO/BC mixed region near the BC layer. As TGO growth continues, the first bulk crack propagates downward along the TGO/BC interface. The TGO/BC interface begins to sustain damage within 500 h. When the TGO thickness at the peak increases to 3.85 μm at 530 h, a crack is initiated at the trough of the TC/TGO interface. The interface crack propagates upward along the TC/TGO interface. When the TGO/BC interface grows below the bulk crack, the bulk crack reaches the TGO layer at 610 h. Subsequently, the first bulk crack becomes stable and ceases to propagate at 860 h. When the TGO

Table 5.4 Material parameters of the model for crack propagation in the TBC [10–12]

Parameter	Value
Material parameters of the TC layer	
Phase-field crack width parameter	$l^{TC} = 0.2\,\mu m$
Critical fracture stress	$\sigma_c^{TC} = 300$ MPa
Slope of the quadratic driving force function	$\zeta^{TC} = 0.5$
Material parameters of the TGO layer	
Phase-field crack width parameter	$l^{TGO} = 0.2\,\mu m$
Critical fracture stress	$\sigma_c^{TGO} = 600$ MPa
Slope of the quadratic driving force function	$\zeta^{TGO} = 0.2$
Material parameters of the BC layer	
Phase-field crack width parameter	$l^{BC} = 0.2\,\mu m$
Critical fracture stress	$\sigma_c^{BC} = 700$ MPa
Slope of the quadratic driving force function	$\zeta^{BC} = 0.1$
Material parameters of the TC/TGO interface	
Initial stiffness	$k_n^0 = k_t^0 = 1 \times 10^{18}$ N/m^3
Initial separation strengths in models I and II	$\sigma_0^0 = \tau_0^0 = 150$ MPa
Critical interfacial energy release rate in model I	$G_{Ic}^i = 30$ N/m
Critical interfacial energy release rate in model II	$G_{IIc}^i = 100$ N/m
Ratio of the initial to ultimate separation strength in models I and II	$\sigma_0^0/\sigma_0^1 = \tau_0^0/\tau_0^1 = 2$

thickness at the peak increases to 6.5 μm, the second bulk crack appears at the TGO/BC interface at 1215 h. Finally, the second bulk crack stabilizes at 1465 h. As seen in Fig. 5.25, at the TGO/BC interface, the thickness of the second bulk crack is approximately twice that of the first crack. In addition, the cracks at the TGO/BC interface ultimately propagate within the TGO layer.

Now, the crack propagation pattern by analyzing the stress distribution in the TBC during the oxidation process is examined. We assume that the principal stresses that account for damage satisfy the condition $\sigma_1 \geq \sigma_2 \geq \sigma_3$. As shown in Fig. 5.26, cracks are initiated and propagate at locations where σ_1 is concentrated, while crack initiation and propagation are primarily driven by σ_{yy}. σ_{yy} is concentrated at the trough of the TC/TGO interface and the peak of the TGO/BC mixed layer near the BC layer. As a result, cracks appear at these two interfaces. As the TGO/BC interface propagates to the lower end of a bulk crack, new TGOs grow at the lower end of the phase-field crack. Due to this part of the volume expansion of the TGO layer, the TGOs above the bulk cracks are forced to withstand an upward force, leading to the spallation of the upper part of the TGO layer and crack propagation into the TGO layer.

Figures 5.27 and 5.28 show cross-sectional SEM images of typical crack propagation in TBC specimens prepared by electron beam-physical vapor deposition (EB-PVD) and atmospheric plasma spraying (APS), respectively. These specimens

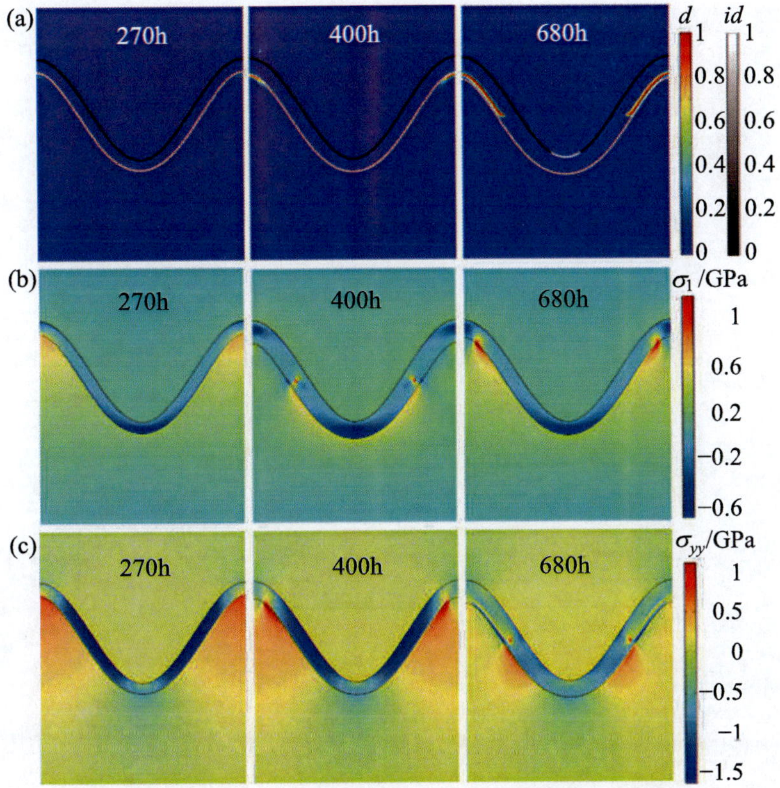

Fig. 5.26 Distribution patterns of the phase field d and stress during crack initiation and propagation. The three columns, from left to right, show the patterns before crack initiation, during bulk crack propagation, and during bulk crack propagation towards the TGO layer and the TC/TGO interface, respectively. **a** Crack phase field d. **b** First principal stress. **c** $\tilde{\sigma}_{yy}$ (In each image in the last two rows (i.e., **b** and **c**), the upper black line represents the TC/TGO interface, and the lower black line represents the TGO/BC interface)

are the same as those discussed in Sect. 5.1.5 and were subjected to no external loading. Cracks appear at the peak of the TGO/BC interface in both the EB-PVD- and APS-prepared TBC specimens. The cracks at the TGO/BC interface propagate to the TGO layer, and cracks appear at the trough of the TC/TGO interface. The results show that TGO growth stress is the primary cause of interfacial failure in the two types of TBCs.

In addition, crack propagation towards the TGO layer is common at the troughs of the TC/TGO interface, the peaks of the TGO/BC interface, and the TGO/BC interface, as shown in Fig. 5.29. Many researchers [11, 14, 16, 41–43] have observed these cracks in experiments. According to our model, cracks at the peaks of the TGO layer are not directly formed in the TGO layer. Instead, they are formed at the peaks of the TGO/BC layer and covered by new TGOs to become internal cracks at the

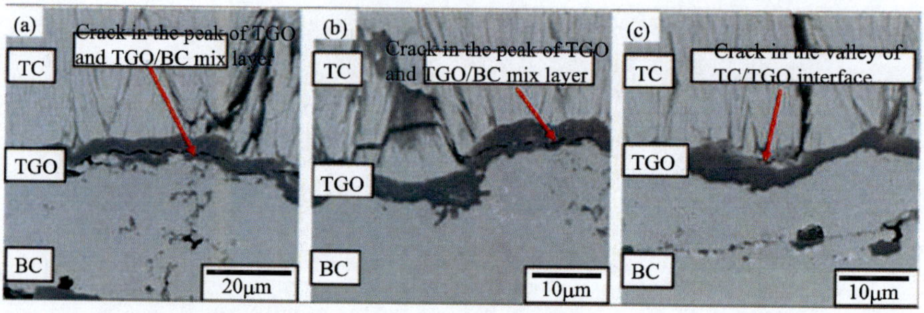

Fig. 5.27 Cross-sectional SEM images of a TBC with an EB-PVD TC layer. **a** and **b** Crack propagation at the peaks of the TGO layer and TGO/BC mixed layer. **c** Crack propagation at the trough of the TC/TGO interface

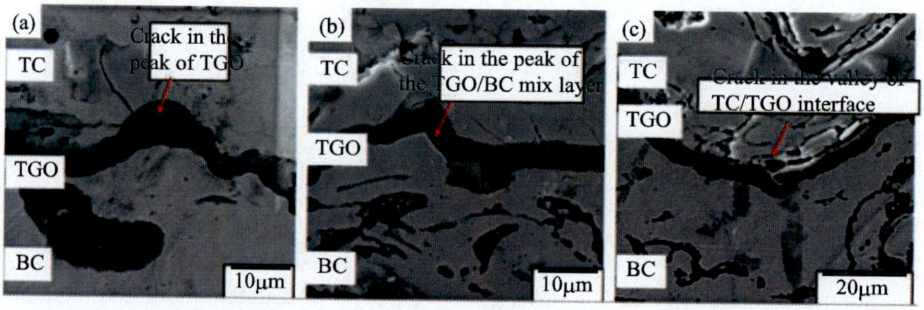

Fig. 5.28 Cross-sectional SEM images of a TBC with an APS TC layer. **a** Crack propagation at the peak of the TGO layer. **b** Crack propagation at the peak of the TGO/BC mixed layer (part of the light-colored BC layer can be seen in the TGO layer at the upper interface of the crack). **c** Crack propagation at the trough of the TC/TGO interface

peaks of the TGO layer. Particularly in Fig. 5.29c, the cracks at the peaks of the TGO layer are morphologically identical to those at the peaks of the TGO/BC interface.

Figure 5.30 shows the variation in the phase-field variable with the TGO thickness. The TGO/BC interface and phase-field variable exhibit a quasiperiodic pattern. Interface cracks are formed when the stress reaches 0.672 GPa. This is because phase-field cracks have a certain width, and consequently, the relation between the interfacial variable and TGO thickness is not completely periodic, and the first crack damages the BC layer. This process can be explained as follows. As Young's modulus of the TC layer is smaller than those of the TGO and BC layers, most of the stress is concentrated at the peak of the TGO/BC interface. Thus, the peak at the TC/TGO interface can be considered to withstand no pressure. The concentrated stress at the TGO/BC interface is released once the first crack at the TGO/BC interface becomes stable. For newly formed TGOs, the first crack is equivalent to the TC/TGO interface. As a result, as TGO growth continues, the first crack is once again subjected to stress concentrations, resulting in the formation of the second crack at the TGO/BC interface. In addition, because the first crack is shorter than the TC/TGO interface,

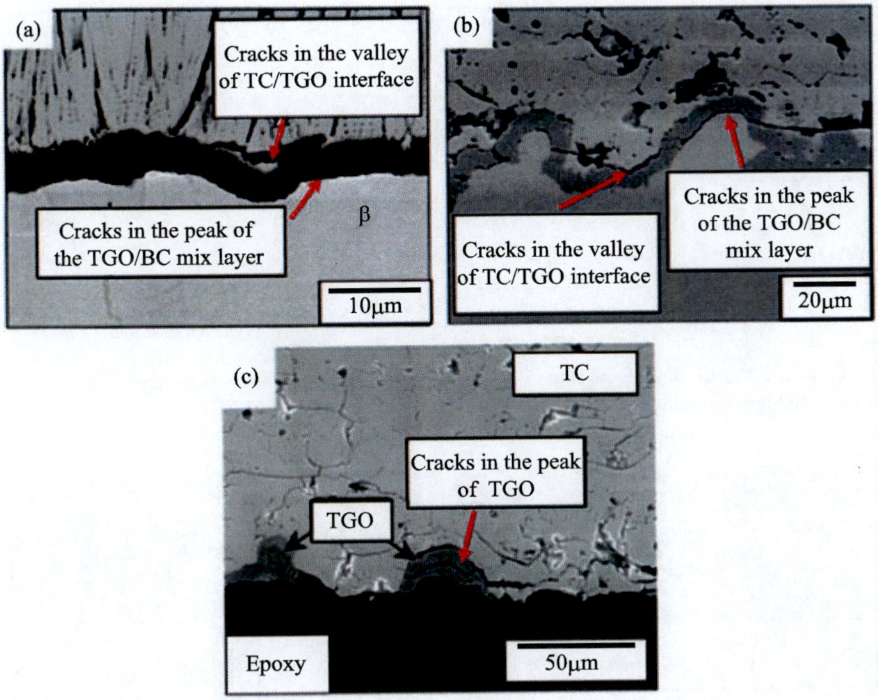

Fig. 5.29 Cross sectional SEM images of TBCs obtained by **a** Mumm et al. [41] **b** Rabici and Evans [11], and **c** Schlichting et al. [14]

the second crack propagates more quickly to the TGO layer. It can be concluded that crack initiation and propagation at the TGO/BC interface depend on the TGO thickness and that equidistant cracks are initiated as TGO growth continues. As shown in Fig. 5.31, it has been experimentally observed that equidistant cracks often appear at a thick TGO peak and that only one crack (i.e., a crack at the TC/TGO interface) appears at a thick TGO trough. This further shows that internal cracks at the peak of the TGO layer are formed from cracks at the TGO/BC interface during the TGO growth process.

Figure 5.32 shows the relations between the crack propagation length and TGO thickness. As TGO growth continues, the crack at the TGO/BC interface becomes stable, whereas the crack at the TC/TGO interface continues to grow. In addition, when the second crack is initiated at the TGO/BC interface, the crack at the TC/TGO interface propagates at an increasing rate. The effects of the cracks at the TGO/BC interface on those at the TC/TGO interface are discussed in the following.

To investigate the effects of the cracks at the TGO/BC interface on those at the TC/TGO interface, the PFM is removed, and only the CZM model is used at the TC/TGO interface.

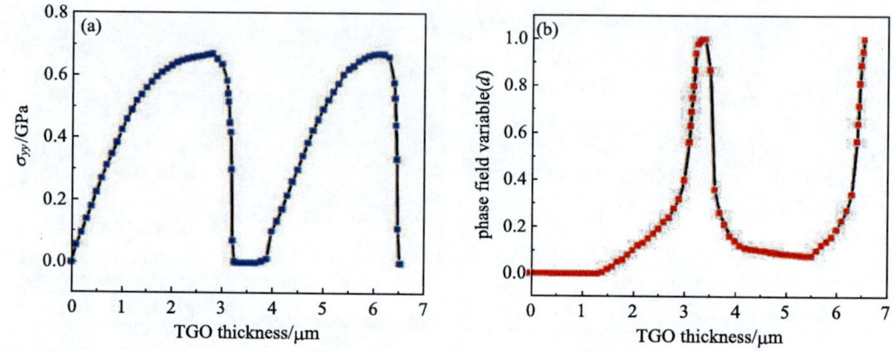

Fig. 5.30 **a** Relation between $\tilde{\sigma}_{yy}$ and the TGO thickness. **b** Relation between the phase field variable and the TGO thickness

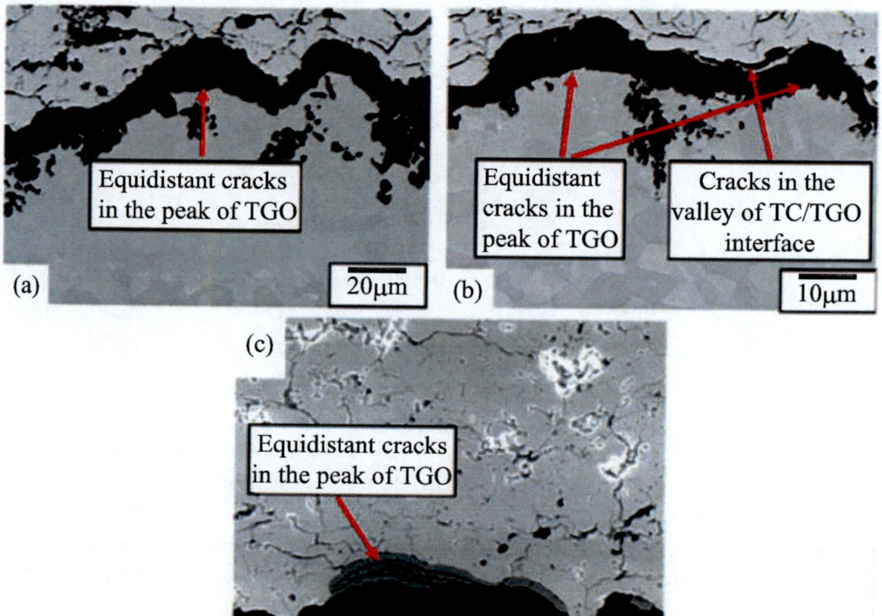

Fig. 5.31 Cross-sectional SEM images of equidistant cracks at the peak of the TGO layer obtained by **a** Schlichting et al. [14] and **b** and **c** Trunova et al. [43]

As shown in Fig. 5.33, a crack appears at a later time at the TC/TGO interface and ultimately propagates over a longer distance when the bulk cracks are not considered than when the bulk cracks are considered. Figure 5.34 shows the increase in the crack length at the TC/TGO interface with the TGO thickness in the two scenarios. When the bulk cracks are considered, a crack is initiated at the TC/TGO interface at a peak

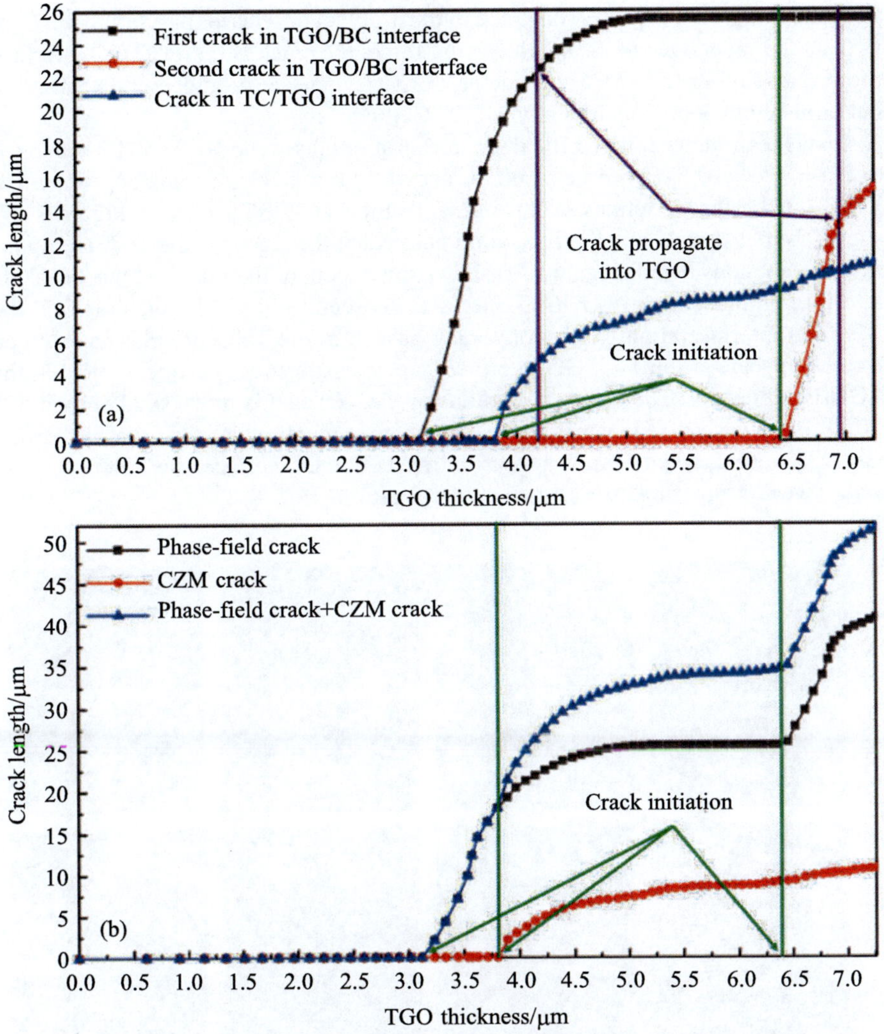

Fig. 5.32 Relations between the crack propagation length and the peak TGO thickness. **a** Relations between the lengths of three cracks and the peak TGO thickness. **b** Relations between the crack length and the peak TGO thickness in different models

TGO thickness of 3.85 μm. In contrast, when the bulk cracks are neglected, a crack begins to appear at the TC/TGO interface at a peak TGO thickness of 5.1 μm. The crack at the TGO/BC interface propagates from the peak to the trough, whereas the peak of the BC layer no longer withstands the tensile stress from the TGO layer. As a result, the TC/TGO interface withstands a higher tensile stress. Consequently, when a crack appears at the TGO/BC interface, the TC/TGO interface is more prone to cracking. As TGO growth continues, the crack at the TC/TGO interface propagates from the trough to the peak. The crack in the TGO layer releases part of the pressure in the TGO layer. Due to the small driving force, the crack at the TC/TGO interface propagates slowly. In short, a crack at the TGO/TC interface promotes crack initiation but inhibits crack propagation at the TC/TGO interface.

In this section, the initial critical separation strengths in models I and II is changed to $\sigma_0^0 = \tau_0^0 = 100$ MPa. As a result, as shown in Fig. 5.35a, a crack appears at the TC/TGO interface before a crack is initiated at the TGO/BC interface. Figure 5.35b shows the phase-field crack propagation path when the cracks at the TC/TGO interface are not considered. Figure 5.36 shows the effects of the cracks at the TC/TGO interface on those at the TGO/BC interface. As seen in Fig. 5.36, the cracks at the TC/TGO interface slightly promote crack growth at the TGO/BC interface. When a crack appears at the TC/TGO interface, more tensile stress is concentrated at the TGO/BC interface. The Young's modulus of the TC layer is much smaller than that of the BC layer, and the stress in the TC layer is much lower than that in the BC layer. Therefore, when a crack appears at the TC/TGO interface, the TGO layer has a very weak promoting effect on crack propagation.

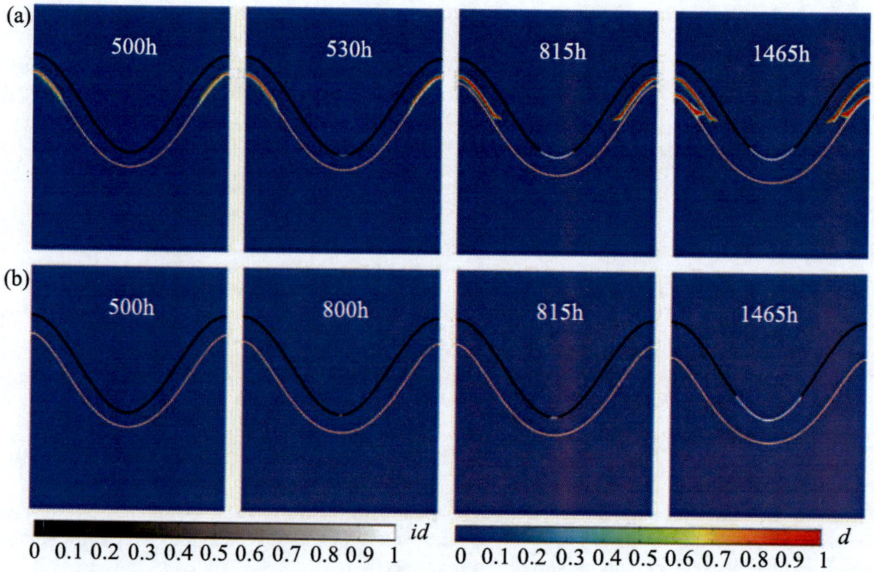

Fig. 5.33 Comparison of the crack growth at the TC/TGO interface simulated in the **a** presence and **b** absence of bulk cracks, respectively

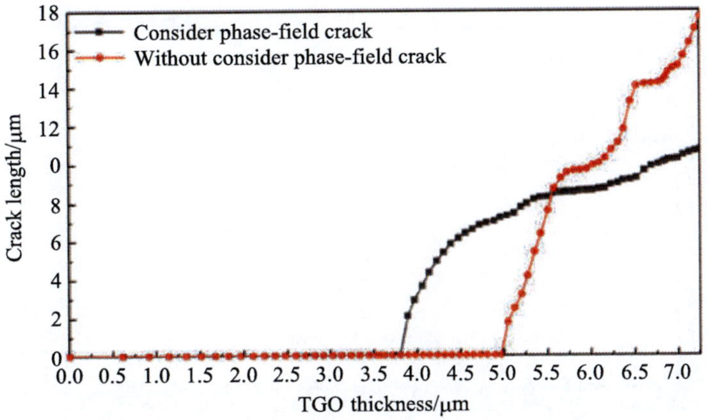

Fig. 5.34 Relations between the crack length at the TC/TGO interface and the peak TGO thickness

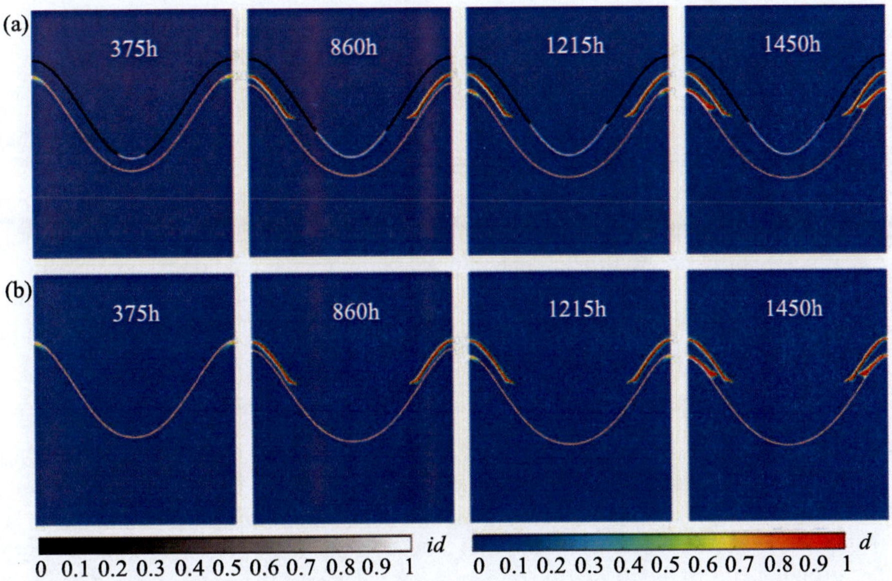

Fig. 5.35 Comparison of the bulk crack growth simulated in the **a** presence and **b** absence of cracks at the TC/TGO interface

5.3 Summary and Outlook

5.3.1 Summary

This chapter focuses on the physically nonlinear coupling oxide growth and failure models for TBCs based on the laws of thermodynamics.

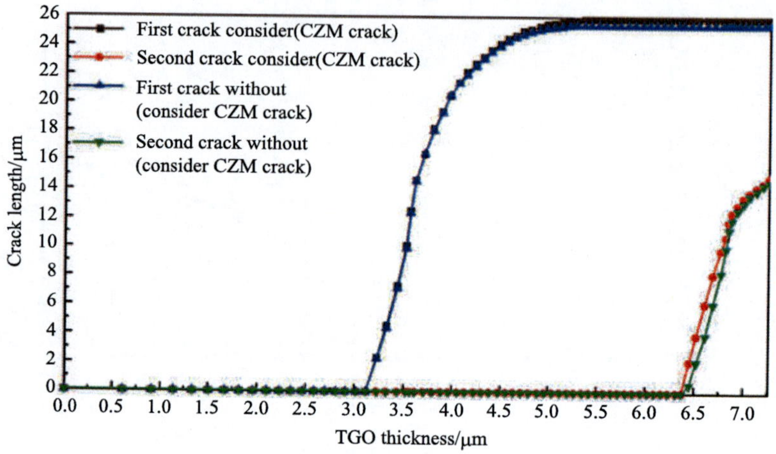

Fig. 5.36 Relations between the bulk crack propagation length and the peak TGO thickness

The main conclusions for the growth model are summarized as follows:

(1) An external tensile load promotes the oxidation reaction, thereby accelerating TGO growth. An external compressive load leads to incomplete interfacial oxidation as well as internal oxidation in the BC layer.
(2) Coupled kinetics states that an external tensile load reduces the critical oxygen concentration required for the oxidation reaction, thereby promoting the oxidation reaction, while an external compressive load does the opposite.
(3) Creep increases the strain in the BC layer and accelerates the oxidation reaction. Under the combined action of creep and diffusion, TGOs grow linearly within a certain period of time.

The main conclusions for the failure model are summarized as follows:

(1) Oxidation in a TBC results in a trend of cracking at the trough of the TC/TGO interface and the peak of the TGO/BC interface.
(2) As the TGO layer grows, newly formed TGOs transform the cracks at the peak of the TGO/BC interface to internal cracks in the TGO layer. Ultimately, TGO growth leads to the formation of equidistant cracks at the peak of the TGO layer and only one crack at the trough of the TC/TGO interface.
(3) Investigations based on coupled kinetics show mutual promotion between the cracks at the TC/TGO and TGO/BC interfaces. The cracks at the TGO/BC interface promote crack initiation but inhibit crack propagation at the TGO/BC interface. The cracks at the TC/TGO interface have a very weak promoting effect on crack initiation and propagation at the TGO/BC interface.

5.3.2 *Outlook*

Future research should focus on the following main areas:

(1) Establishing physically nonlinear coupling failure problems of TBCs that consider a variety of factors (e.g., erosion failure and calcium–magnesium aluminosilicate (CMAS) CMAS corrosion failure).
(2) Developing more experimental methods to determine the coupling coefficients and the effects of external conditions on the coupling coefficients.
(3) Formulating physically nonlinear coupling theories for new TBC failure mechanisms.

References

1. Fox AC, Clyne TW. Oxygen transport by gas permeation through the zirconia layer in plasma sprayed thermal barrier coatings. Surf Coat Technol. 2004;184(2–3):311–21.
2. Evans AG, Mumm D, Hutchinson J, et al. Mechanisms controlling the durability of thermal barrier coatings. Prog Mater Sci. 2001;46(5):505–53.
3. Busso EP, Qian ZQ. A mechanistic study of microcracking in transversely isotropic ceramic–metal systems. Acta Mater. 2006;54(2):325–38.
4. Busso EP, Qian ZQ, Taylor MP, et al. The influence of bondcoat and topcoat mechanical properties on stress development in thermal barrier coating systems. Acta Mater. 2009;57(8):2349–61.
5. Busso EP, Evans HE, Qian ZQ, et al. Effects of breakaway oxidation on local stresses in thermal barrier coatings. Acta Mater. 2010;58(4):1242–51.
6. Loeffel K, Anand L. A chemo-thermo-mechanically coupled theory for elastic–viscoplastic deformation, diffusion, and volumetric swelling due to a chemical reaction. Int J Plast. 2011;27(9):1409–31.
7. Ma K, Schoenung JM. Isothermal oxidation behavior of cryomilled NiCrAlY bond coat: homogeneity and growth rate of TGO. Surf Coat Technol. 2011;205(21–22):5178–85.
8. Chen WR, Wu X, Marple BR, et al. The growth and influence of thermally grown oxide in a thermal barrier coating. Surf Coat Technol. 2006;201(3–4):1074–9.
9. Seo D, Ogawa K, Nakao Y, et al. Influence of high-temperature creep stress on growth of thermally grown oxide in thermal barrier coatings. Surf Coat Technol. 2009;203(14):1979–83.
10. Chen Y, Zhao X, Dang Y, et al. Characterization and understanding of residual stresses in a NiCoCrAlY bond coat for thermal barrier coating application. Acta Mater. 2015;94:1–15.
11. Rabiei A, Evans A. Failure mechanisms associated with the thermally grown oxide in plasma-sprayed thermal barrier coatings. Acta Mater. 2000;48(15):3963–76.
12. Xu T, Faulhaber S, Mercer C, et al. Observations and analyses of failure mechanisms in thermal barrier systems with two phase bond coats based on NiCoCrAlY. Acta Mater. 2004;52(6):1439–50.
13. Singheiser L, Steinbrech R, Quadakkers W, et al. Failure aspects of thermal barrier coatings. Mater High Temp. 2001;18(4):249–59.
14. Schlichting KW, Padture NP, Jordan EH, et al. Failure modes in plasma-sprayed thermal barrier coatings. Mater Sci Eng A. 2003;342(1–2):120–30.
15. Evans AG, He MY, Hutchinson JW. Mechanics-based scaling laws for the durability of thermal barrier coatings. Prog Mater Sci. 2001;46(3–4):249–71.
16. Padture NP, Gell M, Jordan EH. Thermal barrier coatings for gas-turbine engine applications. Science. 2002;296(5566):280–5.

17. Xu GN, Yang L , Zhou YC, et al. A chemo-thermo-mechanically constitutive theory for thermal barrier coatings under CMAS infiltration and corrosion. J Mech Phys Solids. 2019:133.
18. Zhang X, Zhong Z. A coupled theory for chemically active and deformable solids with mass diffusion and heat conduction. J Mech Phys Solids. 2017;107:49–75.
19. Ottosen N, Ristinmaa M. The mechanics of constitutive modeling. Oxford, UK: Elsevier Ltd.; 2005.
20. Ristinmaa M, Ottosen NS. Consequences of dynamic yield surface in viscoplasticity. Int J Solids Struct. 2000;37(33):4601–22.
21. Zhao Y, Chen Y, Ai S, et al. A diffusion, oxidation reaction and large viscoelastic deformation coupled model with applications to SiC fiber oxidation. Int J Plast. 2019;118:173–89.
22. Voigt W. The relation between the two elastic moduli of isotropic materials. 1889;33:573.
23. Hille TS, Turteltaub S, Suiker ASJ. Oxide growth and damage evolution in thermal barrier coatings. Eng Fract Mech. 2011;78(10):2139–52.
24. Bäker M. Finite element simulation of interface cracks in thermal barrier coatings. Comput Mater Sci. 2012;64:79–83.
25. Yang L, Liu QX, Zhou YC, et al. Finite Element Simulation on Thermal Fatigue of a Turbine Blade with Thermal Barrier Coatings. J Mater Sci Technol. 2014;30(4):371–80.
26. Zhu W, Zhang ZB, Yang L, et al. Spallation of thermal barrier coatings with real thermally grown oxide morphology under thermal stress. Mater Des. 2018;146:180–93.
27. Rösler J, Bäker M, Aufzug K. A parametric study of the stress state of thermal barrier coatings part I: creep relaxation. Acta Mater. 2004;52(16):4809–17.
28. Siry CW, Wanzek H, Dau CP. Aspects of TBC service experience in aero engines. Materialwissenschaft und Werkstofftechnik: Mater Sci Eng Technol. 2001;32(8):650–3.
29. Pint B, Wright I, Lee W, et al. Substrate and bond coat compositions: factors affecting alumina scale adhesion. Mater Sci Eng A. 1998;245(2):201–11.
30. Brindley WJ, Miller RA. TBCs for better engine efficiency. Adv Mater Processes. 1989;136(2):29–33.
31. Rösler J, Bäker M, Volgmann M. Stress state and failure mechanisms of thermal barrier coatings: role of creep in thermally grown oxide. Acta Mater. 2001;49(18):3659–70.
32. Kageshima H, Shiraishi K. Relation between oxide growth direction and stress on silicon surfaces and at silicon-oxide/silicon interfaces. Surf Sci. 1999;438(1):102–6.
33. Yata M. External stress-induced chemical reactivity of O_2 on Si(001). Phys Rev B Condens Matter Mater Phys. 2010;81(20): 205402.
34. Miehe C, Welschinger F, Hofacker M. Thermodynamically consistent phase-field models of fracture: variational principles and multi-field FE implementations. Int J Numer Meth Eng. 2010;83(10):1273–311.
35. Zhang X, Vignes C, Sloan SW, et al. Numerical evaluation of the phase-field model for brittle fracture with emphasis on the length scale. Comput Mech. 2017;59(5):737–52.
36. Miehe C, Schänzel L-M, Ulmer H. Phase field modeling of fracture in multi-physics problems. Part I. Balance of crack surface and failure criteria for brittle crack propagation in thermo-elastic solids. Comput Methods Appl Mech Eng. 2015;294:449–85.
37. Paggi M, Reinoso J. Revisiting the problem of a crack impinging on an interface: a modeling framework for the interaction between the phase field approach for brittle fracture and the interface cohesive zone model. Comput Methods Appl Mech Eng. 2017;321:145–72.
38. Camanho PP, Davila CG, De Moura M. Numerical simulation of mixed-mode progressive delamination in composite materials. J Compos Mater. 2003;37(16):1415–38.
39. Cui W, Wisnom M, Jones M. A comparison of failure criteria to predict delamination of unidirectional glass/epoxy specimens waisted through the thickness. Composites. 1992;23(3):158–66.
40. Benzeggagh ML, Kenane M. Measurement of mixed-mode delamination fracture toughness of unidirectional glass/epoxy composites with mixed-mode bending apparatus. Compos Sci Technol. 1996;56(4):439–49.
41. Mumm D, Evans A, Spitsberg I. Characterization of a cyclic displacement instability for a thermally grown oxide in a thermal barrier system. Acta Mater. 2001;49(12):2329–40.

42. Echsler H, Shemet V, Schütze M, et al. Cracking in and around the thermally grown oxide in thermal barrier coatings: a comparison of isothermal and cyclic oxidation. J Mater Sci. 2006;41(4):1047–58.
43. Trunova O, Beck T, Herzog R, et al. Damage mechanisms and lifetime behavior of plasma sprayed thermal barrier coating systems for gas turbines—Part I: experiments. Surf Coat Technol. 2008;202(20):5027–32.

Chapter 6
Thermo–Mechano–Chemical Coupling During CMAS Corrosion in TBCs

Haze and volcanic eruptions increase the sand and dust content in the air, corrosive (e.g., acidic and alkaline) particles are present in marine environments, and fuels may contain some impurities. Oxides in these dust, corrosive particles, and impurities, such as CaO_2, MgO, Al_2O_3, and SiO_2 (collectively referred to as CMAS for short), pose a threat to aeroengine turbines during operation. These oxide particles enter an engine through the internal inlet duct and transform when heated to high temperatures in the compressor and combustion chamber into melts that adhere to the thermal barrier coating (TBC) surface. When the temperature inside the engine exceeds their melting point, CMAS begins to melt and wet the coating surface. As shown in Fig. 6.1, under the action of capillary forces [1, 2], molten CMAS infiltrates the interior of the ceramic layer along the pores and cracks in the coating [3, 4], forming a dense layer. Further, the CMAS remaining in the pores gradually dissolves the coating grains. Ca, Mg, Al, and Si diffuse from the CMAS to the interior of the yttria-stabilized zirconia (YSZ) grains, while Zr and Y diffuse from the interior of the coating grains (i.e., YSZ grains) to the CMAS melts. As the solubility of Y is much higher than that of Zr in CMAS [5, 6], the Y content in the YSZ grains is lower than that before dissolution; that is, the ceramic coating grains lose Y [6, 7]. As the temperature decreases, the YSZ grains that have lost Y transform from a tetragonal phase to a monoclinic phase [6, 7]. From experimental and theoretical perspectives, this chapter systematically analyzes high-temperature CMAS infiltration, deformation due to compositional and microstructural changes, and phase-structure evolution in TBCs. In terms of theoretical modeling, this chapter establishes a model describing the correlations between the CMAS infiltration depth and its key influencing factors, establishes thermo–mechano–chemical decoupling and coupling constitutive theories for high-temperature CMAS corrosion of TBCs, and establishes a thermo–mechano–chemical theory for phase transformation in coatings during cooling based on the phase-field method.

Y. Zhou et al., *Thermal Barrier Coatings: Failure Theory and Evaluation Technology*,
https://doi.org/10.1007/978-981-19-2723-2_6

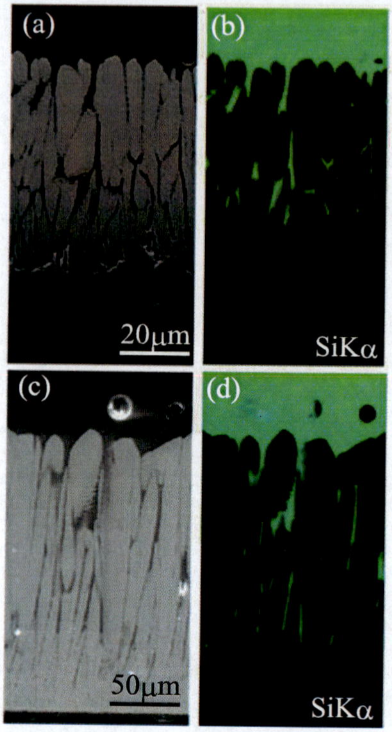

Fig. 6.1 SEM images and Si-elemental maps of a CMAS-infiltrated EB-PVD TBC

6.1 Correlation Analysis of Molten CMAS Infiltration and Its Key Influencing Factors

6.1.1 Theoretical Model for Molten CMAS Infiltration Depth in EB-PVD TBCs

In 2005, Mercer et al. [8] found that there is a critical CMAS infiltration depth for coating failure; that is, when CMAS infiltrates a coating beyond a certain critical depth, cracks are formed in the coating, which then undergoes delamination. This finding shows that the CMAS infiltration depth plays a vital role in the corrosion failure of TBCs. For an electron beam-physical vapor deposition (EB-PVD) coating, molten CMAS infiltrates its interior through the gaps between the columnar crystals under the action of capillary forces and its columnar crystals through voids [9]. In addition, studies [10–12] show that the service temperature, the coated amount of CMAS, and the infiltration time are important environmental factors affecting CMAS infiltration. However, these studies only qualitatively describe the CMAS infiltration process but fail to quantitatively characterize the relationships between the CMAS infiltration depth and its influencing factors with theoretical modeling. Aygun et al. [13] proposed the use of the following equation to describe the infiltration behavior of molten CMAS:

$$t \approx \left[\frac{K_{\mathrm{t}}}{8D_{\mathrm{c}}} \left(\frac{1-\omega}{\omega} \right)^2 L^2 \right] \frac{\mu}{\sigma} \tag{6.1}$$

where t is the CMAS infiltration time, K_{t} and D_{c} are the curvature and diameter of the pores in the TBC, respectively, ω is the effective porosity for the CMAS flow in the TBC, L is the infiltration depth, μ is the viscosity of CMAS, and σ is the surface tension. However, this equation lacks a strict theoretical basis and is unable to satisfactorily describe the process and predict the depth of CMAS infiltration in a TBC.

In view of the above problems, we propose a model that describes the correlations between CMAS infiltration depth and its key influencing factors, including the service temperature, the coated amount of CMAS, infiltration time, and surface roughness of the coating. Subsequently, we verify the reasonableness of the model by comparing results obtained from numerical simulations and experiments. In addition, we discuss the relationship between the microstructure and contact angle of a TBC and, on this basis, the way to control CMAS infiltration depth. This chapter aims to systematically elucidate the factors affecting CMAS infiltration depth, to predict TBC failure locations, and to propose means to mitigate CMAS corrosion.

(1) **1D analytical model without considering tortuosity**

We assume that CMAS covers the surface of the ceramic coating (i.e., the YSZ layer), as shown in Fig. 6.2. An EB-PVD TBC is composed of columnar crystals, the pores between which are perfect channels for molten CMAS as a viscous fluid. CMAS infiltration into a coating is analogous to water infiltration in an unsaturated porous medium [14]. We make the following assumptions about the geometry and hydraulic characteristics of this system: (1) the pores in the TBC are uniform and incompressible; (2) the pores have isotropic properties in the direction perpendicular to the coating surface; (3) the molten CMAS is continuous and incompressible; and (4) the temperature inside the TBC is constant. In addition, research [15] shows that when CMAS reacts with a TBC, $ZrSiO_4$ and $CaAl_2Si_2O_8$ tend to form at the CMAS/ceramic and ceramic/thermal growth oxide (TGO) interfaces, respectively, according to the following chemical equations:

$$SiO_2 + ZrO_2 = ZrSiO_4 \tag{6.2}$$

$$CaO + Al_2O_3 + 2SiO_2 = CaAl_2Si_2O_8 \tag{6.3}$$

In the early infiltration of molten CMAS, only a small amount of molten CMAS reacts with the TGOs before reaching the ceramic/TGO interface. Thus, the chemical reaction in Eq. (6.3) can be ignored. Additionally, the amount of CMAS is small, and it infiltrates for a short period of time, so the chemical reaction in Eq. (6.2) is not notable. Hence, the chemical reactions are neglected in the model.

According to the law of the conservation of mass, the mass of CMAS infiltrating a TBC is equal to the mass of CMAS flowing out of the CMAS layer. Thus, the mass

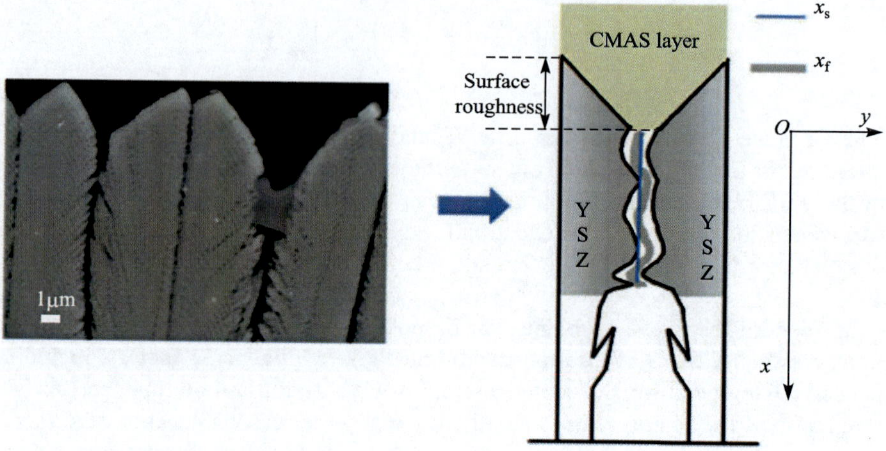

Fig. 6.2 Schematic diagram of a geometric model for the numerical simulation of infiltration based on the true microstructure of a ceramic coating

of CMAS is kept unchanged during the infiltration of molten CMAS, as shown in the following equation [16]:

$$\begin{aligned}\frac{\mathrm{d}m}{\mathrm{d}t} &= \frac{\mathrm{d}}{\mathrm{d}t}\int\limits_V \rho\phi S_{CMAS} dV \\ &= \int\limits_V \left[\frac{\partial(\rho\phi S_{CMAS})}{\partial t} + \nabla\cdot(\rho\phi S_{CMAS}\nu)\right] dV \\ &= \int\limits_V \left[\frac{\partial c}{\partial t} + \rho\phi\nabla\cdot(S_{CMAS}\nu)\right] dV \\ &= 0 \end{aligned} \tag{6.4}$$

where m, V, and ρ are the mass, volume, and density of the molten CMAS, respectively, ϕ is the effective porosity, S_{CMAS} is the degree of saturation of the CMAS and is used to describe the volume fraction of the CMAS in the pores, c is the CMAS concentration in the TBC ($c = \rho\varphi S_{\mathrm{CMAS}}$), and $\boldsymbol{v}$ is the velocity vector of the CMAS. As the law of the conservation of mass holds for any volume, we can obtain the following evolution equation:

$$\frac{\partial c}{\partial t} + \rho\varphi\nabla\cdot(S_{CMAS}\nu) = 0 \tag{6.5}$$

The volume-average method [17] is employed to calculate the average velocity of the CMAS, $\boldsymbol{u}$, in a region with a volume of V_{fluid}:

$$\begin{aligned}\boldsymbol{u} &= \langle S_{\mathrm{CMAS}} \boldsymbol{v} \rangle \\ &= \frac{1}{V_{\mathrm{fluid}}} \int\limits_{V_{\mathrm{fluid}}} S_{\mathrm{CMAS}} \boldsymbol{v} dV\end{aligned} \tag{6.6}$$

where the brackets $\langle\ \rangle$ indicate the average in the fluid domain V_{fluid}. This method overcomes microstructural disorder. Then, Eq. (6.5) can be expressed as

$$\frac{\partial c}{\partial t} + \rho\varphi\nabla \cdot \boldsymbol{u} = 0 \tag{6.7}$$

To simplify this problem, we assume that molten CMAS is a viscous and incompressible fluid and that the pores between the columnar crystals are cylindrical capillary channels. Then, CMAS infiltration in the porous columnar crystals can be regarded as a cylindrical capillary flow [16]. Thus, the Hagen–Poiseuille law can be used to describe the local velocity in the x-axis direction during CMAS infiltration [16, 18]. See Fig. 6.2 for a schematic diagram and the coordinate system.

The Navier–Stokes equation is an important basic equation in fluid dynamics and can be expressed as

$$\begin{aligned}\rho \frac{\mathrm{d}\boldsymbol{v}}{\mathrm{d}t} &= \rho \left[\frac{\partial \boldsymbol{v}}{\partial t} + (\boldsymbol{v} \cdot \nabla)\boldsymbol{v} \right] \\ &= -\nabla p + \mu \nabla^2 \boldsymbol{v} + \rho \boldsymbol{f}\end{aligned} \tag{6.8}$$

where $\boldsymbol{f}$ is the volume force, which is negligible compared to the capillary forces. We treat the molten CMAS as an incompressible, viscous Newtonian fluid, and the whole flow process as steady laminar flow in a circular tube. On this basis, the velocity of the Poiseuille flow is independent of time. Thus, Eq. (6.8) can be written as follows:

$$1 - \nabla p + \mu \nabla^2\ \boldsymbol{v} = \boldsymbol{0} \tag{6.9}$$

We use cylindrical coordinates to describe the capillary flow. As mentioned above, the whole flow process is regarded as a steady laminar flow in a circular tube. Thus, the velocity components and pressure gradient can be expressed as follows:

$$\begin{aligned} v_x &= v(r) \\ v_r &= 0 \\ v_\varphi &= 0 \end{aligned} \tag{6.10}$$

$$\frac{dp}{dx} = \text{Constant}, \ \frac{\partial p}{\partial r} = \frac{\partial p}{\partial \varphi} = 0 \tag{6.11}$$

where x, r, and φ are the axes of the cylindrical coordinate system. Then, the Navier–Stokes equation can be simplified to

$$\frac{1}{r}\frac{\mathrm{d}}{\mathrm{d}r}\left(r\frac{\mathrm{d}v}{\mathrm{d}r}\right)=\frac{\mathrm{d}p}{\mu\mathrm{d}x} \tag{6.12}$$

Molten CMAS is a Newtonian viscous fluid, the velocity of which is zero at the boundaries and is finite. Thus, the boundary conditions can be expressed as

$$\begin{aligned} v_{r=R} &= 0 \\ v_{r=0} &< \infty \end{aligned} \tag{6.13}$$

and it follows that

$$v(r)=\frac{\left(r^2-R^2\right)}{4\mu}\cdot\frac{\mathrm{d}p}{\mathrm{d}x} \tag{6.14}$$

where R is the pore radius, r is the distance between a certain point and the central axis, μ is a temperature-controlled viscosity factor, and p is the capillary force. Compared to p, the gravity of molten CMAS is negligible. As the degree of saturation of Poiseuille flow is 1, the average volume velocity in the fluid domain, u_x, can be calculated as follows:

$$u_x=\frac{dx}{dt}=\frac{1}{A}\int\limits_A 1\cdot v(r)dA=\frac{1}{\pi R^2}\int\limits_0^{2\pi}\int\limits_0^{R} v(r)\cdot r dr d\theta=-\frac{R^2}{8\mu}\cdot\frac{dp}{dx} \tag{6.15}$$

where A is the cross-sectional area of the cylindrical capillary and θ is the apparent contact angle of the CMAS between the columnar crystals controlled by the surface roughness.

Under the assumption of isotropy in the perpendicular direction, substitution of Eq. (6.15) into the 1D form of Eq. (6.7) yields a 1D governing equation. Assuming that dp/dc is a constant, we can derive the following:

$$\frac{\partial c}{\partial t}=\left(\frac{\rho\varphi R^2}{8\mu}\cdot\frac{\mathrm{d}p}{\mathrm{d}c}\right)\frac{\partial^2 c}{\partial x^2} \tag{6.16}$$

where dp/dc is used to describe the relationship between the CMAS concentration and capillary force in the TBC, which can be simplified to a linear relationship as follows:

$$\frac{\mathrm{d}p}{\mathrm{d}c}=\frac{\lambda R\cos\theta}{\rho\varphi\mathrm{d}S_{\mathrm{CMAS}}}=\frac{2\sigma\cos\theta}{\rho\varphi R}\cdot\lambda \tag{6.17}$$

where σ is the surface tension of the molten CMAS on the TBC surface and λ is an approximate slope of the saturation (S)–pressure (P) curve [19, 20]. Then, the 1D governing equation for intercolumnar-crystal gaps or porous media can be expressed as

$$\frac{\partial c}{\partial t} = \left(\frac{\lambda R\sigma \cos\theta}{4\mu}\right) \cdot \frac{\partial^2 c}{\partial x^2} \tag{6.18}$$

(2) **1D analytical model considering tortuosity**

As discussed earlier, the columnar crystals in an EB-PVD TBC have a secondary microstructure at the micro- and nanoscales [21], resulting in a feather-like lateral side of the coating with fractal characteristics. The fractal characteristics of tortuous capillaries in porous media have been extensively studied [22, 23]. As the CMAS infiltration path runs along the fractal interface of columnar crystals, we assume that the actual infiltration length x_{f} and infiltration depth x_{s}, as shown in Fig. 6.2, satisfy the following equation derived from fractal theory:

$$x_{\mathrm{f}} = \varepsilon_{\mathrm{c}}^{1-d} x_{\mathrm{s}}^{d} \tag{6.19}$$

where ε_{c} is the pore size and d is the fractal dimension [17]. Let l be the thickness of the ceramic layer. Then, tortuosity τ can be expressed as

$$\tau = \frac{x_{\mathrm{f}}}{x_{\mathrm{s}}} = \frac{\varepsilon_{\mathrm{c}}^{1-d} l^{d}}{l} = \varepsilon_{\mathrm{c}}^{1-d} l^{d-1} \tag{6.20}$$

Notably, the velocity in Eq. (6.15) refers to the actual distance that the fluid flows per unit time. Thus, the CMAS infiltration velocity in a tortuous capillary is only $1/\tau$ of the velocity in Eq. (6.15) and can be expressed as

$$u_x = \frac{\mathrm{d}x_{\mathrm{s}}}{\mathrm{d}t} = \frac{\mathrm{d}x_{\mathrm{f}}}{\tau \mathrm{d}t} = -\frac{R^2}{8\mu\tau} \cdot \frac{\mathrm{d}p}{\mathrm{d}x_{\mathrm{f}}} = -\frac{R^2}{8\mu\tau^2} \cdot \frac{\mathrm{d}p}{\mathrm{d}x_{\mathrm{s}}} \tag{6.21}$$

As a capillary is tortuous, the actual infiltration length x in Eq. (6.18) cannot describe the coordinates of CMAS infiltration. Thus, we rewrite Eq. (6.18) to

$$\begin{aligned} \frac{\partial c}{\partial t} &= \left(\frac{\lambda R\sigma \cos\theta}{4\mu}\right) \cdot \frac{\partial^2 c}{\partial x_{\mathrm{f}}^2} = \left(\frac{\lambda R\sigma \cos\theta}{4\mu}\right) \cdot \frac{\partial^2 c}{\tau^2 \partial x^2} \\ &= \left(\frac{\lambda R\sigma \cos\theta}{4\mu \varepsilon_{\mathrm{c}}^{2-2d} l^{2d-2}}\right) \cdot \frac{\partial^2 c}{\partial x^2} = D \cdot \frac{\partial^2 c}{\partial x^2} \end{aligned} \tag{6.22}$$

where x is the infiltration depth x_{s}. As the cracks and pores in the TBC are assumed to be uniform and isotropic, the hydrodynamic diffusion coefficient D is isotropic.

Next, let us derive an analytical solution to the 1D model. At the initial time point, the CMAS concentration in the TBC is zero, and the amount of CMAS on the coating surface gradually decreases as the extent of infiltration increases. Thus, the initial and boundary conditions for Eq. (6.22) can be defined as follows:

Initial conditions: $c(x, 0) = 0$

$$\text{Boundary conditions: } c(x = 0, t) = \phi(t) \tag{6.23}$$

$$c(x = l, t) = 0$$

where $\phi(t)$ is the boundary value and is a function of time. Since the coated amount of CMAS is a constant, we can define the following:

$$\int_0^l c(x, t)\mathrm{d}x + \varphi(t) \cdot \frac{M}{\rho} = M \tag{6.24}$$

where M is the mass of the CMAS on the coating surface per unit area, the integral part is the mass of the CMAS infiltrating the coating per unit area, and $\frac{\varphi(t) \cdot M}{\rho}$ is the mass of the CMAS remaining on the coating surface per unit area, as shown in Fig. 6.2. Performing the Laplace transform in the region described in Eq. (6.23) [24], we obtain

$$\begin{aligned} s \cdot \overline{c} &= D \cdot \tfrac{\mathrm{d}^2 \overline{c}}{\mathrm{d}x^2} \\ \overline{c}(x = 0) &= \overline{\varphi}(s) \\ c_x(x = l) &= 0 \end{aligned} \tag{6.25}$$

where $\overline{c}$ is the Laplace transform of c and can be described by the following equation:

$$\overline{c}(x, s) = \int_0^\infty c(x, t) \cdot \exp(-st)\mathrm{d}t \tag{6.26}$$

Similarly, performing the Laplace transform in the region described in Eq. (6.23), we obtain

$$\int_0^l \overline{c}(x, s)\mathrm{d}x + \overline{\varphi}(s) \cdot \frac{M}{\rho} = \frac{M}{S} \tag{6.27}$$

The second equation in Eq. (6.25) is a second-order ordinary differential equation for which the general solution can be expressed as

$$\overline{c}(x,s) = c_1 \cdot \exp\left(\sqrt{\frac{s}{D}}x\right) + c_2 \cdot \exp\left(-\sqrt{\frac{s}{D}}x\right) \tag{6.28}$$

where c_1 and c_2 are constants. The relation between c_1 and c_2 can be obtained based on the last two boundary conditions in Eq. (6.25) as follows:

$$c_2 = c_1 \cdot \exp\left(2\sqrt{\frac{s}{D}} \cdot l\right) = \frac{\overline{\phi}(s)}{1 + \exp(2\sqrt{\frac{s}{D}} \cdot l)} \tag{6.29}$$

Then, the solution in the Laplace transform space can be expressed as

$$\overline{c} = \frac{\overline{\varphi}(s)}{1 + \exp(2\sqrt{\frac{s}{D}} \cdot l)}\left[\exp\left(\sqrt{\frac{s}{D}} \cdot x\right) + \exp\left(\sqrt{\frac{s}{D}} \cdot (2l - x)\right)\right] \tag{6.30}$$

However, $\overline{\varphi}(s)$ in the above equation is a function to be determined. Based on Eqs. (6.27) and (6.30), we have

$$\overline{\varphi}(s) = \frac{\frac{M}{s}}{\frac{M}{\rho} + \sqrt{\frac{D}{s}}\frac{\exp\left(\sqrt{\frac{s}{D}} \cdot l\right) - \exp\left(-\sqrt{\frac{s}{D}} \cdot l\right)}{\exp\left(2\sqrt{\frac{s}{D}} \cdot l\right) + \exp\left(-\sqrt{\frac{s}{D}} \cdot l\right)}} \tag{6.31}$$

Because this is an analytical solution in the Laplace space, we need to perform an inverse transform of Eq. (6.31). $\exp\left(2\sqrt{\frac{s}{D}}\, l\right)$ is in the denominator on the right-hand side of Eq. (6.31), making a direct inverse transform impossible. Then, by performing a Taylor series expansion, we obtain

$$\overline{\phi}(s) = \frac{\frac{M}{s}}{\frac{M}{\rho} + \sqrt{\frac{D}{s}}\frac{\exp\left(\sqrt{\frac{s}{D}} \cdot l\right) - \exp\left(-\sqrt{\frac{s}{D}} \cdot l\right)}{\exp\left(2\sqrt{\frac{s}{D}} \cdot l\right) + \exp\left(-\sqrt{\frac{s}{D}} \cdot l\right)}} \approx \frac{\frac{M}{s}}{\frac{M}{\rho} + \frac{l}{1 + \frac{l}{\frac{s}{D} \cdot l^2}}} \tag{6.32}$$

$$\frac{M}{1 + \exp(-2l\sqrt{\frac{s}{D}})} = \sum_{n=0}^{\infty}(-1)^n \exp\left(-2nl\sqrt{\frac{s}{D}}\right) \tag{6.33}$$

$$\overline{c} = \frac{M}{\frac{M}{\rho} + \frac{l}{1 + \frac{s}{D} \cdot l^2}} \cdot \frac{1}{s}\sum_{n=0}^{\infty}(-1)^n\left\{\exp[x - 2l(n+1)]\sqrt{\frac{s}{D}} + \exp[-x - 2nl]\sqrt{\frac{s}{D}}\right\} \tag{6.34}$$

Performing a convolution transform, we obtain the inverse Laplace transform of $\overline{c}$ as follows:

$$c = L^{-1}\left[\frac{M}{\frac{Ml^2}{\rho D}} \cdot \frac{\left(1 + \frac{l^2}{D} \cdot s\right)}{S + \frac{\frac{M}{\rho}+1}{\frac{\rho}{\left(\frac{Ml^2}{\rho D}\right)}}}\right] \cdot L^{-1}\left\{\frac{1}{s}\sum_{n=0}^{\infty}\left[\begin{array}{c}\exp[x - 2l(n+1)]\sqrt{\frac{s}{D}} \\ + \exp[-x - 2nl]\sqrt{\frac{s}{D}}\end{array}\right]\right\} \tag{6.35}$$

Then, the analytical solution to the 1D theoretical model can be written as follows:

$$\begin{aligned} c(x,t) = {} & \rho \cdot \sum_{n=0}^{\infty}(-1)^n\left\{\operatorname{erfc}\frac{[2l(n+1) - x]}{2\sqrt{Dt}} + \operatorname{erfc}\frac{[2nl + x]}{2\sqrt{Dt}}\right\} \\ & - \frac{\rho^2 D}{Ml}\exp(-kt)\int_0^t\left\{\exp(k\tau) \cdot \sum_{n=0}^{\infty}(-1)^n\left[\begin{array}{c}\operatorname{erfc}\frac{[2l(n+1) - x]}{2\sqrt{D\tau}} \\ + \operatorname{erfc}\frac{[2nl + x]}{2\sqrt{D\tau}}\end{array}\right]\right\}d\tau \end{aligned} \tag{6.36}$$

where erfc is the complementary error function and the expanded form of k is $D(M + \rho l)/Ml^2$, which is a constant value, showing that the CMAS infiltration process follows the Gaussian diffusion process. Then, the evolutionary relationship between the concentration of the CMAS infiltrating the coating, $c(x,t)$, the coated amount of CMAS, M, and the diffusion coefficient D can be obtained by analyzing the above equation. Here, D is related to the coefficient of viscosity, which is related to temperature. Thus, $c(x,t)$ is closely related to the temperature.

(3) **Numerical simulation of CMAS infiltration**

To examine the correctness of the above model, we numerically simulate the extent of the CMAS infiltration in the TBC. Considering CMAS infiltration in the gaps between columnar crystals and the voids in columnar crystals, we can extend Eq. (6.22) into the following 2D form:

$$\begin{gathered} \frac{\partial c}{\partial t} = D\left(\frac{\partial^2 c}{\partial x^2} + \frac{\partial^2 c}{\partial y^2}\right) \\ D = \frac{\lambda R\sigma\cos\theta}{4\mu\tau^2} = \frac{\lambda R\sigma\cos\theta}{4\mu\varepsilon_c^{2-2d}l^{2d-2}} \end{gathered} \tag{6.37}$$

The boundary and initial conditions for numerical simulation are as follows:

$$\text{Boundary conditions}: n \cdot \nabla c = 0 \quad \text{At all external boundaries} \tag{6.38}$$

$$\text{Initial conditions}: c = \rho \quad \text{In the CMAS layer}$$

$$c = 0 \quad \text{In the gaps and coating} \tag{6.39}$$

Table 6.1 Parameters for numerical simulation

Parameter/(unit)	Value	Parameter/(unit)	Value
σ [13]/(N/m)	0.4	*Rc*/(μm)	0.46
$\rho_{\text{CMAS}}/(\text{kg/m}^3)$	2630	*Rp*/(μm)	0.05
φ	0.15	$\cos\theta$	0.17
$\mu_{1100\,°\text{C}}/(\text{Pa}\cdot\text{s})$ [13]	68	$\varepsilon_c/(\mu\text{m})$	0.1
$\mu_{1150\,°\text{C}}/(\text{Pa}\cdot\text{s})$ [13]	31	d_c	1.2
$\mu_{1200\,°\text{C}}/(\text{Pa}\cdot\text{s})$ [13]	15	d_Y	1.6
$D_1/(\text{m}^2/\text{s})$	0.01	$\lambda/(\text{Pa}^{-1})$	6.5×10^{-4}
$D_{\text{YSZ}}/(\text{m}^2/\text{s})$	1.7×10^{-14}	$\tau/(\mu\text{m})$	4.43
$l/(\mu\text{m})$	170		

To better simulate the actual infiltration process, we set the diffusion coefficient in the CMAS layer, D_1, to 0.01 to ensure a uniform concentration in the CMAS layer. Table 6.1 summarizes the values of the diffusion coefficients in the YSZ and porous layers [13]. The surface tension of the CMAS can be calculated using the following equation [13]:

$$\sigma = 271.2 + 3.34[\text{CaO}] + 1.96[\text{MgO}] + 3.47[\text{Al}_2\text{O}_3] + 2.68[\text{FeO}] + 2.92[\text{MnO}] + k \tag{6.40}$$

where σ is the surface tension of the CMAS (unit: mN/m), 271.2 is the value of the surface-tension parameter (unit: mN/m), the value in front of each pair of square brackets is the surface tension of the corresponding component per mole (unit: mN/(m·mol)), the parameter within each pair of square brackets is the molar percentage of the concentration of the corresponding component, and k is a parameter related to the CMAS composition (unit: mN/m). Generally, the fluidity of a flow depends primarily on its viscosity μ, which is related to temperature through the following equation [13, 25]:

$$\mu = A_e \exp\left(\frac{E_B}{k_B T}\right) \tag{6.41}$$

where A_e is a parameter related to the composition of volcanic ash (VA), E_B is the activation energy, k_B is the Boltzmann constant, and T is Kelvin temperature. When the temperature is between the glass transition point and the melting point, heating can significantly reduce the viscosity of molten VA, thereby promoting its infiltration in the TBC. Equations (6.22) and (6.41) show that the infiltration depth increases exponentially with temperature. The CMAS density ρ_{CMAS}, crack radius R_c, pore radius R_P, and $\cos\theta$ can be determined experimentally. The pore size ε_c and the fractal dimensions of the pores and YSZ (d_c and d_Y) can be derived from the microstructure of the coating. ε_c is set to 0.1 μm.

6.1.2 Experiments on the Molten CMAS Infiltration Depth in an EB-PVD TBC and Its Influencing Factors

Based on the above model, we conducted a series of experiments to examine CMAS infiltration in a TBC in different service environments. VA samples collected from the 2009 eruption of Sakurajima, Japan [26], were used as CMAS for the experiments in this section. We selected a TBC substrate consisting mainly of a Ni-based high-temperature alloy and with a thickness of approximately 3.2 mm, as well as a ceramic layer composed of 7YSZ. Compared to an APS YSZ coating, an EB-PVD ceramic coating has intercolumnar gaps that facilitate CMAS infiltration and, consequently, the coating is prone to failure. The experimental results presented in the subsequent section show that the ceramic layer of the EB-PVD TBC had a thickness of approximately 170 μm, a columnar-crystal width of approximately 10 μm, and a porosity of approximately 17%. Table 6.2 summarizes the chemical composition of the VA determined by inductively coupled plasma atomic emission spectroscopy (ICP-AES), showing that the VA is composed primarily of SiO_2, Al_2O_3, CaO, MgO, FeO, TiO_2, Na_2O, K_2O, MnO, and Ta_2O_5.

Evidently, the VA is compositionally more complex than the CMAS prepared in the laboratory. We ground the VA powder in a mortar to obtain uniform particles similar in size to actual VA particles. A NETZSCH-44 was used for differential scanning calorimetry (DSC) to measure the melting point of the VA by heating it from room temperature to 1300 °C at a heating rate of 10 °C/min. Figure 6.3 shows the DSC heating curve of the CMAS. The VA melts at 1065–1282 °C with a peak at approximately 1190 °C; in other words, the VA begins to melt at approximately 1065 °C and is completely melted at approximately 1282 °C. The VA has a lower melting point than the CMAS prepared in the laboratory (melting point: ca. 1230 °C) [27, 28]. The surface of the ceramic layer was ground with SiC abrasive paper of different grit sizes to obtain different surface roughnesses R_a. With microscopic observations, R_a values of 3.9, 5.6, 9.9, and 13.7 μm were obtained. A solution composed of the VA and absolute ethyl alcohol mixed at a ratio of 1 mg:0.01 ml was uniformly sprayed onto the coating surface to obtain a coated amount of VA, M_{VA}, of 1–5 mg/cm^2.

The parameters used in the CMAS corrosion experiments were as follows: temperature, 1100–1250 °C; time, 0.5–4 h; M_{VA}, 1–5 mg/cm^2; and R_a, 3.9–13.7 μm. The purpose of choosing a temperature lower than the melting point of the VA was to determine the critical temperature for VA infiltration. The VA was uniformly applied on the coating surface, and an MSFT-1520P resistance furnace was used to control the temperature. Both the heating and cooling rates were set to 10 °C/min. The obtained cross-sectional TBC specimens were ground with SiC abrasive paper of different

Table 6.2 Main composition of the VA collected from Sakurajima, Japan [29]

Compound	SiO_2	Al_2O_3	CaO	MgO	FeO	TiO_2	Na_2O	K_2O	MnO
Content	56.9	12.48	10.2	3.04	13.1	0.00	4.25	1.72	0.30

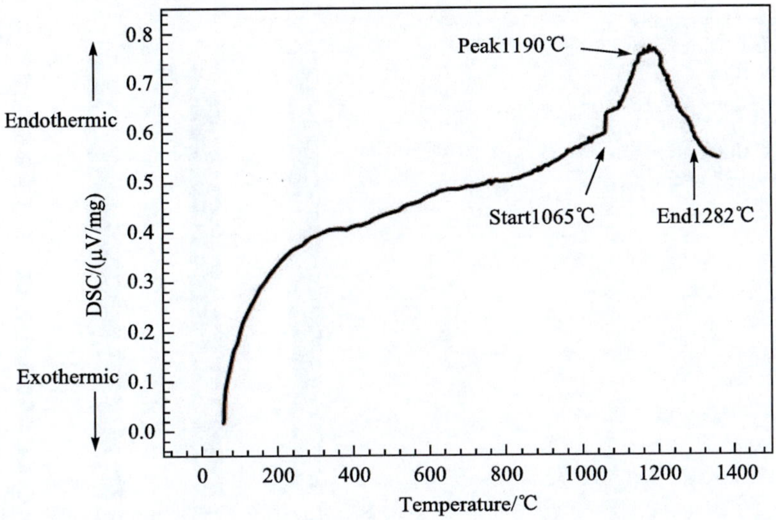

Fig. 6.3 DSC heating curve of the CMAS

grit sizes and subsequently polished with diamond paste. The infiltration behavior of the molten VA in the TBC was investigated using a JSM-7500F field emission scanning electron microscope (FE-SEM), and the elemental distribution was characterized with an energy dispersive spectrometer (EDS). Si is the most abundant element in VA and is absent in the ceramic layer of a TBC. Thus, the distribution of Si is generally used to characterize the VA infiltration depth L_{VA}. To accurately determine L_{VA}, five regions each with an L_{VA} as uniform as possible were selected from each specimen for EDS analysis. Considering instrument errors, an element is generally considered present when its content reaches 2%. Here, we define the infiltration depth as the maximum Si infiltration depth minus the thickness of the VA layer.

6.1.3 CMAS Infiltration Depth in the EB-PVD TBC and Its Influencing Factors

(1) **Correlation between temperature and infiltration depth**

Figure 6.4 shows numerically simulated distributions of the VA concentration in the TBC subjected to infiltration for 2 h at different temperatures [26] (M_{VA}: 3 mg/cm^2; R_a: 13.7 μm). At 1100 °C, the VA concentration on the coating surface is very high because the VA just begins to melt at this temperature. Nevertheless, some of the VA is present in the gaps between the columnar crystals, indicating that once the VA begins to melt, it can rapidly infiltrate the TBC. As the temperature rises to 1150 and 1200 °C, there is a significant increase in the fluidity of the VA, making it easy

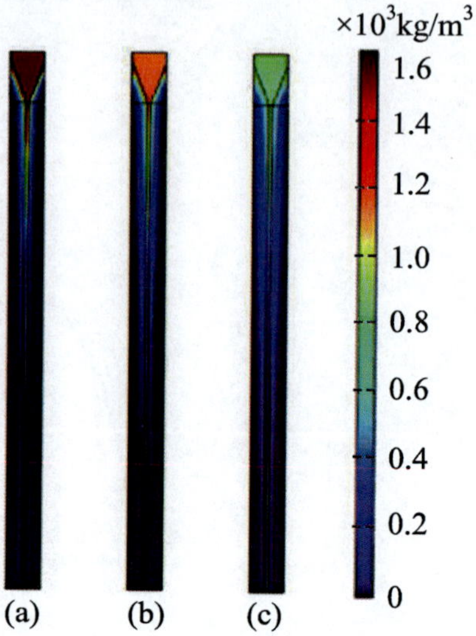

Fig. 6.4 Numerically simulated distributions of the concentration of the VA infiltrating the coatings at **a** 1100 °C, **b** 1150 °C, and **c** 1200 °C (infiltration time, 2 h; M_{VA}, 3 mg/cm^2; R_a, 13.7 μm)

for the VA to infiltrate the YSZ region. Thus, the VA concentrations on the coating surface at 1150 and 1200 °C are lower than that at 1100 °C.

Figure 6.5a, b shows cross-sectional SEM images of the coatings subjected to VA infiltration for 4 h at 1100 and 1200 °C, respectively (M_{VA}: 3 mg/cm^2; R_a: 13.7 μm), and Fig. 6.5c, d shows the corresponding Si distributions [26]. Because the VA just begins to melt at 1100 °C, there is a large amount of VA deposited on the coating surface. As shown in Fig. 6.5c, at 1100 °C, the VA is able to infiltrate the interior of the coating via the large pores between the columnar crystals to a depth of approximately 40.1 μm. As the temperature rises to 1200 °C, the VA content on the surface of the specimen decreases, while L_{VA} increases to 99.9 μm, evidenced by the corresponding Si distribution in Fig. 6.5d.

Figure 6.6 compares the variations in the experimental results and calculated results (including the numerical simulation and analytical results) for L_{VA} with temperature over different periods of time [26]. The mean square errors (MSEs) of the experimental results are also shown in the figure. The experimental results agree well with the numerical simulation and analytical results. In addition, at 1100 °C, there is a small difference between the experimental and calculated results after 1, 2, and 4 h of VA infiltration. This difference increases as the temperature rises. According to the experimental data, at 1250 °C, the VA almost infiltrates to the bottom of the coating within 2 h, and L_{VA} remains at approximately 170 μm after 4 h of infiltration due to the obstruction by the TGO layer.

(2) Correlation between M_{VA} and infiltration depth

Figure 6.7 shows the numerically simulated distributions of the VA concentration in the TBC subjected to infiltration for 2 h at different M_{VA} levels. At an infiltration temperature of 1150 °C, all the specimens have an R_a of 13.7 μm. At an M_{VA} of

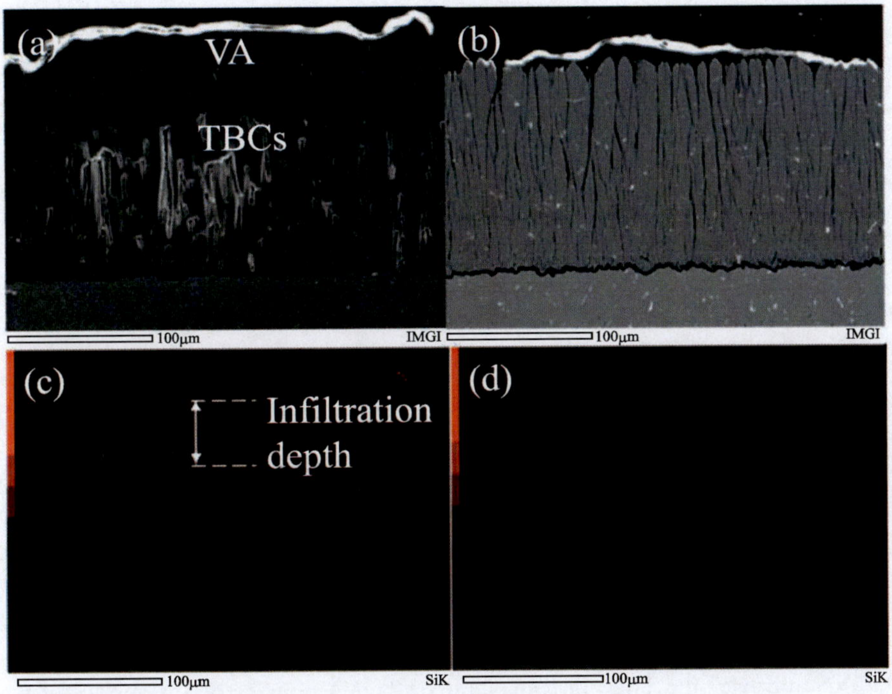

Fig. 6.5 Cross-sectional SEM images of the TBCs infiltrated by the VA for 4 h at **a** 1100 °C and **b** 1200 °C (M_{VA}, 3 mg/cm^2; R_a, 13.7 μm). **c** Si distribution corresponding to **a**. **d** Si distribution corresponding to **b**

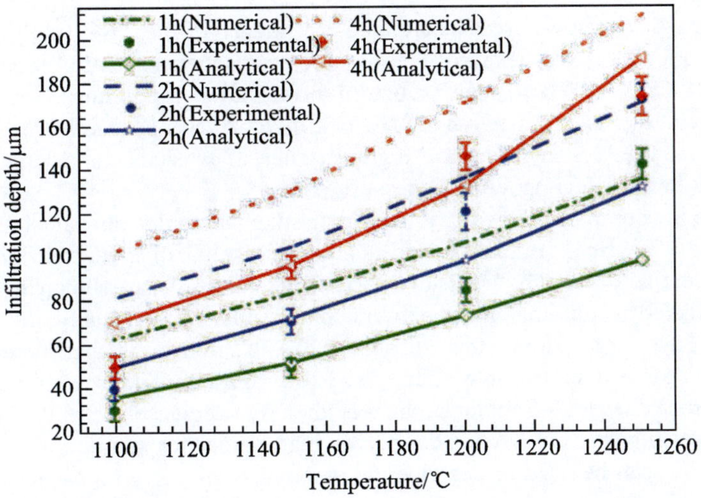

Fig. 6.6 Comparison of the variations in the theoretical and experimental data for L_{VA} with temperature (M_{VA}, 3 mg/cm^2; R_a, 13.7 μm)

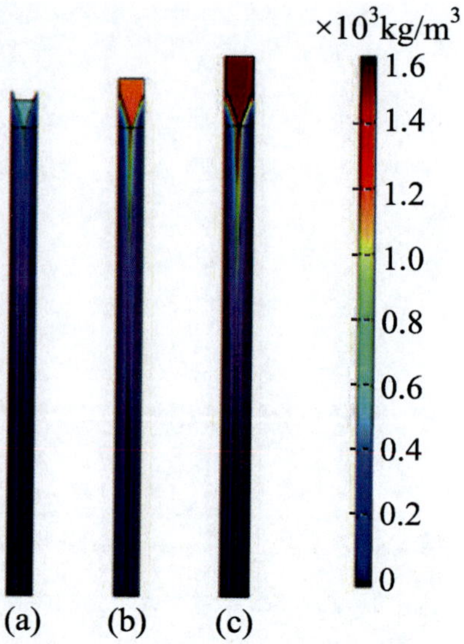

Fig. 6.7 Numerically simulated distributions of the concentration of the VA infiltrating the coatings at different M_{VA} levels: **a** 1 mg/cm^2, **b** 3 mg/cm^2, and **c** 5 mg/cm^2 (infiltration time, 2 h; temperature, 1150 °C; R_a, 13.7 μm)

1 mg/cm^2, all the VA infiltrates the TBC, while the VA concentration in the surface layer is low and L_{VA} is small. In contrast, as the M_{VA} increases to 3 mg/cm^2, not all the VA infiltrates the TBC; instead, some of the VA remains on the surface of the coating, forming a VA layer, and L_{VA} is significantly greater when $M_{VA} = 3$ mg/cm^2 than when $M_{VA} = 1$ mg/cm^2. As the M_{VA} further increases to 5 mg/cm^2, there is an increase in both the thickness and concentration of the VA layer, whereas there is no significant increase in L_{VA}.

Figure 6.8a, b shows cross-sectional SEM images of the TBC subjected to VA infiltration for 8 h at M_{VA} levels of 1 and 3 mg/cm^2, respectively (infiltration temperature, 1150 °C; R_a, 13.7 μm), and Fig. 6.8c, d shows the corresponding Si distributions, respectively. L_{VA} is 64.4 μm when $M_{VA} = 1$ mg/cm^2, and L_{VA} increases rapidly to 113.4 μm as the M_{VA} increases to 3 mg/cm^2. Thus, M_{VA} can be considered as a very important factor affecting VA infiltration.

Figure 6.9 compares the variations in experimental results and calculated results (including numerical simulation and analytical results) of L_{VA} with the M_{VA} at different temperatures. The MSEs of the experimental results are shown in the figure. The calculated results match the experimental results well. Notably, the curves at 1200 °C have larger slopes than those at 1150 °C, indicating that increasing the M_{VA} has a more demonstrable effect on L_{VA} at high temperatures. In addition, as the M_{VA} increases to 4–5 mg/cm^2, there is a relative decrease in the VA infiltration rate, suggesting that the VA inside the coating approaches saturation. The degree of saturation can be used to describe the maximum VA carrying capacity per unit volume of the coating, which can be expressed using λ in Eq. (6.17). Generally, the degree of saturation is considered to be constant. At a low M_{VA}, there is a significant difference between the VA content and λ, and the M_{VA} has a notable effect on L_{VA}.

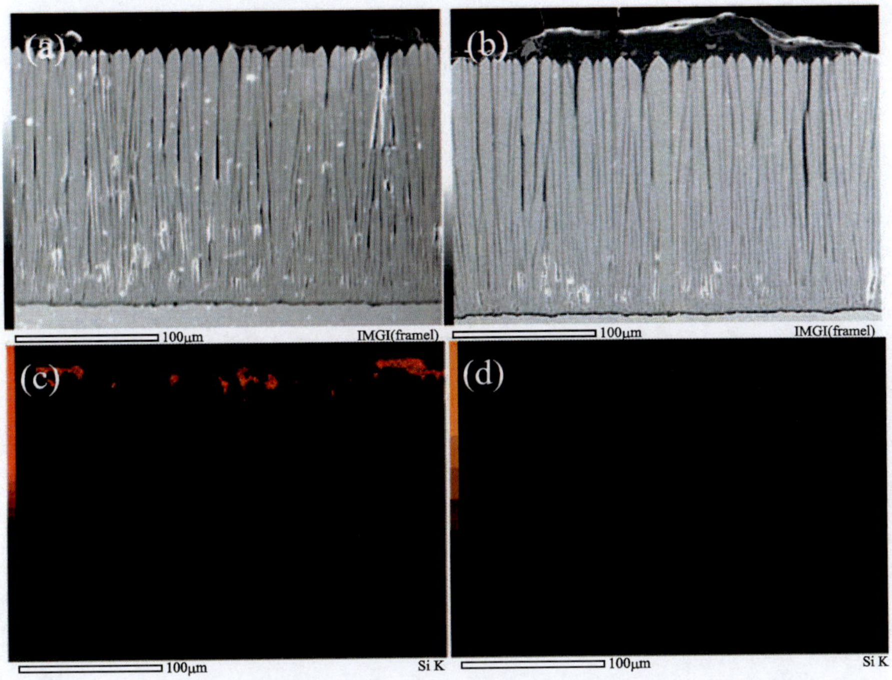

Fig. 6.8 Cross-sectional SEM images of the VA-infiltrated TBC at M_{VA} of **a** 1 mg/cm^2 and **b** 3 mg/cm^2 (infiltration time, 8 h; infiltration temperature, 1150 °C; R_a, 13.7 μm). **c** Si distribution corresponding to **a**. **d** Si distribution corresponding to **b**

In comparison, at a high M_{VA}, there is a small difference between the VA content and λ, and the M_{VA} has an insignificant effect on L_{VA}.

(3) **Correlation between time and infiltration depth**

Figure 6.10 shows numerically simulated distributions of the VA concentration at different infiltration times (temperature: 1150 °C; M_{VA}: 3 mg/cm^2; R_a: 13.7 μm). After 1 h of infiltration, the concentration of the VA layer on the coating surface remains high at approximately 1.6×10^6 kg/m^3. As the infiltration time increases to 2 h, the thickness of the VA layer remains unchanged, whereas there is a considerable decrease in its concentration and a small increase in L_{VA}. As the infiltration time further increases to 3 h, there is a further decrease in the concentration of the VA layer on the coating surface but no notable increase in L_{VA} inside the coating.

Figure 6.11 shows cross-sectional SEM images of the coating at different VA infiltration times at 1250 °C (M_{VA}: 3 mg/cm^2; R_a: 13.7 μm). Figure 6.11a shows an SEM image of the coating subjected to infiltration for 1 h, when L_{VA} is approximately two-thirds of the thickness of the coating, as shown in Fig. 6.11c. After 2 h of VA infiltration, the VA fills all the large pores, and cracks appear in the coating, as shown in Fig. 6.11d and b.

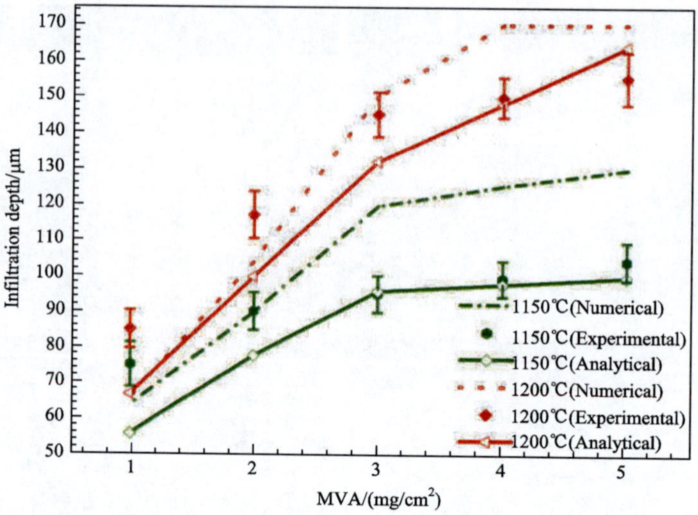

Fig. 6.9 Comparison of the variations in the theoretical and experimental data for L_{VA} with M_{VA} (infiltration time, 4 h; R_a, 13.7 μm)

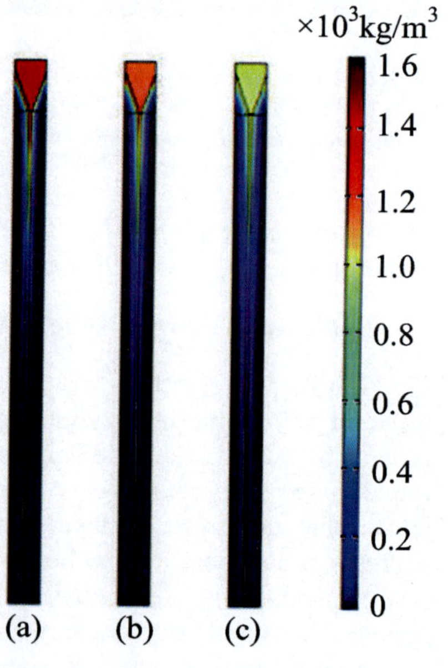

Fig. 6.10 Numerically simulated distributions of the VA concentration after **a** 1 h, **b** 2 h, and **c** 3 h of infiltration at 1150 °C (M_{VA}, 3 mg/cm^2; R_a, 13.7 μm)

Figure 6.12 compares the variations in the experimental results and calculated results (including numerical simulation and analytical results) for L_{VA} with time at different temperatures. The MSEs of the experimental results are also shown in the figure. Within a short time, the VA is able to rapidly infiltrate the interior of

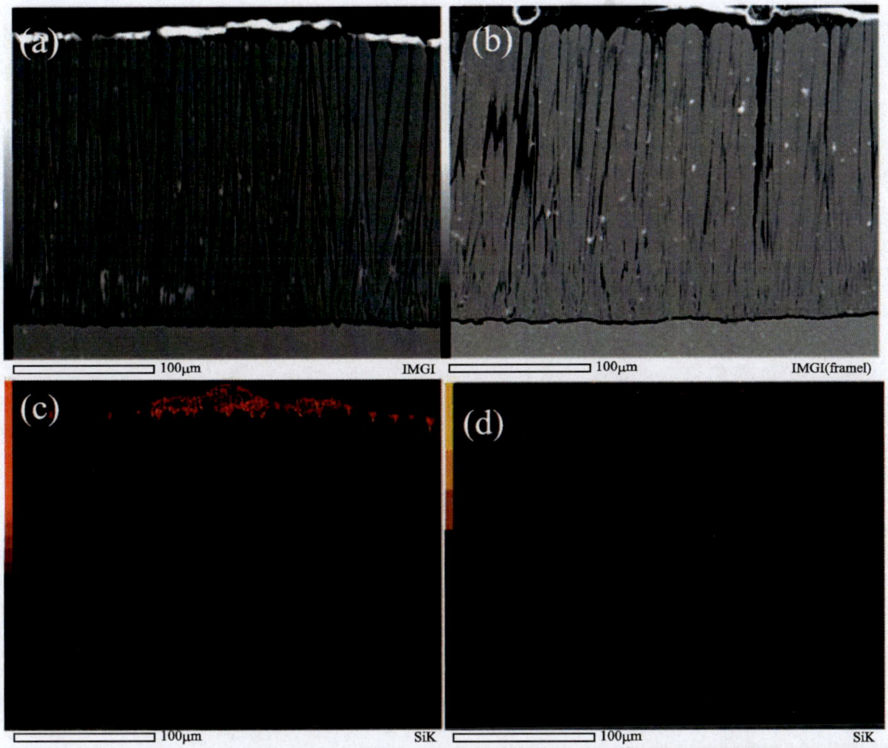

Fig. 6.11 Cross-sectional SEM images of the coatings subjected to VA infiltration for **a** 0.5 h and **b** 2 h at 1250 °C (M_{VA}, 1 mg/cm^2; R_a, 13.7 μm). **c** Si distribution corresponding to **a**. **d** Si distribution corresponding to **b**

the coating. As time passes, there is a gradual decrease in the infiltration rate. The variation in L_{VA} with time follows a parabolic law. According to Eq. (6.1), L_{VA} is directly proportional to the square root of time, which is consistent with Fig. 6.12. A comparison of the experimental, theoretical, and simulation results finds that the difference between the three types of results at 1150 °C is smaller than that at 1250 °C.

(4) **Correlation between R_a and infiltration depth**

Figure 6.13 compares the variations in the experimental results and calculated results (including numerical simulation and analytical values) for L_{VA} with R_a at different temperatures. At 1100 °C, as R_a increases, there are no significant changes in L_{VA}. At 1150 and 1250 °C, as R_a increases, there is a slight increase in L_{VA}. Research [30] shows that R_a mainly affects the fluidity of VA, which in turn affects L_{VA}. In general [30], the surface energy of the ceramic layer is the primary factor obstructing VA infiltration and increases with R_a. At a low R_a, the low surface energy makes it easy for molten VA to spread across the coating surface, resulting in a low VA content on the coating surface per unit area, which in turn leads to a small L_{VA} inside

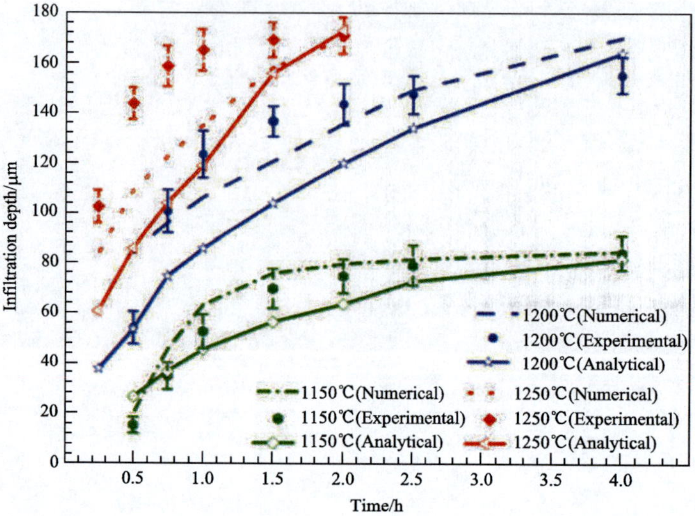

Fig. 6.12 Comparison of the theoretical and experimental data for L_{VA} with time (M_{VA}, 5 mg/cm^2; R_a, 13.7 μm)

the coating. Increasing R_a increases the surface energy, thereby almost completely preventing the VA from spreading across the coating surface, resulting in a relatively high VA content on the coating surface per unit area, which in turn leads to a large L_{VA} inside the YSZ layer. In addition, at low temperatures, VA is not completely melted, and its viscosity, instead of the surface energy of the ceramic layer, is the primary hindrance to its diffusion. Thus, R_a has no significant effect on VA infiltration under this condition.

According to Eq. (6.22), L_{CMAS} is also related to the contact angle of the CMAS in the gaps between columnar crystals. A large contact angle results in a slow infiltration rate. Since the contact angle is related to R_a, we can investigate the correlation between the contact angle and R_a and, on this basis, further control R_a to alter the contact angle and thereby control L_{CMAS}.

(5) **Comprehensive evaluation of influencing service-environment factors**

As discussed above, we analyzed the effects of temperature, M_{VA}, time, and R_a on L_{VA}. Here, we normalize these four factors to evaluate the extents of their effects on L_{VA}. Figure 6.14 shows the variation in the relative rate of change in L_{VA} with the relative rate of increase in each influencing factor. Let x be the relative rate of change in an influencing factor. Then, we have [31]

$$x = \frac{x_i - x_0}{x_0} \times 100\% \tag{6.42}$$

where x_i and x_0 are the current and initial values of an influencing factor, respectively. The change in L_{VA} can be expressed as

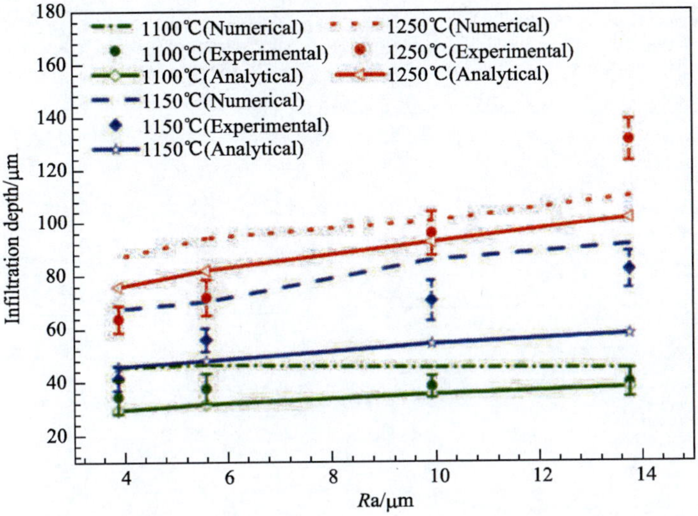

Fig. 6.13 Comparison of the variations in the theoretical and experimental data for L_{VA} with R_a (M_{VA}, 3 mg/cm^2; infiltration time, 0.5 h)

$$y = \frac{y(x_i) - y(x_0)}{y(x_0)} \times 100\% \tag{6.43}$$

where $y(x_0)$ and $y(x_i)$ are the L_{VA} values that correspond to x_0 and x_i, respectively. As shown in Fig. 6.14a, temperature is the primary factor affecting VA infiltration, followed by M_{VA} and time, as evidenced in Fig. 6.14b, c, respectively, while R_a has an insignificant effect on L_{VA}, as shown in Fig. 6.14d. In addition, when $M_{VA} <$ 3 mg/cm^2, M_{VA} has a significant effect on L_{VA}; when $M_{VA} >$ 3 mg/cm^2, M_{VA} has a relatively insignificant effect on L_{VA}. A comparison of the experimental results shows that an increase in temperature of 4.5% results in a 76.8% increase in L_{VA}, whereas an increase in R_a of 251.3% leads to an increase in L_{VA} of only 70.94%. A comprehensive comparison of the influencing factors finds that the theoretical results are in good agreement with the experimental data. In particular, temperature and R_a have the largest and smallest effects on L_{VA}, respectively.

Note that the model introduced in this chapter does not consider the chemical reaction between the VA and the coating. In addition, we make some reasonable assumptions to simplify the model. These factors may all result in some errors in the model. First, we assume uniform pores in the model; however, in an actual TBC, the gaps at the bottom of the TC layer are much smaller than those at the top of the TC layer. Second, in the actual infiltration process, a coating dissolves in the VA and precipitates again in granular form, resulting in a morphological change in the columnar crystals and hence an expansion of the VA flow region [32]. Third, the actions of $ZrSiO_4$ and other products increase the viscosity of the VA. Finally, since it is difficult to measure L_{VA} in real time at high temperatures, the

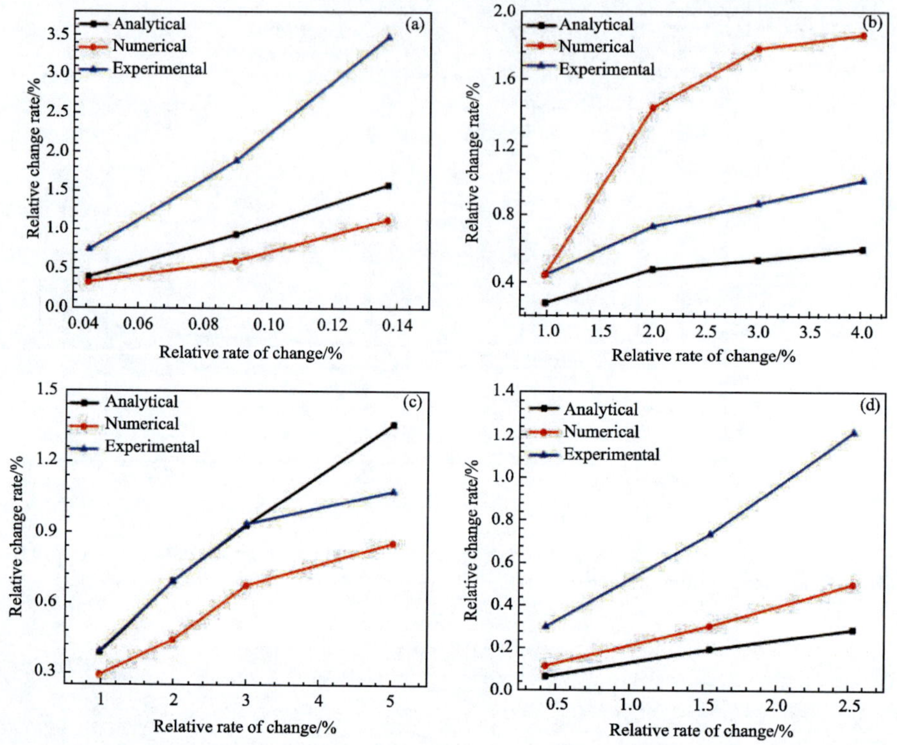

Fig. 6.14 Variations in the relative rate of change in L_{VA} with the relative rates of increase in **a** temperature, **b** M_{VA}, **c** time, and **d** R_a

experimentally measured L_{VA} includes that during heating and cooling stages (even though these contributions are insignificant), whereas the model considers L_{VA} only in the temperature-holding stage. These factors can all affect L_{VA} and therefore result in a difference between theoretical and experimental results.

6.1.4 Infiltration of CMAS Melts in an APS TBC

(1) **Definition and governing equation of CMAS concentration**

To quantitatively describe the CMAS infiltration process in an APS TBC, we denote the concentration of each metal element in the CMAS melts by c_i (i = 1, 2, 3, and 4 representing Ca, Mg, Al, and Si, respectively). According to Eq. (1.29), the conservation equation for the concentration of each element is expressed as follows:

$$\dot{c}_i = -\nabla \cdot \mathbf{j}_i \tag{6.44}$$

The CMAS density ρ_{CMAS} can be expressed as

$$\rho_{\mathrm{CMAS}}(c_i) = \mathrm{M}_{\mathrm{CaO}} \cdot c_1 + \mathrm{M}_{\mathrm{MgO}} \cdot c_2 + \mathrm{M}_{\mathrm{Al_2O_3}} \cdot c_3/2 + \mathrm{M}_{\mathrm{SiO_2}} \cdot c_4 = \sum_{i=1}^{4} \mathrm{M}_i c_i \tag{6.45}$$

According to Fick's second law, we assume that in Eq. (6.44), the diffusion flux of element i, $\mathbf{j}_i$, runs in the opposite direction and is linearly related to the concentration gradient ∇c_i as follows:

$$\boldsymbol{j}_i = -D_j \nabla c_i (i = j) \tag{6.46}$$

Substitution of the above equation into Eq. (6.44) yields

$$\dot{c}_i = D_j \nabla^2 c_i (i = j) \tag{6.47}$$

(2) **Experimental characterization of CMAS concentration**

A TBC with a thickness of 200 μm and cross-sectional dimensions of 1 cm × 0.5 cm was prepared by atmospheric plasma spraying (APS). The prepared TBC was soaked in hydrochloric acid for 48 h to remove the NiCoCrAlY BC layer and to obtain a bare ceramic layer with no substrate. CMAS powder was prepared in the laboratory according to a CaO:MgO:Al_2O_3:SiO_2 molar ratio of 35:10:7:48. The mixed powder was subsequently ground in an agate mortar until a powder particle size of 20–30 μm was obtained. Afterwards, the ground CMAS powder was placed in a crucible and sintered in a holding furnace at 1500 °C. Then, the sintered CMAS block was removed from the furnace and ground again until a powder particle size of 20–30 μm was reached. Thus, the CMAS powder needed for the experiments was obtained.

Four TBC specimens, numbered 1, 2, 3, and 4, were prepared, of which specimens 1, 2, and 3 were used to measure the distributions of the CMAS concentration in the ceramic layer at different corrosion times, and specimen 4 was used to observe and record the evolution of the swelling deformation during the CMAS corrosion of the TBC. The prepared CMAS powder was coated on the TBC surface at 20 mg/cm^2. Specimens 1, 2, and 3 were placed in a holding furnace for 2, 15, and 60 min, respectively, and then cooled with the furnace to room temperature. Next, the amount of CMAS at different depths in the ceramic layer in each specimen was measured by EDS. Specimen 4 was placed in a high-temperature furnace and held for 300 min, during which the lateral morphology of the coating was recorded in real time by a camera.

The SEM image in Fig. 6.15 shows the lateral morphology of the coating subjected to CMAS corrosion for 2 min. The amounts of different elements in the CMAS in different regions selected at different depths were measured using EDS. The atomic percentages of Ca, Mg, Al, Si, Y, and Zr in a region are denoted by $C_{\mathrm{Ca}}\%$, $C_{\mathrm{Mg}}\%$, $C_{\mathrm{Al}}\%$, $C_{\mathrm{Si}}\%$, $C_{\mathrm{Y}}\%$, and $C_{\mathrm{Zr}}\%$, respectively. Considering that it is more difficult for the Zr than for the Y in the coating to dissolve in the CMAS melt during the CMAS corrosion process, we assume that the amount of Zr in each region remains constant. Let ρ_{YSZ} be the mass density of the YSZ coating. The YSZ coating was composed of ZrO_2 stabilized by 8 wt.% Y_2O_3. Let M_{CaO}, M_{MgO}, $M_{Al_2O_3}$, M_{SiO_2}, M_{ZrO_2}, and $M_{Y_2O_3}$ be the molar masses of CaO, MgO, Al_2O_3, SiO_2, ZrO_2, and Y_2O_3, respectively. ρ_{CMAS} can be expressed as

$$\rho_{CMAS} = 0.92\rho_{YSZ}\begin{pmatrix} C_{Ca}\% \cdot M_{CaO} + C_{Mg}\% \cdot M_{MgO} \\ +0.5 \cdot C_{Al}\% \cdot M_{Al_2O_3} + C_{Si}\% \cdot M_{SiO_2} \end{pmatrix} / (M_{ZrO_2} \cdot C_{Zr}\%) \tag{6.48}$$

Note that the above equation calculates the average ρ_{CMAS} in each rectangular box in Fig. 6.15.

(3) **Evolutionary pattern of CMAS concentration**

When using the finite element method (FEM) to solve the governing equations for the CMAS concentration, first we need to transform the differential equations into weak-form integral equations. The weak form of the governing equations for the concentration of an element in the CMAS is expressed as follows:

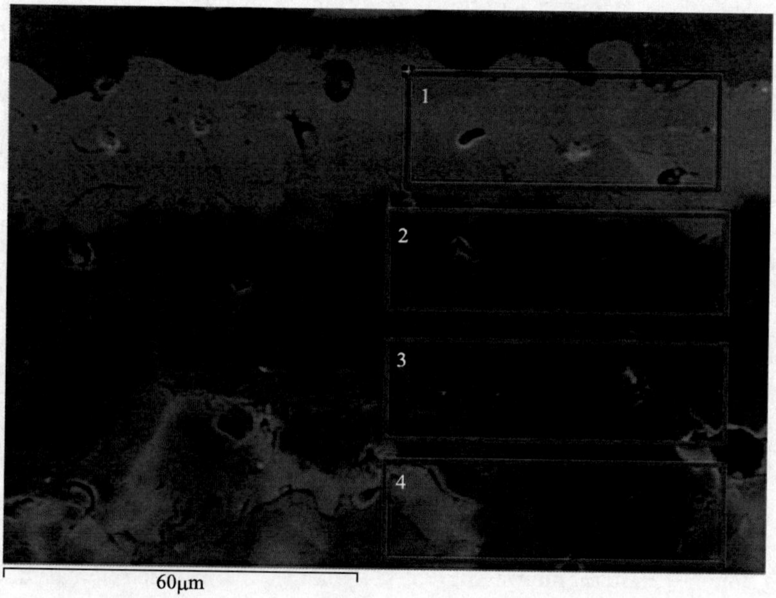

Fig. 6.15 Micromorphology of the coating subjected to high-temperature CMAS corrosion for 2 min

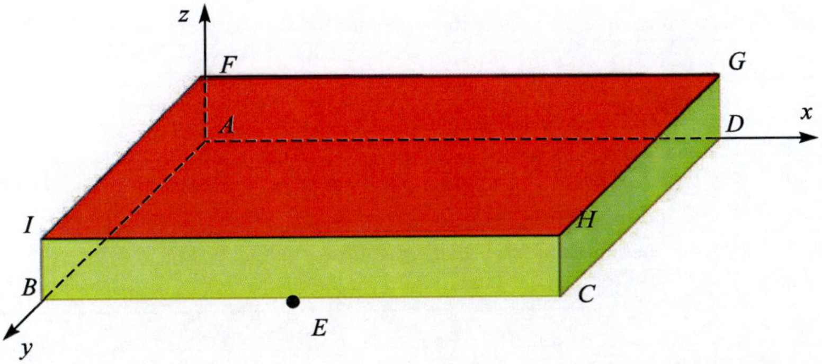

Fig. 6.16 An FE model for the high-temperature CMAS corrosion of a TBC (a bare ceramic layer with no substrate). The red surface region is coated with the CMAS, and the yellow region is the ceramic layer

$$\int_{\Omega} \left(\delta c_j \cdot \dot{c}_i + D_j \delta \nabla c_l \cdot \nabla c_i\right) \mathrm{d}V - \int_{\Gamma} \left(D_j \delta c_l \mathbf{n} \cdot \nabla c_i\right) \mathrm{d}a = 0 (i = j = l)$$

$$\mathbf{n} \cdot \nabla c_i = 0 \tag{6.49}$$

where δc_j is a trial function of the field variable c_j and $\mathbf{n} \cdot \nabla c_i = 0$ indicates that there is no diffusion source at the boundaries of the system under investigation.

The model in Fig. 6.16 has dimensions of 0.5 cm × 1 cm × 200 μm and is discretized with 8000 hexahedral domain elements. During the CMAS corrosion process, the pores in the coating near the ceramic surface are completely filled.

There is no CMAS in the coating at the initial time point. Let $\overline{c}_i^0$ be the CMAS concentration near the ceramic surface (i.e., plane IFGH during the CMAS infiltration process. Considering that the CaO:MgO:Al_2O_3:SiO_2 mole ratio is 35:10:7:48 in the CMAS, we can derive the following expression of $\overline{c}_i^0$:

$$\begin{aligned}
\overline{c}_1^0 &= 35\rho_{\mathrm{CMAS}} \cdot f_{\mathrm{i0}} / \left(35 \cdot \mathrm{M_{CaO}} + 10 \cdot \mathrm{M_{MgO}} + 7 \cdot \mathrm{M_{Al_2O_3}} + 48 \cdot \mathrm{M_{SiO_2}}\right) \\
\overline{c}_2^0 &= 10\rho_{\mathrm{CMAS}} \cdot f_{\mathrm{i0}} / \left(35 \cdot \mathrm{M_{CaO}} + 10 \cdot \mathrm{M_{MgO}} + 7 \cdot \mathrm{M_{Al_2O_3}} + 48 \cdot \mathrm{M_{SiO_2}}\right) \\
\overline{c}_3^0 &= 14\rho_{\mathrm{CMAS}} \cdot f_{\mathrm{i0}} / \left(35 \cdot \mathrm{M_{CaO}} + 10 \cdot \mathrm{M_{MgO}} + 7 \cdot \mathrm{M_{Al_2O_3}} + 48 \cdot \mathrm{M_{SiO_2}}\right) \\
\overline{c}_4^0 &= 48\rho_{\mathrm{CMAS}} \cdot f_{\mathrm{i0}} / \left(35 \cdot \mathrm{M_{CaO}} + 10 \cdot \mathrm{M_{MgO}} + 7 \cdot \mathrm{M_{Al_2O_3}} + 48 \cdot \mathrm{M_{SiO_2}}\right)
\end{aligned} \tag{6.50}$$

where f_{i0} is the porosity of the coating. When the temperature rises to the melting point of the CMAS melts, the pores near the surface in the ceramic layer are completely filled within a short period of time. Then, we define the Dirichlet boundary condition for the concentration of each element of the CMAS that coats the red region in Fig. 6.16 as follows:

$$c_i = \overline{c}_i^0 \tag{6.51}$$

Table 6.3 Simulation parameters used in the model for the high-temperature stage [4, 28–33]

Parameter	Value
Infiltration time (t)	360 min
Temperature of the system in the high-temperature stage (θ_h)	1250 °C
Migration rate of the CMAS (m)	2×10^{-31}s mol^2/kg m^3
Diffusion coefficient of the CMAS (D)	8.84×10^{-13} m^2/s
Corrosion expansion coefficient of the CMAS (β)	0.135
Mass density of the CMAS (ρ_{CMAS})	2.63 g/cm^3
Density of the YSZ	6 g/cm^3
Bulk modulus of the coating (K)	23.81 GPa
Shear modulus of the coating (G)	16.39 GPa
Porosity of the coating (f_{i0})	0.21
Chemical-reaction rate constant (ζ)	1.914×10^{-8}[m^3/s·kg]

The temperature of the system remains at 1250 °C during the simulation process. Table 6.3 summarizes the model parameters used in the numerical calculation process.

Figure 6.17 shows the evolutionary pattern of the distribution of ρ_{CMAS} in the coating along the infiltration depth with the corrosion time. The total amount of CMAS in the coating continues to increase as the corrosion time increases. The pores at the top of the coating are completely filled. Because there are no changes in the pores in the coating during the corrosion process, there are almost no changes in the ρ_{CMAS} in the top region of the coating as the corrosion time increases. As the CMAS continues to infiltrate the interior of the coating, the pores inside the coating gradually transition from a completely empty state to a filled state. Thus, at any time point, the ρ_{CMAS} inside the coating is higher in regions near the surface and lower in regions away from the surface. The experimental results tend to be consistent with the simulation results.

6.2 Microstructural Evolution, Deformation, and Composition Loss of Coatings Due to Corrosion

6.2.1 Microstructural Evolution and Deformation of Coatings

(1) **Microstructural evolution of TBCs**

To compare the cross-sectional micromorphology of APS YSZ coatings before and after corrosion, the original and corroded specimens of coatings were cold embedded

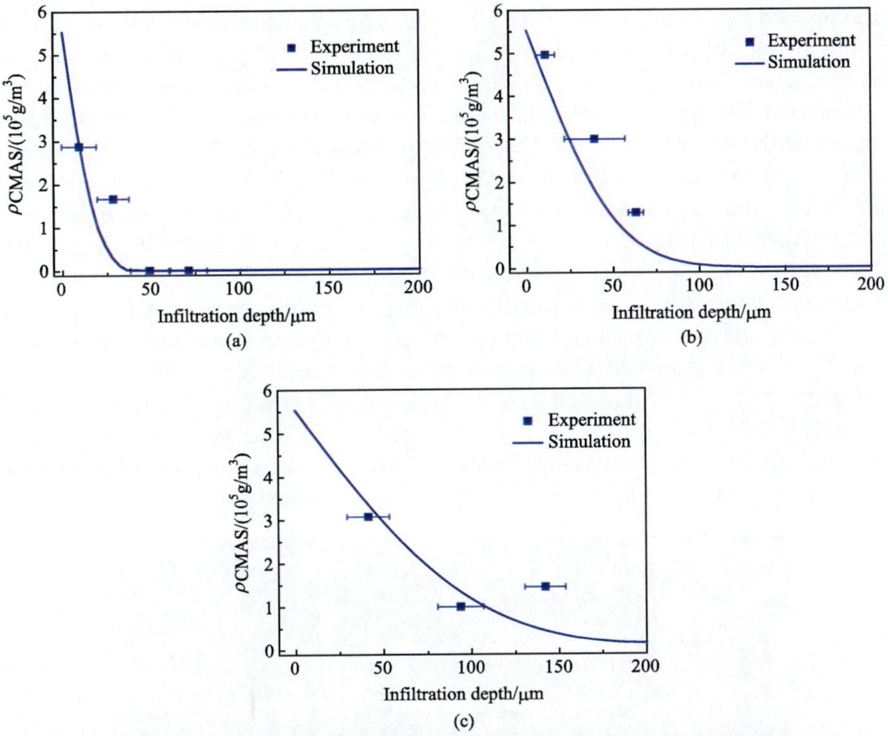

Fig. 6.17 Distributions of the CMAS concentration in the infiltration direction at **a** 2 min, **b** 15 min, and **c** 60 min. The solid lines represent the simulation results; the dots represent the experimental results; and the horizontal error bars indicate the sizes of the rectangles in Fig. 6.15 in the infiltration direction

with epoxy resin. The embedded specimens were then subjected to a metallographic treatment involving two main processes, namely, grinding and polishing. Both processes were performed on an automated Buehler Vector grinder-polisher. In the grinding process, the surface of each specimen was first ground and smoothed on diamond wheels with grit sizes of 45, 15, and 6 μm, respectively, and then polished with diamond metallographic polishing agents with grit sizes of 7, 2.5, and 0.5 μm, respectively. Subsequently, the surface of the specimen was polished with an Al_2O_3 abrasive paste with a grit size of 0.05 μm and then rinsed with water. Ceramic coating materials have low electrical conductivity. Therefore, the surface of each specimen needed to be electroplated with a thin conductive film before SEM observation to ensure that clear images of the specimens could be obtained using FE-SEM (Quanta FEG 250).

Figure 6.18a–d shows the cross-sectional micromorphology of the original APS-YSZ specimen. Figure 6.18a shows the overall morphology of the APS coating. The coating has a uniform thickness of approximately 300 μm, and the ceramic layer is bonded well to the BC layer. Large numbers of microcracks and microvoids are present in the coating. As seen in Fig. 6.18b, there are large voids in the coating.

The layered structure unique to APS coatings and visible microcracks can be seen in Fig. 6.18c. Strip-like structures with grain sizes of 5–10 μm and a large number of microcracks between these strip-like structures can be seen in Fig. 6.18d.

Figure 6.19a–d shows the cross-sectional micromorphology of the coating specimen heat-treated at 1150 °C for 8 h. A layer of residual VA approximately 60–70 μm in thickness on the coating surface can be seen in Fig. 6.19a. As clearly shown in Fig. 6.19b, the VA has infiltrated into the interior of the coating along the pores and microcracks, resulting in a dense region at the top of the coating. The VA has destroyed the original layered structure of the coating specimen and fills the interlayer microcracks, resulting in the formation of spherical grains, as shown in Fig. 6.19c, d.

Figure 6.20a–d shows the micromorphology of the coating specimen heat-treated at 1250 °C for 8 h. Figure 6.20a shows the overall morphology of the coating. The VA almost penetrates through the entire thickness of the coating. There is still some residual VA on the coating surface, but the residual VA layer is thinner than that on the coating subjected to corrosion at 1150 °C, indicating that more VA infiltrated into

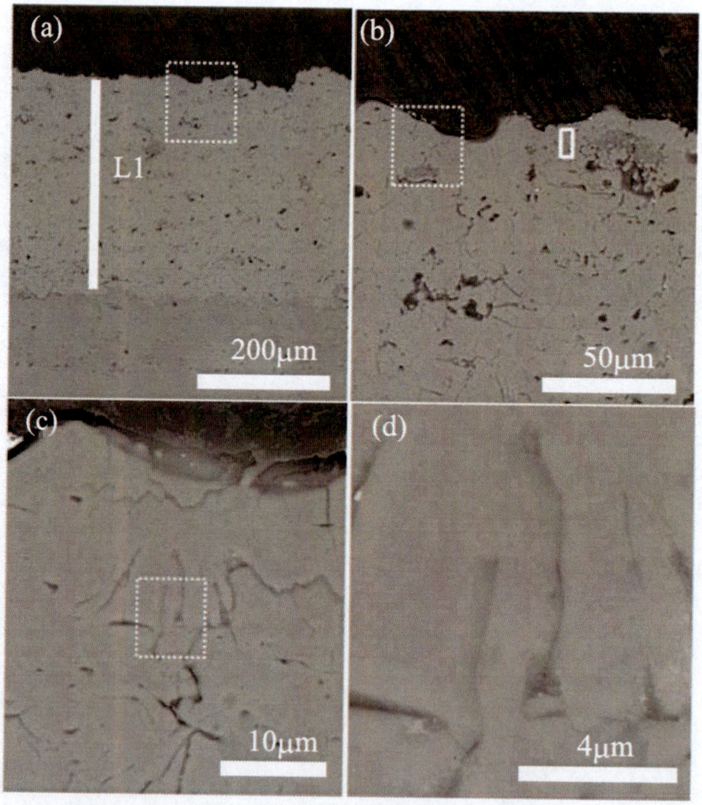

Fig. 6.18 **a** Cross-sectional micromorphology of the original APS-YSZ specimen. **b** Close-up of the structure in the dotted box in **a**. **c** Close-up of the structure in the dotted box in **b**. **d** Close-up of the structure in the dotted box in **c**

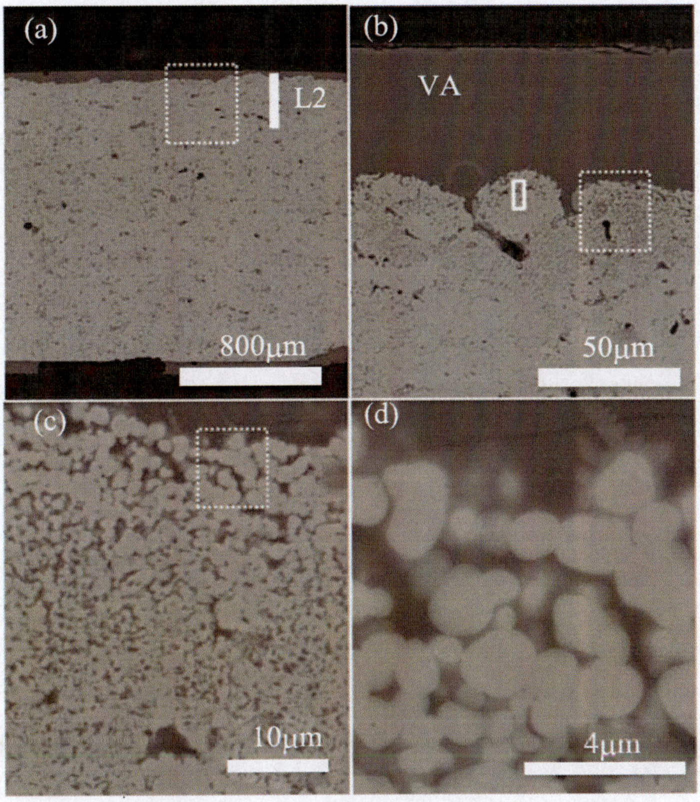

Fig. 6.19 **a** Cross-sectional micromorphology of the specimen coated with VA at 20 mg/cm^2 and heat-treated at 1150 °C for 8 h. **b** Close-up of the structure in the dotted box in **a**. **c** Close-up of the structure in the dotted box in **b**. **d** Close-up of the structure in the dotted box in **c**

the interior of the coating. The enlarged views of local regions in Fig. 6.20b, c show that the original strip-like structures of the coating have completely vanished and that the pores and microcracks are filled with VA, suggesting that the destructive effect of corrosion becomes more prominent as the corrosion temperature rises. Figure 6.20d shows that the coating treated at the higher temperature is more significantly sintered and that its microstructure is considerably degraded.

When corroded by CMAS, an APS coating transforms from a layered structure to a spherical grain structure. To describe the evolutionary pattern of the extent of corrosion in a coating during the CMAS corrosion process, we denote the volume fraction of the CMAS corrosion region in a coating by n, the value of which ranges from 0 to 1. $n = 0$ indicates an intact coating structure, and $n = 1$ means that the coating is completely corroded; i.e., the coating displays a spherical structure. Let us assume that the rate of change in n is directly proportional to the density of the reactant CMAS, ρ_{CMAS}, and the volume fraction of the undissolved coating, $1 - n - f_{i0}$, through the following equation:

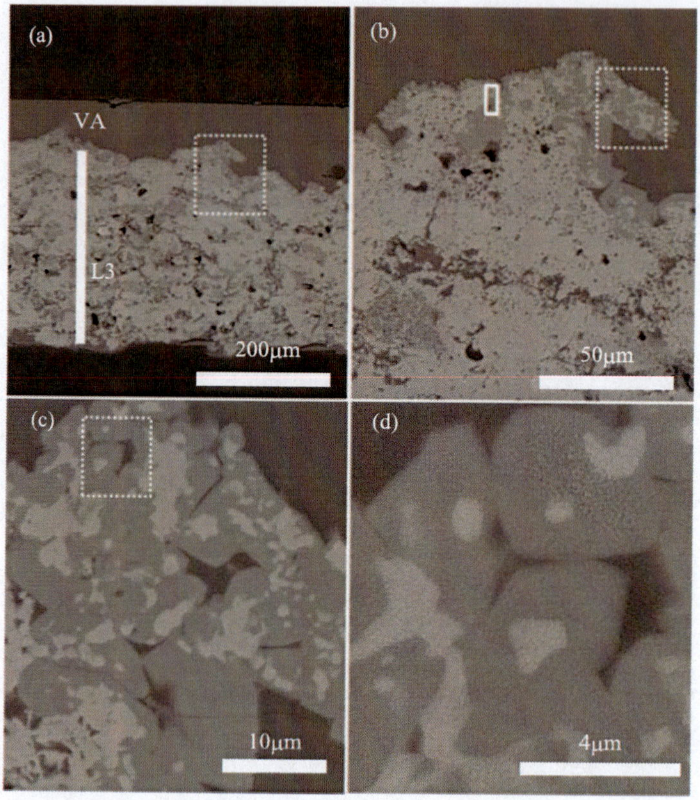

Fig. 6.20 **a** Cross-sectional micromorphology of the specimen coated with VA at 20 mg/cm^2 and heat-treated at 1250 °C for 8 h. **b** Close-up of the structure in the dotted box in **a**. **c** Close-up of the structure in the dotted box in **b**. **d** Close-up of the structure in the dotted box in **c**

$$\dot{n} = \zeta(1 - n - f_{i0})\rho_{CMAS} \tag{6.52}$$

where ζ is the chemical reaction rate constant. The weak form of Eq. (6.52) is written as follows:

$$\int_{\Omega} (\delta n \cdot \dot{n} - \delta n \cdot \zeta(1 - n - f_{i0})\rho_{CMAS})\mathrm{d}V = 0 \tag{6.53}$$

where δn is a trial function of n. With the combination of the above equation with Eq. (6.49) and the use of the FE model from Sect. 5.1.4, the CMAS infiltration and coating corrosion process is simulated with FEM. We set $n = 0$ for the interior of the coating at the initial time point.

Figure 6.21a shows the distributions of the extent of corrosion n along the infiltration depth at different corrosion times, and Fig. 6.21b presents the lateral corrosion morphology of the coating subjected to corrosion for 60 min. Because the amount of CMAS melt is higher in regions close to the surface and lower in regions away from the surface, the corrosion and spheroidization of the coating material become more severe as the distance from the surface decreases. Figure 6.21a shows the evolutionary pattern of the extent of corrosion n along the infiltration depth with the corrosion time. The value of n is large in regions close to the surface and small in regions away from the surface. The FE simulation results agree with the experimental results.

(2) **Corrosion-induced swelling deformation of coatings**

The CMAS that has infiltrated into the pores of a coating gradually dissolves and corrodes the coating material. The separation between the YSZ grains and grain boundaries results in the swelling deformation of the coating.

To describe the swelling deformation of a coating during the corrosion process, we divide the total strain of a coating due to CMAS infiltration and corrosion into two parts:

$$\varepsilon = (\sigma - \boldsymbol{I}\lambda \boldsymbol{tr}\sigma/(2G + 3\lambda))/2G + \beta n\boldsymbol{I} = \varepsilon_e + \varepsilon_n(n) \tag{6.54}$$

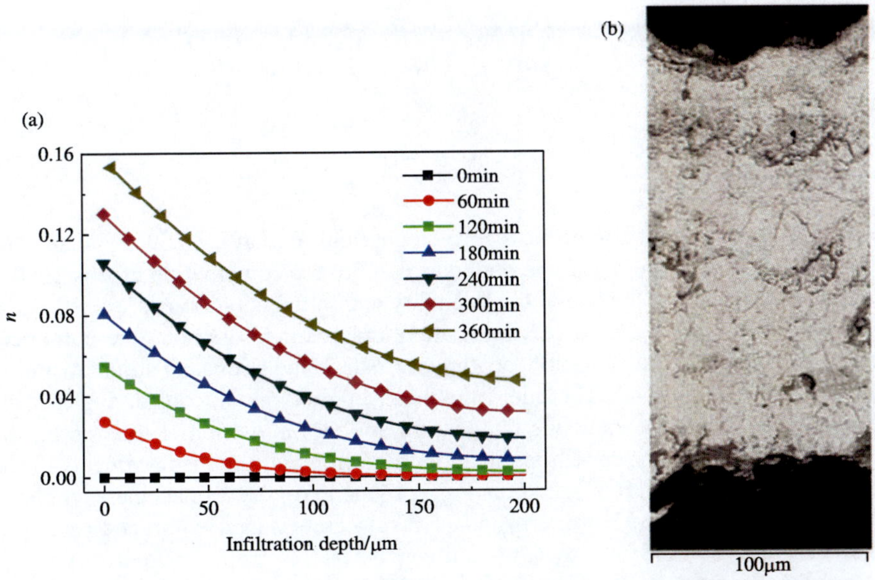

Fig. 6.21 Distributions of the extent of corrosion n along the infiltration depth. **a** Distributions of the extent of the corrosion reaction in the CMAS infiltration direction at different corrosion times. **b** Lateral morphology of the coating subjected to corrosion for 60 min

where $\boldsymbol{\varepsilon}_e$ is the elastic strain, $\boldsymbol{\varepsilon}_n(n)$ is the strain due to the corrosion and dissolution of the coating, $\boldsymbol{I}$ is the second-order unit tensor, λ, G, and K are the Lame's constant, shear modulus, and bulk modulus of the coating, respectively, and β is the swelling coefficient of the coating during CMAS infiltration and corrosion. Assuming that the corrosion-induced strain is directly proportional to the extent of the corrosion reaction, n, we have

$$\varepsilon_n(n) = \beta n \boldsymbol{I} \tag{6.55}$$

Equation (6.54) can be transformed and expressed in terms of stress tensors as follows:

$$\sigma = \boldsymbol{I}\sigma_0/3 + \lambda \boldsymbol{I} \boldsymbol{tr}\varepsilon + 2G\varepsilon - 3K\beta n \boldsymbol{I} = \sigma_e + \sigma_n \tag{6.56}$$

where $\boldsymbol{\sigma}_e$ is the elastic stress and $\boldsymbol{\sigma}_n = -3K\beta n \boldsymbol{I}$ is the compressive stress due to corrosion and swelling.

By neglecting the volume force $\boldsymbol{f}$, we can express the force equilibrium equation as follows:

$$\nabla \cdot \boldsymbol{\sigma} = 0 \tag{6.57}$$

The weak form of the above equation is expressed as follows:

$$-\int_V \nabla \delta \mathbf{u}{:}\sigma dV - \int_\Omega (\delta \mathbf{u} \cdot (\mathbf{n} \cdot \sigma))d\Omega = 0 \tag{6.58}$$

$$\mathbf{n} \cdot \sigma = 0 \tag{6.59}$$

where $\delta\mathbf{u}$ is a trial function of the displacement field. Equation (6.59) indicates that there are no external forces at the boundaries. With the combination of Eqs. (6.49), (6.53), and (6.59) and the use of the boundary and initial conditions provided in the previous two sections for CMAS infiltration and coating corrosion, the numerical simulations of CMAS infiltration, coating corrosion, and stress and strain evolution are carried out using the FE model introduced in Sect. 5.1.4. Considering that the displacement components inside the coating are zero at the initial time point, the displacement boundary conditions include the following: the displacements in the z-direction at points A, B, C, and D in Fig. 6.16 are zero, the displacements in the x-direction at points A and B are zero, and the displacements in the y-direction at points A and D are zero. We assume that the coating material is isotropic and that the temperature of the system remains constant at 1250 °C during the simulation process. Table 6.3 summarizes the model parameters used in the numerical calculation process.

Figure 6.22 shows the evolution of the lateral morphology of bare ceramic-layer specimen 4 with no substrate during the corrosion process. The coating continuously

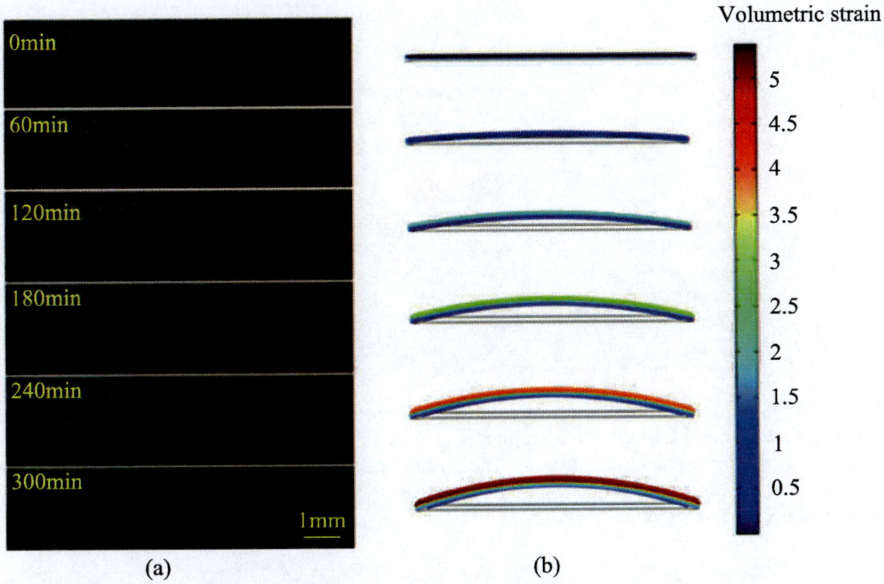

Fig. 6.22 Spatial distribution of the volumetric strain and morphology of the deformed coating at different corrosion times (left: images recorded during the experiment; right: simulation results)

swells upwards as the corrosion time increases. Because the corrosion-induced strain ε_n is directly proportional to n, the volumetric strain inside the coating in the infiltration direction is large in regions close to the surface and small in regions away from the surface; that is, the distribution trend of the volumetric strain is consistent with that of n. In addition, as seen in Fig. 6.22, the experimental results tend to be consistent with the simulation results.

We further establish an FE model for the local corrosion of a ceramic layer with a substrate, as shown in Fig. 6.23. The model has dimensions of 1 cm × 1 cm × 200 μm and is discretized with 69,812 domain elements. During the corrosion process, the pores in the coating close to the red region at the center of the surface of the ceramic layer are completely filled with CMAS. Considering the restraining effect of the substrate on the coating, the displacement on plane ABCD remains constant at zero throughout the process. We assume that the coating material is isotropic and that the temperature of the system remains constant at 1250 °C during the numerical calculations. The model parameters for the numerical calculation are listed in Table 6.3.

After 300 min of high-temperature corrosion, the CMAS infiltrates from the surface of the local coated circular region to the bottom of the coating along the pores while dissolving and corroding the coating beneath the CMAS-coated region. The dissolution of the coating is accompanied by corrosion-induced swelling deformation, while the coating in the uncorroded regions undergoes no deformation. Consequently, an extremely high in-plane compressive stress is formed in the CMAS-coated region of the coating, inducing the coating to buckle. As shown in Fig. 6.24, a tensile

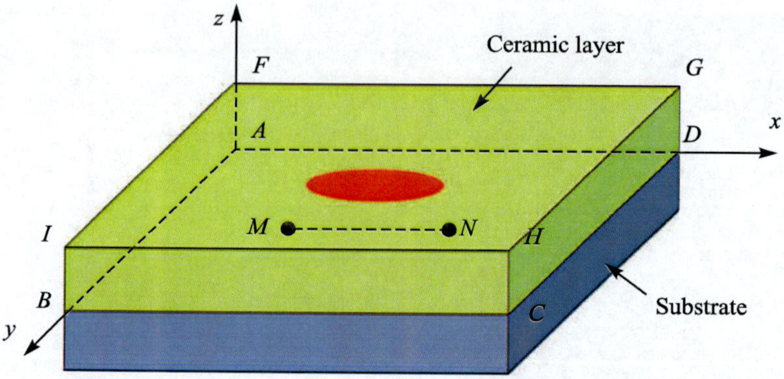

Fig. 6.23 FE model for high-temperature CMAS corrosion of a TBC. Ceramic layer with a substrate. The circular CMAS-coated region is at the center of the surface of the cuboid coating. The blue region represents the substrate. The dotted line is located at the boundary between the ceramic layer and the substrate and is parallel to the *x*-axis, with its center coinciding with the center of the CMAS-coated region on plane *xy*

stress is formed in the out-of-plane direction inside the coating. In addition, out-of-plane stress concentrations appear at the boundaries of the coating, which promote the spalling of the coating from the substrate during the corrosion process, as seen in Fig. 6.25.

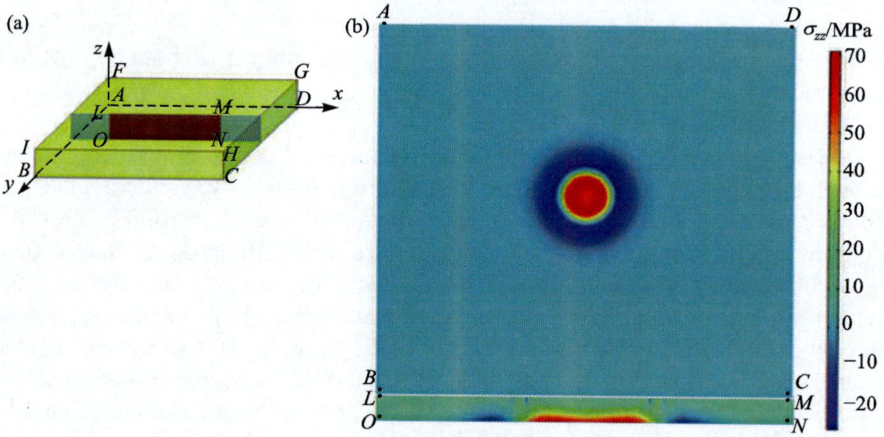

Fig. 6.24 Distribution of the out-of-plane stress (in the *z*-direction) inside the coating subjected to corrosion for 300 min (top: distribution of the stress at the bottom boundary of the coating; bottom: distribution of the stress at a cross-section through the center of the defect)

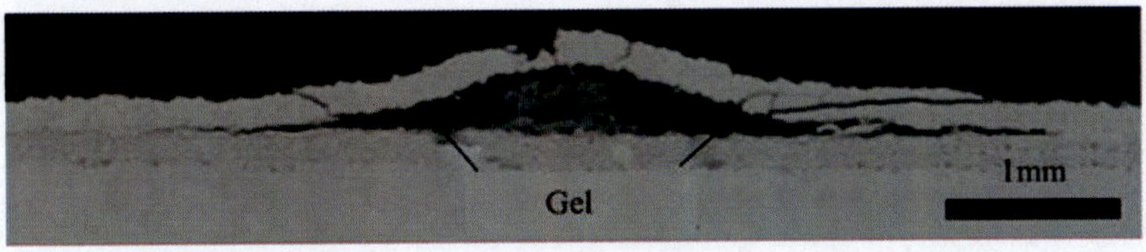

Fig. 6.25 SEM image of an APS TBC subjected to CMAS corrosion at 1250 °C for 4 h [4]

6.2.2 *Thermo–Mechano–Chemical Coupling Theory for CMAS Infiltration and Corrosion in TBCs*

(1) **Theoretical modeling**

Equations (6.47), (6.52), and (6.54) collectively constitute a theoretical decoupling model for CMAS corrosion in TBCs, i.e., a model that does not consider the interactions between deformation, infiltration, and corrosion reactions. Next, let us establish a thermo–mechano–chemical coupling theory for the high-temperature infiltration and corrosion stage in the CMAS corrosion of a TBC from the perspectives of the laws of thermodynamics, as well as analyze and discuss the thermo–mechano–chemical coupling effect.

For the high-temperature corrosion stage, we consider the couplings of the physical processes inside a TBC system, including CMAS infiltration, the dissolution of the TBC, and the swelling deformation in the corroded region. The internal energy E in any domain V is expressed as follows:

$$\dot{E} = \int_V \dot{e} dV = \int_\Omega (\boldsymbol{\sigma} \cdot \boldsymbol{n}) \cdot \boldsymbol{v} d\Omega + \int_V \boldsymbol{f} \cdot \boldsymbol{v} dV - \int_\Omega \boldsymbol{q} \cdot \boldsymbol{n} d\Omega + \int_V r dV - \int_\Omega \sum_{i=1}^{4} \mu_i \boldsymbol{j}_i \cdot \boldsymbol{n} d\Omega \tag{6.60}$$

where μ_i and $\mathbf{j}_i$ are the chemical potential of element i for infiltration and its diffusion flux, respectively, r is the heat absorbed from external sources per unit volume of the material per unit time, and q is the heat absorbed by the material in domain V from boundary Ω per unit time (i.e., the heat flow rate). The differential form of the above equation is as follows:

$$\dot{e} = (\nabla \cdot \boldsymbol{\sigma}) \cdot \boldsymbol{v} + \boldsymbol{\sigma} : \nabla \boldsymbol{v} + \boldsymbol{f} \cdot \boldsymbol{v} - \nabla \cdot \boldsymbol{q} + r - \sum_{i=1}^{4} \nabla \mu_i \cdot \boldsymbol{j}_i - \sum_{i=1}^{4} \mu_i \nabla \cdot \boldsymbol{j}_i \tag{6.61}$$

By substituting the concentration conservation equation in Eq. (6.44) and force equilibrium equation into the above equation, we obtain

$$\dot{e} = \boldsymbol{\sigma} : \dot{\boldsymbol{\varepsilon}} - \nabla \cdot \boldsymbol{q} + r - \sum_{i=1}^{4} \nabla \mu_i \cdot \boldsymbol{j}_i + \sum_{i=1}^{4} \mu_i \dot{c}_i \tag{6.62}$$

According to the second law of thermodynamics, the entropy increase rate is stated as follows:

$$\int_{\Omega} \dot{s} d V \geq - \int_{\Gamma} (\boldsymbol{q} \cdot \boldsymbol{n}/\theta) d a + \int_{\Omega} (r/\theta) d V \tag{6.63}$$

If we transform Eq. (6.63) to an equation, it has the following differential form:

$$\dot{s} = -\nabla \cdot (\boldsymbol{q}/\theta) + r/\theta + \gamma = (\boldsymbol{q} \cdot \nabla\theta)/\theta^2 + (-\nabla \cdot \boldsymbol{q} + r)/\theta + \gamma \tag{6.64}$$

Using Eq. (6.62), we rewrite Eq. (6.64) to

$$\dot{s} = (\boldsymbol{q} \cdot \nabla\theta)/\theta^2 + \left(\dot{e} - \boldsymbol{\sigma} : \dot{\boldsymbol{\varepsilon}} - \sum_{i=1}^{4} \mu_i \dot{c}_i + \sum_{i=1}^{4} \nabla \mu_i \cdot \boldsymbol{j}_i \right) / \theta + \gamma \tag{6.65}$$

By introducing the Helmholtz free energy, we obtain

$$\psi = e - \theta s \tag{6.66}$$

Through Eq. (6.66), we can rewrite Eq. (6.65) to

$$\theta\gamma = -\dot{\psi} - \dot{\theta} s + \boldsymbol{\sigma} : \dot{\boldsymbol{\varepsilon}} + \sum_{i=1}^{4} \mu_i \dot{c}_i - (\boldsymbol{q} \cdot \nabla\theta)/\theta - \sum_{i=1}^{4} \nabla \mu_i \cdot \boldsymbol{j}_i \geq 0 \tag{6.67}$$

Let us assume that the Helmholtz free energy ψ is a function of the temperature θ, the elastic strain $\boldsymbol{\varepsilon}_e$, the molar concentration c_i of element i in the CMAS, and the volume fraction n of the corroded region; i.e.,

$$\psi = \psi(\theta, \boldsymbol{\varepsilon}_e, c_i, n) \tag{6.68}$$

Taking the derivative of the above equation with respect to time, we obtain

$$\dot{\psi} = (\partial\psi/\partial\theta)\dot{\theta} + (\partial\psi/\partial\varepsilon_e) : \dot{\varepsilon}_e + \sum_{i=1}^{4} (\partial\psi/\partial c_i)\dot{c}_i + (\partial\psi/\partial n)\dot{n} \tag{6.69}$$

The total strain is a superposition of the following three parts:

$$\boldsymbol{\varepsilon} = \boldsymbol{\varepsilon}_e + \boldsymbol{\varepsilon}_n(n) + \boldsymbol{\varepsilon}_\theta(\theta) \tag{6.70}$$

where $\boldsymbol{\varepsilon}_e$ is the elastic strain, $\boldsymbol{\varepsilon}_n(n)$ is the deformation of the coating due to corrosion and dissolution, and $\boldsymbol{\varepsilon}_\theta(\theta)$ is the deformation of the coating due to temperature change. Substitution of Eqs. (6.70) and (6.69) into Eq. (6.67) yields

$$(\boldsymbol{\sigma} - \partial\psi/\partial\boldsymbol{\varepsilon}_e) : \dot{\boldsymbol{\varepsilon}}_e + (\boldsymbol{\sigma} : (\partial\boldsymbol{\varepsilon}_\theta/\partial\theta) - \partial\psi/\partial\theta - s)\dot{\theta} + \sum_{i=1}^{4}(\mu_i - \partial\psi/\partial c_i)\dot{c}_i +$$
$$(\boldsymbol{\sigma} : (\partial\boldsymbol{\varepsilon}_n/\partial n) - \partial\psi/\partial n)\dot{n} - (\boldsymbol{q}\cdot\nabla\theta)/\theta - \sum_{i=1}^{4}\nabla\mu_i\cdot\boldsymbol{j}_i = \gamma\theta \geq 0 \tag{6.71}$$

We define the dissipation rates due to chemical reactions, heat conduction, and infiltration as follows:

$$\theta\gamma_n = (\boldsymbol{\sigma} : (\partial\boldsymbol{\varepsilon}_n/\partial n) - \partial\psi/\partial n)\dot{n} \tag{6.72}$$

$$\theta\gamma_\theta = -(\mathbf{q}\cdot\nabla\theta)/\theta \tag{6.73}$$

$$\theta\gamma_c = -\sum_{i=1}^{4}\nabla\mu_i\cdot\mathbf{j}_i \tag{6.74}$$

In Eq. (6.71), $\boldsymbol{\varepsilon}_e$, θ, and c_i can take any values within their respective value ranges. To ensure that Eq. (6.71) is always nonnegative, the coefficients of its first three terms, namely,$\dot{\boldsymbol{\varepsilon}}_e$,$\dot{\theta}$, and$\dot{c}_i$, must be zero. Then, we obtain the following three generalized constitutive relations between the entropy and temperature, between the stress and strain, and between the chemical potential and concentration of element *i*:

$$s = \boldsymbol{\sigma} : (\partial\boldsymbol{\varepsilon}_\theta/\partial\theta) - \partial\psi/\partial\theta \tag{6.75}$$

$$\boldsymbol{\sigma} = \partial\psi/\partial\boldsymbol{\varepsilon}_e \tag{6.76}$$

$$\mu_i = \partial\psi/\partial c_i \tag{6.77}$$

Substituting Eqs. (6.75)–(6.77) into Eq. (6.71), we obtain

$$\gamma\theta = -(\boldsymbol{q}\cdot\nabla\theta)/\theta - \sum_{i=1}^{4}\nabla\mu_i\cdot\boldsymbol{j}_i + (\boldsymbol{\sigma} : (\partial\boldsymbol{\varepsilon}_n/\partial n) - \partial\psi/\partial n)\dot{n}$$
$$= \theta\gamma_\theta + \theta\gamma_c + \theta\gamma_n \tag{6.78}$$

We assume that the volume fraction of the corroded and dissolved region, *n*, is linearly related to the density of the reactant CMAS, ρ_{CMAS}, and the volume fraction of the undissolved coating, $1 - n - f_{i0}$, through the following equation:

$$\dot{n} = \zeta(1 - n - f_{i0})\rho_{CMAS} + A_0 \tag{6.79}$$

where ζ is the chemical reaction rate constant, the value of which depends on the individual corrosion reaction; A_0 represents the effects of other field variables on the dissolution of the coating; and n is the extent of the corrosion and dissolution of the coating.

We define the reference state of the coating as follows:

$$\varepsilon = 0 \quad \sigma = 0 \quad c_i = 0 \quad \theta = \theta_1 \quad n = 0 \tag{6.80}$$

In this state, CMAS infiltration has not occurred, and the coating has not dissolved in the CMAS. To obtain Eq. (6.79), we perform a series expansion of the free energy function ψ with respect to independent variables θ, $\boldsymbol{\varepsilon}$, c_i, and n near the reference state and retain up to the quadratic terms as follows:

$$\begin{aligned}\psi = &\sum_{i=1}^{4} \mu_{0i} c_i + (\sigma_0 \mathrm{tr}\boldsymbol{\varepsilon})/3 + \lambda(\mathrm{tr}\boldsymbol{\varepsilon})^2/2 + G\boldsymbol{\varepsilon} : \boldsymbol{\varepsilon} + \sum_{i=1}^{4} \varsigma_i c_i^2/2 \\ &- 3K\beta n \mathrm{tr}\boldsymbol{\varepsilon} - 2A_1(1 - f_{i0})\rho_{CMAS} n + A_1 \rho_{CMAS} n^2\end{aligned} \tag{6.81}$$

where ς_i, β, and A_1 are constants. The first two terms on the right-hand side of Eq. (6.81) are the first-order expansion of the free energy function with respect to c_i and $\boldsymbol{\varepsilon}$, representing the contributions of the amount of CMAS and strain inside the coating before CMAS infiltration to the free energy. According to Eq. (6.80), the first two terms are zero. The third and fourth terms on the right-hand side of Eq. (6.81) represent the elastic strain energy. The fifth term is the quadratic term of the concentration of element i and represents the main driving force for CMAS infiltration. Considering that the dissolution of the coating is accompanied by a volume expansion, we introduce the sixth term on the right-hand side of Eq. (6.81), i.e., the coupling term between the strain and n. To obtain Eq. (6.79), we introduce the final two terms on the right-hand side of Eq. (6.81), i.e., the coupling term between the CMAS density ρ_{CMAS} and n as well as the quadratic term of n.

By substituting Eq. (6.81) into Eq. (6.76), we obtain

$$\boldsymbol{\sigma} = \frac{\partial \psi}{\partial \varepsilon_e} = \lambda \boldsymbol{I} \mathrm{tr}\boldsymbol{\varepsilon} + 2G\boldsymbol{\varepsilon} - 3K\beta n \boldsymbol{I} = \boldsymbol{\sigma}_e + \boldsymbol{\sigma}_n \tag{6.82}$$

Let us denote the sum of the first two terms ($\lambda I tr\varepsilon + 2G\varepsilon$) and the third term ($-3K\beta n\boldsymbol{I}$) on the penultimate rightmost side of the above equation by the elastic stress $\boldsymbol{\sigma}_e$ and the stress $\boldsymbol{\sigma}_n$ due to the dissolution and swelling of the coating, respectively.

If the temperature of the system remains constant during the corrosion process, we have

$$\theta\gamma_\theta = -(\mathbf{q} \cdot \nabla\theta)/\theta = 0 \tag{6.83}$$

Because both the dissolution of the coating in the CMAS melts and the CMAS infiltration from the surface of the coating are irreversible processes, the dissipation rates due to chemical reactions and infiltration are always nonnegative; i.e., $\theta\gamma_n \geq 0$, and $-\sum_{i=1}^{4} \nabla\mu_i \cdot \mathbf{j}_i \geq 0$. By assuming that the $\dot{n}$ in Eq. (6.72) is directly proportional to its coefficient and considering Eqs. (6.82) and (6.81), we have

$$\dot{n} = \zeta(1 - n - f_{i0})\rho_{\mathrm{CMAS}} + 3\beta K\zeta \mathrm{tr}\boldsymbol{\varepsilon}/(2A_1) \tag{6.84}$$

A comparison of Eqs. (6.84) and (6.79) yields $A_0 = 3\beta K\zeta \mathrm{tr}\boldsymbol{\varepsilon}/(2A_1)$, where A_0 represents the effect of the strain field on the dissolution of the coating (i.e., chemical reactions) and is referred to as the strain coupling term for short.

According to Fick's second law, we assume that the diffusion flux of element i in Eq. (6.74), $\mathbf{j}_i$, is in the opposite direction of and linearly related to the gradient of the chemical potential for infiltration, $\nabla\mu_i$; i.e.,

$$\mathbf{j}_i = -m_j \nabla\mu_i (i = j) \tag{6.85}$$

where m_j is the migration rate of element j_i.

By combining Eqs. (6.85), (6.81), (6.77), and (6.44), we have

$$\dot{c}_i = D_j \nabla^2 c_i - 2m_j A_1(\partial\rho_{CMAS}/\partial c_i)(1 - n - f_{i0})\nabla^2 n \quad (i = j) \tag{6.86}$$

where the second term on the right-hand side, $2m_j A_1(\partial\rho_{CMAS}/\partial c_i)(1 - n - f_{i0})\nabla^2 n$, represents the effect of the chemical field on CMAS infiltration in the coating and is referred to as the chemical coupling term for short and $D_j = \varsigma_i m_j$ $(i = j)$ is the diffusion coefficient of element j.

(2) **Discussion of couplings**

When using the FEM to solve the governing equations for each field variable at the high-temperature stage, we need to transform the differential equations to weak-form integral equations. The weak forms of the governing equations for the concentration of each element in CMAS are expressed as follows:

$$\int_V (\delta c_j \cdot \dot{c}_i + m_j \delta\nabla c_l \cdot \nabla(\partial\psi/\partial c_i))\mathrm{d}V - \int_\Omega (m_j \delta c_l \mathbf{n} \cdot \nabla(\partial\psi/\partial c_i))\mathrm{d}\Omega = 0 \tag{6.87}$$

$$\partial\psi/\partial c_i = \mu_{0i} + \varsigma_p c_i + A_1 n(n - 2(1 - f_{i0}))\partial\rho_{\mathrm{CMAS}}/\partial c_i \ (i = j = p = l) \tag{6.88}$$

$$\mathbf{n} \cdot \nabla(\mu_i) = \mathbf{n} \cdot \nabla(\partial\psi/\partial c_i) = 0 \tag{6.89}$$

where δc_j is a trial function of the field variable c_j. Equation (6.89) indicates that there are no diffusion sources at the boundaries of the system under investigation. The weak form of the governing equation for the extent of the corrosion and dissolution of a coating, *n*, is expressed as follows:

$$\int_V (\delta n \cdot \dot{n} - \delta n \cdot \zeta(1 - n - f_{i0})\rho_{\text{CMAS}} - \delta n \cdot 3\beta K\zeta tr\varepsilon/(2A_1))\text{d}V = 0 \tag{6.90}$$

where δn is a trial function of *n*. By neglecting the volume force $\boldsymbol{f}$, we can express the force equilibrium equation in the following weak form:

$$-\int_V \nabla\delta\mathbf{u}{:}\boldsymbol{\sigma}\text{d}V - \int_\Omega (\delta\mathbf{u} \cdot (\mathbf{n} \cdot \boldsymbol{\sigma}))\text{d}\Omega = 0 \tag{6.91}$$

$$\mathbf{n} \cdot \boldsymbol{\sigma} = 0 \tag{6.92}$$

where $\delta\mathbf{u}$ is a trial function of the displacement $\boldsymbol{u}$, a field variable. Equation (6.92) indicates that there are no external forces at the boundaries of the system under investigation. Here, we again use the FE model introduced in Sect. 5.1.4 as well as the initial and boundary conditions for CMAS infiltration described in Sect. 5.1.4 and those for the extent of the corrosion reaction, *n*, and displacement described in Sect. 5.2.1. In addition, the temperature of the system is kept unchanged at 1250 °C during the numerical calculation process. Table 6.3 summarizes the model parameters used in the numerical calculation process.

Next, let us discuss the effects of the strain coupling term $3\beta K\zeta\text{tr}\varepsilon/(2A_1)$ and the chemical coupling term $2m_j A_1(\partial\rho_{CMAS}/\partial c_i)(1 - n - f_{i0})\nabla^2 n$ in the governing equation on the chemical, strain, and CMAS concentration fields, respectively, by comparing the evolution of the field variables determined based on Eqs. (6.84) and (6.86), which account for the coupling terms, and Eqs. (6.47) and (6.52), which do not consider the coupling effects. Figure 6.26 shows the effects of the coupling parts in the governing equations on the evolution of each field variable. As seen in Fig. 6.26a, the red curve obtained for Eq. (6.84) with the strain coupling removed coincides with the black curve obtained for Eq. (6.84), i.e., the strain coupling has almost no effect on the CMAS infiltration process. A comparison of the black and red curves in Fig. 6.26b shows that at the same infiltration depth, the red curve obtained from Eq. (6.84) with the strain coupling removed is below the black curve obtained from Eq. (6.84), suggesting that strain coupling promotes the corrosion reaction. The strain coupling is represented by the second term in Eq. (6.84) and directly affects the evolution of the extent of the corrosion reaction, *n*. Therefore, the strain coupling has a more significant effect on the chemical field than on the infiltration field. By comparing the black and red curves in Fig. 6.26c, we see that because the strain coupling promotes the corrosion reaction and the corrosion-induced strain is directly proportional to the extent of the corrosion reaction, the strain coupling increases the swelling deformation of the coating. As shown in Fig. 6.25a, at the same infiltration depth, the blue curve obtained from Eq. (6.86) with the chemical coupling term removed is above the black curve obtained from Eq. (6.86), suggesting that chemical coupling inhibits CMAS infiltration and, as a result, reduces the extent of the corrosion and dissolution of the coating and thus its swelling deformation.

In the strain coupling term $3\beta K\zeta \mathrm{tr}\boldsymbol{\varepsilon}/(2A_1)$, $\mathrm{tr}\boldsymbol{\varepsilon}$ has a coefficient of $3\beta K\zeta/(2A_1)$, where β, K, and ζ are all fixed values. In the chemical coupling term $2m_j A_1(\partial\rho_{CMAS}/\partial c_i)(1-n-f_{i0})\nabla^2 n$, $(1-n-f_{i0})\nabla^2 n$ has a coefficient of $2m_j A_1(\partial\rho_{CMAS}/\partial c_i)$, where m_j and $\partial\rho_{CMAS}/\partial c_i$ are fixed values. Currently, only parameter A_1 is unknown, and its value affects both the strain coupling term and the chemical coupling term. Thus, we define A_1 as the coupling parameter. Figure 6.27a–c shows the effects of A_1 on the infiltration, chemical, and deformation fields, respectively. As the value of A_1 is directly proportional to the chemical coupling term $2m_j A_1(\partial\rho_{\mathrm{CMAS}}/\partial c_i)(1-n-f_{i0})\nabla^2 n$ and inversely proportional to the strain coupling term $3\beta K\zeta \mathrm{tr}\boldsymbol{\varepsilon}/(2A_1)$, increasing A_1 amplifies the effects of the chemical

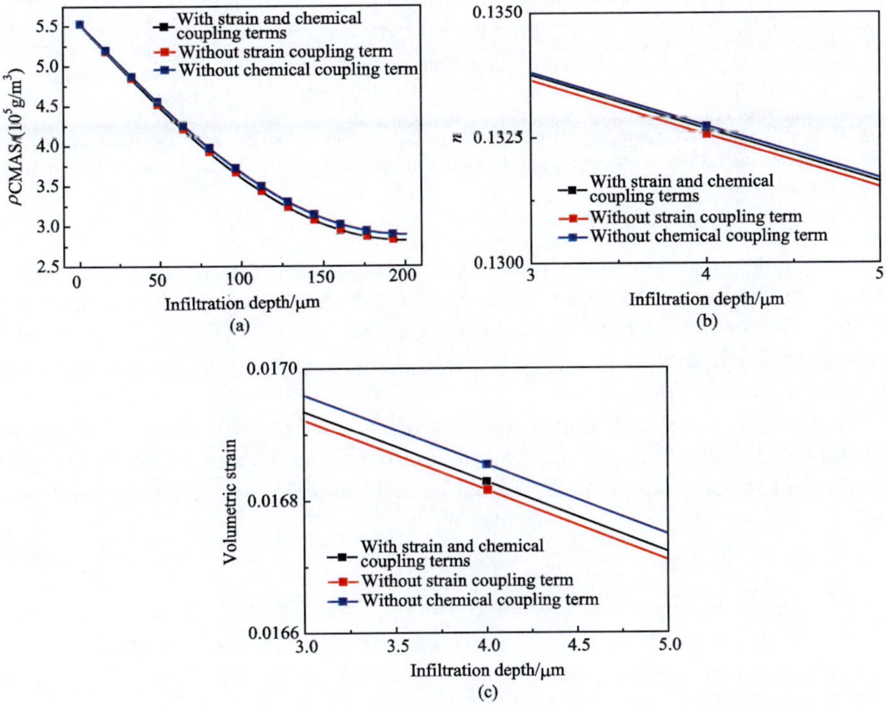

Fig. 6.26 Effects of coupling in the governing equation on the evolution of the field variables inside a coating. **a** ρ_{CMAS}. **b** Extent of the corrosion and dissolution of the coating. **c** Volumetric strain

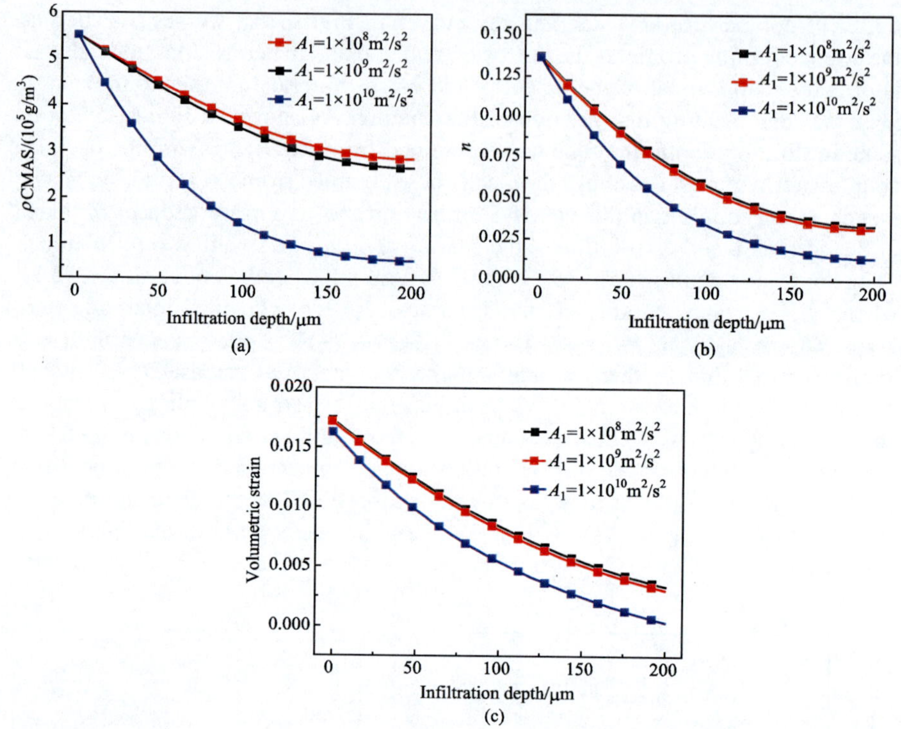

Fig. 6.27 Effects of the coupling parameter A_1 in the governing equation on the evolution of the field variables inside a coating. **a** ρ_{CMAS}. **b** Extent of the corrosion and dissolution of the coating. **c** Volumetric strain

coupling field and reduces the effects of strain coupling. As discussed in the previous paragraph, the chemical coupling term inhibits CMAS infiltration, whereas the strain coupling term promotes the corrosion reaction. Thus, as seen in Fig. 6.27, at the same infiltration depth, increasing A_1 reduces ρ_{CMAS}, the extent of the corrosion reaction, and the volumetric strain.

Figure 6.27a–c shows that as A_1 ranges from 1×10^8 to 1×10^{10} m^2/s^2, there are no significant changes in ρ_{CMAS}, the extent of corrosion reaction, or the volumetric strain. Thus, when A_1 takes a value in this range, the model can yield good predictions.

6.2.3 *Quantitative Characterization of the Distribution Pattern of Y in TBCs Subjected to CMAS Corrosion*

Pure ZrO_2 transitions from the monoclinic phase to the tetragonal phase at approximately 1180 °C. This transition process is accompanied by a volume change of approximately 4%, resulting in an enormous phase-transformation stress inside the coating, which ultimately causes coating failure. We can effectively address this problem by doping ZrO_2 with Y_2O_3 at 7–8%. However, previous research on the ceramic layer was focused primarily on the changes in its thermophysical and thermomechanical properties. For example, Kakuda et al. examined the change in the thermal conductivity of the ceramic layer of a coating after CMAS erosion, and they found that the thermal conductivity of the coating increased after it was filled with glassy CMAS [39]. Cai et al. Explored the evolution law of stress near the elliptical void filled with CMAS under the condition of thermal cycle [40]. From a mechanical perspective, Chen analyzed two failure mechanisms of the ceramic layer after CMAS erosion, namely, cracking of single columnar crystals and spallation of the whole coating [33]. While the above results of thermophysical and thermomechanical analyses are, to a certain extent, convincing, they are not focused on the microstructure of the ceramic layer. Through a microstructural study, Krämer et al. found that high-temperature CMAS corrosion leads to the formation of a new phase but failed to provide compelling evidence or the formation mechanism and quantitative conditions of the new phase [41].

The severe damage caused by CMAS corrosion in an EB-PVD TBC is reflected by structural changes in the ceramic layer, but more importantly, the open structure of the ceramic layer composed of columnar crystals allows the molten CMAS to rapidly infiltrate the boundary between the ceramic layer and the substrate, thereby significantly reducing the strain tolerance and interfacial bonding strength of the ceramic layer and, ultimately, causing spallation of the whole coating or partial spallation of the coating starting from the edges [42–44]. It is difficult to prepare cross-sectional specimens for transmission electron microscopy (TEM) and to employ methods such as SEM and X-ray powder diffraction (XRD) to characterize the interfaces in detail. As a result, the interfacial properties of TBCs have yet to be thoroughly investigated.

Many studies have reported that, during the high-temperature CMAS corrosion of a TBC, as the temperature rises, CMAS infiltrates the ceramic layer, and subsequently, the Y_2O_3 in the ceramic layer dissolves in the CMAS; as a result of the loss of the stabilizer Y_2O_3, the ZrO_2 transforms from the tetragonal phase to the monoclinic phase, resulting in a volume change of approximately 4%. In addition, Y^{3+} is carried into molten CMAS layer. As the CMAS infiltrates the ceramic layer, there is a significant increase in the Ca, Mg, Al, and Si contents but a decrease in the Y content in the ceramic layer as well as an increase in the Y content in the CMAS layer. This suggests that during the high-temperature CMAS corrosion of a TBC, the Y in the ceramic layer is lost to the CMAS layer. Here, we experimentally analyzed the variations in the amount of Y lost to the CMAS layer with the corrosion time and temperature.

(1) **Experimental scheme**

Several coating specimens cut in the dimensions of 1 cm × 1 cm were prepared. Each specimen was uniformly coated with 10 mg of CMAS powder. The specimens were then held at 1250, 1200, or 1150 °C for different periods of time. Subsequently, each sintered specimen was mounted and examined with an electronic probe (EP) to determine the changes in the content of each element in the CMAS and ceramic layers as well as the relevant patterns. The obtained images and data were compared and analyzed.

First, a line scan was performed at a location 5 μm above the CMAS/ceramic interface in each specimen in the CMAS infiltration direction to determine the pattern of change in the Y content along the CMAS infiltration depth, as shown in Fig. 6.28.

Second, we performed point measurements to determine the Y content. Specifically, three different regions parallel to and 5 μm away from the CMAS/ceramic interface were selected on each side of the interface. The Y content was measured at five points selected within each region. The measurements were averaged to prevent large errors. See Fig. 6.29 for details.

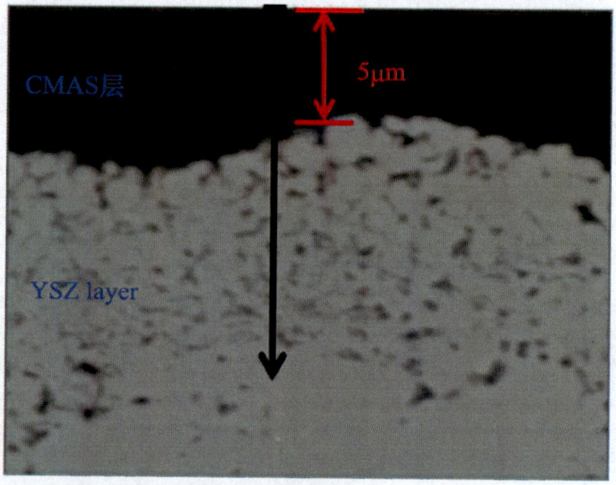

Fig. 6.28 Schematic diagram of a line scan performed to determine the pattern of change in the Y content along the CMAS infiltration depth

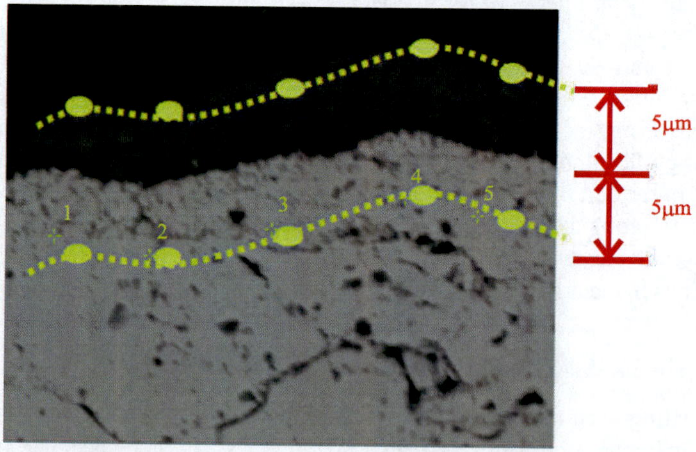

Fig. 6.29 Schematic diagram of point measurements performed to determine the patterns of change in the Y content in the CMAS and ceramic layers

(2) Variation in the Y content in the infiltration direction with corrosion time

Figure 6.30 shows the variation in the content of each element plotted based on the average experimental measurements, with the origin of the coordinates located at 5 μm above the interface between the ceramic and CMAS layers. The curves show that during the CMAS corrosion of the TBC, Ca, Mg, Al, and Si gradually infiltrate the ceramic layer, the Ca, Mg, Al, and Si contents gradually decrease in the CMAS infiltration direction, and the Y content gradually increases in the CMAS infiltration direction until eventually tending to stabilize. In addition, the CMAS infiltration depths at different corrosion temperatures can be seen in Fig. 6.30. As shown in Fig. 6.30a, the CMAS infiltration region ranges from the position of $x = 5$ μm to the position where the Y content becomes essentially constant. Thus, Fig. 6.30 shows that at the same corrosion temperature, the CMAS infiltration depth gradually increases with the corrosion time.

The CMAS remaining in the pores gradually dissolves the coating grains. Ca, Mg, Al, and Si diffuse from the CMAS to the YSZ grains, while the Zr and Y in the YSZ grains diffuse to the CMAS melt. As the solubility of Zr is much higher than that of Y in CMAS, the Y content in the YSZ grains that precipitate from the melts is lower than that in the YSZ grains before dissolution, resulting in a loss of Y in the coating grains that precipitate after cooling [3, 4]. A high amount of CMAS is found to cause large Y losses in the coating. As the amount of CMAS is high in regions near the surface and low in regions away from the surface, the amount of Y in the coating is low in regions near the surface and high in regions away from the surface after corrosion.

Assuming that the concentration of lost Y is directly proportional to ρ_{CMAS}, we can express the Y concentration in the coating after corrosion as follows:

$$c_Y = c_{Y0} - k_Y \rho_{\text{CMAS}} \tag{6.93}$$

where k_Y is a coefficient of proportionality and c_{Y0} is the Y concentration in the original coating. $c_{lost} = k_Y \rho_{CMAS}$ is the concentration of Y lost to the CMAS.

(3) Variation in the Y content of the ceramic and CMAS layers with the corrosion time

The experimental results in the previous section show that the Y content gradually increases along the CMAS infiltration depth during the high-temperature CMAS corrosion of the TBC. In this section, we focus on analyzing and investigating the patterns of change in the element content in the CMAS layer as well as in the corroded and uncorroded regions of the YSZ layer with temperature and time.

Figure 6.31 shows the distribution of the EP-measured content of each element in a TBC subjected to high-temperature CMAS corrosion. Figure 6.32 shows an enlarged view of the area enclosed by the red oval in Fig. 6.31. The distribution of the EP-measured content of each element shows that a small amount of Y appears in the CMAS layer, proving that during the high-temperature CMAS corrosion of the TBC, the Y in the ceramic layer is lost to the CMAS layer.

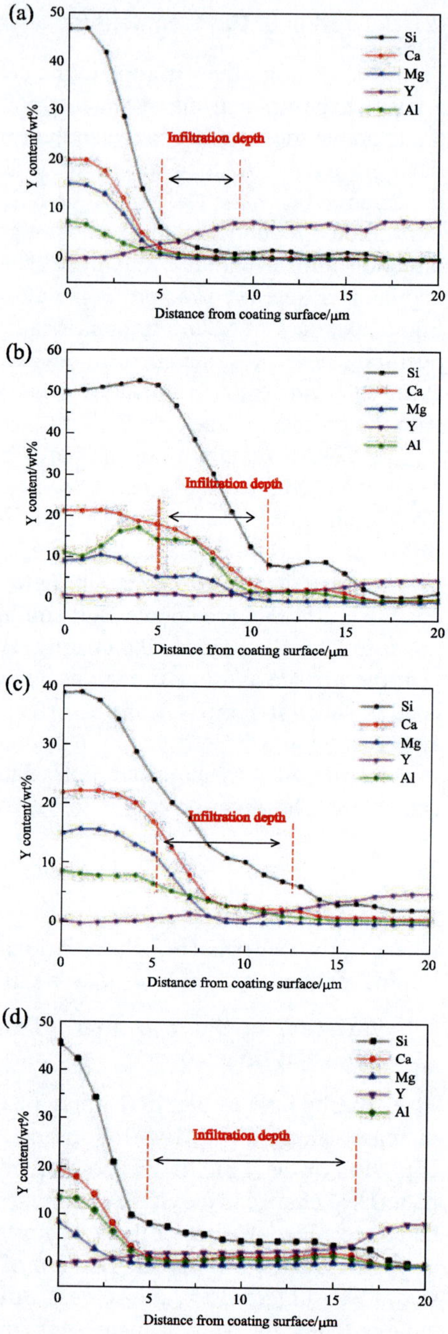

Fig. 6.30 Changes in the Y element content along the infiltration direction with the corrosion time. **a** 1200 °C and 0.5 h. **b** 1200 °C and 1 h. **c** 1200 °C and 2 h. **d** 1200 °C and 4 h.

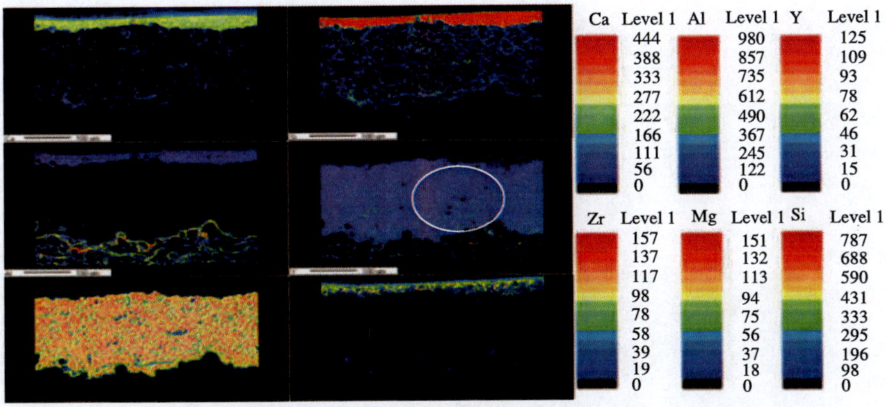

Fig. 6.31 Distribution of the EP-measured content of each element in the TBC subjected to high-temperature CMAS corrosion

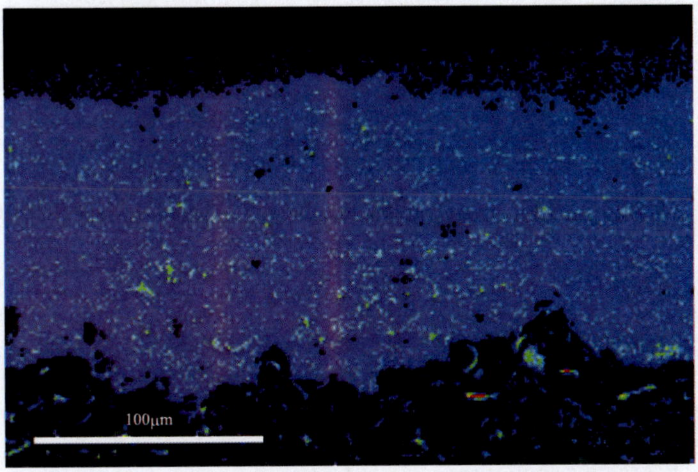

Fig. 6.32 Distribution of the EP-measured Y content in the TBC subjected to high-temperature CMAS corrosion (Close-up of the area enclosed by the red oval in Fig. 6.31)

Table 6.4 summarizes the EP measurements of the Y content in the CMAS layer (i.e., at locations 5 μm above the CMAS/ceramic interface), the corroded region of the YSZ layer (i.e., at locations 5 μm below the CMAS/ceramic interface), and the uncorroded region of the YSZ layer (i.e., the region below the ceramic layer), as shown in Fig. 6.33. The following can be derived from the data in Table 6.4. The Y content in the CMAS layer gradually increases with the corrosion time. At 4 h, the Y content in the CMAS layer increases to 1.68 wt.%. In addition, the Y content in the CMAS-corroded region of the YSZ layer gradually decreases with the corrosion time; specifically, it decreases from 7 wt.% to 3.9 wt.% in 1 h and to 3.4 wt.% in

Table 6.4 Y content in the CMAS and ceramic layers of the TBC subjected to CMAS corrosion at a high temperature of 1200 °C

Time/h	0.5	1	2	4
Y content in the CMAS block (wt.%)	1.0	1.1	1.4	1.7
Y content in the reacted region of the YSZ block (wt.%)	4.5	3.9	3.6	3.4
Y content in the unreacted region of the YSZ block (wt.%)	7.0	6.9	7.0	7.0

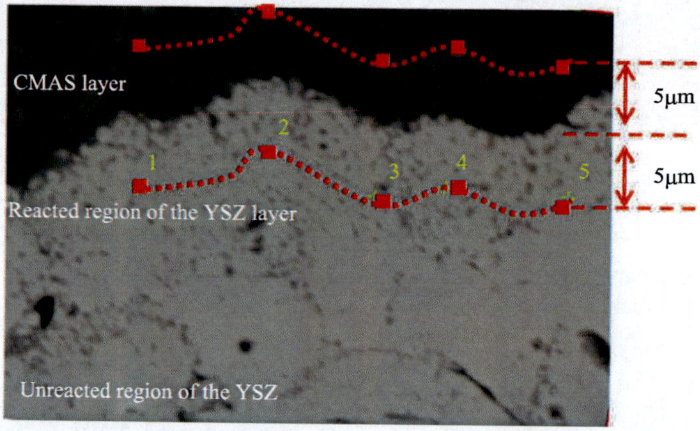

Fig. 6.33 Schematic diagram of the EP measurements of the Y content in each region

4 h. The Y content in the region of the YSZ layer uncorroded by the CMAS remains unchanged.

Figure 6.34 shows the variation in the Y content in each layer. The Y content in the YSZ layer gradually decreases, whereas the Y content in the CMAS layer gradually increases, sufficiently indicating that the Y in the YSZ layer is lost to the CMAS layer during the high-temperature CMAS corrosion of the TBC.

For a TBC, as the temperature rises, CMAS infiltrates the ceramic layer, after which the Y_2O_3 in the ceramic layer becomes dissolved in the CMAS. Due to the loss of the stabilizer Y_2O_3, the ZrO_2 transitions from the tetragonal phase to the monoclinic phase, resulting in a volume change of approximately 4 wt.%. Consequently, tensile stress is formed inside the ceramic layer, causing coating spallation [45–47].

(4) **Variation in the Y content of the CMAS layer with the corrosion temperature**

Now, let us experimentally examine the variations in the Y content of the CMAS layer with the corrosion temperature and time. TBCs were subjected to CMAS corrosion at three different high temperatures, namely, 1150, 1200, and 1250 °C. The specimens of the 1150 °C group were held at 1150 °C for 1, 2, 4, or 8 h. The specimens of the 1200 °C group were held at 1200 °C for 0.5, 1, 1.5, or 2 h. The specimens of the

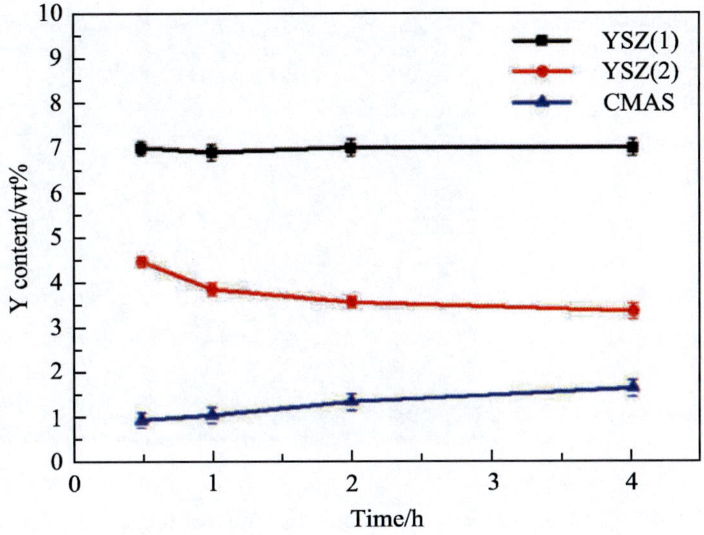

Fig. 6.34 Variation in the EP-measured Y content in each region

1250°C group were held at 1250 °C for 2, 5, 30, 60, or 90 min. After high-temperature corrosion, the specimens were allowed to cool to room temperature, then they were mounted and subsequently ground and polished. Then, the specimens were observed with an EP. Table 6.5 summarizes the EP measurements of the Y content at locations 5 μm above the CMAS/ceramic interface (as shown in Fig. 6.35) for the three groups of specimens subjected to corrosion at different temperatures.

The following can be directly observed in Table 6.5. At the same temperature, the Y content in the CMAS layer of the TBC subjected to high-temperature CMAS corrosion gradually increases with the corrosion time. Because the melting point of CMAS is above 1150 °C, only a small amount of CMAS infiltrates the ceramic layer at 1150 °C and, consequently, there is a small loss of Y. The Y contents in the CMAS layer reach 0.35 wt.% and 0.94 wt.% at 60 and 480 min of corrosion, respectively. The CMAS layer loses 1.04 wt.% and 1.68 wt.% Y after the coating is held at 1200 °C for 30 and 120 min, respectively. Because the melting point of the CMAS is below 1250 °C, at 1250 °C, the CMAS melts and rapidly enters the ceramic layer, resulting in a considerable increase in the loss of Y. The loss of Y reaches 1.61 wt.% and 3.35

Table 6.5 Y content in the CMAS layer of the TBC subjected to CMAS corrosion at different time points

Time/min (°C)	2	5	30	60	90	120	240	480
1150	/	/	/	0.35	/	0.62	0.86	0.94
1200	/	/	1.04	1.16	1.37	1.68	/	/
1250	1.61	1.80	1.88	2.38	3.35	/	/	/

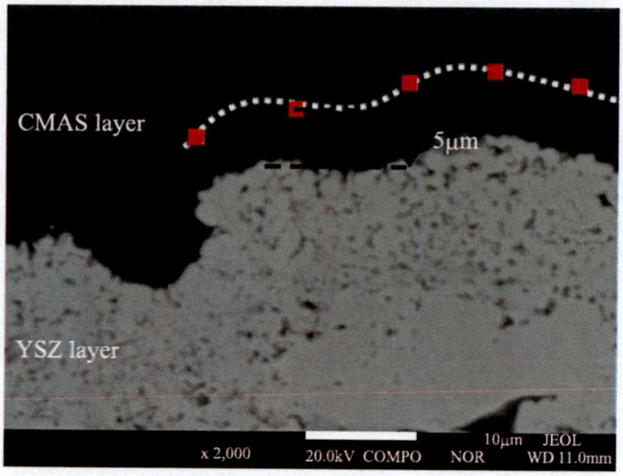

Fig. 6.35 Schematic diagram of the locations where the Y content in the CMAS block was measured with an EP

wt.% at 2 and 90 min of corrosion, respectively. Thus, at the same corrosion time, as the temperature rises, there is a gradual increase in the loss of Y. For example, at 60 min, the Y contents in the CMAS layer reach 0.35 wt.%, 1.16 wt.%, and 2.38 wt.% at 1150, 1200, and 1250 °C, respectively.

Figure 6.36 shows the curves plotted based on the data. We can more directly see the variation in the Y content with the temperature and time from Fig. 6.36. As shown in Fig. 6.36b, the change in temperature more significantly affects the change in the Y content. Thus, we conclude that the corrosion temperature has a more significant effect on the loss of Y in a TBC subjected to high-temperature CMAS corrosion than the corrosion time. An increase in the corrosion temperature results in a considerable change in the infiltrated amount of Y.

6.3 Phase-Structure Characterization and Phase-Field Theory for CMAS Corrosion of Coatings

6.3.1 XRD Characterization of the Evolution of the Coating Phase Structure

To analyze the effects of VA corrosion on the surface phase structure of the coating, we examined the surfaces of uncorroded original and corroded disk specimens with XRD at a scan rate of 5°/min, a scan angle of 10–90°, and a scan step size of 0.02°.

Figure 6.37 shows the surface XRD intensity plots of the original and corroded APS-YSZ specimens. The surface of the original 7YSZ specimen is composed primarily of t'-ZrO_2. A comparison shows that the XRD intensity plot is consistent with the Y_2O_3–ZrO_2 phase diagram [48]. Through a standard XRD phase identification procedure, we find that $ZrSiO_4$ is the primary compositional phase of the surface

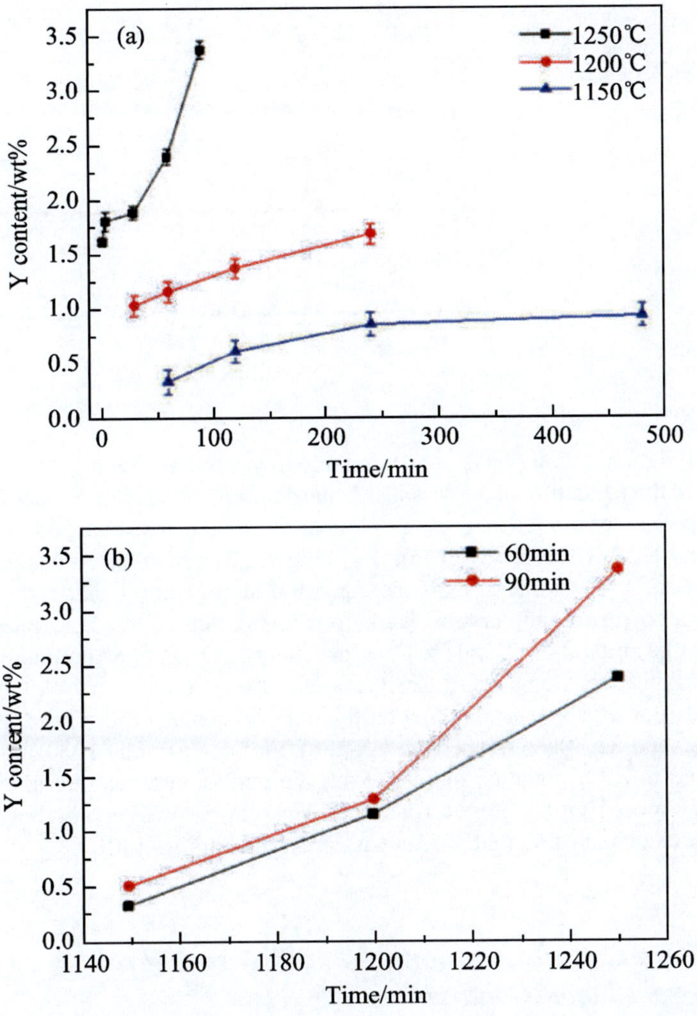

Fig. 6.36 Variation in the EP-measured Y content in the CMAS layer with **a** time and **b** temperature

of the VA-coated coating and that m-ZrO_2 and unreacted or phase-transformed t'-ZrO_2 are present in the surface of the VA-coated coating. A comparison of the XRD intensity plots of the original specimen and the specimen subjected to corrosion at 1150 °C for 8 h shows that the corrosion reaction mainly leads to the formation of $ZrSiO_4$ in the tetragonal phase. This suggests that the VA and YSZ undergo a chemical reaction, that the addition of VA significantly alters the chemical composition of the coating, and that the VA significantly corrodes and damages the coating. In addition, a significant presence of m-ZrO_2 is found among the reaction products. This occurs because after VA corrosion, the Y diffuses to the VA, causing some of t'-ZrO_2 to lose its stability and transform to m-ZrO_2. The transformation from the tetrahedral

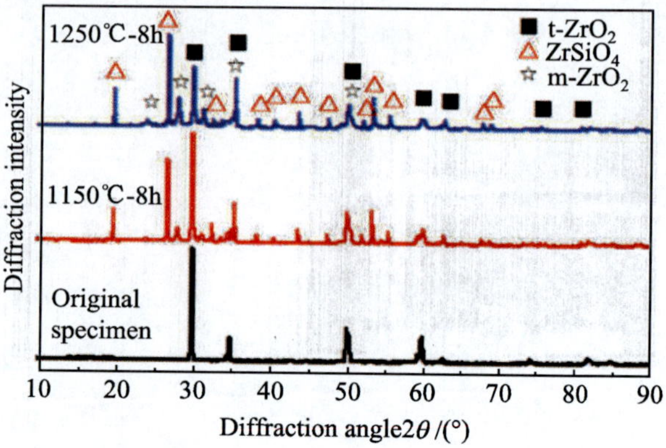

Fig. 6.37 Surface XRD intensity plots of the original and corroded APS-YSZ specimens

phase to the monoclinic phase is accompanied by a volume change of 4–5% [49], resulting in the formation of a stress inside the coating, which causes coating failure. The comparison of the XRD intensity plots of the specimens subjected to corrosion at 1150 and 1250 °C, respectively, for 8 h in Fig. 6.37 shows that the peak intensities of $ZrSiO_4$ and m-ZrO_2 increase as the corrosion temperature increases, suggesting that a high corrosion temperature leads to a high chemical reaction intensity and causes a large amount of t'-ZrO_2 to transform to m-ZrO_2. Rietveld refinement of the XRD data shows that the t-ZrO_2, $ZrSiO_4$, and m-ZrO_2 contents are 49.9 wt.%, 33.7 wt.%, and 16.4 wt.%, respectively, after 8 h of corrosion at 1150 °C and 29.1 wt.%, 38.2 wt.%, and 32.7wt.%, respectively, after 8 h of corrosion at 1250 °C. Both the $ZrSiO_4$ and m-ZrO_2 contents increase with the corrosion temperature, suggesting that the VA more significantly corrodes and damages the YSZ at a higher corrosion temperature, a conclusion that agrees with previous findings [50].

6.3.2 TEM Characterization of the Microstructural Evolution of Coatings

TEM is an important tool for the microstructural analysis and micromorphological observation of materials and is an important means for studying materials. To further investigate the micromorphology and microstructure of the TBC before and after corrosion, we subjected the original and corroded specimens of the TBC to TEM analysis. TEM is performed by transmitting a beam of electrons through a sample and measuring the resulting transmitted and diffracted beams and thus requires that the sample be very thin (typically 50–100 nm in thickness) [51]. The conventional TEM sample preparation technique of mechanical and ion thinning is complex and tedious and significantly damages the sample. In addition, this technique often randomly slices the specimen, making it difficult to find a specific area or location [51]. To address this problem, the focused ion beam (FIB) technique can be employed to directly extract a thin area of a TEM sample from the region that needs to be observed.

It is easy to use a FIB to prepare a TEM sample, and this technique minimally damages the sample and can directly extract a sample from the region of interest. To further analyze the microscopic mechanism of corrosion in coatings, we used a Tescan GAIA3 FIB-SEM system to extract TEM samples from the original coating specimen and the specimens subjected to corrosion at 1150 and 1250 °C, respectively, for 8 h. Subsequently, we observed the morphology of the samples with an FEI Titan G2 60–300 TEM system and used a standard procedure to identify selected-area electron diffraction patterns (SAEDPs).

Figure 6.38 shows TEM images of a sample cut from the original APS-YSZ specimen with a FIB. A micron-level strip-like morphology and interstrip microcracks can be seen in the coating in Fig. 6.38a. The EDS spectrum in Fig. 6.38b shows that Zr and O are the main components in area S1. According to a standard identification procedure, the SAEDP in Fig. 6.38c consists of t-ZrO_2 with a zone axis index of [011]. Figure 6.38d shows a high-resolution image of area S1. The fringe spacing is 0.29 nm, which is consistent with the interplanar crystal spacing in the (011) direction of t-ZrO_2. The scanning TEM (STEM) image in Fig. 6.38e shows an enlarged view of the area enclosed by the white solid-line box in Fig. 6.38a. Nanoscale voids can be seen in Fig. 6.38e. The interstrip microcracks and voids discussed earlier are the cause of a low thermal conductivity in the APS coating. The TEM results are consistent with the aforementioned XRD test results and Raman spectral results as well as previous findings showing that the original YSZ coating is composed mainly of the t-ZrO_2 phase.

Figure 6.39a shows a STEM image of the top reacted region of the sample obtained with a FIB from the coating specimen subjected to VA corrosion at 1150 °C for 8 h. Different from the original specimen, which shows a strip-like grain morphology, the specimen corroded at 1150 °C mainly displays a spherical grain morphology,

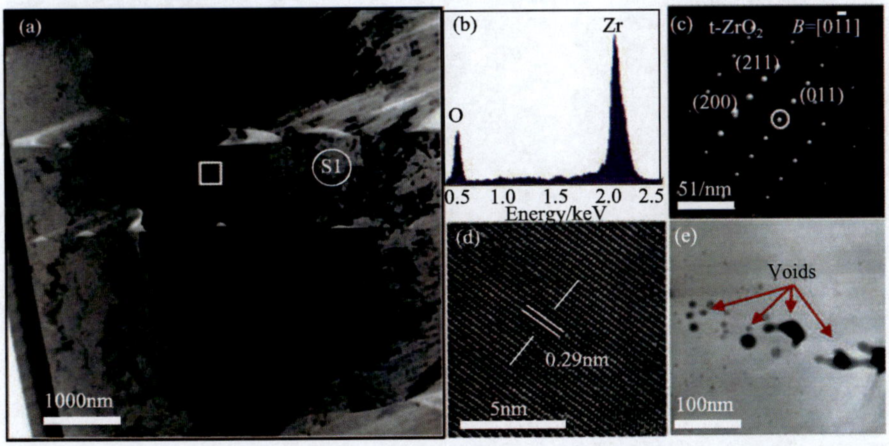

Fig. 6.38 **a** TEM image of a sample obtained with a FIB from the original APS-YSZ specimen. **b** EDS spectrum of area S1 in **a**. **c** SAEDP of area S1. **d** High-resolution image of area S1. **e** STEM image showing a close-up of the area enclosed by the white solid-line box in **a**

consistent with that shown in the SEM image in Fig. 6.19c. Figure 6.39b, c shows the EDS spectrum and SAEDP of area S2, respectively. According to the EDS spectrum, area S2 is composed primarily of Si, Al, Mg, and Fe (the Cu originates from the Cu support rings for the TEM sample), which is consistent with the composition of the VA. In addition, the SAEDP displays an amorphous halo. By comprehensively considering these factors, we determined that the phase in area S2 consists of the residual glassy VA that infiltrated the coating. Based on the EDS spectrum and SAEDP of area S3 shown in Fig. 6.39d, e, respectively, area S3 is composed of t-ZrO_2 with a zone axis index of [011]. This area of the YSZ is undamaged by the VA. Similarly, based on the EDS spectrum and SAEDP, area S4 consists of m-ZrO_2. Zr and Si are the main elements in the EDS spectrum in Fig. 6.39h; these elements originate from the YSZ and VA, respectively. Based on the SAEDP in 5.39(i), area S5 comprises $ZrSiO_4$. The TEM images of the VA corrosion and reaction region show that after infiltrating the coating, the VA reacts with the YSZ to form $ZrSiO_4$, accompanied by a transformation from t-ZrO_2 to m-ZrO_2. This finding is consistent with the XRD and Raman observations.

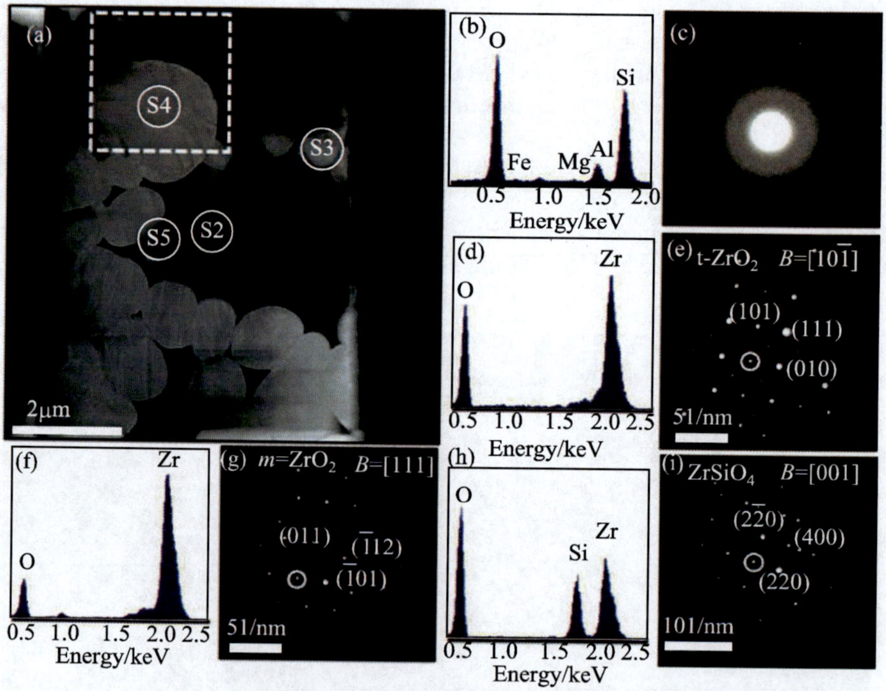

Fig. 6.39 **a** STEM image of the sample obtained with a FIB from the coating specimen subjected to VA corrosion at 1150 °C for 8 h. **b** EDS spectrum of area S2 in **a. c** SAEDP of area S2. **d** EDS spectrum of area S3. **e** SAEDP of area S3. **f** EDS spectrum of area S4. **g** SAEDP of area S4. **h** EDS spectrum of area S5. **i** SAEDP of area S5

6.3.3 Thermo–Mechano–Chemical Coupling Phase-Transformation Theory for Corroded Coatings During the Cooling Process

The SEM and TEM characterization of the corroded YSZ coating in the previous section shows that coating material that loses Y transforms from the tetragonal phase to the monoclinic phase. To study the mechanism of the phase transformation, we establish a thermo–mechano–chemical coupling phase-transformation theory for corroded coatings during the cooling process based on the phase-field method from the perspective of the laws of thermodynamics.

Let us assume that for a coating system, CMAS infiltration and the corrosion, dissolution, and chemical reactions of the coating cease during the cooling process; i.e., c_i and n are field variables that vary only with space. After cooling, in the dissolved region ($n = 1$), the YSZ ceramic grains that have lost Y transform from the tetragonal phase to the monoclinic phase at temperatures lower than the equilibrium temperature. Here, the phase-field method is employed to describe the phase-transformation process of the coating during cooling due to CMAS corrosion. Let η be the phase-transformation order parameter ($\eta = 0$ represents the tetragonal phase; $\eta = 1$ represents the monoclinic phase). η determines whether the dissolved region undergoes a phase transformation. For the sake of simplicity and convenience, we use $\eta \cdot n$ to denote the volume fraction of the monoclinic phase. Next, we derive the Ginzburg–Landau equation for the phase field from the perspective of the laws of thermodynamics.

Here, we introduce the following energy conservation equation for any domain V:

$$\int_V \dot{e}\mathrm{d}V = \int_\Omega (\boldsymbol{\sigma} \cdot \boldsymbol{n}) \cdot \boldsymbol{v}\mathrm{d}\Omega + \int_V \boldsymbol{f} \cdot \boldsymbol{v}\mathrm{d}V - \int_\Omega \boldsymbol{q} \cdot \boldsymbol{n}\mathrm{d}\Omega + \int_V r\mathrm{d}V \tag{6.94}$$

Equation (6.94) can be written in the following differential form:

$$\dot{e} = (\nabla \cdot \boldsymbol{\sigma}) \cdot \boldsymbol{v} + \boldsymbol{\sigma} : \nabla v + \boldsymbol{f} \cdot v - \nabla \cdot \boldsymbol{q} + r \tag{6.95}$$

Substitution of the force equilibrium equation into the above equation gives

$$\dot{e} = \sigma : \nabla v - \nabla \cdot \boldsymbol{q} + r = \boldsymbol{\sigma} : \boldsymbol{\varepsilon} - \nabla \cdot \boldsymbol{q} + r \tag{6.96}$$

The entropy increase rate in the second law of thermodynamics can be expressed as

$$\theta\dot{s} - \dot{e} - \frac{1}{\theta}\mathbf{q} \cdot \nabla\theta + \boldsymbol{\sigma} : \dot{\varepsilon} = \gamma\theta \tag{6.97}$$

By replacing the internal energy e with the Helmholtz free energy ψ, we obtain

$$-\dot{\psi} - \dot{\theta}s - \frac{1}{\theta}\mathbf{q} \cdot \nabla\theta + \boldsymbol{\sigma} : \dot{\boldsymbol{\varepsilon}} = \gamma\theta \tag{6.98}$$

We assume that ψ is a function of temperature θ, elastic strain $\boldsymbol{\varepsilon}_e$, and phase-transformation order parameter η as follows:

$$\psi = \psi(\theta, \boldsymbol{\varepsilon}_e, \eta) \tag{6.99}$$

By taking the derivative of Eq. (6.99) with respect to time, we obtain

$$\dot{\psi} = \frac{\partial\psi}{\partial\theta}\dot{\theta} + \frac{\partial\psi}{\partial\boldsymbol{\varepsilon}_e} : \dot{\varepsilon}_e + \frac{\partial\psi}{\partial\eta}\dot{\eta} \tag{6.100}$$

The total strain is decomposed as follows:

$$\boldsymbol{\varepsilon} = \boldsymbol{\varepsilon}_e + \overline{\varepsilon} + \boldsymbol{\varepsilon}_t(\eta) \tag{6.101}$$

where $\overline{\varepsilon}$ is the high-temperature residual strain and $\boldsymbol{\varepsilon}_t(\eta)$ is the phase-transformation strain. We assume that $\boldsymbol{\varepsilon}_t(\eta)$ has the following expression:

$$\boldsymbol{\varepsilon}_t(\eta) = O\left(3\eta^2 - 2\eta^3\right)\boldsymbol{I} \tag{6.102}$$

Substitution of Eqs. (6.100) and (6.101) into Eq. (6.98) yields

$$-\left(\frac{\partial\psi}{\partial\theta}\dot{\theta} + \frac{\partial\psi}{\partial\boldsymbol{\varepsilon}_e} : \dot{\varepsilon}_e + \frac{\partial\psi}{\partial\eta}\dot{\eta}\right) - \dot{\theta}s + \boldsymbol{\sigma} : [\dot{\varepsilon}_e + \dot{\varepsilon}_t(\eta)] - \frac{1}{\theta}\mathbf{q} \cdot \nabla\theta = \gamma\theta \tag{6.103}$$

The above equation can be further simplified to

$$\left(\boldsymbol{\sigma} - \frac{\partial\psi}{\partial\boldsymbol{\varepsilon}_e}\right) : \dot{\varepsilon}_e + \left(-\frac{\partial\psi}{\partial\theta} - s\right)\dot{\theta} + \left(\boldsymbol{\sigma} : \frac{\partial\boldsymbol{\varepsilon}_t}{\partial\eta} - \frac{\partial\psi}{\partial\eta}\right)\dot{\eta} - \frac{1}{\theta}\mathbf{q} \cdot \nabla\theta = \gamma\theta \tag{6.104}$$

In Eq. (6.104), $\boldsymbol{\varepsilon}_e$, θ, and η can take any values within their respective value ranges. To ensure that Eq. (6.104) is always nonnegative, the coefficients of its first three terms, namely,$\dot{\boldsymbol{\varepsilon}}_e$,$\dot{\theta}$, and η, must be zero. Then, we obtain the following generalized constitutive relations between entropy and temperature and between stress and strain:

$$s = -\frac{\partial\psi}{\partial\theta} \tag{6.105}$$

$$\boldsymbol{\sigma} = \frac{\partial\psi}{\partial\boldsymbol{\varepsilon}_e} \tag{6.106}$$

Substitution of Eqs. (6.105) and (6.106) into Eq. (6.104) yields

$$\gamma\theta = \left(\boldsymbol{\sigma} : \frac{\partial \boldsymbol{\varepsilon}_t}{\partial \eta} - \frac{\partial \psi}{\partial \eta}\right)\dot{\eta} - \frac{1}{\theta}\mathbf{q} \cdot \nabla\theta \tag{6.107}$$

The dissipation rates due to heat conduction and phase transformation are defined as follows:

$$\theta\gamma_\theta = -(\mathbf{q} \cdot \nabla\theta)/\theta \tag{6.108}$$

$$\theta\gamma_\eta = \left(\boldsymbol{\sigma} : \frac{\partial \boldsymbol{\varepsilon}_t}{\partial \eta} - \frac{\partial \psi}{\partial \eta}\right)\dot{\eta} = F_\eta \dot{\eta} \tag{6.109}$$

Because heat flow and phase transformation are both irreversible processes, Eqs. (6.108) and (6.109) are both nonnegative. In Eq. (6.109), $F_\eta = \boldsymbol{\sigma} : \frac{\partial \boldsymbol{\varepsilon}_t}{\partial \eta} - \frac{\partial \psi}{\partial \eta}$ is the driving force for phase transformation. Assuming that $\dot{\eta}$ in Eq. (6.109) is directly proportional to its coefficient F_η, we have

$$\dot{\eta} = LF_\eta = L\left(\boldsymbol{\sigma} : \frac{\partial \boldsymbol{\varepsilon}_t}{\partial \eta} - \frac{\partial \psi}{\partial \eta}\right) \tag{6.110}$$

Equation (6.110) is the evolution equation for the phase-transformation order parameter, i.e., the Ginzburg–Landau equation. We establish the following Helmholtz free energy function:

$$\begin{aligned} \psi \;=\; & \frac{1}{2}\lambda(tr\boldsymbol{\varepsilon})^2 + G\boldsymbol{\varepsilon} : \boldsymbol{\varepsilon} - 2G\overline{\boldsymbol{\varepsilon}} : \boldsymbol{\varepsilon} - \lambda tr\overline{\boldsymbol{\varepsilon}} tr\boldsymbol{\varepsilon} - 3K\boldsymbol{\varepsilon}_t(\eta) : \boldsymbol{\varepsilon} \\ & + \frac{1}{2}\beta'(\nabla\eta)^2 + z(\theta - \theta_c)(3\eta^2 - 2\eta^3) \end{aligned} \tag{6.111}$$

where θ_c is the equilibrium temperature between the tetragonal and monoclinic phases of YSZ, z is a constant of proportionality, and β' is the coefficient of the gradient energy. On the right-hand side of Eq. (6.111), the first two terms represent the elastic strain energy, the third and fourth terms represent the coupling term between the residual strain and total strain during the high-temperature corrosion process, the fifth term represents the coupling term between the phase-transformation strain and total strain, the sixth term represents the gradient energy of the phase-transformation order parameter, and the final term represents the volume free energy in the monoclinic and tetragonal phases.

We define the shear modulus and bulk modulus by the following expressions:

$$G = \left(G_{01} + (G_1 - G_{01})\left(3\eta^2 - 2\eta^3\right)\right)(1 - n) + G_{02}n \tag{6.112}$$

$$K = \left(K_{01} + (K_1 - K_{01})\left(3\eta^2 - 2\eta^3\right)\right)(1 - n) + K_{02}n \tag{6.113}$$

where the subscripts 02, 01, and 1 indicate the CMAS-dissolved region of the coating, the tetragonal phase of the original coating, and the monoclinic phase, respectively.

Substitution of Eq. (6.111) into Eq. (6.106) yields

$$\boldsymbol{\sigma} = \frac{\partial \psi}{\partial \boldsymbol{\varepsilon}_e} = \boldsymbol{\sigma}_0 + \boldsymbol{I}\lambda tr\boldsymbol{\varepsilon} + 2G\boldsymbol{\varepsilon} - 2G\overline{\boldsymbol{\varepsilon}} - \boldsymbol{I}\lambda tr\overline{\boldsymbol{\varepsilon}} - 3K\boldsymbol{\varepsilon}_t(\eta) \tag{6.114}$$

The above equation is the inverse constitutive relation. By taking the inverse of Eq. (6.114), we obtain the following constitutive relation:

$$\boldsymbol{\varepsilon} = \frac{1}{2G}\left((\boldsymbol{\sigma} - \boldsymbol{\sigma}_0) - \frac{\lambda}{2G + 3\lambda}\boldsymbol{I}\mathrm{tr}(\boldsymbol{\sigma} - \boldsymbol{\sigma}_0)\right) + \overline{\boldsymbol{\varepsilon}} + \boldsymbol{\varepsilon}_t(\eta) \tag{6.115}$$

By substituting Eqs. (6.111) and (6.114) into Eq. (6.110) and neglecting the higher-order strain terms, we obtain

$$\dot{\eta} = L\big((3K_0 O\mathrm{tr}\boldsymbol{\varepsilon} - z(\theta - \theta_c))(6\eta - 6\eta^2) + \beta'\nabla^2\eta\big) \tag{6.116}$$

Equation (6.116) is the Ginzburg–Landau equation that governs the evolution of the phase-transformation order parameter.

Figure 6.40 shows the metastable equilibrium curve for the transformation from the tetragonal phase to the monoclinic phase expressed in terms of T_0. Here, we fit the metastable equilibrium curve to a quadratic function as follows:

$$\theta_c = \theta_{c1}c_Y^2 + \theta_{c2} \tag{6.117}$$

where c_Y is the concentration of residual Y in the coating.

By substituting Eq. (6.93) into Eq. (6.117) and then substituting the resulting equation into Eq. (6.116), we obtain

$$\dot{\eta} = L\Big(\big(3K_0 O\mathrm{tr}\boldsymbol{\varepsilon} - z\big(\theta - \big(\theta_{c1}(c_{Y0} - k_Y\rho_{CMAS})^2 + \theta_{c2}\big)\big)\big)(6\eta - 6\eta^2) + \beta'\nabla^2\eta\Big) \tag{6.118}$$

As seen from Eq. (6.118), the evolution of the order parameter is related to the temperature of the system and the amount of CMAS that has infiltrated the coating during the cooling process.

Next, we employ the FEM to solve the evolution equation for each field variable during the phase-transformation process. We divide the simulation process into two stages, namely, a high-temperature corrosion stage and a cooling and phase-transformation stage. During the high-temperature corrosion process, the temperature of the system, θ_h, remains constant at 1250 °C. The field variables needed to be calculated for this process are the CMAS concentration c_i, the extent of the corrosion reaction n, and the concentration of the stabilizer Y c_Y. See Eqs. (6.87) and (6.90)

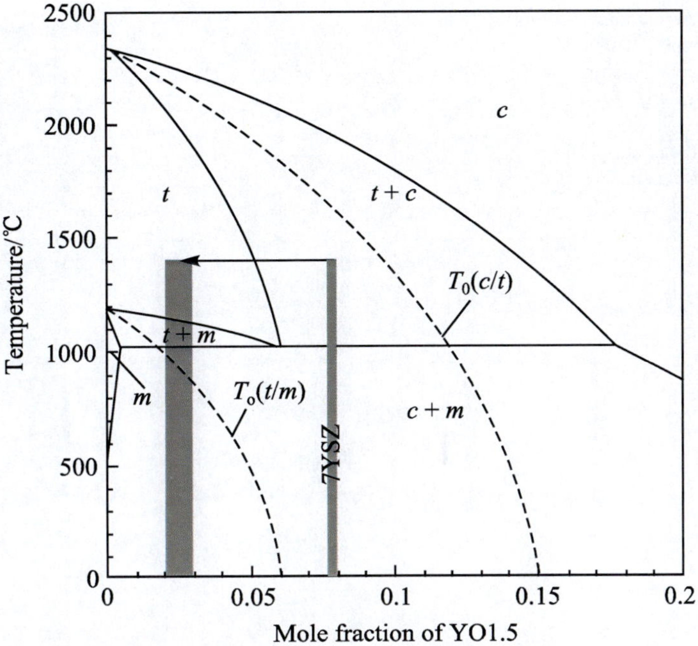

Fig. 6.40 Partial phase diagram of YSZ [3]

for the weak forms of the governing equations for CMAS infiltration and n, respectively, and Eq. (6.93) for the governing equation for Y. The field variables needed to be calculated for the cooling process are the strain, stress, and phase-transformation fields. See Eq. (6.91) for the weak form of the equilibrium equation. The weak form of the governing equation for the phase-transformation order parameter is expressed as follows:

$$\delta\eta \cdot \dot{\eta} - L \cdot \delta\eta\big(3K_0 O\mathrm{tr}\boldsymbol{\varepsilon} - z\big(\theta - \big(\theta_{c1}(c_{YO} - k_Y \rho_{CMAS})^2 + \theta_{c2}\big)\big)\big)(6\eta - 6\eta^2) + \beta' L\delta\nabla\eta \cdot \nabla\eta = 0 \tag{6.119}$$

where $\delta\eta$ is a trial function of the order parameter η. The model in Fig. 6.41 has dimensions of 50 μm × 200 μm.

The boundary condition for c_i during the high-temperature corrosion stage is identical to that shown in Eq. (6.51). At the initial time point, $c_i = 0$, and $n = 0$. The spatial distributions of the c_i and n inside the coating during the cooling and phase-transformation stage are identical to those at the end of the high-temperature corrosion stage. We assume that θ_h is linearly related to the time t through the following equation:

$$\theta = \theta_h - k_m \cdot t \tag{6.120}$$

Fig. 6.41 A corroded coating model. **a** The red solid line represents the CMAS infiltration source, and the grayish-white part represents the coating. **b** Meshing of the coating model

At the initial time point of the cooling stage, both the order parameter and displacement inside the coating are zero. In terms of the boundary conditions, $\eta = 1$ at edge AB, and the displacement at edge DC is fixed at zero.

As shown in Fig. 6.41b, the FE model is discretized with 10,000 domain elements and 500 boundary elements.

See Tables 6.3 and 6.6 for the model parameters for the high-temperature stage and the cooling stage, respectively.

The CMAS that has infiltrated the coating gradually dissolves and corrodes the ceramic layer, while the dissolved stabilizer Y in the coating is lost to the CMAS. Figure 6.42 shows the distributions of the concentration of the Y lost from the coating at different corrosion times. As the corrosion time increases, there is an increase in the amount of Y lost from the coating to the CMAS. According to Eq. (6.93), a higher CMAS concentration leads to a higher concentration of Y lost from the coating. Therefore, the concentration of Y lost from the coating is high at the top and low at the bottom; that is, its distribution is consistent with that of the CMAS concentration.

Figure 6.43a, b shows the distributions of the phase-transformation order parameter η and the volume fraction of the monoclinic phase, respectively, at room temperature after 600 min of corrosion. In Fig. 6.43a, the regions with and without phase transformation are shown in red and blue, respectively; these two regions are separated by a horizontal and straight transition region, i.e., the phase boundaries. Because η determines whether phase transformation occurs in the dissolved region and the extent of the corrosion and dissolution of the coating, n, is high in regions close

Table 6.6 Simulation parameters used in the model at the cooling stage [6, 34–38] (Mamivand et al. 2015)

Parameter	Value
Mass density of the YSZ	6 g/cm^3
Bulk modulus of the tetragonal phase	23.81 GPa
Shear modulus of the tetragonal phase	16.39 GPa
Bulk modulus of the monoclinic phase	187.95 GPa
Shear modulus of the monoclinic phase	94.85 GPa
Molar fraction of Y in the original coating, c_{Y0}	0.08
θ_{c1}	−256,167 K
θ_{c2}	1211.52 K
Coefficient of proportionality, O	0.01
z	5.697×10^5 J/K·m^3
Kinetic coefficient, L	2 m^3/J·s
Coefficient of the gradient energy, β	2.5×10^{-9} J/m
k_Y	1.057×10^{-5} 1/s
Cooling rate of the system, k_m	0.1 K/s

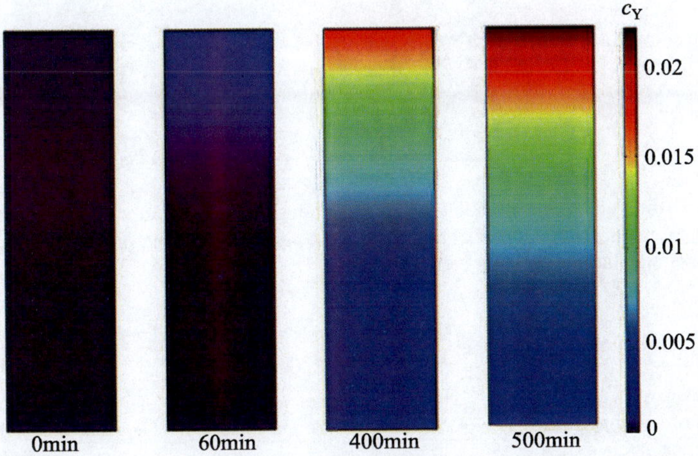

Fig. 6.42 Distributions of the molar fraction of the lost Y precipitating from the coating at different corrosion times

to the surface and low in regions away from the surface, the volume fraction of the monoclinic phase, $\eta \cdot n$, is high in regions close to the surface and low in regions away from the surface, as shown in Fig. 6.43b. During the cooling process, the coating material that has lost Y is distributed from the tetragonal phase to the monoclinic phase. Figure 6.43b shows the distributions of the volume fraction of the monoclinic phase at different cooling times after 600 min of corrosion. As seen in Fig. 6.44, the monoclinic phase begins to extend from the surface of the coating to its interior at 400 °C. As the temperature of the system decreases, there is a continuous increase in the amount of monoclinic phase inside the coating. At a given cooling temperature, the amount of monoclinic phase is high in regions close to the surface of the coating and low in regions away from the surface of the coating.

The coating undergoes a volume expansion when transforming from the tetragonal phase to the monoclinic phase. As shown in Fig. 6.45, the evolutionary pattern of the volumetric strain in the coating with time is consistent with that of the volume

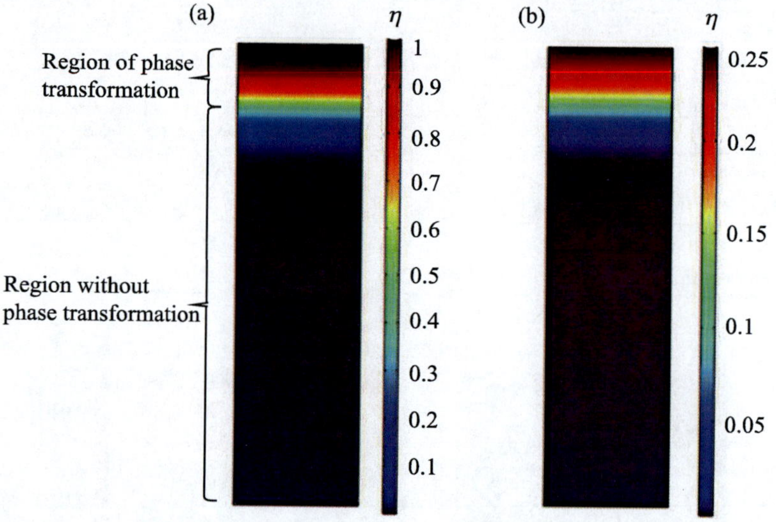

Fig. 6.43 Distributions of the monoclinic phase at room temperature after 600 min of corrosion. **a** Distribution of the order parameter of the monoclinic phase. **b** Distribution of the volume fraction of the monoclinic phase in the ceramic coating

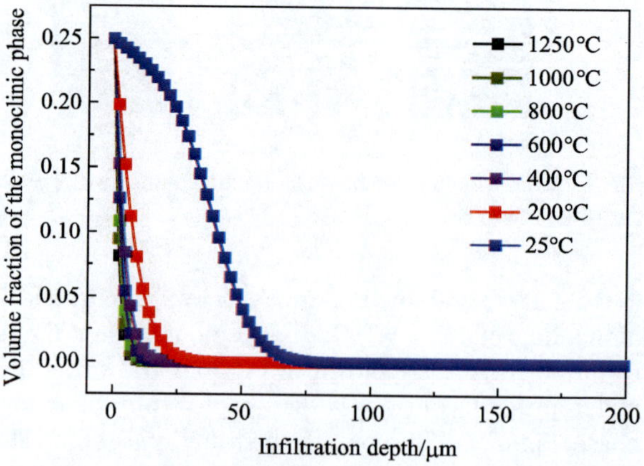

Fig. 6.44 Distributions of the volume fraction of the monoclinic phase at different time points of cooling after 600 min of corrosion

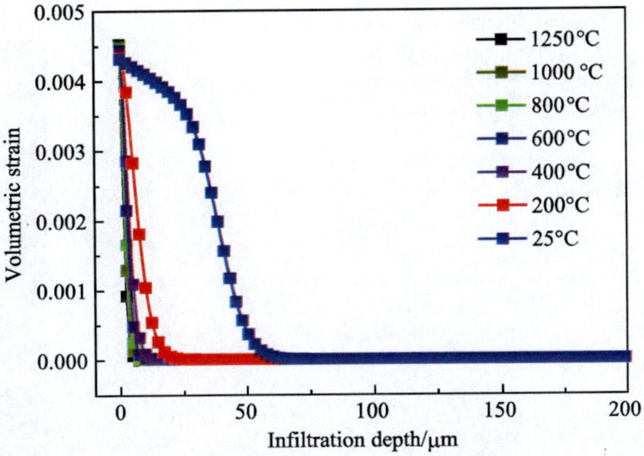

Fig. 6.45 Distributions of the volumetric strain at different time points of cooling after 600 min of corrosion

fraction of the monoclinic phase. At a given cooling temperature, the volumetric strain is distributed in a graded manner along the infiltration depth. As the temperature of the system decreases, the volumetric strain increases continuously in the region where phase transformation occurs (i.e., the red region in Fig. 6.43) but is zero in the region where no phase transformation occurs ($\eta = 0$). During the cooling process, the phase transformation is accompanied by local deformation, which results in stress concentrations in the coating in the region where phase transformation occurs. Figure 6.46 shows the distribution of the stress component σ_{yy} in the coating at room temperature after 600 min of corrosion. An out-of-plane tensile stress is formed at the interface between the region where phase transformation occurs and the region where no phase transformation occurs. The tensile stress, when sufficiently large, promotes coating delamination.

6.4 Summary and Outlook

This chapter investigates the problem of the thermo–mechano–chemical coupling failure of TBCs subjected to CMAS corrosion with the use of a combination of experimental characterization and theoretical modeling. With regard to the theoretical model, we divide the CMAS corrosion process into two stages, namely, a high-temperature corrosion stage and a cooling and phase-transformation stage. For the high-temperature stage, we establish an infiltration–chemical reaction–inelasticity coupling model of CMAS corrosion by combining the force equilibrium equation and the CMAS concentration conservation equation from the perspective of the laws of thermodynamics. For the cooling and phase-transformation stage, we define an order parameter based on the phase-field method to describe the microstructural evolution of a corroded coating that transforms from the tetragonal phase to the monoclinic phase during the cooling process. Based on the effects of the loss of Y at high temperatures, we derive a Ginzburg–Landau equation from the perspective of

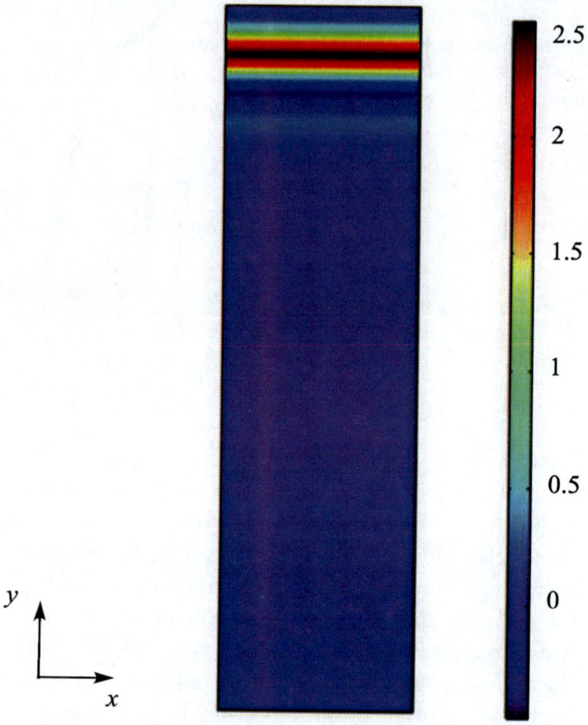

Fig. 6.46 Distribution of the stress component σ_{yy} in the coating at room temperature after 600 min of corrosion

the laws of thermodynamics. Experimentally, we characterize the phase structure and microstructural evolution of coatings subjected to corrosion by SEM, TEM, XRD, and Raman spectroscopy and verify the correctness of the theoretical model.

A key problem in the study of thermo–mechano–chemical coupling failure is the way the field variables affect each other during the evolutionary process. This chapter focuses on discussing the effects of the chemical and strain coupling terms in the governing equations on the evolution of the field variables at the high-temperature stage. In addition, this chapter presents a key coupling parameter A_1 and discusses how it affects each field variable through the coupling terms. However, this parameter remains unknown. In future research, a simple experiment will be performed to determine this coupling parameter.

References

1. Richards LA. Capillary conduction of liquids through porous mediums. Physics. 1931;1:318–33.
2. Pasandideh-Fard M, Qiao YM, Chandra S, Mostaghimi J. Capillary effects during droplet impact on a solid surface. Phys Fluids. 1996;8:650–9.

3. Drexler JM, Shinoda K, Ortiz AL, Li D, Vasiliev AL, Gledhill AD, Sampath S, Padture NP. Air-plasma-sprayed thermal barrier coatings that are resistant to high-temperature attack by glassy deposits. Acta Mater. 2010;58:6835–44.
4. Vidal-Setif MH, Rio C, Boivin D, et al. Microstructural charaif the condition is a consequent (statement cterization of the interaction between 8YPSZ (EB-PVD) thermal barrier coatings and a synthetic CAS. Surf Coat Technol. 2014;239:41–8.
5. Krause AR, Garces HF, Dwivedi G, Ortiz AL, Sampath S, Padture NP. Calcia-magnesia-alumino-silicate (CMAS)-induced degradation and failure of air plasma sprayed yttria-stabilized zirconia thermal barrier coatings. Acta Mater. 2016;105:355–66.
6. Garces HF, Senturk BS, Padture NP. In situ Raman spectroscopy studies of high-temperature degradation of thermal barrier coatings by molten silicate deposits. Scripta Mater. 2014;76:29–32.
7. Shan X, Zou Z, Gu L, Yang L, Guo F, Zhao X, Xiao P. Buckling failure in air-plasma sprayed thermal barrier coatings induced by molten silicate attack. Scripta Mater. 2016;113:71–4.
8. Mercer C, Faulhaber S, Evans AG, Darolia R. A delamination mechanism for thermal barrier coatings subject to calcium–magnesium–alumino-silicate (CMAS) infiltration. Acta Mater. 2005;53(4):1029–39.
9. Vasiliev AL, Padture NP. Coatings of metastable ceramics deposited by solution-precursor plasma spray: II. Ternary ZrO2–Y2O3–Al2O3 system. Acta Materialia. 2006;54(18):4921–28.
10. Wellman R, Whitman G, Nicholls JR. CMAS corrosion of EB PVD TBCs: identifying the minimum level to initiate damage. Int J Refract Metal Hard Mater. 2010;28(1):124–32.
11. Lee K-I, Wu LT, Wu RT, Xiao P. Mechanisms and mitigation of volcanic ash attack on yttria stabilized zirconia thermal barrier coatings. Surf Coat Technol. 2014;260:68–72.
12. Washburn EW. The dynamics of capillary flow. Phys Rev. 1921;17(3):273–83.
13. Kulkarni A, Vaidya A, Goland A, Sampath S, Herman H. Processing effects on porosity-property correlations in plasma sprayed yttria-stabilized zirconia coatings. Mater Sci Eng A. 2003;359(1–2):100–11.
14. Majumdar A. Role of fractal geometry in the study of thermal phenomena. Annu Rev Heat Transfer. 1992;4:51–110.
15. Cai C, Chang S, Zhou Y, Yang L, Zhou G, Wang Y. Microstructure characteristics of EB-PVD YSZ thermal barrier coatings corroded by molten volcanic ash. Surf Coat Technol. 2016;286:49–56.
16. Cai J, Perfect E, Cheng CL, Xu X. Generalized modeling of spontaneous imbibition based on Hagen-Poiseuille flow in tortuous capillaries with variably shaped apertures. Langmuir Acs J Surfaces Colloids. 2014;30(18):5142–51.
17. Lucas R. Ueber das zeitgesetz des kapillaren aufstiegs von flüssigkeiten. Kolloid-Zeitschrift. 1918;23(1):15–22.
18. Fredlund DG, Xing A. Equations for the soil-water characteristic curve. Can Geotech J. 1994;31(4):521–32.
19. Sampath S, Schulz U, Jarligo MO, Kuroda S. Processing science of advanced thermal-barrier systems. MRS Bull. 2012;37(10):903–10.
20. Cai J, Yu B. A discussion of the effect of tortuosity on the capillary imbibition in porous media. Transp Porous Media. 2011;89(2):251–63.
21. Luo L, Yu B, Cai J, Zeng X. Numerical simulation of tortuosity for fluid flow in two-dimensional pore fractal models of porous media. Fractals-Complex Geom Patterns Scaling Nat Soc. 2014;22(4):1450015.
22. Khan Y, Vázquez-Leal H, Faraz N. An auxiliary parameter method using Adomian polynomials and Laplace transformation for nonlinear differential equations. Appl Math Model. 2013;37(5):2702–8.
23. Dyke PPG. An introduction to Laplace transforms and Fourier series. Springer;2001.
24. Wang N, Li C, Yang L, Zhou Y, Zhu W, Cai C. Experimental testing and FEM calculation of impedance spectra of thermal barrier coatings: effect of measuring conditions. Corros Sci. 2016;107:155–71.

25. Yang G, Yang H, Guo X, Li W. Effect of mass ratio of CaO to MgO on crystallization of CaO–MgO–Al_2O_3–SiO_2 glass ceramics. J Chin Ceram Soc. 2010;38(11):2045–9.
26. Yin BB, Liu ZY, Yang L, Wu RT, Zhou YC. Factors influencing the penetration depth of molten volcanic ash in thermal barrier coatings: theoretical calculation and experimental testing. Results Phys. 2019:102169.
27. Kim J, Dunn MG, Baran AJ. Deposition of volcanic materials in the hot sections of two gas turbine engines. J Eng Gas Turbines Power. 1993;115(3):641–51.
28. Nie X. Effects of compositional changes of CMAS on the microstructure of EBPVD thermal barrier coatings. Xiangtan University;2014.
29. Lee KI, Wu LT, Wu RT, Xiao P. Mechanisms and mitigation of volcanic ash attack on yttria stabilized zirconia thermal barrier coatings. Surf Coat Technol. 2014;260:68–72.
30. Zang H. The development of methods about the measurement of contact angle. Fujian Anal Test. 2006;15(2):47–8.
31. Cao L, Hu HH, Gao D. Design and fabrication of micro-textures for inducing a superhydrophobic behavior on hydrophilic materials. Langmuir Acs J Surfaces Colloids. 2007;23(8):4310–4.
32. Liu J, Feng X, Wang G, Yu S. Mechanisms of superhydrophobicity on hydrophilic substrates. J Phys Condens Matter. 2007;19(35):1431–5.
33. Chen X. Calcium–magnesium–alumina–silicate (CMAS) delamination mechanisms in EB-PVD thermal barrier coatings. Surf Coat Technol. 2006;200(11):3418–27.
34. Abubakar AA, Akhtar SS, Arif AFM. Phase field modeling of V_2O_5 hot corrosion kinetics in thermal barrier coatings. Comput Mater Sci. 2015;99:105–16.
35. Wiesner VL, Bansal NP. Mechanical and thermal properties of calcium–magnesium aluminosilicate (CMAS) glass. J Eur Ceram Soc. 2015;35(10):2907–14.
36. Bak T, Nowotny J, Prince K, Rekas M, Sorrell CC. Grain boundary diffusion of magnesium in zirconia. J Am Ceram Soc. 2002;85:2244–50.
37. Chiodelli G, Magistris A, Scagliotti M, Parmigiani F. Electrical properties of plasma-sprayed yttria-stabilized zirconia films. J Mater Sci. 1988;23:1159–63.
38. Ulm F-J, Torrenti J-M, Adenot F. Chemo-poro-plasticity of calcium leaching in concrete. J Eng Mech. 1999;125:1200–11.
39. Drexler JM, Gledhill AD, Shinoda K, et al. Jet engine coatings for resisting volcanic ash damage. Adv Mater. 2011;23(21):2419–24.
40. Peng H, Wang L, Guo L, et al. Progress Nat Sci Mater Int. 2012;22 (5):461–7.
41. Guo HB, Gong SK, Khor KA, Xu HB. Effect of thermal exposure on the microstructure and properties of EB-PVD gradient thermal barrier coatings. Surf Coat Technol. 2003;168(1):23–9.
42. Poirier DR, Geiger GH. Transport phenomena in materials processing. Minerals, Metals & Materials Society;1994.
43. Bai Y, Han ZH, Li HQ, et al. Structure–property differences between supersonic and conventional atmospheric plasma sprayed zirconia thermal barrier coatings. Surf Coat Technol. 2011;205(13):3833–9.
44. Zhao X, Wang X, Xiao P. Sintering and failure behaviour of EB-PVD thermal barrier coating after isothermal treatment. Surf Coat Technol. 2006;200(20):5946–55.
45. Poza P, Gómez-García J, Múnez CJ. TEM analysis of the microstructure of thermal barrier coatings after isothermal oxidation. Acta Mater. 2012;60(20):7197–206.
46. Yu Q, Rauf A, Wang N, et al. Thermal properties of plasma-sprayed thermal barrier coating with bimodal structure. Ceram Int. 2011;37:1093–9.
47. Wellman R, Nicholls J. A review of the erosion of thermal barrier coatings. J Phys D Appl Phys. 2007;40(16):293–305.
48. Scott HG. Phase relationships in the zirconia-yttria system[J]. J Mater Sci. 1975;10(9):1527–35.
49. Cutler RA, Reynolds JR, Jones A. Sintering and characterization of polycrystalline monoclinic, tetragonal, and cubic zirconia. J Am Ceram Soc. 2010;75(8):2173–83.
50. Chen X, Wang R, Yao N, et al. Foreign object damage in a thermal barrier system: mechanisms and simulations. Mater Sci Eng A. 2003;352(1–2):221–31.

51. Zhu H. Research and test methods for materials science. Southeast University Press (Nanjing);2008.
52. Chang HL, Cai CY, Wang Y, et al. Calcium-rich CMAS corrosion induced microstructure development of thermal barrier coatings. Surf Coat Technol. 2017;324:577–84.

Chapter 7
Erosion Failure Mechanisms of TBCs

Erosion failure in a thermal barrier coating (TBC) refers to its thinning, cracking, and spallation under repeated exposure to high-temperature flow carrying hard particles and was once considered the second most important factor inducing coating spallation. However, at an excessively high gas temperature, these particles melt and result in calcium–magnesium aluminosilicate (CMAS) corrosion. As gas temperatures in engines become increasingly high, CMAS corrosion has gradually become the primary risk factor causing TBC spallation. As China is home to a large number of deserts and frequently experiences by wind-borne sand and haze conditions, erosion failure remains an issue that should not be overlooked in airplanes during take-off and low-altitude flight.

This chapter presents erosion failure phenomena in TBCs by focusing on yttria-stabilized zirconia (YSZ) TBC materials and two TBC preparation processes, namely, atmospheric plasma spraying (APS) and electron beam-physical vapor deposition (EB-PVD), in combination with new TBC materials (e.g., rare-earth doped and La-series materials) and new and improved preparation processes (e.g., plasma spraying-physical vapor deposition (PS-PVD) and laser melting) and introduces experimental, numerical simulation, and theoretical methods for studying erosion failure in TBCs, with the goal of facilitating an understanding of the erosion failure mechanisms of TBCs.

7.1 Erosion Failure Phenomena in TBCs

Erosion failure in a TBC is generally understood by using a special wind-tunnel device, a gas spray gun, or an industrial combustion device to spray one or multiple types of hard particles onto its ceramic surface at a certain angle, velocity, and temperature while observing the failure mode of its ceramic layer and analyzing the factors affecting the erosion performance of its ceramic layer.

Y. Zhou et al., *Thermal Barrier Coatings: Failure Theory and Evaluation Technology*,
https://doi.org/10.1007/978-981-19-2723-2_7

7.1.1 Failure Phenomena in TBCs

As shown in Fig. 7.1, erosion failure in a TBC refers to the phenomenon in which under repeated impacts from flow carrying hard particles, the ceramic layer in the area of action first becomes dense, followed by thinning, crack formation, or even coating spallation. These hard particles originate mainly from two sources. (1) Particles formed within the engine. These particles can be either carbon particles formed during the combustion process or particles formed from the erosion of the inner walls of the combustion chamber and the turbine blades of the engine. (2) External objects sucked into the gas turbine, such as sand particles, dust, and metallic (e.g., Al) particles. In some studies, TBC failures caused by particles of the first and second sources are referred to as erosion and impact damage by external objects, respectively [1–6]. However, most studies refer to both types of damage as erosion and view the differences in damage caused by particles of different natures as different failure mechanisms [7–10].

7.1.2 TBC Erosion Rate

Generally, the ratio of the decrease in the mass of the ceramic layer, Δw_{TBC}, to the mass of the particles involved in erosion, w_P, is defined as the erosion rate of the TBC, w_e:

$$w_e = \Delta w_{TBC}/w_P$$

The erosion rate depends on the properties of both the impacting particles and ceramic layer and is affected primarily by the mass (size and density), velocity, and angle of the particles as well as the physical and mechanical parameters (e.g., density, elastic modulus, strength, hardness, and fracture toughness) of the ceramic layer. In addition, the erosion rate is related to the temperatures of the TBC and impacting particles during erosion [1–6].

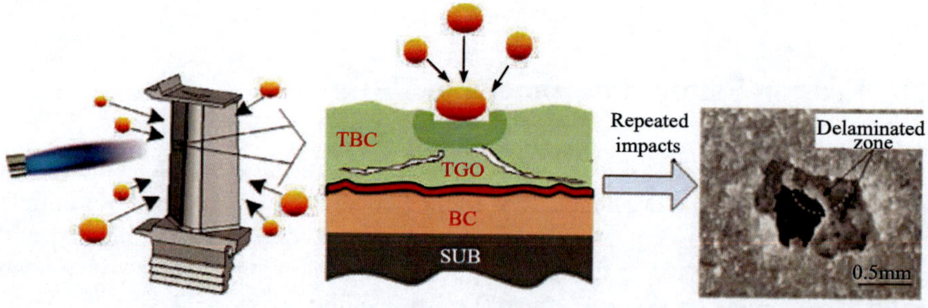

Fig. 7.1 Erosion damage and spallation of a TBC [11]

7.1.3 Comparison of the Erosion Performance of Various Coatings

YSZ TBCs, which are the currently most commonly used TBC systems, and EB-PVD [3–6, 13, 14] and APS [11, 14–22], which are two typical TBC preparation processes, are the focus of research on erosion failure in TBCs. Research has shown that owing to their columnar-crystal structure, EB-PVD TBCs have a higher strain tolerance and better erosion performance than APS TBCs [1, 11]. However, to address the shortcomings (i.e., inadequate temperature capabilities and susceptibility to CMAS corrosion) of YSZ, various new TBC materials, such as rare-earth modified YSZ coatings, La- and Ta-series coatings, functionally graded coatings, segmented, surface-crack-containing APS coatings (Fig. 7.2a) [23], and PS-PVD coatings, which combine the advantages of PS and PVD coatings (Fig. 7.2b) [24, 25], have successively emerged, and their erosion performance has garnered attention.

Cernuschi et al. [15], Wellman and Nicholls [1], Steinberg et al. [26], Shin and Hamed [27], and Tian et al. [28] compared the effects of the preparation process, the state (velocity and angle) of erosive particles, and the erosion temperature on the erosion performance of YSZ coatings, as shown in Fig. 7.3. As seen in the figure, when the erosion angle, velocity, and temperature remain unchanged, the EB-PVD TBC has the best erosion performance, followed sequentially by the APS TBC containing vertical cracks, the PS-PVD TBC, and the APS TBC. In addition, the erosion rate of the APS TBC is 10 times or even more than that of the EB-PVD TBC.

Cernuschi et al. [29] compared the erosion performance of a YSZ TBC and new TBCs, including an Al-modified YSZ/YSZ (YAG) two-layer TBC and a Gd-Zr/YSZ (Gd) two-layer TBC, all of which were prepared by PS. As shown by the results in Fig. 7.4, at the same erosion angle and velocity, the YAG and Gd coatings resistant to CMAS corrosion were actually inferior to the YSZ coating in terms of erosion performance. This finding was later validated by Steinberg et al. [30]. In addition, it can be seen from Fig. 7.4 that the APS coating containing multiple vertical cracks had the best erosion performance. This, to some extent, suggests that APS TBCs with vertical cracks have good application prospects.

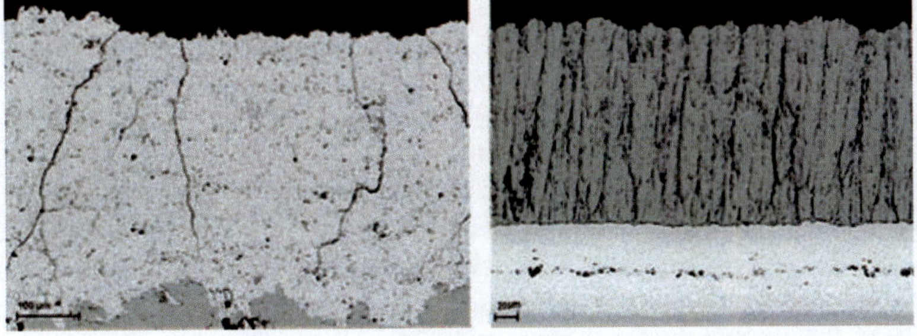

(a) (b)

Fig. 7.2 **a** APS TBC with vertical surface microcracks; **b** PS-PVD TBC [16]

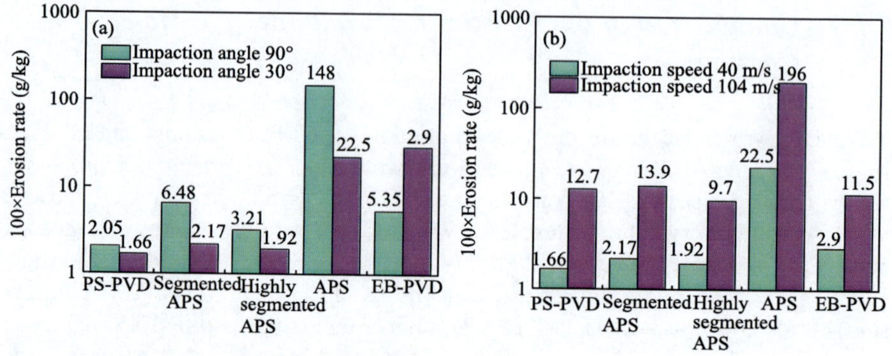

Fig. 7.3 Comparison of the erosion rates of TBCs at 700 °C: **a** under an erosion velocity of 40 m/s and various erosion angles; **b** under an erosion angle of 30° and various erosion velocities [16]

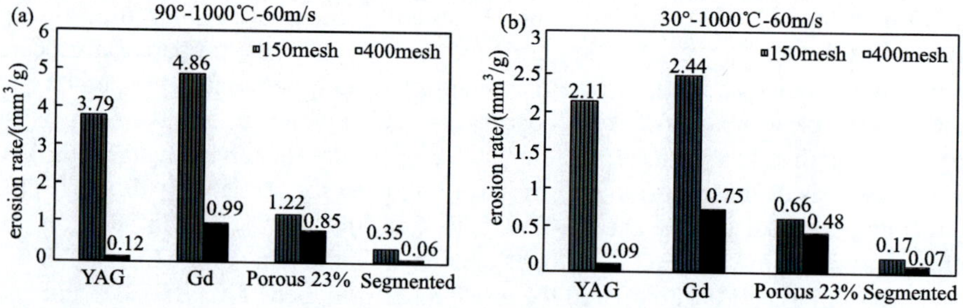

Fig. 7.4 Comparison of the erosion rates of TBCs at 1000 °C under an erosion angle of **a** 90°; **b** 30° [29]

7.1.4 General Pattern of the Erosion Performance of TBCs

For each type of TBC, the erosion rate increases as the erosion angle (i.e., the angle between the direction of motion of the particles and the ceramic surface) and particle velocity increase, as shown in Fig. 7.3a and b. The erosion rate w_e is related to the erosion angle ϕ and particle velocity v through the following empirical equation: [15]

$$w_e \propto (v \sin \phi)^n \tag{7.1}$$

where n is a constant. $n \approx 3$ for ZrO_2 ceramic blocks. Of the five TBCs in Fig. 7.3a and b, the EB-PVD TBC has the smallest n at 1.4, and the APS TBC has the largest n at 2.3 [15]. Research has shown that temperature has an insignificant effect on the erosion rate of a TBC, suggesting that the erosion temperature does not alter the erosion failure mechanism. It has also been found that the erosion rate of a TBC decreases slightly as the temperature increases, which may be related to the sintering or increased inelastic deformation capacity of the ceramic layer at high temperatures [4].

A multitude of mechanical parameters affect the erosion performance of a TBC, of which hardness can often reflect the comprehensive mechanical properties of the material. Therefore, studying the relation between the erosion performance and hardness of a TBC can, to a certain extent, reveal the effects of the mechanical properties of the TBC material on the erosion performance of the TBC. Cernusch et al. [15] and Janos et al. [20] found that the relation between the erosion rate w_e of an APS TBC and the Vickers microhardness H_t of its ceramic layer can be fitted to the following equation:

$$w_e = aH_t^b + c \tag{7.2}$$

where *a*, *b*, and *c* are constants related to experimental conditions. Generally, coatings with high hardness have excellent erosion performance. Thus, *a* is generally negative.

Hassani et al. [31], Chen et al. [32], and Wellman et al. [4] noted that the erosion performance of a TBC is related to its hardness H and elastic modulus E or their relation. Thus, the erosion rate w_e can be described as a function of H, E, H/E, or H^3/E^2:

$$w_e = f\left(H, E, H/E, H^3/E^2\right) \tag{7.3}$$

The specific functional relation can be determined by fitting experimental data.

Recently, Algenaid et al. [33] analyzed the correlation between the erosion performance of a TBC prepared by suspension PS by adjusting its microstructure (e.g., pores). They experimentally found that the erosion performance of a coating can be more accurately evaluated based on its fracture toughness than its hardness and elastic modulus. The erosion rate is related to the fracture toughness through the following equation:

$$w_e = aK_I^C + c \tag{7.4}$$

where K_I^C is the fracture toughness of the coating and *a*, and *c* are constants related to experimental conditions. Fracture toughness represents the capacity of a material to withstand loading and accounts for both the capacity to withstand mechanical loading and the effects of defects. Thus, it is more reasonable to use the fracture toughness of a coating to evaluate its erosion performance.

Turbine blades with TBCs have very complex geometries and microstructures, for example, different surface structures and microstructures (e.g., different sizes of and spacings between columnar-crystal grains in EB-PVD coatings). In addition, for a TBC with an APS ceramic layer, the porosity of the ceramic layer has a significant effect on the erosion performance of the TBC. Wellman and Nicholls [1], Nicholls and Wellman [11], and Wellman et al. [12] found that the erosion rate of a ceramic layer with a convex geometrical structure is much higher than that of a ceramic layer with a concave geometrical structure. An EB-PVD TBC exhibits good erosion performance when the angle between the columnar crystals and the surface of the eroded ceramic layer is greater than 70°. As the angle decreases in the range of 30–70°, there is a linear increase in the erosion performance of an EB-PVD TBC. At an angle of 50°, the erosion rate of an EB-PVD TBC is higher than that of an APS TBC under the same erosion conditions. The erosion rate of an EB-PVD TBC at an angle of 30° is approximately 10 times that at an angle of 70°. At an angle of less than 7.5°, the erosion rate of an EB-PVD TBC can reach as high as 6000 g/kg. Moreover, the erosion performance of an EB-PVD TBC depends heavily on the diameter of the columnar crystals. An EB-PVD TBC with columnar crystals large in diameter has a high erosion rate. The erosion failure mode, to a large extent, depends on the ratio of the diameter of the erosive particles, *D*, to the diameter of the columnar crystals, *d*, i.e., *D*/*d* [5, 11, 12]. To ensure that an EB-PVD ceramic layer exhibits good erosion performance during service, it is necessary to keep the angle between the columnar crystals and the surface of the eroded ceramic layer between 75 and 90° and the diameter of the columnar crystals as small as possible. To improve the erosion performance of an APS TBC, it is necessary to minimize voids and cracks in the ceramic layer and to appropriately prefabricate cracks perpendicular to the interface [11, 15, 18–21].

7.2 Erosion Failure Modes of Typical TBCs

Different types of TBCs differ in preparation process and microstructure; as a result, they have different erosion failure modes. This section focuses on introducing the erosion failure modes of EB-PVD, APS, and PS-PVD coatings and the relevant influencing factors.

7.2.1 Erosion Failure Modes of EB-PVD TBCs

For an EB-PVD TBC with a columnar-crystal structure, the collisions between the erosive particles and its ceramic layer can be viewed as the actions between the particles and one or multiple columnar crystals, and its failure mode depends on the particle size (mass) and velocity, the collision angle, and the yield strength of the columnar crystals, and, particularly, the *D*/*d* ratio. According to the findings of Nicholls and Wellman et al., the erosion failure of EB-PVD TBCs can be grouped into three modes, namely, erosion due to densification, erosion due to compressive damage, and erosion due to external particles [1, 3–7, 11–13].

(a) Mode I: erosion due to densification. This failure mode generally occurs at a *D*/*d* ratio below 2. Upon impact on the surface of the ceramic layer, small, low-velocity particles interact with only one or a very small number of columnar crystals, and the ceramic layer densifies only in the near surface within a depth

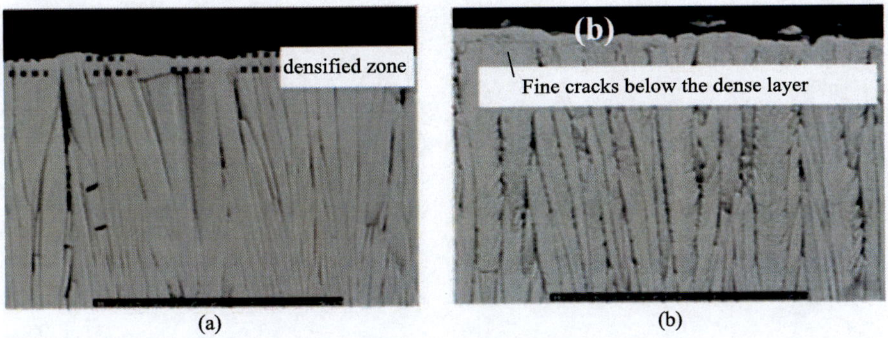

Fig. 7.5 Erosion failure mode I of EB-PVD TBCs: **a** formation of a densified layer; **b** formation of small interlayer cracks beneath the densified layer [3]

of less than 20 μm, forming a densified layer, as shown in Fig. 7.5a. Under this condition, the columnar-crystal structure of the ceramic layer remains intact, and a small number of cracks may be formed between the densified layer (a plastic layer) and the undeformed ceramic layer (an elastic layer), as shown in Fig. 7.5b. This failure mode can occur at both low and high temperatures. However, the thickness of the densified layer formed at high temperatures does not increase as a result of the "softening" of the ceramic layer. In this mode, the erosion performance of a TBC is related to Young's modulus, hardness, and fracture toughness of the ceramic layer. As these parameters vary with temperature, the erosion rate of a TBC layer in this mode varies with temperature.

(b) Mode II: erosion due to compressive damage. This is a failure mode between erosion due to densification and erosion due to damage from external particles and generally occurs at *D*/*d* ratios of 2–12. When large particles impact the surface of the ceramic layer at a high velocity, a small amount of coating delamination occurs or a densified layer is formed in the impact zone; however, cracks are not formed inside the ceramic layer, nor does the ceramic layer undergo plastic deformation as a whole, as shown in Fig. 7.6. Compared to mode I, erosion occurs independently in single or multiple columnar crystals and more columnar-crystal units are involved in the erosion process in mode II. Therefore, while more energy is absorbed during impact than in mode I, no cracks are formed at the interface between the densified layer and the insignificantly deformed ceramic layer due to the participation of more units in mode II. In addition, compared to mode III, no shear zone or cracks are formed in mode II. Thus, the condition for this failure mode is that the instantaneous stress to which each columnar-crystal unit is subjected is less than its fracture strength. Under this condition, the densification is completely attributed to the voids in the ceramic layer. This is a result of the fatigue caused by repeated impacts from particles. This failure mode can occur at both room and high temperatures.

(c) Mode III: erosion due to damage from external particles. At a *D*/*d* ratio greater than 12, when large, high-velocity particles impact the surface of the ceramic

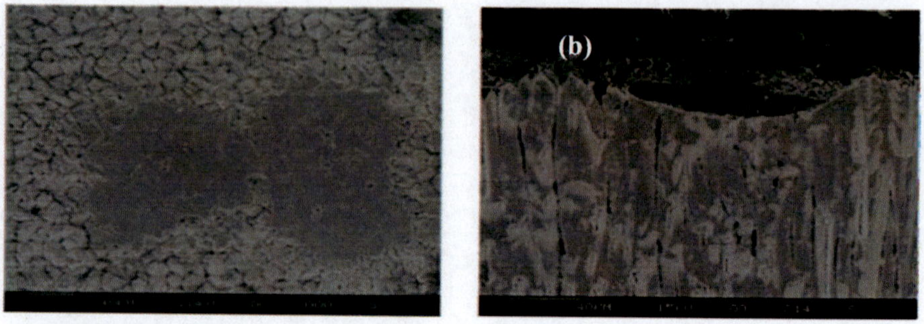

(a) (b)

Fig. 7.6 Erosion failure mode II of EB-PVD TBCs: **a** surface morphology; **b** cross-sectional image [1]

layer, the ceramic layer undergoes large deformation over a very large area, which can extend to the transition layer or even the substrate. Under this condition, the columnar-crystal units even bend and fracture, and macroscopic cracks are formed when multiple columnar-crystal units fracture and connect with each other. Under normal circumstances, cracks are first formed in the collision zone and then propagate outwardly. As the fracture locations of the columnar-crystal units are not on the same horizontal line, the cracks do not propagate in parallel to the interfaces of the TBC; instead, the propagating cracks form a crack zone that "kinks," which is referred to as the "kink band," as shown in Fig. 7.7 [3, 10]. This failure mode can occur at both room and high temperatures. However, under the same loading conditions, high temperature results in large deformation in a ceramic coating. Thus, the pits left by erosive particles are larger at high temperatures than at room temperature, as shown in Fig. 7.7. At high temperatures, another erosion failure mode may occur, as shown in Fig. 7.7d. In this mode, the coating does not deform significantly as a whole, and no visible cracks are formed, even though the columnar-crystal units undergo buckling. This erosion failure mode occurs only at temperatures above 800 °C and basically does not occur at room temperature [12].

7.2.2 Erosion Failure Mode of APS TBCs [1, 15, 21]

An APS TBC has a typical layered structure, and its entire ceramic layer can be viewed as a structure formed by strip units stacked layer-by-layer. During spraying and cooling, voids with a volume of 10–15% are formed in the ceramic layer. The Y_2O_3 added to increase the stability of the ZrO_2 leads to the formation of a large number of microcracks parallel to the interfaces in the coating. As a result of the presence of voids and cracks, an APS ceramic layer has a low elastic modulus and,

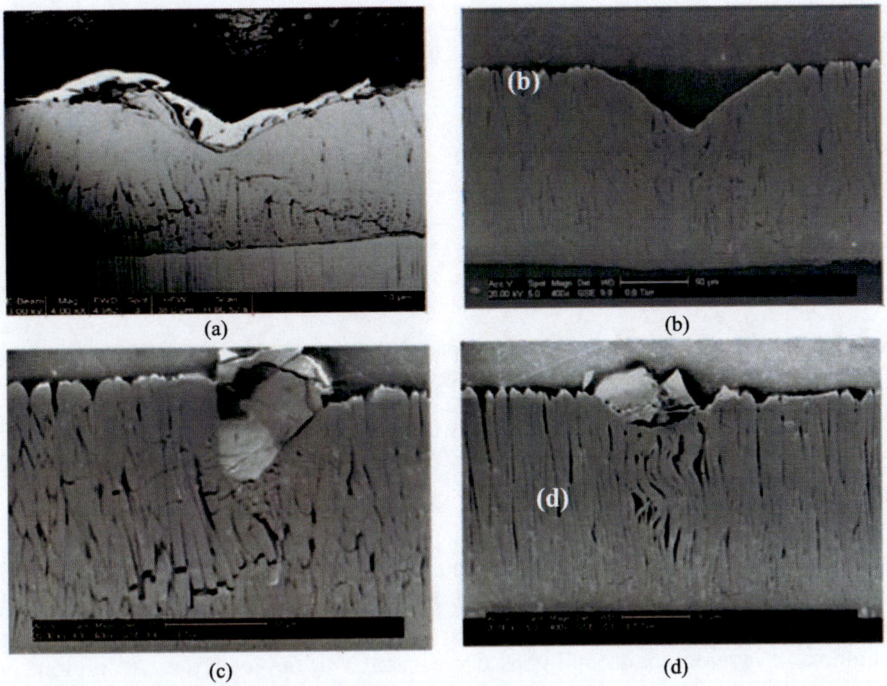

Fig. 7.7 Erosion failure mode III of EB-PVD TBCs: **a** damage at 1200 °C; **b** damage at 800 °C; **c** damage at room temperature; **d** damage at 800 °C [3, 12]

therefore, a low load-carrying capacity. APS TBCs are inferior to EB-PVD TBCs in terms of erosion performance. The collisions between particles and the ceramic layer can be viewed as the actions between particles and one or multiple layered units in this zone, as shown in Fig. 7.8. The erosion failure of an APS TBC is characterized by the formation and propagation of cracks at the boundaries of the layered units as well as the spallation of the ceramic layer in blocks and patches along these boundaries. Unlike the case of an EB-PVD TBC, no visible densified layer appears in an APS TBC during erosion failure. The erosion performance of an APS TBC, to a large extent, depends on the distribution of the pores and cracks in the ceramic layer. An increase in the number of pores and parallel cracks in the ceramic layer provides more initiation sources for the spallation of the layered units, making the coating more susceptible to spallation. Conversely, a coating with a dense ceramic layer (i.e., a ceramic layer with a small number of pores and cracks) exhibits good erosion performance.

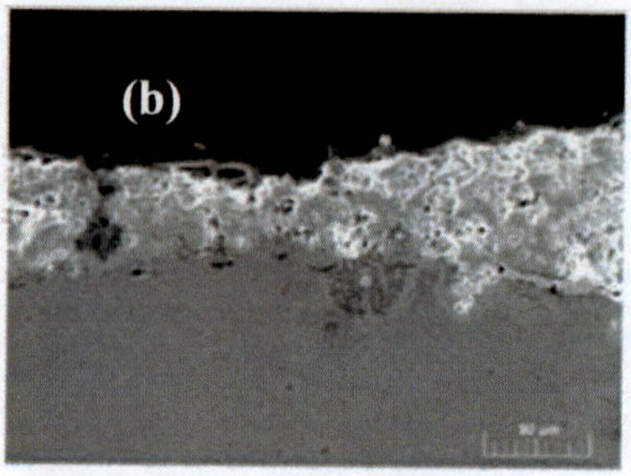

(a) (b)

Fig. 7.8 Images of an APS TBC eroded by Al particles at 30° and 140 m/s: **a** optical image of the eroded zone; **b** cross-sectional SEM image of the eroded zone [15]

7.2.3 Erosion Failure Mode of PS-PVD TBCs [15, 26–28, 34, 35]

A PS-PVD TBC possesses both the porous, lamellar structure of an APS TBC and the gap-rich two-layer structure of an EB-PVD TBC, as shown in Fig. 7.2. The co-existence of the two structures results in a porosity of 19–29%. In addition, large numbers of layered cracks and gaps perpendicular to the interfaces are present in a PS-PVD TBC. The presence of gaps reduces the damage to the coating outside the zone of action of erosive particles. Consequently, PS-PVD TBCs display better erosion performance than APS TBCs and even EB-PVD TBCs. As shown in Fig. 7.9, three zones, namely, a significantly spalled zone (R1), an erosion-affected zone (R2), and an undamaged zone (R3), are formed when a PS-PVD TBC is eroded. Layered cracks appear in zone R1 (i.e., the damaged zone), resulting in layer-by-layer coating spallation. Coin-shaped cracks are formed in zone R2, accompanied by partial coating spallation often in an incline direction. While no visible erosion-induced deformation appears in zone R3, regions prone to crack formation and spallation are also formed in this zone. Generally, as shown in Fig. 7.9b, cracks are formed at weak locations (e.g., pores and microcracks) in the zone of direct action of erosive particles (i.e., zone R1) and propagate under repeated erosion by particles. A crack ceases to propagate upon reaching the nearby gap. The formation, propagation, and connection of more cracks in zone R1 lead to coating spallation in this zone. Moreover, stress transfer results in the formation of coin-shaped cracks in the nearby zone R2. The erosion performance of a PS-PVD TBC depends on the gap distribution and the crystal grain size in the ceramic layer as well as its porosity. A PS-PVD TBC displays good erosion performance when its ceramic layer contains small pores, fine crystal grains, and a large number of gaps evenly distributed at a given porosity.

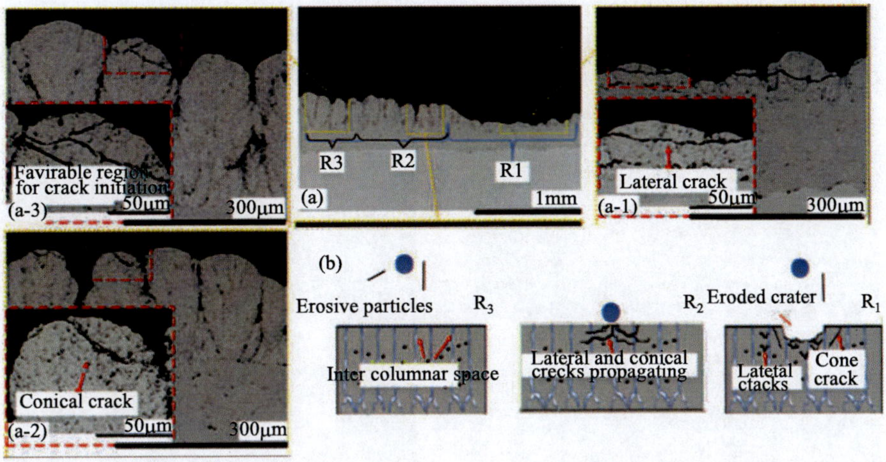

Fig. 7.9 Erosion failure mode of PS-PVD TBCs: **a** three damaged zones; **b** schematic diagram of the failure mode [35]

7.2.4 CMAS Erosion Failure of TBCs

When the surface temperature of the TBC is lower than the melting point of CMAS (i.e., mixed CaO, MgO, Al_2O_3, and SiO_2 particles), CMAS are the particles causing erosion failure in the coating. When the surface temperature (e.g., above 1240 °C) of the TBC is higher than the melting point of CMAS, erosive CMAS particles melt. In fact, accompanied by the erosion and CMAS corrosion processes, the erosion process can still cause coating spallation, as shown in Fig. 7.10d. In addition, molten CMAS is extremely prone to diffusion in the voids and cracks in the ceramic layer, forming a glass phase with a triclinic structure (i.e., anorthite ($CaAl_2Si_2O_8$) [1, 16, 17, 36]), as shown in Fig. 7.10a–c, during the subsequent cooling process. Drexler et al. [17] proved that an engine could be eroded by molten CMAS through erosion by volcanic ash (VA) erupted from the Eyjafjallajökull volcano in Iceland. Under the action of molten CMAS, no visible densified zone is formed in the ceramic layer, and the ceramic layer does not undergo significant spallation, as shown in Fig. 7.10a and b. However, the formation of the glass-phase anorthite with a triclinic structure significantly reduces the strain tolerance (i.e., deformation resistance) of the TBC [16, 17, 36]. More importantly, the coefficient of thermal expansion (CTE) of CMAS is considerably lower than that of a ceramic coating. As a result, an additional compressive stress is generated in a ceramic layer with CMAS-infiltrated voids and cracks during cooling. When this compressive stress accumulates to a sufficiently high level, it leads to the formation of cracks at the interface between the CMAS-infiltrated ceramic layer and the ceramic layer beneath it, causing coating spallation.

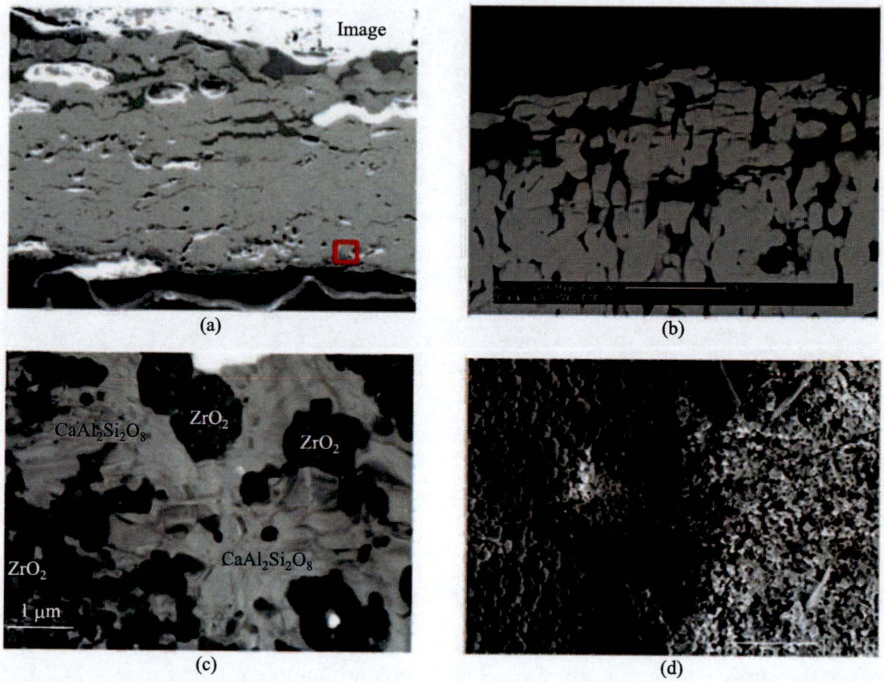

Fig. 7.10 **a** Image of an APS coating subjected to erosion from Eyjafjallajökull VA at 1200 °C for 24 h [16, 17]; **b** cross-sectional SEM image of an EB-PVD TBC subjected to CMAS erosion [1]; **c** high-resolution bright-field TEM image of the red boxed area in (**a**) [16]; **d** surface morphology of a CMAS-eroded area [37]

7.2.5 Factors Affecting the Erosion Performance of TBCs

According to the above discussion on the differences in the erosion rates and failure modes of TBCs prepared by different processes, the erosion performance of an EB-PVD TBC depends heavily on the *D*/*d* ratio, the erosion performance of an APS TBC depends primarily on the porosity of the ceramic layer and the bond strength between the layered units, and the erosion performance of a PS-PVD TBC depends on its microstructure (e.g., crystal grains and gaps) and composition. This section provides a brief introduction to other important factors affecting the erosion performance of TBCs, including the material composition of the ceramic layer, high-temperature aging, and high-temperature corrosion.

(1) **Effects of the material composition of the ceramic layer**

To improve the thermal insulation capacity and phase stability of TBCs, researchers across the world have extensively studied TBC materials and found that adding a small amount of rare-earth oxides (e.g., Gd_2O_3, Dy_2O_3, and Yb_2O_3) [1, 4, 7, 14] or Hf [4] to the YSZ ceramic layer can effectively reduce the heat transfer coefficient (HTC) of an EB-PVD TBC. However, currently, any doping agents capable of reducing the HTC of a coating reduce the erosion performance of the ceramic layer. For example, Wellman and Nicholls [1] found that adding Dy_2O_3 or Gd_2O_3 at 2 mol% to the conventional ZrO_2-8wt.%Y_2O_3 ceramic layer approximately doubled the erosion rate of the TBC, while adding Gd_2O_3 at 4 mol% increased the erosion rate by 20-fold. In another example, Wellman et al. [4] found that adding a small amount of Hf to the conventional ZrO_2-8wt.%Y_2O_3 ceramic layer increased its erosion rate at room temperature from 10.92 g/kg (for conventional YSZ) to 16.51 g/kg. Under normal circumstances, an as-prepared EB-PVD TBC is composed mainly of a metastable tetragonal phase (t' phase). Under thermal cycling, the t' phase tends to transform to the monoclinic phase (m phase). In addition, the phase structure of the ceramic layer can be varied by doping or improving the process parameters. Zhu et al. [14] investigated the differences in the erosion performance of the conventional t'-phase ZrO_2-7wt%Y_2O_3, cubic-phase (c-phase) ZrO_2-4mol%Y_2O_3-3mol%Gd2O_3-3mol%Yb_2O_3, c-phase ZrO_2-4mol%Y_2O_3-3mol%Gd2O_3-3mol%Yb_2O_3-TiO_2-Ta_2O_5, c-phase ZrO_2-4mol%Y_2O_3-4mol%Gd2O_3-4mol%Yb_2O_3, and $Zr_2Gd_2O_7$ coatings impacted by particles at high velocities (above 170 m/s) and temperatures (above 1000 °C). They found that both the doped c-phase ceramic layers and $Zr_2Gd_2O_7$ coating were inferior to the t'-phase ZrO_2-7wt%Y_2O_3 ceramic layer in terms of erosion performance. Despite this, lightly doped EB-PVD TBCs are still superior to APS TBCs in terms of erosion performance, and the doping of Hf can improve the erosion performance of a TBC at high temperatures (i.e., in real-world service environments) [4]. Ranmanujam and Nakamura also noted that doping the ceramic layer with a small amount of a ductile metal phase (e.g., CoNiCrAlY alloy) could improve the erosion performance of an APS TBC [21]. These results indicate that doping may help to produce TBC materials with ideal comprehensive thermal insulation and erosion performance.

(2) **Effects of high-temperature aging**

During service at high temperatures, TBCs undergo aging processes such as sintering, phase transformation, and oxidation, the effects of which on erosion performance are an important area of research on erosion mechanisms. For an EB-PVD TBC, high-temperature aging (e.g., sintering) reduces the spacing between columnar crystals or even causes two or more columnar crystals to connect with each other, forming one columnar crystal, thereby resulting in a columnar-crystal size larger than that before aging and, therefore, a higher erosion rate [1, 13]. For example, Wellman and Nicholls [13] found that aging an EB-PVD TBC at 1100 °C for 30 and 100 h increased its erosion rate at room temperature by approximately 2.6 and 9.76 g/kg, respectively, and that aging the TBC at 1500 °C for 24 h was triple that quadrupled its erosion rate at room temperature. This is because before aging, the gaps between the columnar crystals can be viewed as free surfaces, and it is not easy for the fracture (i.e., microcrack) of a single columnar crystal to propagate to the nearby columnar-crystal unit. However, aging results in connection between the columnar crystals and thus increases the columnar-crystal size, thereby increasing the size of fracture-induced microcracks. In addition, as a result of the shortening of the gaps, the microcracks become more prone to propagation toward the nearby columnar crystals, thereby causing coating spallation over a large area. The opposite is true for an APS TBC. High-temperature aging reduces the numbers of voids and microcracks in the ceramic layer and improves the interlayer bond strength of the layered structure, thereby enhancing the erosion performance of the TBC. For example, Janos et al. [20] found that aging APS TBCs at 1260, 1371, and 1482 °C for 16 h reduced their erosion rates by approximately 53.6%, 64.3%, and 69.5%, respectively. Evidently, high-temperature aging has completely different effects on the two typical types of TBCs. Specifically, high-temperature aging weakens the erosion performance of EB-PVD TBCs but improves that of APS TBCs.

(3) **Effects of high-temperature CMAS corrosion**

As the environmental factors causing CMAS corrosion become increasingly severe, TBCs are inevitably subjected to concurrent CMAS corrosion and erosion. However, research examining the evolutionary pattern of the erosion performance of TBCs subjected to CMAS corrosion remains lacking. Recently, Steinberg et al. [26] used the suspension PS process to prepare a sacrificial Al layer on the surface of an EB-PVD coating and studied the evolution of the erosion performance of the coating after CMAS corrosion. They compared five specimens of the EB-PVD coating equipped with a sacrificial Al layer (i.e., an as-coated specimen), a specimen aged at a high temperature of 1250 °C for 30 min, a specimen corroded by CMAS (C1) with a high Si content at 1250 °C for 30 min, a specimen corroded by CMAS (C1) with a low Si content at 1250 °C for 30 min, and a specimen corroded by Iceland VA (IVA) at 1250 °C for 30 min. They found that high-temperature aging alone reduced the erosion performance of the TBC, corrosion by CMAS with a low Si content or IVA actually improved the erosion performance of the TBC, and corrosion by CMAS with a high Si content reduced the erosion performance of the TBC. The erosion rates of the five TBC specimens at 90° and room temperature were found to be 2.57, 3.67, 4.36, 1.25, and 1.65 mm^3/g, respectively. This may be related to the increase in the hardness and modulus of the coating after CMAS infiltration.

7.3 Numerical Simulation of the Correlations Between the Erosion Parameters of TBCs

The erosion failure modes of various types of TBCs and various factors affecting their erosion performance have been experimentally identified. However, it is very difficult to experimentally determine the correlations between various intricate and complex factors as well as the extent of their effects on erosion performance. In addition, establishing a theoretical model of erosion failure that comprehensively accounts for various factors is a formidable task. Hence, dimensional analysis combined with numerical simulation is an important means for determining empirical relations in the investigation of the erosion failure mechanisms of TBCs. The finite element method (FEM) [3, 7–13, 37–40] and the Monte Carlo method [41, 42] are predominantly used to simulate the erosion process of TBCs, with the goals of determining the key factors affecting the erosion performance of the TBCs and verifying experimental and theoretical results. In the following, we use the FEM as an example to briefly introduce the process of numerically simulating the erosion of a TBC and present the main results.

7.3.1 Dimensional Analysis Theory

The earliest concept of the "dimension" can be traced back to 1822 when Fourier [43] applied this geometrical concept in the realm of physics. According to Fourier, if the unit of a parameter changes, not only the value of the parameter but the values of all the quantities related to the parameter change. In 1914, Buckingham [44] proposed the use of several quantities with a dimensionless power (referred to as π) to describe each physical law, which forms the common fundamental rule that numerical simulations follow. In 1922, Bridgman [45] referred to the concept proposed by Buckingham as the "Π theorem," and this theorem constitutes the core idea of dimensional analysis. Any physical law can always be expressed by a certain functional relation. For a certain type of physical problem, let a be a dimensional quantity that is a function of some dimensional quantities that are independent of each other $(a_1, a_2, \ldots, a_n)$; i.e.,

$$a = f(a_1, a_2, \ldots, a_k, a_{k+1}, \ldots, a_n) \tag{7.5}$$

Basic dimensional quantities with independent dimensions can be found among the independent variables. Let us assume that the first k $(k \leq n)$ independent variables in dimensional quantities $a_1, a_2, \ldots, a_n$ have independent dimensions. Dimensional independence means that the dimension of any quantity among the independent variables cannot be obtained by combining the dimensions of other independent variables in any manner. Let us treat the k dimensionally independent quantities (i.e., $a_1, a_2, \ldots, a_k$) as basic quantities and express their dimensions as follows [46, 47]:

$$[a_1] = A_1, [a_2] = A_2, \cdots, [a_k] = A_k$$

Then, the dimensions of the remaining quantities can be expressed as follows:

$$[a] = A_1^{m_1} A_2^{m_2} \ldots A_k^{m_k},$$

$$[a_{k+1}] = A_1^{p_1} A_2^{p_2} \ldots A_k^{p_k},$$

$$\ldots$$

$$[a_n] = A_1^{q_1} A_2^{q_2} ... A_k^{q_k}$$

If we increase the units of measurement of the first k basic dimensional quantities (i.e., $a_1, a_2, \ldots, a_k$) in Eq. (7.5) by factors of $a_1, a_2, \ldots, a_k$, respectively, then the first k independent variables of function f each become a constant of 1, and the values of the remaining $(n - k)$ parameters (i.e., $a, a_{k+1}, \ldots, a_n$) can be expressed as follows [46, 47]:

$$\Pi = \frac{a}{a_1^{m_1} a_2^{m_2} \cdots a_k^{m_k}},$$

$$\Pi_1 = \frac{a_{k+1}}{a_1^{p_1} a_2^{p_2} \cdots a_k^{p_k}}$$

$$\cdots \quad \cdots \quad \cdots$$

$$\Pi_{n-k} = \frac{a_n}{a_1^{q_1} a_2^{q_2} \cdots a_k^{q_k}}$$

We can see that the values of $\Pi, \Pi_1, \cdots, \Pi_{n-k}$ are independent of the originally selected units of measurement because they are dimensionless relative to units of measurement $A_1, A_2, \ldots, A_k$. In addition, the values of $\Pi, \Pi_1, \cdots, \Pi_{n-k}$ are independent of the k units of measurement selected for basic dimensional quantities $a_1, a_2, \ldots, a_k$. We refer to these $(n - k)$ parameters as dimensionless parameters.

Then, Eq. (7.5) can be written in the following form:

$$\Pi = f(1, 1, \ldots, \Pi_1, \ldots, \Pi_{n-k}) \tag{7.6}$$

where $\Pi_1, \Pi_2, \ldots, \Pi_{n-k}$ correspond to the values of $a_{k+1}, a_{k+2}, \ldots, a_n$, respectively. Of all the variables of function f, the first k variables all have a constant value of 1, and only the latter $(n - k)$ variables (i.e., $\Pi_1, \Pi_2, \ldots, \Pi_{n-k}$) are dimensionless variables that play a role in function f. Thus, the functional relation in Eq. (7.6) can be written as follows:

$$\Pi = f(\Pi_1, \Pi_2, \ldots, \Pi_{n-k}) \tag{7.7}$$

Clearly, the relation between n dimensional quantities $(a_1, a_2, \ldots, a_n)$ independent of the selection of the system of units of measurement can be transformed to the relation between $(n + 1 - k)$ dimensionless quantities $(\Pi, \Pi_2 \ldots, \Pi_{n-k})$ composed of $(n + 1)$ dimensional quantities. This is the well-known Π theorem [46, 47], which provides an effective method and means for correlating and analyzing multiple physical parameters.

7.3.2 Dimensional Analysis of Erosion in TBCs

In essence, the erosion of a TBC is a process of collision between particles and the surface of the ceramic layer and can be viewed as a dynamic indentation process. Figure 7.11 schematically shows various types of damage caused by erosion of a TBC. There are three types of parameters that determine the extent of damage. (1) Parameters of the erosive particles, including the density ρ_p, diameter D (the particles are simplified into spheres), initial velocity v_0, and kinetic energy $U = \frac{1}{2}mv_0^2 = \frac{\pi}{12}\rho_P D^3 v_0^2$ of the particles. (2) Parameters of the TBC, namely, the density ρ_{tbc}, initial porosity f_0, thickness h_{tbc}, Young's modulus E_{tbc}, and Poisson's ratio ν_{tbc} of the ceramic layer, the critical stress or critical strain corresponding to the densification zone and pits formed during erosion due to the presence of defects (e.g., voids and cracks) in the ceramic layer (the critical stress and critical strain of the ceramic layer are often described by researchers with the concept of "yield" used to describe metallic materials and are referred to as the yield strength σ_{tbc}^{Y} and yield strain ε_{tbc}^{Y} of the ceramic layer [3, 7, 8, 10], respectively, with $\varepsilon_{tbc}^{Y} = \sigma_{tbc}^{Y}/E$), as well as the thickness h_{bc}, Young's modulus E_{bc}, yield strain ε_{bc}^{Y}, density ρ_{bc}, and Poisson's ratio ν_{bc} of the transition layer. Assuming that the action of erosion is limited to the ceramic layer, the effects of the transition layer and substrate are negligible. (3) Damage characterization parameters, namely, quantities related to time t and location r, including the indentation depth $\delta(r, t)$, the indentation width on the surface of the ceramic layer $w(t)$, the radius of the densification zone $R(t)$, the stress state $\sigma_{ij}(r, t)$, and the instantaneous velocity of the particles $v(t)$, as well as time independent residual quantities, e.g., the residual depth at the cumulative indentation site δ_P^R, the residual indentation depth δ^R, the indentation width on the surface of the ceramic layer w^R, the rebound velocity of the particles v^R, the stress inside the TBC after the completion of the erosion process $\sigma_{ij}(r)$, and the crack length in the event that cracks appear in the ceramic layer.

During erosion, the damage parameters (e.g., stress field $\sigma_{ij}(r, t)$, indentation depth $\delta(r, t)$, and particle velocity $v(t)$) inside a TBC should be functions of the parameters of the particles and the TBC. Thus, we can establish their functional relations as follows:

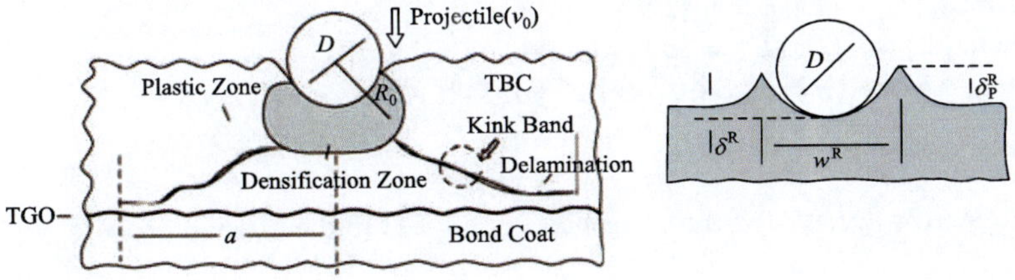

Fig. 7.11 Schematic diagram of the erosion failure of a TBC

Table 7.1 Dimensions of the parameters related to the erosion of a TBC

Parameter	Symbol	Dimension
Particle density	ρ_p	ML^{-3}
Particle diameter	D	L
Particle mass	m_p	M
Initial particle velocity	v_0	LT^{-1}
Density of the ceramic layer	ρ_{TBC}	ML^{-3}
Yield strength of the ceramic layer	σ_{TBC}^{y}	$L^{-1}MT^{-2}$
Elastic modulus of the ceramic layer	E_{TBC}	$L^{-1}MT^{-2}$
Thickness of the ceramic layer	H_{TBC}	L
Porosity	f_0	1
Stress	σ	$L^{-1}MT^{-2}$
Energy	U	L^2MT^{-2}

$$\begin{aligned} \sigma_{ij}(r,t) &= A\left(\rho_P, m, D, v_0, \rho_{TBC}, f_0, h_{TBC}, \sigma_{TBC}^{Y}, \varepsilon_{TBC}^{Y}, v_{TBC}, t, r\right) \\ \delta(t) &= B\left(\rho_P, m, D, v_0, \rho_{TBC}, f_0, h_{TBC}, \sigma_{TBC}^{Y}, \varepsilon_{TBC}^{Y}, v_{TBC}, t, r\right) \end{aligned} \tag{7.8}$$

As seen in Eq. (7.8), damage parameters $\sigma_{ij}(r,t)$ and $\delta(r,t)$ are each a function of 12 independent variables. As a result, it is very difficult to establish their relations. This problem can be simplified by dimensional analysis. Three basic dimensions, namely, mass M, length unit L, and time unit T, are selected as the units of measurement. The dimensions of the parameters related to the erosion of a TBC can be derived using combinations of the basic units of measurement, as shown in Table 7.1.

To facilitate analysis, the particle density ρ_p, particle diameter D, the yield strength of the ceramic layer σ_{tbc}^{Y}, and time t are selected as the dimensions of the basic quantities. Dimensional analysis of various types of parameters related to the erosion process can be carried out based on the dimensions in Table 7.1.

Example 7.1: Derive the dimensionless time $\bar{t}$ in the erosion process.

Solution: Based on the propagation velocity of elastic waves in materials, $v = \sqrt{E/\rho}$, and $t = L/v$,
we have $[t] = [L][E]^{-0.5}[\rho]^{0.5}$,
where $[L] = [D]$, $[E] = [\sigma_{TBC}^{Y}]$, and $[\rho] = [\rho_P]$.
Then, we have $[t] = [D][\sigma_{TBC}^{Y}]^{-0.5}[\rho_P]^{0.5}$.
Thus, $\bar{t} = t/[t] = \frac{t\sqrt{\sigma_{tbc}^{Y}/\rho_P}}{D}$.

Example 7.2: Derive the dimensionless energy Ω in the erosion process.

Solution: Based on the work done by external forces, $[W] = [F][L] = [\sigma][L]^3$, and particle energy $[U] = [W]$.
we have $\Omega = \frac{U}{\sigma_{tbc}^{Y} D^3}$.

Based on the abovementioned dimensional analysis method in combination with the practice in Examples 7.1 and 7.2, readers can perform dimensional analysis on other parameters. The following dimensionless geometric quantities are obtained:

$$\bar{R}(t) = \frac{R(t)}{D}, \bar{\delta}(t) = \frac{\delta(t)}{D}, \bar{\delta}_P^R = \frac{\delta_P^R}{D}, \bar{\delta}^R = \frac{\delta^R}{D}, \bar{w}(t) = \frac{w(t)}{D}, \text{time}\bar{t} = \frac{t\sqrt{\sigma_{tbc}^Y/\rho_P}}{D},$$
$$\text{velocity}\bar{v}(t) = \frac{v(t)}{v_0}, \text{and}\Omega = \frac{\text{U}}{\sigma_{tbc}^Y D^3}, \frac{\sigma_{ij}(r,t)}{\sigma_{tbc}^y}, f_0, \frac{\rho_p}{\rho_{tbc}}, \frac{D}{h_{tbc}}$$

Evidently, the dimensionless stress state $\frac{\sigma_{ij}(r,t)}{\sigma_{\text{tbc}}^Y}$, indentation depth $\overline{\delta}(t)$, cumulative indentation height $\overline{\delta}_p(t)$, indentation width $\overline{w}(t)$, and velocity $\overline{v}(t)$ are each a function of $\frac{r}{D}, \frac{t\sqrt{\sigma_{tbc}^Y}}{D}, f_0, \frac{U}{\sigma_{tbc}^Y D^3}, \frac{D}{h_{tbc}}, \varepsilon_{tbc}^Y$, and $\frac{\rho_P}{\rho_{tbc}}$. For example,

$$\begin{aligned} \frac{\sigma_{ij}(r,t)}{\sigma_{tbc}^y} &= F_{ij}\left[\frac{r}{D}, \frac{t\sqrt{\sigma_{tbc}^Y}}{D}, f_0, \frac{U}{\sigma_{tbc}^Y D^3}, \frac{D}{h_{tbc}}, \varepsilon_{tbc}^Y, \frac{\rho_P}{\rho_{tbc}}\right] \\ \overline{\delta}(t) &= h\left[\frac{r}{D}, \frac{t\sqrt{\sigma_{tbc}^Y}}{D}, f_0, \frac{U}{\sigma_{tbc}^Y D^3}, \frac{D}{h_{tbc}}, \varepsilon_{tbc}^Y, \frac{\rho_P}{\rho_{tbc}}\right] \end{aligned} \quad (7.9)$$

For a specific TBC system, f_0 and $\frac{D}{h_{tbc}}$ are known parameters.

The FEM can be employed to analyze the erosion process of the TBC in detail to determine the effects of the other five parameters on the erosion performance of the TBC as well as the functional forms of the damage parameters [3, 6, 8, 37–39]. In the following, we present the main methods and conclusions with the EB-PVD coating as an example.

7.3.3 *Numerical Simulation Analysis of the Correlations Between Erosion Parameters*

(1) **Geometric model, basic equations, and boundary conditions**

The TBC erosion process is dynamically simulated in ABAQUS. Figure 7.12 shows the FE analysis model. The ceramic layer with a density of ρ_{tbc} and a thickness of h_{tbc} can be viewed as an axisymmetric structure composed of columnar-crystal units (cylinders) with a diameter of d that are spaced at a distance of g apart. The coefficient of friction between adjacent cylinders during collisions can be set to 0 or a certain constant. A spherical particle with a density of ρ_p, a diameter of D, and an initial velocity of v_0 impacts the cylindrical units composing the ceramic layer and causes it to deform and subsequently rebounds.

Fig. 7.12 A simplified FE model for the erosion of an EB-PVD TBC

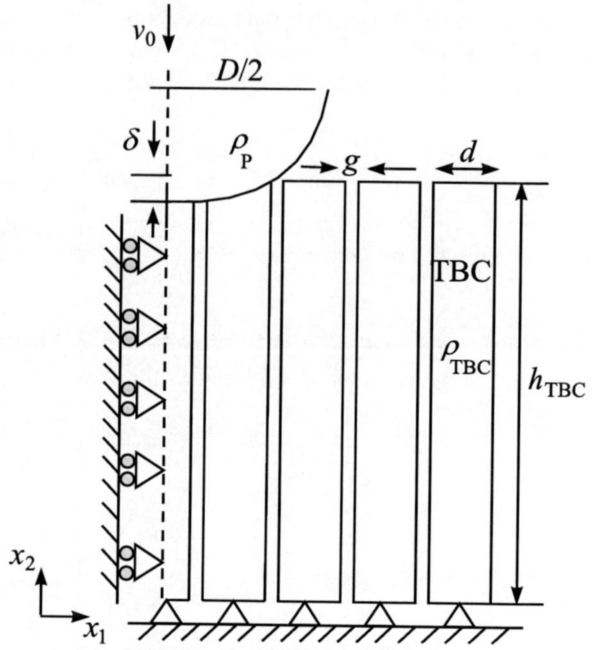

Upon erosion (dynamic indentation) by the particle, the ceramic layer becomes densified and thus undergoes irreversible deformation due to the presence of defects (e.g., pores). Under these conditions, the erosive particle can be simplified into a rigid body relative to the ceramic layer, while the ceramic layer can be regarded as an ideal elastoplastic material, the constitutive equation of which for the elastic deformation stage is as follows:

$$\varepsilon_{ij} = \frac{1+\nu_{TBC}}{E_{TBC}}\sigma_{ij} - \frac{\nu_{TBC}}{E_{TBC}}\delta_{ij}\sigma_{kk} \tag{7.10}$$

By applying the von Mises yield criterion, we have

$$\left(\sigma_1^2 - \sigma_2^2\right)^2 + \left(\sigma_2^2 - \sigma_3^2\right)^2 + \left(\sigma_1^2 - \sigma_3^2\right)^2 = 2\left(\sigma_{TBC}^Y\right)^2 \tag{7.11}$$

where σ_1, σ_2, and σ_3 are the principal stresses. In the actual simulation process, the effect of the width of the coating can be ignored; that is, the coating can be considered to be in a plane strain state. In addition, as the particle velocity can reach as high as 300 m/s during erosion, the strain rate $\dot{\varepsilon}$ inside the ceramic layer is very large. When $\dot{\varepsilon} > 10^3$–10^4 s^{-1}, its effect on the stress should be considered. Chen [38] introduced the following strain-hardening model in the form of a power function:

Table 7.2 Material parameters of each layer of the TBC subjected to erosion

Material	E (GPa)	ν	Density ρ (kg/m^3)	Yield strength σ_y (MPa)	Tangent modulus E_T (GPa)
YSZ	200	0.25	3610	600	_
Foam	0.5	0.1	295	_	_
TGO	320	0.25	4000	_	_
BC	110	0.33	7380	200	5

$$\dot{\varepsilon} = \dot{\varepsilon}_r \left(\frac{\sigma^Y(\dot{\varepsilon})}{\sigma^Y(0)} - 1 \right)^n \tag{7.12}$$

where $\sigma^Y(\dot{\varepsilon})$ is the yield strength that varies with $\dot{\varepsilon}$, $\sigma^Y(0)$ is the yield strength when $\dot{\varepsilon} = 0$, and $\dot{\varepsilon}_r$ and n are material constants ($\dot{\varepsilon}_r = 2 \times 10^4\ \text{s}^{-1}$ and $n = 3$ for TBCs).

The displacement at the axis of symmetry of the ceramic layer in the x_1-direction is zero; i.e., $u_1|_{x_1=0,t} = 0$. The boundary conditions for the displacement at the bottom of the ceramic layer are $u_2|_{x_2=0,t} = 0$ and $u_1|_{x_2=0,t} = 0$. See Table 7.2 for the material parameters in detail [10, 17, 36, 37].

(2) **Correlation analysis of the erosion parameters**

Figure 7.13 shows the results of the FE simulation. It can be found that the variations in the dimensionless indentation depth $\delta(t)/D$ and velocity $v(t)/v_0$ with the dimensionless time $t\sqrt{\sigma^Y_{tbc}/\rho_P}/D$ are independent of parameters ε^Y_{tbc} and $\frac{\rho_P}{\rho_{tbc}}$, suggesting that the variations in $\delta(t)/D$ and $v(t)/v_0$ with time depend solely on the dimensionless particle kinetic energy, Ω. As seen in Fig. 7.13b, when $\Omega = 0.2$, the particle velocity $v(t)/v_0$ is close to zero at $t\sqrt{\sigma^Y_{tbc}/\rho_P}/D \cong 0.5$, indicating that the particle penetrates to the maximum possible depth. Subsequently, $v(t)/v_0$ turns negative, suggesting that the particle rebounds. At $t\sqrt{\sigma^Y_{tbc}/\rho_P}/D \cong 0.6$, the particle separates from the impacted body. Because both the indentation depth $\delta(t)/D$ and velocity $v(t)/v_0$ are independent of ε^Y_{tbc} and $\frac{\rho_P}{\rho_{tbc}}$ and are the basic quantities that determine the stress field $\frac{\sigma_{ij}(r,t)}{\sigma^Y_{tbc}}$ and other geometric quantities (e.g., the cumulative indentation height $\overline{\delta}_p(t)$ and indentation width $\overline{w}(t)$) of the ceramic layer, all of these quantities of the ceramic layer are independent of ε^Y_{tbc} and $\frac{\rho_P}{\rho_{tbc}}$. Thus, the functional equations of these quantities can be simplified to functions of $\frac{r}{D}$, $\frac{t\sqrt{\sigma^Y_{tbc}}}{D}$ and $\frac{U}{\sigma^Y_{tbc}D^3}$, while the residual quantities independent of time are functions of $\frac{r}{D}$, $\frac{U}{\sigma^Y_{tbc}D^3}$ or $\frac{U}{\sigma^Y_{tbc}D^3}$. For example,

$$\frac{\sigma_{ij}(r,t)}{\sigma^y_{tbc}} = F_{ij}\left[\frac{r}{D}, \frac{t\sqrt{\sigma^Y_{tbc}}}{D}, \frac{U}{\sigma^Y_{tbc}D^3}\right] \tag{7.13}$$

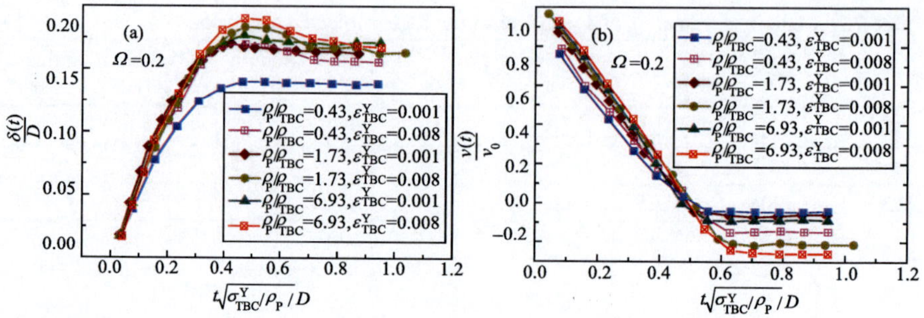

Fig. 7.13 **a** Variation in the dimensionless indentation depth $\delta(t)/D$; **b** relations of the variation in the dimensionless particle velocity $v(t)/v_0$ with ε_{tbc}^{Y} and ρ_P/ρ_{tbc} (the variation in $v(t)/v_0$ is independent of ε_{tbc}^{Y} and ρ_P/ρ_{tbc} except when they are very small) [39]

$$\frac{\sigma_{ij}^{R}(r)}{\sigma_{tbc}^{y}} = F_{ij}^{R}\left[\frac{r}{D}, \frac{U}{\sigma_{tbc}^{Y} D^3}\right] \tag{7.14}$$

Thus, the function of the variation in each geometric quantity (e.g., δ^R/D and δ_P^R/D) characterizing the erosion damage with the particle kinetic energy, Ω, can be obtained by fitting its values with the corresponding different values of Ω. Note that the thickness h_{tbc} and porosity f_0 of the ceramic layer are fixed when simulating the key influencing factors for δ^R/D and δ_P^R/D, and consequently, these quantities are related to h_{tbc} and f_0 despite being independent of ε_{tbc}^{Y} and ρ_P/ρ_{tbc}. Chen et al. [3] obtained the following at $h_{tbc} = 100$ μm and $f_0 = 0.1$:

(1) During the collision between a particle and a ceramic layer when erosion occurs, if the ceramic layer undergoes significant deformation and a densified zone is formed, a notable pit (or referred to as an indentation) appears. Under these conditions, the dimensionless collision energy can be expressed as follows:

$$\Omega = \frac{\mathrm{U}}{\sigma_{tbc}^{Y} D^3} = \frac{\pi}{12}\left[\frac{\rho_P}{\sigma_{tbc}^{Y}(0)}\right] v_0^2$$

where $\sigma_{tbc}^{Y}(0)$ is the yield strength of the ceramic layer when $\dot{\varepsilon} = 0$. The relations of the residual indentation depth δ^R, the residual depth at the cumulative indentation site δ_P^R, the indentation width on the surface of the ceramic layer w^R, and the rebound velocity of the particle v^R with Ω can be expressed as follows [3, 39]:

$$\frac{\delta^R}{D} = 0.38\sqrt{\Omega}\left(0.1 + 0.84\sqrt{\Omega}\right) \tag{7.15}$$

$$\overline{\delta}_p^R = \frac{\delta_P^R}{D} = 0.011\sqrt{\Omega}\left(1 + 3.1\sqrt{\Omega} + 27\Omega\right) \tag{7.16}$$

$$\overline{w}^R = \frac{w^R}{D} = 2\sqrt{\overline{\delta}^R + \overline{\delta}_p^R - \left(\overline{\delta}^R + \overline{\delta}_p^R\right)^2}\left(\text{when } \overline{\delta}^R + \overline{\delta}_p^R < \frac{D}{2}\right)$$
$$= D\left(when\overline{\delta}^R + \overline{\delta}_p^R > \frac{D}{2}\right) \quad (7.17)$$

$$\frac{v^R}{v} \approx -58\varepsilon_{tbc}^y\left(1 - 2.1\sqrt{\frac{v_0}{v_{tbc}}} - 17\varepsilon_{tbc}^y\right)\left(v_{tbc} = \sqrt{E_{tbc}/\rho_{tbc}}\right) \quad (7.18)$$

The relation between the thickness of the densified zone formed in the ceramic layer and δ^R can be expressed as follows [3]:

$$h_D = 4\sqrt{\delta^R D} = 2.4D\sqrt{\sqrt{\Omega}\left(0.1 + 0.84\sqrt{\Omega}\right)} \quad (7.19)$$

(2) If the velocity of the erosive particle is low, the deformation of the ceramic layer is very small, and no visible residual pit is formed. In this case, the duration of the action of the particle on the ceramic layer, referred to as the duration of indentation under loading t_0, can be expressed as follows [3, 9]:

$$\frac{t_0 v_0}{D} = 2.54\left(\frac{\rho_P v_0^2}{E_{tbc}}\right)^{2/5} \quad (7.20)$$

The indentation depth caused by particle impact is as follows:

$$\frac{\delta^R}{D} = 0.86\left(\frac{\rho_P v_0^2}{E_{tbc}}\right)^{2/5} \quad (7.21)$$

Under these conditions, the stress state inside the ceramic layer can be expressed in the following form:

$$\frac{\sigma_{ij}(t, r, z)}{\left(\rho_P v_0^2\right)^{1/5} E_{tbc}^{4/5}} = f_{ij}\left(\frac{t v_0}{D}, \frac{r}{D}, \frac{z}{D}\right) \quad (7.22)$$

where the stress function f_{ij} can also be determined by numerical simulation.

7.4 Erosion Failure Behavior Analysis Considering Microstructural Effects

There are several questions concerning EB-PVD TBCs. Is it acceptable to simplify their columnar-crystal structure to a homogeneous system composed of columns and gaps? In particular, the intercolumnar gaps are connected via dendritic and columnar crystals; thus, is it appropriate to simplify the gaps to completely unstressed voids? In addition, the columnar crystals themselves contain pores, which is why they can become densified under compression. Thus, is it acceptable to neglect the intrinsic pores in the columnar crystals and their effects on the mechanical performance of the coating in the simplified model? To answer these questions, the author of this book proposes an FE modeling method for TBCs based on their true microstructure (see Chap. 2 for details) and introduces a Gurson model for porous media and a foam yield model to describe the yielding of columnar crystals and pores, respectively, and, on this basis, investigates the erosion behavior and characteristics of EB-PVD TBCs.

7.4.1 Numerical Model of the True Microstructure of an EB-PVD TBC

The method for establishing an FE geometric model of a TBC that accounts for its microstructural effects based on its real microstructure from scanning electron microscopy (SEM) images (i.e., the image-based FEM) is described in detail in Chap. 2 and, therefore, is not described here. Instead, a geometric model of the coating is directly provided in Fig. 7.14 (the columnar crystals and gaps are shown in blue and yellow, respectively). To compare the real and simplified structures, a simplified model is also established. During erosion, the elastic wave generated by particle impact also affects the zone near the contact zone. Due to the limitations in the SEM system in terms of imaging resolution and scale, the length of the real coating microstructure is approximately 370 μm in the x-direction. We continue to use simplified structures for other zones. The columnar crystals and gaps are 12 and 3 μm in size, respectively, and the particle diameter is 100 μm. We make the following five assumptions: (1) the columnar crystals as well as the thermal growth oxide (TGO) and transition layers are isotropic materials; (2) all layers are well bonded, and there is no delamination; (3) the phase transformation and sintering effects of the ceramic layer are not considered; (4) the formation of kink bands is the main erosion failure mode; and (5) the TBC is in a plane strain state. The boundary conditions described in Sect. 6.2.2 are used again here.

Baither et al. [48] measured the yield strength of the columnar crystals σ_{TBC}^{y} and the yield stress of the gap structure σ_{Foam}^{y} in an EB-PVD TBC at 1000 °C as 600 and 20 MPa, respectively. The density of the spherical erosive particle is 3900 kg·m^{-3}, and its modulus and strength are much higher than those of the coating. Thus, the particle is simplified to a rigid body. See Table 7.3 for the material parameters of other layers, including the transition and TGO layers [49].

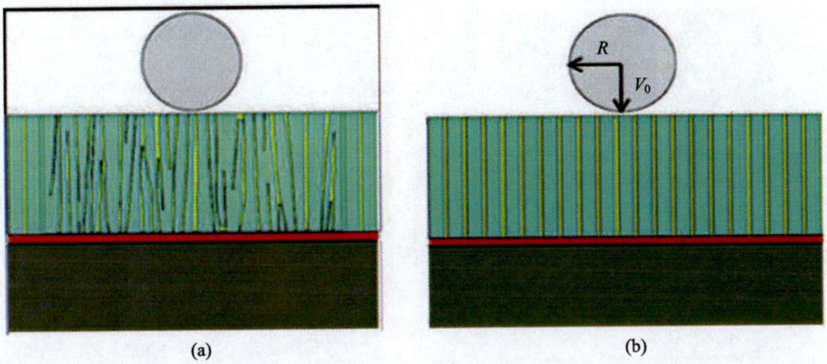

Fig. 7.14 A simplified FE model of the erosion of an EB-PVD TBC: **a** real microstructure, **b** simplified structure

Table 7.3 Material parameters of various layers of the TBC subjected to erosion

Material	E (GPa)	ν	Density ρ (kg/m^3)	Yield strength σ_y (MPa)	Tangent modulus E_T (GPa)
YSZ	200	0.25	3610	600	–
Foam	0.5	0.1	295	–	–
TGO	320	0.25	4000	–	–
BC	110	0.33	7380	200	5

7.4.2 Yield Conditions Considering the Microstructure

As shown in Fig. 7.15a, the columnar crystals in a TBC are a porous material, which is the primary cause of the plastic deformation of and kink-band formation in the coating during high-temperature erosion. Evans et al. [50] obtained stress–strain curves for a YSZ coating at high temperatures from isostatic pressure tests, as shown in Fig. 7.15b. The coating significantly yielded when the stress exceeded a certain value, after which stress and strain hardening was insignificant and the coating was close to an ideal elastoplastic material. Thus, an EB-PVD ceramic layer subjected to high-temperature erosion can be simplified to an ideal elastoplastic material.

In 1981, based on the Gurson model, Tvergaard [51] proposed the following yield model for porous media (also known as the Gurson–Tvergaard–Needleman model):

$$F=\left(\frac{\sigma_e}{\sigma_y}\right)^2+2q_1 f\cosh\left(q_2\frac{3P}{\sigma_{TBC}^y}\right)-q_3 f^2-1=0 \tag{7.23}$$

This yield condition was written by Crowell into the commercial FE software ABAQUS in 2012 and can be very easily numerically solved. In Eq. (7.23), σ_{TBC}^y is the yield strength of the columnar crystals, $\sigma_e=\sqrt{(3/2)\mathbf{S}:\mathbf{S}}$ is the equivalent von Mises stress, **S** is the Cauchy stress deviator, $P=-1/3\sigma:\mathbf{I}$ is the hydrostatic pressure, f is the porosity of the columnar crystals ($f=0.1$), and parameters $q_1=q_2=q_3=1$.

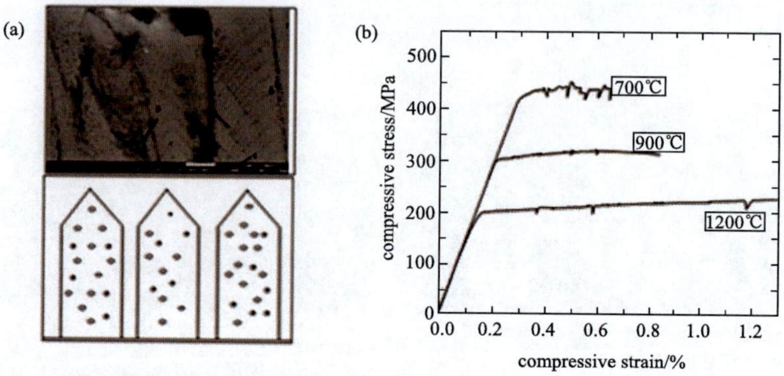

Fig. 7.15 **a** Schematic diagram of porous columnar crystals; **b** isostatic compressive stress–strain curves of YSZ

The gaps between columnar crystals are often connected via feather-like dendritic crystals to the columnar crystals and are similar to a foam material. Thus, the following yield condition for low-density foam materials [52] can be used to describe the yield of the gaps due to erosion:

$$F=\sqrt{(\sigma_e)^2+\alpha^2P^2}-\alpha P_c=0 \tag{7.24}$$

where P_c is the yield strength under hydrostatic compression and $\alpha = 3k/\sqrt{9-k^2}$ is the shape factor of the elliptic yield surface ($k = 1$ is the ratio of the yield strength under hydrostatic tension to that under hydrostatic compression).

7.4.3 Correlation Analysis of Various Parameters in the Erosion Process

Figure 7.16 shows the evolution of the indentation depth with time at different particle energy levels. While the real microstructure geometrically differs from the simplified structure, the variations in the dimensionless indentation depths in both structures with time display similar patterns, and the maximum dimensionless depths in both structures are very close. This is because an ideal elastoplastic model is used as the columnar-crystal material in both structures. In each structure, when the particle collides with the coating, all the material near the collision zone instantaneously reaches the yield state. In addition, Fig. 7.16 compares the variations in the maximum dimensionless indentation depths ($\bar{\delta}_{max} = \delta_{max}/H_{TBC}$) in the real microstructure and simplified structure with the dimensionless kinetic energy (0–0.32). The variations in the maximum dimensionless indentation depths in both structures with the dimensionless kinetic energy also display similar patterns. Specifically, a high particle

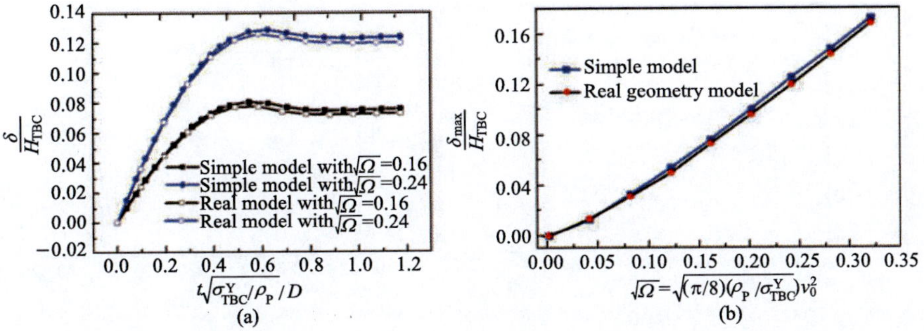

Fig. 7.16 **a** Relations between the indentation depth and time and **b** relations between the maximum indentation depth and particle energy for the two structures

energy level leads to a large maximum indentation depth; at a given particle energy level, the maximum dimensionless indentation depths in the real microstructure and simplified structure are very close. The fitted equations for the real and simplified models, as shown in Fig. 7.16, are obtained by fitting with the dimensionless kinetic energy as an independent variable and the maximum dimensionless indentation depth as a dependent variable. The fitted equation for the simplified structure is as follows:

$$\overline{\delta}_{\max} \approx 0.32\sqrt{\Omega} + 0.35\left(\sqrt{\Omega}\right)^2 \tag{7.25}$$

The fitted equation for the real microstructure is as follows:

$$\overline{\delta}_{\max} \approx 0.32\sqrt{\Omega} + 0.29\left(\sqrt{\Omega}\right)^2 \tag{7.26}$$

The coefficients in Eqs. (7.27) and (7.28) reflect that the extent of the dependence of the indentation depth on the particle kinetic energy is related to the material parameters of the TBC. Because the real microstructure of the TBC is identical to the simplified structure in terms of yield strength, Young's modulus, density, and coating, their fitted relations between the maximum indentation depth and the kinetic energy of the particle as well as the coefficients are very similar. The small difference in the coefficient of proportionality is considered to be derived from the difference between the parameters of the two structures. As mentioned in Sect. 7.1, the elastic modulus, to a certain extent, reflects the erosion resistance of a coating. Thus, different structures differ in the equivalent elastic modulus and, therefore, the indentation depth. Here, let us take low-density gaps as an example. As the elastic modulus of low-density gaps is much lower than that of columnar crystals, we can introduce the overall equivalent elastic modulus of the coating as follows[aa]:

$$E_{eq} = E_{eq0}\exp(-Bp) \tag{7.27}$$

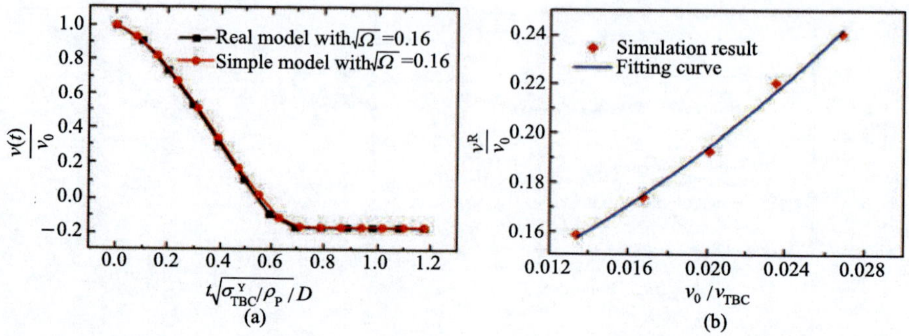

Fig. 7.17 **a** Relation between the particle velocity and the erosion time; **b** relation between the residual and initial velocities of the particle

where E_{eq0} is the elastic modulus of gapless columnar crystals, B is a constant greater than zero, and p is the ratio of the gap area to the columnar-crystal area ($p = 0.241$ and 0.25 for the real and simplified structures, respectively). The small difference between the p values of the two leads to an insignificant difference between their maximum indentation depths.

Figure 7.17 shows the variation in the dimensionless velocity of the particle, $v(t)/v_0$, for both the real microstructure and simplified structure with dimensionless time with a particle kinetic energy fixed at 0.16. The particle velocity continuously decreases and reaches zero at an indentation time of 0.59, at which time the particle rebounds. Because the ceramic layer is an ideal elastoplastic material, the stress in the indented and deformed zone should be equal to the yield strength of the coating. Thus, the contact pressure can be described as follows:

$$F_q = C\pi a^2 \sigma_{TBC}^y \tag{7.28}$$

where C is a plastic constraint factor that relies on the indentation depth (generally, $C = 1$–2 for TBCs) [50] and a is the contact radius, which can be expressed as follows:

$$a = \sqrt{R^2 - (R - \delta)^2} = \sqrt{2R\delta - \delta^2} \tag{7.29}$$

where $R = D/2$ is the radius of the particle. Evidently, when the real microstructure is close to the simplified structure in terms of both the yield strength of the coating and the indentation depth, their contact radii and pressures are identical. According to Newton's second law, the particle velocity is related to the pressure through the following equation:

$$F_q \Delta t = m\Delta v \tag{7.30}$$

Because the real microstructure and simplified structure are identical in terms of the contact pressure F_q and the particle mass, their particle velocities are also identical, as shown in Fig. 7.17. Therefore, because the real microstructure and simplified structure are essentially similar in terms of material parameters and geometry during the erosion process, the deformation behaviors of the particle and coating after erosion in the two structures are basically identical. Figure 7.17 shows the following fitted relation between the residual and initial velocities of the particle:

$$v_R/v_0 \approx 0.092\exp\left(50v_0/\sqrt{E_{TBC}/\rho_{TBC}}\right) + 0.012 \tag{7.31}$$

7.4.4 Analysis of Typical Erosion Failure Modes

The formation of kink bands and their cracks is a notable and dangerous failure mode of EB-PVD TBCs and has been found in TBCs on the blades of turbine engines in service, as shown in Fig. 7.18. Kink bands refer to the bending of columnar crystals along the densified zone where high compressive stresses are present. Because the columnar crystals in a brittle YSZ coating have a low strain tolerance, the formation of kink bands is accompanied by the formation, propagation, and transmission of cracks, which cause the columnar crystals to fracture and, ultimately, macrocracks to form inside the coating and the coating to spall. In addition, as indicated in Fig. 7.18b, kink bands are formed only in the vicinity of the plastic zone and begin to grow only after the indentation depth in the columnar crystals exceeds a certain value. Thus, studying the evolutionary characteristics, formation conditions (e.g., the critical particle velocity and energy), and influencing factors of kink bands is the key to understanding the erosion failure mechanisms of TBCs.

Figure 7.19 shows the whole process of the erosion of both the real and simplified coating structures by a particle at a velocity of 150 m/s and an angle of 90° as

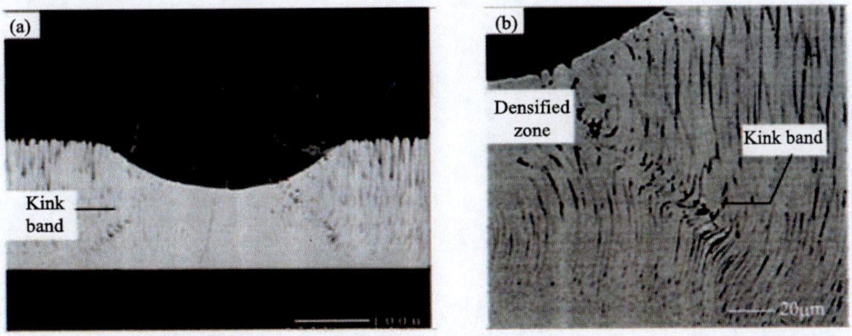

Fig. 7.18 Kink bands and microcracks in an EB-PVD coating caused by erosion [53]

well as the corresponding formation of kink bands. There is indeed a critical indentation depth for kink-band formation, with values of 0.047 and 0.051 in the real microstructure and simplified structure, respectively. In each case, after the critical depth is exceeded, the kink bands intensify as the indentation depth of the particle increases and reaches its maximum level as the maximum indentation depth of the particle is reached.

Figure 7.20 shows the effects of particle energy on kink bands. In both the real microstructure and the simplified structure, the kink bands become increasingly prominent as the particle energy increases. In addition, no kink bands are formed at a low particle energy level. For example, at a particle energy of 0.128, basically no kink bands are formed in the simplified structure, whereas the formation of kink bands can be observed in the real microstructure. As the particle energy increases slightly (e.g., to 0.132), notable kink bands can be observed. Thus, 0.128 can be treated as the particle energy threshold for kink-band formation in the real coating microstructure.

As shown in Figs. 7.19 and 7.20, the formed kink bands are basically symmetrical about the impact zone, which is consistent with the experimental results in Fig. 7.18. In addition, some findings show that kink bands are related to the width of columnar

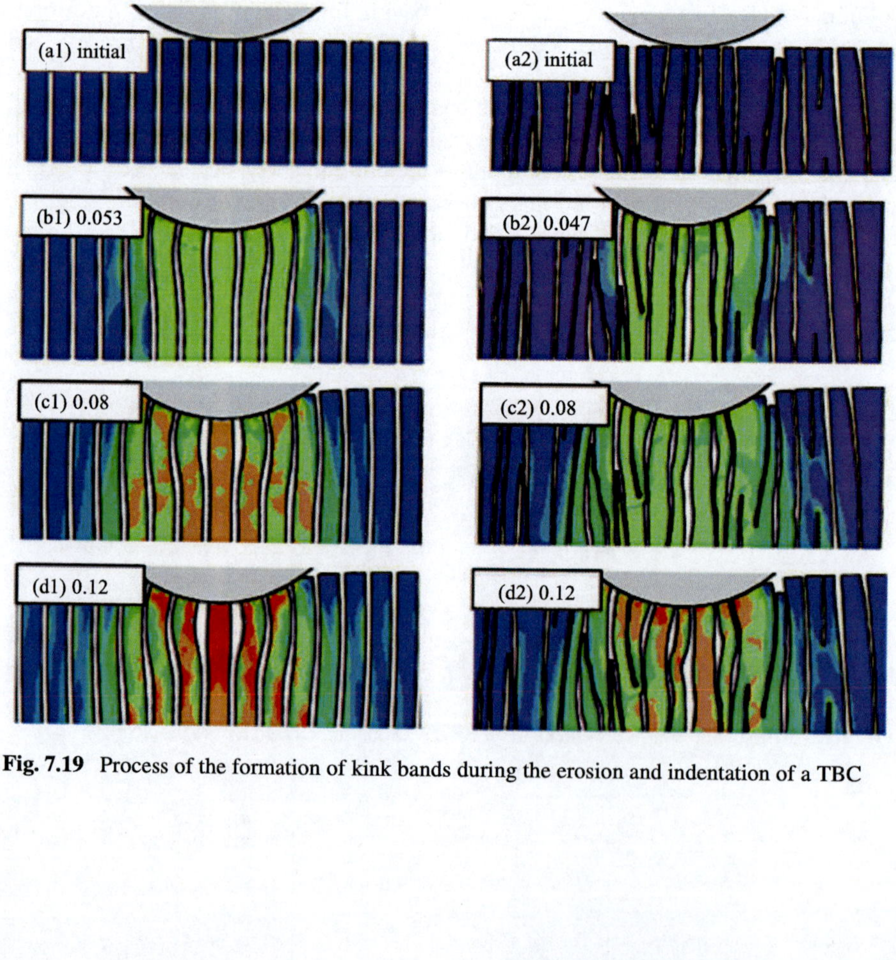

Fig. 7.19 Process of the formation of kink bands during the erosion and indentation of a TBC

(a) $\sqrt{\Omega}$=0.128

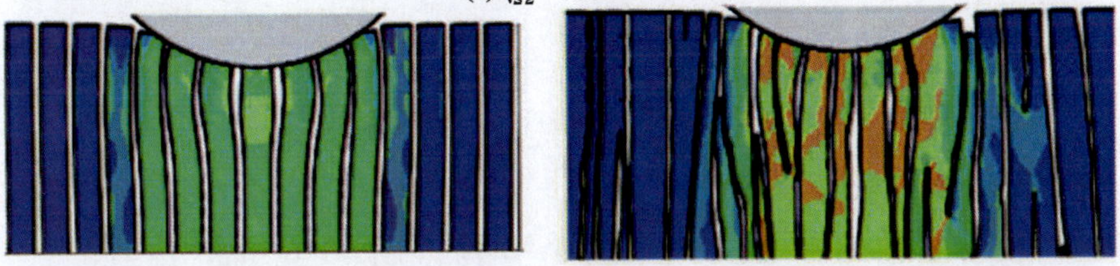

(b) $\sqrt{\Omega}$=0.132

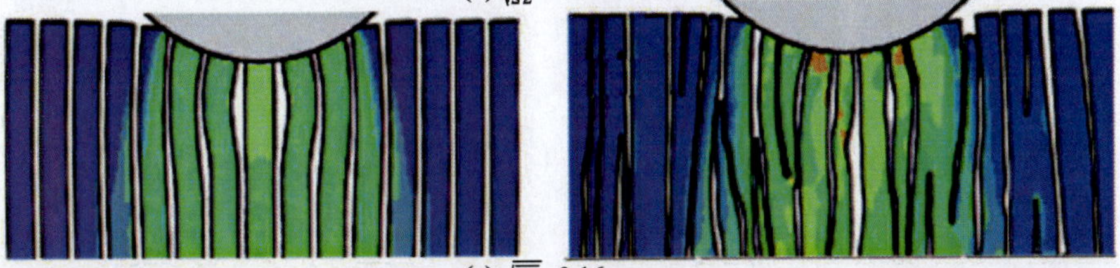

(c) $\sqrt{\Omega}$=0.16

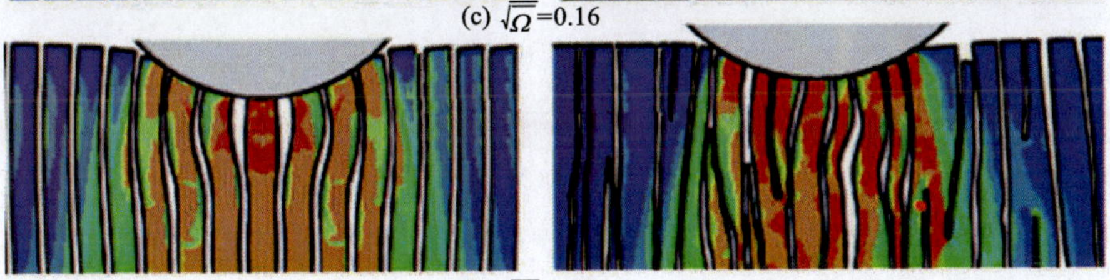

(d) $\sqrt{\Omega}$=0.28

Fig. 7.20 Effects of erosive-particle energy on kink bands

crystals and that thinner columns are more prone to developing kink bands. For example, as seen in Fig. 7.20, at a particle kinetic energy of 0.16, kink bands have already been formed, whereas the wide columns do not bend until the energy reaches 0.28. This phenomenon was experimentally observed by Vaughn and Hutchinsons. Moreover, it can be seen from Fig. 7.20 that while displaying a tendency toward symmetry, the kink bands in the real microstructure form irregular patterns in local zones due to the inconsistent microstructural characteristics (e.g., widths and gaps) of the columns. As a result, the simulated kink bands in the simplified structure differ from the experimental results.

The formation of kink bands in a TBC is a result of the sudden release of the intercolumn constraint due to plastic deformation. Drawing on Hutchinson's plastic bifurcation model for columns [54], Wei et al. [55] proposed a microbuckling and bridging kink-band model (Fig. 7.21), the characteristics of which are similar to the structural characteristics of the kink bands formed in the columnar crystals in a TBC. The indentation depth δ is related to the kink-band angle of columnar crystals, θ, through the following equation:

$$\frac{\delta}{2h_f} = 2s\sin\theta + 2w(1-\cos\theta) + 2w\varepsilon_{TBC}^{fc} \tag{7.32}$$

where

$$s = S/h_f = 2\sqrt{1+2w\lambda\tan\theta + w^2\tan^2\theta}\cos\left(\psi - \frac{2}{3}\pi\right) - w\tan\theta \tag{7.33}$$

$$\psi = \frac{1}{3}\cos^{-1}\left\{\frac{-\left|\frac{\omega\hat{\sigma}_{TBC}^{fc}}{2\cos\theta} - (1+w\tan\theta)^3 - (\lambda-1)\left(1+3w^2\tan^2\theta\right)\right|}{\left(1+2w\lambda\tan\theta + w^2\tan^2\theta\right)^{3/2}}\right\} \tag{7.34}$$

$\hat{\sigma}_{TBC}^{fc} = \frac{\sigma_{TBC}^{fc}}{E_f}, \sigma_{TBC}^{fc} = E_{TBC}\varepsilon_{TBC}^{y} + E_{TBC}^{t}\left(\varepsilon_{TBC}^{fc} - \varepsilon_{TBC}^{y}\right), \varepsilon_{TBC}^{y} = \frac{\sigma_{TBC}^{y}}{E_{TBC}}, r = \frac{E_{TBC}^{t}}{E_{TBC}},$
$\lambda = \frac{1+r}{1-r}, \omega = \frac{12w^2}{1-r}, w = \frac{W}{h_f}.$

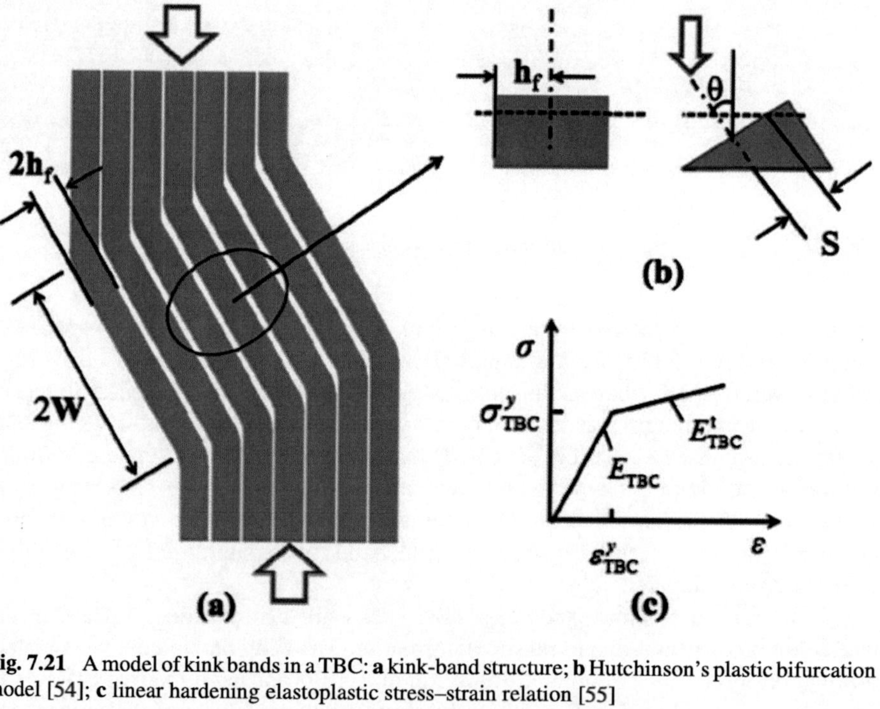

Fig. 7.21 A model of kink bands in a TBC: **a** kink-band structure; **b** Hutchinson's plastic bifurcation model [54]; **c** linear hardening elastoplastic stress–strain relation [55]

In Fig. 7.21, W and h_f are the half-widths of the kink bands and columnar crystals, respectively, and ε_{TBC}^{y} and ε_{TBC}^{fc} are the yield strain of the columnar crystals and the critical strain for kink-band formation, respectively.

Substitution of Eq. (7.26) into Eq. (7.32) yields the following relation between the critical particle energy for kink-band formation and the kink-band angle θ:

$$0.32\sqrt{\Omega} + 0.29\left(\sqrt{\Omega}\right)^2 - \frac{4h_f}{H_{TBC}}\left(s\sin\theta + w(1-\cos\theta) + w\varepsilon_{TBC}^{fc}\right) = 0 \quad (7.35)$$

The following values are derived from the finite element (FE) simulation results: $W = 18$ μm, $h_f = 4$ μm, and $\varepsilon_{TBC}^{fc} = 0.5\%$. In addition, we set the yield strength σ_{TBC}^{y}, Young's modulus E_{TBC}, and linear hardening modulus E_{TBC}^{t} of the coating to 600 MPa, 200 GPa, and 50 GPa, respectively. By substituting these values into Eq. (7.35), we find, as shown in Fig. 7.22, that the critical energy for kink-band formation varies from 0.11 to 0.13 as θ varies in the range of 16°–20°. This conclusion is consistent with the numerical simulation results. As shown in Fig. 7.20, no kink bands are formed when the particle energy is lower than 0.128.

The kink bands and cracks near the interface experimentally observed in the EB-PVD coating during erosion are closely related to the tensile stress σ_{22} in the erosion process. Figure 7.23 shows the evolution of the stresses in the real coating microstructure with time at a particle kinetic energy of 0.32. During the initial stage of particle indentation ($t = 4$ ns), compressive stresses are generated directly beneath the impact zone and propagate downward along the columnar crystals, while tensile stresses are generated at the edges of the columnar crystals near the collision zone. As shown in Fig. 7.18a, the maximum σ_{22} at point A reaches 136 MPa. As the particle continues to indent the coating downward ($t = 24$ ns), the compressive stresses propagate continuously, while the maximum tensile stresses at the edges of the columnar crystals propagate outward away from the collision zone. As shown in Fig. 7.18b, the maximum tensile stress at point B increases to 409 MPa. These stresses

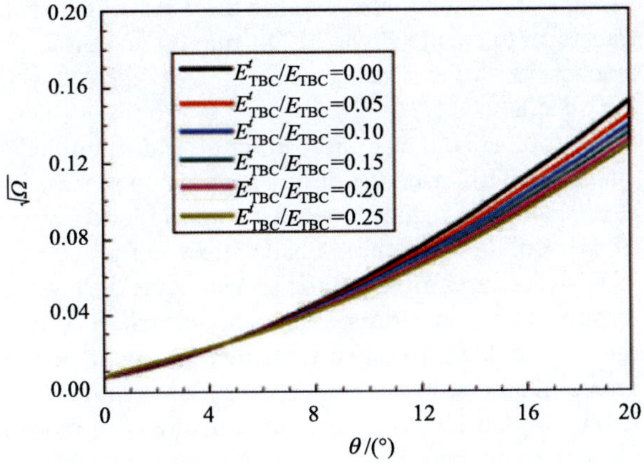

Fig. 7.22 Relation between the critical particle energy for kink-band formation and the kink-band angle

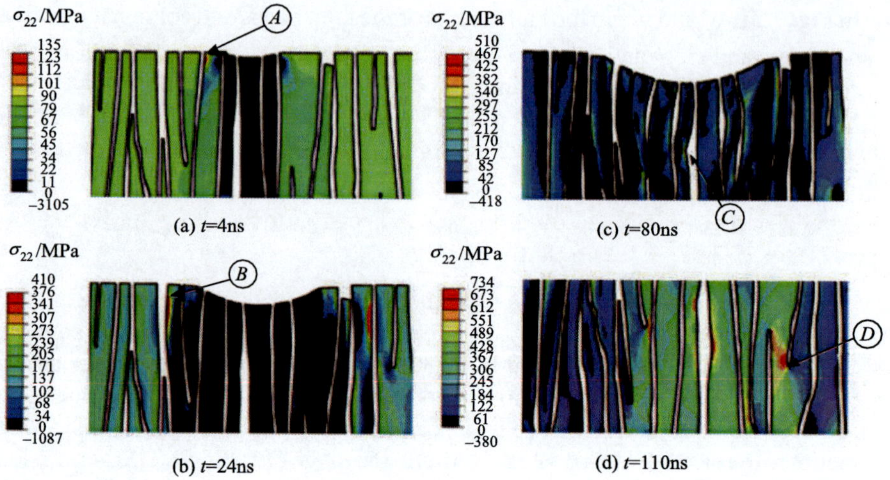

Fig. 7.23 Evolution of the stresses in the real coating microstructure with time at a particle kinetic energy of 0.32

may lead to the formation of cracks on the coating surface similar to those in Fig. 7.5. As the particle further indents the coating, as shown in Fig. 7.18c, a large area of the coating is subjected to compressive stresses as the contact radius increases, resulting in the bending of the columns (i.e., the formation of kink bands) and the formation of tensile stresses in local zones, which, in turn, leads to the formation of cracks within the kink bands. As the indentation depth of the particle reaches its maximum ($t =$ 80 ns), the kink bands are compressed to the extreme, and then the particle begins to rebound, resulting in the formation of rebound tensile stresses within the kink bands. For example, the maximum tensile stress at point C reaches 500 MPa; this point represents a location that is often the origin of cracking in the kink bands. When the compressive stresses propagate to the ceramic-layer/transition-layer interface ($t =$ 110 ns), all the internal stresses in the zones of the coating in contact with the particle transform to tensile stresses. The maximum tensile stress, which occurs at point D, reaches 735 MPa. These tensile stresses lead to the formation of cracks within the kink bands.

To compare the real microstructure and simplified structure, the stress field in the simplified TBC structure during the erosion process is obtained numerically. Similar stress evolution patterns are observed in the two structures. During the initial stage of erosion, the maximum tensile stress is formed near the erosion zone. Compressive stresses are formed in the erosion zone during the indentation process, whereas tensile stresses are formed after the particle rebounds. However, the magnitude of the stress field in the real microstructure, due to its irregular structure and defects, is greater than that in the simplified structure. Figure 7.24 shows the evolution of the maximum tensile stress in each structure with erosion time under the same conditions (i.e., the particle velocity and radius as well as material parameters). During

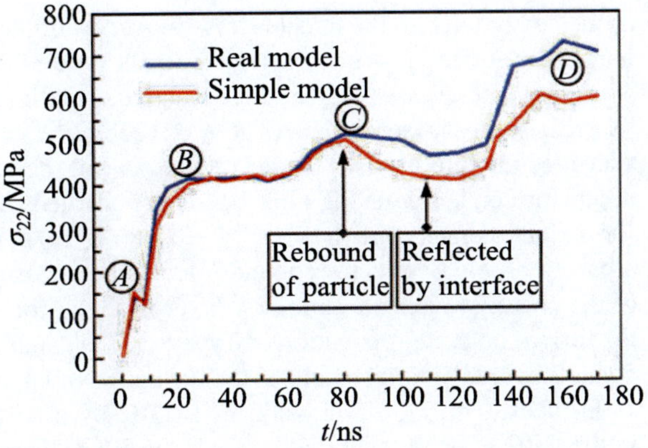

Fig. 7.24 Evolution of the maximum tensile stress in each structure with time

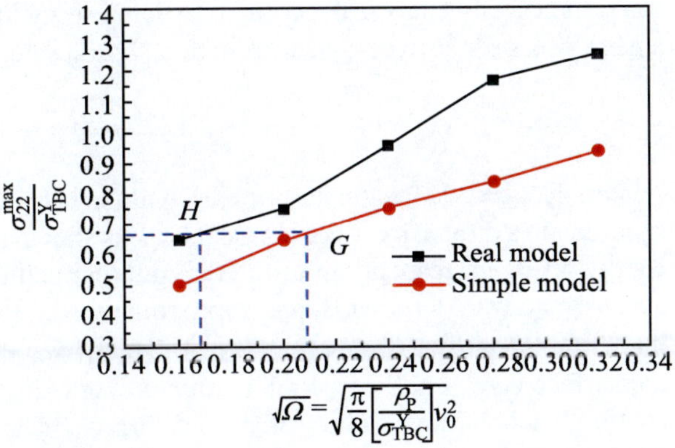

Fig. 7.25 Relation between the maximum residual stress in each structure and the particle energy

the initial stage of erosion, the location and magnitude of the maximum tensile stress in the simplified structure are basically consistent with those in the real microstructure model. After the particle rebounds, the maximum tensile stress appears near the TGO/BC interface in each structure; however, the maximum tensile stress in the real microstructure remains higher than that in the simplified model. When the dimensionless kinetic energy of the impacting particle is 0.32, after the particle rebounds, the maximum tensile stress values inside the coating are approximately 600 and 735 MPa in the simplified structure and real microstructure, respectively. Figure 7.25 shows the relations between the maximum residual stress inside the coating and the particle energy after the completion of the erosion process. Evidently, as the particle energy increases, the residual stress increases and is higher in the real microstructure than in the simplified structure. Thus, the real microstructure model of the TBC is in fact more prone to erosion than the simplified model.

The experimental results for the erosion of an EB-PVD TBC in Fig. 7.5 show that while no kink bands are formed and the coating does not undergo plastic deformation over a large area, a small number of surface cracks are formed in the near-surface zone of the coating due to the tensile stresses. According to the analysis of the stress field during the erosion process, the tensile stress at point B in Fig. 7.23 reaches 410 MPa (Fig. 7.24). Let us assume that there exists a 2-μm-long microcrack at point B. Then, its stress intensity factor (SIF) of $K = 1.122\sigma\sqrt{\pi a} = 1.12\,\text{MPa}\sqrt{\text{m}}\sqrt{\text{m}}$ has already exceeded the fracture toughness (1 MPa $\sqrt{m}$) of the coating [56]. As the indentation depth further increases, if kink bands are formed, the tensile stresses lead to the formation of cracks with a SIF of $K = 1.28$ MPa $\sqrt{m}$ near the kink bands. When the erosive particle begins to rebound, the tensile stresses in a considerably large area of the coating under the action of the erosive particle exceed 400 MPa, resulting in the formation and propagation of Type I cracks capable of penetrating the columnar crystals, which, in turn, results in the formation of large cracks, as shown in Fig. 7.6d.

Evidently, during the erosion of a TBC by a particle, the large cracks formed in the TBC after the rebound of the particle are dangerous, and their driving force depends, in essence, on the coating parameters and the particle energy. Figure 7.25 can satisfactorily analyze the critical particle energy for crack formation. The critical tensile stress for surface-crack formation should satisfy

$$K_{IC} - 1.122\sigma_c\sqrt{\pi a} = 0 \tag{7.36}$$

When $K_{IC} = 1$ MPa $\sqrt{m}$ the critical tensile stress is approximately 366 MPa (corresponding to a dimensionless stress of 0.61). As shown in Fig. 7.25, the critical particle energy levels corresponding to the real microstructure and simplified structure are approximately 0.16 (point H) and approximately 0.195 (point G), respectively. When the particle energy exceeds the critical value (e.g., when $\sqrt{\Omega} > 0.32$), the SIFs of the surface cracks in the real microstructure and simplified structure models of the coating reach 2.06 $\text{MPa}\sqrt{\text{m}}\sqrt{\text{m}}$ and 1.56 $\text{MPa}\sqrt{\text{m}}\sqrt{\text{m}}$, respectively, both of which exceed the fracture toughness of Type I cracks in the coating, thereby forming cracks. In addition, these data suggest that the real microstructure model is more prone to erosion failure than the simplified model.

7.5 Erosion Failure Mechanisms and Resistance Indices of TBCs

The theoretical analysis of the erosion failure mechanisms of TBCs includes the establishment of local failure models, the provision of theoretical analytical solutions to the criteria and key influencing factors for the erosion failure of TBCs, and reasonable explanations to experimental phenomena and numerical simulation results, based on which the optimum design of TBC preparation processes is obtained. In this regard, Chen et al., Fleck et al., and the author of this book have fully investigated the theoretical erosion failure models of EB-PVD TBCs [3, 7–13, 37–40].

As EB-PVD TBCs and APS TBCs differ significantly in failure modes, their theoretical erosion analysis models also differ considerably and are therefore discussed separately in this section.

7.5.1 Erosion Resistance Index of EB-PVD TBCs

By summarizing the experimental results obtained by Nicholls et al. and based on the erosion failure modes of EB-PVD TBCs, Chen et al. and Fleck et al. summarized the following three failure mechanisms of EB-PVD TBCs [3, 7, 10].

Mechanism I: lack of notable residual deformation, also referred to as formation of a dynamic elastic zone (Fig. 7.26a). This mechanism occurs when the particle is both small in size and low in kinetic energy. Under these conditions, no notable densified zone is formed in the ceramic layer. The interactions between the particle and the ceramic layer are completely elastic. In the shallow surface zone of the ceramic layer, local columnar units close to the erosive particle are subjected to tension, which may result in the fracture of the columnar units in this zone. The fracture of a large number of columnar units leads to the formation of transverse cracks, as shown in Fig. 7.26a.

Mechanism II: formation of a densified and deformed zone, also known as formation of a quasistatic elastic indentation. This mechanism occurs when both the size and kinetic energy of the erosive particle are moderate. Under these conditions, the ceramic layer undergoes significant deformation, resulting in the formation of a notable densified layer. The densified layer may also delaminate from the ceramic layer underneath, as shown in Figs. 7.26b and 7.5, corresponding to erosion failure mode I (i.e., erosion due to densification).

Mechanism III: formation of a notable residual pit, also referred to as formation of a quasistatic plastic indentation. This mechanism occurs when the particle has both a high velocity and a large size. Under this condition, a notable densified layer, a prominent residual pit, pronounced delamination cracks, and visible kink bands are formed in the ceramic layer, as shown in Figs. 7.26c and 7.6. This mechanism corresponds to erosion failure mode III (i.e., damage by an external particle).

By summarizing FE numerical simulation results, Chen et al. found that the stress and deformation leading to the erosion failure of a TBC are mainly functions of the dimensionless kinetic energy of the particle Ω, time $t\sqrt{\sigma_{tbc}^{Y}/\rho_P}/D$, and location r/D, that the residual quantity at a specific location is a function of only Ω, and that the specific stress function can be determined by numerical simulation. On this basis, failure criteria were established for the above three failure mechanisms [10].

Under failure mechanism I, the main interaction between the particle and the ceramic layer is the elastic energy exchange. The response time and impact/indentation depth of the particle as well as the stress form under this mechanism are given by Eqs. (7.22)–(7.24), respectively. The cracks formed in the ceramic layer consist primarily of transverse cracks in single columnar-crystal units (these cracks are perpendicular to the axes of the columnar units), as shown in Fig. 7.26a.

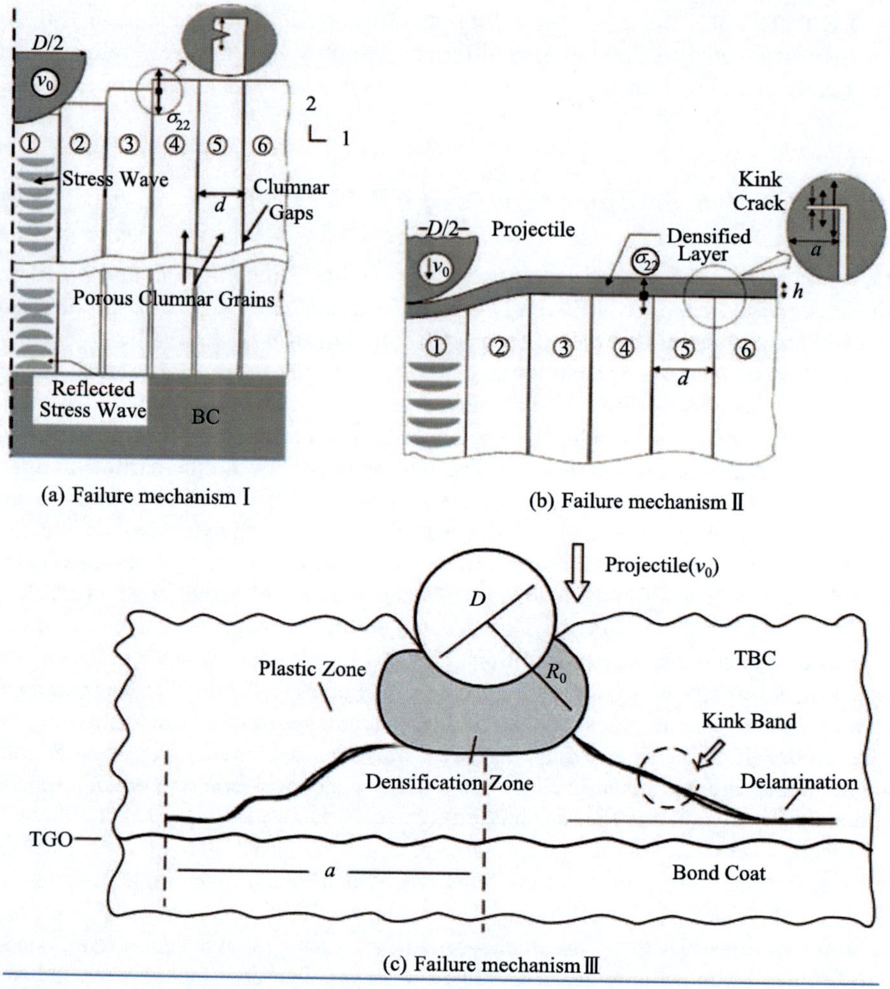

Fig. 7.26 Schematic diagrams of the three failure mechanisms of EB-PVD TBCs: **a** dynamic elastic zone; **b** quasistatic elastic indentation; **c** quasistatic plastic indentation [10]

The stress σ_{22} induces the propagation of the transverse cracks and results in Type I crack propagation mode. The amplitude of the change in dimensionless σ_{22} and its mean, $\Delta\sum = \Delta\sigma_{22}/\left(\left(\rho_P v_0^2\right)^{1/5} E_{tbc}^{4/5}\right)$ and $\overline{\sum} = \overline{\sigma}_{22}/\left(\left(\rho_P v_0^2\right)^{1/5} E_{tbc}^{4/5}\right)$, respectively, during the collision process can be determined from the FE simulation results. Under the action of σ_{22}, the SIF and energy-release rate of the crack propagation can be expressed as follows:

$$K_I \approx \sqrt{\pi a_0}[\Delta\sigma_{22} + \overline{\sigma}_{22}],\ G \approx \frac{\pi a_0}{E_{tbc}}[\Delta\sigma_{22} + \overline{\sigma}_{22}]^2 \tag{7.37}$$

The criterion for the separation of a single columnar-crystal unit is that the energy-release rate is equal to the fracture toughness of the ceramic layer; i.e., $G = \Gamma_{tbc}$. Under this condition, the crack length is comparable to the diameter of the columnar-crystal unit and thus can be expressed by d. With this failure mechanism, the following crack propagation criterion can be established based on Eq. (7.37):

$$\pi\left[\sum + \Delta\sum\right]^2 \geq \frac{\Gamma_{tbc}}{\left(\rho_P v_0^2\right)^{2/5} E_{tbc}^{3/5} d} \tag{7.38}$$

where υ_P and v depend on the erosive particle (i.e., external conditions). Then, we can define the erosion failure resistance index of a TBC by this failure mechanism as $\Xi_1 = \Gamma_{tbc}/E_{tbc}^{3/5} d$. Evidently, a high Ξ_1 results in a low erosion rate and, correspondingly, good erosion performance by the TBC.

Under failure mechanism II, erosion leads to the formation of a densified layer in the ceramic layer. The gaps between the columnar structural units in the ceramic layer can be viewed as vertical cracks, as shown by the inset in the upper right corner of Fig. 7.26b. Under the action of σ_{22}, the vertical cracks kink toward the position with a low strength, i.e., the interface between the densified and ceramic layers. During this process, due to the presence of transverse cracks and vertical cracks (i.e., gaps), σ_{22} results in both opening (Type I) and dislocation (Type II) propagation of the cracks. According to the FE simulation results, the SIF values can be expressed as follows:

$$\frac{K_I}{\Delta\sigma_{22}\sqrt{a}} = 1.76 - \frac{2.24a}{d} \quad \text{and} \quad \frac{K_{II}}{\Delta\sigma_{22}\sqrt{a}} = 0.20 - \frac{0.17a}{d} \tag{7.39}$$

Assuming that the delamination does not propagate, $K_I \to 0$ and $a/d = 0.75$ based on Eq. (7.39). If the cracks propagate only in single columnar-crystal units, the model is equivalent to a simple tensile stress model. In this case, $K_I/\overline{\sigma}_{22}\sqrt{a} = 1.76$. The densified layer can separate completely from the ceramic layer underneath only when $a/d \to 1$. According to the FE simulation result of $\sigma_{22} \sim 1/h_D$, the erosion failure resistance index of a TBC by this failure mechanism is defined by $\Xi_2 = \Gamma_{tbc}/E_{tbc}^{3/5} d\sigma_{tbc}^{Y}$. Clearly, a high Ξ_2 results in a low erosion rate and, correspondingly, good erosion performance by the TBC.

Under failure mechanism III, erosion results in large deformation and a notable residual pit in the ceramic layer, as shown in Fig. 7.26c. It has been shown that in this case, kink bands lead to delamination, which is strongly related to the residual stress in the eroded ceramic layer. The following equation must be satisfied for delamination [42].

$$\Delta_{th} = \frac{\sigma_{tbc}^{Y}(\delta_{\max} D)^{1/4}}{\sqrt{E_{tbc}\Gamma_{tbc}}} \tag{7.40}$$

When large cracks are formed, $\Delta_{th} \geq 0.3\times10^{-2}$. On this basis, a residual pit depth $\delta_{\max} \geq 25\ \mu\text{m}$ is derived as the precondition for a TBC to fail by this mechanism at 1150 °C, which is consistent with the experimental results in Figs. 7.5a and 7.6b. When the depth of the densified layer or the pit is less than 20 μm, as shown in Figs. 7.5a and 7.6b, no cracks are formed over a large area.

As shown in Fig. 7.26c, when the quasistatic plastic indentation zone is large enough to approach the TGO layer, the kink bands extend to the TGO layer and form delamination at the ceramic/TGO interface. The delamination propagation satisfies the following equation [3]:

$$\frac{a_{delam}}{h_{tbc}} = \left[\frac{\left(\sigma_{tbc}^{Y}\right)^{2} h_{tbc}}{E_{tbc}\Gamma_{tbc}}\right]\left[\frac{\sigma_{tbc}^{Y}}{E_{tbc}}\right]^{\alpha}\left[\frac{\rho_{P} v_{0}^{2}}{\sigma_{tbc}^{Y}}\right]^{\beta} \tag{7.41}$$

where $\alpha = 1$ and $\beta = 1/4$. As suggested by this conclusion, the index that determines the erosion performance of a TBC by this failure mechanism is $\Xi_3 = \Gamma_{tbc}E_{tbc}^{2}/\left(\sigma_{tbc}^{Y}\right)^{3}$. This index indicates that after the delamination, the TBC has a high resistance to erosion and spallation owing to its low yield strength and high fracture toughness.

7.5.2 Resistance Index of APS TBCs

As the erosion failure of an APS TBC is mainly manifested by the delamination of the layered units, its erosion performance depends primarily on the thickness and bonding of the layered units. On this basis, Li et al. proposed a simple model to evaluate the erosion performance of an APS TBC [22]. In this model, the erosion rate R_c of an APS TBC is defined as follows:

$$R_c = C\frac{h_l}{\lambda} \tag{7.42}$$

where h_l is the unit-layer thickness, λ is the inter-unit-layer bond ratio, and C is a constant related to the erosion test conditions.

Assuming that a proportion K_e of the total energy of the erosive particles causes unit-layer delamination, the total energy of N_p impacting particles causing the delamination of unit layers is as follows:

$$E_n = \frac{1}{2}K_e N_p m v_0^2 \tag{7.43}$$

where m and v_0 are the particle mass and velocity, respectively.

Let us assume that the ceramic layer comprises layered units composed of connected ideal particles. Let h_l be the thickness of each layer and S_0 be the contact surface area between each particle and the ceramic layer underneath. According to the law of conservation of energy, if the action of N_p impacting particles leads to the separation of N_c particles in the unit layers of the ceramic layer, we have

$$E_n = 2\lambda\gamma_c N_c S_0 \tag{7.44}$$

where γ_c is the surface energy of a layered unit.

According to Eqs. (7.44) and (7.45), we have

$$\frac{N_p m}{N_c} = \frac{4\lambda\gamma_c S_0}{k_e v_0^2} \tag{7.45}$$

It is experimentally found that the mass of a coating lost due to erosion, W_c, is directly proportional to the total mass of the erosive particles, W_p. In addition, we have the following:

$$N_c = \frac{W_c}{S_0 h_l \rho_{tbc}} \text{ and } N_p = \frac{W_a}{m} \tag{7.46}$$

Let us define the erosion resistance as $A_c = W_a \rho_{tbc}/W_c$, then substitution of Eq. (7.46) into Eq. (7.45) yields

$$A_c = \frac{2\gamma_c\lambda}{E_{eff} h_l} \tag{7.47}$$

where $E_{eff} = 0.5k_e v^2$ is the effective particle energy per unit mass used to delaminate the coating and $2\gamma_c\lambda$ is the fracture toughness of the ceramic layer. Clearly, the erosion performance of an APS TBC depends on the fracture toughness of the ceramic layer, the kinetic energy of the particles, and the thickness of the layered units.

7.6 Erosion Failure Mechanism Diagrams of TBCs

Myriad parameters affect the performance of TBCs. As a result, it is very difficult to determine the relations between the erosion damage and the parameters of particles and coatings by theoretical analysis. In most cases, their relations are empirically determined by numerical simulations combined with experiments. In addition, the empirical relations vary with different types of environments, coatings, and modes, making it difficult to use them to guide practical applications. Establishing a failure mechanism diagram capable of determining whether erosion occurs in TBCs in various service environments and, if so, the main failure modes can provide a direct basis for understanding their failure mechanisms and guiding their applications. Hence, researchers have attempted to establish erosion failure mechanism diagrams of TBCs based on theoretical and numerical simulation results. Using EB-PVD TBCs as an example, we introduce in the following the main approach for establishing erosion failure mechanism diagrams of TBCs and the main results.

7.6.1 Establishment of a Failure Mechanism Diagram from a Theoretical Perspective

Fleck and Zisis provided analytical solutions for the three erosion failure mechanisms of EB-PVD TBCs and, on this basis, produced their failure mechanism diagram [7].

First, several other dimensionless parameters need to be defined, namely, $\overline{b} = b/R$ (b is the surface contact radius of the pit, and R is the particle radius), load $\overline{P} = P/(E_{tbc}R^2)$, pressure $\overline{p}_{av} = P/(\pi b^2 E_{tbc})$, velocity $\overline{v} = v\sqrt{\rho_P/E_{tbc}}$, and $\overline{\delta} = \delta/h_{tbc}$. As described earlier, the dimensionless particle velocity can be derived from $\overline{v} = v\sqrt{\rho_P}/(\sqrt{\rho_{tbc}}c)$. $c = \sqrt{E_{tbc}/\rho_{tbc}}$ is the velocity of sound in the ceramic layer.

Under failure mechanism I (i.e., when a dynamic elastic zone is formed), there is an elastic collision between the particle and the ceramic layer, which can be viewed as the propagation of an elastic wave in the ceramic layer. As shown in Fig. 7.12, the indentation follows the geometric relationship below

$$b^2 = 2R\delta \tag{7.48}$$

When propagating at a velocity of $c = \sqrt{E_{tbc}/\rho_{tbc}}$ in the columnar structure units in the ceramic layer, the elastic wave generates a sound pressure of $p_{ed} = \rho_{tbc}cv = v\sqrt{E_{tbc}\rho_{tbc}}$, where v is the instantaneous velocity of the particle. Evidently, the maximum pressure generated by the elastic wave is $p_{\max} = v_0\sqrt{E_{tbc}\rho_{tbc}}$, where v_0 is the initial velocity of the particle. The instantaneous contact pressure can be expressed as

$$P_{ed} = \pi b^2 p_{ed} = 2\pi R\delta v\sqrt{E_{tbc}\rho_{tbc}} \tag{7.49}$$

or in the following dimensionless form:

$$\overline{P}_{ed} = \frac{P_{ed}}{R^2 E_{tbc}} = 2\pi\sqrt{\frac{\rho_{tbc}}{\rho_P}\frac{h_{tbc}}{R}}\overline{v}\overline{\delta} \tag{7.50}$$

By applying Newton's second law, we have

$$\Delta\overline{v}_{ed} = -\frac{3}{4}\sqrt{\frac{\rho_{TBC}}{\rho_p}}\left(\frac{h_{tbc}}{R}\right)^2\left[\Delta\overline{\delta}^2\right] \tag{7.51}$$

Equation (7.51) shows the relation between the velocity and the impact/indentation depth of the particle. The maximum impact/indentation depth is achieved at $v = 0$ and can be expressed as

$$\Delta\overline{\delta}^2_{ed,\max} = \frac{4}{3}\sqrt{\frac{\rho_p}{\rho_{tbc}}}\left(\frac{R}{h_{tbc}}\right)^2 \overline{v}_0 \tag{7.52}$$

The elastic wave reflects upon propagating to the interface of the TBC. Then, the maximum propagation duration of the elastic wave in the ceramic layer is as follows:

$$t_{ed,end} = \frac{2h_{tbc}}{c} \tag{7.53}$$

Under failure mechanism II (i.e., when a quasistatic elastic indentation is formed), each columnar-crystal unit in the ceramic layer undergoes elastic deformation. As shown in Fig. 7.11, the displacement in the x_2-direction at a distance of r from the center can be expressed as follows:

$$u = \delta - \frac{r^2}{D} \tag{7.54}$$

Then, the contact load is given by

$$\left.\begin{array}{l} P_{es} - \int_0^b 2\pi r p_{es}(r)dr \\ p_{es}(r) = E_{tbc}\dfrac{u}{h_{tbc}} \end{array}\right\} \Rightarrow P_{es} = \frac{2\pi E_{tbc}}{h_{tbc}}\left[\frac{\delta b^2}{2} - \frac{b^4}{4D}\right]$$

By substituting the geometric relationship in Eq. (7.44) into the above equation, we obtain the contact load and its dimensionless form as follows:

$$P_{es} = \frac{\pi E_{tbc} R\delta^2}{h_{tbc}} \text{ and } \overline{P}_{es} = \pi\left(\frac{h_{tbc}}{R}\right)\overline{\delta}^2 \tag{7.55}$$

The mean stress and its dimensionless form are expressed as follows:

$$p_{es} = \frac{E_{tbc}\delta}{2h_{tbc}}, \overline{p}_{es} = \frac{p_{es}}{E_{tbc}} = \left(\frac{E_{tbc}\delta}{2h_{tbc}}\right)/E_{TBC} = \frac{1}{2}\overline{\delta} \tag{7.56}$$

By applying Newton's second law, we have

$$\Delta\left(\overline{v}^2_{es}\right) = -\frac{1}{2}\left(\frac{h_{tbc}}{R}\right)^2 \Delta\left(\overline{\delta}\right)^3 \tag{7.57}$$

Equation (7.57) provides the relation between the impact/indentation depth and instantaneous velocity of the particle under the quasistatic elastic indentation mechanism. At $v = 0$, the maximum impact/indentation depth of the particle under the quasistatic elastic indentation mechanism can be calculated as follows:

$$\overline{\delta}^3_{es,\max} = 2\overline{v}_0^2\left(\frac{R}{h_{tbc}}\right)^2 \tag{7.58}$$

Under failure mechanism III (i.e., when a quasistatic plastic indentation is formed), the columnar-crystal units undergo large deformations, and the mean contact stress is related to the yield strength of the ceramic layer through the following empirical relation:

$$p_{ps} = C\sigma^Y_{tbc} \tag{7.59}$$

where C is a constant related to the shape of the indenter and the yield strain ($C = 3$ when a spherical indenter indents the surface of an ideal rigid-plastic material). Fleck and Zisis proved that C ranges from 1 to 2 for ideal elastoplastic TBCs [7]. Then, at an indentation depth of δ, the contact load and its dimensionless form can be expressed as follows:

$$P_{ps} = C\sigma^Y_{tbc}\pi b^2 \text{ and } \overline{P}_{ps} = \frac{P_{ps}}{R^2 E_{tbc}} = 2\pi C\varepsilon^Y_{tbc}\left(\frac{h_{tbc}}{R}\right)\overline{\delta} \tag{7.60}$$

By applying Newton's second law, $\int P_{ps}dt = \int madt$, we have

$$\Delta\left(\overline{v}^2_{ps}\right) = -\frac{3}{2}C\varepsilon^Y_{tbc}\left(\frac{h_{tbc}}{R}\right)^2\Delta\left(\overline{\delta}\right)^2 \tag{7.61}$$

Through similar analysis, we obtain the maximum impact/indentation depth of the particle as follows:

$$\overline{\delta}^2_{ps,\max} = \frac{2}{3}\frac{1}{C\varepsilon^Y_{tbc}}\left(\frac{R}{h_{tbc}}\right)^2\left(\overline{v}_0^2\right) \tag{7.62}$$

As such, we can see that the relation between the velocity and impact/indentation depth of the particle under each failure mechanism can be determined based on Eqs. (7.51), (7.57), or (7.61) and that the relation between the maximum impact/indentation depth and initial velocity of the particle can be determined based on Eqs. (7.52), (7.58), or (7.62). To examine the correctness of the abovementioned analytical expressions, Fleck et al. analyzed and computed the stress and velocity inside a TBC during the erosion process by FE simulations. They found a good match between the two, thereby validating the correctness of the theoretical analytical expressions [7]. On this basis, Fleck et al. established a failure mechanism diagram

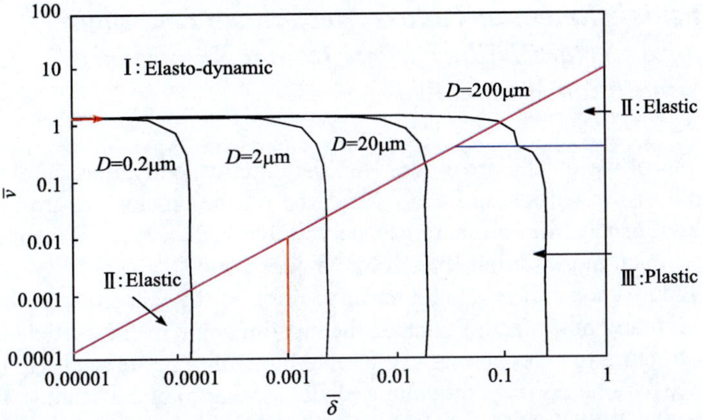

Fig. 7.27 Failure mechanism diagram of TBCs subjected to erosion by particles of different diameters

[7] for TBCs for different erosive particle diameters D and with $\overline{v} = v\sqrt{\rho_P/E_{tbc}}$ and $\overline{\delta} = \delta/h_{tbc}$ as parameters, as shown in Fig. 7.27. The failure mechanism diagram is summarized as follows:

(1) Under failure mechanism I (i.e., when a dynamic elastic zone is formed), $\overline{v} = \overline{v}_{ed}$, and its relation with $\overline{\delta}$ is given by Eq. (7.51). Under failure mechanism II (i.e., when a quasistatic elastic zone is formed), $\overline{v} = \overline{v}_{es}$, and its relation with $\overline{\delta}$ is given by Eq. (7.57). Under failure mechanism III (i.e., when a quasistatic plastic zone is formed), $\overline{v} = \overline{v}_{ps}$, and its relation with $\overline{\delta}$ is given by Eq. (7.61); $C = 2$.

(2) Boundary values under various failure mechanisms. Under failure mechanism I, $t_{ed,end} = \frac{2h_{tbc}}{c}$ and is evidently very short. Thus, the decrease in the particle velocity given by Eq. (7.53) is very small relative to the change in the displacement. Hence, the particle velocity can be approximately considered to be directly proportional to the displacement during this process; i.e., $\delta = vt$. Then, we have $\overline{\delta} = 2\overline{v}\sqrt{\rho_{tbc}/\rho_p}$, as shown by the purple line in Fig. 7.27. Under failure mechanism II, the quasistatic elastic indentation zone stops at the location where the strain reaches the yield strain of the ceramic layer; i.e., $\overline{\delta}_{es/ps} = \delta/h_{tbc} = \varepsilon_{TBC}^{Y}$, as shown by the red line in Fig. 7.27. If the particle velocity is so high that it results in a very large strain rate (e.g., $\dot{\varepsilon}_c = v/h_{tbc} = 10^{-6}s^{-1}$), the strain rate effect should be considered. Under this condition, another type of boundary appears between failure mechanisms II and III, as shown by the blue line in Fig. 7.27.

7.6.2 Establishment of Failure Mechanism Diagrams for a Certain Failure Mode from a Numerical Simulation Perspective

The relations of various failure modes with the initial velocity, diameter, and indentation depth of the particle are analyzed based on the erosion failure mechanism diagram established from a theoretical perspective in Fig. 7.27. Evidently, failure mode III is much more serious than failure modes I and II. According to the earlier analysis, failure mode III is related to, in addition to the velocity and diameter of the particle, many other factors such as the erosion angle of the particle as well as the microstructure (e.g., columnar-crystal and intercolumnar-crystal-gap sizes) and performance parameters (e.g., modulus and yield strength) of the coating. Therefore, it is extremely difficult to theoretically establish the correlations between various influencing parameters and the erosion performance. Fortunately, the numerical calculation can be conveniently used to predict the erosion failure of the coating. Here, using the erosion failure mode III as an example, we introduce a method for analyzing failure mechanism diagrams [57].

(1) Basic model and equations

Figure 7.28 shows the basic structure of an 8YSZ EB-PVD TBC that contains a 100-μm-thick ceramic layer, a 100-μm-thick transition layer, and a 100-μm-thick substrate material, and a coating length of 0.98 mm is used for numerical simulations. The columnar-crystal and intercolumnar-crystal-gap widths are 6 and 2 μm, respectively. The diameter of the particle is 100 μm and is varied for analysis. In addition, we assume that the ceramic and transition layers and the substrate are all isotropic materials and are well bonded. Moreover, other factors (e.g., TGOs and phase transformation in the coating) are not considered.

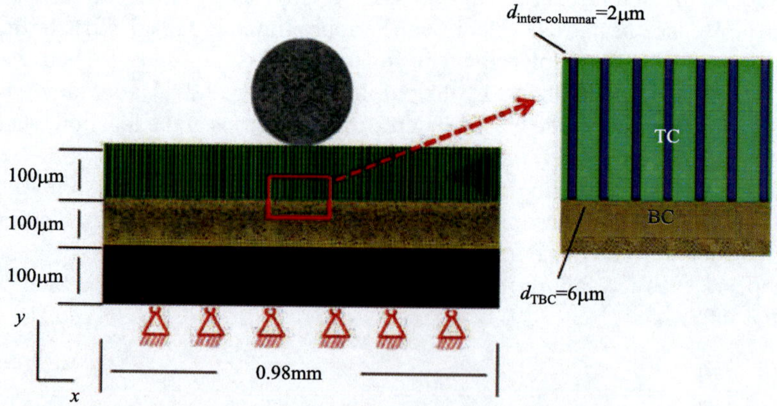

Fig. 7.28 Geometric model, boundary conditions, and mesh distribution of an EB-PVD TBC used for numerical simulations

Figure 7.28 shows the FE mesh and boundary conditions. The mesh is generated using four-node plane-strain reduced integration elements (CPE4R) and contains a total of 101,508 cells. To prevent computational nonconvergence due to the excessive deformation of the mesh, the arbitrary Lagrangian–Eulerian (ALE) algorithm is implemented in the collision zone. The boundary conditions of the model are defined as follows: the displacements in the x- and y-directions at the bottom of the substrate are zero; i.e., $u = 0$, and $v = 0$.

The yield criteria for porous media and low-density foam materials described earlier are selected for the columnar crystals and intercolumnar-crystal gaps, respectively. The yield condition for the columnar crystals is as follows:

$$F = (\frac{\sigma_e}{\sigma_y})^2 + 2q_1 f \cosh(q_2 \frac{3p}{2\sigma_y}) - (q_3 f^2 + 1) = 0 \tag{7.63}$$

where σ_{TBC}^{y} is the yield strength of the columnar crystals, $\sigma_e = \sqrt{(3/2)\mathbf{S} : \mathbf{S}}$ is the equivalent von Mises stress, $\mathbf{S}$ is the Cauchy stress deviator, $P = -1/3\sigma : \mathbf{I}$ is the hydrostatic pressure, f is the porosity of the columnar crystals ($f = 0.1$), and parameters $q_1 = q_2 = q_3 = 1$.

The yield condition for the intercolumnar-crystal gaps is as follows:

$$F = \sqrt{(\sigma_e)^2 + \alpha^2 P^2} - \alpha P_c = 0 \tag{7.64}$$

where P_c is the hydrostatic compressive yield strength and $\alpha = 3k/\sqrt{9 - k^2}$ is the shape factor of the elliptical yield surface ($k = 1$ is the ratio of the hydrostatic tensile yield strength to the hydrostatic compressive yield strength).

To analyze the effects of the particle as well as the microstructure and performance of the coating on the erosion failure of the coating, we consider changes in the following three types of parameters: (1) the parameters of the particle, namely, its radius (25–125 μm), velocity (50–200 m/s), and erosion angle (30–90°); (2) the microstructural parameters of the coating, namely, the columnar-crystal size (2–20 μm) and the intercolumnar-crystal-gap size (1–4 μm); and (3) the performance of the coating, namely, the Young's modulus (22–200 GPa) and yield strength (400–600 MPa) of the columnar crystals.

If only mode III is considered, the crack formation criterion established earlier based on Chen et al.'s theoretical analysis for crack formation under this mechanism can be used as the failure criterion for erosion mode III; i.e.,

$$\Delta_{th} = \frac{\sigma_{\mathrm{TBC}}^{y} (2\delta_{\max} R)^{1/4}}{\sqrt{E_{\mathrm{TBC}} \Gamma_{\mathrm{TBC}}}} \tag{7.65}$$

where $\delta_{\max}$ is the maximum indentation depth of the erosive particle, Γ_{TBC} is the fracture toughness of the ceramic layer, and R is the radius of the erosive particle. It is generally agreed that at $\Delta_{th} \geq 3 \times 10^{-3}$, plastic cracks begin to form, i.e., failure mode III occurs. Equation (7.65) shows that when the properties of the erosive particle and coating are determined, the radius of the particle as well as the yield strength, Young's modulus, and fracture toughness of the coating are all fixed values. Then, crack formation is related only to the maximum indentation depth in the coating. When erosion failure mechanism diagrams are established, if there is no particular need to analyze the effects of the performance parameters of the coating, the following parameter values are generally used: $\sigma_{\mathrm{TBC}}^{y} = 400$ MPa, $E_{\mathrm{TBC}} = 200$ GPa, and $\Gamma_{\mathrm{TBC}} = 50$ J/m^2 [28].

(2) Erosion failure mechanism diagrams of the TBC

Figure 7.29 shows the failure mechanism diagrams of the TBC eroded by a particle at a certain angle and with different radii and velocities. The black curve is the boundary of crack formation in columnar crystals, which is obtained from Eq. (7.63). The columnar crystals and the intercolumnar-crystal gaps are 6 and 2 μm wide, respectively. The green zone means that no cracks are formed; i.e., the coating is safe. The yellow zone means that cracks may be formed, i.e., the coating is at risk. When the particle impacts the coating at a velocity of 100 m/s, as the radius of the particle increases from 25 to 100 μm, the indentation depth increases from 6.14 to 20.52 μm. Thus, at a given particle velocity, there is a positive correlation between the indentation depth and the radius of the erosive particle. In other words, a particle with a large radius is very likely to cause crack formation. Similarly, when the radius of the impacting particle is 50 μm, as the impact velocity of the particle increases from 50 to 200 m/s, the indentation depth increases from 4.58 to 16.51 μm. When the radius of the erosive particle is fixed, a high particle velocity results in a large maximum indentation depth and is very likely to cause crack formation. At an erosion velocity of 100 m/s, the critical particle radius for crack formation in the coating is approximately 58 μm. A comparison of the erosion at angles of 60° and 30° in Fig. 7.29 shows that when the particle impacts the coating at a velocity of 50 m/s, the critical values of the particle radius for crack formation in the coating are approximately 85, 95, and 125 μm at incident angles of 90°, 60°, and 30°, respectively. This comparison reveals that the incident angle of the erosive particle, to a certain extent, affects coating failure. A large incident angle of the particle is very likely to cause coating failure, which is in good agreement with the experimental results.

Wang et al. [58] and Wellman and Nicholls [1] found experimentally that the erosion rate of a TBC with a columnar-crystal structure was related to the columnar-crystal structure (e.g., column and intercolumn-gap widths) in addition to external conditions. Thus, we analyze the erosion behavior of the coating at different columnar-crystal widths and intercolumnar-crystal-gap widths under impact by a particle with a radius of 50 μm at a velocity of 100 m/s, as shown in Fig. 7.30. A comparison of all the points in the figure at columnar-crystal widths of 6, 10, and 20 μm shows that a small columnar-crystal width leads to a large maximum indentation depth and is highly likely to cause crack formation in the coating. However, when the velocity and radius of the particle are fixed, the maximum indentation depths at the three columnar-crystal widths are close to each other. A comparison

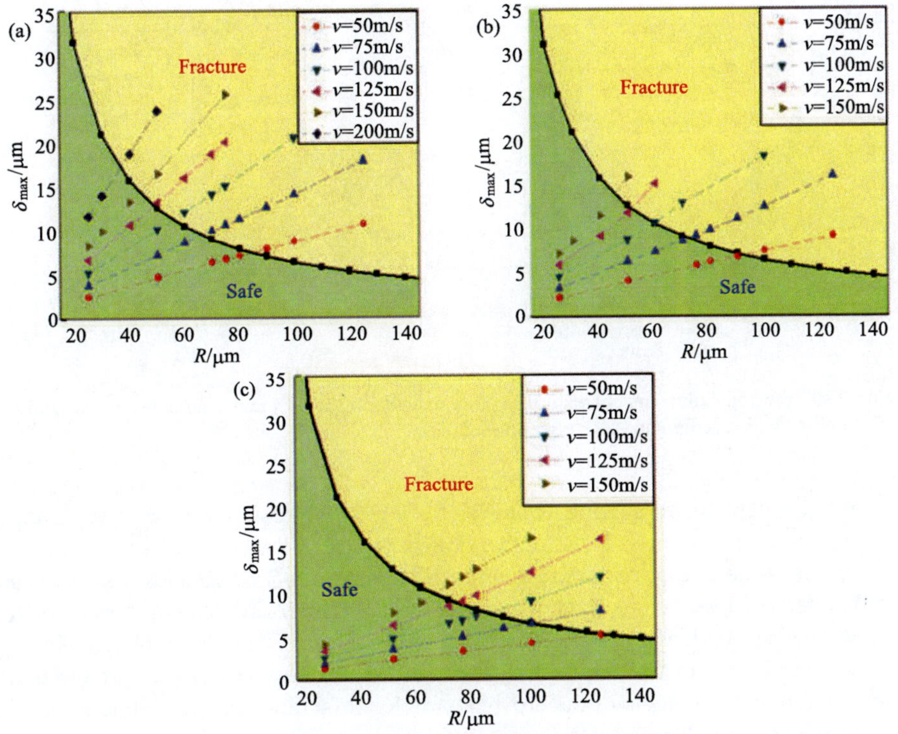

Fig. 7.29 Failure mechanism diagrams considering different radii and velocities of the particle: **a**; **b** $\theta = 60°$; **c** $\theta = 30°$

of the maximum indentation depths at intercolumnar-crystal-gap widths of 1, 2, and 4 μm, respectively, shows the following. Similar to the pattern found at different columnar-crystal widths, while a small intercolumnar-crystal-gap width results in a large maximum indentation depth and is very likely to cause crack formation in the coating, the maximum indentation depths at the three intercolumnar-crystal-gap widths are similar when the velocity and radius of the particle are fixed. Thus, it can be concluded that compared to the particle parameters, the coating microstructure (columnar-crystal and intercolumnar-crystal-gap widths) has a relatively weak effect on erosion failure behavior, particularly in mode III.

Figure 7.31 shows the erosion failure mechanism diagrams corresponding to different values of Young's modulus and yield strength of the columnar crystals when the radius, incident velocity, and angle of the particle are 25–150 μm, 100 m/s, and 90°, respectively, and the columnar-crystal and intercolumnar-crystal-gap widths are 6 and 2 μm, respectively. The effects of Young's modulus of the columnar crystals on the failure threshold are basically negligible, as shown in Fig. 7.31a, whereas the indentation depth increases as Young's modulus increases. The effect of the yield strength of the coating on its erosion failure behavior is essentially similar to that of its Young's modulus. Specifically, a low yield strength leads to a large maximum

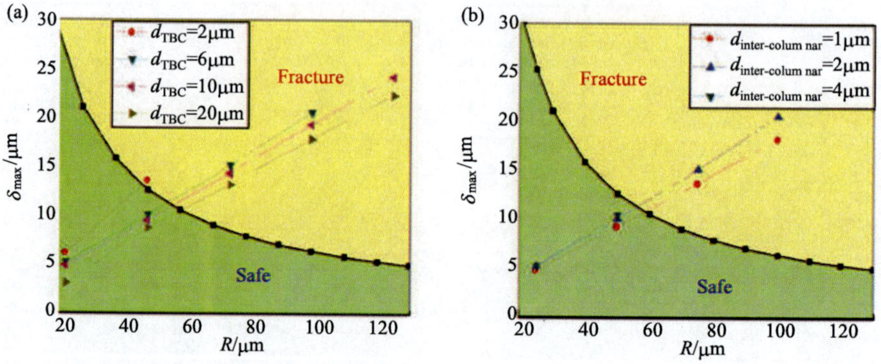

Fig. 7.30 Erosion failure mechanism diagrams considering **a** different columnar-crystal widths and **b** different intercolumnar-crystal-gap widths

indentation depth in the coating. However, the yield strength of the coating has a more significant effect on the failure threshold than its Young's modulus. In addition, the coating is prone to erosion failure at a low yield strength. This conclusion is similar to that derived by Chen et al. and suggests that a high erosion temperature leads to a low yield strength and therefore facilitates crack formation in the kink bands.

A combination of the above analyses shows that a high velocity, a large radius, and a large incident angle of the erosive particle all facilitate crack formation in the coating, that a small columnar-crystal width and a small intercolumnar-crystal-gap width both lead to a large maximum indentation depth and facilitate crack formation, and that the coating is prone to cracking at low Young's modulus and low yield strength.

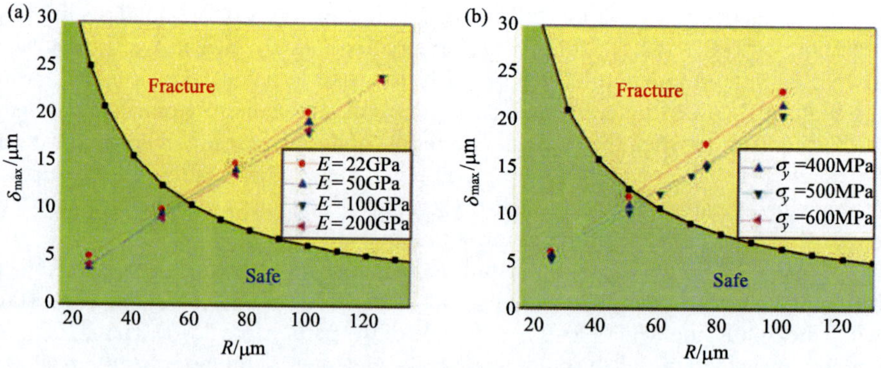

Fig. 7.31 Erosion failure mechanism diagrams of the coating corresponding to different values of **a** Young's modulus and **b** yield strength of the coating

7.7 Summary and Outlook

7.7.1 Summary

This chapter elucidates the erosion failure phenomena, patterns, and mechanisms of TBCs while focusing on their erosion failure behavior, parameter correlations, and failure mechanism diagrams based on numerical simulations. This chapter is summarized as follows:

(1) The erosion failure phenomena and patterns as well as typical failure modes of TBCs prepared by PS, PVD, and new processes (e.g., PS-PVD) are analyzed and summarized.
(2) The correlations of the parameters of an erosive particle (loading) with the performance and failure behavior of a TBC are established based on dimensionless parametric analysis in combination with numerical simulations of the erosion process, and the effects of the real microstructure of a TBC on its erosion failure behavior are further analyzed.
(3) The erosion resistance indices of TBCs are theoretically analyzed, the erosion failure mechanisms of TBCs are revealed based on numerical simulations combined with experiments, and erosion failure mechanism diagrams are constructed, thereby providing a basis for optimizing and applying TBC preparation processes.

7.7.2 Outlook

Future research can be performed in the following main areas:

(1) Experimentally investigating the erosion failure process of TBCs in key environments (e.g., thermal shock by high-temperature fuel gas and high-speed rotation) is necessary. In particular, developing real-time, in situ detection methods for erosion damage to accurately evaluate and predict the thickness of coatings, the time points and locations of crack formation and propagation, and the extent of the damage sustained by coatings is necessary.
(2) Establishing erosion failure models for TBCs, particularly APS TBCs, is important to provide reasonable and accurate explanations of erosion test results and, on this basis, to determine the correlations of the erosion performance of TBCs with their material and preparation process parameters.
(3) The effects of interfacial oxidation, CMAS corrosion, and erosion in TBCs need to be comprehensively considered to reveal their multimodal, multiparametric failure modes and mechanisms as well as to obtain performance evaluation and optimization indices.

References

1. Wellman RG, Nicholls JR. A review of the erosion of thermal barrier coatings. J Phys D Appl Phys. 2007;40:293–305.
2. Zhou Y, Liu Q, Yang L, Wu D, Mao W. Failure mechanisms and life prediction of thermal barrier coatings. Acta Mech Solida Sin. 2010;31(5):504–31.
3. Chen X, Wang R, Yao N, Evans AG, Hutchinson JW, Bruce RW. Foreign object damage in a thermal barrier system: mechanisms and simulations. Mater Sci Eng, A. 2003;352:221–31.
4. Wellman RG, Nicholls JR, Murphy K. Effect of microstructure and temperature on the erosion rates and mechanisms of modified EB PVD TBCs. Wear. 2009;267:1927–34.
5. Wellman RG, Nicholls JR. Erosion, corrosion and erosion-corrosion of EB PVD thermal barrier coatings. Tribol Int. 2008;41:657–62.
6. Steenbakker RJL, Wellman RG, Nicholls JR. Erosion of gadolinia doped EB PVD TBCs. Surf Coat Technol. 2006;201:2140–6.
7. Fleck NA, Zisis TH. The erosion of EB PVD thermal barrier coatings: The competition between mechanisms. Wear. 2010;268:1214–24.
8. Zisis TH, Fleck NA. The elastic-plastic indentation response of a columnar thermal barrier coating. Wear. 2010;268:443–54.
9. Evans AG, Fleck NA, Faulheber S, Vermaak N, Maloney M, Darolia R. Scaling laws governing the erosion and impact resistance of thermal barrier coatings. Wear. 2006;260:886–94.
10. Chen X, He MY, Spitsberg I, Fleck NA, Hutchinson JW, Evans AG. Mechanisms governing the high temperature erosion of thermal barrier coatings. Wear. 2004;256:735–46.
11. Nicholls JR, Wellman RG. A comparison between the erosion behavior of thermal spray and electron beam physical vapor deposition thermal barrier coatings. Wear. 1999;233–235:352–61.
12. Wellman RG, Deaakin MJ, Nicholls JR. The effect of TBC morphology on the erosion rate of EB PVD TBCs. Wear. 2005;258:349–56.
13. Wellman RG, Nicholls JR. On the effect of ageing on the erosion of EB PVD TBCs. Surf Coat Technol. 2004;177–178:80–8.
14. Zhu DM, Miller RA, Kuczmarski MA. Development and life prediction of erosion resistant turbine low conductivity thermal barrier coatings. NASA/TM;2010. p. 215669
15. Cernuschi F, Lorenzonia L, Capelli S, Guardamagnaa C, Karger M, Vaβen R, von Niessenc K, Markocsand N, Menueye J, Giolli C. Solid particle erosion of thermal spray and physical vapour deposition thermal barrier coatings. Wear. 2011;271:2909–18.
16. Drexler JM, Aygun A, Li DS, Vaßen R, Steinke T, Padture NP. Thermal-gradient testing of thermal barrier coatings under simultaneous attack by molten glassy deposits and its mitigation. Surf Coat Technol. 2010;204:2683–8.
17. Drexler JM, Gledhill AD, Shinoda K, Vasiliev AL, Reddy KM, Sampath S, Padture NP. Jet engine coatings for resisting volcanic ash damage. Adv Mater. 2011;23:2419–24.
18. Mishra SB, Prakash S, Chandra K. Studies on erosion behavior of plasma sprayed coatings on a Ni-based superalloy. Wear. 2006;260:422–32.
19. Westergård R, Axén N, Wiklund U, Hogmark S. An evaluation of plasma sprayed ceramic coatings by erosion, abrasion and bend testing. Wear. 2000;246:12–9.
20. Janos BZ, Lugscheider E, Remer P. Effect of thermal aging on the erosion resistance of air plasma sprayed zirconia thermal barrier coating. Surf Coat Technol. 1999;113:278–85.
21. Ramanujam R, Nakamura T. Erosion mechanisms of thermally sprayed coatings with multiple phases. Surf Coat Technol. 2009;204:42–53.
22. Li CJ, Yang GJ, Ohmori A. Relationship between particle erosion and lamellar microstructure for plasma-sprayed alumina coatings. Wear. 2006;260:1166–72.
23. Guo HB, Vaßen R, Stöver D. Atmospheric plasma sprayed thick thermal barrier coatings with high segmentation crack density. Surf Coat Technol. 2004;186:353–63.
24. Ambühl P, Meyer P. Thermal coating technology in controlled atmospheres (ChamProTM), In: Lugscheider E, Kammer PA, editors. Proceedings of the ITSC. Düsseldorf, Germany: DVS-Verlag;1999. p. 291–92.

25. Rezanka S, Mack DE, Mauer G, Sebold D, Guillon O, Vaßen R. Investigation of the resistance of open-column-structured PS-PVD TBCs to erosive and high-temperature corrosive attack. Surf Coat Technol. 2017;324:222–35.
26. Steinberg L, Mikulla C, Naraparaju R, Toma FL, Großmann H, Schulz U, Leyens C. Erosion resistance of CMAS infiltrated sacrificial suspension sprayed alumina top layer on EB-PVD 7YSZ coatings. Wear. 2019;438–439: 203064.
27. Shin DY, Hamed A. Influence of micro–structure on erosion resistance of plasma sprayed 7YSZ thermal barrier coating under gas turbine operating conditions. Wear. 2018;396–397:34–47.
28. Tian HL, Wang CL, Guo MQ, Gao JG, Cui YJ, Wang FY, Liu EB, Jin G. Erosion resistance and toughening mechanism of AlBO and BNw whiskers modified thermal barrier coatings. Ceram Int. 2020;46:4573–80.
29. Cernuschi F, Guardamagna C, Capelli S, Lorenzoni L, Mack DE, Moscatelli A. Solid particle erosion of standard and advanced thermal barrier coatings. Wear. 2016;348–349:43–51.
30. Steinberg L, Naraparaju R, Heckert M, Mikulla C, Schulz U, Leyens C. Erosion behavior of EB-PVD 7YSZ coatings under corrosion/erosion regime: Effect of TBC microstructure and the CMAS chemistry. J Eur Ceram Soc. 2018;38:5101–12.
31. Hassani S, Bielawski M, Beres W, Martinu L, Balazinski M, Klemberg-Sapieha JE. Predictive tools for the design of erosion resistant coatings. Surf Coat Technol. 2008;203:204–10.
32. Chen J, Beake BD, Wellman RG, Nicholls JR, Dong HS. An investigation into the correlation between nano-impact resistance and erosion performance of EB-PVD thermal barrier coatings on thermal ageing. Surf Coat Technol. 2012;206:4992–8.
33. Algenaid W, Ganvir A, Calinas RF, Varghese J, Rajulapati KV, Joshi S. Influence of microstructure on the erosion behaviour of suspension plasma sprayed thermal barrier coatings. Surf Coat Technol. 2019;375:86–99.
34. Satyapal M, Céline R, Nicholas C, Jonas H, Stefan B, Nicolaie M, Per N. Understanding the effect of material composition and microstructural design on the erosion behavior of plasma sprayed thermal barrier coatings. Appl Surf Sci. 2019;488:170–84.
35. Schmitt MP, Harder BJ, Wolfe DE. Process-structure-property relations for the erosion durability of plasma spray-physical vapor deposition (PS-PVD) thermal barrier coatings. Surf Coat Technol. 2016;297:11–8.
36. Mercer C, Faulhaber S, Evans AG, Darolia R. A delamination mechanism for thermal barrier coatings subjected to calcium-magnesium-alumino-silicate (CMAS) infiltration. Acta Mater. 2005;53:1029–39.
37. Xie XY, Guo HB, Gong SK, Xu HB. Lanthanum–titanium–aluminum oxide: A novel thermal barrier coating material for applications at 1300 °C. J Eur Ceram Soc. 2011;31:1677–83.
38. Chen X. Calcium-magnesium-alumina-silicate (CMAS) delamination mechanism in EB PVD thermal barrier coating. Surf Coat Technol. 2006;200:3418–27.
39. Chen X, Hutchinson JW, Evans AG. Simulation of the high temperature impression of thermal barrier coatings with columnar microstructure. Acta Mater. 2004;52:565–71.
40. Chen X, Hutchinson JW. Particle impact on metal substrates with application to foreign object damage to aircraft engines. J Mech Phys Solids. 2002;50:2269–690.
41. Wellman RG, Nicholls JR. A monte carlo model for predicting the erosion rate of EB PVD TBCs. Wear. 2004;256:889–99.
42. Nicholls JR, Stephenson DJ. Monte carlo modeling of erosion processes. Wear. 1995;186–187:64–77.
43. Fourier JBJ. Analytic theory of heat. New York: Cambridge University Press; 1955.
44. Bukingham E. On physically similar systems: illutrations of the use dimensional analysis [J]. Phys Rev. 1914;4:345–76.
45. Bridgman PW. Dimensional analysis. New Haven: Yale University Press; 1922.
46. Huang Y. Characterization of the stress–strain relations in thin films with nanoindentation. Xiangtan: Xiangtan University; 2006.
47. Ma Z. Characterization of the mechanical properties of metallic thin-film materials with nanoindentation. Xiangtan: Xiangtan University; 2011.

48. Baither D, Bartsch M, Baufeld B, et al. Ferroelastic and plastic deformation of t′-Zirconia single crystals. J Am Ceram Soc. 2004;84(8):1755–62.
49. Yang L, Li HL, Zhou YC, Zhu W, Wei YG, Zhang JP. Erosion failure mechanism of EB-PVD thermal barrier coatings with real morphology. Wear. 2017;392–393:99–108.
50. Evans AG, Mumm DR, Hutchinson JW, et al. Mechanisms controlling the durability of thermal barrier coatings[J]. Prog Mater Sci. 2001;46(5):505–53.
51. Tvergaard V. Influence of voids on shear band instabilities under plane strain conditions[J]. Int J Fract. 1981;17(4):389–407.
52. Deshpande VS, Fleck NA. Isotropic constitutive models for metallic foams[J]. J Mech Phys Solids. 2000;48(6–7):1253–83.
53. Fleck NA, Zisis T. The erosion of EB-PVD thermal barrier coatings: the competition between mechanisms. Wear. 2010;268(11):1214–24.
54. Hutchinson J. Post-bifurcation behavior in the plastic range. J Mech Phys Solids. 1973;21:163–90.
55. Y.G. Wei, W. Yang, K.-Z. Huang, Theoretical and experimental researches of postmicrobuckling for fiber-reinforced composites, Sci. China Ser. A-Math. Phys. Astron. Technol. Sci. 37 (1994) 1077–87
56. Mao WG, Dai CY, Yang L, Zhou YC. Interfacial fracture characteristic and crack propagation of thermal barrier coatings under tensile conditions at elevated temperatures. Int J Fract. 2008;151:107–20.
57. Zhu W, Jin YJ, Yang L, Pi ZP, Zhou YC. Fracture mechanism maps for thermal barrier coatings subjected to single foreign object impact. Wear. 2018;414–415:303–9.
58. Wang M, Fleck NA, Evans AG. Elastodynamic erosion of thermal barrier coatings. J Am Ceram Soc. 2011;94:160–7.

Chapter 8
Basic Mechanical Properties of TBCs and Their Characterization

During service, crack initiation and propagation occur on the surface and at the interfaces of the thermal barrier coating (TBC) on an aeroengine turbine blade, eventually leading to TBC spallation, which rapidly exposes the metallic substrate to high-temperature gas and causes the blade to fracture and, in severe cases, the engine to explode. TBC spallation is affected by a multitude of factors that directly or indirectly promote stress formation inside the TBC. Of the many factors, the mechanical properties of the TBC on a turbine blade are the most crucial factor directly affecting the service life of the TBC. Therefore, the characterization of the mechanical properties of TBCs on turbine blades is vitally important.

The basic mechanical properties of a TBC involve many aspects, including the hardness, elastic modulus, Poisson's ratio, fracture toughness, residual stress, creep, and fatigue. Since properties such as fracture toughness and residual stress have been individually discussed in relevant chapters, this chapter focuses on discussing the characterization of Young's modulus, nanoindentation hardness, and high-temperature creep behavior of ceramic coatings.

The elastic modulus, also known as Young's modulus, is the main measure of how difficult it is for a material to undergo elastic deformation. A high elastic modulus means that a high stress is required for a material to undergo certain elastic deformation. In other words, a material with a high stiffness experiences small elastic deformation under a certain stress. The elastic modulus is the core parameter of the constitutive relations of a material. A TBC is a very thin layer of brittle material with many pores and scattered mechanical properties. In particular, an electron beam-physical vapor deposition (EB-PVD) coating comprises a series of columnar crystals. Thus, it is a daunting task to accurately characterize the elastic modulus of a TBC.

Creep refers to the slow, permanent deformation of a solid material under stress. Creep is more severe in a material subjected to high temperatures or temperatures near its melting point for an extended period of time. The rate of change in the creep increases continuously as the temperature rises. Long-term exposure to high

Y. Zhou et al., *Thermal Barrier Coatings: Failure Theory and Evaluation Technology*,
https://doi.org/10.1007/978-981-19-2723-2_8

temperatures leads to both large creep in the metallic material of an engine blade and a certain amount of creep in its ceramic component. Therefore, it is very important to thoroughly understand and characterize the creep deformation behavior of TBCs.

This chapter consists of three sections. Section 8.1 mainly describes the macroscopic elastic modulus of EB-PVD TBCs and its characterization methods. Section 8.2 focuses on the presentation of the temporal and spatial correlations between the mechanical properties and microstructure of TBCs. Section 8.3 is mainly dedicated to the discussion on the characterization methods for the creep behavior of TBCs as well as the effects of creep behavior on interfacial stresses.

8.1 In Situ Measurement of the Elastic Behavior of EB-PVD TBCs

EB-PVD coatings are predominantly applied on the hottest-end components of turbine blades, with the main goal of protecting directional or single-crystal Ni-based metallic blades. Due to its columnar structure, as shown in Fig. 1.7 in the Introduction of the book, an EB-PVD coating is anisotropic, and its service life is more significantly affected by its in-plane Young's modulus. This special structure makes the direct measurement of the in-plane Young's modulus very difficult. Removing a coating from its substrate prior to measurement causes the coating to fragment. Here, we introduce a microbending measurement method for in-plane Young's moduli—digital image correlation (DIC) [1–3]. For the sake of convenience, the in-plane Young's modulus is referred to as Young's modulus unless stated otherwise.

8.1.1 DIC-Based Microbending Testing

The DIC technique (see Chap. 12 for details) is capable of accurately measuring surface deformation and indirectly determining Young's modulus of an EB-PVD TBC based on the constitutive relations through measurement of deformation and external loading. Eberl et al. [1, 2] and He et al. [3] cleverly designed a microbending test method, as shown in Fig. 8.1. Specifically, the metallic disk shown in Fig. 8.1 is in fact the substrate material. The circumferential side of the disk is sprayed with a transition layer, on which an EB-PVD ceramic layer is deposited. Table 8.1 summarizes the dimensional parameters of each specimen. In other words, the radial direction of the disk is equivalent to the thickness direction of the coating that we commonly refer to. The columnar crystals run along the radial direction. We define Young's modulus in the radial direction or thickness direction as the out-of-plane Young's modulus and Young's modulus in the circumferential direction as the in-plane Young's modulus. One approximately trapezoidal hole is prepared at locations A, B, C, and D on the disk, as shown in Fig. 8.1, using electric sparks to facilitate the subsequent application of mechanical loading. Thereby, the preparation of specimens for elastic modulus measurement is completed.

Now, we introduce the loading process. As shown in Fig. 8.2, the brown disk is the specimen in Fig. 8.1 and is loaded at trapezoidal holes A, B, C, and D using loading pins. In Fig. 8.2, the specimen is loaded only at hole B in the direction indicated by the white arrow. This loading method is in fact the blister loading discussed in detail in Sect. 9.5 of Chap. 9.

Here, the method for measuring deformation is presented. Under mechanical loading along the thickness of the coating, the coating protrudes along its thickness or the radial direction of the disk. During this process, the deformation of the coating is accurately measured by the DIC technique. Speckles are created on the surface of the brown disk in Fig. 8.2. In particular, the speckles near the four holes are the measurement zones, as shown in Fig. 8.3. Then, the variation process of the speckles is measured by imaging. Here, the raster and speckle-zone sizes are set to 1 μm

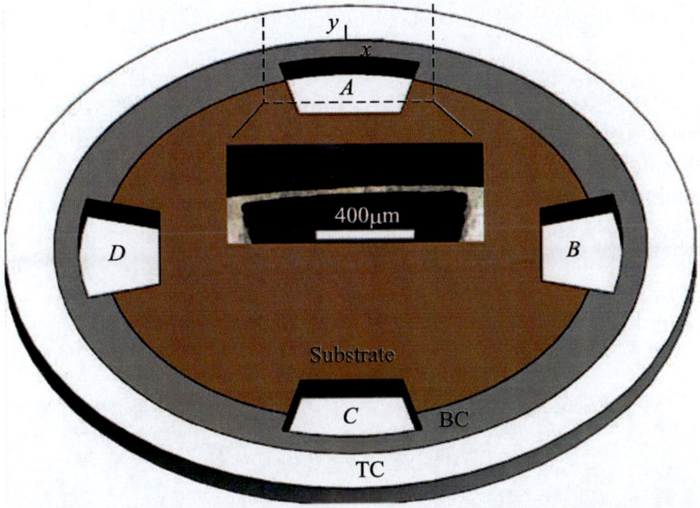

Fig. 8.1 A specimen for elastic modulus measurement and its preparation method. The disk is a metallic substrate circumferentially sprayed with a transition layer and subsequently coated with an EB-PVD ceramic layer. One approximately trapezoidal hole is prepared at locations A, B, C, and D using electric sparks to facilitate the application of mechanical loading

Table 8.1 Parameters of microbending test specimens and measured and FE-simulated values of their Young's moduli

Specimen no	Width (μm)	h_{na} (μm)	h_{TBC} (μm)	h_{BC} (μm)	E_{BC} (GPa)	ETBC (GPa) by Eq. (8.1)	ETBC (GPa) by simulation
a1	550	120	110	50	155	16	20
a2	550	107	110	40	155	25	30
b1	400	145	150	40	155	15	15–20

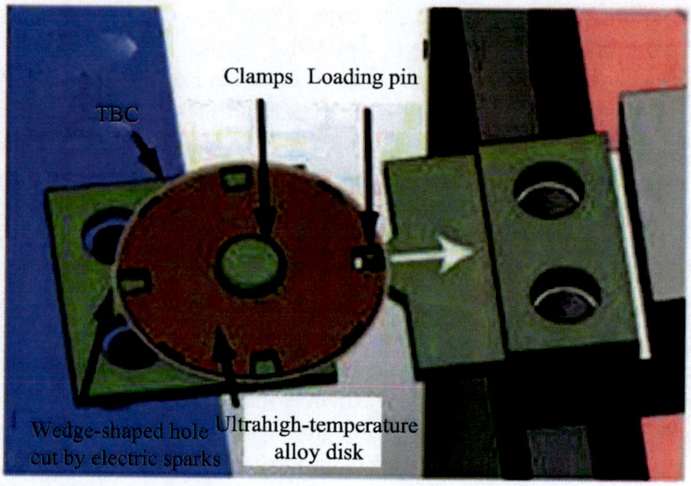

Fig. 8.2 Specimen loading method for elastic modulus measurement

and 400 μm × 290 μm, respectively. As long as the deformation and applied load are measured, Young's modulus of the EB-PVD ceramic coating can be obtained by numerical simulation.

The loading process is numerically simulated using the finite element method (FEM). Figure 8.4a shows the finite element (FE) mesh. We assume that the bond coat (BC) layer is isotropic and has elastic–plastic constitutive relations, Young's modulus E of 155 GPa, a yield strength of 750 MPa, and a strain hardening exponent n of 0.2, Hemker et al. [4] and that the EB-PVD ceramic coating has an in-plane

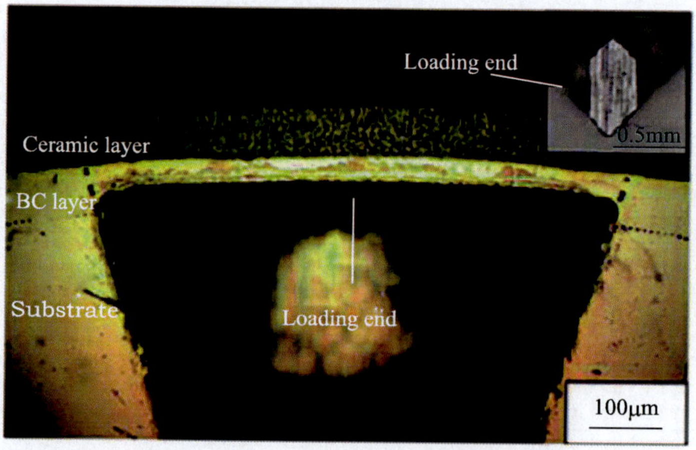

Fig. 8.3 Enlarged view of a local area of a speckle deformation measurement zone

Young's modulus E_p of 10–50 MPa and an out-of-plane Young's modulus E of 160 GPa. The numerically simulated results based on different selected values of E_p are compared with the experimental result, and the selected E_p that corresponds to the smallest error between the simulated and experimental results is considered to be the actual E_p in the experiment. However, 10–50 MPa is an excessively wide range of E_p. In addition, elastic–plastic constitutive relations are associated with nonlinear problems, the FE calculation of which may produce large errors. Therefore, the value of Young's modulus obtained through comparison of the numerically simulated and experimental results alone may include a large error. Hence, we need to first estimate Young's modulus through analysis based on a simple theoretical model.

An analysis of Fig. 8.4 finds that it is fully possible to simplify this diagram to a beam with a multilayer structure. As early as 1990, Suo and Hutchinson [5] put forward an excellent theoretical model for this problem (see Fig. 9.14 in Chap. 9 for its schematic diagram). Let h_{TBC} and h_{BC} be the thicknesses of the ceramic layer (i.e., the TC layer) and the transition layer (i.e., the BC layer), respectively, and E_{TBC} and E_{BC} be their Young's moduli, respectively. The position of the neutral axis, h_0, under a bending moment is expressed by the following equation [2, 5]:

$$\frac{h_0}{h_{\text{BC}}} = \frac{1 + \frac{2E_{\text{TBC}}h_{\text{TBC}}}{E_1 h_1} + \frac{E_{\text{TBC}}h_{\text{TBC}}^2}{E_1 h_1^2}}{2\left(1 + \frac{E_{\text{TBC}}h_{\text{TBC}}}{E_{\text{BC}}h_{\text{BC}}}\right)} \tag{8.1}$$

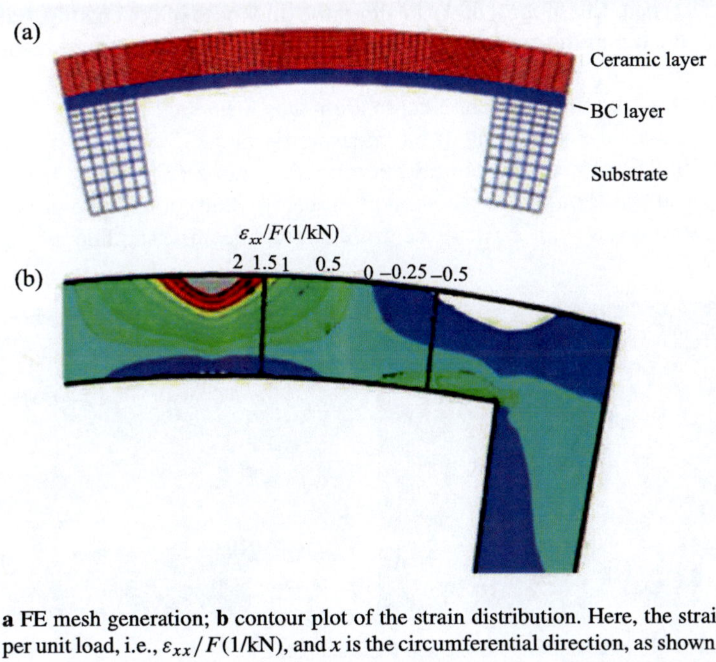

Fig. 8.4 **a** FE mesh generation; **b** contour plot of the strain distribution. Here, the strain refers to the strain per unit load, i.e., ε_{xx}/F(1/kN), and x is the circumferential direction, as shown in Fig. 8.1

In Eberl et al.'s experiment, [2] $h_{\mathrm{IBC}} = 110\ \mu\mathrm{m}$, $h_{\mathrm{BC}} = 50\ \mu\mathrm{m}$, and $E_{\mathrm{BC}} = 155$ GPa [4]. The neutral axis is located at the position where the DIC-measured strain is zero. The following presents a detailed analysis. Then, E_{TBC} is estimated to be 20 $\pm$ 10 GPa.

Figure 8.4b shows the contour plot of the circumferential strain ε_{xx} obtained by FE calculation (here, the strain refers to the strain per unit load, i.e., ε_{xx}/F(1/kN). The red zone in the figure is the central zone of symmetry where the maximum tensile strain appears, and x is in the circumferential direction. Clearly, the strains are concentrated and vary significantly (i.e., there is a large strain gradient) near the loading line. Similarly, both the strains and strain gradient are large near the edge (shoulder) of the support end.

8.1.2 *Analysis of the Experimental and Numerical Simulation Results*

Let us first analyze the strain measurements. According to the preliminary FE calculation results, as shown in Fig. 8.4b, the strains are found to be large near the central zone and the support end. Therefore, we focus on analyzing the strain measurements in these two zones. As shown in Fig. 8.5a, paths AA' and BB' in the central zone run along the thickness of the coating, with points A and B at the top of the coating and points A' and B' at the bottom of the transition layer. As shown in Fig. 8.5b, paths CC' and DD' near the support end run along the thickness of the coating, with points C and D at the top of the coating and points C' and D' at the bottom of the transition layer.

The deformation under mechanical loading was measured by the DIC technique. Figure 8.6 shows the variations in the measured strain ε_{xx} (in the x-direction, i.e., the radial direction) near the central microbending zone (Fig. 8.6a), the measured strain ε_{xx} near the shoulder of the support end (Fig. 8.6b), and the load-normalized measured strains (i.e., ε_{xx}/F) in the central zone and near the shoulder of the support

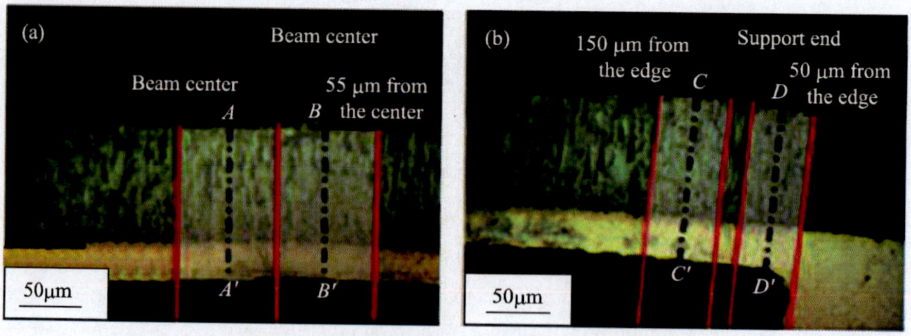

Fig. 8.5 Vicinities of **a** the central zone and **b** the support end

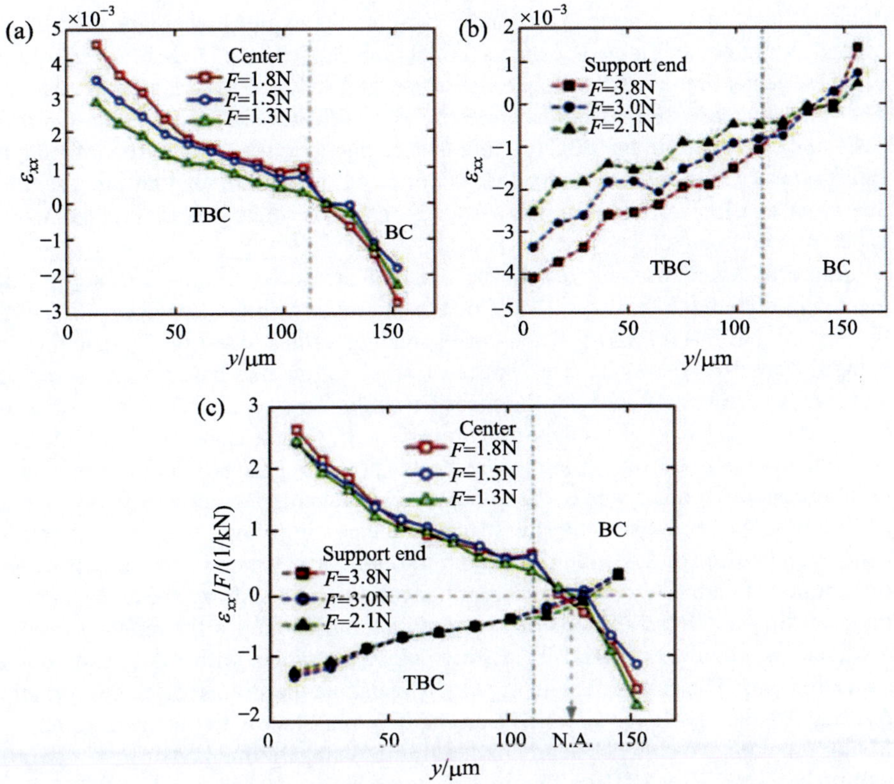

Fig. 8.6 Variations in normal strain ε_{xx} in the x-direction (i.e., radial direction) with the positions in the thickness direction (i.e., AA' and CC') under different loads: **a** ε_{xx} near the central microbending zone; **b** ε_{xx} near the shoulder of the support end; **c** load-normalized ε_{xx} (i.e., ε_{xx}/F) in the central zone and the shoulder of the support end

end (Fig. 8.6c) with the positions in the y-direction (i.e., AA' and CC'). A qualitative comparison of the experimental measurements reveals that the central zone can be basically analyzed using beam theory. A tensile strain is formed in the ceramic layer, whereas a compressive strain is formed in the transition layer. In addition, as the load changes, there is a significant change in the tensile strain in the ceramic layer but little change in the compressive strain in the transition layer. A similar phenomenon is observed near the shoulder of the support end. However, Fig. 8.6c shows that the above strain–load relations no longer exist after load normalization, suggesting that there is a linear relation between the strain and the load and that the deformation of the coating is still within the elastic range.

The elastic strain at the shoulder of the specimen (Fig. 8.5b) is smaller than the tensile strain at the central cross-section and is difficult to measure. In addition, as the load exceeds a certain critical value, cracks perpendicular to the loading direction are formed inside the ceramic layer of the TBC. As the load increases, the cracks run through the BC and ceramic layers and appear at the positions where the maximum normal strains occur. Figure 8.7 shows the relation between the DIC-measured strain ε_{xx} at the center of the top of the coating (i.e., point A in Fig. 8.5) and the applied load. The sharp increase in the strain seen in Fig. 8.7 indicates crack formation. On this basis, the critical strain for TBC fracture is determined to be 3.5–5 × 10^{-3}. Through careful observation of the in situ images collected during the loading process, the exact load causing crack initiation is identified, and the critical load for this specimen is determined to be 2.0 N.

Figure 8.8a and b compares the normalized measured strains (i.e., ε_{xx}/F) at the center and shoulder, respectively, of the microbending test specimen a1 (from the experiment in Fig. 8.5) and the corresponding values based on FE calculation using different E_{TBC} values. The values of ε_{xx}/F simulated based on $E_{\mathrm{TBC}} = 20$ and 50 GPa are consistent with the measurements. Thus, it remains impossible to determine E_{TBC} by relying solely on comparison of the measured and numerically simulated values of the overall strain. Hence, the problem must be simplified to obtain reliable results. In other words, the number of influencing factors should be kept at a minimum. An analysis of the stress state of the coating shows that only the stress distribution near path AA' in Fig. 8.5 is monotonic. Thus, we now focus on the stress distribution at path AA' inside the ceramic layer. Figure 8.9 shows the experimental data for different cross-sections of specimen a2 and compares the relevant results obtained by calculation based on a series of E_{TBC} values with the experimental measurements. The comparison in Fig. 8.9 reveals that the strains inside the ceramic coating calculated at $E_{\mathrm{TBC}} = 30$ GPa are highly consistent with the measurements. Young's moduli of different specimens determined using this method differ but are all within the range of 10–30 GPa. Table 8.1 summarizes the detailed results for different specimens, including Young's modulus determined based on the simple beam theory, the position of the neural axis, and Young's modulus determined through comparison of the simulated and measured values of the strain in the central zone.

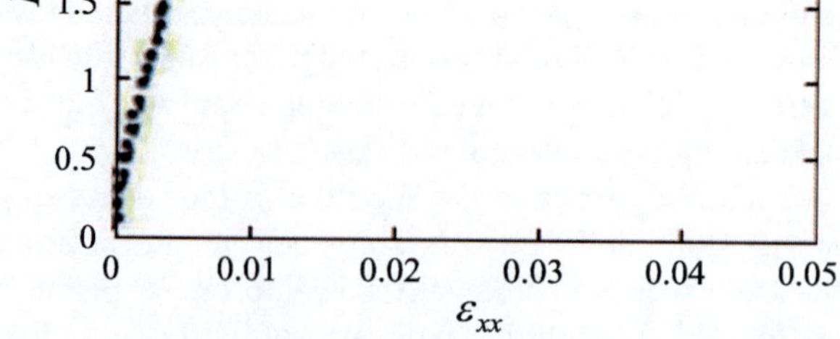

Fig. 8.7 Relation between the measured strain at the center of the top of the coating (i.e., point A in Fig. 8.5) and the applied load. The sharp increase in the strain indicates crack formation

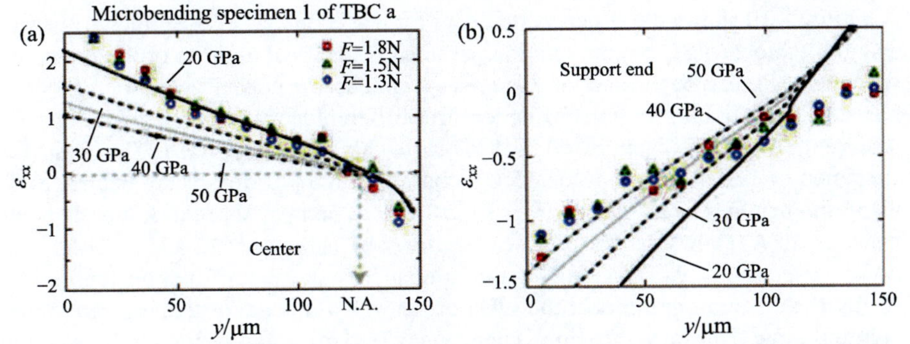

Fig. 8.8 Comparison of the normalized measured strains (ε_{xx}/F) from the experiment in Fig. 8.5 and the values obtained by FE calculation using different E_{TBC} values: **a** normalized strain at the center of microbending specimen a1; **b** normalized strain at the shoulder of specimen a1

Fig. 8.9 Experimental data for different cross-sections of specimen a2 and comparison of the results obtained by FE simulation using a series of E_{TBC} values and the experimental measurements

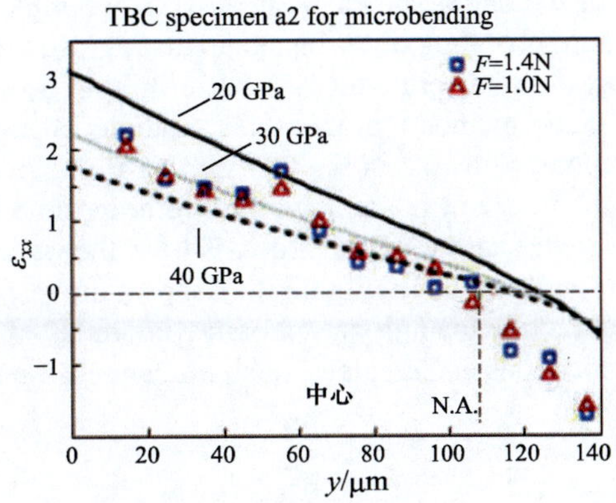

8.1.3 Factors Affecting the Elastic Modulus Measurement Accuracy

An analysis from a mathematical perspective shows that the abovementioned microbending characterization method for Young's modulus of a TBC is in fact an inverse problem. Inverse problems usually face the issue of whether they have a unique solution. Hence, how reliable is this microbending method? Here, let us compare the experimental data in Fig. 8.6 and the series of FE simulation results for microbending in Fig. 8.8. It is found that the E_{TBC} near the center of specimen a1 is approximately 20 GPa, which is comparable to the predicted value for the position of the neural axis (see Table 8.1), and that E_{TBC} near the shoulder of the specimen is 50 GPa, which is much higher than the value measured at the center of the specimen. When the strain in the shoulder was measured, cracks formed at the center of the specimen under the tensile stress. The local strain changes related to these cracks may not be properly captured in the FE simulation.

Figure 8.10 shows the experimental data for a specimen (i.e., specimen b1) with a slightly thicker TBC but preparation parameters identical to those of the previous specimens. Similar to the case of previous specimens, the normalized curves plotted based on the data for specimen b1 under three different loads converge to one curve. The neutral axis of specimen b1 is 145 μm away from its surface. The E_{TBC} of specimen b1 is determined to be 15 GPa based on the position of the neutral axis using the bending beam theory (i.e., Eq. (8.1)). In comparison, the finite element analysis (FEA) shows that the E_{TBC} of specimen b1 is between 15 and 20 GPa.

However, an EB-PVD coating has a columnar crystal structure, making it possible to observe the finest grains near the substrate as well as the grains that thicken as the columns grow. This microstructural change may lead to a significant difference in the E_{TBC}. In a numerical experiment on a TBC with Young's moduli of 50 GPa near the transition layer and 10 GPa at the top of the ceramic layer, [1, 2] Eberl et al. found that the calculated strains were surprisingly consistent with measurements. This simulation result is almost identical to that obtained in this study based on Young's modulus of 20 GPa, as shown in Fig. 8.10. This simple numerical experiment indicates that the microbending test is unable to distinguish the two cases. Therefore, it is necessary to resort to other methods for determining E_{TBC}, of which the simplest is a nanoindentation method that measures modulus changes. An indentation test was performed along path AA' in Fig. 8.5 using a Berkovich indenter with an indentation depth of 1 μm. Readers are advised to refer to the original literature [6] and the monograph by the author of this book [7] for the nanoindentation method for characterizing Young's moduli. Figure 8.11 shows the results of Young's modulus obtained based on the unloading data. Young's modulus of the TBC is basically 25 GPa at all the locations and displays no significant changes across the TBC.

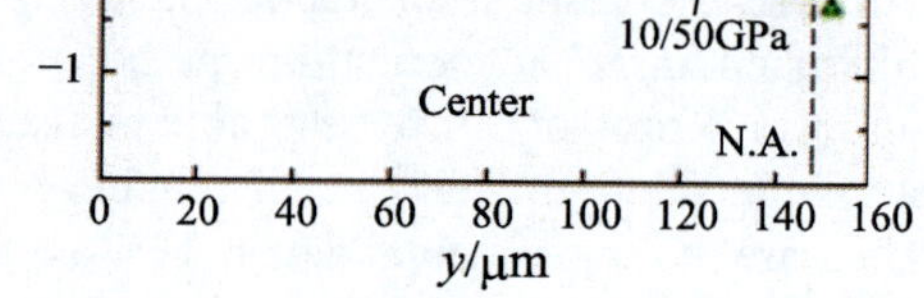

Fig. 8.10 Comparison of experimental data and FE simulation results for different specimens

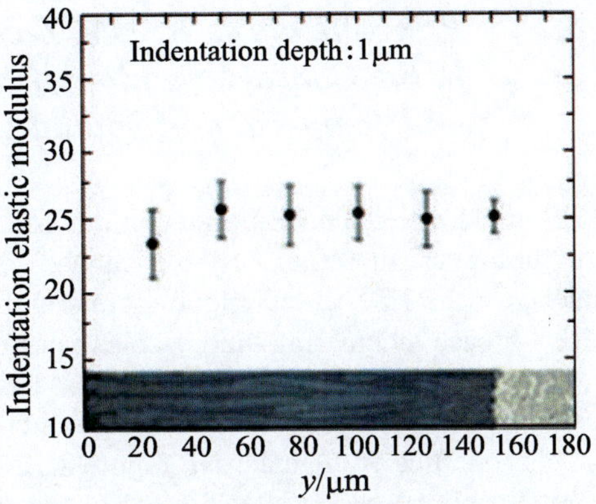

Fig. 8.11 Young's modulus of the TBC measured by the nanoindentation method

8.2 Temporal and Spatial Correlations Between the Mechanical Properties and Microstructure of TBCs

The properties of a material depend on its microstructure. Establishing the correlations between the microstructure and macroscopic properties of materials is an ambition pursued by the academic and industrial communities. To this end, it is vitally necessary to develop theories and methods to characterize these correlations. A TBC has a multilayer structure composed of layers with significantly different properties, and these properties vary with service time. Therefore, developing theories, methods, and experimental equipment for characterizing the temporal and spatial correlations between the mechanical properties and microstructure of TBCs is both a necessity and a challenge. This section focuses on introducing characterization methods based on the high-speed nanoindentation mapping (HSNM) technique and deconvolution theory [8]. The HSNM technique maps the feature distribution of the mechanical properties of each layer through extensive nanoindentation testing [9, 10] and, on this basis, establishes the correlations between the microstructure and mechanical properties. Further, deconvolution of experimental data [11] yields the distribution of various microphases (e.g., the β-NiAl and γ'-Ni_3Al phases in the thermal growth oxide (TGO) layer) and produces the evolutionary pattern of the mechanical properties with thermal cycling, based on which the failure mechanisms at different numbers of thermal cycles are analyzed. Thus, the HSNM technique is very important for the FE modeling of the microstructure of TBCs, the investigation of the temporal and spatial evolutionary patterns of the mechanical properties of TBCs, and the development of data models for predicting the service life of TBCs [12].

8.2.1 Principle of the HSNM Technique in the Characterization of Microstructure and Temporal and Spatial Correlations

HSNM is a characterization technique for studying the local mechanical properties of multiphase materials. Using advanced electronic techniques and novel mechanical designs, the HSNM technique increases the indentation test speed (approximately 1 s for each indentation) and is able to perform extensive testing on a specific region within a short period of time and, on this basis, spatially map the features of the material. During testing, each indentation point provides local information on the mechanical properties of the material. Figure 8.12 shows a schematic diagram depicting the principle of HSNM [12]. In Fig. 8.12a, the indenter producing the Vickers indentation is large and therefore masks some information related to phase 1 or phase 2, thereby resulting in mean field information. In comparison, in Fig. 8.12b, a nanoindenter can both distinguish the information related to phase 1 or phase 2 of the microstructure and characterize the macroscopic mechanical properties related to phase 1 or phase 2. On the one hand, a large database of such highly localized micron- or nanoscale mechanical properties can be used to study the relations between the microstructure and the corresponding mechanical properties (e.g., hardness and elastic modulus). On the other hand, advanced statistical analysis of large volumes of experimental data can be carried out to produce precise results and information. The HSNM technique has already yielded some excellent results for materials such as cementing agents [9, 10]. However, the mapping of TBC systems faces some unique difficulties, such as local damage sustained by coatings, the interactions between the layers of multilayer coatings, and interfacial problems, which make it more difficult to use the HSNM technique to characterize the mechanical properties of TBCs [13].

Two key problems facing the use of the HSNM technique to study the correlations between the properties and microstructure of TBCs need to be addressed, namely, (1) rapid, extensive indentation testing within a short period of time and (2) the rapid analysis of large volumes of experimental data and obtaining of quantitative microscopic material information through data deconvolution analysis. There are a variety of methods for data deconvolution. For example, Constantinides et al. [12] simulated the deconvolution of some data by fitting their probability distribution function under the assumption that the data follow a multimodal Gaussian distribution. By fitting to cumulative distributions, Ulm et al. [14] addressed the limitation of fitting probability distribution functions. Recently, Veytskin et al. [15] performed analysis using the Gaussian mixture deconvolution technique by maximizing the expected value. While these techniques can be used to quantitatively analyze the feature distribution of each microphase, they have two main disadvantages, namely, the required assumption that the data follow a certain distribution (e.g., Gaussian and Lorentz) and the loss of some spatial information. Similarly, most studies on the mapping of the local mechanical properties of TBCs have been limited to only the mean effect [16] but failed to conduct any rigorous statistical processing. The HSNM technique and deconvolution method developed by Vignesh et al. [8] is capable of mapping different

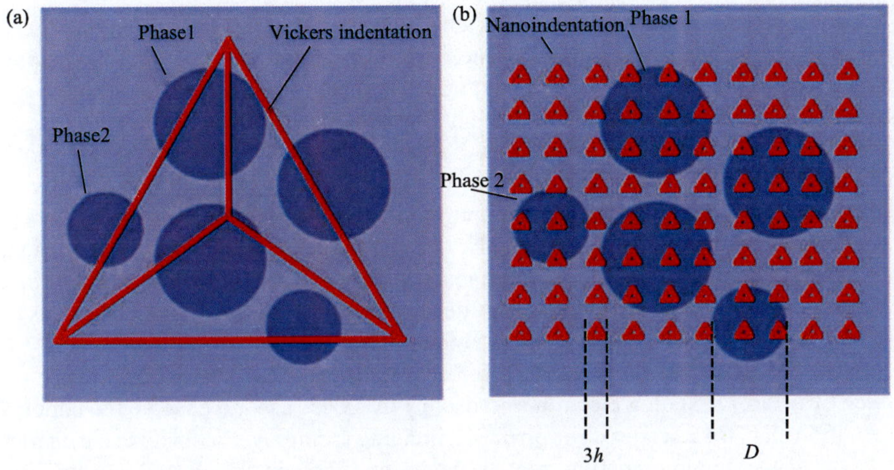

Fig. 8.12 Schematic diagram of the principle of HSNM. **a** A large indenter completely fails to extract local information. **b** A nanoindenter is able to extract local information

positions in the BC and ceramic layers as well as interfacial zones of a TBC after different numbers of thermal cycles and subsequently comparing and correlating the obtained feature distribution map and microstructural map. To determine the properties of the constitutive phases while ensuring that positional information is not lost, new methods based on k means clustering can be employed to perform deconvolutional mapping on materials. During thermal cycling, the failure mechanisms of TBC systems can be comprehensively understood based on their microstructural evolution patterns and the resulting mechanical property distribution patterns.

8.2.2 HSNM and Deconvolution Techniques

Diamond Berkovich indenters ($E = 1141$ GPa; $\nu = 0.07$) are used in all nanoindentation tests. The HSNM process involves surface approaching, surface inspection, loading, unloading, and specimen placing (for the next indentation). With this method, it takes less than 1 s to test each indentation specimen. High-speed indentation testers, an outcome of the tremendous progress made in speed and minimum indentation spacing, have promoted the rapid realization of several thousand of indentation tests over large areas (usually 1000 μm^2) and are capable of rapidly producing high-resolution mechanical property distribution maps [17]. This is the so-called HSNM technique. Maps of different sizes (usually containing more than 10,000 indentations) are drawn on the BC layer and the ceramic coating and at their interface before and after thermal cycling. A molten quartz specimen is used to calibrate the area function and to correct the load correlations. The hardness and elastic modulus of each indentation are calculated using the standard the Oliver–Pharr method [6].

According to the latest work of Phani and Oliver, Phani and Oliver [17] an indentation load is selected to ensure that the ratio of the minimum indentation spacing to the indentation depth is 10. The maximum indentation depth depends, to a large extent, on the signal-to-noise ratio at shallow depths and the image resolution at large depths (or indentation spacings). To capture the inherent features of various characteristics of TBCs, Vignesh et al. [8] experimentally studied the reliability of simple tests and data to determine an optimum maximum indentation depth. Hence, in all tests, the maximum load is 2 mN, and the indentation spacing is 1 μm (i.e., at least 10 times the depth). The mechanical property distribution maps of original and thermally cycled specimens are produced based on 212,500 indentation tests to obtain structure–property correlations at the micron scale.

Now, we introduce the procedure for obtaining the phase distribution map of a microstructure. Scanning electron microscopy (SEM) is used to observe the nanoindentation map. Backscatter electron (BSE) imaging is employed to analyze the energy spectra of the ceramic coating, the BC layer, and their interface to determine the phase composition. In addition, an optical microscope is used to image the phase distribution of the coating.

In the above, we introduced the use of the nanoindentation technique to rapidly obtain a high-resolution mechanical property distribution map and the use of SEM and BSE imaging techniques to produce a microstructural phase distribution map. If the correlation between the two maps can be determined, the HSNM technique can then be used to obtain microstructural information for a coating. Determining this correlation requires a modern mathematical tool—the convolution theorem. Now, we briefly introduce the basic concepts of the convolution theorem. Convolution and deconvolution are a mathematical method of integral transformations that has been extensively applied in many areas. The use of convolution to address the problems encountered in testing has already yielded excellent results. In comparison, the application of deconvolution has been lagging, and it has not drawn wide attention from the engineering and academic communities until recently when Schroeter, Hollaender, and Gringarten et al. solved the stability problem of its computational method. Some experts believe that the application of deconvolution is another major leap in the history of the testing field. With the development and application of new testing tools and techniques and their closer integration with research results in other fields, the theory of convolution and deconvolution will inevitably play an increasingly important role in the testing field. Convolution is an important operation in analytical mathematics. Let $f(x)$ and $g(x)$ be two integrable functions in R1 (a 1D domain of real numbers). Integration of $f(x)$ and $g(x)$ yields

$$\int_{-\infty}^{\infty} f(\tau)g(x-\tau)d\tau \tag{8.2}$$

It can be proven that the above integral exists with respect to almost any real number x. Then, for each selected value of x, this integral defines a new function $h(x)$, which is referred to as the convolution of f and g and is denoted by

$$h(x) = (f * g)(x) \tag{8.3}$$

It is easy to prove that

$$(f * g)(x) = (g * f)(x) \tag{8.4}$$

In addition, $h(x) = (f * g)(x)$ remains an integrable function. That is, to say, through substitution of multiplication by convolution, the L1 (R1) space is an algebra or even a Banach algebra. Convolution is closely related to the Fourier transform. By taking advantage of this property (i.e., the product of the Fourier transforms of two functions is equal to the Fourier transform of their convolution), we can simplify the solving of many problems in Fourier analysis. The function obtained by convolution, $h(x) = (f * g)(x)$, is generally smoother than both f and g. In particular, when g is a smooth function with a compact set and f is a locally integrable function, their convolution $h(x) = (f * g)(x)$ is also a smooth function. By exploiting this property, for any integrable function f, we can easily construct a sequence of smooth functions f_S approaching f. This method is called function smoothing or regularization. The concept of convolution can also be extended to sequences, measures, and generalized functions.

When an instrument is used to observe and record a physical phenomenon and process, the obtained observations and records reflect not only the physical phenomenon and process but also the features of the instrument. The nonideal features of the instrument can distort the obtained observations and records, and this distortion can be expressed by the following mathematical convolution [11]:

$$Y(t) = S\left\{ \int_{-\infty}^{\infty} H(t;\tau)X(\tau)d\tau \right\} + \xi(t) \tag{8.5}$$

where $X(\tau)$ is the real physical quantity, $Y(t)$ is the measured data, $H(t;\tau)$ is the impulse response of the instrument at time τ, $S\{\cdot\}$ represents the nonlinearity of the recording medium or sensing element, and $\xi(t)$ is the noise. Deconvolution is a technique that estimates the original physical quantity $X(t)$ based on the observation $Y(t)$ and the knowledge on the statistical features of noise. In the current work, the data deconvolution technique is used to determine the distribution of each phase of a multiphase specimen from its spatial feature map. To this end, a so-called k-means clustering algorithm is adopted that is similar to that described elsewhere [18], and this algorithm is presented in Chap. 11 of this book. The main function of k-means clustering is to divide n observation results into k clusters. k-means clustering is an iterative optimization technique that randomly initializes k cluster centers by moving them each to a new point and minimizes the sum of the squares of the intracluster distances during each iteration. The iterative process converges when the intracluster distances are no longer reduced, thus dividing the data into a predetermined number of clusters. The number of clusters into which data are merged must be decided through the determination of the number of phases existing in the mapping area based on prior knowledge. Alternatively, optimum parameters can be selected by iteratively running the algorithm on different numbers of clusters in combination with error minimization. When the algorithm partitions the data in this manner, the datapoints in the same cluster are more similar to each other than to those in other clusters. Therefore, it is expected that each cluster represents a different phase or feature on the spatial map. After clustering, the mean and standard deviation of the datapoints in each cluster can be used to quantitatively measure the properties of the corresponding phase. This deconvolution method has been applied to all the spatial feature maps obtained from the BC layer, the ceramic coating, and their interface to obtain the distribution map of each phase or characteristic (β-NiAl, γ/γ′-Ni, YSZ, and TGOs). Note that, unlike the Gaussian deconvolution method, this method retains the spatial information of each constituent phase, which is vitally important for studying the evolution of local mechanical properties [8].

8.2.3 *Characterization of the Mechanical Properties of BC Layers and Ceramic Coatings by HSNM*

Now, we introduce the use of the characterization methods based on the HSNM technique and deconvolution theory to study a TBC with a NiCoCrAlY BC layer and a YSZ coating [8]. Both the BC and ceramic layers of the TBC were deposited by atmospheric plasma spraying (APS) equipment. The substrate was made of Ni-based alloy C263. Each specimen had dimensions of 10 × 10 × 5 mm. To enhance the adhesion of the coating, the substrate was sandblasted before the coating was deposited. Subsequently, the coarse substrate was covered with a BC layer prepared with 47.05Ni-23Co-17Cr-12.5Al-0.45Y(wt%) raw powder with particle sizes ranging from 22 to 45 μm. Then, a layer of 7 wt% Yttria partially Stabilized Zirconia (YSZ) with particle sizes ranging from 10 to 45 μm was deposited on the BC layer. The average thicknesses of the BC layer and the coating were 105 and 175 μm, respectively.

To analyze the evolution of its mechanical properties, the coating was subjected to thermal cycling. Specifically, the specimen was heated to 1100 °C for 20 min and held at 1100 °C for 40 min, after which the specimen was cooled to 300 °C within 10 min. The required numbers of thermal cycles were automatically completed in a thermal cycling furnace. The specimen was removed from the furnace after 5, 10, 50, and 100 cycles, respectively, to characterize its mechanical properties and microstructure.

To prevent the cutting process from potentially damaging the thermal-cycling specimen, the specimen was first cold-mounted in epoxy resin and then cut by an abrasive wheel cutting machine to expose its cross-section. The cross-section was fixed with Bakelite powder for further polishing using SiC sandpaper with grit sizes of 1000, 2000, and 4000, successively. Finally, the specimen was subjected to vibratory polishing for 2 h in a 60-nm silica gel solution to obtain a smooth polished surface.

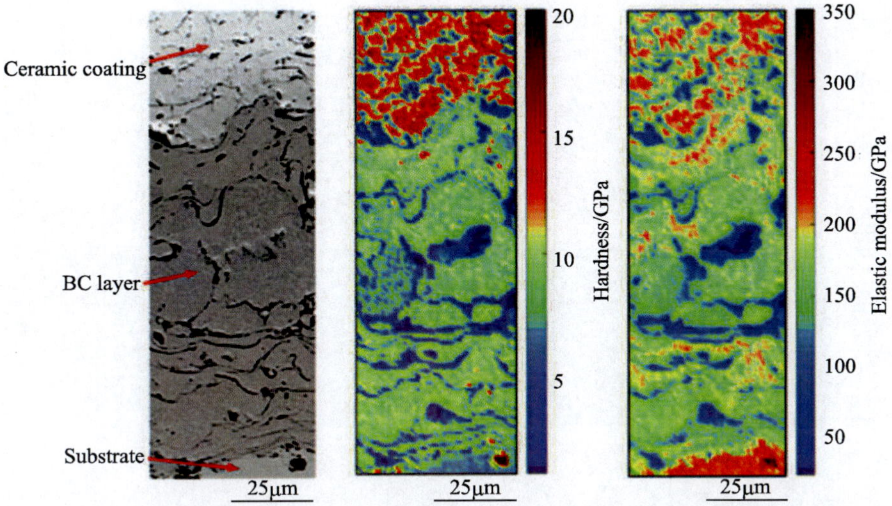

Fig. 8.13 Cross-sectional SEM image of the TBC-coated substrate and its corresponding hardness and elastic modulus maps

Figure 8.13 shows the cross-sectional microstructure as well as hardness and elastic modulus maps of the TBC. A total of 7500 point indentations were made in an area of $50 \times 150\ \mu m^2$. The interlayer and intralayer changes in hardness and elastic modulus were obtained by mapping, which reflected that the indentation spacing was sufficient to capture the key microstructural features. This also proves that the technique is capable of handling sudden feature and morphological changes, which were drawn through mapping at the cross-layer, cross-interface, and indentation locations in the coating. In the following section, we analyze in detail the changes in the spatial features of each layer and their evolution with thermal cycling.

Figure 8.14 shows the microstructure, hardness, and elastic modulus of the BC layer during thermal cycling. The interfaces of the coating are very discernible, which is a typical characteristic of an APS coating. The hardness of the BC layer is higher before thermal cycling than after thermal cycling. The opposite phenomenon is observed in the elastic modulus of the BC layer. The difference between the changes in the hardness and elastic modulus with the number of thermal cycles can be ascribed to the microstructural changes observed in the SEM image.

As shown in Fig. 8.14b, a two-phase microstructure is observed in the specimen after five thermal cycles. The dark and light gray shadows correspond to the β-NiAl phase and the γ/γ'-Ni substrate, respectively. This observation is confirmed by the energy dispersive X-ray spectroscopy (EDS) measurement of the Al content and is similar to that made by Hemker et al. [4]. The β-NiAl zones are relatively large and interconnected. A comparison shows an extremely strong correlation between the microstructure and hardness map of the specimen after five thermal cycles. The changes in the features of the two phases and their morphologies are accurately captured. The β-NiAl phase is harder than the γ/γ'-Ni phase, whereas the opposite is true for their elastic moduli. Compared to the γ/γ'-Ni phase in a face-centered cubic

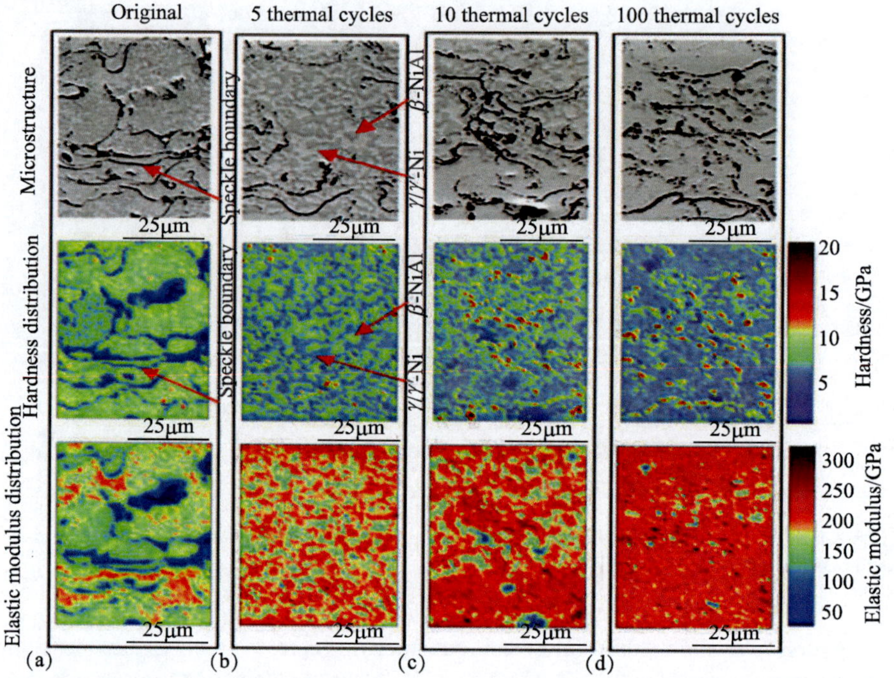

Fig. 8.14 Cross-sectional SEM images of the NiCoCrAlY BC layer **a** in the original specimen and after **b** 5, **c** 10, and **d** 100 thermal cycles and its corresponding hardness and elastic modulus maps

(FCC) crystal structure, the presence of a body-centered cubic (BCC)-like crystal structure in β-NiAl can improve the hardness of the material. A comparison of the microstructure before and after five thermal cycles shows that the crack boundaries have healed significantly after five thermal cycles. A scrutiny of the hardness map after five thermal cycles reveals some zones or spots with very high hardness. An EDS analysis of these spots shows that they are likely Al-rich oxides formed inside the BC layer. Figure 8.14c shows the results for the specimen obtained after 10 thermal cycles. A comparison of the microstructure of the specimen after five and 10 thermal cycles shows a decrease in the relative proportion of the β-NiAl phase. The hardness map also shows an increase in the area fraction of the high-hardness zones, suggesting that the area fraction of the oxides increases as the number of thermal cycles increases. As the number of thermal cycles continues to increase to 100 (Fig. 8.14d), the β-NiAl phase is almost completely depleted, while the proportion of the oxides further increases. Interestingly, as shown in the hardness map, the oxides are first discovered at the splashing boundary. The feature distribution maps after 50 thermal cycles are similar to those after 100 thermal cycles and are therefore not shown here. Overall, the hardness and elastic modulus maps of the BC layer each, to a certain extent, display an extremely strong correlation with the microstructure and accurately capture the microstructural changes at the micron scale.

The BC-layer/ceramic-coating interface is one of the most critical zones of a TBC and directly affects its thermal cycling life. In this zone, the changes in the microstructure and the corresponding mechanical properties of a TBC can be observed. Considering that the TGO layer between the BC layer and ceramic layer is generally formed on the order of microns in thickness, there are no reliable methods for measuring its local mechanical properties, while high-speed mapping at the micron scale can produce accurate measurements of its mechanical properties. Figure 8.15 shows the microstructure surrounding the BC-layer/ceramic-coating interface and the corresponding hardness and elastic modulus maps before and after thermal cycling. An undulating interface, which is a typical characteristic of an APS coating, can be observed. A comparison of the microstructure of the specimen before and after thermal cycling shows that the formation of TGOs (the dark zones at the interface) began during the fifth thermal cycle and that the thickness of the TGO layer increased as the number of thermal cycles increased. The corresponding hardness and elastic modulus maps show similar trends. The TGOs can be identified based on the interfacial zones with high hardness and elastic modulus. Interestingly, the thickness of the TGO layer measured from these maps exhibits a parabolic growth trend, which has been reported in previous microstructural studies [19]. Apart from the gradual growth of TGOs with an increasing number of thermal cycles, the hardness maps and microstructural images clearly show that the β-NiAl phase that acts as a source of Al on the BC-layer side becomes depleted as the number of thermal cycles increases. The difference in the elastic modulus due to TGO growth is one of the main driving forces for thermal stress mismatch. As shown in Fig. 8.15, after the specimen is subjected to 100 thermal cycles, microcracks are formed in the ceramic layer directly above the interface, and the hardness and elastic modulus in the crack zones decrease. Overall, the high-speed mapping technique can be employed to measure the local mechanical properties of coatings at the interfaces, particularly the mechanical properties of the BC-layer/ceramic-coating interface of a TBC. In FE simulation analysis, the local elasticity of different layers of a TBC subjected to thermal cycling can be easily measured.

Figure 8.16 shows the evolution of the micromorphology, hardness, and elastic modulus of the ceramic coating subjected to thermal cycling testing. The microstructural images clearly show that the coating is loose and porous, which might cause difficulty in high-speed mapping testing. However, the distribution maps reveal that it is feasible to implement the mapping technique to reflect large deformation and feature changes caused by pores. An extremely strong correlation can be found between the microstructure and properties of the specimen in each case. For example, the porous zones exhibit a low hardness and elastic modulus. Within the thermal cycling temperature and time ranges investigated in this work, the hardness and elastic modulus of the ceramic coating remain stable as the number of thermal cycles increases. The pores form because the deposition of the coating is mixed with those formed due to the polishing of the specimen. Thus, it is not easy to quantify the volume fraction of pores based on the microscopic images.

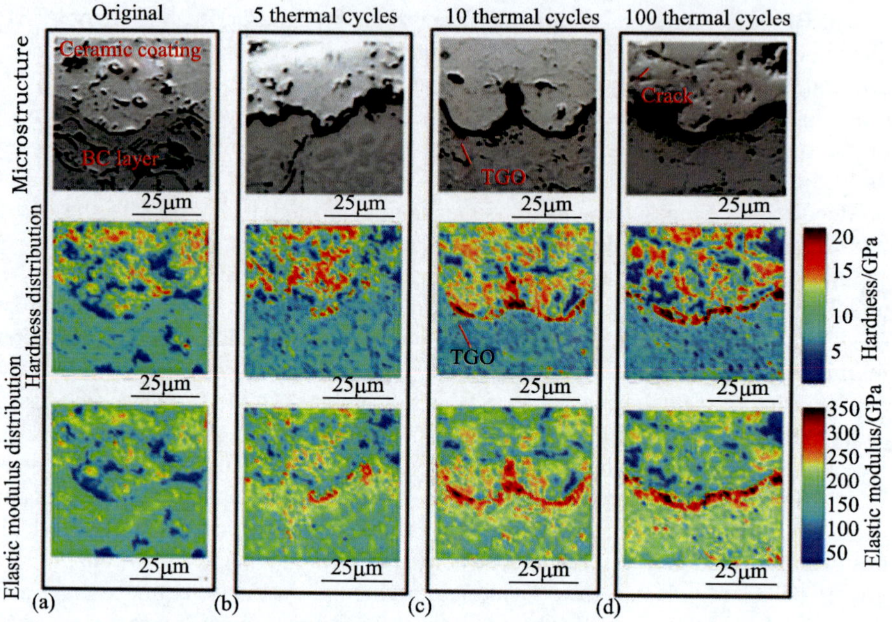

Fig. 8.15 Cross-sectional SEM images of the BC-layer/ceramic-coating interface **a** in the original specimen and after **b** 5, **c** 10, and **d** 100 thermal cycles and its corresponding hardness and elastic modulus maps

8.2.4 Characterization of the Phase Distribution of the Microstructure of the TBC Based on Deconvolution

The correlations between the microstructure and the feature distribution maps of the mechanical properties can be clearly seen in Fig. 8.16. However, these correlations are only qualitative results instead of quantitative analysis results. The determination of the phase structure of each layer of a TBC based on extensive mechanical property data requires deconvolution analysis of high-speed mapping data. Conventional methods often need to assume that the data can be curve-fitted to a Gaussian or Lorentz distribution. In contrast, the clustering algorithm used in this work, which was developed by Vignesh et al. [8], does not require any assumptions and retains relevant post-deconvolution data and can thus reconstruct a phase distribution map from the feature distribution map of mechanical properties. In addition, the clustering algorithm does not require any initial estimate or an expected value with a certain output range during the curve-fitting process. The deconvolution method has been briefly introduced in Sect. 8.2.2. Readers interested in the use of the deconvolution theory to perform deconvolution analysis are advised to attentively study relevant monographs or textbooks [11]. Figure 8.17a and b shows the microstructural image and hardness map, respectively, of the BC layer after five thermal cycles,

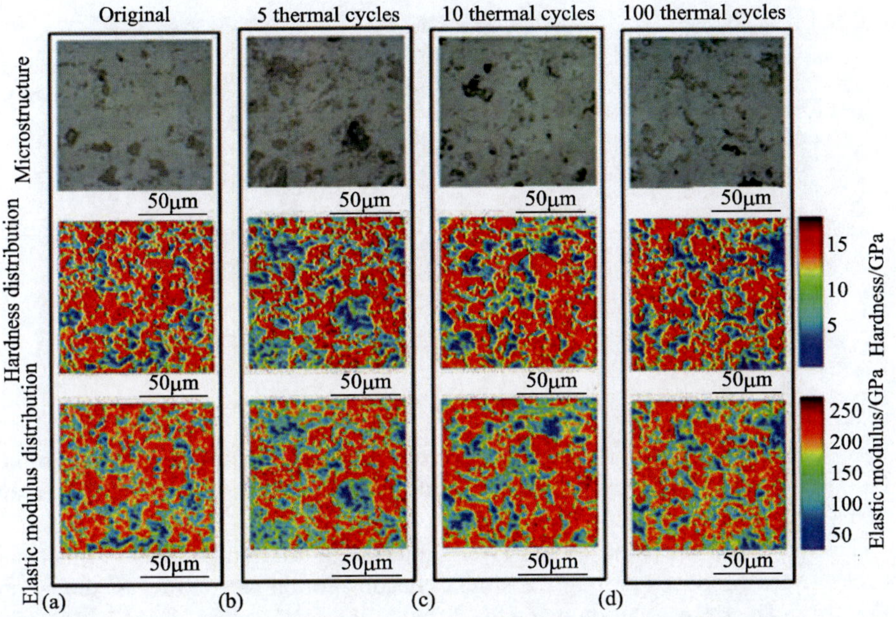

Fig. 8.16 Cross-sectional SEM images of the YSZ coating **a** in the original specimen and after **b** 5, **c** 10, and **d** 100 thermal cycles and its corresponding hardness and elastic modulus maps

which clearly display a one-to-one correspondence between them. In fact, there are only three phases in the microstructural image in Fig. 8.17a, which correspond to the β-NiAl phase, the γ/γ′-Ni phase, and the TGO phase formed due to internal oxidation, respectively. Similarly, there are three types of color—blue, light yellow, and red—in Fig. 8.17b. Based on the hardness of the bulk material, the red, light yellow, and blue in Fig. 8.17b correspond to the hardnesses of the TGOs, the β-NiAl phase, and the γ/γ′-Ni phase, respectively. Here, through detailed comparison with different crack patterns corresponding to acoustic emissions at different frequencies discussed in Chap. 11, we find completely identical relations between the hardness and microphases, which are, in fact, simpler and do not require a wavelet transform. In other words, the deconvolution introduced in Sect. 8.2.2 is, in fact, not needed, and cluster analysis alone suffices. The cluster analysis method introduced in Sect. 11.2.3 of Chap. 11 can be fully adopted here. Hence, a detailed introduction to cluster analysis is neglected here. Based on the hardness data, the map in Fig. 8.17b is divided into three clusters that correspond to the β-NiAl phase, the γ/γ′-Ni phase, and the TGO phase formed due to internal oxidation. As discussed in Sect. 8.2.2, we assume that the mean and standard deviation of the datapoints in each cluster represent those of the corresponding phase. Figure 8.17c shows the phase map of the microstructure obtained through cluster analysis of Fig. 8.17b. A comparison of the microstructure (Fig. 8.17a) and the corresponding deconvolution map (Fig. 8.17c) shows that the microstructure is basically consistent with that shown in the deconvolution map in

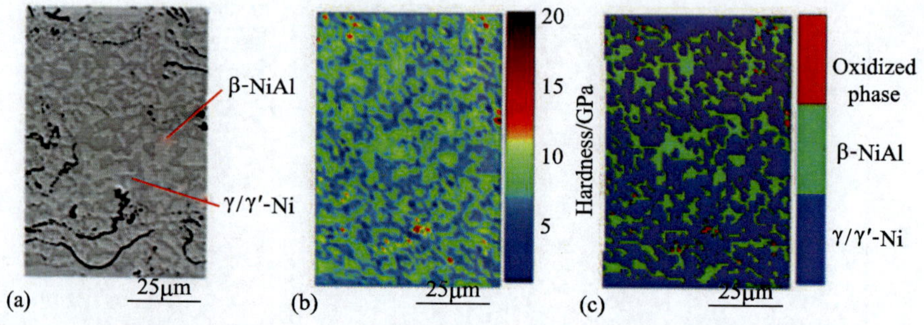

Fig. 8.17 **a** Microstructure, **b** hardness map, and **c** post-deconvolution phase map of the BC layer after five thermal cycles

terms of size and shape. The area fraction of each phase is estimated based on the ratio of the number of intracluster datapoints obtained from the deconvolution map to the total number of datapoints.

Deconvolution maps of the BC layer subjected to different numbers of thermal cycles were produced through a similar deconvolution operation, as shown in Fig. 8.18. The deconvolution maps for different thermal cycling conditions clearly show the evolution of the microstructure. β-NiAl is the main phase in the original coating. Due to the depletion of Al, γ/γ′-Ni becomes the main phase after 100 thermal cycles. As shown in Fig. 8.19, the proportion of the oxidized phase increases as the number of thermal cycles increases. While these results can also be obtained from the hardness maps, the deconvolution technique is able to produce quantitative analysis results in addition to visual results.

The hardness and elastic modulus of each phase in the BC layer and ceramic coating (including TGOs) were determined through deconvolution, as shown in Fig. 8.20. As expected, the γ/γ′-Ni phase has the lowest hardness of all the phases. In addition, both the hardness and elastic modulus of the β-NiAl phase decrease at the 50th thermal cycle and remain stable afterwards. This is because the change in the β-NiAl phase with thermal cycling leads to Ni enrichment, which, in turn, lowers

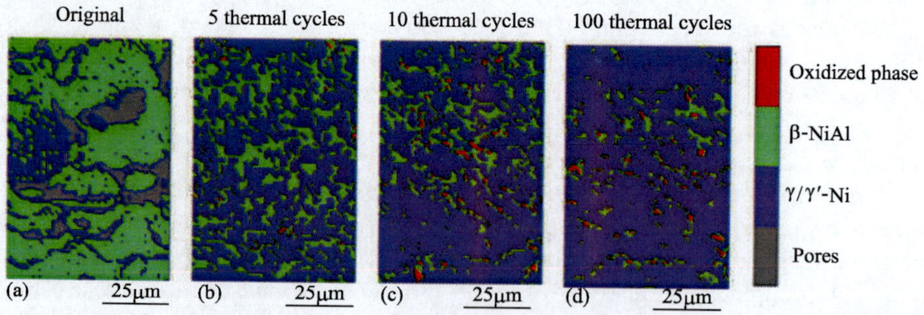

Fig. 8.18 Deconvolution phase maps of the BC layer after different numbers of thermal cycles

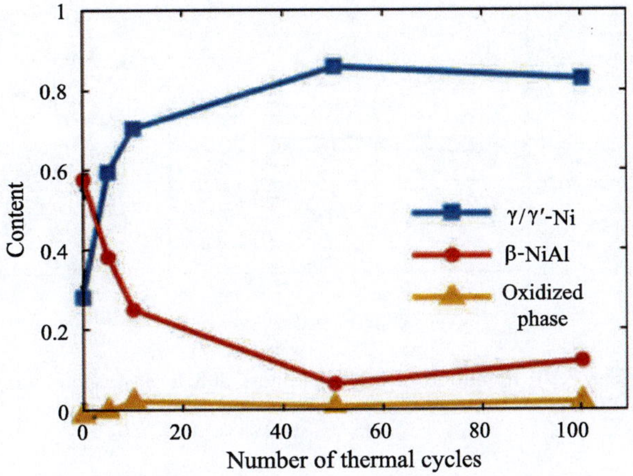

Fig. 8.19 Evolution of the content of each phase with the number of thermal cycles determined by deconvolution

the hardness. Similarly, as the number of thermal cycles increases, the Ni_3Al nano-precipitates distributed in the γ/γ'-Ni phase coarsen [4]. Interestingly, this does not lead to any notable changes in the hardness. Within the thermal cycling temperature and time ranges tested in this work, as the number of thermal cycles increases, the hardness of the YSZ coating remains stable, Jamali et al. [20] whereas the hardness of the TGOs slightly increases. However, the increase in the mean hardness is insignificant compared to the standard deviation.

Opposite the results for the hardness, the elastic modulus of each phase in the transition layer (Fig. 8.20b) is higher after five cycles than in the original state and subsequently undergoes no significant changes. While the elastic modulus of the YSZ coating remains constant (similar to its hardness) during thermal cycling, the elastic modulus of the TGOs increases as the number of thermal cycles increases, which may be attributed to the compositional changes in the oxides, particularly the changes in the Al content. In addition, in the beginning, the TGOs were relatively small in size, and the indenter might have been simultaneously pressed into the TGOs and the surrounding transition layer or coating when the hardness was measured, resulting in measurement errors. Thus, the features of TGOs can be viewed as apparent features instead of inherent features. The hardness and elastic modulus of each phase determined in this manner are approximately consistent with the values reported previously based on a small number of indentation tests [21]. Table 8.2 summarizes the hardness and elastic modulus of each phase determined through high-speed mapping and deconvolution. A comparison shows that the experimental results obtained from the conventional nanoindentation test for each layer after different numbers of thermal cycles are consistent with those obtained from mapping. There are some differences within the expected experimental dispersion ranges for this system, probably because the conventional test is not designed for specific functions. These results validate the accuracy of the high-speed mapping technique for measurements of the complex TBC material system.

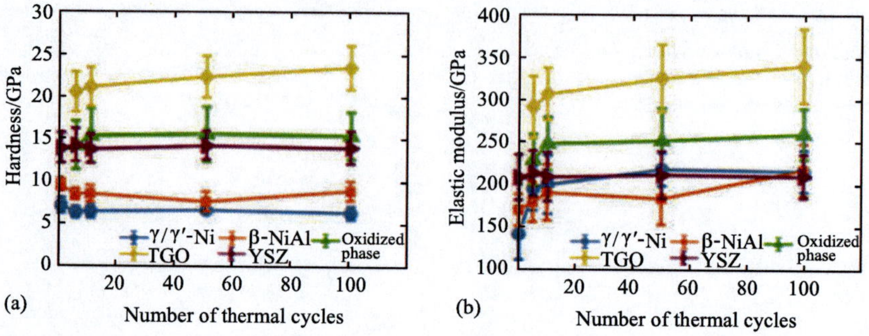

Fig. 8.20 Evolution of the **a** hardness and **b** elastic modulus of each phase and layer with the number of thermal cycles

8.3 Creep Behavior of TBCs

Creep deformation is a major failure mechanism of TBCs due to their long-term exposure to high temperatures. The creep rate often increases as the temperature rises. In addition, the creep rate is related to material properties, loading time and duration, loading temperature, and the stress of the loaded structure.

8.3.1 High-Temperature Creep Behavior of EB-PVD TBCs

Compared to conventional creep characterization methods (tensile and compressive methods), the indentation method for measuring creep requires a smaller specimen size and only a flat surface to be measured that is perpendicular to the indenter. Compared to the conventional Vickers indentation and nanoindentation methods, the indentation method used here is applicable for higher temperatures and more suitable for characterizing the creep behavior of TBCs (see Fig. 8.21 for its principle) [22].

A high-temperature atmosphere furnace and a flat cylindrical indenter with a diameter of 2 mm were used in an indentation creep test. The support for the indenter and specimen was made from a whole block of SiC. A displacement sensor with an accuracy of 1 μm was connected to a computer to measure the indentation depth of the indenter. Since the creep resistance of the SiC was much higher than that of the layers deposited by EB-PVD, the measured creep originated from only the layers deposited by EB-PVD. During the test, the specimen was first heated in an air furnace to 1100–1300 °C and then subjected to an indentation load of 30, 50, or 70 MPa. The load was kept unchanged during the test. After the test was completed, the specimen was unloaded and then allowed to cool in the high-temperature furnace.

During the test, only the YSZ ceramic coating was indented to eliminate the effect of the substrate. Here, we introduce the work of Jana et al. on an EB-PVD YSZ ceramic coating [22]. Due to the absence of the effect of the substrate, the theoretical

Table 8.2 Hardness (H) and elastic modulus (E) values determined by high-speed nanoindentation and conventional nanoindentation

Specimen conditions				Original	5 thermal cycles	10 thermal cycles	50 thermal cycles	100 thermal cycles
High-speed nanoindentation	$\gamma/\gamma' - Ni$	H	GPa	7.2 ± 0.9	6.4 ± 0.7	6.5 ± 0.8	6.7 ± 0.4	6.4 ± 0.8
		E		141.2 ± 30.6	195.0 ± 25.6	199.7 ± 34.2	218.0 ± 19.4	215.8 ± 24.4
	β − NiAl	H		9.6 ± 0.7	8.5 ± 0.7	8.5 ± 1.0	7.7 ± 1.2	9.0 ± 1.1
		E		169.6 ± 18.5	178.8 ± 22.9	190.1 ± 32.7	183.5 ± 30.7	216.9 ± 30.3
	Internal oxidized phase	H		–	14.2 ± 2.8	15.4 ± 3.1	15.6 ± 3.2	15.5 ± 2.7
		E		–	195.0 ± 25.6	199.7 ± 34.2	218.0 ± 19.4	215.8 ± 24.4
	TGO	H		–	20.5 ± 2.4	21.2 ± 2.3	22.4 ± 2.5	23.5 ± 2.6
		E		–	290.4 ± 36.6	305.7 ± 30.9	325.1 ± 39.8	339.6 ± 43.9
	YSZ	H		13.9 ± 1.8	14.3 ± 1.9	13.8 ± 1.7	14.3 ± 1.6	14.1 ± 1.9
		E		207.7 ± 26.5	212.8 ± 26.4	208.4 ± 28.5	211.1 ± 27.4	209.6 ± 25.3
Conventional nanoindentation	BC	H		9.9 ± 1.0	7.4 ± 0.4	8.3 ± 0.9	7.7 ± 0.6	7.5 ± 0.4
		E		165.9 ± 18.7	218.5 ± 19.3	200.8 ± 25.6	213.1 ± 34.4	222.7 ± 27.9
	TC	H		14.4 ± 3.6	13.1 ± 4.7	15.6 ± 2.6	15.9 ± 2.1	16.5 ± 2.6
		E		206.9 ± 38.1	193.0 ± 47.6	207.4 ± 32.7	227.9 ± 32.5	230.9 ± 14.7

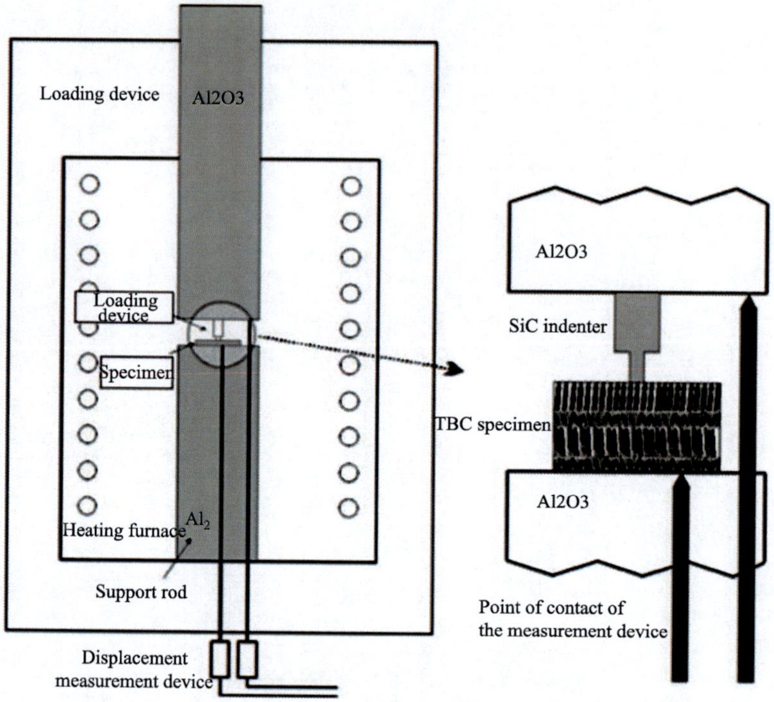

Fig. 8.21 Schematic diagram of a high-temperature indentation device

modeling of an indentation creep test on a bare coating involves the establishment of the correlation between the indentation pressure and the strain rate [22]:

$$\sigma = \frac{p}{k_1} \quad p = \frac{4F_\mathrm{i}}{\pi d^2}\dot{\varepsilon} = \frac{v_\mathrm{i}}{k_2 d} \tag{8.6}$$

where v_i is the displacement rate during the indentation process (unit: μm/s), $\dot{\varepsilon}$ is the strain rate (unit: s^{-1}), d is the diameter of the indenter (unit: μm), F_i is the indenter load (unit: N), p is the indenter pressure (unit: MPa), and k_1 and k_2 are specific constants of the material.

Figure 8.22 shows the indentation depth versus time curves at different temperatures. Specifically, Fig. 8.22a shows the indentation depth versus time curves corresponding to different indentation loads and indentation displacement rates at the same temperature of 1300 °C, and Fig. 8.22b shows the indentation depth versus time curves corresponding to different temperatures and indentation displacement rates at the same load of 50 MPa. The results show that the indentation creep deformation is small within the temperature range of 1100–1200 °C, that notable creep deformation occurs only at 1250 °C and above, and that the strain rate increases as the temperature increases.

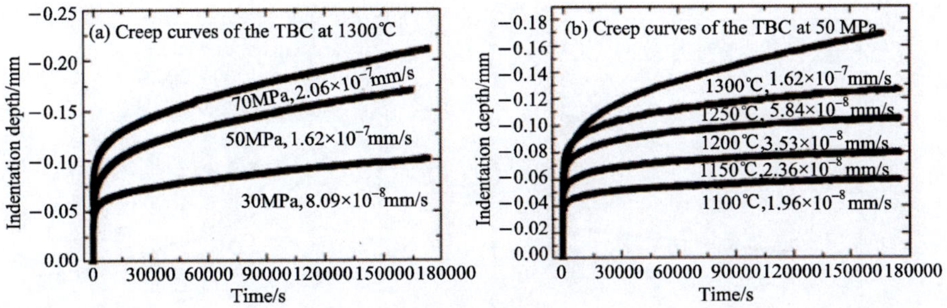

Fig. 8.22 Indentation depth versus time curves. **a** Temperature = 1300 °C, load = 30 MPa, and displacement rate = 8.09×10^{-8} mm/s; temperature = 1300 °C, load = 50 MPa, and displacement rate = 1.62×10^{-7} mm/s; temperature = 1300 °C, load = 70 MPa, and displacement rate = 2.06×10^{-7} mm/s. **b** Load = 50 MPa, temperature = 1100 °C, and displacement rate = 1.96×10^{-8} mm/s; load = 50 MPa, temperature = 1150 °C, and displacement rate = 2.36×10^{-8} mm/s; load = 50 MPa, temperature = 1200 °C, and displacement rate = 3.53×10^{-8} mm/s; load = 50 MPa, temperature = 1250 °C, and displacement rate = 5.84×10^{-8} mm/s; load = 50 MPa, temperature = 1300 °C, and displacement rate = 1.62×10^{-7} mm/s

The creep strain rate of a polycrystalline ceramic is related to the load and temperature through the following equation [22]:

$$\dot{\varepsilon} \sim \sigma^n \exp\left(-\frac{Q_c}{RT}\right) \tag{8.7}$$

where n is the stress exponent, Q_c is the heat activation energy of the creep, R is the gas constant, and T is the temperature. n and Q_c can be obtained from the experimental results in Fig. 8.22 and Eq. (8.7). Figure 8.23 shows the relations of the creep strain rate with stress and temperature. Specifically, Fig. 8.23a shows the relations between the creep strain rate and stress at different temperatures, and Fig. 8.23b shows the relations between the creep strain rate and 1/RT at different loads. n and Q_c are determined from Fig. 8.23a to be 0.05–1.12 and 215–329 kJ/mol, respectively.

8.3.2 Creep Behavior of TGOs Under Tensile Stress

The creep behavior of TGOs is another important factor affecting the service life of a coating. Kang and Mercer designed the experimental method shown in Fig. 8.24 for characterizing the creep of TGOs [23, 24]. A mechanical balance is used to control the load accuracy within ±10 μN. The loading path is supported by an air bearing, which also serves as a medium for cooling the specimen. In addition, the specimen is heated by a direct current (DC) power source via stainless steel clamps. Subsequently, the displacement is measured using an infrared (IR) laser position sensor placed at the topmost location. The clamps below are air-cooled by the connected fins and fan,

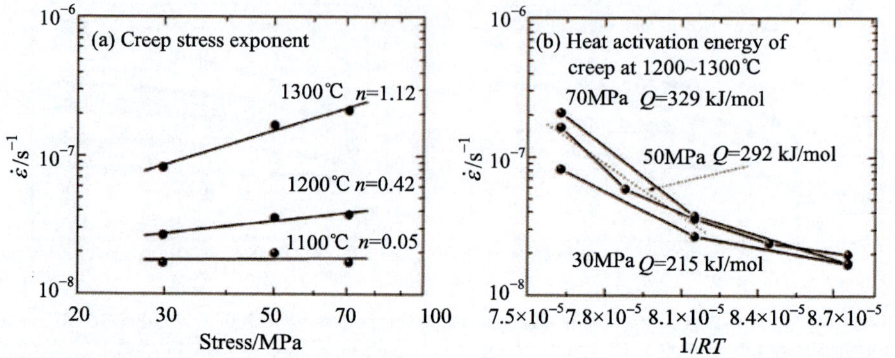

Fig. 8.23 Relations between the creep strain rate of a ceramic coating with stress and temperature: **a** Creep strain rate–stress relations at different temperatures; **b** creep strain rate–1/RT relations at different loads (R: gas constant; T: temperature)

while an IR pyrometer is used to monitor the temperature to prevent the weighing sensor from overheating. The thickness of the TGO layer is measured in real time through analysis of the different temperatures monitored by two temperature sensors. This method is referred to as the emissivity difference method (EDM) [25].

In this work, the TGOs were composed of single-phase α-Al_2O_3 that thermally grew on commercial high-temperature FeCrAlY alloy thin strips with a thickness of 50–1000 μm (see Table 8.3 for their composition). The main difference between the first and second batches of specimens was the difference in the Y and Ti concentrations. The first batch of specimens contained 510 ppm Ti and 560 ppm Y, compared to the 800 ppm Ti and 280 ppm Y in the second batch of specimens. Each specimen was cut into a strip-like specimen with transverse dimensions of 5 mm × 50 mm. To eliminate the residual stresses, each specimen was annealed in vacuum at 1100 °C for 24 h, after which the surface of the specimen was ground and polished to a

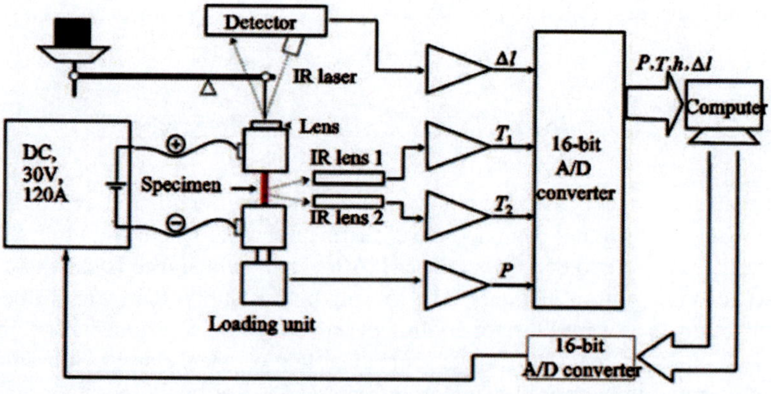

Fig. 8.24 Schematic diagram of the experimental setup

Table 8.3 Composition of the two batches of specimens

	Fe (%)	Cr (%)	Al (%)	Mn (ppm)	Si (ppm)	Zr (ppm)	Ti (ppm)	Y (ppm)
First batch of specimens	73.3	21.4	5.1	1450	440	800	510	560
Second batch of specimens	72.9	21.7	5.3	1560	560	720	800	280

roughness of 1 μm. Before testing, each specimen was cleaned with acetone, and the measurement accuracies for its size and weight were ensured to be ±10 μm and ±10 μg, respectively.

Since the TGOs in each specimen were a very thin layer (only several microns in thickness) growing on an FeCrAlY metallic substrate, it was impossible to remove the substrate and then separately measure the creep of the TGOs, as is the case with the measurement of the creep of a ceramic coating in Fig. 8.21; instead, the TGO layer and substrate had to be subjected to tensile testing together. Then, to determine the creep behavior of the TGOs, it was necessary to separately know their stress and strain. The creep of the TGOs could be determined only through the establishment of the relations of their strain rate with stress and temperature. Thus, the substrate needed to be first subjected to creep testing to (1) provide basic data for separately obtaining strain and stress data for the TGOs and to (2) verify the accuracy of experimental method and equipment. An FeCrAlY substrate specimen was heated in Ar to 1200 °C, after which the specimen was loaded and its displacement was measured to determine its creep performance. The thickness of the TGO layer formed during this test process was less than 0.3 μm. Such a thin TGO layer had absolutely no effect on the stress in the FeCrAlY substrate, that is, on the creep performance of the FeCrAlY substrate. Thus, the minimum thickness h_{TGO} was determined to be 0.3 μm. Due to the very small h_{TGO}, the tensile creep test results were considered to be the creep test results for the substrate. The stable creep rate of the metallic substrate, $\dot{\varepsilon}_{ss}$, could be expressed as follows:

$$\frac{\dot{\varepsilon}_{sS}}{\dot{\varepsilon}_0} = \left(\frac{\sigma}{\sigma_0}\right)^n \tag{8.8}$$

where $\dot{\varepsilon}_0$ is the reference strain rate ($\dot{\varepsilon}_0 = 10^{-6}\text{s}^{-1}$) and σ_0 is the reference stress ($\sigma_0 = 0.55$ MPa for the first batch of specimens; $\sigma_0 = 1.85$ MPa for the second batch of specimens). The values of the stress exponent n for the first and second batches of specimens were determined using Eq. (8.8) based on the experimental data to be 2 and 4, respectively. Notable TGO growth occurs at temperatures above 1200 °C. Under this condition, the tensile creep test results include the deformation of both the substrate and TGOs. In the case where their interface is intact, the substrate and the TGOs have identical strains and strain rates. It is assumed that at such a high temperature, the substrate is completely softened and basically unable to withstand loading. Thus, the stress within the TGOs is expressed as follows:

$$\sigma_{TGO} = \frac{P}{2h_{TGO}w} \tag{8.9}$$

where P is the tensile load, w is the width of the specimen, and the "2" in the denominator indicates that both sides of the metal are oxidized to TGOs. The creep of the TGOs is measured through the following steps: (1) Pre-oxidize the specimen in air at 1200 °C under the minimum load ($P_{min} \approx 0.1$ N) until the TGO layer reaches the predetermined thickness (h_{TGO}=1 μm). (2) Apply a fixed load P, and measure the creep displacement. (3) After a certain period of time when a steady-state creep rate is reached, reduce the load to $P_{min} \approx 0.1$ N, and continue the oxidation process; continually monitor the displacement during the TGO growth period. (4) Change the applied load to 2P, and determine the new creep rate when $h_{TGO} = 2$ μm. (5) Repeat Step (3). (6) Change the applied load to 3P, and determine the new creep rate when $h_{TGO} = 3$ μm. (7) Repeat Step (3). (8) Change the applied load to 4P, and determine the new creep rate when $h_{TGO} = 4$ μm. (9) Terminate the test after a certain period of time when a steady-state creep rate is reached.

Figure 8.25 shows the relations of the test load, displacement, and h_{TGO} with time. The creep test of the TGO layer was carried out at $h_{TGO} = 1$, 2, 3, and 4 μm, respectively.

The relations between the strain rate and stress in the TGOs on the alloy at four different values of h_{TGO}, as shown in Fig. 8.26, are derived from the load and displacement data in Fig. 8.25. The data for different substrate thicknesses are scattered in Fig. 8.26a and concentrated in Fig. 8.26b–d. In addition, there are no signs of cracking in the TGOs when $\sigma_{TGO} < 200$ MPa. At a high σ_{TGO}, many cracks perpendicular to the loading direction appear on the surface of the TGOs, and the displacement of the TGOs increases at an increasing rate after a small creep strain occurs.

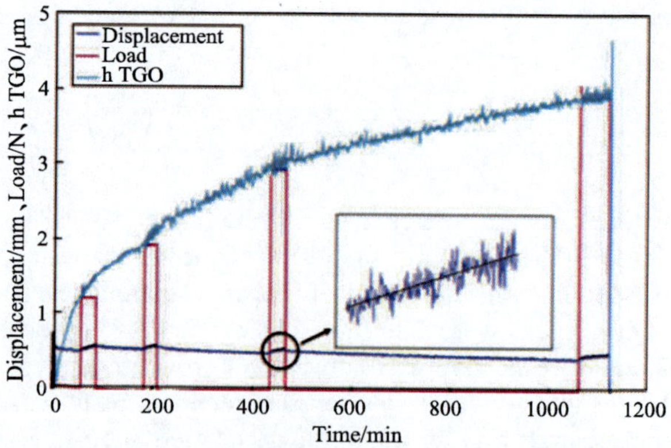

Fig. 8.25 Relations of the load, displacement, and h_{TGO} with time during the creep test

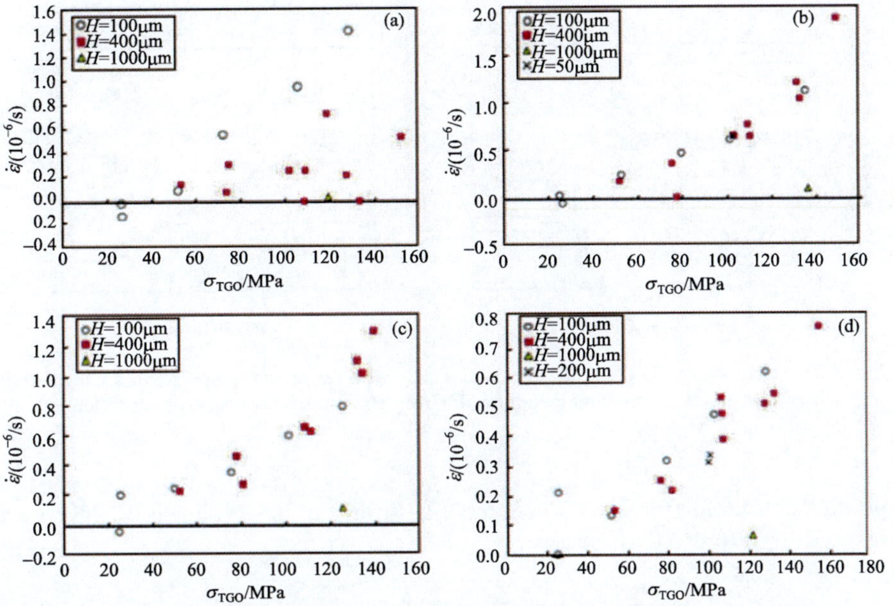

Fig. 8.26 Relations between the strain rate and stress at four different values of h_{TGO} (H is the thickness of the metallic substrate and ranges from 50 to 1000 μm). **a** $h_{TGO} = 1$ μm. **b** $h_{TGO} = 2$ μm. **c** $h_{TGO} = 3$ μm. **d** $h_{TGO} = 4$ μm

The relation of the creep strain rate of a polycrystalline TGO ceramic with stress and temperature can also be represented by Eq. (8.7). The stress exponent n and creep activation energy Q_c can be obtained from the experimental results in Fig. 8.26 in combination with Eq. (8.7). Figure 8.27 shows the relations of the creep strain rate with stress and temperature. Specifically, Fig. 8.27a shows the relations between the creep strain rate and stress at different temperatures, and Fig. 8.27b shows the relations between the creep strain rate and 1/RT at different loads. Based on Fig. 8.27a, $n = 1.5$–2.0. A thick TGO layer creeps slowly, and a high Y content results in a low creep rate. A comparison with the work of Cho et al. [26] yields a Q_c of 685 kJ/mol.

8.3.3 Effects of the Creep Behavior of TBCs on Interfacial Stresses

Interfacial stress evolution causes damage to a TBC and results in its spallation. The mechanical properties of the TC, TGO, and BC layers remain the principal factor affecting interfacial stresses. Here, the effects of creep on interfacial stresses are analyzed using FE numerical simulation [27]. We assume that the TGO and BC layers undergo both creep deformation and plastic deformation, or creep +

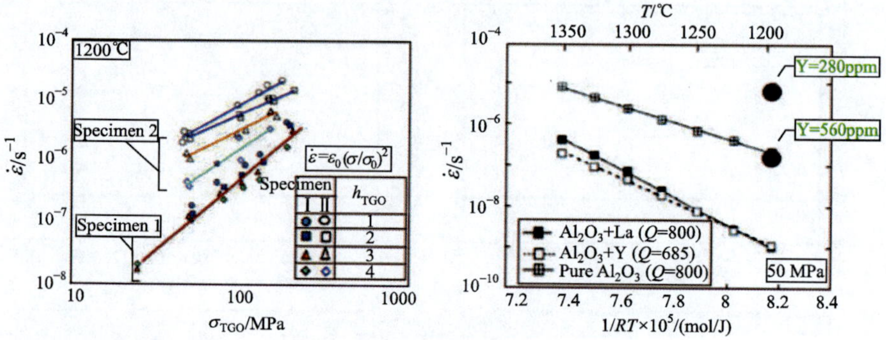

Fig. 8.27 Relations of the creep strain rate of TGOs with stress and temperature. **a** Creep strain rate–stress relations at different temperatures. **b** Creep strain rate–1/RT relations at different loads (R: gas constant; T: temperature)

plastic deformation for short. Then, the total strains of the TGO and BC layers are respectively expressed as follows:

$$\boldsymbol{\varepsilon}_{TGO} = \boldsymbol{\varepsilon}^{th}_{TGO} + \boldsymbol{\varepsilon}^{e}_{TGO} + \boldsymbol{\varepsilon}^{p}_{TGO} + \boldsymbol{\varepsilon}^{c}_{TGO} + \boldsymbol{\varepsilon}^{g}_{TGO} \tag{8.10}$$

$$\boldsymbol{\varepsilon}_{BC} = \boldsymbol{\varepsilon}^{th}_{BC} + \boldsymbol{\varepsilon}^{e}_{BC} + \boldsymbol{\varepsilon}^{p}_{BC} + \boldsymbol{\varepsilon}^{c}_{BC} \tag{8.11}$$

where the subscripts TGO and BC indicate the TGOs and BC material, respectively, $\boldsymbol{\varepsilon}^{th}$, $\boldsymbol{\varepsilon}^{e}$, $\boldsymbol{\varepsilon}^{p}$, and $\boldsymbol{\varepsilon}^{c}$ are the thermal, elastic, plastic, and creep strains, respectively, and $\boldsymbol{\varepsilon}^{g}$ is the TGO growth strain, which is expressed by

$$\boldsymbol{\varepsilon}^{g}_{TGO} = \varepsilon^{g}_{v}\mathbf{I} \tag{8.12}$$

where the growth strain is described in Eq. (4.15) of Chap. 4, $\varepsilon^{g}_{v} = 0.08$, and **I** is the unit tensor. Figure 8.28 shows the contour plots of tensile stress σ_{22} after 300 h of oxidation at 1100 °C. Here, the TGO and BC layers are each modeled to be elastic (Fig. 8.28a), ideal elastic–plastic (Fig. 8.28b), and creep + plastic (Fig. 8.28c).

Figure 8.29a and b shows the evolutionary curves of the tensile stress σ_{22} at the trough of the TC layer and the peak of the BC layer, respectively. When elastic and ideal elastic–plastic models of the TGO and BC layers are used, the considerably high compressive growth stress caused by oxidation is mainly concentrated within the TGO layer and results in an out-of-plane displacement of the TGO layer. Such an out-of-plane displacement exerts a tensile stress on the trough of the TC layer and the peak of the BC layer. Therefore, the high tensile stress increases cumulatively as the oxidation time extends.

When the TGO and BC layers undergo creep + plastic deformation, large creep deformation rapidly relaxes the growth stress, resulting in a decrease in the out-of-plane displacement, thereby inhibiting the tensile stress from accumulating at the

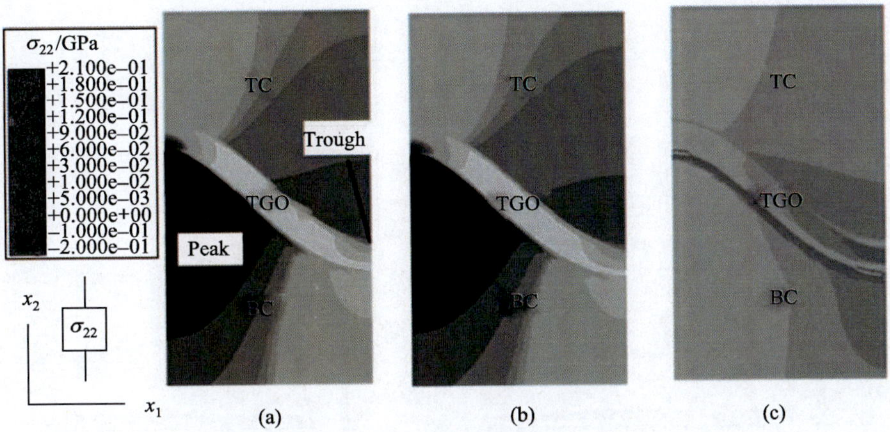

Fig. 8.28 Contour plots of stress σ_{22} after 300 h of oxidation at 1100 °C, with TGO and BC layers modeled as **a** elastic, **b** ideal elastic–plastic, and **c** creep + plastic materials

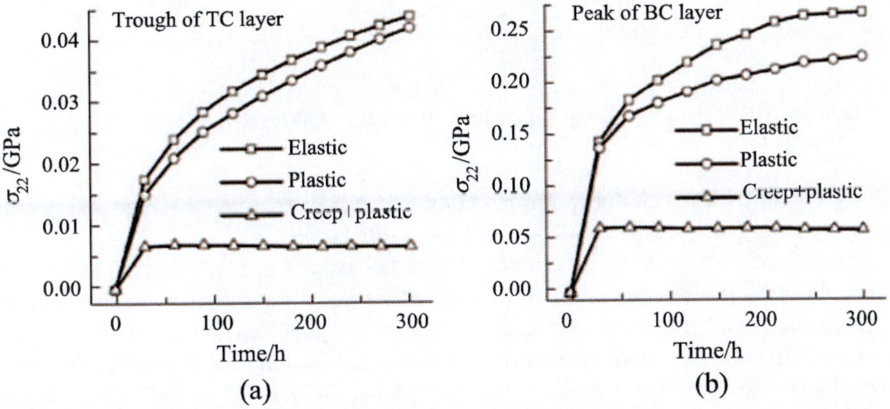

Fig. 8.29 Variation in the tensile stress σ_{22} at **a** the trough of the TC layer and **b** the peak of the BC layer with the oxidation time

trough of the TC layer and the peak of the BC layer, which means that the creep deformation of the TGO and BC layers significantly affects the interfacial stresses during the oxidation process. However, the creep behavior of the TGO and BC layers is expected to vary with the particle size or the proportions of the components. For example, as a result of the change in the particle size or the proportions of the components, the creep strain rate may span several orders of magnitude [28, 29]. To reflect the difference in the creep rate, the parameters for three types of creep in Table 8.4 are used. The creep model is expressed by the following equation:

$$\dot{\varepsilon} = A\sigma^n \exp\left(-\frac{Q_c}{RT}\right) \tag{8.13}$$

Table 8.4 Parameters for fast-, medium-, and slow-rate creep

	$A\ (\text{MPa}^{-1} \times \text{s}^{-1})$	n	Q_c (J/mol)
Slow-rate creep	10^{-6}	1	162,800
Medium-rate creep	10^{-4}	1	162,800
Fast-rate creep	10^{3}	1	162,800

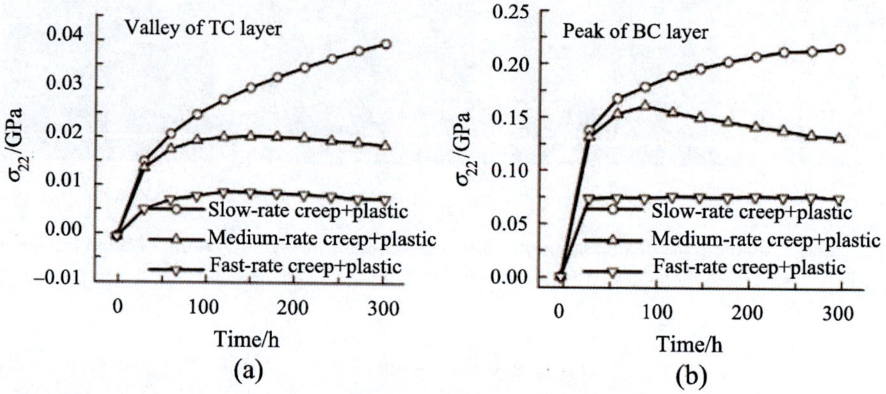

Fig. 8.30 Evolution of the stress σ_{22} at **a** the trough of the TC layer and **b** the peak of the BC layer at three creep rates by using a creep + plastic model for the TGO layer

While the values of the parameters in Table 8.4 are not obtained from real experimental data, they are still within reasonable ranges.

Let us assume that the TGO layer can be studied using a model with the parameters of each of the three types of creep and simultaneously undergoes creep and plastic deformation and that an ideal elastic–plastic model can be used for the BC layer. Figure 8.30 shows the evolution of the tensile stress σ_{22} at the trough of the TC layer and the peak of the BC layer, respectively. Clearly, creep has a notable effect on the stresses.

8.4 Summary and Outlook

8.4.1 Summary

The mechanical properties of TBCs are a key measure for both quality evaluation and quality enhancement. However, since a coating is a thin layer, its property characterization differs completely from that of a bulk material. In addition, the mechanical properties of TBCs vary with service time. TBCs operate in adverse environments (e.g., high temperatures, high-speed rotation, and calcium–magnesium aluminosilicate (CMAS) corrosion), which make it very difficult to investigate and analyze their high-temperature mechanical properties. First, this chapter introduces the microbending method for characterizing Young's modulus of TBCs, the method for correlating the temporal and spatial evolution and microstructure of TBCs, and the high-temperature creep behavior of TBCs and TGOs. The microbending method for characterizing Young's modulus of TBCs involves a clever microbending design, in situ high-accuracy characterization based on DIC, FE numerical simulation, and bending beam theory. The reliability of the microbending method is verified by the nanoindentation method. Young's modulus of EB-PVD TBCs is characterized by the microbending method to be between 15 and 25 GPa. Second, this chapter presents in detail the temporal and spatial correlations between the mechanical properties and microstructure of TBCs. A NiCoCrAlY BC layer and a YSZ ceramic coating were prepared by APS. The evolutionary patterns of the mechanical properties of each layer and interface of the TBC material subjected to different numbers of thermal cycles were determined through HSNM. The results show an extremely strong correlation between the micron-scale microstructure and the local mechanical properties of the TBC. The mechanical properties of the phase structure in each layer were determined. Specifically, the distribution of the phase structure in each layer was determined through deconvolution analysis of the phase composition data using the *k*-means clustering algorithm. On this basis, the evolutionary patterns of different phases in each layer and their local mechanical properties during thermal cycling testing were analyzed. Third, this chapter introduces the methods for characterizing the high-temperature creep of coatings as well as the effects of creep on the interfacial stresses in a coating. A high-temperature indentation method, including a characterization model and the strain rate measurement, was designed to characterize the creep properties of ceramic coatings. The creep stress exponent n and creep activation energy Q_c of ceramic coatings are determined to be 0.05–1.12 and 215–329 kJ/mol, respectively. Another method was designed to characterize the creep of TGOs formed due to the oxidation of the transition layer under high-temperature tension. The n and Q_c of TGOs are determined to be 1.5–2.0 and 685 kJ/mol, respectively. A thick TGO layer creeps slowly, and a high Y content results in a low creep rate. A numerical analysis of interfacial stresses based on the creep of ceramic coatings and TGOs finds that creep has a considerable effect.

8.4.2 Outlook

Note that research conducted around the world in the past three decades on the mechanical properties of TBCs has achieved great results in areas such as theories and equipment and made tremendous contributions to the development of aeroengines and gas turbines. From the perspectives of both engineering applications and frontier scientific research, the academic and industrial communities will attach great importance and continue to pay attention to the investigation of the mechanical properties of TBCs. In particular, with the development of aeroengines and gas turbines, coatings will operate at increasingly high temperatures and in increasingly adverse environments. As a result, the research on their mechanical properties will face increasingly serious challenges, including the development of theoretical characterization models, the realization of the coupling and decoupling of thermo–mechano–chemical environments, the research and development of characterization equipment, the obtaining of macro- and micromechanical information, fatigue behavior, creep–fatigue interactions, the spatial and temporal correlations between mechanical properties and microstructure, the ferroelasticity of YSZ, Li et al. [30] and new coatings (e.g., $Hf_6Ta_2O_{17}$) [31]. These challenges are well worth the efforts of researchers.

References

1. Eberl C, Gianola DS, Wang X, He MY, Evans AG, Hemker KJ. A method for in situ measurement of the elastic behavior of a columnar thermal barrier coating. Acta Mater. 2011;59:3612–20.
2. Eberl C, Gianola DS, Hemker KH. Mechanical characterization of coatings using microbeam bending and digital image correlation techniques. Exp Mech. 2010;50:85–97.
3. He MY, Hutchinson JW, Evans AG. A stretch/bend method for in situ measurement of the delamination toughness of coatings and films attached to substrates, J Appl Mech. 78(2011);011009–1–011009–5
4. Hemker KJ, Mendis BG, Eberl C. Characterizing the microstructure and mechanical behavior of a two-phase NiCoCrAlY bond coat for thermal barrier systems. Mater Sci Eng A. 2008;483:727–30.
5. Suo Z, Hutchinson JW. Interface crack between two elastic layers. Int J Fractures. 1990;43:1–18.
6. Oliver WC, Pharr GM. An improved technique for determining hardness and elastic modulus using load and displacement sensing indentation experiments. J Mater Res. 1992;7:1564–83.
7. Zhou YC, Yang L, Huang YL. Micro-and macromechanical properties of materials. Taylor & Francis Group: CRC Press; 2013.
8. Vignesh B, Oliver WC, Kumar GS, Phani PS. Critical assessment of high speed nanoindentation mapping technique and data deconvolution on thermal barrier coatings. Mater Des. 2019;181: 108084.
9. Sebastiani M, Moscatelli R, Ridi F, Baglioni P, Carassiti F. High-resolution high speed nanoindentation mapping of cement pastes: unravelling the effect of microstructure on the mechanical properties of hydrated phases. Mater Des. 2016;97:372–80.
10. Brown L, Allison PG, Sanchez F. Use of nanoindentation phase characterization and homogenization to estimate the elastic modulus of heterogeneously decalcified cement pastes. Mater Des. 2018;142:308–18.
11. Zou M. Deconvolution and signal recovery. National Defense Industry Press;2001.
12. Constantinides G, Chandran KSR, Ulm FJ, et al. Grid indentation analysis of composite microstructure and mechanics: principles and validation. Mater Sci Eng A Struct Mater Prop Microstruct Process. 2006;430:189–202.
13. Bolelli G, Righi MG, Mughal MZ, Moscatelli R, Ligabue O, Antolotti N, Sebastiani M, Lusvarghi L, Bemporad E. Damage progression in thermal barrier coating systems during thermal cycling: a nano-mechanical assessment. Mater Des. 2019;166: 107615.
14. Ulm FJ, Vandamme M, Bobko C, et al. Statistical indentation techniques for hydrated nanocomposites: concrete, bone, and shale. J Am Ceram Soc. 2010;90(9):2677–92.
15. Veytskin YB, Tammina VK, Bobko CP, et al. Micromechanical characterization of shales through nanoindentation and energy dispersive x-ray spectrometry. Geomech Energy Environ. 2017;9:21–35.

16. Nath S, Manna I, Majumdar JD. Nanomechanical behavior of yttria stabilized zirconia (YSZ) based thermal barrier coating. Ceram Int. 2015;41:5247–56.
17. Phani PS, Oliver WC. A critical assessment of the effect of indentation spacing on the measurement of hardness and modulus using instrumented indentation testing. Mater Des. 2019;164: 107563.
18. Hartigan JA, Wong MA. A K-means clustering algorithm. Appl Stat. 1979;28:100–8.
19. Yun J, Wee S, Park S, et al. Method for predicting thermal fatigue life of thermal barrier coatings using TGO interface stress. Int J Precis Eng Manuf. 2020;21:1677–85.
20. Jamali H, Mozafarinia R, Shoja-Razavi R, et al. Comparison of hot corrosion behaviors of plasma-sprayed nanostructured and conventional YSZ thermal barrier coatings exposure to molten vanadium pentoxide and sodium sulfate. J Eur Ceram Soc. 2014;34:485–92.
21. Deng SJ, Wang P, He YD, et al. Surface microstructure and high temperature oxidation resistance of thermal sprayed NiCoCrAlY bond-coat modified by cathode plasma electrolysis. J Mater Sci Technol. 2017;33:1055–60.
22. Jana V, Dorčáková F, Dusza J, Bartsch M. Indentation creep of free-standing EB-PVD thermal barrier coatings. J Eur Ceram Soc. 2008;28:241–6.
23. Kang KJ, Mercer C. Creep properties of a thermally grown alumina. Mater Sci Eng, A. 2008;478:154–62.
24. Sharma SK, Ko GD, Kang KJ. High temperature creep and tensile properties of alumina formed on FeCr alloy foils doped with yttrium. J Eur Ceram Soc. 2009;29:355–62.
25. Lee SS, Sun SK, Kang KJ. In-situ measurement of the thickness of aluminum oxide scales at high temperature. Oxid Met. 2005;63:73–85.
26. Cho J, Harmer MP, et al. Effect of yttrium and lanthanum on the tensile creep behavior of aluminum oxide. J Am Ceram Soc. 1997;80(4):1013–7.
27. Lin C, Li YM. Interface stress evolution considering the combined creep–plastic behavior in thermal barrier coatings. Mater Des. 2016;89:245–54.
28. Pan D, Chen MW, Wright PK, Hemker KJ. Evolution of a diffusion aluminide bond coat for thermal barrier coatings during thermal cycling. Acta Mater. 2003;51:2205–17.
29. Taylor MP, Evans HE, Busso EP, Qian ZQ. Creep properties of a Pt-aluminide coating. Acta Mater. 2006;54:3241–52.
30. Li JB, Yang L, Zhou YC, Li ZP, Liu ZJ, Tsai MF. Preparation of epitaxial yttria-stabilized zirconia with non-equilibrium t' phase and ferroelastic domain by pulsed laser deposition. Appl Surf Sci. 2021;537: 147790.
31. Tan ZY, Yang ZH, Zhu W, Yang L, Zhou YC, Hu XP. Mechanical properties and calcium-magnesium-alumino-silicate (CMAS) corrosion behavior of a promising $Hf_6Ta_2O_{17}$ ceramic for thermal barrier coatings. Ceram Int. 2020;46:25242–8.

Chapter 9
Fracture Toughness Characterization of TBCs

Fracture is an issue worthy of attention during the design and use of all engineering materials. Based on the test results for different materials, the fracture can be approximately categorized into three types, namely, brittle fracture, quasibrittle fracture, and ductile fracture. The inherent brittleness of ceramic materials is their most essential mechanical property, and their fracture process is basically a brittle one. Thermal barrier coatings (TBCs) are coated on the surfaces of turbine blades in an aeroengine to reduce the surface temperature of the blades, thereby enhancing the thermal efficiency of the engine. However, in practical application, thermal stresses cause brittle fractures in the coatings and, as a result, affect the normal operation of the whole component. In the early 1920s, Griffith [1] proposed the classical theory of brittle fracture. On this basis, through decades of development, researchers around the world in the field of the mechanics of ceramic materials have preliminarily established a set of theories of fracture mechanics for ceramic materials. Griffith noted that fracture strength is not a characteristic constant of material but instead depends on the geometrical characteristics (e.g., sizes, shapes, and locations) of its internal defects or cracks. With the development of fracture mechanics, another more effective parameter—fracture toughness—is used to describe the ability of a material to resist crack instability and propagation or fracture.

This chapter focuses primarily on introducing the methods for characterizing the surface and interfacial fracture toughness of TBCs, mainly including the methods for characterizing surface fracture toughness, conventional methods for testing interfacial fracture toughness, new methods (e.g., indentation, buckling, and blister tests) for characterizing surface and interfacial fracture toughness, and methods for characterizing the high-temperature fracture toughness of TBCs.

Y. Zhou et al., *Thermal Barrier Coatings: Failure Theory and Evaluation Technology*,
https://doi.org/10.1007/978-981-19-2723-2_9

9.1 Surface Fracture Toughness K_{IC} Characterization of TBCs

9.1.1 Definition of Fracture Toughness

According to Irwin's theory [2], the fracture toughness K_{IC} represents the extreme applied stress-field strength that a Type I crack can withstand. Crack propagation occurs when the numerical value of the applied stress-field strength K_I reaches or exceeds the K_{IC} of a material. Irwin considered crack systems in isotropic, uniform, and continuous materials. For a crack system with a defined geometry, the fracture toughness K_{IC} defined by Irwin should be a constant. To obtain specific values of fracture toughness by testing, ASTM International formulated a test standard, ASTM E-24 [3], to give an authoritative definition for fracture toughness from a testing perspective. It states that fracture toughness is "the resistance to crack propagation under plane-strain conditions" and that "under Type 1 loading conditions in which the loading rate is low and plastic-zone correction is negligible, the plane-strain fracture toughness, denoted by K_{IC}, is obtained by the operable test technique stipulated in the E-399 method and is equivalent to the resistance encountered by cracks at the start of propagation..." [3]. To date, some techniques for testing the K_{IC} of ceramic materials have emerged around the world [4], including the test methods for different geometric components (e.g., the single-edge notched beam (SENB) method, the double torsion (DT) method, the double cantilever beam (DCB) method, the chevron notch (CHV) method, and the indentation method) and the preparation methods for ideal crack tips (e.g., the bridge indentation (BI) method, the wedge indentation (Pw) method, and the making-cracks-before-sintering (MBS) method). This section investigates the surface K_{IC} of TBCs mainly using the SENB method.

9.1.2 Surface K_{IC} Characterization of the Pure Ceramic Surface of TBCs Using the SENB Method

The SENB method is similar to the three-point bending method used to test the K_{IC} of metal specimens except that the precracks are replaced by a single-edge notch and the K_{IC} of material is tested using the three- or four-point bending method, as shown in Fig. 9.1. To date, a series of reports are available on the effects of the notch width and depth of the SENB on K_{IC} [5, 6]. In the following, we introduce the use of the SENB method to test the K_{IC} of pure ceramics [7, 8]. During the loading process, the load increases with time and the ceramic coating fractures when the load reaches the maximum, which is the fracture point of the ceramic coating. K_{IC} can be calculated using the following equation [9]:

$$K_{IC} = Y \frac{3Fl}{2Bw^2} \sqrt{a} \tag{9.1}$$

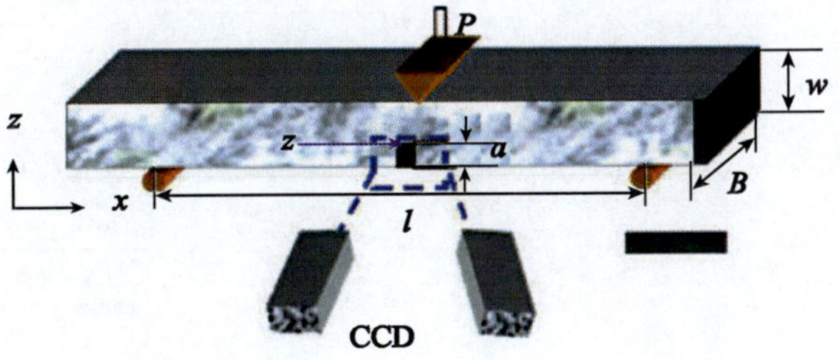

Fig. 9.1 Schematic diagram of the SENB test of a pure ceramic TBC

Table 9.1 Values of coefficient A_i

	A_0	A_1	A_2	A_3	A_4
Three-point bending $l/w = 8$	1.96	−2.75	13.66	−23.98	25.22
Three-point bending $l/w = 4$	1.93	−3.07	14.53	−25.11	25.8

where F is the applied load, l is the span, B is the width of the specimen, w is the thickness of the specimen, a is the height of the notch, and Y is a dimensionless coefficient [10] related to a/w and the loading rate and can be expressed by the polynomial of a/w in Eq. (9.2) with $0 \le a/w \le 0.6$.

$$Y = A_0 + A_1\frac{a}{w} + A_2\left(\frac{a}{w}\right)^2 + A_3\left(\frac{a}{w}\right)^3 + A_4\left(\frac{a}{w}\right)^4 \tag{9.2}$$

Table 9.1 summarizes the values of coefficient A_i (i = 1, 2, 3, 4) in Eq. (9.2).

TBC specimens were prepared through the following procedure. First, the surface of a large Al substrate was treated to remove grease, surface stresses, sharp corners, and burrs. Then, 8 wt.% Y_2O_3-ZrO_2 powder was sprayed onto the surface of the Al substrate to obtain a thickness of approximately 4 mm. Test specimens with the required dimensions (2 mm × 4 mm × 20 mm) were then prepared by cutting. Subsequently, the Al substrate was removed using KOH through the chemical reaction of $2Al + 2KOH + 2H_2O = 2KAlO_2 + 3H_2\uparrow$, followed by the polishing of each specimen. Finally, a notch was prepared on each specimen using a precision diamond wire cutting machine. The notch had a depth 0.4 times that of the whole specimen and a width of approximately 250 μm. Figure 9.2 shows the procedure in detail. The specimens were subjected to 0, 50, 100, 150, 200, and 300 thermal cycles. Figure 9.3 shows a photograph of the final specimens.

Prior to the SENB test, a layer of matte black paint was manually sprayed onto the surface of the pure ceramic coating to create speckles. A charge-coupled device (CCD) image acquisition system was used to record the changes in the bit values of the pixels on the surface of the tested object in real time during two consecutive stages before and after loading at different times, based on which the changes in the strain field on the tested surface in different states were obtained. This test was conducted

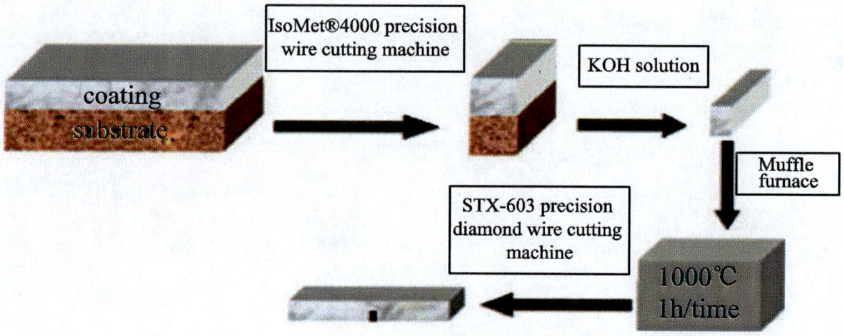

Fig. 9.2 Flowchart of the preparation of a pure ceramic TBC

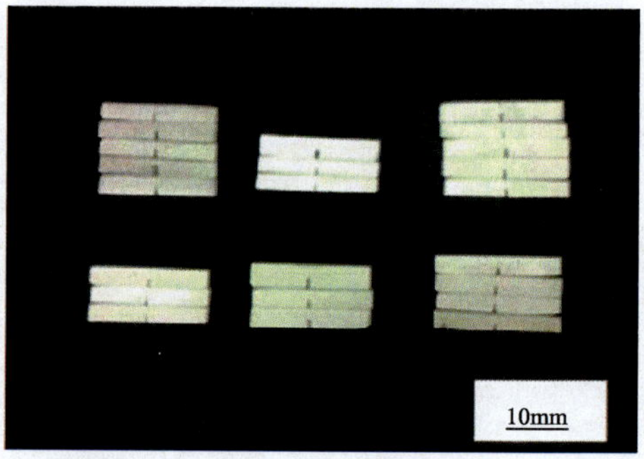

Fig. 9.3 Photograph of notched pure ceramic specimens obtained after removing the substrate

using the bending test method on a WDTI-5 testing machine at a bending test rate of 0.01 mm/min. During the bending process, the CCD camera was used to collect real-time data on the changes in the strain above the tested notch and to observe the changes in the strain at the notch, based on which the time when the ceramic coating began to crack as well as the strain at fracture were determined. Figure 9.1 shows the test zone. See Chap. 11 for the details of the measurement of the strain based on digital image correlation (DIC). Before the bending test, a calibration panel 10 mm × 8 mm in size was used to perform calibration to control the error within 0.05%. The acquisition rate was set to one image per second. During the test process, the influence of factors such as vibration, noise, and light on the test results had to be prevented.

Figure 9.4 shows the bending load *P*–time *t* relations obtained from testing on a universal tensile testing machine. When *P* increases to the bending strength of the ceramic (i.e., the highest point of the curve), the coating completely fractures, after which the applied *P* gradually decreases with time. The data represented by each red point (point C) are the values corresponding to the fracture of the ceramic during the test on the tensile testing machine. For the pure ceramic specimen in the original state, when the deflection ω is 0.0185 mm, the corresponding *P* is 7.74 N. Under this condition, the coating is in the elastic stage. The contour plot for point A shows that the strains in the whole ceramic and at the notch are relatively uniform. As *P* further increases to 12.57 N (corresponding to a ω of 0.0271 mm), the strains further increase. As seen from the contour plot for point C, the strain at the notch of the ceramic undergoes significant localization, suggesting that the ceramic might have cracked and that the cracks primarily propagate vertically upward along the notch. At a P of 13.51 N (corresponding to a ω of 0.0317 mm), the slope of the P–t curve is significantly smaller than that in the original state, the ceramic has already completely fractured, and the P begins to rapidly decrease. As seen in the contour plot for point D, which corresponds to a ω of 0.0371 mm and a P of 11.67 N, the contour plot for the strain in the zone above the notch is even more nonuniform. Under this condition, the strain is much larger than that when the coating undergoes brittle fracture under the largest load, suggesting that the cracks have already begun to propagate upward. An analysis of the effects of the number of thermal cycles shows that the P under which the ceramic completely fails increases from 13.51 N in the original state to 25.83 N after 150 thermal cycles; as the number of thermal cycles increases further, the P under which the ceramic fractures decreases to 20.99 N. It takes 180, 181, 219, 245, 208, and 201 s for the pure ceramic coating in the original state to completely fracture when subjected to 50, 100, 150, 200, and 300 thermal cycles, respectively.

Figure 9.5 shows the values of K_{IC} of the pure ceramic coatings subjected to different numbers of thermal cycles. As the number of thermal cycles increases, the K_{IC} of the pure ceramic first increases and then gradually decreases. After heat treatment in the early stage, the K_{IC} of the ceramic coating increases from 1.12 $\pm$ 0.12 to 2.45 $\pm$ 0.2 MPa·m$^{1/2}$. As the number of thermal cycles increases, the K_{IC} of the ceramic coating decreases to 1.84 $\pm$ 0.05 MPa·m$^{1/2}$. The main cause of the increase in the K_{IC} during the early stage is that the ceramic coating is sintered and becomes denser. The gradual decrease in the K_{IC} during the later stage may be ascribed to the appearance of microcracks and pores in the ceramic coating, which lowers its performance. There are two reasons for the decrease in the performance of the ceramic: (1) during the sintering process, the ceramic coating becomes denser, resulting in a decrease in its strain tolerance; and (2) under cold/thermal shocks, there exists a temperature gradient field in the ceramic, producing thermal stresses that cause some cracks to propagate or lead to stress concentration in the ceramic. Figure 9.6 shows that cracks (mostly transcrystalline cracks) and pores appear in the ceramic. Under thermal cycling, the cracks continuously propagate and coalesce. By using the three-point bending method, Choi et al. [11] found that K_{IC} increased from 1.15 to 2.38 MPa·m$^{1/2}$ before 100 thermal cycles, a result consistent with that obtained by the SENB method.

The strain field in the notch zone was monitored in real time using the DIC technique. An analysis of the inflection point of the strain yielded the time required for the strain at the fracture of the coating to occur. Then, the substitution of the time measured by the DIC technique into the three-point P–t–ω curves in Fig. 9.4 gave the times required for the ceramic to fracture after being subjected to different numbers of thermal cycles, which are 160, 160, 182, 219, 177, and 162 s, corresponding to P values of 12.57, 19.62, 19.92, 23.28, 19.23, and 19.97 N, respectively. The coating fractures earlier than the time measured by the tensile testing machine. This is because the tensile testing machine evaluates the overall load-bearing capacity of a specimen,

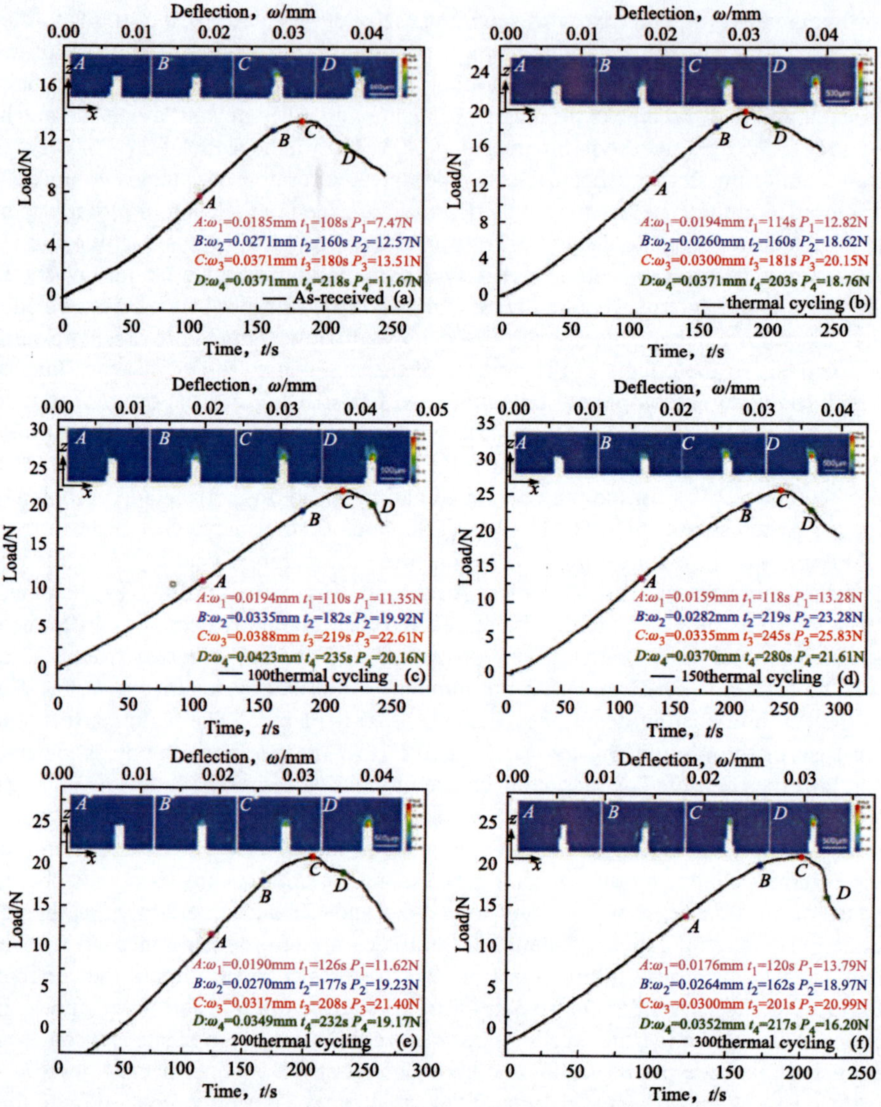

Fig. 9.4 Three-point bending load-time-deflection (P–t–ω) curves of the ceramic coating subjected to different numbers of thermal cycles: **a** original specimen and specimens subjected to **b** 50, **c** 100, **d** 150, **e** 200, and **f** 300 thermal cycles

whereas the DIC technique characterizes the local fracture of a specimen. During the loading process, the ceramic coating cracks at the notch but still has enough capacity to resist its fracture. As the applied P gradually increases, the cracks in the ceramic further propagate vertically upward. When the maximum P required for failure is reached, the ceramic coating completely loses its resistance and therefore

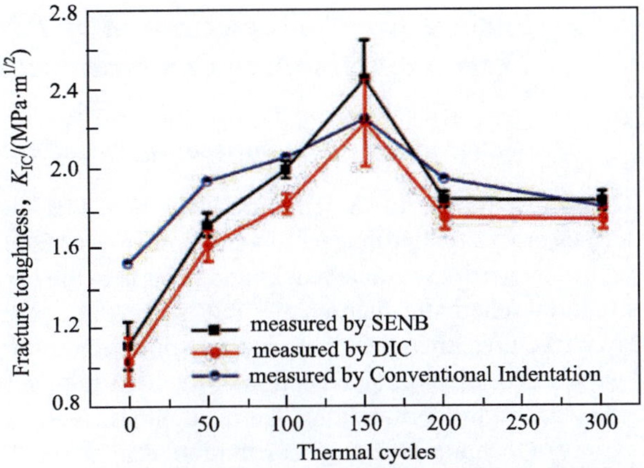

Fig. 9.5 Evolution of the K_{IC} of the ceramic coating under thermal cycling

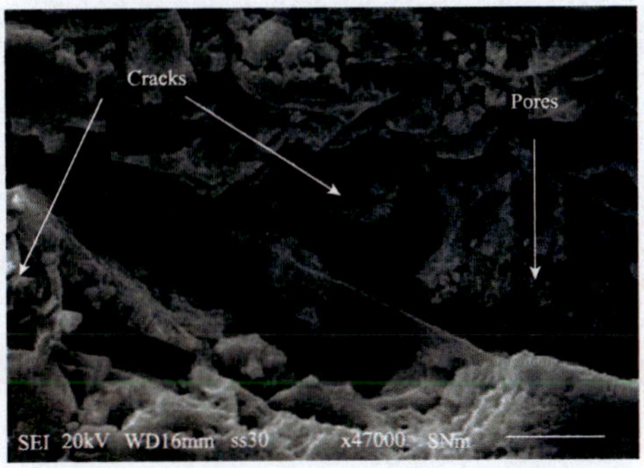

Fig. 9.6 SEM image of the fracture of the ceramic coating subjected to 300 thermal cycles

fractures. As seen from the contour plots of strain in Fig. 9.4, after being subjected to different numbers of thermal cycles and prior to fracture (i.e., at point A), the strain at the notch of the ceramic displays no significant difference from the strains at other locations, suggesting that the ceramic remains in the elastic stage during the bending process. As P gradually increases to point B, the contour plots of strain for the notch of the ceramic change, and the strain is larger than the original value, which might be attributed to crack formation at the notch of the ceramic. The strain–time curve also shows that fracture occurs under this condition. As P further increases to point C (i.e., the fracture load required for the pure ceramic to completely fail), the strain becomes even larger, and the cracks have already propagated upward. We can observe in Fig. 9.4 that the time and deflection at the fracture of the ceramic first increases and then gradually decreases as the number of thermal cycles increases. After being subjected to different numbers of thermal cycles, the increase in the fracture load indicates an increase in the fracture resistance of the ceramic. The K_{IC} of the ceramic increases from 1.04 ± 0.12 to 2.23 ± 0.22 MPa·m$^{1/2}$ and then decreases to 1.75 ± 0.05 MPa·m$^{1/2}$.

9.1.3 Surface K_{IC} Characterization of TBCs Using Three-Point Bending Combined with Acoustic Emission

(1) Theoretical model for the surface K_{IC} characterization of TBCs

The failure modes of the ceramic layer in a TBC system mainly include vertical surface cracks that propagate inward along the surface and internal transverse cracks caused by vertical surface cracks and defects in the ceramic layer. Figure 9.7 shows a structural schematic diagram of a TBC system under bending load. When the ceramic layer fractures, the stress state in the x-direction in the coating can be considered to be a result of the combined action of the load-induced tensile stress σ_{xx} and residual stress σ_{xx}^r in the x-direction. The residual stress in a coating can be measured by a variety of techniques such as indentation, Raman spectroscopy, and X-ray diffraction. The elastic properties of the bond coat (BC) layer in a TBC system are similar to those of the substrate, and thus the BC layer and substrate can be collectively viewed as one layer. Hence, in this chapter, a TBC system is regarded as a two-layer system composed of a ceramic layer and a substrate. In other words, a TBC system under a three-point bending load can be viewed as a two-layer composite beam model. Since the plane assumption for bending holds, the cross-sectional normal strain remains linearly distributed along the y-axis; i.e., $\varepsilon = y/\rho$. The magnitude of the normal stress σ_M in a coating in the x-direction is obtained based on Hooke's law

$$\sigma_M = E_c y/\rho \tag{9.3}$$

where E_c is the elastic modulus of the ceramic layer, y is the distance between the coating and the neutral axis of the composite beam, and ρ is the radius of curvature. Figure 9.8 shows a schematic diagram depicting the calculation of the curvature. Straight-line AB indicates the position of the neutral axis before loading, and curved line AB shows the shape of the neutral axis after loading. By approximating the shape of the neutral axis of a bent TBC as an arc, we have

$$\rho = [(L/2)^2 + \delta^2]/2\delta \tag{9.4}$$

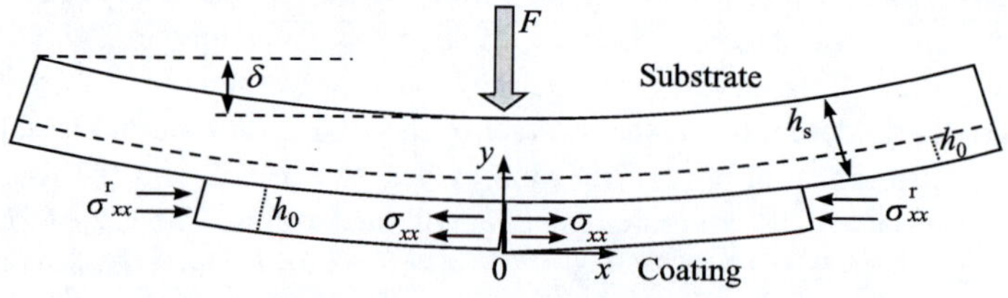

Fig. 9.7 Three-point bending composite beam model of a TBC

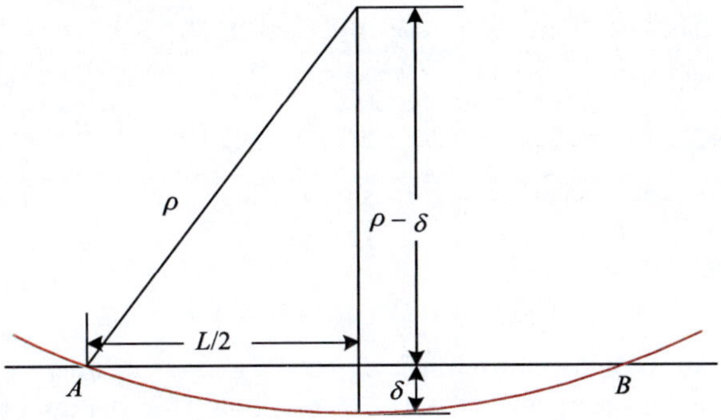

Fig. 9.8 Schematic diagram of the calculation of the curvature of a TBC specimen

$$P = P_1 - C_1 P_3 - C_2 \frac{M_3}{h} \tag{9.5}$$

where h_0 is the distance between the ceramic/BC interface and the neutral axis. The expression of h_0 can be derived from the static equilibrium conditions as follows [12]:

$$h_0 = \left(E_s h_s^2 - E_c h_c^2\right)/(2E_s h_s + 2E_c h_c) \tag{9.6}$$

where E_c and E_s are the elastic moduli of the ceramic layer and the substrate, respectively, δ is the bending deflection, and h_c and h_s are the thicknesses of the ceramic layer and the substrate, respectively.

Under a three-point bending load, a TBC fractures, generating Type I cracks that propagate in the direction perpendicular to the direction of σ_{xx}. According to Griffith's criterion for brittle fracture ($K_I = K_{IC}$) in combination with the theory of linear elastic fracture mechanics, K_{IC} can be expressed as follows:

$$K_{IC} = \sigma \sqrt{\pi a_0} Y_I \tag{9.7}$$

where a_0 is the critical crack length and σ is the critical fracture stress of the coating, which consists of the normal stress formed due to bending and the residual stress; i.e.

$$\sigma = \sigma_r + \sigma_M \tag{9.8}$$

where σ_r is the residual stress in the coating and σ_M is the normal stress in the coating in the x-direction formed under the bending load. The K_{IC} of the ceramic layer can be calculated through the measurement of the critical load thresholds for a coating (e.g., the critical deflection δ for coating cracking and the critical crack length a_0) during the three-point bending test process.

Table 9.2 Spraying parameters of the APS TBC

Powder	Current (A)	Voltage (V)	Feed rate (g/min)	Spraying distance (mm)	Carrier gas flow rate (l/min)
NiCrAlY	450	40–50	10	150–200	5.0
8YSZ	450	40–50	5	100–120	4.5

(2) Surface K_{IC} characterization of TBCs

Atmospheric plasma spraying (APS) TBC specimens were prepared for testing. Some of the specimens were oxidized at a high temperature for 300 h. The specimens were prepared by the following procedure:

I. Specimen preparation and treatment. (1) Treatment of the substrate. A 2-mm -thick GH3030 Ni-based high-temperature alloy was selected as the substrate material. The alloy plate was cut into pieces with dimensions of 80 mm × 9 mm × 2 mm using an IsoMet®4000 wire cutting machine. Before it was sprayed with a coating, the substrate needed to be subjected to a series of treatments, namely, cleaning, surface softening, sandblasting or coarsening with high-pressure water, and low-stress treatment. Surface coarsening is a necessary step that creates an uneven substrate surface to facilitate molten particles to become "hooked" onto the substrate as they spread, coagulate, and shrink over the substrate, thereby enhancing the bonding strength of the coating. (2) Spraying of a BC layer. A layer of NiCrAlY powder approximately 150 μm in thickness was sprayed onto the surface of the substrate by the vacuum plasma spraying (VPS) technique. To ensure sufficient diffusion between the component of the BC layer and the substrate material, the coated substrate was held at 1000 °C in a vacuum for 2 h after spraying. (3) Spraying of a ceramic layer. A layer of 8 wt.% Y_2O_3-ZrO_2 powder approximately 400 μm in thickness was APS-sprayed onto the surface of the BC layer. See Table 9.2 for the spray parameters in detail.

II. High-temperature oxidation of the specimens. Some of the prepared TBC specimens were placed in a high-temperature resistance furnace and heated to 1000 °C and kept at this temperature for 300 h, after which the furnace was turned off and the specimens were cooled to room temperature in the furnace.

III. Metallographic treatment of the specimens. Before marking the cross-section of the TBC with speckles, the sides of the TBC were first subjected to metallographic treatment. The surface was polished to reduce the effects of burrs and substrate oxides on acoustic emission (AE) signals and sensor–specimen coupling. Specifically, each specimen was polished using sandpaper with grit sizes of 800 and 1000 to remove the burrs and oxides from each side before and after high-temperature oxidation. Subsequently, the specimen was polished using sandpaper with a grit size of 1500 to create a smooth surface. Finally, the cross-section of the specimen was polished using metallographic sandpaper with grit sizes of 1000 and 3000. Figure 9.9 schematically shows the dimensions of a specimen.

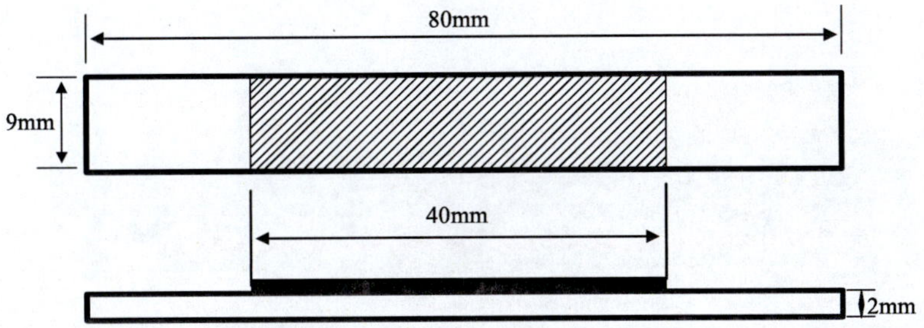

Fig. 9.9 Schematic diagram of the TBC specimen dimensions

IV. Preparation of speckle marks. After the metallographic treatment, a uniform layer of matte white paint was sprayed onto the cross-section of the TBC, followed by a light layer of matte black paint, to prepare a randomly distributed artificial speckle field.

It has been shown that [13] under bending load, cracks are first initiated on the surface of the ceramic layer and propagate along the loading direction, forming vertical surface cracks. As the load increases, the number of cracks gradually increases until saturation, and the cracks continually propagate toward the ceramic/BC or BC/substrate interface. Eventually, interface cracks are initiated and propagate at the ceramic/BC or BC/substrate interface. Thus, during the test process, it is necessary to accurately measure the starting point of surface crack initiation, a critical threshold that is used to estimate the surface K_{IC} of the TBC before and after oxidation.

An RG2000-10 universal material testing machine was used in the test. The speed-loading mode was selected, and the loading rate was set to 0.2 mm/min. In addition, a PCI-2 AE data acquisition system (Physical Acoustics Corporation, USA) was used in the test. Two sensors were employed during the test to acquire AE signals from the designated zone. Prior to the test, the wave velocity in the material was measured using the pencil core method to be 5630 ± 732 m/s. The response frequency of the sensors ranged from 70 to 400 kHz. The AE threshold and sampling frequency were set to 38 dB and 2 MHz, respectively. The timing parameters were set as follows: a peak definition time (PDT) of 50 μs, hit definition time (HDT) of 200 μs, and hit lockout time (HLT) of 300 μs. An ARAMIS digital image processing system (GOM, Germany) with a maximum measurement accuracy of 0.01% was used to perform digital speckle correlation (i.e., DIC) [14, 15]. The image acquisition frequency was set to one image per two seconds. White light was used for image acquisition during the test. First, the light intensity was adjusted to ensure the acquisition of clear speckle images. Before the test, the system was calibrated, and the parameters of the two cameras were determined. The calibrated instrument was capable of accurately measuring strains within an area of 10×8 to 100×75 mm^2. In this test, the error of DIC strain measurement was controlled by 0.05%. See Chap. 12 for the detailed

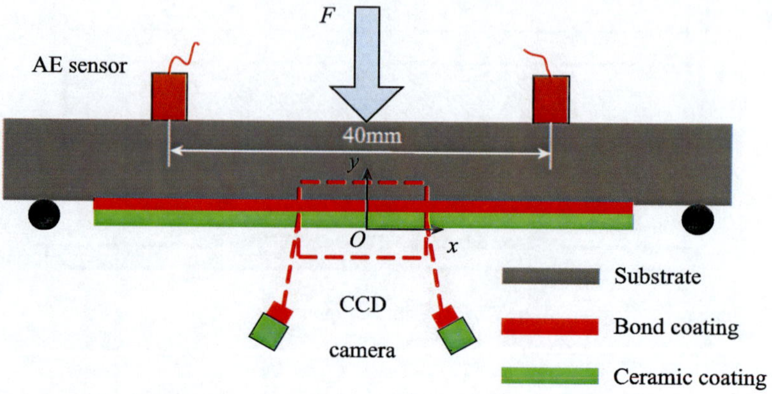

Fig. 9.10 Schematic diagram of the three-point bending test setup for the TBC

principle and procedure of DIC strain measurement. The ceramic surface of each TBC specimen was placed facing down, and its normal stress was tensile, as shown in Fig. 9.10.

The failure process of the TBC under bending load is found to involve two stages: (1) when the strain in the coating exceeds its fracture failure threshold, vertical surface cracks are initiated and then continually propagate and proliferate with the increase in strain; and (2) when the interfacial strain exceeds the interfacial fracture failure threshold, interface cracks are initiated and subsequently propagate and coalesce. This finding is consistent with that reported previously [16]. Figure 9.11 shows cross-sectional images of the unoxidized and oxidized TBC specimens after bending failure. We also observe from the test that there is only one vertical surface crack in the ceramic layer of the oxidized specimen. Figure 9.12a shows the contour plot of the cross-sectional strain along the direction of the tensile stress (i.e., x-direction) in an unoxidized TBC specimen under bending loading for 300 s. It is observed that several strain-concentration zones with strains of approximately 0.7% are distributed in the ceramic coating along the x-axis. Figure 9.12b shows the contour plot of the cross-sectional interfacial strain in the vertical direction (i.e., y-direction) in a TBC specimen oxidized for 300 h and subsequently subjected to bending loading for 550 s. There is only one strain-concentration zone with strains of approximately 3.0%, and the interface cracks propagate from this zone to both ends of the coating/substrate interface.

Figure 9.13 shows the variation in P on the TBC as well as the cumulative number of AE events (N_{AE}), the strain at the strain-concentration point in the ceramic layer in the x-direction, and the strain at the interfacial strain-concentration point in the y-direction in the TBC, subjected to the bending load, with time before and after oxidation. A large number of AE signals are generated during the crack initiation and propagation processes. As a result, there is a substantial change in the N_{AE} during crack initiation. When vertical surface cracks are initiated in the coating, there is an increase in the rate of the strain ε_{xx} at the strain-concentration point in the x-direction,

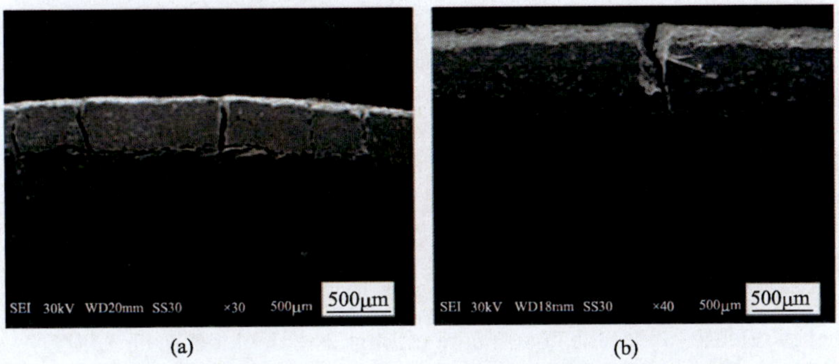

Fig. 9.11 Cross-sectional SEM images of the TBC after bending failure: **a** an unoxidized specimen; **b** a specimen subjected to 300 h of oxidation

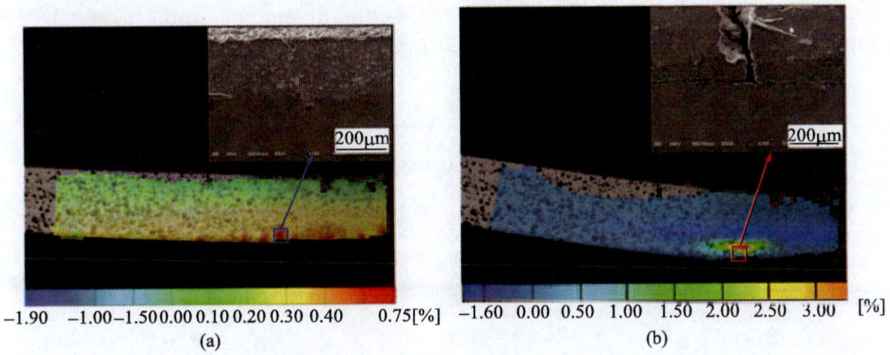

Fig. 9.12 Strain contour plots of the TBC under bending loading: **a** strain ε_{xx} in an unoxidized coating specimen; and **b** interfacial strain ε_{yy} in a specimen after 300 h of oxidation

corresponding to a notable inflection point [17]. Similarly, when interface cracks are initiated, there is an increase in the rate of the strain ε_{yy} at the interfacial strain-concentration point along the *y*-direction, corresponding to another notable inflection point [17]. As seen in Fig. 9.13a, at approximately 200 s, there is an increase in the N_{AE} in the unoxidized specimen as well as a sharp increase and a notable inflection in the strain ε_{xx} in the ceramic layer. At this time point, the corresponding strain in the coating is approximately 0.5%, and the corresponding load and deflection are 300 N and 0.667 mm, respectively, as shown at point A in Fig. 9.13a. A comparison with the cross-sectional SEM image shows that a surface crack is formed in the ceramic layer under this load. Subsequently, as the load and deflection increase, the number of cracks increases until saturation, followed by interface crack initiation. At point B (at approximately 720 s) in Fig. 9.13a, there is a change in the N_{AE} and an increase in the interfacial strain rate. A comparison with the corresponding scanning electron microscopy (SEM) image of the interface shows that cracks are formed at the ceramic/BC interface. At this time point, the corresponding interfacial strain, load, and deflection are approximately 0.27%, 460 N, and 2.4 mm, respectively. Figure 9.13b shows the results obtained for an oxidized TBC specimen using a similar analysis method. Point A is the critical starting point of vertical surface crack initiation, corresponding to a strain of approximately 0.75% in the coating, a load of 180 N, and a deflection of 0.833 mm. Point B is the starting point of interface crack initiation, corresponding to an interfacial strain of approximately 0.71%, a load of 142 N, and a deflection of 1.166 mm.

Based on the above critical thresholds for vertical surface crack initiation in the unoxidized specimen (i.e., load = 300 N and deflection = 0.667 mm), the surface K_{IC} and energy release rate of the coating are ultimately calculated to be 1.005 MPa·m$^{1/2}$ and 20.2 J·m^{-2}, respectively. During the calculation process, the Young's moduli of the ceramic layer and the metallic substrate are set to 49 and 220 GPa, respectively [18]. By using the single-edge V-notched beam method, Choi et al. [19] calculated the surface K_{IC} of a ZrO_2-8%Y_2O_3 TBC prepared by PS at room temperature to be 0.95 ± 0.09 MPa·m$^{1/2}$. By using the four-point bending method, Yamazaki et al. [20] calculated the critical energy release rate of the ceramic layer of a PS TBC without

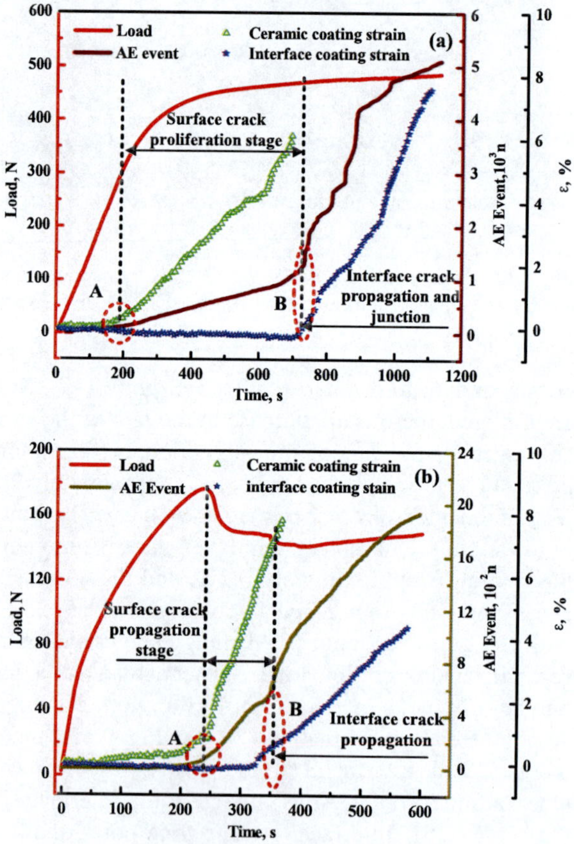

Fig. 9.13 Variation in the load on the TBC as well as the cumulative number of AE events, coating strain, and interfacial strain in the TBC under bending loading with time: **a** an unoxidized specimen; and **b** a specimen after 300 h of oxidation

heat treatment to be 25 J·m^{-2}. Through an indentation test based on the Lawn–Evans–Marshall model, Chen [18] measured a surface K_{IC} of a TBC of 0.52 MPa·m$^{1/2}$. Evidently, the surface K_{IC} values obtained in this chapter are consistent with those reported in the above studies.

The critical load and deflection for surface crack initiation in the ceramic layer of the oxidized TBC are 180 N and 0.833 mm, respectively. On this basis, the surface K_{IC} and energy release rate of the coating are ultimately calculated to be 3.531 MPa·m$^{1/2}$ and 83.120 J·m^{-2}, respectively. During the calculation process, the elastic modulus of the ceramic layer is set to 150 GPa [21], and the residual stress in the ceramic layer is set to –200 MPa [22]. By using the four-point bending method, Yamazaki et al. [20] calculated the critical energy release rate of the ceramic layer of a PS TBC subjected to 2000 h of oxidation to be 75 J·m^{-2}. Through an indentation test, Chen [18] calculated the K_{IC} of a TBC subjected to 100 thermal cycles at 1000 °C to be 3.06 MPa·m$^{1/2}$. Clearly, the surface K_{IC} of the oxidized TBC obtained in this chapter agrees well with those in the above studies.

9.2 Conventional Methods for Characterizing the Interfacial K_{IC} of TBCs

9.2.1 *Theoretical Model for the Interfacial K_{IC} Characterization of TBCs*

Research [23] has shown that the durability of TBCs depends on a series of events, namely, crack nucleation, propagation, and coalescence, and that the cumulative nucleation, propagation, and coalescence of interface cracks eventually results in a large area of coating debonding or even spallation. Therefore, compared to the vertical surface cracks, the formation of interface cracks is more detrimental to the use of TBCs. Mao [13] conducted extensive research based on the three-point bending test and found that there are two failure modes of a TBC under bending load. In failure mode I, vertical surface cracks directly extend to the BC/substrate interface, followed by the rapid propagation of interface cracks at the BC/substrate interface. In failure mode II, after vertical surface cracks propagate to the ceramic/BC interface, interface cracks are initiated and propagate at the ceramic/BC interface.

Interfacial K_{IC} is an important measure of the interfacial mechanical properties of a coating. Suo and Hutchinson [24] and Hutchinson et al. [25] extensively studied the interfacial mechanical properties of coatings and developed a Suo–Hutchinson model for calculating the cracking energy release rate and interfacial K_{IC} of two-layer systems. The interface cracking behavior of a coating/substrate system under loading can be equivalent to the composite superposition of two equilibrium states in Fig. 9.14a, b. The equilibrium equations of the system in Fig. 9.14a can be written as follows:

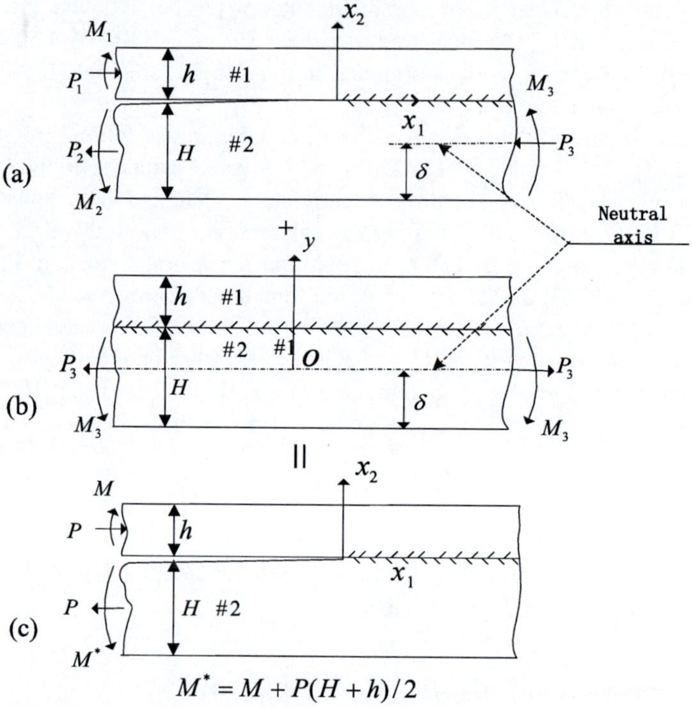

Fig. 9.14 Suo–Hutchinson model and its boundary loads

$$P_1 - P_2 - P_3 = 0 \tag{9.9}$$

$$M_1 - M_2 + P_1\left(\frac{h}{2} + H - \delta\right) + P_2\left(\delta - \frac{H}{2}\right) - M_3 = 0 \tag{9.10}$$

where M and P are the bending moment and load per unit width, respectively, h and H are the thicknesses of the coating and substrate, respectively, and δ is the distance between the free surface of the substrate and the neutral axis of the system. Of the six load parameters in Eqs. (9.9) and (9.10), P_2 and M_2 are independent. After superposition, P and M are the only two parameters controlling the crack-tip singularity.

$$P = P_1 - C_1 P_3 - C_2 \frac{M_3}{h} \tag{9.11}$$

$$M = M_1 - C_3 M_3 \tag{9.12}$$

where $C_i\,(i = 1, 2, 3)$ is a dimensionless parameter and is expressed by Eq. (9.15).

The expressions of the interfacial energy release rate and the interfacial stress intensity factor (SIF) are ultimately obtained as follows:

$$G = \frac{c_1}{16}\left[\frac{P^2}{Ah} + \frac{M^2}{Ih^3} + 2\frac{PM}{\sqrt{AI}h^2}\sin\gamma\right] \tag{9.13}$$

$$K = \sqrt{\frac{p^2}{2}\left[\frac{P^2}{Ah} + \frac{M^2}{Ih^3} + 2\frac{PM}{\sqrt{AI}h^2}\sin\gamma\right]} \tag{9.14}$$

The following shows the expressions of the relevant parameters in Eqs. (9.15)–(9.21):

$$C_1 = \frac{\Sigma}{A_0} \quad C_2 = \frac{\Sigma}{I_0}\left(\frac{1}{\eta} - \Delta + \frac{1}{2}\right) \quad C_3 = \frac{\Sigma}{12I_0} \tag{9.15}$$

$$A_0 = \frac{1}{\eta} + \Sigma \quad I_0 = \frac{1}{3}\left\{\Sigma\left[3\left(\Delta - \frac{1}{\eta}\right)^2 - 3\left(\Delta - \frac{1}{\eta}\right) + 1\right] + 3\frac{\Delta}{\eta}\left(\Delta - \frac{1}{\eta}\right) + \frac{1}{\eta^3}\right\} \tag{9.16}$$

$$\Delta = \frac{1 + 2\Sigma\eta + \Sigma\eta^2}{2\eta(1 + \Sigma\eta)} \quad \Sigma = \frac{1 + \alpha}{1 - \alpha} \quad \eta = \frac{h}{H} \tag{9.17}$$

$$\alpha = \frac{\Gamma(\kappa_2 + 1) - (\kappa_1 + 1)}{\Gamma(\kappa_2 + 1) + (\kappa_1 + 1)} \quad \beta = \frac{\Gamma(\kappa_2 - 1) - (\kappa_1 - 1)}{\Gamma(\kappa_2 + 1) + (\kappa_1 + 1)} \quad \delta = \frac{1 + 2\Sigma\eta + \Sigma\eta^2}{2\eta(1 + \Sigma\eta)}h \tag{9.18}$$

$$c_1 = \frac{\kappa_1 + 1}{\mu_1} \quad \kappa = \frac{3 - \nu}{1 + \nu} \quad \mu = \frac{E}{2(1 + \nu)} \tag{9.19}$$

$$p = \sqrt{\frac{1 - \alpha}{1 - \beta^2}} \quad A = \frac{1}{1 + \Sigma(4\eta + 6\eta^2 + 3\eta^3)} \quad I = \frac{1}{12(1 + \Sigma\eta^3)} \tag{9.20}$$

$$\sin\gamma = 6\Sigma\eta^2(1 + \eta)\sqrt{AI} \tag{9.21}$$

The cracking behavior of a coating under the three-point bending load can be considered to be a result of the following equivalent bending moments and equivalent loads: the load P_1 and bending moment M_1 in the coating as well as the load P_3 and bending moment M_3 on the substrate.

9.2.2 *Interfacial K_{IC} Characterization of TBCs Using the Three-Point Bending Method*

The specimens used in the test were identical to those used in Sect. 9.1.3 in terms of geometric dimensions as well as the materials, process, and procedure used for preparation. The speckle marking method and procedure were identical to those in Sect. 9.1.3. The critical thresholds for interface crack initiation are obtained in Sect. 9.1.3. Based on the critical loads for interface crack initiation in the two types of specimens, four loads were selected for each type of specimen: 470, 475, 480, and 485 N for unoxidized specimens and 144, 146, 148, and 150 N for oxidized specimens. After bending, the half-length a of the interface cracks in the specimens under each load was determined by SEM. During the test process, an AE instrument was used to record the AE signals from each specimen, which were subsequently used to quantitatively analyze the interfacial damage of the specimen. The loading mode and rate for the universal material testing machine in the test were identical to those in Sect. 9.1.3. In addition, the settings for the AE acquisition system were identical to those in Chap. 2. See Fig. 9.10 in Sect. 9.1.3 for a schematic diagram of the test setup.

The failure modes of a TBC under a three-point bending load and a tensile load share a common characteristic: in both modes, vertical surface cracks perpendicular to the tensile stress direction are formed in the ceramic layer, followed by the formation of interface cracks in the coating along the shear stress direction [26]. This phenomenon is closely related to the stress state in the coating under the bending and tensile loads. In both the bending and tensile processes, there exist tensile stresses in the coating and shear stresses at its interfaces. Due to the spraying technique, the material properties of each layer, and the stress state in each layer, interface cracks can be initiated and propagate at different locations in a coating. Research [13] has shown that TBCs have two interfacial bending failure modes. In failure mode I, vertical surface cracks propagate to the BC/substrate interface, followed by the rapid initiation and propagation of interface cracks at the BC/substrate interface; due to the weakened stress field, other surface microcracks lack sufficient energy to propagate. In failure mode II, surface microcracks propagate to the ceramic/BC interface and form macrocracks, followed by the initiation and propagation of interface cracks at the ceramic/BC interface; during this process, a small number of cracks cross the ceramic/BC interface and penetrate the BC layer.

According to the three-point bending test results for the unoxidized TBC, after extending to the ceramic/BC interface, the vertical surface cracks in the TBC propagate in three directions: (1) the surface cracks continue to propagate to the BC layer; (2) the cracks change direction and propagate toward the two sides along the ceramic/BC interface; and (3) the cracks coalesce to form interface cracks, as shown in Fig. 9.15a. This failure mode is in line with failure mode II mentioned above. The three-point bending test results for the oxidized TBC show that after extending to the ceramic/BC interface, the surface cracks do not propagate along this interface;

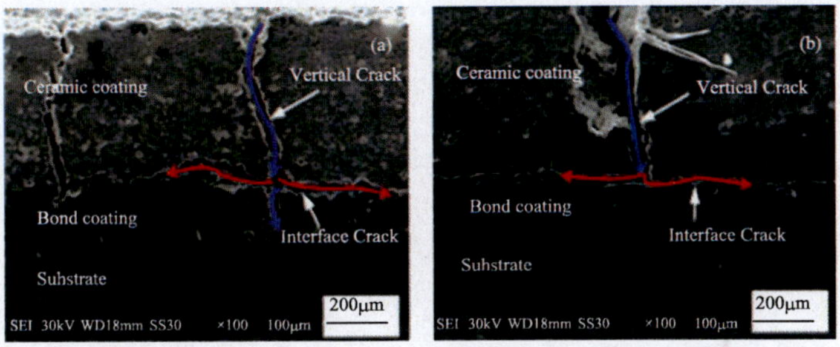

Fig. 9.15 Cross-sectional SEM images of **a** an unoxidized specimen and **b** a 300-h-oxidized specimen

instead, they penetrate through the BC layer and extend to the BC/substrate interface and subsequently propagate and extend in two directions at this interface, as shown in Fig. 9.15b. This failure mode is the failure mode I mentioned above. In the test, only one vertical surface crack forms in the oxidized TBC under bending loading. Figure 9.15 shows SEM images of (a) an unoxidized specimen and (b) a 300-h-oxidized specimen after bending.

In the test discussed in Sect. 9.1.3, the AE and digital image correlation techniques were used to investigate the critical thresholds for interface crack initiation in the TBC under bending loading in detail. The critical loads for interface crack initiation in the unoxidized and oxidized TBC specimens are determined to be 460 and 142 N, respectively. Figure 9.16 shows the analysis model and equivalent boundary loads for interface cracks in a coating under bending load. In the figure, σ_r is the residual stress in the coating, l is the span, a is the half-length of the interface crack, $P(t)$ is the magnitude of a time-varying load, M and M_3 are the bending moments on the coating and substrate, respectively, and Q is the shear force. The a of the interface crack formed under bending loading is much smaller than the l of the specimen. We can thus deduce that the length of the interface crack, M_3 varies only with P and is independent of l. Therefore, M_3 is a function of $P(t)$. In addition, we assume that $Q = 0$. Finally, in a three-point bending test, we can conclude that the interfacial SIF is formed by the following equivalent bending moments and equivalent loads [24]:

$$M_1 = 0 \;\; M_3 = -\frac{P(t)}{4b} l \;\; P_1 = \sigma_r h \;\; P_3 = 0 \tag{9.22}$$

The interfacial K_{IC} is estimated by substituting Eq. (9.22) into Eqs. (9.11) and (9.12) in Sect. 9.2 and then substituting the resulting P and M into Eq. (9.14). Other parameters of the material of each layer of the TBC used in the calculation are obtained from the literature [18]. The residual stress in the ceramic layer of the unoxidized TBC is measured to be –57 MPa using the indentation method based on the theory in Chap. 2. The residual stress in the ceramic layer of the oxidized TBC is –200 MPa [22]. The critical loads $P(t)$ for the unoxidized and oxidized TBC specimens are 460 and 142 N, respectively. The width b of each specimen is 9 mm.

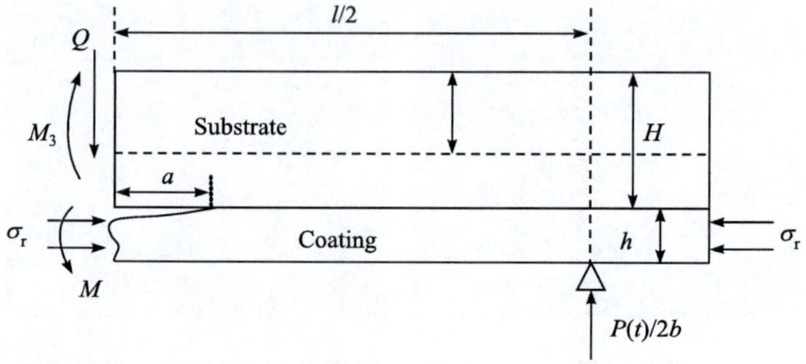

Fig. 9.16 Analysis model and equivalent loads suitable for interface cracks formed under three-point bending load

We measure the critical load for interface crack initiation in the unoxidized TBC and its K_{IC} and energy release rate to be 460 N, 2.13 MPa·m$^{1/2}$, and 51.82 J·m^{-2}, respectively. Employing the blister test, Zhou et al. [27] measured the interfacial K_{IC} of a TBC to be 0.52–0.61 MPa·m$^{1/2}$. Using the four-point bending method, Choi et al. [28] measured the interfacial K_{IC} of a TBC to be 1.15 ± 0.07 MPa·m$^{1/2}$. Using the three-point bending method, Wu [17] measured the critical energy release rate at the ceramic/BC interface of an unoxidized TBC to be 64.5–91.2 J·m^{-2}.

We measure the critical load for interface crack initiation in the TBC subjected to 300 h of oxidation and its K_{IC} and energy release rate to be 142 N, 2.39 MPa·m$^{1/2}$, and 27.24 J·m^{-2}, respectively. Using the three-point bending method, Mao [13] measured the K_{IC} at the BC/substrate interface of an oxidized TBC to be 2.10 MPa·m$^{1/2}$. Thurn et al. [29] measured the K_{IC} at the BC/substrate interface of a TBC heated to 1200 °C to be 2.0 MPa·m$^{1/2}$. Evidently, our test results are basically consistent with those reported in the abovementioned studies.

9.3 Surface and Interfacial K_{IC} Characterization of TBCs Using the Indentation Method

9.3.1 Surface K_{IC} Characterization of TBCs Using the Indentation Method

When used to apply a load to the surface of a brittle coating, a rigid indenter creates an indentation on the surface of the coating as well as cracks surrounding the indentation, as schematically shown in Fig. 9.17. Crack formation and propagation, as seen in this figure, suggest that the SIF resulting from the applied load reaches or exceeds the resistance (i.e., K_{IC}) needed for crack formation and propagation. Thus, the K_{IC} of a material can be determined based on the relation between the magnitude of the applied load and the crack length. During the indentation process, the SIF of a coating

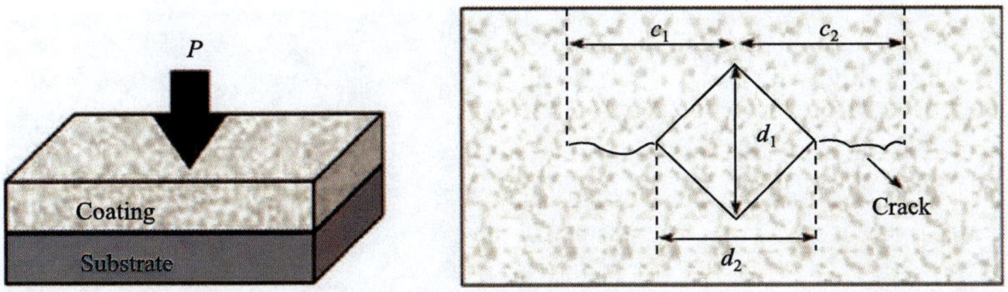

Fig. 9.17 Indentation and cracks on the surface of a TBC

consists of two parts, namely, the SIF resulting from the applied load, K_p, and that from the residual stress, K_r. By ignoring the plastic behavior of brittle materials, we can obtain the K_{IC} of and the residual stress in a TBC by superposing the two parts of the SIF [30–34]:

$$K_{IC}^{sur} = K_P + K_r = \chi \frac{P}{c^{3/2}} + \frac{4}{\sqrt{\pi}} \sigma_r h_t^{1/2} - \frac{2}{\sqrt{\pi}} \sigma_r h_t / c^{1/2} \quad (9.23)$$

where K_{IC}^{sur} is the surface fracture toughness, σ_r is the residual stress, P is the indentation load, c is the crack half-length, h_t is the thickness of the coating, and χ is a constant related to the properties of the indenter and coating. The above equation can be rewritten as

$$\frac{P}{c^{3/2}} = \frac{K_{IC}^{sur} - \frac{4}{\sqrt{\pi}} \sigma_r h_t^{1/2}}{\chi} + \frac{2}{\chi \sqrt{\pi}} \sigma_r h_t c^{-1/2} \quad (9.24)$$

By treating $\frac{P}{c^{3/2}}$ as a function of $c^{-1/2}$ and linear fitting, we obtain the slope and intercept of the linear fitting equation. In the equation, $\chi = \delta \left(\frac{E}{H}\right)^{1/2}$, where δ is a parameter related to the geometry of the indenter (δ= 0.016 for a Vickers indenter), and E and H are the Young's modulus and hardness of the coating, respectively, $\left(K_{IC}^{sur} - \frac{4}{\sqrt{\pi}} \sigma_r h_t^{1/2}\right)/\chi$ and $\frac{2}{\sqrt{\pi}} \sigma_r h_t$ are the intercept and slope of the linear fitting equation, respectively. When the thickness of the coating is known, the residual stress in the material can be obtained based on the slope, and substitution of the residual stress into the intercept yields the surface K_{IC} of the coating.

Five loads (29.4, 49, 98, 196, and 294 N) were selected for the indentation test. Each load was held for 10 s and then removed within 15 s. The test results for the surface of the ceramic layer of the TBC show that a set of sharp radial cracks is formed along the extension of each diagonal of the square plastic indentation, that cracks of different lengths are formed under different applied indentation loads, and that a higher load leads to the formation of longer cracks. Figure 9.18a, b shows metallographic micrographs of the Vickers indentations on the surface of the ceramic layer of the original TBC under an applied load of 49 N and on the surface of the

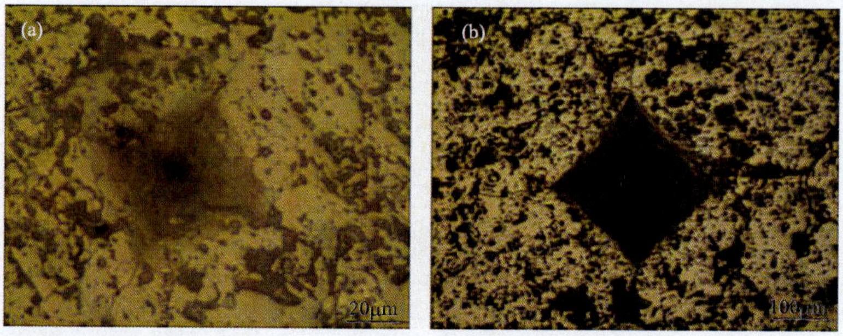

Fig. 9.18 Metallographs of Vickers indentations: **a** graph of the indentation on the surface of the ceramic layer of the original TBC under an applied load of 49 N and **b** graph of the indentation on the surface of the ceramic layer of the TBC subjected to 100 thermal cycles under an applied load of 196 N

ceramic layer of the TBC after 100 thermal cycles under an applied load of 196 N, respectively.

According to the model equation in Eq. (9.24) that considers the residual stress effect, we can calculate the K_{IC} of and the residual stress in the TBC. Figure 9.19 shows the variation in the surface K_{IC} of the TBC with the number of thermal cycles. The test results show that under an indentation load of 98 N, K_{IC} increases from 0.52 to 3.06 MPa·m$^{1/2}$ as the number of thermal cycles increases from 0 to 50. The K_{IC} of the TBC first increases and then decreases as the number of thermal cycles increases. This may be because, during the early stage of the thermal cycling process, the sintering of the ceramic coating heals the microcracks, thereby increasing its K_{IC}. However, during the later stage of the thermal cycling process, the K_{IC} of the TBC begins to decrease, possibly due to thermal growth oxide (TGO) growth. Through a tensile test at room temperature, Qian et al. [35] measured the K_{IC} of a TBC to be 0.228 MPa·m$^{1/2}$, which is in good agreement with the result calculated using the model equation in Eq. (9.24).

9.3.2 Interfacial K_{IC} Characterization of TBCs Using the Indentation Method

When used to apply a load to the interface between the ceramic layer and the transition layer, a rigid indenter creates an indentation at the interface as well as cracks near the interface that propagate along the interface, as shown in Fig. 9.20. Thus, we can calculate the interfacial K_{IC} based on the SIF at the cracks. The SIF at the indentation consists of two parts, namely, the SIF caused by the applied load, K_p, and the contribution of the residual stress in the coating near the interface, K_r. Therefore, the interfacial K_{IC} can be expressed as follows [30]:

Fig. 9.19 Variation in the surface K_{IC} of the TBC with the number of thermal cycles

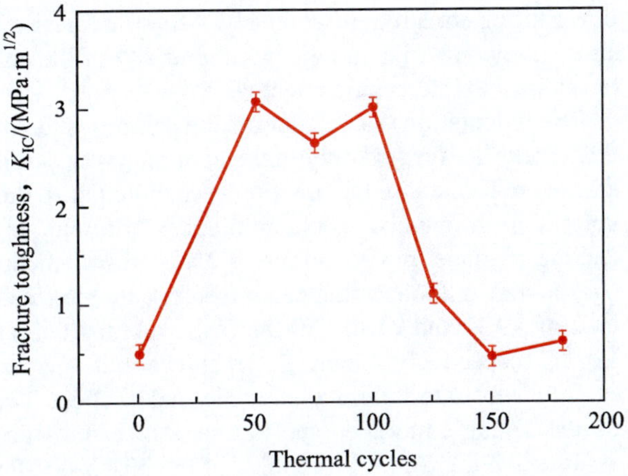

Fig. 9.20 Schematic diagram of an indentation and cracks at the interface of a TBC

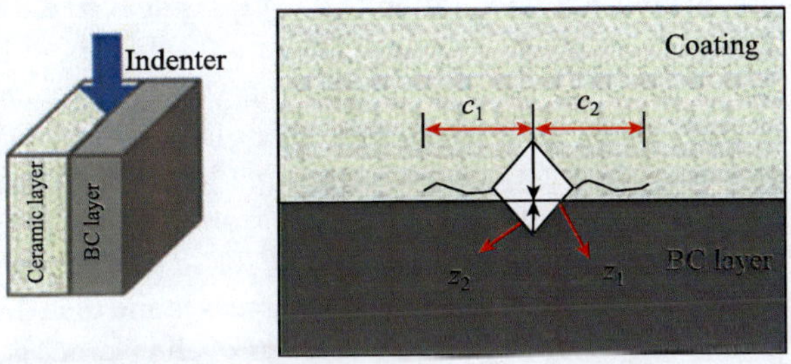

$$K_{IC}^{\text{int}} = K_P + K_r = \chi \frac{P}{c^{3/2}} + \frac{2}{\sqrt{\pi}} \sigma_r c^{1/2} \tag{9.25}$$

where $K_{\text{IC}}^{\text{int}}$ is the interfacial fracture toughness, σ_r is the residual stress, P is the indentation load, c is the crack half-length, h_t is the thickness of the coating, and χ is a constant related to the properties of the indenter and coating. The above equation can be rewritten as.

$$\frac{P}{c^{3/2}} = \frac{K_{IC}^{\text{int}}}{\chi} - \frac{2\sigma_r}{\sqrt{\pi}\chi} c^{1/2} \tag{9.26}$$

By treating $\frac{P}{c^{3/2}}$ as a function of $c^{1/2}$ and linear fitting, the slope and intercept of the linear fitting equation can be obtained. In this equation, $\chi = \delta\left(\frac{E}{H}\right)_i^{1/2} = \delta\left[\frac{z_1}{z_1+z_2}\left(\frac{E}{H}\right)_c^{1/2} + \frac{z_2}{z_1+z_2}\left(\frac{E}{H}\right)_s^{1/2}\right]$, where δ is a parameter related to the geometry of the indenter ($\delta = 0.016$ for a Vickers indenter), E and H are the Young's modulus and hardness of the coating, respectively, and the subscripts c and s indicate the coating and the transition layer, respectively; $\frac{K_{IC}^{\text{int}}}{\chi}$ and $\frac{2\sigma_r}{\sqrt{\pi}\chi}$ are the intercept and slope of the linear fitting equation, respectively. After the linear relation is obtained, the residual stress in the material and the interfacial K_{IC} of the coating can be determined using the slope and intercept, respectively.

The indentation test results for the ceramic/substrate interface of the TBC show that cracks are formed only in the ceramic layer parallel to the interface and that no cracks are formed in the direction perpendicular to the interface. Cracks of different lengths are formed by indentation under different applied loads, with a higher load leading to longer cracks. Figure 9.21a, b shows metallographic micrographs of the Vickers indentations at the ceramic/substrate interface of the TBC under an applied load of 49 N and of the 50-thermal-cycle-treated TBC under an applied load of 196 N, respectively. Figure 9.21b shows that new oxides (i.e., TGOs) form at the ceramic/substrate interface after thermal cycling. Thermal cycling has little effect on the Young's modulus and hardness of a metallic material. In this chapter, we assume that the Young's modulus of the Ni-based alloy substrate is a constant; i.e., $E = 220$ GPa [36], and $H = 1.67$ GPa. The variation in the interfacial K_{IC} of the TBC with the number of thermal cycles is analyzed using the model equation based on the conventional theory of fracture mechanics in Eq. (9.26). Figure 9.22 shows the variations in the surface K_{IC} (marked by dots) and interfacial K_{IC} (marked by squares) of the TBC with the number of thermal cycles. The results calculated using the theoretical model equation in Eq. (9.26) show an increase in the interfacial K_{IC} under an indentation load of 98 N from 0.54 to 2.8 MPa·m$^{1/2}$ with an increasing number of thermal cycles from 0 to 70. As the number of thermal cycles increases, the interfacial K_{IC} of the TBC first increases and then decreases. This may be because, during the early stage of the thermal cycling process, the sintering of the ceramic coating heals some microcracks, thereby increasing its K_{IC}. However, during the later stage of the thermal cycling process, the K_{IC} of the TBC begins to decrease, which might be a result of TGO growth. Using the blister test, Zhou et al. [27] measured the interfacial K_{IC} values of original TBCs to be 0.52–0.61 MPa·m$^{1/2}$. By using the four-point bending method, Choi et al. [28] measured the K_{IC} values of TBCs at room and high temperatures, respectively. Specifically, they found the K_{IC} at room temperature to be 1.15 ± 0.07 MPa·m$^{1/2}$, which is in good agreement with our measured interfacial K_{IC} (0.54 MPa·m$^{1/2}$) of the original TBC. In addition, Fig. 9.22 shows that overall, the surface K_{IC} of the TBC is higher than its interfacial K_{IC}.

9.4 Interfacial K_{IC} Characterization of TBCs Using the Buckling Method

9.4.1 Buckling Test for Determining the Interfacial K_{IC} of TBCs

A GH3030 Ni-based alloy was selected as the substrate material for the TBC used in the test. The APS process was used to prepare a ceramic coating. The TBC system

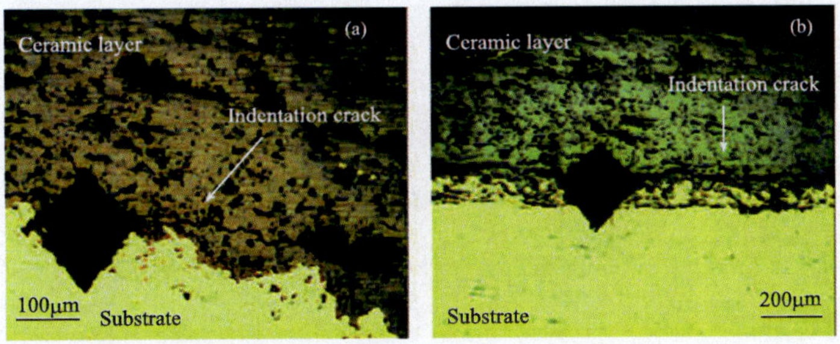

Fig. 9.21 Metallographs of the indentations at the interface of the TBC: **a** an original specimen under an applied load of 49 N and **b** a specimen subjected to 50 thermal cycles and then an applied load of 196 N

Fig. 9.22 Variations in the surface and interfacial K_{IC} values of the TBC determined by the indentation method with the number of thermal cycles

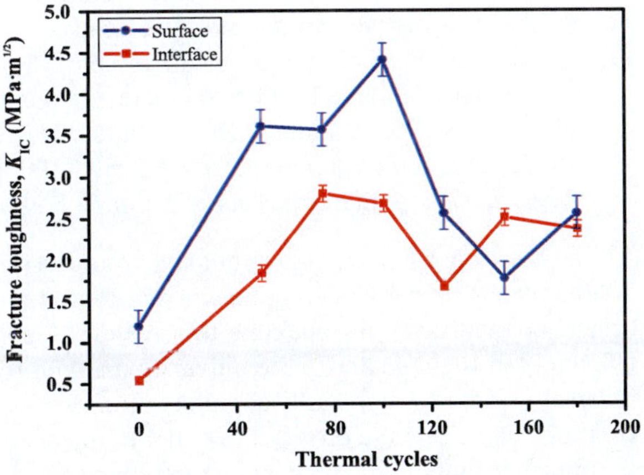

consisted of three layers, namely, a ceramic layer, a BC layer, and a superalloy substrate. Before spraying, we first used a wire cut electric discharge machine to cut TBC specimens into cuboids with dimensions of 20 mm × 5 mm × 5 mm. If no preexisting penetrating interface cracks are prepared in a specimen but instead the specimen is directly subjected to uniaxial compression, it is very difficult to accurately predict the location and time at which the coating will buckle. In addition, since coatings exhibit a variety of failure modes, it is not possible to summarize patterns, and more importantly, theoretical models suitable for describing their failure modes are lacking. Hence, this section presents the design and preparation of an interfacial defect between the ceramic and BC layers. Specifically, this interfacial defect is prepared by placing a Ni film with the same width as the substrate between the ceramic and BC layers. This way, a penetrating interfacial defect is approximated as a preexisting penetrating interface crack.

Figure 9.23 shows the procedure for preparing a TBC specimen with a preexisting crack, which mainly involves the following steps:

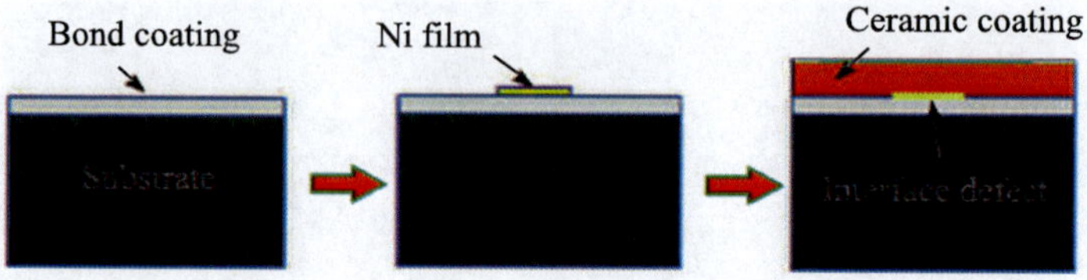

Fig. 9.23 Schematic diagram of the preparation of a TBC specimen with a preexisting crack

(1) First, clean the surface of the substrate material to be sprayed by degreasing, heat treating, shot blasting, and ultrasonic cleaning. Then, spray a BC layer composed of $NiCr_{22}Al_7Y_{0.2}$ powder onto the surface of the substrate using the PS technique.
(2) Place a Ni film with a thickness of approximately 40 μm on top of the BC layer as a preexisting penetrating interface crack.
(3) Employ the APS technique again to spray 8 wt% Y_2O_3-ZrO_2 powder onto the BC layer and the Ni film to create a ceramic layer. Figure 9.24 shows the dimensions of the obtained TBC specimen. The thicknesses of the substrate, the BC layer, and the ceramic layer are 5 mm, 100 μm, and 350 μm, respectively. The preexisting interface crack is 5 mm in both length and width.

The buckling test here was a uniaxial compression test performed on a REGER 2000 universal material testing machine at room temperature. Figure 9.25 shows a schematic diagram of the buckling test method. A cuboidal specimen was placed at the center of the indenter of the universal material testing machine. A disk-shaped indenter was used to uniaxially compress the specimen at a loading rate of 250 N/min from one end along the direction parallel to the coating/substrate interface. During the uniaxial compression process, we employed a PCI-2 AE nondestructive detection system and an AVT-Manta G504 CCD camera to monitor the buckling delamination processes of the TBC specimen in real time. The threshold, preamplifier gain, and sampling frequency of the AE system were set to 38 dB, 40 dB, and 2 MHz, respectively. An AE sensor was placed on the surface of the substrate to collect the AE signals during the buckling delamination processes of the coating. A CCD camera equipped with a 50-mm focal lens and a resolution of 2452 × 2056 was placed directly in front of the cross-section of the specimen to record the buckling delamination processes in real time. The sampling frequency was set to one image per second.

Figure 9.26 shows the variations in the stress σ and N_{AE} in the TBC specimen with the strain ε during the strain loading process. Clearly, due to the very large difference between the thicknesses of the coating and the substrate, the σ–ε curve of the TBC specimen is completely insensitive to the coating buckling delamination processes during loading. In other words, the buckling and interfacial delamination of the coating cannot be observed from the σ–ε curve. The AE detection technique has the unique advantage of being able to continuously detect signals related to microcrack formation in materials in various testing environments. Thus, based on the N_{AE} and

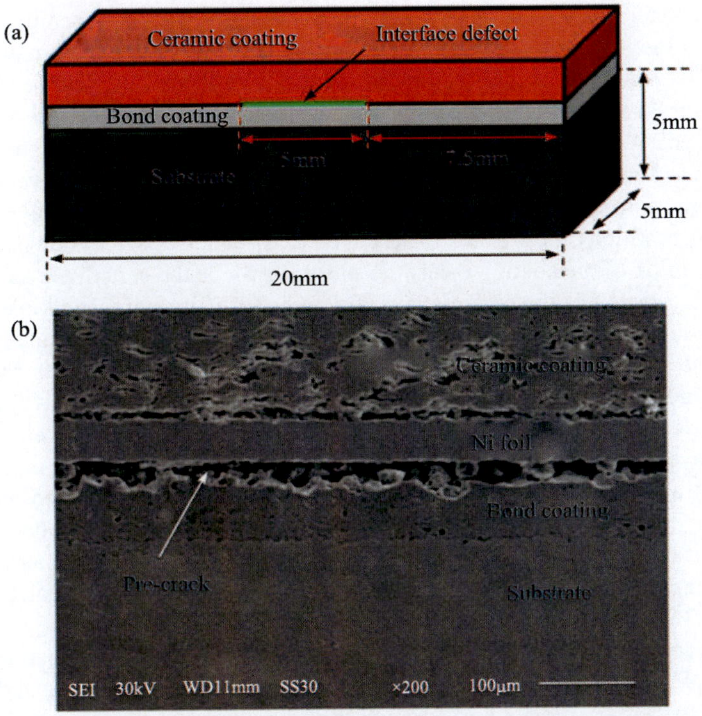

Fig. 9.24 **a** Schematic diagram of a TBC specimen with a preexisting crack. **b** Cross-sectional SEM image of the area with the preexisting crack

Fig. 9.25 Schematic diagram of the buckling test method combined with the AE detection technique and a CCD monitoring system

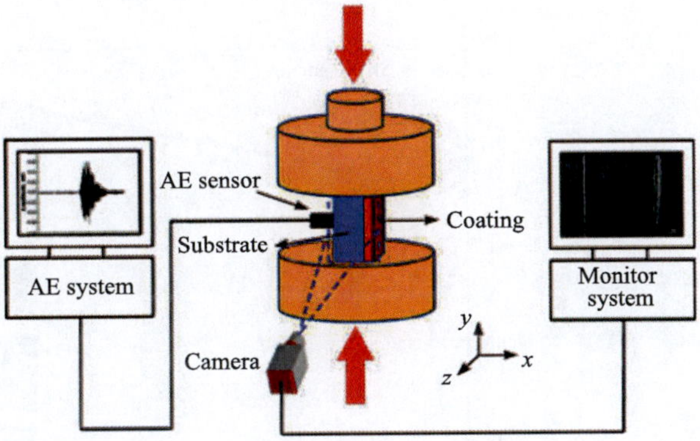

the in situ typical features detected by the CCD camera in real time, the buckling delamination processes of the TBC are divided into four stages: (I) The relation between the N_{AE} and the applied ε shows that as ε increases, N_{AE} increases steadily until reaching an inflection point (point A) at $\varepsilon = -0.17\%$. However, no buckling or notable macrocracks are observed by the CCD camera. The phenomenon of the increase in the N_{AE} with increasing ε is attributed to microcrack formation in the coating. (II) As ε increases, N_{AE} further increases until reaching another inflection point (point B) at $\varepsilon = -0.578\%$, at which time notable buckling is observed by the CCD camera. As ε further increases, there is a gradual increase in the amplitude of the buckling deflection of the coating. (III) As ε increases to –0.58% (very close to point B), the amplitude of the buckling deflection of the ceramic coating reaches its maximum, at which time kinked cracks appear at the crack tips in the areas where the buckled portion of the coating meets the substrate. Since the ceramic layer is brittle, the buckled portion of the ceramic layer fractures at the instant when kinked cracks appear, resulting in its spallation. At this time, N_{AE} is found to basically remain constant as ε increases until it reaches –0.95%, when there is a sudden increase in N_{AE} and another inflection point (i.e., point C). (IV) During Stage III, as ε increases, N_{AE} almost remains constant, but there is a continual accumulation of energy in the ceramic layer. Interfacial delamination begins when its required energy is reached by the energy accumulated in the ceramic layer, accompanied by a sudden increase in N_{AE} at the inflection point C. At this time, interface cracks appear at inflection point C, which is thus the starting point of interfacial delamination. As ε further increases, there is a rapid increase in N_{AE} until another inflection point (point D, corresponding to $\varepsilon = -1.36\%$) occurs. As the compressive ε continues to increase, N_{AE} is found to increase very slowly, indicating the spallation of the whole ceramic coating.

Figure 9.27 shows typical photographs of the buckling of the coating and its spallation due to the formation of kinked cracks in its buckled portion. Figure 9.27a shows that at $\varepsilon = -0.578\%$, the coating undergoes a typical buckling process. The maximum amplitude of the buckling deflection reaches 200 μm, and kinked microcracks appear at the tips where the buckled portion of the coating and the substrate come in contact.

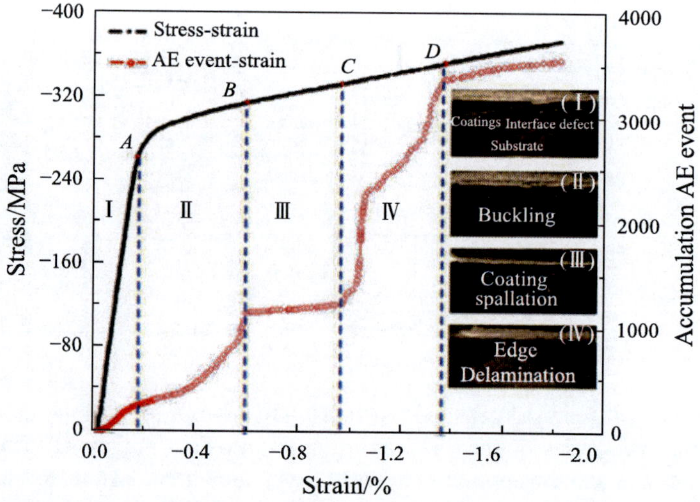

Fig. 9.26 Variations in the stress and number of AE events in the TBC specimen with strain

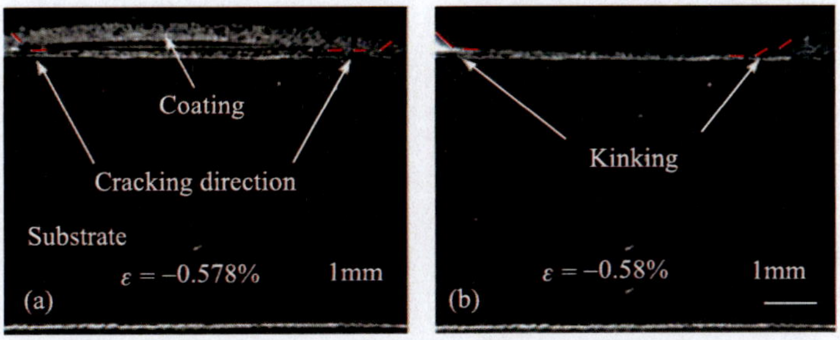

Fig. 9.27 **a** Buckling and **b** spallation processes of the coating

The initiation of microcracks eventually causes the buckled portion of the ceramic coating to spall. As seen from Fig. 9.27b, the buckled portion of the coating spalls at $\varepsilon = -0.58\%$, which, according to an analysis of the fracture surface of the coating, is caused by the propagation of the kinked cracks. This phenomenon occurs because the K_{IC} of the coating is lower than that of the coating/substrate interface, reflecting the typical mechanism of competition between interfacial buckling delamination and surface fracture in a coating. When the inherent K_{IC} of a coating is lower than its interfacial K_{IC} (i.e., the interfacial bond is stronger than the coating material), cracks propagate toward the interior of the coating instead of along the interfaces, forming so-called kinked cracks [37, 38]. Kim et al. [39] observed a similar phenomenon in an experiment on a diamond-like thin film on a steel substrate. When the inherent K_{IC} of a coating is higher than its interfacial K_{IC} (i.e., it is difficult for the coating itself to fracture), cracks propagate along the interfaces.

Figure 9.28 shows that after the spallation of the buckled portion of the coating, interfacial delamination begins when the energy accumulated over a period of time reaches the critical energy required for interface crack propagation. At $\varepsilon = -0.95\%$, interface crack initiation begins. As ε increases, interfacial delamination suddenly occurs, and the length of interface cracks continually increases. At $\varepsilon = -1.36\%$, the whole coating delaminates, which is the ultimate failure mode. Figure 9.29 presents the relation between the interface crack propagation length and the compressive ε under external loading. We find that as the compressive ε increases, the interface crack length continually increases until the whole coating spalls. Hence, we hope to characterize the interfacial bonding performance of a TBC through the relation between the interface crack length and the applied ε.

Figure 9.30 shows a schematic diagram of the buckling delamination failure mechanism of a TBC specimen with a preexisting crack based on the phenomena observed during the test (see Figs. 9.27 and 9.28). Under an applied compressive load, the TBC specimen with a preexisting crack first buckles when the critical buckling conditions are met (Fig. 9.30b). As the compressive load further increases, kinked cracks are initiated in areas where the buckled portion of the coating and the substrate meet (Fig. 9.30c). Kinked cracks propagate very rapidly once initiated. When the cracks propagate to the surface of the ceramic coating, the buckled portion of the coating

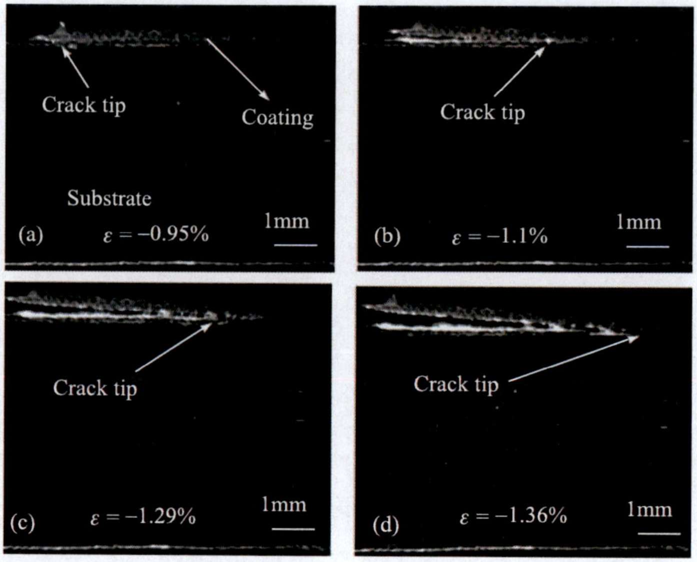

Fig. 9.28 Interface crack propagation at different strains after the spallation of the buckled portion of the coating: **a** $\varepsilon = -0.95\%$, **b** $\varepsilon = -1.1\%$, **c** $\varepsilon = -1.29\%$, and **d** $\varepsilon = -1.36\%$

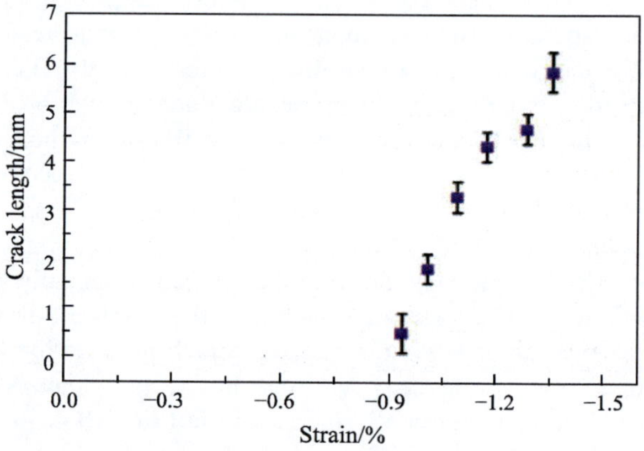

Fig. 9.29 Relation between the interface crack propagation length and the compressive strain under external loading

spalls (Fig. 9.30d). With a further increase in compressive strain, interfacial delamination occurs when the energy accumulated over a period of time reaches the critical energy required for interface crack propagation (Fig. 9.30e). As the compressive strain continues to increase, the range of interfacial delamination increases rapidly, and the length of interface cracks continually increases, eventually resulting in the spallation of the whole ceramic coating (Fig. 9.30f).

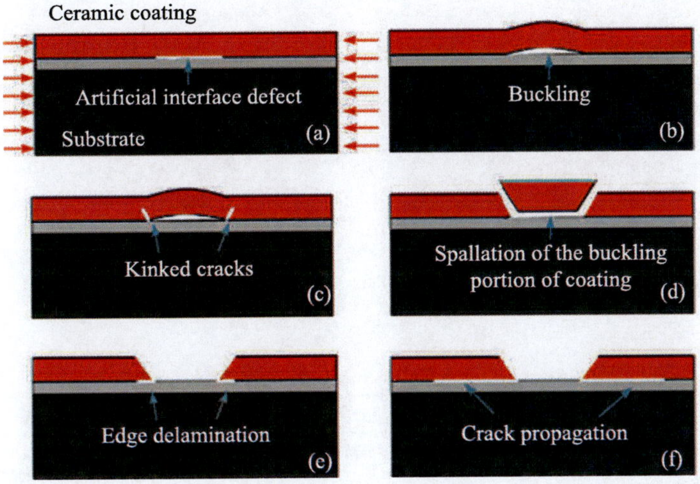

Fig. 9.30 Schematic diagram of the buckling delamination failure mechanism of a TBC specimen with a preexisting crack

9.4.2 FE Simulation for the Buckling of TBC Interfacial K_{IC}

Figure 9.31 shows the geometry of the 2D FE model established using the commercial FE software ABAQUS 6.10. The model consists of a ceramic layer, a BC layer, and a substrate, which are 350 μm, 100 μm, and 5 mm in thickness, respectively. The model has a total length of 20 mm, and the spallation length of the buckled portion of the coating is 5 mm. We need to establish only half of the model considering its symmetry. Notably, due to its porosity, growth orientation, and crystal structure, the Young's modulus of a PS ceramic layer is often anisotropic. However, the anisotropy of Young's modulus becomes prominent only after thermal cycling for a period of time, while for a coating that is prepared by spraying and has yet to undergo thermal cycling, the anisotropy of its Young's modulus is not pronounced. Thus, to simplify the Young's modulus of the ceramic layer, we do not consider the effect of the anisotropy of Young's modulus on the results of FE calculation and treat each layer as a uniform, isotropic material. Specifically, in the FE calculation, we consider the ceramic layer as a linear elastic material and the BC layer and substrate each as an elastoplastic material. Their constitutive relations are as follows:

$$\varepsilon = \begin{cases} \sigma/E & \sigma \le \sigma_y \\ (\sigma_y/E)/(\sigma/\sigma_y)^{1/n} & \sigma > \sigma_y \end{cases} \tag{9.27}$$

where σ_y is the yield stress, n is the power hardening exponent, E is the Young's modulus, and v is the Poisson's ratio. We assume that the material parameters of the

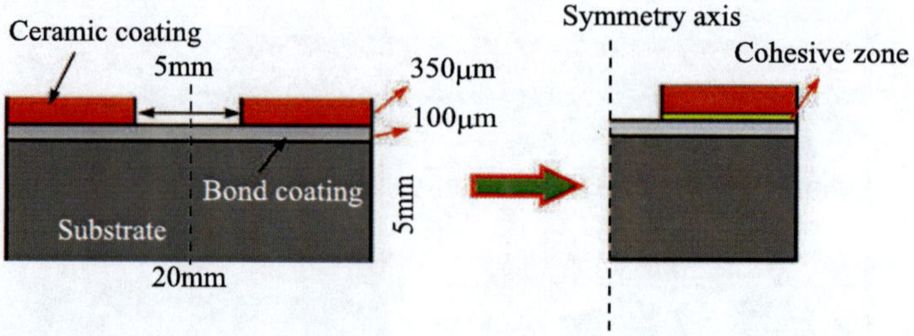

Fig. 9.31 Schematic diagram of the geometry of the FE model

Table 9.3 Material parameters used in the FE model

Layer	*E* (GPa)	σ_y (MPa)	ν	n
Ceramic coating	48	–	0.22	–
Bond coating	175	280	0.3	0.12
Substrate	175	280 280	0.3	0.12

BC layer are identical to those of the substrate. Table 9.3 summarizes the material parameters in detail [40].

Figure 9.32 shows the generated mesh and boundary conditions of the FE model. The ceramic layer, BC layer, and substrate materials are each discretized using CPE4R elements, while the cohesive zone at the ceramic/BC interface is simulated with four-node plane-strain reduced integration cohesive elements (COH2D4). The mesh is gradually refined along the zone close to the interface (see the inset in Fig. 9.32). Due to the symmetry of the model, symmetrical boundary conditions are applied to the left side of the model to prevent it from moving in the horizontal direction. A vertical displacement constraint is applied to the bottom edge of the substrate so that it cannot move in the vertical direction. A uniform axial displacement is applied to the right side of the model. Based on the magnitude of the axial displacement, the applied strain can be defined as $\varepsilon = \Delta x/L$, where L is the length of the geometric model and Δx is the magnitude of the applied axial displacement, which is obtained from the strain values applied in the buckling test.

The cohesive zone model (CZM) can be used to simulate interface cracks in different structures [41–44]. Thus, we select the CZM to simulate the interface crack initiation and propagation during the compression process. The CZM can be characterized by the cohesive force–opening displacement relation under normal stress or shear stress, which involves a total of six parameters, namely, the interfacial tensile strength σ_n^0 and shear strength σ_s^0, the critical opening displacement δ_n^0 and slip displacement δ_s^0, and the areas enclosed by the cohesive force–opening displacement curve, G_{nc} and G_{sc}. In this section, the bilinear CZM in Fig. 9.33 is used as the failure criterion for the cohesive elements.

The CZM can be described by the following expressions:

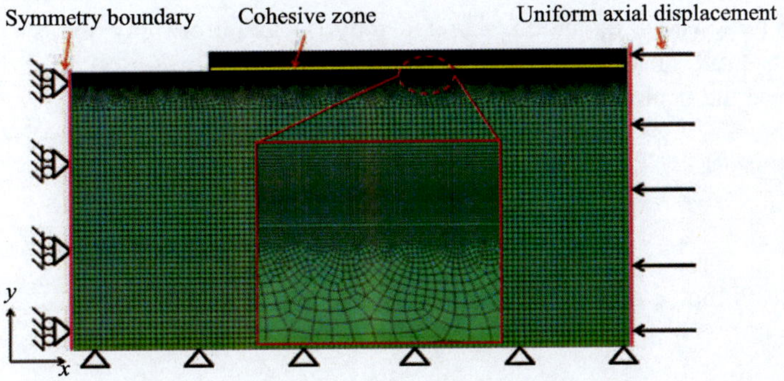

Fig. 9.32 Meshing and boundary conditions of the FE model (the yellow line shows the zone of cohesive elements, and the inset is a close-up of the refined mesh near the ceramic/BC interface)

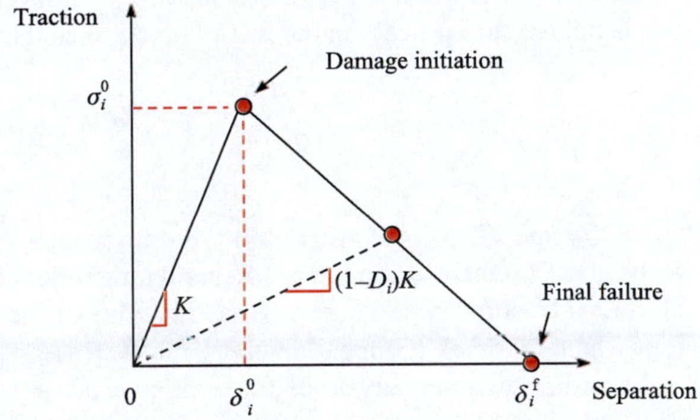

Fig. 9.33 Traction–separation curve of cohesive elements (cohesive zone model)

$$\sigma_i = \begin{cases} K\delta_i, & \delta_i \le \delta_i^0 \\ (1 - D_i)K\delta_i, & \delta_i^0 < \delta_i < \delta_i^f \quad (i = n,\ s) \\ 0, & \delta_i \ge \delta_i^f \end{cases} \tag{9.28}$$

$$D_i = \begin{cases} 0, & \delta_i \le \delta_i^0 \\ \frac{\delta_i^f(\delta_i - \delta_i^0)}{\delta_i(\delta_i^f - \delta_i^0)}, & \delta_i^0 < \delta_i < \delta_i^f \quad (i = n,\ s) \\ 1, & \delta_i \ge \delta_i^f \end{cases} \tag{9.29}$$

where σ is the stress, δ is the displacement, D is the damage variable, K is the initial interfacial stiffness, the subscripts n and s indicate the normal and tangential direction, respectively, and the superscripts 0 and f indicate the critical displacement when the element is first damaged and the displacement when the element completely fails, respectively. The normal and tangential critical fracture energies are expressed as follows:

$$G_{nc} = \frac{1}{2}\sigma_n^0\delta_n^f \text{ and } G_{sc} = \frac{1}{2}\sigma_s^0\delta_s^f \tag{9.30}$$

In the mixed loading mode, damage may occur before any stress component reaches its limit. In this section, the quadratic nominal stress criterion is used to characterize the occurrence of interfacial damage [45, 46]:

$$\left(\frac{<\sigma_n>}{\sigma_n^0}\right)^2 + \left(\frac{\sigma_s}{\sigma_s^0}\right)^2 = 1 \tag{9.31}$$

where the symbol $< >$ is the Macaulay bracket, which is defined as follows:

$$<\sigma_n> = \begin{cases} \sigma_n, & \sigma_n \geq 0 \\ 0, & \sigma_n < 0 \end{cases} \tag{9.32}$$

The Macaulay bracket means that the element is not damaged under pure compressive stress. We adopt the following damage evolution relationship to describe the interfacial fracture process in the mixed mode once the damage criterion is met [47]:

$$\left(\frac{G_n}{G_{nc}}\right)^\alpha + \left(\frac{G_s}{G_{sc}}\right)^\alpha = 1 \tag{9.33}$$

where G_n and G_s are the work done by the normal and tangential forces, respectively, and G_{nc} and G_{sc} are the normal and tangential critical fracture energies at the interfacial fracture, respectively. Here, the value of the power exponent α is set to 1. Then, the total crack propagation energy is $G = G_n + G_s$.

According to the analysis of the available test results, we find that σ_n^0 and σ_s^0 are within the ranges of 100–350 and 10–30 MPa, respectively [48–50]. In the FE simulation, we set the σ_n^0 and σ_s^0 of the interfacial cohesive elements to 200 and 20 MPa, respectively. We assume that $G_{nc} = G_{sc}$ and that they vary in the range of 70 to 150 J/m^2 [51]. The shape function of the CZM is defined as $\delta_{ratio} = \delta_i^0/\delta_i^f$, with its value set to 0.25.

Notably, the CZM considers material property degradation caused by damage evolution, which leads to a severe convergence problem in FE calculations. Thus, to improve the computational convergence, the viscous regularization options in ABAQUS are used for the interfacial cohesive elements. This method has been validated by Faou et al. [52] and Gao and Bower [53], who found that the convergence problem can be effectively addressed by adjusting the viscous parameters. The introduction of viscous parameters ensures that the tangential stiffness matrix of a degraded material always remains positive at sufficiently small time increments [52]. However, caution should be exercised when introducing viscous parameters, because unreasonable values of viscous parameters may substantially reduce the reliability of the computational results even though convergence can be achieved. Therefore, viscous parameters act as a double-edged sword—they can improve the convergence of computational results on the one hand, but produce unreliable computational results on the other hand. Therefore, the selection of viscous parameters should be based on the following basic principles for the validity of their values:

(1) The viscosity adjustment time (VAT) should be slightly shorter than the time increment.
(2) The viscous dissipation energy generated by the introduction of viscous parameters should be less than 5% of the total strain energy.
(3) The ratio of the viscous dissipation energy to the total strain energy should be independent of the selected VAT.
(4) The VAT should not vary with the stress distribution in the model.

Extensive FE calculations show that the above basic principles can be satisfied well in the VAT range of 0.0001–0.0005 and that a VAT of 0.0005 leads to a stable computational process and results that converge to quasistatic values.

Figure 9.34 shows the initiation and propagation of interface cracks in the TBC under compressive loading. The degradation process of the interfacial cohesive elements depends on the scalar damage variable (SDEG). When the SDEG = 0, the interfacial cohesive elements are intact and undamaged, and when SDEG = 1, the interfacial cohesive elements fail, and cracks are formed. Note that the damage to the ceramic layer, the BC layer, or the substrate is not considered in the model. Consequently, the contour plot in Fig. 9.34 shows only the magnitude and distribution of the damage variable within the interfacial cohesive zone. We find that the initial interfacial damage occurs when the compressive strain increases to a certain level and that once the damage evolution relationship is satisfied, the cohesive elements begin to degrade until they eventually fail completely, at which time interface cracks are formed. As seen in Fig. 9.34, interfacial delamination begins at a compressive strain of –0.5%. As the strain increases, there is a rapid increase in the interface crack length. As the strain increases to –1.01%, the interface crack length reaches 2.5 mm. The interface crack initiation and propagation processes simulated using the finite element method (FEM) are consistent with those observed during the test in Sect. 9.4.1 (see Fig. 9.28).

Figure 9.35 compares the interface crack length–strain relations observed and measured experimentally with those simulated using the FEM. In the FE simulation process, a large number of repeated calculations are performed by changing the interfacial bonding energy Γ until the FE simulation results best agree with the experimental results, and the corresponding Γ value is that of the TBC system to be solved. As seen in Fig. 9.35, there is a continuous increase in the interface crack length with increasing compressive strain, and the FE and experimental results display similar trends. Moreover, the FE simulation results at Γ = 100 and 130 J/m^2 are very consistent with the experimental results, except for the magnitude of the initial strain at interfacial delamination. We attribute this discrepancy to the accuracy of the CCD camera used in the experiment; microcracks may have been formed at strains in the range of –0.58% to –0.95% (see Stage III in Fig. 9.26) but could not be distinguished in the CCD camera. The trends of our FE and experimental results are similar to the results of Soulignac et al. [54], thus validating the feasibility of this method.

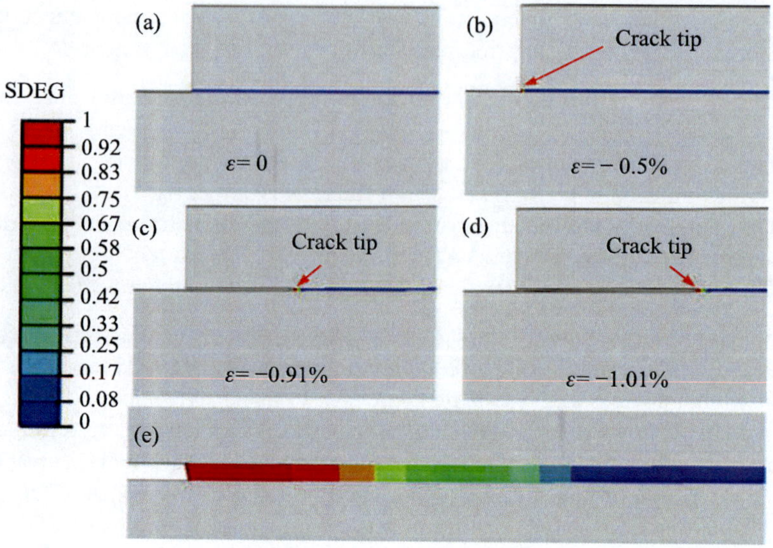

Fig. 9.34 Contour plot of the variation in the damage variable (i.e., SDEG) of the interfacial cohesive elements with the compressive strain ε: **a** $\varepsilon = 0$, **b** $\varepsilon = -0.5\%$, **c** $\varepsilon = -0.91\%$, and **d** $\varepsilon = -1.01\%$. **e** Enlarged view of the contour plot of the interface crack-tip region

Fig. 9.35 Variation in the interface crack propagation length with strain (the pink squares represent the experimental results, and the red, green, and blue dashed lines represent the FEM calculation results at interfacial bonding energies $\Gamma = 70$, 100, and 130 J/m^2, respectively)

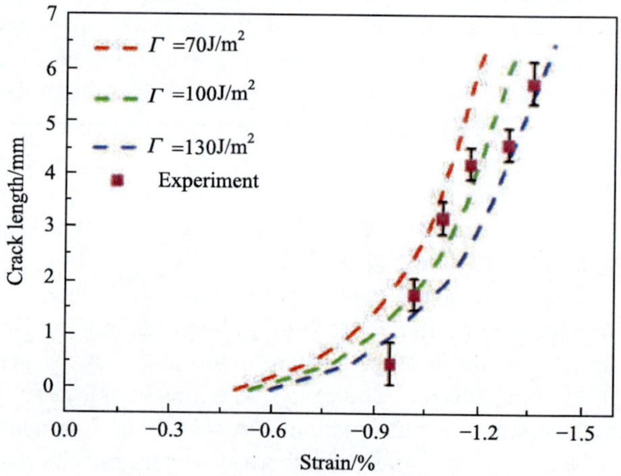

9.4.3 Theoretical Model for the TBC Interfacial K_{IC} Characterization Based on Buckling Delamination

According to nonlinear delamination theory, the coating on a substrate may undergo interfacial delamination when subjected to a residual compressive stress or an applied compressive load. Figure 9.36 shows two common and typical failure modes: (a)

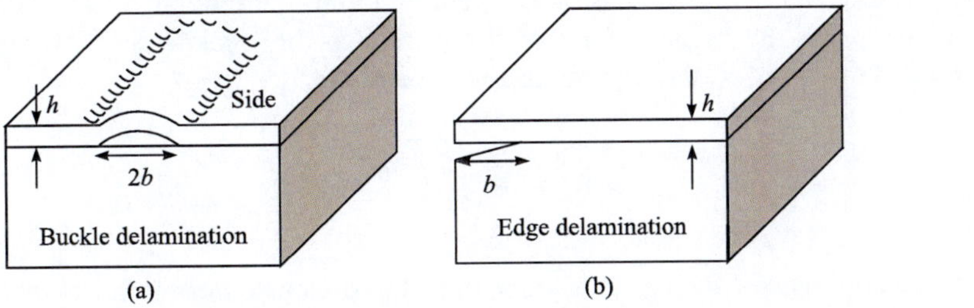

Fig. 9.36 Delamination modes of coatings under compression: **a** buckling delamination (straight-side buckling) and **b** edge delamination

buckling delamination and (b) edge delamination. Generally, there are three typical modes of buckling delamination, namely, straight-side buckling, circle blistering, and telephone-cord buckling.

For straight-side buckling delamination, the interfacial energy release rate G and the mixed-mode phase angle ψ can be calculated using the following model proposed by Hutchinson and Suo [37]:

$$G/G_0 = \left(1 - \frac{\sigma_c}{\sigma_0}\right)\left(1 + 3\frac{\sigma_c}{\sigma_0}\right) \tag{9.34}$$

and

$$\tan\psi = \frac{4\cos\omega + \sqrt{3}\xi\sin\omega}{-4\sin\omega + \sqrt{3}\xi\cos\omega} \tag{9.35}$$

where G_0 is the energy stored in the coating per unit area and is expressed by the following equation:

$$G_0 = \frac{(1 - \upsilon^2)h\sigma_0^2}{2E} \tag{9.36}$$

where σ_c is the critical buckling stress, which is related to the Young's modulus E, Poisson's ratio v, and thickness h of the coating as well as the interface crack length $2b$. σ_c is expressed as follows:

$$\sigma_c = \frac{\pi^2}{12}\frac{E}{1 - \upsilon^2}\left(\frac{h}{b}\right)^2 \tag{9.37}$$

A coating buckles only when its internal stress σ_0 exceeds its critical buckling stress σ_c. In Eq. (9.35), ω is the phase-angle shift caused by the elastic mismatch between the coating and substrate materials (generally, $\omega = 52°$ for the ceramic/BC interface) and ξ is a dimensionless buckling amplitude defined as the ratio of the deflection at the highest point of the buckle to the thickness of the coating and related to σ_0 and σ_c through the following equation [37]:

$$\xi = \sqrt{\frac{4}{3}\left(\frac{\sigma_0}{\sigma_c} - 1\right)} \tag{9.38}$$

According to Moon et al.'s work [55], the dimensionless stress is related to the interface crack half-length b through the following equation:

$$\frac{\sigma_0}{\sigma_c} = \left(\frac{b}{b_0}\right)^2 \tag{9.39}$$

where b_0 is the interface crack half-length at the start of buckling. Substitution of Eq. (9.39) into Eqs. (9.34) and (9.35) gives the following relations of G and ψ with b:

$$G = G_0\left[1 - \left(\frac{b_0}{b}\right)^2\right]\left[1 + 3\left(\frac{b_0}{b}\right)^2\right] \tag{9.40}$$

and

$$\psi = \arctan\left[\frac{2\cos 52^o + \sqrt{\left[\left(\frac{b}{b_0}\right)^2 - 1\right]}\sin 52^o}{-2\sin 52^\circ + \sqrt{\left[\left(\frac{b}{b_0}\right)^2 - 1\right]}\cos 52^\circ}\right] \tag{9.41}$$

For edge delamination, when the interface crack length (i.e., $2b$) increases to several times the coating thickness, G becomes independent of $2b$ and tends to a stable value (i.e., G_0). The mode of interfacial delamination in a steady state tends to pure delamination mode II.

According to Eq. (9.40), we can obtain the variation in G with b. As shown in Fig. 9.37, as b increases, G continuously increases and eventually tends to a stable value. In the buckling test, $b_0 = 2.5$ mm. At the time when the coating begins to delaminate as a result of interface crack propagation, the b observed from the CCD camera is 3 mm, and σ_0 is –186 MPa. According to Eq. (9.36), the critical G at the start of interfacial delamination, G_{cr}, is calculated to be 120 J/m^2. As b increases, G tends to a steady-state value of $G_{ss} = 150$ J/m^2. Note that the theoretical model used to calculate G does not consider plastic deformation, while the FE simulation in Sect. 9.4.2 considers the effect of plastic deformation and it is found that the critical interfacial bonding energy at interfacial delamination ranges from 100 to 130 J/m^2. Based on nonlinear buckling delamination theory, G_{cr} is calculated to be

120 J/m^2, which is consistent with the FE simulation result, indicating that the effect of plastic deformation is negligible. In addition, a comparison of the FE-simulated initial and propagated interface crack lengths in the absence and presence of plastic deformation shows no significant changes, suggesting that in this section, the effect of plastic deformation on the formation of interface cracks, crack propagation length, and interfacial bonding performance are negligible.

Based on Eq. (9.41), the variation in the interfacial ψ with b is obtained. As shown in Fig. 9.38, as b increases, ψ increases from –40° to –85°. At the start of interfacial delamination, the b and critical ψ (ψ_{cr}) observed experimentally are 3 mm and –56°, respectively. As b increases, ψ tends to a stable value close to –90°, suggesting that the delamination process is close to pure type II crack.

Based on nonlinear buckling delamination theory, G and the corresponding ψ are obtained for a TBC that undergoes interfacial delamination under compressive loading. Next, we compare these results with previously reported results corresponding to different deposition methods, treatment methods, failure modes, and test methods, as shown in Table 9.4.

As seen in Table 9.4, the energy release rate G of a TBC varies with the deposition method, treatment method, failure mode, and test method, and its value ranges from 49 to 260 J/m^2. The APS TBC in this section has a G value in the range of 120 to 150 J/m^2 and is dominated by fracture mode II. The G values obtained in this section are comparable to the data in the literature but are higher than those of most PS TBCs reported in the literature, mainly due to the difference in the fracture modes and the lack of consideration of the TGO layer. It is well known that G is significantly related to ψ. Hutchinson and Suo [37] formulated an empirical equation based on a summary of phenomena to describe the mode dependence of G

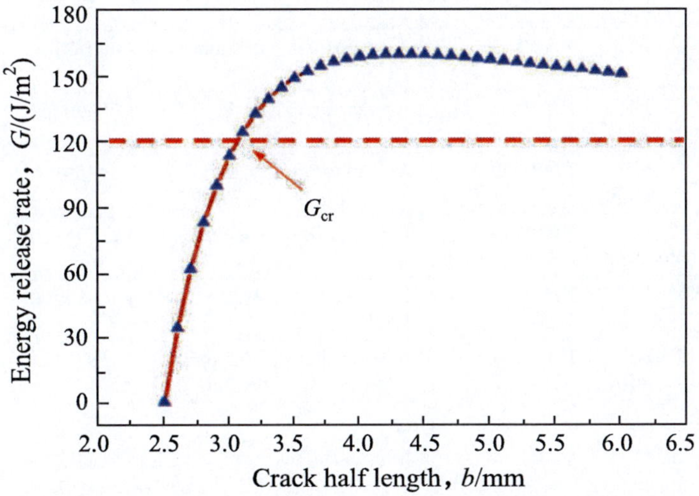

Fig. 9.37 Variation in the interfacial energy release G with the interface crack half-length b (the red dashed line shows the critical interfacial bonding energy)

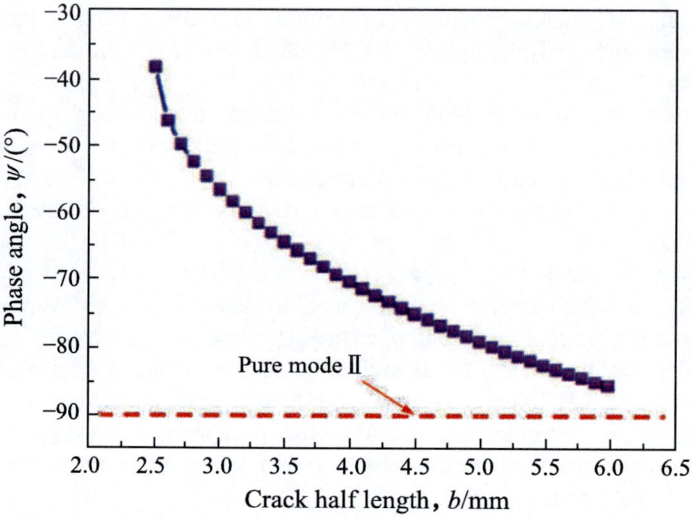

Fig. 9.38 Evolution of the interfacial mixed-mode phase angle ψ with the crack half-length b (the red dashed line shows pure fracture mode II)

Table 9.4 Comparison of energy release rate for TBCs corresponding to different fracture modes, preparation processes, treatment methods, and test methods

Materials bond coating	substrate	Deposition method	Treatment	Test method	G (J/m^2)	Fracture mode
NiCrAlY	Ni alloy	APS	As-deposited	Compression	120–150*	II
NiCrAl	Cr alloy	Flame-sprayed	As-deposited	Shear	260 [56]	II
PtAl	Ni alloy	EB-PVD	As-deposited	Indentation	49 [57]	Mixed
NiCoCrAlY	Ni alloy	EB-PVD	As-deposited	Barb	60–90 [58]	II
NiCoCrAlY	Ni alloy	EB-PVD	As-deposited	Push-out	84 [59]	I
NiCoCrAlY	Ni alloy	EB-PVD	1150 °C, 10 h	Push-out	115 [59]	Mixed
NiCoCrAlY	Ni alloy	EB-PVD	1100 °C, 100 h	Wedge	56–80 [60]	II
β–(Ni, Pt)Al /β–NiAl	Ni alloy	EB-PVD	As-deposited	Four-point bending	110 [61]	Mixed

*Indicates the results obtained in this section

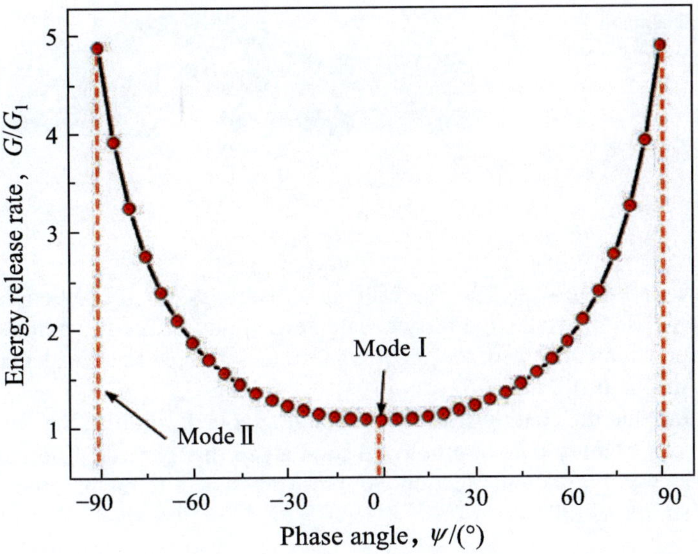

Fig. 9.39 Relation between the mode-dependent energy release rate and mixed-mode phase angle

$$G(\psi) = G_{\mathrm{I}}[1 + \tan^2(1 - \lambda)\psi] \quad (9.42)$$

where $G(\psi)$ is the mode-dependent interfacial energy release rate, G_{I} is the interfacial energy release rate in pure mode I ($\psi = 0°$), and λ is the mode sensitivity factor (generally, $\lambda = 0.3$) [55]. Equation (9.42) shows that G increases with ψ. The G in pure mode II ($\psi = \pm 90°$) is approximately 4–5 times that in pure mode I ($\psi = 0°$), as shown in Fig. 9.39. In addition, TGO formation adversely affects the interfacial bonding performance of a TBC and generally lowers its G.

9.5 Interfacial K_{IC} Characterization of TBCs by the Blister Test

Figure 9.40 shows the principle of the blister test for characterizing the interfacial K_{IC} of a TBC. During the preliminary specimen preparation stage, a penetrating hole is first prepared by microfabrication in a flat, smooth substrate, onto which a TBC is subsequently deposited. During the test process, an air or hydraulic pressure p is applied through the hole to the coating with a thickness of h. The coating blisters under p. When the separation pressure (i.e., p) reaches the critical value p_c, the coating begins to debond along the interface and, as a result, becomes separated from the substrate. As p increases, cracks continuously propagate along the interface, eventually causing the coating to gradually delaminate. This test can be performed

Fig. 9.40 Schematic diagram of the blister test

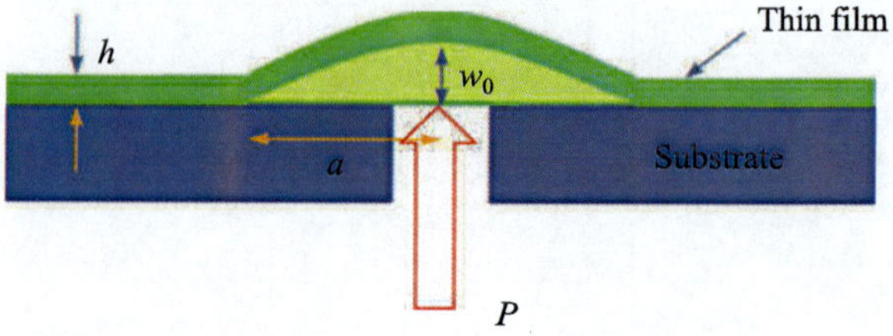

to measure parameters such as the critical pressure p_c for the debonding of the coating/substrate interface, the blister diameter $2a$, and the center-point deflection w_0, based on which the K_{IC} of the coating/substrate interface can be calculated using a suitable mechanical model [62].

Assuming that the coating/substrate debonding zone is circular, the von Karman nonlinear plate theory can be employed to analyze this process. The energy of a circular plate with a large deflection can be expressed in terms of deflections and midplane strains as follows:

$$U_s^{\mathrm{plate}} = \pi D \int_0^a r dr \left\{ (\nabla^2 w)^2 - \frac{2}{r}(1 - \upsilon_1) w' w'' + \frac{12}{h^2}[\varepsilon_1^2 + 2(\upsilon_1 - 1)\varepsilon_2] - \frac{2wq}{D} \right\} \tag{9.43}$$

where u and w are the displacements of any point on the midplane in the r- and z-directions, respectively, D is the bending stiffness ($D = E_1 h^3/12(1 - \upsilon_1^2)$), $\nabla^2 = \mathrm{d}^2/\mathrm{d}r^2 + 1/r(\mathrm{d}/\mathrm{d}r)$ is the Laplace operator, E_1, υ_1^2, and h are the Young's modulus, Poisson's ratio, and thickness of the coating, respectively, and ε_1 and ε_2 are the first and second strain invariants, respectively. By eliminating ε_2 and using the Euler–Lagrange variational principle to find the minimum potential energy, Zhou et al. [62] transformed the Euler–Lagrange differential equation to a nonlinear coupling governing equation for u and w with the following boundary conditions:

$$w'(0) = 0, w(a) = 0, w'(a) = 0, \ u(0) = 0, \text{ and } u(a) = 0 \tag{9.44}$$

The following dimensionless parameters are introduced:

$$\xi = \frac{r}{a}, \overline{w} = \frac{w}{h}, \overline{u} = \frac{u}{h}, p = c_1^0 q, \eta = \frac{h}{a} \tag{9.45}$$

where $c_1^0 = 8(1 - \upsilon_1^2)/E_1$ and $\overline{w}$ can be easily solved using the following equation:

$$\overline{w} = \frac{3p}{8\eta^4 k^2}\left[\frac{2[I_0(k\xi) - I_0(k)]}{I_1(k)k} + 1 - \xi^2\right] \tag{9.46}$$

where $I_m(x)$ is the mth-order modified Bessel function and $k^2 = 12\varepsilon_1/\eta^2$ is determined using the following transcendental equation:

$$\frac{1}{3}\left(\frac{2\eta^4 k^3}{3p}\right)^2 = \left[\frac{3}{4} + \frac{4}{k^2} - \frac{1}{2}\left[\frac{I_0(k)}{I_1(k)}\right]^2 - \frac{1}{k}\frac{I_0(k)}{I_1(k)}\right] \tag{9.47}$$

The stress at the boundary $N_{r|r=1}$ of the coating can be expressed in terms of the following dimensionless parameter $S_r{}^0$:

$$S_r^0 = \frac{c_1^0 N_{r|r=a}}{h} = \left(\frac{3p}{2\eta^3 k^2}\right)^2 \left[\frac{8}{k^2}\frac{3}{2}\left[\frac{I_0(k)}{I_1(k)}\right]^2 - \frac{2}{k}\frac{I_0(k)}{I_1(k)}\right] \tag{9.48}$$

The dimensionless bending moment at the boundary $r = a$ can be expressed as follows:

$$M_r^0 = \frac{c_1^0 M_{r|r=a}}{h^2} = \frac{p}{2\eta^2 k^2}\left[k\frac{I_0(k)}{I_1(k)} - 2\right] = \frac{p}{2\eta^2} f(k) \tag{9.49}$$

where $f(k) = 1/k^2[kI_0(k)/I_1(k) - 2]$. Based on Eq. (9.46), the center-point deflection under a small load q can be easily obtained

$$w(0) = q\frac{3(1-\upsilon_1^2)a^4}{16E_1h^3} \tag{9.50}$$

The energy release rate of the coating/substrate interface, G_a, proposed by Suo and Hutchinson [24], can be written in a dimensionless form as $\overline{G_a} = 1/3(16\eta^2/p)^2 c_1^0 G_a h$; i.e.

$$\begin{aligned}\overline{G_a} &= \frac{1}{3}\left(\frac{4\eta^2}{p}\right)^2\left[\frac{(S_r^0)^2}{A} + \frac{(M_r^0)^2}{I} - 2\frac{S_r^0 M_r^0}{\sqrt{AI}}\sin\gamma\right] \\ &= \frac{4}{3}\frac{1}{I} f^2(k)(\lambda^2 + 1 - 2\lambda\sin\gamma)\end{aligned} \tag{9.51}$$

where the parameter $\lambda = \sqrt{I/A}S_r^0/M_r^0$ and the dimensionless parameters A, I, Σ, and $\sin\gamma$ are given by the following equations:

$$A = \frac{1}{1 + \sum(4\eta_0 + 6\eta_0^2 + 3\eta_0^3)}$$

$$I = \frac{1}{12(1 + \sum\eta_0^3)} \sin\gamma = 6\sum\eta_0^3(1 + \eta_0)\sqrt{AI}$$

$$\sum = \frac{c_2}{c_1} = \frac{1+\alpha}{1-\alpha} \quad \eta_0 = \frac{h}{H}$$

where H is the thickness of the substrate and $c_j = (\Gamma_j + 1)/\mu_j$, in which $\mu_j (j = 1, 2)$ is the shear modulus and $\Gamma_j = 3 - 4\upsilon_j$ for plane strain cases and $\Gamma_j = (3 - \upsilon_j)/(1 + \upsilon_j)$ for plane stress cases (where υ_j is the Poisson's ratio), and the subscripts 1 and 2 indicate the coating and the substrate, respectively. The test method presented in this chapter is intended for plane strain cases (i.e., $c_1 = c_1^0$). Generally, the dimensionless energy release rate $\overline{G_a}$ in Eq. (9.51) depends on four dimensionless parameters, namely, p, η, η_0, and $\sum$. For cases where the coating is much thinner than the substrate, numerical calculation results show that $\overline{G_a}$ is almost independent of parameters η_0 and $\sum$ (or α, which is a Dundurs' parameter). Let us further consider Eq. (9.47). $\overline{G_a}$ which depends only on a combined parameter p/η^4. The deformation of the substrate is not considered in the calculation of the displacement of the debonded coating; once this displacement is obtained, so are the stress and bending moment of the coating, as shown in Eqs. (9.48) and (9.49). Suo and Hutchinson [24] studied the problem of interface cracks between two elastic layers based on composite beam theory and experimentally determined the energy release rate of interface cracks between two elastic layers, which can be expressed in terms of membrane stresses and bending moments. In this chapter, Eq. (9.51) is the expression of the energy release rate obtained by Suo and Hutchinson, which also includes the deformation of the substrate. In addition, an examination of the dimensionless parameters defined in Eq. (9.45) reveals that the combined parameter p/η^4 can be written as follows:

$$p/\eta^4 = \frac{8(1-\upsilon_1^2)q}{E_1}\left(\frac{a}{h}\right)^4 \tag{9.52}$$

It can be seen that $\overline{G_a}$ depends on q as well as the E_1, υ_1, h, and a (the debond radius) of the coating. In view of the above situation, the energy release rate in the theory of elasticity can be fitted as follows:

$$\overline{G_a} = \exp\left[-\sum_{i=1}^{9} a_i \left(\frac{3p}{128\eta^4} - \overline{a}\right)^{i-1}\right] \tag{9.53}$$

The values of the coefficients in the above equation are given in Table 9.5. On the other hand, the parameter can also be fitted as follows:

$$\lambda = \sqrt{\frac{I}{A}\left[\sum_{i=1}^{7} b_i \left(\frac{3p}{128\eta^4} - \overline{b}\right)^{i-1}\right]} \tag{9.54}$$

The values of the relevant coefficients in Eq. (9.54) are also provided in Table 9.5. The phase angle ψ can be expressed in the following analytical form [24]:

Table 9.5 Fitting coefficients

I	a_i	b_i
1	9.21E-01	1.0647
2	1.36E-01	9.79E-02
3	−1.6259E-2	−2.2795E-2
4	4.53E-05	2.65E-03
5	6.31E-04	−8.1940E-4
6	4.2S18E-4	1.05E-03
7	−1.4939E-4	−2.4690E-4
8	−1.9844e-5	
9	6.70E-06	
	$\bar{a} = 4.0150$	$\bar{b} = 3.1050$

$$\psi = \tan^{-1}\left[\frac{-\lambda \sin\omega - \cos(\omega + \gamma)}{-\lambda \sin\omega + \cos(\omega + \gamma)}\right] \tag{9.55}$$

where the angular quantity ω depends on geometric parameter η_0 as well as Dundurs' parameters α and β [63].

To test the interfacial K_{IC}, hydraulic pressure was applied through a manual pump to the TBC through a hole drilled at the center of the substrate, as shown in Fig. 9.40. The pressure in the hole was measured using a 5 MPa pressure sensor. The deflection at the center of the coating was measured using a linear variable differential transformer (LVDT) sensor. To measure the interface crack length in the coating, the specimens were subjected to multiple loading/unloading cycles at loading and unloading rates of 0.2 MPa/min. After the macroscopic test, each specimen was cut into four pieces along the midline. An SEM system was used to observe the characteristics of the interface crack and to measure its length, as shown in Fig. 9.41.

The debond radius a of a coating is an important parameter for determining its interfacial K_{IC} and can be determined using the following three methods: (1) SEM observation after testing the macroscopic parameters, (2) the compliance method,

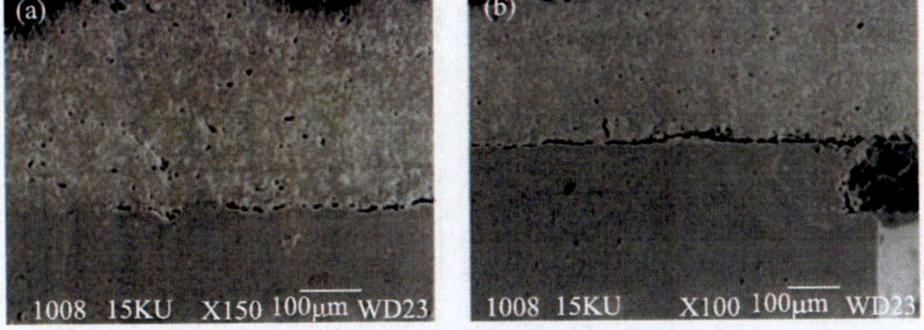

Fig. 9.41 SEM images of the interface crack formed during the blister test: **a** crack tip and **b** overall view

which determines a mainly based on the slope of the load–deflection curve, and (3) ultrasonic testing. The Commission on Testing Methods of the International Society for Rock Mechanics recommends that the compliance method be used to test the K_{IC} of rocks [64]. The main idea of the compliance method lies in the linearization of cyclic loading–unloading curves. Figure 9.42 shows the loading–unloading curves during a loading–unloading process. For the curve corresponding to the initial load, when the maximum deflection $\omega(0)$ is less than 5 μm, the loading curve is a straight line, based on the slope of which the elastic parameters of the coating material can be obtained; when $\omega(0) > 5$ μm, the loading curve is not a straight line, and the nonlinear relation between $\omega(0)$ and the hydraulic pressure q can be predicted using the analytical expression in Eq. (9.46) by setting $\xi = 0$. Figure 9.43 illustrates the basic principle of the compliance method. Specifically, a straight line is drawn through each cyclic loading–unloading curve, and two points, H and L, are defined on each line. The highest point H is the location where the deflection begins to decrease during the unloading phase of the cycle, and the corresponding pressure is denoted by q_H. The lowest point L is located at the reloading phase of the cycle and is defined by the pressure level $q_L = 0.5q_H$. Then, a straight line is drawn between points H and L, q and $\omega(0)$ follow a linear relation, and the a of the coating can be determined using Eq. (9.50).

Ultrasonic micrography is a nondestructive test technique that obtains material damage information through analysis of reflected ultrasonic waves. In the test of this section, an ultrasonic transducer with a frequency of 10 MHz, a focal distance of 10 mm, and a sensor radius of 6.4 mm was used. Both the specimen and ultrasonic transducer were immersed in a water tank. Figure 9.44 compares the measurements by the three methods, which yield consistent measurements of the interface crack length,

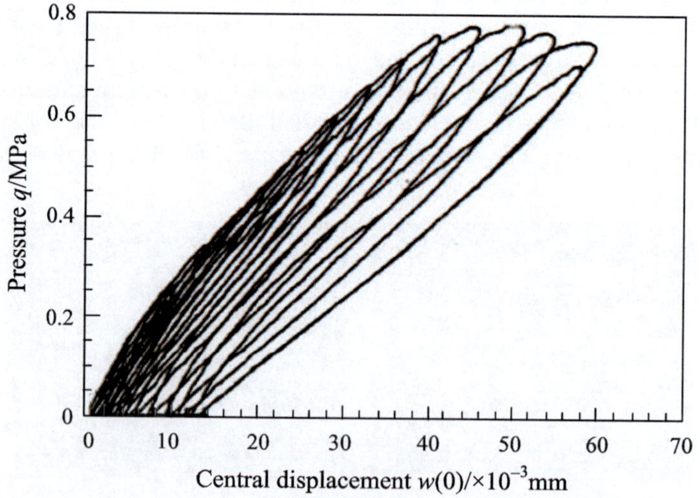

Fig. 9.42 Load–deflection curves of the coating during the debonding process

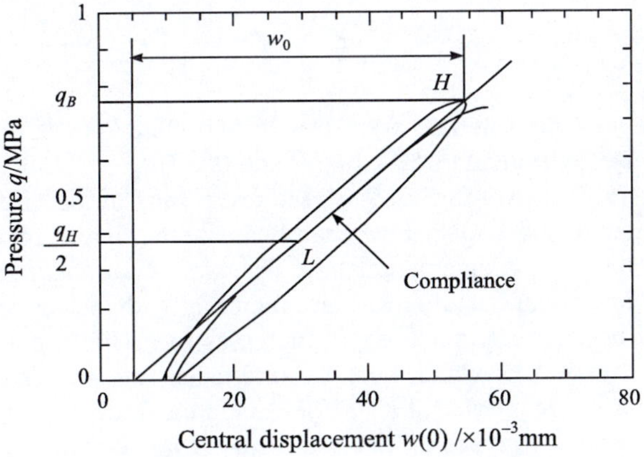

Fig. 9.43 Principle of the compliance method in the linearization of the cyclic loading–unloading curves

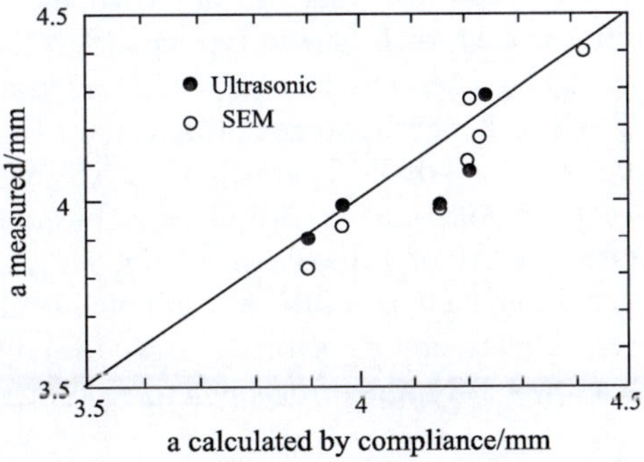

Fig. 9.44 Comparison of the measurements of the interface crack length by the compliance method and SEM

suggesting that the compliance method is effective in determining the interface crack length.

Jensen and Thouless [65] investigated the effect of residual stress σ_{res} on the interfacial K_{IC} of coatings with the blister test. They found that σ_{res} has a significant effect when it is much higher than the critical stress σ_c. Here, σ_c is the in-plane buckling stress of the clamped circular plate

$$\sigma_c = \frac{14.68D}{ha^2} \tag{9.56}$$

Then, for a coating with an h of 0.35 mm and an a of 4 mm, $\sigma_c = 90.9$ MPa. The σ_{res} in a yttria-stabilized zirconia (YSZ) ceramic coating deposited on a NiCrAlY BC layer by PS was measured to be approximately –11 ± 10 MPa. According to Jensen and Thouless's work [65], at a very low σ_{res}/σ_c, σ_{res} has little effect on the interfacial K_{IC} measured by the blister test.

For a plane strain problem, the effective interfacial K_{IC} can be expressed analytically as follows [62]:

$$K_i = \sqrt{3}\frac{\sqrt{h}}{c_1^0}\left[\frac{p\cosh(\pi\varepsilon)}{4\eta^2}\right](G_a)^{1/2} \tag{9.57}$$

If the material parameters, q, and a are known, the effective interfacial K_{IC} can be determined by Eqs. (9.53) and (9.57).

As shown in Fig. 9.45, the test results of the effective interfacial K_{IC} of TBCs are relatively dispersed, mainly due to the significant effect of the microstructure on the macroscopic parameters. The variation in the interfacial fracture shows that the interfacial fracture resistance increases with increasing interface crack length. When the interface crack length increases to its limit, the interfacial fracture resistance saturates, at which time the interfacial fracture resistance is considered to be the effective interfacial K_{IC}. The data in the figure reveal that the interfacial K_{IC} values of the two types of TBCs are 0.9–1.2 and 0.6–0.7 MPa·m$^{1/2}$, respectively, and that the corresponding G values are 20–35.6 and 15.9–21.7 J/m^2, respectively.

It is well known that the fractures of multilayer material systems at the interface are often in the form of mixed-type cracks. If the material parameters, q, and a are known, ψ can be determined by Eq. (9.55). Figure 9.46 shows the variation in the ψ of the TBCs with interface crack propagation. The maximum differences in the ψ values of the two types of TBCs are 1.0° and 0.5°, respectively. The average ψ of the coatings changes from –26° to –30°. A crack propagation length of 0.3 mm corresponds to an effective SIF of 1.0 MPa·m$^{1/2}$ (Fig. 9.45) and a ψ of –26.5° (Fig. 9.46). Under this condition, the type I SIF (K_I) and type II SIF (K_{II}) are 0.0915 and 0.0404 MPa·m$^{1/2}$, respectively, and K_I is related to K_{II} through $K_{II} = -0.4421\ K_I$. A crack propagation length of 1.2 mm corresponds to an effective SIF of 1.1 MPa·m$^{1/2}$ (Fig. 9.45) and

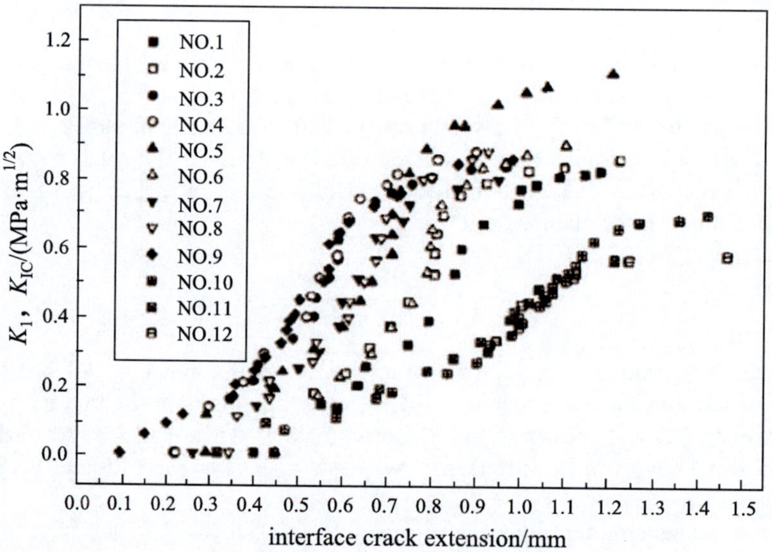

Fig. 9.45 Relation between interface crack propagation length and interfacial K_{IC}

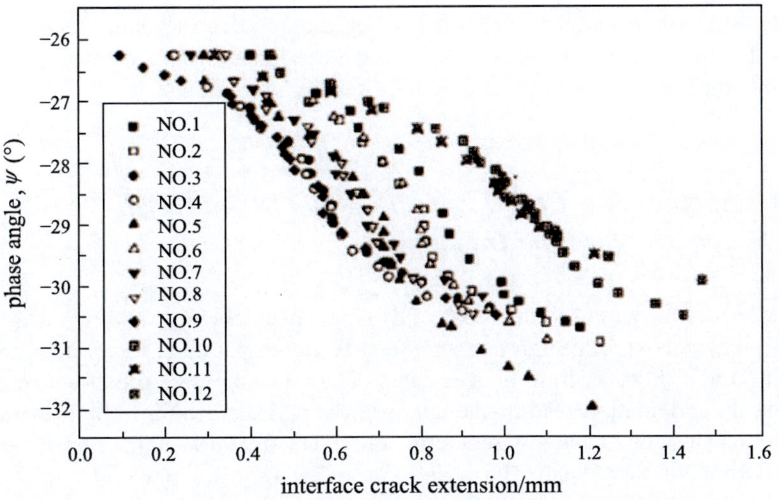

Fig. 9.46 Relation between the interface crack propagation length and phase angle

a ψ of –32° (Fig. 9.46). Under this condition, $K_{\text{I}} = 0.09639$ MPa·m$^{1/2}$ and $K_{\text{II}} =$ 0.053 MPa·m$^{1/2}$, and K_{I} is related to K_{II} through $K_{\text{II}} = -0.5498\, K_{\text{I}}$. Evidently, type II crack propagation is the dominant crack propagation mechanism. According to the findings of Suo and Hutchinson [24], ψ depends heavily on the loading mode. This dependence can be seen from Eq. (9.54), which shows that ψ depends strongly on the parameter $\lambda = \sqrt{I/AS_r^0}/M_r^0$, that is, the loading mode. Evan and Hutchinson [66] found that the energy release rate depends on the mixed mode

$$G_i = G_0(1 + \tan^2\psi) \tag{9.58}$$

where G_0 is the energy release rate at $\psi = 0$, i.e., the energy release rate of type I cracks. In the above discussion, the energy release rate refers to G_i, meaning that the energy release rate depends on the mixed-mode phase angle ψ. Thus, when $\psi = 0$, G_0 is calculated using Eq. (9.57) to be 11.88–25.88 J/m^2, the steady-state ψ is –31°, and the corresponding K_{IC} is 0.52–1.02 MPa·m$^{1/2}$.

9.6 In Situ K_{IC} Characterization of TBCs at High Temperatures

The previous sections show that currently, the K_{IC} of TBCs is characterized mostly at room temperature or after heat treatment, making it impossible to determine their high-temperature stress state and mechanical properties. However, TBCs usually operate at high temperatures, and changes in the mechanical properties of TBCs in high-temperature environments are a key factor affecting their failure mechanisms. For this reason, this section focuses on the indentation and three-point bending methods for characterizing the K_{IC} of TBCs at high temperatures.

9.6.1 Surface K_{IC} Characterization of TBCs at High Temperatures by Indentation

The APS process was used to prepare TBC specimens, the K_{IC} of which was subsequently characterized at high temperatures. High-temperature alloys may undergo oxidation and creep at high temperatures. Thus, to eliminate the effect of oxidation on the indentation results, the coating was deposited on a flat polycrystalline Al_2O_3 plate instead of a high-temperature alloy. The thickness of the ceramic coating ranged from 300 to 400 μm. The dimensions of each specimen were 15 mm (length) × 15 mm (width) × 3 mm (thickness). Prior to the indentation test, the surface of each specimen was ground and polished using a diamond abrasive paste to a finish of 1 μm. Subsequently, each polished specimen was carefully cleaned in an ultrasonic oscillator and then dried.

Figure 9.47 shows the high-temperature in situ indentation test system. Each specimen was heated and held at a high temperature in a muffle furnace. The temperature of the specimen was measured using a thermocouple placed near the specimen stage in the uniform temperature zone. Before the indentation test, each specimen was placed in a high-temperature furnace in a laboratory environment and then heated to and held at the target test temperature to allow it to reach thermal equilibrium. Three target test temperatures (25, 800, and 1000 °C) were selected. Under a given load, a cube-corner indenter can more easily produce cracks than Vickers and Berkovich indenters. Therefore, a cube-corner indenter was used to perform an in situ high-temperature indentation test on the surface of the coating. The load-control mode was employed to accurately control the maximum indentation load during the indentation test process. Three maximum indentation loads (80, 100, and 120 N) were selected, each of which was held for 20 s during the test. After the in situ high-temperature indentation test, an SEM image of residual impression was taken to measure the length of each indentation crack. Note that the high-temperature K_{IC} of the TBC was measured by the cube-corner indenter, whereas the Young's modulus and hardness of the TBC were measured by the standard indentation technique with a Berkovich indenter. Corresponding to the previous indentation test, the test temperature was set to 25, 800, and 1000 °C, respectively. The indentation test was performed under a maximum load of 3 N. The nominal constant loading rate was set to 0.1 N/s during the loading–unloading process, and the test was repeated at least five times to determine the average value at each selected test temperature [67].

The theoretical model for the characterization of the surface K_{IC} of a TBC at high temperatures by indentation is consistent with Eqs. (9.23) and (9.24) in Sect. 9.3,

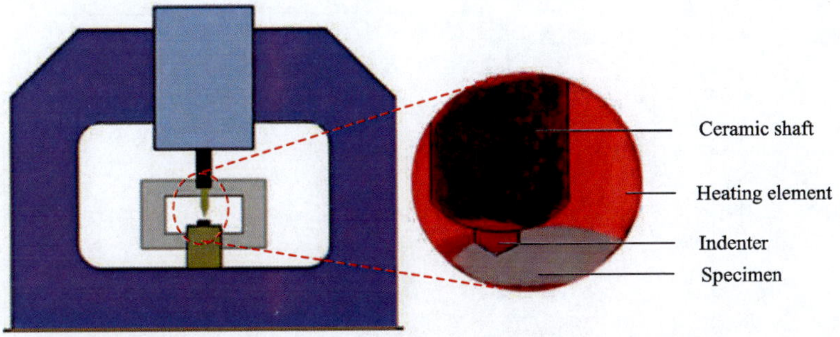

Fig. 9.47 Schematic diagram of a high-temperature indentation test system [67]

except that the values of the material parameters are replaced by those for the corresponding temperature. Equation (9.24) shows that there is a linear relation between $\frac{P}{c^{3/2}}$ and $c^{-1/2}$ and that the residual stress and K_{IC} can be determined based on the slope and intercept, respectively. To obtain a better linear relation, it is necessary to perform a series of indentation tests under different maximum indentation loads and to measure the corresponding crack lengths (c_1, c_2, and c_3). In this section, the surface of the TBC was subjected to indentation tests under different loads (80, 100, and 120 N) at various temperatures. An SEM system was then used to observe the residual impressions and measure the indentation crack lengths. Figure 9.48 shows SEM images of typical residual impressions at different temperatures. The shape of the residual impressions is usually triangular. One crack is formed at each vertex of the triangular residual impression. The lengths of the three cracks, denoted by c_1, c_2, and c_3, respectively, are measured. The average crack length $c = (c_1 + c_2 + c_3)/3$ is calculated. Through the above indentation test, the relations between $\frac{P}{c^{3/2}}$ and $c^{-1/2}$ at different temperatures are obtained, as shown in Fig. 9.49. All the indentation data for each given temperature are fitted to a linear equation. The residual stress in and surface K_{IC} of the TBC at each temperature are then determined through the calculation of the slope and intercept of the corresponding equation, respectively.

An analysis of Eq. (9.24) shows that the values of the parameter χ at different temperatures are needed to determine the K_{IC} of and residual stress in the TBC at

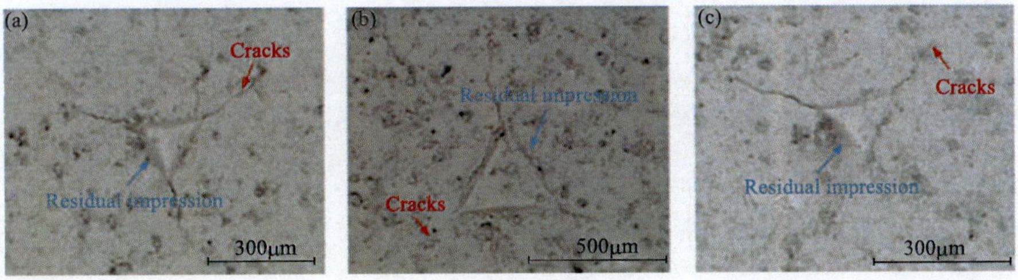

Fig. 9.48 SEM images of typical residual impressions at different temperatures: **a** 25 °C, **b** 800 °C, and **c** 1000 °C

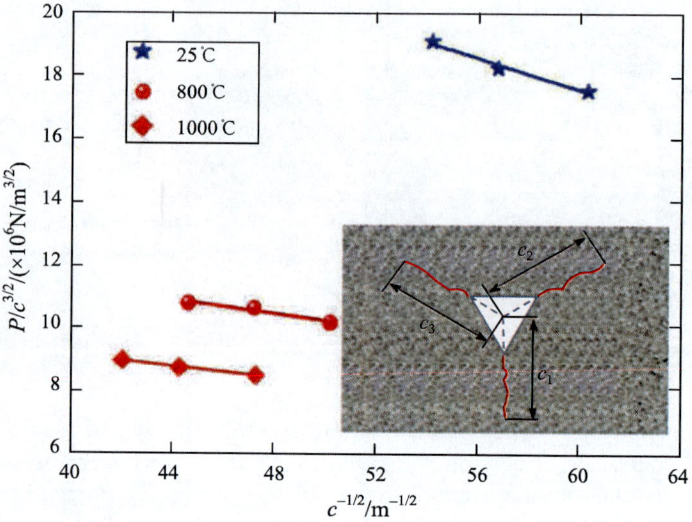

Fig. 9.49 The relations between $\frac{P}{c^{3/2}}$ and $c^{-1/2}$ at different temperatures: 25 °C, 800 °C, and 1000 °C

the corresponding temperatures. According to the expression $\chi = \delta(E/H)^{1/2}$, the values of the parameter χ at different temperatures can be determined only when the values of elastic modulus and hardness of the TBC at the corresponding temperatures are available. Based on the Oliver–Pharr method, the elastic modulus and hardness of a TBC can be determined using the load–displacement curve from the indentation test under a low load [68]. Thus, the elastic modulus is calculated to be 59.34 ± 4.32, 45.46 ± 3.11, and 39.47 ± 3.59 GPa at 25, 800, and 1000 °C, respectively. Correspondingly, the hardness is calculated to be 1.72 ± 0.34, 1.32 ± 0.17, and 1.10 ± 0.23 GPa, respectively. We can find that both the elastic modulus and hardness of the TBC decrease as the test temperature increases. Then, the surface K_{IC} of and residual stress in the TBC at the corresponding temperatures can be calculated using Eq. (9.24). Figure 9.50 shows the surface K_{IC} of and residual stress in the TBC at each selected temperature. At 25, 800, and 1000 °C, the K_{IC} values of the TBCs are 1.25, 0.91, and 0.75 MPa·m$^{1/2}$, respectively, and the residual stresses in the TBCs are –131.3, –55.5, and –45.5 MPa, respectively. Figure 9.50 shows that the surface K_{IC} of and residual stress in the TBC are highly sensitive to temperature. Both the surface K_{IC} and the absolute value of the residual stress decrease as the temperature increases, where the negative sign of the residual stress means that it is compressive.

Without considering the effect of the residual stress on K_{IC}, the K_{IC} characterized by indentation can be expressed as follows:

$$K_{IC} = \chi \frac{P}{c^{3/2}} \tag{9.59}$$

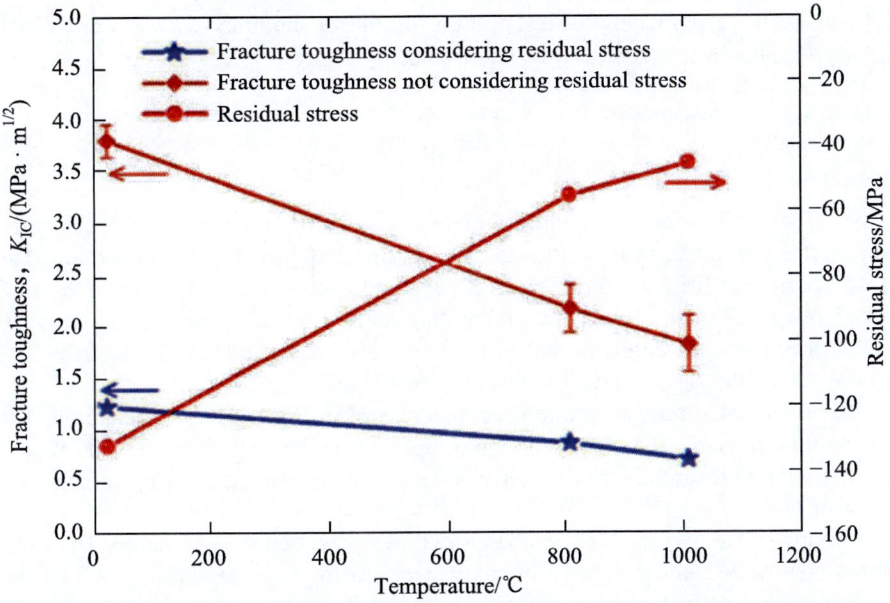

Fig. 9.50 Variations in the surface K_{IC} of and residual stress in the TBC with temperature

Figure 9.50 also shows the variation in the surface K_{IC} of the TBC with temperature without considering the effect of the residual stress. At 25, 800, and 1000 °C, the K_{IC} values of the TBCs are 3.84 ± 0.16, 2.21 ± 0.23, and 1.87 ± 0.28 MPa·m$^{1/2}$, respectively. The K_{IC} of the TBC without considering the residual stress also decreases with temperature and is higher than that determined with the consideration of the residual stress at the same temperature. This is because tensile stress can promote crack propagation, whereas compressive stress has an inhibiting effect on crack propagation [69, 70]. Therefore, when a material is in a tensile stress state, the K_{IC} determined without the consideration of the residual stress is lower than the inherent K_{IC}; when the material is subjected to a compressive stress, the K_{IC} determined without the consideration of the residual stress is higher than the inherent K_{IC}. Hence, in the case where the residual stress is not considered, the increase in K_{IC} can be attributed to the compressive stress in the TBC.

9.6.2 K_{IC} Characterization of TBCs at High Temperatures by the Three-Point Bending Test

(1) Elastic modulus characterization of the substrate of a TBC at high temperatures by the three-point bending test

During the three-point bending test process, the elastic modulus of a substrate, which is a typical beam structure, can be determined by the following equation [71]:

$$E_s = \frac{FL^3}{4bh^3\delta} \tag{9.60}$$

where b, h, and E_s are the width, thickness, and elastic modulus of the substrate, respectively, L is the span of the clamp, F is the load, and δ is the deflection. The displacement at the central point of the substrate can be determined using the DIC technique (see Chap. 11 for details). We know from Eq. (9.60) that E_s can be calculated once the dimensions (b and h) of the specimen, L, and the slope of the F–δ curve for the three-point bending test are determined.

Here, we use a pure substrate specimen shown in Fig. 9.51 as an example. The substrate was made of a Ni-based high-temperature alloy (GH536) and was 50 mm long, 6 mm wide, and 2 mm thick. The span of the three-point bending support was 30 mm.

Figure 9.52 shows the high-temperature three-point bending test setup. The specimen was heated and held at a high temperature in a muffle furnace. A thermocouple placed near the specimen stage in the uniform temperature zone was used to measure the temperature. Prior to the bending test, the specimen was placed in a high-temperature furnace in a laboratory environment and then heated to and held at the target test temperature to allow it to reach thermal equilibrium. Four target test temperatures (30, 400, 600, and 800 °C) were selected. A CCD high-speed camera was placed in front of the cross-section of the specimen to record its bending process. To eliminate the effect of high-temperature IR radiation on imaging, a blue filter was placed in front of the CCD camera. The image sampling frequency was set to one image per second. The three-point bending loading rate was set to 0.4 mm/min. Figure 9.53 shows the load–deflection curves of the substrate during the three-point bending process. The deflection of the substrate increases linearly with increasing load. The slope of the load–deflection curve decreases with increasing temperature, suggesting that the substrate softens as the temperature increases. Figure 9.54 shows that the substrate elastic modulus E_s decreases as the temperature increases. As the temperature increases from 30 °C to 600 °C, E_s decreases from 125 to 98 GPa; when the temperature exceeds 600 °C, E_s drops sharply, reaching 68 GPa at 800 °C [72].

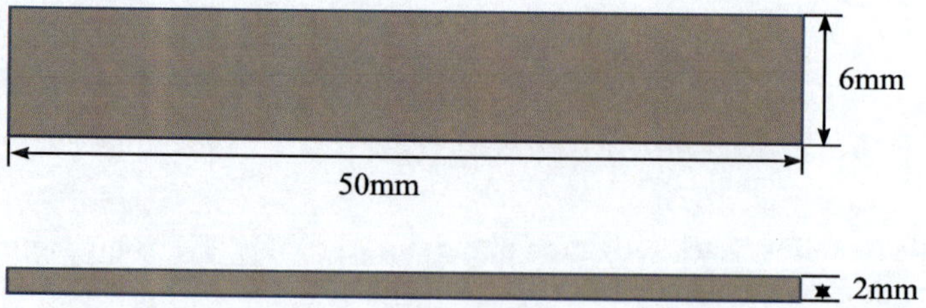

Fig. 9.51 Dimensions of a specimen for a high-temperature three-point bending test

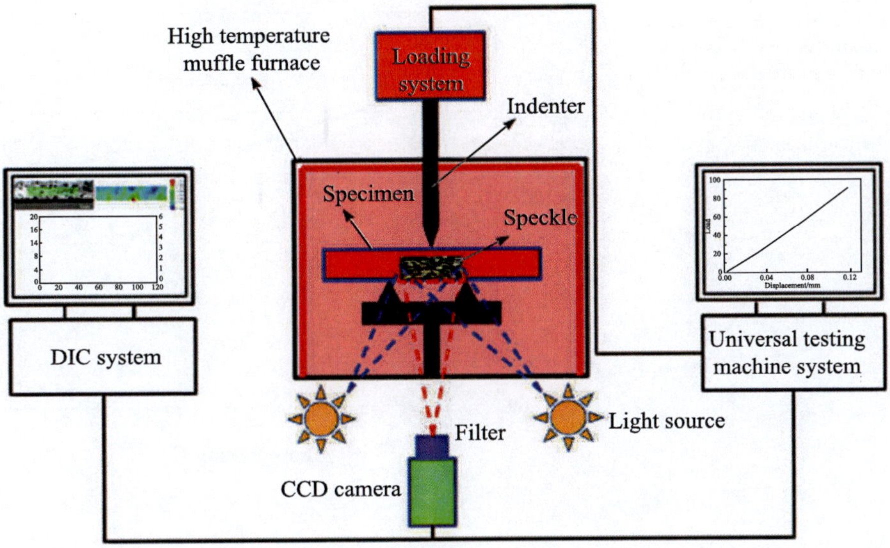

Fig. 9.52 Schematic diagram of the loading process for a high-temperature three-point bending test

Fig. 9.53 Load–deflection curves of the substrate obtained in the three-point bending test

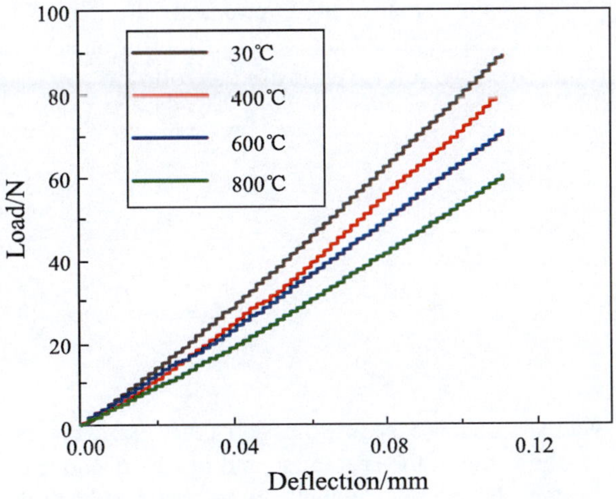

(2) **Elastic modulus characterization of the ceramic layer of a TBC at high temperatures by the three-point bending test**

A TBC system can be viewed as a two-layer composite beam structure. According to the composite beam theory, the elastic modulus of the ceramic layer is given by the following expression [72]:

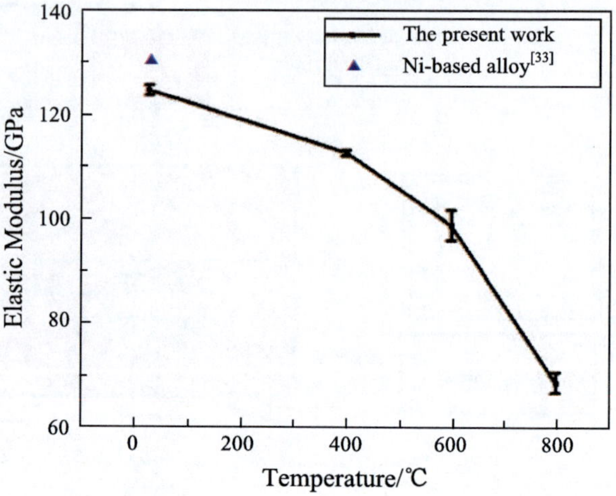

Fig. 9.54 Variation in the substrate elastic modulus with temperature

$$E_c I_c + E_s I_s = \frac{FL^3}{48\delta} \tag{9.61}$$

where E_c and E_s are the elastic moduli of the ceramic layer and the substrate, respectively, L is the span of the clamp, F is the load, δ is the deflection, and I_c and I_s are the cross-sectional moments of inertia of the ceramic layer and the substrate, respectively, which are expressed as follows:

$$I_c = b \int_{-l-l_c}^{-l} y^2 dy \tag{9.62}$$

$$I_s = b \int_{-l}^{l_s - l} y^2 dy \tag{9.63}$$

where l_c and l_s are the thicknesses of the ceramic layer and the substrate, respectively, b is the width of the specimen, and l is the distance between the neutral axis and the ceramic/BC interface, which can be expressed as follows:

$$l = \frac{E_s l_s^2 - E_c l_c^2}{2E_s l_s + 2E_c l_c} \tag{9.64}$$

From Eqs. (9.61) to (9.64), we know that once the dimensions (b and h) of the three-point bending test specimen, E_s, L, and the slope of the F–δ curve are determined, E_c can be calculated.

A 50-mm-long, 6-mm-wide APS TBC specimen was prepared. The substrate of the coating was made of a Ni-based high-temperature alloy (GH536) and was 2 mm thick. An 80-μm-thick NiCoCrAlY BC layer was sprayed onto the surface of the substrate by VPS. A 1-mm-thick top ceramic layer was sprayed onto the surface of the BC layer by APS. The span of the three-point bending support was 30 mm.

Figure 9.52 shows the high-temperature three-point bending test process. Figure 9.55 shows the load–deflection curves of the TBC specimen obtained from the three-point bending test at different temperatures. The variation in these curves is consistent with that of the pure substrate specimen. The deflection increases linearly with increasing load, and the slope of the curve decreases as the temperature increases. Based on Eq. (9.61), the elastic modulus of the ceramic layer of the TBC specimen, E_c, at different temperatures can be determined. Figure 9.56 shows the variation in E_c with temperature. Evidently, E_c gradually decreases with increasing temperature. Specifically, as the temperature increases from 30 to 800 °C, E_c decreases from 21 to 13 GPa. The E_c at room temperature (30 °C) is consistent with the values (21.3–25.5 GPa) previously obtained from curvature monitoring [73].

(3) **Surface K_{IC} characterization of TBCs at high temperatures by the three-point bending test**

During the three-point bending test process, to allow a single surface crack and a steadily propagating interface crack to form in a TBC specimen, it is necessary to cut a penetrating preexisting crack with a certain depth on the surface of its ceramic layer. Under tensile stress on the surface of the ceramic layer, the vertical surface crack in the ceramic layer begins to propagate along the preexisting crack. Crack initiation and propagation can be detected by the DIC technique (see Chap. 11 for details), based on which the critical stress for crack propagation can be obtained. The surface K_{IC} (K_{TC}) of the TBC can be determined using the following equation:

$$K_{TC} = Y\sigma_{cr}\sqrt{\pi a_0} \tag{9.65}$$

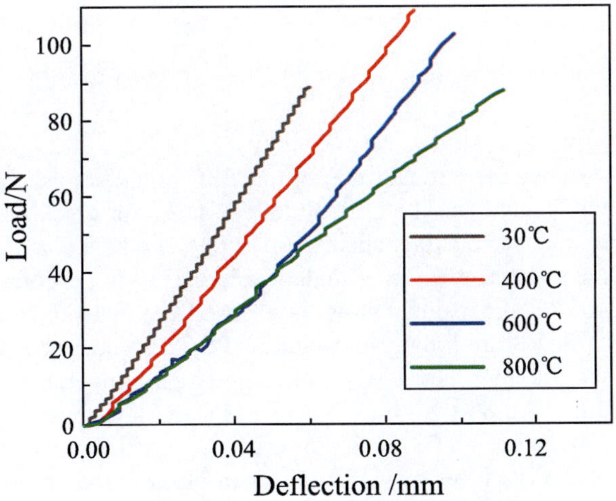

Fig. 9.55 Load–deflection curves of the TBC obtained from the three-point bending test at different temperatures

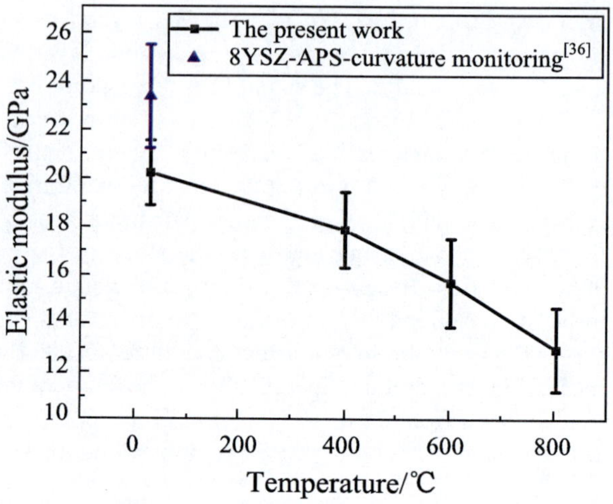

Fig. 9.56 Variation in the elastic modulus of the ceramic layer with temperature

where K_{TC} is the surface fracture toughness of the TBC, σ_{cr} is the critical stress for crack propagation, a_0 is the length of the preexisting crack, and Y is a geometrical factor. σ_{cr} can be expressed as follows:

$$\sigma_{cr} = \frac{E_c y}{\rho} + \sigma_r \quad (9.66)$$

where σ_r is the residual stress in the TBC, E_c is the elastic modulus of the ceramic layer, y is the distance between the crack tip and the neutral axis, and ρ is the radius of curvature of the neutral axis. y and ρ are given by the following equations:

$$y = l + \frac{l_c}{2}$$

$$\rho = \frac{[(L/2)^2 + \delta_{cr}^2]}{2\delta_{cr}}$$

Once the critical time for the TBC to undergo surface fracture is obtained by the DIC technique, the critical deflection δ_{cr} for crack propagation is determined, then σ_{cr} can be calculated using Eq. (9.66), and hence, K_{TC} is determined. Note that only the residual stress at room temperature (30 °C) is considered. At temperatures above 400 °C, the residual stress is released and tends to zero.

A 50-mm-long, 6-mm-wide APS TBC specimen was prepared. The substrate of the coating was made of a Ni-based high-temperature alloy (GH536) and was 2 mm thick. An 80-μm-thick NiCoCrAlY BC layer was sprayed onto the surface of the substrate by VPS. A 1-mm-thick top ceramic layer was sprayed onto the surface of the BC layer by APS. The span of the three-point bending support was 30 mm, and the length of the preexisting surface crack, a_0, was 0.5 mm. The residual stress near the crack tip was measured by Raman spectroscopy. The characteristic peak of 8YSZ powder appeared at 639.46 cm^{-1}, and a Raman frequency shift of 641.04 cm^{-1} occurred near the crack tip in the ceramic layer. On this basis and combined with the Raman piezospectroscopic coefficient, the residual stress near the crack tip in the ceramic layer at room temperature (30 °C) was calculated to be –31.2 MPa.

To obtain the critical deflection δ_{cr} for crack propagation, the variations in the strain and deflection at the crack tip with time were extracted from the DIC system and are shown in Fig. 9.57. The strain increases slowly with time. At 37 s, there is a sharp increase in the strain at the crack tip, suggesting that crack initiation and propagation begin at the crack tip and correspond to a δ_{cr} of 0.21 mm (Fig. 9.57a). The δ_{cr} values at 400, 600, and 800 °C are similarly determined to be 0.12, 0.13, and 0.15 mm, respectively. Based on Eqs. (9.65) and (9.66) in combination with the δ_{cr} value, the surface K_{IC} of the TBC can be determined.

Figure 9.58 shows the evolution of the surface K_{IC} of the TBC with temperature. Evidently, the surface K_{IC} of the TBC decreases as the temperature increases. Without considering the effect of the residual stress at room temperature, the surface K_{IC} values of the TBC at 30, 400, 600, and 800 °C are calculated to be 2.54 ± 0.04, 1.28 ± 0.03, 1.23 ± 0.04, and 1.16 ± 0.03 MPa·$m^{1/2}$, respectively (Fig. 9.58a). The variation in the surface K_{IC} of the TBC with temperature determined considering the effect of the residual stress is similar to that determined without considering the

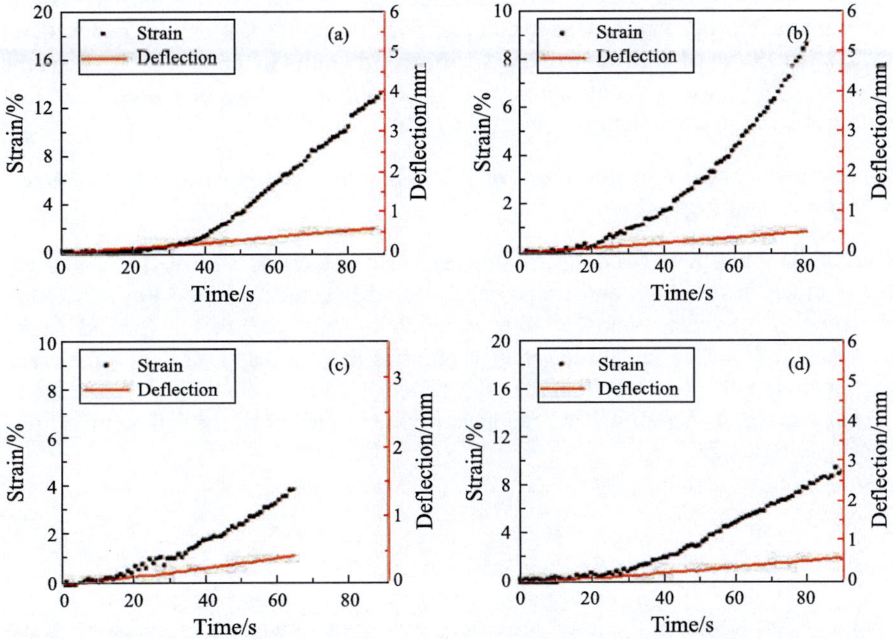

Fig. 9.57 Evolution of the strain and deflection at the tip of the surface crack in the ceramic layer with time at different temperatures: **a** 30 °C, **b** 400 °C, **c** 600 °C, and **d** 800 °C

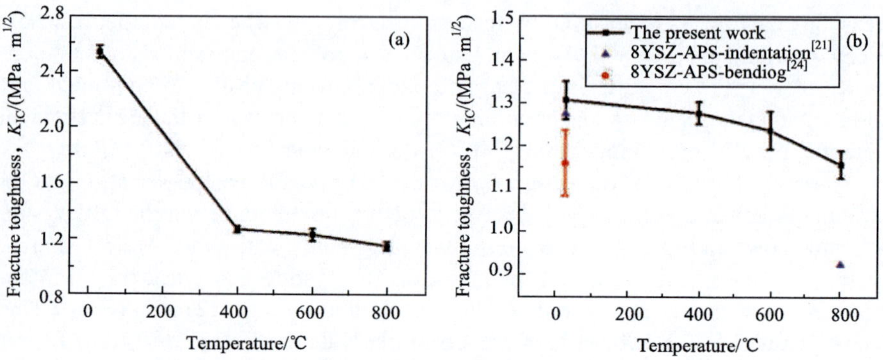

Fig. 9.58 Evolution of the surface K_{IC} of the TBC with temperature: **a** without the consideration of the effect of the residual stress at room temperature and **b** with the consideration of the effect of the residual stress at room temperature

effect of the residual stress. Considering the effect of the residual stress, the surface K_{IC} values of the TBC at 30, 400, 600, and 800 °C are calculated to be 1.31 ± 0.04, 1.28 ± 0.03, 1.23 ± 0.04, and 1.16 ± 0.03 MPa·m$^{1/2}$, respectively (Fig. 9.58b). The surface K_{IC} of the TBC determined with the consideration of the effect of the residual stress is lower than that determined without the consideration of the effect of the residual stress, which can be ascribed to the fact that compressive stress inhibits crack propagation. Under compressive stress, the K_{IC} determined without the consideration of the residual stress is higher than the inherent K_{IC}, whereas the K_{IC} determined with the consideration of the residual stress is consistent with that obtained by high-temperature indentation [67].

(4) **Interfacial K_{IC} characterization of TBCs at high temperatures by the three-point bending test**

During the three-point bending test process, a surface crack is formed at the preexisting crack in the TBC and propagates toward the interface, forming a steadily propagating interface crack. The time at which the interface crack is formed can be determined based on the abrupt-change point of the normal strain along the crack propagation path extracted from the DIC system.

The interfacial K_{IC} of a TBC can be expressed in terms of the following critical energy release rate G_c:

$$G_c = \left(\frac{P^2}{2W}\right)\left(\frac{\partial C}{\partial l}\right) \tag{9.67}$$

where l is the length of the interface crack, W is the width of the specimen, P is the critical load for interface crack propagation, and C is the compliance of the specimen and is expressed as follows:

$$C = \delta / P \tag{9.68}$$

Figure 9.59a shows the load–deflection relations during interface crack initiation and propagation in the TBC at different temperatures. Basically, the deflection increased linearly with increasing load, suggesting that the specimen was at the elastic stage. According to Eq. (9.68), the compliances of the TBC specimen at different temperatures can be determined based on the slope of the load–deflection curve. Figure 9.59b presents the relations between the interface crack propagation length in the TBC and deflection at different temperatures, showing that the interface crack length increased with increasing deflection. Figure 9.60 shows the relations between the interface crack length and the compliance at different temperatures; it is found that the two are nearly linearly related and that a higher temperature leads to lower goodness of linear fit. At 30, 400, 600, and 800 °C, $\partial C/\partial l$ values are calculated to be 1×10^{-5}, 2×10^{-5}, 3.8×10^{-5}, and 5.8×10^{-5} N^{-1}, respectively. Based on Eq. (9.68) and combined with the average load F and the $\partial C/\partial l$ value, the interfacial K_{IC} values of the TBC at different temperatures can be calculated.

Figure 9.61 shows the variation in the interfacial K_{IC} of the TBC with temperature. At temperatures below 600 °C, the interfacial K_{IC} increases almost linearly with increasing temperature, a trend that is consistent with the theoretical result of Essa et al. [74]. When the temperature is higher than 600 °C, the interfacial K_{IC} increases significantly, which may be related to the brittle–ductile transition temperature (approximately 650 °C) of the BC layer. As the temperature increases beyond 650 °C, the fracture strain in the BC layer increases substantially, indicating that the BC layer exhibits ductility [75]. This finding is also corroborated by the interface crack length–compliance relations at different temperatures (Fig. 9.60). Evidently, the compliance of the TBC specimen at 800 °C is much higher than those at 30, 400, and 600 °C. As the temperature increases from room temperature (30 °C) to 800 °C, the interfacial K_{IC} of the TBC increases from 83.7 to 156.3 J/m^2.

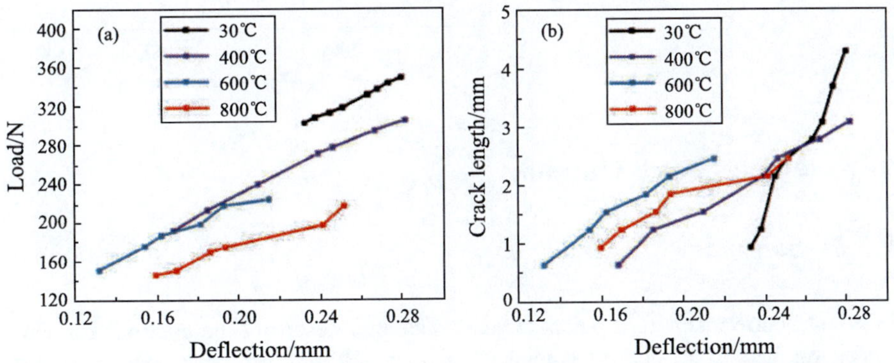

Fig. 9.59 **a** Load–deflection relations and **b** interface crack length–deflection relations at different temperatures

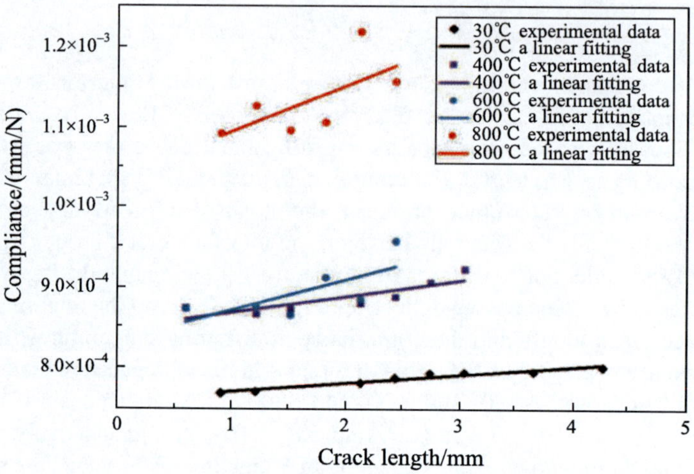

Fig. 9.60 Interface crack length–compliance relations at different temperatures

Fig. 9.61 Variation in the interfacial K_{IC} of the TBC with temperature

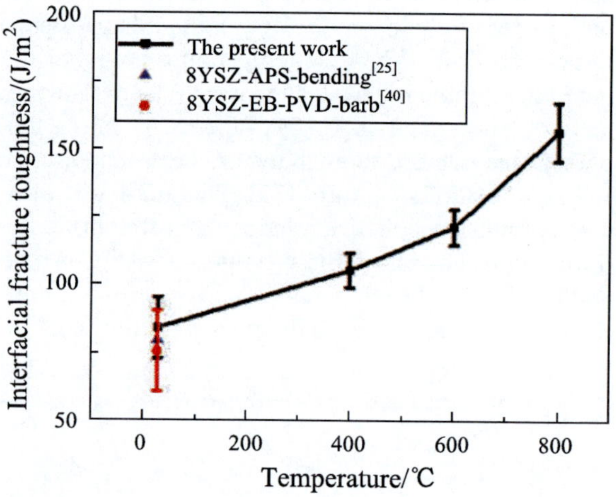

9.7 Summary and Outlook

9.7.1 Summary

From a fracture mechanics perspective, this chapter describes the methods for characterizing the surface and interfacial K_{IC} values of TBCs at room and high temperatures and the characterization of the surface and interfacial K_{IC} values of ceramic blocks and TBCs using nondestructive testing techniques (e.g., AE and digital speckle correlation). The following provides a summary of this chapter:

(1) Based on the three-point bending test in combination with nondestructive testing techniques such as AE and digital speckle correlation, methods for quantitatively characterizing the surface and interfacial K_{IC} of TBCs are developed and extended to the measurement and characterization of high-temperature K_{IC}. The evolutionary pattern of K_{IC} with temperature is obtained, which provides test data support for analyzing the failure mechanisms of TBCs.
(2) Based on the theory of fracture mechanics, indentation methods for quantitatively characterizing the surface and interfacial K_{IC} values of TBCs are established and applied to the quantitative characterization of in situ indentations at high temperatures.
(3) The interfacial bonding performance of TBCs is characterized using the buckling method and the blister test from three perspectives, namely, experimental research, numerical simulation and calculation, and theoretical analysis. The relations of the evolution of the buckling morphology with the interfacial energy release rate and phase angle are obtained.

9.7.2 Outlook

Future research will be focused on the following areas:

(1) Surface and interfacial K_{IC} characterization of electron beam-physical vapor deposition (EB-PVD) TBCs. The indentation methods presented in this chapter for characterizing the surface and interfacial K_{IC} of TBCs are applicable only to APS TBCs with a layered, stacked structure. Further research is needed for EB-PVD TBCs with a columnar crystal structure.
(2) In situ K_{IC} characterization at high temperatures. The high-temperature K_{IC} characterization methods mentioned in this chapter are capable of characterizing K_{IC} at temperatures up to 1000 °C, which is due to the limitations of the test equipment and instruments. In theory, high-temperature K_{IC} characterization methods can be used to measure K_{IC} at even higher temperatures, thereby providing effective test data for the development and application of new coatings operating at 1400 °C and above.

References

1. Griffith AA. The phenomena of rupture and flow in solids. Philos Trans R Soc Lond. 1921;A221:163.
2. Irwin GR. Fracture. In Handbuck der Physik 6. Edited by Fluegge S (Springer, Berlin), 1958, 551.
3. Gong J. Fracture mechanics of ceramic materials. Beijing: Tsinghua University Press, 2001.
4. Gong J, Guan Z. Advancements in techniques for measuring fracture toughness of ceramics in China. Bullet Chin Ceramic Soc. 1996;1:53–7.

5. Xu S, Shen G, Tyson WR. Effect of crack-tip plasticity on crack length estimation methods for SENB sample. Eng Fract Mech. 2005;72(9):1454–9.
6. Xu J, Zhang ZL, Nyhus B, Sun DB. Effects of crack depth and specimen size on ductile crack growth of SENT and SENB specimens for fracture mechanics evaluation of pipeline steels. Int J Press Vessels Pip. 2009;86(12):787–97.
7. Wan J, Zhou M, Yang XS, Dai CY, Zhang Y, Mao WG, Lu C. Fracture characteristics of freestanding 8wt% Y_2O_3–ZrO_2 coatings by single edge notched beam and Vickers indentation tests. Mater Sci Eng A. 2013;581:140–4.
8. Wan J. A study on the fracture toughness of and residual stress in thermal barrier coating systems based on indentation tests. Xiangtan University, 2012.
9. Peng LM, Cao JW, Noda K, Han KS. Mechanical properties of ceramic–metal composites by pressure infiltration of metal into porous ceramics. Mater Sci Eng A. 2004;374(1–2):1–9.
10. Roberts RJ, Rowe RC, York P. The measurement of the critical stress intensity factor (KIC) of pharmaceutical powders using three point single edge notched beam (SENB) testing. Int J Pharm. 1993;91(2–3):173–82.
11. Choi SR, Zhu DM, Miller RA. Effect of sintering on mechanical properties of plasma-sprayed zirconia-based thermal barrier coatings. J Am Ceram Soc. 2005;88(10):2859–67.
12. Ray AK, Steinbrech RW. Crack propagation studies of thermal barrier coatings under bending. J Eur Ceram Soc. 1999;19:2097–109.
13. Mao W. Interfacial failure analysis of thermal barrier coatings under combined thermal and mechanical loadings. Xiangtan University, 2006.
14. Florando JN, LeBlanc MM, Lassila DH. Multiple slip in copper single crystals deformed in compression under uniaxial stress. Scripta Mater. 2007;57(6):537–40.
15. Kang J, Wilkinson DS, Jain M, Embury JD, Beaudoin AJ, Kim S, Mishira R, Sachdev AK. On the sequence of inhomogeneous deformation processes occurring during tensile deformation of strip cast AA5754. Acta Mater. 2006;54(1):209–19.
16. Zhou Q. Plasma spray techniques. Jiangsu Science Press (Jiangsu), 1982.
17. Wu D. Real-time testing, analysis, and experimental simulation system development for the interfacial failure of thermal barrier coatings. Xiangtan University, 2011.
18. Chen Q. Testing and analysis of the residual stress in and fracture toughness of thermal barrier coatings under thermal cycling. Xiangtan University, 2010.
19. Choi SR. High-temperature slow crack growth, fracture toughness and room-temperature deformation behavior of plasma-sprayed ZrO_2–8 wt.% Y_2O_3. Ceramic Eng Sci Proc.1998;19:293–301.
20. Yamazaki Y, Schmidt A, Scholz A. The determination of the delamination resistance in thermal barrier coating system by four-point bending tests. Surf Coat Technol. 2006;201:744–54.
21. Guo SQ, Kagawa Y. Young's moduli of zirconia top-coat and thermally gorwn oxide in a plasma-sprayed thermal barrier coating system. Scripta Mater. 2004;50:1401–6.
22. Freborg AM, Ferguson BL, Brindley WJ. Modeling oxidation induced stresses in thermal barrier coatings. Mater Sci Eng A. 1998;245(2):182–90.
23. Evans AG, Mumm DR, Hutchinson JW, Meier GH, Pettit FS. Mechanisms controlling the durability of thermal barrier coatings. Prog Mater Sci. 2001;46:505–53.
24. Suo Z, Hutchinson JW. Interface crack between two elastic layers. Int J Fract. 1990;43:1–19.
25. Hutchinson JW, He MY, Evans AG. The influence of imperfections on the nucleation and propagation of bucking driven delaminations. J Mech Phys Solids. 2000;48:709–43.
26. Chen ZB, Wang ZG, Zhu SJ. Tensile fracture behavior of thermal barrier coatings on superalloy. Surf Coat Technol. 2011;205:3931–9.
27. Zhou YC, Hashida T, Jian CY. Determination of interface fracture toughness in thermal barrier coating system by blister tests. J Eng Mater Technol. 2003;125:176–83.
28. Choi SR, Zhu D, Miller RA. Fracture behavior under mixed-mode loading of ceramic plasma-sprayed thermal barrier coatings at ambient and elevated temperatures. Eng Fract Mech. 2005;72:2144–59.
29. Thurn G, Schneider GA, Bahr HA, Aldinger F. Toughness anisotropy and damage behavior of plasma sprayed ZrO2 thermal barrier coatings. Surf Coat Technol. 2000;123:147–59.

30. Kim SS, Chae YH, Choi SY. Characteristics evaluation of plasma sprayed ceramic coatings by nano/micro-indentation test. Tribol Lett. 2004;17(3):663–9.
31. Marshall DB, Lawn BR. Residual stress effects in sharp contact cracking: I, Indentation fracture mechanics. J Mater Sci. 1979;14:2001–11.
32. Evans AG, Charles AE. Fracture toughness determination by indentation. J Amer Ceramic Soc. 1976;59:371–2.
33. Anstis GR, Chantikul P, Lawn BR, Marshall DB. A critical evaluation of indentation techniques for measuring fracture toughness: I, direct crack measurement. J Amer Ceramic Soc. 1981;64:533–9.
34. Chantikul P, Anstis GR, Lawn BR, Marshall DB. A critical evaluation of indentation techniques for measuring fracture toughness: II, strength method. J Amer Ceramic Soc. 1981;64:539–43.
35. Qian G, Nakamura T, Berndt CC, Leigh SH. Tensile toughness test and high temperature fracture analysis of thermal barrier coatings. Acta Mater. 1997;45(4):1767–84.
36. Lawn BR, Fuller ER. Measurement of thin-layer surface stresses by indentation fracture. J Mater Sci. 1984;19:4061–7.
37. Hutchinson JW, Suo Z. Mixed mode cracking in layered materials. Adv Appl Mech. 1991;29:63–191.
38. He MY, Bartlett A, Evans AG. Kinking of a crack out of an interface: role of in-plane stress. J Am Ceram Soc. 1991;74(4):767–71.
39. Kim HJ, Moon MW, Kim DI, Lee KR, Oh KH. Observation of the failure mechanism for diamond-like carbon film on stainless steel under tensile loading. Scripta Mater. 2007;57(11):1016–9.
40. Yang L, Zhong ZC, You J, Zhang QM, Zhou YC, Tang WZ. Acoustic emission evaluation of fracture characteristics in thermal barrier coatings under bending. Surf Coat Technol. 2013;232:710–9.
41. Jiang LM, Zhou YC, Hao HX, Liao YG, Lu C. Characterization of the interface adhesion of elastic-plastic thin film/rigid substrate systems using a pressurized blister test numerical model. Mech Mater. 2010;42(10):909.915.
42. Tvergaard V, Hutchinson JW. The relation between crack growth resistance and fracture process parameters in elastic-plastic solids. J Mech Phys Solids. 1992;40(6):1377–97.
43. Needleman A. An analysis of decohesion along an imperfect interface. Int J Fract. 1990;42(1):21–40.
44. Chandra N, Li H, Shet C, Ghonem H. Some issues in the application of cohesive zone models for metal-ceramic interfaces. Int J Solids Struct. 2002;39(10):2827–55.
45. Camanho PP, Davila CG, De Moura MF. Numerical simulation of mixed-mode progressive delamination in composite materials. J Compos Mater. 2003;37(16):1415–39.
46. Xu W, Wei YG. Strength analysis of metallic bonded joints containing defects. Comput Mater Sci. 2012;53(1):444–50.
47. Mi Y, Crisfield MA, Davies GAO, Hellweg HB. Progressive delamination using interface elements. J Compos Mater. 1998;32(14):1246–72.
48. Zhao PF, Sun CA, Zhu XY, Shang FL, Li CJ. Fracture toughness measurements of plasma-sprayed thermal barrier coatings using a modified four-point bending method. Surf Coat Technol. 2010;204(24):4066–74.
49. Zhou YC, Tonomori T, Yoshida A, Liu L, Bignall G, Hashida T. Fracture characteristics of thermal barrier coatings after tensile and bending tests. Surface Coatings Technol. 2002;157(2):119.127.
50. Wu DJ, Mao WG, Zhou YC, Lu C. Digital image correlation approach to cracking and decohesion in a brittle coating/ductile substrate system. Appl Surf Sci. 2011;257(14):6040–3.
51. Zhu W, Yang L, Guo JW, Zhou YC, Lu C. Determination of interfacial adhesion energies of thermal barrier coatings by compression test combined with a cohesive zone finite element model. Int J Plast. 2015;64:76–87.
52. Faou JY, Parry G, Grachev S, Barthel E. How does adhesion induce the formation of telephone cord buckles?. Phys Rev Lett. 2012;108(11): 116102.

53. Gao YF, Bower AF. A simple technique for avoiding convergence problems in finite element simulations of crack nucleation and growth on cohesive interfaces. Modell Simul Mater Sci Eng. 2004;12(3):453.
54. Soulignac R, Maurel V, Rémy L, Köster A. Cohesive zone modelling of thermal barrier coatings interfacial properties based on three-dimensional observations and mechanical testing. Surf Coat Technol. 2013;237:95–104.
55. Moon MW, Jensen HM, Hutchinson JW, Oh KH, Evans AG. The characterization of telephone cord buckling of compressed thin films on substrates. J Mech Phys Solids. 2002;50(11):2355–77.
56. Xu ZH, Yang Y, Huang P, Li X. Determination of interfacial properties of thermal barrier coatings by shear test and inverse finite element method. Acta Mater. 2010;58(18):5972–9.
57. Vasinonta A, Beuth JL. Measurement of interfacial toughness in thermal barrier coating systems by indentation. Eng Fract Mech. 2001;68(7):843–60.
58. Guo SQ, Mumm DR, Karlsson AM, Kagawa Y. Measurement of interfacial shear mechanical properties in thermal barrier coating systems by a barb pullout method. Scripta Mater. 2005;53(9):1043–9.
59. Kim SS, Liu YF, Kagawa Y. Evaluation of interfacial mechanical properties under shear loading in EB-PVD TBCs by the pushout method. Acta Mater. 2007;55(11):3771–81.
60. Mumm DR, Evans AG. On the role of imperfections in the failure of a thermal barrier coating made by electron beam deposition. Acta Mater. 2000;48(8):1815–27.
61. Théry PY, Poulain M, Dupeux M, Braccini M. Spallation of two thermal barrier coating systems: experimental study of adhesion and energetic approach to lifetime during cyclic oxidation. J Mater Sci. 2009;44(7):1726–33.
62. Zhou YC, Hashida T, Jian CY. Determination of interface fracture toughness in thermal barrier coating system by blister tests. J Eng Mater Technol. 2003;125(2):176–82.
63. Dundurs J. Edge-bonded dissimilar orthogonal elastic wedges under normal and shear loading. J Appl Mech. 1996;36:650–2.
64. International society for rock mechanics. Suggested methods for determining the fracture toughness of rock. Int J Rock Mech Min Sci. 1988;25:71–96.
65. Jensen HM, Thouless MD. Effects of residual stresses in the blister test. Int J Solids Struct. 1993;30:779–95.
66. Evans AG, Hutchinson JW. Effects of non-planarity on the mixed mode fracture resistance of biomaterial interfaces. Acta Metall. 1989;37:909–16.
67. Qu ZL, Wei K, He Q, He RJ, Pei YM, Wang SX, Fang DN. High temperature fracture toughness and residual stress in thermal barrier coatings evaluated by an in-situ indentation method. Ceramics Int. 2018;44:7926–7929.
68. He RJ, Qu ZL, Pei YM, Fang DN. High temperature indentation tests of YSZ coatings in air up to 1200 °C. Mater Lett. 2017;209:5–7.
69. Taskonak B, Mecholsky JJ, Anusavice KJ. Residual stresses in bilayer dental ceramics. Biomaterials. 2005;26:3235–41.
70. Ewart L, Suresh S. Crack propagation in ceramics under cyclic loads. J Mater Sci Technol. 1987;22:1173–92.
71. Hibbeler RC, Bending, Mech. Mater, seventh ed., Prentice Hall, New Jersey, 2008, pp. 254–361.
72. Zhu W, Wu Q, Yang L, Zhou YC. In situ characterization of high temperature elastic modulus and fracture toughness in air plasma sprayed thermal barrier coatings under bending by using digital image correlation. Ceramics Int. 2020;46:18526–18533.
73. Okajima Y, Sakaguchi M, Inoue H. A finite element assessment of influential factors in evaluating interfacial fracture toughness of thermal barrier coating. Surf Coat Technol. 2017;313:184–90.
74. Essa SK, Liu R, Yao MX. Temperature and exposure-dependent interfacial fracture toughness model for thermal barrier coatings. Surf Coat Technol. 2019;358:505–10.
75. Ray AK, Das DK, Venkataraman B, Roy PK, Goswami B, Roy N, Das SK, Parida N, Tarafder S, Chaudhuri S, Sahay SK, Ghosh RN. Characterization of rupture and fatigue resistance of TBC superalloy for combustion liners. Mater Sci Eng A. 2005;405:194–200.

Chapter 10
Residual Stresses in TBCs

Thermal barrier coatings (TBCs) are widely used in fields such as aviation, aerospace, and large-scale thermal power generation due to their ability to effectively lower the operating temperature of high-temperature heat-resistant alloy substrates and prolong the service life of high-temperature components. However, crack initiation and propagation and, in more serious cases, premature spallation cause TBC failure, which is a key issue that attracts the attention of researchers. Residual stresses are a primary factor causing TBC failure, and consequently, studying their influence on the durability of TBCs is vitally important.

This chapter focuses on describing the causes, simulation, and prediction ofdestructive and nondestructive characterization methods for residual stresses in TBCs, as well as elucidating the effects of residual stresses on TBC failure and the relevant mechanisms.

10.1 Formation of Residual Stresses in TBCs

10.1.1 Causes of Residual Stresses in TBCs

In the absence of external forces, the stresses that exist in equilibrium inside an object are referred to as residual stresses. The theoretical basis of residual stresses is described by Eqs. (2.172) and (2.271) in Chap. 2. Residual stresses can be formed during the preparation process (i.e., they exist before service) or during service (i.e., they are formed after the object has been in service for a certain period). The formation of residual stresses during the TBC preparation process is mainly attributed to the following factors:

(1) The quenching stress σ_q formed due to the rapid contraction of a sheet-like coating. During the atmospheric plasma spraying (APS) process, molten particles impact and spread across the surface of the substrate and are immediately

Y. Zhou et al., *Thermal Barrier Coatings: Failure Theory and Evaluation Technology*,
https://doi.org/10.1007/978-981-19-2723-2_10

quenched to the substrate temperature T_s. When cooling from the processing temperature to T_s, a coating with a lamellar structure sprayed onto a substrate contracts rapidly, resulting in the formation of residual stress known as the quenching stress σ_q, [1–3] which is expressed mathematically as [4]

$$\sigma_q = \alpha_c(T_m - T_s)E_c \tag{10.1}$$

where T_m is the melting point of the sprayed material and E_c and α_c are the elastic modulus and coefficient of thermal expansion (CTE) of the coating, respectively. Generally, the measured value of σ_q in a brittle ceramic coating is lower than its calculated value because the residual stresses in the coating are released by microcracks formed in the coating.

(2) Thermal mismatch stress σ_t. This is a thermal stress formed at high temperatures and during cooling in the spraying process due to the difference between the CTEs of the coating and the substrate material [5]. σ_t is expressed mathematically as [4]

$$\sigma_t = E_c\Delta\alpha(T_m - T_s)(1 + \nu)/(1 - \nu^2) \tag{10.2}$$

where ν is the Poisson's ratio of the coating and $\Delta\alpha = \alpha_c - \alpha_s$ (α_s is the CTE of the substrate).

(3) Phase-transformation stress σ_p. This is the stress formed in a coating when molten (liquid) particles solidify or undergo solid-state phase transformation during the thermal spraying process. σ_p can also be formed during service. For example, the stress corresponding to the structural changes caused by calcium–magnesium aluminosilicate (CMAS) discussed in Chap. 6 is a typical σ_p. If we consider only the σ_p during the preparation process, the σ_p in a TBC is generally not high. When σ_p is negligible, [4, 6–8] there are only two types of residual stress in a TBC, σ_q and σ_t. Then, the total residual stress can be expressed as follows:

$$\sigma = \sigma_q + \sigma_t \tag{10.3}$$

10.1.2 Influencing Factors of the Residual Stresses in TBCs

Residual stresses in TBCs are influenced by many factors, including the grain morphology, [9, 10] cooling rate, [1] liquid-to-solid phase transformation, and thermal growth oxides (TGOs) [11, 12].

(1) **Effect of the grain morphology on the residual stresses in TBCs**

Grain morphology has a significant impact on the service life of TBCs. Compared to an APS TBC with an equiaxed crystal structure, an electron beam-physical vapor deposition (EB-PVD) TBC has a columnar crystal structure with high strain tolerance and hence can resist the formation of surface cracks and extend the service life of the TBC. As another process capable of producing TBCs with a columnar crystal structure, plasma spraying-physical vapor deposition (PS-PVD) combines the features of APS and EB-PVD and can be employed to prepare TBCs with a mixed layered and columnar structure. In addition, PS-PVD can be used to improve the strain tolerance of coatings, and therefore, prolong their service life.

Research shows that grain morphology has a significant impact on the residual stresses in TBCs [9]. Using electron backscatter diffraction (EBSD), Song et al. [9] investigated the effect of grain morphology on the residual stresses/strains in APS TBCs. They found that the strains of equiaxed grains were greater than those of columnar grains and that the residual stresses in a TBC composed of a large number of equiaxed grains were much higher than those in a TBC composed of columnar grains. Large temperature gradients occur in equiaxed grains as a result of high cooling rates. Therefore, a coating composed of columnar grains in a high proportion has a long service life due to the low residual stresses in columnar grains. Figure 10.1 shows two different types of grains in sprayed coatings, namely, columnar grains (within the dashed rectangles and denoted by "1") and equiaxed grains (within the dashed rectangles and denoted by "2"). Coating N1 has many more equiaxed grains than coating N2. Through a statistical analysis of 200 zones in each APS coating, the relative percentages of equiaxed grains with an aspect ratio < 2 in coatings N1 and N2 are determined to be 80.5% and 46.4%, respectively. Notably, the statistical content of equiaxed grains includes the unmelted particles.

To compare the residual stresses in columnar and equiaxed grains in a sprayed yttria-stabilized zirconia (YSZ) coating, 4.2 million grains were measured and analyzed to determine their low-angle ($\Phi < 5^{o}$) integrated angular misorientation density (IMD). Figure 10.2 shows the orientation map and corresponding IMD distribution map of a typical YSZ coating. The IMD varies from zone to zone. Figure 10.2c shows the quantitative data for the correlation between the aspect ratio and IMD of grains. The IMD of equiaxed grains with an aspect ratio < 2 is much higher than that of columnar grains. Compared to coatings comprising columnar grains, YSZ coatings comprising equiaxed grains display a higher strain capacity. To verify this, the residual stresses in two YSZ coatings with different contents of equiaxed grains, N1 and N2, were tested by means of X-ray diffraction (XRD), with the results summarized in Table 10.1. Clearly, the residual stresses in coating N1, which had a higher equiaxed grain content, are much higher than those in coating N2.

(2) **Effects of the cooling rate and substrate temperature on the residual stresses in TBCs**

Preparation conditions have an important impact on residual stresses. Here, we discuss only the effects of the cooling rate R_c and substrate temperature T_s on the residual stresses in TBCs. The R_c in the preparation process is a key influencing factor, and the residual stresses increase with increasing R_c [1, 13]. Widjaja et al. [7] examined the effect of R_c on the residual stress distribution at the ceramic/BC interface in an APS TBC and simulated the R_c by varying the heat convection from the specimen

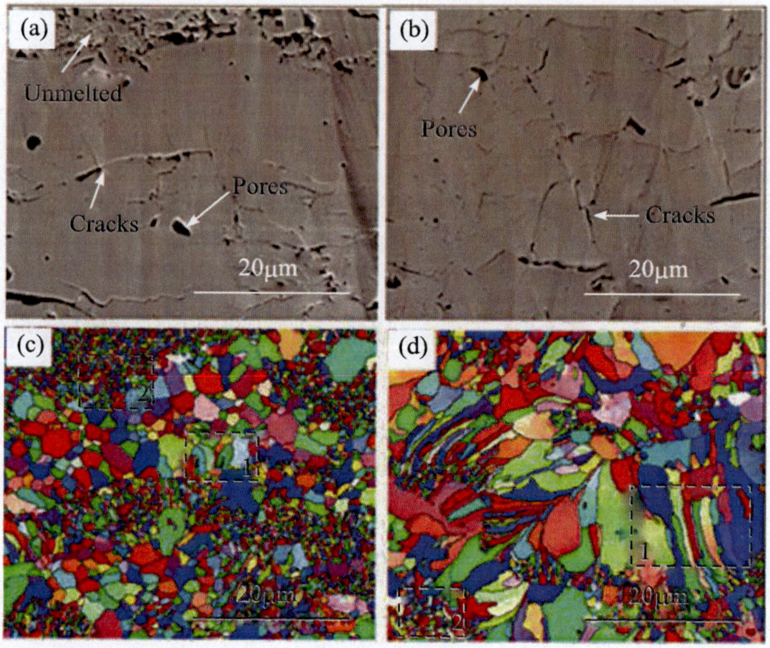

Fig. 10.1 **a**, **b** Cross-sectional SEM and **c**, **d** EBSD grain orientation maps of APS TBCs: **a**, **c** coating N1; **b**, **d** coating N2

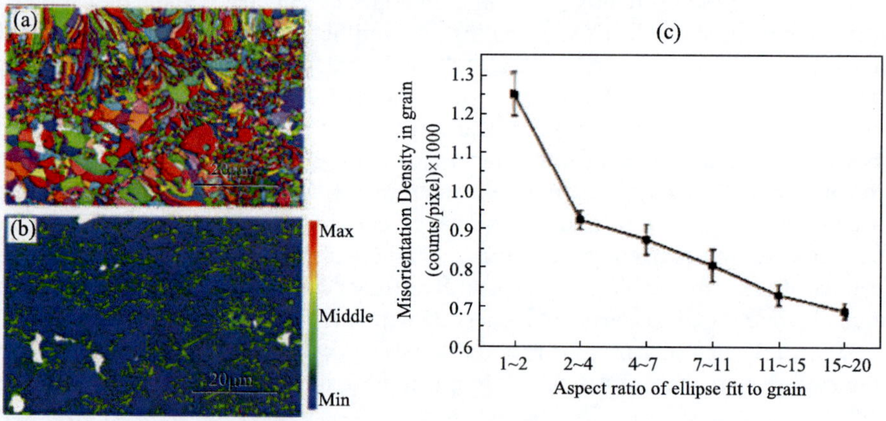

Fig. 10.2 **a** Orientation map of an APS YSZ TBC, **b** cross-sectional low-angle IMD distribution of the YSZ coating, and **c** mean IMDs of grains with different aspect ratios

Table 10.1 Spraying parameters and residual stresses of APS coatings

		N1 coating	N2 coating
Parameters	Current/A	560	680
	Ar/(L·min^{-1})	25	15
	H_2/(L·min^{-1})	12	14
	Spray distance/mm	120	120
	Content of equiaxed grains	80.5%	46.4%
	(Deviation)	(3.66%)	(4.46%)
Residual stresses (MPa)	As sprayed	43.75	34.98
	After 30 thermal cycles	– 48.1	– 25.4
CTE ($\times 10^{-6}$/°C)	Ceramic topcoat	9.37	10.64
	Ni-based superalloy	13.30	13.30

to the environment. Figure 10.3 shows the effect of R_c on the residual radial stress at the BC/TC interface. A high R_c leads to a high residual stress. At a high convection coefficient, a large amount of heat is dissipated into the air, resulting in a high R_c. Figure 10.4 shows the effect of the preheating temperature of the substrate, $T_{0,\text{substrate}}$, during the preparation process on the residual radial stress at the bond coat/top coat (BC/TC) interface in a TBC. The residual stress formed during the thermal spraying process is much lower when $T_{0,\text{substrate}} = 300$ °C than when $T_{0,\text{substrate}} = 250$ °C and further decreases as $T_{0,\text{substrate}}$ increases to 627 °C. Thus, increasing T_s during preheating can significantly reduce the quenching stress formed in a TBC during the preparation process [1].

Relevant research on the formation mechanism of equiaxed grains in YSZ coatings shows that a high R_c inhibits grain growth, thereby resulting in the formation

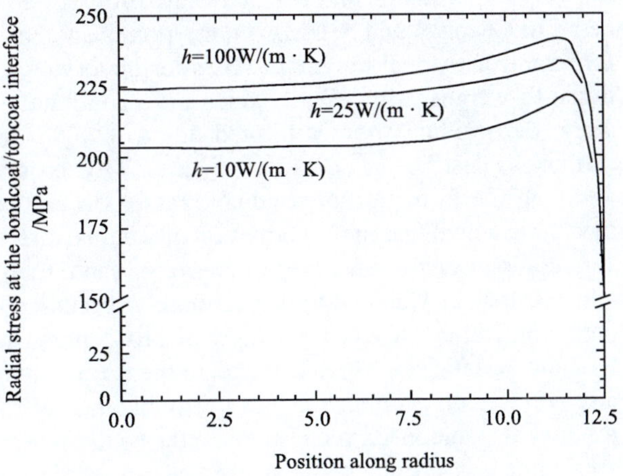

Fig. 10.3 Effect of the cooling rate on the residual radial stress at the BC/TC interface

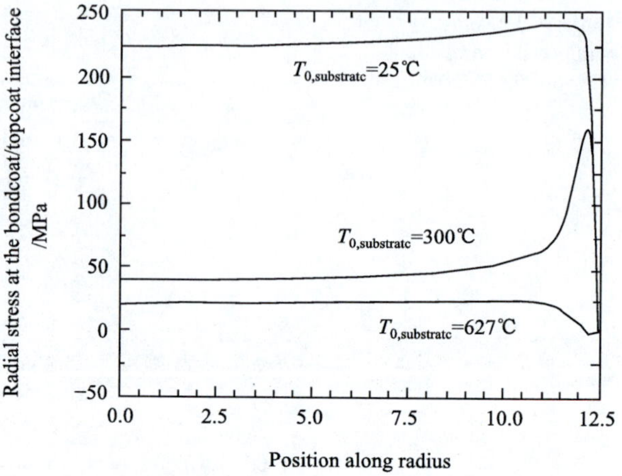

Fig. 10.4 Effect of $T_{0,\text{substrate}}$ on the residual radial stress at the BC/TC interface

of a large number of equiaxed grains [14]. A high R_c means a high convective heat transfer coefficient (CHTC), which allows more heat to be dissipated to the surrounding environment and thus less heat to be transferred to the neighboring grains through conduction. Therefore, there are small temperature changes in the zones near equiaxed grains, leading to a high-temperature gradient. The large residual strains in equiaxed grains are a result of the high-temperature gradient caused by a high R_c, which also explains why a high R_c results in high residual stresses [11].

10.2 Simulation and Prediction of the Residual Stresses in TBCs

Now, let us discuss the residual stresses formed during service. Interfacial oxidation is a very important factor causing residual stresses during service and is analyzed in detail in Chaps. 4 and 5. CMAS corrosion is another key influencing factor for the formation of residual stresses in TBCs during service. In particular, with an increasing thrust-to-weight ratio in aeroengines and a continual rise in the turbine inlet temperature, the residual stresses formed due to CMAS corrosion become increasingly serious, as described in detail in Chap. 6. TBC erosion is another important factor affecting the formation of residual stresses. In particular, there is a very complex mechanism involved in the formation of residual stresses in TBCs due to erosion by very complex environmental particles (e.g., sand, fuel, and haze particles) across the vast territory of China, a topic thoroughly discussed in Chap. 7. During service at high temperatures, the ceramic layer of TBC sinter, causes changes in both the grain size and orientation, which can lead to the formation of large residual stresses. This area of research is not covered in this book and has rarely been studied by Chinese researchers. Interested readers can refer to the relevant literature [15, 16]. In this section, to analyze the residual stresses formed during service, we select an actual turbine blade with complex geometry as an object of analysis. Evidently, it is impossible to obtain an analytical solution for a complex geometry, and hence numerical simulation is needed.

10.2.1 Stress Field Evolution and Danger Zone Prediction of a Turbine Blade with a TBC

A turbine blade with a TBC has a very complex geometry, and each of its layers has physically nonlinear constitutive relations. To qualitatively understand the stress field in a turbine blade with a TBC, we simplify the problem in the numerical finite element (FE) simulation process by making the following assumptions: (a) there is only one cooling channel; (b) the tenon has a rectangular shape; (c) each layer of the coating has a uniform thickness; (d) convection and radiation are not considered; (e) phase transformation is excluded; (f) the interfacial failure of the coating is not taken into account; and (g) the substrate, the transition layer, and the TGOs are ideal elastoplastic materials, and the ceramic coating is an elastic material.

A turbine blade with a TBC was modeled using a combination of CATIA and ABAQUS software. Specifically, the ceramic layer, the TGO layer, the BC layer, the substrate, and the tenon were first modeled separately in CATIA and then combined in ABAQUS to generate an overall model of a turbine blade with a TBC [17]. Each layer of the coating uniformly covers the surface of the blade (including the blade body and the tenon surface). The ceramic layer, the BC layer, the blade substrate, and the tenon are 0.3, 0.1, approximately 1.0, and 8 mm in thickness, respectively. The blade body is 37 mm in length. The TGO layer is generally in an elastic state during thermal cycling. Thus, the influence of the thickness of the TGO layer on the stress distribution in the ceramic and BC layers is neglected. Hence, to reduce the workload in FE analysis, we set the thickness of the TGO layer to 20 μm. The thickness of the TGO layer and nonhomogeneous TGO growth have a tremendous but local effect on the stress field. Thus, while the selected thickness of the TGO layer differs significantly from the actual thickness, it has little influence on the stress distribution in the coating. Figure 10.5 shows the geometrical model of the turbine blade with TBC.

We set the boundary conditions of the model of the turbine blade with a TBC as follows. The axial displacement of the lower surface of the base is constrained, i.e., the displacement component of all the points on this surface in the z-direction, w, is set to zero. To prevent the model from translating or rotating as a whole during the computational process, the midpoint of the lower surface of the base is also fixed.

The first kind of temperature boundary condition is selected as the temperature boundary of the model. The convection and radiation effects between the system and the environment are neglected. The temperature boundary is implemented as shown in the thermal cycling profiles in Fig. 10.6. Considering the initial residual stress during the preparation process, the thermal cycling process of the system

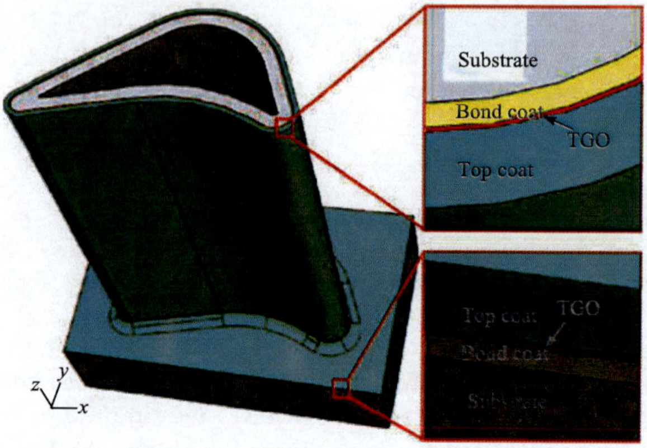

Fig. 10.5 An FE model of a turbine blade with a TBC

involves two steps. The whole system is first cooled as a whole from the preparation temperature (set to 600 °C) to room temperature (25 °C) and then subjected to thermal cycling. We assume that the system is in a stress-free state at the completion of the preparation. The residual stress formed due to the mismatch between the CTEs of the layers within the system after cooling to room temperature is referred to as the initial residual stress and plays an important role in the stress distribution and evolution of the system. The specific thermal cycling procedure is as follows. The outer surface of the ceramic layer is increased from room temperature to the maximum temperature (1200 °C) within 2 min, this temperature is held for 5 min, and then the surface is cooled to room temperature within 5 min. Similarly, the inner surface of the cooling channel is increased from room temperature to the maximum temperature (700 °C) within 2 min, this temperature is held for 5 min, followed by cooling to room temperature within 5 min. Table 10.2 summarizes the material parameters used in the simulation. Figure 10.7 shows the mesh generated for the system. Most zones of the blade body and the base are simulated using hexahedral element C3D8RT, and the zone connecting the blade body and the base is simulated using tetrahedral element C3D4T. A total of approximately 210,000 elements are used.

In the FE simulation, we ignore the interfacial damage and cracks between the coating layers and between the coating and the substrate; that is, we consider the bond at each interface to be intact. Considering that the ceramic layer is a brittle material, we use the first strength theory to examine the stress evolution in the ceramic layer and to predict danger zones in the coating based on the maximum principal stress distribution.

To facilitate the analysis, we separate the ceramic layer from the model of the system of the turbine blade with a TBC. Figure 10.8 shows the contour plot of the maximum principal stress distribution in the ceramic layer at the end of 50 thermal cycles. The stress concentration zones are marked with ovals and denoted A–H. Specifically, Zones A and B are distributed on the pressure side, Zone C is located at the connection (chamfer) between the trailing edge and the base, Zone D is located

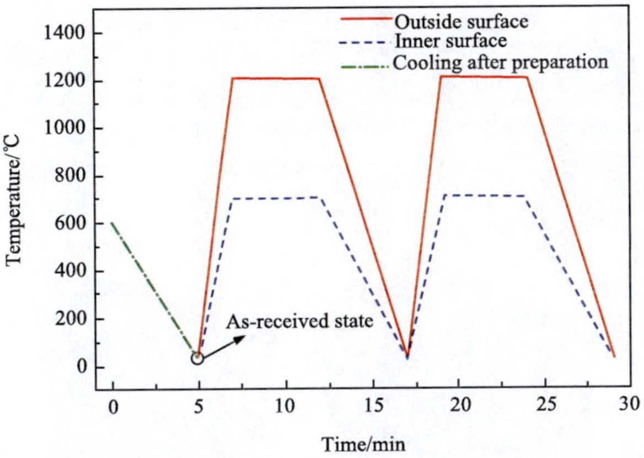

Fig. 10.6 Thermal cycling profiles for the turbine blade with a TBC

Table 10.2 Material parameters of the different layers of the TBC [18–21]

	Substrate	BC	TGO	TC
Temperature range (K)	300 K–1500 K	300 K–1500 K	300 K–1500 K	300 K–1500 K
Young's modulus (GPa)	220–120	200–110	400–320	48–22
Poisson's ratio	0.31–0.35	0.30–0.33	0.23–0.25	0.10–0.12
Thermal expansion coefficient (10^{-6}/K)	14.8–18.0	13.6–17.6	8.0–9.6	9.0–12.2
Thermal conductivity (W/(m K))	88–69	5.8–17.0	10–4	2.0–1.7
Density (kg/m3)	8500	7380	3984	3610
Specific heat (J/(kg K))	440	450	755	505
Yield strength (MPa)	800	426–114	10,000 –1000	–

at the connection (chamfer) between the leading edge and the base, Zone E is located at the connection (chamfer) between the pressure side and the base, Zones F and G are distributed on the suction side, and Zone H is located at the connection (chamfer) between the suction side and the base. The stresses are relatively concentrated in the abovementioned zones, which are referred to as "danger zones" since failure within the ceramic layer first occurs in these zones. The predicted results are basically in line with the results reported in the literature [22], in which the failure zones are distributed at the connection between the pressure side and the base and gradually expand to the surrounding area with the connection as the center. Figure 10.8 shows the evolution of the residual stress in each danger zone with the number of thermal cycles. At the beginning of thermal cycling, the residual stress in the ceramic layer

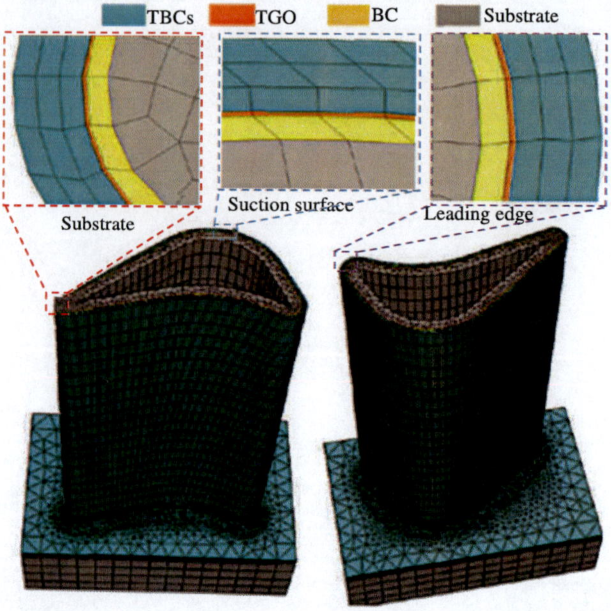

Fig. 10.7 Meshing of the system

is low. As the number of thermal cycles increases, the residual stress in each zone gradually increases mainly due to the accumulation of plastic deformation in the system. However, there is a decrease in the stress increase rate after approximately 30 thermal cycles. This is because stress relaxation becomes increasingly prominent with increasing residual stress in the system, while during the first few thermal cycles, the stress in the system is low, and the creep plays a minor role.

10.2.2 Analysis of the Stress Field in the Turbine Blade with a TBC Using the Fluid–Solid Coupling Method

(1) **Mathematical model for fluid–solid coupling**

The governing equations for the fluid domain mainly include the continuity equation, the Reynolds-averaged Navier–Stokes equation, and the energy equation, which are described in a Cartesian coordinate system as follows.

To ensure a constant mass flow, a fluid should satisfy the following mass conservation equation [24]:

$$\frac{\partial \rho_f}{\partial t} + div(\rho_f \boldsymbol{v}) = 0 \tag{10.4}$$

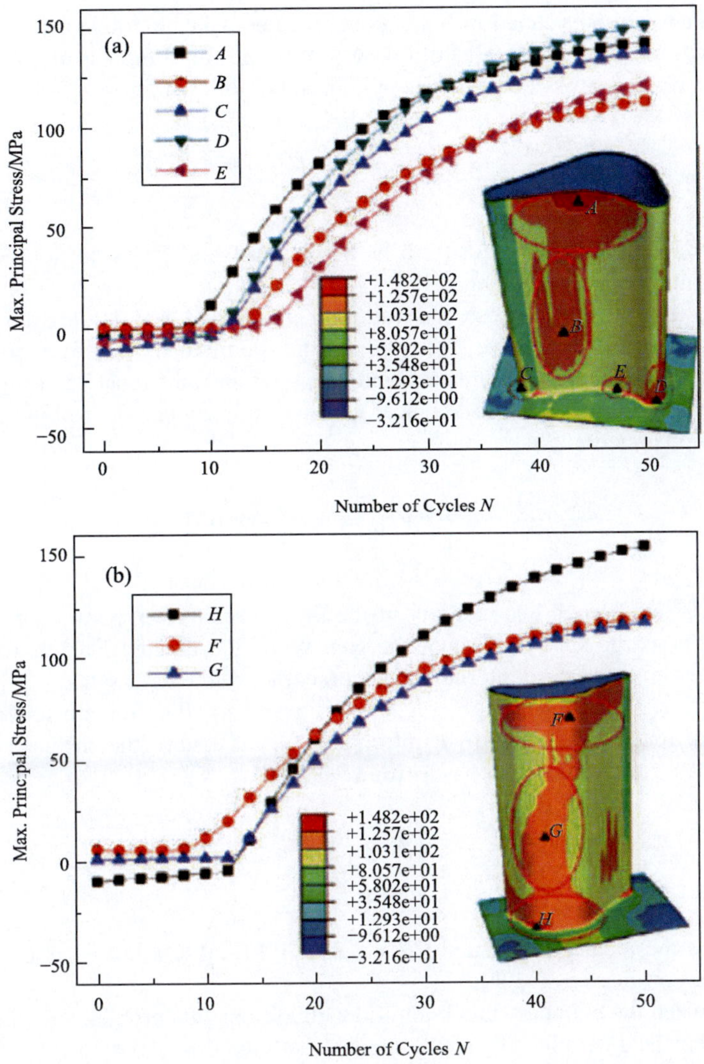

Fig. 10.8 Distribution of the maximum principal stress on the **a** front side and **b** back side of the ceramic layer [21]

where ρ_f is the fluid density, which varies both temporally and spatially for a compressible fluid, and $\boldsymbol{v}$ is the velocity vector.

The Navier–Stokes equation is the most fundamental equation in fluid mechanics and is used to solve nearly all fluid flow problems. For compressible, nonviscous fluids, the Navier–Stokes equation is written in the following form [24]:

$$\frac{D(\rho_f v)}{Dt} = -\text{grad}p + \text{div}(\mu_f \,\text{grad}\, v) + S_M \tag{10.5}$$

where μ_f is the dynamic viscosity, p is the pressure, and S_M is the source term of the momentum conservation equation.

During the coupling process, the wall temperature and heat flux are the coupling variables that transfer between the solid and fluid domains. Therefore, coupling is a process involving energy transport, and an energy balance must be ensured in the process of solving the problem. Hence, we need to introduce the following energy conservation equation [24];

$$\frac{\partial(\rho_f T_f)}{\partial t} + \text{div}(\rho_f v T_f) = \text{div}(\frac{k_f}{c_p}\text{grad}\, T_f) + S_\phi \tag{10.6}$$

where k_f is the thermal conductivity of the fluid, T_f is the temperature of the flow field, and c_p is the specific heat of the gas. When the chemical reactions are not considered and there is no internal heat source, the source term $s_\phi = 0$.

When Eqs. (10.4)–(10.6) are solved simultaneously, they remain unclosed. An equation of state for gas needs to be introduced. Let us assume that the gas is an ideal gas, the state equation of which is as follows [24]:

$$\begin{cases} p = R\rho_f T_f \\ e_f = c_p T_f \end{cases} \tag{10.7}$$

where R is the ideal gas constant ($R \approx 8.31441$ J/(mol·K)) and e_f is the internal energy per unit mass of the fluid.

Thus, when the boundary and inlet/outlet conditions of a problem are known, the fluid part can be determined by simultaneously solving Eqs. (10.4)–(10.7). Then, the basic physical quantities (i.e., velocity, pressure, and temperature) within the fluid domain are obtained. Here, the ideal k–ε model is selected as the turbulence model.

In a turbine blade system with a TBC, there is heat transfer between the layers of the TBC as well as between the TBC and the substrate. In addition, these components have the same constitutive relations in problems involving thermal stress. In the whole solid domain, their governing equations mainly include heat transfer equations and thermoelastoplastic constitutive equations.

The heat transfer between the interior of the TBC and the substrate follows Fourier's law of heat transfer, i.e.

$$\rho_s c \frac{\partial T}{\partial t} = \nabla \cdot (k_s \nabla T) \tag{10.8}$$

where c is the specific heat of the solid, k_s is the corresponding thermal conductivity and ρ_s is the density of the solid.

In the analysis of the failure mechanism of TBCs, stress is an important criterion for determining failure. At high temperatures, the thermal stress in a turbine blade with a TBC may cause local plastic strains in the coating but generally does not lead to large-scale plastic dissipation. Therefore, the thermal stress is calculated using the decoupling method while the influence of plastic dissipation on the temperature is ignored. Through the realization of the conjugate heat transfer between the fluid and solid domains using the fluid–solid coupling method, the distribution of the temperature field in the turbine blade is first calculated, based on which the stress field in the turbine blade is then calculated. To numerically calculate the thermal stress, it is assumed that each layer has a uniform thickness and is a continuous, isotropic material and that the interlayer connection is intact.

The stress field is calculated using the decoupling method. The stress field satisfies the following force equilibrium equation:

$$\nabla \cdot \boldsymbol{\sigma} + \boldsymbol{F} = \mathbf{0} \tag{10.9}$$

When the volume force is not considered, the equilibrium equation can be written as follows:

$$\nabla \cdot \boldsymbol{\sigma} = \mathbf{0} \tag{10.10}$$

The material is assumed to be an isotropic ideal elastoplastic material, and the rate form of its thermo–elastoplastic constitutive relations is used in the calculation. Then, the strain rate can be expressed as follows:

$$\dot{\boldsymbol{\varepsilon}} = \dot{\boldsymbol{\varepsilon}}^{\mathrm{e}} + \dot{\boldsymbol{\varepsilon}}^{p} + \dot{\boldsymbol{\varepsilon}}^{therm} \tag{10.11}$$

where $\dot{\boldsymbol{\varepsilon}}^{\mathrm{e}}$ is the elastic strain rate, with the following expression:

$$\dot{\boldsymbol{\varepsilon}}^{\mathrm{e}} = \frac{1}{E}[(1+\nu)\bar{\mathbf{I}} - \nu\mathbf{I} \otimes \mathbf{I}] : \dot{\boldsymbol{\sigma}} \tag{10.12}$$

and $\dot{\boldsymbol{\varepsilon}}^{therm}$ is the strain rate due to temperature change and is expressed as follows:

$$\dot{\boldsymbol{\varepsilon}}^{therm} = \alpha \Delta \dot{T} \mathbf{I} \tag{10.13}$$

In Eqs. (10.12) and (10.13), E is Young's modulus, ν is Poisson's ratio, $\bar{\mathbf{I}}$ is the fourth-order symmetric unit tensor, $\mathbf{I}$ is the second-order symmetric unit tensor, α is the CTE, and $\Delta \dot{T}$ is the rate of temperature change. The plastic strain rate $\dot{\boldsymbol{\varepsilon}}^{p}$ is expressed as follows:

$$\dot{\boldsymbol{\varepsilon}}^{p} = \gamma \frac{\partial f}{\partial \boldsymbol{\sigma}} \tag{10.14}$$

where γ is the plastic flow factor and f is the von Mises yield function, which can be expressed as follows:

$$f = \sqrt{\boldsymbol{\sigma} : \boldsymbol{\sigma} - \frac{1}{3}(\mathrm{tr}(\boldsymbol{\sigma}))^2} - \sqrt{\frac{2}{3}} Y = 0 \tag{10.15}$$

where Y is the yield strength. As the material is assumed to be an ideal elastoplastic material, Y is a constant. Then, according to the flow rule, $\dot{\boldsymbol{\varepsilon}}^{p}$ can be obtained from Eqs. (10.14) and (10.15):

$$\dot{\boldsymbol{\varepsilon}}^{\mathrm{p}} = \gamma \frac{\partial f}{\partial \boldsymbol{\sigma}} = \gamma \frac{\boldsymbol{\sigma} - \frac{1}{3}\mathrm{tr}(\boldsymbol{\sigma})\mathbf{I}}{\sqrt{\boldsymbol{\sigma} : \boldsymbol{\sigma} - \frac{1}{3}(\mathrm{tr}\boldsymbol{\sigma})^2}} = \gamma \frac{\boldsymbol{S}}{|\boldsymbol{\sigma}'|} \tag{10.16}$$

where $\boldsymbol{S} = \boldsymbol{\sigma} - 1/3(\mathrm{tr}\boldsymbol{\sigma})\mathbf{I}$ is the deviatoric stress tensor. In addition, tensor $\boldsymbol{n}$ is defined as follows:

$$\frac{\mathbf{S}}{|\mathbf{S}|} = \boldsymbol{n} \tag{10.17}$$

To calculate γ, according to the consistency condition, we have

$$\dot{f} = \frac{\partial f}{\partial \boldsymbol{\sigma}} : \dot{\boldsymbol{\sigma}} = 0 \tag{10.18}$$

Here, the condition that Y is a constant is used. The above equation is combined with the following equation:

$$\dot{\boldsymbol{\sigma}} = \mathbf{C} : (\dot{\boldsymbol{\varepsilon}} - \dot{\boldsymbol{\varepsilon}}^{therm} - \dot{\boldsymbol{\varepsilon}}^{p}) \tag{10.19}$$

Here,

$$\mathbf{C} = \lambda \mathbf{I} \otimes \mathbf{I} + 2\mu \bar{\mathbf{I}} \tag{10.20}$$

where λ is a Lame constant and μ is the shear modulus. Then, the following equation can be obtained:

$$0 = \frac{\partial f}{\partial \boldsymbol{\sigma}} : \mathbf{C} : (\dot{\boldsymbol{\varepsilon}} - \dot{\boldsymbol{\varepsilon}}^{therm} - \gamma \boldsymbol{n}) \tag{10.21}$$

Based on the above equation, the following expression of γ is obtained [21]:

$$\gamma = \frac{\frac{\partial f}{\partial \sigma} : \boldsymbol{C} : (\dot{\boldsymbol{\varepsilon}} - \dot{\boldsymbol{\varepsilon}}^{therm})}{\frac{\partial f}{\partial \sigma} : \boldsymbol{C} : \boldsymbol{n}} \tag{10.22}$$

Through a combination of the above equation and the following expression of the strain rate:

$$\dot{\boldsymbol{\sigma}} = \mathbf{C} : (\dot{\boldsymbol{\varepsilon}} - \dot{\boldsymbol{\varepsilon}}^{therm} - \dot{\boldsymbol{\varepsilon}}^{p}) = \boldsymbol{L} : (\dot{\boldsymbol{\varepsilon}} - \dot{\boldsymbol{\varepsilon}}^{therm}) \tag{10.23}$$

we can obtain the expression of the ideal elastoplastic stiffness tensor $\boldsymbol{L}$ as follows:

$$\boldsymbol{L} = \boldsymbol{C} - \frac{\boldsymbol{C} : \boldsymbol{n} \otimes \frac{\partial f}{\partial \sigma} : \boldsymbol{C}}{\frac{\partial f}{\partial \sigma} : \boldsymbol{C} : \boldsymbol{n}} \tag{10.24}$$

Simplification of Eq. (10.24) yields

$$\boldsymbol{L} = \lambda \mathbf{I} \otimes \mathbf{I} + 2\mu(\bar{\mathbf{I}} - \beta \boldsymbol{n} \otimes \boldsymbol{n}) \tag{10.25}$$

Finally, the following thermo–ideal elastoplastic constitutive relation is obtained [21]:

$$\dot{\boldsymbol{\sigma}} = \boldsymbol{L} : (\dot{\boldsymbol{\varepsilon}} - \dot{\boldsymbol{\varepsilon}}^{therm}) = \mathbf{L} : [\dot{\boldsymbol{\varepsilon}} - \alpha \Delta \dot{T} \mathbf{I}] \tag{10.26}$$

$$\boldsymbol{L} = \lambda \mathbf{I} \otimes \mathbf{I} + 2\mu(\mathbf{I} - \beta \boldsymbol{n} \otimes \boldsymbol{n}) \qquad \beta = \begin{cases} 1 & \text{When } \boldsymbol{\sigma} \text{ is on the yield surface} \\ 0 & \text{When } \boldsymbol{\sigma} \text{ is in the yield surface} \end{cases} \tag{10.27}$$

where $\lambda = \frac{\nu E}{(1+\nu)(1-2\nu)}$ $\mu = \frac{E}{2(1+\nu)}$. When the plane strain assumption is used, the ideal thermoelastoplastic constitutive relation is expressed as follows:

$$\dot{\boldsymbol{\sigma}} = \overline{\boldsymbol{L}} : (\dot{\boldsymbol{\varepsilon}} - \dot{\boldsymbol{\varepsilon}}^{therm}) = \overline{\mathbf{L}} : [\dot{\boldsymbol{\varepsilon}} - \overline{\alpha} \Delta \dot{T} \mathbf{I}] \tag{10.28}$$

$$\boldsymbol{L} = \bar{\lambda} \mathbf{I} \otimes \mathbf{I} + 2\mu(\bar{\mathbf{I}} - \beta \boldsymbol{n} \otimes \boldsymbol{n}) \qquad \beta = \begin{cases} 1 & \text{When } \boldsymbol{\sigma} \text{ is on the yield surface} \\ 0 & \text{When } \boldsymbol{\sigma} \text{ is in the yield surface} \end{cases} \tag{10.29}$$

where $\bar{\lambda} = \frac{(1-\nu)\nu E}{(1+\nu)(1-3\nu)}$ $\bar{\mu} = \frac{E}{2(1+\nu)}$ $\bar{\alpha} = (1+\nu)\alpha$.

When $\boldsymbol{\sigma}$ is on the yield surface Conjugated heat transfer analysis [25] is applicable to thermal structures containing multiple modes of heat transfer and is suitable for investigating the heat transfer process of a turbine blade that requires the simultaneous consideration of internal and external temperature fields. Conjugated heat transfer is neglected in many analytical and numerical models. The most common approach is to separately analyze and compute the fluid and heat transfer and to provide a CHTC in advance to calculate the wall temperature instead of obtaining the CHTC based on the actual gas flow rate, temperature, and pressure. Conjugated heat transfer analysis does not require that a specific CHTC be given in advance. Generally, there are two methods for solving multifield coupling problems. (1) Direct coupling. This method establishes and solves a whole set of governing equations for the fluid and solid domains to realize full fluid–solid coupling. The direct coupling method is mainly used to solve physical processes in which the time scale significantly affects the physical variables. However, this method has a very low computational efficiency in practice due to the mismatch between the coefficient matrices of many computational approaches. (2) Weak coupling. This method establishes separate governing equations for the fluid and solid domains, respectively, and uses different computational approaches to solve them. In addition, this method meets the temperature and heat flow continuity conditions only at the coupling boundary, which drastically improves the computational efficiency. Thus, we have the following continuity conditions [25]:

$$T_{\mathrm{f}} = T_S \tag{10.30}$$

$$-k_{\mathrm{s}}\frac{\partial T_S}{\partial \boldsymbol{n}} = k_{\mathrm{f}}\frac{\partial T_{\mathrm{f}}}{\partial \boldsymbol{n}} \tag{10.31}$$

where k_s and k_f are the heat transfer coefficients of the solid and fluid, respectively, and $\boldsymbol{n}$ is the unit vector in the outer normal direction of the wall.

Based on the above governing equations and interfacial coupling conditions, the temperature field of a turbine blade can be calculated to determine the temperature distribution in different zones of the turbine blade at different stages. Then, with the temperature field as the temperature boundary, the constitutive equations for the solid domain can be solved to obtain the thermal stress distribution.

(2) **FE model for fluid–solid coupling**

For a complex system that includes a flow field and a turbine blade with a TBC that has a multilayer structure, theoretical analysis methods are no longer applicable, while experimental methods require multiple repetitions and are very costly. Hence, we obtain the temperature and stress fields in a turbine blade using numerical methods. A numerical calculation and analysis scheme for fluid–solid coupling heat transfer is established based on the aforementioned mathematical model for fluid–solid coupling, as shown in Fig. 10.9. The fluid domain is simulated using FLUENT, with computational fluid dynamics (CFD) as the theoretical basis. The solid domain is realized using the large-scale commercial software ABAQUS. The coupling between the two domains is achieved through mutual transfer between two variables, the wall temperature and the CHTC, in the third-party software MpCCI.

A turbine blade with a TBC has a very complex structure due to both the complex shape of the turbine blade and, more importantly, the multilayer structure of the TBC, which leads to a problem similar to cross-scale problems, posing a series of difficulties for FE calculation. Specifically, FE modeling is extremely difficult, and so is mesh generation. The TGO layer in the TBC is much thinner than the turbine

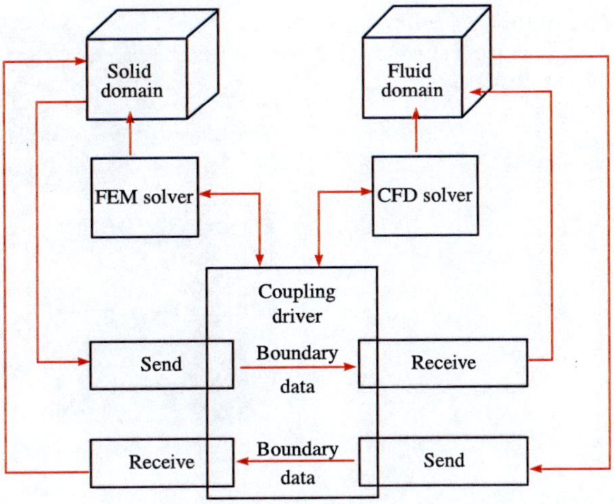

Fig. 10.9 A numerical analysis scheme for fluid–solid coupling

blade but is also the most important layer. To ensure computational accuracy, it is necessary to ensure a certain aspect ratio for the mesh of the TGO layer. This leads to an extremely large number of mesh cells, which, in turn, results in a long computational cycle and high requirements for the computing hardware.

A 3D model of a turbine blade with TBC consisting of a substrate, a BC layer, a TGO layer, and a TC layer is established in this chapter. To facilitate the investigation, the base is not considered, and the following assumptions are made: (1) internal cooling is achieved through a single cooling channel; (2) the TBC has a uniform thickness and is an isotropic material; and (3) the interlayer interfaces are intact. Figure 10.10 shows the established geometric model. The BC layer is 100 μm thick. The TGO layer is generally 5 μm thick in the initial stage of service and thickens with time to a maximum of approximately 10 μm. Here, we set the thickness of the TGO layer to 10 μm, the thickness of the TC layer to 250 μm, and the height of the whole blade to 60 mm. Figure 10.10 shows a mesh generated with hexahedral cells for the whole 3D model of the turbine blade with a TBC. During the preprocessing (mesh generation) process in ABAQUS, meshes are generated separately for the substrate (SUB), the transition layer (BC), the TGO layer (TGO), and the ceramic coating (TC). The stress distribution in the TGO layer is an important factor causing TBC failure. To better extract the stress information of the TGO layer, regular hexahedral cells are used for the whole TGO layer, and a total of six layers of cells are set along its thickness. In addition, to ensure that the aspect ratio of the cells in the TGO layer is distributed within a reasonable range, the mesh for the TGO layer is refined in the radial direction to produce a density 1.5 times that of the radial cells in the ceramic layer and twice that in the substrate. The mesh generated for the whole turbine blade and TBC system contains a total of 7,527,000 cells, of which 5,188,150 (68.92%) are in the TBC part (the BC, TGO, and TC layers). Moreover, eight-node linear heat transfer brick (DC3D8) and eight-node reduced integration (C3D8R) elements are used for the fluid–solid coupling and thermal stress analysis stages, respectively.

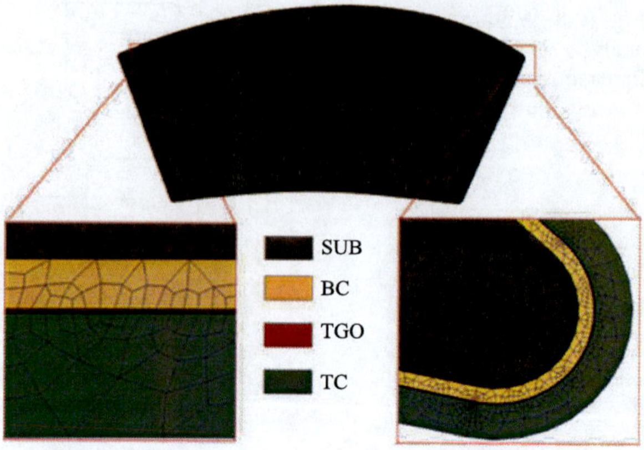

Fig. 10.10 Geometric model and mesh of a turbine blade with a TBC

The TBC material exhibits nonlinear behavior at high temperatures. Except for the ceramic layer, which is a pure elastic material, the plasticity of all the other layers is considered. Particular attention is paid to the phenomenon of stress-induced yielding of the TGO layer at high temperatures. The yield strength of the TGO layer is very high at room temperature but decreases significantly at high temperatures. The TGO layer yields under thermal stress. Tables 10.2 and 10.2 summarize the needed material parameters and the yield strengths, respectively [26–28] (Table 10.3).

In FE analysis, the determination of the load and boundary conditions plays a decisive role in the accuracy of the calculation results. In this chapter, we need to determine the thermal load and force boundary conditions. Analysis shows that the thermal boundary conditions for the TBC and the turbine blade consist of three parts, namely, the convective heat transfer between the high-temperature gas in the external flow field and the turbine blade, the heat transfer inside the blade, and the convective heat transfer between the cooling gas in the internal cooling channel and the turbine blade. To facilitate analysis, we make the following simplifications to the cooling channel inside the turbine blade when selecting the thermal load boundary conditions: the fluid behavior inside the cooling channel is not considered; the CHTC of the wall,

Table 10.3 Material yield strengths

Temperature (°C)	Substrate (MPa)	Transition layer (MPa)	TGO layer (MPa)
20	800	426	10,000
200	800	412	10,000
400	800	396	10,000
600	800	362	10,000
800	800	284	10,000
1000	800	202	1000
1100	800	114	1000

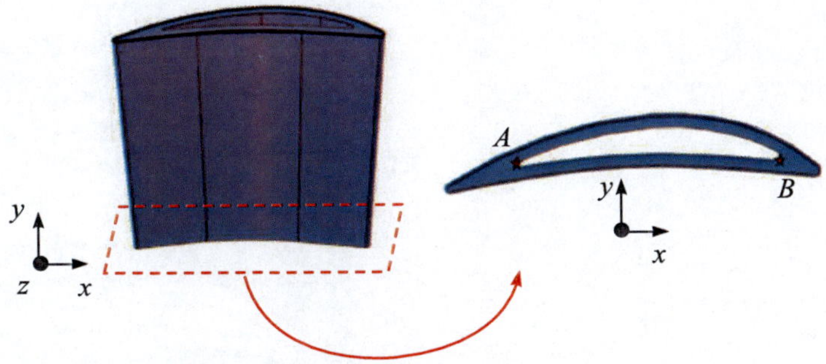

Fig. 10.11 Boundary conditions for the turbine blade with a TBC

h_f, is fixed at 300 W/m^2·K; and the temperature of the cooling gas is 873 K. In the coupling process, the convective heat transfer between the high-temperature gas and the blade provides a temperature boundary and a CHTC for the blade.

In the thermal stress analysis, we need to ensure that the blade is able to expand freely and does not undergo rigid-body displacement and rotation under loading. Thus, the following boundary conditions, as shown in Fig. 10.11, are selected: (1) point A is fixed in the x- and y-directions, i.e., $u_A = v_A = 0$; (2) point B is fixed in the y-direction, i.e., $v_B = 0$; and (3) the bottom surface of the blade does not undergo displacement in the z-direction, i.e., $w(x, y, 0) = 0$. This way, the boundary conditions selected based on points A and B ensure that the blade expands freely but does not rotate. Boundary condition (3) ensures that the blade does not undergo rigid-body displacement.

The blades are arranged in a periodic array on a turbine disk. High-temperature gas enters through the inlet and is discharged through the outlet under turbocharging. This provides a basis for simplifying the external flow-field model. There is no longer a need to establish the internal flow field in the whole engine, and instead, the flow field within a period can be selected based on the periodicity condition, thereby significantly reducing the modeling difficulty and computational load. Figure 10.12 shows the FE model of the external flow field, which contains an inlet, an outlet, a periodic boundary, and a heat transfer wall. The specific modeling procedure is as follows. Step 1: Establish a 3D external flow-field model that contains the blade based on the established model of the turbine blade with TBC and the actual external flow field in ABAQUS, and subsequently, elongate it to create a 3D external flow field with the same height as that of the blade. Step 2: Remove the turbine blade part and retain the flow-field part through Boolean subtraction of the whole model. Step 3: Export the 3D flow-field part from ABAQUS, and name it 3D-flow-file.STP for use as the mesh generation file in the next step. When the periodic boundary is determined, it is necessary to ensure that the throat is constricted and has consistent geometric features at the corresponding locations along the boundary.

The temperature and velocity gradients of the fluid are very large near the blade wall. To more accurately simulate the boundary-layer effect, we use the structured mesh generation technique to improve the mesh quality for the boundary layer. The

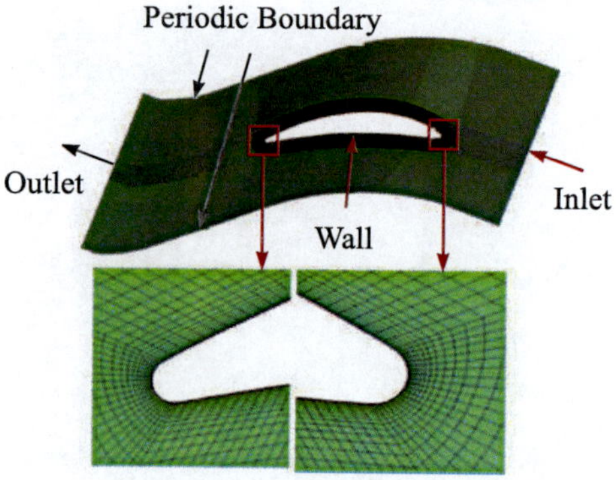

Fig. 10.12 A 3D model and mesh of the external flow field surrounding the turbine blade

commercial software ICEM CFD has good preprocessing capabilities, with structured mesh generation as an outstanding feature. Thus, in this chapter, ICEM CFD is used to generate a structured mesh for the fluid domain through the following steps: (1) Importation of the geometry exported from ABAQUS: Establish the geometric features, including an inlet, an outlet, a periodic boundary, and a wall; (2) Geometry partitioning: Generate a regular block (rectangular in this chapter) structure that contains the whole flow field based on the imported geometry model, correlate the geometric features (e.g., points and lines) of the block to the corresponding locations in the geometry, divide the block into regularly partitioned structures based on the geometric features, and remove the subblocks that do not contain the geometry; (3) Block mapping: Create one-to-one mapping relations between the geometry and the corresponding subblocks and connect them; and (4) Mesh generation: Set an overall mesh density for the block structure, refine the mesh for the boundary layer, set the thickness of the first layer of the mesh to 0.01 mm and set 20 layers based on a ratio of 1.05, generate a mesh for the block, and directly export the mesh for the geometry. Figure 10.12 shows the 3D mesh generated using the structured mesh generation technique for the external flow field surrounding the turbine blade. The mesh far from the wall is relatively sparse. A type O refined mesh layer, as mentioned earlier, is generated for the near-wall zone to obtain more accurate information (e.g., the heat transfer coefficient and temperature) about the fluid.

In the established model, high-temperature gas enters through the inlet, undergoes convective heat transfer with the wall and the blade, and is discharged through the outlet after being compressed by the flow field. Generally, the velocity at the inlet and the total pressure are known conditions. In this chapter, we provide the boundary conditions for the flow field, as shown in Table 10.4, and make the following assumptions for the fluid: (1) the external flow field is a single flow field and is isolated from the cooling gas inside the blade; (2) under stable working conditions, the blade operates stably, and both the pressure and velocity of the high-temperature gas remain stable; and (3) in the calculation, the high-temperature gas satisfies the equation of state of an ideal gas.

Table 10.4 Boundary conditions for the external flow field

Total pressure of the incoming flow (atm)	Static pressure of the incoming flow (atm)	Total temperature of the incoming flow (K)	Turbulence intensity (%)
14.22	14	1350	3

(3) **Stress field in the turbine blade with a TBC**

Thermal stresses are a primary factor causing TBC failure. In particular, during stages (i.e., heating and cooling stages) with a large temperature gradient, there is a significant temperature change within a short period of time, resulting in large thermal mismatch stress and microcrack propagation. As early as 2001, Zhou and Hashida predicted that thermal stress-induced crack initiation and propagation in a TBC occur during the cooling stage [18]. Generally, it is difficult to understand the phenomenon in which a coating does not spall during heating (i.e., loading) but does during cooling (i.e., unloading). Using acoustic emission (AE), a nondestructive testing technique, Yang et al. [29, 30] found that TBC failure indeed occurs during the cooling stage. This phenomenon is mechanistically explained as follows. During the cooling stage, as a TBC cools from a very high temperature to room temperature within a short period of time, the CTEs of the substrate, the BC layer, and the ceramic layer change inconsistently, resulting in the formation of a large compressive thermal mismatch stress between the BC and ceramic layers, which, in turn, causes the coating to buckle and cracks to form at the interface. Further propagation of the interface cracks causes the coating to spall. In the study of crack propagation problems, the maximum principal stress is often used as the critical value for crack propagation. In the discussion on the thermal stresses in a TBC, to facilitate the prediction of danger zones for TBC spallation failure, the maximum principal stresses in the BC, TGO, and TC layers of the TBC are mainly analyzed.

Figure 10.13 shows the temperature field in the turbine blade with a TBC. Specifically, Fig. 10.13a and b show the temperature distributions in the TBC and the substrate of the turbine blade, respectively. The maximum temperature of the ceramic layer is 1296 K, located at the leading edge of the surface, but that of the substrate of the turbine blade decreases to 1224 K, located at the leading edge. In other words, the 350 μm thick TBC reduces the surface temperature at the leading edge of the turbine blade by 72 K. At other locations, the surface temperature of the TBC is significantly higher than the temperature of the substrate. Evidently, the TBC satisfactorily reduces the surface temperature of the alloy substrate. Figure 10.14 shows a contour plot of the maximum principal stress distribution in the BC layer. The maximum principal stress is negative (i.e., compressive) at the leading and trailing edges, with a maximum value of –88 MPa. In comparison, the maximum principal stress is tensile on both the pressure side and suction side, with a maximum value of 118 MPa. At temperatures above 1000 °C, the yield strength of the BC layer is only 110 MPa. The temperature field distribution (Fig. 10.13) shows that during the initial stage of cooling, there is a small plastic strain in the high-temperature zone of the leading edge. The presence of a plastic strain partially releases the stress, and hence, the maximum stress does not occur at the leading edge despite a high temperature in this zone. The maximum temperatures occur near both sides of the leading and trailing edges, and the temperature gradients in these zones are relatively large. As a result, the mismatch caused by the material parameters is more pronounced in these zones during cooling, thereby leading to the formation of higher thermal mismatch stresses.

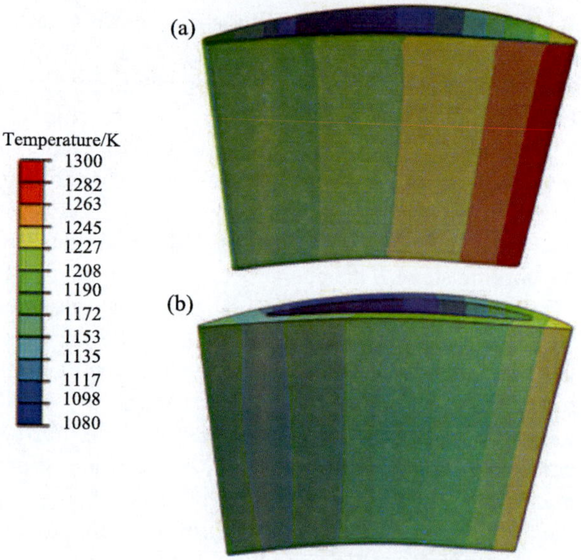

Fig. 10.13 Temperature field in the turbine blade with a TBC: **a** temperature distribution in the TBC and **b** temperature distribution in the substrate of the turbine blade

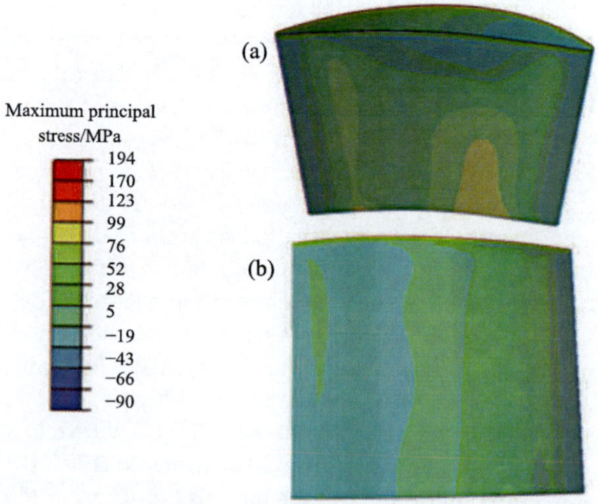

Fig. 10.14 Contour plot of the maximum principal stress distribution in the BC layer during the cooling stage: **a** concave side of the blade and **b** convex side of the blade

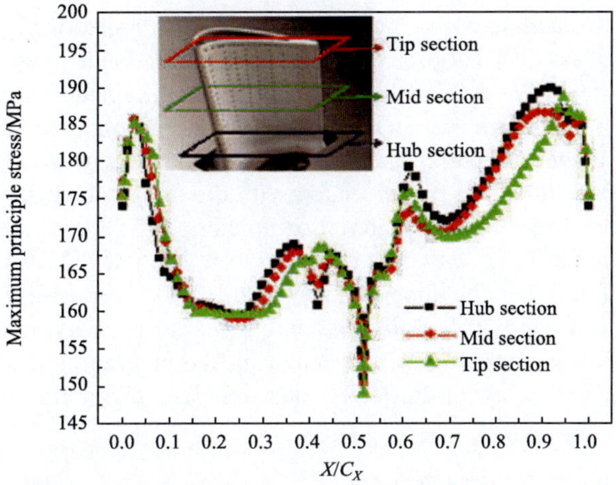

Fig. 10.15 Distribution of the maximum principal stress in the BC layer along the height of the blade during the cooling stage (X is the coordinate along the surface of the TBC; C_X is the axial chord length; dimensionless lengths $X/C_X = 0$ and $X/C_X = 0.5$ are the locations of the leading and trailing edges of the blade, respectively; the suction and pressure sides are located at $0.3 < X/C_X < 0.42$ and $0.72 < X/C_X < 0.9$, respectively)

Figure 10.15 shows the maximum principal stress distribution in the BC layer along the height of the blade during the cooling stage. The stress distribution curves corresponding to three selected cross-sections basically coincide with each other except for the locations at $0.3 < X/C_X < 0.42$ and $0.72 < X/C_X < 0.9$. This result suggests that the stress distribution is directly related to the temperature distribution along the height of the blade; that is, there is little difference in the temperature distribution along the height of the blade. As a result, the stress distribution is basically consistent along the height of the blade. In addition, the stresses in the abovementioned two areas are higher in the base than at the other two cross-sections. The primary cause of the notable difference in the abovementioned two areas is that the locations corresponding to these two areas are the closest to points A and B in the selected boundary conditions in Fig. 10.11. Due to the boundary conditions, stress concentration occurs at points A and B, resulting in large stresses in the TBC in nearby areas. In contrast, the midsection and tip section of the blade are far from the bottom surface and are thus minimally affected by the boundary conditions.

Figure 10.16 shows the contour plot of the maximum principal stress distribution in the TGO layer during the cooling stage. The TGO layer is made of α-Al_2O_3. As the Young's modulus of the TGO layer is much higher than that of any other layer, the stress is much higher in the TGO layer than in the BC layer. In addition, compared to the BC layer, the stress is positive across the TGO layer and is notably lower at the leading edge than in the other zones, which is a result of the variation in the yield strength with temperature. The yield strength of the TGO layer is low (only 1 GPa) at high temperatures but increases to 10 GPa as the temperature decreases to below 800 °C. Consequently, in the initial stage of cooling, both the temperature and stress are high at the leading edge, and the material undergoes certain inelastic deformation. As the temperature continues to drop, the yield strength increases and no plastic stress is formed anymore. The plastic deformation sustained during the initial stage is irreversible and stores some strain energy, resulting in lower stress at the leading edge than in any other zone. Figure 10.17 shows the maximum principal stress distribution in the TGO layer along the height of the blade during the cooling stage. Except for the leading edge and its nearby zones where the stress remains basically consistent, the stress varies significantly from zone to zone. This result suggests that of the stresses in different layers of the TBC, the stress in the TGO layer is relatively little correlated with the temperature distribution but is more sensitive to changes in the material parameters. In addition, the stress level is notably higher in the TGO layer than in the adjacent BC and TC layers. Interface cracks are usually formed on the upper surface of the TGO layer, which is analyzed in detail later.

Figure 10.18 shows a contour plot of the maximum principal stress distribution in the ceramic (TC) layer during the cooling stage. The stress distribution in the ceramic layer is basically consistent with that in the BC layer. That is, the maximum

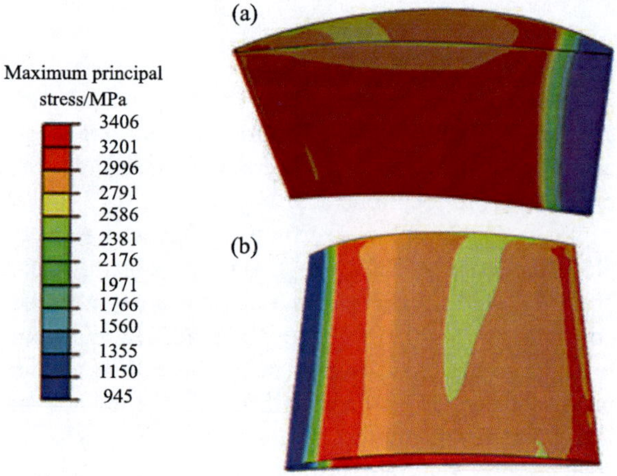

Fig. 10.16 Contour plot of the maximum principal stress distribution in the TGO layer during the cooling stage: **a** concave side of the blade and **b** convex side of the blade

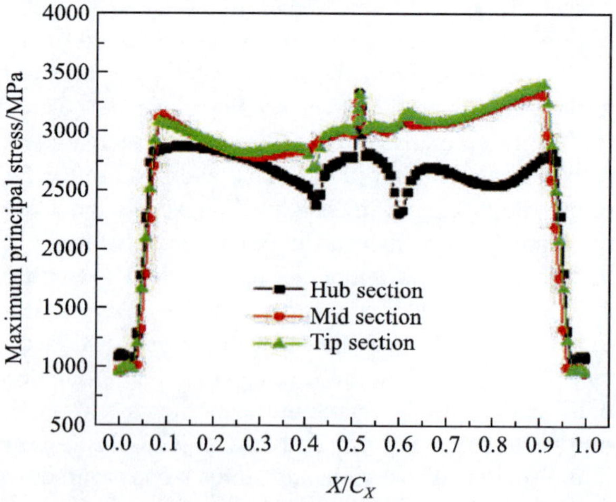

Fig. 10.17 Distribution of the maximum principal stress in the TGO layer along the height of the blade during the cooling stage

stress values occur on both sides of the leading and trailing edges, but there is no compressive stress in the ceramic layer, and the stress in the ceramic layer is slightly higher than that in the BC layer and has a maximum value of 194 MPa, located near the pressure side of the leading edge. Similar to the analysis of the BC and TGO layers, Fig. 10.19 shows the maximum principal stress distribution in the ceramic (TC) layer along the height of the blade during the cooling stage. Due to the boundary conditions, the stresses at $0.3 < X/C_X < 0.42$ and $0.72 < X/C_X < 0.9$ are slightly higher at the cross-section near the tenon at the bottom than at the midsection and the blade tip section. In addition, the stresses are basically consistent at all other locations.

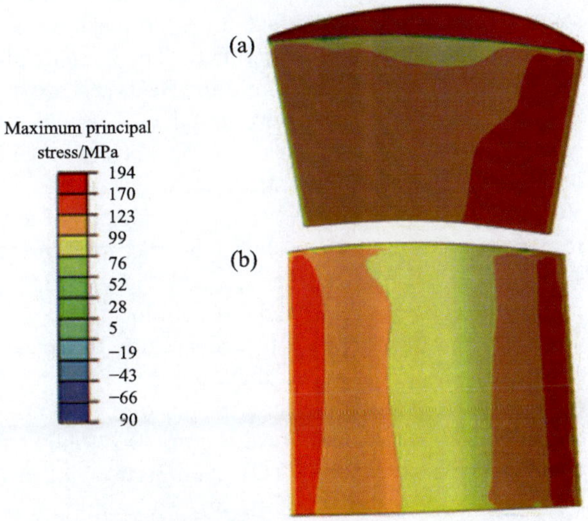

Fig. 10.18 Contour plot of the maximum principal stress distribution in the ceramic (TC) layer during the cooling stage: **a** concave side of the blade and **b** convex side of the blade

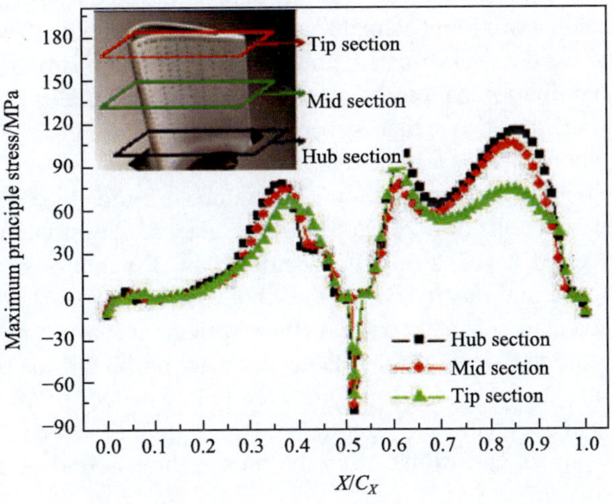

Fig. 10.19 Distribution of the maximum principal stress in the ceramic (TC) layer along the height of the blade during the cooling stage

(4) **Prediction of danger zones in the turbine blade with a TBC**

Danger zones in a TBC are usually predicted based on the maximum stress zones. In most studies on crack initiation and propagation, the maximum principal stress is used as the criterion for crack propagation. In addition, the maximum principal stress is independent of the coordinates. Hence, the maximum principal stress is selected in this section as the object of discussion.

Extensive research [31] shows that TBC spallation occurs primarily at the TC/TGO and TGO/BC interfaces. The ceramic and transition layers on the two sides of the TGO layer are the focus of the analysis. While the stresses in the TGO layer are very high, the available research results show that crack propagation in the cooling stage is dominated by interface crack propagation. Hence, the stresses in the TGO layer are not discussed in detail here. The contour plots of the maximum principal stress distributions in the ceramic and BC layers are obtained in the previous section. To analyze their failure danger zones, the zones where the highest maximum principal stresses occur in the ceramic and BC layers are marked. As shown in Fig. 10.20a, the highest maximum principal stress occurs in zones A and B on the pressure side of the BC layer, which is close to the leading edge and the blade root at the trailing edge, respectively. Figure 10.20b shows the maximum principal stress distribution on the suction side. The marked zone C where the stress is high is at the blade root near the trailing edge. Zones A, B, and C are all located in the areas of the blade root where there is a significant change in temperature. In fact, the thermal strain rate equation in Eq. (10.13) clearly shows that a large change in temperature (ΔT) leads to high thermal stress. In addition, the presence of the boundary conditions at the blade root results in stress concentrations. Figure 10.21 shows the maximum principal stress distribution in the ceramic (TC) layer. Figure 10.21a shows the maximum principal stress on the pressure side. The highest maximum principal stress occurs in zone D, which is located in the part of the blade root close to the pressure side at the leading edge and corresponds to zone B in the transition layer. Figure 10.21b shows the maximum principal stress on the suction side. The highest maximum principal stress occurs in zones E and F, of which zone F similarly corresponds to zone C. The distribution pattern of zones D, E, and F resembles that of zones A, B, and C in the transition layer; that is, zones D, E, and F are located in the parts of the blade root where there is a large temperature gradient.

Previous research on TBC failure has mostly considered the stress distribution patterns in local zones of turbine blades, the temperature boundaries of which are fixed. Liu [17] similarly considered the distribution of the maximum principal stress in the multilayer TBC on the whole turbine blade. However, this researcher similarly selected a constant outer surface temperature for the TBC and determined the temperature field solely from the heat transfer of the turbine blade and, on this basis, calculated the stress field and predicted failure danger zones in the TBC. Another approach is to obtain the temperature field through fluid–solid coupling conjugate heat transfer analysis and to use it as the temperature boundary to determine the stress field.

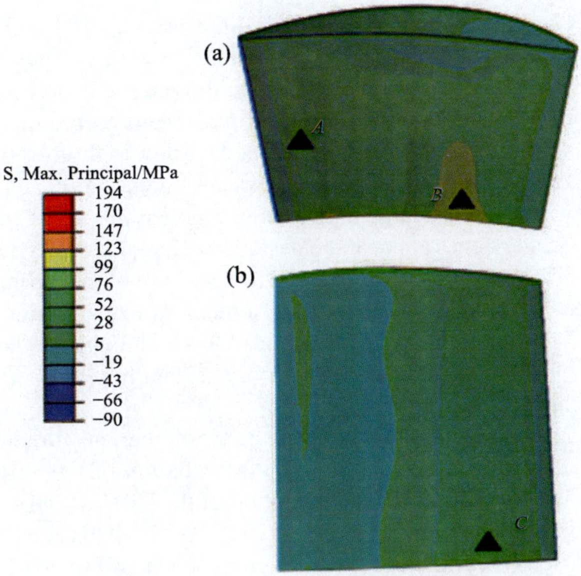

Fig. 10.20 Distribution of the maximum principal stress in the transition (BC) layer (A, B, and C are the predicted failure danger zones): **a** concave side of the blade and **b** convex side of the blade

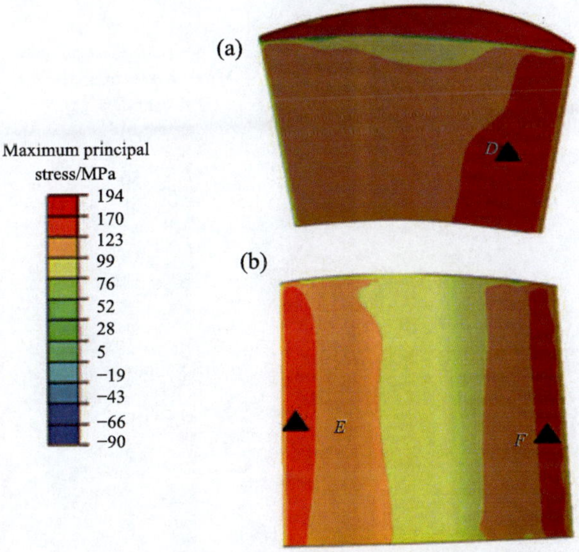

Fig. 10.21 Distribution of the maximum principal stress in the ceramic (TC) layer (D, E, and F are the predicted failure danger zones): **a** concave side of the blade, **b** convex side of the blade

Figure 10.22 compares the danger zones predicted by the two approaches to illustrate the difference between the results obtained by these approaches. Figure 10.22a shows the maximum principal stress distribution and failure danger zones in the ceramic layer obtained based on a fixed temperature boundary [31]. It can be seen that on the pressure side, zones A and B are located at the tip section and midsection of the blade, respectively, whereas zones C, D, and E are all located at the connecting chamfer between the blade and the base. Zones F, G, and H are located in the part of the suction side with the highest curvature. Thus, it is determined that there are significant geometric changes in these three zones. Sudden geometric changes lead to a high curvature, which, in turn, leads to high stress. Figure 10.22b shows the stress field in the blade obtained using the fluid–solid coupling method. Zones D, E, and F do not exhibit significant geometric changes. According to earlier analysis, there is a large temperature gradient in each of these three zones. Similarly, the danger zones are near the base. Therefore, the stress distribution in the TBC is jointly affected by the geometry and temperature of the turbine blade. Under a fixed temperature boundary condition, the temperature distribution no longer has a notable effect. Therefore, the use of the fluid–solid coupling method to obtain a temperature field closer to the actual one is very important for ensuring the accuracy of the calculated stress field.

For the verification of the above numerical simulation results, Fig. 10.23 shows the experimental results for the spallation of the TBC on a turbine blade [17, 32]. The location of the delamination of the TBC on the suction side of the turbine blade is indicated by an arrow in Fig. 10.23a. This location corresponds to zone E in Fig. 10.22. On the pressure side shown in Fig. 10.23b, Rudder [32] observed the spallation zones of the TBC on a turbine blade after multiple cycles in service. There

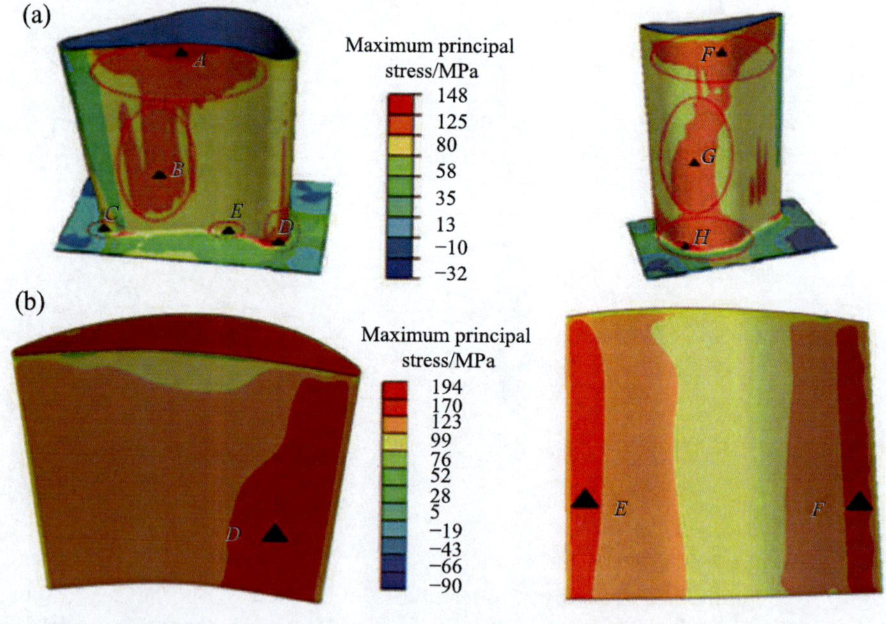

Fig. 10.22 Distribution of the maximum principal stress in the ceramic layer: **a** stress distribution on the concave and convex sides of the blade obtained using a fixed temperature boundary; **b** stress distribution on the concave and convex sides of the blade obtained using the actual temperature boundary determined by the fluid–solid coupling method

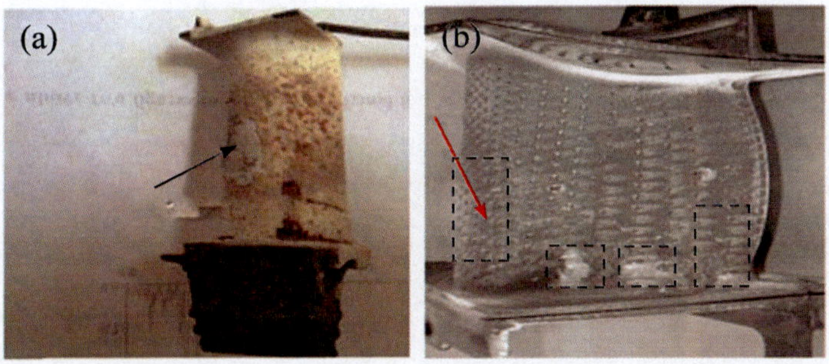

Fig. 10.23 Experimental results of TBC spallation failure

are two notable spallation zones near the base in the middle of the pressure side of the blade, and there are large spallation zones near the leading and trailing edges. The zones enclosed by the dashed boxes as well as those indicated by an arrow match zones A, B, and D in Figs. 10.20 and 10.21 very well.

In short, the TBC failure danger zones predicted based on the stress field calculated with the temperature field, which is obtained using the fluid–solid coupling method and considering the heat transfer between the high-temperature gas and the turbine blade, as the temperature boundary condition for the turbine blade is in good agreement with the experimental results. This approach provides a new direction for studying TBC spallation as a failure mode, that is, the consideration of stresses and crack propagation in a TBC in the actual temperature field.

10.3 Destructive Characterization of the Residual Stresses in TBCs

10.3.1 Characterization by the Curvature Method

The curvature method, first proposed by Hobson and Reiter, [33] is advantageous because it has a simple testing setup and can directly measure residual stresses in TBCs. This method relies on the principle that the residual stresses formed during the formation of a coating on a substrate cause curvature changes, based on which the residual stresses can be calculated. Mismatch deformation during the cooling process following thermal spraying is the primary source of residual stresses. Figure 10.24 shows a schematic diagram of the change in the curvature of a TBC during the thermal spraying process. Since the CTE of the metallic substrate is higher than that of the ceramic layer, the substrate contracts more than the ceramic layer during the cooling stage after spraying, resulting in mismatch deformation ΔL, as shown in Fig. 10.24a. Since the substrate and the ceramic layer actually form a whole, the deformation of the substrate should be geometrically compatible with that of the ceramic layer. Figure 10.24b shows the stress state in the substrate and the ceramic layer. In this stress state, the TBC undergoes compatible deformation to "release" some of the stress, as shown in Fig. 10.24c. Let κ_1 and κ_0 be the curvatures of the specimen before and after it is sprayed with a ceramic layer, respectively. Then, the difference in the curvature of the specimen is $\Delta\kappa = \kappa_0 - \kappa_1$. The TBC specimen is simplified to a two-layer composite beam. In the absence of external forces, the residual stresses are self-balanced in an object. Thus, the condition that the axial force and the bending moment at any cross-section are zero is met.

The axial force equilibrium condition is expressed as follows:

$$\int \sigma_c \, dA_c + \int \sigma_s \, dA_s = 0 \quad (10.32)$$

where A_c and A_s are the cross-sectional areas of the coating and substrate of the composite beam, respectively, and σ_c and σ_s are the residual stresses in the coating and the substrate, respectively. Due to the presence of a mismatched strain in the coating, we have the following expressions:

Fig. 10.24 Schematic diagram of the change in the curvature of a TBC during the thermal spraying process

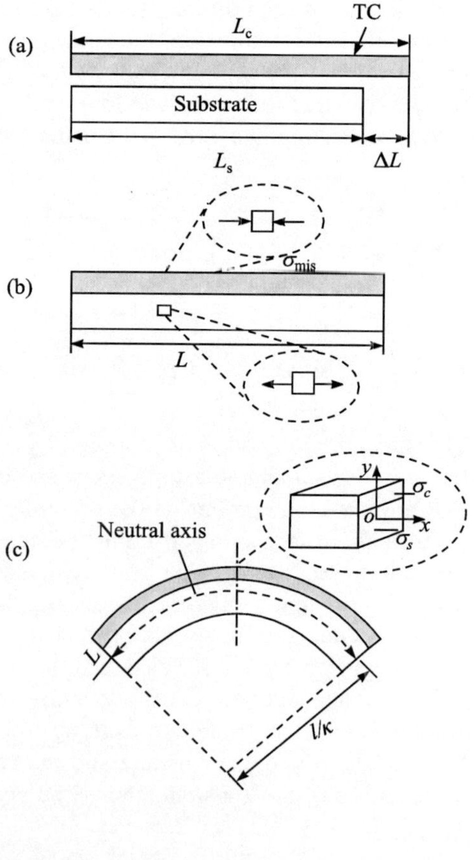

$$\sigma_c = -\overline{E_c}\Delta\kappa y + \sigma_{mis} \tag{10.33}$$

$$\sigma_s = -\overline{E_s}\Delta\kappa y \tag{10.34}$$

where $\overline{E_i} = E_i/(1 - \nu_i)$ (E_i and ν are the elastic modulus and Poisson's ratio of the corresponding material, respectively) and σ_{mis} is the mismatch stress in the coating. Substitution of Eqs. (10.33) and (10.34) into Eq. (10.32) gives

$$\overline{E_c}\int_{h_1}^{h_1+h_c} (-\Delta\kappa y + \frac{\sigma_{mis}}{\overline{E_c}})\,dy - \overline{E_s}\int_{-h_0}^{h_1} \Delta\kappa y dy = 0 \tag{10.35}$$

where h_1 is the distance between the neutral plane of the beam and the ceramic-layer/substrate interface and h_0 is the distance between the lower surface of the substrate and the neutral plane of the beam.

The bending-moment equilibrium condition states that

$$\int_{h_1}^{h_1+h_c} \sigma_c y\,dy + \int_{-h_0}^{h_1} \sigma_s y dy = 0 \tag{10.36}$$

Substitution of Eqs. (10.33) and (10.34) into Eq. (10.36) yield the following bending-moment equilibrium equation of the beam before and after thermal spraying:

$$\overline{E_c}\int_{h_1}^{h_1+h_c} (-\Delta\kappa y + \frac{\sigma_{mis}}{\overline{E_c}})\,y dy - \overline{E_s}\int_{-h_0}^{h_1} \Delta\kappa y^2 dy = 0 \tag{10.37}$$

Simultaneously solving Eqs. (10.35) and (10.37) yield the following expression for σ_{mis} in the coating [34, 35]:

$$\sigma_{mis} = \frac{\Delta\kappa(\overline{E_c}^2 h_c^4 + 4\overline{E_c} h_c^3 \overline{E_s} h_s + 6\overline{E_c} h_c^2 \overline{E_s} h_s^2 + 4\overline{E_c} h_c \overline{E_s} h_s^3 + \overline{E_s}^2 h_s^4)}{6h_c h_s \overline{E_s}(h_c + h_s)} \tag{10.38}$$

Figure 10.25a and b plot the morphologies of a TBC specimen before and after the spraying of the ceramic layer, respectively, obtained using the 3D digital image correlation (DIC) technique [34]. Figure 10.25c gives the curves obtained by fitting the centerlines extracted from the morphological plots. The curvature before the spraying of the ceramic layer, κ_1, is –0.98 ± 0.1 m^{-1}, while the curvature after the spraying of the ceramic layer, κ_0, is –1.56 ± 0.2 m^{-1}. The elastic moduli of the ceramic layer and the substrate, E_c and E_s, are set to 34 and 143 GPa, respectively.

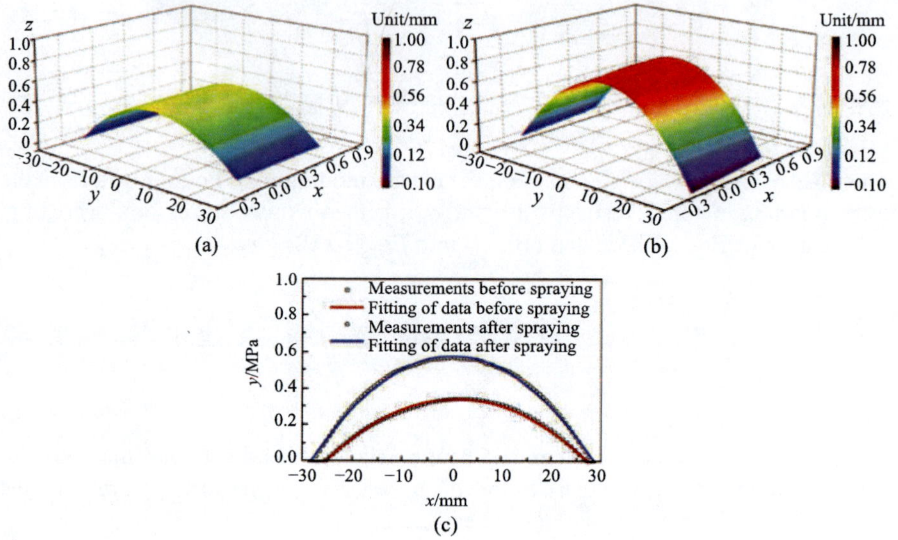

Fig. 10.25 Morphological plots and fitted curves of a TBC specimen: **a** morphology before spraying; **b** morphology after spraying; **c** fitted curves for the specimen before and after spraying [34]

The σ_{mis} in the ceramic layer is calculated to be −105.8 MPa using Eq. (10.38), where the negative sign indicates compressive residual stress. Many experimental and theoretical analyses show that residual stresses in TBCs are generally compressive and range from 20 to 200 MPa [36–38]. This is mainly because the CTEs of porous coatings are lower than those of metallic substrate materials, and changes in process parameters (e.g., spray flame temperature) lead to different values of residual stresses in coatings.

The solid line in Fig. 10.26 shows the residual stress distribution in the coating, ranging from −86 to −70 MPa. According to Eq. (10.33), the residual stress consists of two parts, namely, the mismatch stress σ_{mis} due to the mismatch strain between the ceramic coating and the metallic substrate and the part of stress "released" from the bending deformation of the specimen, which ranges from 20 to 35 MPa. Generally, a thick substrate undergoes small bending deformation and "releases" a small amount of stress, thus resulting in a high residual stress in the coating.

10.3.2 Characterization by the Drilling Method

The drilling method, also known as the blind-hole method and the trepanning method, is currently widely used to measure in-plane residual stresses in coatings. The drilling method for measuring residual stresses was first proposed by German researcher Mathar [39] in 1934 and later gradually developed and improved by other researchers

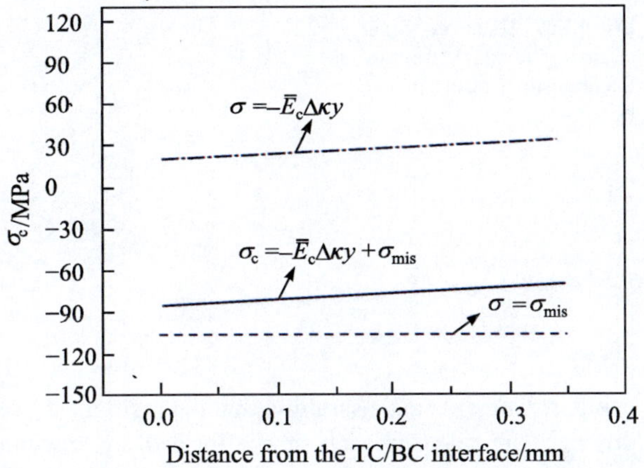

Fig. 10.26 Distribution of the residual stress in a coating with the thickness of the ceramic layer [34]

such as Soete and Vancrombrugge, [40] culminating a whole set of theories. The procedure of the drilling test is described as follows. A specifically designed foil strain rosette is adhered to the surface of a coating. A small hole is drilled at the center of the strain rosette to a depth close to the diameter of the hole. This process leads to a local stress release. The released strain can be read from the strain indicator connected to each strain gauge. Based on an elastic model, the average principal stress and the principal stress direction angle within the depth of the hole can be determined. This is a simple, low-cost, high-accuracy measurement technique and has become a standard measurement method that is widely used in engineering practice. The ASTM E837 Standard Test Method for Determining Residual Stresses by the Hole-Drilling Strain-Gage Method stipulates the method, requirements, and corresponding drilling procedure for determining residual stresses.

In recent years, test analysis by the drilling method based on DIC (an optical technique) has been developed. DIC, as a substitute for strain gauges, can be used in conjunction with the drilling method to determine the in-plane residual stresses in materials. DIC is a typical noncontact optical measurement technique that has advantages such as good accuracy and high reliability and uses a standardized test procedure, which makes it easier to use in practice. In addition, damage caused by testing using the DIC technique to a specimen is often negligible or repairable. Therefore, this technique is sometimes described as "semidestructive." A combination of the new step-by-step drilling method and DIC is currently the latest research tool for residual stress measurement.

The layer-by-layer drilling method (see Fig. 10.27 for its model) was developed to determine the variation in residual stress in the vertical direction. The basic concept of this method is described as follows [41–43]. A small blind hole is prepared at an arbitrary location on the surface of an elastic or isotropic material. Let H be the total depth of the coating on the specimen, h the depth drilled each time, D_0 the radius of the borehole, D the distance between the measuring point and the center of the borehole, and ε_1, ε_2, and ε_3 the strains released in the 0°, 45°, and 90° directions in the area surrounding the borehole, respectively. By measuring the residual strains

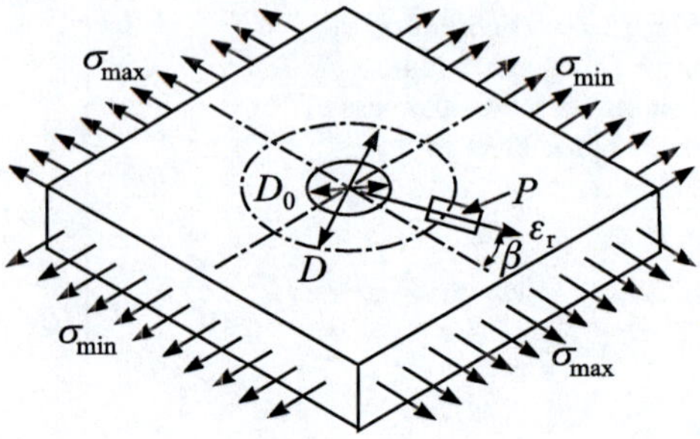

Fig. 10.27 Relation of the strain and residual stress in the area surrounding the borehole

using instruments and assuming that the residual stresses in the area surrounding the borehole are released each time after drilling, we can calculate the three principal stresses, σ_1, σ_2, and σ_3, (i.e., residual stresses) using the constitutive relations.

The step-by-step drilling method assumes that the residual stresses in each micro-layer are uniformly distributed. The radially distributed residual strain generated by each hole in a single-layer structure can be expressed as follows:

$$\varepsilon_r = A(\sigma_1 + \sigma_3) + B(\sigma_1 - \sigma_3)\cos 2\varphi + C\tau_{13}\sin 2\varphi \tag{10.39}$$

$$A = -\frac{1}{2E}(1+\nu)\left(\frac{D_0}{D}\right)^2 \quad \text{and} \quad B = -\frac{1}{2E}\left[4\left(\frac{D_0}{D}\right)^2 - 3(1+\nu)\left(\frac{D_0}{D}\right)^4\right]$$

where τ_{13} is the shear stress, φ is the angle of the radial location in the clockwise direction relative to the principal stress axis, and A and B are correction coefficients.

Equation (10.39) shows that the influences of material parameters on the residual surface stresses are reflected only in coefficients A and B. Thus, in a multilayer structure, the influences of the material parameters of each layer on the residual surface stresses can be calculated based on the correction coefficients, which can be obtained from the literature [44]. Now, let us consider the following case. Using the inverse method, certain deformation stresses are applied to a specimen, and then the stress–strain relations at 0° and 45° are measured. Let u^i and v^i be the displacements in the x- and y-directions at the ith depth increment in the rectangular coordinate system, respectively, and σ_x^{ij} be the stress in the x-direction. Then, the influences of the multilayer structure of the specimen on the stress at the drilling depth can be obtained as follows:

$$A_{\mathrm{ij}} = \frac{u^i_{(\theta=45^\circ)} + v^i_{(\theta=45^\circ)}}{\sqrt{2}\sigma_{\mathrm{x}}^{\mathrm{ij}}} \quad B_{\mathrm{ij}} = \frac{\sqrt{2}u^i_{(\theta=0^\circ)} - \left(u^i_{(\theta=45^\circ)} + v^i_{(\theta=45^\circ)}\right)}{\sqrt{2}\sigma_{\mathrm{x}}^{\mathrm{ij}}} \tag{10.40}$$

Based on the constitutive model of a multilayer structure subjected to the drilling method, the residual strains for a suitable reference system in the three directions can be determined from the following equations [41]:

$$\begin{cases} \varepsilon_1 = A(\sigma_1 + \sigma_3) + B(\sigma_1 - \sigma_3) \\ \varepsilon_2 = A(\sigma_1 + \sigma_3) + C\tau_{13} \\ \varepsilon_3 = A(\sigma_1 + \sigma_3) - B(\sigma_1 - \sigma_3) \end{cases} \tag{10.41}$$

By reorganizing the above equations, we have

$$\begin{cases} (A + B)\sigma_1 + (A - B)\sigma_3 = \varepsilon_1 \\ A\sigma_1 + C\tau_{13} + A\sigma_3 = \varepsilon_2 \\ (A - B)\sigma_1 + (A + B)\sigma_3 = \varepsilon_3 \end{cases} \tag{10.42}$$

Equation (10.42) can be written in matrix form as follows:

$$\begin{bmatrix} A + B & 0 & A - B \\ A & C & A \\ A - B & 0 & A + B \end{bmatrix} \begin{bmatrix} \sigma_1 \\ \tau_{13} \\ \sigma_3 \end{bmatrix} = \begin{bmatrix} \varepsilon_1 \\ \varepsilon_2 \\ \varepsilon_3 \end{bmatrix} \tag{10.43}$$

By defining the following parameters:

$$P = (\sigma_1 + \sigma_3)/2 \quad Q = (\sigma_3 - \sigma_1)/2 \quad T = \tau_{13} \tag{10.44}$$

$$p = (\varepsilon_1 + \varepsilon_3)/2 \quad q = (\varepsilon_3 - \varepsilon_1)/2 \quad t = (\varepsilon_3 + \varepsilon_1 - 2\varepsilon_2)/2 \tag{10.45}$$

we can simplify Eq. (10.43) to

$$\begin{cases} AP = p \\ BQ = q \\ BT = t \end{cases} \tag{10.46}$$

Correction coefficients A and B can be used to analyze only the case of through-holes, not cases where stresses are nonuniformly distributed along the thickness direction, for which A and B must be calibrated experimentally. The integration method is widely used to analyze these cases and assumes that residual stresses are uniformly distributed in a microlayer. At a drilling depth h, the residual strain $\varepsilon_i(h)$ represents the integral of the stress generated by an infinitesimal strain component in the depth direction ($0 < h < H$)

$$\begin{cases} p(h) = \int_0^h \overline{A}(H, h)P(H)\mathrm{d}H \\ q(h) = \int_0^h \overline{B}(H, h)Q(H)\mathrm{d}H \quad 0 \le h \le H \\ t(h) = \int_0^h \overline{B}(H, h)T(H)\mathrm{d}H \end{cases} \tag{10.47}$$

where

$$\begin{cases} P(H) = [\sigma_1(H) + \sigma_3(H)]/2 \\ Q(H) = [\sigma_3(H) - \sigma_1(H)]/2 \\ T(H) = \tau_{13}(H) \end{cases} \quad (10.48)$$

$$\begin{cases} p(h) = [\varepsilon_1(h) + \varepsilon_3(h)]/2 \\ q(h) = [\varepsilon_3(h) - \varepsilon_1(h)]/2 \\ t(h) = [\varepsilon_3(h) + \varepsilon_1(h) - 2\varepsilon_2(h)]/2 \end{cases} \quad (10.49)$$

In Eq. (10.47), the influence functions $\overline{A}(H, h)$ and $\overline{B}(H, h)$ represent the stress released per unit depth along the total depth H due to the drilling depth h, and their specific forms cannot be determined and analyzed. Therefore, Eq. (10.46) is broken up into i discrete steps to perform the calculation.

$$\begin{cases} \sum_{j=1}^{j=i} \overline{A}_{ij} P_j = p_i \\ \sum_{j=1}^{j=i} \overline{B}_{ij} Q_j = q_i \\ \sum_{j=1}^{j=i} \overline{B}_{ij} T_j = t_i \end{cases} \quad (10.50)$$

i.e.

$$\begin{cases} \sum_{j=1}^{j=i} \overline{A}_{ij} [[\sigma_1(H) + \sigma_3(H)]/2]_j = [[\varepsilon_1(h) + \varepsilon_3(h)]/2]_i \\ \sum_{j=1}^{j=i} \overline{B}_{ij} [[\sigma_3(H) - \sigma_1(H)]/2]_j = [[\varepsilon_3(h) - \varepsilon_1(h)]/2]_i \\ \sum_{j=1}^{j=i} \overline{B}_{ij} [\tau_{13}(H)]_j = [[\varepsilon_3(h) + \varepsilon_1(h) - 2\varepsilon_2(h)]/2]_i \end{cases} \quad (10.51)$$

Under the same test conditions, TBC specimens with three different ceramic-layer thicknesses were separately tested and analyzed using the step-by-step drilling method. To eliminate the influence of the nonuniformity of residual stresses, three collection points were selected at each radial location in each specimen for testing. Three small holes were drilled in each specimen, and the obtained residual stress values were averaged. The material parameters in the model originated from the mechanical property parameters (elastic modulus $E = 48$ GPa; Poisson's ratio $\nu = 0.1$) of the TBC investigated by Mao et al. [45] at room temperature. A high-speed drilling machine capable of ensuring the accuracy of the hole size and center position was used in the step-by-step drilling test. To obtain an accurate stress distribution, a depth of 50 μm was drilled each time. Due to the inherent vibrations of the drilling machine and other factors, the prepared boreholes were not completely cylindrical. Figure 10.28 is a contour plot showing the variation in the residual stress in the original state [41] (i.e., at a drilling depth of 0). Since the original thermally cycled specimens were all in a stress-free state, their contour plots are essentially unchanged. Therefore, for the sake of convenience, only one contour plot is used for all the original specimens. Accordingly, all the subsequent contour plots are obtained from specimens starting from the drilling of the first hole. Figures 10.29, 10.30 and 10.31 show the contour plots of the residual stress distribution in the original TBC specimens with ceramic-layer thicknesses of 200, 300, and 400 μm, respectively, corresponding to the contour plots of the residual stress at each stage point in Fig. 10.32a, b, c, d, e, and f. The curves clearly show that there is peak stress in each specimen, which can be attributed to two causes: (1) the plastic deformation of the substrate leads to stress relaxation near the interface; and (2) the nonlinear interactions of the residual stresses differ in nature (temperature and quenching) (Fig. 10.30).

Figure 10.29a, b, c, and d shows the contour plots of the residual stress in the original APS TBC specimen with a ceramic-layer thickness h_{TC} of 200 μm, corresponding to the points on the curves in Fig. 10.32, obtained by the drilling of the original and thermally cycled TBC specimens with h_{TC} of 200, 300, and 400 μm, respectively. At $N = 0$, the residual stress in the TBC decreases as h_{TC} increases. In each specimen, the residual stress first increases and then decreases with increasing depth and tends to vary more slowly near the interface, which is a result of the mismatch between the physical parameters of the materials. The experimental results show that the residual stress in the ceramic layer of the original APS TBC with h_{TC} = 200 μm increases from 230.43 to 264.58 MPa and then decreases to 185.69 MPa; the residual stress in the ceramic layer of the original APS TBC with $h_{TC} = 300$ μm increases from 163.58 to 211.34 MPa and then decreases to 161.05 MPa; and the residual stress in the ceramic layer of the original APS TBC with $h_{TC} = 400$ μm increases from 150.61 to 206.53 MPa and then decreases to 152.88 MPa. These residual stress test results are consistent with those obtained by Liu et al. [46] using

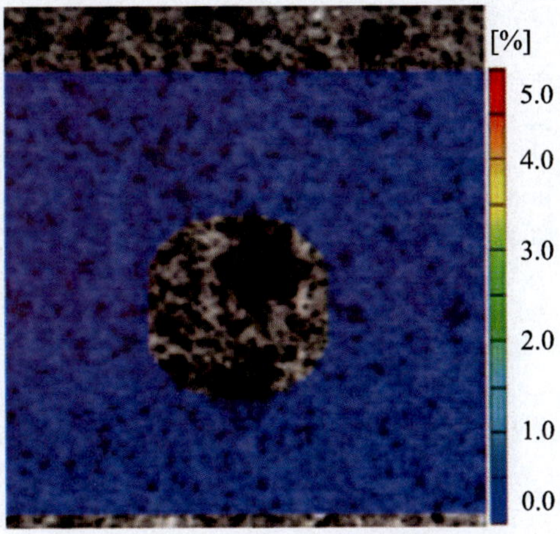

Fig. 10.28 Contour plot of the residual stress in the original TBC specimens [41]

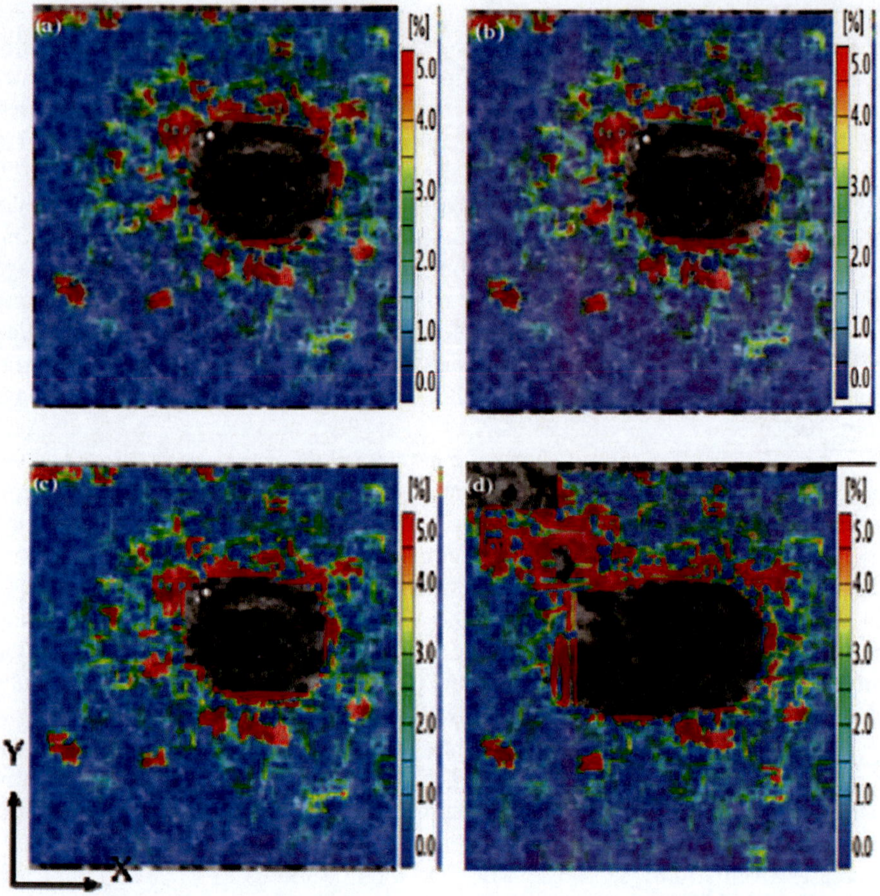

Fig. 10.29 Contour plots of the residual stress in the original APS TBC with a ceramic-layer thickness of 200 μm varying at different depths: [41] **a** 50 μm, **b** 100 μm, **c** 150 μm, and **d** 200 μm

the curvature method and by Mao et al. [45] using a new Vickers hardness model in terms of value and trend. That is, the residual stress in the original APS TBC decreases as h_{TC} increases from 200 to 300 and further to 400 μm. This trend is in line with that of high-velocity oxygen fuel (HVOF) metallic coatings with different thicknesses measured by Clyne and Gill [47] using XRD. Based on the data obtained by Godoy et al. [48] and the available findings, increasing the ratio of the thickness of the coating to the thickness of the substrate can effectively reduce residual stresses.

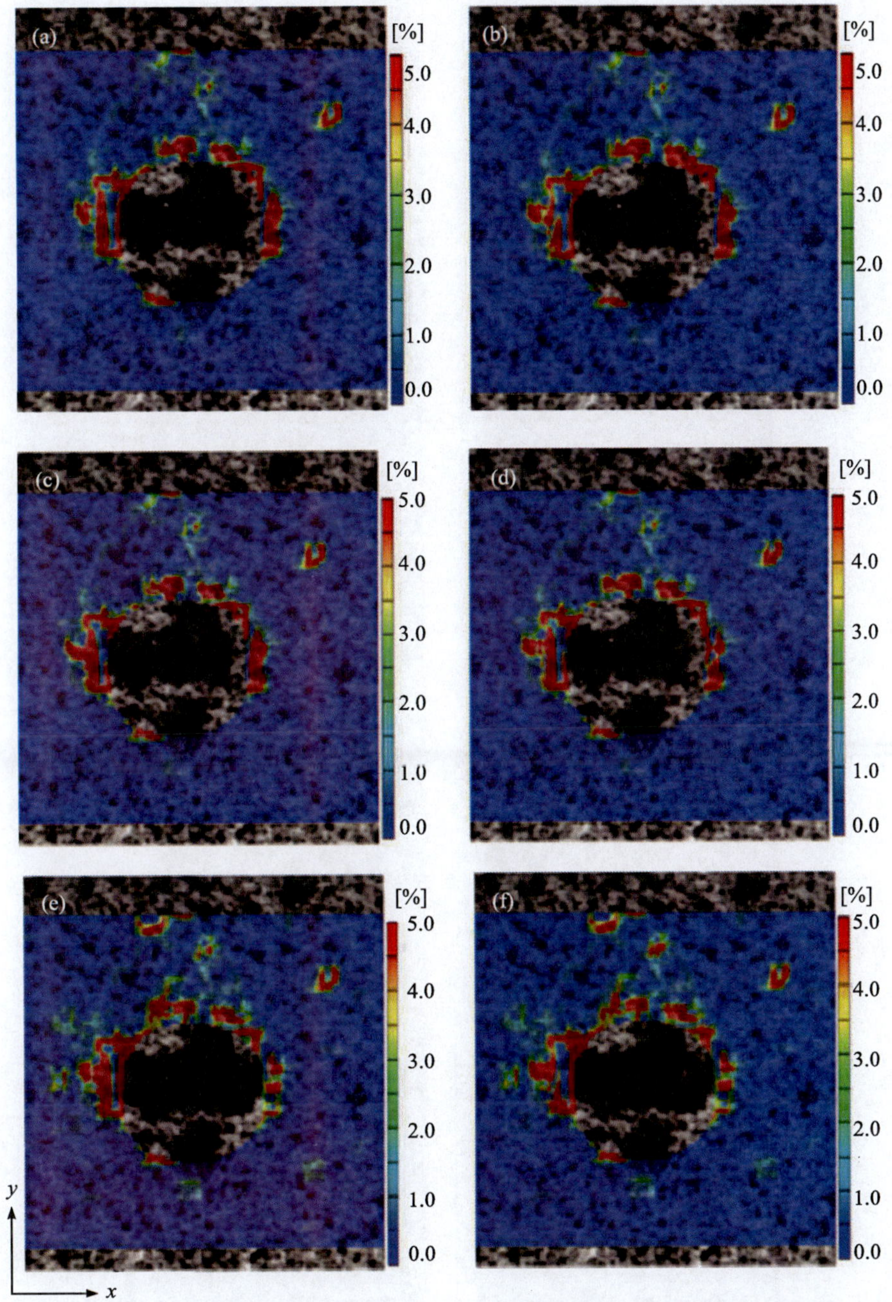

Fig. 10.30 Contour plots of the residual stress in the original APS TBC with a ceramic-layer thickness of 300 μm varying at different depths: [41] **a** 50 μm, **b** 100 μm, **c** 150 μm, **d** 200 μm, **e** 250 μm, and **f** 300 μm

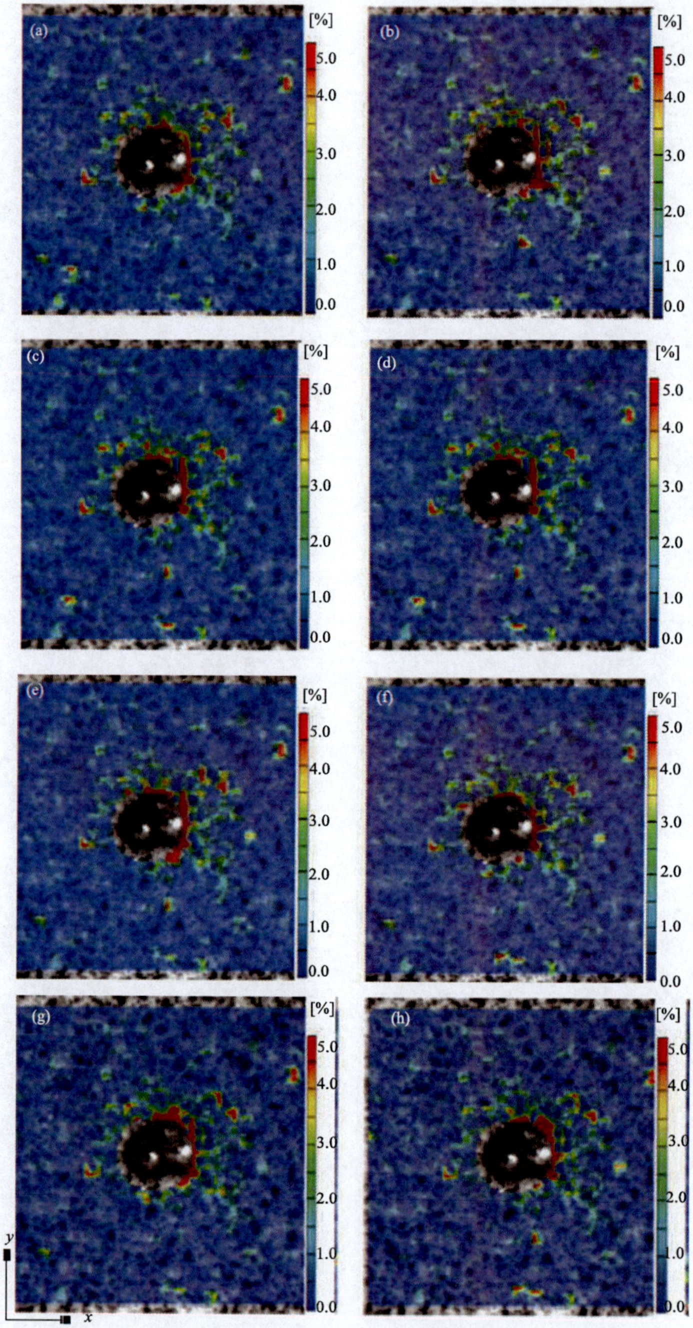

Fig. 10.31 Contour plots of the residual stress in the original APS TBC with a ceramic-layer thickness of 400 μm varying at different depths: [41] **a** 50 μm, **b** 100 μm, **c** 150 μm, **d** 200 μm, **e** 250 μm, **f** 300 μm, **g** 350 μm, and **h** 400 μm

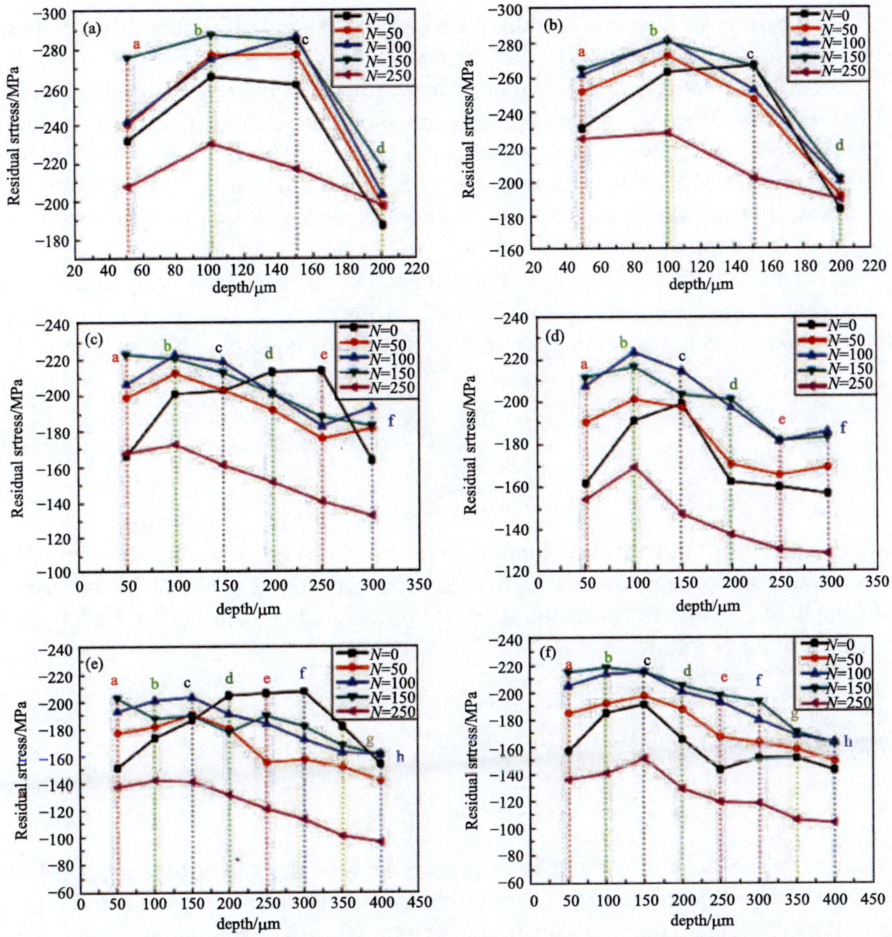

Fig. 10.32 Variation in the maximum (left) and minimum (right) values of the residual stress in each APS TBC specimen with depth before and after different numbers of thermal cycles N [41]. The ceramic-layer thickness is **a b** 200 μm, **c d** 300 μm, and **e f** 400 μm

10.3.3 Characterization by the RCM

The basic testing principle of the ring-core method (RCM) is that a circular groove is prepared on the surface of a coating and that the ring core is then separated from the main body of the specimen, during which the residual stresses in the ring core are released. This method measures the stresses in a coating based on the elastic deformation of the material (i.e., the stress release effect) [49]. In recent years, test analysis using a DIC (an optical technique)-based RCM has been developed. DIC, as a substitute for strain gauges, can be used in combination with the RCM to determine in-plane residual stresses in materials. This method is sometimes described as "semidestructive." A combination of the new RCM and DIC is currently the latest research tool for residual stress measurement.

A circular groove is often prepared by mechanical cutting or a focused ion beam (FIB). However, the measuring depth is limited by the milling capacity of the FIB, which is generally several tens of microns, much smaller than the depth (250–500 μm) of typical YSZ ceramic layers in APS TBCs. Zhang et al. [49] cut a circular groove on the surface of an APS TBC using a picosecond laser and measured the local residual stresses in the coating using DIC.

When an APS TBC in an equibiaxial stress state is subjected to ring coring, the relaxation strain $\varepsilon(h)$ measured at the depth of the circular groove, h, is related to the residual stress in the coating through the following equation [49]:

$$\varepsilon(h) = \int_0^h A(H,h) \cdot \sigma(H,h)\mathrm{d}H \tag{10.52}$$

where $\sigma(H,h)$ is the stress at a depth of H and $A(H,h)$ is a kernel function used to describe the surface strain at the depth of the circular groove, h, caused by unit stress at a depth of H. Assuming that the stress follows a gradual distribution, Eq. (10.52) can be written as follows:

$$[\varepsilon] = \begin{bmatrix} \varepsilon_1 \\ \vdots \\ \varepsilon_\mathrm{n} \end{bmatrix} = [A][\sigma] = \begin{bmatrix} a_{11} & \cdots & 0 \\ \vdots & \ddots & \vdots \\ a_{n1} & \cdots & a_{nn} \end{bmatrix} \begin{bmatrix} \sigma_1 \\ \vdots \\ \sigma_n \end{bmatrix} \tag{10.53}$$

where A (a coefficient matrix) is the discrete form of the kernel function $A(H,h)$ in Eq. (10.52). Equation (10.53) shows that the coefficient matrix needs to be first determined to obtain the stress distribution from the surface relaxation strain.

Generally, the coefficient matrix is obtained through FE modeling [50]. The ceramic layer of a TBC is often characterized by a porous microstructure with many defects. In addition, the coating/substrate interface has a certain roughness. Both factors affect the values of elements in the coefficient matrix. Thus, the local cross-sectional microstructure of a TBC is characterized using micro-computerized tomography (CT), and on this basis, an image-based FE model is established to obtain an accurate coefficient matrix.

The TBC specimen [49] mentioned in this section had dimensions of 50 mm × 25 mm × 5.5 mm and consisted of a Hastelloy X alloy substrate, a 150 μm thick NiCoCrAlY BC layer prepared by high-velocity air–fuel (HVAF) thermal spraying, and a 250-μm-thick APS 8YSZ ceramic layer. The APS TBC specimen was subjected to isothermal heat treatment at 1150 °C for 190 h in a high-temperature furnace. To place it in a laser microprocessor, the thermally treated specimen was cut into smaller pieces (5 mm × 5 mm), with the substrate thinned to 2 mm. Figure 10.33 shows the evolutionary behavior of the surface strain relaxation during the progressive preparation of a circular groove with a diameter of 500 μm. The distribution of the displacement field shows that the surface displacement of the ceramic layer remains consistent with the compressive stress in the coating. The displacement field is distributed axisymmetrically, and the residual stress in the surface coating is approximately an equibiaxial residual stress.

A comparison of Fig. 10.34a and b shows that using a laser to prepare a groove on the surface of an 8YSZ coating does not cause notable damage to the coating. The local cross-sectional microstructure of the TBC was characterized using micro-CT. Grayscale threshold segmentation was carried out on the obtained micrograph of the local microstructure. On this basis, a 2D axisymmetric image-based FE model was established (Fig. 10.34c). The mesh generated for the image-based FE model has good properties and contains approximately 110,000 three-node axisymmetric elements (CAX3). A mesh sensitivity analysis confirms that the mesh is sufficiently refined and can ensure computational convergence.

Table 10.5 summarizes the values of the Young's moduli and CTEs of the ceramic layer used in the FE simulation. According to the reports in the relevant literature, [51–55] the value range of the Young's moduli for the ceramic layer is wide (2–190 GPa). Hence, particular caution should be exercised in the selection of the value of the Young's modulus of the ceramic layer. The maximum values of Young's modulus reported in the abovementioned literature [51–55] were obtained from indentation testing, which yields the Young's modulus of a local microzone of the coating. Thus, a large number of typical representative values of the Young's modulus of the microstructure of a coating at the 100 μm scale are required. A beam bending test is often used to measure the overall Young's modulus of a specimen. Reportedly, the Young's modulus of the ceramic layer of a TBC ranges from 2 to 30 GPa [51–53].

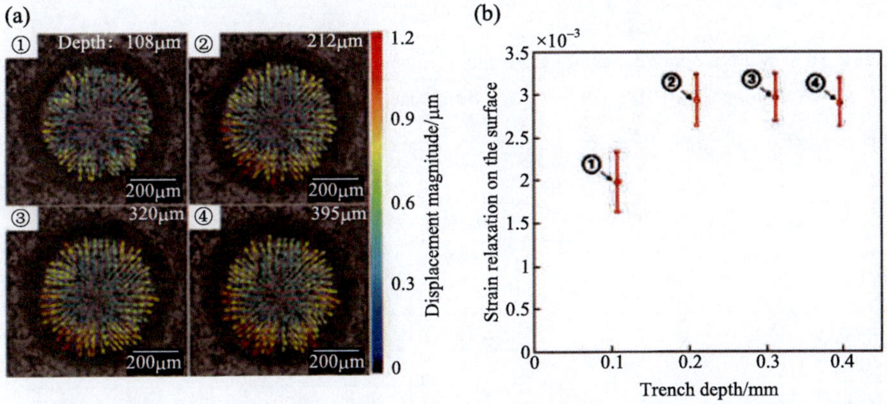

Fig. 10.33 Displacement and strain results obtained by testing the APS TBC specimen thermally treated at 1150 °C for 190 h with the RCM in combination with DIC: **a** surface displacement maps obtained by measuring the groove at different depths using DIC; **b** variation in the average surface relaxation strain (in the radial direction) with the groove depth [49]

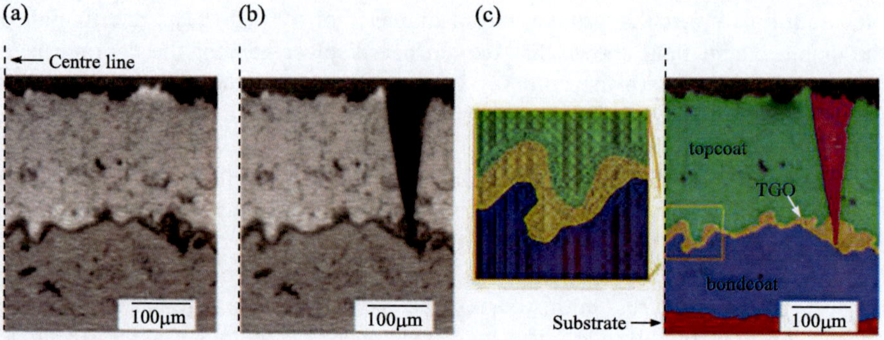

Fig. 10.34 Micro-CT graphs of the centerline section of the column: **a** before the preparation of a circular groove; **b** after the preparation of a circular groove on the surface of the coating; **c** the segmented image of different cross-sectional regions of the specimen (the inset shows the FE mesh generated from the micrograph) [49]

Thus, in this section, the values of the Young's modulus at room temperature and 900 °C are set to 17.5 and 12.4 GPa, respectively [56].

The coefficient matrix ***A*** in Eq. (10.53) is obtained using a previously reported method [57]. The ceramic layer is evenly divided into nine sublayers. The *i*th sublayer is groove-cut and hence removed (the pink part in Fig. 10.34c). A calibration pressure is applied to each of the remaining sublayers of the ceramic layer. The surface strain caused by the pressure in the *j*th ($j < i$) sublayer is recorded. A Python script is developed to recursively remove this sublayer, a calibration pressure is applied, and the resulting surface strain is recorded. This way, the coefficient matrix is reconstructed. According to Eq. (10.53), once the coefficient matrix and strain are determined, so

Table 10.5 Material parameters in the FE model

T (°C)	YSZ topcoat		TGO		Bondcoat				Substrate	
	E	α × 10^{-6}	E	α × 10^{-6}	E	α × 10^{-6}	Stress	Plastic strain	E	α × 10^{-6}
	(GPa)	(°C^{-1})	(GPA)	(°C^{-1})	(GPa)	(°C^{-1})	(MPa)		(GPa)	(°C^{-1})
25	17.5	9.68	380	5.1	183		1000	0	211	12.6
100						10.34				
200		9.7				11.3			201	13.6
400					152	12.5	2500	0.23	188	
600			353				2200	0.3	173	14.0
700		9.88								
800			338		109	14.2	375	0.022	157	15.4
900	12.4					16.0	60	0.02		
1000		10.34	312	9.8			19	0.01	139	16.3

is the residual stress. Based on the variation in the displacement and strain with the groove depth in Fig. 10.29, the average residual stress in the ceramic layer is obtained as –94 ± 8 MPa. Figure 10.35 shows the evolution of the residual stress with the groove depth. The results show a notable residual stress gradient in the coating. The distribution of the residual stress in the coating obtained using the RCM agrees well with that obtained using synchrotron radiation XRD [58]. Notably, the residual stress measured by synchrotron radiation XRD is the average stress on the X-ray path, whereas the residual stress measured by the RCM is the local residual stress in the test zone.

Considering that residual stresses are formed during the preparation of a coating when it is cooled from a high temperature to room temperature, the relaxation strain and stress values measured by the RCM are compared with the theoretically predicted values with an elastic thermal mismatch stress model (Fig. 10.36a). The residual stress is formed due to cooling from the stress-free temperature (970 °C), which is slightly lower than the holding temperature (1150 °C). Figure 10.36b compares the evolution of the calculated and RCM-measured values of the residual stress with the groove depth. Both results show that the in-plane residual stress in the coating becomes pronounced only at a depth of approximately 50 μm beneath the surface due to the surface roughness of the coating. As shown in Fig. 10.34a, the surface roughness of the coating at said cross-section was measured by micro-CT to be approximately 45 μm (the peak–valley distance).

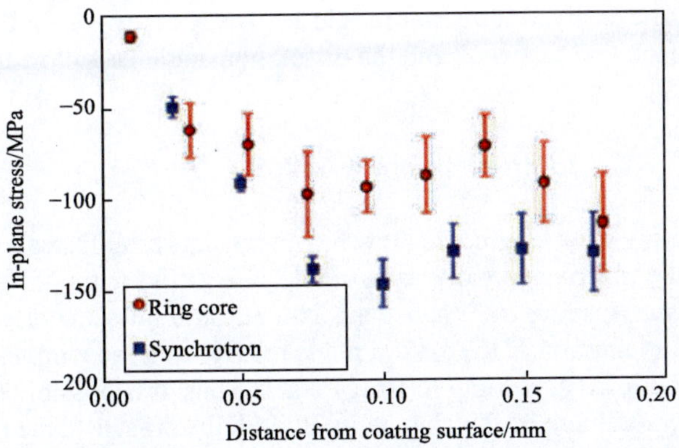

Fig. 10.35 Variation in the residual stress in the coating with depth (the red dots show the residual stress values obtained using the RCM; the blue squares show the residual stress values obtained using synchrotron radiation XRD; the error bars in the measurements by the core-ring method are obtained from the strain uncertainties determined by the Monte Carlo error propagation method) [49]

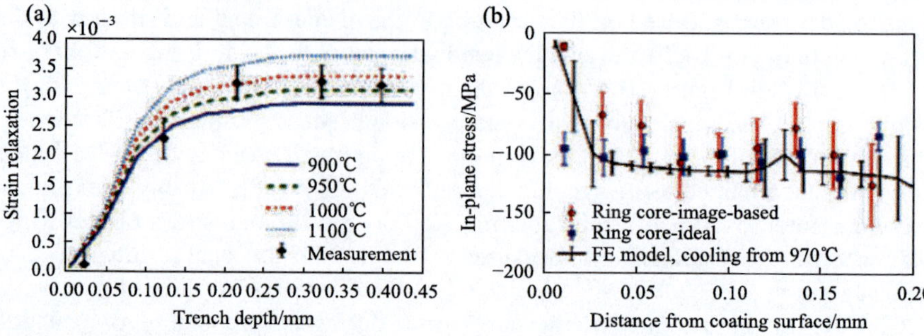

Fig. 10.36 Validation of the evolution of the image-based FEM-determined residual stress in the APS TBC against the results obtained using the RCM: **a** variation in the strain with the groove depth; **b** comparison of the FEM-predicted and RCM-determined residual thermal mismatch stress curves [49]

10.4 Nondestructive Characterization of the Residual Stresses in TBCs

10.4.1 XRD Characterization

When an X-ray beam with a certain wavelength irradiates a polycrystal, the diffracted X-ray beam can be received at a certain angle 2θ. The X-ray wavelength λ, the spacing between the diffraction planes d, and the diffraction angle 2θ follow the well-known Bragg's law [59]

$$2d\sin\theta = n\lambda\ (n = 1, 2, 3, 4, 5\ldots\ldots) \tag{10.54}$$

When the X-ray wavelength λ is known, Bragg's law establishes a definitive relation between the macroscopically measurable diffraction angle 2θ and the microscopic interplanar spacing d. When there is a stress $\boldsymbol{\sigma}$ in a material, d inevitably varies with the orientation of the crystal plane relative to $\boldsymbol{\sigma}$, accompanied by a corresponding change in 2θ according to Bragg's law. Thus, it is possible to determine $\boldsymbol{\sigma}$ through the measurement of the change in 2θ with the crystal-plane orientation.

X-rays are able to penetrate a depth of approximately 10 μm. Therefore, we can approximately consider the surface stress in a material measured by XRD to be 2D; i.e., the stress in the normal direction $\sigma_3 = 0$. As shown in Fig. 10.37, we assume that the longitudinal and transverse directions of a plate specimen are the two principal directions of the in-plane residual stress. Let ϕ and ψ be two azimuth angles of an arbitrary spatial direction OP, σ_ϕ be the residual stress to be determined in the coating in direction OA, $\varepsilon_{\phi\psi}$ be the normal strain along direction OP (direction OP is within the vertical plane formed by the direction (OA) of the stress to be determined and the normal direction of the coating surface), and σ_1 and σ_2 be the principal stresses in the x- and y-directions, respectively. According to the theory of elasticity, we have

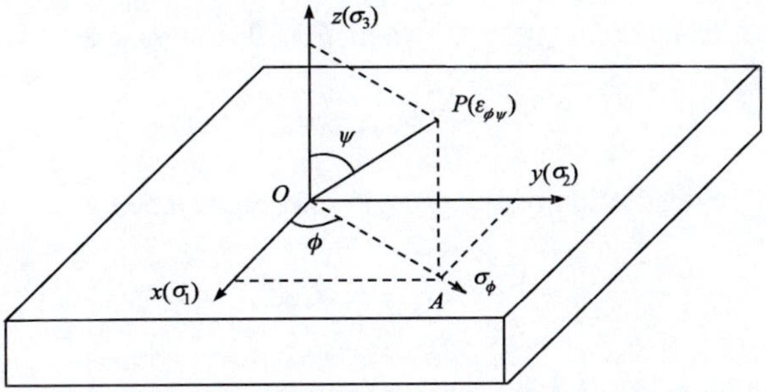

Fig. 10.37 Relations between the stress to be determined and the principal stresses (σ_1, σ_2, and σ_3)

$$\varepsilon_{\phi\psi} = \alpha_1^2\varepsilon_1 + \alpha_2^2\varepsilon_2 + \alpha_3^2\varepsilon_3 \tag{10.55}$$

In Eq. (10.55), ε_1, ε_2, and ε_3 are the principal strains, and α_1, α_2, and α_3 are the direction cosines of OP relative to the directions of the principal strains and are expressed as follows:

$$\begin{cases} \alpha_1 = \sin\psi\cos\phi \\ \alpha_2 = \sin\psi\sin\phi \\ \alpha_3 = \cos\psi = \sqrt{1-\sin^2\psi} \end{cases} \tag{10.56}$$

Then, we have

$$\varepsilon_{\psi\phi} = (\sin\psi\cos\phi)^2\varepsilon_1 + (\sin\psi\sin\phi)^2\varepsilon_2 + (1-\sin^2\psi)\varepsilon_3 \tag{10.57}$$

The linear elastic constitutive relations of the ceramic layer of a TBC are as follows:

$$\begin{cases} \varepsilon_1 = \frac{1}{E}[\sigma_1 - \nu(\sigma_2+\sigma_3)] \\ \varepsilon_2 = \frac{1}{E}[\sigma_2 - \nu(\sigma_1+\sigma_3)] \\ \varepsilon_3 = \frac{1}{E}[\sigma_3 - \nu(\sigma_1+\sigma_2)] \end{cases} \tag{10.58}$$

where E and ν are the elastic modulus and Poisson's ratio of the ceramic layer of the TBC, respectively. By substituting Eq. (10.58) into Eq. (10.57) and letting $\sigma_3 = 0$, we have

$$\varepsilon_{\psi\phi} = \frac{1+\nu}{E}(\sigma_1\cos^2\phi + \sigma_2\sin^2\phi)\sin^2\psi - \frac{\nu}{E}(\sigma_1+\sigma_2) \tag{10.59}$$

The stress along direction OP, $\sigma_{\phi\psi}$, is related to the principal stresses (σ_1, σ_2, and σ_3) through the following equation:

$$\sigma_{\phi\psi} = (\sin\psi\cos\phi)^2\sigma_1 + (\sin\psi\sin\phi)^2\sigma_2 + (1-\sin^2\psi)\sigma_3 \tag{10.60}$$

Considering $\sigma_3 = 0$, when $\psi = 90°$, we have

$$\sigma_{\phi\psi} = \sigma_\phi = \sigma_1 \cos^2\phi + \sigma_2 \sin^2\phi \tag{10.61}$$

Substitution of Eq. (10.61) into Eq. (10.59) gives

$$\varepsilon_{\psi\phi} = \frac{1+\nu}{E}\sigma_\phi \sin^2\psi - \frac{\nu}{E}(\sigma_1 + \sigma_2) \tag{10.62}$$

Equation (10.62) shows the basic relation of the stress to be determined on the surface of a specimen in a specific direction. We can see from Eq. (10.62) that the strain in direction OP is contributed by two parts, namely, the stress σ_ϕ in direction OA and the in-plane principal stresses $(\sigma_1 + \sigma_2)$, and that the contribution of the principal stresses $(\sigma_1 + \sigma_2)$ to σ_ϕ remains constant when there is a change in the angle between the normal directions of the diffraction plane and the specimen surface, ψ. The strain is only linearly related to $\sin^2\psi$. Then, by taking the partial derivative of Eq. (10.62), we have

$$\frac{\partial \varepsilon_{\phi\psi}}{\partial \sin^2\psi} = \frac{1+\nu}{E}\sigma_\phi \tag{10.63}$$

Thus

$$\sigma_\phi = \frac{E}{1+\nu}\frac{\partial \varepsilon_{\phi\psi}}{\partial \sin^2\psi} \tag{10.64}$$

According to Bragg's relation, the strain can be expressed in terms of the relative change in the spacing between the diffraction planes and linked with the displacement of the diffraction peak; i.e.

$$\varepsilon_{\phi\psi} = \frac{\Delta d}{d} = \frac{d_\psi - d_0}{d_0} = -\frac{\cot\theta_\psi}{2}\Delta 2\theta_\psi \tag{10.65}$$

Since $\theta_\psi \approx \theta_0$, we have

$$\varepsilon_{\phi\psi} = -\frac{\cot\theta_0}{2}(2\theta_\psi - 2\theta_0) \tag{10.66}$$

where θ_0 and θ_ψ are the Bragg angles of the diffraction peak in a stress-free state and in a stressed state, respectively. By substituting Eq. (10.66) into Eq. (10.64) and using the degree as the unit of 2θ, we have

$$\sigma_\phi = -\frac{E}{2(1+\nu)}\cot\theta_0 \frac{\pi}{180}\left(\frac{\partial 2\theta}{\partial \sin^2\psi}\right) \tag{10.67}$$

The above equation is commonly referred to as the XRD-based $\sin^2\psi$ method, [59] with $-\frac{E}{2(1+\nu)}\cot\theta_0\frac{\pi}{180}$ defined as the stress constant of the $\sin^2\psi$ method. We can see that this stress constant is related to the elastic modulus and Poisson's ratio of the ceramic layer of the TBC. Residual stresses are measured using a texture goniometer or an XRD system with stress attachments. An X-ray beam with a certain wavelength successively irradiates a specimen at different angles of incidence ψ to scan specific crystal planes, as shown in Fig. 10.38, and each $2\theta_\psi$ is measured. Because each reflection is generated by planes of the same type (hkl) but with different orientations relative to the surface of the specimen, changes in $2\theta_\psi$ reflect changes in the spacing between the planes of the same type (hkl) that are in different azimuths from that of the specimen surface. To construct the relational graph of $2\theta_\psi \sim \sin^2\psi$, measurements are often first taken at four points, with $\psi = 0°, 15°, 30°,$ and $45°$, respectively. Subsequently, the peak is determined through fitting using the Lorentz method or the Gaussian method. On this basis, the slope $\partial 2\theta/\partial \sin^2\psi$ is determined. $2\theta_0$ is set to the value of 2θ at $\psi = 0$. As long as the stress constant is known, the residual stress in a specimen can be determined using Eq. (10.65).

Based on the abovementioned principle of using X-rays to measure residual stresses, the four-point method was employed to measure the residual surface stress in the ceramic layer of a thermally cycled 8YSZ TBC. To reduce measurement errors, high-angle diffraction peaks are usually used during the selection of a plane for scanning [60]. As seen from the XRD intensity spectrum of the ceramic layer of the 8YSZ TBC in Fig. 10.39, the (312) plane has a single diffraction peak with a diffraction angle greater than 90° and was thus selected for XRD scanning. A D-max2500X diffractometer was employed with four ψ values (0°, 15°, 30°, and 45°) and the following parameters: scanning rate, 0.2°/min; scanning step angle, 0.02°; scanning range, 93.5–96.5°; high voltage and current of the X-ray tube, 40 kV and 250 mA, respectively.

Figure 10.40 shows the diffraction peak patterns of the (312) plane of the surface of the 8YSZ TBC subjected to different numbers of thermal cycles under different

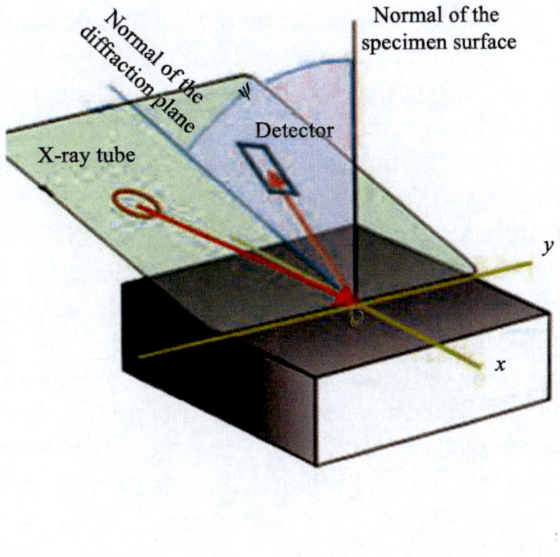

Fig. 10.38 Schematic diagram of XRD

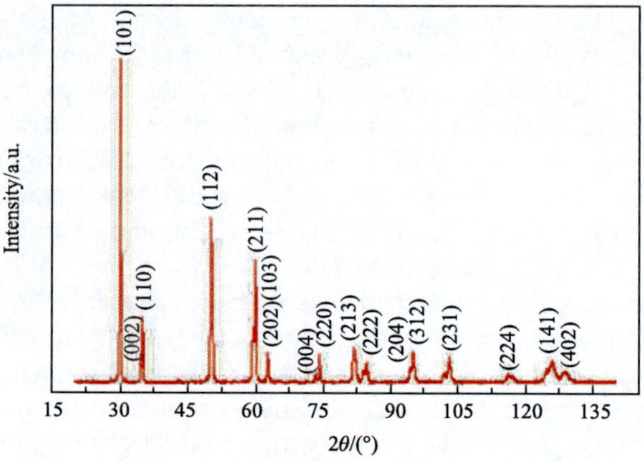

Fig. 10.39 XRD intensity spectrum of the raw 8YSZ material

ψ angles (0°, 15°, 30°, and 45°). The 2θ value of each peak was determined using Gaussian fitting. Subsequently, the value of $\partial 2\theta/\partial \sin^2 \psi$ was obtained through linear fitting using the least squares method, as shown in Table 10.6. We assume that the Poisson's ratio of the 8YSZ material does not vary with thermal cycling, that $\nu = 0.1$, [45] and that the residual surface stress in the ceramic layer of the TBC is equiaxial and 2D. Then, by substituting the results in Table 10.6 into Eq. (10.67) and based on the elastic modulus results, we can calculate the residual stress σ_ϕ.

Figure 10.41 shows the relation between the residual surface stress in the ceramic layer of the 8YSZ TBC and the number of thermal cycles. We assume that the Young's modulus E of the surface of the ceramic layer during thermal cycling remained constant at 50 GPa. The value of the residual stress calculated under this assumption first increases from –82 to –124 MPa and then decreases to –57 MPa. These results are marked with solid circles in Fig. 10.41. However, in actual applications, Young's modulus is found to vary with the number of thermal cycles, as shown in Table 10.7. The value of the residual stress calculated by substituting the results in Table 10.7 into Eq. (10.67) first increases in magnitude from –82 to –212 MPa and then decreases to –72 MPa. These results are marked with solid squares in Fig. 10.41. The experimental results show that after thermal cycling, the residual stress in the ceramic layer of the TBC is compressive and that the residual stress increases as the number of thermal cycles increases before reaching 180. These test results for the residual stress are consistent with those obtained by Teixeira et al. [61] Hamacha et al. [62] and Jordan and Faber [63] by XRD in terms of value and trend. The increase in the residual stress is mainly a result of the mismatch between the CTEs of the ceramic layer and the metallic substrate, leading to an accumulation of residual stress after thermal cycling. In addition, during thermal cycling, the elastic/plastic deformation, high-temperature creep, interfacial oxidation, and sintering of the TBC affect the change in the residual stress [64]. In the subsequent thermal cycling process, the residual stress begins to decrease. Scanning electron microscopy (SEM) observations of the surface of these specimens show that a large number of microcracks are formed on the surface of the coating specimens, leading to the release of residual stress [63, 64]. A comparison of

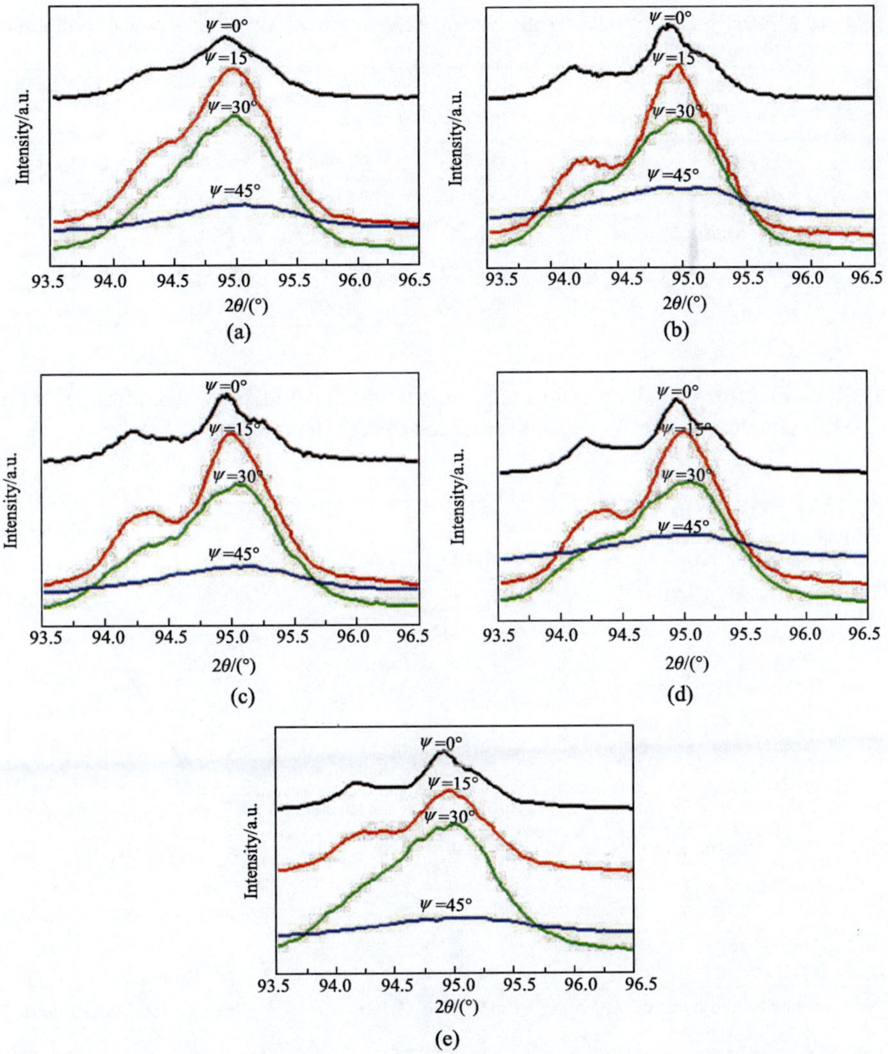

Fig. 10.40 Diffraction peaks of the (312) plane of the ceramic layer of the 8YSZ TBC at $\psi = 0°$, 15°, 30°, and 45°: **a** original specimen and the specimen after **b** 50, **c** 100, **d** 180, and **e** 250 thermal cycles, respectively [60]

the two types of results in Fig. 10.41 shows that the residual stress values obtained with the consideration of the change in the Young's modulus are higher than those obtained without such consideration, with the maximum difference between the two reaching 90 MPa. Thus, when XRD is used to measure the residual stresses in a TBC, it is necessary to consider the change in Young's modulus to ensure more accurate results. Figure 10.41 shows that the measurements of the residual surface stress in the TBC by Vickers indentation tests (marked with triangles) are consistent with those

Table 10.6 Surface XRD results for the ceramic layer of the 8YSZ TBC subjected to different numbers of thermal cycles [60]

ψ		0°	15°	30°	45°	$\frac{\partial 2\theta}{\partial \sin^2 \psi}$
$\sin^2\psi$		0	0.067	0.250	0.500	
$2\theta_\psi$	0 cycle	94.886°	94.933°	94.978°	95.022°	0.225
	50 cycles	94.893°	94.909°	94.986°	95.026°	0.247
	100 cycles	94.916°	94.951°	95.011°	95.063°	0.256
	180 cycles	94.873°	94.901°	94.960°	95.046°	0.342
	250 cycles	94.951°	94.975°	94.997°	95.034°	0.156

by XRD in terms of values and trends, suggesting that Vickers indentation tests are also effective in measuring residual surface stresses in coatings.

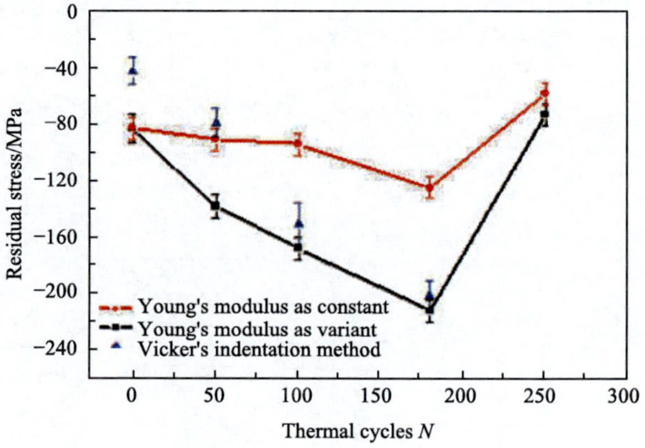

Fig. 10.41 Variation in the residual stress in the ceramic layer of the 8YSZ TBC with the number of thermal cycles [60]

Table 10.7 Young's moduli of the surface of the ceramic layer of the TBC [60]

Number of thermal cycles	Young's modulus E (GPa)			Weibull parameters	
	Maximum	Minimum	Average	m	E_0
0	52.94	38.12	47.99	13.52	49.81
30	82.06	46.39	62.03	8.12	65.76
50	95.53	57.27	73.15	8.69	77.29
100	107.87	45.72	80.28	5.01	89.93
150	111.89	62.68	88.29	9.08	93.29
180	102.64	51.56	78.42	6.33	84.58

10.4.2 Characterization by Raman Spectroscopy

While studying light scattering in benzene in 1928, Indian physicist Raman discovered the presence of a spectral line with a frequency different from (higher or lower than) that of the incident light with an extremely low intensity, apart from the spectral line of the Rayleigh-scattered light with the same frequency as the incident light in the spectral line of the scattered light. Spectral lines with lower and higher frequencies are referred to as Stokes and anti-Stokes lines, respectively. This phenomenon is called the Raman scattering effect, [65] as shown in Fig. 10.42.

According to the Raman scattering effect, when a monochromatic light beam is incident on an object, the photons and the molecules of the object collide with each other, causing light scattering, and the inelastically scattered beam forms a Raman spectrum after splitting. The Raman scattering spectrum is related to the intensity of molecular vibrations in an object, and molecular vibrations can interact with the excitation light and result in Raman scattering only when accompanied by a change in polarizability. The presence of stresses in an object causes some stress-sensitive characteristic spectra to shift or deform. Raman peak frequency shifts are briefly explained in the following. When an object is subjected to compressive stress, its molecular bond length often decreases, which, according to the relation between the force constant and the bond length, is accompanied by an increase in the force constant. As a result, there is an increase in the vibration frequency and a shift of the spectral band toward high frequencies. Conversely, when a solid object is subjected to tensile stress, its spectral band shifts toward low frequencies, as shown in Fig. 10.43.

The shift of the characteristic Raman peak of the 8YSZ material near 640 cm^{-1} is directly proportional to the stress to which it is subjected. We assume that the stress in the ceramic layer is an equiaxial plane stress. Then, the residual stress is related to the Raman frequency shift through the following equation:

$$\sigma_{xx} = \sigma_{yy} = \Delta\omega/(2\Pi_u) \tag{10.68}$$

where $\Delta\omega$ is the Raman frequency shift (unit: cm^{-1}) and Π_u is the piezospectroscopic coefficient (PSC) ($\Pi_u = 25cm^{-1}/GPa$). The Raman frequency shift in the ceramic layer of a TBC can be measured using a Raman spectrometer. On this basis, the residual stress can be calculated.

Figure 10.44a, b, and c shows the micro-Raman frequency shift patterns of APS TBC specimens with a ceramic-layer thicknesses h_{TC} of 200, 300, and 400 μm, respectively, based on the position ω_0 of the laser Raman peak of ceramic powder

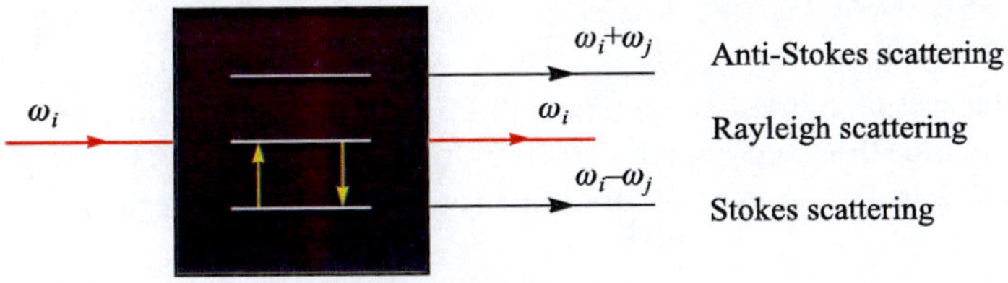

Fig. 10.42 Schematic diagram of the Raman scattering process

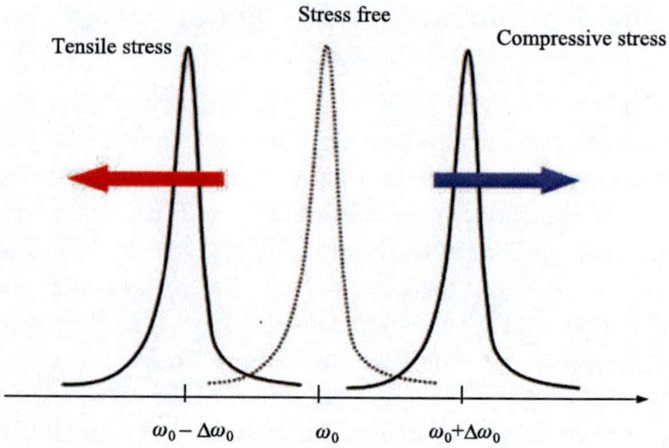

Fig. 10.43 Relation between the Raman frequency shift and stress

particles at 635.56 cm^{-1}. The residual surface stress in the TBC specimens can be calculated using Eq. (10.68). Table 10.8 summarizes the values of the residual surface stress in the APS TBC specimens with $h_{TC} = 200$, 300, and 400 μm. The results show that as h_{TC} increases, there is a gradual decrease in the residual surface stress in the ceramic layer. When the APS process is used to prepare a TBC, the first ceramic power particles deposited on the substrate are in direct contact with the substrate metal, leading to a pronounced CTE-induced thermal mismatch effect. As the spray thickness continuously increases, the subsequently sprayed ceramic powder is directly deposited on the ceramic layer already sprayed onto the substrate. In this case, the thermal mismatch effect due to the CTE of the substrate material gradually weakens. A comparison of the values of residual stress measured by Raman spectroscopy and XRD shows that the measurements obtained by the two methods are basically consistent with each other and that the data are reliable [41]. In addition, as seen in Table 10.8, the residual surface stress in the material gradually decreases with increasing h_{TC}. This is because there is a better match between the layers of a thicker coating, resulting in a lower residual stress. The values of the residual surface stress in the specimens with $h_{TC} = 200$, 300, and 400 μm are measured by XRD to be 82, 60, and 44 MPa, respectively, compared to the Raman spectroscopy-measured values of 79, 64, and 33 MPa, respectively. The values of the residual surface stress within a depth of 10 μm in the original specimens measured by XRD and micro-Raman spectroscopy are basically in good agreement.

10.4.3 Characterization of Residual Stresses in the TGO Layer by PLPS

Photoluminescence piezospectroscopy (PLPS), a type of Raman spectroscopy, is a mature method for studying the stresses in the TGO layer. The frequencies of some characteristic Raman peaks or fluorescence peaks of a material are sensitive to

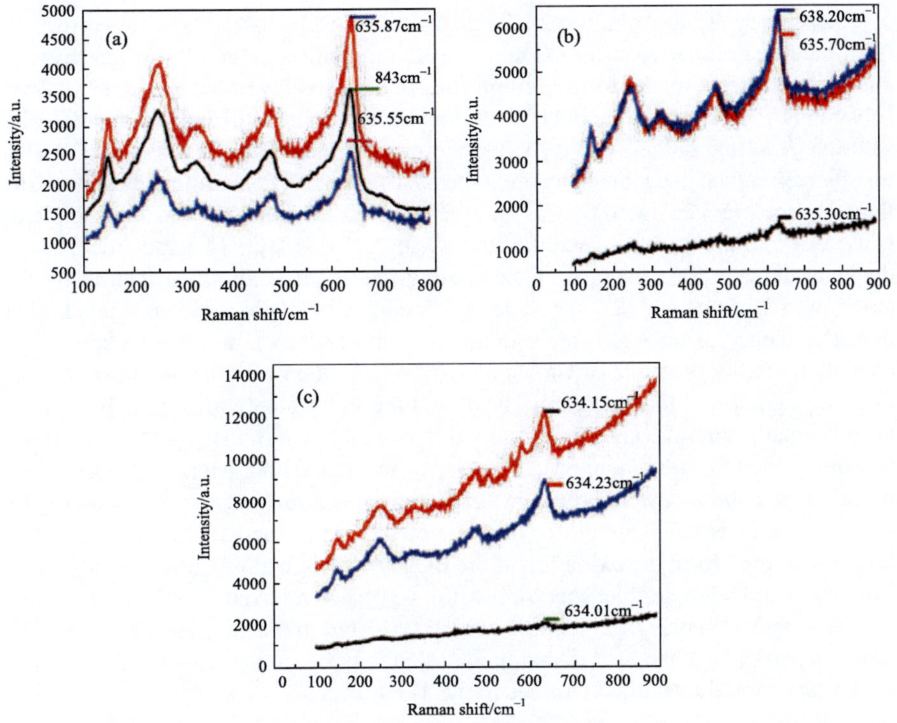

Fig. 10.44 Raman spectra of the TBC specimens with different ceramic layer thicknesses: **a** 200 μm, **b** 300 μm, and **c** 400 μm [41]

Table 10.8 Residual surface stresses in the TBC specimens with different ceramic-layer thicknesses [41]

	200 μm	300 μm	400 μm
	3.74 ± 0.2	2.93 ± 0.2	2.19 ± 0.2
$\Delta\omega$ (cm^{-1})	5.29 ± 0.2	2.93 ± 0.2	2.19 ± 0.2
	2.93 ± 0.2	3.74 ± 0.2	0.64 ± 0.2
	74.8 ± 4	58.60 ± 4	43.80 ± 4
σ_{xx} (*MPa*)	105.8 ± 4	58.60 ± 4	43.80 ± 4
	58.60 ± 4	74.80 ± 4	12.80 ± 4
$\overline{\sigma}_{xx}$ (*MPa*)	79.73 ± 4	64.00 ± 4	33.47 ± 4

the strain (stress) in the material, and there is a correspondence, which is generally linear, between the two. Based on this feature, Raman spectroscopy can be easily used to measure residual stresses in materials. In the past decade, Raman spectroscopy has evolved into an important field of research in the micromechanical testing of materials. PLPS primarily determines the residual stresses in the TGO layer based on the shifts of the characteristic spectral peaks of the fluorescence emitted by Cr^{3+} ions under the illumination of a laser that penetrates through the ceramic layer.

The primary product of interfacial oxidation in a TBC is α-Al_2O_3, which is formed from the reaction between the Al and O in the transition layer. Cr^{3+} often diffuses and exists in Al_2O_3 in the form of impurities in the TGO layer, the BC layer, and the substrate. In addition, Cr^{3+} and Al^{3+} have very similar ionic radii and thus form a solid solution, resulting in the TGO layer containing Cr^{3+} ions. The stress to which Al_2O_3 is subjected can be determined through the measurement of the frequency shift of the fluorescence line generated by Cr^{3+} ions. Figure 10.45 shows the basic principle of using PLPS to measure the residual stresses in the TGO layer of a specimen. PLPS does not damage the structure of the coating specimen at all when measuring the stress in its TGO layer. YSZ has a spectral band gap of 12 eV, which is much higher than the energy of an Ar^+ laser source (514 nm, 2.41 eV), and the energy of the characteristic fluorescence of Cr^{3+} ions (693 nm, 1.78 eV) in PLPS. Except for the defects (e.g., voids) in the ceramic layer and the grain boundaries, which scatter a laser beam, the ceramic layer is partially transparent to an irradiating Ar^+ laser beam and the excited fluorescence [66]. Therefore, the excited characteristic fluorescence of Cr^{3+} ions can be detected through irradiation of the surface of a YSZ coating by a suitable laser beam. This fluorescence is formed from the excitation of the TGO layer (generated from the oxidation of the BC layer at high temperatures) under laser irradiation and has a doublet spectrum with two peaks (referred to as R_1 and R_2) and a main body at 14,300–14,500 cm^{-1}, which is related to the stress level in the TGO layer. In particular, there is a strong linear relation between the wavenumber shift of the R_2 peak and the residual stresses in the TGO layer.

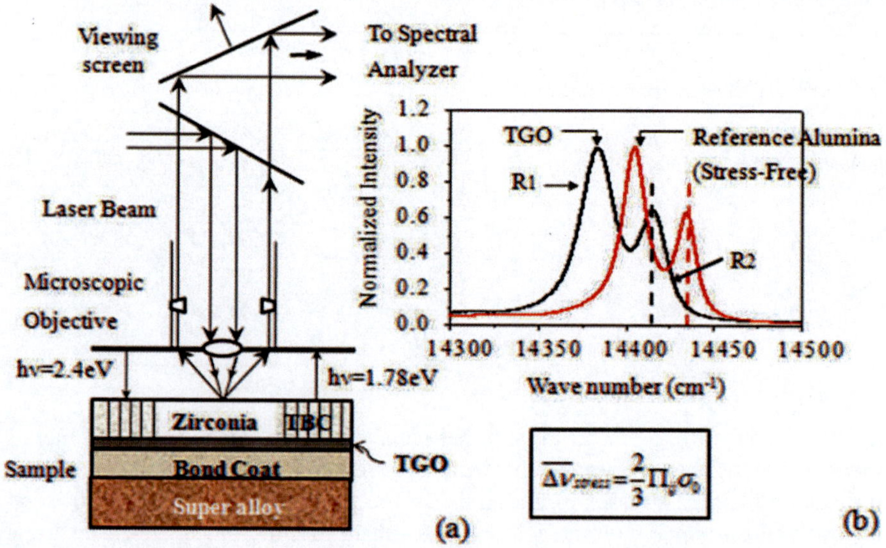

Fig. 10.45 **a** Schematic diagram of the principle for PLPS; **b** typical R_1/R_2 fluorescence spectra of a Cr^{3+}-containing α-Al_2O_3 layer with and without stresses, respectively [67]

$$\Delta\nu = \frac{1}{3}\Pi_{ii}\sigma_{jj} \quad (10.69)$$

where ν is the frequency shift of the R_2 fluorescence peak (unit: cm^{-1}), Π_{ii} is the stress coefficient of α-Al_2O_3 (unit: cm^{-1}/GPa), and σ_{jj} is the stress in a certain direction in the reference standard (unit: GPa). Assuming that the TGO layer is isotropic, the frequency shift can be expressed by the following equation:

$$\Delta\nu = \frac{1}{3}(\Pi_{11} + \Pi_{22} + \Pi_{33})(\sigma_{11} + \sigma_{22} + \sigma_{33}) \quad (10.70)$$

Let us assume that the coating is in a plane stress state and that $\sigma_{xx} = \sigma_{yy} = \sigma$ and $\sigma_{zz} = 0$. Then, we have

$$\Delta\nu_{stress} = \frac{2}{3}\Pi_{ij}\sigma \quad (10.71)$$

The precise wavenumber position of the R_2 peak can be determined through testing. On this basis, the stress in the coating can be determined. In Eq. (10.71), $\Delta\nu_{stress}$ is the wavenumber shift of the R_2 peak, $\sigma = \sigma_{xx} = \sigma_{yy}$ is the residual stress in the α-Al_2O_3 layer, and the value of the stress coefficient Π_{ij} is 7.61 cm^{-1}/GPa [68]. In the stress-free state, the wavenumbers of the R_1 and R_2 peaks of the fluorescence spectrum are 14,402 and 14,432 cm^{-1}, respectively. The residual stress is compressive when the characteristic peaks shift to the right (i.e., the wavenumbers increase) and tensile when characteristic peaks shift to the left.

Figure 10.46 shows the frequency spectra of the characteristic peaks of TBC specimens subjected to 1, 10, and 100 thermal cycles, respectively. With an increasing number of thermal cycles, the wavenumber positions of the two characteristic peaks of the fluorescence generated by Cr^{3+} ions both remain on the left side of the wavenumber position (14,432 cm^{-1}) corresponding to a stress-free state, indicating a compressive residual stress in the TGO layer at this time. The wavenumber positions of the peaks of the specimen subjected to 10 thermal cycles remain notably skewed to the left, indicating that the residual stress in the TGO layer continues to increase as the number of thermal cycles increases from 1 to 10. However, the measured position of the R_2 peak of the specimen subjected to 100 thermal cycles is skewed to the right, suggesting a change in the stress state in the coating and a decrease in the residual stress.

The Raman shift of the R_2 peak is directly proportional to the residual stresses in the TGO layer. On this basis, each specimen was measured at five different locations to determine the mean and deviation of the in-plane equibiaxial residual stress in the TGO layer. Figure 10.47 shows the evolution of the residual stress in the TGO layer with the number of thermal cycles. The evolution of the residual stress in the TBC involves four stages, i.e., the residual stress increases first rapidly and then stably, after which it decreases first rapidly and then slowly. Specifically, (a) as shown in Fig. 10.47, the residual stress in the TGO layer of the TBC is 2.35 GPa after one cycle

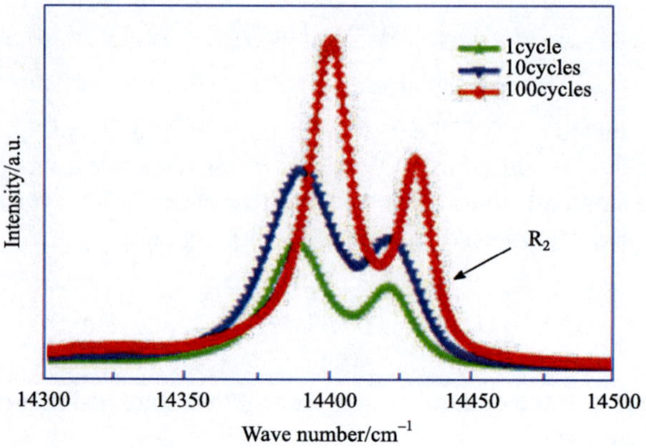

Fig. 10.46 PLPS-determined wavenumber positions of the R_2 fluorescence peak under different numbers of thermal cycles [67]

of high-temperature oxidation (approximately 50 min) and rapidly increases to 2.65 GPa after 10 cycles of oxidation; (b) subsequently, the stress increases at a slower rate, and the maximum residual stress (approximately 2.8 GPa) is reached after 20 cycles, at which point the residual stress in the TGO layer begins to decrease (i.e., an inflection point occurs); (c) the following cyclic oxidation (20–50 cycles) leads to a sharp decrease in the residual stress in the TGO layer to 0.6 GPa; and (d) the residual stress begins to decrease slowly during the subsequent cycles and is close to zero after 260 cycles.

Figure 10.48 shows SEM images of TBC specimens subjected to 1, 24, 28, 50, 180, and 260 thermal cycles at 1150 °C, respectively. The ceramic layer in the PVD TBC displays a columnar crystal structure. As the oxidation process proceeds, black

Fig. 10.47 Variation in the residual stress in the TGO layer with the number of thermal cycles [67]

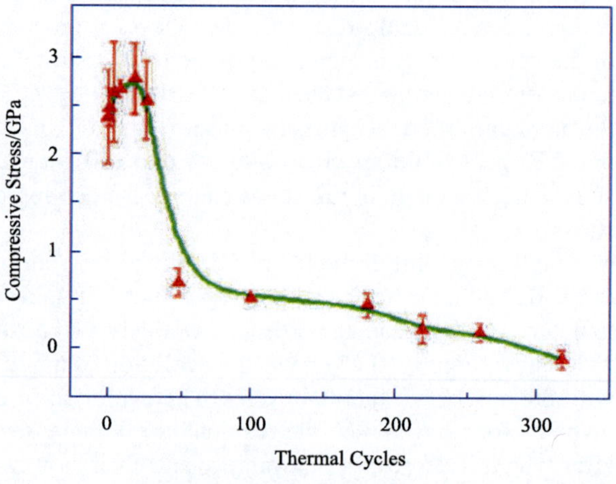

products are formed at the grain boundaries in the ceramic layer, suggesting that high-temperature oxidation also occurs in the ceramic layer. In Fig. 10.48, clear interfaces can be seen between the ceramic layer, the intermediate transition layer, and the substrate in the TBC before oxidation. After one cycle of oxidation, a continuous TGO layer with a thickness of approximately 1 μm is formed between the ceramic layer and the intermediate transition layer. According to energy spectrum analysis, the TGO layer is basically composed of Al_2O_3. This is because Al has a high oxidation capacity and is prone to selective oxidation. Generally, this oxide film is very dense and can reduce the oxidation rate of the TBC. In addition, Al_2O_3 is highly stable at high temperatures and can prevent the BC layer from being further oxidized. After 24 cycles, the TGO layer reaches a thickness of approximately 4 μm and is continually distributed along the interface, and some gray oxide particles are distributed in the BC layer. After 28 cycles, cracks begin to form in the TGO layer. After 50 cycles, the TGO layer reaches a thickness of 5.6 μm but becomes loose, accompanied by the formation of cracks. It is indicated that as the thickness of the TGO layer increases, the TBC undergoes interfacial damage, which gradually expands with further TGO growth until the coating spalls and fails. This phenomenon is consistent with previous findings in the literature [69] and further shows that TGO growth is a key factor causing TBC failure. After 180 and 260 cycles, notable cracks appear in the TGO layer, and its thickness reaches approximately 8 μm.

The PLPS measurements of the residual stress in the TGO layer show that the cyclic oxidation of the TBC is accompanied by a stress release after reaching a certain stage (after 28 cycles). We know that the stresses in a material are released upon crack formation or propagation. Therefore, it can be inferred from the decrease in the residual stress in the TGO layer that cracks form or propagate in the coating. The inflection point of the stress likely corresponds to the initial time point at which cracks are formed. To determine the correlation between stress evolution and interfacial failure in the TBC, we analyze the evolution of the residual stress in the TGO layer as well as the evolution of the interfacial microstructure. According to Figs. 10.47 and 10.48, an intact, dense TGO layer with no visible defects is formed on the TBC after one thermal cycle. After 24 cycles, the TGO layer thickens to a large extent but remains free of notable cracks and other defects. In this stage, the residual stress in the TGO layer increases significantly as the TGO layer thickens. Cracks are formed once the residual stress in the TGO layer accumulates to a certain extent and exceeds the fracture strength of the TGO layer or the interfacial bond strength. Hence, we can infer that the sharp decrease in the residual stress in the TGO layer during the subsequent cycles is a result of crack formation or propagation. This inference can be supported by the microstructure of the TBC. After 50 cycles, there is a significant morphological change in the TGO layer. Specifically, the TGO layer fragments, with cracks parallel to the interface formed in the middle. In addition, interface cracks appear between the ceramic and TGO layers, and embedded oxides that had grown into the BC layer are found in the TGO layer. The stress in the TBC decreased sharply to 0.6 GPa, suggesting that crack formation and TGO-layer fragmentation substantially reduced the residual stress in the TGO layer. After 180 cycles, the ceramic layer delaminates from the surface of the TBC. Moreover, the SEM image

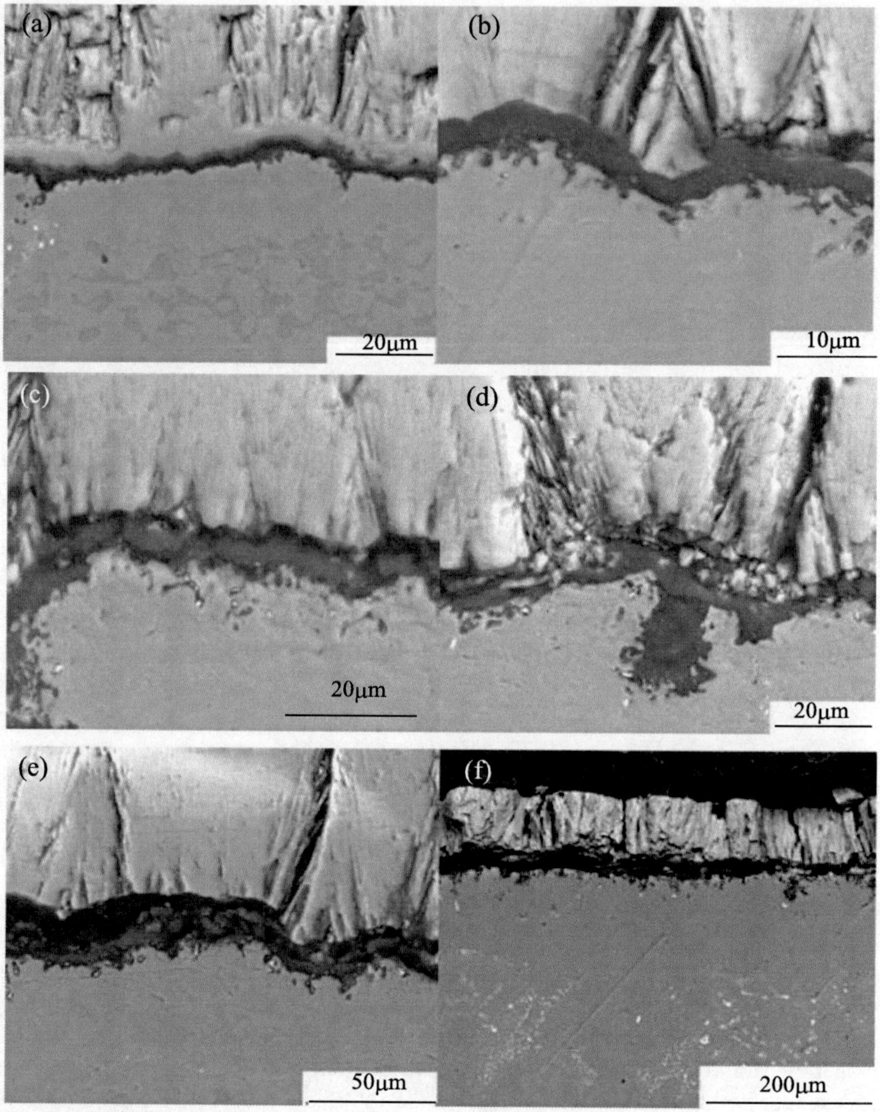

Fig. 10.48 Microstructure of PVD TBC specimens after **a** 1, **b** 24, **c** 28, **d** 50, **e** 180, and **f** 260 cycles of high-temperature oxidation [67]

shows the presence of large interface cracks between the ceramic and TGO layers and even interface cracking. After 260 cycles, the ceramic layer spalls in a large area. The SEM image shows that the Al_2O_3 in the TGO layer forms individual blocks or falls off due to the grinding and polishing process and that there is a notable gap between the ceramic and BC layers. In this stage, the residual stress decreases to zero, indicating that the whole TGO layer is in a free, unconstrained state. This conclusion indicates that we can determine the crack initiation time based on the evolution of the residual stress in the TGO layer, and this time corresponds to the number of cycles when there is an inflection point on the residual stress evolution curve of the TGO layer at which the residual stress sharply decreases from the maximum value. Furthermore, we can determine the time from crack propagation to coating spallation, corresponding to the number of cycles during which the residual stress in the TGO layer decreases to zero, as shown in Fig. 10.47. Therefore, the service life of a TBC can be estimated based on the above two times, which can then be used to provide guidance on its application and optimization.

When PLPS is used to measure the residual stresses in the TGO layer of a TBC, the laser beam needs to penetrate through the ceramic layer and excite the Cr^{3+} ions in the TGO layer to generate fluorescence signals. The morphology of the ceramic layer varies significantly with the preparation process. Sprayed ceramic layers have a stacked, layered structure, ceramic layers prepared by deposition have a columnar crystal structure, and PS-PVD coatings display a dendritic or featherlike structure [70]. The photoelectric effect between a ceramic layer and a laser beam varies with the morphology of the ceramic layer. When PLPS is used to measure the residual stresses in the TGO layer of a TBC, the ceramic-layer preparation process has a significant impact on the measurement.

Liu et al. [71] studied the feasibility of using PLPS to measure the residual stresses in the TGO layers of APS and EB-PVD TBCs. The PLPS technique employs a laser beam to penetrate the ceramic layer of a TBC to excite the Cr^{3+} ions in the TGO layer to generate a fluorescence spectrum and calculates the residual stresses in the TGO layer of the TBC based on the shift of the peaks of the fluorescence spectrum of the Cr^{3+} ions. When the PLPS nondestructive detection technique is applied to an APS TBC, the unique scaly structure of the APS ceramic layer causes the laser beam to diffuse. As a result, the intensity of the laser beam in the TGO layer is reduced, and the detection range is increased (i.e., the diameter of the laser beam in the TGO layer is transversely increased), thereby reducing the fluorescence signal intensity and lowering the spatial resolution, as shown in Fig. 10.49a [71]. Currently, PLPS is suitable for detecting residual stresses only in the TGO layers of 20–50-μm-thick APS TBCs [72]. An EB-PVD ceramic layer has a unique columnar crystal structure that can guide laser-beam propagation and does not cause severe laser-beam diffusion, [73] as shown in Fig. 10.49b [71]. Therefore, PLPS can be used to satisfactorily detect residual stresses in the TGO layers of EB-PVD TBCs.

While an APS ceramic layer significantly reduces the laser-beam signal intensity, it does not alter the detection results obtained from the laser beam-excited fluorescence spectrum (peak shifts), as shown in Fig. 10.49a [71]. Hence, we hope that through appropriate instrument calibration and settings, higher-intensity laser beams can irradiate APS TBCs to excite sufficient fluorescence signals that can then be collected to calculate the residual stresses in the TGO layer. The laser power, exposure time, and cumulative number are the primary parameters that need to be adjusted for this technique, which, however, in no way affects the evaluation of the peak shifts. Lima et al. [74] conducted a relevant study and successfully measured the residual stresses in the TGO layer of an APS TBC with a 250-μm-thick ceramic

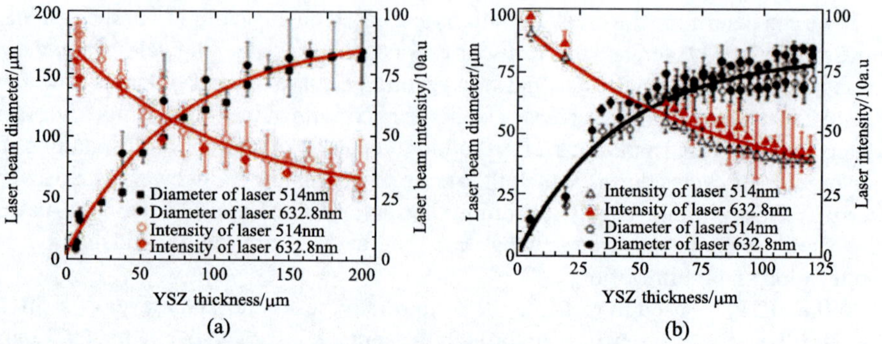

Fig. 10.49 Variation in the diameter of a red laser beam (k = 632.8 nm) and a green laser beam (k = 514 nm) with the YSZ thickness: **a** APS YSZ and **b** EB-PVD YSZ [71]

layer, proving the feasibility of using PLPS to measure residual stresses in the TGO layer of an APS TBC.

Rossmann et al. [75] and Tao et al. [76] investigated the use of PLPS systems to nondestructively detect residual stresses in the TGO layers of PS-PVD TBCs. Their test results show that the use of PLPS to detect residual stresses in the TGO layer of PS-PVD TBCs is completely feasible. While the feasibility of using PLPS to detect residual stresses in the TGO layers of PS-PVD TBCs has been proven through testing, the photoelectric effect between the laser beam and the microstructure of the coating during the detection process requires further investigation. Since a PS-PVD TBC is microstructurally similar to an EB-PVD TBC, it is preliminarily inferred that the voids in the dendritic or featherlike microstructure [70] of the ceramic layer in a PS-PVD TBC may guide laser-beam propagation and that a PS-PVD TBC exhibits a photoelectric effect similar to that of an EB-PVD TBC and has good versatility.

10.5 Summary and Outlook

10.5.1 Summary

This chapter discusses three aspects of residual stresses in TBCs, namely, causes and influencing factors, simulation and prediction, and destructive and nondestructive characterization methods, as well as illustrates the impacts of residual stresses on TBC failure and the relevant mechanisms. The following is a summary of this chapter:

(1) There are three main types of residual stress in a TBC, namely, quenching stress, thermal mismatch stress, and phase-transformation stress. The grain morphology, cooling rate, and TGO formation each have a significant impact on residual stresses in TBCs.
(2) The residual stresses in a TBC on a turbine blade are simulated and predicted using two approaches based on the first boundary condition (temperature boundary) and fluid–solid coupling, respectively. The residual stresses obtained

by the fluid–solid coupling approach are more consistent with the stress-field distribution than those obtained by the fixed temperature approach.

(3) Methods for quantitative characterization of residual stresses in TBCs are established based on three destructive detection methods, namely, the curvature method, the drilling method, and the RCM. On this basis, the residual stress evolution pattern is obtained.

(4) The methods for quantitative characterization of residual stresses in TBCs are established based on nondestructive detection methods such as XRD and Raman spectroscopy. On this basis, the influences of residual stress on TBC failure and the relevant mechanisms are determined.

10.5.2 Outlook

The currently available methods for destructive and nondestructive characterization of residual stresses are suitable for testing only TBC test specimens. The global detection of the residual stresses in the TBC of a turbine blade faces tremendous challenges. The curvature of a blade has an enormous impact on the measuring accuracy for the residual stresses in the TBC. Therefore, the global detection of residual stresses in the TBC of a turbine blade using PLPS and the testing and production of contour plots of stresses in large-area curved specimens through the linkage and coordination of optical focusing and four-axis mechanical loading platforms are important directions for future development for realizing nondestructive testing and life prediction of TBCs.

References

1. Das B, Gopinath M, Nath AK, Bandyopadhyay PP. Effect of cooling rate on residual stress and mechanical properties of laser remelted ceramic coating. J Eur Ceram Soc. 2018;38:3932–44.
2. Li GR, Lei J, Yang GJ, Li CX, Li CJ. Substrate-constrained effect on the stiffening behavior of lamellar thermal barrier coatings. J Eur Ceram Soc. 2018;38:2579–87.
3. Zhang X, Watanabe M, Kuroda S. Effects of processing conditions on the mechanical properties and deformation behaviors of plasma-sprayed thermal barrier coatings: evaluation of residual stresses and mechanical properties of thermal barrier coatings on the basis of in situ curvature measurement under a wide range of spray parameters. Acta Mater. 2013;61:1037–47.
4. Lee MJ, Lee BC, Lim JG, Kim MK. Residual stress analysis of the thermal barrier coating system by considering the plasma spraying process. J Mater Sci Technol. 2014;28:2161–8.
5. Keyvani A, Bahamirian M, Kobayashi A. Effect of sintering rate on the porous microstructural, mechanical and thermomechanical properties of YSZ and CSZ TBC coatings undergoing thermal cycling. J Alloy Compd. 2017;727:1057–66.
6. Wang Y, Tian W, Yang Y, Li CG, Wang L. Investigation of stress field and failure mode of plasma sprayed Al_2O_3–13%TiO_2 coatings under thermal shock. Mater Sci Eng, A. 2009;516:103–10.
7. Widjaja S, Limarga AM, Yip TH. Modeling of residual stresses in a plasma sprayed zirconia/alumina functionally graded-thermal barrier coating. Thin Solid Films. 2003;434:216–27.

8. Boley BA, Weiner JH. Theory of Thermal Stresses. Courier Corporation;2012.
9. Song X, Zhang J, Lin C, Liu Z, Jiang C, Kong M, Zeng Y. Microstructures and residual strain/stresses of YSZ coatings prepared by plasma spraying. Mater Lett. 2019;240:217–20.
10. Nicholls JR, Lawson KJ, Johnstone A, Rickerby DS. Methods to reduce the thermal conductivity of EB-PVD TBCs. Surf Coat Technol. 2002;151:383–91.
11. Zhang XC, Xu BS, Wang HD, Wu YX. Modeling of the residual stresses in plasma-spraying functionally graded ZrO_2/NiCoCrAlY coatings using finite element method. Mater Des. 2006;27:308–15.
12. Lima CRC, Cinca N, Guilemany JM. Study of the high temperature oxidation performance of thermal barrier coatings with HVOF sprayed bond coat and incorporating a PVD ceramic interlayer. Ceram Int. 2012;38:6423–9.
13. Mehboob G, Liu MJ, Xu T, Hussain S, Mehboob G, Tahir A. A review on failure mechanism of thermal barrier coatings and strategies to extend their lifetime. Ceram Int. 2020;46:8497–521.
14. Jung IH, Bae KK, Song KC, Yang MS, Ihm SK. Columnar grain growth of yttria-stabilized-zirconia in inductively coupled plasma spraying. J Therm Spray Technol. 2004;13:544–53.
15. Hutchinson RG, Fleck NA, Cocks ACF. A sintering model for thermal barrier coatings. Acta Mater. 2006;54:1297–306.
16. Kumar S, Cocks ACF. Sintering and mud cracking in EB-PVD thermal barrier coatings. J Mech Phys Solids. 2012;60:723–49.
17. Liu Q. Finite element simulation of the failure of turbine blades with TBCs. Xiangtan University;2012.
18. Zhou YC, Hashida T. Coupled effects of temperature gradient and oxidation on thermal stress in thermal barrier coating system. Int J Solids Struct. 2001;38(24):4235–64.
19. Mao WG, Zhou YC, Yang L, et al. Modeling of residual stresses variation with thermal cycling in thermal barrier coatings. Mech Mater. 2006;38(12):1118–27.
20. Xu ZH, Yang Y, Huang P, et al. Determination of interfacial properties of thermal barrier coatings by shear test and inverse finite element method. Acta Mater. 2010;58(18):5972–9.
21. Busso EP, Qian ZQ, Taylor MP, et al. The influence of bondcoat and topcoat mechanical properties on stress development in thermal barrier coating systems. Acta Mater. 2009;57(8):2349–61.
22. Tamarin Y, International A. Protective coatings for turbine blades. Asm Intl;2002.
23. Yang L, Liu QX, Zhou YC, Mao WG, Lu C. Finite element simulation on thermal fatigue of a turbine blade with thermal barrier coatings. J Mater Sci Technol. 2014;30:371–80.
24. Wang F. Computational fluid dynamics analysis—principles and applications of CFD software. Tsinghua University Press;2004.
25. Zhu W, Wang JW, Yang L, Zhou YC, Wei YG, Wu RT. Modeling and simulation of the temperature and stress fields in a 3D turbine blade coated with thermal barrier coatings. Surf Coat Technol. 2017;315:443–53.
26. Bai Y, Ding C, Li H. Isothermal oxidation behavior of supersonic atmospheric plasma-sprayed thermal barrier coating system. J Therm Spray Technol. 2013;22(7):1201–9.
27. Moridi A, Azadi M, Farrahi GH. Numerical simulation of thermal barrier coating system under thermo-mechanical loadings. Proceed World Congress Eng. 2011;3:1959–64.
28. Busso EP, Evans HE, Qian ZQ, Taylor MP. Effects of breakaway oxidation on local stresses in thermal barrier coatings. Acta Mater. 2010;58(4):1242–51.
29. Yang L, Zhou Y C, Lu C. Damage evolution and rupture time prediction in thermal barrier coatings subjected to cyclic heating and cooling: an acoustic emission method. Acta Materialia. 2011;59:6519.6529.
30. Yang L, Zhou YC, Mao WG, Lu C. Real-time acoustic emission testing based on wavelet transform for the failure process of thermal barrier coatings. Appl Phys Lett. 2008;93:231906.
31. Zhou Y, Liu Q, Yang L, Wu D, Mao W. Failure mechanisms and life prediction of thermal barrier coatings. Acta Mech Solida Sin. 2010;31(5):504–31.
32. Wu RT, Osawa M, Yokokawa T. Degradation mechanisms of an advanced jet engine service-retired TBC component. J Solid Mech Mater Eng. 2010;4(2):119.130.

33. Hobson MK, Reiter H. Residual stress in ZrO_2-8% Y_2O_3 plasma-sprayed thermal barrier coatings. Surf Coat Technol. 1988;33:33–42.
34. Zeng Y. A study on the measurement of the Young's moduli of and residual stresses in thermal barrier coatings. Jiangsu University;2019.
35. Forschelen PJJ, Suiker ASJ, Sluis OVD. Effect of residual stress on the delamination response of film-substrate systems under bending. Int J Solids Struct. 2016;97–98:284–99.
36. Khan AN, Lu J, Liao H. Effect of residual stresses on air plasma sprayed thermal barrier coatings. Surf Coat Technol. 2003;168:291–9.
37. Limarga AM, Vaßen R, Clarke DR. Stress distributions in plasma-sprayed thermal barrier coatings under thermal cycling in a temperature gradient. J Appl Mech. 2011;78(1):011003.
38. Yang LX, Yang F, Long Y, Zhao Y, Xiong X, Zhao XF, Xiao P. Evolution of residual stress in air plasma sprayed yttria stabilised zirconia thermal barrier coatings after isothermal treatment. Surf Coat Technol. 2014;251:98–105.
39. Mathar J. Determination of initial stresses by measuring the deformation around drilled holes. Trans ASME. 1934;56(4):249.254.
40. Soete W, Vancrombrugge R. An industrial method for the determination of residual stresses. Proc SESA. 1950;8(1):17–28.
41. Chi G. Testing of the residual stresses in thermal barrier coating materials based on the drilling method and digital image correlation. Xiangtan University;2015.
42. Wu Z, Lu J, Han B. Study of residual stress distribution by a combined method of Moire interferometry and incremental hole drilling, part II: implementation. J Appl Mech. 1998;65(4):844–50.
43. Woo W, An GB, Kingston EJ, De Wald AT, Smith J, Hill MR. Through-thickness distributions of residual stresses in two extreme heat-input thick welds: a neutron diffraction, contour method and deep hole drilling study. Acta Mater. 2013;61(10):3564–74.
44. Chen Y-H, Chen X, Xu N, Yang L. The digital image correlation technique applied to hole drilling residual stress measurement: SAE Technical Paper;2014.
45. Mao WG, Wan J, Dai CY, Ding J, Zhang Y, Zhou YC, Lu C. Evaluation of microhardness, fracture toughness and residual stress in a thermal barrier coating system: a modified Vickers indentation technique. Surf Coat Technol. 2012;206(21):4455–61.
46. Liu D, Seraffon M, Flewitt PEJ, Simms NJ, Nicholls JR, Rickerby DS. Effect of substrate curvature on residual stresses and failure modes of an air plasma sprayed thermal barrier coating system. J Eur Ceram Soc. 2013;33(15–16):3345–57.
47. Clyne T, Gill S. Residual stresses in thermal spray coatings and their effect on interfacial adhesion: a review of recent work. J Therm Spray Technol. 1996;5(4):401–18.
48. Godoy C, Souza E, Lima M, Batista J. Correlation between residual stresses and adhesion of plasma sprayed coatings: effects of a post-annealing treatment. Thin Solid Films. 2002;420:438–45.
49. Zhang X, Li C, Withers PJ, Markocsan N, Xiao P. Determination of local residual stress in an air plasma spray thermal barrier coating (APS-TBC) by microscale ring coring using a picosecond laser. Scripta Mater. 2019;167:126–30.
50. Sebastiani M, Eberl C, Bemporad E, Pharr GM. Depth-resolved residual stress analysis of thin coatings by a new FIB–DIC method. Mater Sci Eng, A. 2011;528:7901–8.
51. Wakui T, Malzbender J, Steinbrech RW. Strain analysis of plasma sprayed thermal barrier coatings under mechanical stress. J Therm Spray Technol. 2004;13:390–5.
52. Tang F, Schoenung JM. Evolution of Young's modulus of air plasma sprayed yttria-stabilized zirconia in thermally cycled thermal barrier coatings. Scripta Mater. 2006;54:1587–92.
53. Harok V, Neufuss K. Elastic and inelastic effects in compression in plasma-sprayed ceramic coatings. J Therm Spray Technol. 2001;10:126–32.
54. Thompson JA, Clyne TW. The effect of heat treatment on the stiffness of zirconia top coats in plasma-sprayed TBCs. Acta Mater. 2001;49:1565–75.
55. Zhu JG, Xie HM, Hu ZX, Chen PW, Zhang QM. Cross-sectional residual stresses in thermal spray coatings measured by moire´ interferometry and nanoindentation technique. J Therm Spray Technol. 2012;21:810–7.

56. Ranjbar-Far M, Absi J, Mariaux G, Dubois F. Simulation of the effect of material properties and interface roughness on the stress distribution in thermal barrier coatings using finite element method. Mater Des. 2010;31:772–81.
57. Korsunsky AM, Sebastiani M, Bemporad E. Residual stress evaluation at the micrometer scale: analysis of thin coatings by FIB milling and digital image correlation. Surf Coat Technol. 2010;205:2393–403.
58. Li C, Jacques SDM, Chen Y, Xiao P, Beale AM, Michiel MD, Markossan N, Nylen P, Cernik RJ. Precise strain profile measurement as a function of depth in thermal barrier coatings using high energy synchrotron X-rays. Scripta Mater. 2016;113:122–6.
59. Zhou S. Principles, methods, and applications of X-ray diffraction. Chongqing University Press;1991, 29.55.
60. Chen Q. Test analysis of the fracture toughness and residual stresses in thermal barrier coatings under thermal cycling conditions. Xiangtan University;2010.
61. Teixeira V, Andritschky M, Fischer W, Buchkremer HP, Stöver D. Analysis of residual stress in thermal barrier coatings. J Mater Process Technol. 1999;92–93:209.216.
62. Hamacha R, Dionnet B, Grimaud A, Nardou F. Residual stress evolution during the thermal cycling of plasma-sprayed zirconia coatings. Surf Coat Technol. 1996;80:295–303.
63. Jordan DW, Faber KT. X-ray residual stress analysis of a ceramic thermal barrier coating undergoing thermal cycling. Thin Solid Films. 1993;235:137–41.
64. Dong ZL, Khor KA. Microstructure formation in plasma sprayed functionally graded NiCoCrA1Y/Yttria-stabilized Zirconia coatings. Surf Coat Technol. 1999;11:181–6.
65. Wang D, Xu B, Dong S. Research progress on practical measurement of residual stress in coatings. Heat Treat Met. 2006;31(5):48–53.
66. Gell M, Sridharan S, Wen M, et al. Photoluminescence Piezospectroscopy: a multi-purpose quality control and NDI technique for thermal barrier coatings. Int J Appl Ceram Technol. 2004;1(4):316–29.
67. Zhou C. A study on the correlation between interfacial failure and stress evolution in the oxidation layer of thermal barrier coatings. Xiangtan University, 2014.
68. Wu F, Jordan EH, Ma X, Gell M. Thermally grown oxide growth behavior and spallation lives of solution precursor plasma spray thermal barrier coatings. Surf Coat Technol. 2008;202:1628–35.
69. Xu H, Gong S, Liu F. Recent development in materials design of thermal barrier coatings for gas turbine. Acta Aeronautica Et Astronautica Sinica. 2000;1:7–12.
70. Niessen KV, Gindrat M, Refke A. Vapor phase deposition using plasma spray-PVD. J Therm Spray Technol. 2010;19(1–2):502–9.
71. Liu D, Lord O, Stevens O, et al. The role of beam dispersion in Raman and photo-stimulated luminescence piezo-spectroscopy of yttria-stabilized zirconia in multi-layered coatings. Acta Mater. 2013;61(1):12–21.
72. Schlichting KW, Vaidyanathan K, Sohn YH, et al. Application of Cr3+ photoluminescence piezo-spectroscopy to plasma-sprayed thermal barrier coatings for residual stress measurement. Mater Sci Eng A. 2000;291(1–2):68–77.
73. Tolpygo VK, Clarke DR, Murphy KS. Evaluation of interface degradation during cyclic oxidation of EB-PVD thermal barrier coatings and correlation with TGO luminescence. Surf Coat Technol. 2004;188:62–70.
74. Lima CRC, Dosta S, Guilemany JM, et al. The application of photoluminescence piezospectroscopy for residual stresses measurement in thermally sprayed TBCs. Surf Coat Technol. 2017;318:147–56.
75. Rossmann L, Northam M, Sarley B, et al. Investigation of TGO stress in thermally cycled plasma-spray physical vapor deposition and electron-beam physical vapor deposition thermal barrier coatings via photoluminescence spectroscopy. Surface Coating Technol. 378:125047.
76. Tao S, Yang J, Zhao H, et al. Evolution of residual stresses in PS-PVD TBCs with thermal cycling. Thermal Barrier Coatings, 2017;1.

Chapter 11
Real-Time Acoustic Emission Characterization of Cracks in TBCs

During long-term service in extreme conditions (e.g., gas thermal shock, air-film cooling, centrifugal forces, particle erosion, calcium–magnesium aluminosilicate (CMAS) corrosion, fatigue, creep, and temperature gradients, accompanied by chemical reactions), the thermal barrier coating (TBC) systems on the turbine blades in aeroengines sustain various types of damage (e.g., vertical, horizontal, and oblique cracks as well as opening and sliding interface cracks) due to a variety of mechanisms (e.g., thermal mismatch, interfacial oxidation, corrosion and infiltration, and phase transformation). These cracks are the root cause of the eventual spallation but differ completely in terms of the contribution to spallation and the mechanism. Real-time nondestructive testing of the crack initiation and evolution processes is an effective means for determining coating spallation mechanisms in various environments.

Acoustic emission (AE) testing is a nondestructive technique for detecting the damage process from strain energy (stress wave) signals based on the physical phenomenon in which the formation of damage is inevitably accompanied by a release of strain energy. In addition, AE signals vary with material properties, the form of damage, and external loading. Hence, the AE technique has become an important tool for detecting crack initiation and evolution. However, for TBCs, the AE technique faces three tremendous challenges: (1) the detection of crack signals in complex environments (e.g., high-temperature gas), (2) the identification of various complex crack patterns, and (3) the quantitative evaluation of the extent of the damage.

This chapter focuses on the AE detection, pattern recognition, and quantitative evaluation methods for complex high-temperature environments as well as the major failure mechanisms of TBCs revealed by real-time AE characterizations.

Y. Zhou et al., *Thermal Barrier Coatings: Failure Theory and Evaluation Technology*,
https://doi.org/10.1007/978-981-19-2723-2_11

11.1 High-Temperature AE Detection Method

11.1.1 Basic Principle of AE Detection

The AE technique originated from Kaiser's research work in Germany in the 1950s [1] and was later extensively studied in industrial countries (e.g., the United States and Japan) and gradually applied to various fields [2–4]. AE refers to the phenomenon in which a material emits transient elastic waves due to an instantaneous energy release in local areas under the action of external factors and is therefore also called stress-wave emission. AE is a very common physical phenomenon that accompanies the deformation and fracture of most materials. A release of a sufficient amount of energy can generate audible sounds. However, for most materials, the AE signals generated by deformation or fracture are too weak to be directly detected by the human ear and thus require some sensitive electronic devices for detection. The AE technique detects, records, and analyzes AE signals using instruments and, on this basis, infers the AE source. As a dynamic nondestructive detection technique, the AE technique involves a number of basic concepts, including the AE source, wave propagation, acoustic/electric energy conversion, signal processing, data display and recording, and analysis and evaluation. Figure 11.1 shows the basic principle of the AE technique.

11.1.2 Waveguide Rod/Wire Transmission Technique for Complex High-Temperature Environments

Complex high-temperature conditions (e.g., high-temperature thermal cycling, gas thermal shock, high-temperature oxidation, high-temperature CMAS corrosion, thermomechanical combination, and thermo–mechano–chemical coupling) are inevitable service loads to which TBCs are subjected [5]. Therefore, signal detection under these complex conditions is necessary for understanding the failure process and mechanism of TBCs [6, 7]. However, the available AE sensors can only be used at room temperature, which therefore necessitates the use of devices such as waveguide rods (WRs), waveguide wires (WWs), and wireless AE sensors to achieve

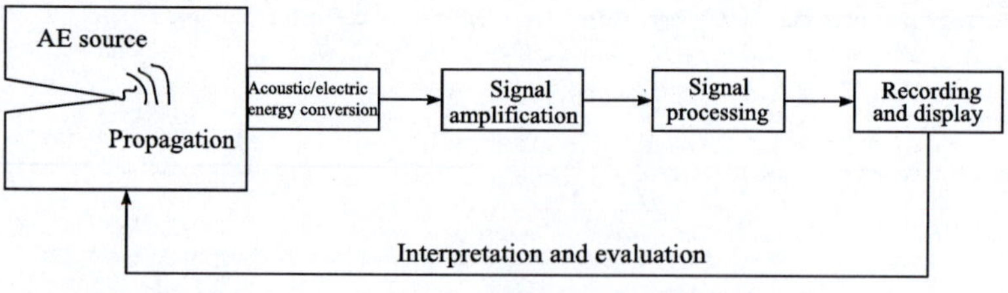

Fig. 11.1 Schematic diagram of the basic principle of the AE technique

real-time detection of crack signals in TBCs in complex high-temperature environments [8–10]. However, the wireless AE sensor technique has yet to mature, and its sensitivity and effectiveness for signals requires further investigation. Hence, this chapter primarily discusses the AE signal detection method based on the WR/WW technique.

When used in signal transmission, one end of a WR (or WW) is closely connected to the specimen via a sealing agent and mechanical devices to ensure good contact between the WR and the specimen, while the other end of the WR is coupled with an AE sensor. Connecting a test specimen and a sensor with a WR prevents direct contact between the sensor and the high-temperature specimen. Generally, a WR is a metal rod with a low acoustic impedance and a high-temperature resistance (e.g., a stainless-steel tube with a diameter of 5 mm). It is difficult to connect large-diameter WRs to TBC specimens, particularly those on turbine blades with a complex curved structure. Therefore, an ideal WR (or WW) for TBCs should have the following properties and capabilities: (1) High flexibility, which allows the WR to easily move with the TBC specimen and prevents the WR itself from generating vibration or stress waves that lead to the formation of an additional AE source. (2) Capability to realize rapid, low-attenuation transmission of stress waves, which prevents the WR from losing weak AE signals. (3) Capability to minimize waveform distortion. In fact, the difference between a WR and a specimen in material properties and geometric dimensions causes waveform changes in the AE signals during propagation in the WR, thereby making the pattern recognition of the AE source inconvenient. Therefore, a WR that causes the least possible waveform distortion should be selected. (4) A high melting point. This property ensures that the WR does not melt in a high-temperature furnace. On this basis, we present a WW transmission technique specifically developed for the complex service conditions (e.g., high-temperatures and gas thermal shock) and complex curved structure of TBCs on turbine blades [10, 11]. In this technique, metal wires capable of moving freely and with good acoustic transmission capabilities are used to collect signals.

On this basis, we analyzed the acoustic transmission characteristics of several types of WWs, including steel, Fe, Al, and Pt wires, as well as the transmitted acoustic waves in an Al rod to determine the difference between WWs and WRs. The WW technique was implemented as follows. An ultrasonic machine was used to emit ultrasonic waves with the same amplitude, frequency, and energy via a sensor, which were then allowed to propagate in the selected WWs. Subsequently, the waves were received by another sensor, and their waveforms were stored. Except for the Al rod, each WW had a diameter of 0.5 mm and a length of 1 m. As shown in Fig. 11.2, the Pt wire received the ultrasonic wave signal the earliest; in other words, the acoustic wave traveled the fastest and had the largest amplitude in the Pt wire, suggesting that this wire exhibited the least ultrasonic wave attenuation. Thus, Pt wires can be used as a WW material. A test on the Al wire and rod, with a low melting point, was conducted to compare the influence of the diameter on the acoustic transmission characteristics. As seen in Fig. 11.2, the ultrasonic signal transmitting in the Al rod was received earlier and attenuated less significantly than that in the Al wire, suggesting that a large diameter is favorable to ultrasonic wave propagation.

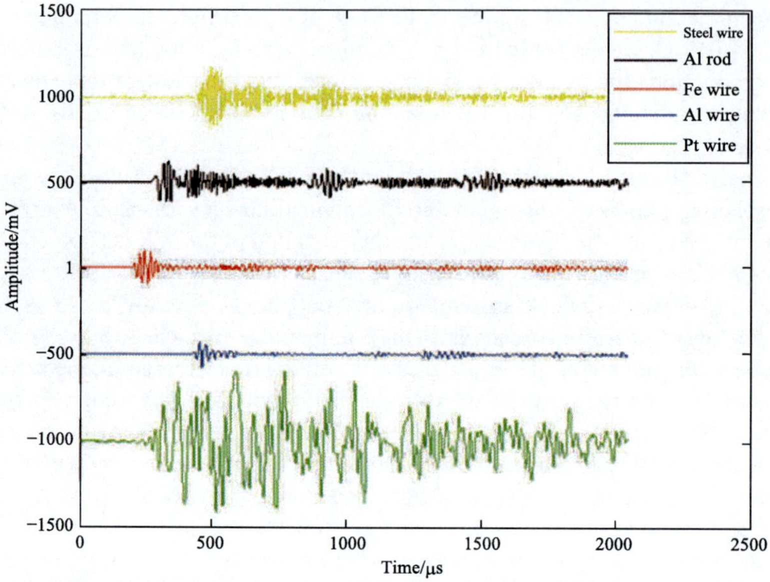

Fig. 11.2 Comparison of the acoustic transmission characteristics of different WWs and a WR

Exercise 11.1 Design and describe a real-time AE signal detection method for TBCs under high-temperature thermal cycling.

High-temperature thermal cycling is a common test method for simulating the heating and cooling service conditions for TBCs in aeroengines at take-off and landing. During heating and cooling, defects are formed in TBCs due to factors such as thermal mismatch and oxidation. Nondestructive testing for this process can provide a direct basis for determining TBC spallation mechanisms (e.g., whether cracks are formed during the heating or cooling stage).

Real-time detection of the damage sustained by a TBC during this process involves the following basic steps:

(1) Select a suitable Pt WW based on Sect. 11.1.2 or the actual test conditions.
(2) Weld one end of the WW to the bottom surface of the substrate of the TBC, and connect the other end of the WW to an AE sensor through mechanical forces. Subsequently, perform an active ultrasonic wave test and determine the connectivity of the WW based on the waveform, energy, and amplitude of the signal.
(3) Based on the schematic diagram in Fig. 11.3, heat the resistance furnace to the test temperature, then quickly open the door of the furnace and place the clamp holding the TBC specimen in the furnace for heating. Due to its excellent flexibility, the Pt wire is able to easily move with the specimen and does not vibrate. The AE wave emitted by the damage activity (AE source) in the specimen is transmitted through the WW, received by the AE sensor and subsequently reaches an AE instrument where it is amplified, recorded, and stored.

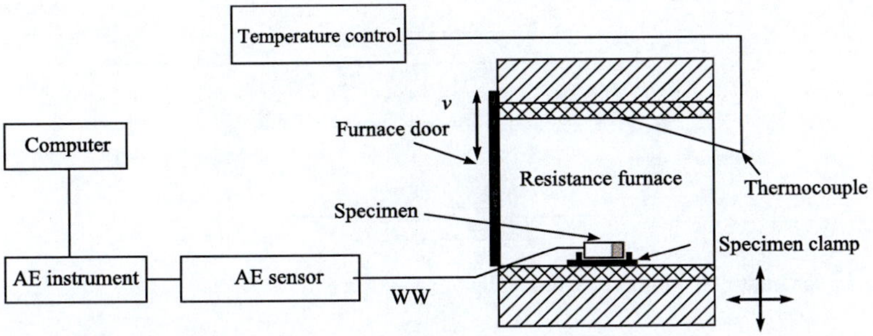

Fig. 11.3 An AE detection test setup for the thermal fatigue of a TBC

(4) During the cooling stage, quickly remove the specimen from the furnace and allow it to cool in air. During this stage, AE detection can also be achieved because the WW and the specimen are welded together.
(5) After the test is completed, analyze, and evaluate the whole AE signal. The relevant method is described after this section.

11.1.3 AE Signal Detection Method Based on Regional Signal Selection

Noise in complex environments has always been hindrance to AE detection. Here, we present a denoising method based on regional signal selection [11]. The basic concept of this method is that the maximum and minimum AE signal response times are set based on the acoustic transmission velocity in a specimen to detect the AE signal only from the region of interest, thereby eliminating the noise from other regions (e.g., the loading device and the environment).

Here, we use a 1D TBC rod specimen with a length of L (Fig. 11.4) as an example. The maximum distance that the AE signal generated by damage (i.e., a crack) that can be received by sensor A can travel is L. If sensor A receives an AE signal that travels a distance of L, it means that the crack is located at the other end of the specimen (i.e., the location of sensor B). Similarly, the maximum distance that the AE signal generated by damage that can be received by sensor B can travel is also L. If sensor B receives an AE signal that travels a distance of L, it means that the crack is located at the other end of the specimen (i.e., the location of sensor A). Let v be the acoustic transmission velocity in the TBC specimen. Then, the maximum response time of the signal from the damage (AE) source received by sensors A and B is

$$t_{\max} = \frac{L}{v} \tag{11.1}$$

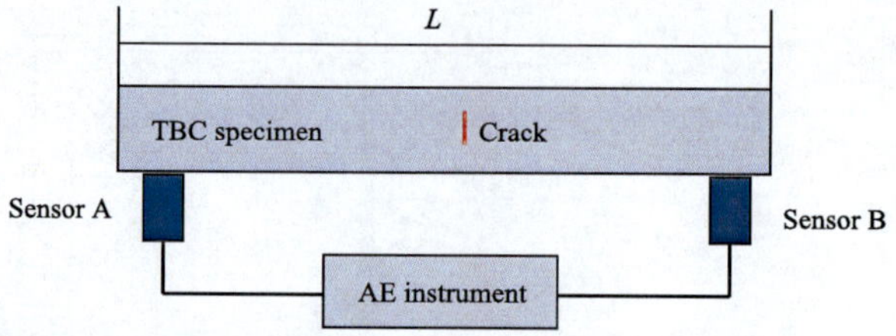

Fig. 11.4 Schematic diagram of regional signal detection

Thus, we can set signals that reach sensors A and B in no longer than t_{max} as effective signals and other signals as noise signals.

The regional signal detection method is highly advantageous in noise elimination. Here, a tensile test on a 1D TBC specimen is used as an example. In addition to those generated by the TBC due to damage (i.e., the effective signal), there are, in fact, AE signals caused by the noise from the tension clamps, the mechanical sound from the gearing during the movement of the clamps, the ambient noise in the laboratory, and the electrical noise. These types of noise can also be received and collected by the AE sensor. However, setting the maximum response time of effective AE signals based on the gauge length of the tensile test specimen (i.e., the effective test length) can shield noise from the clamps and mechanical gearing. For example, the distance between the clamps and the nonadjacent sensor is definitely greater than the gauge length. Therefore, the time taken for the AE signal generated due to the movement of the clamps or the damage sustained by the clamps to reach one of the two sensors must be longer than the maximum response time. Therefore, this AE signal can be categorized as a noise signal and subsequently shielded.

As an exercise, readers can design a maximum response time for a planar specimen (e.g., a plate-like test specimen). Hint: at least three sensors are required for a planar specimen.

Note that while it is impossible to simplify complex components such as turbine blades to 1D or 2D components and it is very difficult to design a probe to accurately calculate the maximum response time, it remains possible to set the maximum response time based on the maximum 3D size of a blade during actual detection to shield most of the noise from outside the specimen region. The AE signals collected due to an inaccurate response time (e.g., mechanical noise and electromagnetic interference) can be further eliminated through analysis of AE signal characteristics (e.g., frequency and energy), as is discussed in Sect. 11.3.

11.2 Analysis of the Key Parameters for Crack Pattern Recognition

11.2.1 *Key Failure Modes of TBCs and the Time-Domain Characteristics of the Relevant AE Signals*

Extensive studies [9–16] have shown that there are four failure modes, namely, surface cracks, sliding interface cracks, opening interface cracks, and plastic deformation of the substrate of TBCs under thermal loading, mechanical loading, and a combination of these. As shown in Fig. 11.5, surface cracks refer to cracks perpendicular to the interface in the ceramic coating and are often caused by the tensile stress σ_{11} in the ceramic layer due to the thermal mismatch between the layers. Sliding interface cracks refer to cracks caused by the shear stress σ_{21} at the ceramic/BC interface. The tensile stress σ_{22} at the ceramic/BC interface is the primary mechanism causing the delamination of the ceramic layer, i.e., the formation of opening interface cracks. Generally, σ_{22} is caused by the formation of nonuniform thermal growth oxides (TGOs) at a rough interface. Thermal, tensile, and compressive stresses can all lead to the plastic deformation of the substrate.

However, AE signals associated with various failure modes of TBCs cannot be distinguished in the time domain. As shown in Fig. 11.6, a vertical surface crack and an opening interface crack display no notable differences in waveform parameters such as the rise time, duration, energy, amplitude, and ring count (see Fig. 11.6c for the physical meaning of each parameter). While Ma et al. [17] note that the waveforms of vertical surface cracks and interface cracks differ in the sequential order of the longitudinal and transverse waves, these differences are extremely difficult to differentiate in practical analysis. It is exceedingly difficult to carry out pattern recognition based on time-domain AE signal waveforms. This is mainly because signals vary with the propagation distance, ambient noise, and measurement system and can even undergo waveform changes. Therefore, it is infeasible to perform pattern recognition based on AE signal waveforms (i.e., time-domain information).

11.2.2 *Pattern Recognition of TBC Failure Modes Based on Characteristic Frequencies*

We have established that it is impossible to identify damage patterns based on AE signals in the time domain (i.e., physical space), but is it possible to analyze them in another space (e.g., image space)? Is it possible to determine the key parameters for the recognition of damage patterns in image space?

Fourier transform (FT) shows that any continuously measured time series or signal can be expressed in terms of an infinite superposition of sinusoidal wave signals with

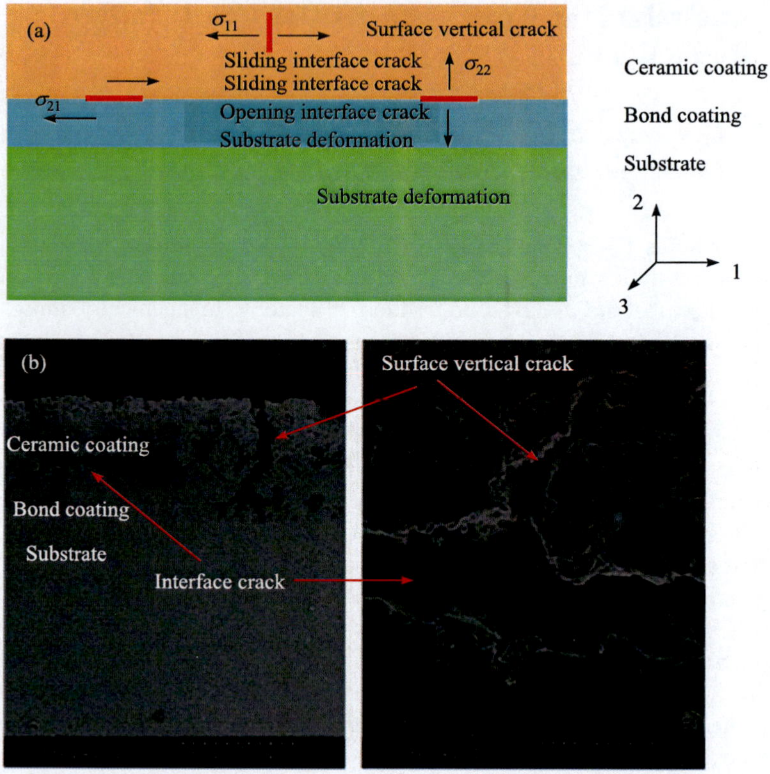

Fig. 11.5 **a** Four damage mechanisms: surface cracks, sliding interface cracks, opening interface cracks, and plastic deformation of the substrate. **b** SEM images of the surface and interface cracks

different frequencies, i.e., the frequencies, amplitudes, and phases of different sinusoidal signals in a directly measured original signal can be calculated through integration. From a physical perspective, the FT can be understood as a linear transform of an original function with a set of special functions (trigonometric functions) as the orthogonal basis, the physical meaning of which is the projection of the original function on each set of basis functions. The basic FT equation is as follows:

$$F(\omega) = F[f(t)] = \int_{-\infty}^{\infty} f(t)e^{-i\omega t}dt \tag{11.2}$$

where ω is the frequency, t is time, and $e^{-i\omega t}$ is the complex function.

Simple tensile and compressive tests can be performed on TBC specimens to analyze the AE frequency characteristics of various crack patterns in TBCs. Scanning electron microscopy (SEM) observations of the microstructure of a coating during different stages of the failure process show that under tensile loading, early-stage

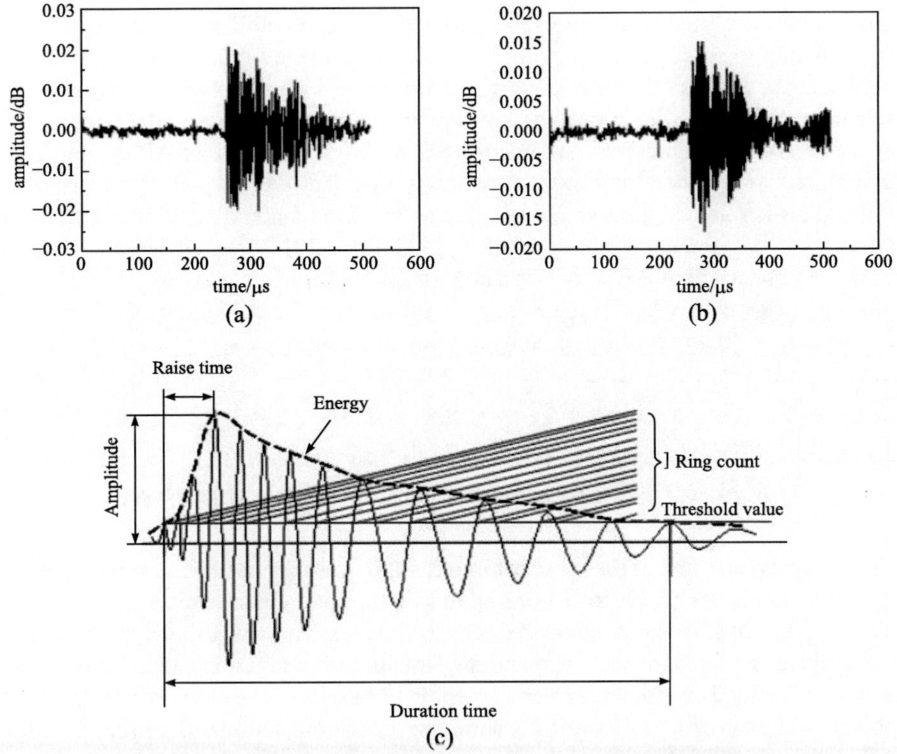

Fig. 11.6 **a** Waveform of the AE signal from a vertical surface crack. **b** Waveform of the AE signal from an opening interface crack. **c** Definition of the simplified waveform parameters for AE signals

cracks originate from the surface cracks in the coating, followed by the formation of sliding interface cracks. When the tensile stress reaches the yield strength of the substrate material, the substrate begins to undergo significant plastic deformation. In a compressive test, opening interface cracks are formed due to Poisson's ratio effect. Extensive statistical analysis of AE signals generated during each stage reveals an extremely important phenomenon—the frequency of early-stage AE signals (Fig. 11.7a) during the tensile test process is concentrated at 0.21 MHz (Fig. 11.7b). When the surface cracks are saturated and numerous interface cracks appear, a large number of AE signals (Fig. 11.7c) have a frequency of 0.29 MHz (Fig. 11.7d). The AE signals caused by the plastic deformation of the substrate (Fig. 11.7e) have a frequency of 0.14 MHz (Fig. 11.7f). In a compressive test, there are AE signals with a fourth frequency (Fig. 11.7g)—0.44 MHz (Fig. 11.7h)—in addition to those with the three frequencies that appear under tensile loading. While the frequencies of AE signals of each pattern fluctuate near their concentrated frequency, there are significant differences between those of different patterns. Hence, we conclude that "there is a correspondence between the AE signals generated by different types of cracks or deformation in TBCs and frequencies". In fact, our research article reporting this finding was quickly accepted and published in Applied Physics Letters in 2008 [18]. In the same year, Benson et al. of University College London noted in a report published in Science that the spectral characteristics of AE signals generated by seismic activity in volcanic structures vary with the seismic source model [19], which bears a striking resemblance to our research findings. In addition, AE detection in particle-enhanced metal-matrix composites shows that the frequency spectrum of AE signals is related only to the damage pattern of the material and independent of the crack size and the magnitude of the applied load [20]. Thus, we believe that the frequency spectrum of AE signals is a key parameter for identifying the failure mode of a TBC. However, whether the frequency spectrum is the only parameter and what its mechanism is require further analysis and verification.

11.2.3 Extraction of the Characteristic Parameters for Pattern Recognition Based on Cluster Analysis

Cluster analysis is one of the most commonly used unsupervised pattern recognition (UPR) methods. Its goal is to, according to a certain rule, divide objects to be classified into several clusters based on their characteristics. There is no need to determine these clusters in advance or to make any assumptions regarding their number or structure. Instead, these clusters are determined based on the characteristics of the objects to be classified. Figure 11.8 summarizes the cluster analysis procedure for AE signals [21–23]. First, an AE instrument is used to collect the AE signals generated by the specimen due to damage. Then, the simplified characteristic waveform parameters of the AE signals are extracted. Subsequently, the characteristic parameters are subjected to a cluster analysis based on a certain similarity measure and clustering criteria to group AE signals of the same or similar type. Finally, based on the cluster analysis results, relations between the types of damage and the characteristic parameters of the AE signals are established to determine the AE signals that can best characterize different damage patterns.

The similarity measure (e.g., the Euclidean distance) refers to a measure of the similarity between two signals. There are several types of clustering methods (also called clustering algorithms), including partitioning, hierarchical, density, grid, and model methods. *k*-means clustering (a hierarchical algorithm) can be used to carry out pattern recognition on the AE signals of TBCs. The main concept of this algorithm is described as follows.

k-means clustering is a dynamic clustering algorithm with the following basic principle [24]: First, the number of clusters, k, is predefined, and *k* samples are selected randomly or according to a certain rule as the initial cluster centers. Then, the remaining samples are classified based on the principle of proximity. Thus, an initial clustering scheme is obtained. In addition, the mean of each cluster is calculated to update its center. Subsequently, the samples are reclassified based on the new cluster centers. This process is repeated until the cluster centers converge, at which point clustering is completed. The basic procedure is described as follows:

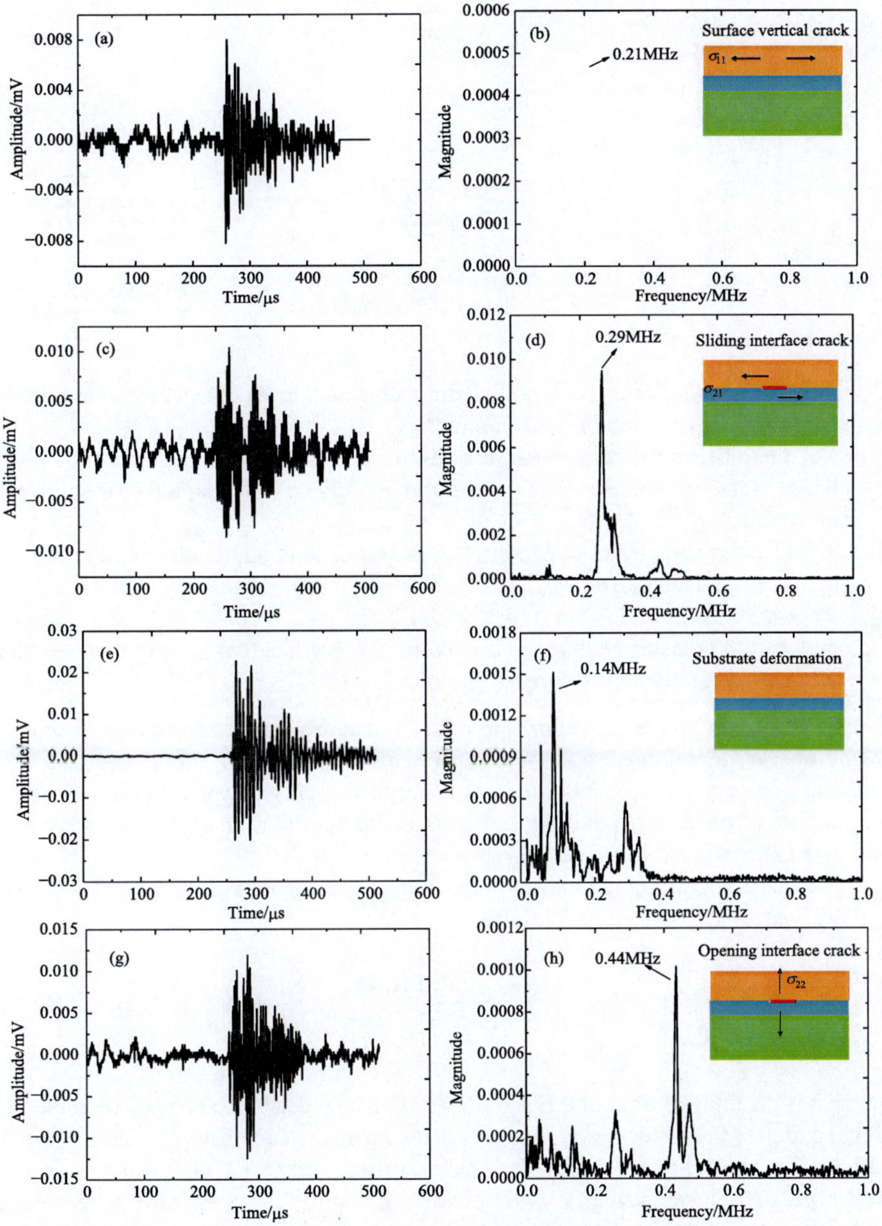

Fig. 11.7 Typical AE signals associated with TBC failure. **a**, **c**, **e**, and **g** are waveforms, and **b**, **d**, **f**, and **h** are FT spectra corresponding to the respective waveforms

Fig. 11.8 Flowchart for the cluster analysis of AE signals

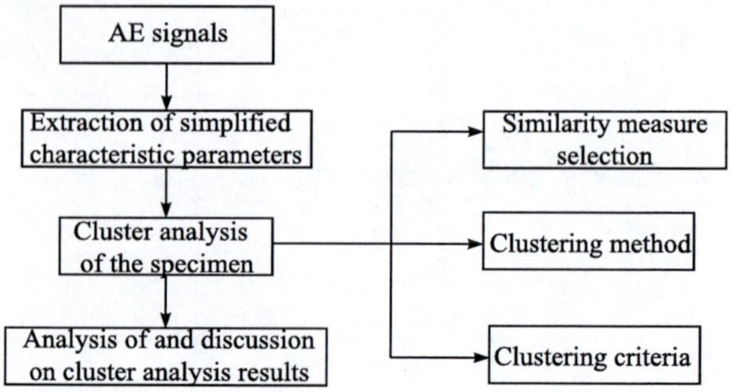

(1) Select k initial cluster centers C_i from a cluster sample consisting of n signals based on one or some of their parameters.
(2) Calculate the distance between each remaining signal and each cluster center based on the parameter values of the signal and assign the signal to the nearest cluster center.
(3) After all the signals are clustered, the cluster centers are recalculated based on the mean of each cluster.
(4) Repeat Steps (2) and (3) until the cluster locations are converged (i.e., the error between the cluster centers calculated in two consecutive iterations meets the requirements), then clustering is completed.

Evidently, the above procedure produces clustering results for cases in which the number of signal clusters, k, is determined or preselected. There exist two key problems, namely, the determination of the number of clusters for classification (i.e., the number of patterns, k) and the selection and determination of the characteristic parameters for signal classification.

k is generally determined based on the analysis of the silhouette value (s-value), which is calculated as follows [23–25]:

$$s(k) = \frac{1}{n}\sum_{1}^{n}\frac{[\min(b(i,k) - a(i))]}{\max[a(i), \min(b(i,k))]} \tag{11.3}$$

where n is the total number of AE signals, $b(i,k)$ is the mean distance between the ith signal and the kth cluster center, and $a(i)$ is the mean distance between the ith signal and the signals of the same type. The s-value varies with k. A high s-value indicates well-separated clusters, that is, good clustering results. It is generally agreed that when the s-value >0.6, clustering is highly effective, and the corresponding k is reasonable.

In terms of the selection and determination of parameters, a large number of signal parameters generally contain a large amount of information on damage patterns. Therefore, when the characteristic parameters are not determined, the signal parameter values should be extracted as completely as possible. However, if there is a strong correlation between the signal parameters (i.e., the parameters reflect similar information when used in pattern recognition), these parameters, as invalid parameters, can instead increase the computational and analysis workload [17]. Thus, for AE signals, we select the five most representative simplified characteristic waveform parameters, namely, the amplitude, peak frequency, energy, rise time, and duration, as clustering variables (i.e., the parameters for pattern recognition).

These five parameters differ in dimension, and their numerical values cannot be compared. It is therefore necessary to normalize the clustering variables, that is, to transform all the parameters to values in the range of [0, 1]. Mean–variance standardization is the most commonly used method for data standardization, the principle of which is to transform data to a standard normal distribution with a mean of 0 and a standard deviation of 1. Here, a specific parameter, amplitude, is used as an example. Let us assume that there are *n* AE signals. Then, the mean and standard deviation of the amplitudes of these *n* signals can be calculated. The normalized amplitude of the *i*th signal is expressed as follows [26, 27]:

$$\overline{A_i} = \frac{A_i - A_\mu}{A_\sigma} \tag{11.4}$$

where A_i and $\overline{A_i}$ are the amplitude and normalized amplitude of the signal, respectively, and A_μ and A_σ are the mean and standard deviation of the amplitudes of the *n* signals, respectively.

In specific analysis, one, multiple, or all of these five parameters can be selected for clustering. An optimal number of clusters, *k*, can be determined based on the variation in the *s*-value. Subsequently, the clustering results are displayed based on different parameters, and parameters capable of distinguishing different types of signals are extracted as the characteristic parameters for pattern recognition.

(1) **Identification of the TBC failure modes under tension and the extraction of the relevant characteristic parameters**

Using the abovementioned method, the amplitude, peak frequency, energy, rise time, and duration of 1457 AE signals from a TBC specimen during the tensile test process are selected as clustering variables for cluster analysis [23]. First, the number of patterns, *k*, is determined based on the *s*-value, as shown by the *s*–*k* curve in Fig. 11.9. The *s*-value reaches its maximum and is >0.6 when the AE signals are classified into four types, suggesting that the AE signals from the TBC under tension can be classified into four types.

Figure 11.10 shows the distributions of the clustered amplitudes and peak frequencies of the AE signals generated by the tensile test specimen due to damage. Figure 11.10 shows that the type A, B, and C signals are basically similar in terms of amplitude distribution, which ranges from 38 to 65 dB. However, their frequencies are distributed in significantly different bands (250–350, 170–250, and 40–150 kHz, respectively), indicating that frequency is an effective parameter (i.e., characteristic parameter) for differentiating them. Compared to the type A, B, and C signals, the

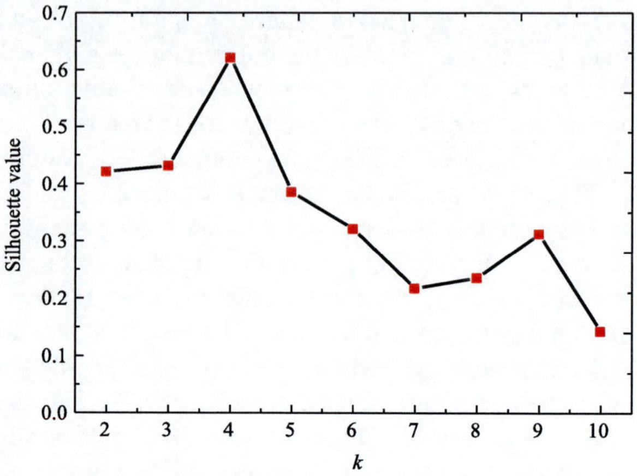

Fig. 11.9 *s*–*k* curve of the AE signals generated by a TBC specimen due to damage during the tensile failure process

type D signals have relatively large amplitudes but dispersed frequencies. It can then be concluded that type D is different from types A, B, and C and can be further analyzed based on the failure characteristics.

Figure 11.11a and b show the relationship between the distributions of the clustered peak frequencies and durations and the relation between the distributions of the clustered amplitudes and durations, respectively. As seen in Fig. 11.11a, the distributions of the durations of all but the type D signals coincide with each other and range from 0 to 500 μs, while their frequencies are distinguishable, which again suggests that frequency is a characteristic parameter for crack pattern recognition. Figure 11.11b clearly shows that there is a significant overlap between the distributions of the amplitudes and durations of type A, B, and C signals, which thus cannot be distinguished at all, suggesting that it is very difficult to successfully carry out pattern recognition in the analysis of AE signals from TBCs without choosing frequency as a characteristic parameter.

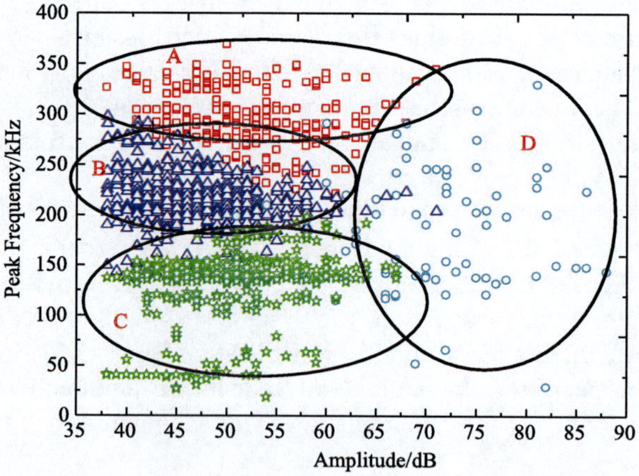

Fig. 11.10 Distributions of the clustered amplitudes and peak frequencies of the AE signals from the tensile test specimen

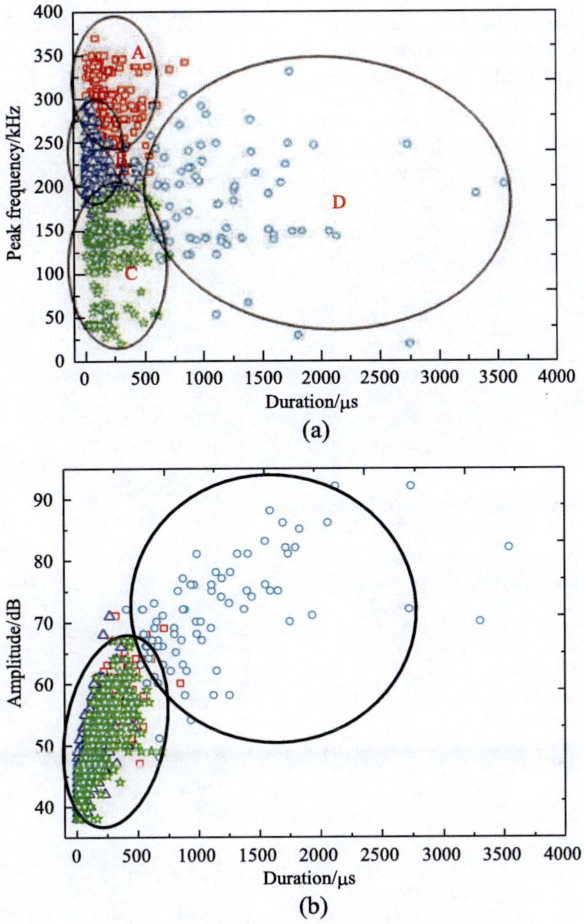

Fig. 11.11 **a** Distributions of the clustered peak frequencies and durations of the AE signals from the tensile test specimen; **b** Distributions of the clustered amplitudes and durations of the AE signals from the tensile test specimen

Figure 11.12 shows the evolution of different types of AE signals with time in the tensile test process. The type B signals are the first to appear during the tensile failure process, followed mainly by the type A signals. In contrast, there are relatively few type C and D signals. Research [10, 11, 13] shows that surface cracks are mainly formed during the early stage of the tensile loading process due to the low strain tolerance of the ceramic layer. The low tensile deformation capacity of the ceramic layer leads to rapid propagation of the surface cracks toward the ceramic/BC interface, and upon arrival, the cracks then grow and propagate along this interface. The frequencies of the type A and B signals are in the ranges of 0.26–0.30 and 0.20–0.23 MHz, respectively. On this basis, it can be inferred that the type B and A signals are from surface cracks and sliding interface cracks, respectively. The type C signals have a frequency ranging from 0.13 to 0.16 MHz and consist primarily of signals generated by the plastic deformation of the substrate and some noise signals. The

Fig. 11.12 Variation in the different failure modes of a TBC under tension with time

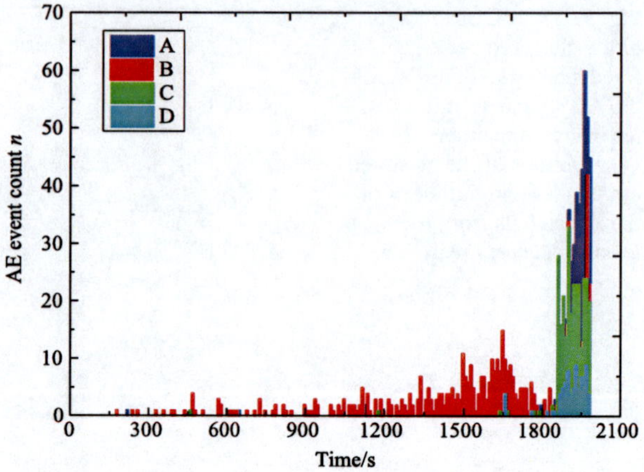

Fig. 11.13 *s*–*k* curve for the AE signals from a TBC specimen during the compressive failure process

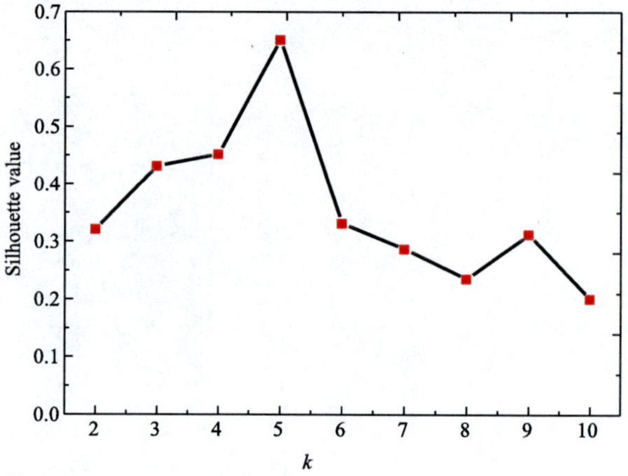

type D signals are considerably high in both amplitude and energy and are generally believed to originate from high-energy macrocracks [14].

(2) **Identification of TBC failure modes under compression and the extraction of the relevant characteristic parameters**

To further determine the validity of the pattern recognition of failure modes and the relevant characteristic frequency parameter, 3122 AE signals from a TBC specimen during the failure process under compressive loading are subjected to a cluster analysis using the same method [23]. Figure 11.13 shows the s–k curve for the clustering of the AE signals generated under compressive loading. When the damage-induced AE signals are classified into five types, a maximum *s*-value of 0.6512 is obtained, indicating that the damage-induced AE signals from the TBC specimen under compressive loading can be classified into five types.

Figure 11.14 shows the distributions of the clustered amplitudes and peak frequencies of the AE signals generated by the compressive test specimen due to damage. The amplitudes of the type A′, B′, C′, and E′ signals are basically identical and range from 38 to 55 dB, indicating that these signals are indistinguishable based on their amplitudes. However, the frequencies of the type A′, B′, C′, and E′ signals are distributed in significantly different bands (250–350, 170–250, 0–150, and 400–450 kHz, respectively), suggesting that these signals can be classified well based on their frequencies. The type D′ signals have large amplitudes and a wide frequency range that basically cover the aforementioned four bands. On this basis, it can be found that the type A′, B′, and C′ signals are respectively identical to the type A, B, and C signals generated during the tensile failure process in terms of frequency distribution, suggesting that they correspond to the same damage pattern and that the pattern of AE signals generated by cracks is independent of the loading mode.

Figures 11.15 and 11.16 show the relationship between the distributions of the clustered peak frequencies and durations and the relation between the distributions of the clustered amplitudes and durations of the damage-induced AE signals from the TBC specimen under compression, respectively. There is a significant overlap between the distributions of both the amplitudes and durations of the type A′, B′, C′, and E′ signals, and hence, it is difficult to differentiate these signals based on their amplitudes and durations. The type D′ signals have both relatively large amplitudes and long durations.

Figure 11.17 shows the variation in the different types of AE signals with time in the compressive failure process. The type E′ signals are the first to appear in the compressive failure process, followed mainly by the type A′ and C′ signals. In comparison, there are relatively few type B′ and D′ signals. Research [28] shows that under compressive loading, opening, and sliding interface cracks are the dominant types of interface cracks in a coating at compressive strains less than and >0.37%, respectively. The frequencies of type E′, A′, and C′ signals are in the ranges of 0.4–0.5, 0.26–0.30, and 0.13–0.16 MHz, respectively. Thus, it can be determined

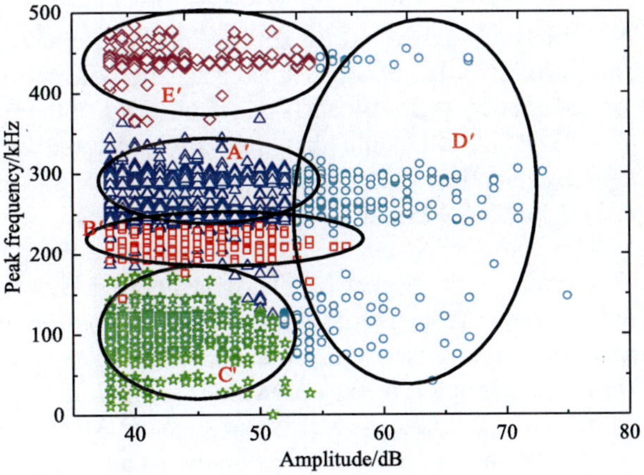

Fig. 11.14 Distributions of the clustered amplitudes and peak frequencies of the AE signals from the compressive test specimen

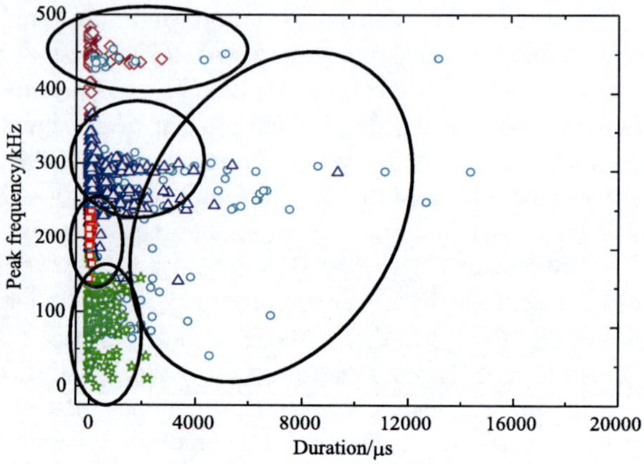

Fig. 11.15 Distributions of the clustered peak frequencies and durations of the AE signals from the compressive test specimen

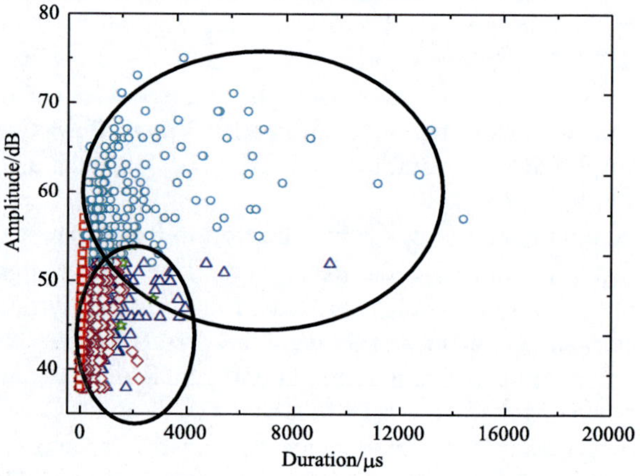

Fig. 11.16 Distributions of the clustered amplitudes and durations of the AE signals from the compressive test specimen

that the type E' signals are generated by opening interface cracks, that the type A′ signals are generated by sliding interface cracks, and that the type C' signals consist of signals generated by the plastic deformation of the substrate and some low-frequency noise signals. The frequencies of the type B′ signals range from 0.20 to 0.23 MHz, indicating that they are generated by surface cracks. The type D′ signals are considerably high in both amplitude and energy, indicating that they are AE signals caused by high-energy macrocracks.

The above cluster analysis shows that the peak frequency is most capable of characterizing the pattern of AE signals generated by a TBC due to damage and that there is a significant overlap between the distributions of the values of each remaining parameter, and consequently, these parameters cannot be used to classify AE signals. Hence, frequency is the key parameter for identifying the crack patterns of TBCs. In addition, the frequency-spectrum analysis of AE signals (Figs. 11.7, 11.9, and 11.14) shows that the frequencies of AE signals caused by surface cracks range from 0.20

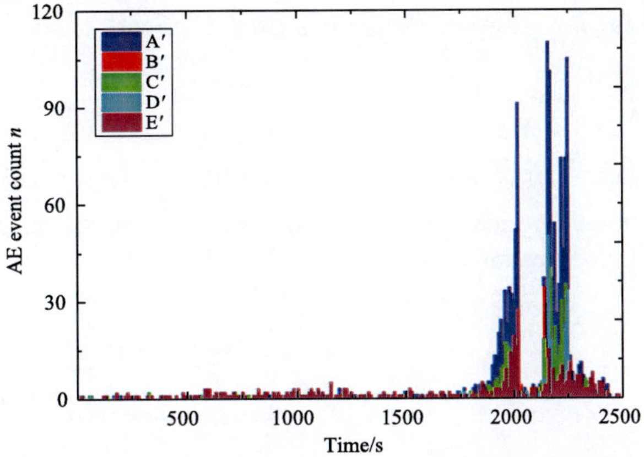

Fig. 11.17 Variation in the different failure modes of the TBC specimen with time

to 0.23 MHz, with a dominant frequency of 0.21 MHz; the frequencies of AE signals caused by sliding interfacial signals range from 0.26 to 0.30 MHz, with a dominant frequency of 0.29 MHz; the frequencies of AE signals caused by opening interfacial signals range from 0.42 to 0.46 MHz, with a dominant frequency of 0.44 MHz; and the frequencies of AE signals caused by the plastic deformation of the substrate range from 0.13 to 0.16 MHz, with a dominant frequency of 0.15 MHz.

11.3 Intelligent Crack Pattern Recognition Methods Based on Wavelets and Neural Networks

The analysis in the previous section shows that frequency is the characteristic parameter for identifying crack patterns and that the frequency for each failure mode is not fixed but instead varies within a certain range. In addition, thousands of AE signals are generated during the failure process of a TBC. Manual identification of each signal through frequency-spectrum analysis involves a very large workload. Therefore, effective extraction of the characteristic frequency bands of AE signals and intelligent pattern recognition are desired to significantly improve efficiency. Wavelet transform (WT) is a method for the time–scale (frequency) analysis of signals and is capable of displaying local signal features in both the time and frequency domains [29–32]. Neural networks (NNs) are a mathematical method established through the simulation of biological NNs from an information processing perspective. As an effective tool for the intelligent analysis of big data, this method formulates an algorithm through training on the signals of known patterns and uses it to identify other unknown signals [33, 34]. Establishing an NN training algorithm based on key parameters for pattern recognition (e.g., characteristic frequency-band information) extracted through the WT may realize intelligent identification of crack patterns based on AE signals. To this end, this section presents an intelligent method for crack pattern recognition based on the combination of WT and NN [35].

11.3.1 Basic Principle and Method of WT

(1) **WT**

Let $\widehat{\psi}(\omega)$ be the FT of function $\psi(t) \in L^2(R)$ ($L^2(R)$ is the square-integrable real-number space, i.e., the signal space with limited energy. When $\widehat{\psi}(\omega)$ satisfies the following admissibility condition [29–32]:

$$C_{\omega} = \int_{-\infty}^{+\infty} \frac{\left|\widehat{\psi}(\omega)\right|^2}{|\omega|} d\omega < \infty \tag{11.5}$$

$\psi(t)$ is called a wavelet function or mother wavelet.

By scaling and translating $\psi(t)$, we have

$$\psi_{a,b}(t) = \frac{1}{\sqrt{a}} \psi\left(\frac{t-b}{a}\right) \tag{11.6}$$

where a is the scaling factor (or scale factor) and b is the translation factor (both a and b are real numbers, and $a > 0$). $\psi_{a,b}(t)$ is referred to as a continuous wavelet basis function (WBF) dependent on a and b.

Let t^* and $\Delta\psi$ be the time-domain center and radius of the WBF $\psi(t)$, respectively, and ω^* and $\Delta\widehat{\psi}$ be the frequency center and radius of $\widehat{\psi}(\omega)$, respectively. Then, the wavelet scaling function $\psi_{a,b}(t)$ is a window function with a time-domain center at $b + at^*$ and a radius of $a\Delta\psi$ as follows:

$$\left[b + at^* - a\Delta\psi, b + at^* + a\Delta\psi\right]$$

The corresponding frequency-domain window is as follows:

$$\left[\frac{\omega^*}{a} - \frac{\Delta\widehat{\psi}}{a}, \frac{\omega^*}{a} + \frac{\Delta\widehat{\psi}}{a}\right]$$

Thus, the WT reflects the signal information within the following time–frequency window:

$$\left[b + at^* - a\Delta\psi, b + at^* + a\Delta\psi\right] \times \left[\frac{\omega^*}{a} - \frac{\Delta\widehat{\psi}}{a}, \frac{\omega^*}{a} + \frac{\Delta\widehat{\psi}}{a}\right] \tag{11.7}$$

Therefore, a WT analysis of a signal can be understood as follows: the wavelet window shifts along the whole time-axis through the WT translation factor *b* to gradually analyze the local time-domain signal, and the local time-domain signal within the time window is analyzed at different frequency-domain intervals by varying the scaling factor *a*. Thus, the WT can also be understood as a series of filter functions used to analyze a signal in different frequency ranges. The time window in Eq. (11.7) has the following important property. As *a* decreases, the width of the time window decreases, and the width of the frequency window increases, under which the time-domain resolution is very high but the frequency-domain resolution is reduced, and vice versa. Whether the frequency-domain and time-domain windows are adjustable is the fundamental difference between the WT and the FT.

The continuous WT of a given square-integrable signal $f(t)$ (i.e., $f(t) \in L^2(R)$) is expressed as follows:

$$W_f(a,b) = \langle f, \psi_{a,b} \rangle = \frac{1}{\sqrt{a}} \int_{-\infty}^{\infty} f(t)\overline{\psi\left(\frac{t-b}{a}\right)}dt \tag{11.8}$$

where $\overline{\psi\left(\frac{t-b}{a}\right)}$ is the conjugate of $\psi\left(\frac{t-b}{a}\right)$.

The inverse WT is defined as follows:

$$f(t) = \frac{1}{C_\omega} \int_{0}^{\infty} a^{-2} \int_{-\infty}^{\infty} W_f(a,b)dadb \tag{11.9}$$

In practical applications, AE signals are generally discretized data with discontinuous time intervals. Therefore, it is necessary to discretize a continuous wavelet and its transform. The scaling factor *a* and translation factor *b* are discretized as follows:

$$\begin{cases} a = a_0^j, & a_0 > 0 \\ b = ka_0^j b_0 & a_0 > 0 \end{cases} \quad j,k\text{为整数} \tag{11.10}$$

Then, a discrete wavelet basis (DWB) can be defined as follows:

$$\psi_{j,k}(t) = a_0^{-j/2}\psi\left(a_0^{-j}t - kb_0\right) \tag{11.11}$$

The discrete WT is expressed as follows:

$$W_f(j,k) = \langle f(t), \psi_{j,k}(t) \rangle = \int f(t)\overline{\psi_{j,k}(t)}dt \tag{11.12}$$

Then, for any square-integrable signal $f(t)$, we have

$$f(t) = \sum_{-\infty}^{\infty}\sum_{-\infty}^{\infty} W_f(j,k)\psi_{j,k}(t) \tag{11.13}$$

To facilitate calculation, the constants a_0 and b_0 in a DWB are generally set to 2 and 1, respectively, to constitute a binary wavelet and a binary WT. All the WTs in this chapter refer to binary WTs with the following discrete wavelet series:

$$\psi_{j,k}(t) = 2^{-j/2}\psi(2^{-j}t - k) \tag{11.14}$$

where j and k are integers and are also called the scale factor and the translation factor, respectively.

(2) **Multiresolution analysis (MRA)**

The WT scaling parameter, in essence, describes the range (i.e., scale) of the observed signal and is also generally called the resolution in signal processing. Thus, the WT can also be understood as a multiscale signal analysis, i.e., a coarse-to-fine, level-by-level signal analysis method [29–32]. The concept of the WT can be explained using the analogy of the local and global relations between the focal length of a camera and a scene. A camera is used to observe target $f(t)$. Then, the WBF $\psi_{a,b}(t)$ represents the role of the camera (the role of b is equivalent to a translation scan by the camera from left to right, and the role of the scaling factor a is equivalent to the camera zooming in on the object or the camera zooming-out from the object). At a large a, the field of view is wide and thus allows general observation of the signal. Conversely, at a small a, the field of view is narrow and thus allows a detailed observation of the signal. Here, a binary WT is used as an example. The scale is a^j (for the sake of convenience, j is generally directly called the scale), the time-domain resolution is a^j, and the resolution in the frequency space is $f_s/2^j$ (f_s is the effective sampling frequency).

Figure 11.18 shows a visual description of a multiresolution analysis (MRA). It involves the establishment of a nested sequence of closed subspaces $\{V_j\}_{j\in Z}$, each with a closure approximating L_2(R), where j is the scale and V_j is the approximation of L_2(R) at a resolution of $f_s/2^j$ (a high resolution results in a high level of approximation). Another analysis function, namely, the scale function $\phi(t)$, closely related to the wavelet function $\psi(t)$, is introduced in the MRA, and a sequence of orthonormal wavelet subspaces $\{W_j\}_{j\in Z}$ of L_2(R) is established.

The subspace sequence $\{V_j\}_{j\in Z}$ of an MRA has the following properties:

(1) Nesting property: $V_{j+1} \subset V_j$ $(j \in Z)$.
(2) Approximability and isolatability: $\bigcup_{j\in Z} V_j = L^2(R)$, $\bigcap_{j\notin Z} V_j = \{0\}$.
(3) Scalability: $f(t) \in V_0 \Leftrightarrow f(2^{-j}t) \in V_j$.
(4) Translation invariance: $f(t) \in V_0 \Leftrightarrow f(t-k) \in V_0$ holds for any integer k.
(5) Orthogonality: If there exists a function $\phi(t) \in V_0$ such that $\{\phi(t-k), k \in Z\}$ is the orthonormal basis of V_0, then the function system

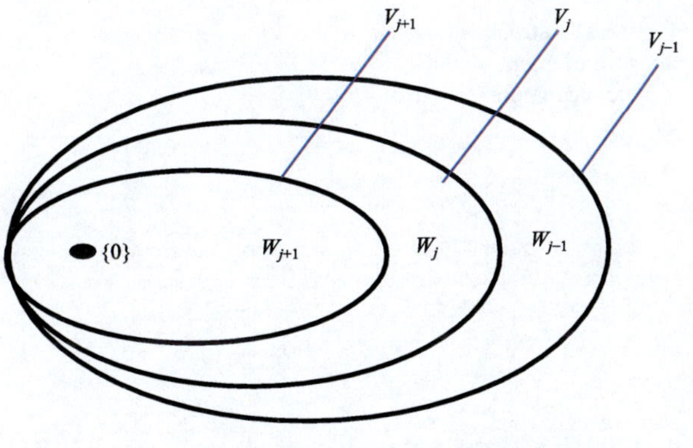

Fig. 11.18 Approximation of the sequence of linear function spaces $\{V_j\}_{j\in Z}$ of the space L_2(R)

$\{\phi_{j,k}(t) = 2^{-j/2}\phi(2^{-j}t - k), j, k \in Z\}$ $(j, k \in Z)$ constitutes the orthonormal basis of subspace V_j. $\phi(t)$ and V_j are called the scale function and scale space of the MRA, respectively.

Based on the above analysis, the sequence of the scale spaces in an MRA is generated through the scaling and translation of the same scale function at different scales, i.e., one MRA $\{V_j\}_{j\in Z}$ corresponds to one scale function. Thus, sometimes $\{V_j\}_{j\in Z}$ is also called the MRA generated from scale function $\phi(t)$, and $\phi(t)$ is called the generator of $\{V_j\}_{j\in Z}$. For a fixed j, the function system $\{\phi_{j,k}(t), k \in Z\}$ is an orthonormal basis of V_j. However, due to their nested relation, there is a spatial overlap between $\{V_j\}_{j\in Z}$. Thus, $\{V_j\}_{j\in Z}$ is not the orthogonal decomposition of L_2(R). Therefore, it is necessary to establish an orthogonal complement space W_j for V_j. As shown in Fig. 11.18, we have

$$V_j \perp W_j \quad V_{j-1} = V_j \oplus W_j \tag{11.15}$$

For subspace sequence $\{W_j\}_{j\in Z}$, we have

(1) For any j and j', subspaces W_j and $W_{j'}$ are mutually orthogonal.
(2) $\ldots, V_0 = V_1 \oplus W_1 = (V_2 \oplus W_2) \oplus W_1 = \cdots = V_j \oplus W_1 \oplus W_2 \oplus \cdots \oplus W_j$. When $j \to +\infty$, one orthogonal decomposition of L^2(R) is obtained as follows:

$$L^2(\mathrm{R}) = \bigoplus_{j\in Z} W_j \tag{11.16}$$

The MRA described in Fig. 11.18 can be further vividly described using a wavelet decomposition tree structure, e.g., the three-level wavelet decomposition diagram in Fig. 11.19, where S represents the original signal and can be considered as the subspace undecomposed by the WT or the subspace at decomposition level 0, i.e., V_0. At decomposition level 1, S can be decomposed into a low-frequency part C1 and a high-frequency part D1, which represent the subspace and the orthogonal complement W_1 of V_1, respectively. At decomposition level 2, the subspace C1 of V_1 is further decomposed into a low-frequency part C2 and a high-frequency

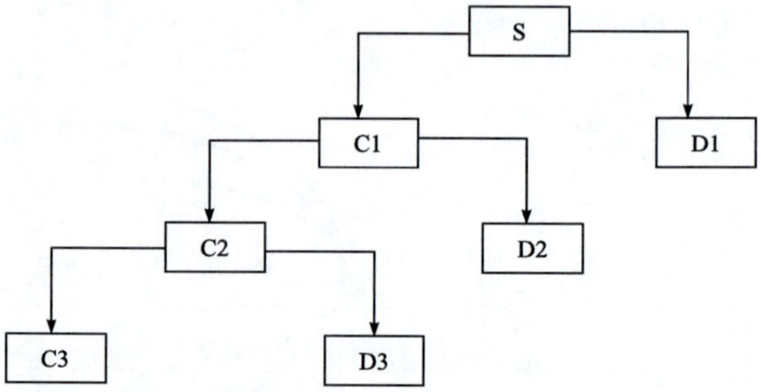

Fig. 11.19 Structural diagram of a three-level wavelet decomposition tree

part D2. This process is then similarly repeated. We can see from Fig. 11.19 that there is an inclusion relation between spaces V_j, i.e., C3 ∈ C2 ∈ C1. At each level, $V_{j-1} = V_j + W_j$; e.g., C1 = C2 + D2. The orthogonal complement spaces of V_j (i.e., the wavelet spaces W_j) are complementary in frequency and are not in an inclusion relation; i.e., D1, D2, and D3 represent detail signals in different frequency bands. In addition, the original signal S is the sum of C3, D1, D2, and D3 (i.e., S = C3 + D1 + D2 + D3). On this basis, we conclude that an MRA establishes an orthogonal wavelet basis (OWB) that acts as a bandpass filter highly approximating the space L^2(R) in frequency [36, 37].

If $f(t) \in W_0$, then $f(t) \in V_{-1} - V_0$. As an MRA is scalable, $f\left(2^{-j}t\right) \in V_{j-1} - V_j$, i.e., W_j. Let the orthonormal basis of W_0 be composed of the integer translation $\{\psi(t-k)\}_{k \in Z}$ of function $\psi(t)$. Then, for all the scales $j \in Z$, $\psi_{j,k}(t) = 2^{-j/2}\psi(2^{-j}t - k), k \in Z$ must be the orthonormal basis of subspace W_j. Based on the orthogonality of $\left\{W_j\right\}_{j \in Z}$, it can be inferred that $\left\{\psi_{j,k}(t) = 2^{-j/2}\psi(2^{-j}t - k), k \in Z\right\}$ constitutes the orthonormal basis of the space L_2(R).

According to the definition of the DWB, $\psi_{j,k}(t)$ is exactly the OWB obtained from the scaling and translation of the same generating function. Thus, $\psi(t)$ is called the wavelet function, and correspondingly, W_j is called the wavelet space with a scale of j.

For any signal $f(t) \in L^2(R)$, let $C_j f(t)$ and $D_j f(t)$ be its smooth approximation obtained through projection to space W_j at a certain limited scale of j and its detailed information, respectively. Due to the orthogonality between spaces W_j, the signal at a certain scale (e.g., j) can be expressed as follows:

$$f(t) = C_j f(t) + \sum_{i=0}^{j} D_i f(t) \tag{11.17}$$

(3) **Two-scale equations and orthogonal WT**

A multiscale analysis essentially establishes a spatial level-by-level 2D subdivision framework to perform an MRA. The core of a multiscale analysis is to find the orthonormal basis scale function $\phi(t)$ and wavelet function $\psi(t)$ of spaces V_0 and

W_0. As long as $\phi(t)$ and $\psi(t)$ are known, the analysis can be performed step-by-step. Thus, it is vitally important to understand the basic properties of $\phi(t)$ and $\psi(t)$.

Two-scale equations are the basic properties given by multiscale analysis to $\phi(t)$ and $\psi(t)$ and describe the inherent relation between the basis functions of two adjacent scales V_j and V_{j-1}. Let $\phi(t)$ be an orthonormal basis function (OBF) of scale space V_0 and $\psi(t)$ be an OBF of wavelet space W_0. Because $V_0 \subset V_{-1}$ and $W_0 \subset V_{-1}$, $\phi(t)$ and $\psi(t)$ must belong to space V_{-1}. Then, $\phi(t)$ and $\psi(t)$ can be expanded in terms of the normal basis of space V_{-1} as follows [29, 37]:

$$\begin{cases} \phi(t) = \sum\limits_{l \in Z} h(l)\phi_{-1,l}(t) = \sqrt{2}\sum\limits_{l \in Z} h(l)\phi(2t - l) \\ \psi(t) = \sum\limits_{l \in Z} g(l)\phi_{-1,l}(t) = \sqrt{2}\sum\limits_{l \in Z} g(l)\phi(2t - l) \end{cases} \tag{11.18}$$

where the coefficients $h(l)$ and $g(l)$ can be very easily obtained through inner-product operations as follows:

$$\begin{cases} h(l) = \langle \phi(t), \phi_{-1,l}(t) \rangle \\ g(l) = \langle \psi(t), \phi_{-1,l}(t) \rangle \end{cases} \tag{11.19}$$

Equation (11.18) describes the relation between basis functions at adjacent scales and is referred to as a two-scale equation. Equations (11.18) and (11.19) show that once $\psi(t)$ and $\phi(t)$ are given (see Sect. 11.3.2 for their determination), $h(l)$ and $g(l)$ can be determined. In addition, because $W_0 \perp V_0$, $h(l)$ and $g(l)$ need to be mutually orthogonal. On this basis, it can be derived that $g(l) = (-1)^l h(1 - l)$ [29].

In fact, two-scale equations exist in any adjacent scale spaces, i.e.,

$$\begin{cases} \phi_{j,k}(t) = \sum\limits_{l \in Z} h(l - 2k)\phi_{j-1,l}(t) \\ \psi_{j,k}(t) = \sum\limits_{l \in Z} g(l - 2k)\phi_{j-1,l}(t) \end{cases} \tag{11.20}$$

The two-scale coefficients $h(l)$ and $g(l)$ depend on $\phi(t)$ and $\psi(t)$ but are independent of j. A low-frequency smooth approximation $\phi_{j,k}(t)$ is obtained by passing $\phi_{j-1,l}(t)$ through $h(l)$, which is thus also called the low-pass filter coefficient. Similarly, the high-frequency details $\psi_{j,k}(t)$ are obtained by passing $\phi_{j-1,l}(t)$ through $g(l)$, which is thus called the high-pass filter coefficient. $h(l)$ and $g(l)$ are often referred to as filter coefficients. The FTs of $h(l)$ and $g(l)$ are called a low-pass wavelet filter and a high-pass filter, respectively.

The scale and wavelet coefficients at scale j are defined as follows:

$$c_{j,k} = \langle f(t), \phi_{j,k} \rangle = \int_{-\infty}^{\infty} f(t)\phi_{j,k}dt$$
$$d_{j,k} = \langle f(t), \psi_{j,k} \rangle = \int_{-\infty}^{\infty} f(t)\psi_{j,k}dt \tag{11.21}$$

Then, it is possible to construct the smooth approximation $C_j f(t)$ and detailed decomposition $D_j f(t)$ (the notations signify a high-frequency approximation of the signal) of a square-integrable function or signal $f(t)$ from L^2(R) to V_j and W_j, respectively:

$$C_j f(t) = \sum_{k \in Z} c_{j,k}\phi_{j,k}(t)$$
$$D_j f(t) = \sum_{k \in Z} d_{j,k}\psi_{j,k}(t) \tag{11.22}$$

where $C_j f(t)$ and $D_j f(t)$ are both notations and represent the approximation of the low-frequency part and the high-frequency approximation of the signal, respectively.

The signal $f(t)$ reconstructed after the wavelet decomposition can be described as follows:

$$f(t)=C_j f(t) + \sum_{i=1}^{j} D_i f(t) = \sum_{k \in Z} c_{j,k}\psi_{j,k}(t) + \sum_{i=1}^{j}\sum_{k \in Z} d_{i,k}\psi_{j,k}(t) \tag{11.23}$$

For the sake of convenience, $C_j f(t)$ and $D_j f(t)$ are denoted as $f_0(t)$ and $f_j(t)$, respectively. Then, the signal $f(t)$ reconstructed after the wavelet decomposition can be described as follows:

$$f(t)=\sum_{i=1}^{j} f_i(t) \tag{11.24}$$

Note that in Eqs. (11.21)–(11.23), j is a specific scale, k and i are integers, and no summation is carried out even if they are repeated unless the summation symbol is explicitly written. This MRA concept, which involves the decomposition of a signal into components in different frequency ranges, is the basis for damage pattern recognition based on AE signal patterns. Various damage patterns can be identified only if they can be differentiated in different frequency intervals.

(3) **Wavelet packet transform**

Wavelet packet (WP) analysis is a finer signal analysis method than wavelet analysis. Wavelet analysis decomposes only the scale space V_j (i.e., the low-frequency part), whereas WP analysis further decomposes the wavelet space W_j (i.e., the

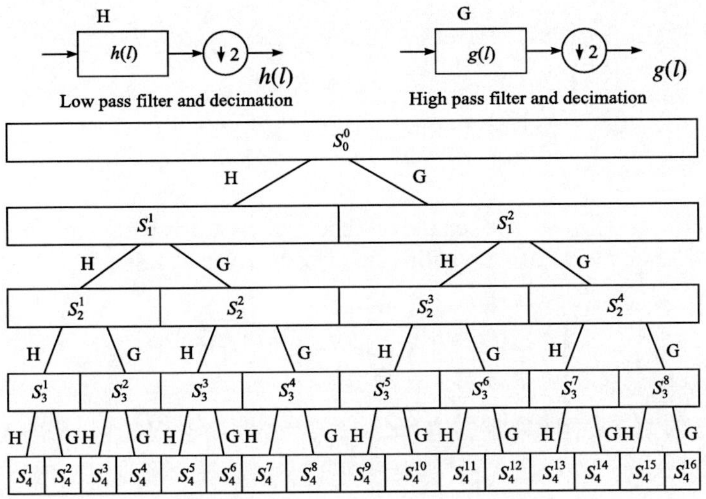

Fig. 11.20 Structural diagram of a four-level WP decomposition tree

high-frequency part) while decomposing the scale space V_j [38] to obtain a higher frequency-spectrum resolution and better frequency-spectrum characteristics for a signal.

Here, we use a four-level decomposition to illustrate the wavelet packet transform (WPT). Figure 11.20 shows the WP decomposition tree. In the figure, $h(k)$ and $g(k)$ are the low- and high-pass filter coefficients, respectively, and the low- and high-frequency parts are simultaneously further decomposed at each level. For a four-level decomposition (i.e., decomposition scale $j = 4$), the frequency space is divided into 16 frequency bands.

To binarily subdivide the wavelet subspace W_j in terms of frequency, the natural approach is to unify the scale subspace V_j and wavelet subspace W_j with a new subspace U_j^n. Let

$$U_j^1 = V_j, \quad U_j^2 = W_j, \quad j \in Z \tag{11.25}$$

Then, the orthogonal decomposition of $V_j = V_{j+1} + W_{j+1}$ can be unified through the decomposition of U_j^n as follows:

$$U_j^1 = U_{j+1}^1 + U_{j+1}^2 \tag{11.26}$$

Subspaces U_j^n and U_j^{2n} are defined as the WP spaces of functions $u_{j,k}^n(t)$ and $u_{j,k}^{2n}(t)$, respectively. In addition, let $u_{j,k}^n(t)$ satisfy the following two-scale equations:

$$\begin{cases} u_{j,k}^{2n-1}(t) = \sqrt{2}\sum\limits_{l\in Z} h(l-2k)u_{j,l}^{n}(2t-l) \\ u_{j,k}^{2n}(t) = \sqrt{2}\sum\limits_{l\in Z} g(l-2k)u_{j,l}^{n}(2t-l) \end{cases} \tag{11.27}$$

where $g(l) = (-1)^l h(1-l)$ (i.e., the two coefficients are also mutually orthogonal). When $n = 1$, the above two equations are directly given as follows:

$$\begin{cases} u_{j,k}^{1}(t) = \sqrt{2}\sum\limits_{l\in Z} h(l-2k)u_{j,l}^{1}(2t-l) \\ u_{j,k}^{2}(t) = \sqrt{2}\sum\limits_{k\in Z} g(l-2k)u_{j,l}^{1}(2t-l) \end{cases} \tag{11.28}$$

A comparison with the two-scale equations of the $\phi(t)$ and $\psi(t)$ of an MRA shows that $u_{j,k}^{1}(t)$ and $u_{j,k}^{2}(t)$ are the scale function and WBF of the WPT, respectively. Equation (11.28) is equivalent to Eq. (11.26). By expanding this equivalent expression to cases where n is a nonnegative integer, we obtain the following equivalent expression of Eq. (11.27):

$$U_j^n = U_{j+1}^{2n-1} + U_{j+1}^{2n} \tag{11.29}$$

Low-pass $h(k)$ and high-pass $g(k)$ filtering of the signal $f(t)$ in the subspace U_j^n yields the approximate signal $d_{j,k}^{2n}(t)$ and detailed signal $d_{j,k}^{2n+1}(t)$ at the next level of decomposition [36, 38]:

$$\begin{aligned} d_{j,k}^{2n-1}(t) &= \sum_{k} h(l-2k)d_{j-1,l}^{n}(t) \\ d_{j,k}^{2n}(t) &= \sum_{k} g(l-2k)d_{j-1,l}^{n}(t) \end{aligned} \tag{11.30}$$

For a given AE signal $f(t)$, its WPT coefficient $d_{j,k}^{n}$ can be obtained from the following equation:

$$d_{j,k}^{n} = \langle f(t), u_{j,k}^{n}\rangle = \int_{-\infty}^{\infty} f(t)u_{j,k}^{n}dt, \tag{11.31}$$

Then, the WP component of the signal at each node can be reconstructed using the following equation:

$$f_j^n(t) = \sum_{k=-\infty}^{\infty} d_{j,k}^{n}u_{j,k}^{n}(t), \tag{11.32}$$

Again, note that the scale j and number n in the above equation are only integers and that no summation is performed even though they are repeated.

Let us revisit the four-level (i.e., $j = 4$) WP decomposition in Fig. 11.21. A signal $f(t)$ (represented by S) with an effective sampling frequency f_s (generally half of the signal sampling frequency) is decomposed into $2^j = 16$ subsignals in different frequency bands, with the frequency range of each band as follows:

$$\left[\frac{(n-1)f_s}{2^j}, \frac{nf_s}{2^j}\right], \quad n = 1, 2, \ldots 2^j \tag{11.33}$$

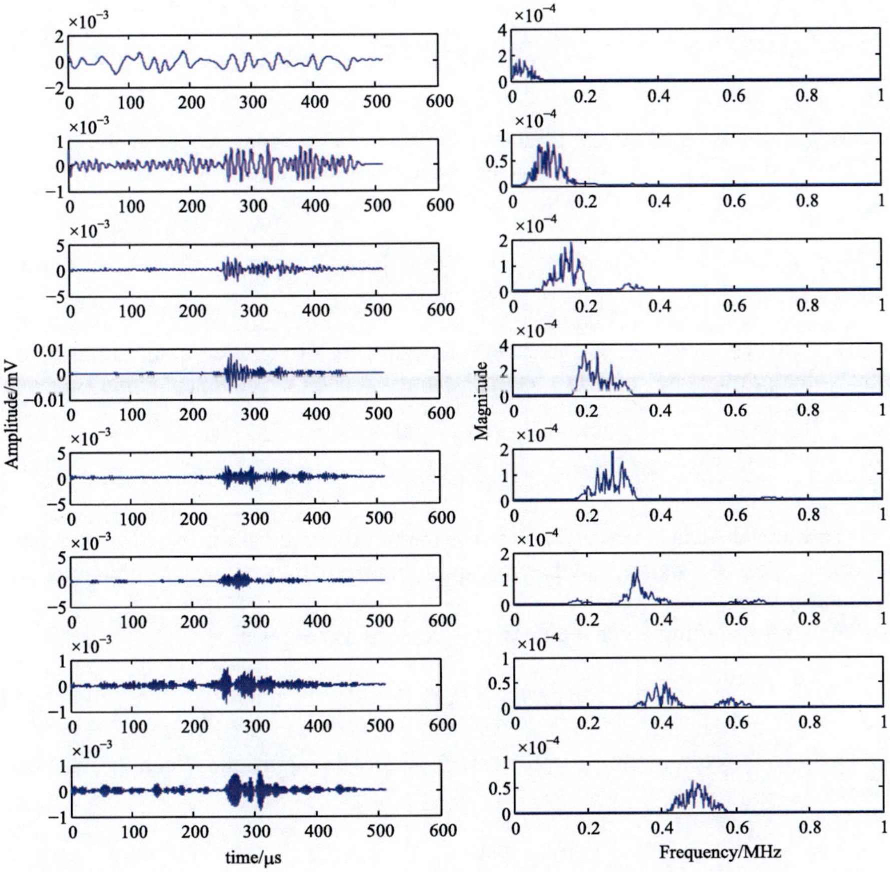

Fig. 11.21 WT results for an AE signal generated by a surface crack and the frequency spectrum in each frequency band

(4) **Wavelet energy spectrum coefficient**

Based on the earlier analysis, the reconstruction of a signal subjected to a WT or WPT at scale j can be described as the sum of the signals in all the subspaces at scale j, as is expressed in the following equation (using the case of a WPT as an example):

$$f(t) = \sum_{n=1}^{2^j} f_j^n(t) \tag{11.34}$$

The nodal energy at each scale is calculated as follows:

$$E_j^n(t) = \sum_{\tau=t_0}^{t} (f_j^n(\tau))^2 \tag{11.35}$$

where t_0 and t are the starting time and duration of the signal, respectively.

The total energy of the signal is calculated using the following equation [35, 38]:

$$E(t) = \sum_{n=1}^{2^j} E_j^n(t) \tag{11.36}$$

The wavelet energy spectrum coefficient (WESC) is defined as the ratio of the nodal energy of an AE signal to its total energy at each wavelet scale, that is,

$$R_n = \frac{E_j^n}{E(t)}, \quad n = 1, 2 \cdots 2^j \tag{11.37}$$

The wavelet scale corresponding to the maximum value of the WESC is the characteristic scale of a signal, and the corresponding frequency band is the characteristic frequency band of the signal.

The corresponding eigenvector $\mathbf{R}$ is expressed as follows:

$$\mathbf{R} = [R_1, R_2, \cdots R_n], \quad n = 1, 2, \ldots 2^j \tag{11.38}$$

This eigenvector is used for NN-based pattern recognition, which is discussed later.

11.3.2 Wavelet Analysis of AE Signals from TBCs Due to Damage

(1) **Wavelet basis selection**

There is a wide range of varieties of wavelet bases (WBs). Any function that satisfies the admissibility condition in Eq. (11.5) can constitute a WB. The analysis results for an AE signal vary with the WB. Many researchers have studied the selection of WBFs for AE signal analysis and noted that a DWB that is similar to the AE signal in question, that is compactly supported in the time domain, that decays rapidly in the frequency domain, and that has a vanishing moment of a certain order should be selected [37, 39, 40]. Commonly used WBFs include the Daubechies, symlet, and coiflet wavelets. Repeated tests and analyses show that the Daubechies wavelets are suitable for AE signals produced by TBCs due to damage. The Daubechies WBs have a vanishing moment of 4–8. Considering that low vanishing moments are unable to highlight the singularity of signals and that overly high vanishing moments increase the computational load of wavelet analysis, the "db8" wavelet is recommended [23, 35].

(2) **WPT-based time-domain AE signal analysis**

AE signals are subjected to a four-level WP decomposition using the db8 wavelet. Each signal is decomposed into 16 nodal subsignals with a frequency width of 0.0625 MHz. The frequency range at each node can be determined using Eq. (11.33). Figures 11.21, 11.22, 11.23, 11.24, 11.25, 11.26, 11.27 and 11.28 show the frequency spectra and the reconstructed time–frequency plots of AE signals generated by a surface crack, a sliding interface crack, an opening interface crack, and the plastic deformation of the substrate, respectively. The frequency spectra of the AE signals produced by TBCs due to damage are basically distributed between 0 and 0.5 MHz. Thus, only the waveforms and frequency spectra of the first eight nodal signals are presented here. The energy of the signal generated by the surface crack is concentrated mainly at node S_4^4, corresponding to a frequency band of 0.1875–0.25 MHz; the energy of the signal generated by the sliding interface crack is concentrated mainly at node S_4^5, corresponding to a frequency band of 0.25–0.3125 MHz; the energy of the signal generated by the opening interface crack is concentrated mainly at node S_4^8, corresponding to a frequency band of 0.4375–0.5 MHz; and the energy of the signal generated by the plastic deformation of the substrate is concentrated mainly at node S_4^3, corresponding to a frequency band of 0.125–0.1875 MHz. These findings are consistent with the results in Figs. 11.10 and 11.14 (0.20–0.23, 0.26–0.30, 0.4–0.5, and 0.13–0.16 MHz for AE signals generated by surface cracks, sliding interface cracks, opening interface cracks, and plastic deformation of the substrate, respectively).

(3) **Extraction of the characteristics of the WESC of AE signals**

Based on Eq. (11.37), the values of the WESC of the AE signals generated by the surface crack, the sliding interface crack, the opening interface crack, and the

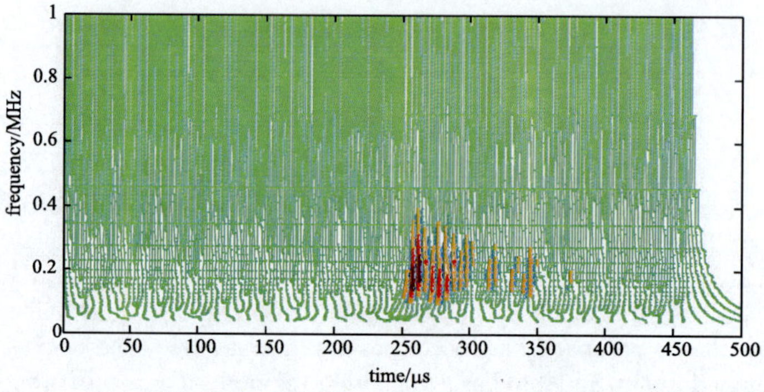

Fig. 11.22 WPT-derived time–frequency plot of an AE signal generated by a surface crack

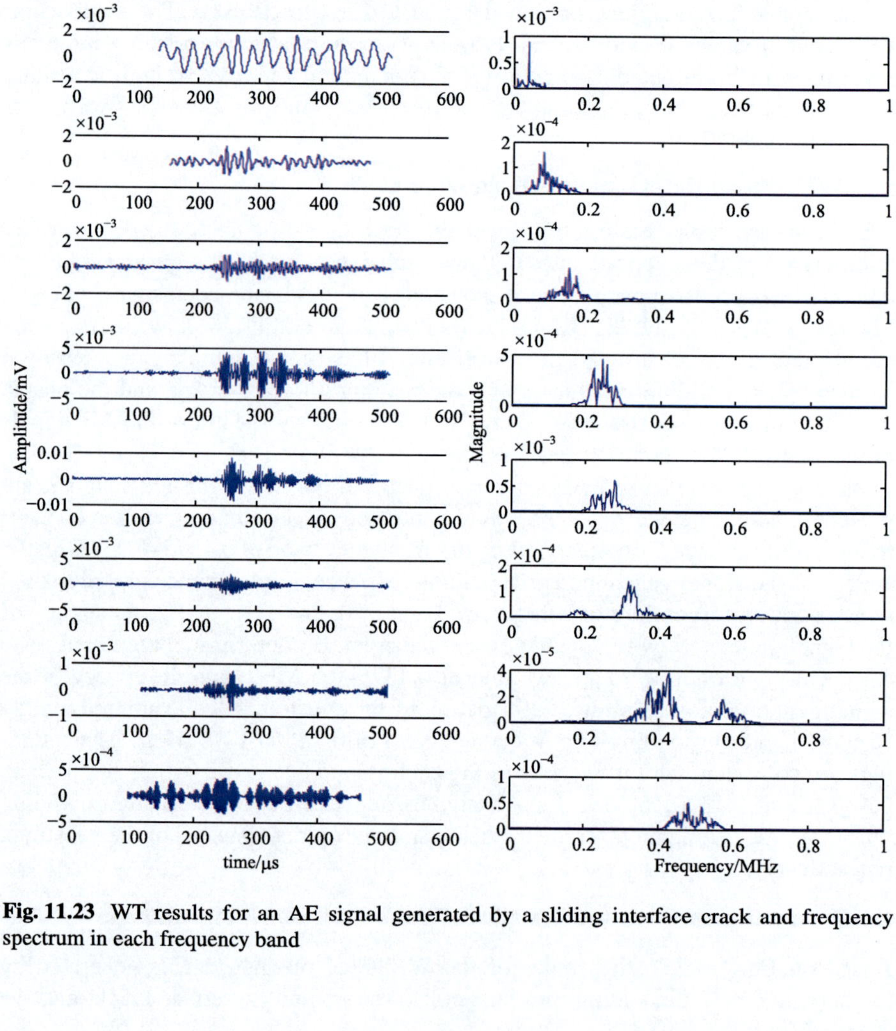

Fig. 11.23 WT results for an AE signal generated by a sliding interface crack and frequency spectrum in each frequency band

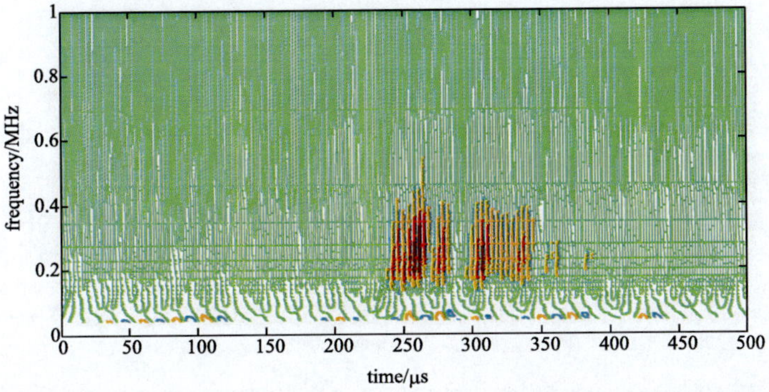

Fig. 11.24 WPT-derived time–frequency plot of an AE signal generated by a sliding interface crack

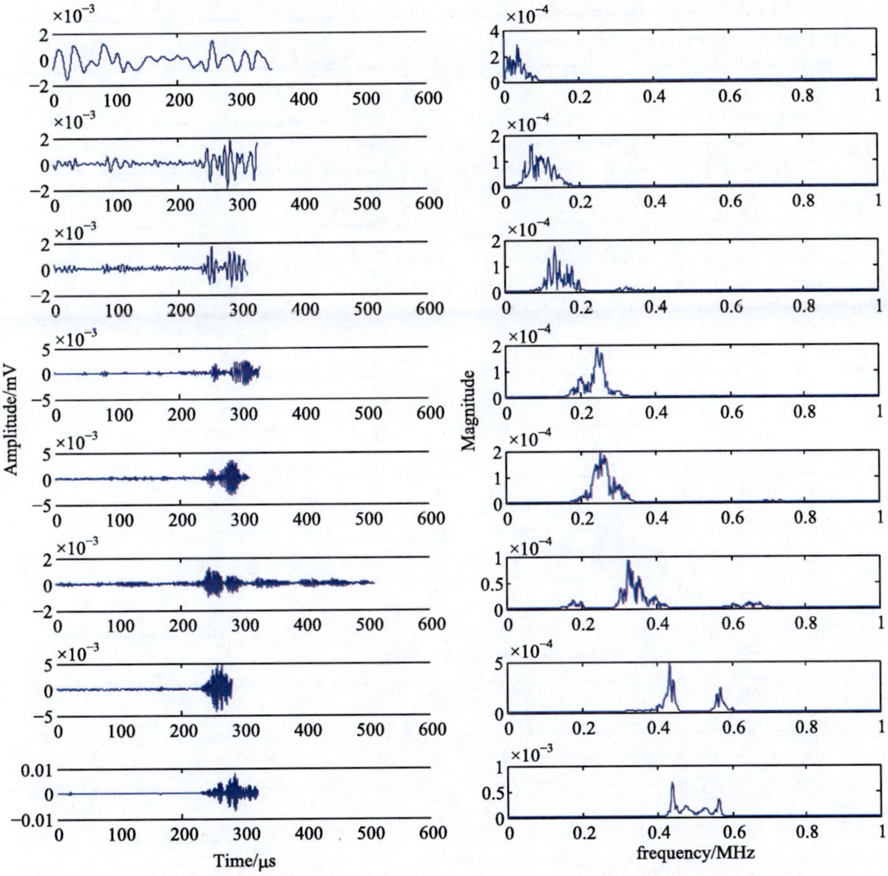

Fig. 11.25 WT results for an AE signal generated by an opening interface crack and frequency spectrum in each frequency band

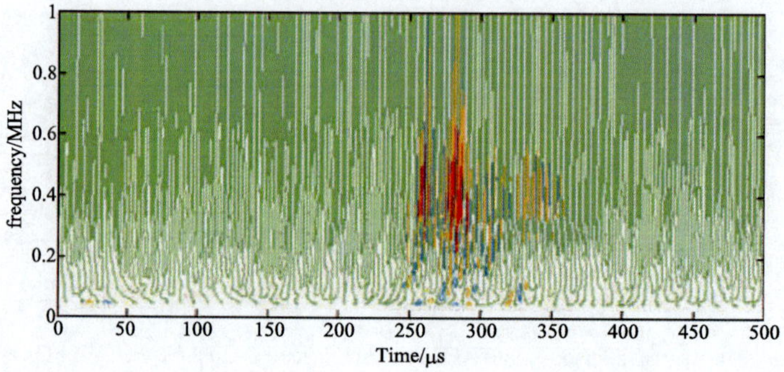

Fig. 11.26 WPT-derived time–frequency plot of an AE signal generated by an opening interface crack

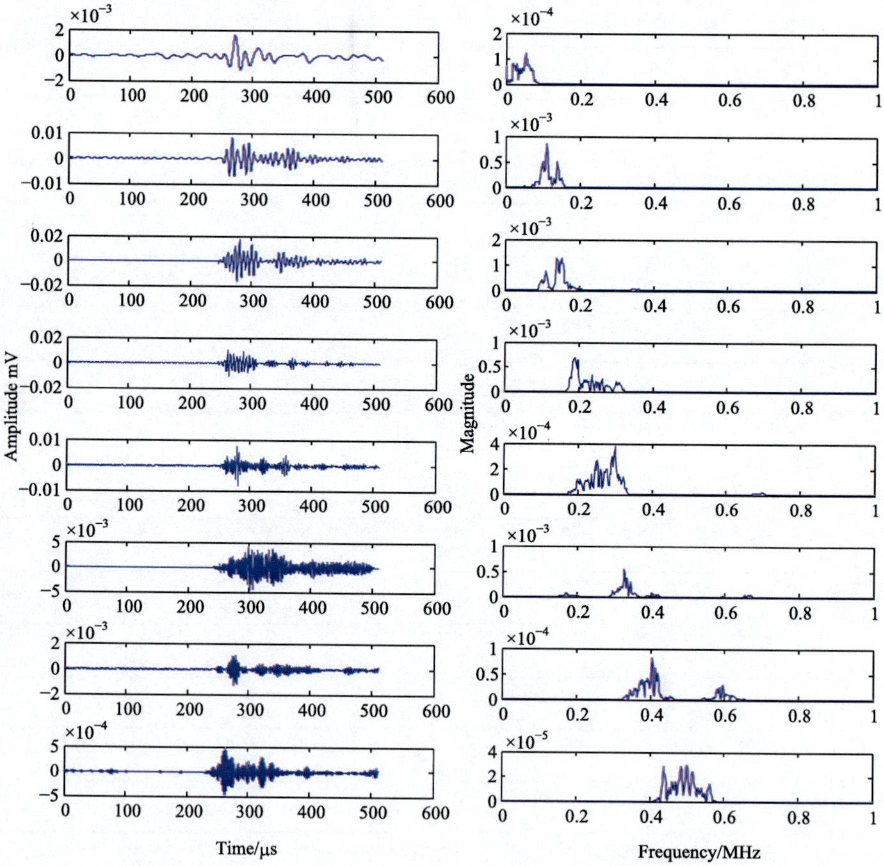

Fig. 11.27 WT results for an AE signal generated by the plastic deformation of the substrate and the frequency spectrum in each frequency band

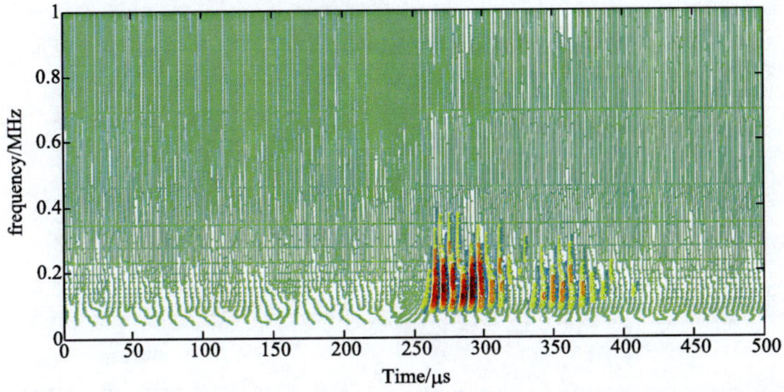

Fig. 11.28 WPT-derived time–frequency plot of an AE signal generated by the plastic deformation of the substrate

plastic deformation of the substrate are extracted. On this basis, each corresponding eigenvector **R** is obtained. As shown in Fig. 11.29a–d, the energy is concentrated in the frequency bands of 4(S_4^4), 5(S_4^5), 8(S_4^8), and 3(S_4^3), respectively, which is identical to the results shown in Figs. 11.21, 11.22, 11.23, 11.24, 11.25, 11.26, 11.27 and 11.28. Figure 11.29 gives a visual distribution of each signal in different frequency bands, suggesting that the use of WESC as the eigenvector of AE signals can satisfactorily facilitate the pattern classification and recognition of these signals.

11.3.3 NN-Based Intelligent Method for Pattern Recognition of AE Signals

(1) **Backpropagation artificial NNs**

As a type of mathematical model established through the simulation of biological NNs using mathematical and physical methods from an information processing perspective [33, 34], artificial NNs (ANNs) are widely used in fields such as pattern recognition, image processing, and intelligent control due to their self-learning, self-organization, and self-adaptation capabilities. Backpropagation (BP) means backward error propagation. Backpropagation neural networks (BPNNs) are currently the most widely used NNs and are a typical multilayer forward network. In a BPNN, the layers are mostly fully connected, and there are no connections in the same layer. Figure 11.30 shows the structure of a three-layer BP network, which consists of an input layer, a hidden layer, and an output layer.

Here, we use a three-layer BP network as an example to discuss the basic principle of the BP algorithm. To facilitate analysis, a three-layer network as shown in Fig. 11.30 is first defined. In the figure, R_n is an input neuron (also called an input vector), n is the number of neuron nodes in the input layer and depends on the dimension of the input vector for pattern recognition, p_i is the hidden-layer unit (i is the

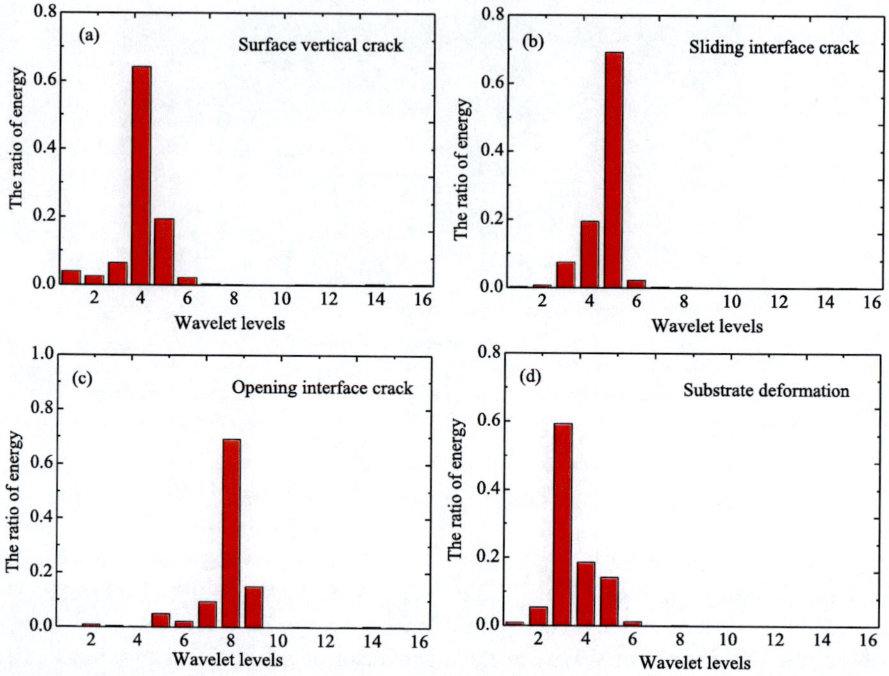

Fig. 11.29 Distribution of the values of the WESC. **a** Surface crack. **b** Sliding interface crack. **c** Opening interface crack. **d** Plastic deformation of the substrate

Fig. 11.30 The topology of a three-layer BP network

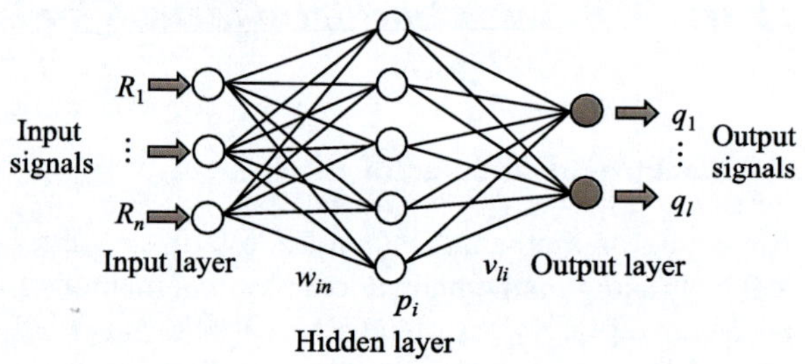

number of neuron nodes in the hidden layer), and q_l is a neuron in the output layer (l is the number of neuron nodes in the output layer and depends on the number of patterns identified from pattern recognition). The BP algorithm consists of forward signal transfer and backward error propagation [33, 34].

(1) Forward information transfer

Forward transfer is a process in which a signal is transferred from the input layer to the hidden layer and further to the output layer. The output p_i of the ith neuron in the hidden layer is related to the input neuron R_n through the following equation:

$$p_i = g\left(\sum_{n=1}^{N} w_{in} R_n\right) \quad (11.39)$$

The following shows the transfer relation between the output q_l of the lth (unrelated to the l in $h(l)$) neuron in the output layer and the hidden-layer neuron p_i:

$$q_l = g\left(\sum_{i=1}^{I} v_{li} p_i\right) \quad (11.40)$$

where w_{in} and v_{li} are the weight matrices from the hidden layer to the output layer and from the output layer to the hidden layer, respectively, N is the number of nodes in the input layer, and I is the number of nodes in the hidden layer. $g(x)$ is a transfer function, which can generally be an S-type tangent function, a logarithmic function, or a linear function, with the logarithmic function being the most common choice [41, 42]:

$$g(x) = \frac{1}{1 + \exp(-x)} \quad (11.41)$$

(2) Backward error propagation process

First, the transfer error is defined as the sum of the squared errors between the ideal output vector Q_l and the output neuron q_l of the trained network as follows:

$$e = \frac{1}{2}\sum_{l=1}^{L} (Q_l - q_l)^2 \quad (11.42)$$

where L is the number of output neuron nodes.

The weight of the output layer is corrected as follows [33, 41]:

$$\Delta v_{li} = -\eta \frac{\partial e}{\partial v_{li}} = -\eta \frac{\partial e}{\partial q_l} \cdot \frac{\partial q_l}{\partial v_{li}} \quad (11.43)$$

By combining Eqs. (11.40)–(11.42), we have

$$\Delta v_{li} = \eta q_l (1 - q_l)(Q_l - q_l) p_i \quad (11.44)$$

Similarly, the weight of the input layer can be corrected as follows:

$$\Delta w_{in} = -\eta \frac{\partial e}{\partial w_{in}} = -\eta \frac{\partial e}{\partial p_i} \cdot \frac{\partial p_i}{\partial w_{in}} = -\eta \frac{\partial e}{\partial q_l} \cdot \frac{\partial q_l}{\partial p_i} \cdot \frac{\partial p_i}{\partial w_{in}} \quad (11.45)$$

By combining Eqs. (11.39)–(11.42), we have

$$\Delta w_{in} = \eta p_i(1 - p_i)\sum_l q_l(1 - q_l)(Q_l - q_l)v_{li} R_n \tag{11.46}$$

In Eqs. (11.43)–(11.46), η is the learning factor. The weights of the input and output layers of a BP network are continuously adjusted until the standard error or the condition for terminating the iteration is met, and then the training of the network is completed.

To increase the learning efficiency, η should be set to a large value. However, an overly large η value can easily lead to oscillations. Therefore, to set η to a sufficiently large value that does not cause oscillations, another inertia term (i.e., the contribution of the previous weight) can often be introduced to the weight correction process. The weights of the output and input layers corrected with the consideration of the inertia term are as follows [33, 41]:

$$\Delta w'_{in}(m)=\Delta w_{in}(m) + \beta[w_{in}(m) - w_{in}(m - 1)] \tag{11.47}$$

$$\Delta v'_{li}(m)=\Delta v_{li}(m) + \beta[v_{li}(m) - v_{li}(m - 1)] \tag{11.48}$$

where β is the momentum factor and $\Delta w_{in}(m)$ and $\Delta v_{li}(m)$ are determined using Eqs. (11.44) and (11.46), respectively.

(1) **BP network-based pattern recognition of AE signals produced by TBCs due to damage**

The application of a BPNN algorithm to TBCs requires the consideration of the following aspects:

(1) The numbers of nodes in the input, output, and hidden layers (N, L, and I, respectively). For a TBC, the WESC contains 16 components. Thus, $N = 16$. In addition, a TBC has four failure modes. Hence, $L = 4$. Let [1 0 0 0], [0 1 0 0], [0 0 1 0], and [0 0 0 1] represent a vertical crack, a sliding interface crack, an opening interface crack in a TBC, and the deformation of the substrate, respectively. I is determined using the following equation [33, 35, 36]:

$$I = \sqrt{N + L} + a \tag{11.49}$$

where a is a constant between 1 and 10. Because N and L are determined to be 16 and 4, respectively, it can be preliminarily determined that I can be set to a number between 5 and 14. Subsequently, the I value corresponding to the smallest network error can be determined by trial and error. Thus, for a TBC, $I = 10$.

(2) Selection of the initial weights w_{in} and v_{li}. The BP algorithm is a gradient algorithm. Excessively large initial weights can extremely easily cause the weighted input and output to fall in the saturation zone and result in very small values of Δv_{li} and Δw_{in} (i.e., stagnation in the process of adjusting the weights of the

network). Based on long-standing experience, the initial weights w_{in} and v_{li} are generally set to random numbers in the range of (–1, 1) [33, 36].

(3) Determination of η, β, the maximum number of cycles, and the expected error. η determines the changes in the weights of the network during each training cycle. A large η value may lead to unstable network training, whereas a small η value may result in overly small weight adjustments during the system training process and therefore a too low rate of convergence. Generally, η is set to a value between 0.01 and 1 [36]. For a TBC, η, β, the maximum number of iterations of the network, and the expected error are set to 0.8, 0.95, 5000, and 0.001, respectively [35].

Here, 400 AE signals of known patterns (100 signals produced by surface cracks, 100 by sliding interface cracks, 100 by opening interface cracks, and 100 by the plastic deformation of the substrate) are used as a sample for network training and pattern recognition. Specifically, 60, 20, and 20% of the signals are randomly selected to form a training set, a test set, and a validation set, respectively. The training set is responsible for training the NN and adjusting its weights. The validation set is involved in the NN training and monitors the network error. The test set is not involved in the training of the NN but is instead used to examine the effectiveness of the NN in recognizing unknown AE signals [43]. Figure 11.31 shows the variation curves of the error of the BP network during the training process. The error curves obtained for the training, validation, and test sets display basically consistent trends: the error gradually decreases as the number of epochs of training increases and reaches the minimum after 175 epochs of training, at which point the network meets the design requirements. Table 11.1 summarizes the effectiveness of the trained BP network in identifying AE signals. The table shows that the network can identify the four types of AE signal at a rate above 98%. Hence, this NN is able to meet the goal of automatically identifying AE signals produced by a TBC due to damage.

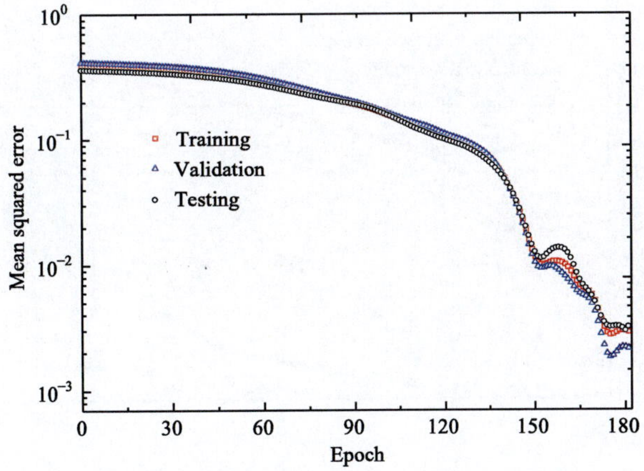

Fig. 11.31 Training error curves of the BP network

Table 11.1 Recognition results of AE signals produced by a TBC due to damage using the BP network

Crack pattern	Signal sample size	Number of recognized signals	Recognition rate (%)
Vertical surface crack	100	98	98
Sliding interface crack	100	99	99
Opening interface crack	100	99	99
Plastic deformation of the substrate	100	100	100

Further, the BP network is used for the pattern recognition of AE signals produced by a TBC due to damage under tensile and compressive loading. Figure 11.32 shows the variation in the number of AE signals produced by different patterns of cracks under tensile loading with time. In the early stage (i.e., the first 100 s of loading), the substrate yields, and surface cracks are the primary damage (crack) pattern in the TBC [23]. However, the surface cracks tend to saturate at 1600–1800s. At 1850s, sliding interface cracks begin to appear and grow rapidly, and the surface of the ceramic layer undergoes notable spallation. Compared to those due to surface and interface cracks, very few signals are produced by the plastic deformation of the substrate and hence can be neglected, indicating that the TBC has low-deformation capacity. In addition, the small number of signals produced by the opening interface cracks suggests that the spallation of the coating under tension is caused by the interfacial shear stress.

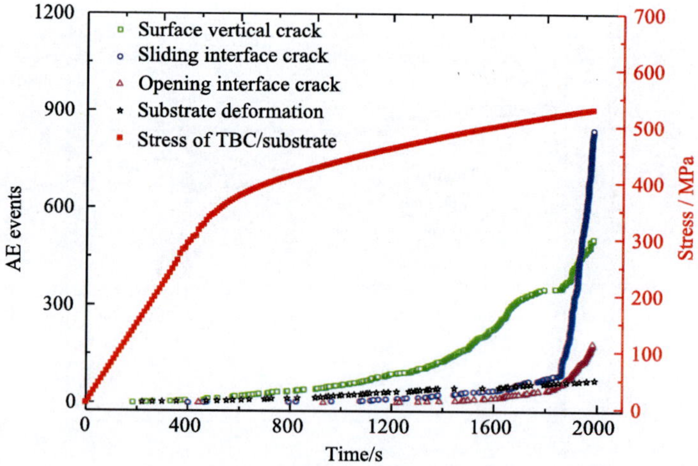

Fig. 11.32 Evolution of the number of AE signals produced by various crack patterns under tensile loading

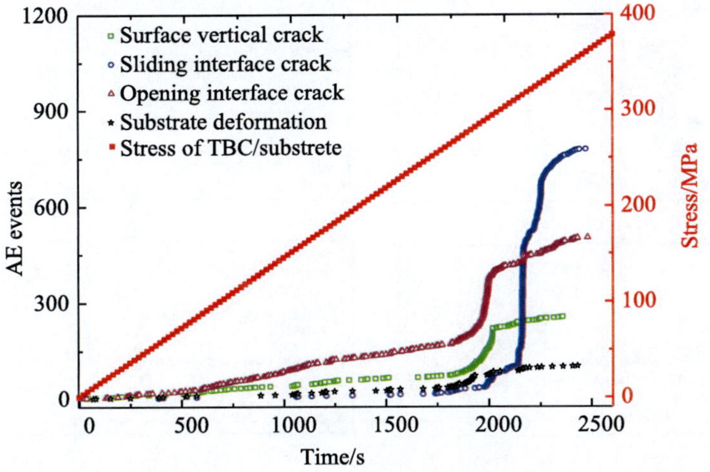

Fig. 11.33 Evolution of the number of AE signals produced by various crack patterns under compressive loading

As shown in Fig. 11.33, there are three main damage patterns (i.e., surface cracks, sliding interface cracks, and opening interface cracks) under compressing loading, and the signals produced by the plastic deformation of the substrate can be basically neglected. During the first 2000s of compressive loading, the AE signals are mainly produced by opening interface cracks. After the first 2000s, the number of signals produced by sliding interface cracks increases rapidly and becomes much larger than that of signals produced by opening interface cracks, indicating that sliding interface cracks are the primary failure mode of the TBC spallation under compressive loading.

Based on the above analysis, the procedure of the WPT-NN-based intelligent method for the damage pattern recognition of a TBC based on AE signals is summarized in Fig. 11.34. First, AE signals produced due to damage are collected. Then, the AE signals are subjected to a four-level WP decomposition using the "db8" wavelet function to extract the WESC at each node as their characteristic parameter, which is subsequently used as the input vector of the BPNN. Afterward, the damage pattern generating AE signals is identified based on the output vector of the BP network. Finally, the failure mode of the TBC is determined based on the distributions of the AE signals generated by different damage patterns. Note that if a new crack pattern appears during the failure process of a coating, the above method can be used to redetermine the WESC and the NN structure to realize intelligent identification of the failure mode of the coating based on AE signals.

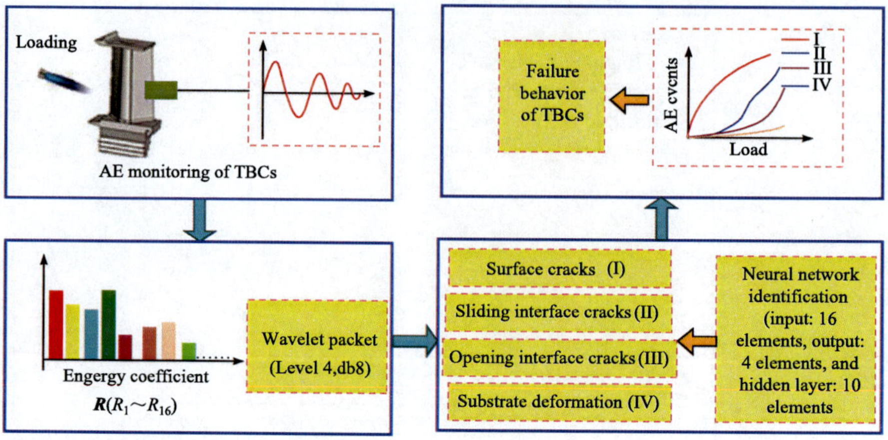

Fig. 11.34 A flowchart for the identification of crack patterns in a TBC based on AE signals

11.4 Quantitative Evaluation of the Key Damage in TBCs

11.4.1 Basic Approach for Damage Quantification

There are two issues that need to be addressed in the quantitative characterization of AE signals in the damage process of a TBC; one is the determination of the parameters descriptive of the damage of the TBC and the AE parameters characteristic of the damage of the TBC, and the other is the establishment of the quantitative relations between the AE and damage parameters of the TBC. According to the above analysis, vertical surface cracks and interface cracks (including both sliding and opening interface cracks) are the key crack patterns in TBCs. If the quantitative relations between various crack patterns and the parameters of AE signals can be established through simple loading, they can then be used in combination with the characteristics of AE signals generated in complex environments to predict damage evolution in complex environments.

The analysis in Sect. 11.3 shows that the failure of a TBC under axial tensile loading involves the following three processes: (1) vertical surface cracks are initiated and propagate in the ceramic layer and increase rapidly in number with increasing load until a saturated crack density is reached; (2) the vertical surface cracks extend to and subsequently propagate along the ceramic/BC interface, forming interface cracks; during this process, a small number of vertical surface cracks even extend to the BC/substrate interface; and (3) the connection of interface cracks eventually causes the coating to spall. Therefore, the vertical surface crack density is an important parameter for quantitatively characterizing the damage sustained by a TBC during the tensile failure process. Under axial compressive loading, the ceramic layer is prone to buckling. The buckling of the ceramic layer from the substrate leads to interface cracking, which first appears at the locations with a relatively weak bond at the TC/BC interface [28]. The failure mainly involves three macroscopic process: (1) the onset of ceramic layer buckling and the initiation of interface cracks, (2) the propagation of the buckling-induced interface cracks and the increase in the displacement in the buckled coating, and (3) the fracture and interfacial delamination and spallation of the buckled coating. Therefore, interface crack length is an important parameter for characterizing the buckling damage of a TBC.

The energy of AE signals produced by both vertical surface cracks and interface cracks originates from the energy released during crack formation or propagation. Thus, the following must hold:

$$E_{el} = \alpha E_{AE} \tag{11.50}$$

where E_{el} is the strain energy released during crack propagation, E_{AE} is the AE signal energy, and α is a proportionality coefficient related to the AE instrument. Thus, once the correlations between E_{el} and the quantitative parameters of surface and interface cracks are established, so are the correlations between the quantitative parameters of surface and interface cracks and E_{AE}; i.e., a quantitative evaluation can be realized [28, 45].

11.4.2 Quantitative Analysis of the Surface Crack Density

When a system composed of a brittle coating and a ductile substrate is subjected to tensile loading, surface cracks are gradually formed in the brittle coating and eventually saturate. Let $2a$ be the distance between two cracks. Then, the crack density can be defined as $\rho = 1/(2a)$. Based on McGuigan et al.'s shear-lag model for the crack density of brittle films [45] and considering the nonuniform distribution of stresses along the thickness of the TBC, we present a model to analyze the surface crack density of a TBC under tensile loading.

As shown in Fig. 11.35, the BC layer and the substrate are viewed as one layer, and the ceramic layer is divided into three layers, namely, d_1, d_2 (where the stresses are linearly distributed along the thickness), and d_3 (i.e., the interfacial layer) [44]. $u_1(x)$, $u_2(x)$, and $u_3(x)$ are the displacements in layers d_1, d_2, and d_3, respectively, and $\tau_c(x)$ and $\tau_s(x)$ are the shear stresses in layers d_2 and d_3, respectively. The ceramic coating and the substrate are viewed as an ideal elastic material and an ideal elastoplastic material, respectively. The normal stress in the thin layer d_1 can be considered to be uniformly distributed. The normal stress $\sigma(x)$ in layer d_1 in the separated small coating block and the shear stress $\tau_s(x)$ in the interfacial layer satisfy the following equilibrium equations:

$$\frac{d\sigma(x)}{dx} + \frac{\tau_c(x)}{d_1} = 0 \tag{11.51}$$

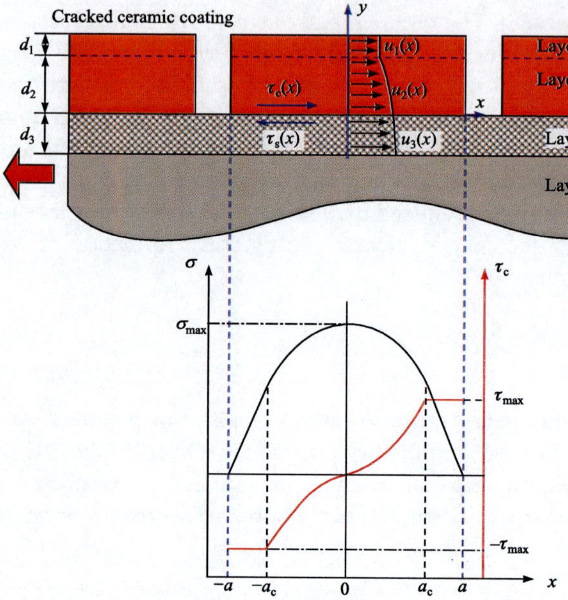

Fig. 11.35 A shear-lag model for the surface crack density of a TBC and stress distribution curves

$$\tau_c(x) = -\tau_s(x) \tag{11.52}$$

The geometric equations of the strains in the coating are expressed as follows:

$$\varepsilon_c = \frac{du_1(x)}{dx} + \varepsilon_r \tag{11.53}$$

$$\varepsilon_s = \frac{du_3(x)}{dx} \tag{11.54}$$

where ε_s and ε_c are the strains in layer d_3 (i.e., the interfacial layer) and layer d_1, respectively, and ε_r is the residual strain in the ceramic layer. On this basis, the following three equilibrium equations are obtained:

$$\sigma(x) = E_c\left[\frac{du_1(x)}{dx} + \varepsilon_r\right] \tag{11.55}$$

$$\tau_c(x) = \frac{G_c[u_2(x) - u_1(x)]}{d_2} \tag{11.56}$$

$$\tau_s(x) = \frac{G_s[u_3(x) - u_2(x)]}{d_3} \tag{11.57}$$

where E_c and G_c are the elastic and shear moduli of the ceramic layer, respectively, and G_s is the shear modulus of the substrate. By substituting Eqs. (11.55) and (11.56) into Eq. (11.51), we have

$$\frac{d^2\sigma(x)}{dx^2} + \frac{G_c}{d_1 d_2}\left[\frac{du_2(x)}{dx} - \frac{du_1(x)}{dx}\right] = 0 \tag{11.58}$$

$$\frac{d^2\sigma(x)}{dx^2} + \frac{G_s}{d_1 d_3}\left[\frac{du_2(x)}{dx} - \frac{du_3(x)}{dx}\right] = 0 \tag{11.59}$$

Substituting Eqs. (11.53) and (11.55) into Eq. (11.58) and substituting Eq. (11.54) into Eq. (11.59) gives

$$\frac{d^2\sigma(x)}{dx^2} + \frac{G_c}{d_1 d_2}\left[\frac{du_2(x)}{dx} + \frac{\sigma(x)}{E_c} - \varepsilon_r\right] = 0 \tag{11.60}$$

$$\frac{d^2\sigma(x)}{dx^2} + \frac{G_s}{d_1 d_3}\left[\frac{du_2(x)}{dx} - \varepsilon_s\right] = 0 \tag{11.61}$$

By simultaneously solving Eqs. (11.60) and (11.61), we have

$$\frac{d^2\sigma(x)}{dx^2} - K^2\sigma(x) = -K^2 E_c(\varepsilon_s + \varepsilon_r) \tag{11.62}$$

where $K = \sqrt{\frac{G_s G_c}{d_1 E_c (G_s d_2 - G_c d_3)}}$. The general solution of Eq. (11.62) is as follows:

$$\sigma(x) = c_1 e^{Kx} + c_2 e^{-Kx} + E_c(\varepsilon_s + \varepsilon_r) \tag{11.63}$$

Substitution of Eq. (11.63) into Eq. (11.51) yields

$$\tau_c(x) = -d_1 K(c_1 e^{Kx} - c_2 e^{Kx}) \tag{11.64}$$

In Eqs. (11.63) and (11.64), both c_1 and c_2 are coefficients to be determined. When the substrate undergoes plastic deformation, the normal stress $\sigma(x)$ in layer d_1 and the shear stress in layer d_3 follow distributions as shown by the curves in Fig. 11.36. c_1 and c_2 can be determined according to the boundary conditions.

When $x < a_c$, the normal stress at the center of the fractured coating block is the highest, corresponding to a shear stress of 0 at the center of the interfacial layer, i.e., $\sigma(0) = \sigma_b$, and $\tau_s(0) = 0$. c_1 and c_2 are determined based on these two known conditions in combination with Eqs. (11.63) and (11.64). The normal stress in the coating and the shear stress in the interfacial layer are expressed as follows:

$$\sigma(x) = [\sigma_b - E_c(\varepsilon_s + \varepsilon_r)]\cosh(Kx) + E_c(\varepsilon_s + \varepsilon_r) \tag{11.65}$$

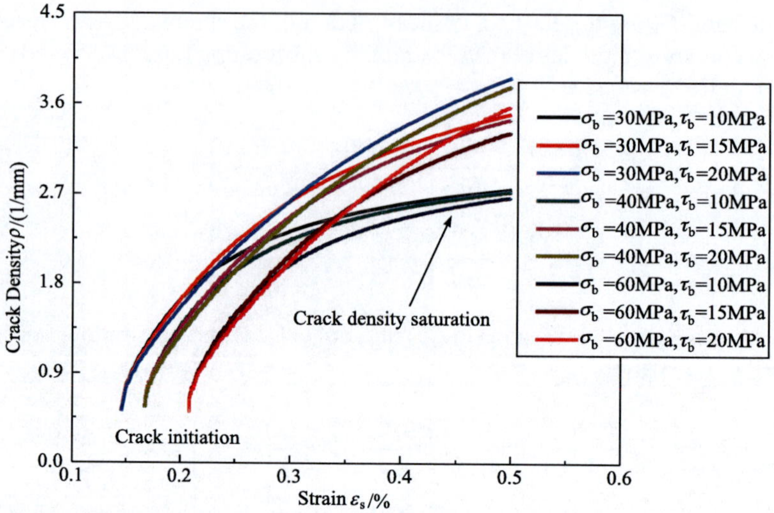

Fig. 11.36 Variation in the surface crack density ρ with the substrate strain ε_s at different values of fracture strength σ_b and shear strength τ_b

$$\tau_c(x) = d_1 K[E_c(\varepsilon_s + \varepsilon_r) - \sigma_b] \sinh(Kx) \tag{11.66}$$

When $a_c \le x \le a$, the normal stress at the free boundary of the fractured coating block is zero, and the shear stress in the interfacial layer reaches the maximum and does not vary with the location, i.e., $\sigma(a) = 0$, and $\tau_s(x) = \tau_b$. Based on these known conditions, the following expression of the normal stress in the coating is obtained:

$$\sigma(x) = \frac{\tau_b}{d_1}(a - x) \tag{11.67}$$

$$\tau_c(x) = \tau_b \tag{11.68}$$

The maximum normal stress σ_b and shear stress $\tau_c(x)$ in the coating satisfy the following equilibrium relation:

$$\sigma_b = \frac{1}{d_1}\int_0^a \tau_c(x)dx \tag{11.69}$$

The stress distribution curves in Fig. 11.35 show that the function is continuous at $x = a_c$, i.e.,

$$\tau_b = d_1 K[E_c(\varepsilon_s + \varepsilon_r) - \sigma_b] \sinh(K a_c) \tag{11.70}$$

By substituting Eqs. (11.66) and (11.68) into Eq. (11.69) and simultaneously solving the resulting equation and Eq. (11.70) with the consideration of $\cosh^2(Ka_c)-\sinh^2(Ka_c)=1$, we obtain the following analytical expression of the surface crack density of the coating:

$$\rho=\frac{1}{2a}=\frac{\tau_b K}{2}\left[\sigma_b K d_1 - M + \sqrt{M^2+\tau_b^2}+\tau_b \operatorname{arcsin} h\frac{\tau_b}{M}\right]^{-1} \quad (11.71)$$

where $M = d_1 K[E_c(\varepsilon_s+\varepsilon_r)-\sigma_b]$.

Equation (11.71) shows that the surface crack density of a TBC is related to the fracture strength of the ceramic layer (σ_b), the interfacial shear strength (τ_b), the residual stress formed during the preparation of the coating (σ_r), and the strain of the substrate (ε_s). Figure 11.36 shows the crack densities calculated based on different values of σ_b and τ_b. Clearly, the crack density ρ increases with ε_s and tends to saturate as ε_s reaches a certain value, a high σ_b leads to a large ε_s corresponding to the crack initiation point, and a low τ_b leads to a small saturation crack density (ρ_{satu}).

To verify the correctness of the surface crack density model for coatings, we prepared three groups of specimens through the adjustment of the preparation process. For each group of specimens, the Young's modulus of the ceramic layer, E_c, and the Young's modulus of the substrate, E_s, were 48 and 165 GPa, respectively, and the Poisson's ratios of the ceramic layer and the substrate were set to 0.22 [46] and 0.3, respectively. The three groups of specimens differed in σ_r, σ_b, and τ_b, as shown in Table 11.2. See elsewhere [44] for the detailed testing method.

Figure 11.37 shows the theoretical and experimental results of the surface crack densities of the three groups of APS TBC specimens. Surface cracks begin to appear in the ceramic layer when ε_s reaches a certain value, and as ε_s increases, the number of surface cracks in the ceramic layer increases and eventually saturates. The ε_s corresponding to the initiation of surface cracks in the three types of TBC specimens varies in the range of 0.3 to 0.375%, which is consistent with the experimental results. The ε_s values corresponding to crack saturation in the three types of TBC specimens are 0.5%, 0.75%, and 0.90%, respectively, which are in good agreement with the experimental results. An analysis of both the experimental and calculation results for the three types of specimens shows that a high σ_b leads to a high tensile fracture resistance of the coating and a high ε_s corresponding to the surface crack initiation

Table 11.2 Mechanical property parameters of three types of TBC specimens

	Residual stress σ_r/MPa	Fracture strength σ_b/MPa	Shear strength τ_b/MPa
Group 1	−510.8 ± 1.7	123.3 ± 10.9	23.8 ± 2.4
Group 2	−48.0 ± 4.5	107.0 ± 10.9	15.4 ± 1.3
Group 3	−74.7 ± 3.1	94.3 ± 5.7	10.2 ± 1.6

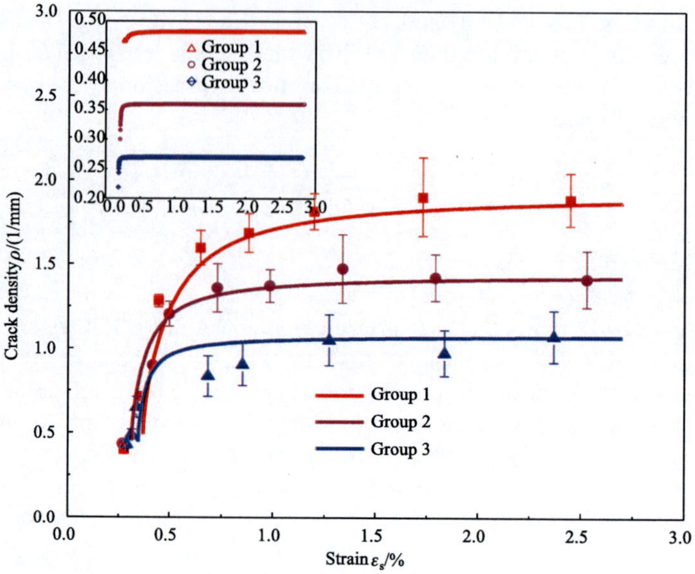

Fig. 11.37 Variation in the surface crack density ρ with the substrate tensile stress ε_s

and that a high τ_b leads to a pronounced tensile stress transfer effect in the coating and a large ρ_{satu}.

Now, we can quantify the relation between the surface crack density and the cumulative energy of the AE signals generated by the surface cracks. Assuming that the vertical surface cracks penetrate the ceramic layer in the thickness and width directions and are uniformly distributed, the total surface crack area S is as follows:

$$S = 2whL\rho \tag{11.72}$$

where L is the effective signal sampling distance (see Fig. 11.2) and w and h are the width and thickness of the ceramic layer, respectively. Under tensile loading, the surface cracks in the coating can be viewed as Type I cracks in a plane stress state with the following energy release rate:

$$G_s = \frac{K_{IC}^2}{E_c} \tag{11.73}$$

where K_{IC} is the fracture toughness of the ceramic layer. Then, the total energy released by the surface cracks in the ceramic layer, E_{total}^s, is expressed as follows:

$$E_{total}^S = G_s \cdot S = 2whL\rho\frac{K_{IC}^2}{E_c} \tag{11.74}$$

Fig. 11.38 Curves of the AE energy E_{AE} versus the surface crack density ρ

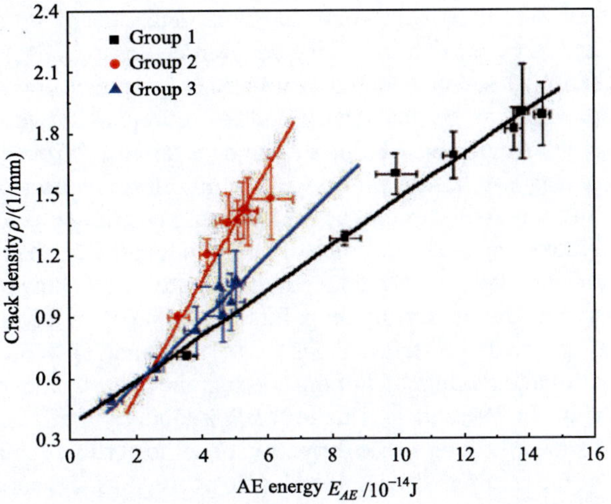

The final expression of the total AE energy released by surface cracks is as follows:

$$E_{AE}^{S} = 2whL\rho \frac{K_{IC}^{2}}{\alpha E_{c}} \tag{11.75}$$

The energy of a single AE signal is calculated using the following equation [44]:

$$E_{AE} = \int_{0}^{T} \frac{U^{2}(t)}{R} dt \tag{11.76}$$

where T and U(t) are the duration and voltage amplitude of the signal, respectively, and R is the internal resistance of the AE instrument (here, $R = 1 \times 10^{21}\ \Omega$).

Figure 11.38 shows the quantitative relations between surface crack density and E_{AE} for the three groups of TBC specimens. A notable linear relation can be observed for each group of TBC specimens. An analysis based on Eq. (11.75) reveals that the slope of the linear relation depends on K_{IC} and the conversion coefficient α, both of which are related to the preparation process of the ceramic coating and the microstructure (e.g., distribution of pores and defects) of its coating. In addition, α is related to the AE instrument.

11.4.3 Quantitative Analysis of the Interface Cracks

A coating can buckle and spall under compressive loading, as it is much thinner than its substrate. The whole buckling process of a TBC under compressive loading can be recorded through real-time detection of its deformation using techniques and devices such as charge-coupled devices (CCDs) and digital image correlation (DIC), based on which the formation of interface cracks in the TBC and its spallation process can be analyzed. In addition, a quantitative relation between the interface crack length and AE energy can be established from recorded AE signals generated during the failure process using the aforementioned pattern recognition method. To accurately capture the buckling and cracking phenomenon in the ceramic layer, a preexisting interface crack of a certain length can be prepared (see elsewhere [28] for the detailed preparation method). Figure 11.39 shows the buckling process of a TBC under compressive loading. First, a compressive load is applied to a TBC containing a preexisting interface crack (Fig. 11.39a and f). When the compression exceeds the critical buckling load of the coating, the coating above the interfacial defect buckles (Fig. 11.39b and g). Due to the Poisson's ratio effect, an opening interface crack is formed, leading to the formation of some vertical surface cracks in the nearby zones of the coating. The buckled brittle coating, as shown in Fig. 11.39c and h, suddenly fractures at both ends along a direction deviating at a certain angle from the vertical direction of the coating and therefore spalls, indicating the kinking of the interface cracks. This is consistent with the oxidation and buckling phenomenon discovered by Wright and Evans [47]. Further, a sliding interface crack is formed (i.e., delamination occurs) near the interface due to shear stress, as shown in Fig. 11.39d and i. Under external loading, the interface cracks gradually propagate and eventually cause the coating to spall.

Here, we consider that an axially compressed TBC is in a plane stress state and assume that the coating is in an elastic stage under a compressive stress σ. Then, the elastic energy G_0 stored in the ceramic layer with a thickness of h is expressed as follows:

$$G_0 = \frac{1 - \nu_c^2}{2E_c}\sigma^2 h \tag{11.77}$$

where E_c and ν_c are the elastic modulus and Poisson's ratio of the ceramic layer, respectively. Considering the residual compressive stress σ_0 due to the preparation of the ceramic layer, the stress in the coating can be expressed as follows:

$$\sigma = \sigma_0 + E_c\varepsilon \tag{11.78}$$

where ε is the strain applied to the ceramic layer.

Here, we simplify the buckling of the ceramic layer to a stability problem of a column with two fixed ends under compression. The critical buckling stress of a coating that contains an interfacial defect with a length of $2b$ is as follows:

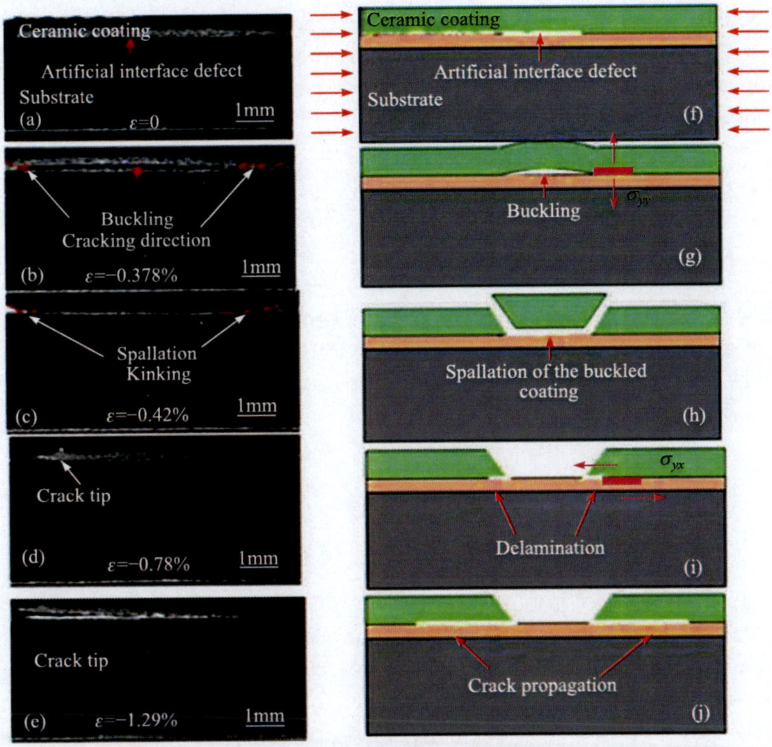

Fig. 11.39 Schematic diagram of the buckling process of a TBC

$$\sigma_{cr} = \frac{\pi^2}{12}\frac{E_c}{1-\nu_c^2}\left(\frac{h}{b}\right)^2 \tag{11.79}$$

The strain energy release rate G is a function related to the half-length of the crack at the buckling interface, b, with the following expression [48, 49]:

$$G(b) = G_0\left(1-\frac{\sigma_{cr}}{\sigma}\right)\left(1+3\frac{\sigma_{cr}}{\sigma}\right) \tag{11.80}$$

Substitution of Eq. (11.79) into (11.80) yields

$$G(b) = G_0\left[1+\frac{\pi^2 E_c h^2}{6\left(1-\nu_c^2\right)\sigma}b^{-2}-\frac{\pi^4 E_c^4 h^4}{48\left(1-\nu_c^2\right)^2\sigma^2}b^{-4}\right] \tag{11.81}$$

Thus, the energy released during interface crack propagation can be determined based on G. The total energy E_{total}^{I} released by the interface crack with a half-length of b and a width of w formed from the propagation of a preexisting crack with a length of b_0 is as follows:

$$E_{total}^{I} = 2w\int_{b_0}^{b} G(b)db \tag{11.82}$$

Based on the linear relation between the AE energy (E_{AE}^{I}) and the corresponding E_{total}^{I} of an interface crack in combination with Eqs. (11.77), (11.81), and (11.82), the following correlation between the interface crack and the AE energy is obtained:

$$\begin{aligned} E_{AE}^{I} &= \frac{1}{\alpha}E_{total}^{I} \\ &= \frac{2w}{\alpha}\left[\frac{(1-\nu_c^2)\sigma^2 h}{2E_c}(b-b_0) + \frac{\pi^4 E_c h^5}{288(1-\nu_c^2)}\left(\frac{1}{b^3}-\frac{1}{b_0^3}\right) - \frac{\pi^2 h^3 \sigma}{12}\left(\frac{1}{b}-\frac{1}{b_0}\right)\right] \end{aligned} \tag{11.83}$$

Figure 11.40 shows the relation between the crack half-length b and the AE energy E_{AE}^{I} of a TBC under compressive loading. As b increases, the corresponding E_{AE}^{I} increases in an overall linear fashion. A comparison shows a good agreement between the experimental and calculation results.

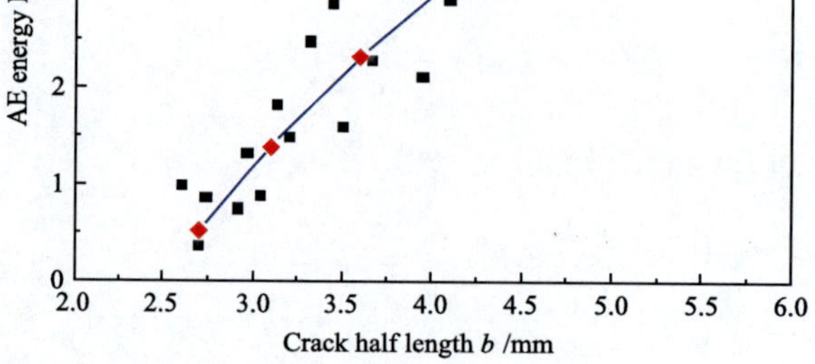

Fig. 11.40 Relation between the crack half-length and AE energy

11.5 Determination of TBC Failure Mechanisms Based on AE Detection

11.5.1 Failure Mechanisms Under Thermal Cycling

Thermal cycling is the easiest, most effective experimental means for simulating the service state of TBCs during take-off, flight, and landing. However, TBC failure mechanisms under thermal cycling are not thoroughly understood, and whether spallation occurs during take-off, flight, or landing remains unclear. Therefore, it is particularly important to examine this process in real time. Because sensors for AE detection cannot be used at high-temperatures, AE detection is generally performed only at room temperature. Ma et al. [17] were the first to use the AE technique to investigate the failure processes of TBCs under four- and three-point bending at room temperature. Fu et al. [8] and Ma et al. [17] employed the AE technique to test the cooling process of TBCs under thermal shock or thermal fatigue conditions. Using the WW transmission technique, pattern recognition method, and quantitative evaluation method introduced in this chapter, we detected the AE signals generated by a TBC during a thermal fatigue process at high-temperatures and analyzed its failure mechanisms [10].

A physical vapor deposition (PVD) TBC specimen composed of a DZ125 alloy substrate with dimensions of 40 mm × 6 mm × 3 mm, a 50-μm NiCrAlY transition layer, and a 100-μm yttria-stabilized zirconia (YSZ) ceramic layer was subjected to a thermal cycling test in a high-temperature resistance furnace. First, the resistance furnace was heated to the required temperature (e.g., 800 °C). Then, the TBC was placed in the furnace and heated for 5 min, after which the TBC was removed from the furnace and allowed to cool naturally in air for 10 min. The whole thermal cycling process was carried out in stationary air. The AE signals generated during both the heating and cooling processes were detected using the WW technique. In the experiment, one end of the Pt wire was welded to the substrate of the TBC and the other end was connected to an AE sensor through a mechanical device. The AE signal sampling frequency was set to 1 MHz.

The damage process and mechanism of the TBC under thermal cycling were analyzed based on the AE signals as follows:

(1) **Evolution and pattern recognition of cracks induced by thermal cycling**

The AE signals were subjected to a four-level decomposition with the db8 wavelet using the aforementioned wavelet analysis method. The corresponding frequency bands are D1: 0.25–0.50 MHz; D2: 0.125–0.25 MHz; D3: 0.0625–0.125 MHz; D4: 0.03125–0.0625 MHz; and C4: 0–0.03125 MHz. Based on the analysis in Sect. 11.2, AE signals generated by vertical surface cracks have a frequency 0.23 MHz, and consequently these cracks are in the frequency band D2; AE signals generated by opening and sliding interface cracks have dominant frequencies of 0.29 and 0.43 MHz, respectively, and therefore, these cracks are both in the frequency band

D1. An analysis of the damage signals generated by the TBC during the thermal cycling process shows that most of the signals are in the frequency band D1 or D2; i.e., they were generated by vertical surface cracks and interface cracks.

Further, the pattern recognition was applied to AE signals during the heating and cooling stages of each thermal cycle using the characteristic WESC. As clearly seen from the results in Fig. 11.41, most of the AE signals detected during the heating process are generated by vertical surface cracks, whereas the AE signals are predominantly produced by interface cracks during the cooling process. This suggests that, in each thermal cycle, the damage to the TBC is primarily reflected by the vertical cracks in the ceramic layer during the heating stage and the cracks at the interface between the ceramic layer and the intermediate transition layer during the cooling stage. In fact, as early as in 2002, the author of this book carried out a theoretical analysis of the compressive buckling of coatings during the cooling stage and, on this basis, hypothesized that coatings spall during the cooling stage [50]. However, this hypothesis had not been verified experimentally until tests were conducted to detect the AE signals generated by coatings during the whole failure process under thermal cycling in real time and the corresponding crack pattern recognition was carried out.

(2) **Quantitative characterization of the surface cracks in a coating**

Here, we regard a TBC system as an elastic coating and its transition layer and substrate as one layer. Considering the residual compressive stress and thermal mismatch stress formed during the preparation process, the thermal stress in the ceramic layer can be expressed as follows [51, 52]:

$$\sigma = \frac{E_t^*\left[E_s^* h_s(\alpha_s - \alpha_t)(T - T_r)\right]}{E_s^* h_s + E_t^* h_t} + \sigma_r \tag{11.84}$$

The thermal strain in the ceramic layer can be expressed as follows [51, 52]:

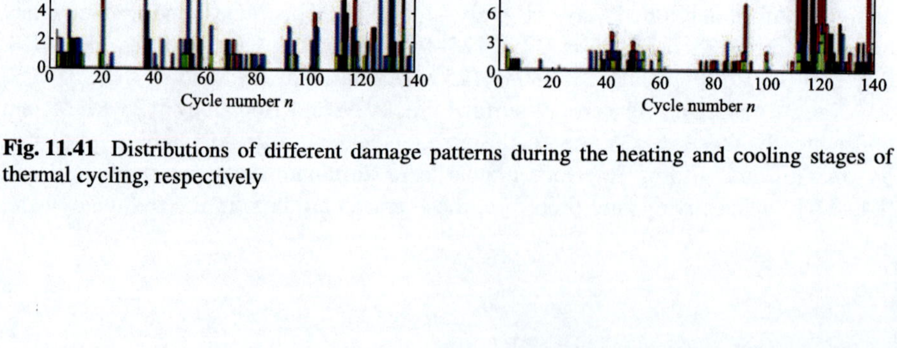

Fig. 11.41 Distributions of different damage patterns during the heating and cooling stages of thermal cycling, respectively

$$\varepsilon = \frac{E_s^* h_s(\alpha_s - \alpha_t)(T - T_r)}{E_s^* h_s + E_t^* h_t} + \frac{\sigma_r}{E_t^*} \tag{11.85}$$

where $E_i^* = \frac{E_i}{1-\nu_i}$ ($i = t$ or s), E is Young's modulus ($E = 48$ and 200 GPa for the ceramic layer and the transition layer, respectively), ν is Poisson's ratio ($\nu = 0.1$ and 0.3 for the ceramic layer and the transition layer, respectively), α is the coefficient of thermal expansion (CTE) ($\alpha = 9 \times 10^{-6}$ and 14.8×10^{-6} for the ceramic layer and the transition layer, respectively), h is the thickness, t and s signify the coating and the substrate, respectively, T is temperature, T_r is room temperature, and σ_r is the initial residual stress in the coating ($\sigma_r = -70$ MPa) [56]. Based on the above equation, the stress in the coating is 155.9 MPa at $T = 800$ °C. The stress in the coating returns to σ_r after it completely cools.

Evidently, during a thermal cycle, the stress in the ceramic layer changes from tensile stress in the heating stage to compressive stress in the cooling stage. The tensile stress in the heating stage also plays a vital role in the failure of the coating and can induce the formation of vertical surface cracks in the coating. While the compressive stress can close the cracks in the coating, it promotes the buckling of the coating and therefore leads to interfacial delamination (i.e., interface cracking). This analysis agrees very well with the AE detection results in that vertical surface cracks were formed in the coating during the heating process whereas interface cracks were generated during the cooling process.

According to the conventional shear-lag model, when the substrate of a coating is in an elastic stage, the crack density ρ of the coating is expressed as follows [10]:

$$\rho = \frac{k}{2}\operatorname{arccosh}\left(\frac{-G\varepsilon}{\sigma_b d_1 d_2 k^2 - G\varepsilon}\right)^{-1} \tag{11.86}$$

When the substrate is in a plastic stage, ρ is expressed as follows:

$$\rho = \frac{\tau_b k^2 d_2}{2G}(A + B + C)^{-1} \tag{11.87}$$

where $A = \frac{\sigma_b k^2 d_1 d_2 - G\varepsilon}{G}\cosh\left(\operatorname{arcsinh}\left(\frac{\tau_b k d_2}{G\varepsilon - \sigma_b k^2 d_1 d_2}\right)\right)$,

$$B = \frac{\tau_b k d_2}{G}\operatorname{arcsinh}\left(\frac{\tau_b k d_2}{G\varepsilon - \sigma_b k^2 d_1 d_2}\right) \text{ and}$$

$$C = \varepsilon.$$

In Eqs. (11.86) and (11.87), $k = \sqrt{\frac{G}{d_1 d_2 E}}$, G is the shear modulus of the interfacial layer, E is the elastic modulus of the coating, d_1 and d_2 are the thicknesses of the coating and the interfacial layer, respectively, and k is the fracture toughness of the coating. Evidently, when ε_s is known, the crack density can be determined using Eq. (11.86) or (11.87).

During the heating process, the tensile stress in the ceramic layer gradually increases and reaches the maximum at the furnace temperature. According to the principle of strain compatibility, the substrate strain ε_s can be considered to be the same as that in Eq. (11.85). Assuming that the damage increases linearly with the number of thermal cycles during the thermal cycling process, the final crack density can be determined using Eq. (11.86) or (11.87). Assuming that $\sigma_r = -70$ MPa, ε_s is calculated to be 0.29% using Eq. (11.85), meaning that the substrate undergoes plastic deformation. Thus, the crack density is calculated using Eq. (11.87). Here, the fracture strength σ_b of the ceramic layer is set to the range of 40–80 MPa [50, 57], and the shear strength τ_b of the interfacial layer is set to the range of 6–20 MPa [58, 59]. Figure 11.42 shows the correlation between the surface crack density and the AE event counts N_{AE}. Clearly, under the assumption of linear damage accumulation, the crack density calculated using the shear-lag model increases linearly with the number of thermal cycles. At $\sigma_b = 80$ MPa and $\tau_b = 12$ MPa, the theoretically calculated values of the crack density are in good agreement with the experimental results. In addition, all the test points fall between the curve corresponding to $\sigma_b =$ 80 MPa and $\tau_b = 12$ MPa and the curve corresponding to $\sigma_b = 60$ MPa and $\tau_b =$ 12 MPa, indicating that the quantitative linear relation between the crack density and N_{AE} (Fig. 11.42) obtained using the shear-lag model is reliable.

(3) **Quantitative characterization of interface cracks**

The interface crack propagation under cyclic thermal loading follows the classical fatigue crack growth model:

$$da/dN = C\Delta K^m \tag{11.88}$$

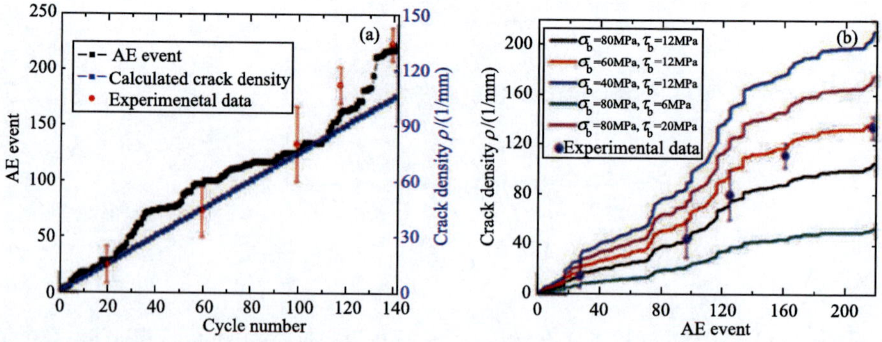

Fig. 11.42 **a** Evolution of the surface crack density and the cumulative AE event counts of a TBC under cyclic thermal loading with number of thermal cycles. **b** Relation between the surface crack density and the cumulative AE event counts

where C and m are constants and ΔK is the amplitude of the change in the stress intensity factor (SIF) generated by the compressive stress during cooling and can be expressed as follows [60]:

$$\begin{aligned}\Delta K &= F\left(\frac{a}{W}\right)\Delta\sigma\sqrt{\pi a}\\ &= \frac{a/W}{\sqrt{\pi(1-a/W)}}\left[7.264-9.37\left(\frac{a}{W}\right)+2.74\left(\frac{a}{W}\right)^2+\cdots\right]\Delta\sigma\sqrt{\pi a}\end{aligned} \tag{11.89}$$

where $\Delta\sigma$ is the amplitude of the compressive stress during cooling (in this example, the initial residual stress is set as $\Delta\sigma$), a is the interface crack length, W is the width of the specimen ($0 \le \frac{a}{W} \le 1$), and $F\left(\frac{a}{W}\right)$ is a geometric factor. In this example, $W =$ 3 mm, and a is on the order of microns, so a/W is close to zero. Hence, the geometric factor is selected as $F(a/W) = 7.264\frac{a/W}{\sqrt{\pi}}$.

Here, the dimensionless crack length $\overline{a} = a/a_0$ and dimensionless SIF $\overline{\Delta K_{\mathrm{II}}} = \frac{\Delta K}{K_{\mathrm{IIC}}}$ are introduced. By substituting them into Eqs. (11.88) and (11.89), we can express a as follows:

$$a = a_0\left[\left(1-\frac{3}{2}m\right)D\left(\frac{7.264a_0^{3/2}\Delta\sigma}{K_{\mathrm{IIC}}W}\right)^m N+1\right]^{\frac{2}{2-3m}} \tag{11.90}$$

where a_0 is the initial crack length, D is a constant, K_{IIC} is the interfacial fracture toughness, and N is the number of cycles.

Based on experimental observations, the value of interface crack length a after 140 cycles and the value of a_0 are set to 18 and 1 μm, respectively, the exponent m is set to the range of 0–2, and K_{IIC} is set to 0.73 MPa [60]. Figure 11.46a shows the evolution of a and N_{AE} with the number of thermal cycles at $m = 0.6$ and $\sigma_r = -70$ MPa. While there is a certain deviation between the experimental and calculated values of a, they display similar upward trends. In addition, while N_{AE} gradually increases with the number of thermal cycles, there is no linear relation between them in Fig. 11.43a. Thus, the correlation between a and N_{AE} is described in Fig. 11.43b. The evolution of a with N_{AE} involves two stages. During approximately the first 70 cycles, due to the random initiation of interface cracks, there is no notable regular relation between the value of a calculated simply based on the fatigue crack growth model and N_{AE}. With the progression of crack propagation, a increases linearly with N_{AE}. Figure 11.44 also shows that interface crack propagation is significantly affected by m. A large m leads to a high crack growth rate and can easily cause unstable interface crack propagation, which is macroscopically reflected by coating spallation.

(4) **Prediction of critical damage events**

Coating spallation is caused by the gradual evolution of microdamage such as vertical surface cracks and interface cracks. Whether the final spallation can be predicted

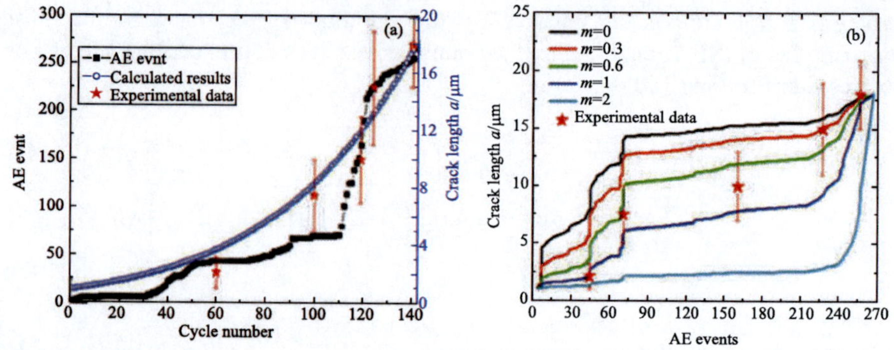

Fig. 11.43 **a** Evolution of the interface crack length and AE event counts. **b** Correlation between interface crack length and AE event counts

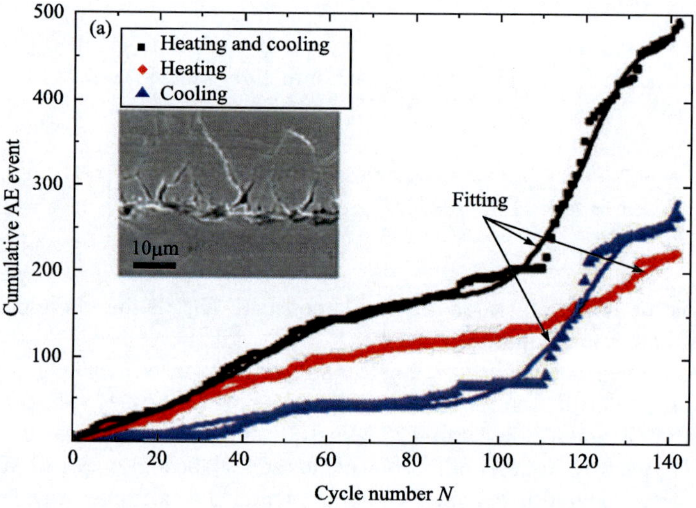

Fig. 11.44 Critical number of cycles predicted based on the statistical analysis of the cumulative number of AE events

based on early-stage damage signals is an issue of vital interest in engineering applications. Here, based on renormalization theory, we present a model that predicts critical damage points from the statistical pattern of early-stage damage signals. Critical points can refer to generalized loads such as the time of damage, critical load, and number of cycles. The specific physical concept involved in this model is that macroscopic failure results from the gradual evolution of damage and that various types of damage are intercorrelated; hence, early-stage development trends can be used to predict critical sudden-change points. The model is mathematically expressed as follows:

$$f(N) = A + B\left(N_f - N\right)^m \left[1 + C\cos\left(2\pi \frac{\log\left(N_f - N\right)}{\log \lambda} + \phi\right)\right] \tag{11.91}$$

where $f(N)$ is the cumulative number of AE events, N is the number of cycles, A, B, and C are all dimensionless fitting parameters, m is the power exponent, λ is the scale parameter, and φ is the conversion parameter. See elsewhere [61, 62] for the meaning of the above equation and the seven fitting parameters.

Figure 11.44 shows the results of a statistical analysis of all the AE signals and those generated during the heating and cooling stages of thermal cycling, respectively. All the fitted curves match the experimental results. The signals generated during the cooling stage become scattered near the point of damage, corresponding to the sudden change in crack propagation. Based on the statistical analysis of the signals generated during the heating stage, the signals generated during the cooling stage, and all the signals, the critical numbers of cycles are predicted to be 221, 151, and 164, respectively. The cycle number predicted based on the signals generated during the heating stage is larger than that predicted based on the signals generated during the cooling stage, which is, however, closer to the experimental results. A large number of cracks occur at the interface and in the coating after 170 cycles, as seen in the microstructure shown in Fig. 11.44. This result shows two important phenomena: (1) for TBCs, interface cracks are a more dangerous damage pattern than vertical surface cracks, and (2) real-time detection and prediction of the evolution of interface cracks can facilitate a more accurate prediction of the spallation life of a coating.

11.5.2 Failure Mechanism Under High-Temperature CMAS Corrosion

CMAS corrosion is considered to be the most dangerous service condition for TBCs. Real-time detection and quantitative analysis of the spallation process of coatings under CMAS corrosion provides key data and a direct basis for understanding their failure mechanism under CMAS corrosion. Hence, we detected the AE signals generated by a TBC subjected to CMAS corrosion at high-temperatures in real time using the high-temperature WW technique and an independently developed high-temperature gradient furnace, as well as analyzed its failure mechanism and extent using methods such as wavelet and quantitative analysis [63].

An APS TBC composed of a DZ125 alloy substrate with dimensions of 15 mm × 10 mm × 3 mm, a NiCrAlY transition layer with a thickness of 100 μm, and a YSZ ceramic layer with a thickness of approximately 200 μm was used in the experiment. Based on the analysis of the main chemical composition of sediments (e.g., volcanic ash), CMAS powder with a SiO_2:Al_2O_3:CaO:MgO ratio of 48.5:11.8:33.2:6.5 (wt.%) was selected (see elsewhere [63] for its preparation method). As shown in Fig. 11.45, the TBC was subjected to a CMAS corrosion test in a high-temperature gradient furnace. Specifically, the TBC was heated with the furnace to 1250 °C within 30 min

Fig. 11.45 Detection of the AE signals generated by a TBC subjected to high-temperature CMAS corrosion and the temperature distribution of the specimen

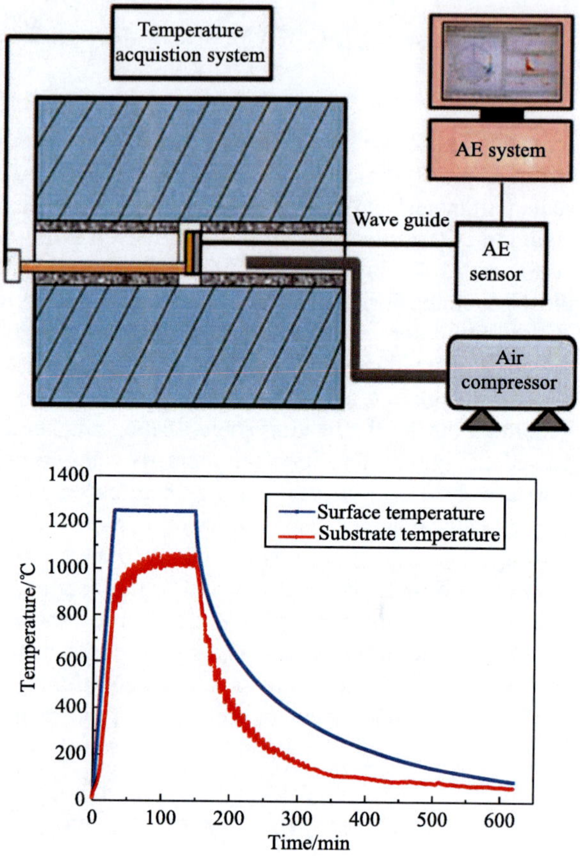

and was subsequently held at this temperature for 120 min, after which the TBC was allowed to cool with the furnace to 100 °C. During this process, the AE signals generated by the TBC due to damage were detected in real time using the WW technique. The surface of the substrate was cooled by airflow. The whole process was carried out in stationary air. The AE signals generated due to damage were detected using the WW technique. Here, one end of the Pt wire was electrically welded to the substrate of the TBC, and the other end was connected to an AE sensor through a mechanical device. The parameters used in the experiment included an AE signal sampling frequency of 2 MHz, a preamplifier gain of 40 dB, a hit length of 2 k, a frequency range of 0.1–1 MHz, and a sampling threshold of 40 dB.

(1) **Characteristics and pattern recognition of the AE signals of a TBC during failure under high-temperature CMAS corrosion**

Figure 11.46 shows the variation in the AE event counts N_{AE} for a CMAS-covered coating specimen, a non-CMAS-covered coating specimen, and a pure substrate specimen with the corrosion time. During the heating stage (stage I), the N_{AE} value

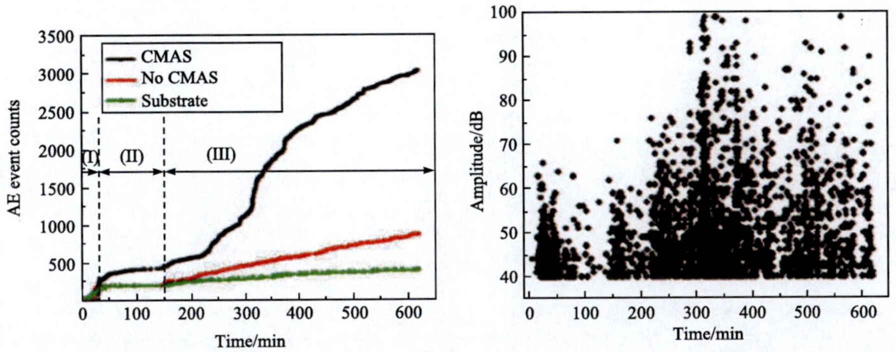

Fig. 11.46 The AE event counts and AE signal amplitude distribution for the TBC subjected to high-temperature CMAS corrosion

for each of the three specimens gradually increases with temperature. The specimens were held at a test temperature of 1250 °C for 120 min. During stage II, the N_{AE} value for each specimen increases at a lower rate or does not increase at all. However, as the temperature decreases (i.e., during the cooling stage (stage III)), there is a significant increase in N_{AE}. While the N_{AE} trends for the three specimens are similar, the changes in the N_{AE} values for the non-CMAS-covered coating specimen and the pure metallic substrate specimen are basically consistent with each other during stages I and II, and the N_{AE} value for the non-CMAS-covered coating specimen is significantly larger during stage III. This suggests that the AE signals generated during stages I and II are related only to the substrate and that, owing to the generation of AE signals due to damage, the N_{AE} value for each coating specimen is larger than that for the metallic substrate specimen during stage III. In addition, throughout the test process, the N_{AE} value for the CMAS-covered coating specimen is significantly larger than that for each of the other specimens, indicating that CMAS corrosion causes more damage to the coating. Moreover, as shown in Fig. 11.46, while the AE signal amplitudes are relatively low during stage I of the CMAS corrosion process, they are distributed in the same interval during stages II and III. As a result, it is very difficult to identify the crack patterns in the coating based on the amplitude.

The AE signals generated during the high-temperature CMAS corrosion process were subjected to a cluster analysis using the method discussed in Sect. 11.3 with six characteristic parameters of AE signals, namely, the amplitude, peak frequency, duration, rise time, counts, and energy. The results in Fig. 11.47 show that there are five damage patterns. Very similar to the results in Sect. 11.3, the five types of signals can be differentiated based on only their frequencies, again proving that characteristic frequency is a key parameter for identifying the failure mode of a TBC. As shown in Fig. 11.48, the peak frequencies of the type A, B, C, and D signals rarely coincide with each other, but their amplitudes are all distributed between 40 and 75 dB. It is basically impossible to classify the AE signals with their amplitudes as the characteristic quantity. In contrast, the AE signals are best classified based on their frequencies. In addition, four typical frequency bands can be found in Fig. 11.48.

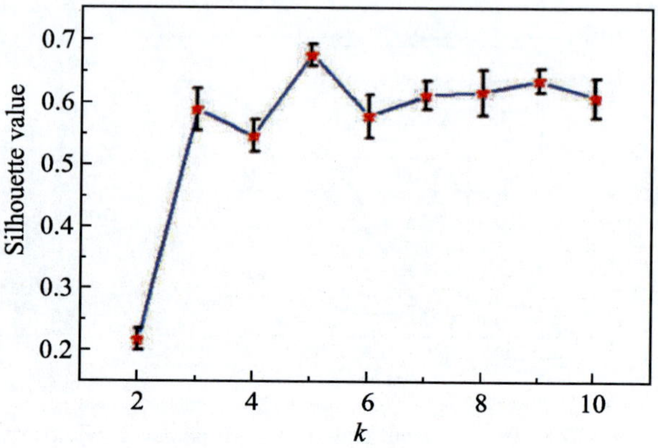

Fig. 11.47 Cluster analysis of the AE signals generated during high-temperature CMAS corrosion

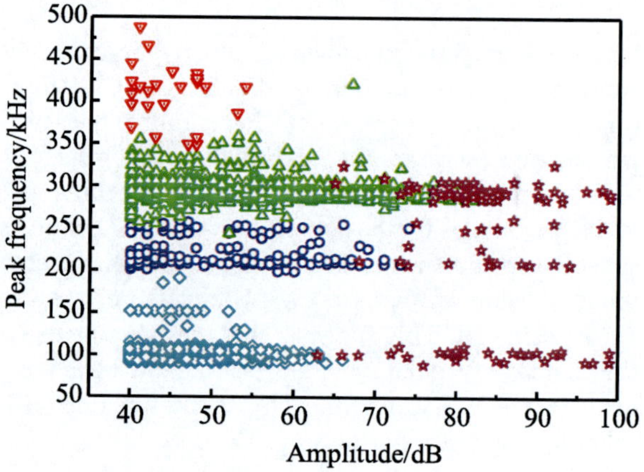

Fig. 11.48 Clustered AE signals

The type A, B, C, and D signals are in the frequency bands of 90–180, 200–250, 250–350, and 350–500 kHz, respectively, corresponding to plastic deformation of the substrate, sliding interface cracks, vertical surface cracks, and opening interface cracks, respectively, as identified earlier. In addition, the type E signals have large amplitudes and exist in each frequency band.

To analyze the failure mode of the TBC during the high-temperature CMAS corrosion process, we observed the micromorphology of a CMAS-covered coating specimen and a non-CMAS-covered coating specimen using SEM. As shown in Fig. 11.49a, the APS TBC contained a metallic substrate, a BC layer, a ceramic layer, and a non-CMAS-infiltrated CMAS layer. After CMAS corrosion, vertical surface cracks (Fig. 11.49b) and interface cracks (Fig. 11.49c) appear in the ceramic layer. Under the same test conditions, only surface cracks are formed in the non-CMAS-covered coating specimen while no visible interface cracks are found.

Further, the AE signals were subjected to wavelet decomposition, and pattern recognition was carried out on the AE signals generated during the heating (I),

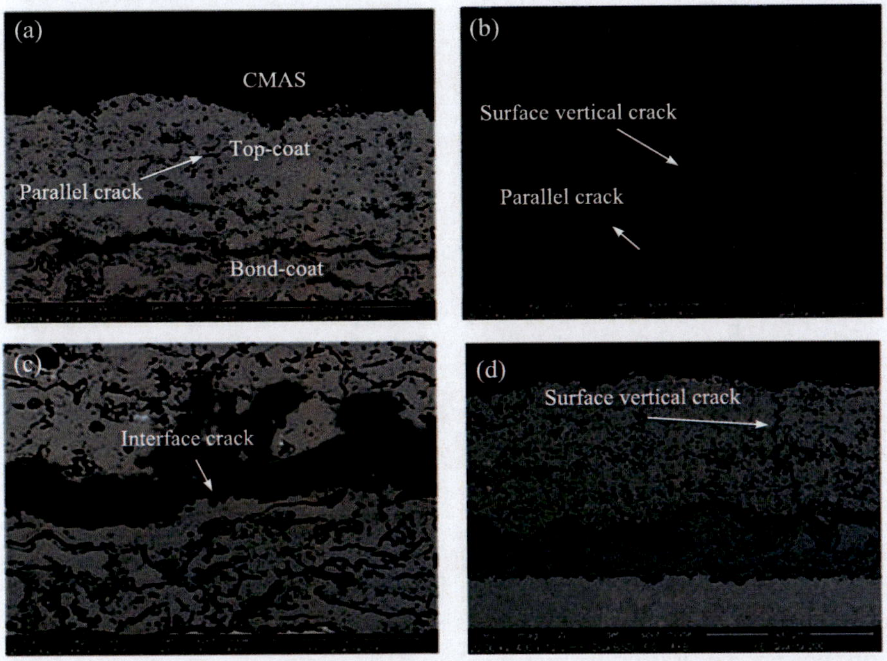

Fig. 11.49 Cross-sectional SEM images of TBC specimens: **a** the CMAS-covered coating specimen; **b** vertical and horizontal surface cracks; **c** interface cracks; **d** the non-CMAS-covered coating specimen

temperature-holding (II), and cooling (III) stages of the CMAS corrosion process. As seen from the results in Fig. 11.50, during stages I and II, there are few AE signals, all of which are generated by the plastic deformation of the substrate, whereas there are a large number of AE signals generated by sliding interface cracks and a small number of AE signals generated by opening interface cracks and vertical surface cracks during stage III. In addition, significantly more AE signals are generated by sliding interface cracks than opening interface cracks. This finding is consistent with the result obtained by Fan et al. [64] using the finite element method (FEM), in which the phase angle of cracks at the ceramic/BC interface is >45°; that is, sliding cracks are the predominant type of cracks at the ceramic/BC interface.

(2) Thermodynamic analysis of CMAS corrosion and mechanism diagrams of CMAS corrosion-induced delamination

Evans and Hutchinson proposed a theoretical model for delamination cracks in TBCs subjected to high-temperature CMAS corrosion under temperature gradients, which is referred to as the E–H model [65]. Krause et al. [66] revised the E–H model by considering the effect of the strain generated by the *t*-to-*m* phase transformation of ZrO_2, ε_{transf}, during the cooling process. Here, we analyze the delamination cracks in a TBC subjected to high-temperature CMAS corrosion using the improved

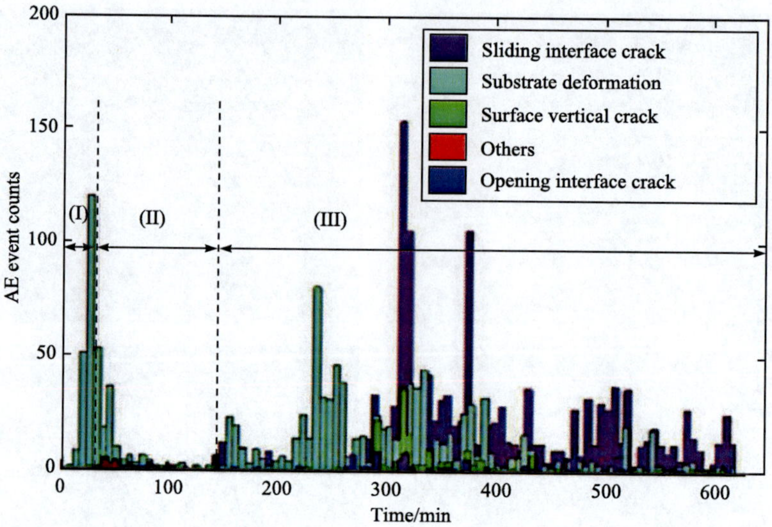

Fig. 11.50 Distribution of different failure modes of the TBC subjected to CMAS corrosion with time

delamination model [63]. As shown in Fig. 11.51, the BC layer and substrate of a TBC subjected to high-temperature CMAS corrosion are simplified as one layer due to their similar CTEs. CMAS infiltrates the coating with a thickness of H to form a CMAS layer with a thickness of h. It is assumed that a crack is generated at a depth of d. Let E_1 and E_2 be the elastic moduli of the CMAS-infiltrated layer and the non-CMAS-infiltrated layer, respectively, and v be their Poisson's ratio.

The phase transformation strain $\varepsilon_{transf} = p\Delta V/3V$, where $\Delta V/V$ is the volume change ($\Delta V/V = 0.03$) [66] and p is the volume coefficient of ZrO_2 during the t-to-m phase transformation. Let us assume that CMAS infiltration alters the Young's modulus E of the coating but does not affect its CTE. By considering only the thermal mismatch between the coating and the metallic substrate, we can obtain the following expression of the stress in the TBC at y, $\sigma(y)$, during the cooling process under the assumption that the coating is linearly elastic and isotropic [63, 65, 66]:

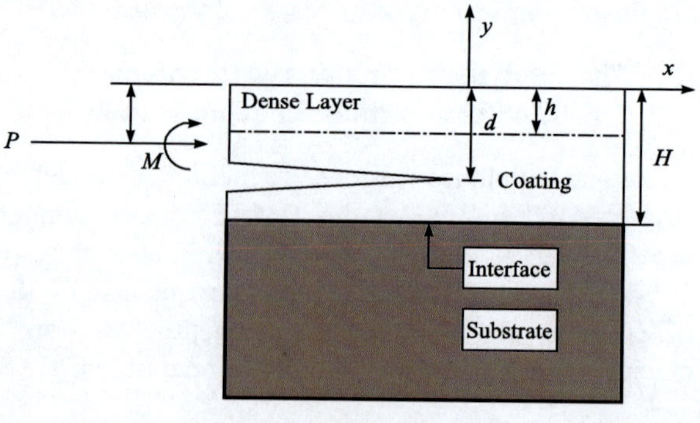

Fig. 11.51 Schematic diagram of a two-layer coating and analysis of the distribution of different CMAS corrosion-induced failure modes with time

$$\sigma(y) = \begin{cases} \frac{E_1\alpha_{TBC}\Delta T_{surf\ sub}}{(1-\nu)}\left\{1+\frac{y}{H}-\frac{\Delta\alpha\Delta T_{sub}+\varepsilon_{transf}}{\alpha_{TBC}\Delta T_{surf/sub}}\right\}, & 0>y>-h \\ \frac{E_2\alpha_{TBC}\Delta T_{surf\ sub}}{(1-\nu)}\left\{1+\frac{y}{H}-\frac{\Delta\alpha\Delta T_{sub}}{\alpha_{TBC}\Delta T_{surf/sub}}\right\}, & -h>y>-H \end{cases} \tag{11.92}$$

where $\Delta T_{sur} = T^i_{sur} - T_{sur}$ is the difference in the surface temperature of the specimen (T^i_{sur} and T^i_{sub} are the initial surface temperature of the specimen and the initial backside surface temperature of the substrate, respectively), $\Delta T_{sub} = T^i_{sub} - T_{sub}$ is the difference in the temperature of the substrate, $\Delta T_{sur/sub} = \Delta T_{sur} - \Delta T_{sub}$ is the instantaneous difference between the difference in the surface temperature of the specimen and the difference in the temperature of the substrate, α_{TBC} is the CTE of the ceramic layer, and $\Delta\alpha = \alpha_{\text{sub}} - \alpha_{TBC}$ (α_{sub} is the CTE of the metallic substrate).

The force P and the moment M can be determined based on $\sigma(y)$.

$$P = \int_{-H}^{0} \sigma(y)dy \tag{11.93}$$

$$M = \int_{-H}^{0} \sigma(y)(D+y)dy \tag{11.94}$$

where D is the distance between the surface and neutral axis of the bent coating [65]. The stress intensity factors (SIFs) of Type I and II crack tips (denoted K_{I} and K_{II}, respectively) are expressed as follows [65]:

$$K_I = \frac{P}{\sqrt{2AH}}\cos\omega + \frac{M}{\sqrt{2Ih^3}}\sin\omega \tag{11.95}$$

$$K_{II} = \frac{P}{\sqrt{2AH}}\sin\omega + \frac{M}{\sqrt{2Ih^3}}\cos\omega \tag{11.96}$$

The strain energy release rate G is as follows:

$$G = \frac{1}{2\overline{E_2}}\left(\frac{P^2}{Ah} + \frac{M^2}{Ih^3}\right) \tag{11.97}$$

In the above equations, A is the dimensionless effective cross-sectional area, I is the moment of inertia, $\omega = 52.1°$, and $\overline{E_2} = E_2 / (1-\nu^2)$ [65].

The delamination cracks at or near the interface are of the mixed type. Thus, their phase angle ψ and mixed fracture toughness Γ^i_C are expressed as follows [65]:

$$\psi = \tan^{-1}\left(\frac{K_{II}}{K_I}\right) \tag{11.98}$$

$$\Gamma_C^i = \Gamma_{IC}^i\left[1 + \tan^2(1 - \lambda)\psi\right] \tag{11.99}$$

where λ is the mode-mixing coefficient and Γ_{IC}^i is the fracture toughness of Type I cracks. Based on $G = \Gamma_C^i$, we can draw the boundary of delamination cracking.

Figure 11.52a shows the relation between the boundary of delamination cracking and P ($H = 200$ μm; $d = 200$ μm; infiltration depth $h = 40$ μm) [66, 67]. The red dashed curve is the fitted curve for the cooling stage (from $T_{sur}^i = 1250$ °C to $T_{\mathrm{sur}} = 79$ °C and from $T_{sub}^i = 1050$ °C to $T_{\mathrm{sur}} = 61$ °C) obtained from the test. The zone between straight lines of the same color is the nondelamination zone, while the other zone is the delamination zone. Straight lines of different colors represent the boundaries of delamination at different p values. Figure 11.52a shows that the curve is in the nondelamination zone at the start of the cooling process and enters the delamination zone when the temperature decreases to a certain level. During the cooling process, at a high p value, the coating enters the delamination zone at a high-temperature and at an early time. This is because at a high p value, ε_{transf} is large, and the coating is prone to delamination. When the surface temperature of the coating, $\mathrm{T_{sur}}$, decreases to approximately 400 °C, the temperature curve is in the delamination zone and is the furthest away from the boundary of delamination, suggesting that delamination cracks are most likely to appear in the coating at this temperature. The black star symbols in the delamination zone signify the temperatures ($\mathrm{T_{sur}} =$ ca. 400 °C, which completely matches the result produced by the model) at which the AE signals detected in the test are the strongest. Considering that the test lasted for a short period of time, p is set to 0.05. Figure 11.52b shows the delamination mechanism diagram drawn with a p of 0.05 and different values of crack depth d. As can be found in the figure, during the cooling process, cracks appear first near the interface between the CMAS-infiltrated layer and the non-CMAS-infiltrated layer and then at the interface between the ceramic layer and the metallic substrate. These results show that the spallation of a TBC with a temperature gradient subjected to CMAS corrosion during the cooling process occurs primarily when the temperature decreases to approximately 400 °C. Delamination cracks are formed mainly near the boundary of the CMAS-infiltrated layer or the BC/ceramic interface.

11.5.3 Failure Mechanism Under Gas Thermal Shock

Gas thermal shock is an inevitable service condition for TBCs on engine turbine blades. Currently, focus is placed mostly on the failure mechanisms under stationary high-temperature conditions, whereas the failure mechanisms under gas thermal shock receive little attention due to the limitations regarding test simulation devices and real-time detection in complex environments. In addition, it is more difficult to theoretically analyze the failure mechanism of TBCs on turbine blades due to their structural complexity. Real-time detection of AE signals generated by TBCs on turbine blades under gas thermal shock can provide a key basis for understanding

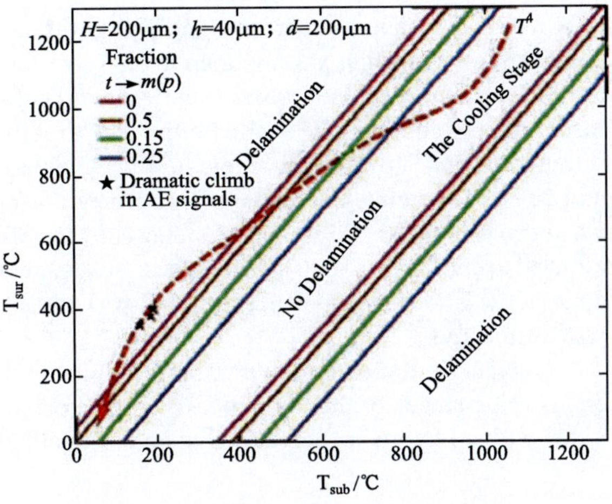

(a)

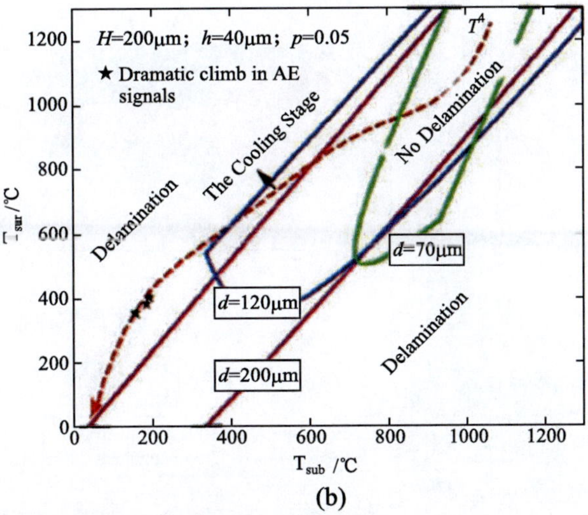

(b)

Fig. 11.52 Delamination mechanism diagrams of a TBC at **a** $p = 0$–0.25 and **b** $p = 0.05$

their failure mechanisms. Hence, we detected the AE signals generated by a TBC on a turbine blade during the failure process under gas thermal shock using the high-temperature WW technique and an independently developed system for testing TBCs on turbine blades in stationary service conditions [68, 69]; in addition, we analyzed its failure mechanism and extent using methods such as wavelet and quantitative analysis [70].

A PVD TBC on a triple turbine blade of a certain type was used in the experiment. The TBC consisted of a CMSX-4 substrate, a NiCoCrAlY transition layer with a thickness of approximately 80 μm, and a YSZ ceramic layer with a thickness of 150 μm. The thermal shock test was conducted in an independently developed test system for TBCs on turbine blades in stationary service conditions, as schematically shown in Fig. 11.53. A supersonic spray gun was used to mix and burn oxygen and a combustion-supporting gas and then to jet-spray the resulting gas at a high speed to the TBC on a turbine blade while cooling air was passed through the cooling channels inside the turbine blade to simulate the temperature gradient service conditions of turbine blades. The gas flow rate could be adjusted between Mach numbers of 0.3 and 1. The following thermal shock parameters were used in the test. The coating was heated to 1100 °C within 10 s and held for 30 s, after which the coating was allowed to cool in air for 40 s. When the crack length reached 10 mm or the spallation zone of the TBC reached 10%, the TBC was considered to have failed, and the test was terminated.

The crack initiation and evolution throughout the heating, temperature-holding, and cooling stages of the test process were detected in real time using the AE technique with a Pt wire as the WW. One end of the Pt wire was welded to the substrate

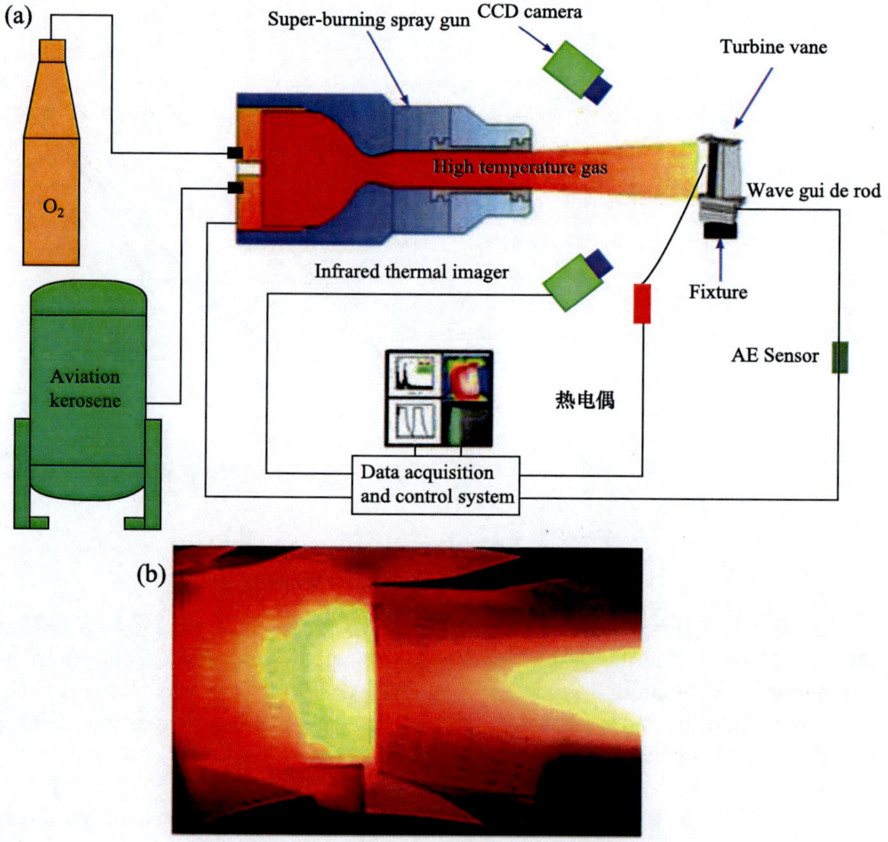

Fig. 11.53 **a** Schematic diagram of the setup for testing the TBC on a turbine blade under gas thermal shock and detecting AE signals in real time. **b** TBC on the turbine blade under thermal shock

of the TBC, and the other end was connected to an AE sensor through a mechanical device. The following parameters were used in the test: an AE signal sampling frequency of 2 MHz, a preamplifier gain of 40 dB, a hit length of 2 k, a frequency range of 0.1–1 MHz, and a sampling threshold of 40 dB. In addition, the distribution of the temperature field was detected using the infrared technique in real time in combination with AE signal detection to analyze the failure mechanism of the TBC on the turbine blade.

(1) **Distribution of the temperature field on the TBC on the turbine blade**

Figure 11.54 shows the distribution of the temperature field on the surface of the TBC in the temperature-holding stage of the thermal shock. The temperature is higher at the leading edge than at other locations, and the temperature is higher on the pressure side than on the suction side. This is because the high-temperature, high-speed gas is jet-sprayed to the leading edge, resulting in a poor cooling effect. Figure 11.55 shows the detailed temperature distribution along the height of the turbine blade. The temperature decreases from the leading edge to the trailing edge. The temperature on the pressure side (points 1–4) is significantly higher than that on the suction side (points 5–9), validating the finding from the thermal image sequence. The temperature at the middle section (cross-section B) is higher than those at other sections (cross-sections A and C). The temperature at the front end of the middle section is higher than that at cross-sections A and C by approximately 80 °C, which is attributed to concentration of the flame spot with a diameter of 30 mm at the middle section.

Figure 11.56 shows the variation in the temperature at cross-section B with time during thermal shock cycling. The temperature at the leading edge (point 1) increases within 10 s from room temperature to 1100 °C (which translates to a heating rate of 100 °C/s), then is held at 1100 °C for 30 s, and finally decreases within 40 s to room temperature, thus satisfying the thermal shock cycling condition. A similar temperature trend is found at point 2, where the highest temperature is approximately 1000 °C. Other locations (points 3 and 4) on the pressure side do not reach a thermal equilibrium state, with the temperatures at points 3 and 4 increasing from room

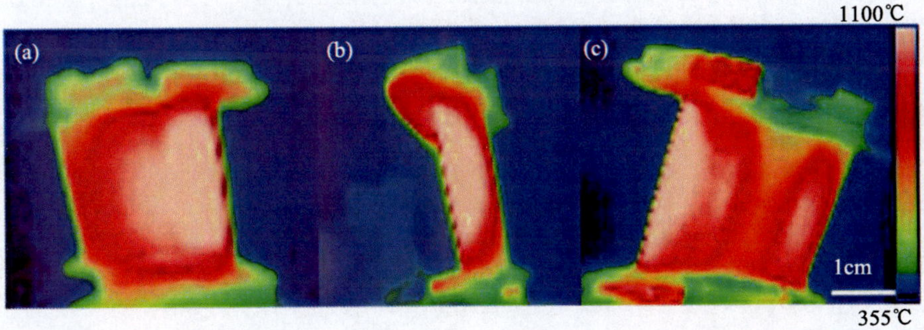

Fig. 11.54 Temperature distribution in the TBC in the temperature-holding stage of the thermal shock: **a** pressure side, **b** leading edge, and **c** suction side

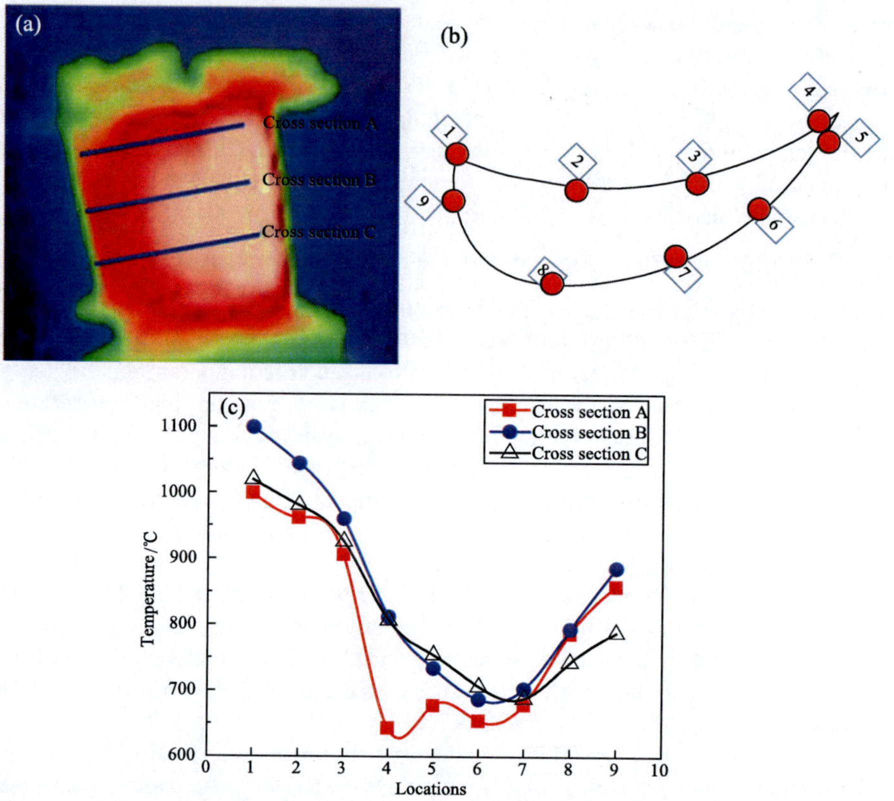

Fig. 11.55 Temperature distribution at various cross-sections of the TBC on the turbine blade in the temperature-holding stage: **a** three cross-sections along the height direction, **b** distribution of the selected points at the cross-sections, and **c** temperatures at different interfacial checkpoints

temperature to 900 and 883 °C, respectively. In addition, the heating rates on the suction side (points 5–9) are significantly lower than those on the pressure side. No locations on the suction side reach a thermal equilibrium state. The temperature on the suction side increased with time, with the highest temperatures at points 5–9 being 755, 790, 791, 825, and 896 °C, respectively.

(2) Evolution of the damage to the TBC on the turbine blade

The AE signals generated by the TBC on the turbine blade under gas thermal shock were analyzed using the aforementioned cluster analysis and wavelet frequency-spectrum analysis. Figure 11.57 shows the cluster analysis results for the AE signals. Figure 11.57a shows that the *s*-value is the highest when $k = 5$, suggesting the presence of five types of AE signals. Based on the peak frequencies shown in Fig. 11.57b, these types of AE signals correspond to four damage patterns, namely, substrate

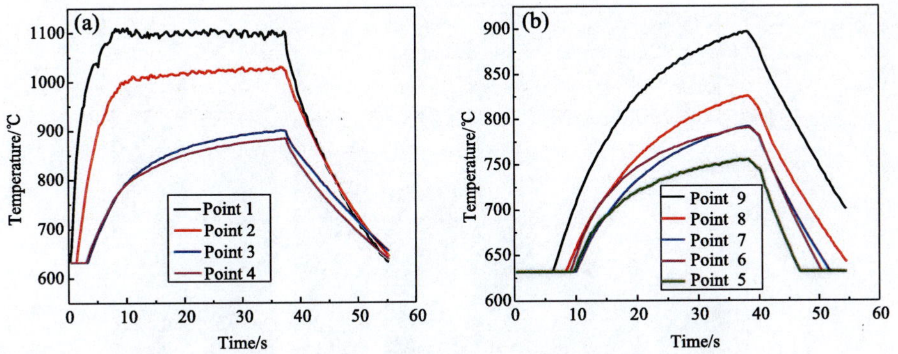

Fig. 11.56 Changes in the temperature at each checkpoint during the thermal shock cycling: **a** points 1–4 on the pressure side; **b** points 5–9 on the suction side

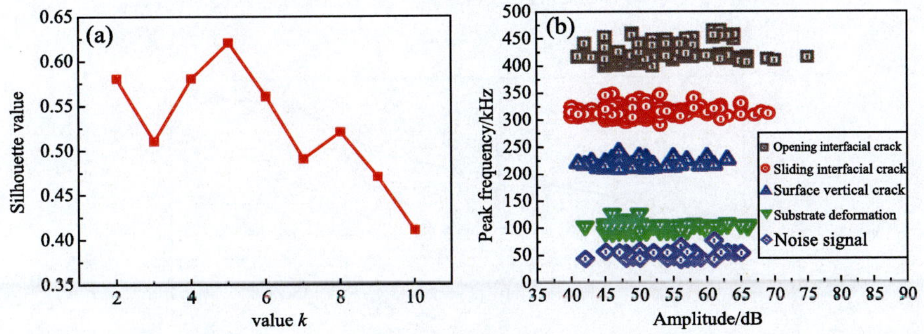

Fig. 11.57 **a** Cluster analysis of the AE signals generated by the TBC under thermal shock; **b** distributions of the amplitudes and peak frequencies of five types of signals

deformation (90–110 kHz), vertical surface cracks (200–220 kHz), sliding interface cracks (280–325 kHz), and opening interface cracks (400–450 kHz), as well as noise (20–60 kHz).

Note that similar to the thermal cycling test results, there are very few AE signals in the heating and temperature-holding stages, most of which are generated by vertical surface cracks, and the AE signals detected during the cooling stage are generated by interface cracks [aa]. Detailed results are not shown here. Figure 11.58a and b shows the evolution of the number of AE events N_{AE} and cumulative N_{AE} corresponding to each damage pattern with the number of cycles N. N_{AE} increases with N. When $N = 100$, vertical surface cracks and sliding interface cracks are the primary damage patterns. When $N = 300$, the N_{AE} corresponding to opening interface cracks increases rapidly. This reveals the following failure mechanism of the coating under thermal shock. Under the thermal mismatch stress, vertical surface cracks appear in a coating when heated due to the formation of a tensile stress. These surface cracks gradually propagate toward the interface and, upon arrival, along the interface due

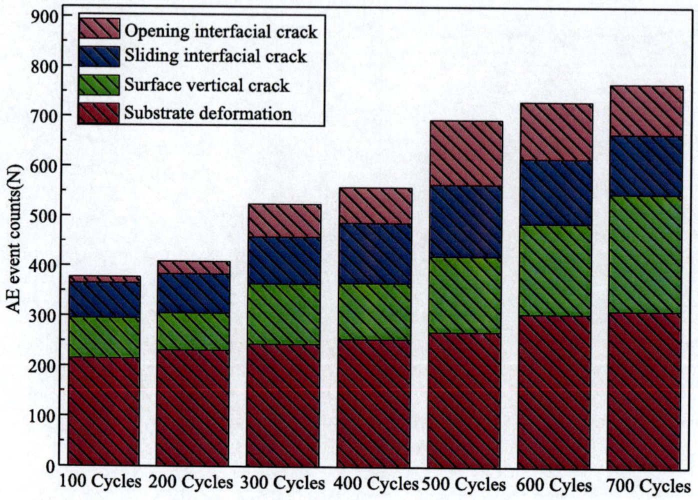

(a)

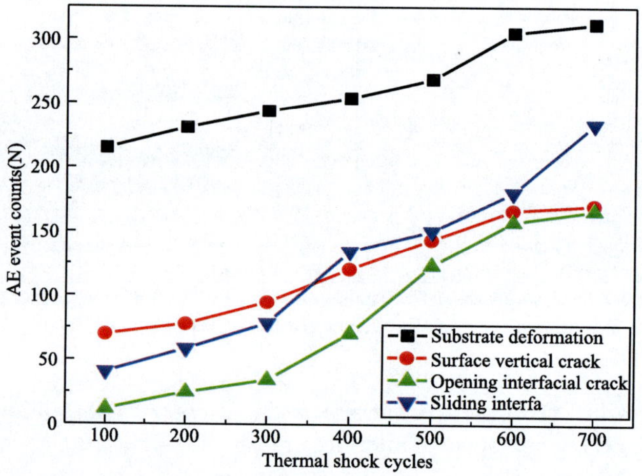

(b)

Fig. 11.58 Variations in **a** the number of AE events and **b** the cumulative number of AE events corresponding to different damage patterns with the number of thermal shock cycles

to the interfacial tensile stress formed as a result of the buckling of the coating under compressive stress during cooling as well as the interfacial shear stress. The propagation of the cracks gradually causes the coating to spall. This mechanism can be verified based on the macroscopic spallation of the coating, particularly its microstructural evolution. See Figs. 11.59 and 11.60 for the detailed results.

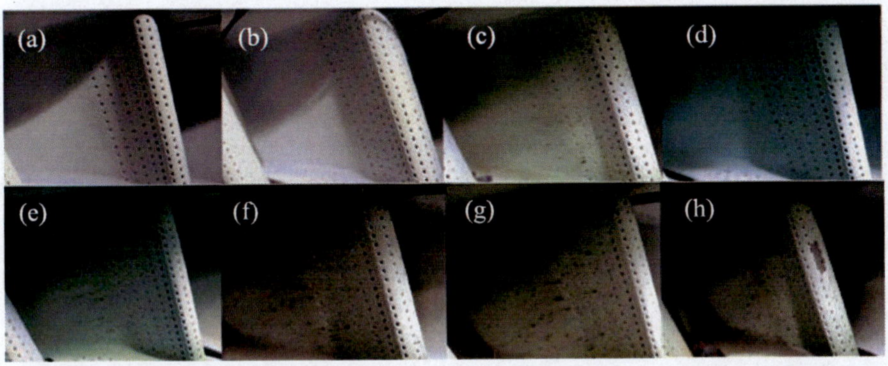

Fig. 11.59 Surface morphology of the TBC after different numbers of thermal shock cycles: **a** 0, **b** 100, **c** 200, **d** 300, **e** 400, **f** 500, **g** 600, and **h** 710

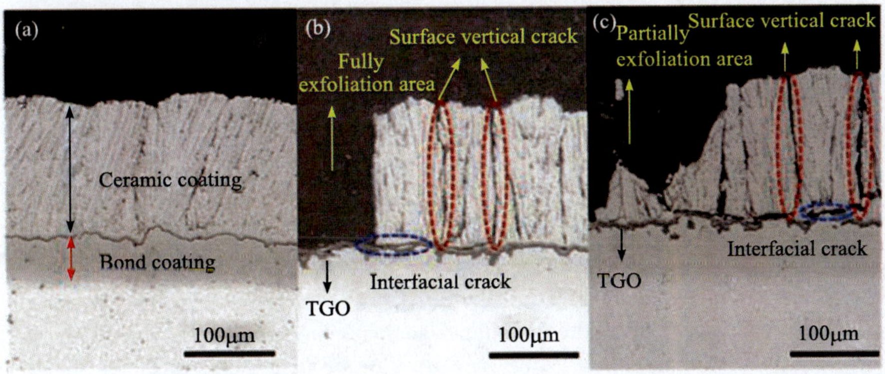

Fig. 11.60 Microstructural observations of the TBC: **a** nonspalled zone at the leading edge, **b** completely spalled zone, and **c** partially spalled zone

11.6 Summary and Outlook

11.6.1 Summary

This chapter describes the real-time AE detection, pattern recognition, and quantitative evaluation methods for crack evolution in TBCs as well as their failure mechanisms determined using the AE technique. The following is a summary:

(1) The WW technique and regional signal detection method can be used to effectively detect AE signals generated due to crack evolution in TBCs in various environments (e.g., under high-temperatures, gas thermal shock, and high-temperature CMAS corrosion).

(2) AE signal frequency is a key parameter for identifying crack patterns in TBCs. This finding can be applied to the analysis of the failure mechanisms of TBCs under various loading conditions and expanded to other materials or structures.
(3) An intelligent method for crack pattern recognition in TBCs is established based on frequency (a key pattern recognition parameter) and through the exploitation of the advantages of the WT in the characteristic frequency-spectrum window and the preservation of time-domain information in combination with the NN training method.
(4) Quantitative relations of the surface crack density and the interface crack length with the AE event counts are established based on the mechanical analysis of crack initiation and propagation in TBCs under single loading (tensile or compressive) conditions and are used to quantitatively characterize the failure mechanisms of TBCs under high-temperature thermal cycling and high-temperature CMAS corrosion.
(5) The failure mechanisms of TBCs under various loading conditions (e.g., thermal cycling, high-temperature CMAS, and high-temperature gas thermal shock) are determined using the AE technique. In addition, an important phenomenon—that coatings spall during the cooling process—is discovered.

11.6.2 Outlook

Future research should be focused on the following areas:

(1) Theoretical validation of the correlation between the failure mode and frequency. While various test results and signal analyses have shown that frequency is a key parameter for identifying the failure mode of a TBC, further research is required to theoretically validate this point.
(2) Detection of AE signals generated by TBCs during the failure process under high-temperature, high-speed rotation conditions. High-speed rotation is the primary load to which the TBC on a turbine blade in operation is subjected. However, detection is currently limited to stationary high-temperature environments. Developing AE signal detection and analysis methods for high-speed rotation conditions and elucidating the failure mechanism of TBCs under high-temperature, high-speed rotation conditions are important development trends.
(3) Detection of AE signals generated by TBCs during the failure process under high-temperature vibration conditions. High-temperature vibration may be another key load causing the TBC on a turbine blade to spall but has yet to receive considerable attention. Both the vibration and inherent frequency of specimens under vibration conditions pose new difficulties in signal detection and pattern recognition.
(4) AE characterization of the failure mechanism of new types of TBC. Currently, YSZ coatings can operate only at temperatures below 1200 °C and are therefore unable to meet the requirements for engine development. Studying the failure

mechanism of new-generation TBCs using the AE characterization technique can provide valid data to facilitate the development and application of new coatings.

References

1. Yuan Z, Ma Y, He Z. Acoustic emission technique and its applications. China Machine Press (Beijing);1985.
2. Ying SP. The use of acoustic emission for assessing the integrity of a nuclear reactor pressure vessel. NDT Int. 1979;12:175–1710.
3. Hamstad MA, Bianchetti R, Mukherjee AK. A correlation between acoustic emission and the fracture toughness of 2124–T851 aluminum. Eng Fract Mech. 1977;9:663–74.
4. Nomura H, Takahisa K, Koyama K, Sakai T. Acoustic emission from superconducting magnets. Cryogenics. 1977;17:471–81.
5. Miller RA. Thermal barrier coatings for aircraft engines-history and directions. In: Proceedings of thermal barrier coating workshop, NASA conference publication, 3312;1995. p. 17–34.
6. Padture NP, Gell M, Jordan EH. Thermal barrier coatings for gas-turbine engine applications. Science. 2002;296(5566):280–4.
7. Hutchinson JW. The role of mechanics in advancing thermal barrier coatings. In: The 22th international congress of theoretical and applied mechanics, 24–29 August 2008.
8. Fu L, Khor KA, Ng HW, Teo TN. Non-destructive evaluation of interface failure of functionally graded thermal barrier coatings. Surf Coat Tech. 2000;130(2–3):233–9.
9. Wang L, Ming C, Zhong XH, Ni JX, Tao SY, Zhou FF, Wang Y. Prediction of critical rupture of plasma-sprayed yttria stabilized zirconia thermal barrier coatings under burner rig test via finite element simulation and in-situ acoustic emission technique. Surf Coat Tech. 2019;367:58–74.
10. Yang L, Zhou YC, Lu C. Damage evolution and rupture time prediction in thermal barrier coatings subjected to cyclic heating and cooling: an acoustic emission method. Acta Mater. 2011;59:6519–210.
11. Yang L. A nondestructive evaluation of the oxidation, damage, and fracture of thermal barrier coatings. Doctoral disseration, Xiangtan University;2007.
12. Sun YL, Li JG, Zhang WX, Wang TJ. Local stress evolution in thermal barrier coating system during isothermal growth of irregular oxide layer. Surf Coat Tech. 2013;216:237–50.
13. Kumar V, Balasubramanian K. Progress update on failure mechanisms of advanced thermal barrier coatings: a review. Prog Org Coat. 2016;90:54–82.
14. Liang LH, Li XN, Liu HY, Wang YB, Wei YG. Power-law characteristics of damage and failure of ceramic coating systems under three-point bending. Surf Coat Tech. 2016;285:113–1110.
15. Shan X, Chen WF, Yang LX, Guo FW, Zha XF, Xiao P. Pore filling behavior of air plasma spray thermal barrier coatings under CMAS attack. Corros Sci. 2020;167: 108478.
16. Guo HB, Gong SK, Zhou CG, Xu HB. Investigation on hot-fatigue behaviors of gradient thermal barrier coatings by EB-PVD. Surf Coat Tech. 2001;148:110–6.
17. Ma XQ, Cho S, Takemoto M. Acoustic emission source analysis of plasma sprayed thermal barrier coatings during four-point bend tests. Surf Coat Tech. 2001;139:55–62.
18. Yang L, Zhou YC, Mao WG, Lu C. Real-time acoustic emission testing based on wavelet transform for the failure process of thermal barrier coatings. Appl Phys Lett. 2008;93: 231906.
19. Benson PM, Vinciguerra S, Meredith PG, Young RP. Laboratory simulation of volcano seismicity. Science. 2008;322:248–52.
20. Woo SC, Goo NS. Analysis of the bending fracture process for piezoelectric composite actuators using dominant frequency bands by acoustic emission. Compos Sci Tech. 2007;67(7–8):1499–508.

21. Marec A, Thomas JH, Guerjouma R. Damage characterization of polymer-based composite materials: Multivariable analysis and wavelet transform for clustering acoustic emission data. Mech Syst Signal Process. 2008;22(6):1441–64.
22. Refahi OA, Heidary H, Ahmad M. Unsupervised acoustic emission data clustering for the analysis of damage mechanisms in glass/polyester composites. Mater Des. 2012;37:416–22.
23. Yang L, Kang HS, Zhou YC, Zhu W, Lu C. Frequency as a key parameter in discriminating the crack modes of thermal barrier coatings: cluster analysis of acoustic emission signals. Surf Coat Technol. 2015;264:97–104.
24. Xie Z. Statistical Analysis and Applications with MATLAB—40 Case Studies. Beihang University Press, Beijing;2010.
25. Gutkin R, Green C, Vangrattanachai S, et al. On acoustic emission for failure investigation in CFRP: pattern recognition and peak frequency analyses. Mech Syst Signal Process. 2011;25(4):1393–407.
26. Sause M, Gribov A, Unwin A. Pattern recognition approach to identify natural clusters of acoustic emission signals. Pattern Recogn Lett. 2012;33(1):17–23.
27. Moevus M, Godin N, Mili MR, Rouby D, Reynaud P, Fantozzi G, Farizy G. Analysis of damage mechanisms and associated acoustic emission in two SiC/[Si–B–C] composites exhibiting different tensile behaviours. Part II: Unsupervised acoustic emission data clustering. Comp Sci Technol. 2008;68(6):1258–65.
28. Yang L, Zhong ZC, Zhou YC, Zhu W, Zhang ZB, Cai CY, Lu C. Acoustic emission assessment of interface cracking in thermal barrier coatings. Acta Mech Sin. 2016;32:342–8.
29. Fan Q. Wavelet analysis. Wuhan University Press, Wuhan;2008.
30. Chui CK. An introduction to wavelets. San Diego: Academic Press; 1992.
31. Newland DE. An introduction to random vibrations, spectral and wavelet analysis. Longman Scientific & Technical, Essex;1993.
32. Daubechies I. The wavelet transform, time frequency localization and signal analysis. IEEE T Inform Theory. 1990;36(5):961–1005.
33. Tian Y. Hybrid neural network techniques. Science Press, Beijing;2010.
34. Zhang D. Application and design of neural networks in MATLAB. China Machine Press, Beijing;2009.
35. Yang L, Kang HS, Zhou YC, He LM, Lu C. Intelligent discrimination of failure modes in thermal barrier coatings: wavelet transform and neural network analysis of acoustic emission signals. Exp Mech. 2015;55:321–30.
36. Feng L. A study on an electromagnetic acoustic emission signal processing platform based on wavelet analysis and neural network identification. Master's thesis, Hebei University of Technology;2010.
37. Chen Z. A study on wavelet basis and wavelet transform analyzer systems in engineering signal processing. Doctoral dissertation, Chongqing University;1998.
38. Khamedi R, Fallahi OAR. Effect of martensite phase volume fraction on acoustic emission signals using wavelet packet analysis during tensile loading of dual phase steels. Mater Des. 2010;31(6):2752–9.
39. Wu JD, Wang YH, Chiang PH, Bai MR. A study of fault diagnosis in a scooter using adaptive order tracking technique and neural network. Expert Syst Appl. 2009;36(1):49–56.
40. Zhang P. Research on integrated acoustic emission signal processing platforms. Tsinghua University, Beijing;2002.
41. Kim EY, Lee YJ, Lee SK. Heath monitoring of a glass transfer robot in the mass production line of liquid crystal display using abnormal operating sounds based on wavelet packet transform and artificial neural network. J Sound Vib. 2012;331(14):3412–27.
42. Kuo CC. Artificial recognition system for defective types of transformers by acoustic emission. Expert Syst Appl. 2009;36(7):10304–11.
43. Kandaswamy A, Kumar CS, Ramanathan RP. Neural classification of lung sounds using wavelet coefficients. Comput Biol Med. 2004;34(6):523–37.
44. Yang L, Zhong ZC, Zhou YC, Lu C. Quantitative assessment of the surface crack density in thermal barrier coatings. Acta Mech Sin. 2014;30:167–74.

45. McGuigan AP, Briggs GAD, Burlakov VM, Yanaka M, Tsukahara Y. An elastic-plastic shear lag model for fracture of layered coatings. Thin Solid Films. 2003;424(2):219–23.
46. Cao XQ, Vassen R, Stoever D. Ceramic materials for thermal barrier coatings. J Eur Ceram Soc. 2004;24:1–10.
47. Wright PK, Evans AG. Mechanisms governing the performance of thermal barrier coatings. Curr Opin Solid State Mater Sci. 1999;4(3):255–65.
48. Choi SR, Hutchionson JW, Evans AG. Delamination of multilayer thermal barrier coatings. Mech Mater. 1999;31(7):431–47.
49. Hutchionson JW, Suo Z. Mixed mode cracking in layered. Adv Appl Mech. 1992;29:63–163.
50. Zhou YC, Hashida T. Thermal fatigue failure induced by delamination in thermal barrier coating[J]. Int J Fatigue. 2002;24:407–17.
51. Zhang XC, Xu BS, Wang HD, Wu YX. J Appl Phys. 2007;101: 083530.
52. Hsueh CH. Thin Solid Films. 2002;418:182.
53. Mao WG, Chen Q, Dai CY, Yang L, Zhou YC, Lu C. Effects of piezo-spectroscopic coefficients of 8 wt.% Y_2O_3 stabilized ZrO_2 on residual stress measurement of thermal barrier coatings by Raman spectroscopy. Surf Coat Technol. 2010; 204:3573–7.
54. Teixeira V, Andritschky M, Fischer W, Buchkremer HP, Stöver D. Surf Coat Technol. 1999;120–121:103.
55. Suzuki K, Shobu T. Mater Sci Forum. 2010;638–642:906.
56. Vasinonta A, Beuth JL. Eng Fract Mech. 2001;68:843.
57. Schwingel D, Taylor R, Haubold T, Wigren J, Gualco C. Surf Coat Technol. 1998;108–109:910.
58. Xu ZH, Yang YC, Huang P, Li XD. Acta Mater. 2010;58:5972.
59. Cruse TA, Dommarco RC, Bastias PC. J Eng Mater Technol. 1998;120:26.
60. Choi SR, Zhu D, Miller RA. Eng Fract Mech. 2005;72:2144.
61. Bufe CG, Varnes DJ. Predictive modeling of the seismic cycle of the greater san francisco bay region. J Geophys Res. 1993;98:9871–83.
62. Sornette D, Sammis CG. Complex critical exponents from renormalization group theory of earthquakes: Implications for earthquake predictions. J Phy I France. 1995;5:607–9.
63. Yang L, Yang TT, Zhou YC, Wei YG, Wu RT, Wang NG. Acoustic emission monitoring and damage mode discrimination of APS thermal barrier coatings under high temperature CMAS corrosion. Surf Coat Technol. 2016;304:272–82.
64. Fan X, Xu R, Wang TJ. Interfacial delamination of double-ceramic-layer thermal barrier coating system. Ceram Int. 2014;40(9):13793–802.
65. Evans AG, Hutchinson JW. The mechanics of coating delamination in thermal gradients. Surf Coat Technol. 2007;201(18):7905–16.
66. Krause AR, Garces HF, Dwivedi G. Calcia-magnesia-alumino-silicate (CMAS) induced degradation and failure of air plasma sprayed yttria-stabilized zirconia thermal barrier coatings. Acta Mater. 2016;105:355–66.
67. Pujol G, Ansart F, Bonino JP. Step-by-step investigation of degradation mechanisms induced by CMAS attack on YSZ materials for TBC applications. Surf Coat Tech. 2013;237:71–8.
68. Zhou YC, Yang L, Zhong v, Cai CY. Type of testing equipment for detecting the failure process of thermal barrier coating in a simulated working environment, USA, 14/760, 444.
69. Zhou Y, Yang L, Zhong Z, Cai C. A test device for simulating the service conditions of thermal barrier coatings and detecting their failure in real time. China National Invention Patent (ZL201310009293.5).
70. Zhu W, Zhang CX, Yang L, Zhou YC, Liu ZY. Real-time detection of damage evolution and fracture of EB-PVD thermal barrier coatings under thermal shock: An acoustic emission combined with digital image correlation method. Surf Coat Tech. 2020;399: 126151.

Chapter 12
Characterization of the Microstructural Evolution of TBCs by Complex Impedance Spectroscopy

When subjected to gas thermal shock, oxidation, or calcium–magnesium aluminosilicate (CMAS) corrosion, a thermal barrier coating (TBC) undergoes complex microstructural evolution (e.g., the contraction or filling of the pores in the coating due to sintering and CMAS infiltration as well as the formation of an interfacial oxide layer due to oxidation) before macroscopical spallation or even cracking. Currently, microstructural evolution is evaluated using destructive and/or non-real-time characterization methods (e.g., scanning electron microscopy (SEM) and transmission electron microscopy (TEM)) that are time-consuming and require the destruction of the specimen. Real-time detection of microstructural evolution can provide an important reference for understanding the failure mechanism of TBCs and, particularly, for establishing key microstructural criteria.

Of the parameters descriptive of the microstructure of a TBC, the porosity and the thickness of the interfacial oxide layer (i.e., the thermal growth oxide (TGO) layer), h_{TGO}, are particularly important due to the following reasons. Porosity is a key microstructural feature of a TBC related to its thermal insulation performance. Sintering, particularly the filling of pores due to CMAS infiltration, reduces the thermal insulation performance of TBCs, which is an issue of great concern to those tasked with the design and application of TBCs. Interfacial oxidation has long been regarded as the primary key factor causing coating spallation. Many studies have shown that in the absence of CMAS, a coating spalls when its h_{TGO} reaches a certain value. Therefore, real-time determination of h_{TGO} and porosity is highly effective for evaluating the application performance of TBCs.

This chapter focuses on introducing the detection of the microstructural evolution (e.g., interfacial oxidation and porosity changes) of TBCs by complex impedance spectroscopy (CIS), including the detection principle, suitable detection conditions for TBCs, and quantitative microstructural evaluations.

Y. Zhou et al., *Thermal Barrier Coatings: Failure Theory and Evaluation Technology*,
https://doi.org/10.1007/978-981-19-2723-2_12

12.1 Basic Principle of Characterization by CIS

12.1.1 Principle of CIS

CIS is used to measure the electrochemical response of a system to an applied small-amplitude perturbation [1, 2]. For example, the alternating current (AC) impedance of a tested system to which a small-amplitude AC signal voltage (current) is applied is determined through the measurement of its current (voltage) response. This impedance is represented by a complex number and is referred to as complex impedance. The impedance parameters are related to each other through the following equations:

$$|Z| = \sqrt{\text{Re}(Z)^2 + \text{Im}(Z)^2} \quad \phi = \tan^{-1}\frac{\text{Im}[Z]}{\text{Re}[Z]} \tag{12.1}$$

where $|Z|$, ϕ, Re(Z), and Im(Z) are the modulus, phase angle, real part, and imaginary part of the impedance Z, respectively.

Complex impedance spectra can be graphed using two types of plots in different forms, namely, a Nyquist plot, which displays the real and imaginary parts of the impedance, and a Bode plot, which displays the phase angle, modulus, and frequency of the impedance. Figure 12.1 shows the Nyquist and Bode plots for a parallel resistor and capacitor (*RC*) circuit.

The impedance of a parallel *RC* circuit is as follows:

$$\frac{1}{Z} = \frac{1}{R} + j\omega C \tag{12.2}$$

Here, the resistance *R* and capacitance *C* are as follows:

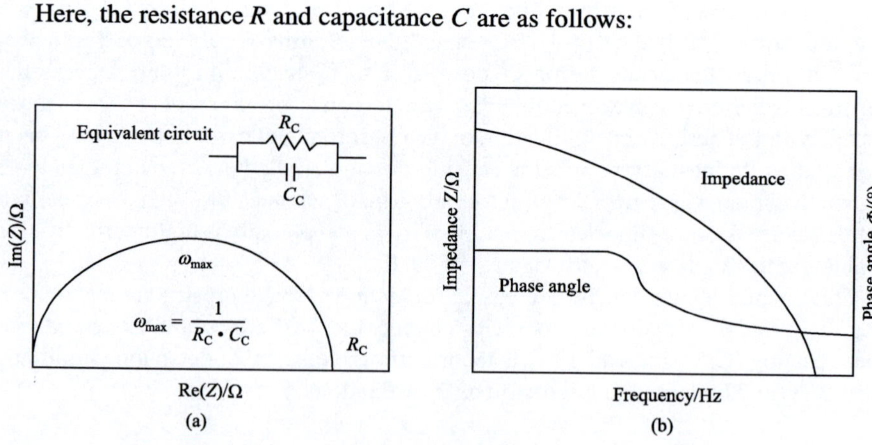

Fig. 12.1 **a** Nyquist plot and **b** Bode plot of a parallel *RC* circuit

$$R = \frac{\rho h}{A} \quad \text{and} \quad C = \frac{\varepsilon_0 \varepsilon A}{h} \tag{12.3}$$

In Eqs. (12.2) and (12.3), j is the imaginary unit $\left(j^2 = -1\right)$, ω (rad/s) is the angular velocity ($\omega = 2\pi f$, f(Hz) is the frequency), ρ is the resistivity, ε is the relative permittivity, ε_0 is the vacuum permittivity, A (mm^2) is the electrode area, and h (mm) is the thickness of the specimen. Evidently, Z changes as R and C change as a result of changes in the thickness or area of the specimen. Similarly, we can analyze the changes in the values of these element parameters based on the changes in Z and, on this basis, analyze the microstructural changes in the specimen. Thus, it is necessary to establish an equivalent circuit for a system examined by CIS.

The impedance Z of a multilayer system can be expressed as follows:

$$Z = Z_1 + Z_2 + \ldots + Z_n \tag{12.4}$$

where $Z_1, Z_2, \ldots, Z_n$ are the complex impedances of the layers.

12.1.2 Analysis of the Impedance Responses of TBCs

The substrate and intermediate transition layer of a TBC have excellent conductivity, and consequently, their complex impedance is negligible. Thus, the impedance of an unoxidized TBC system primarily reflects that of its ceramic layer. After a certain period of oxidation, the Al ions in the transition layer and the O ions in the ceramic layer react to form TGOs that are composed mainly of Al_2O_3 [3–5]. In this case, the complex impedance spectrum of the system is mainly characterized by the impedances of two layers, namely, the ceramic layer and the TGO layer. After a long period of oxidation, the number of Al ions becomes insufficient for oxidation. Under this condition, the Co, Cr, and Ni ions in the transition layer undergo oxidation to form mixed oxides (MOs) [6, 7]. As a result, the impedance spectrum of the system is characterized by the impedances of three layers, namely, the ceramic layer, the TGO layer, and the MO layer.

By simplifying each layer to a simple parallel *RC* circuit, we obtain an equivalent circuit for the TBC system, as shown in Fig. 12.2 (R_S is the contact resistance of the electrodes, and the subscripts C, T, and M signify the ceramic layer, the TGO layer, and the MO layer, respectively). The Nyquist plot of the complex impedance spectrum of an ideal parallel *RC* circuit is a standard semicircle. However, the Nyquist plot obtained from the measurement of the complex impedance spectrum of a TBC does not comprise one or multiple standard semicircles due to material anisotropy; this is often called the dispersion effect. A constant phase element (CPE) can be used instead of an ideal capacitance to reflect the dispersion effect of the material [2, 6–8].

The complex impedance of a CPE, Z_{CPE}, is defined as follows [2, 6–8]:

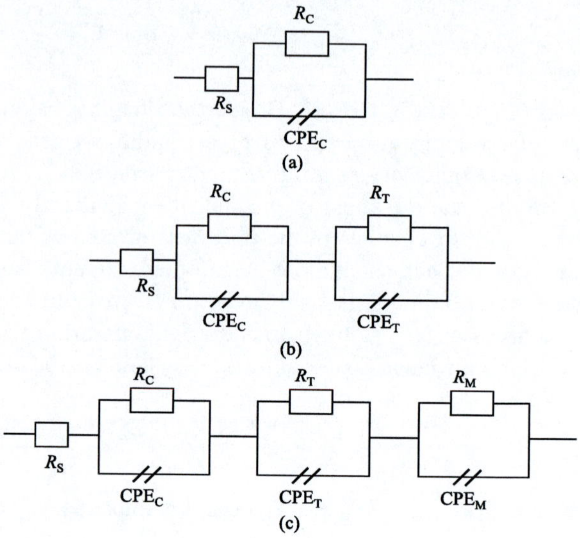

Fig. 12.2 Equivalent circuits for a TBC at different oxidation stages: **a** the unoxidized stage; **b** after a short period of oxidation; **c** after a long period of oxidation

$$Z_{CPE} = \frac{1}{Q(j\omega)^{\beta}} \tag{12.5}$$

where $Q(\Omega^{-1}(\text{rad/s})^{-\beta})$ is a constant and β is a fitting constant between –1 and 1 (also known as the dispersion effect coefficient) that reflects the anisotropy of the material (when $\beta = -1$, the CPE is a pure inductor; when $\beta = 0$, the CPE is a pure resistor, and the Q value is the reciprocal of the resistance R; when $\beta = +1$, the CPE is a pure capacitor, and the Q value is the capacitance C).

Equations (12.2)–(12.4) show that the complex impedance of a system is related to the R and C of each layer and that changes in the R and C are closely related to the microstructure and thickness of the material as well as the electrode area. Once the changes in the element parameters (i.e., the R and CPE of each layer) of the equivalent circuit for a TBC are detected, the changes in the microstructure (e.g., thickness) of the TBC can be analyzed. Therefore, accurate calculation of these parameters is vitally important. In the following, we use Fig. 12.2 as an example to analyze the relation between impedance spectra and circuit element parameters.

(1) **Equivalent circuit for a single-layer system**

Figure 12.2a shows the equivalent circuit for the impedance spectrum of a single-layer TBC that accounts for the dispersion effect. The corresponding complex impedance Z is as follows:

$$Z = R_s + \frac{R_C}{1 + Q_C(j\omega)^{\beta_C} R_C} \tag{12.6}$$

Letting $\tau_C = (R_C Q_C)^{\frac{1}{\beta_C}}$, we obtain the following:

$$\frac{R_C}{Z - R_s} = 1 + (\omega\tau_C)^{\beta_C}\left(\cos\frac{\pi}{2}\beta_C + j\sin\frac{\pi}{2}\beta_C\right) \tag{12.7}$$

Letting $D = \mathrm{Re}[Z]$, $B = -\mathrm{Im}[Z]$, we have

$$\begin{aligned} \frac{R_C(D - R_S)}{(D - R_s)^2 + B^2} &= 1 + (\omega\tau_C)^{\beta_C}\cos\frac{\pi}{2}\beta_C \\ \frac{R_C B}{(D - R_s)^2 + B^2} &= (\omega\tau_C)^{\beta_C}\sin\frac{\pi}{2}\beta_C \end{aligned} \tag{12.8}$$

Then, we obtain the following:

$$(\omega\tau_C)^{\beta_C} = \frac{B}{(D - R_s)\sin\frac{\pi}{2}\beta_C - B\cdot\cos\frac{\pi}{2}\beta_C} \tag{12.9}$$

Based on Eqs. (12.8) and (12.9), we have

$$(D - R_s)^2 + B^2 = (D - R_s)\cdot R_C - R_C B\cdot\cot\frac{\pi}{2}\beta_C \tag{12.10}$$

$$\left(D - R_S - \frac{R_C}{2}\right)^2 + \left(B + \frac{R_C}{2}\cdot\cot\frac{\pi}{2}\beta_C\right)^2 = \frac{R_C^2}{4\sin^2\frac{\pi}{2}\beta_C} \tag{12.11}$$

As seen in Eq. (12.11), the real and imaginary parts of the complex impedance constitute an arc shaped curve with a diameter of $R_C/(\sin\frac{\pi}{2}\beta_C)$ that intersects the real coordinate axis at $(R_S, 0)$ and $(R_S + R_C, 0)$. When $\beta_C = 1$ (i.e., without the dispersion effect), Eq. (12.11) can be simplified to

$$\left(D - R_S - \frac{R_C}{2}\right)^2 + B^2 = \frac{R_C^2}{4} \tag{12.12}$$

The above equation shows that the real and imaginary parts of the complex impedance constitute an arc with a diameter of R_C that intersects with the real coordinate axis at $(R_S, 0)$ and $(R_S + R_C, 0)$. Figure 12.3 reveals the physical meaning of Eqs. (12.11) and (12.12) (–B and D are the imaginary and real parts of the impedance, respectively). The two arcs intersect the real coordinate axis at two identical points, but the arc with the dispersion effect (i.e., arc 2) has a larger radius than the arc without the dispersion effect (i.e., arc 1).

(2) **Equivalent circuit for a two-layer system**

For the sake of simplicity, both the ceramic and TGO layers are regarded as simple parallel *RC* circuits, and neither the dispersion effect nor R_S is considered. Then, the equivalent circuit in Fig. 12.2b is transformed into that in Fig. 12.4 (the subscripts 1 and 2 signify the ceramic layer and the TGO layer, respectively). Then, the complex impedance of the system is as follows:

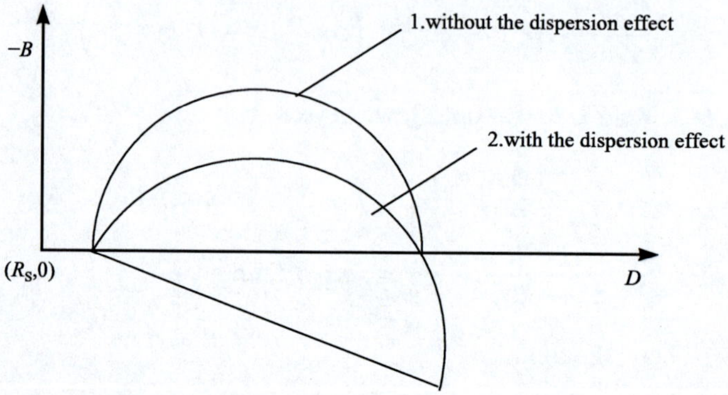

Fig. 12.3 Impedance spectra of a single capacitive reactance arc

Fig. 12.4 The simplified double capacitive reactance equivalent circuit for a TBC

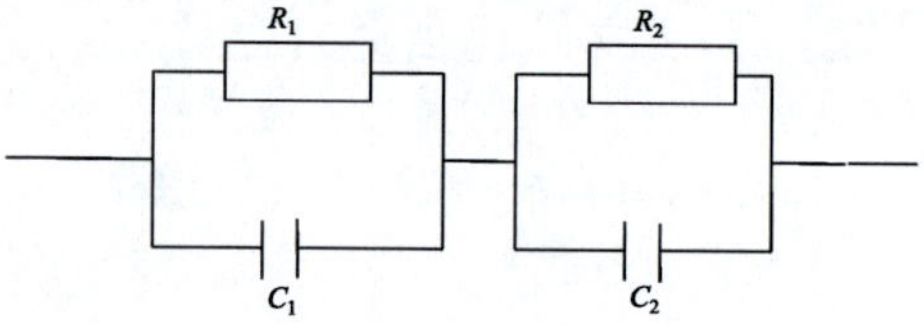

$$Z = \frac{R_1}{1 + j\omega R_1 C_1} + \frac{R_2}{1 + j\omega R_2 C_2}$$
$$= \frac{R_1 + j\omega R_2 R_1 C_1 + R_2 + j\omega R_1 R_2 \cdot C_2}{1 + j\omega R_1 C_1 + j\omega R_2 C_2 + (j\omega)^2 \cdot R_1 R_2 \cdot C_1 \cdot C_2} \tag{12.13}$$

Letting $\tau_1 = R_1 C_1 = \rho_1 \varepsilon_1$, $\tau_2 = R_2 C_2 = \rho_2 \varepsilon_2$, we obtain the following parameter values for the TBC [6]:

$$\rho_C = 1 \times 10^9 \Omega/m, \rho_T = 1 \times 10^{12} \Omega/m, \varepsilon_C = 35, \varepsilon_T = 6 \tag{12.14}$$

Thus, $\tau_2 >> \tau_1$. Then, Eq. (12.13) is simplified to

$$Z = \frac{R_1 + R_2 + j\omega R_1 R_2 \cdot C_2}{1 + j\omega R_2 C_2 + (j\omega)^2 \cdot R_1 R_2 C_1 C_2} \tag{12.15}$$

At high frequencies (i.e., $\omega >> 1$), by neglecting the terms that do not contain ω and subsequently substituting Z_{Hf} for the Z in the above equation, we obtain the following:

$$Z_{Hf} = \frac{R_1}{1 + j\omega R_1 C_1} \tag{12.16}$$

At low frequencies, by neglecting the term that contains ω^2 and substituting Z_{Lf} for the Z in Eq. (12.15), we have

$$Z_{Lf} = R_1 + \frac{R_2}{1 + j\omega R_2 C_2} \tag{12.17}$$

When $\beta_C = 1$, Eq. (12.6) has the same form as that of Eqs. (12.16) and (12.17). According to the earlier discussion, the complex impedance spectrum of a two-layer model consists of two semicircles. The real and imaginary parts of the complex impedance in Eq. (12.16) constitute the high-frequency arc that intersects the real axis at points (0, 0) and (R_1, 0), while the real and imaginary parts of the complex impedance in Eq. (12.17) constitute the low-frequency arc that intersects the real axis at points (R_1, 0) and ($R_1 + R_2$, 0). Figure 12.5 shows the physical meaning of Eqs. (12.16) and (12.17) (–B and D are the imaginary and real parts of the impedance, respectively). We find in Fig. 12.3 that the arcs corresponding to the complex impedance of a single-layer system with and without the dispersion effect intersect the real axis at identical points; however, they have different radii. According to Eq. (12.14), $\tau_T >> \tau_C$ for a TBC. We have reason to believe that when the dispersion effect is considered, the impedance in the high-frequency range depends on the first layer (i.e., the ceramic layer), while the impedance in the low-frequency range depends on the second layer (i.e., the TGO layer). In other words, the high-frequency arc on the left represents the ceramic layer, while the low-frequency arc on the right represents the TGO layer. Based on Fig. 12.5, the complex impedance spectrum of a two-layer TBC system can be simply regarded as a linear superposition of the impedance values of both layers. The following shows the complex impedance of a two-layer TBC system in different frequency ranges:

$$Z_{Hf} = \frac{R_C}{1 + R_C Q_C (j\omega)^{\beta_C}} \tag{12.18}$$

$$Z_{Lf} = R_S + R_C + \frac{R_T}{1 + R_T Q_T (j\omega)^{\beta_T}} \tag{12.19}$$

(3) **Equivalent circuit for a three-layer system**

Fig. 12.5 The ideal double capacitive reactance arc impedance spectrum of a TBC

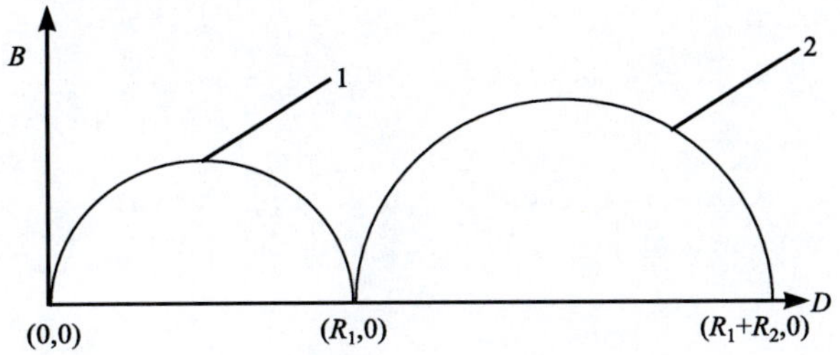

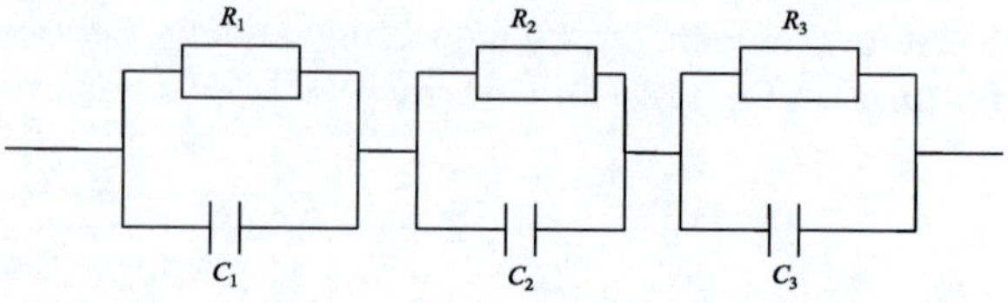

Fig. 12.6 The simplified triple capacitive reactance equivalent circuit for a TBC

Here, we use Fig. 12.6 to analyze the three-layer impedance system of a TBC without considering the dispersion effect or R_S (the subscripts 1, 2, and 3 signify the ceramic layer, the TGO layer, and the MO layer, respectively). The complex impedance of the equivalent circuit in Fig. 12.6 is as follows:

$$\begin{aligned} Z &= Z_1 + Z_2 + Z_3 \\ &= \frac{R_1}{1 + j\omega R_1 C_1} + \frac{R_2}{1 + j\omega R_2 C_2} + \frac{R_3}{1 + j\omega R_3 C_3} \end{aligned} \tag{12.20}$$

Letting $\tau_1 = R_1 C_1 = \rho_1 \varepsilon_1, \tau_2 = R_2 C_2 = \rho_2 \varepsilon, \tau_3 = R_3 C_3$, we obtain the following for the TBC [6]:

$$\begin{aligned} &\rho_C = 1 \times 10^9\ \Omega/m,\ \rho_T = 1 \times 10^{12}\ \Omega/m,\ \rho_M = 1 \times 10^{11}\ \Omega/m \\ &\varepsilon_C = 35,\ \varepsilon_T = 6,\ \varepsilon_M = 0.9 \end{aligned} \tag{12.21}$$

Thus, $\tau_2 >> \tau_3 >> \tau_1$. Then, Eq. (12.20) can be simplified to

$$Z = \frac{R_1 + R_2 + R_3 + j\omega R_2 R_3 C_3 + j\omega R_1 R_3 C_3 + (j\omega)^2 R_1 R_2 R_3 C_2 C_3}{1 + j\omega R_3 C_3 + (j\omega)^2 R_2 R_3 C_2 C_3 + (j\omega)^3 R_1 R_2 R_3 C_1 C_2 C_3} \tag{12.22}$$

At high frequencies, by neglecting the zero- and first-order terms that contain ω and subsequently substituting Z_{Hf} for the Z in the above equation, we obtain the following:

$$Z_{Hf} = \frac{R_1}{1 + j\omega R_1 C_1} \tag{12.23}$$

At intermediate frequencies, by neglecting the zero-order term and terms of order higher than 2 that contain ω and substituting Z_{Mf} for the Z in Eq. (12.22), we have

$$Z_{Mf} = R_1 + \frac{R_2}{1 + j\omega R_2 C_2} \tag{12.24}$$

At low frequencies, by neglecting the terms of second and higher orders that contain ω and substituting Z_{Lf} for the Z in Eq. (12.22), we obtain the following:

$$Z_{Lf} = R_1 + R_2 + \frac{R_3}{1 + j\omega R_3 C_3} \tag{12.25}$$

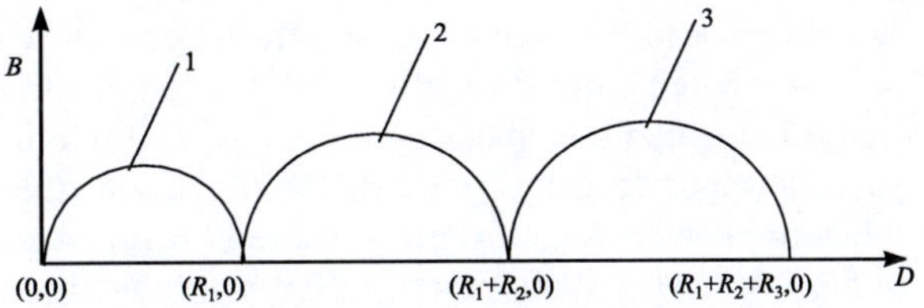

Fig. 12.7 The ideal triple capacitive reactance arc impedance spectrum of a TBC

Similarly, Eqs. (12.23)–(12.25) are reflected by three semicircles in the impedance spectrum, as shown in Fig. 12.7. According to Eq. (12.21), $\tau_T >> \tau_M >> \tau_C$ for a TBC. Thus, the high-frequency arc on the left of the spectrum represents the complex impedance spectrum of the first layer (i.e., the ceramic layer), the intermediate-frequency arc in the middle represents the complex impedance spectrum of the second layer (i.e., the MO layer), and the low-frequency arc on the right represents the complex impedance spectrum of the third layer (i.e., the TGO layer).

Similar to the derivation process for a two-layer system, when the dispersion effect is considered, we have

$$Z_{Hf} = R_S + \frac{R_C}{1 + R_C Q_C (j\omega)^{\beta_C}} \tag{12.26}$$

$$Z_{Mf} = R_S + R_C + \frac{R_M}{1 + R_M Q_M (j\omega)^{\beta_M}} \tag{12.27}$$

$$Z_{Lf} = R_S + R_C + R_M + \frac{R_T}{1 + R_T Q_T (j\omega)^{\beta_T}} \tag{12.28}$$

The above analysis shows that the complex impedance response of a TBC can be divided into three zones, namely, a high-frequency zone (corresponding to the ceramic layer, which has the lowest resistivity of the three layers), an intermediate-frequency zone (corresponding to the MO layer), and a low-frequency zone (corresponding to the TGO layer). Note that in the actual oxidation process, the MO layer occurs rarely, is often discontinuous, and is generally mixed with the TGO layer to form a single layer. In addition, the ceramic layer is a structure with multiple gaps or voids and may produce lower-frequency responses.

12.2 Numerical Simulation of the Complex Impedance Spectral Characteristics of TBCs

12.2.1 FE Principle of CIS

The finite element method (FEM) was first introduced by Fleig et al. [9] to the CIS analysis of materials and was used to analyze the distribution of the electric potential and complex impedance at the interface between two solid electrolytes in contact with each other. Later, through the FEM in combination with experiments, Deng et al. [10] investigated the changes in the h_{TGO} and microstructure of an electron beam-physical vapor deposition (EB-PVD) TBC during the sintering process and drew the following conclusions: (1) an increase in h_{TGO} causes the peak for the TGO layer of the TBC in the Bode plot to shift toward the high-frequency range and increases its intensity; (2) an increase in the conductivity of the TGO layer (σ_{TGO}) causes the frequency at the peak for the TGO layer to shift toward the high-frequency range, and σ_{TGO} increases with the sintering time and temperature. The FEM is employed in the CIS characterization of the microstructural evolution of TBCs because it is difficult to theoretically and experimentally determine the propagation process of an electric field in the extremely high impedance ceramic and TGO layers. The basic principle of the microstructural testing of TBCs is given in Sect. 12.1.2 but is generally based on the assumption of ideal elements. Therefore, it is necessary to use the FEM to determine the characteristics of and characterization method for the complex impedance spectra of TBCs.

FEM-based complex impedance simulation mainly involves the solving of frequency-domain dynamic equations that consider conduction and displacement currents. Its theoretical basis is described as follows.

Generally, the current density $\boldsymbol{J}$, electric displacement $\boldsymbol{D}$, and magnetic field intensity $\boldsymbol{H}$ satisfy the Maxwell–Ampere law

$$\nabla \times \boldsymbol{H} = \boldsymbol{J} + \mathrm{j}\omega \boldsymbol{D} \tag{12.29}$$

where j is the imaginary unit and ω is the angular frequency ($\omega = 2\pi f$, where f is the frequency). Since a magnetic field has no sources, we take the divergence of the left- and right-hand sides of the equal sign in Eq. (12.29) to obtain [11]

$$\nabla \cdot (\nabla \times \boldsymbol{H}) = \nabla \cdot (\boldsymbol{J} + \mathrm{j}\omega \boldsymbol{D}) = 0 \tag{12.30}$$

$\boldsymbol{D}$ in a homogeneous medium can be described as follows:

$$\boldsymbol{D} = \varepsilon_{\mathrm{r}}\varepsilon_0 \boldsymbol{E} \tag{12.31}$$

where ε_r and ε_0 are the relative permittivity and vacuum permittivity, respectively, and $\boldsymbol{E}$ is the electric field strength. According to Ohm's law, $\boldsymbol{J}$ is expressed as follows:

$$\boldsymbol{J} = \sigma \boldsymbol{E} \tag{12.32}$$

where σ is the conductivity. Substitution of Eqs. (12.32) and (12.31) into Eq. (12.30) yields

$$\nabla[\sigma E + j\omega\varepsilon_r\varepsilon_0 E] = 0 \tag{12.33}$$

Because an electric field has no rotation, $\boldsymbol{E}$ and the electric potential V are related through the following equation:

$$\boldsymbol{E} = -\nabla V \tag{12.34}$$

Then, Eq. (12.33) can be expressed as follows:

$$\nabla[(\sigma + j\omega\varepsilon_r\varepsilon_0)\nabla V] = 0 \tag{12.35}$$

The above equation is the governing equation for calculating the complex impedance spectrum of a TBC using the FEM. Based on Eq. (12.33), the expression within the parentheses in Eq. (12.35) is a constant. Then, Eq. (12.35) can be expressed as follows:

$$\nabla^2 V = 0 \tag{12.36}$$

For a 3D model, its governing equation is as follows:

$$\frac{\partial^2 V}{\partial X^2} + \frac{\partial^2 V}{\partial Y^2} + \frac{\partial^2 V}{\partial Z^2} = 0 \tag{12.37}$$

When the boundary conditions are determined, the distribution of V can be obtained based on Eq. (12.37). On this basis, the complex current density $\boldsymbol{J}$ can be determined using Eqs. (12.34) and (12.32), and $\boldsymbol{D}$ can be calculated using Eq. (12.31). Thus, the complex current density $\hat{J}$ can be written as follows:

$$J = \mathbf{J} + \mathrm{j}\omega\mathbf{D} \tag{12.38}$$

The current in the complex form, $\hat{I}$, can be obtained through the integration of $\hat{J}$:

$$I = \int (\boldsymbol{J} + \mathrm{j}\omega\boldsymbol{D}) \cdot d\boldsymbol{n} \tag{12.39}$$

where $\boldsymbol{n}$ is the direction perpendicular to the surface of the TBC and Ω is the electrode area. Thus, we can determine the complex impedance Z by calculating the ratio of the AC voltage U to the current I in the complex form; i.e.

$$Z = U/I = U/\int (\boldsymbol{J} + \mathrm{j}\omega\boldsymbol{D}) \cdot d\boldsymbol{n} \tag{12.40}$$

12.2.2 FE Model for the Complex Impedance Spectrum of a TBC

As shown in Fig. 12.8, the TBC specimen selected for testing or calculation has a square shape with dimensions of 10 cm × 10 cm and a Pt electrode deposited on the surface of its ceramic layer. In reality, the TBC comprises a yttria-stabilized zirconia (YSZ) layer, a TGO layer, a BC layer, and a substrate (Fig. 12.8b). Figure 12.9a shows the FE model established based on the geometry of the TBC. Because both the BC layer and substrate are conductors, only the YSZ and TGO layers need to be considered in the 3D FE model in the calculation of the complex impedance spectrum. The Pt electrode is located at the center of the specimen and has a diameter of *D* (*D* is a variable parameter with a value range of 1–9 mm in the calculation; note that *D* differs from the electric displacement ***D***). To ensure comparability between the numerical simulation and experimental results, the TGO and BC layers were completely bonded during the measurement process. Thus, the whole bottom surface of the TGO layer can be treated as another electrode. The thickness of the YSZ layer, h_{YSZ}, ranges from 150 to 300 μm, and h_{TGO} ranges from 1 to 3 μm.

Here, we assume that the ceramic layer is a homogeneous medium, neglect the boundary effect, [12–14] and select an impedance response temperature of 400 °C. Then, the conductivity of the YSZ layer σ_{YSZ} is 1×10^{-3} S/m, [15] and its relative permittivity ε_{rYSZ} is 28 [9, 10]. At a small h_{TGO}, the effect of the MO layer is

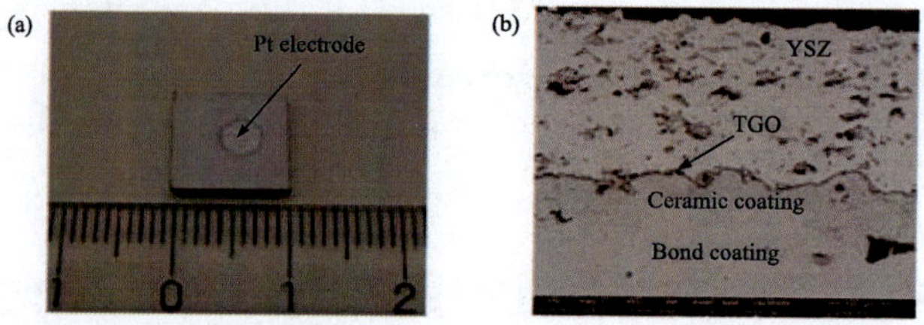

Fig. 12.8 **a** A TBC specimen for complex impedance measurement; **b** microstructure of the TBC

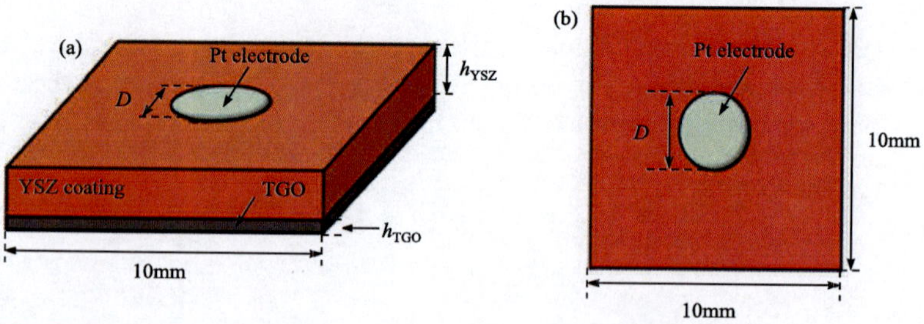

Fig. 12.9 **a** FE model of the TBC; **b** electrode on the surface of the coating

not considered. Under this condition, the TGO layer has a conductivity of $\sigma_{TGO} = 3 \times 10^{-7}$ S/m [16] and a relative permittivity of $\varepsilon_{rTGO} = 10$ [10].

The three boundary conditions used in the finite element (FE) calculation are described as follows: (1) There is no potential drop at the Pt/YSZ or YSZ/TGO interface, an AC voltage of 1 V is applied to the Pt electrode, and the bottom surface of the TGO layer is grounded. (2) The TBC is insulated from the external environment, meaning that there is no current between the outer surface of the TBC and the external environment (i.e., $(\boldsymbol{J} + \mathrm{j}\omega \boldsymbol{D}) \cdot \boldsymbol{n} = 0$). (3) There is no charge accumulation in the YSZ or TGO layer, that is, the current density is continuous at both the Pt/YSZ and YSZ/TGO interfaces.

12.2.3 Complex Impedance Spectral Characteristics of TBCs

(1) Electric-field divergence and its influencing factors

Figure 12.10 shows the effect of electrode symmetry on electric field lines (EFLs). Here, symmetric electrodes mean that the size of the Pt electrode on the surface of the ceramic layer is consistent with that of the electrode on the surface of the TGO layer. Conversely, asymmetric electrodes mean that there is an inconsistency between the sizes of the two electrodes. Figure 12.10a, b shows that when the size of the Pt electrode is identical to that of the electrode on the surface of the TGO layer, all the EFLs are uniformly distributed and parallel to each other. When the electrodes on the two surfaces are inconsistent in size, the EFLs are divergent from the electrode on the surface of the ceramic layer to the electrode on the surface of the TGO layer, as shown in Fig. 12.10c, d. Electric-field divergence caused by asymmetric electrodes leads to errors in the measurement of the complex impedance spectrum, thereby affecting the accuracy of the analysis of the microstructural evolution.

Figure 12.11 shows the evolution of the electrode divergence with the frequency. For the complex impedance spectrum of a TBC with symmetric electrodes (Fig. 12.11a), the EFLs are uniform and nondivergent at various frequencies (from 0.1 Hz to 100 kHz). In comparison, the EFLs from asymmetric electrodes are divergent, particularly at low frequencies, as shown in Fig. 12.11b. Based on the frequency response characteristics of a TBC, frequencies can be divided into three ranges, namely, low frequencies (below 100 Hz), corresponding to the response of the pores in the coating, intermediate frequencies (100 Hz–100 kHz), corresponding to the response of the TGO layer, and high frequencies (above 100 kHz), corresponding to the response of the coating. Figure 12.11b shows that the EFLs are the most divergent at low frequencies. The divergence of the EFLs decreases significantly at intermediate frequencies but remains pronounced, as shown in Fig. 12.11c. The divergence of the EFLs is insignificant at high frequencies, as shown in Fig. 12.11d, and changes little as the frequency further increases. This also suggests that the errors caused by

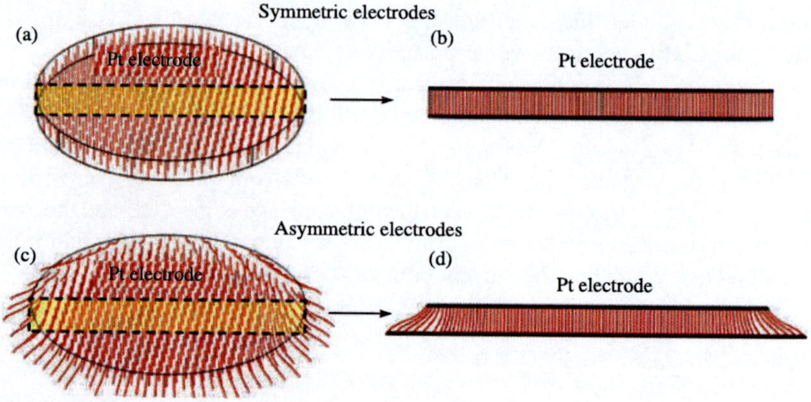

Fig. 12.10 Effect of electrode symmetry on the distribution of EFLs: **a**, **b** distribution of EFLs from symmetric electrodes; **c**, **d** distribution of EFLs from asymmetric electrodes

asymmetric electrodes originate mainly from the pores in the TGO layer and the coating.

Figure 12.12 shows the Nyquist plot (showing the real and imaginary parts of the impedance) and Bode plot of the complex impedance spectrum of the TBC simulated based on a Pt electrode with a diameter of 3 mm. The Nyquist plot shows two typical semicircles, corresponding to the two characteristic frequency spectrum ranges (100–10 kHz and above 100 kHz) in the Bode plot, respectively. We already know from the analysis in Sect. 12.1 that the low and high frequencies originate from the responses of the TGO and YSZ layers, respectively [17, 18]. In addition, Fig. 12.12 shows that the impedance frequency response corresponding to asymmetric electrodes is identical to that corresponding to symmetric electrodes, but the impedance amplitude (i.e., the diameter of the semicircle) corresponding to asymmetric electrodes is much smaller than that corresponding to symmetric electrodes. Moreover, because the response

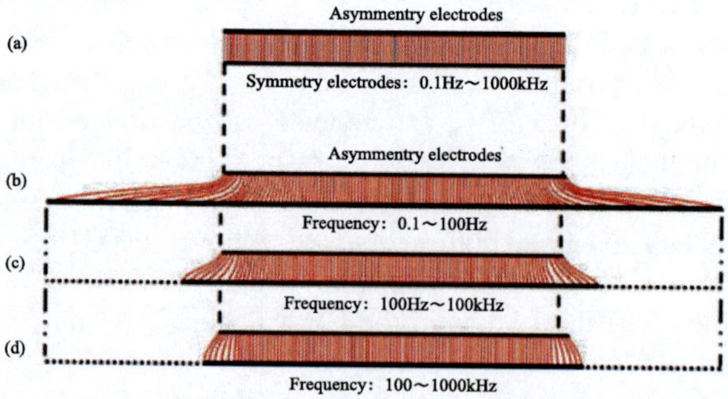

Fig. 12.11 Variation in EFLs with frequency: **a** symmetric electrodes; **b** low-frequency range for asymmetric electrodes; **c** intermediate-frequency range for asymmetric electrodes; **d** high-frequency range for asymmetric electrodes

frequency of the TGO layer ranges from 100 Hz to 100 kHz, asymmetric electrodes have a much more significant effect on the impedance amplitude of the TGO layer than on that of the YSZ layer. There are two characteristic peaks in the Bode plot in Fig. 12.12b, located at 10^3–10^4 and 10^7 Hz, corresponding to the characteristic frequencies of the TGO and YSZ layers, respectively. Asymmetric electrodes affect the Bode plot mainly by reducing the intermediate-frequency range (i.e., the phase angle at the characteristic frequency of the TGO layer) but do not affect the peak for the YSZ layer. This finding is consistent with that obtained from Fig. 12.11.

The resistance R values of the YSZ and TGO layers can be determined through the fitting of the impedance spectra obtained from Fig. 12.12 using the double-layer parallel R–CPE model (Fig. 12.5) in the ZView impedance analysis software. According to the following relation of R with the thickness h of the material, conductivity σ, and electrode area A

$$R = h/(\sigma A) \tag{12.41}$$

When σ and A are known, once R is determined, so are the YSZ and TGO thicknesses h_{YSZ} and h_{TGO}. As a result of the divergence of the EFLs produced by asymmetric electrodes, the area where the EFLs pass through the TBC is larger than A. Under this condition, there is an error in the thicknesses calculated based on A. Table 12.1 compares the h_{YSZ} and h_{TGO} obtained through the fitting of the FE-simulated complex impedance spectra to the equivalent circuit in Fig. 12.5 using ZView with those set for the FE model. The thicknesses determined through simulation based on symmetric electrodes are identical to those set for the model. There is an error in both h_{YSZ} and h_{TGO} determined through simulation based on asymmetric electrodes. Because the divergence of the EFLs increases as the frequency decreases and the frequency response range of the TGO layer is lower than that of the YSZ layer, the error (63.8%) in the simulated value of h_{TGO} is greater than that (11.9%) in the simulated value of h_{YSZ}.

Figure 12.13 shows the variation in the divergence of EFLs with electrode diameter D. The extent of the divergence of EFLs, d, is almost independent of D. As D increases from 1 to 9 mm, d remains almost the same. In the Nyquist plot in Fig. 12.14, the semicircles corresponding to the TGO and YSZ layers decrease in size as D increases. This is because the divergence of EFLs increases parameter A in Eq. (12.41) and, therefore, reduces R. In the Bode plot, an increase in D significantly increases the peak value for the TGO layer but has little effect on the peak value for the YSZ layer. Experimental studies have shown that the divergent effect of asymmetric electrodes affects the results [14, 19]. The experimental results in Fig. 12.15 [19] are similar to the results of the FE analysis: as D increases, the peak value for the TGO layer increases, whereas the peak value for the YSZ layer changes little.

(2) Effects of electrode divergence on h_{YSZ} and h_{TGO}

To study the effects of electrode divergence on the microstructure of the YSZ layer and the measurement of h_{TGO}, the complex impedance values of the YSZ and TGO layers of different thicknesses (Z_{YSZ} and Z_{TGO}, respectively) are simulated. According to

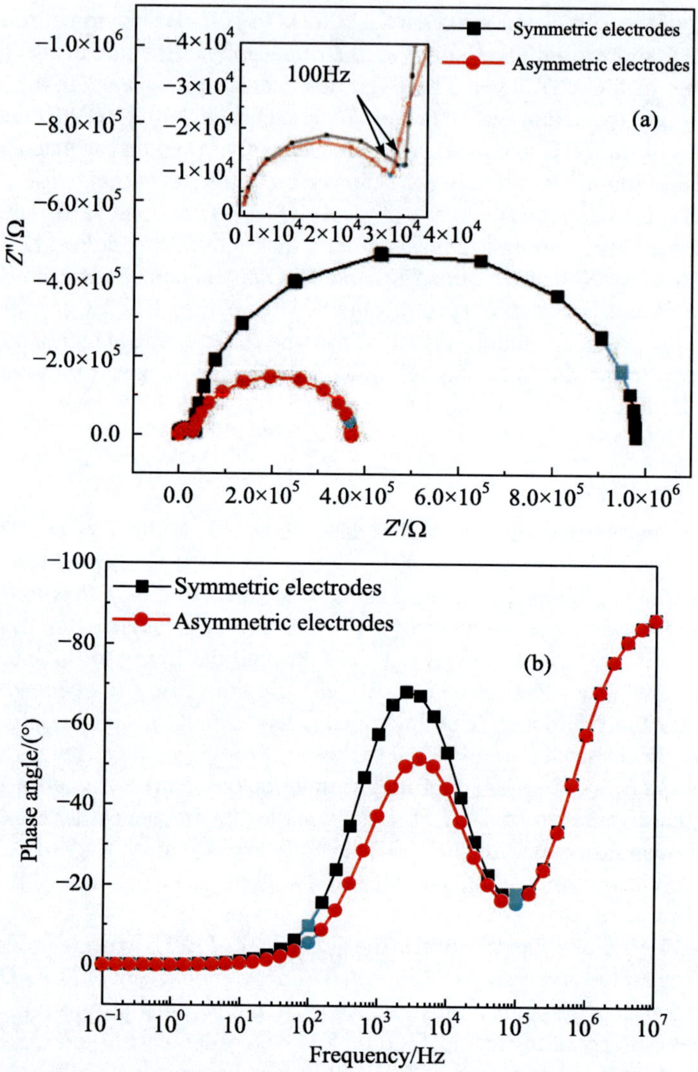

Fig. 12.12 **a** Nyquist plot and **b** Bode plot of the complex impedance of the TBC simulated based on a Pt electrode with a diameter of 3 mm

the earlier complex impedance analysis (i.e., Eq. (12.13)), an increase in either h_{YSZ} or h_{TGO} increases the impedance. Specifically, according to Eqs. (12.16) and (12.17), the two semicircles corresponding to the TGO and YSZ layers in the Nyquist plot intersect with the real part at points (0, 0), (R_1, 0), and ($R_1 + R_2$, 0). Thus, an increase in h_{YSZ} results in an increase in R_1, which, in turn, alters the sizes of both semicircles corresponding to the YSZ and TGO layers. However, if h_{YSZ} remains unchanged, a

Table 12.1 Errors in h_{YSZ} and h_{TGO} were simulated based on a Pt electrode with a diameter of 3 mm

Electrode type	Model-calculated h_{YSZ} (μm)	CIS-determined h_{YSZ} (μm)	$Error_{YSZ}$ (%)	Model-calculated h_{TGO} (μm)	CIS-determined h_{TGO} (μm)	$Error_{TGO}$ (%)
Symmetric electrodes	250	250	0.00	2.00	2.00	0.00
Asymmetric electrodes	250	222	11.2	2.00	0.725	63.8

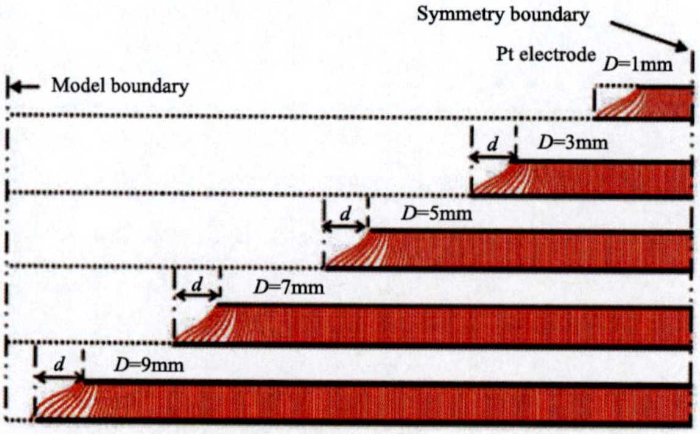

Fig. 12.13 Effect of the electrode diameter D on the electric-field divergence: **a** $D = 1$ mm; **b** $D = 3$ mm; **c** $D = 5$ mm; **d** $D = 7$ mm; **e** $D = 9$ mm

change in h_{TGO} does not affect the response of the YSZ layer because R_1 remains unchanged.

Here, we analyze the effect of electric-field divergence on thickness characterization. We start with the analysis of the effect of h_{YSZ} at a given electrode size and h_{TGO}. Figure 12.16 shows the results obtained for a certain frequency (e.g., 10^4 Hz) selected from the frequency range in which electrode divergence has an impact. The extent of the divergence of EFLs increases as h_{YSZ} increases. As expected, in the Nyquist plot in Fig. 12.17a, the diameters of both semicircles corresponding to the YSZ and TGO layers increase accordingly as h_{YSZ} increases. In the Bode plot in Fig. 12.17b, as h_{YSZ} increases, the value of the characteristic peak increases, whereas the frequency at the characteristic peak remains unchanged. This result suggests that, in essence, electrode divergence does not alter the impedance response properties of the material but affects the measurement accuracy for resistance or thicknesses as a result of the increase in the electric-field propagation area.

Table 12.2 compares the values of h_{YSZ} and h_{TGO} obtained through the fitting of the FE-simulated complex impedance spectra to the equivalent circuit in Fig. 12.5 using ZView with those actually set for the FE model. While only h_{YSZ} changes, the values of both h_{YSZ} and h_{TGO} derived from the fitting of the complex impedance spectra contain large errors that increase as h_{YSZ} increases. Specifically, the errors in the values of h_{YSZ} and h_{TGO} are as high as 8.7–14.5% and 54.2–69.6%, respectively. Thus, error correction is needed when asymmetric electrodes are used to determine the h_{YSZ} or h_{TGO} of a TBC by CIS [20].

The divergence of the EFLs from the TGO layer is similarly analyzed through FE simulations at different values of h_{TGO}. Figure 12.18 shows the results. Similarly, the extent of the divergence of EFLs increases as h_{TGO} increases. The results obtained from the analysis of the values of h_{YSZ} and h_{TGO} determined by CIS based on different values of h_{TGO} are similar to those obtained based on different values of h_{YSZ}. As shown in Fig. 12.18, the electrode divergence increases as h_{TGO} increases. In the Nyquist plot of the impedance spectra in Fig. 12.8b, as h_{TGO} increases, the diameter

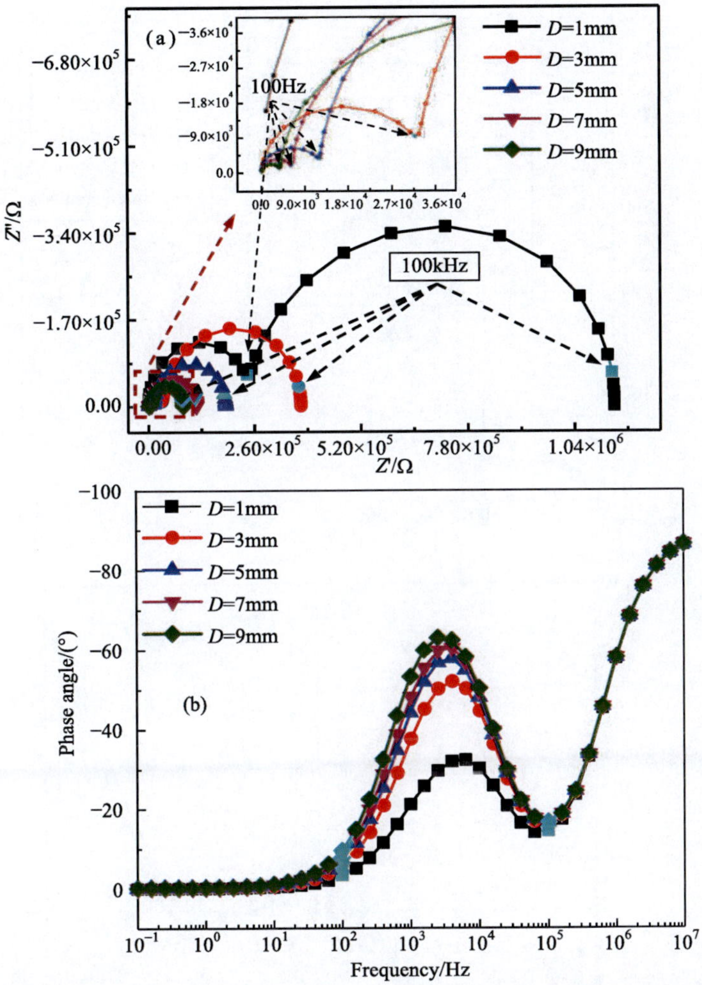

Fig. 12.14 Complex impedance spectra of the TBC simulated using the FEM with different values of electrode diameter *D*: **a** Nyquist plot; **b** Bode plot

of the semicircle corresponding to the YSZ layer remains unchanged, whereas the diameter of the semicircle corresponding to the TGO layer increases. The Bode plot in Fig. 12.18c is similar to that in Fig. 12.17b: as h_{TGO} increases, the value of the characteristic peak increases, whereas its frequency remains unchanged. This result again suggests that in essence, electrode divergence does not alter the impedance response properties of the material but affects the measurement accuracy for resistance or thicknesses due to the increase in the electric-field propagation area.

Table 12.3 summarizes the values of h_{YSZ} and h_{TGO} determined through the fitting of the FE-simulated complex impedance spectra at different values of h_{TGO} to the

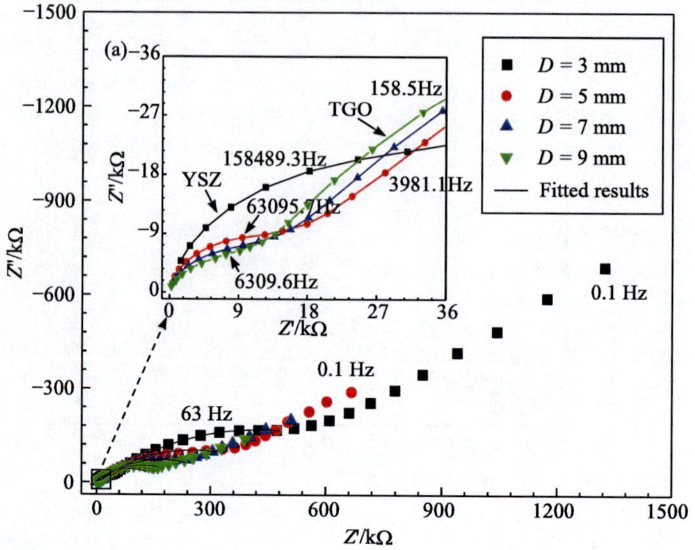

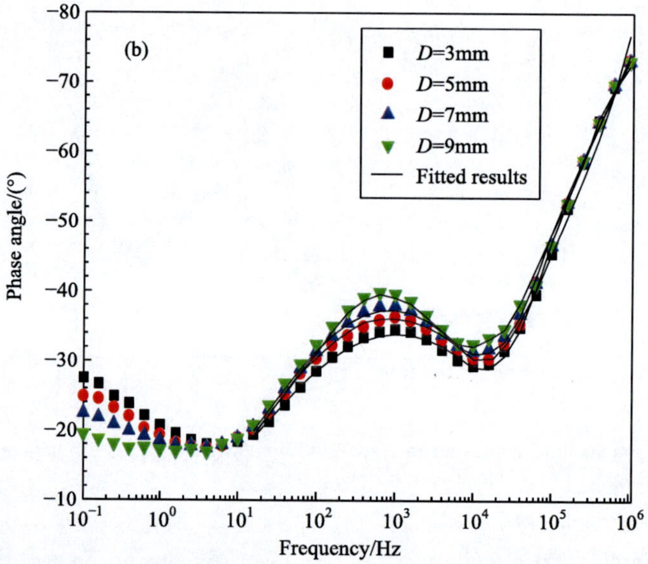

Fig. 12.15 Measured complex impedance spectra of the TBC at different values of electrode diameter *D*: **a** Nyquist plot; **b** Bode plot

equivalent circuit in Fig. 12.5. As h_{TGO} increases from 1 to 3 μm, the errors in the CIS-determined values of h_{YSZ} and h_{TGO} increase from 9.4% to 12.6% and from 53.7% to 63.9%, respectively. This result suggests that the errors in the CIS-determined values of both thicknesses increase as h_{TGO} increases. The response frequency of the TGO layer is lower than that of the YSZ layer. Therefore, a comparison with Table 12.2

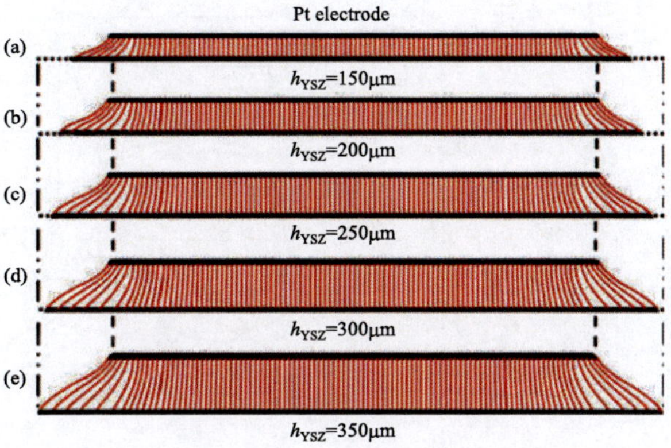

Fig. 12.16 Variation in the electrode-divergence phenomenon with h_{YSZ}

shows that h_{TGO} affects the error in the CIS-determined value of h_{YSZ} less than it affects that of its own CIS-determined value. Thus, electrode divergence affects the CIS-determined value of h_{TGO} to a larger extent than the CIS-determined value of h_{YSZ}.

12.2.4 Asymmetric Electrode Error Correction Models

Table 12.4 summarizes the errors in the CIS-determined values of h_{YSZ} and h_{TGO} at different electrode sizes. The errors in the CIS-determined values of both h_{YSZ} and h_{TGO} decrease as the electrode diameter D increases. As D increases from 1 to 9 mm, the errors in the CIS-determined values of h_{YSZ} and h_{TGO} decrease from 26.2% to 1.5% and from 89.3% to 23.6%, respectively. Thus, the electrode size is extremely important in the characterization of TBCs by CIS. This result is discussed in detail in the following section.

Now, let us first analyze the thickness measurement errors. Because the area through which the current flows is larger than the actual electrode area due to the presence of asymmetric electrodes, the value of the resistance R determined based on the relation between h, R, σ, and A described in Eq. (12.41) is lower than the actual value. Considering the effect of electric-field divergence, we can write Eq. (12.41) as follows:

$$h^* = R^* \sigma A \tag{12.42}$$

where $R^* = h/(\sigma A^*)$ and $A^* = \pi(D/2 + d)^2$ (d is the electrode divergence length (see Fig. 12.13)). According to the definition of relative errors, the relative error Δ in h can be described as follows:

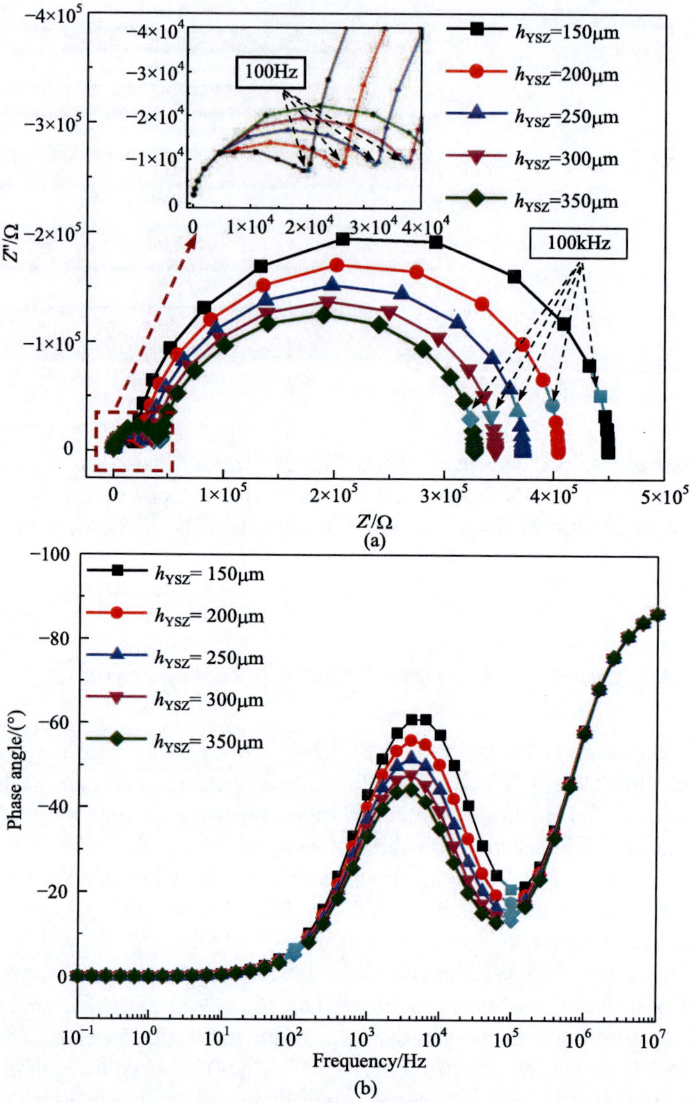

Fig. 12.17 Complex impedance spectra of the TBC at different values of h_{YSZ}: **a** Nyquist plot; **b** Bode plot

$$\Delta = \frac{h - h^*}{h} = \frac{4}{\frac{D^2}{Dd + d^2} + 4} \tag{12.43}$$

According to Eq. (12.43), when the divergence diameter of the electric field is proven to be d in each case (as shown in Fig. 12.13), Δ decreases as D increases. Note that the divergence diameter of the electric field from the TGO layer is larger

Table 12.2 Errors in the CIS-determined values of h_{YSZ} and h_{TGO} at different actual values of h_{YSZ}

Actual h_{YSZ} (μm)	CIS-determined h_{YSZ} (μm)	$\text{Error}_{\text{YSZ}}$ (%)	Actual h_{TGO} (μm)	CIS-determined h_{TGO} (μm)	$\text{Error}_{\text{TGO}}$ (%)
150	137	8.65	2	0.916	54.20
200	182	9.21	2	0.806	59.70
250	222	11.90	2	0.725	63.80
300	261	12.80	2	0.661	66.90
350	299	14.50	2	0.608	69.60

than that of the electric field from the YSZ layer, and consequently, the error in the CIS-determined value of h_{TGO} is larger than that in the CIS-determined value of h_{YSZ}.

Since asymmetric electrodes are generally used to determine complex impedance spectra, particularly those of TBCs, the electric-field divergence phenomenon leads to inevitable errors in the CIS-determined thicknesses. Hence, based on the FE and experimental results in Table 12.4 and Fig. 12.19, the following empirical error correction models for h_{YSZ} and h_{TGO} are established [21]:

$$\begin{aligned} \Delta_{h_{YSZ}} &= -0.089 + \frac{0.111}{\sqrt{D/L}} \\ \Delta_{h_{TGO}} &= 1.21 - 1.026\sqrt{D/L} \end{aligned} \tag{12.44}$$

where D is the electrode diameter and L is the dimension parameter of the TBC specimen (e.g., L is the side length for a square TBC specimen and the diameter for a circular TBC specimen).

The corrected h_{YSZ} and h_{TGO} (denoted $h_{\text{YSZ}}{}^{**}$ and $h_{\text{TGO}}{}^{**}$, respectively) are expressed as follows [21]:

$$\begin{aligned} h_{\text{YSZ}}^{**} &= \frac{h_{\text{YSZ}}^{*}}{1 - \Delta_{h_{\text{YSZ}}}} = \frac{h_{\text{YSZ}}^{*}}{1 - (-0.089 + \frac{0.111}{\sqrt{D/L}})} \\ h_{\text{TGO}}^{**} &= \frac{h_{\text{TGO}}^{*}}{1 - \Delta_{h_{\text{TGO}}}} = \frac{h_{\text{TGO}}^{*}}{1 - (1.21 - 1.026\sqrt{D/L})} \end{aligned} \tag{12.45}$$

In summary, the FE-simulated complex impedance spectra of the TBC show that the use of asymmetric electrodes in CIS characterization leads to an electrode-divergence phenomenon that causes the EFLs to propagate in the TBC over an area larger than the actual area of the electrode and further induces errors in CIS-determined thicknesses or microstructure. The simulations show that the electrode divergence increases as the frequency decreases. For a TBC, this phenomenon is more pronounced in the low-frequency response range of the TGO layer. In addition, the electrode divergence increases as the electrode size decrease or as h_{YSZ} and h_{TGO} increase. For a TBC, the main objective of CIS is to determine its h_{TGO}. Hence,

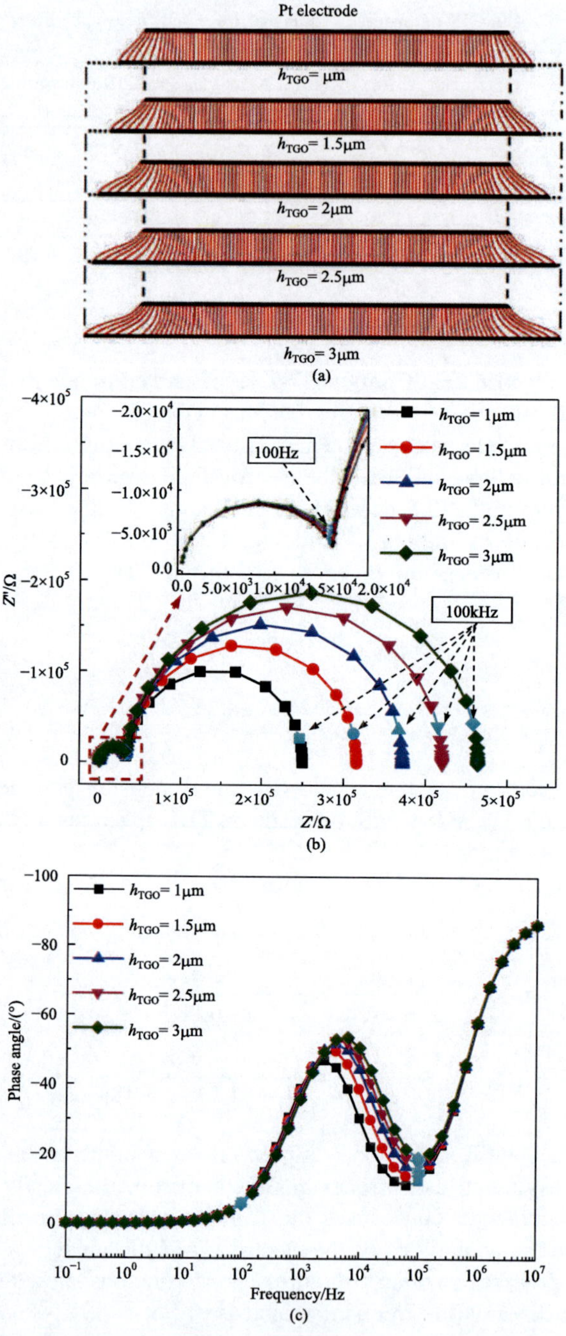

Fig. 12.18 FE-simulated complex impedance characteristics of the TBC at different values of h_{TGO}: **a** distribution of EFLs; **b** Nyquist plot; **c** Bode plot

Table 12.3 Errors in the CIS-determined values of h_{YSZ} and h_{TGO} at different actual values of h_{TGO}

Actual h_{TGO} (μm)	CIS-determined h_{TGO} (μm)	$Error_{TGO}$ (%)	Actual h_{YSZ} (μm)	CIS-determined h_{YSZ} (μm)	$Error_{YSZ}$ (%)
1	0.463	53.70	250	226	9.35
1.5	0.605	59.70	250	224	11.10
2	0.725	63.80	250	222	11.90
2.5	0.828	66.90	250	221	11.70
3	0.922	69.30	250	218	12.60

Table 12.4 Errors in the CIS-determined values of h_{YSZ} and h_{TGO} at different electrode diameters D

D (mm)	h_{YSZ} in model (μm)	h_{YSZ} from CIS (μm)	$Error_{YSZ}$ (%)	h_{TGO} in model (μm)	h_{TGO} from CIS (μm)	$Error_{TGO}$ (%)
1	250	184	26.20	2	0.214	89.30
3	250	222	11.90	2	0.725	63.80
5	250	232	7.15	2	1.04	47.80
7	250	236	5.73	2	1.27	36.30
9	250	246	1.52	2	1.53	23.60

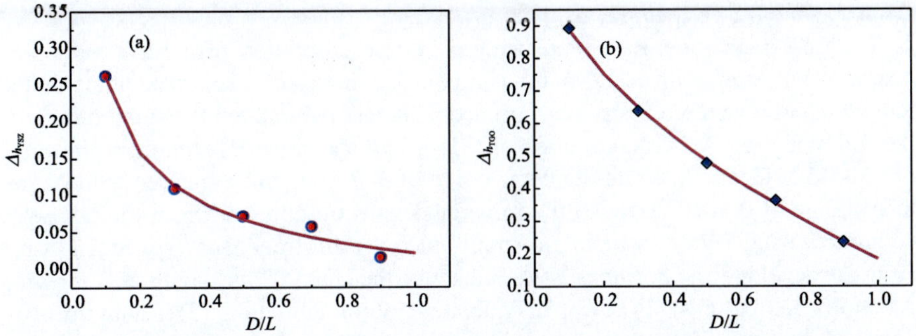

Fig. 12.19 **a** Variation in the errors in the CIS-determined values of (**a**) h_{YSZ} and **b** h_{TGO} with the electrode diameter (the points represent the experimental data, and the curves are obtained through the fitting of the experimental data based on Eq. (12.43))

this chapter presents correction models for h_{YSZ} and h_{TGO} that consider electrode divergence.

12.3 Parametric Optimization of CIS for TBCs

The analysis in the above section shows that the electric-field divergence phenomenon exists in the characterization of TBCs by CIS and that electric-field divergence is related to factors such as frequency, electrode size, and thickness. Apart from these factors, is the characterization of the microstructural evolution of TBCs by CIS related to other factors? The answer is yes. For example, resistivity varies with temperature. At an impedance of $10^6\ \Omega$, does the current response to a certain voltage exceed the measurable range? What is the suitable electrode area? With these questions in mind, we discuss and optimize the parameters of CIS for TBCs using the FEM combined with experiments in this section [19].

12.3.1 FE Simulation and Impedance Measurement

The atmospheric plasma spraying (APS) YSZ TBC specimens used in this section were identical to those used in Sect. 12.2. Specifically, each specimen had dimensions of 10 mm × 10 mm and consisted of a 250-μm-thick ceramic layer, a 1-μm-thick TGO layer, and a 90-μm-thick BC layer. A Ag electrode layer, the size of which varied with the experimental requirements, was deposited at the center of the outer surface of the ceramic layer by vacuum sputtering, and then the TBC was annealed in a high-temperature furnace at 300 °C for 30 min to enhance the bond between the electrode and the ceramic layer.

The impedance was measured using a system composed of a 1260 frequency response analyzer with a 1296 high-impedance interface (Solartron, the United Kingdom) in a vacuum resistance furnace. The test parameters considered include the following: an AC voltage range of 0.1–1 V, a specimen test temperature range of 250–450 °C, an electrode diameter range of 3–7 mm, and a frequency measurement range of 0.1–10^6 Hz. All the impedance spectra obtained from the test were simulated using ZView, which is a small but powerful impedance spectrum fitting software developed by Scribner Associates Inc., the United States, with the following functions and features: (1) powerful equivalent circuit modeling; (2) instant fitting of common circuits; (3) data processing and plotting; (4) batch data processing; (5) data analysis and fitting; and (6) compatibility with Windows 7/8/10. The FE simulation method used in this section is identical to that introduced in Sect. 12.2 and is thus not repeated here.

12.3.2 Optimal AC Voltage Amplitude

The complex impedance is the ratio of the AC voltage to the current. According to this definition, as long as the AC current can be accurately measured, the impedance spectrum should be independent of the amplitude of the applied voltage. Generally, an impedance measurement system applies a voltage of approximately 1–2 V. For a TBC system containing a YSZ layer and a TGO layer, considering that its impedance can reach as high as 10^5 Ω or even 10^6 Ω, the current is on the order of 10^{-2} mA or even 10^{-3} mA; in addition, its impedance increases as the frequency decreases (readers can analyze this trend based on the derivation in Sect. 12.2). This means the test instruments must have extremely high accuracy. Thus, it is recommended that the voltage selected for measuring complex impedance spectra be sufficient to ensure a current above 10^{-2} mA or be close to the upper voltage limit of the impedance measurement system.

First, the effect of the voltage on the test results was analyzed while other measurement parameters (e.g., temperature and electrode size) were set to fixed values. The complex impedance of the specimen was measured at a temperature of 400 °C (the specimen was placed in a constant-temperature furnace until its temperature stabilized) and an electrode diameter D of 3 mm. The Solartron 1296 + 1260 impedance measurement system has a maximum AC voltage amplitude of 2 V. Thus, three voltage amplitudes (0.1, 0.5, and 1.0 V) were selected to analyze the effect of voltage on complex impedance spectrum measurement. Figure 12.21 shows the results. As expected, the voltage has almost no effect on the complex impedance spectrum in the high- and intermediate-frequency ranges, which are the response ranges of the YSZ and TGO layers, respectively. However, in the low-frequency range, the impedance increases as the voltage decrease, suggesting that the magnitude of the current might have approached or exceeded the measuring range of the instrument or that the current measurement accuracy decreased. Since the influence of neither pores in the ceramic layer nor the electrode effect is considered in the FE calculation, two complete semicircles are available in the FE-simulated Nyquist plot in Fig. 12.22a. In addition, because the measurement by the instrument is not limited by the current measurement accuracy, the impedance, as shown in the Nyquist plot in Fig. 12.22a and the Bode plot in Fig. 12.22b, does not vary with the voltage. On this basis, it is concluded that within the measurement range of the instrument, the amplitude of the AC voltage does not affect the measurement accuracy and thus can take any value. However, considering the high-impedance characteristics of TBCs, it is recommended that a voltage above 0.5 V be selected to ensure the reliability and accuracy of the measurement.

12.3.3 *Effects of the Test Temperature and Optimal Test Temperature*

The electrical properties of materials are generally temperature dependent. However, the evolution of the electrical properties with temperature varies from material to material. Here, we take conductivity as an example. In general, the conductivity of a metallic material decreases as the temperature increases, while the opposite trend is observed in semiconductor materials, and materials exist that are insensitive to temperature. For a TBC, the properties of its ceramic and TGO layers are closer to those of a semiconductor. As a result, the increase in the conductivity σ of a TBC with temperature can be described using the classical Arrhenius equation as follows [13]:

$$\sigma = A\exp\left(-\frac{E}{kT}\right) \tag{12.46}$$

where k is the Boltzmann constant, T is the absolute temperature, E is the activation energy, and A is the conductivity at room temperature. Therefore, increasing the test temperature can help to increase conductivity, thereby reducing the resistances of the YSZ and TGO layers (R_{YSZ} and R_{TGO}, respectively) and the diameters of the semicircles corresponding to their impedance responses. This result shows that compared to AC voltage, temperature has a more significant effect on the impedance response of a TBC. Figure 12.23 shows the complex impedance of the TBC at different test temperatures. The Nyquist plot shows that as the temperature increases, the semicircles corresponding to both the YSZ (on the left) and TGO layers decrease in size. In addition, the Bode plot shows that as the temperature increases, the frequencies corresponding to the peaks for both the TGO and YSZ layers in the Bode plot shift to the high-frequency range, and the peak phase angle for the YSZ layer decreases. These observations suggest that the impedance of a TBC decreases as the temperature increases. Moreover, we can see that as the temperature increases from 350 to 450 °C, the impedance responses of both the YSZ and TGO layers tend to stabilize. Thus, to ensure comparability in the impedance spectra of TBCs, selecting a test temperature above 350 °C is recommended. Furthermore, considering factors such as the interfacial oxidation in a TBC, the oxidation of its substrate, and the increase in the resistivity of its metallic substrate with temperature, selecting a temperature between 350 and 400 °C is recommended for measuring the complex impedance spectrum of a TBC.

Next, let us briefly analyze the trends of the impedance and the frequency at the peak phase angle, using the YSZ layer (which responds in the high-frequency range) as an example. According to Eq. (12.16), in the high-frequency range, the effect of the YSZ layer plays a dominant role, whereas the effect of the TGO layer is negligible. The impedance of the YSZ layer, Z_{YSZ}, in Eq. (12.16) is further decomposed into a real part and an imaginary part as follows:

$$Z'_{YSZ} = \frac{R_{YSZ}}{1 + \omega^2 R^2_{YSZ} C^2_{YSZ}} \tag{12.47}$$

$$Z'''_{YSZ} = -\frac{\omega R^2_{YSZ} C_{YSZ}}{1 + \omega^2 R^2_{YSZ} C^2_{YSZ}} \tag{12.48}$$

The tangent of the phase angle θ is calculated using the following equation:

$$\tan(\theta) = \frac{Z''}{Z'} \tag{12.49}$$

The phase angle of the YSZ layer, θ_{YSZ}, can be written as follows:

$$\theta_{YSZ} = \tan^{-1}\left(\frac{-\frac{\omega R_{YSZ}^2 C_{YSZ}}{1+\omega^2 R_{YSZ}^2 C_{YSZ}^2}}{\frac{R_{YSZ}}{1+\omega^2 R_{YSZ}^2 C_{YSZ}^2}}\right) \tag{12.50}$$

Substitution of $R = \rho l/A$, $C = \varepsilon_r \varepsilon_0 A/h$ into Eq. (12.50) simplifies it to

$$\theta_{YSZ} = \tan^{-1}(-\omega \rho_{YSZ} \varepsilon_{YSZ}) \tag{12.51}$$

Equation (12.46) shows that a temperature increase leads to an increase in the conductivity of the YSZ layer, σ_{YSZ}. Thus, as the temperature increases, the resistivity of the YSZ layer, ρ_{YSZ}, decreases, and consequently, the height of the peak for the YSZ layer decreases. Anderson et al. [22] noted that the angular frequency at the peak ω_p for the TGO layer can be expressed as follows:

$$\omega_p \approx \frac{1}{R_{TGO} C_{TGO}} \sqrt{\left(\frac{R_{YSZ} + R_{TGO}}{R_{YSZ}}\right)} \tag{12.52}$$

By substituting the expressions of R and C of each layer, we can further simplify the above equation to

$$\omega_p \approx \frac{1}{\rho_{TGO} \varepsilon_{TGO}} \sqrt{\left(1 + \frac{\rho_{TGO} \delta_{TGO}}{\rho_{YSZ} \delta_{YSZ}}\right)} \tag{12.53}$$

A temperature increase reduces the ρ_{YSZ} and ρ_{TGO} (the resistivity of the TGO layer), thereby increasing ω_p and shifting the peak for the TGO layer to the high-frequency range. In addition, temperature affects the complex impedance response of the electrode (near approximately 0.1 Hz). As the temperature increases from 250 to 350 °C, the peak intensity of the electrode response increases. A further temperature increase from 350 to 450 °C reduces the height of the peak.

Here, we analyze the effect of temperature on the complex impedance spectrum of the TBC from a theoretical perspective using the FEM. Table 12.5 summarizes the variation in σ_{TGO} and σ_{YSZ} with temperature at a voltage amplitude of 1 V and an electrode diameter of 3 mm. Figure 12.24a, b shows Nyquist and Bode plots at different temperatures, respectively. Both the semicircles corresponding to the impedance responses of the YSZ and TGO layers decrease in size as the temperature increases. In the Bode plot, as the temperature increases, the frequency at the peak for the TGO layer shifts to the right, and the phase angle corresponding to the peak for the YSZ layer decreases. This phenomenon is more pronounced at temperatures between 250 and 300 °C, which is consistent with the experimental results. This result again suggests that test temperature plays a vital role in the microstructural characterization of TBCs by CIS.

Figure 12.25 shows the measured complex impedance spectra of the TBC at different temperatures fitted using ZView. R_{YSZ} and R_{TGO} decrease significantly as the test temperature increases from 250 to 300 °C, but tends to stabilize as the temperature increases beyond 350 °C. This observation again indicates that the microstructural characterization of a TBC does not vary with the test temperature when it exceeds 350 °C, which is, therefore, selected as the temperature for measuring the complex impedance spectra of TBCs in many cases.

12.3.4 Effect of the Electrode Size

As mentioned earlier, when the complex impedance spectrum of a TBC is measured, a Ag or Pt electrode needs to be prepared on the surface of the nonconductive ceramic layer, while the metallic substrate, which is a conductor, can be directly used as another electrode. Therefore, the characterization of TBCs by CIS mostly involves asymmetric electrodes. Section 12.2 details that asymmetric electrodes can introduce the electric-field divergence effect, which, in turn, affects the measurement of h_{YSZ} and h_{TGO}. In this section, we further analyze the effect of electrode size on the measurement of the complex impedance spectrum of the TBC based on the FE simulations in Sect. 12.2 in combination with experimental testing. The complex impedance spectrum was measured at a temperature of 400 °C and a voltage amplitude of 1 V.

Figure 12.13 shows the variation in the divergence of EFLs with the electrode size. The extent of the divergence of EFLs, *d*, is almost independent of the electrode size. As the diameter of the Pt electrode, *D*, increases from 1 to 9 mm, *d* remains almost the same. Both the FE results in Fig. 12.14 and the test results in Fig. 12.26 show that the impedance decreases as the electrode size increases. In particular, the electrode diameter, when less than 5 mm, has a significant effect. According to the definition of the resistance *R* and impedance (e.g., Eq. (12.16)), increasing the electrode size decreases R_{YSZ} and R_{TGO} while reducing the proportion of the complex impedance response of the electrode (approximately 0.1 Hz) in the whole impedance response. Both FE and experimental results show that the electrode size has no effect on the frequency spectrum of the YSZ layer. As shown in Fig. 12.26b, as the electrode size increases, there is no shift in the frequency at the peak and no change in the phase

Table 12.5 Conductivities of YSZ and TGO layers (σ_{YSZ} and σ_{TGO}) at different temperatures[23]

Temperature/°C	250	300	350	400	450
σ_{YSZ}/(S m^{-1})	7×10^{-6}	5×10^{-5}	1×10^{-4}	4×10^{-4}	1×10^{-3}
σ_{TGO}/(S m^{-1})	3×10^{-9}	2×10^{-8}	6×10^{-8}	2×10^{-7}	1×10^{-6}

angle. In addition, as the electrode size increases, there is similarly no shift in the frequency at the peak for the TGO layer but a slight increase in its phase angle.

Despite having no significant effect on the frequency characteristics, the electrode size affects the thickness and microstructure measurement accuracies for a coating as R_{YSZ} and R_{TGO} decrease (see the effects of divergence on h_{YSZ} and h_{TGO} in Sect. 12.2 for a detailed analysis). Therefore, an electrode diameter of 5 mm or larger is recommended.

12.3.5 Summary

In this section, based on FE simulations and experimental tests, we investigate the effects of AC voltage amplitude, temperature, and electrode size on the complex impedance spectrum of a TBC and optimize the test parameters. The results show the following:

(1) The AC voltage amplitude has no effect on test results. However, to ensure high current measurement accuracy, selecting a voltage amplitude of 1 V is recommended.
(2) The test temperature is an important parameter affecting the characterization of a TBC by CIS. The suitable temperature range is 350–400 °C.
(3) Increasing the electrode size reduces the impedance but does not affect the frequency spectrum characteristics. Considering the effect of asymmetric electrodes on the thickness measurement accuracy, selecting an electrode diameter ≥5 mm is recommended.

12.4 Characterization of Interfacial Oxidation in TBCs by CIS

CIS has been widely used to characterize the thickness of the interfacial oxide layer in a TBC [8, 17, 24–30]. The basic procedure is described as follows. First, a suitable equivalent circuit is established based on the microstructural characteristics of the TBC. Then, according to the evolutionary pattern, the complex impedance spectrum of the TBC is measured under various conditions. The value of each element parameter is determined through the simulation of the impedance spectrum with an equivalent circuit. Finally, the microstructure (e.g., thickness) and its evolutionary pattern are determined based on the values of the element parameters. In this section, we introduce a method to characterize TBCs by CIS based on the following basic procedure: equivalent circuit → complex impedance spectrum measurement → impedance spectrum simulation → microstructural evolution.

12.4.1 Equivalent Circuit for Interfacial Oxidation in TBCs

TBCs prepared by two typical processes—EB-PVD and APS—display certain differences in microstructure and oxidation characteristics, and consequently, their equivalent circuits also differ. An EB-PVD TBC, as shown in Fig. 12.27a, contains a ceramic layer that exhibits a columnar crystal structure with multiple gaps and a smooth interface between its ceramic and transition layers. Figure 12.27b shows the morphology of the TBC after oxidization. A layer of continuous, dense oxides (i.e., TGOs) can be seen at the interface between the ceramic layer and the transition layer of the oxidized TBC. Because both the transition layer and the substrate are conductive metals, their impedances are negligible compared to those of ceramic and TGO layers. On this basis, an equivalent circuit for an EB-PVD TBC is established, as shown in Fig. 12.27c, [10, 22, 31] that consists of two parallel *RC* circuits, (corresponding to the TGO and ceramic layers, respectively) connected in series (R_1 and C_1 are the resistance and capacitance of the ceramic layer, respectively, and R_2 and C_2 are the resistance and capacitance of the TGO layer, respectively). Note that the formation of a dense TGO layer on the surface of the transition layer through preoxidation before the preparation of the ceramic layer can reduce the TGO growth rate during service. Thus, as-prepared and oxidized TBC systems each display two microstructural layers, and their corresponding equivalent circuits each consist of two layers. In addition, while the ceramic layer with a columnar crystal structure contains gaps, these gaps are not completely isolated from the crystals in the ceramic layer, and hence, the current can be transferred in the coating at the gaps. Therefore, the gap-containing ceramic layer is also represented by a parallel *RC* circuit. This indirectly suggests that it is difficult to directly characterize the vertical surface cracks and gaps of EB-PVD TBCs by CIS.

The ceramic layer of an APS TBC, as shown in Fig. 12.28a, has a lamellar structure that contains multiple pores and cracks, and the roughness of its ceramic-layer/transition-layer interface is much higher than that in an EB-PVD coating. Figure 12.28b shows the morphology of the TBC after oxidization. A continuous TGO layer is formed at the interface and may be accompanied by microcracks. In addition, due to sintering, the number of small pores in the ceramic layer decreases. However, as time elapses, large pores and cracks are gradually formed in the ceramic layer. Particular attention should be directed to a possible oxidation phenomenon during the preparation of an APS TBC, as shown in Fig. 12.28d. Specifically, an oxide layer is not only formed at the ceramic-layer/transition-layer interface but also formed throughout the transition layer and at the transition-layer/substrate interface. This oxidation phenomenon is mainly related to the transition-layer preparation process, and its mechanism and effect on the performance of a TBC can be found elsewhere [32]. Figure 12.28e shows the equivalent circuit for the complex impedance spectrum of an APS TBC [23, 29, 30, 33–35] (C_{YSZ} and R_{YSZ} are the CPE and resistance of the grains in the ceramic layer, respectively, C_P and R_P are the CPE and resistance of the pores in the ceramic layer, respectively, and C_{TT} and R_{TT} are the CPE and resistance of the TGO layer, respectively). While the transition layer is oxidized as a whole, the TGO layer can still be considered a parallel circuit because the transition layer itself is a conductor.

12.4.2 *Measurement of the Complex Impedance Spectrum of a TBC*

Figure 12.20 shows a schematic diagram of the measurement of the complex impedance spectrum of a TBC. A specimen equipped with asymmetric electrodes is placed in a high-temperature furnace. The complex impedance spectrum of the TBC is measured after the temperature reaches the set value and becomes stable. The complex impedance spectrum data are recorded and then fitted using impedance analysis software to determine the parameter value of each element in the equivalent circuit, followed by a microstructural analysis. In the following, we briefly introduce the impedance measurement system, electrode preparation, connection between the high-temperature furnace and electrodes, and simulation software.

(1) *Z* measurement system

The calculated or measured impedance spectra at various temperatures and electrode sizes presented earlier show that the impedance of a TBC is as high as above 10^6 Ω, placing high requirements on the performance of complex impedance measurement instruments. Currently, the impedance spectra of TBCs are mostly measured using frequency response analyzers (e.g., the 1260 series) equipped with high-impedance interfaces (e.g., the 1296 series) manufactured by Solartron, a former United Kingdom-based company that has been acquired by Princeton Applied Research in the United States. An impedance measurement system capable of measuring impedances greater than 10^8 Ω should be selected in practice.

(2) Electrode preparation

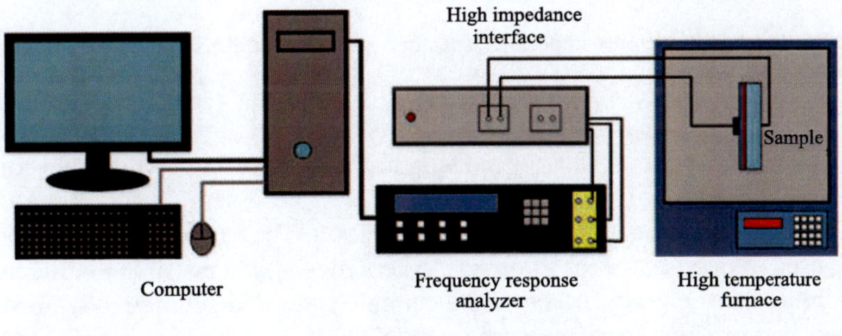

Fig. 12.20 Schematic diagram of the measurement of the impedance of the TBC

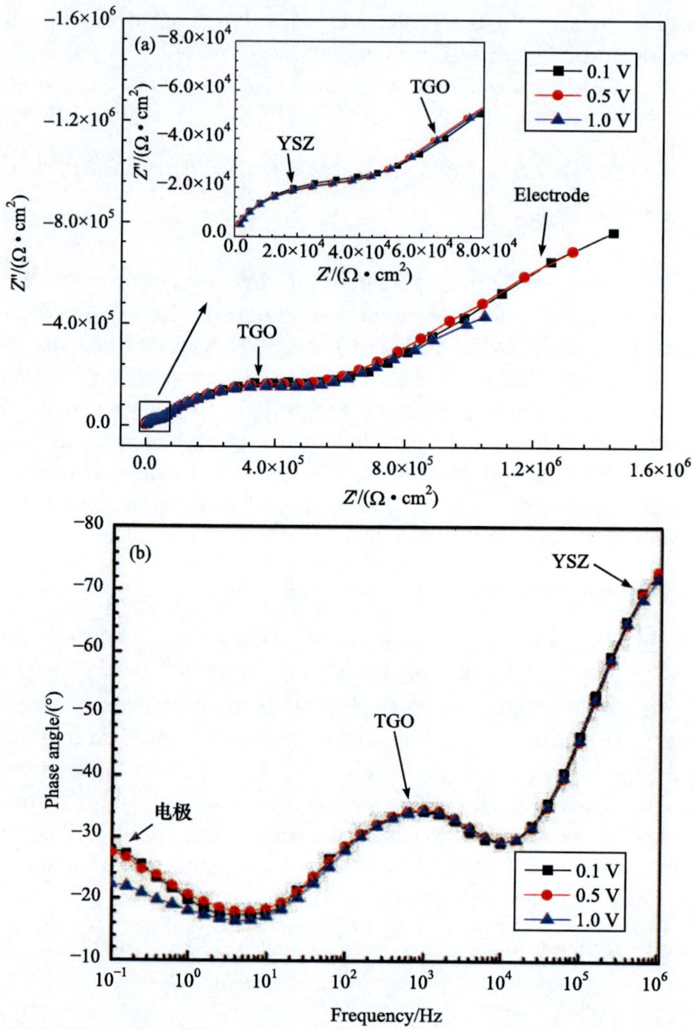

Fig. 12.21 Measured complex impedance spectra of the TBC at different voltage amplitudes: **a** Nyquist plot; **b** Bode plot

The ceramic layer of a TBC is nonconductive. Therefore, when the complex impedance spectrum of a TBC is measured, it is necessary to spray a Ag or Pt electrode onto the outer surface of the ceramic layer. The vacuum sputtering process can be used to deposit Au (or Pt or Ag). Alternatively, it is possible to directly coat Ag in liquid form on the surface of the ceramic layer and subsequently anneal the ceramic layer in a high-temperature furnace at 300 °C for 30 min to enhance the bond between the electrode and the ceramic layer. The selection of the electrode size is discussed in detail in Sect. 12.3, where the selection of an electrode with a diameter

Fig. 12.22 FE-calculated complex impedance spectra of the TBC at different voltage amplitudes: **a** Nyquist plot; **b** Bode plot

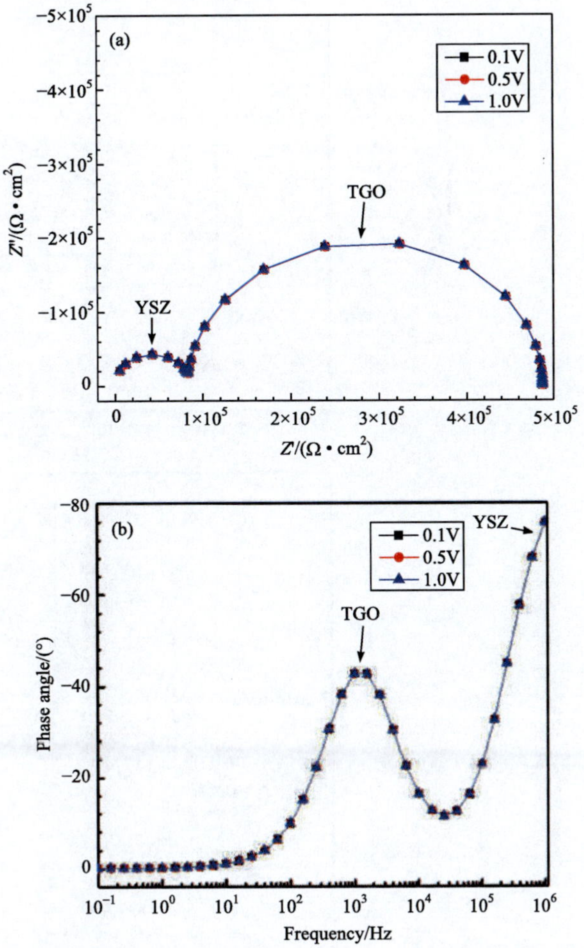

greater than 5 mm is recommended. However, to reflect the local characteristics, it is not recommended to coat a specimen in its entirety with an electrode.

(3) Temperature control and electrode connection for the specimen

The impedance of a TBC decreases as the temperature increases. To improve the impedance measurement accuracy and to relax the high-impedance measurement requirements for instruments, the complex impedance spectrum of a TBC is generally measured at a certain temperature. The effect of test temperature on impedance spectra is analyzed in detail in Sect. 12.3. In the range of room temperature to 300 °C, the temperature has a significant effect on impedance measurements. At higher temperatures, a TBC may undergo further oxidation. Thus, the complex impedance spectrum of a TBC is usually measured in a vacuum furnace at 350–400 °C. Under

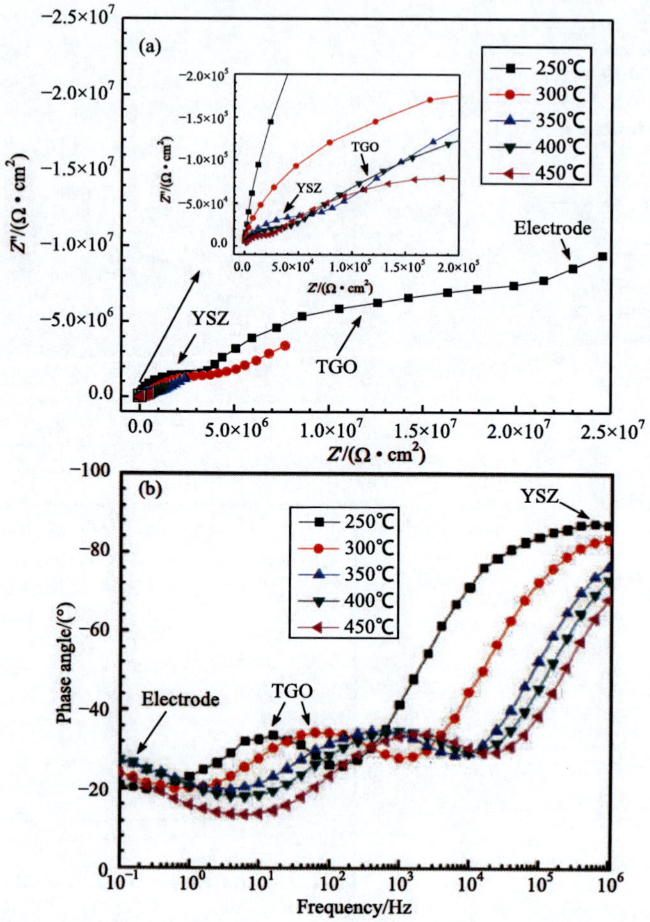

Fig. 12.23 Measured complex impedance spectra of the TBC at different temperatures: **a** Nyquist plot; **b** Bode plot

this condition, as shown in Fig. 12.20, it is necessary to place a carbon electrode resistant to high temperatures between the electrodes of the specimen and the complex impedance measurement system. When the impedance of a specimen is measured in a furnace, special attention should be paid to ensure insulation between the specimen and the furnace, between the specimen and the test bench, and between the electrodes and the furnace so that the current can be transferred only in the specimen.

(4) Impedance spectrum simulation

An impedance spectrum represented by Nyquist and Bode plots reflects the information of the complex impedance spectrum of the whole specimen but fails to determine the information and evolution of each layer and each characteristic structure. Therefore, as a crucial step, an equivalent circuit is used to simulate an impedance spectrum

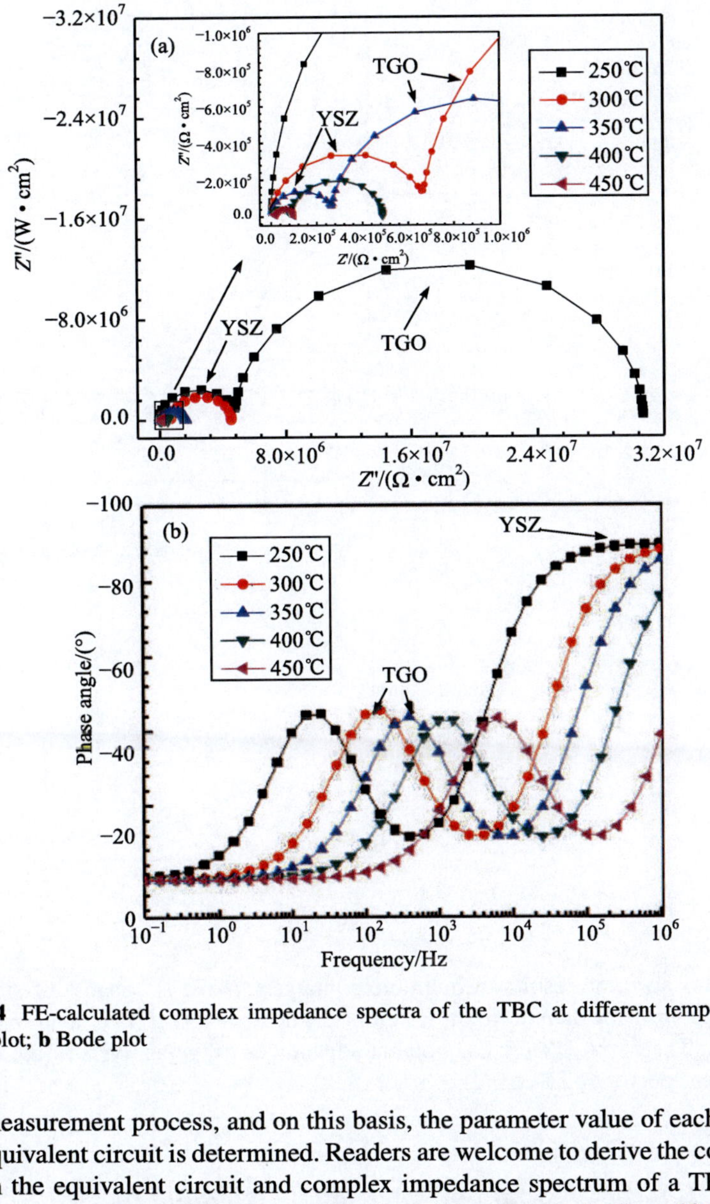

Fig. 12.24 FE-calculated complex impedance spectra of the TBC at different temperatures: **a** Nyquist plot; **b** Bode plot

in the measurement process, and on this basis, the parameter value of each element in the equivalent circuit is determined. Readers are welcome to derive the correlation between the equivalent circuit and complex impedance spectrum of a TBC based on the basic testing principle for complex impedance spectra in Sect. 12.2 or to fit the data using impedance simulation software (e.g., ZView and EQU). Of the available software, ZView, developed by Scribner Associates Inc., United States, is more convenient for use due to the following functions and features: (1) It has powerful equivalent circuit modeling functions. (2) It obtains the parameter value ranges of the elements of key modules in different frequency ranges and regions and, using them as initial values, gradually corrects parameter values through multiple fitting processes

Fig. 12.25 Calculated and measured values of the resistances of the **a** YSZ layer and **b** TGO layer at different temperatures

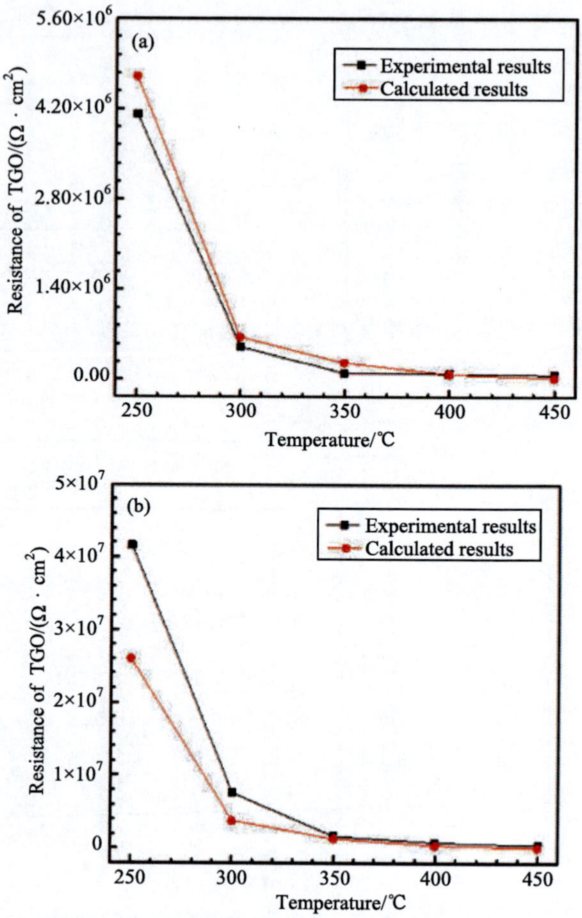

and obtains accurate results through error analysis. (3) It is capable of simultaneously simulating the data of multiple complex impedance spectra to analyze the data evolution. Therefore, ZView has notable advantages in the analysis of the complex impedance spectra of TBCs.

12.4.3 Characterization of Interfacial Oxidation by CIS

In the following, we introduce a method for characterizing TGO growth and the microstructural evolution of the ceramic layer by CIS using an APS TBC as an example.

(1) **Measurement and equivalent circuit simulation of *Z* spectra**

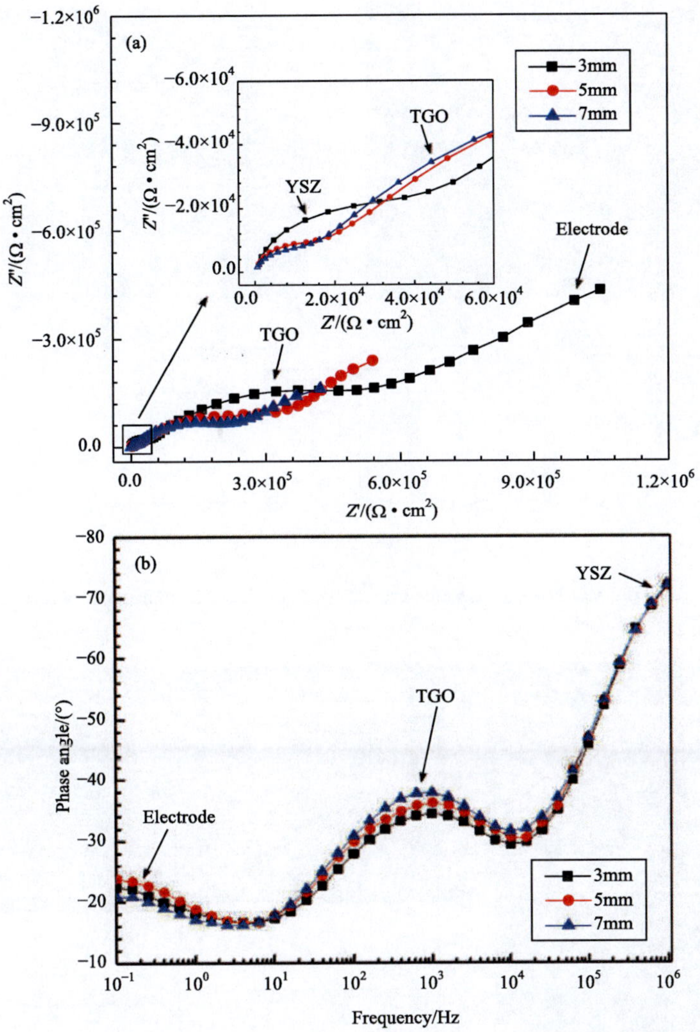

Fig. 12.26 Complex impedance spectra of the TBC at different electrode diameters: **a** Nyquist plot; **b** Bode plot

Figure 12.29a, b shows the variation in the modulus and phase angle of the impedance of an APS TBC with frequency after oxidation at 800 °C for 0, 50, 100, 200, 400, 500, 1000, and 2000 h. Evidently, there is a significant increase in the impedance of the TBC system after oxidation. Based on the phase angle plot in Fig. 12.29b, the frequency response of the TBC system can be divided into three ranges, namely, (1) a high-frequency range (10^4–5×10^6 Hz) with a peak phase angle appearing near 1 MHz, (2) an intermediate-frequency range (10–10^3 Hz) with a peak phase angle

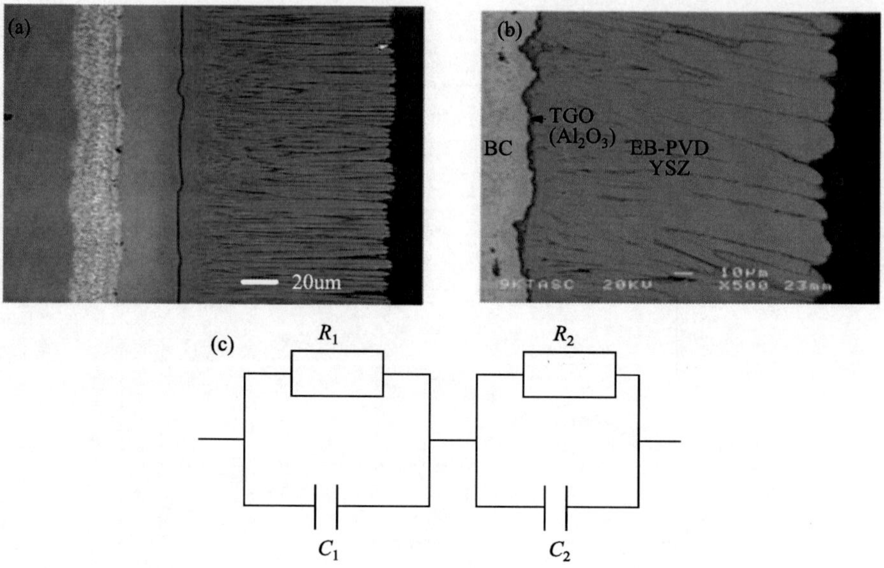

Fig. 12.27 An EB-PVD TBC: **a** columnar crystal structure; **b** oxidized microstructure; **c** equivalent circuit

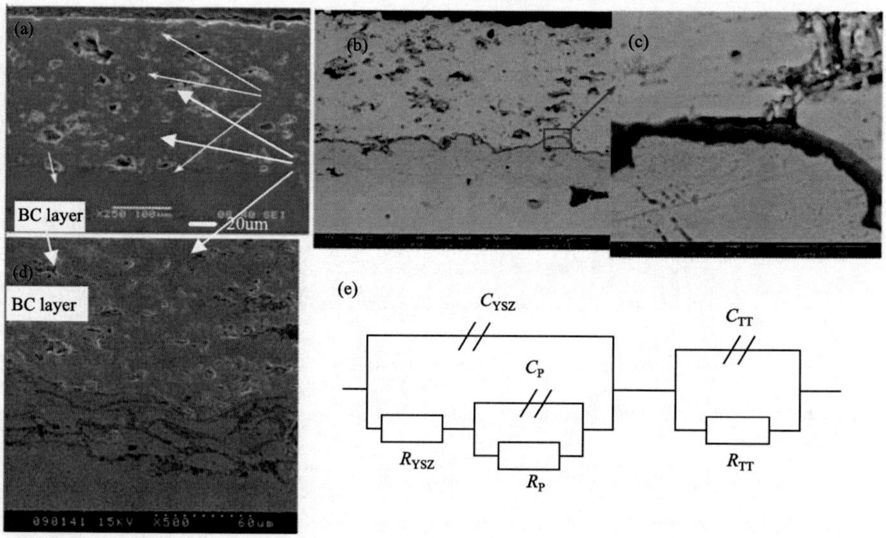

Fig. 12.28 An APS TBC: **a** porous coating structure; **b–d** oxidized microstructure; **e** equivalent circuit

occurring near 10 Hz, and (3) a low-frequency range (0.01–10 Hz). Originating from the YSZ layer, the peak phase angle in the high-frequency range (near 1 MHz) is highly discernible and does not change significantly, regardless of whether the TBC is oxidized. The peak phase angle in the intermediate-frequency range (i.e., the phase angle of the TGO layer) increases as the oxidation duration increases. Due to the presence of a large number of pores and cracks, which obviously have the highest impedance, in the ceramic layer, the characteristic frequency of its impedance response is the lowest [38, 39]. In addition, the pores and cracks in an APS TBC undergo changes as the oxidation process proceeds. As a result, the phase angle of the impedance in the low-frequency range varies greatly depending on the occurrence of high-temperature oxidation of TBCs and, if so, the oxidation duration, as evidenced in Fig. 12.29b.

The impedance responses of the YSZ and TGO layers as well as various characteristic microstructures can be clearly seen from the impedance spectra in Fig. 12.29. However, the values of parameters (e.g., resistance) of the impedance of each layer cannot be clearly determined from these spectra. Figure 12.30 shows the impedance spectra of the TBC before oxidation and after 400 h of oxidation, respectively, simulated with the equivalent circuit in Fig. 12.28e in ZView. The measured impedance spectra match the spectra simulated with the equivalent circuit very well, suggesting that the equivalent circuit is correct and that it is reliable to characterize the experimentally measured data using the simulated parameter values of the elements in the equivalent circuit. Note that only the impedance spectra simulated with the equivalent circuit are given here. In fact, the parameter values of the circuit elements can also be obtained based on the simulation results. Here, we present only the method, and the data are provided in figures through subsequent analysis.

(2) Determination of the thickness h_{TGO} or content C_{TGO} of the TGO layer

For a continuous, homogeneous TGO layer (e.g., an Al_2O_3 layer), h_{TGO} can be easily determined based on the following relation between the resistance R, resistivity ρ, and electrode area A:

$$R = \rho h / A \tag{12.54}$$

where R is given based on the impedance spectrum simulated with an equivalent circuit, and ρ and A are known. Thus, h_{TGO} and its evolution with the oxidation time can be obtained.

If the TGO layer is discontinuous and the transition layer is oxidized as a whole (as shown in Fig. 12.28d), h_{TGO} can be quantitatively analyzed by calculating the C_{TGO}. When the area of the gilded electrode used for complex impedance measurement, A, is a fixed value, h_{TGO} is directly proportional to the volume of the TGO layer. Assuming that the volume of the intermediate transition layer remains unchanged, the extent of the oxidation of a TBC can be expressed in terms of the ratio of the volume of the oxides to the volume of the intermediate transition layer, i.e., the volume fraction of the TGOs. Thus, we have

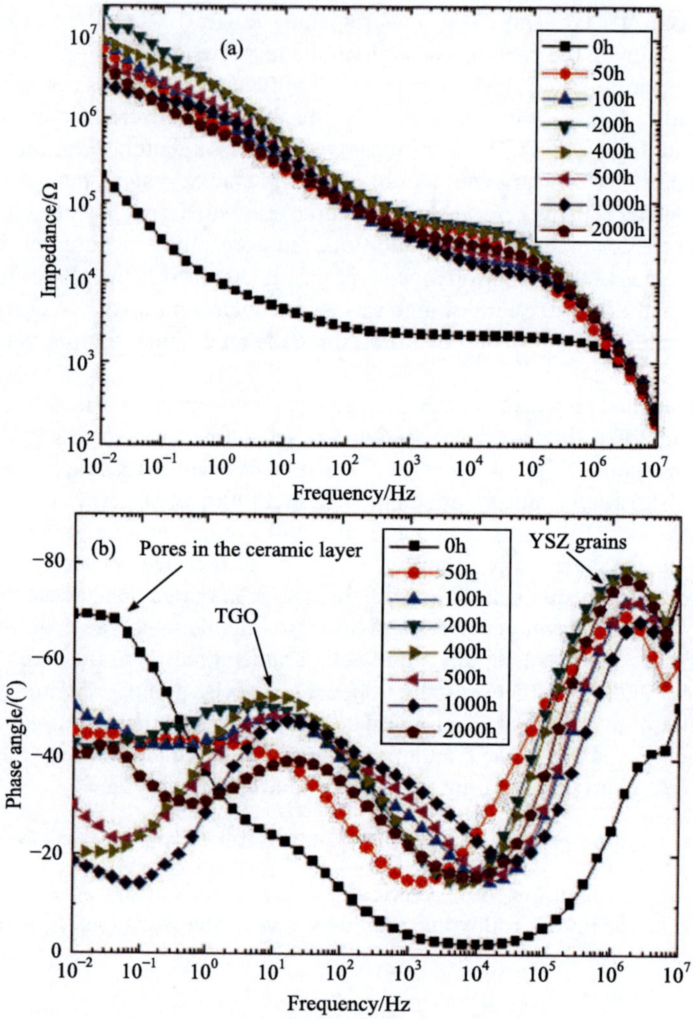

Fig. 12.29 Complex impedance spectra of a TBC before and after oxidation: **a** impedance; **b** phase angle

$$V_{TGO}\% = kR_{TGO} \tag{12.55}$$

where k is a constant. To measure the extent of the oxidation of a TBC by CIS, we need only to first determine the resistance of the TGO layer, R_{TGO}, by simulating the measured complex impedance spectrum with an equivalent circuit and then calculate h_{TGO} using Eq. (12.54) or C_{TGO} using (12.55).

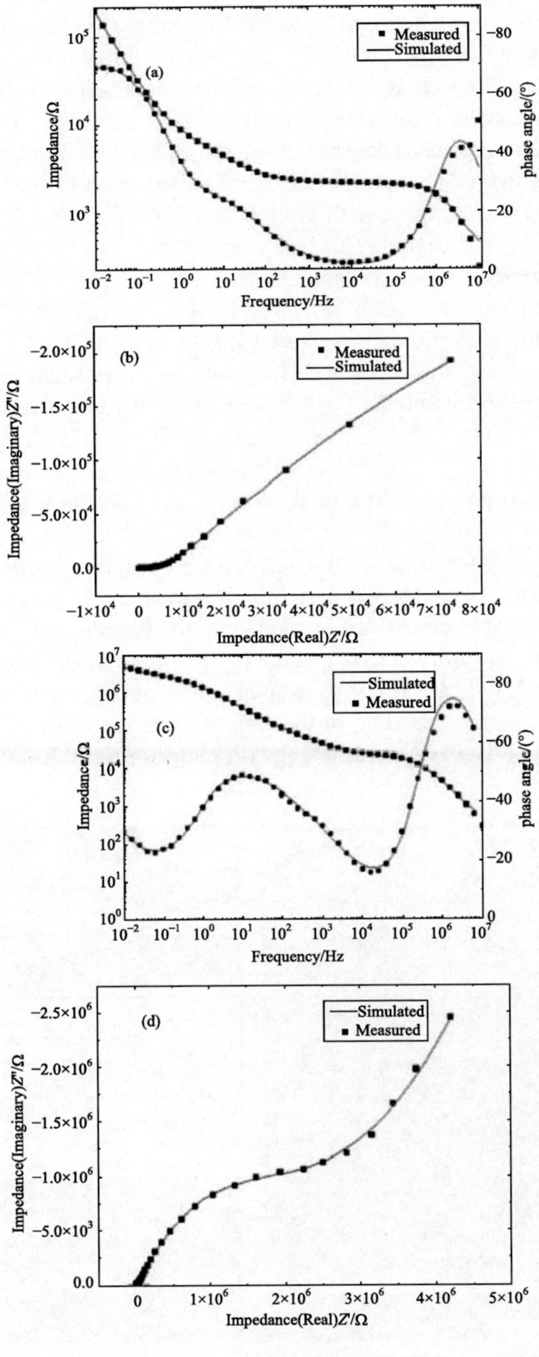

Fig. 12.30 Simulated complex impedance spectra of an APS TBC: **a** and **b** before oxidation; **c**, **d** after 400 h of oxidation

C_{TGO} can be analyzed through image recognition. We calculate C_{TGO} through image processing in MATLAB with an example to illustrate the basic idea and procedure as follows: (1) Read in an SEM image using the imrad function (Fig. 12.31a); (2) extract the image of the transition layer region using the imcrop function (Fig. 12.31b); (3) set thresholds based on the different colors of the TGOs and NiCrAlY; detect the TGO region (Fig. 12.31c), set its color to a value of 0, and calculate the area of the region with a color value of 0 (i.e., the TGO region), S_{TGO}; (4) extract the NiCrAlY region (12.31(d)), set its color to a value of 1, and calculate the area of the region with a color value of 1 (i.e., the NiCrAlY region), $S_{NiCrAlY}$; and (5) calculate C_{TGO} as $TGO\% = S_{TGO}/(S_{TGO} + S_{NiCrAlY}) \times 100\%$. The detailed calculation method is available elsewhere [32].

Figure 12.32a shows the relation between the R_{TGO} obtained through simulation with an equivalent circuit and the C_{TGO} calculated using MATLAB. Evidently, R_{TGO} is directly proportional to C_{TGO}. The straight line does not pass through the origin. This is because the complex impedance spectrum in the intermediate-frequency range is a result of the combined action of the TGO layer and the interface between the ceramic layer and the intermediate transition layer, and consequently, the simulated value of R_{TGO} is the sum of the actual value of R_{TGO} and the value of the resistance of the interface and is thus greater than the actual value of R_{TGO}. During the oxidation process, no notable cracks are formed, and no clear delamination occurs at the interface between the ceramic layer and the intermediate transition layer. As a result, there are no notable changes in the parameter values of the resistor or the CPE of the interface. Consequently, the trend of the resistance reflects that of the TGO growth and microstructure. The last point in Fig. 12.32a does not fall on the straight line,

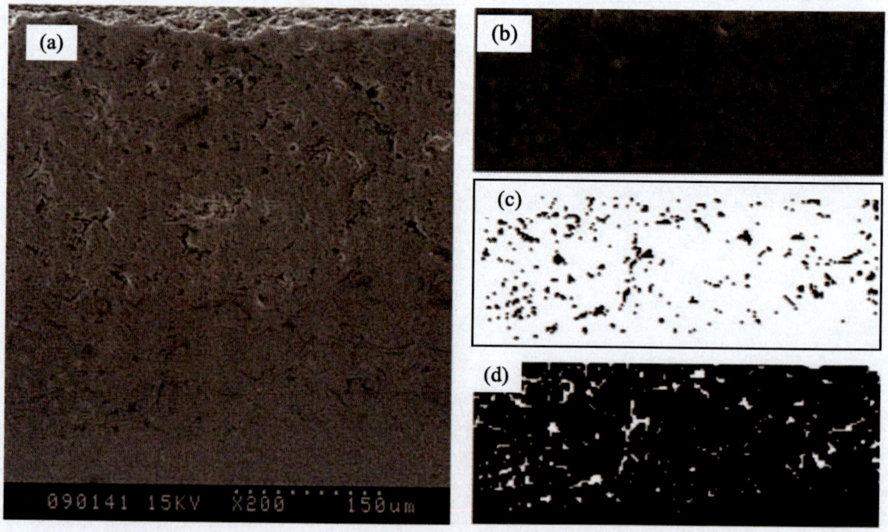

Fig. 12.31 Calculation of the C_{TGO} of an APS TBC: **a** cross-sectional SEM image of the TBC; **b** extracted intermediate transition layer; **c** TGO area; **d** NiCrAlY region

which may be due to the formation of microcracks in the TGO layer or the decrease in the resistivity of the TGO layer [7]. Figure 12.32b shows the evolution of C_{TGO} with the square root of the oxidation time, SR_{OT}. C_{TGO} is directly proportional to SR_{OT}. The slope of the straight line in Fig. 12.32b reflects the oxidation rate of the TBC. There is a notable decrease in the slope of the straight line after 500 h of oxidation, suggesting a decrease in the oxidation rate. On the one hand, the dense and thermally stable Al_2O_3 formed from the reaction between Al and O inhibits the TGO growth. On the other hand, because the TGO layer is composed primarily of Al_2O_3, as the oxidation process proceeds, there is an inevitable decrease in the Al concentration, which, in turn, reduces both the Al diffusion and the TGO growth rate. This result agrees with the theoretical calculation results of Zou [36]. In their investigation of the oxidation kinetics of an APS TBC at 950 °C, Song and Xiao [7] similarly found that the oxidation rate was directly proportional to SR_{OT} and that the bilinear curve representing the relation between the oxidation rate and SR_{OT} had two slopes that corresponded to different oxidation characteristics. The first point in Fig. 12.32b is not on the straight line because the corresponding C_{TGO} depends on the extent of oxidation during the preparation process and is independent of the high-temperature oxidation time.

(3) Evaluation of the variation in the microstructure of a TBC with the high-temperature oxidation time

Now, let us recall the physical meaning of a CPE—the dispersion effect on the complex impedance spectrum caused by the material inhomogeneity of the TBC. In the definition equation for the complex impedance of a CPE ($Z_{CPE} = 1/Q(j\omega)^{\beta}$), β reflects the heterogeneity of the material, and its variation reflects the variation in the composition or microstructure of the material. When β is close to 1, Q can be approximated as the capacitance value. Then, the variation in the microstructure of a TBC during the oxidation process can be analyzed and evaluated based on the variation in the values of parameters β and Q of the CPE in the equivalent circuit.

Figure 12.33 shows the variation in the values of CPE parameters A_{TT} and n_{TT} (i.e., Q and β) of the TGO layer of a TBC with oxidation time. The exponent n_{TT}

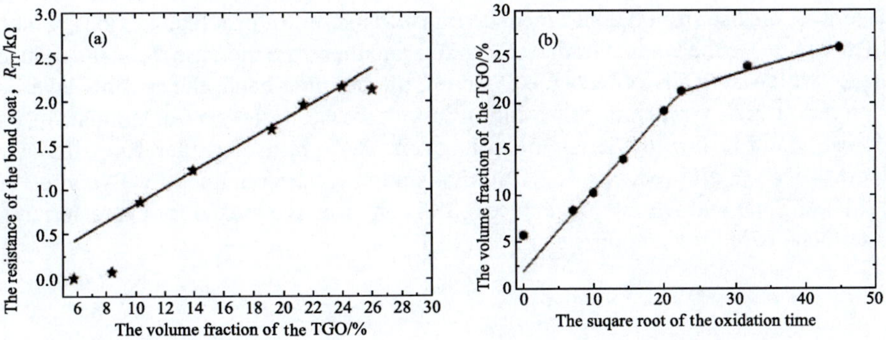

Fig. 12.32 **a** R_{TGO}–C_{TGO} relation. **b** C_{TGO}–SR_{OT} curve

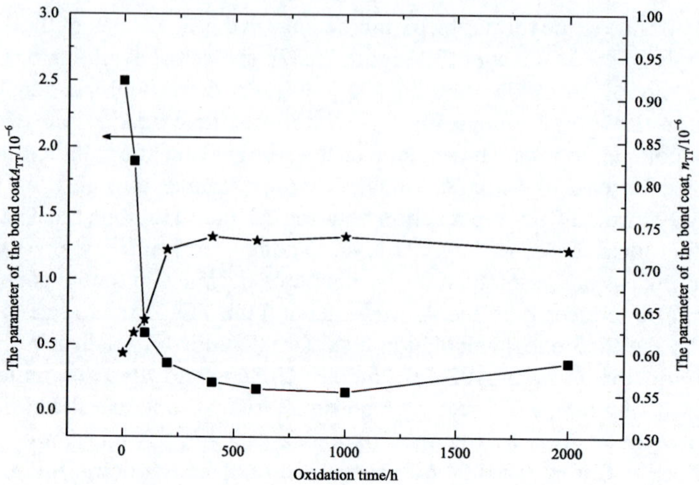

Fig. 12.33 Variation in the values of the CPE parameters of the TGO layer of a TBC with oxidation time

remains basically unchanged as the oxidation time increases from 200 to 2000 h, suggesting that there are no significant changes in the composition of the TGO layer during this stage. In contrast, A_{TT} decreases as the oxidation time increases, which can be attributed to the increase in C_{TGO} (which is equivalent to the increase in h_{TGO}).

Figure 12.34 shows the variation in the values of the resistance of the YSZ grains (i.e., R_{YSZ}) in a TBC and the values of its CPE parameters A_{YSZ} and n_{YSZ} with oxidation time. The value of n_{YSZ} is close to 1 and remains basically unchanged during the oxidation process, suggesting that the YSZ layer does not undergo significant structural changes during the oxidation process and is a dielectric material close to a pure capacitor. The values of R_{YSZ} and A_{YSZ} do not change significantly as the oxidation time increases beyond 400 h. However, as the oxidation time increases before reaching 200 h, there is a significant increase in the value of R_{YSZ} and a considerable decrease in the value of A_{YSZ}. During this stage, on the one hand, the abundant O ions in the YSZ layer rapidly react with the metallic ions in the ceramic layer that diffuse from the intermediate transition layer to form black oxides, which in turn increase the volume of the YSZ grains and therefore increase R_{YSZ} and reduce A_{YSZ} (which basically reflects the C value); on the other hand, the ceramic layer is sintered at high temperatures, resulting in a decrease in the size or number of its pores, which in turn increases the volume of YSZ grains and therefore similarly increases R_{YSZ} and reduces A_{YSZ}. Similar findings were obtained by Byeon et al. [29] and Zhan and Desai [37]. Hence, CIS is an effective means for characterizing interfacial oxidation and pore evolution in TBCs.

oxidation time

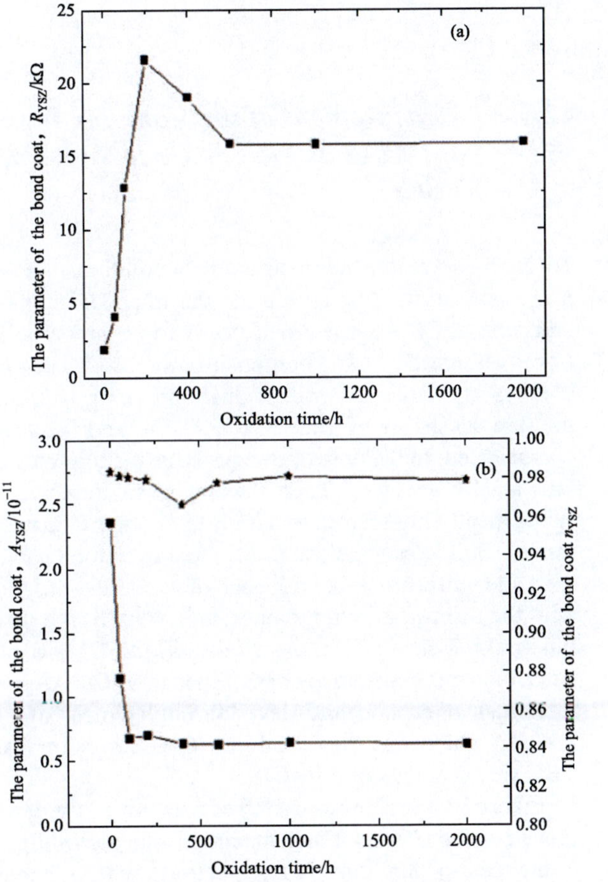

Fig. 12.34 Variation in the values of **a** the resistance of the YSZ grains R_{YSZ} and **b** CPE parameters A_{YSZ} and n_{YSZ} of a TBC with oxidation time

12.5 Characterization of CMAS Corrosion in TBCs with CIS

CMAS corrosion is currently the most hazardous service condition facing TBCs. CMAS melts not only fill the pores in a coating but also induce phase transformations, chemical reactions, and mismatch stresses and cause defects such as cracks and spallation. Currently, destructive methods (e.g., SEM and X-ray diffraction (XRD)) are primarily used to examine CMAS-corroded TBCs and to analyze the microstructural evolution and failure mechanism of TBCs before and after corrosion. The use of nondestructive testing methods (e.g., CIS) to analyze the microstructural evolution of CMAS-corroded TBCs is vitally important for determining the corrosion mechanism of CMAS and providing guidance on engineering projects. Researchers across the world have conducted studies in relevant areas [17, 39–42]. In this section, we focus on introducing methods for characterizing CMAS melts and CMAS-corroded coatings by CIS.

12.5.1 Measurement of the Complex Impedance Spectra of CMAS as Well as Uncorroded and CMAS-Corroded TBCs

To characterize the microstructural evolution of CMAS-corroded TBCs by CIS, it is necessary to first determine the impedance characteristics of CMAS. To this end, first, a CMAS specimen needs to be prepared for complex impedance spectrum measurement. The composition of CMAS can be determined based on actual engines in service as well as dust and smog in the environment. Here, a powder mixture consisting of CaO, MgO, Al_2O_3, and SiO_2 at an atomic ratio of 33:9:13:45 was selected. In the preparation process, the four oxides were mixed according to the ratio and blended well. Then, the powder mixture was placed in a crucible, which was subsequently transferred to a high-temperature furnace, heated from room temperature to 1300 °C, and held at this temperature for 8 h to allow the oxides to sufficiently melt and uniformly mix with each other. Afterward, the mixture was allowed to cool with the furnace to room temperature. Finally, the mixture condensed into a CMAS block. After its surface was polished, the CMAS block was ready for testing to determine the impedance characteristics of CMAS melts. Further, the CMAS block was crushed, ground, and sieved through a mesh of a certain size to prepare CMAS powder, which was then used in a TBC corrosion test in which a CMAS-corroded TBC was characterized by CIS.

Figure 12.35a shows a sketch of a specimen used for testing the impedance performance of the CMAS. Each specimen was a thin film with a diameter of 5 mm and a thickness of 500 μm. A Ag electrode with a diameter of 3 mm was prepared at the center of each of the upper and lower surfaces of each specimen. To facilitate a comparison with the impedance performance of uncorroded and CMAS-corroded TBCs, pure YSZ specimens were also prepared. The melting point of the CMAS was high and exceeded the temperature that the substrates of all the specimens used in the test could withstand. Consequently, when the CMAS was used to corrode a TBC with a substrate composed of a Ni-based alloy in a high-temperature furnace, the metallic substrate first underwent significant deformation, causing coating spallation and making it impossible to measure the complex impedance spectrum of the TBC. Thus, an independent ceramic layer needed to be used in the test; that is, the ceramic layer needed to be separated from the metallic substrate. The TBC used in the test was prepared by APS. To facilitate the preparation of an independent ceramic layer, we directly sprayed a 250-μm-thick ceramic layer onto the surface of an Al plate and then cut the Al plate covered with a ceramic layer into a square shape with dimensions of 10 mm × 10 mm. Subsequently, the Al plate was corroded in a diluted hydrochloric acid solution until the ceramic layer completely separated

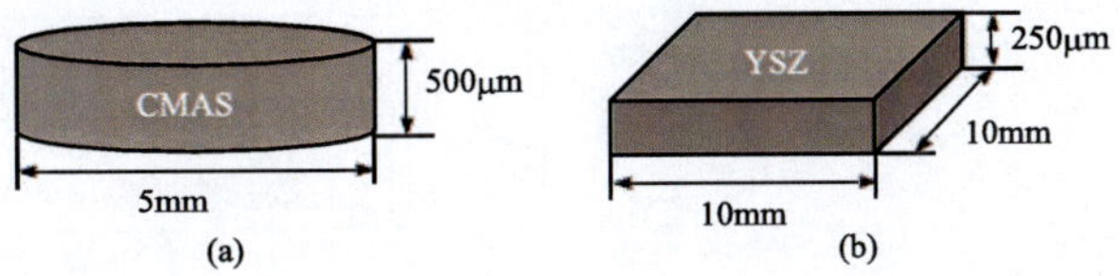

Fig. 12.35 **a** A CMAS specimen used for complex impedance spectrum testing; **b** a YSZ layer specimen

from the Al plate. Thus, the independent ceramic layer needed for the test was obtained [41]. Figure 12.35b shows a sketch of an independent ceramic layer specimen. A 1260 frequency response analyzer equipped with a 1296 high-impedance interface (Solartron, the United Kingdom) was used as the complex impedance spectrum measurement system. The measurement was performed at room temperature, a voltage of 500 mV, and a frequency of 1–10^6 Hz.

During the high-temperature corrosion test, the CMAS powder was first weighed according to the required density—5 mg of CMAS powder on every cm^2 of the TBC. Then, the surface of the ceramic layer was evenly coated with the weighed CMAS powder. Subsequently, specimens coated with the CMAS powder and uncoated specimens were placed in a high-temperature furnace at the same time and subjected to 0, 2, 4, 8, and 16 h of oxidation. The melting point of the CMAS powder used in the test was approximately 1240 °C. To allow the CMAS to sufficiently melt, a corrosion temperature of 1250 °C was selected for the test. After the high-temperature corrosion process was completed, first, electrodes with a diameter of 5 mm needed to be prepared on the surfaces of the ceramic-layer specimens using the method introduced earlier, with the only difference being the need for preparing electrodes on both surfaces of each ceramic-layer specimen due to the lack of a BC layer and a substrate. In addition, the microstructural evolution of the coating was observed by SEM to analyze the measured complex impedance spectra.

12.5.2 Complex Impedance Spectrum Characteristics of CMAS

Figure 12.36 compares the complex impedance spectra of the YSZ layer and the pure CMAS specimen. As seen from Fig. 12.36a, there is a semicircle in the high-frequency range and a tail arc in the low-frequency range for both the CMAS specimen and the YSZ layer. Evidently, the semicircle for the CMAS specimen in the high-frequency range represents its own impedance response, while the semicircle for the YSZ layer in the high-frequency range represents the impedance response of the YSZ grains and the pores in the YSZ layer. As shown in Fig. 12.36b, the peak phase angles of the CMAS specimen and the YSZ layer appear near 10^6 Hz in the high-frequency range, and the peak phase angle for the CMAS specimen is greater

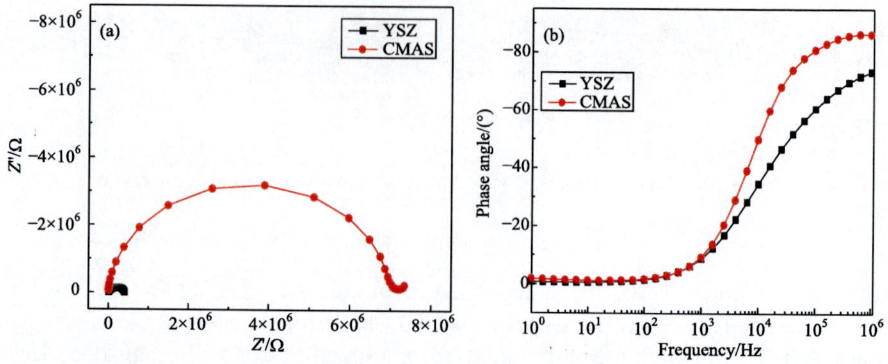

Fig. 12.36 Complex impedance spectra of YSZ and CMAS: **a** Nyquist plot; **b** Bode plot

than that for the YSZ layer, suggesting a difference between the resistivities of the CMAS and the YSZ layer.

To determine the resistivities of YSZ and CMAS (ρ_{YSZ} and ρ_{CMAS}), the measured complex impedance spectra were fitted to circuits using ZView to obtain the approximate complex impedances of YSZ and CMAS (Z_{YSZ} and Z_{CMAS}). Figure 12.37 shows the fitted curves and the fitting circuits. The Nyquist plots show that there is good agreement between the fitted and measured curves, suggesting the reliability of the selected fitting circuits. In the circuits, the subscripts YSZ, CMAS, pores, and electrode signify the ceramic layer, the CMAS layer, the pores, and the electrode response, respectively. The resistance R is related to the resistivity ρ through the following equation:

$$R = \frac{\rho h}{A} \tag{12.56}$$

where h is the thickness of the tested specimen and A is the electrode area. ρ_{YSZ} and ρ_{CMAS} are calculated to be approximately 7.0×10^5 and 2.7×10^6 Ω·m, respectively, which provides data support for testing the CMAS-corroded TBC by CIS and lays a foundation for establishing its equivalent circuit.

12.5.3 *Complex Impedance Response of the CMAS-Corroded TBC*

Even when uncorroded by CMAS, a YSZ coating undergoes microstructural changes at high temperatures due to sintering. To analyze the effect of CMAS corrosion, let us first characterize the complex impedance response of the TBC sintered at 1250 °C. Figure 12.38a, b shows the Nyquist and Bode plots for the independent ceramic-layer specimens that were not corroded by the CMAS and that were held at 1250 °C

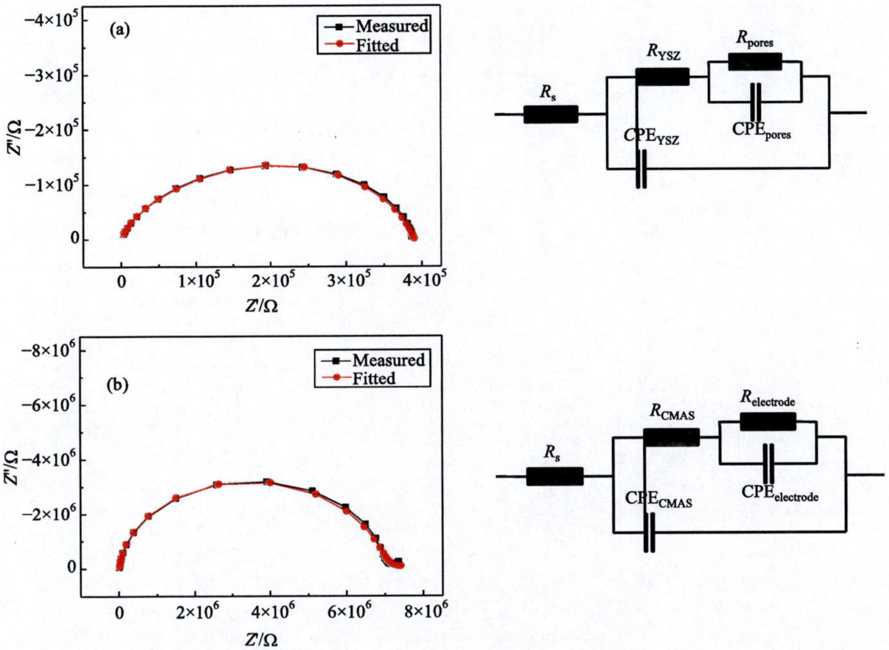

Fig. 12.37 Fitted curves and equivalent circuits for **a** YSZ and **b** CMAS

for different periods of time, respectively. As shown in Fig. 12.38a, there is only one semicircle for the ceramic-layer specimen held at 1250 °C for 0 h (hereinafter referred to as the 0 h specimen). In contrast, in addition to a semicircle circle in the high-frequency range, the diameter of which is significantly smaller than that of the semicircle corresponding to the 0 h specimen, there is a tail arc in the low-frequency range for each of the ceramic-layer specimens held at 1250 °C for 2–16 h, respectively. This is because a large number of pores and cracks shrank or even vanished during sintering, resulting in a decrease in the impedance of the ceramic layer. Moreover, it is indicated that the impedance in the high-frequency range is affected by the pores in a coating. Simulations with equivalent circuits find that, as seen in Fig. 12.38c, there is a sharp decrease in the resistance of the YSZ coating, R_{YSZ}, during the initial stage of sintering (e.g., the first 2 h). As the YSZ sintering time increases, there is a further decrease in the resistance of the pores but at a lower rate. This is because a large number of pores melt and sinter at a high temperature within a short period of time; furthermore, in the absence of the thermal mismatch effect of the transition layer and substrate, no cracks or pores are formed in the YSZ coating due to mismatch stresses.

The CMAS-coated YSZ coating corrodes at 1250 °C. Specifically, the CMAS first melts and then infiltrates into the YSZ coating, fills its pores, and undergoes chemical reactions with it, thereby altering its impedance performance. Figure 12.39a, b shows the Nyquist and Bode plots of the impedance spectra of the YSZ coating corroded

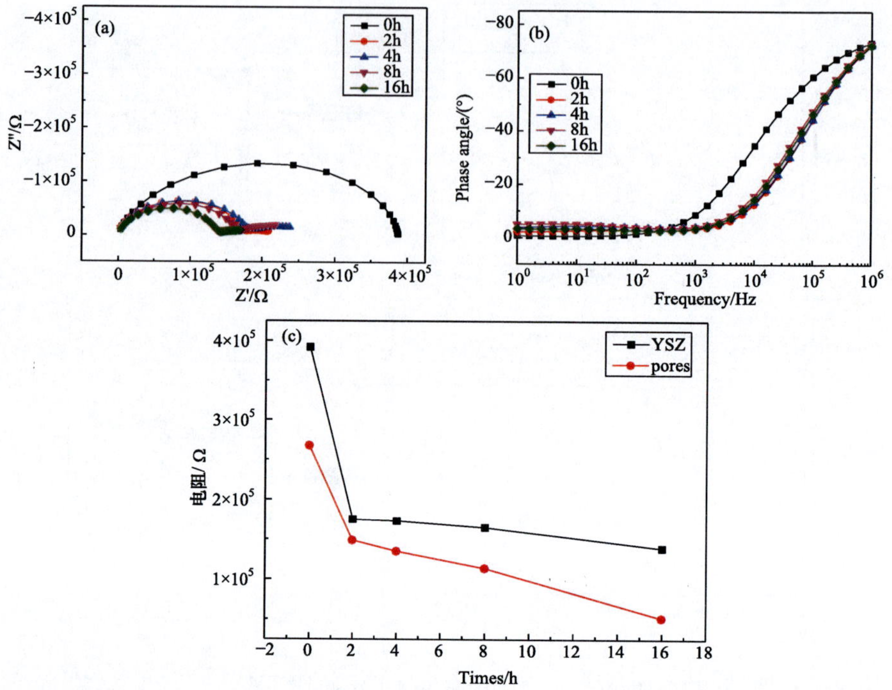

Fig. 12.38 **a** Nyquist plot and **b** Bode plot of the impedance of the YSZ coating after sintering for different periods of time; **c** variation in the resistances of the YSZ coating and the pores with sintering time

at 1250 °C for different periods of time, respectively. As shown in Fig. 12.39a, the impedance response of the uncorroded ceramic-layer specimen is represented by a semicircle and that of each CMAS-corroded ceramic-layer specimen is similarly represented by a semicircle, the diameter of which is, however, significantly larger than that of the semicircle corresponding to the uncorroded specimen. As the temperature-holding time increases from 2 to 16 h, there is an increase in the impedance response of the ceramic layer, which is reflected by a gradual increase in the diameter of the semicircle. From Fig. 12.38, we know that the Z_{YSZ} gradually decreases for the YSZ coating subjected only to sintering, indicating that CMAS corrosion significantly increases Z_{YSZ} and that CMAS can continuously melt and infiltrate into a coating as long as there is a sufficient amount of CMAS. Figure 12.39b shows that the peak phase angle for each corroded specimen in the high-frequency range is significantly greater than that of the uncorroded specimen. Specifically, the peak phase angle for the YSZ coating increases from 70° to 80° after CMAS corrosion. A comparison with Fig. 12.38b shows consistency between the peak phase angles of the CMAS-corroded specimens and the CMAS. Thus, it can be inferred that CMAS corrosion has a significant impact on the electrical performance of YSZ. Moreover, a new peak is observed near 10^3 Hz in Fig. 12.39b, suggesting that a new

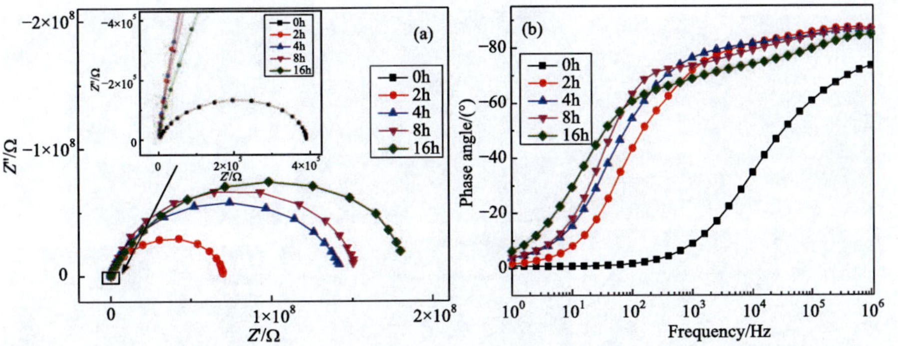

Fig. 12.39 **a** Nyquist plot and **b** Bode plot of the impedance response of the CMAS-corroded YSZ coating

material is formed in the YSZ layer. The frequency at the peak shifts toward the low-frequency range with an increase in oxidation time. Deng et al. [10] found that the shift of the left side of the peak phase angle for a coating to the low-frequency range is caused by the microstructural changes and is accompanied by a change in its conductivity and that the shift of the right side of the peak phase angle for a coating to the high-frequency range is due to the effect of its thickness and is not accompanied by a change in its conductivity. Hence, we can infer that for each CMAS-corroded YSZ-coating specimen, the appearance of a new peak phase angle at 10^3 Hz is a result of the formation of new material from the chemical reaction between the YSZ coating and the CMAS that infiltrated into the interior of the YSZ coating. The resistivity of the newly formed material differs from those of CMAS and YSZ.

Figure 12.40 shows microstructural images of the CMAS-corroded TBC. As shown in Fig. 12.40a, after 0.5 h, the CMAS infiltrates into the coating to a depth of 70 μm, and the pores in the CMAS-infiltrated part of the coating notably decrease in number and become dense. During the infiltration process, the molten CMAS and the YSZ coating undergo chemical reactions in the high-temperature environment to form a mixed layer with an impedance performance different from that of the YSZ coating and CMAS. Figure 12.40b, c show microstructural images of the coatings subjected to 2 and 16 h of CMAS corrosion, respectively (note that the left side is the upper surface of the coating). No clear CMAS infiltration boundary can be seen in these two images, indicating that the CMAS infiltrates into the whole coating, as verified by the energy spectra in Fig. 12.40d–f. Despite a more pronounced elemental accumulation on the upper surface, the elements are distributed across the coating. However, as the corrosion time increases, there is a continual increase in the element concentration, again indicating that at a sufficient amount, CMAS can continually infiltrate into a coating. Moreover, Fig. 12.40 shows that notable cracks are formed in the coating after 2 and 16 h of corrosion, indicating that despite the absence of the thermal mismatch between the transition layer and the substrate, whether CMAS infiltrated into the coating and the difference in the CMAS concentration can both cause mismatch stresses in the coating, and therefore, cause it to spall.

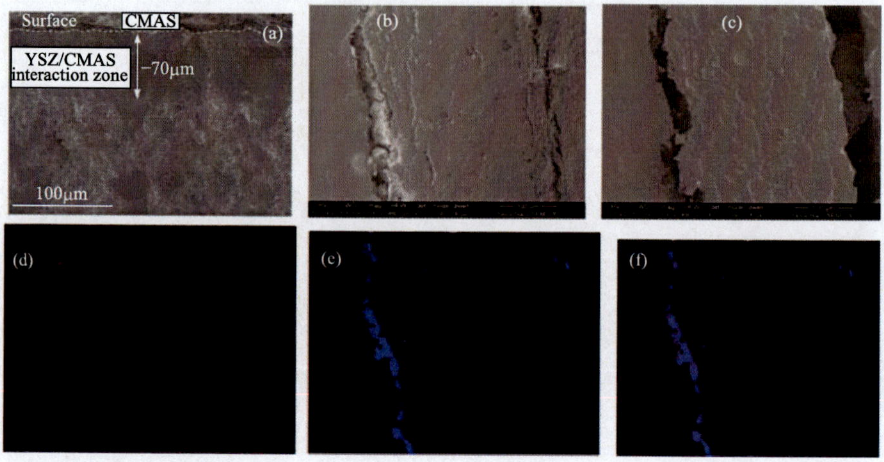

Fig. 12.40 Microstructural evolution and composition distribution of the YSZ coating subjected to high-temperature CMAS corrosion for **a** 0.5 h; **b** 2 h; **c** 16 h; **d** Distribution of Ca after 16 h; **e** Distribution of Si after 2 h; and **f** Distribution of Si after 16 h

To determine the impedances of the YSZ layer and the mixed layer (Z_{YSZ} and Z_{Mix}) after CMAS corrosion, the complex impedance spectrum of the YSZ coating was fitted to a circuit using ZView. Considering that the impedance spectrum mainly reflects the impedance characteristics of the ceramic and mixed layers, an equivalent circuit was established for the CMAS-corroded TBC, as shown in Fig. 12.41 (where the subscripts YSZ and Mix signify the YSZ layer and the mixed layer, respectively).

Figure 12.42 shows the fitting results. Both Z_{YSZ} and Z_{Mix} increase rapidly as the corrosion time increases from 0 to 4 h and they tend to stabilize as the corrosion time increases beyond 4 h. This phenomenon occurs for the following reason. During the initial stage of oxidation, the porosity of the ceramic layer is relatively large. As a result, CMAS is able to rapidly infiltrate into the interior of the ceramic layer through the pores and reacts with the YSZ, leading to a rapid increase in Z_{YSZ}. After 4 h, the porosity of the ceramic layer further decreases, which is accompanied by a decrease in the amount of CMAS that infiltrated into the coating, the phase transformation of the YSZ, and a decrease in the rate of the chemical reaction between the YSZ and CMAS. As a result, Z_{YSZ} tends to stabilize.

In addition, CIS has been employed to characterize the oxidation performance of new TBCs (e.g., Gd_2O_3-Yb_2O_3 [43] and $Sm_2SrAl_2O_7$ [44]) and to determine

Fig. 12.41 An equivalent circuit for the CMAS-corroded TBC

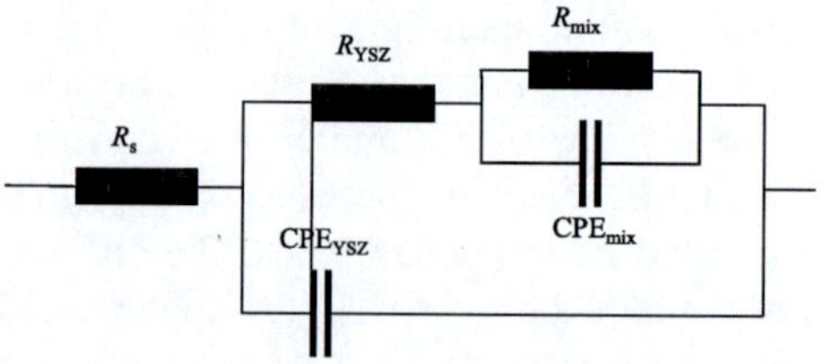

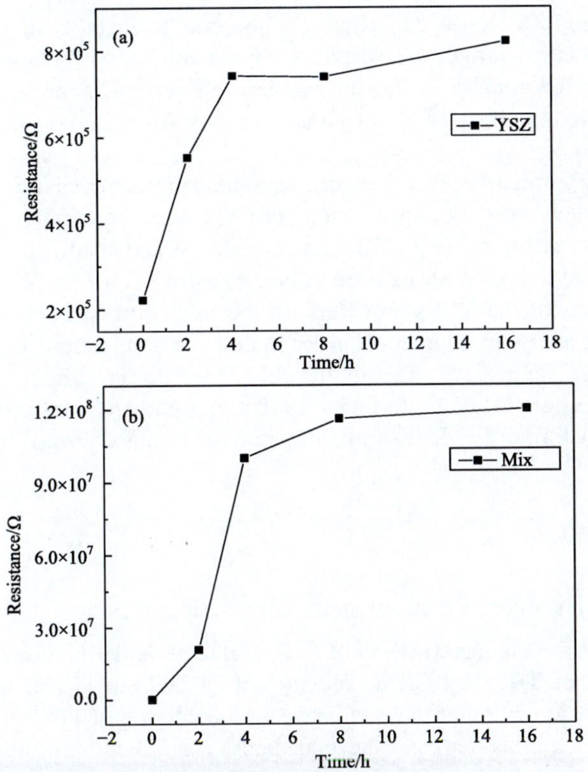

Fig. 12.42 Variation in **a** Z_{YSZ} and **b** Z_{Mix} with CMAS corrosion time

the mechanical performance [45, 46] and damage [47, 48] of TBCs. Readers are encouraged to refer to relevant literature.

12.6 Summary and Outlook

12.6.1 Summary

This chapter describes real-time methods for characterizing the interfacial oxidation and microstructural evolution of TBCs by CIS. The following is a summary of this chapter:

(1) For a TBC, the impedance response frequencies of the interfacial oxide layer and the ceramic layer can be clearly differentiated. Therefore, CIS is an effective tool for characterizing the interfacial oxidation and microstructural evolution of TBCs.
(2) The complex impedance response characteristics of the microstructure of a TBC before and after interfacial oxidation are theoretically derived. The response

of the microstructural evolution of the ceramic layer occurs mainly in the high-frequency range, the response of the thickness evolution of the TGO layer appears mainly in the intermediate-frequency range, and the response in the low-frequency range originates mainly from factors such as pores and electrodes.

(3) FE and experimental tests show that asymmetric electrodes (used for measuring the complex impedance spectrum of TBCs) cause errors in the characterization of the microstructure (e.g., TGO layer thickness and coating porosity) of TBCs by CIS and that such errors need to be corrected.

(4) The effects of the CIS parameters on the microstructural characterization of TBCs are analyzed in detail. It is found that test temperature and electrode size are the key parameters affecting the characterization results.

(5) The evolution of the interfacial oxidation, porosity, and CMAS-corroded microstructure of EB-PVD and APS coatings is characterized by CIS.

12.6.2 Outlook

Future research should be focused on the following areas:

(1) Currently, the characterization of the interfacial oxidation and microstructural evolution of TBCs by CIS is mostly not carried out in real time. Real-time CIS characterization of the microstructural evolution of TBCs under simulated service conditions (e.g., gas thermal shock) is a direction of future research.

(2) The structure of turbine blades is a key factor to be considered in the characterization of the interfacial oxidation and microstructural evolution of TBCs. Currently, research is focused mostly on the characterization of coatings at a specimen level. CIS characterization methods that consider the turbine blade structure and the effect of curvature on characterization results along with the correction of relevant errors are important development trends.

(3) It is also important to characterize the interfacial oxidation and microstructure evolution of new TBCs by CIS. The currently available YSZ coatings can operate only at temperatures below 1200 °C, and therefore, are unable to meet the requirements of engine development. Studying the oxidation performance and CMAS corrosion resistance of next-generation TBCs through CIS characterization is expected to provide valid data for the development and application of new coatings.

References

1. M. Shi. Principle and application of alternating current impedance spectroscopy. National Defense Industry Press (Beijing), 2000.
2. Ali MS, Song SH, Xiao P. Evaluation of degradation of thermal barrier coatings using impedance spectroscopy. J Eur Ceram Soc. 2002;22:101–7.
3. Nitin PP, Gell M, Jordan EH. Thermal barrier coatings for gas-turbine engine applications. Science. 2002;296:280–4.

4. Quadakkers WJ, Shemet V, Sebold D, Anton R, Wessel E, Singheiser L. Oxidation characteristics of a platinized MCrAlY bond coat for TBC systems during cyclic oxidation at 1000 °C. Surf Coat Technol. 2005;199:77–82.
5. Czech N, Fietzek H, Juez-Lorenzo M, Kolarik V, Stamm W. Studies of the bond-coat oxidation and phase structure of TBCs. Surf Coat Technol. 1999;113:157–64.
6. Ogawa K, Minkov D, Shoji T, Sato M. Hashimoto H. NDE of degradation of thermal barrier coating by means of impedance spectroscopy. NDE&E Int. 1999,32:177–185.
7. Song SH, Xiao P. An impedance spectroscopy study of high-temperature oxidation of thermal barrier coatings. Mater Sci Eng B. 2003;97:46–53.
8. Wang X, Mei J, Xiao P. Non-destructive evaluation of thermal barrier coatings using impedance spectroscopy. Surf Coat Technol. 2001;21:855–9.
9. Fleig J, Rodewald S, Maier J. Microcontact impedance measurements of individual highly resistive grain boundaries: General aspects and application to acceptor-doped $SrTiO_3$. J Appl Phy. 2000;87:2372–81.
10. Deng LF, Xiong YS, Xiao P. Modelling and experimental study of impedance spectra of electron beam physical vapour deposition thermal barrier coatings. Surf Coat Technol. 2007;201:7755–63.
11. COMSOL Multiphysics 4.3 User's Manual.
12. Joo JH, Choi GM. Electrical conductivity of YSZ film grown by pulsed laser deposition. Solid State Ionics. 2006;177:1053–7.
13. Zhao W, Kim IJ, Gong J. Influence of thickness on the electrical conductivity of YSZ electrolytes. J Ceram Soc Jpn. 2010;118:550–4.
14. Tanno M, Ogawa K, Shoji T. Influence of asymmetric electrode geometry on an impedance spectrum of a plasma-sprayed thermal barrier coating system. Surf Coat Technol. 2010;204:2504–9.
15. Henn FEG, Buchanan RM, Jiang N, Stevenson DA. Permittivity and AC conductivity in yttria-stabilized zirconia. Appl Phy A. 1995;60:515–9.
16. Yang F, Shinmi A, Xiao P. Electrical and dielectric properties of thermally grown oxide (TGO) on fecralloy substrate studied by impedance spectroscopy. Adv Ceram Coat Interfaces IV, (2010) 87–95.
17. Li C. Complex impedance spectroscopy measurement technology and application for high temperature performance evaluation of thermal barrier coatings. Master's thesis, Xiangtan University, 2015.
18. Wang X, Mei J, Xiao P. Determining oxide growth in thermal barrier coatings (TBCs) non-destructively using impedance spectroscopy. J Mater Sci Lett. 2001;20:47–9.
19. Wang NG, Li CF, Yang L, Zhou YC, Zhu W, Cai CY. Experimental testing and FEM calculation of impedance spectra of thermal barrier coatings: Effect of measuring conditions. Corros Sci. 2016;107:155–71.
20. Wang Z, Zhao M, Wu R, Xing Y, Yang J, Wang M, Li Z, Song N, Wan C, Pan W. Non-destructive evaluation of thermally grown oxides in thermal barrier coatings using impedance spectroscopy. J Europ Ceramic Soc. 2019;39:5048–5058.
21. Yang L, Zhu W, Li CF, Zhou YC, Wang NG, Wei YG. Error and modification in thermal barrier coatings measurement using impedance spectroscopy. Ceram Int. 2017;43:4976–83.
22. Anderson P, Wang X, Xiao P. Impedance spectroscopy study of plasma sprayed and EB-PVD thermal barrier coatings. Surf Coat Technol. 2004;185:106–19.
23. Chen W. Complex impedance spectrum measurement technique for evaluating the high-temperature performance of thermal barrier coatings and its applications. Master's thesis, Xiangtan University, 2014.
24. Yang X, Weia L, Lia J, Zhang B, Wang S, Guo H. Microstructural evolution of plasma spray physical vapor deposited thermal barrier coatings at 1150 °C studied by impedance spectroscopy. Ceram Int. 2018;44:10797–805.
25. Chen Q, Liu M, Zhang J, Deng Z, Mao J. High-temperature oxidation behavior and analysis of impedance spectroscopy of 7YSZ thermal barrier coating prepared by plasma spray-physical vapor deposition. Chinese J Aeronaut. 2018;31(8):1764–1773.
26. Heung R, Wang X, Xiao P. Characterisation of PSZ/Al2O3 composite coatings using electrochemical impedance spectroscopy. Electrochim Acta. 2006;51:1789–96.
27. Song SH, Xiao P, Weng LQ. Evaluation of microstructural evolution in thermal barrier coatings during thermal cycling using impedance spectroscopy. J Eur Ceram Soc. 2005;25:1167–73.

28. Jayaraj B, Desai VH, Lee CK, Sohn YH. Electrochemical impedance spectroscopy of porous ZrO2–8 wt.% Y2O3 and thermally grown oxide on nickel aluminide. Mater Sci Eng A. 2004;372:278–286.
29. Byeon JW, Jayaraj B, Vishweswaraiah S, Rhee S, Desai VH, Sohn YH. Non-destructive evaluation of degradation in multi-layered thermal barrier coatings by electrochemical impedance spectroscopy. Mater Sci Eng A. 2005;407:213–225.
30. Steil MC, Thevenot F. Densification of Yttria-Stabilized Zirconia impedance specfroscopy analysis. J Electrochem Soc. 1997;144(1):390–8.
31. Zhang C, Zhou C, Gong S, Li H, Xu H. Evaluation of thermal barrier coating exposed to different oxygen partial pressure environments by impedance spectroscopy. SurfaceCoatings Technol. 2006;201:446–451.
32. Younshonics TM. Overview of thermal barrier coatings in diesel engines. J Therm Spray Technol. 1997;616:50–6.
33. Yang F, Xiao P. Nondestructive evaluation of thermal barrier coatings using impedance spectroscopy. Int J Appl Ceram Technol. 2009;6(3):381–99.
34. Anderson S, Wang X, Xiao P. Effect of isothermal heat treatment on Plasma-Sprayed Yttria stabilized Zirconia studied by impedance spectroscopy. J Am Ceram Soc. 2005;88(2):324–330 (2005)
35. Yang L. A study on nondestructive evaluation of the oxidation, damage, and fracture of thermal barrier coatings. Doctoral dissertation, Xiangtan University, 2007.
36. Zou J. Theoretical thermodynamic analysis of the interfacial oxidation of thermal barrier coatings. Xiangtan: Xiangtan University; 2004.
37. Zhan J, Desai V. Evaluation of thickness, porosity and pore shape of plasma sprayed TBC by electrochemical impedance spectroscopy. Surf Coat Technol. 2005;190:98–109.
38. Huang H, Liu C, Nl L, Zhou C. Evaluation of TGO growth in thermal barrier coatings using impedance spectroscopy. Rare Metals. 2011;30(Spec. Issue):643.
39. Zhang J, Desai V. Determining thermal conductivity of plasma sprayed TBC by electrochemical impedance spectroscopy. Surf Coat Technol. 2005;190:90–7.
40. Liu C, Huang H, NI L, Zhou C. Evaluation of thermal barrier coatings exposed to hot corrosion environment by impedance spectroscopy. Chinese J Aeronaut. 2011;24:514–519.
41. Huang H, Liu C, Ni L, Zhou C. Evaluation of microstructural evolution of thermal barrier coatings exposed to Na2SO4 using impedance spectroscopy. Corros Sci. 2011;53:1369–74.
42. Wu J, Guo HB, Abbas M, Gong SK. Evaluation of plasma sprayed YSZ thermal barrier coatings with the CMAS deposits infiltration using impedance spectroscopy. Progress Nat Sci Mater Int. 2012;22(1):40–7.
43. Zhang D-H, Guo H-B, Gong S-K. Impedance spectroscopy study of high-temperature oxidation of Gd2O3-Yb2O3 codoped zirconia thermal barrier coatings. Trans Nonferrous Met Soc China. 2011;21:1061–1067.
44. Baskaran T, Arya SB. Influence of ceramic top coat and thermally grown oxide microstructures of air plasma sprayed Sm2SrAl2O7 thermal barrier coatings on the electrochemical impedance behavior. Surface Coat Technol. 2018;344:601–613.
45. Gómez-García J, Rico A, Garrido-Maneiro MA, Múnez CJ, Poza P, Utrilla V. Correlation of mechanical properties and electrochemical impedance spectroscopy analysis of thermal barrier coatings. Surface Coat Technol. 2009;204:812–815.
46. Hilpert T, Ivers-Tiffe´e E. Correlation of electrical and mechanical properties of zirconia based thermal barrier coatings. Solid State Ionics 175 (2004) 471–476.
47. Jayaraj B, Vishweswaraiah S, Desai VH, Sohn YH. Electrochemical impedance spectroscopy of thermal barrier coatings as a function of isothermal and cyclic thermal exposure. Surf Coat Technol. 2004;177–178:140–51.
48. Wu NQ, Ogawa K, Chyu M, Mao SX. Failure detection of thermal barrier coatings using impedance spectroscopy. Thin Solid Films. 2004;457:301–306.

Chapter 13
Nondestructive Testing of the Surface and Interfacial Damage and Internal Pores of TBCs

Interfacial defects, strain fields, and porosity are key parameters affecting the thermal insulation performance and service life of thermal barrier coatings (TBCs). During service, the continual extension of interfacial defects, sudden changes in porosity, and increases in stress can all cause coating spallation and failure. Seeking an effective method for examining the internal state of TBCs is vitally important for preventing accidents and losses due to sudden TBC spallation and failure. Conventionally, coating quality evaluation and property characterization are achieved mostly through destructive testing, with stringent requirements for the testing environment. Ongoing efforts have been devoted to finding a simple, economical, and rapid nondestructive method for the quality control of as-prepared coatings and state diagnosis of in-service coatings to provide a direct basis for correctly understanding the failure mechanism of coatings and accurately predicting their service life. To date, a multitude of advanced testing techniques (e.g., acoustic emission (AE), X-ray diffraction (XRD), thermography, AC complex impedance spectroscopy (CIS), Cr^{3+} piezoelectric spectroscopy, infrared (IR) spectroscopy, and synchrotron radiation scanning) have been successfully employed in research to locate damage in TBCs, to determine their interfacial oxidation behavior, and to detect their interfacial delamination and stress evolution [1–6]. However, due to the complex failure mechanisms of TBCs, the difficulty in the prediction of their service life, and their increasingly wide application, nondestructive testing (NDT) techniques remain an important topic of research in the TBC field in the future. The reliability evaluation of TBCs depends, to a significant extent, on the type and reliability of NDT data. Therefore, developing economic, convenient, effective, and highly reliable testing schemes is a crucial step in TBC research.

This chapter focuses on analyzing the application of digital image correlation (DIC), X-ray computed tomography (CT), and IR thermography (IRT) in TBC failure monitoring; characterizing the deformation field, porosity, interfacial defects, and thickness of TBC specimens during testing; and providing experimental means for analyzing the failure mechanism of TBCs.

Y. Zhou et al., *Thermal Barrier Coatings: Failure Theory and Evaluation Technology*,
https://doi.org/10.1007/978-981-19-2723-2_13

13.1 Characterization of the Strain Fields of TBCs Using DIC

This section focuses on analyzing the application of DIC in TBC failure monitoring. Through the characterization of the deformation field of TBC specimens under tensile and three-point bending conditions in conjunction with NDT methods (e.g., AE), a digital speckle/AE-combined failure criterion analysis method for TBCs is formulated. The evolution of the deformation and strain fields of TBC specimens at room and high temperatures and under high-temperature calcium–magnesium aluminosilicate (CMAS) corrosion is determined. Experimental methods for analyzing the failure mechanism of TBCs are provided.

13.1.1 Basic Principle of DIC Characterization of the Strain Field

The testing principle of DIC is the processing of digital images of the surface of a test object at adjacent stages collected under different conditions to obtain pixel deformation information. DIC is an optical testing technique established based on digital image processing and numerical calculation [7–12]. In DIC, the distribution of the grayscale values of an image of an object contains displacement and deformation information. In the computer science field, each pixel in a grayscale digital image has only one sampling color. This type of image is often displayed in grayscale from the darkest black to the brightest white, even though in theory, this sampling can be of different shades of any color or even different colors at different brightness levels. A color image contains a large amount of information, which a computer processes slowly. In comparison, a grayscale image has only one channel. As a result, it is possible to completely extract the feature information from a target image while increasing the processing speed. Generally, digital images of a test object before and after deformation are referred to as the reference and target images, respectively. The testing principle is specifically described as follows. An arbitrary square region with the pixel to be solved, P(x_0, y_0), as the center is selected from the reference image as the subregion of the reference image. To calculate the displacement of point P, a square reference subset with P as the center and a total of (2 M + 1) × (2 M + 1) pixels are selected from the reference image and used to track the the corresponding position of point P in the deformed image. Then, calculations are performed on the target image based on the relevant predefined correlation function. A certain search method is used to find the subregion with P'(x_0', y_0') as the center that has the largest or smallest correlation coefficient with the subregion of the reference image. On this basis, the displacement components of point P in the subregion of the reference image in the *x*- and *y*-directions, *u* and *v*, respectively, are determined. In the calculation process, the region to be calculated in the reference image is generally divided into many virtual mesh cells, and the displacement of the whole region is then

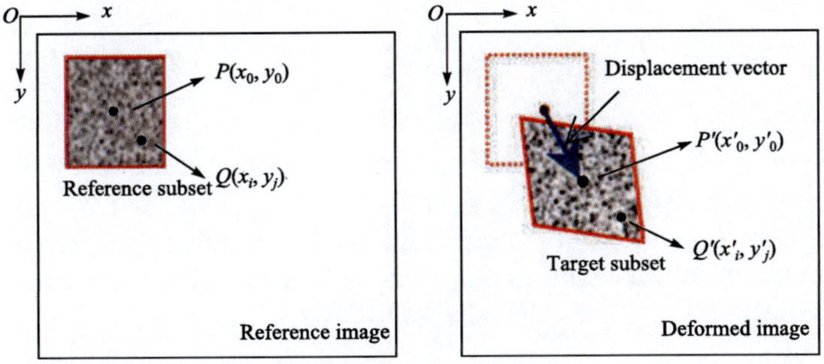

Fig. 13.1 Schematic diagram of the basic principle of DIC [14]

determined based on the displacement of each mesh node [13]. If there are only rigid-body translations between the reference and deformation subsets (i.e., all the points in the subset undergo the same displacement), a zero-order function can be used [7]. Figure 13.1 shows that the subregion of the target image differs from the subregion of the source image in both the location of the center (i.e., displacement) and the shape (i.e., deformation). When the transformation from the subregion of the reference image to the subregion of the target image involves tensile, compressive, or shear deformation or rigid-body rotation, a first-order function is used in the calculation. There is a one-to-one functional relation between point Q(x, y) in the subregion of the reference image and the corresponding point Q'(x', y') in the subregion of the target image as follows [14, 15]:

$$\begin{aligned} x' &= x_0 + \Delta x + u + u_x \Delta x + u_y \Delta y \\ y' &= y_0 + \Delta y + v + v_x \Delta x + v_y \Delta y \end{aligned} \tag{13.1}$$

where Δx and Δy are the distances between point Q(x, y) and the center of the subregion of the reference image (i.e., P(x_0, y_0)) in the x- and y-directions, respectively; u and v are the displacements of the center of the subregion of the reference image in the x- and y-directions, respectively; and u_x, u_y, v_x, and v_y are the displacement gradients of the subregion of the image.

To evaluate the similarity between the reference and deformation subsets, it is necessary to first define a cross-correlation (CC) criterion or a sum-of-squared-difference (SSD) correlation criterion and then complete the matching process by searching for the relevant correlation coefficient. Research [7, 16] shows that the zero-mean normalized SSD (ZNSSD) correlation function is equivalent to the commonly used zero-mean normalized CC function. However, the ZNSSD function is little affected by linear changes in the grayscale of the subregion of the target image, is highly resistant to interference, and can be easily optimized. As a result, the ZNSSD correlation function is often used to evaluate the similarity between image subregions before and after deformation. The specific form of the ZNSSD function is as follows:

$$C_{ZNSSD}(\mathrm{p}) = \sum_{x=-M}^{M}\sum_{y=-M}^{M}\left[\frac{f(x,y)-f_m}{\sqrt{\sum_{x=-M}^{M}\sum_{y=-M}^{M}[f(x,y)-f_m]^2}} - \frac{g(x',y')-g_m}{\sqrt{\sum_{x=-M}^{M}\sum_{y=-M}^{M}[g(x',y')-g_m]^2}}\right]^2 \quad (13.2)$$

where $\mathbf{p} = [u, u_x, u_y, v, v_x, v_y]^T$ is the parameter vector that describes the deformation state of the subregion of the image, $f(x, y)$ and $g(x', y')$ are the grayscale values of the pixels in the subregions of the reference and target images, respectively, and f_m and g_m are the mean grayscale values of the subregions of the reference and target images, respectively, and are calculated as follows:

$$f_m = \frac{1}{(2M+1)^2}\sum_{x=-M}^{M}\sum_{y=-M}^{M}[f(x,y)]^2 \quad g_m = \frac{1}{(2M+1)^2}\sum_{x=-M}^{M}\sum_{y=-M}^{M}[g(x',y')]^2 \quad (13.3)$$

Finally, the displacement field in the subregion of the image is determined through calculation, and the full-field displacement field and strain field are obtained successively.

13.1.2 Preparation of Digital Speckles

Speckles are the carrier of the surface deformation information of objects. Speckles can generally be categorized into three types, namely laser speckles, artificial speckles, and natural speckles, as shown in Fig. 13.2. Laser speckles refer to speckles with different random brightness levels formed from diffuse reflection interference when a laser beam irradiates the rough surface of an object. This speckle preparation method is suitable for measuring small deformation but is expensive. Natural speckles, as their name implies, refer to some textures on the surface of an object that can serve as random speckles. This type of speckle does not require time or effort to prepare but is present only in certain objects (e.g., wood).

Artificial speckles are therefore the most commonly used and can be further divided into two types, namely room-temperature speckles and high-temperature speckles. Two methods are available for producing room-temperature speckles. (1) In the first method, a pen is used to manually draw random black dots, which serve as speckles, on the surface of an object. This method is crude with significant limitations and thus is not recommended. (2) The second method is paint spraying. In this method, a specimen is first ground with 1000-grit sandpaper and then ultrasonically cleaned. Subsequently, a white or black matte paint is sprayed onto the surface of the specimen. This speckle preparation method is fast, convenient, and effective. However, the sprayed paint is prone to decoloration, delamination, and spallation during tests conducted at temperatures above 200 °C. Therefore, room-temperature speckles are no longer suitable. Instead, high-temperature speckles are needed.

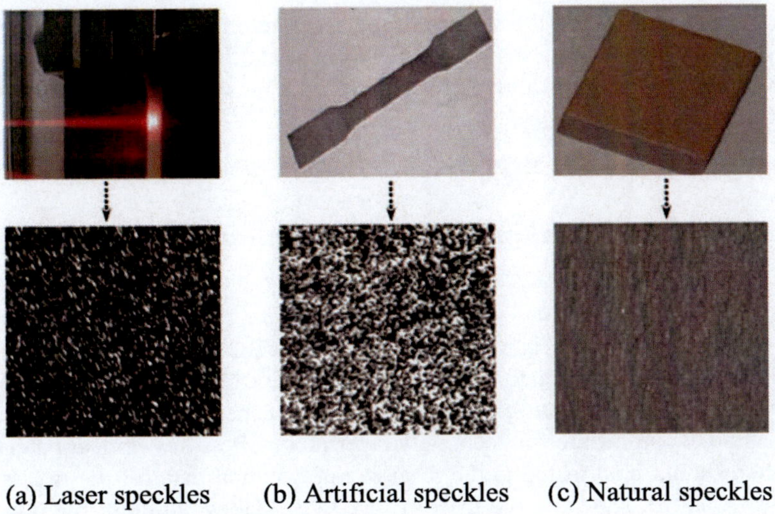

Fig. 13.2 Common speckle types [17]

There are a variety of high-temperature speckle preparation methods, which cannot be clearly classified, including polishing the surface of an object, [16] coating the surface of a material with Ag, [16] and using refractory coatings [18] and high-temperature-resistant adhesives [19]. In the surface polishing method, the surface of a specimen is polished with an abrasive cloth, knife grinders, and files, producing speckles with randomly distributed grayscale values on the surface. This method is easy to implement but difficult to control. The Ag plating method produces elemental Ag through the reactions between $AgNO_3$, a strong ammonia solution, glucose, and purified water. This method is easy to use and effective in producing speckles. However, Ag is unable to withstand temperatures higher than 950 °C, which is its melting point. Refractory coatings are a type of material with Al_2O_3 as the aggregate capable of resisting temperatures as high as 1730 °C. However, because CMAS contains Al_2O_3 and has a viscous texture, speckles can be produced only through splashing using a brush, resulting in speckle particles that are too large and uneven. Therefore, Ag plating is not a suitable method. Speckles can also be prepared by spraying a high-temperature-resistant adhesive. This method is easy to administer and is stable. The high-temperature speckles mentioned in this chapter were produced using this method as follows.

The liquid component of an inorganic adhesive (DB5012) resistant to high temperatures (no lower than 1700 °C) was diluted and mixed well with alcohol at a certain ratio in a dish or a container. Then, CoO powder was weighed according to a weight ratio of 10% and added to the above diluted liquid component. Subsequently, the solution containing CoO powder was transferred to an Iwata W71 spray gun and sprayed under high-pressure air conditions to the surface of a specimen. Afterward, the specimen rested at room temperature for 12–24 h. Finally, to enhance the bond between the speckles and the surface of the specimen, the speckled specimen was placed and held for 2 h in a vacuum drying oven at 150 °C. Then, the specimen was removed from the oven and allowed to cool in the air, thus completing the preparation of high-temperature speckles.

13.1.3 DIC/AE-Combined Method for Failure Criterion Analysis

The AE technique and DIC can be used to analyze a TBC under loading to determine its failure mode, critical fracture time, and fracture locations. Here, an atmospheric plasma spraying (APS) coating subjected to three-point bending is used as an example to illustrate the procedure for accurately determining these key critical parameters. Attention should be directed to the local strain evolution data for the two fracture regions, A and B, shown in the inset of Fig. 13.3. The evolution of the transverse and longitudinal strains (ε_{xx} and ε_{zz}, respectively) in regions A and B with time was extracted using the strain processing software of the ARAMIS system. In addition, the variation in the AE event counts corresponding to different crack patterns with loading time throughout the entire loading process was determined using wavelet analysis (refer to Chap. 10 of the book for details on the AE technique), and the results are shown in Fig. 13.3. In region A, ε_{zz} changes relatively little, whereas ε_{xx} shows a regular variation trend throughout the loading process. During the initial stage of bending loading, the ε_{xx} in region A increases linearly as the deflection increases. As a typical brittle material at room temperature, the ceramic displays notable elasticity under loading. However, there is a clear turning point in ε_{xx} in region A at $\varepsilon_{xx} = 0.75\%$ (i.e., loading time $t = 152$ s). Comparison with the image shows that a noticeable crack appears in region A at this point. During the initial stage of loading, region B is affected by the compressive stress, and ε_{zz} in region B increases linearly as the loading process proceeds. After crack initiation occurs on the inner surface of the ceramic layer, region B is subjected to longitudinal tensile stress, reflecting that the appearance of surface cracks promotes the initiation of interface cracks. At $t = 312$ s, the vertical surface crack reaches the interface and begins to propagate along it. There is a notable turning point in the corresponding ε_{zz} in region B. After the AE signals are subjected to wavelet decomposition transformation, the damage mode of the TBC during the bending failure process is then determined based on the distribution of the maximum value of the wavelet characteristic energy spectrum coefficient of the signals and whether the coefficient value is greater than 0.4. On this basis, the variation in the AE event counts corresponding to various damage modes with loading time is determined, as shown in Fig. 13.3b. At the turning point of ε_{xx} in region A (i.e., $t = 152$ s), there is a sharp increase in the AE event count corresponding to vertical surface cracks, indicating the formation of surface cracks in region A. At the turning point of ε_{zz} in region B (i.e., $t = 312$ s), there is a sharp increase in the AE event count corresponding to interface cracks,

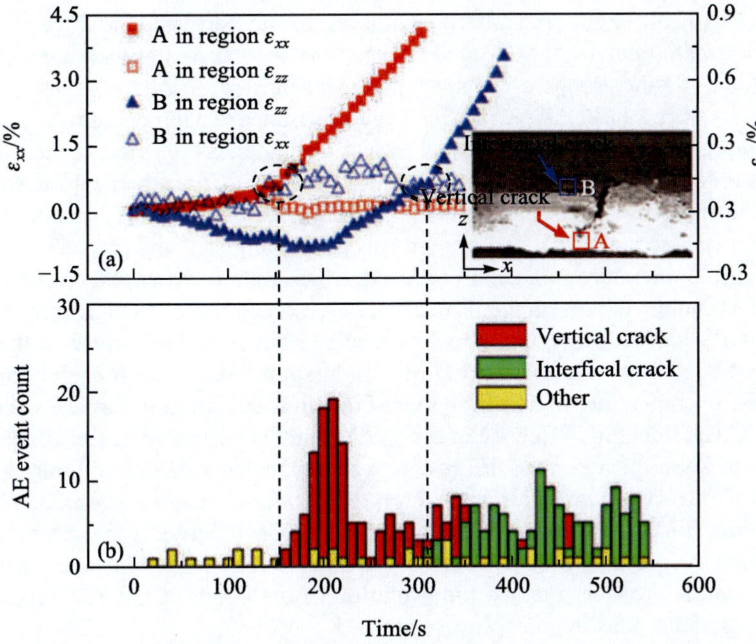

Fig. 13.3 Variation in local strains (**a**) and AE signals (**b**) of a TBC under bending loading with time

suggesting the formation of interface cracks in region B. The critical surface and interfacial fracture times are consistent with those using the DIC test. Thus, key parameters (e.g., the initiation time of cracks of different types, the critical fracture strain, and the evolution of the strain field in the system) during the bending failure process can be accurately determined based on the turning points of the strain.

13.1.4 DIC Characterization of the High-Temperature Strain Field in TBCs

(1) Evolution of the strain field on the TBC surface during the high-temperature tensile process

The surfaces of I-shaped TBC specimens were speckled using the spraying method for preparing high-temperature speckles introduced in Sect. 13.1.2. Each TBC specimen covered with high-temperature speckles was fixed inside a high-temperature furnace using a specially designed high-temperature-resistant fixture. Subsequently, the specimen was loaded using a tensile testing machine. Small holes were prepared in the sidewall of the high-temperature furnace to facilitate real-time, simultaneous DIC measurement of the deformation field on the surface of the specimen, as shown in Fig. 13.4 [20]. High-temperature tensile tests were performed at three temperatures (800, 900, and 1000 °C). Three specimens were tested at each temperature, and a total of four groups of 12 specimens were tested. Prior to the test, the test parameters were adjusted. Specifically, a predetermined test temperature was set, the position-tension mode of the universal testing machine was selected, the tensile rate was set to 0.3 mm/min, an average heating rate of 20 °C/min was selected for the high-temperature furnace, and the imaging frequency of the DIC system was set to one image per second. The DIC system was height-adjusted and placed on the side of the high-temperature furnace. In addition, a blue light filter was placed in front of each CCD camera to prevent the IR thermal radiation generated at high temperatures from interfering with imaging. The lenses were pointed at the surface of the TBC to be tested. A monochromatic LED blue light source was used for light compensation. The temperature and heating rate of the high-temperature furnace were then adjusted. The imaging frequency of the CCD cameras was set after the temperature stabilized. Subsequently, the DIC system was used to capture digital images of the surface of the test specimen during the high-temperature tensile process. The ZNSSD correlation function was used to evaluate the similarity between the subregions of the images before and after deformation to reduce the measurement error caused by thermal disturbance. Finally, the evolution of the strain of the TBC during the high-temperature tensile process was obtained.

We can clearly see from Fig. 13.5 that the strain on the surface of the ceramic layer gradually changes from the initial small compressive strain to a tensile strain. As the tensile test proceeds, strain concentration areas gradually form in the ceramic layer, where cracks eventually initiate and propagate [21]. Notable strain concentration occurs in the strain field within the red circle, where a crack is initiated. Later, as the test proceeds, the crack continually grows and propagates, accompanied by the initiation and growth of new cracks. Thus, this crack can be considered to be the

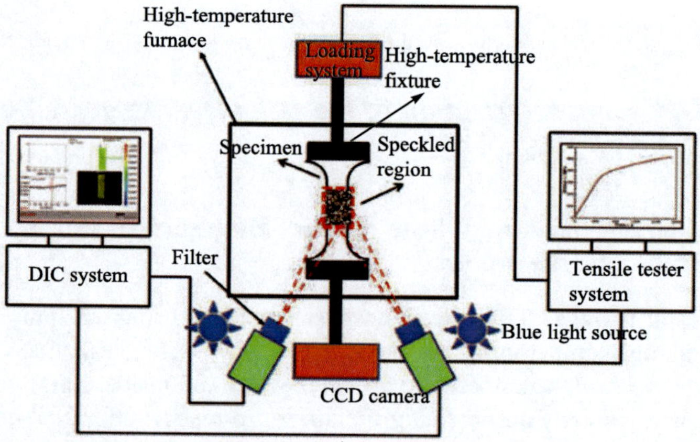

Fig. 13.4 Schematic diagram of a real-time, in situ strain measurement setup for a TBC during the high-temperature tensile process

initial crack in this specimen during the test process. On this basis, three reference points within the strain concentration area where the initial crack had been formed were selected, and their time and strain data were extracted. Then, the time and mean strain corresponding to the initial crack were calculated and fitted to obtain the time–strain curve for the initial crack, as shown in Fig. 13.6 [21]. We can see a notable sudden-change point (i.e., point A) in the strain from the time–strain curve for each initial crack in Fig. 13.6. As the load increases, the ceramic layer releases its internal stress through the formation of cracks. Thus, point A corresponds to the onset of the initiation of the initial crack, and the corresponding strain is the critical fracture strain at the onset of the initiation of the initial crack. The critical fracture strains for the ceramic layer of the electron beam-physical vapor deposition (EB-PVD) TBC specimens at 800, 900, and 1000 °C are found to be 0.40%, 0.33%, and 0.12%, respectively.

(2) **Evolution of the strain field in a TBC during high-temperature three-point bending**

In this section, we introduce the characterization of the cross-sectional deformation and strain fields in an APS TBC during the high-temperature three-point bending process through the high-temperature three-point bending test combined with DIC, as well as analyze the initiation and propagation of surface and interface cracks in the coating based on the contour plots of its cross-sectional strain [20]. The substrate of the APS TBC specimen was made of a Ni-based high-temperature alloy (GH536) and had a thickness of 2 mm. An 80-μm-thick NiCoCrAlY BC layer was sprayed onto the surface of the substrate using the vacuum plasma spraying (VPS) technique. A 1-mm-thick ceramic TC layer was sprayed onto the surface of the BC layer using the APS technique. To allow a single surface crack and a steadily propagating interface crack to form in the coating during the three-point bending process, a 0.5-mm-deep through precrack was cut in advance into its surface. Figure 13.7 shows the dimension of the TBC specimen and its cross-sectional high-temperature speckle pattern. Cross-sectional high-temperature speckles were produced in the APS TBC using the spraying method introduced in Sect. 13.1.2. Then, the TBC specimen with high-temperature speckles was fixed in a high-temperature furnace with a specially designed high-temperature-resistant fixture and subjected to three-point bending loading using a universal material testing machine. Small holes were prepared on the sidewall of the high-temperature furnace to facilitate real-time, simultaneous DIC measurement of the cross-sectional deformation field in the TBC, as shown in Fig. 13.8 [20]. High-temperature three-point bending tests were conducted at four temperatures (30, 400, 600, and 800 °C). The position-loading mode of the universal testing machine was selected. The tensile rate was set to 0.4 mm/min, an average heating rate of 20 °C/min was selected for the high-temperature furnace, and the imaging frequency of the DIC system was set to one image per second.

Figure 13.9 shows the evolution of the cross-sectional strain field in the TBC during the three-point bending process [20]. The tensile ε_{xx} at the tip of the precrack increases continuously as the loading time t increases. At room temperature (T = 30 °C) and t = 37 s, continual strain concentration occurs at the tip of the precrack,

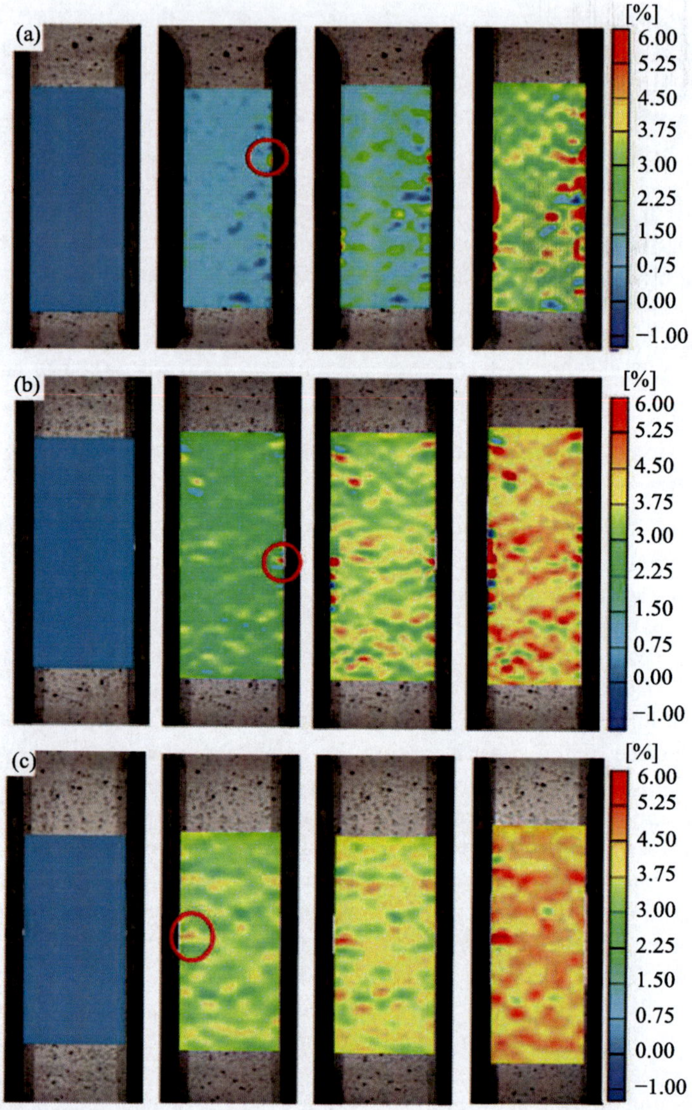

Fig. 13.5 Contour plots showing the evolution of the strain of EB-PVD TBC specimens at **a** 800 °C, **b** 900 °C, and **c** 1000 °C

and ε_{xx} increases to 0.89%, as shown in Fig. 13.9a. Beyond this time point, a crack is initiated at the tip of the precrack and continuously propagates along the precrack. Similar strain evolution is observed at 400, 600, and 800 °C. At $T = 400$ °C and $t = 20$ s, continual strain concentration occurs at the tip of the precrack, and a crack is initiated at $\varepsilon_{xx} = 0.62\%$, as shown in Fig. 13.9b. At $T = 600$ °C and $t = 19$ s, continual strain concentration occurs at the tip of the precrack, and a crack is initiated

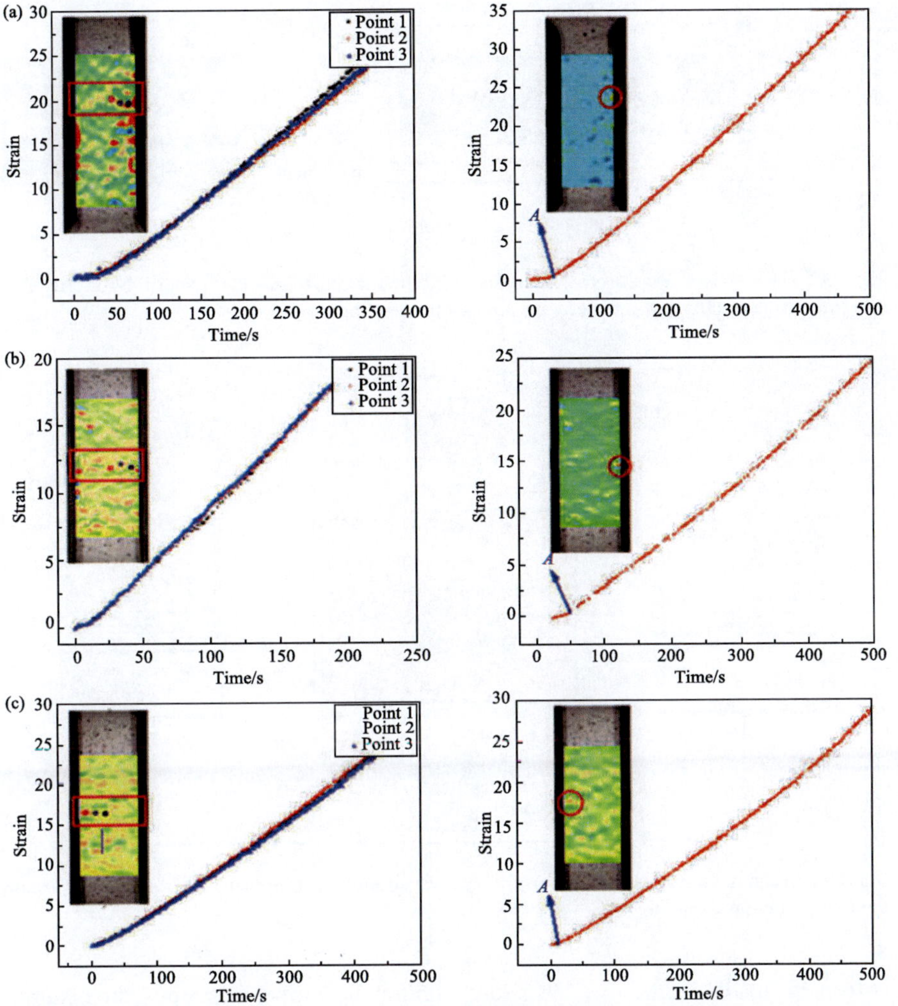

Fig. 13.6 Time–strain curves for the initial cracks in the EB-PVD TBC specimens at **a** 800 °C, **b** 900 °C, and **c** 1000 °C

at $\varepsilon_{xx} = 0.62\%$, as shown in Fig. 13.9c. At $T = 800$ °C and $t = 27$ s, continual strain concentration occurs at the tip of the precrack, and a crack is initiated at $\varepsilon_{xx} = 0.99\%$, as shown in Fig. 13.9d.

Figure 13.10 shows the initiation and propagation of the surface and interface cracks in the TBC during the three-point bending process at room temperature [20]. A surface crack is formed at the tip of the precrack in the precrack-containing TBC specimen after its fracture toughness is reached under external loading ($t = 90$ s). As the load increases, the surface crack begins to propagate toward the interface starting from the tip of the precrack. As the load further increases, an interface crack

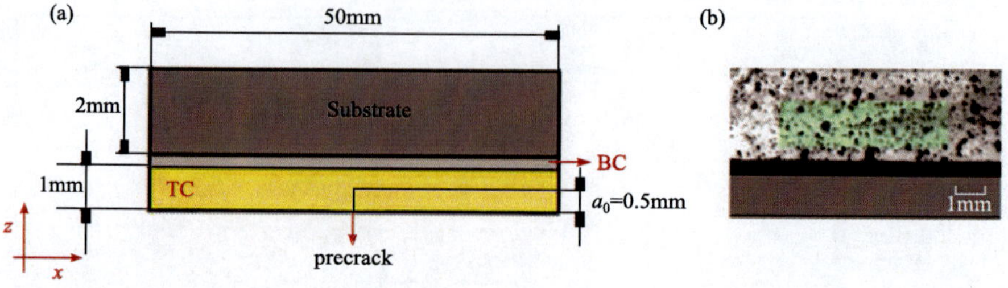

Fig. 13.7 Dimension **a** and cross-sectional high-temperature speckle pattern **b** of the TBC specimen subjected to high-temperature three-point bending testing

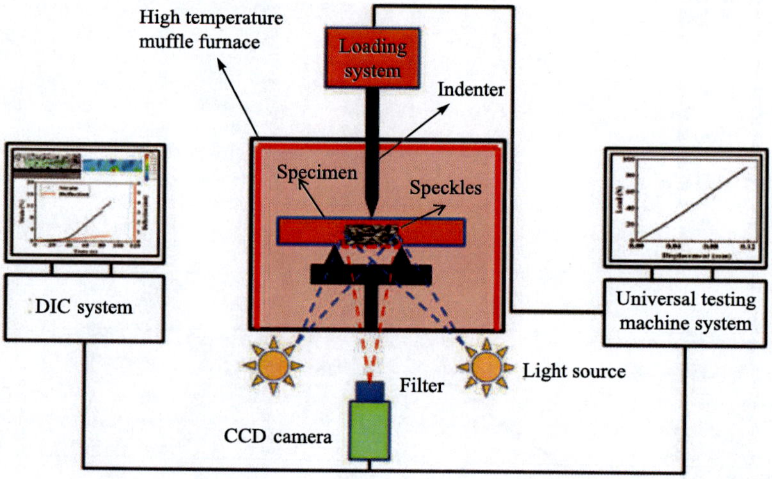

Fig. 13.8 Schematic diagram of a real-time, in situ strain measurement setup for the TBC under three-point bending loading

is formed (loading time $t = 240$ s) and continually propagates along the ceramic-layer/BC-layer interface ($t = 390$ s). Figure 13.11 gives the contour plots showing the evolution of the principal strain $\varepsilon_{\text{principal}}$ during the three-point bending process [20]. $\varepsilon_{\text{principal}}$ increases with loading time. The $\varepsilon_{\text{principal}}$ at each location where a crack appears or propagates is higher than that when there are no cracks. When the surface crack starts from the tip of the precrack, the opening displacement of the surface crack increases as the deflection increases, resulting in an increase in $\varepsilon_{\text{principal}}$, as shown in Fig. 13.11b. When the surface crack propagates to the ceramic-layer/BC-layer interface, an interface crack is formed and subsequently propagates as the deflection increases, resulting in a sharp increase in $\varepsilon_{\text{principal}}$ at the interface, as shown in Fig. 13.11b, c. To determine the location and loading time corresponding to interface crack formation, the variation in the normal strain in the z-direction along the interface crack propagation path, ε_{zz}, with loading time was extracted from the DIC system. As shown in Fig. 13.12, ε_{zz} increases first slowly and then sharply as

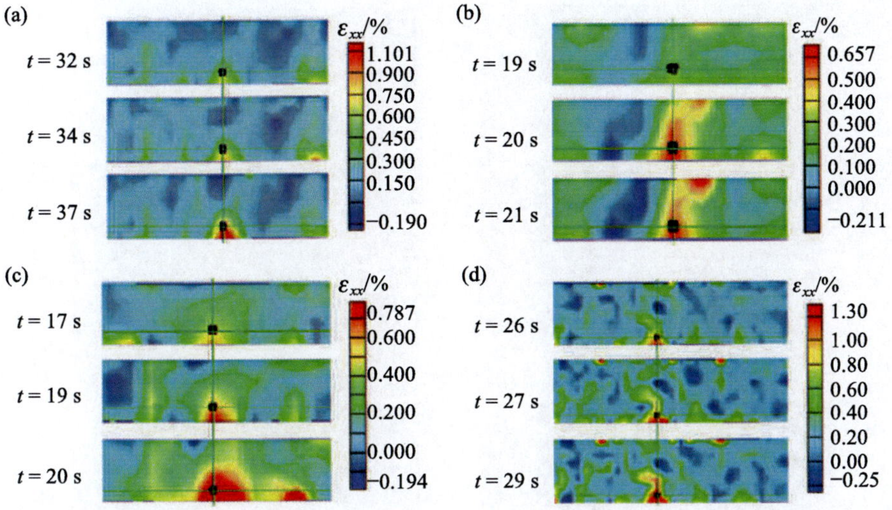

Fig. 13.9 Contour plots showing the evolution of the cross-sectional strain of the TBC during the high-temperature three-point bending process at **a** 30 °C, **b** 400 °C, **c** 600 °C, and **d** 800 °C

the loading time increases [20]. The strain at the inflection point can be viewed as the fracture strain for interface crack formation. On this basis, the location and loading time corresponding to interface crack formation can be determined. Note that the fracture strain and loading time corresponding to interface crack formation at 400, 600, and 800 °C are determined in the same way as in the case of room temperature (30 °C).

13.1.5 DIC Characterization of the High-Temperature CMAS Corrosion-Induced Strain Field in a TBC

In this section, we introduce the evolution of the deformation and strain fields in an APS TBC during the high-temperature CMAS corrosion process, as well as analyze the initiation and propagation of surface and interface cracks in the coating based on the strain contour plots for the specimens [22]. See Chap. 5 of the book for thermo–mechano-chemical coupling tests and theoretical studies relating to CMAS. A specimen with artificially sprayed high-temperature speckles was placed on the specimen stage inside the chamber of a high-temperature furnace. The CCD cameras of the DIC system were adjusted to obtain the clearest image on the computer screen, with a magnification of 200x, a resolution of 1624 × 1236, and a focal length varying approximately from 80 to 400 mm. The temperature of the high-temperature furnace was set to 1250 °C, and the specimen was held at this temperature for 30 min. Figure 13.13 shows a schematic diagram of the test setup. The imaging frequency

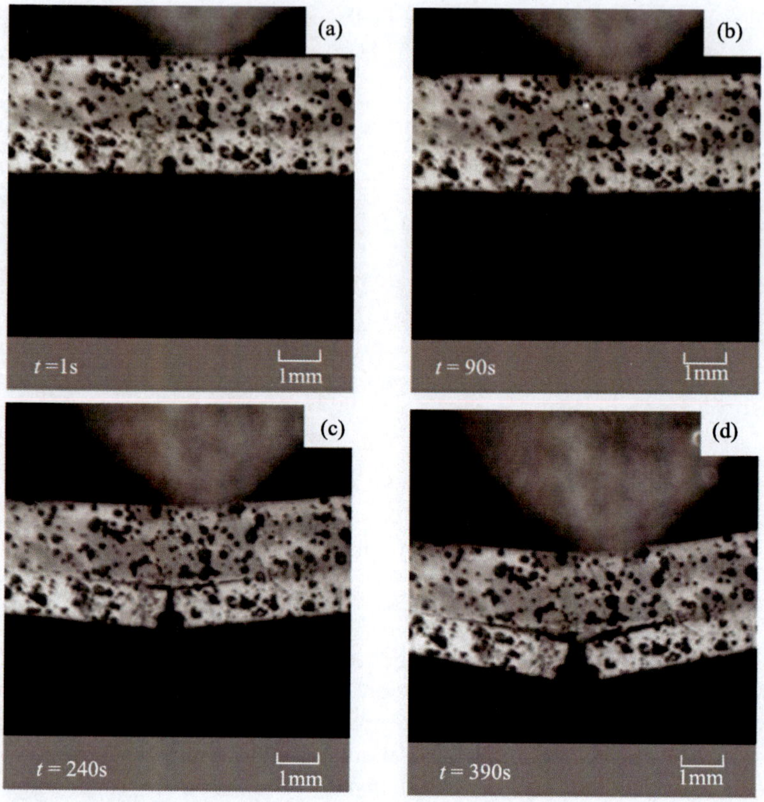

Fig. 13.10 Initiation and propagation of surface and interface cracks in the TBC during the high-temperature three-point bending process: **a** initial state; **b** surface crack initiation; **c** interface crack initiation; **d** interface crack propagation

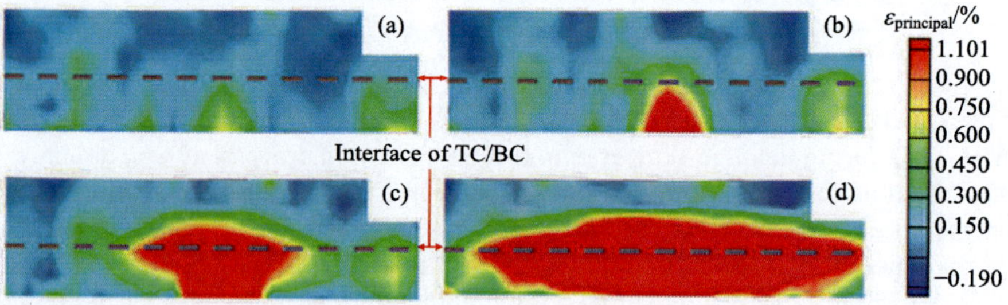

Fig. 13.11 Contour plots showing the evolution of $\varepsilon_{\text{principal}}$ of the TBC during the three-point bending process: **a** $t = 32$ s; **b** $t = 40$ s; **c** $t = 47$ s; **d** $t = 55$ s

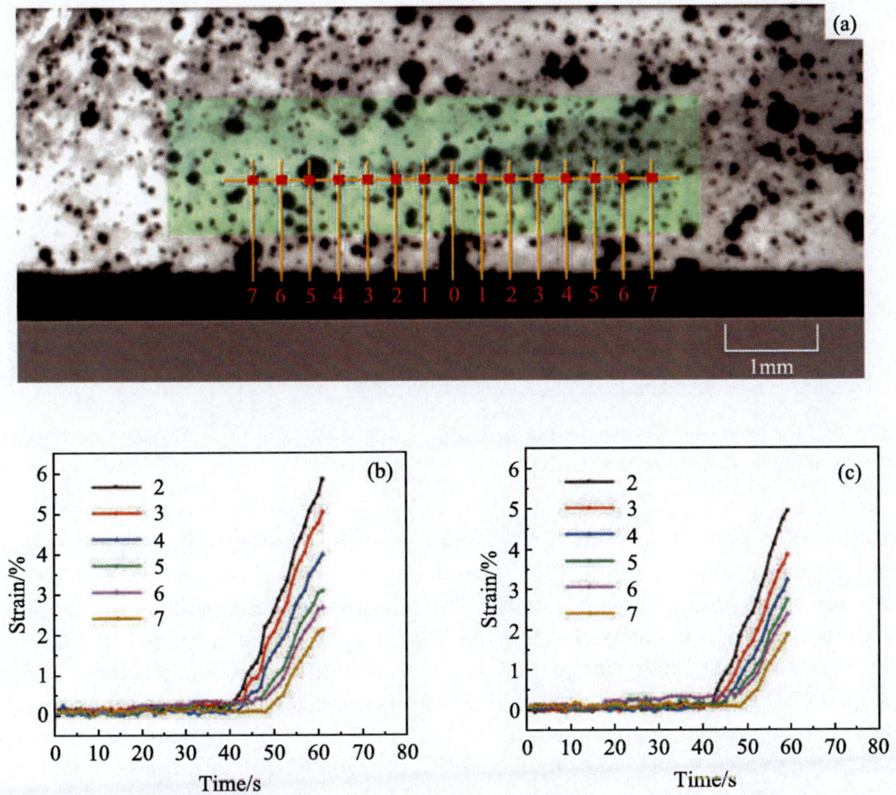

Fig. 13.12 Evolution of ε_{zz} at different locations on the interface crack propagation path with loading time: **a** numbered locations along the interface crack propagation path; **b** evolution of ε_{zz} along the left side of the interface crack propagation path; **c** evolution of ε_{zz} along the right side of the interface crack propagation path. (Numbers 1–7 indicate different locations along the interface crack propagation path.)

was set to one image per 40 s. The collected digital images were analyzed and postprocessed using the ARAMIS software (GOM, Germany). A rectangular region within the "reference image" was selected as the computational region of interest. During calculation, the image subregion size was set to 29 pixels × 29 pixels, the distance between adjacent calculation points was set to 7 pixels × 7 pixels, the scale factor was set to 0.1 mm/pixels, and the strain measurement error was 0.05%.

The wavelength and energy of the electromagnetic waves emitted by an object to the surrounding environment vary with temperature. This is the so-called thermal radiation. An object can both emit and absorb electromagnetic waves. A higher radiation intensity leads to a higher absorption intensity. Thus, as the temperature in the high-temperature furnace increased continuously, the temperature of the specimen increased, and the specimen emitted an increasing intensity of electromagnetic waves, making it increasingly difficult for the cameras to distinguish the specimen

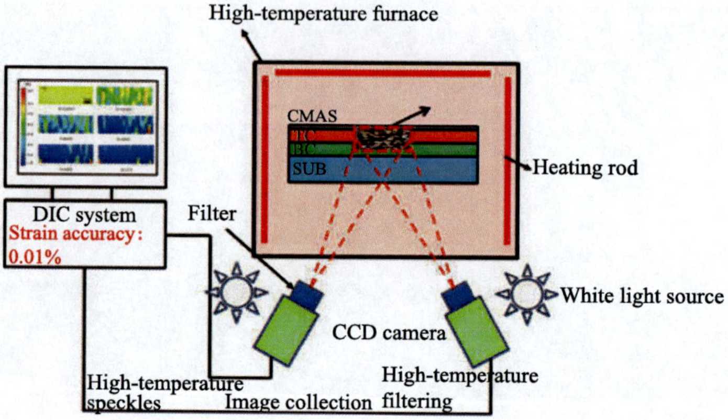

Fig. 13.13 Schematic diagram of the test setup for measuring the high-temperature CMAS corrosion-induced strain field in a TBC

surface. Sutton noted in 1996 that filtering is necessary for collecting digital images at temperatures above 750 °C [20]. Therefore, filtering was required for the CMAS corrosion test, which was conducted at 1250 °C. However, adding filters significantly weakens the light intensity. Thus, an additional light source is needed to intensify the light beam. The light source used in the test comprised two white halogen lamps (Fig. 13.13) that had a long service life, an ability to produce sufficient light beams, and could operate at high temperatures. To comprehensively measure the influence of CMAS corrosion on the strains in the TBC, the evolution of the cross-sectional and surface strains in a specimen coated with CMAS at 10 mg/cm^2 was also measured. For the measurement of the cross-sectional strain field, because CMAS occurs on the surface of the coating and the speckles are prepared on a cross-section, these two processes do not affect each other, and the specimen can be monitored in real time. Figure 13.14 shows the test procedure. For the measurement of the surface strain field, because CMAS occurs on the surface of the coating and the speckles also need to be prepared on the surface of the coating, it is necessary to first preheat the CMAS and allow it to infiltrate into the coating prior to preparing speckles on the surface and measuring the strain using DIC. Figure 13.15 shows the test procedure.

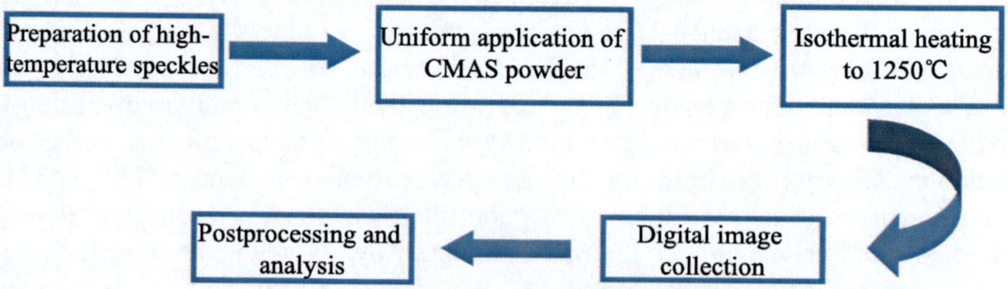

Fig. 13.14 Test procedure for measuring the cross-sectional strain in the coating using DIC

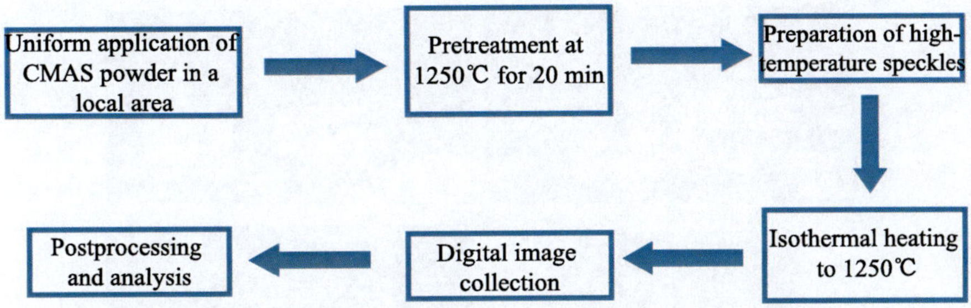

Fig. 13.15 Test procedure for measuring the surface strain in the coating using DIC

(1) Evolution of the cross-sectional strain field in a TBC without CMAS corrosion

To analyze the extent of the effect of CMAS corrosion on the strain in a TBC, we first analyze the cross-sectional strain field in a TBC without CMAS corrosion [22]. From the perspective of microstructural changes, the bond between the ceramic layer and substrate of the non-CMAS-corroded specimen remains intact at a high temperature of 1250 °C. In addition, the substrate turns black due to high-temperature oxidation, but the speckles on the cross-section of the coating remain present, as shown in Fig. 13.16a. When the temperature decreases to approximately 100 °C, the ceramic-layer/substrate interface cracks, but there are no significant changes in the speckle field, as shown in Fig. 13.16b. When the specimen cools to room temperature, as shown in Fig. 13.16c, the ceramic layer spalls completely from the substrate. There are two possible causes of the TBC failure: (1) during the cooling process, the difference between the coefficients of thermal expansion (CTEs) of the ceramic and substrate materials leads to thermal mismatch; (2) the specimen used in the test might contain defects formed during the preparation process, resulting in a weak bond between the ceramic and BC layers, which, in turn, causes premature coating spallation and failure.

To further analyze the trend of the strain field in the non-CMAS-corroded TBC during the cooling process, the digital images collected during the test were post-processed using the software ARAMIS [22]. Of the obtained digital images, the last digital image collected at the beginning of the cooling process was selected as the "reference image." Then, a rectangular area 8 mm × 1.8 mm in size within the reference image was selected as the computational region of interest, as shown in Fig. 13.16d. During calculation, the image subregion size was set to 29 pixels × 29 pixels, the distance between adjacent calculation points was set to 7 pixels × 7 pixels, and the scale factor was set to 0.1 mm/pixels. The calculation error was 0.05%. On this basis, the strain field in each direction in the TBC was calculated. The thermal mismatch effect was reflected primarily in the evolution of the strain in the *x*-direction, ε_{xx}. Now, let us first analyze the changes in ε_{xx}. Figure 13.17 shows the ε_{xx} contour plots for the non-CMAS-corroded coating. As the temperature decreases,

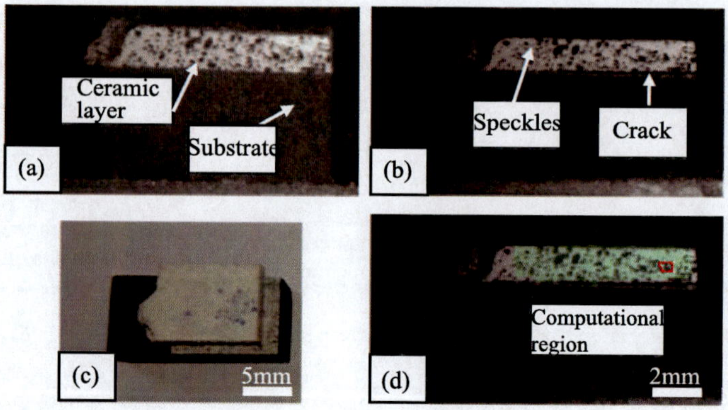

Fig. 13.16 Morphology of the non-CMAS-corroded specimen at **a** 1250 °C, **b** 100 °C, and **c** room temperature (25 °C); **d** computational region

ε_{xx} gradually changes from the initial value of 0 to a compressive strain, which then continually increases in magnitude, accompanied by an increase in the compressive strain region. Above 800 °C, there is a small compressive strain in the coating as a whole and no significant tensile strain on its upper side. Below 400 °C, there is a large and uniform compressive strain in the whole coating. As the temperature decreases to 100 °C, a large tensile strain appears at the coating/substrate interface, suggesting that the thermal mismatch effect between the ceramic layer and substrate begins to become intense and that the coating may have cracked. As shown in Fig. 13.16c, the ultimate failure of the coating is manifested in the form of interface cracks. Therefore, it is necessary to analyze the evolution of the strain in the *y*-direction, ε_{yy}, in the coating with different amounts of CMAS. Figure 13.18 gives contour plots showing the variation in ε_{yy} for the non-CMAS-corroded coating during the cooling process. As the temperature decreases, the ε_{yy} in the coating gradually turns into a compressive strain, which then continually increases in magnitude, accompanied by an increase in the compressive strain region. In addition, as the temperature decreases, there is a gradual increase in the ε_{yy} at the boundary of the lower side of the coating (i.e., at the interface between the coating and the substrate) and an increase in the tensile strain region. When the temperature decreases to approximately 100 °C, the tensile strain at the interface becomes prominent, suggesting that cracking occurs between the coating and the substrate and that the coating separates from the substrate; that is, cracks are formed at the juncture of the coating and the substrate, which, in turn, lead to coating spallation and failure. The trend of ε_{yy} is consistent with that of ε_{xx}. The evolution curves of the local strains of the coating are analyzed in the following section.

(2) **Evolution of the cross-sectional strain field in the bare ceramic layer subjected to CMAS corrosion**

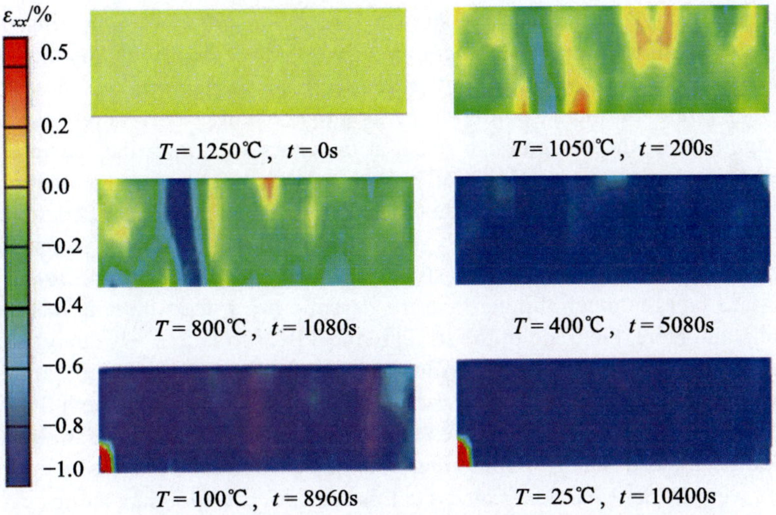

Fig. 13.17 Evolution of the ε_{xx} field in the non-CMAS-corroded TBC with temperature

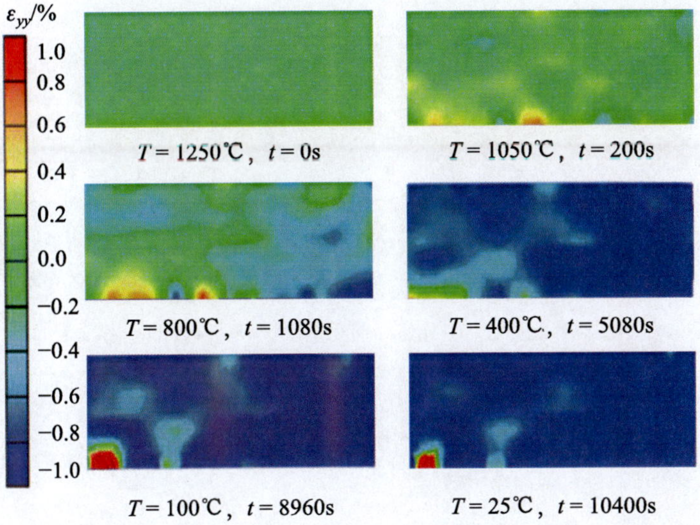

Fig. 13.18 Evolution of the ε_{yy} field in the non-CMAS-corroded TBC with temperature

To facilitate further investigation of the evolution of the strain field in the CMAS-corroded TBC system, it is necessary to understand the evolution of the strain in an independent ceramic layer subjected to CMAS corrosion [22]. To this end, a substrate-free TBC specimen coated with CMAS at 10 mg/cm^2 was subjected to DIC testing using the aforementioned method. Digital images of the specimen were collected during the cooling process and subsequently postprocessed using ARAMIS. Figure 13.19 shows the evolution of ε_{xx} in the specimen. When the specimen is cooled to 1050 °C, a compressive strain is present in most of the coating. As the temperature further decreases, the compressive strain region gradually expands. A uniform compressive strain is present in the specimen when it further cools to 800 °C. There is a continual increase in both the area of the compressive strain region and the magnitude of the compressive strain at the cross-section of the specimen until the temperature decreases to 25 °C. Throughout this process, no significant tensile strain region appears. This is because the specimen is an independent ceramic layer with no substrate, and, consequently, there is no thermal mismatch in the system, leading to no formation of a tensile strain on the lower side of the specimen. Since the CMAS layer is much thinner than the ceramic layer, their thermal mismatch is insignificant. Here, the trend of the strain in the TBC is qualitatively analyzed based on the contour plots showing the evolution of ε_{xx} in Fig. 13.19, which, however, is far from sufficient for determining the magnitude of the strain in the TBC caused by CMAS corrosion. Thus, it is necessary to quantify the trend of the strain. To this end, the strain values at 15 points randomly selected from the cross-section of the coating were extracted and averaged. On this basis, a corresponding strain curve was produced, as shown in Fig. 13.20. During the cooling process, there is a continual but gentle increase in the ε_{xx} in the coating. The maximum ε_{xx} is only –0.75%, with a test error of 0.05%.

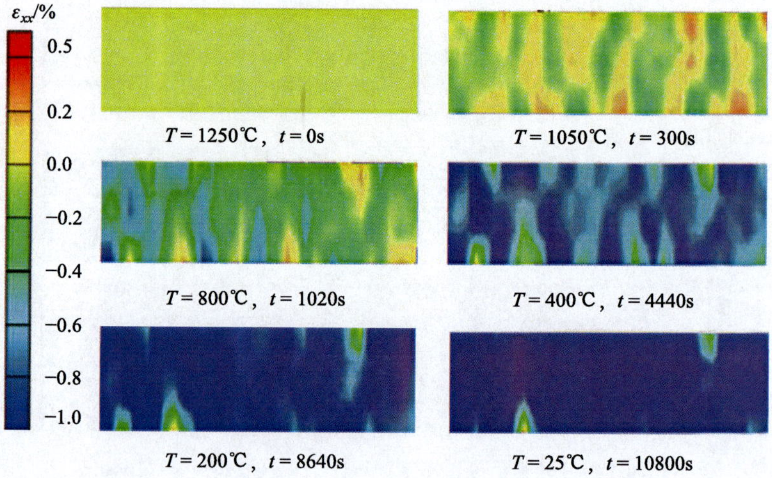

Fig. 13.19 Evolution of the ε_{xx} field in a substrate-free TBC specimen coated with CMAS at 10 mg/cm^2 with temperature

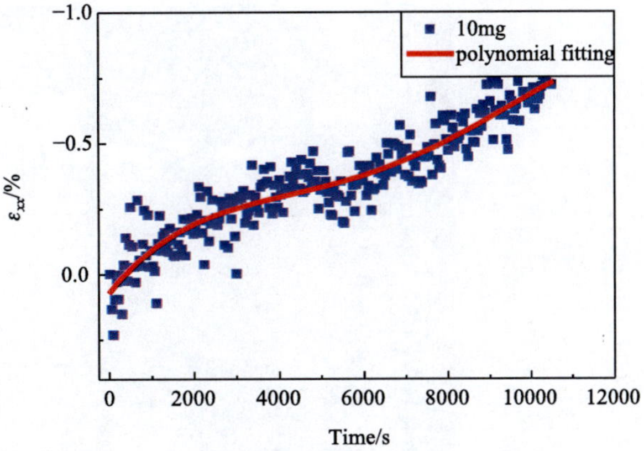

Fig. 13.20 ε_{xx}–time curve for the substrate-free coating specimen coated with CMAS at 10 mg/cm^2

13.1.6 Evolution of the Cross-Sectional Strain Field in TBCs Coated with Different Amounts of CMAS

TBC specimens coated with CMAS at 5, 10, and 20 mg/cm^2 were each subjected to a corrosion test under the same test conditions (i.e., each specimen was held at 1250 °C for 30 min and subsequently allowed to cool to room temperature). Figures 13.21 and 13.22 show the test results [23]. First, let us macroscopically analyze the variation in the strain field in the APS TBC specimens coated with different amounts of CMAS from the collected digital images (Fig. 13.21). Figure 13.21a shows the morphology of the specimen coated with CMAS at 20 mg/cm^2 at 1250 °C. The interfacial bond between the coating and the substrate is intact at 1250 °C. During the cooling process, as the temperature decreases to 400 °C, notable cracking appears at the coating/substrate interface of the specimen coated with CMAS at 20 mg/cm^2 (Fig. 13.21b). As the temperature further decreases to approximately 300 °C, cracking begins to occur in the specimens coated with CMAS at 10 and 5 mg/cm^2, respectively. After cooling to room temperature (25 °C), the coating spalls from the substrate in each specimen (Fig. 13.21c). A comparison with the non-CMAS-corroded specimen discussed earlier shows the following. When a CMAS-corroded TBC specimen cools to 300–400 °C, cracks tend to form at the interface between the coating and the substrate, causing the interface to crack and, ultimately, the coating to spall and fail. In contrast, the interface of the non-CMAS-corroded specimen cracks after the specimen cools to approximately 100 °C, which eventually leads to spallation and failure, suggesting that CMAS corrosion caused the TBC to spall earlier.

The corrosion of a TBC by CMAS at high temperatures is a complex process. At high temperatures, CMAS melts and infiltrates into the interior of the TBC and undergoes chemical reactions with the TBC, causing a phase transformation. The CMAS powder that does not participate in the reactions fills the pores of the coating. During the cooling stage, the CMAS solidifies and thereby alters the structure of the coating, substantially reducing its strain tolerance and leading to a more pronounced

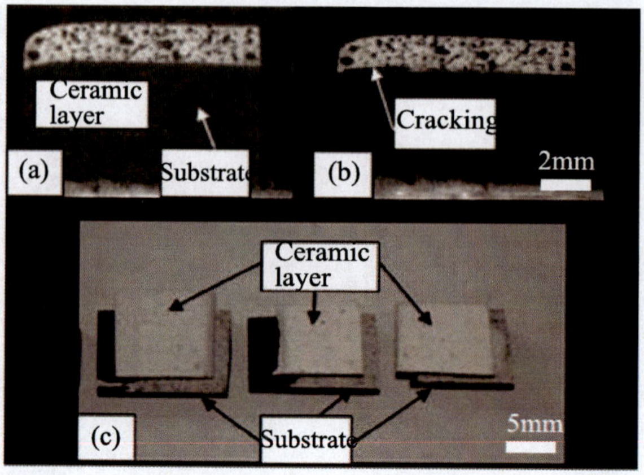

Fig. 13.21 Morphologies of CMAS-coated APS TBC specimens at different temperatures and with different amounts of CMAS: **a** 1250 °C and 20 mg/cm^2; **b** 400 °C and 20 mg/cm^2; **c** room temperature and 5, 10, and 20 mg/cm^2

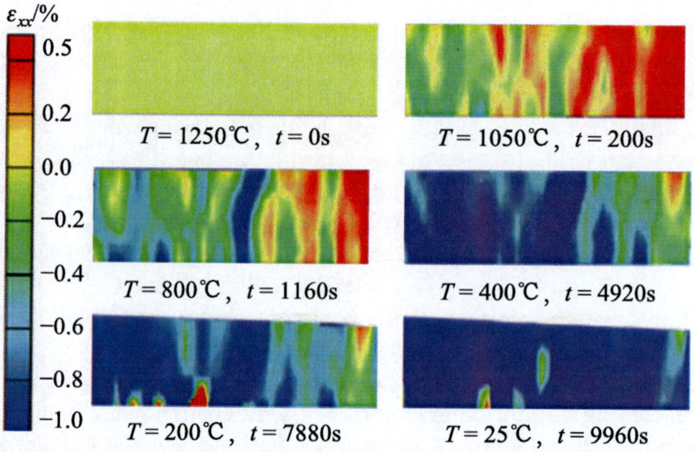

Fig. 13.22 Evolution of the ε_{xx} field in a TBC specimen coated with CMAS at 5 mg/cm^2 with temperature

thermal mismatch effect. The earlier analysis of the strain field in the non-CMAS-corroded specimen shows a significant change in the strain in the coating during the cooling stage due to the thermal mismatch. Therefore, we consider that the thermal mismatch has an extremely significant impact on a coating during the cooling stage and is highly likely to cause the coating to spall and fail. Thus, to measure the strain induced by the thermal mismatch, the first digital image collected at the start of the cooling stage was similarly used as the reference image for the postprocessing of each of the specimens coated with different amounts of CMAS. A rectangular area of the same size as that of the area selected for the analysis of the non-CMAS-corroded specimen was selected as the computational region to calculate the variation in the strain in each CMAS-corroded specimen during the cooling process. As the thermal mismatch effect was mainly reflected in the evolution of ε_{xx}, we first analyzed the variation in ε_{xx} in the specimens coated with different amounts of CMAS.

The thermal mismatch between the coating and the substrate was the driving force for the cracking and spallation of the TBC subjected to high-temperature CMAS corrosion during the cooling stage. Figure 13.22 shows the evolution of ε_{xx} in the specimen coated with CMAS at 5 mg/cm^2. During the cooling process, there is a continual increase in the compressive strain and a gradual increase in the compressive strain region. At 200 °C, a tensile strain appears at the coating/substrate interface. However, when the coating cools to room temperature, the tensile strain almost ceases to exist, and a compressive strain is present at the entire cross-section of the coating. The trend of ε_{xx} in the specimens coated with CMAS at 10 and 20 mg/cm^2 is similar to that seen in Fig. 13.22, with the only difference being that a tensile strain has already appeared at the coating/substrate interface at approximately 400 °C. Similarly, as each of the two specimens coated with more CMAS cool to room temperature, the tensile strain gradually decreases, and a compressive strain is present through the whole cross-section of the coating. Clearly, the ε_{xx} contour plots for the TBC specimens coated with CMAS at 5, 10, and 20 mg/cm^2 exhibit basically similar trends during the cooling process, with the main difference being the temperature and time at which a tensile strain appears at the interface.

As shown in Fig. 13.21, the ultimate coating failure is manifested in the form of interface cracks. Thus, it is necessary to analyze the evolution of ε_{yy} in the TBCs coated with different amounts of CMAS. Figure 13.23 shows the evolution of ε_{yy} in the TBC specimen coated with CMAS at 10 mg/cm^2 during the cooling process. As the temperature decreases from 1250 to 1050 °C, the ε_{yy} in the coating is a small tensile strain. This is because the CMAS melts and infiltrates into the coating and subsequently fills its pores, thereby altering its structure and thermophysical properties. As the temperature further decreases to 400 °C, the ε_{yy} in the coating turns into a notable compressive strain, and a tensile strain appears at the coating/substrate interface due to the thermal mismatch caused by the difference between the CTEs of the coating and the substrate. The presence of the tensile strain leads to the initiation and formation of cracks at the coating/substrate interface. As the temperature continues to decrease, there is a continual increase in both the compressive strain region in the coating and the tensile strain at the interface. It is in this process that the cracks continually propagate and eventually lead to coating spallation and failure. The evolutionary trend of ε_{yy} in each of the specimens coated with CMAS at 5 and 20 mg/cm^2, respectively, is consistent with that of ε_{yy} in the specimen coated with CMAS at 10 mg/cm^2. A comparison of the evolution of the ε_{xx} and ε_{yy} contour plots for the specimens coated with different amounts of CMAS shows the following. During the cooling process, both the ε_{xx} and ε_{yy} in each specimen are compressive and increase continuously. In a certain stage, a tensile strain appears at the coating/substrate interface in each specimen. However, the time and temperature at which this tensile strain appears vary between the specimens coated with different amounts of CMAS. It is easily found that a larger amount of coated CMAS leads to

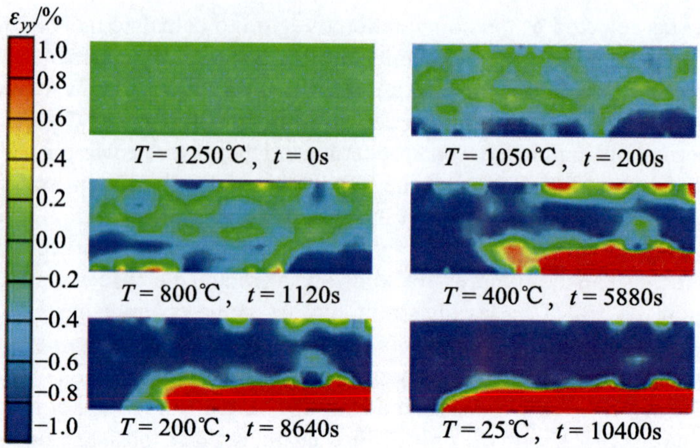

Fig. 13.23 Evolution of the ε_{yy} field in a TBC specimen coated with CMAS at 10 mg/cm^2 with temperature

the formation of a tensile strain at the interface earlier and at a higher temperature. A tensile strain has already appeared at the interface when the temperature decreases to approximately 400 °C, suggesting that the thermal mismatch effect has already become prominent at this temperature. In comparison, a tensile strain appears in the non-CMAS-corroded specimen (Fig. 13.20) at a later time; specifically, a tensile strain appears in this specimen only when the temperature decreases to approximately 100 °C. Evidently, the CMAS increases the thermal mismatch between the coating and the substrate and thus leads to premature coating spallation and failure.

The failure of the CMAS-corroded TBC is ultimately reflected by the interface cracks caused by the shear strain. To determine the strain in the coating at failure and the time at which the coating spalled, it is necessary to analyze the strain curves for the *y*- and *x*-directions. First, five points were randomly selected from the red tensile strain region at the bottom of the ε_{yy} contour plot for each of the specimens coated with different amounts of CMAS to calculate the mean ε_{yy}. Figure 13.24 shows the ε_{yy}–cooling time curves. A common characteristic can be seen in the trends of ε_{yy} in the specimens coated with CMAS at 0–20 mg/cm^2—during the cooling process, there is a sudden change in ε_{yy} in each specimen, after which ε_{yy} remains large, indicating that the coating has cracked. In addition, we can clearly see that the time of the sudden change in ε_{yy} for the specimens coated with different amounts of CMAS differs. The specimens coated with CMAS at 10 and 20 mg/cm^2 undergo a sudden change in ε_{yy} at $t = 5000$ s, the earliest of all the specimens. In other words, the earliest cracking and failure of the coating at the interface with the metallic substrate occur in these two specimens. As the cooling process proceeds, at $t = 7000$ s, there is a sudden change in the ε_{yy} in the specimen coated with CMAS at 5 mg/cm^2, that is, this specimen fails at $t = 7000$ s. In contrast, at $t = 9000$ s, there is a sudden change in ε_{yy} in the non-CMAS-corroded specimen, and the interface between the coating and the metallic substrate in this specimen cracks. Evidently, high-temperature CMAS

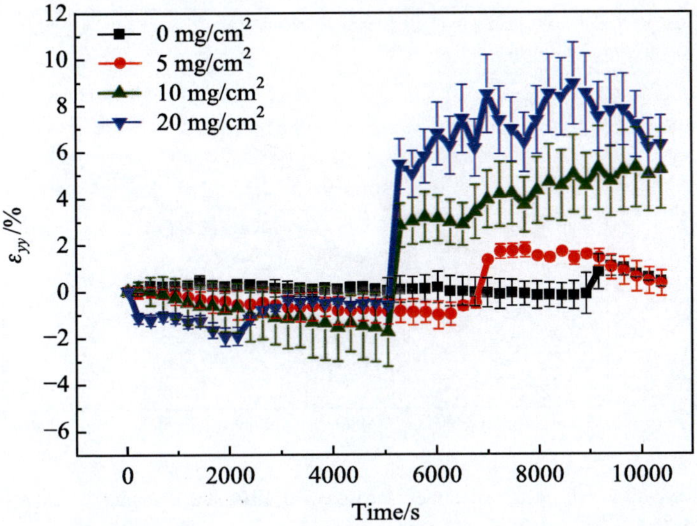

Fig. 13.24 ε_{yy}–time curves for specimens coated with different amounts of CMAS during the cooling process

corrosion accelerates the spallation of the ceramic layer and thus causes it to fail prematurely. After a large amount of CMAS powder infiltrates into the ceramic layer and fills its porous structure, the failure of the coating is no longer affected by the amount of coated CMAS.

To analyze the critical ε_{xx} at which the TBC system failed during the cooling process, we randomly extracted 15 ε_{xx} data points from the digital images of each of the specimens coated with different amounts of CMAS and calculated the mean ε_{xx}. Figure 13.25 shows the variation in ε_{xx} in coatings with different amounts of CMAS with the cooling time throughout the cooling process. A comparison of ε_{xx} values in the specimens coated with different amounts of CMAS shows the following. At the same time point, the ε_{xx} in each CMAS-corroded specimen is much smaller than that in the non-CMAS-corroded specimen. As the amount of coated CMAS increases, ε_{xx} tends to decrease but to an insignificant extent. For each specimen, the horizontal coordinate in Fig. 13.25 is the cooling time corresponding to the sudden change in ε_{yy} in Fig. 13.24, and the corresponding vertical coordinate ε_{xx} in Fig. 13.25 is the critical strain at which the coating/substrate interface cracks. Figure 13.25 clearly shows that as the amount of coated CMAS increases, there is a gradual decrease in the critical strain at which the coating cracks. The critical strain at which the non-CMAS-corroded specimen cracks (corresponding to point A in Fig. 13.25) is –1.6%, compared to approximately –1.0%, –0.7%, and –0.55% (corresponding to points B, C, and D in Fig. 13.25) for the specimens coated with CMAS at 5, 10, and 20 mg/cm^2, respectively. This finding is consistent with the previously reported measurement of the critical strain (approximately 0.7%) at which a TBC fails under thermal shock. [19, 24, 25] These results show that the critical strain at which each

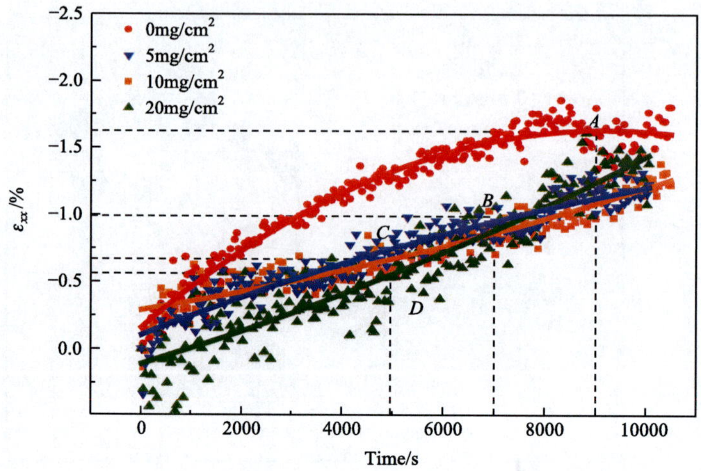

Fig. 13.25 ε_{xx}–time curves for specimens coated with different amounts of CMAS during the cooling process

CMAS-corroded specimen cracks is much smaller than that at which the non-CMAS-corroded specimen cracks and that CMAS corrosion accelerates the spallation failure of the TBC, shortens its life, and alters its structure, thereby resulting in a significant change in its strain. The test results obtained through the variation in the amount of coated CMAS show that the critical strain varies little with the amount of coated CMAS. However, as the amount of coated CMAS increases, the critical strain still displays a downward trend, suggesting that CMAS powder corrosion has a significant impact on the life of the coating. Preventing the harm caused by CMAS corrosion is indeed a top priority and an urgent task.

13.1.7 Evolution of the Surface Strain Field in a TBC Subjected to CMAS Corrosion

Tests on the corrosion of the TBC with CMAS were conducted at high temperatures. With the available equipment, it is impossible to conduct a 3D DIC measurement at high temperatures. Therefore, it is necessary to separately measure the evolution of the surface strain in a TBC subjected to CMAS corrosion. The melting point of CMAS is approximately 1230 °C. Thus, the variation in the surface strain field during the temperature-holding stage was primarily measured while keeping the other test conditions unchanged [22] After the test, the specimen was removed from the furnace, and the coating had already spalled from the substrate. Here, the strain field is analyzed to determine the cause of the TBC failure. Changes in the cross-sectional and surface strains in the TBC specimen coated with CMAS at 10 mg/cm^2 for high-temperature corrosion were calculated using the aforementioned ARAMIS strain analysis software. Figure 13.26 shows the evolution of the cross-sectional $\varepsilon_{principal}$ field in the TBC specimen subjected to high-temperature CMAS corrosion. In the temperature-holding stage of the CMAS corrosion process, there is a tensile strain in most of the coating, while a small compressive strain appears in the upper layer of the coating, which is caused by the infiltration of the CMAS powder that had melted at high temperatures into the interior of the coating. At high temperatures, the specimen undergoes thermal expansion. However, due to the large difference between the CTEs of the ceramic and metallic substrate materials, the ceramic layer is pulled by the metallic substrate. Consequently, a large tensile strain occurs at the bottom of the ceramic layer. In the cooling stage, a compressive strain appears in the entire cross-section of the coating. In addition, as the temperature decreases, there is a decrease in the tensile strain but an increase in the compressive strain region. This is the so-called compression of the coating due to its contraction during the cooling process. However, the ceramic layer contracts slowly due to its small CTE, whereas the metallic substrate contracts rapidly due to its large CTE. As a result, the substrate deforms faster than the ceramic layer and thus exerts a tensile stress on the ceramic layer, which is the cause of the tensile strain at the bottom of the ceramic layer.

To study the evolution of the surface strain field in a TBC corroded by CMAS powder, a TBC specimen coated with CMAS powder on only a 5 mm × 5 mm area at the center of its surface was subjected to a DIC test. Figure 13.27 shows the variation in the surface $\varepsilon_{principal}$ in the TBC specimen determined using ARAMIS. The following can be directly seen from the $\varepsilon_{principal}$ contour plots. In the temperature-holding stage, as time passes, the surface $\varepsilon_{principal}$ of the coating, which is a tensile strain, increases continually, a maximum tensile strain appears at the boundary between the powder-coated and powder-free regions, and the tensile strain concentration region expands continually. In the cooling stage, the surface tensile strain of the coating decreases gradually. In particular, a compressive strain appears in the powder-free region because of the thermal expansion/contraction of the coating. To

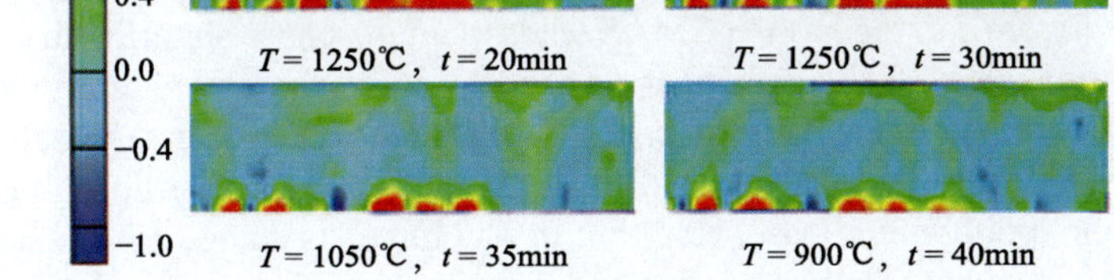

Fig. 13.26 Evolution of the surface $\varepsilon_{principal}$ field in a TBC specimen coated with CMAS at 10 mg/cm^2 with temperature

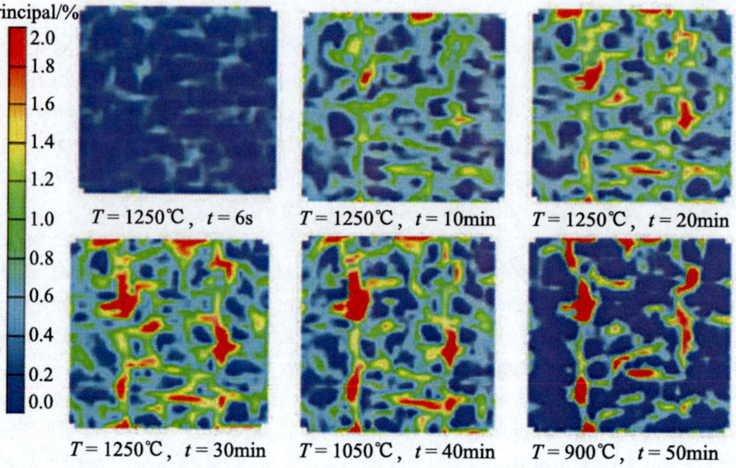

Fig. 13.27 Evolution of the surface $\varepsilon_{principal}$ field in a CMAS-corroded TBC specimen with time

accurately determine the evolutionary pattern of the surface strain in local regions of the coating, $\varepsilon_{principal}$–time curves for local surface regions of the coating were extracted, as shown in Fig. 13.28 (points a, b, and c are located at the boundary between the powder-coated and powder-free regions, in the powder-free region, and in the powder-coated region, respectively). The fitting curves show that $\varepsilon_{principal}$ is the largest at point a, followed by that at point c, and that at point b is the smallest. In the first 2000s, as time passes, the $\varepsilon_{principal}$ at points a, b, and c gradually increases and reaches maximum values of 1%, 0.3%, and 0.5%, respectively. After the first 2000s (i.e., during the cooling stage), as time passes, $\varepsilon_{principal}$ at each point decreases fast due to thermal expansion/contraction and gradually turns compressive. $\varepsilon_{principal}$ at point c is larger than that at point b but smaller than that at point a. This is because CMAS powder transforms into a molten state and becomes fluid at 1250 °C.

In summary, the high-temperature CMAS corrosion of a TBC is divided into three processes, as shown in Fig. 13.29. First, the CMAS melts at high temperatures and fills the pores of the TBC. The pores and microcrystalline structure of the coating undergo sintering, resulting in a dense coating structure and the slow formation of transverse cracks on the surface and in the interior of the coating. Then, the molten CMAS infiltrates into the coating along its thickness under the capillary action [26]. The Y in the coating structure is dissolved in the CMAS, forming $Ca_2Y_8(SiO_4)_6O_2$ [27]. The presence of the molten CMAS leads to liquid-phase sintering in the interior of the coating and thus the densification of the coating, thereby altering its physical properties (e.g., thermal conductivity, CTE, and fracture toughness). Finally, CMAS corrosion causes the coating to transform from the stable nonequilibrium tetragonal phase to the monoclinic phase, resulting in a volume shrinkage of 3–5%. Thermal mismatch occurs between the coating and the substrate. Continuous thermal cycling

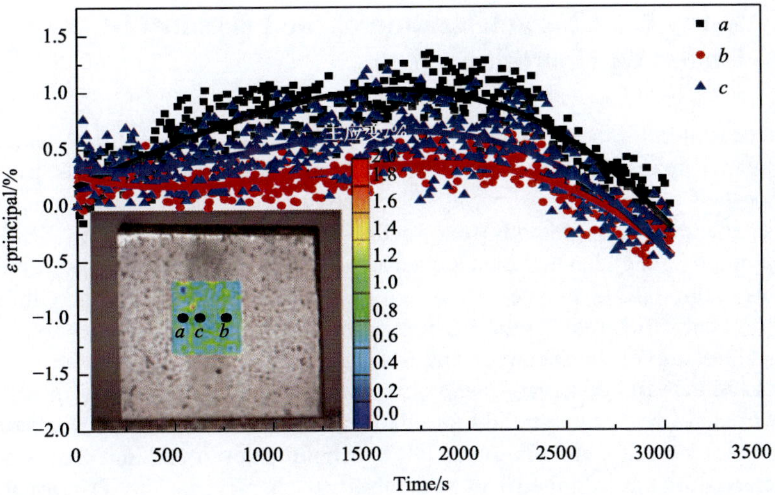

Fig. 13.28 Evolution of the surface $\varepsilon_{\text{principal}}$ field in a CMAS-corroded TBC specimen with time

and alternation of tensile and compressive strains lead to crack propagation, which, in turn, causes coating failure.

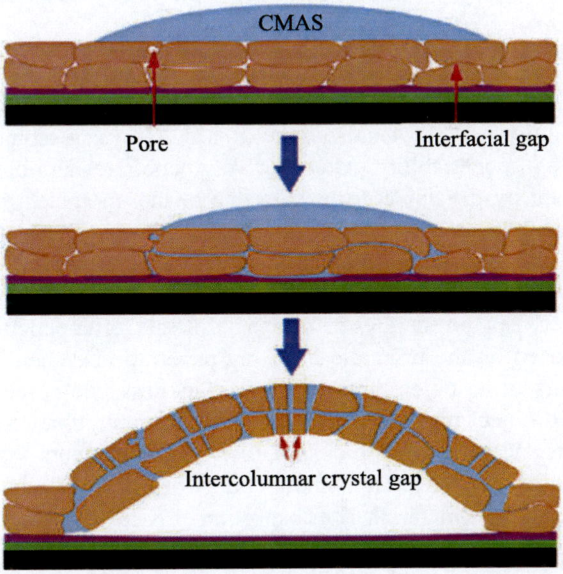

Fig. 13.29 Schematic diagram of the CMAS corrosion of a TBC

13.2 X-ray CT Characterization of the Pores in TBCs Under VA Corrosion

Characterization of pores and defects in TBCs helps us to further understand their internal microstructure and their evolution with service conditions, providing an important basis for investigating their performance. Conventional microstructural characterization methods (e.g., scanning electron microscopy (SEM) and optical microscopy) have shown powerful capabilities. However, these methods share a common disadvantage—they require specimen treatments (e.g., cutting and polishing) before characterization. These preparation processes damage the original internal structure of the specimen and introduce artificial defects, thereby resulting in errors in material observation and characterization. Therefore, we urgently need a type of characterization method for nondestructively observing and analyzing the internal structure of materials. As an NDT technique developed in recent years, CT characterization has exhibited a powerful ability to characterize the 3D internal structure of materials. This technique can be employed to nondestructively examine the internal structure of a material, to generate digital images with different grayscale values, and to determine the density distribution in each internal region of the material. The obtained 2D reconstructed images can then be restructured in 3D to determine the 3D internal spatial structure of the material. Hence, the application of the CT characterization technique in the TBC field is of great importance.

13.2.1 Principle of the CT Characterization of the Internal Structure of an Object

CT, or X-ray CT, is used in many material fields because it is nondestructive and offers high resolution and 3D visualization [28–32]. X-rays are a type of high-energy radiation with high penetrability. After an X-ray penetrates an object, its energy is partially absorbed by the object, primarily due to the photoelectric and Compton effects caused by the collisions between X-rays and the electrons in the object. These collisions lead to energy loss, and the X-ray absorption capacity (i.e., X-ray attenuation coefficient) of an object can be determined through the measurement of the difference between the energies of X-rays before and after penetrating the object. CT testing is based on the principle of X-ray penetration. When X-rays are used to scan the layers of an object with a certain thickness, the computer divides the X-ray-penetrated object into cuboids of the same volume, which are referred to as volume elements. With the same thickness, the volume elements absorb some of the energy of the X-rays, while the remaining energy of the X-rays passes through the interatomic spaces in the material. The absorption of X-rays by objects plays a major role in CT imaging. In a homogeneous object, the attenuation of X-rays follows an exponential trend. In the case where X-rays penetrate a nonhomogeneous object (i.e., an object composed of multiple components with different densities), the attenuation coefficient varies from point to point. In essence, a CT image is an image of the attenuation coefficient. The X-ray attenuation coefficient or absorption coefficient of each volume element is calculated using a computer based on the obtained projected value. The coefficient values are then arranged into a digital matrix. Each number in the digital matrix is converted using a digital/analog converter into a small square (i.e., a pixel) in grayscale ranging from black to white. These pixels are arranged according to the matrix to obtain a CT image. The attenuation coefficient is calculated using the following procedure [33].

The intensity of X-rays decreases from I_0 to I after they penetrate a volume element with a thickness of d. The attention of the X-ray follows the Beer–Lambert law. The projected value P is expressed as follows:

$$\mu d = \ln\left(\frac{I_0}{I}\right) = P \tag{13.4}$$

If the medium in the X-ray scan path is nonhomogeneous, the medium distributed along the path can be divided into n blocks, each of which is a volume element with a thickness of d. The projected value P is given by the following equation:

$$d \cdot \sum_{i=1}^{n} \mu_i = \ln\left(\frac{I_0}{I}\right) = P \tag{13.5}$$

If the medium in the X-ray scan path is nonhomogeneous and the attenuation coefficient μ changes continuously (i.e., μ is a function of the path l), the projected value P is expressed as follows:

$$P = \int_{-\infty}^{\infty} \mu(l)dl = \ln\left(\frac{I_0}{I}\right) \tag{13.6}$$

Based on the definition of the CT value, the calculated value of the attenuation coefficient of each volume element is converted to a CT value for the corresponding pixel. Thus, the distribution of CT values is obtained. Conversion of the CT value of each pixel on the image plane into grayscale yields a grayscale distribution on the image plane, thereby forming a CT image.

The penetrability of X-rays is directly proportional to their energy and inversely proportional to the density of the penetrated material. Based on this property, it is possible to distinguish materials with different atomic numbers as well as different internal densities and thicknesses of the same type of material and, therefore, to produce images with different grayscale values. In 3D imaging, a "virtual knife" is used to cut a specimen into virtual slices for observation, and then these images with different grayscale values are reconstructed to obtain the 3D structural information of the whole specimen. The CT technique can be used to determine the 3D internal structure (e.g., pores, cracks, and the distribution of different components) of a specimen without the need for special treatment of the specimen or a vacuum environment.

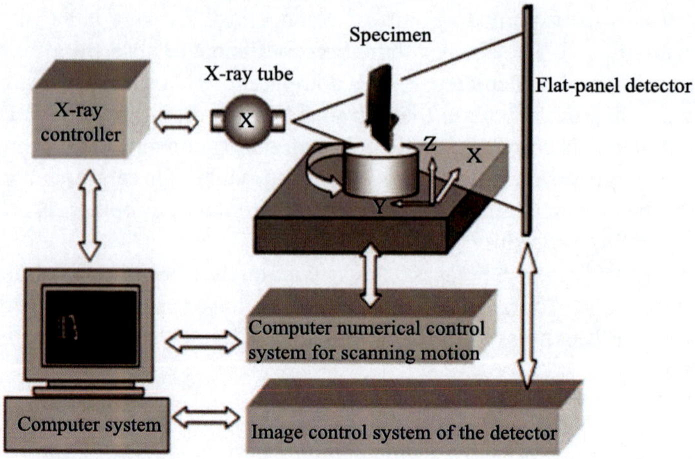

Fig. 13.30 Composition of a CT system

In addition, with a high detection resolution, the CT technique can produce high-quality images of the interior of a specimen [28]. Moreover, as an NDT technique, CT can effectively prevent unknown damage incurred during surface and cross-sectional treatments of specimens.

The working system of CT consists primarily of an X-ray source, a detector, a data acquisition system, a computer numerical control, and a computer system. Figure 13.30 shows the composition of the system.

X-ray source system. This system is used to emit and control X-rays. When heated by the tube voltage to a white-hot state, the tungsten wire at the cathode produces a large number of electrons, which are accelerated by the electric field and then impact the anode (usually made of a metal) at high speed. During the impact process, the electrons slowly decelerate until they come to rest. As a result, the X-ray spectrum is continuously distributed [28]. The X-ray energy depends on the tube voltage. As high-speed electrons impact the target material, a small portion of them are converted to X-rays radiating outward, and most of the kinetic energy is converted into thermal energy. Therefore, tungsten, a metal with a high melting point, is generally selected as the target material.

Detector system. The main function of the detector is to receive the X-rays that penetrate through the specimen and convert these signals with different degrees of attenuation into electrical signals [28, 30]. There are two types of CT detectors, namely linear-array and area-array detectors. The detection units of a linear-array detector have independent processing circuits. As a result, linear-array detectors can produce high-quality images but have a complex structure and are expensive to fabricate. In comparison, area-array detectors have a higher X-ray utilization efficiency, receive and convert X-ray signals at faster rates, and are less costly to manufacture, making them the first choice.

13.2.2 *Extraction of Pores in TBCs and 3D Reconstruction of CT Images*

(1) **Extraction of pores based on image morphology**

Image morphology is a computational means for image processing, originating from a branch of biology that explores the morphology and structure of plants and animals. Image morphology produces an output by applying structuring elements with specific shapes to the input image [34].

Erosion and dilation are the two fundamental operations in image morphology [34]. Erosion uses a 3 × 3 structuring element to approximate the original image and performs an "AND" operation. If all the results are 1, then the pixel of the structural image has a value of 1; otherwise, the pixel has a value of 0. The erosion operation removes some of the pixels at the boundary of an object, thereby reducing the size of the binary image by removing all original edge pixels. Figure 13.31a shows a schematic diagram of the erosion operation. Opposite to erosion, the dilation algorithm uses a 3 × 3 structuring element to scan each pixel of a binary image and performs an "AND" operation on the structuring element and the part of the binary image covered by it. If all the results are 0, then the pixel of the structural image has a value of 0; otherwise, the pixel has a value of 1. The dilation operation adds pixels to the boundary of the object in the binary image, thereby increasing the size of the binary image by adding a new complete set of edge pixels. Figure 13.31b shows a schematic diagram of the dilation operation.

Top hat is an advanced morphological transform based on the two fundamental morphological operations (i.e., dilation and erosion) and is the difference between the original image and the opened image. The opening operation first erodes the original image and then dilates the resulting image with a structuring element. Opposite to the opening operation, the closing operation performs a dilation operation first, followed by an erosion operation. Figure 13.32 shows the computational procedure of the opening operation. We can see that the details of the original image are preserved after the opened image is subtracted from the original image. Therefore, the top-hat operation is effective in detecting the details of background images in image segmentation and is suitable for extracting micropores in TBCs [35].

Region growing is a "point-radiation"-type object identification method. First, a region (pixel) is selected as the starting point of identification. Then, all the nearby pixels within the user-defined grayscale range are grouped together. The newly identified region is treated as a new starting point of identification, and the growing process continues. This way, a region can be generated [36]. In region growing, most micropores are manually identified and selected. Hence, this method is usually used as a supplement in image segmentation.

Figure 13.32 shows the process of extracting the pores in a TBC, involving a combination of top hat and region growing. As shown in Fig. 13.32, the use of the top hat for image morphology processing is able to identify most of the pores, while region growing complements the top hat by identifying the individual pores that the top hat has missed. Hence, the two methods can be used in conjunction to identify and extract the majority of the pores in a TBC specimen.

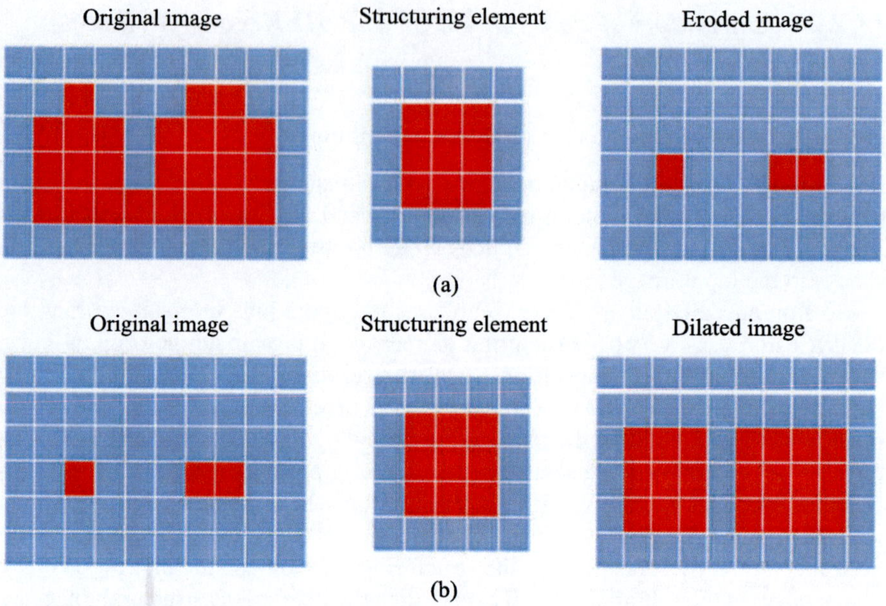

Fig. 13.31 Schematic diagram of image operations: **a** erosion; **b** dilation

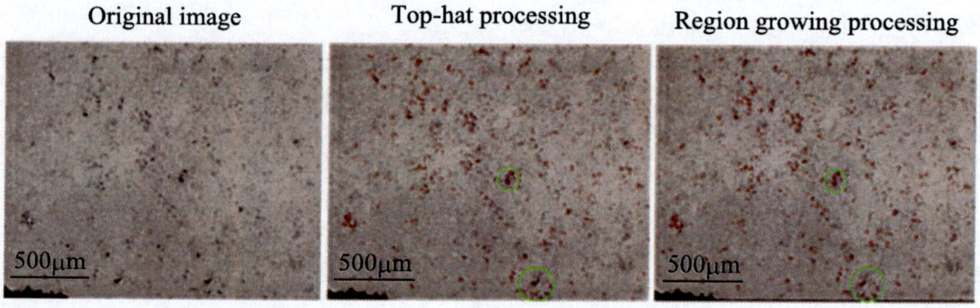

Fig. 13.32 Pore extraction process

(2) **3D image reconstruction**

A 3D image of a specimen can be produced through reconstruction of its CT scan slice images using the software Avizo. Specifically, in Avizo, the slice images of different locations of a specimen are first subjected to processing such as denoising and then reconstructed using the volume rendering module to produce a 3D structural image of the specimen. Figure 13.33 shows the sectional information for a certain location in the TBC specimen in the XY, YZ, and XZ directions. A 3D image of the TBC specimen is produced through the stacking of all its slice images for different locations and directions, forming a basis for analyzing its internal pores. Figure 13.33 shows that the coating has a rough surface and a thickness of approximately 250 μm.

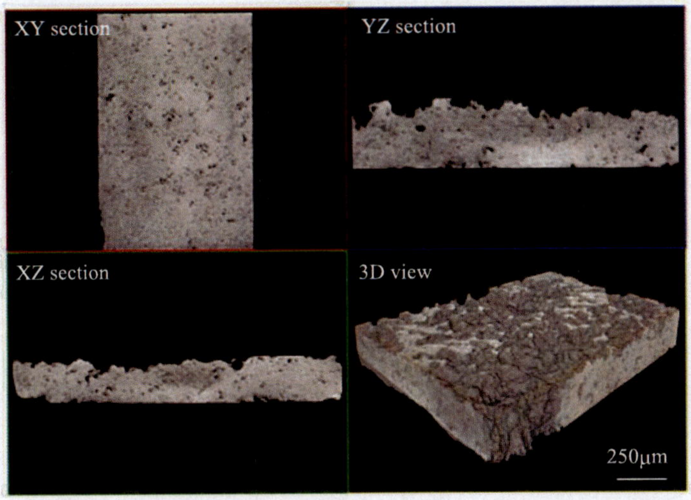

Fig. 13.33 Schematic diagram of 3D reconstruction

13.2.3 *CT Characterization of the Evolution of Pores in TBCs Under VA Corrosion*

CT is an imaging technique that is completely different from ordinary radiation-based techniques. This technique divides a test object into slices for scanning and separately images each slice, thereby preventing the slices from interfering with each other. With high detection sensitivity and spatial resolution, CT can produce high-quality images of a test object; clearly and accurately determine its internal structural relations, material composition, and defect conditions; and locate and measure its defects, therefore playing a vital role in NDT. In the following, the influence of high-temperature volcanic ash (VA) corrosion on the pore-related properties (i.e., pore size, pore shape, and porosity) of coatings is examined by CT using APS TBC as an example [37].

(1) **Influence of high-temperature VA corrosion on the size of pores in TBC**

The pores were first segmented from an APS TBC specimen through image morphology processing involving a combination of top hat and region growing. Then, in Avizo, these segmented pores formed a new label, based on which the volume of each pore was determined through analysis and computation. The volume unit was set to μm when the image was imported. The radius of a pore was calculated based on its volume as follows:

$$r = \frac{1}{2}\sqrt[3]{\frac{6V}{\pi}} \tag{13.7}$$

where r and V are the radius and volume of the pore, respectively.

Next, the pores were visualized in 3D. Figure 13.34 shows the 3D distribution of pores in the whole ceramic coating, with the images generated based on the principle that adjacent pores were colored differently. Using some standard programs (including reconstruction, segmentation, labeling, and screening) in Avizo, the 3D distribution of the pores within different radius ranges was obtained, and the corresponding proportions of the total volume of the pores were statistically quantified. The radii were calculated using Eq. (13.7).

The pores were segmented from the TBC specimen using image morphology processing involving a combination of top hat and region growing [36]. Because the minimum scan size was 0.89 μm, we set the minimum value to 0.89 μm in the pore radius classification and considered any identified pores with radii smaller than 0.89 μm to be noise. Specifically, the pore radius was classified into four ranges, namely 0.9–5, 5–10, 10–20, and >20 μm. The pores with radii in these four ranges in the original APS TBC account for 26.83%, 35.29%, 27.36%, and 10.33% of the total pore volume, respectively. Clearly, most of the pores in the APS TBC are smaller than 20 μm in radius. The 3D extraction of the pores in a specimen not only allows obtaining the size of each pore but also makes it possible to determine the specific location of each pore based on its corresponding X-, Y-, and Z-coordinates. In addition, this method yields some volume-related information (e.g., surface area and volume) that is unattainable using other characterization methods.

Figure 13.35 shows the variation in the relative frequency of the pores in the ceramic layer of the original APS TBC with the relative radius. Here, relative x- and y-values are used here for nondimensionalization. The relative radius refers to

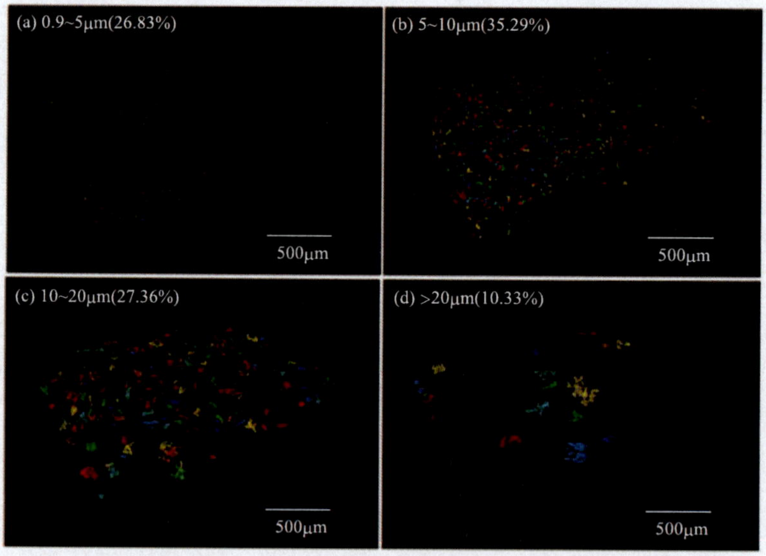

Fig. 13.34. 3D distribution of the pores with different radius ranges in the original coating specimen: **a** 0.9–5 μm; **b** 5–10 μm; **c** 10–20 μm; **d** >20 μm

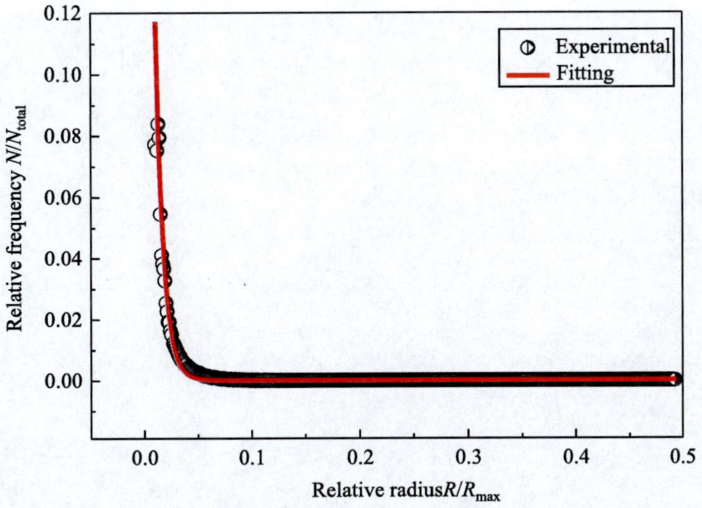

Fig. 13.35 Distribution of the relative radius and relative frequency of the pores in the original coating specimen

the radius of a pore divided by the maximum radius of the identified pores, and the relative frequency refers to the number of pores with a certain radius divided by the total number of identified pores. In Fig. 13.35, the black data points represent the radius of each pore and the number of pores detected by CT, while the red curve shows the functional relation between the relative frequency and relative radius obtained through fitting the statistical data for the pores.

Figure 13.35 clearly shows that in the ceramic layer of the original APS TBC, the number of pores decreases as the radius increases, that is, there are a large number of small pores and a small number of large pores, which might be caused by the APS preparation process. The relation of the relative frequency and relative radius of the pores follows the exponential distribution expressed in the following equation: $y = 0.45e^{-135.13x}$, with a goodness-of-fit of 91%. The results of the quantitative analysis of the size and number of pores can be used as a basis for subsequent analysis of the internal microstructure of the TBC. The pores were segmented and extracted from a specimen under VA corrosion using the same method. Figure 13.36 shows the 3D distribution of the pores in the corroded specimen.

The pores with radius ranges of 0.9–5, 5–10, 10–20, and >20 μm in the TBC specimen corroded by VA (coated at 20 mg/cm^2) and held at 1250 °C for 24 h account for 17.05%, 27.4%, 26.97%, and 27.83% of the total pore volume, respectively. A comparison of the original and corroded specimens shows that the proportion of the pores with radius ranges of 0.9–5, 5–10, and 10–20 μm in the TBC decreases by 9.78%, 7.89%, and 0.39%, respectively, after corrosion, suggesting that high-temperature VA corrosion primarily affects the pores smaller than 20 μm in radius. This is because the VA transformed into a molten state at high temperatures and subsequently infiltrated into the ceramic layer and filled its small pores, whereas some

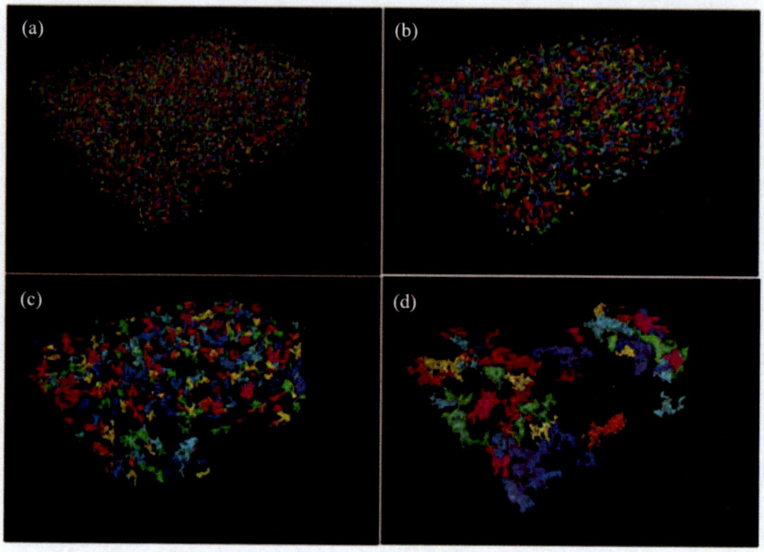

Fig. 13.36. 3D distribution of the pores with different radius ranges in a coating specimen under high-temperature VA corrosion: **a** 0.9–5 μm, **b** 5–10 μm, **c** 10–20 μm, and **d** >20 μm

adjacent or close large pores grew and coalesced, resulting in an increase in the pore size. This finding is consistent with that obtained by Pelissari et al. [38] According to the earlier analysis, high-temperature VA corrosion predominantly affected small pores (i.e., those smaller than 20 μm in radius), but the specific distribution remains unclear. To analyze in detail the influence of high-temperature VA corrosion on the pore size in the APS TBC, the distribution of the radius and number of pores in an original specimen and a specimen subjected to high-temperature VA corrosion (i.e., a specimen coated with VA at 20 mg/cm^2 and held at 1250 °C for 24 h) was determined, as shown in Fig. 13.37 [37].

In Fig. 13.37, the *x*- and *y*-axes represent the pore radius and the corresponding relative pore frequency, respectively. The relative frequency refers to the ratio of the number of pores with a certain radius to the total number of pores. Because the two curves gradually coincide with each other as the radius increases, only the part of each curve with a pore radius less than 12 μm is presented. The black dashed-dotted curve and red solid curve show the distribution of the radius and relative frequency of the pores in the original and high-temperature VA-corroded APS TBC specimens, respectively. Figure 13.37 shows that the red solid curve is below the black dashed-dotted curve in the radius range of 0.89–3 μm and slightly above the black dashed-dotted curve in the radius range above 3 μm, suggesting that the high-temperature VA-corroded specimen contains fewer pores smaller than 3 μm in radius and slightly more pores larger than 3 μm in radius than the original specimen. After melting, the VA rapidly filled the small pores, thereby densifying the ceramic layer and causing the coating more prone to spallation failure. This result is consistent with that obtained by Pelissari et al. [38].

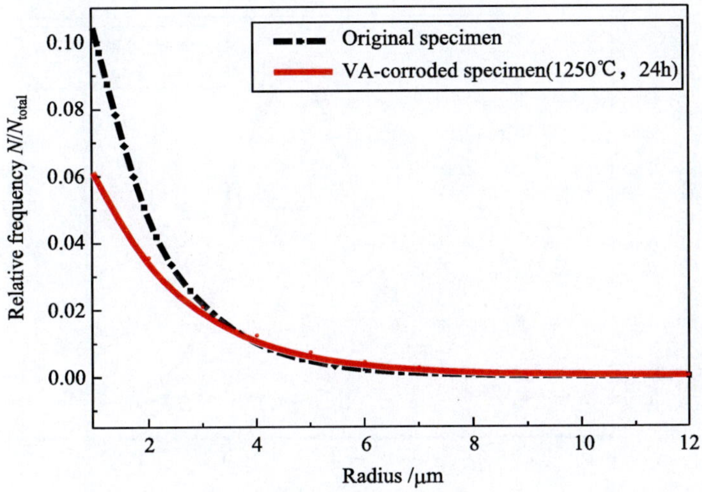

Fig. 13.37 Distribution of the radius and relative frequency of the pores in an original TBC specimen and a VA-corroded TBC specimen

(2) **Influence of high-temperature VA corrosion on the shape of the pores in TBCs**

The sphericity ψ is an important parameter describing the shape of a pore and represents the closeness of the shape of a pore to a sphere. A high level of closeness means that the shape of the pore is close to that of a sphere, which has a ψ of 1. After pores are segmented from a TBC, Avizo can be used to identify the position coordinates of each pixel comprising a pore. On this basis, 3D information (e.g., the spatial location, volume, and shape) can be obtained for each pore. In the Avizo software, the shape of a pore is reflected by its ψ determined through the calculation of the shape factor S_f. S_f can be directly calculated in the software using the following equation [39]:

$$S_f = \frac{A^3}{36\pi V^2} \tag{13.8}$$

where A is the surface area of a pore (unit: μm^2) and V is the volume of a pore (unit: μm^3). The ψ and S_f of a pore are related through the following conversion equation:

$$\psi = \sqrt[3]{\frac{1}{S_f}} \tag{13.9}$$

Substitution of Eq. (13.8) into Eq. (13.9) yields the following expression of ψ:

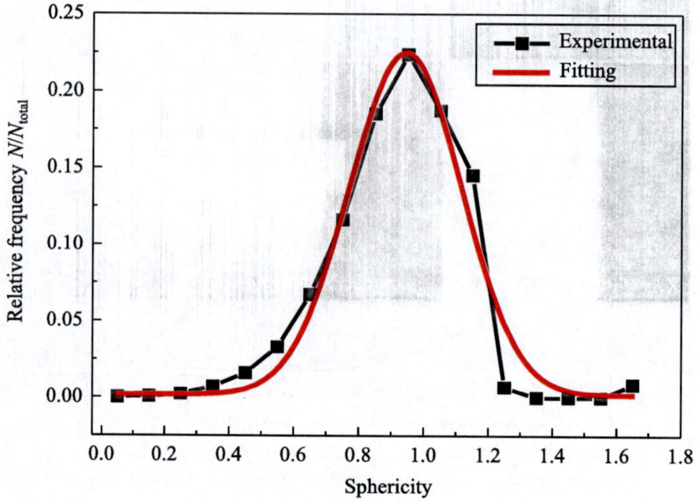

Fig. 13.38 Distribution of the sphericity and relative frequency of the pores in the original APS TBC specimen

$$\psi = \frac{\pi^{1/3}(6V)^{2/3}}{A} \tag{13.10}$$

After all the pores were extracted from the original APS TBC specimen, the ψ of each pore was calculated using Eq. (13.10). Figure 13.38 shows the distribution of the ψ values of all the pores in the original specimen [37].

In Fig. 13.38, the x- and y-axes show the ψ value calculated using Avizo and the corresponding relative frequency, respectively. The relative frequency refers to the ratio of the number of pores with a certain ψ value to the total number of pores. The black curve was obtained through statistical calculation in the software Origin based on the ψ values of all the pores in the specimen, while the red curve shows the function fitted to the statistical data. The fitting results show that the ψ and relative frequency of the pores in the original specimen follow a normal distribution centered at 0.94 and expressed as $y = 0.22e^{-16.69(x-0.94)^2}$ and have a goodness-of-fit of 95%.

According to the previously reported classified ranges of ψ values, pores with ψ values of 0.8–1, 0.4–0.6, and 0–0.2 are referred to as spherical or spheroidal pores, intermediate spherical pores, and flat pores, respectively [29]. Based on this classification, we can see that the majority of the pores in the original specimen are spherical or spheroidal pores with regular shapes. The presence of regularly shaped pores results in approximately similar structural, thermal, and mechanical properties between different parts of a TBC and thus helps to improve its working performance, prolong its service life, and enhance its stability.

The shape of the pores in a TBC is a factor affecting its performance. Here, to determine the influence of high-temperature VA corrosion on the shape of the pores in the APS TBC, the ψ values of the original specimen and specimens corroded by VA under different conditions are analyzed side-by-side, as shown in Fig. 13.29 [37].

In Fig. 13.39, the x- and y-axes show the amount of coated VA (0, 5, 10, 15, and 20 mg/cm^2) and the corresponding ψ value, respectively. The broken lines connecting solid red circles and connecting solid black squares show the variation in the ψ of the TBC corroded by VA at 1250 and 1150 °C, respectively. Each data point represents the mean ψ of all the pores in a certain specimen, with the error bars showing the corresponding standard deviation. The radii of the pores in the TBC range widely from 0.89 to 100 μm, which results in a significant fluctuation in ψ and thus large error bars.

As shown in Fig. 13.39, the VA corrosion temperature significantly affects the ψ of the pores. A higher VA corrosion temperature leads to a higher ψ, i.e., a higher level of irregularity in the shape of the pores. Such pore shape irregularity leads to inconsistent heat transfer in a TBC or even causes a thermal mismatch between the components of a TBC and eventually threatens its working performance and service life in an engine. The influence of the pore shape on thermal conductivity has been reported previously [40]. Compared to the corrosion temperature, the amount of coated VA has little influence on the ψ of the pores. At a corrosion temperature of 1250 °C, the ψ of the pores fluctuates little with the amount of coated VA. This is because at this temperature, the VA melts to a large degree and subsequently infiltrates into and corrodes the TBC to a similarly large extent.

(3) **Influence of high-temperature VA corrosion on the porosity of TBCs**

The porosity of the TBC was determined by processing the pore data obtained from CT scans. Specifically, the ratio of the volume of the pores in each "sliced" specimen to the volume of the "sliced" specimen was used as the area porosity of this layer. The area porosities of all the slices comprising a specimen were averaged as the

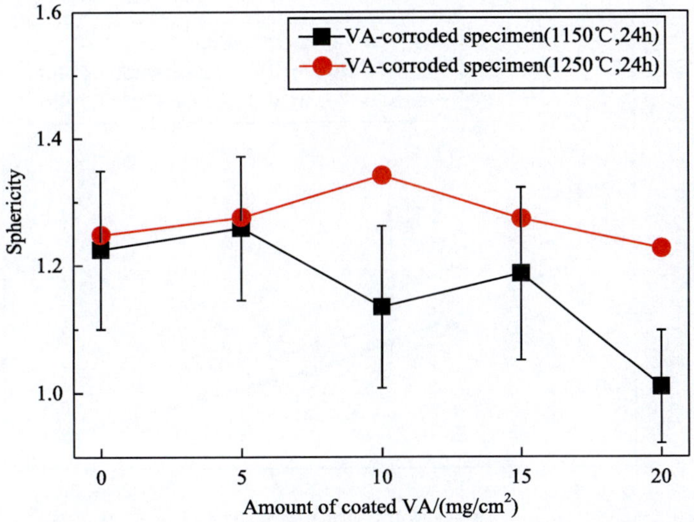

Fig. 13.39 Variation in the shape of the pores in the TBC with the VA corrosion conditions

porosity of the specimen. The standard deviation of all the area porosities was used as the error bar. Here, a "slice" refers to a layer with a thickness of 0.89 μm in the thickness direction of a TBC specimen.

To analyze the influence of high-temperature VA corrosion on the porosity of the APS TBC at different temperatures and coated with different amounts of VA, the porosities of an original APS TBC specimen and APS TBC specimens corroded by VA under different conditions (corrosion temperature: 1150 and 1250 °C; the amount of coated VA: 0, 5, 10, 15, and 20 mg/cm^2) are compared, as shown in Fig. 13.40 [37].

In Fig. 13.40, the *x*- and *y*-axes represent the amount of coated VA (0, 5, 10, 15, and 20 mg/cm^2) and the corresponding porosity, respectively. The broken lines connecting solid red dots and connecting solid blue triangles show the variation in the porosity of the APS TBC after VA corrosion at 1150 and 1250 °C, respectively. The single solid black square data point shows the porosity of the original APS TBC. The trends of the curves show that the amount of coated VA has a significant influence on the porosity of the TBC. The internal porosity of the TBC decreases as the amount of coated VA increases. This is because the VA infiltrates into the interior of the coating through the adjacent or interconnected pores and cracks and subsequently fills the microdefects in the coating and, as a result, accelerates the sintering of the coating under long-term high-temperature service conditions. In addition, the porosity of each VA-corroded APS TBC specimen is lower than that of the original APS TBC specimen. This finding is consistent with the conclusion obtained by Wu et al. [41] The porosity decreases as the corrosion temperature increases. However, compared to the amount of coated VA, the VA corrosion temperature has a relatively weak influence on the porosity. This is because at 1150 °C, the VA is in a molten glassy state and has mostly infiltrated into the ceramic layer with a thickness of

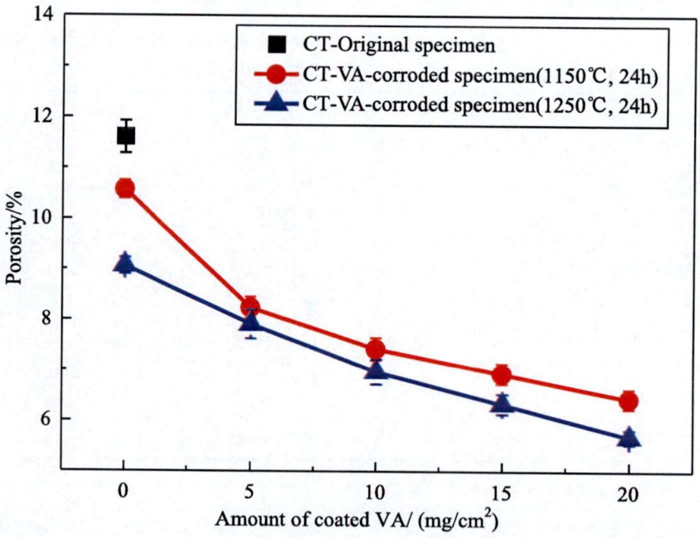

Fig. 13.40 Variation in the porosity of the TBC with the VA corrosion conditions

approximately 250 μm [42]. A decrease in the porosity leads to an increase in the thermal conductivity of a TBC, thereby affecting its thermal insulation performance and, consequently, the service life of the turbine blade to which it is applied [43].

13.3 IRT NDT Technique and Its Current Application Status

Over the years, numerous methods have been developed for characterizing the internal defect states of materials. Destructive testing methods (e.g., SEM and metallographic microscopy [44]) are able to detect defects only within local planes. In addition, these methods can irreversibly damage specimens; thus, the specimens are not reusable. As the most common NDT techniques, ultrasonic (UT), eddy-current, and X-ray testing can be employed to detect internal defects in materials. However, the detection probes used in these techniques are incapable of sufficiently covering components with complex geometries (e.g., gas turbine blades) [39]. Moreover, coupling agents are often required in the testing of these components. However, a coupling agent fills the surface pores of a coating on a turbine blade and, thereby, damages the key characteristic structure of the coating that provides thermal insulation functions. Consequently, these testing techniques are unsuitable for characterizing internal defects in TBCs. In recent years, IRT NDT has become an important technique for characterizing the internal defect state of materials. IRT is an NDT technique based on transient heat transfer. The internal defects in a specimen affect heat transfer, resulting in a nonuniform surface temperature distribution. IRT identifies the interfacial damage of a specimen based on its surface thermogram recorded by an IR thermal imager [46]. IRT has the advantages of a large single detection area, visual detection results, high detection efficiency, and noncontact detection. The detection of the internal defects (e.g., cracks and voids) in a specimen can be achieved through the collection of surface temperature distribution data for the specimen with an IR thermal imager, the processing of the temperature image sequence, the extraction of the characteristics of temperature changes at the defects, and the analysis of the internal state of the specimen.

13.3.1 Principle of IRT

Any object with a temperature higher than 0 K emits energy (radiant energy) in the form of electromagnetic waves, a process referred to as thermal radiation [47]. An object with a higher surface temperature radiates a larger amount of energy. While radiating energy, the surface of an object also inevitably absorbs the energy radiated from other objects. The radiant thermal energy from excitation sources received by the surface of an object consists of three components, namely the absorbed energy, reflected energy, and transmitted energy, and the ratios of their dissipated portions are referred to as the absorptivity α_λ, reflectivity ρ_λ, and transmissivity τ_λ, respectively, which are related to each other through the following equation:

$$\alpha_\lambda + \rho_\lambda + \tau_\lambda = 1 \tag{13.11}$$

For an object with an extremely low τ_λ, the radiant energy reaching its surface can only be reflected or absorbed. For a material with a τ_λ of 0, the above equation can be written as follows:

$$\alpha_\lambda = 1 - \rho_\lambda \tag{13.12}$$

An object unable to transmit or reflect radiant energy on its surface is called a blackbody ($\alpha_\lambda = 1$). The radiant energy emitted by an actual object to the surrounding environment is always lower than that emitted by a blackbody.

The energy radiated by an actual object is related to its temperature through the following equation:

$$Q = \varepsilon \sigma A T^4 \tag{13.13}$$

where Q is the amount of energy radiated by an actual object (unit: W), A is the radiation area (unit: m^2), σ is the blackbody radiation constant or the Stefan–Boltzmann constant ($\sigma = 5.67 \times 10^{-8} W/(m^2 \cdot K^4)$), and ε is the surface emissivity of the object, which is defined as the ratio of the amount of energy radiated by the object to that radiated by a blackbody at the same temperature.

An object with a wavelength-independent ε is referred to as a graybody [47]. Generally, the ε of an object is a wavelength-dependent variable. The ε of a solid varies slowly with the wavelength and can be usually considered a constant within a narrow waveband.

An IR thermal imager measures the temperature of an object by converting the received IR radiant energy emitted by its surface to temperature, as shown in Fig. 13.41. During the temperature measurement process, to accurately measure the surface temperature of an object, the IR radiant energy received by the thermal imager from the surrounding environment or atmosphere is eliminated through compensation [48].

The total IR radiant energy W_{tot} received by an IR thermal imager consists of three parts, [48] namely the radiant energy emitted by the test object, E^{at}_{obj}; the radiant energy reflected by the object, E^{at}_{refl}; and the atmospheric radiant energy E_{atm}, i.e.,

$$W_{tot} = E^{at}_{obj} + E^{at}_{refl} + E_{atm} \tag{13.14}$$

The amount of IR radiant energy emitted by a target object to the surrounding environment can be calculated using the following equation:

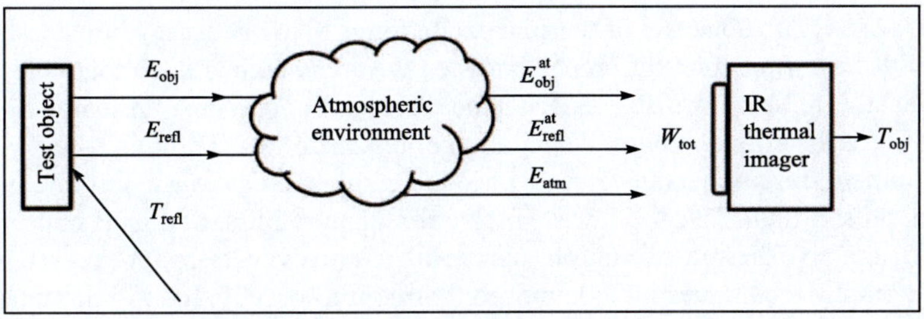

Fig. 13.41 Schematic diagram of the principle of temperature measurement by an IR thermal imager (E_{obj} and E_{refl} are the radiant energy emitted and reflected by the object, respectively)

$$E_{obj}^{at} = \varepsilon_{obj} \cdot \tau_{atm} \cdot \sigma \cdot \left(T_{obj}\right)^4 \tag{13.15}$$

where T_{obj} is the surface temperature of the test object (unit: K) and ε_{obj} is the emissivity of the test object.

The amount of IR radiant energy reflected by a target object can be calculated as follows:

$$E_{refl}^{at} = \left(1 - \varepsilon_{obj}\right) \cdot \tau_{atm} \cdot \sigma \cdot \left(T_{refl}\right)^4 \tag{13.16}$$

The amount of atmospheric radiant energy is expressed as follows:

$$E_{atm} = (1 - \tau_{atm}) \cdot \sigma \cdot (T_{atm})^4 \tag{13.17}$$

Based on Eqs. (13.15), (13.16), and (13.17), (13.14) can be organized as follows:

$$W_{tot} = \varepsilon_{obj} \cdot \tau_{atm} \cdot \sigma \cdot \left(T_{obj}\right)^4 + \left(1 - \varepsilon_{obj}\right) \cdot \tau_{atm} \cdot \sigma \cdot \left(T_{refl}\right)^4 + (1 - \tau_{atm}) \cdot \sigma \cdot (T_{atm})^4$$

$$T_{obj} = \sqrt[4]{\frac{W_{tot} - \left(1 - \varepsilon_{obj}\right) \cdot \tau_{atm} \cdot \sigma \cdot \left(T_{refl}\right)^4 - (1 - \tau_{atm}) \cdot \sigma \cdot (T_{atm})^4}{\varepsilon_{obj} \cdot \tau_{atm} \cdot \sigma}} \tag{13.18}$$

where T_{refl} is the temperature of the reflection surface of the target object, ε_{obj} is the emissivity, τ_{atm} is the atmospheric transmissivity, and T_{atm} is the atmospheric temperature.

13.3.2 IRT-Based Damage Detection

There are two types of IRT detection, namely active IR detection and passive IR detection, which differ principally in whether an actively controlled thermal excitation mode is used. The objective of thermal excitation is to inject energy into a test object to produce a temperature difference between the areas with and without defects and to reflect this temperature difference on the surface. A core topic of research on active IRT NDT is the efficient thermal excitation of test materials. Through recent years of development, researchers have used various energy sources (e.g., sunlight, halogen lamps, pulsed light, laser, UT waves, electromagnetic induction, and microwaves) [49, 50] as active thermal excitation sources in their investigations. The most common thermography techniques include pulsed thermography (PT), lock-in thermography (LIT), modulated thermography, and laser thermography [51–54]. In this section, PT and LIT are discussed in detail.

In photoexcited heating systems, data acquired by IR cameras are often contaminated by various noise sources (e.g., external reflection). Noise (e.g., the difference in the optical properties between areas of a specimen and uneven heating by the excitation source) leads to abnormal thermal modes in thermography, which further complicates damage detection. As a result of noise and uneven heating, IR images contain a small amount of defect information with low reliability. In addition, the microcrack information in IR images is partially covered by noise. To address these problems, it is necessary to process IR image sequences. This is also a key technique in the IRT NDT field for eliminating the interference from adverse factors, improving the signal-to-noise ratio (SNR) of IR images, and enhancing defect visibility. There are four common techniques for processing IR image sequences [55]. (1) Thermal signal reconstruction: an IR image sequence is reconstructed by taking its double logarithm and fitting it to a polynomial, followed by taking the derivative of the reconstructed sequence with respect to time to obtain first- and second-order images. (2) Principal component analysis (PCA): based on the concept of dimensionality reduction, an image sequence is decomposed into a small number of signal space-varying orthogonal eigenfunctions and time-varying principal components (PCs); these PCs are capable of reflecting most of the information of the original image sequence, and the information contained in each PC is not duplicative of that contained in any other PC. (3) LIT: an image sequence is collected by locking in the phase angle of the periodic IR radiation from the test specimen to improve the SNR of IR images. (4) Pulse phase thermography: the obtained IR image sequence is subjected to Fourier transform to produce amplitude and phase images that reflect thermal information at different depths, thus overcoming the time-consuming problem of LIT NDT.

(1) PT

PT is a rapid NDT technique that involves heating the surface of a test specimen using an intense pulsed light through the photothermal effect. If there is a defect in the specimen, heating changes the thermal energy transfer mode at the defect, resulting in a change in the surface temperature field of the specimen [49]. An IR camera and a computer system can be used to monitor the changes in the surface temperature of a specimen to quantify the extent of its damage. The principle of PT is described in detail as follows. An intense pulsed heat flow irradiates a specimen. The amount of thermal energy received by the surface of the specimen per unit area can be expressed using the following 1D heat transfer equation [56]:

$$\frac{\partial^2 T(x,t)}{\partial x^2} - \frac{1}{\alpha}\frac{\partial T(x,t)}{\partial t} = -\frac{q(t)\delta(x)}{k} \tag{13.19}$$

where k is the thermal conductivity of the specimen, $\alpha = k/\rho c$ is the thermal diffusivity (ρ is the density and c is the specific heat), T is the temperature, t is the heating time, and the $\delta(x)$ function signifies that the heat source is present only on the surface. For a 1D semi-infinitely long rod subjected to pulsed heating, it is very easy to determine the temperature when there are no defects:

$$T_n(x,t) = \frac{J_0}{\sqrt{\pi\rho c k t}} \exp\left(-\frac{x^2}{4\alpha t}\right) \tag{13.20}$$

When there exists a defect in the specimen, the surface temperature $T_d(0, t)$ of the specimen varies with the defect below the surface and is expressed as follows:

$$T_d(0,t) = \frac{J_0}{\sqrt{\pi\rho c k t}} \left[1 + 2\sum_{n=1}^{\infty} R^n \exp\left(-\frac{n^2 d^2}{\alpha t}\right)\right] \tag{13.21}$$

where J_0 is the magnitude of the pulse energy, d is the depth of the defect, and R is the effective thermal reflectivity of the solid/air interface at the defect (R is usually approximately taken as 1). R is related to the thermal resistivity Z through $R+Z=1$.

On this basis, the difference between the surface temperatures of the area directly above the defect and the area with no defects below it is determined as follows:

$$\Delta T = T_d(0,t) - T_n(0,t) = \frac{2J_0}{\sqrt{\pi\rho c k t}} \exp\left(-\frac{d^2}{\alpha t}\right) \tag{13.22}$$

Thus, after a specimen is subjected to pulsed heating, whether defects exist in the specimen can be determined by monitoring its surface temperature field using IRT. In addition, the depth of each defect below the surface can be calculated based on the difference between the periods of time needed for the area above the defect and the area with no defects below it to reach the peak surface temperature.

The heat transfer equation is a parabolic one. Therefore, the speed of heat propagation would be infinite if temperature propagation were treated as wave propagation. In fact, the temperature rises very little beyond a certain distance of heat transfer under thermal pulse action. Therefore, in engineering practice, the distance between the starting point of thermal pulse transfer and the point at which the temperature rises to a certain value (e.g., the point at which the temperature rises by $1/\mathrm{e}$ due to the pulsed heat source) is approximated as the distance of thermal pulse transfer. Particular attention should be directed to the fact that thermal pulse transfer is not heat wave propagation. In many research papers and textbooks, the heat transfer from a pulsed heat source is referred to as heat wave propagation, which is inappropriate. Interested readers can refer to the relevant literature [57, 58].

Tang et al. [59] quantified the diameter and depth of the debonding defects in TBCs using pulsed IRT in combination with principal component thermography and neural network theory (as shown in Fig. 13.42). The depth and diameter prediction errors for defects with a diameter-to-depth ratio of 1.2–4.0 and a depth of 1.0–2.5 mm were approximately 4–10%, demonstrating the effectiveness of this method in the quantitative detection of defects. During the detection process, the pulse energy, sampling frequency, and sampling time were set to 2280 J, 90 Hz, and 1 s, respectively. A total of 28 TBC specimens with flat-bottomed hole defects, which mimicked delamination defects, were tested using pulsed IRT. An SC7000 IR camera was used to capture 320 × 256 × 90 thermal image sequences. Therefore, each pixel of an image should be described with 90 quantities (components), resulting in a very large volume of data. The contribution and cumulative contribution of each image pixel can be calculated using Eqs. (13.23) and (13.24), respectively.

$$y_k = \lambda_k / \sum_{i=1}^{n} \lambda_i \tag{13.23}$$

$$\sum_{k=1}^{m} y_k = \sum_{k=1}^{m} \lambda_k / \sum_{i=1}^{n} \lambda_i \tag{13.24}$$

where y_k is the contribution of the kth component Z_k and $\sum_{k=1}^{m} y_k$ is the cumulative contribution of $Z_1, Z_2, \ldots, Z_m$.

Based on the principle of data dimensionality reduction and reconstruction, data containing defect features were extracted from the complex and lengthy multiframe IR thermal image sequence using PCA through singular value decomposition (SVD). The principle of SVD is expressed as $[A] = [U][R][V]^T$. Each column vector in matrix $[A]$ records the spatial location of a pixel in the IR thermal images, while

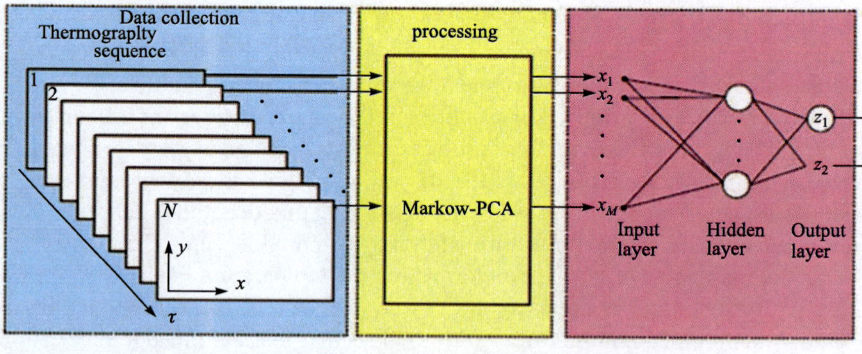

Fig. 13.42 Test results for the carbon fiber-reinforced plastic components of a space shuttle obtained using an IR thermal imager [59]

each row vector records the variation in the temperature with time at a certain location or pixel in the thermal images. [R] is a 90 × 90 diagonal matrix, where all the nondiagonal elements are equal to zero and the diagonal elements are the singular eigenvalues of [A]. [U] is the left singular matrix of [A], with the column vectors in [U] being the singular eigenvectors of [A], i.e., the column vectors 1, 2, 3,... in [U] correspond to PC1, PC2, PC3,... in [A]. The first several PCs contain the vast majority of the information contained in [A]. Through separate extraction and analysis of the first few PCs, complex and extensive data can be significantly simplified without removing the characteristic defect information from the thermal images. With the filtering of redundant and unimportant information, PCs that describe the main information features can be extracted. Figure 13.43 shows the feature maps of PC1–9.

Figure 13.44 shows the contributions of PC1–9. The cumulative contribution of PC1–5 reaches 97.4%, suggesting that these PCs can represent the majority of the thermal image sequences. Thus, the first five PCs can be used in the subsequent analysis of the diameter and depth of the debonding defects in the TBC. During the establishment of a neural network model, PC1–5 corresponding to a total of 100 points (including all the center points of the defects and some defect-free dots in the selection zone) (Fig. 13.45) were collected as sample data for analysis. Sample S4 in the "prediction zone" (Fig. 13.46) was selected as an object of analysis. The established neural network model was used to predict the depths and diameters of the defects, as shown in Fig. 13.46. As seen in Fig. 13.46a, four defects (#2, #4, #5, and #7) were identified, with the diameter-to-depth ratio of the smallest defect (i.e., #4) being 1.2. Figure 13.46b shows the distribution of the defect sizes on a transverse line passing through the center of each defect in Fig. 13.46a.

We sprayed TBCs with different thicknesses on metallic substrates [60] and prepared interfacial defects with different sizes at the ceramic-layer/BC-layer interface. Subsequently, the correlations of the coating thickness and transverse defect size with the surface temperature distribution of the coating were investigated using long-pulse IRT. Figure 13.47 shows the prefabricated interfacial defects. Surface temperature signals corresponding to different defect sizes were collected using a cooled focal-plane-array IR thermal imager (FLIR, USA), as shown in Fig. 13.48. The IR thermal images show that the temperature in the defect zones is significantly higher than that in the defect-free zones. In addition, the boundary of the defects in the IR thermal images becomes increasingly distinguishable as the interfacial defect sizes increase, while the outlines of the defects with small transverse sizes are blurry.

PCA is often used to denoise a thermal image sequence containing multiple images with noise caused by uneven heating, to accentuate the difference between temperature data for defect and defect-free zones, and to improve the ability to identify defects in thermal images [60, 61]. The temperature data (T) contained in the sequence of IR thermal images of a specimen surface collected by the IR thermal imager during the cooling stage were nondimensionalized and subsequently processed using the PCA algorithm to produce a reconstructed IR thermal image, as shown in Fig. 13.49. The defects are clearly distinguishable with complete shapes in the reconstructed IR thermal image.

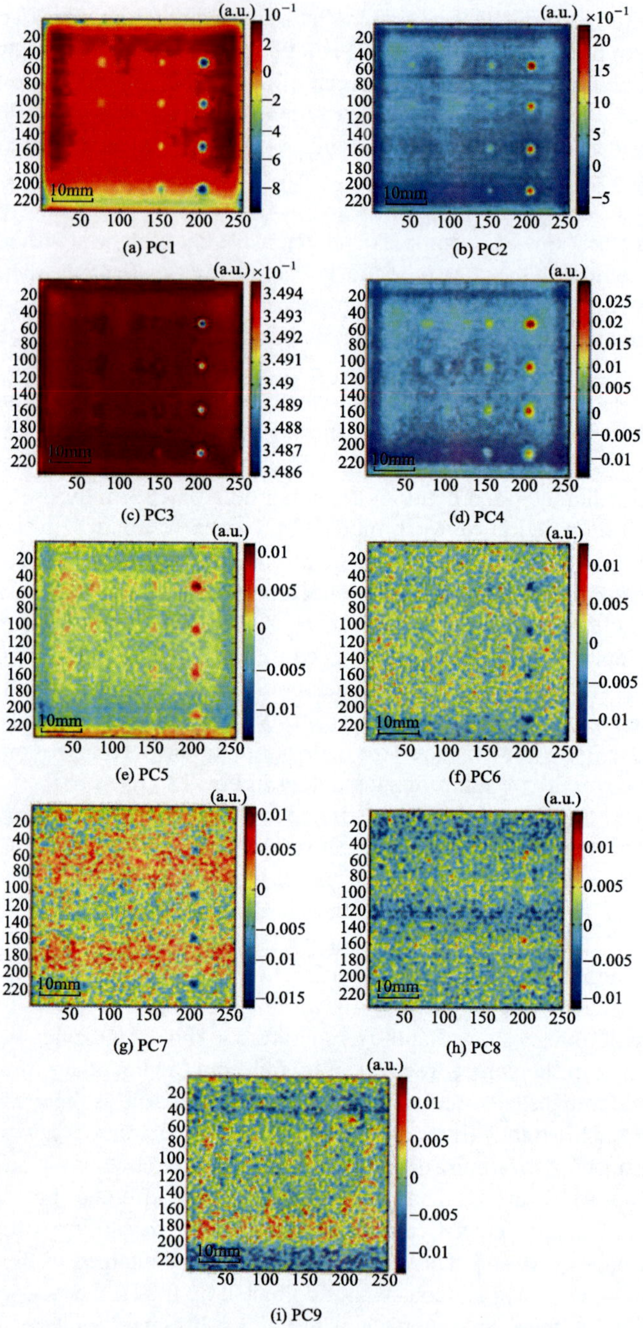

Fig. 13.43 Feature maps of PCs PC1–9 [59]

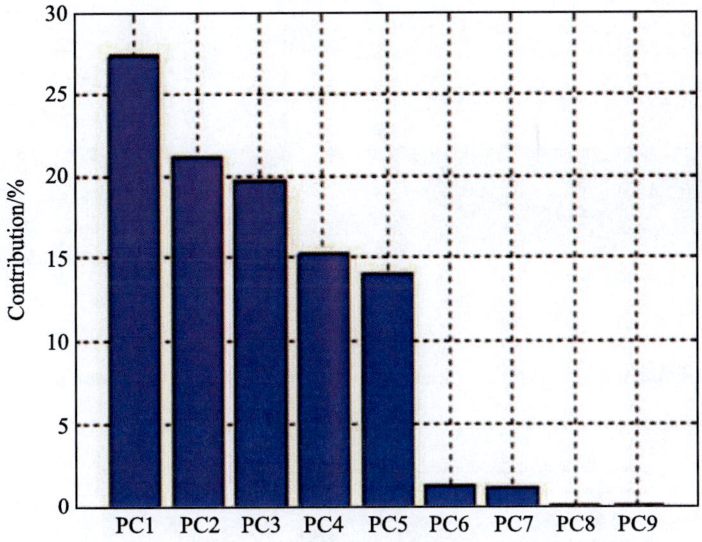

Fig. 13.44 Contributions of PC1–9 [59]

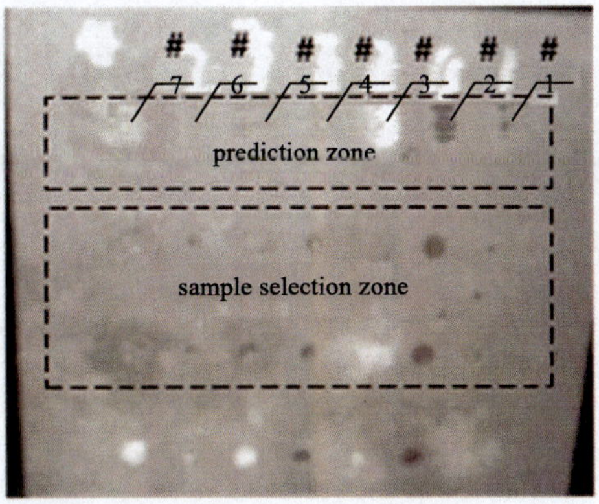

Fig. 13.45 The selection zone and prediction zone of a sample [59]

Uneven transient illumination by an excitation source on the surface of a specimen leads to an uneven temperature distribution in its IR thermal image. The influence of background thermal noise can be eliminated through background subtraction and fitting. Figure 13.50a shows a 3D thermal image of an arbitrary defect (the *x*- and *y*-axes show the spatial position coordinates, and the *z*-axis shows the temperature of the defect). The temperature is unevenly distributed in the defect-free zone. Overall, the temperature is high on the left and low on the right. The temperature of the defect zone is higher than that of the surrounding nondefect zone. Figure 13.50b shows the thermal image obtained after background subtraction and fitting. In this image, the temperature is evenly distributed in the defect-free zone, and the original temperature distribution features are preserved, demonstrating that the influence of

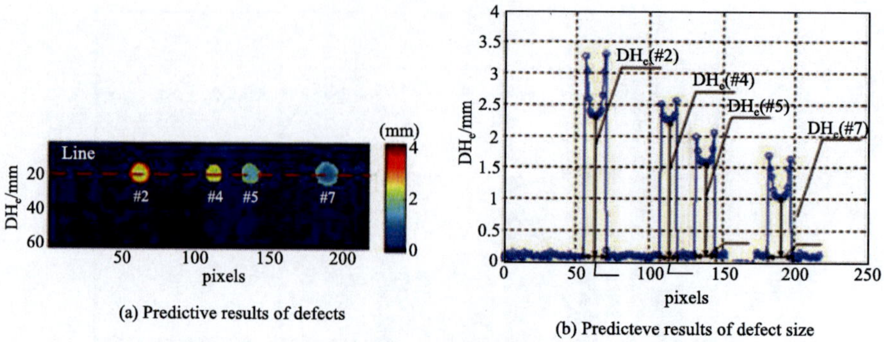

Fig. 13.46 Prediction results of the neural network: **a** Predicted defect locations. **b** Predicted defect sizes [59]

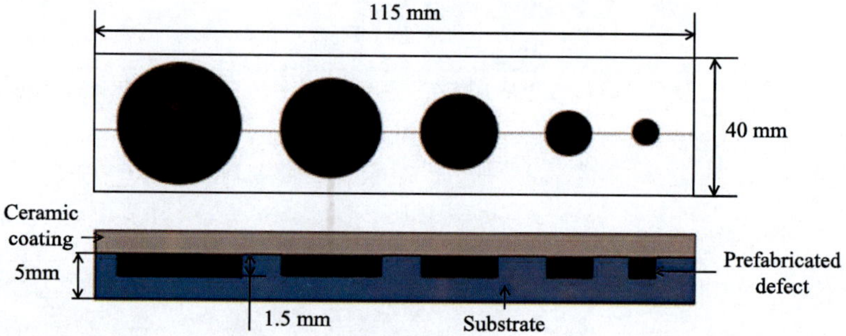

Fig. 13.47 Structural schematic diagram of interfacial defects [60]

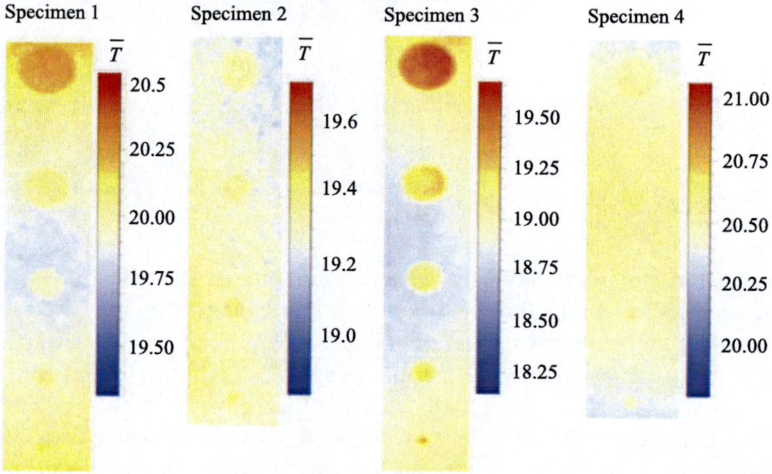

Fig. 13.48 Original thermal image sequence in a certain frame [60]

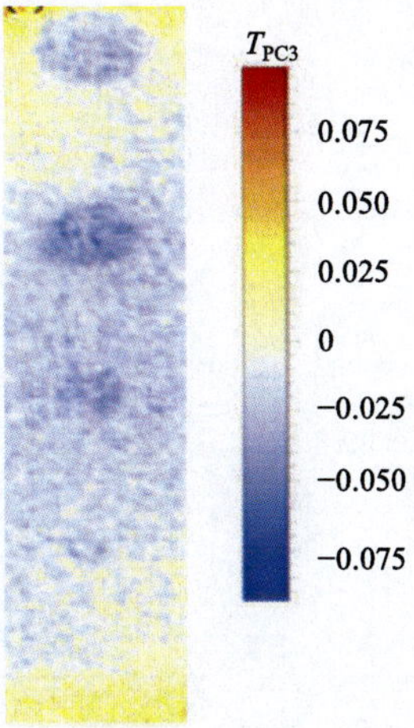

Fig. 13.49 The reconstructed IR thermal image of a sample processed using the PCA algorithm [60]

uneven background temperature on defect identification is eliminated by background subtraction and fitting.

From the perspectives of finite element (FE) numerical simulation and experimentation, Shrestha and Kim [62, 63] nondestructively evaluated the thickness of the ceramic layer of a TBC by thermal wave imaging (TWI) and compared PT- and LIT-based characterizations. To simulate the features of pulsed heating, a rectangular heat flow is applied to the surface of a coating specimen, as expressed in the following:

$$Q = \frac{Q_{\max}}{2}\left(1 - \mathrm{sign}(t - t_{\mathrm{p}})\right) \tag{13.25}$$

$$\mathrm{sign}(t - t_{\mathrm{p}}) = \begin{cases} -1 & 0 < t \leq t_{\mathrm{p}} \\ 0 & t = t_{\mathrm{p}} \\ 1 & t > t_{\mathrm{p}} \end{cases} \tag{13.26}$$

where Q is the incident heat flow, $Q_{\max}$ is the peak pulsed heat flow, t_{p} is the pulsed heating time, and t is the total time. The Fourier transform is an interesting data processing technique that is capable of transforming data from the time-domain space to the frequency-domain space. Using the Fourier transform, phase angles can be extracted from IR thermal image sequences obtained from PT and LIT tests.

The 1D discrete Fourier transform is expressed as follows:

$$F_{\mathrm{n}} = \Delta t \sum_{k=0}^{N-1} T(k\Delta t)\exp^{\frac{j2\pi nk}{N}} = \mathrm{Re}_n + \mathrm{Im}_n \tag{13.27}$$

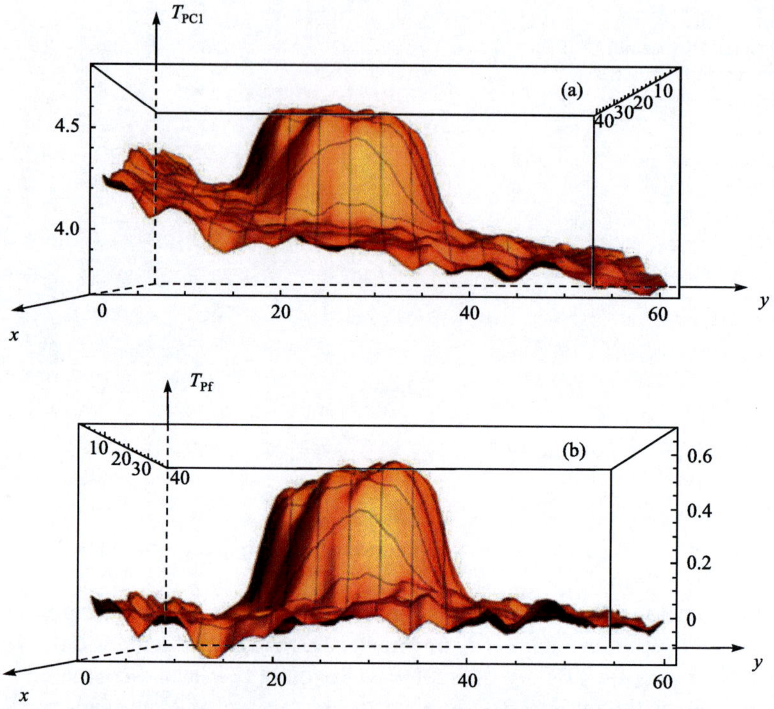

Fig. 13.50 3D thermal images of a defect in a sample: **a** original 3D distribution of the temperature of the defect; **b** 3D distribution of the temperature of the defect after background subtraction and fitting [60]

where j is the imaginary unit ($j^2 = -1$), n is the frequency increment ($n = 0, 1, \ldots, N$), Δt is the sampling time interval, and *Re* and *Im* are the real and imaginary parts of the transform, respectively. The phase angle can be obtained from the real and imaginary parts of the complex transform, as expressed below:

$$\varphi_n = \tan^{-1}\left(\frac{\mathrm{Im}_n}{\mathrm{Re}_n}\right) \tag{13.28}$$

Figure 13.51 illustrates the basic principle of coating thickness evaluation based on thermography. When thermal energy is applied to the surface of a coating, the difference in its thickness interrupts heat transfer, thereby affecting the variation in its temperature distribution with time. The difference in the temperature distribution is a key parameter for evaluating the thickness of a coating. This quantitative information can be obtained by processing IR thermal image data. Figure 13.52 shows the geometric size distribution of the specimen (180 mm × 180 mm) used by Shrestha and Kim [62]. The in-plane thickness of the ceramic layer showed a stepped distribution ranging from 0.6 to 0.1 mm at an interval of 0.1 mm. In the FE simulation, the energy and duration of the single-pulse heat flow were set to 9 kJ and 10 ms, respectively, and the time step size and duration were set to 0.01 and 5 s, respectively.

The response of the surface of the coating to the applied heat flow within 5 s was determined by transient analysis from computational and experimental perspectives. Figure 13.53 shows the surface temperature decay curves corresponding to different coating thicknesses. Figures 13.54 and 13.55 show the FE-simulated and experimental PT images at different time intervals, respectively. Figures 13.53, 13.54 and 13.55 show that because the pulse duration is very short, the temperature increases faster than it decays. The surface temperature reaches equilibrium within 3 s due to

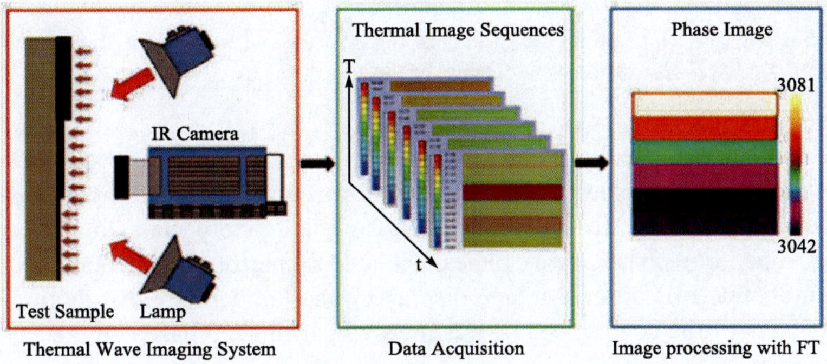

Fig. 13.51 Basic principle of coating thickness evaluation based on TWI [62]

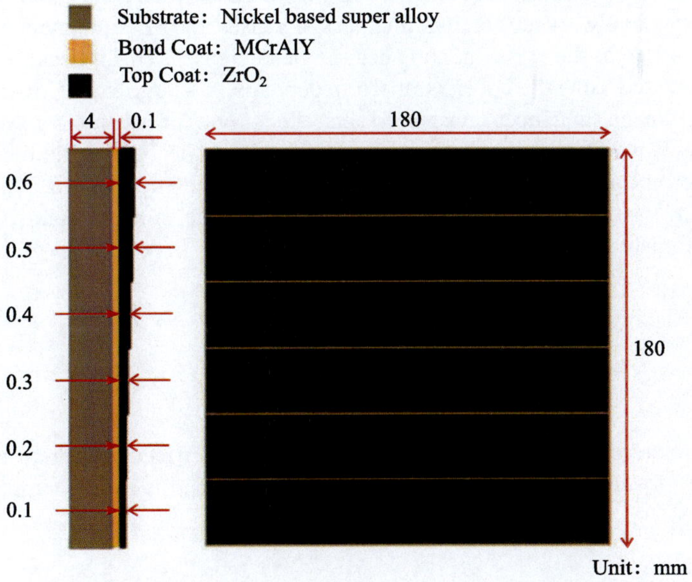

Fig. 13.52 Dimensions of the sample and distribution of the thickness of the ceramic layer [62]

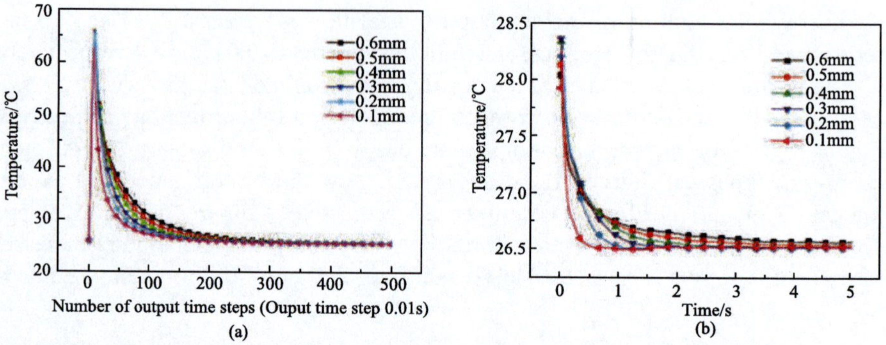

Fig. 13.53 Surface temperature decay curves for coatings with different thicknesses: **a** FE simulation results; [62] **b** experimental results [63]

the thermal diffusion in each direction. In addition, a large coating thickness leads to a relatively high surface temperature. The variation in the surface temperature distribution provides sufficient information for qualitatively evaluating the coating. The corresponding phase angle can be calculated by performing a Fourier transform on each obtained IR thermal image displaying the temperature distribution. Then, the thickness of the coating can be quantitatively evaluated based on the calculated phase angle.

Figure 13.56 shows the phase angle images corresponding to different coating thicknesses obtained by the Fourier transform. Figure 13.57 gives the relations between the coating thickness and phase angle obtained by FE simulation and experiments, respectively. As the coating thickness increases, the FE-simulated phase angle increases, whereas the experimental phase angle decreases. This is because the phase angle calculated using the Fourier transform depends, to a large extent, on the number of thermal image sequences considered (i.e., the number of harmonic waves) during processing. Two harmonic waves were considered in the FE calculation, whereas all the thermal image sequences were considered in the experiments. The relation between the coating thickness and phase angle can be determined as follows by fitting the FE-simulated and experimental data, respectively:

$$T = 0.14295 + 1.11883\varphi + 0.1415\varphi^2 \tag{13.29}$$

$$T = 5.5418 - 3.2137\varphi + 0.4522\varphi^2 \tag{13.30}$$

The thickness of a coating specimen with a known surface temperature distribution can be inverted from Eqs. (13.29) and (13.30), with the results summarized in Table 13.1.

(2) LIT

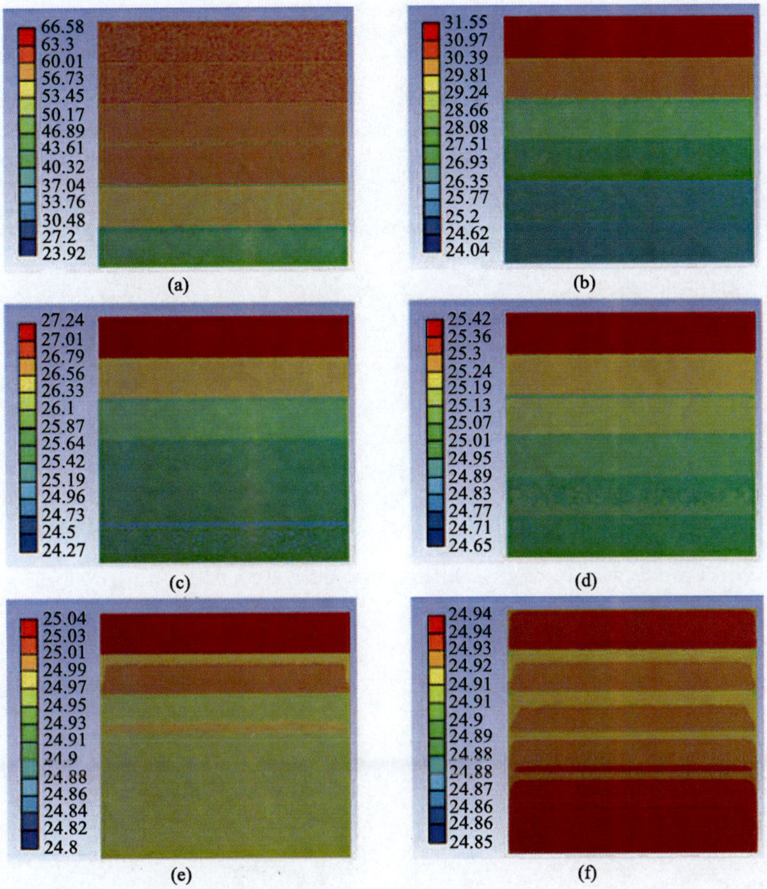

Fig. 13.54 FE-simulated PT images at different time intervals: **a** 0.01 s; **b** 1 s; **c** 2 s; **d** 3 s; **e** 4 s; **f** 5 s [62]

LIT was proposed by Prof. Busse of the University of Stuttgart, Germany [64] The basic principle of LIT detection is described as follows. A modulation signal generator is used to control the intensity of a light source in such a way that it varies sinusoidally. Thermal radiation from the light source is used to heat a test object. Thermal-image collection and heating are conducted on the same side of the test object. The presence of a defect in the surface or interior of the test object leads to uneven diffusion of the heat flow generated by the external excitation heat source in the object, thereby causing a temperature difference between the defect and defect-free zones. An IR thermal imager is then used to detect the thermal radiation from the test object. Multiple thermal images are collected using the digital lock-in technique at specific time points of the heating cycle. Subsequent reconstruction of the signals in the thermal images produces signals of the temperature changes in the surface and interior of the test object. Phase and amplitude images for the temperature changes

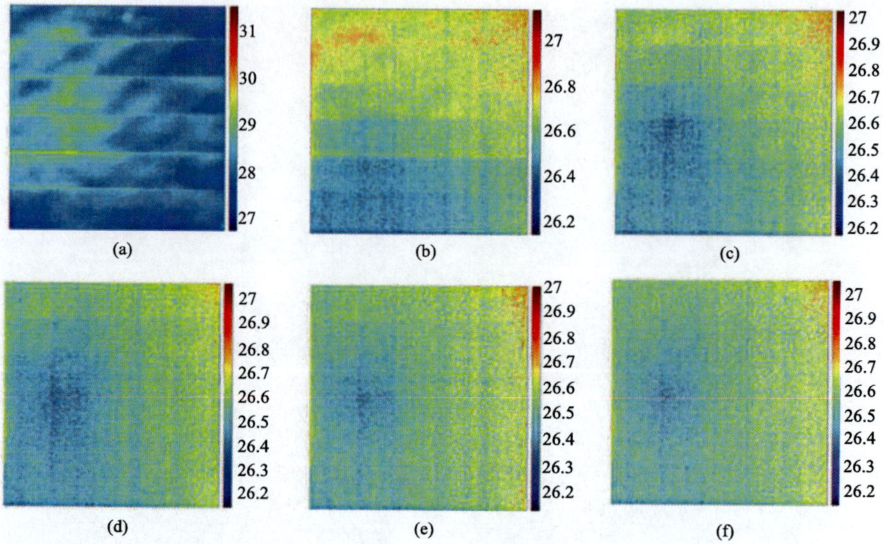

Fig. 13.55 Experimental PT images at different time intervals: **a** 0.01 s; **b** 1 s; **c** 2 s; **d** 3 s; **e** 4 s; **f** 5 s [63]

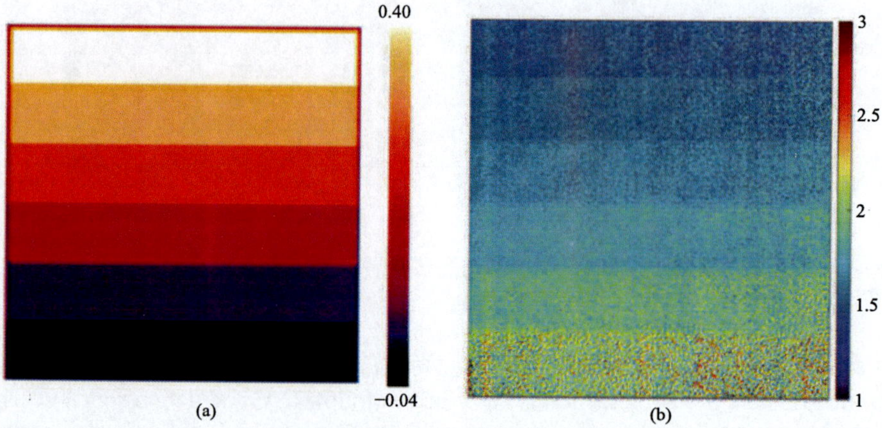

Fig. 13.56 Phase angle images corresponding to different coating thicknesses derived from PT images: **a** FE-simulated images; [62] **b** experimental images [63]

at each point on the surface of the test object are extracted, based on which whether a defect exists and, if so, its features are determined.

The 1D thermal response of an isotropic, homogeneous medium when excited by a uniform heat source with a modulation frequency of ω is expressed as follows [56]:

$$T(x, t) = T_0 e^{-x/\mu} \cos\left(\frac{2\pi x}{\lambda} - \omega t\right) \tag{13.31}$$

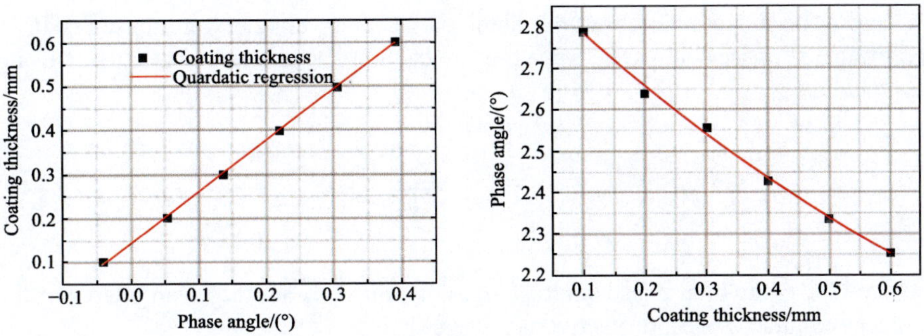

Fig. 13.57 Relations between the coating thickness and phase angle determined by PT: **a** FE-simulated results; [62] **b** experimental results [63]

Table 13.1 PT- and LIT-predicted coating thicknesses and the corresponding errors [62]

Actual thickness (T) (mm)	PT			LIT		
	Calculated (Radian)	Predictive T (mm)	Error (%)	Calculated (Radian)	Predictive T (mm)	Error (%)
0.1	−0.0407	0.0977	2.35	3.081	0.1067	6.73
0.2	0.0546	0.2045	2.23	3.076	0.1911	4.43
0.3	0.1382	0.303	0.09	3.069	0.2988	0.39
0.4	0.2212	0.3974	0.66	3.061	0.4069	1.74
0.5	0.3061	0.4987	0.26	3.053	0.4991	0.19
0.6	0.3906	0.6016	0.26	3.042	0.5997	0.06

where $\mu = \sqrt{2a/\omega}$ is the thermal diffusion length. The defect depth is directly proportional to μ. Evidently, a lower heating modulation frequency allows the detection of greater defect depths. In the detection process, a high modulation frequency is often first selected to facilitate the detection of defects in deep zones, followed by gradually decreasing the frequency. Compared to conventional PT, LIT has the following advantages: (a) phase maps are independent of the surface emissivity; (b) detection is unaffected by uneven heating; (c) the low heating temperature causes no damage to the surface of the material. For a thermal imager with a certain accuracy, compared to amplitude detection, phase detection can improve the measurement accuracy, providing the possibility of accurately measuring defect information.

Tang et al. [65] characterized the interfacial defects in a TBC using LIT and analyzed the influences of the output power of the light source, *P*; the number of analysis periods, *N*; and the modulation frequency f_e on the detection results. The TBC specimens used in their study had dimensions of 140 mm × 100 mm × 5 mm, and each contained 24 artificial flat-bottomed hole defects (produced using electric sparks) 10–2 mm in diameter and 0.3 mm in depth. To analyze the influences of *P*,

N, and f_e on defect resolution, the obtained IR phase images were normalized. In addition, the phase contrast between the defect and intact zones was calculated using Eq. (13.32).

$$C_{Ph} = \left| \frac{\overline{Ph_d} - \overline{Ph_s}}{\overline{Ph_s}} \right| \tag{13.32}$$

where C_{Ph} is the total phase contrast and $\overline{Ph_d}$ and $\overline{Ph_s}$ are the mean phases of the defect and intact zones, respectively.

First, the influence of P on the total phase contrast in IR thermal images was analyzed. The f_e, sampling frequency f_s, and N were set to 0.2 Hz, 30 Hz, and 4, respectively. Four P values (800, 1000, 1400, and 1800 W) were used. Figure 13.58 shows IR phase images at different P values. When the defect size in the test specimen was fixed, at a low P, the thermal excitation energy of the test specimen was low, and there was very little total phase contrast between the defect and intact zones. The largest total phase contrast was obtained at a P of 1000 W.

On this basis, the influence of N on the total phase contrast in IR thermal images was analyzed. The f_e, f_s, and P were set to 0.2 Hz, 30 Hz, and 1000 W, respectively. Six N values (1, 2, 3, 4, 5, and 6) were used. Figures 13.59 and 13.60 show the

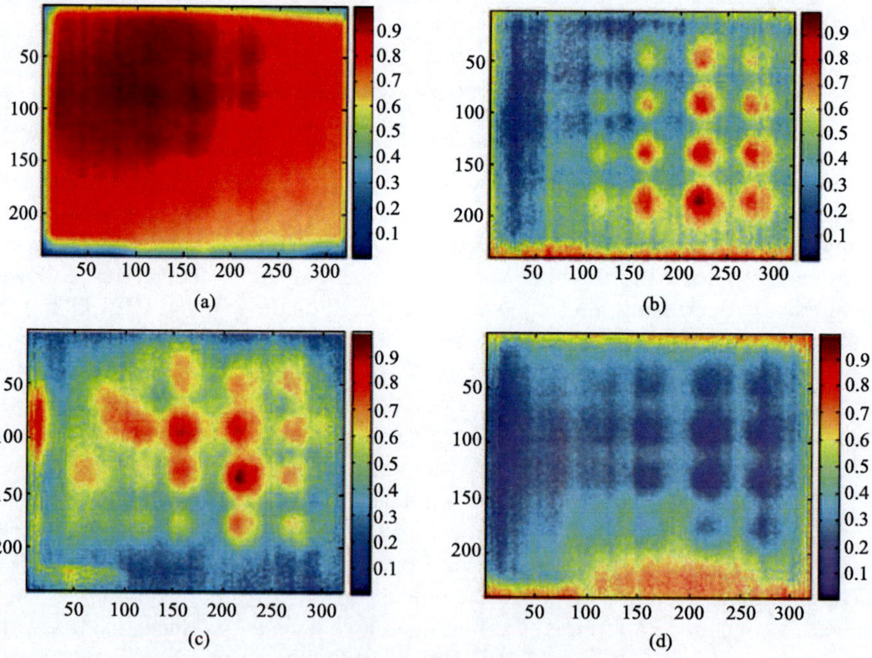

Fig. 13.58 IR phase images obtained at different output powers of the light source: [65] **a** 800 W-30 Hz-0.2 Hz-4; **b** 1000 W-30 Hz-0.2 Hz-4; **c** 1400 W-30 Hz-0.2 Hz-4; **d** 1800 W-30 Hz-0.2 Hz-4

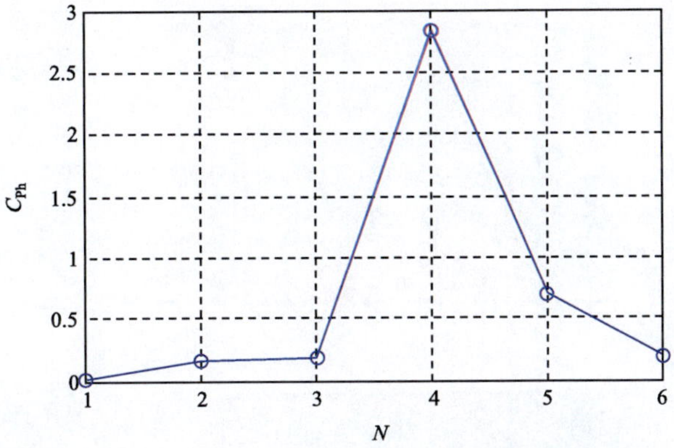

Fig. 13.59 Influence of the number of analysis periods N on the total phase contrast C_{Ph} in IR thermal images [65]

influence of N on IR phase images. C_{Ph} first increased and then decreased as N increased. This is because at a low N, the heat transfer process in the specimen was not yet complete, and there was no significant difference between the thermal signals of the defect and intact zones. At a high N, too much information was collected in the steady state, which, to some extent, masked the difference between the signals of the defect and intact zones. At $N = 4$, the maximum total phase contrast between the defect and intact zones occurred, and thus, the defect zone could be satisfactorily distinguished.

Finally, the influence of f_e on the total phase contrast in IR thermal images was analyzed. f_s, P, and N were set to 30 Hz, 1000 W, and 4, respectively. Six f_e values (0.1, 0.16, 0.18, 0.2, 0.22, and 0.32 Hz) were used. Figures 13.61 and 13.62 show the influence of f_e on IR phase images. f_e significantly influenced the detection results because it directly affected the thermal diffusion length, as shown in Eq. (13.33).

$$\Lambda = \sqrt{\alpha/(\pi f_e)} \tag{13.33}$$

where Λ is the thermal diffusion distance and α is the coefficient of thermal diffusion. As f_e increased, C_{Ph} first increased and then decreased. For the given defect size range, the maximum C_{Ph} was obtained at an f_e of approximately 0.2 Hz.

From the perspectives of FE numerical simulation and experimentation, Shrestha and Kim [62, 63] nondestructively evaluated the thickness of the ceramic layer of a TBC by thermography and compared PT- and LIT-based characterizations. During the LIT process, the surface of a specimen was heated cyclically. A thermal imaging system was used to collect a series of IR images of the surface of the specimen. The modulated heating was compared with the temperature measured through the extraction of the sinusoidal wave mode at each point in the images. To simulate the characteristics of heating by a sinusoidal heat flow, the heat flow excited by a thermal excitation source can be expressed as follows:

$$Q = \frac{Q_0}{2}(1 + \cos(2\pi f t)) \tag{13.34}$$

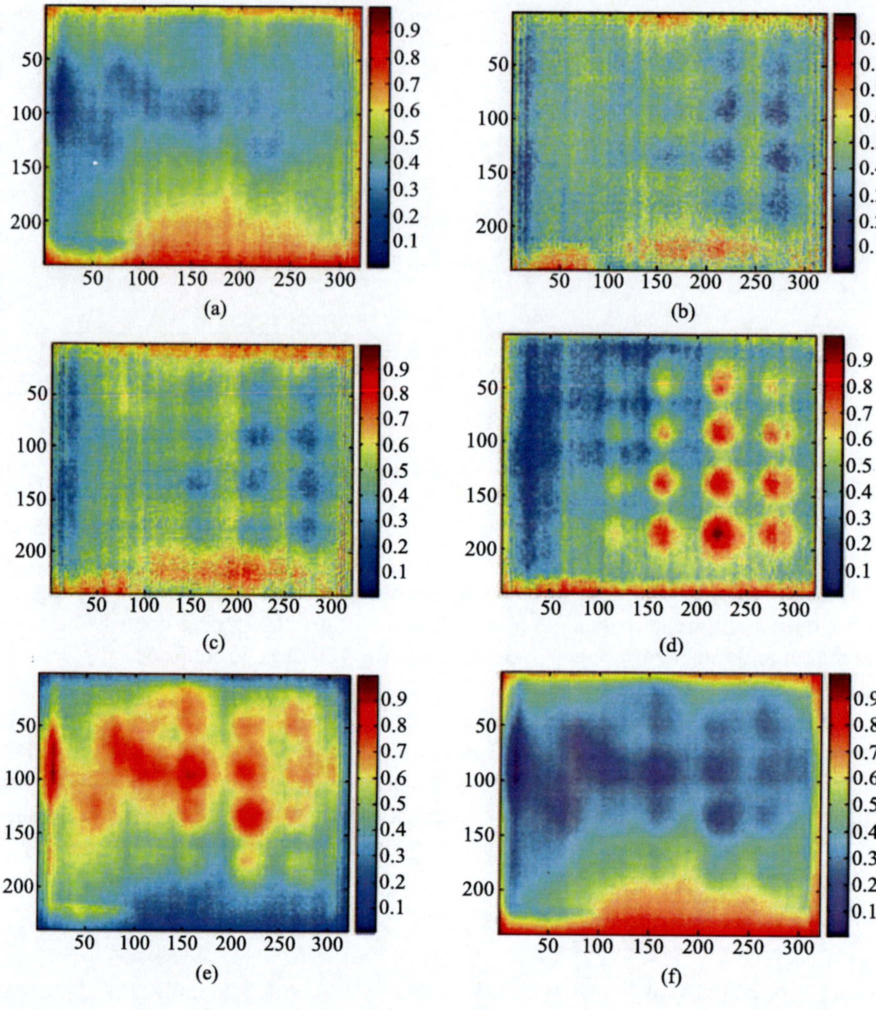

Fig. 13.60 Influence of the number of analysis periods on the total phase contrast in IR thermal images: [65] **a** 1000 W-30 Hz-0.2 Hz-1; **b** 1000 W-30 Hz-0.2 Hz-2; **c** 1000 W-30 Hz-0.2 Hz-3; **d** 1000 W-30 Hz-0.2 Hz-4; **e** 1000 W-30 Hz-0.2 Hz-5; **f** 1000 W-30 Hz-0.2 Hz-6

where Q is the power density of the heat flow, Q_0 is the intensity of the heat source, f is the modulation frequency, and t is the time.

The responses of the surface of the coating to the heat flow during 15 complete excitation cycles at modulation frequencies of 1, 0.5, 0.2, 0.1, 0.05, 0.02, and 0.01 Hz were obtained using FE simulation. Figure 13.63 shows the LIT images at different excitation cycles and time intervals. Figure 13.64 shows the cyclic variation in the surface temperature at a modulation frequency of 0.2 Hz. Figures 13.63 and 13.64 show that under cyclic heating over a sufficiently long time, the surface temperature cyclically evolves from a transient state to a steady state in the form of a sinusoidal

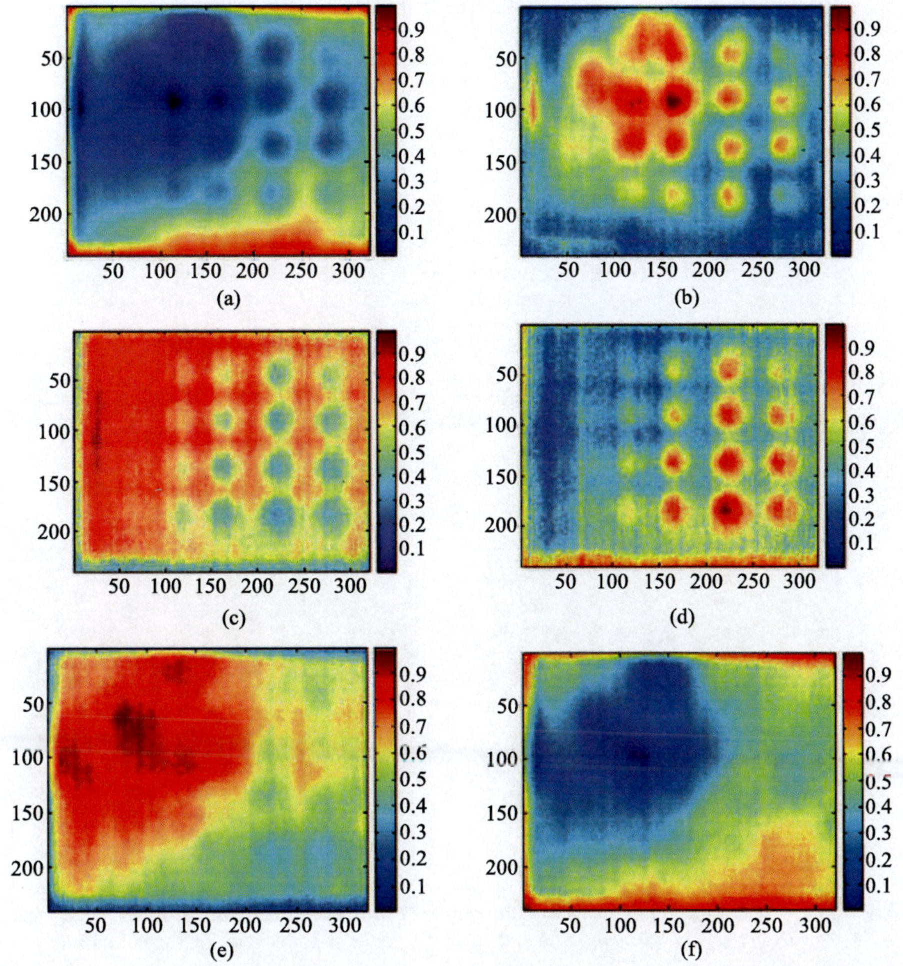

Fig. 13.61 Influence of the modulation frequency on the total phase contrast in IR thermal images: [65] **a** 1000 W-30 Hz-0.1 Hz-4; **b** 1000 W-30 Hz-0.16 Hz-4; **c** 1000 W-30 Hz-0.18 Hz-4; **d** 1000 W-30 Hz-0.2 Hz-4; **e** 1000 W-30 Hz-0.22 Hz-4; **f** 1000 W-30 Hz-0.32 Hz-4

curve. After 12 cycles, the surface temperature reaches a steady state from a transient state. Figure 13.66 shows the FE-simulated and experimental relations between the coating thickness and phase angle at different modulation frequencies. The phase angle decreases as the coating thickness increases. A decrease in the modulation frequency results in a decrease in the difference between the phase angles of thick and thin coatings. At a certain frequency, there is almost no difference between the phase angles of the coatings. Thus, high modulation frequencies are more sensitive to the difference in phase angles than low modulation frequencies (Fig. 13.65).

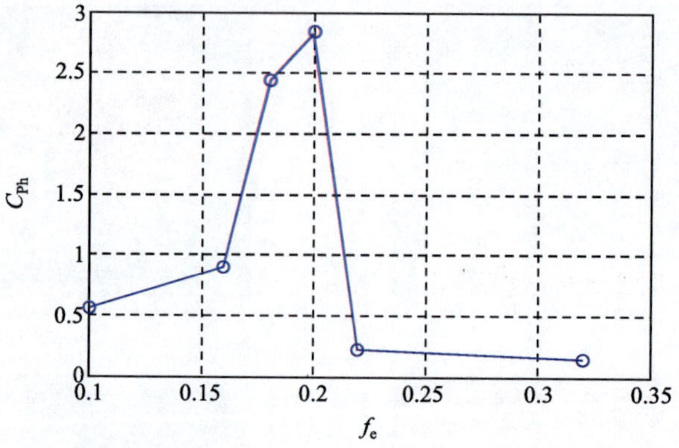

Fig. 13.62 Influence of the modulation frequency f_e on the total phase contrast C_{Ph} in IR thermal images [65]

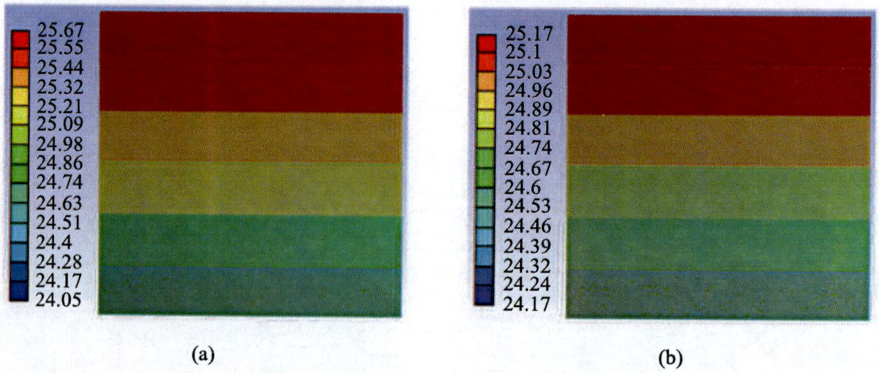

Fig. 13.63 FE-simulated LIT images at different excitation cycles and time intervals: **a** 1st cycle, 2.5 s; **b** 1st cycle, 4.9 s; **c** 8th cycle, 37.5 s; **d** 8th cycle, 39.3 s; **e** 15th cycle, 72.5 s; and **f** 15th cycle, 75 s [62]

Table 13.1 summarizes the PT- and LIT-predicted coating thicknesses and the corresponding errors. The PT-predicted phase angle increases as the coating thickness increases, whereas the LIT-predicted phase angle decreases as the coating thickness increases. This is because the phase angle determined by the Fourier transform depends on the heating method and the number of IR thermal image sequences analyzed (i.e., the number of harmonic waves). IR thermal images obtained by PT provide noncyclic and decaying thermal data, whereas IR thermal images obtained by LIT give cyclic and modulated thermal data. Useful information is contained in 2–5 harmonic waves in PT and LIT, depending on the modulation frequency.

Equations (13.29) and (13.30), obtained by PT and LIT, respectively, were used to predict coating thickness. The predicted and calculated values were then compared to determine the percentage errors of the predictions. The percentage errors of the coating thickness predictions by both PT and LIT decrease as the coating thickness

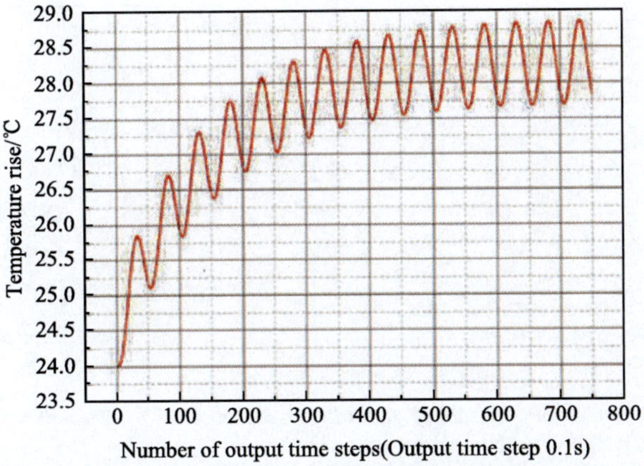

Fig. 13.64 FE-simulated surface temperature profile at a modulation frequency of 0.2 Hz by LIT [62]

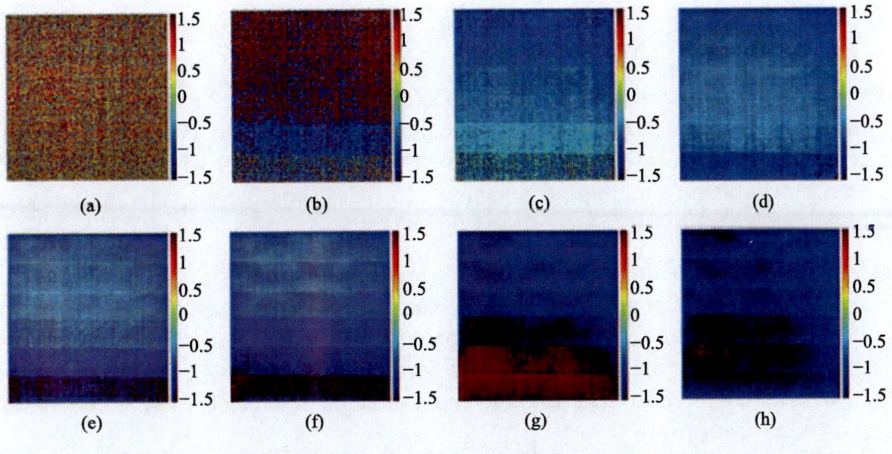

Fig. 13.65 Experimental LIT phase images at different modulation frequencies: **a** 2 Hz; **b** 1 Hz; **c** 0.5 Hz; **d** 0.2 Hz; **e** 0.1 Hz; **f** 0.05 Hz; **g** 0.02 Hz; **h**0.01 Hz [63]

increases. Between the two techniques, the error for thin coating thickness predicted by PT is lower, while the error for thick coating thickness predicted by LIT is lower. The prediction errors of both techniques are satisfactory and within acceptable limits.

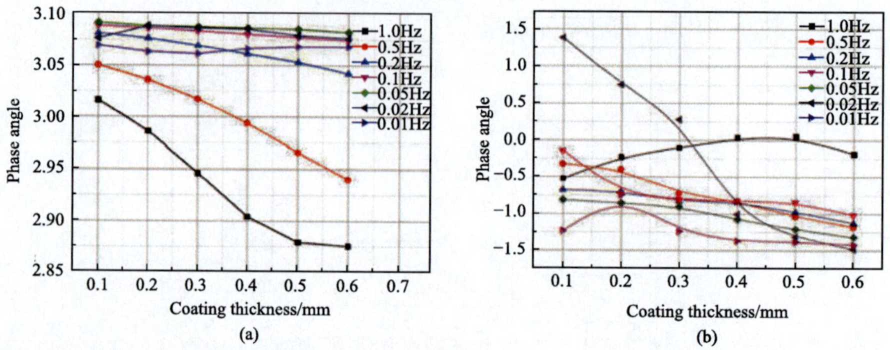

Fig. 13.66 Surface temperature decay curves for coatings with different thicknesses: **a** FE-simulated results; [62] **b** experimental results [63]

13.4 Summary and Outlook

13.4.1 Summary

This chapter describes the application of DIC, X-ray CT, and IRT in the monitoring of TBC failure; characterizes and analyzes the deformation field, porosity, and interfacial defects in TBC specimens during the test process; and provides experimental means for analyzing the failure mechanism of TBCs. The following is a summary of this chapter:

(1) The evolution of the surface and cross-sectional strain fields in TBCs with time during the high-temperature tension, high-temperature three-point bending, and high-temperature CMAS corrosion processes is determined based on DIC. For a coating, the critical time for crack formation and cracking locations can be satisfactorily determined based on the time point at which the strain undergoes sudden changes.

(2) An image segmentation processing method is proposed based on the X-ray CT characterization technique. The distribution pattern of the pores in an APS TBC is determined. The influences of VA corrosion on the evolution of pores in coatings are analyzed. In addition, the experimental results are verified using several common pore detection methods.

(3) The basic principle of IRT in the detection of defects is elucidated. The interfacial and surface defects in coatings are detected using PT and LIT in conjunction with image processing algorithms. The influences of the output power of the light source, the number of analysis periods, and modulation frequency on the test results are analyzed.

13.4.2 Outlook

Currently, DIC, X-ray CT, and IRT are used to characterize only strains, pores, and defects in flat TBC specimens. At higher temperatures, thermal radiation, thermal disturbance, oxidation, and ablation interfere with the DIC-based characterization of strains. In addition, speckles undergo significant decorrelation at high temperatures, resulting in a lower measurement accuracy. These problems pose significant challenges to strain measurement at high temperatures. Investigating anti-interference algorithms for strain measurement at high temperatures and high-temperature speckles is the future development direction for DIC. When X-ray CT is used to characterize the pores and defects in a TBC, the substrate material of the TBC absorbs a significant portion of the X-rays, and the attainable minimum pixel size is only 0.89 μm. Currently, this technique cannot be used in the 3D reconstruction of pores and defects in TBCs on turbine blades. Improving the technical parameters of X-ray CT, the penetrability of X-rays, and CT scan resolution is the future development direction for X-ray CT. Eliminating interference from adverse factors, improving the SNR of IR images, and enhancing defect visibility are the future development directions for IRT.

References

1. Yang L, Zhou YC, Mao WG, Lu C. Real-time acoustic emission testing based on wavelet transform for the failure process of thermal barrier coatings. Appl Phys Lett. 2008;93(23): 231906.
2. Veal BW, Paulikas AP, Hou PY. Tensile stress and creep in thermally grown oxide. Nature Material. 2006;5(5):349–51.
3. Zhang C, Zhou C, Gong S, Li H, Xu H. Evaluation of thermal barrier coating exposed to different oxygen partial pressure environments by impedance spectroscopy. Surf Coat Technol. 2006;201(1–2):446–51.
4. Christensen RJ, Lipkin DM, Clarke DR, Murphy K. Nondestructive evaluation of the oxidation stresses through thermal barrier coatings using Cr3+ piezospectroscopy. Appl Phys Lett. 1996;69(24):3754–6.
5. Yu F, Bennett TD. Phase of thermal emission spectroscopy for properties measurements of delaminating thermal barrier coatings. J Appl Phys. 2005;98(10): 103501.
6. Thornton J, Cookson D, Pescott E. The measurement of strains within the bulk of aged and as-sprayed thermal barrier coatings using synchrotron radiation. Surf Coat Technol. 1999;120–121(10):96–102.
7. Pan B, Qian KM, Xie HM, et al. Two-dimensional digital image correlation for in-plane displacement and strain measurement: a review. Meas Sci Technol. 2009;20(6): 062001.
8. Pan B, Xie H, Xu B, et al. Development of sub-pixel displacements registration algorithms in digital image correlation. Adv Mech. 2005;34(3):345–52.
9. Pan B, Xie HM, Xu BQ, et al. Performance of sub-pixel registration algorithms in digital image correlation. Meas Sci Technol. 2006;17(6):1615–21.
10. Pan B, Xie H. Full-field strain measurement based on least-square fitting of local displacement for digital image correlation method. Acta Optica Sinica. 2007;27(11):1980–6.
11. Pan B, Xie H, Xia Y, et al. Large-deformation measurement based on reliable initial guess in digital image correlation method. Acta Optica Sinica. 2009;29(2):400–6.

12. Wu D, Fang Y, Zhang M. Experimental simulation system of transient aerodynamic heating for high-speed flight vehicle. Aviat Metrol Measur Technol. 2003;(23):9–11, 14.
13. Pan B, Wu D, Xia Y. High-temperature deformation field measurement by combining transient aerodynamic heating simulation system and reliability-guided digital image correlation. Opt Lasers Eng. 2010;48(9):841–8.
14. Pan B, Wu D, Gao Z. A non-contact high-temperature deformation measuring system based on digital image correlation technique. Acta Aeronautica et Astronautica Sinica. 2010;10:1960–7.
15. Zhang D, Zhang X, Cheng G. Compression strain measurement by digital speckle correlation. Exp Mech. 1999;39(1):62–5.
16. Pan B, Xie HM, Guo ZQ, et al. Full-field strain measurement using a two-dimensional Savitzky-Golay digital differentiator in digital image correlation. Opt Eng. 2007;46(3): 033601.
17. Gao H, Liu H, Qi Z, et al. Research on variation law of fatigue crack-tip displacement and strain fields based on high-speed digital image correlation method. Acta Armamentarii. 2015;36(9):1772–81.
18. Wang W. Application of digital image correlation method to the high temperature deformation testing of thermal structure materials. Harbin Institute of Technology, 2014.
19. Wu D. Real-time testing and analysis of interfacial damage in thermal barrier coatings and development of a test simulation system. Xiangtan University, 2011.
20. Zhu W, Wu Q, Yang L et al. In situ characterization of high temperature elastic modulus and fracture toughness in air plasma sprayed thermal barrier coatings under bending by using digital image correlation. Ceramics Int. 2020.
21. Si M. Tensile failure analysis of EB-PVD thermal barrier coating systems based on digital image correlation. Xiangtan University, 2016.
22. Yang J. DIC characterization and analysis of the strain field in thermal barrier coatings subjected to high-temperature CMAS corrosion. Xiangtan University, 2016.
23. Yang L, Yang J, Xia J, Zhu W, Zhou YC, Wei YG, Wu RT. Characterization of the strain in the thermal barrier coatings caused by molten CaO-MgO-Al_2O_3-SiO_2 using a digital image correlation technique. Surf Coat Technol. 2017;322:1–9.
24. Lyons JS, Liu J, Sutton MA. High-temperature deformation measurements using digital-image correlation. Exp Mech. 1996;3:64–70.
25. Maurel V, De Bodman P, Rémy L. Influence of substrate strain anisotropy in TBC system failure. Surf Coat Technol. 2011;206(7):1634–9.
26. Mercer C, Faulhaber S, Evans AG, et al. A delamination mechanism for thermal barrier coatings subject to calcium-magnesium-alumino-silicate (CMAS) infiltration. Acta Mater. 2005;53(4):1029–39.
27. He Q, Liu XJ, Liu B, et al. Influence of CMAS infiltration on microstructure of plasma-sprayed YSZ thermal barrier coating. China Surface Eng. 2012;25(4):42–7.
28. Mathews JP, Campbell QP, Xu H, et al. A review of the application of X-ray computed tomography to the study of coal. Fuel. 2017;209:10–24.
29. Vagnon A, Lame O, Bouvard D, et al. Deformation of steel powder compacts during sintering: correlation between macroscopic measurement and in situ microtomography analysis. Acta Mater. 2006;54(2):513–22.
30. Scheck C, Zupan M. Ductile fracture evaluated through micro-computed X-ray tomography. Scripta Mater. 2011;65(12):1041–4.
31. Kurumlu D, Payton EJ, Young ML, et al. High-temperature strength and damage evolution in short fiber reinforced aluminum alloys studied by miniature creep testing and synchrotron microtomography. Acta Mater. 2012;60(1):67–78.
32. Amanat N, Tsafnat N, Loo BCE, et al. Metallurgical coke: an investigation into compression properties and microstructure using X-ray microtomography. Scripta Mater. 2009;60(2):92–5.
33. Zhang X. Methods for correcting beam hardening artifacts in CT images. Shenyang University of Technology, 2017.
34. Wirjadi O. Survey of 3d image segmentation methods. Institut Techno-und Wirtschaftsmathematik, 2007.

35. Soille P, Morphological image analysis: principles and applications. Springer Science & Business Media, 2013.
36. Iassonov P, Gebrenegus T, Tuller M. Segmentation of X-ray computed tomography images of porous materials: a crucial step for characterization and quantitative analysis of pore structures. Water Resou Res. 2009;45(9).
37. Cai X. CT characterization of the evolution pattern of pores in APS thermal barrier coatings subjected to volcanic ash corrosion. Xiangtan University, 2018.
38. Pelissari PIBGB, Angélico RA, Salvini VR, et al. Analysis and modeling of the pore size effect on the thermal conductivity of alumina foams for high temperature applications. Ceram Int. 2017;43(16):13356–63.
39. Zhu W, Cai XN, Yang L, et al. The evolution of pores in thermal barrier coatings under volcanic ash corrosion using X-ray computed tomography. Surf Coat Technol. 2019;357:372–8.
40. Li H, Zeng Q, Xu S. Effect of pore shape on the thermal conductivity of partially saturated cement-based porous composites. Cement Concrete Compos. 2017;81:87–96.
41. Wu J, Guo HB, Gao YZ, et al. Microstructure and thermo-physical properties of yttria stabilized zirconia coatings with CMAS deposits. J Eur Ceram Soc. 2011;31(10):1881–8.
42. Xia J, Yang L, Wu RT, et al. On the resistance of rare earth oxide-doped YSZ to high temperature volcanic ash attack. Surf Coat Technol. 2016;307:534–41.
43. Pia G, Casnedi L, Sanna U. Porosity and pore size distribution influence on thermal conductivity of yttria-stabilized zirconia: experimental findings and model predictions. Ceram Int. 2016;42(5):5802–9.
44. Goldstein JI, Newbury DE, Michael JR et al. Scanning electron microscopy and X-ray microanalysis. Springer, 2017.
45. Heida J H. Nondestructive evaluation of superalloy specimens with a thermal barrier coating. 16th world conference on nondestructive testing, montreal, August 30-September 3, 2004.
46. Montesano J, Fawaz Z, Bougherara H. Use of infrared thermography to investigate the fatigue behavior of a carbon fiber reinforced polymer composite. Compos Struct. 2013;97:76–83.
47. Mollmann K, Karstadt D, Pinno F, Vollmer M. Selected critical applications for thermography: Convections in fluids, selective emitters and highly reflecting materials. Proceedings of the infrared camera calibration conference, USA (Las Vegas, NV), 2005: 161–173.
48. Michalski L, Eckersdorf K, Kucharski J, McGhee J. Temperature measurement. UK: Wiley West Sussex, 2001.
49. Bates D, Smith G, Lu D, et al. Rapid thermal non-destructive testing of aircraft components. Compos B Eng. 2000;31(3):175–85.
50. Vavilov VP, Pawar SS. A novel approach for one-sided thermal nondestructive testing of composites by using infrared thermography. Polym Testing. 2015;44:224–33.
51. Handbook of technical diagnostics: fundamentals and application to structures and systems. Springer Science & Business Media, 2013.
52. Fazeli H, Mirzaei M. Shape identification problems on detecting of defects in a solid body using inverse heat conduction approach. J Mech Sci Technol. 2012;26(11):3681–90.
53. Junyan L, Liqiang L, Yang W. Experimental study on active infrared thermography as a NDI tool for carbon–carbon composites. Compos B Eng. 2013;45(1):138–47.
54. Li Y, Yang Z, Zhu J, et al. Investigation on the damage evolution in the impacted composite material based on active infrared thermography. NDT and E Int. 2016;83:114–22.
55. Vavilov VP, Burleigh DD. Review of pulsed thermal NDT: Physical principles, theory and data processing. NDT and E Int. 2015;73:28–52.
56. Almond DP, Pickering SG. An analytical study of the pulsed thermography defect detection limit. J Appl Phys. 2012;111: 093510.
57. Zhou Y. Solid mechanics of materials (II). Sciences Press, 2005.
58. Duan Z, Zhou Y, Bo Y, Wang W. Research advances in the study of heat waves. Adv Mech. 1992;22(4):433–48.
59. Tang Q, Dai J, Liu J, et al. Quantitative detection of defects based on Markov–PCA–BP algorithm using pulsed infrared thermography technology. Infrared Phys Technol. 2016;77:144–8.

60. Wei J. NDT and quantitative characterization of interfacial defects in thermal barrier coatings based on infrared thermography. Xiangtan University, 2020.
61. Rajic N. Principal component thermography for flaw contrast enhancement and flaw depth characterization in composite structures. Compos Struct. 2002;58(4):521–8.
62. Shrestha R, Kim W. Evaluation of coating thickness by thermal wave imaging: a comparative study of pulsed and lock-in infrared thermography—Part I: simulation. Infrared Phys Technol. 2017;83:124–131.
63. Shrestha R, Kim W. Evaluation of coating thickness by thermal wave imaging: a comparative study of pulsed and lock-in infrared thermography—Part II: experimental investigation. Infrared Phys Technol. 2018;92:24–29.
64. Wang Z, Zhu J, Tian GY, et al. Comparative analysis of eddy current pulsed thermography and long pulse thermography for damage detection in metals and composites. NDT and E Int. 2019;107: 102155.
65. Tang QJ, Dai JM, Bu CW, Qi LT, Li DY. Experimental study on debonding defects detection in thermal barrier coating structure using infrared lock-in thermographic technique. Infrared Phys Technol. 2016;107:463–8.

Chapter 14
Thermal Insulation Effect of TBCs on Turbine Blades

Comprehensive factors in areas such as aerodynamics, structure, strength, and cooling must be considered in the turbine blade design for advanced aeroengines. As a key thermal protection technology, thermal barrier coatings (TBCs) are vitally important for the optimum design of turbine blades. The thermal insulation effect of TBCs, its influencing factors, and the availability of the corresponding mature evaluation methods all limit the development of aeroengines. Mastering evaluation and testing techniques for the thermal insulation effect of TBCs on turbine blades is urgently needed to improve the performance of advanced aeroengines.

Based on the research progress made in recent years across the world on the thermal insulation effect of TBCs on turbinc blades, this chapter describes in detail theoretical analysis methods and experimental testing techniques for the thermal insulation effect of TBCs and, on this basis, further discusses its key influencing factors.

14.1 Theoretical Analysis of the Thermal Insulation Effect

In the mid-1950s, the maximum front inlet temperature of turbines did not exceed 1200 K and was lower than the melting point of the blade materials. As a result, turbine blades could be directly exposed to high-temperature combustion gases without the need for cooling. In the 1960s, as the temperature at the leading edge of turbine blades increased, hollow turbine blades began to be cast, and cooling air was guided into their interior for cooling. This technology is referred to as internal cooling. In the 1970–80s, the front inlet temperature of turbines rose to 1600–1700 K. To improve the temperature capacity of turbine blades to meet the demand for high-performance engines, three thermal protection technologies, namely single-crystal superalloys (SCSAs), TBCs, and film cooling, were used successively. Since the 1990s, the front inlet temperature of advanced turbine engines has generally surpassed 2000 K [1], and the three thermal protection technologies have been further integrated and

Y. Zhou et al., *Thermal Barrier Coatings: Failure Theory and Evaluation Technology*,
https://doi.org/10.1007/978-981-19-2723-2_14

applied to turbine blades to ensure the safe operation of the engines. Consequently, the heat transfer process in turbine blades with TBCs is extremely complex and involves myriad mutually interacting factors. TBC technology has been introduced in detail in earlier chapters. The following is a brief introduction to the other two thermal protection technologies.

(1) **SCSAs**

SCSAs were first successfully developed and used in the 1980s and have since developed from the first generation to the fourth generation. This technology increases the temperature of an alloy primarily by increasing the mass fraction of the precious metal elements (Re and Ru). As a result, SCSAs became increasingly costly. For example, the Re contents in the second- and third-generation SCSAs were 3% and 6%, respectively. In comparison, in addition to a Re content of 6%, the fourth-generation SCSAs contain a Ru content of 3%. To reduce the Re content in the second-generation SCSAs, General Electric developed a René N515 alloy (Re content: 1.5%) with a performance close to that of René N5 and gradually used it to replace René N5 in aeroengine turbine blades [2]. Similarly, Cannon-Muskegon developed an SCSA (called CMSX-8) with a Re content of 1.5%. Below 1300 K, the creep life of CMSX-8 is comparable to that of CMSX-4. In addition, CMSX-8 has good structural stability with exceptional fatigue resistance, oxidation resistance, and castability. However, at super high temperatures (i.e., above 1094 °C), the creep life of CMSX-8 is considerably shorter than that of CMSX-4 [3]. In addition, due to their high melting point and γ/γ' two-phase structure similar to that of Ni-based alloys, Co-based SCSAs have the potential to be developed into new-generation SCSAs with higher working temperatures [4]. The published performance data have shown that the creep performance of Co-based alloys is comparable to that of the first-generation Ni-based SCSAs [5]. However, Co–Al–W system alloys still face many challenges. For example, they have a low γ'-phase dissolution temperature and a very narrow γ/γ' two-phase structural zone [6]. As another example, when exposed to high temperatures, these alloys mainly form CoO, Co_3O_4, and mixed oxides and, as a result, have poor high-temperature oxidation resistance [7]. Overall, there is a year-to-year increase in the research investment in the relevant fields across the world. Work related to high-throughput design, preparation, and characterization of SCSAs has been included in multiple key research and development programs launched in China in recent years. However, there is a slow improvement in the temperature capabilities of alloys, at a rate of only 20–30 °C per generation. It remains difficult to rely solely on SCSAs to meet the demand for the development of advanced aeroengines.

(2) **Cooling technology**

The research on cooling technology started early, and tremendous progress has been made through decades of development. Cooling technology can be divided into internal cooling and external film cooling. In internal cooling, cooling air is guided into the internal cooling channels of a turbine blade to achieve enhanced convective heat transfer with the blade and to thereby absorb the heat from its inner wall. In external film cooling, cooling air is discharged from the cooling-air film holes on the wall of a blade and subsequently forms a uniform low-temperature cooling-air film on the surface of the blade under the combined action of the pressure of the high-temperature combustion gas and friction. This uniform film separates the high-temperature combustion gas and the surface of the blade, thereby reducing the surface temperature of the blade and achieving cooling.

To enhance the internal heat transfer intensity and to improve the cooling efficiency, ribs or trailing-edge pin fins are placed in the cooling channels to increase the flow turbulence. Webb et al. [8] and Jubran et al. [9] found that the convective heat transfer coefficient (CHTC) increases with increasing rib height and spacing and that staggered ribs often lead to a higher heat transfer efficiency than inline ribs. Upon passing flow-disturbing ribs, the flow of a cooling-air current against the wall becomes separated and subsequently reattaches. The change in its flow conditions increases the local heat transfer coefficient (HTC), thereby enhancing the convective heat transfer effect. The use of ribs and pin fins can increase the heat transfer area by 6–10% and more than double the HTC.

The research on film cooling began in the 1970s and has advanced considerably through more than four decades of development. Bogard and Thole [10] reviewed film cooling in gas turbines, noting that the efficiency of films is primarily affected by factors such as the blowing ratio, the geometry and structure of the holes, and the extent of the turbulence. Chinese researchers Dai and Lin [11] published a detailed review on film cooling and listed the following factors that affect the cooling performance of a film: (1) the geometric parameters of the film holes (including their ejection angle, size, length, aperture ratio, spacing, and outlet shape), (2) the geometric parameters of the blade (including the leading-edge shape and flow-direction surface curvature of the blade as well as the geometry and surface roughness of the cooling-medium transport channel), (3) the aerodynamic parameters of the holes (including the mainstream velocity, blowing ratio, the cooling-flow/mainstream momentum and density ratios, the turbulence intensity of the mainstream, the development of the boundary layer in front of the film holes, the pressure gradient, and the unsteady wake flow), and (4) other factors (including the presence of gaps downstream of the film holes and gap leakage). With the development of the turbine blade manufacturing process, the temperature capability of turbine blades can be improved by approximately 400 K using the current cooling technology. However, the extensive extraction of the air from the air compressor reduces the thermal efficiency of an engine. Moreover, the complex structure reduces the blade strength and increases the processing difficulty. Hence, little room is left for further advancing this technology.

14.1.1 Heat Transfer Modes of Turbine Blades

Currently, the three major thermal protection technologies—TBCs, SCSAs, and film cooling—are applied in combination to the turbine blades of advanced aeroengines to ensure their safe operation. Therefore, the heat transfer process in a turbine blade during service is extremely complex and involves multiple heat transfer modes. The

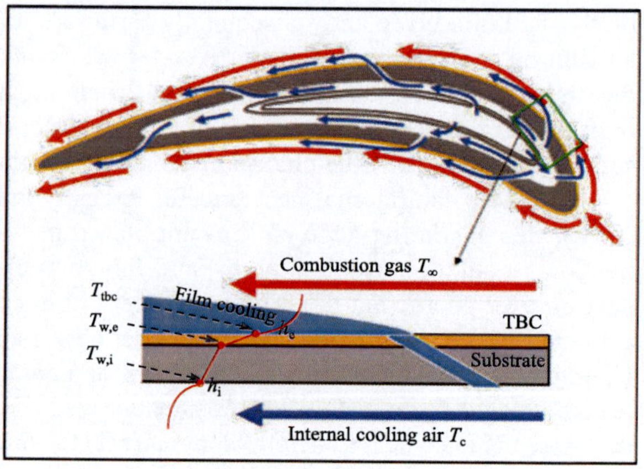

Fig. 14.1 Cooling technique for a TBC on a turbine blade in service [10, 12]

abovementioned factors must be considered simultaneously in the investigation and analysis of the thermal insulation effect of TBCs. Figure 14.1 shows a schematic diagram of the cooling of a turbine blade with a TBC in the presence of a high-temperature combustion gas. The surface of the TBC on the hollow turbine blade with a curved surface structure is subjected to the impact of the high-temperature, high-speed combustion gas. The heat flow is transferred to the surface of the blade through convection and radiation. In addition, there are multiple columns of film holes several hundred microns in diameter on the blade. The internal cooling air forms a cooling-air film on the surface of the coating through the film holes, which separates the external surface of the coating from the high-temperature combustion gas. Inside the TBC on the turbine blade, the heat flow is transferred from high-temperature regions to low-temperature regions. The whole heat transfer process involves three basic modes, namely heat conduction, heat convection, and thermal radiation.

Heat conduction. Heat conduction is a mode of thermal energy transfer that relies on the thermal motion of microscopic particles (e.g., molecules, atoms, and free electrons) while the components of the object undergo no relative displacement. The heat conduction in a solid is described with Fourier's law, which states that the heat conduction per unit time is directly proportional to the temperature gradient and that the heat transfer direction is opposite to the temperature gradient direction:

$$\boldsymbol{q} = -k\nabla T \tag{14.1}$$

where $\boldsymbol{q}$ is the heat flow vector, T is the temperature, and k is the thermal conductivity of the material.

Heat convection. Heat convection refers to the heat transfer caused by the mixing of the cold and hot parts of a fluid due to their relative displacement resulting from the macroscopic motion of the fluid. The heat transfer between a fluid and the surface of an object that occurs when the fluid flows over the surface of the object, which is referred to as convective heat transfer, receives considerable attention and is extensively used in engineering. Based on the cause of the flow, convective heat transfer can be divided into natural convection and forced convection. Natural convection is caused by the difference between the densities of the cold and hot parts of a fluid, whereas forced convection is generated by pumps, fans, or other differential pressure actions. Newton's law of cooling is the basic equation for calculating convective heat transfer:

$$-k\frac{\partial T}{\partial n}\bigg|_{w} = h(T_{\mathrm{w}}-T_{\mathrm{f}}) \tag{14.2}$$

where $\boldsymbol{n}$ is the outer normal direction of the object, w signifies the wall, T_{w} and T_{f} are the wall temperature and the environmental fluid temperature, respectively, and h is the interfacial CHTC, which not only depends on both the physical properties of the fluid and the shape and size of the heat transfer surface but also is closely related to the flow rate. According to the above equation, the heat transfer flux at the fluid/solid interface is directly proportional to the interfacial CHTC and the temperature difference at the fluid/solid interface. Research on convective heat transfer is focused on the theoretical and experimental determination of the CHTC under the given conditions.

Thermal radiation. Thermal radiation refers to the emission of radiant energy by an object due to heat. In nature, every object continuously emits thermal radiation to the surrounding space and absorbs thermal radiation. Heat transfer via this mode is radiant heat transfer. In contrast to heat conduction and heat convection, thermal radiation does not require a heat transfer medium and can occur in a vacuum. The radiant heat flux of an actual object can be calculated using the following empirically modified form of the Stefan–Boltzmann law:

$$-k\frac{\partial T}{\partial n}\bigg|_{w} = \varepsilon A\sigma T^{4} \tag{14.3}$$

where A is the surface area of the radiating object, ε is the emissivity of the object, and σ is the Boltzmann constant (i.e., the radiation constant of a blackbody). Here, the radiant heat flux of an object refers to the heat radiated outward by the object. For an actual object, its radiant heat transfer flow consists of the heat that it absorbs and radiates.

14.1.2 Definition of the Thermal Insulation Effect of TBCs on Turbine Blades

As mentioned earlier, the thermal insulation effect of a TBC is a combined result of the blade structure, the coating, the cooling film, and the combustion gas. Due to the complexity of the blade structure and the heat transfer between the combustion gas, cooling film, and coating, it is difficult to consider all the factors during the establishment of a thermal insulation effect model for a TBC. The most common and most direct definition of the thermal insulation effect of a TBC is the difference between the external surface temperature of the coating and the temperature at the coating/metal interface. Because metals have high thermal conductivity and allow fast heat transfer, researchers consider the temperature at the interface to be close to the temperature of the free surface of the substrate. More importantly, it is extremely difficult to measure the interfacial temperature. Thus, the thermal insulation effect of a TBC is often defined as the difference between the free surface temperatures of the coating and the substrate. According to the Fourier heat transfer equation, this temperature difference is as follows:

$$\Delta T = T_{\mathrm{tbc}} - T_{\mathrm{w,e}} = \frac{q}{k} \tag{14.4}$$

where ΔT is the thermal insulation effect of the TBC, T_{tbc} is the surface temperature of the coating, $T_{\mathrm{w,e}}$ is the surface temperature of the substrate, q is the heat flow that passes through the coating, and k is the thermal conductivity of the coating. Based on this definition, the thermal insulation effect of a TBC depends completely on the thermal conductivity of its compositional material. A low thermal conductivity leads to a high thermal insulation effect. Hence, thermal conductivity-based composition and process design has always been a popular research focus in the TBC field [13–17].

However, in reality, engine designers and engineers have discovered that the use of TBCs fails to produce a good thermal insulation effect and that TBCs spall and block the film holes. In other words, TBCs actually play a negative instead of a positive role. For example, in their experimental investigation into the cooling efficiency of the film at the leading edge of a turbine blade with a TBC, Maikell et al. [18] found that under the same cooling-air conditions, the application of a TBC led to a significant decrease in the temperature of the substrate, but the surface temperature of the coating was 3 °C higher than that of the blade without the coating, making it difficult to consider the contribution of the TBC. According to a report by Harrison [19], using the definition in Eq. (14.4) to design a turbine blade may lead to overestimation of the service life of the turbine blade by approximately 10–15%, thus significantly increasing the unreliability of the engine.

This is primarily because the thermophysical properties of ceramic coatings and metallic substrates substantially differ. A thin ceramic TBC significantly alters the flow and temperature fields near the surface of a blade. In other words, Eq. (14.4) fails to consider the contribution of the film and combustion gas, i.e., the environmental effects. In fact, the thermal insulation effect is extremely vulnerable to the effects of these environmental factors. As a consequence, the thermal insulation effect of TBCs varies significantly for different structures under different settings of combustion gas and cooling air, making it impossible to truly understand the contribution of TBCs. Hence, thermal insulation effect test results obtained based on specimens do not reflect the true conditions of turbine blades in actual combustion chambers. Blades designed by engineers based solely on Eq. (14.4) would encounter problems, possibly serious ones.

In view of this issue, Dees et al. [20] defined the thermal insulation effect of a TBC as the difference between the surface temperatures of the substrate of the blade before and after the application of the TBC, respectively:

$$\Delta T = T_{\mathrm{w,e,notbc}} - T_{\mathrm{w,e}} \tag{14.5}$$

where $T_{\mathrm{w,e,notbc}}$ is the temperature of the substrate surface in the absence of a coating (i.e., the surface in contact with the combustion gas) and $T_{\mathrm{w,e}}$ is the substrate surface in the presence of a coating (i.e., the coating/substrate interface). This definition directly reflects the comprehensive effect of the application of a TBC on the surface of the substrate of the turbine blade, including the temperature gradient caused by the coating, the effect of the coating on the heat flow, and the disturbance caused by the coating to the heat flow.

14.1.3 Nondimensionalization of the Thermal Insulation Effect

While Eq. (14.5) gives the overall change in the temperature field of a turbine blade after the application of a TBC, the thermal insulation effect of the TBC is still affected by numerous factors such as the structure of the turbine blade, combustion gas, cooling air, and film holes. In addition, factors to be considered in the thermal insulation effect analysis and design of TBCs remain intricate and complex. In view of this problem, Davidson et al. [21] proposed a dimensionless overall cooling efficiency ϕ as follows:

$$\phi = \frac{T_\infty - T_{\mathrm{w,e}}}{T_\infty - T_\mathrm{c}} \tag{14.6}$$

where T_∞ and T_c are the inlet temperatures of the combustion gas and cooling air, respectively, and $T_{\mathrm{w,e}}$ is the wall temperature of the external surface of the blade. When a TBC is applied, the overall cooling efficiency ϕ' is as follows:

$$\phi' = \frac{T_\infty - T'_{\mathrm{w,e}}}{T_\infty - T_c} \tag{14.7}$$

where $T'_{w,e}$ is the wall temperature of the external surface of a blade with a TBC.

A comparison of Eqs. (14.6) and (14.7) yields the dimensionless thermal insulation effect of a TBC:

$$\Delta\phi = \phi' - \phi \tag{14.8}$$

Equations (14.6) and (14.7) show that theoretical estimates of $T'_{w,e}$ and $T_{w,e}$ are required to theoretically determine $\Delta\phi$. $T'_{w,e}$ and $T_{w,e}$ depend on the heat transfer process between the combustion gas, cooling air, coating, and substrate, a process primarily affected by the following factors: the inlet temperature of the combustion gas, T_∞; the inlet temperature of the cooling air, T_c; the CHTC of the coating, h_e; the CHTC of the free surface of the substrate, h_i; the thermal conductivity of the coating, k_{tbc}; the thermal conductivity of the substrate, k; the thickness of the coating, d_{tbc}; and the thickness of the substrate, d. The effects of the blade with a complex curved structure on the high-velocity combustion gas and cooling air lead to changes in the parameters (e.g., velocity, pressure, temperature, and direction) of the airflow on the free surfaces of the coating and substrate. These parameters are all different from those of the input combustion gas and cooling air, and all are nonuniform. To address this problem, we assume that the heat transfer between the flow-field domain (the combustion gas and cooling air) and the solid domain (the coating and substrate) occurs at the thermal boundary layer. The temperatures of the combustion gas and cooling air in the thermal boundary layer are denoted $T_{e,conv}$ and $T_{i,conv}$, respectively. On this basis, the following functional relation between the thermal insulation effect of a TBC, $\Delta\phi$, and the nine parameters is established [12]:

$$\Delta\phi = f(k, d, k_{tbc}, d_{tbc}, h_e, h_i, T_\infty - T_c, T_\infty - T_{e,conv}, T_\infty - T_{i,conv}) \tag{14.9}$$

Further, an analysis of the nine influencing parameters based on the π theorem for dimensionless analysis yields the following five mutually independent dimensionless parameters:

$$Bi = \frac{d \cdot h_e}{k} \tag{14.10}$$

$$Bi_{tbc} = \frac{d_{tbc} \cdot h_e}{k_{tbc}} \tag{14.11}$$

$$\alpha = \frac{T_\infty - T_{,conv}}{T_\infty - T_c} \tag{14.12}$$

$$\eta = \frac{T_\infty - T_{e,conv}}{T_\infty - T_c} \tag{14.13}$$

$$R = \frac{h_e}{h_i} \tag{14.14}$$

Then, Eq. (14.9) can be expressed with the following dimensionless functional relation:

$$\Delta\phi = f(Bi, Bi_{\text{tbc}}, \alpha, \eta, R) \tag{14.15}$$

where *Bi* is the Biot number of the substrate (i.e., the ratio of the thermal resistance of the convective heat transfer boundary layer to the thermal resistance of the material comprising the substrate of the blade), Bi_{tbc} is the Biot number of the TBC (i.e., the ratio of the thermal resistance of the convective heat transfer boundary layer to the thermal resistance of the TBC), *R* is the ratio of the external CHTC to the internal CHTC, and α and η are the dimensionless temperatures of the combustion gas and cooling air in the thermal boundary layer, respectively.

It has already been proven by the π theorem that as long as the dimensionless parameters in Eq. (14.15) remain unchanged, the result remains the same, regardless of the changes in the physical quantities that comprise the dimensionless parameters. For example, when *Bi* in Eq. (14.10) is fixed, $\Delta\phi$ remains the same, regardless of the changes in *d*, h_e, and *k*. According to Eq. (14.9), if *Bi* remains unchanged, the use of three sets of data for *d*, h_e, and *k* requires $3 \times 3 \times 3 = 27$ tests each, while the use of 50 sets of data for *d*, h_e, and *k* requires 125,000 tests each. In comparison, the use of the π theorem (i.e., Eq. (14.15)) requires only one test, thereby significantly reducing the number of tests. A substantial increase in the number of tests significantly wastes both manpower and resources and results in very large test errors. Thus, reducing the number of influencing factors through the π theorem (i.e., Eq. (14.15)) has important significance for the investigation of the thermal insulation effect.

To determine the detailed functional relation between $\Delta\phi$ and the five dimensionless parameters, based on Fourier's law of heat transfer and Newton's law of cooling, we analyze in detail the heat transfer along the thickness direction in a turbine blade equipped with a cooling film before and after the application of a TBC. The following overall cooling efficiency of a turbine blade before and after the application of a TBC is obtained as follows:

$$\phi = \frac{1}{1 + Bi + R}(\alpha - \eta) + \eta \tag{14.16}$$

$$\phi' = \frac{1 + Bi_{\text{tbc}}}{1 + Bi + Bi_{\text{tbc}} + R}(\alpha' - \eta') + \eta' \tag{14.17}$$

where $\alpha' = \alpha + \Delta\alpha$ ($\Delta\alpha$ is the extent to which the TBC affects α) and $\eta' = \eta + \Delta\eta$ ($\Delta\eta$ is the extent to which the TBC affects η).

Based on Eqs. (14.8), (14.16), and (14.17), $\Delta\phi$ is determined as follows:

$$\Delta\phi = (a - b)(a - \eta) + a\Delta\alpha \tag{14.18}$$

where

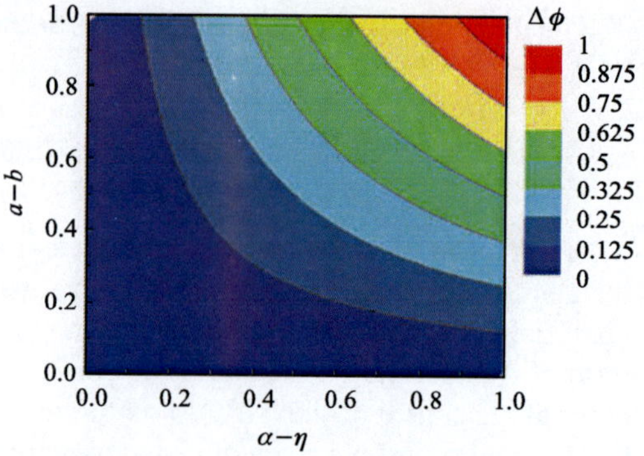

Fig. 14.2 Variation in the TBC performance $\Delta\phi$ with dimensionless parameters [12]

$$a = \frac{1 + Bi_{\mathrm{tbc}}}{1 + Bi + Bi_{\mathrm{tbc}} + R} \quad b = \frac{1}{1 + Bi + R} \tag{14.19}$$

In Eq. (4.18), $(a - b)$ represents the effect of the TBC on the thermal resistance ratio of the blade, and $(\alpha - \eta)$ represents the difference between the temperatures of the internal and external thermal boundary layers of the blade. The TBC provides heat insulation by altering the thermal resistance ratio. In addition, in Eq. (4.18), $a\Delta\alpha$ represents the change in the cooling efficiency of the blade resulting from the influence of the TBC on the temperature of the cooling air.

Figure 14.2 shows the evolution of the thermal insulation effect of the TBC, $\Delta\phi$, with $(a - b)$ and $(a - \eta)$. $\Delta\phi$ increases as $(a - b)$ and $(a - \eta)$ increase. $(a - \eta)$ represents the difference between the temperatures of the internal and external thermal boundary layers of the blade. Figure 14.3 also shows that $\Delta\phi$ decreases as η increases, suggesting that the thermal insulation effect of the TBC is insignificant in zones with a large η (e.g., film holes and trailing-edge slots). $(a - b)$ increases as $d_{\mathrm{tbc}}/k_{\mathrm{tbc}}$ increases, indicating that increasing the thickness of the TBC or reducing its thermal conductivity helps to enhance its thermal insulation effect. $(a - b)$ increases as h_{e} and h_{i} increase, suggesting that the TBC exhibits a higher thermal insulation effect in high-velocity, high-turbulence-intensity zones and that increasing the internal cooling rate and internal turbulence intensity is conducive to enhancing the thermal insulation effect of the TBC.

14.2 Numerical Simulation of the Thermal Insulation Effect

The theoretical model for the thermal insulation effect introduced above focuses on the heat transfer in the turbine blade along its thickness direction under the combined action of the TBC and cooling film. It is relatively easy to obtain theoretical solutions for simple objects of analysis such as panel-like and cylindrical specimens. In comparison, it is extremely difficult to obtain analytical solutions for turbine blades

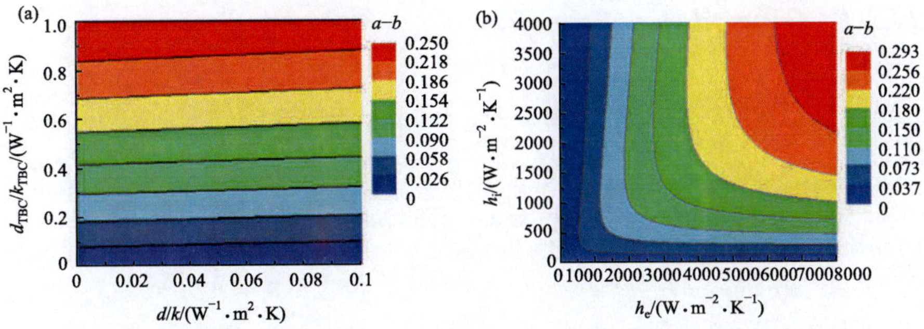

Fig. 14.3 Variation in $(a - b)$ with **a** d/k and d_{tbc}/k_{tbc} as well as with **b** h_e and h_i

due to their complex structure. In addition, the heat exchange, heat convection, and heat transfer of the combustion gas and cooling air vary with the structure (e.g., curved surface and film holes) of a turbine blade. As a result, the 3D temperature field in a turbine blade is distributed unevenly, and the heat flow is transferred both along the thickness direction and from high-temperature zones to low-temperature zones. Consequently, it is difficult to obtain analytical solutions for the temperature field of a turbine blade and the thermal insulation effect of its TBC. With the development of numerical modeling of complex structures (e.g., turbine blades), methods for calculating fluid–solid heat transfer and its coupling, and numerical methods for solid deformation, numerical simulation has become an important tool for analyzing the thermal insulation effect of TBCs on turbine blades.

Based on the theoretical model, both the overall cooling efficiency of a turbine blade and the thermal insulation effect of its TBC depend heavily on the heat transfer of the TBC with the combustion gas and cooling air. These effects all need to be manifested through specific forms of the temperature field to reflect the thermal insulation effect. The temperature field of the TBC on a turbine blade depends on factors such as the temperature and flow rate of the high-temperature combustion gas and cooling air. In turn, the temperature field of a turbine blade affects the temperature and heat transfer of the combustion gas and cooling air. Here, the combustion gas and cooling air are collectively referred to as the flow field, while the turbine blade is referred to as the solid. This way, the interaction between the flow field and solid (i.e., coupled heat transfer) can be determined. In the early days, due to the limitations of the calculation methods and the capability of coupled heat transfer, the heat transfer between the flow field and solid was generally determined through decoupling. The demand for a further improved cooling design for turbine blades and the application of curved blade, film cooling, and coating technologies led to a higher degree of coupling between the flow field and solid. Under these conditions, the temperature field obtained through decoupling differs significantly from the actual temperature field. To address this problem, researchers have proposed various implementation methods for coupled heat transfer, which are described below in three modes, namely uncoupled, weakly coupled, and strongly coupled.

14.2.1 Coupled Heat Transfer

Uncoupled. When used to calculate the temperature field and thermal insulation efficiency of the TBC on a turbine blade, this method neglects the changes in the flow field of the combustion gas and cooling gas, uses the heat flow from the fluid to the solid as the boundary condition at the fluid/solid interface, and determines the temperature field in the TBC on the turbine blade based on the Fourier heat transfer equation. The heat transfer equation for the solid domain is as follows:

$$\rho c \frac{\partial T}{\partial t} = \nabla \cdot (k \nabla T) \tag{14.20}$$

where c is the specific heat. The heat flow from a given fluid to a blade at the fluid/solid interface is difficult to measure and therefore is often expressed in terms of the temperature and CHTC of the fluid in the thermal boundary layer as follows:

$$-k \frac{\partial T}{\partial n}\bigg|_w = h_{\mathrm{e}}(T_{\mathrm{e,conv}} - T_{\mathrm{w,e}}) \tag{14.21}$$

This type of method has the advantages of fast solving speed and good convergence. Using the uncoupled method, Ziaeiasl and Ramezanlou [22] studied the temperature field in a turbine blade equipped with film cooling and a TBC. They found that the coating could reduce the surface temperature of the substrate by up to 100 °C and its thermal insulation effect increased as its thickness increased. The accuracy of the thermal insulation effect of a TBC calculated using the uncoupled method depends on the accuracy of the temperature and CHTC of the fluid on the surface of the coating. However, it is exceedingly difficult to measure the temperature and CHTC of the combustion gas in various zones of a turbine engine blade under working conditions. As a result, for complex turbine blades equipped with film cooling and TBCs, it is difficult to use this method to analyze the influences of the film-cooling structure, nonnormal temperature cooling air, radiation, and particularly the temperature field of a nonuniform air current on the thermal insulation effect of TBCs.

Weakly coupled. This method solves the flow and temperature field of a fluid based on the Navier–Stokes equations and then calculates the interfacial heat flux using the obtained interfacial temperature of the fluid domain as the boundary condition of the solid domain, followed by the calculation of the temperature field of the fluid using the heat flux as the boundary of the fluid domain. The iteration is repeated to ensure temperature continuity and heat flux conservation at the interface. Figure 14.4 shows the calculation principle of this method [23]. The temperature field in the solid domain is calculated based on Eq. (14.20), while the temperature field in the fluid domain is solved based on the Navier–Stokes equations under Stokes's assumption [24]:

$$\frac{\partial \rho}{\partial t} + \nabla \cdot (\rho \boldsymbol{v}) = 0 \tag{14.22}$$

$$\frac{\partial (\rho v)}{\partial t} + \nabla \cdot (\rho v v) = -\nabla p + \nabla \cdot \tau + f \tag{14.23}$$

$$\frac{\partial (\rho h)}{\partial t} - \frac{\partial p}{\partial t} + \nabla \cdot (\rho \boldsymbol{v} h) = \nabla \cdot (\lambda \nabla T) + \mu \mathbf{I} \cdot \nabla p + \boldsymbol{\tau}:\nabla \boldsymbol{v} + \dot{Q} \tag{14.24}$$

where ρ is the density of the fluid, t is the time, $\boldsymbol{v}$ is the velocity, p is the pressure, μ is the viscosity, h is the total entropy per unit mass, λ is the thermal conductivity of the fluid, T is the temperature, f is the volume force, $\dot{Q}$ is the heat source, and $\boldsymbol{\tau}$ is the viscous stress tensor and can be expressed as follows:

$$\tau_{ij} = 2\mu \left(s_{ij} - \frac{1}{3} \frac{\partial v_k}{\partial x_k} \delta_{ij} \right) \tag{14.25}$$

where s_{ij} is the component of the deformation velocity tensor, ν_k is the turbulent viscosity of the fluid, and δ_{ij} is the Kronecker delta function. In addition, the ideal gas state equation is as follows:

$$p = \rho RT \tag{14.26}$$

where R is the Avogadro constant.

Temperature and heat flux continuity are the basic conditions for coupled heat transfer at the fluid/solid interface:

$$T_s = T_l \tag{14.27}$$

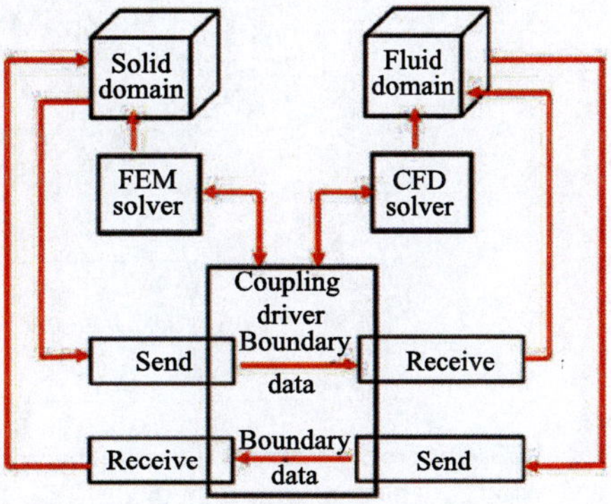

Fig. 14.4 Computational procedure of the weakly coupled method [23]

$$k\frac{\partial T}{\partial n}\bigg|_s = \lambda\frac{\partial T}{\partial n}\bigg|_l \tag{14.28}$$

where $\boldsymbol{n}$ is the normal direction and the subscripts s and l signify the solid and fluid, respectively. Heselhaus and Vogel [25] analyzed the temperature distribution in turbine blades with cooling structures using uncoupled and coupled numerical simulation methods, respectively, as shown in Fig. 14.5. They found a maximum difference of 73 °C between the surface temperatures of coupled and uncoupled blades. In addition, Heselhaus and Vogel [25] and Sondak and Dorney [26] independently examined the problem of heat transfer in turbine rotors and 3D blades using the weakly coupled method (i.e., by satisfying the boundary conditions in Eqs. (14.27) and (14.28) at the fluid/solid interface). By comparing with the temperature field in the blades simulated under adiabatic conditions, they verified the necessity of fluid/solid coupling for ensuring solution accuracy. Using the weak fluid/solid coupling-based numerical simulation method, Zhu et al. [23] studied the temperature distribution in a turbine blade with a multilayer TBC. They found that the TBC exhibited an excellent thermal insulation effect at the leading and trailing edges of the blade, consistent with the experimental results. The weak coupling-based numerical simulation requires repeated iterations at the interface, resulting in a low computational speed and poor convergence. However, because the weak-coupling method uses different solvers for the computational domains of the fluid and solid and the solvers can be flexibly modified based on regional features, it is, to a certain extent, advantageous for studying the problem of heat transfer in TBCs with microstructures such as multiple layers and pores.

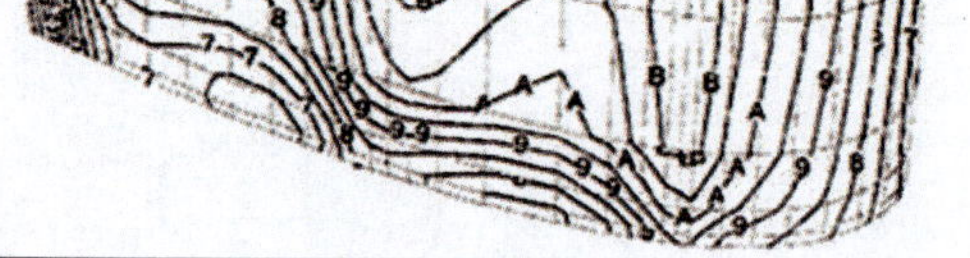

$T_{coupled}/$	1	2	3	4	5	6	7	8	9	A	B
$T_{uncoupled}$	0.94	0.95	0.96	0.97	0.98	0.99	1.00	1.01	1.02	1.03	1.04
$T_{coupled}-$ $T_{uncoupled}$	−73K										+43K

Fig. 14.5 Difference between the temperatures of a turbine blade calculated using the uncoupled and weakly coupled numerical methods [25]

Strongly coupled. This method extends the governing equation for the fluid domain to the solid domain, calculates the heat transfer in the solid domain using a degenerate energy equation, and solves the two domains in a unified manner. The heat transfer in the fluid domain satisfies Eq. (14.24). The following degenerate energy equation is the governing equation for the heat flow that has moved to the solid domain:

$$\frac{\partial(\rho h)}{\partial t} + \nabla \cdot (\rho v h) = \nabla \cdot (k \nabla T) + \dot{Q} \tag{14.29}$$

where h is the entropy per unit mass (the same as h in Eq. (14.24)). Because the fluid and solid domains are solved in a unified manner, the conditions in Eqs. (14.27) and (14.28) are automatically satisfied at the interface. Through experimentation in conjunction with the strong-coupling-based numerical calculation method, Eifel et al. [27] analyzed the effects of the cooling structure of a blade on its cooling performance. In addition, they calculated the heat transfer in the internal flow channels of the blade using the commercial software CFX. They found that the flow-disturbing ribs produced a better cooling effect when arranged in a staggered pattern than when arranged in a parallel pattern and that the staggered arrangement improved the cooling efficiency of the blade by 12.5% and reduced the maximum surface temperature of the blade by 33.5% compared to that of the parallel arrangement. Moritz et al. [28] calculated the thermal load in a blade with cooling channels and film holes at the leading edge using the commercial software CHTflow and found that the results agreed well with the experimental measurements. Overall, the strongly and weakly coupled methods have their respective advantages. The weakly coupled method allows flexible calculations based on regional features but performs poorly in terms of convergence. In particular, when used to study turbine blades with film holes, pin fins, and TBCs, this method involves multiple complex coupling interfaces and experiences tremendous difficulties in data transfer between the two computational domains, program implementation, and computational convergence. The strong-coupling method calculates the fluid and solid together with good coupling performance and is more convenient for numerically simulating TBCs. However, this method requires an extraordinarily long computational time.

14.2.2 Turbulence Models

In the flow field surrounding a turbine blade, due to its high velocity, the combustion gas disturbs the laminar flow regime of the fluid, resulting in the occurrence of small vortices and flow disturbance, a phenomenon that is referred to as turbulence. Turbulence at the fluid/solid boundary layer directly affects the intensity of the heat transfer between the fluid and solid. Turbulence simulation is the key to improving the accuracy of numerical simulation. However, turbulence is a difficult research problem that has challenged mathematicians and mechanicians over the world for over two centuries and has yet to be solved. In the engineering field, there are currently three main types of methods for numerically simulating turbulence, namely direct numerical simulation (DNS), large-eddy simulation (LES), and Reynolds-averaged Navier–Stokes equations (RANS). Due to their very large computational loads, DNS and LES are rarely used in engineering. In particular, it is difficult to apply these two methods to turbine blades with film structures and multilayer TBC systems. In the RANS method, the variables in the Navier–Stokes equations are decomposed into time-averaged and fluctuating quantities, and a Reynolds stress term and turbulent kinetic energy are introduced into the equations. In addition, the Reynolds stress is assumed to be proportional to the strain with the turbulent viscosity coefficient as the coefficient of proportionality, which is determined based on the new equations. Because the objective is to calculate the mean turbulent motion, the turbulent motion can be reflected without the need for a very small computational step size and mesh size. Thus, the RANS method has a considerably lower computational load than the other two methods. In addition, the RANS method is able to meet the accuracy requirements of engineering tasks, has high computational cost-effectiveness, and, as a result, is widely used in engineering.

To calculate the turbulent dynamic viscosity in a flow field using the RANS method, researchers have established various turbulence models, of which the $k-\varepsilon$, $k-\omega$, and shear stress transport (SST) models are extensively used. The SST model [29] combines the advantages of the $k-\varepsilon$ and $k-\omega$ models. Specifically, the transport equation related to the dissipation rate ε in the $k-\varepsilon$ model is first written in the form of ω, and then the expression of the SST model is obtained by taking the weighted sum of the $k-\omega$ model and the transformed $k-\varepsilon$ model based on the mixed function of $\phi_3 = F_1\,\phi_1 + (1-F_1)\,\phi_2$ as follows:

$$\frac{\partial}{\partial t}(\rho k') + \frac{\partial}{\partial x_j}(\rho k' v_j) = \frac{\partial}{\partial x_j}\left[(\mu + \frac{\mu_t}{\sigma_{k3}})\frac{\partial k'}{\partial x_j}\right] + P_k - \beta' \rho k' \omega \tag{14.30}$$

$$\begin{aligned}\frac{\partial}{\partial t}(\rho\omega) + \frac{\partial}{\partial x_j}(\rho\omega v_j) = &\frac{\partial}{\partial x_j}\left[\left(\mu + \frac{\mu_t}{\sigma_{\omega 3}}_j\right)\frac{\partial \omega}{\partial x_j}\right] + (1-F_1)\frac{2\rho}{\sigma_{w2}\omega}\frac{\partial k'}{\partial x_j}\frac{\partial \omega}{\partial x_j} \\ &+ \alpha_3 \frac{\omega}{k'} P_k - \beta_3 \rho \omega^2\end{aligned} \tag{14.31}$$

where ρ is the density, k' is the turbulent kinetic energy, ω is the specific dissipation rate, v_j is the velocity, μ is the viscosity of the fluid, $\mu_t = \rho\frac{k'}{w}$ is the turbulent viscosity, $\sigma_{k3}, \sigma_{\omega 2}, \sigma_{\omega 3}, \alpha_3, \beta_3, F_1, \beta'$ are constant coefficients, and P_k is the turbulent viscosity term with the following specific form:

$$P_k = \mu_t\left(\frac{\partial u_i}{\partial x_j} + \frac{\partial u_j}{\partial x_i}\right)\frac{\partial u_i}{\partial x_j} - \frac{2}{3}\frac{\partial u_k}{\partial x_k}\left(3\mu_t\frac{\partial u_k}{\partial x_k} + \rho k'\right) \tag{14.32}$$

Yoshiara et al. [30] studied the problem of heat transfer in Mark II and C3X blades using the unstructured mesh solver TAS-code and computationally compared the

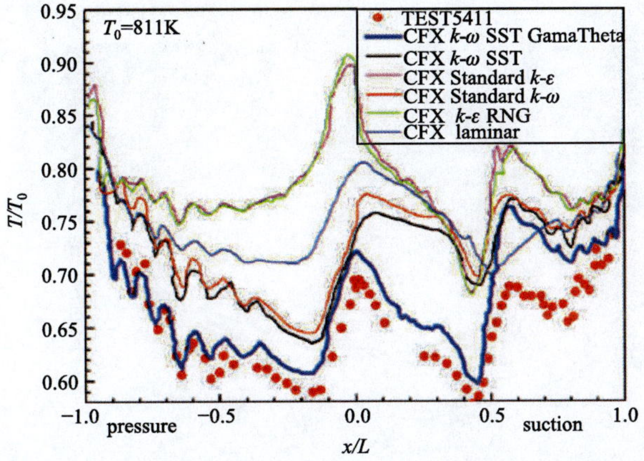

Fig. 14.6 Distribution of the temperature at the middle section of the blade simulated using different turbulence models (T_0 is the inlet temperature of the combustion gas; X is the location along the chord length of the blade; L is the chord length; the pressure-side zone, suction-side zone, leading edge, and trailing edge of the blade are located at $-1 < X/L < 0$, $0 < X/L < 1$, $X/L = 0$, and $X/L = -1$ and 1, respectively)

three turbulence models. They found that the SST turbulence model produced slight errors in the pressure distribution but was the most accurate in capturing the locations of the transition points. Chinese researcher Dong [31] investigated the characteristics of the transition flow in the boundary layer in an air-cooled turbine blade and its effects on temperature and compared multiple turbulence models (SST models) in terms of the ability to identify transition flow. Fluid–heat coupling in multiple cases was calculated using the commercial software FLUENT and CFX, as shown in Fig. 14.6. The numerical results for the complex flow field and heat transfer in the turbine blade produced by different turbulence models differ significantly. The numerical results produced by the CFX SST model are more consistent with the experimental results, proving that the SST turbulence model exhibits a higher accuracy for calculating the flow and heat transfer in turbine blades. The SST turbulence model has also been validated in multiple studies examining the cooling efficiency of turbine blades [32–35].

14.2.3 Numerical Simulation of the Thermal Insulation Effect of TBCs

Based on the above strong-coupling-based numerical simulation method, the thermal insulation effect of TBCs is studied in this section using the TBC on a certain 3D guide vane developed by the U.S. National Aeronautics and Space Administration (NASA) as an example.

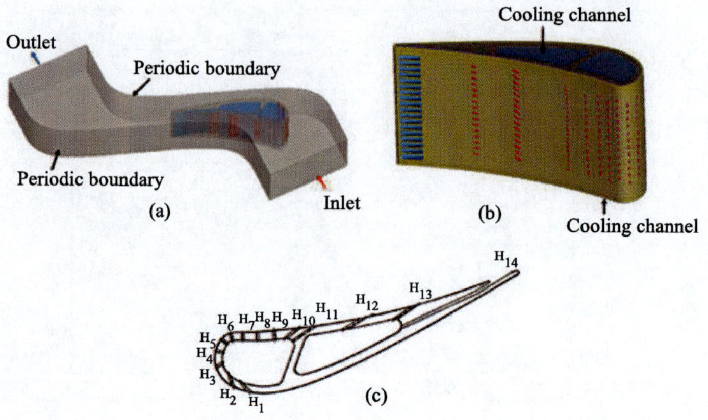

Fig. 14.7 Turbine vane models: **a** fluid-domain model, **b** solid-domain model, **c** cross-section of the vane

(1) Modeling

A geometric model can be established using common geometric modeling software (e.g., UG, SOLIDWORKS, and PEOE). The geometric model established in this section consists of a solid part and a fluid part, as shown in Fig. 14.7. A guide vane composed of a front cooling chamber and a back cooling chamber is present in the fluid channel. The surface of the vane is coated with a 300-μm-thick single-layer TBC. There are 14 rows of film holes (one row (H_{14}) of trailing-edge slots and 13 rows (H_1–H_{13}) of film holes) on the vane. Table 14.1 summarizes the material parameters of the vane and ceramic layer.

(2) Mesh generation

Figure 14.8 shows the computational mesh (an unstructured tetrahedral mesh) generated using ANSYS ICEM 16.0. The mesh is refined at the film holes and trailing edge. In addition, multiple boundary layers are generated in the computational domains for the fluid and solid to improve the computational accuracy of the coupled heat transfer at the interface. The mesh contains a total of 9.21 million cells. To further examine the stability of the mesh, the calculation results and experimental data are compared in Fig. 14.9. The numerical calculation results are basically consistent with the measured Mach numbers, thus verifying the accuracy of the calculation results.

Table 14.1 Material parameters of the vane and TBC

Material parameter	Vane	Coating
Density (kg·m^{-3})	8700	4930
Specific heat capacity (J·g^{-1}·K^{-1})	434	418
Thermal conductivity (W·m^{-1}·K^{-1})	60.5	1.02

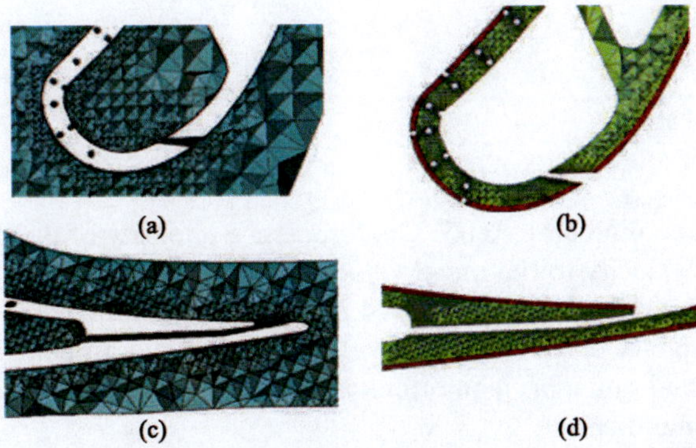

Fig. 14.8 Computational mesh: **a** leading edge of the fluid domain, **b** leading edge of the solid domain, **c** trailing edge of the fluid domain, **d** trailing edge of the solid domain

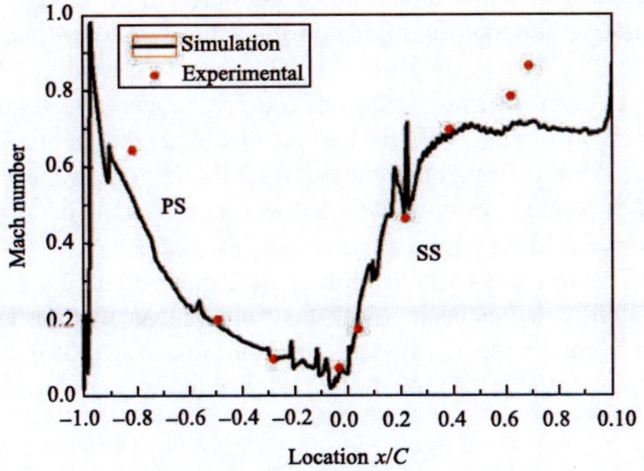

Fig. 14.9 Comparison of the calculated and measured Mach numbers (x represents the location along the chord; C is the chord length; PS is the pressure side of the blade ($-1 < x/C < 0$); SS is the suction side of the blade ($0 < x/C < 1$); the leading and trailing edges of the blade are located at $x/C = 0$ and $x/C = -1$ and 1, respectively; x/C, similar to s/L in Fig. 14.6, represents the relative location along the chord length of the blade)

(3) Material parameters and boundary conditions

In the fluid domain, we assume that the combustion gas is an ideal gas and do not consider the deformation of the solid. Let $T_{t,inlet}$ and $P_{t,inlet}$ be the total temperature and total pressure of the fluid at the inlet boundary, respectively, $P_{s,outlet}$ be the mean static pressure of the fluid at the outlet boundary, and $T_{t,inlet,coolant}$ and $P_{t,inlet,coolant}$ be the total temperature and total pressure of the cooling air at the inlet boundary, respectively. Table 14.2 summarizes the fluid boundary conditions for simulating coupled heat transfer. We set the variables at the periodic boundaries of the fluid domain to be equivalent and set the turbulence intensity of the fluid at the inlet of the

Table 14.2 Fluid boundary conditions

$P_{t,\text{ inlet}}$ (atm)	$P_{t,\text{ inlet, coolant}}$ (atm)	$T_{t,\text{ inlet}}$ (K)	$T_{t,\text{ inlet, coolant}}$ (K)	$P_{s,\text{ outlet}}$ (atm)
3.447	3.509	709	339	2.064

mainstream and cooling air to 0.05. The turbulence intensity of the fluid is the ratio of the turbulent velocity to the mean velocity and can be converted to the value of the turbulent kinetic energy k' in Eq. (14.30) at the boundaries. In addition, we set the velocity boundary conditions of the solid wall as non-slip boundary conditions and set the thermal boundary conditions at the fluid/solid interface as temperature and heat flux continuity.

(4) Result analysis

Figure 14.10 shows the overall cooling efficiency of the 3D vane. Figure 14.10a and b shows the results obtained without the consideration of the effect of the TBC. The cooling efficiency of the vane is higher in the middle section than at the top and bottom, because the cooling-air film primarily covers the middle of the vane. The overall cooling efficiency of the surface of the TBC-covered turbine vane is higher than that of the case where the TBC is not considered, as shown in Fig. 14.10c and d. In addition, the overall cooling efficiency of the TBC-covered vane is more uniform, suggesting that the temperature gradient in the vane is relatively small, which helps to reduce the thermal stress over a longer time of use.

Figure 14.11 shows the overall cooling performance of the vane at different heights. The cooling efficiencies in the absence and presence of the TBC, ϕ and ϕ_{tbc}, respectively, are higher in the middle section of the vane (mid, 50%) than at its bottom (hub, 5%) and top (tip, 95%). At the trailing edge, ϕ and ϕ_{tbc} are lower at the top than

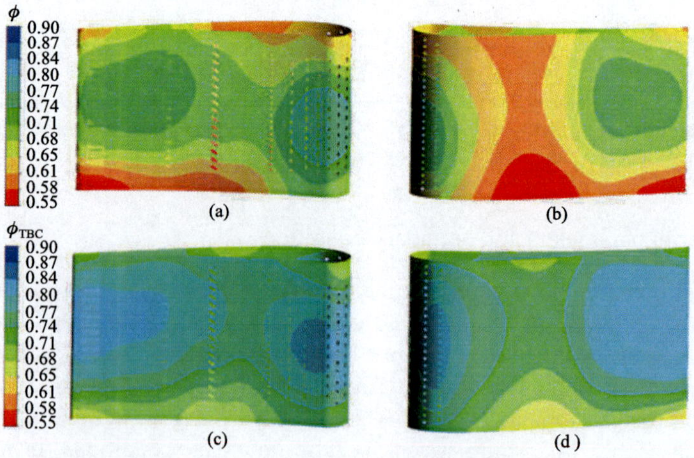

Fig. 14.10 Overall cooling efficiency of the vane: **a** TBC-free pressure side, **b** TBC-free suction side, **c** TBC-covered pressure side, and **d** TBC-covered suction side

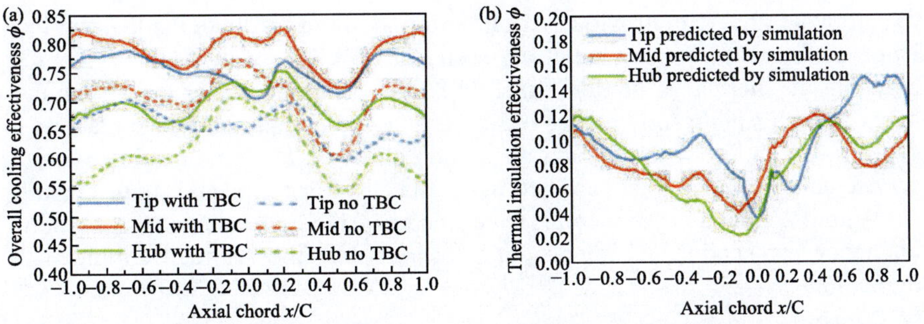

Fig. 14.11 Cross-sectional cooling efficiency of the vane: **a** overall cooling efficiency and **b** thermal insulation efficiency of the TBC

at the bottom, because the temperature of the cooling air gradually increases from the top to the bottom and the internal convective heat transfer efficiency decreases as the temperature of the internal cooling air increases. As shown in Fig. 14.11b, the thermal insulation efficiency of the TBC, $\Delta\phi$, is low at the leading edge but higher at the trailing edge, because the flow rate on the pressure side increases from the leading edge to the trailing edge. An analysis based on Eq. (14.18) shows that the CHTC h_e increases as the flow rate increases. The inadequate film cooling on the suction side leads to a decrease in the dimensionless temperature of the thermal boundary layer, η, at the trailing edge, which is a cause of the increase in $\Delta\phi$. There are two peak values of $\Delta\phi$ in the front cooling channel ($x/C = -0.3$) and back cooling channel ($x/C = 0.4$), suggesting a good internal cooling efficiency, which is favorable to the improvement in the thermal insulation efficiency of the TBC. This result is consistent with that obtained by Davidson et al. [21]—the efficiency of a TBC increases with increasing internal cooling efficiency achieved through an increase in the Reynolds number. Thus, a high internal HTC can effectively improve the thermal insulation efficiency of TBCs in cooling systems.

14.3 Testing Methods for the Thermal Insulation Effect

The problem of convective heat transfer is solved through numerical simulation based on semi-empirical equations. The accuracy of the numerical solution depends on the quality of the numerical mesh and the accuracy of the turbulence model. A complex blade under the action of a high-velocity combustion gas exhibits complex flow characteristics such as turbulent transition and fluctuation. It is difficult to obtain a true solution for the convective heat transfer in the thermal boundary layer through numerical simulation. In addition, there are large errors in numerically simulating thermal insulation effects. To address these deficiencies, corresponding experimental studies can be conducted. However, the overall tests are associated with problems such as high costs, great difficulties, and an inability to examine specific areas and therefore are unable to provide accurate information to facilitate the thermal insulation performance improvement and design of TBCs. In view of this issue, tests targeting the thermal insulation effect of TBCs on turbine blades are essential, as they can both significantly reduce test costs and, more importantly, facilitate an in-depth investigation of the thermal insulation effect of TBCs on turbine blades in service environments. There are two main aspects of experimental studies, namely (1) setups for simulating the service environment of TBCs on turbine blades (cooling efficiency tests) and (2) techniques for measuring the temperature of turbine blades in real time.

14.3.1 Setups for Simulating the Service Environment of Turbine Blades

Simulation of the service environment of a turbine blade refers to the simulation of its flow and heat transfer conditions. Thus, a service environment simulation setup must satisfy the following conditions: (a) high-temperature, high-velocity combustion gas; (b) high-pressure cooling air; (c) high-speed rotation; (d) dynamic and thermodynamic similarity; and (e) data measurability. These stringent requirements increase the difficulty in developing service simulation setups and pose enormous challenges regarding materials and measurement techniques. Earlier, to study the cooling efficiency of turbine blades, NASA [36] developed a low-temperature, low-pressure test under conditions satisfying geometric, kinetic, dynamic, and thermal similarities and noted, through derivation, that the test should satisfy similarities in pressure coefficient *P*, Reynolds number *Re*, and Prandtl number *Pr*:

$$\frac{p}{\rho v^2} = \mathrm{P} \tag{14.33}$$

$$\frac{\rho v l}{\mu} = \mathrm{Re} \tag{14.34}$$

$$\frac{c_p \mu}{k} = \mathrm{Pr} \tag{14.35}$$

This is a specific application of the π theorem in the flow-field simulation of thermal insulation effect model tests. Similar to the analysis of Eq. (14.15), as long as P, Re, and Pr remain unchanged, the results remain the same, regardless of the changes in the physical quantities (e.g., flow rate and pressure).

Based on this principle, NASA [37] established a test setup for guide vanes. In this setup, a wind tunnel and compressed air were used to simulate equivalent combustion-air and cooling-air environments for guide vanes under low-temperature, low-pressure conditions. This setup was further used to study the cooling performance of turbine blades. The results verified the reliability, feasibility, and superiority of tests based on the π theorem (i.e., physical similarity principle). A cooling simulation setup was also developed at the Turbulence and Turbine Cooling Laboratory of the University of Texas at Austin. Based on this setup, Dees et al. [20] and Davidson et al. [21] studied the thermal insulation effect of internal cooling, film cooling, and TBCs. Figure 14.12 shows the basic structure of this setup. In this setup, a combustion-gas flow with a certain temperature is generated through resistance heating. In addition, a high-pressure compressor is used to generate cooling air and to guide it into the interior of the blade. To produce similar flow conditions for the blade, a small section of a wind tunnel is used as the simulated section, and the test section is designed to be circular. The turning corner of the circular wind tunnel is modified as the simulated section of the turbine blade in such a way that it can accommodate three vanes that form two flow channels. Three simulated vanes are placed in the test section of the wind tunnel to form two flow channels. To produce similar flow conditions for the blade, the parameters are adjusted to allow the P, Re, Bi, and Pr of the simulated wind tunnel to be similar to those of the real-world service environment. To facilitate the installation of thermocouples and data measurement, the simulated vanes are 10 times as large as the actual ones. Similarly, Lynch et al. [38] of the Pennsylvania State University established a cooling performance test section containing seven turbine blades produced by enlarging the Pack-B low-pressure turbine blade while maintaining its geometry.

In China, Zhang et al. [39] of North China Electric Power University experimentally investigated the cooling efficiency at the leading edge of a blade based on a stationary-cascade wind-tunnel test setup. Specifically, they analyzed the effects of the blowing ratio and Reynolds number of the mainstream on the cooling efficiency of the blade at the leading edge. Lu et al. [40] established a small-scale low-velocity simulation setup for analyzing cooling efficiency with a gas heating net and cooling air and, on this basis, studied the cooling efficiency of cylindrical film holes embedded in transverse trenches. Based on the cooling efficiency test bench shown in Fig. 14.13,

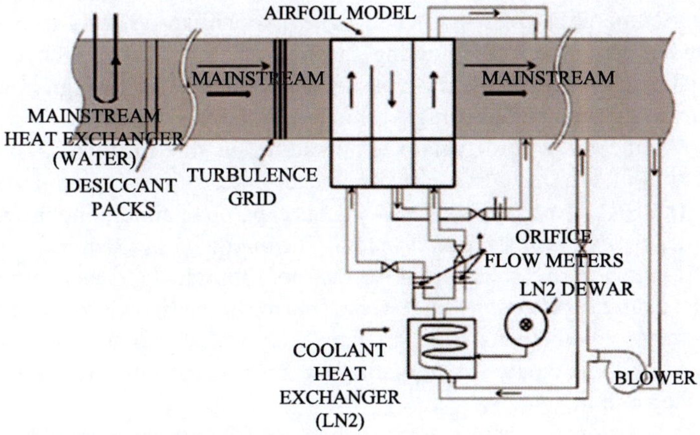

Fig. 14.12 Turbine cooling simulation test setup

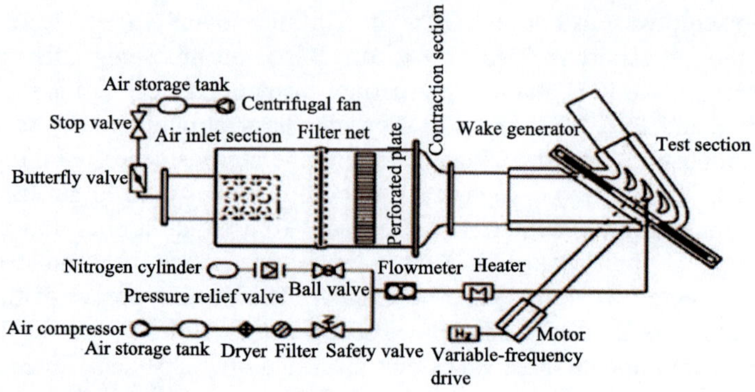

Fig. 14.13 Turbine cooling simulation test setup

Li et al. [41] of Xi'an Jiaotong University examined the cooling efficiency of the film on the surface of a turbine blade with compound-angle fan-shaped holes under the wake effect. They found that the wake significantly reduced the cooling efficiency of the film on the surface of the blade. At a Strouhal number of the wake of 0.36 (the Strouhal number is a dimensionless parameter that characterizes the nonstationarity of flow and is defined in Li et al.'s study as $Sr = \frac{2\pi NnD}{60u}$, where N is the rotational velocity, n is the number of pin fins in the turbulence generator, D is the diameter of each pin fin, and u is the velocity of the fluid), the cooling efficiency of the film on the surface of the blade decreased, on average, by 35 and 26% at low and high mass flow ratios, respectively, showing that the cooling efficiency of the film decreased at a decreasing rate with increasing mass flow ratio. The above model test setups provide an important basis for designing film cooling for turbine blades.

With continually increasing service temperature, it becomes difficult to simulate the effects of turbulence, hot spots, radiation, and rotation on the heat transfer in the TBCs on turbine blades under low-temperature, low-pressure conditions. Developing high-temperature, high-pressure, high-rotational-speed service environment simulation setups to study the cooling and thermal insulation effect of TBCs on moving/stationary blades is an inevitable necessity for developing high-performance aeroengines. High-temperature high-velocity combustion gases, high-speed rotation, and cooling air with a certain temperature are three difficult issues facing the simulation of extreme service environments. Internationally, the high-pressure burner rig (HPBR) developed by NASA [42] is capable of simulating the real-world high-temperature high-pressure environment surrounding combustion gases in aeroengines. Researchers are able to observe the morphological changes in a specimen through the quartz observation window and image acquisition system of the HPBR. The high-velocity combustion-gas simulator with a Mach number of 0.3–1 developed by NASA [43] is capable of simulating supersonic high-temperature combustion gases, as shown in Fig. 14.14.

To study the failure mechanism and reliability of TBCs in extreme service environments, we developed a setup for testing TBCs on turbine guide vanes in simulated

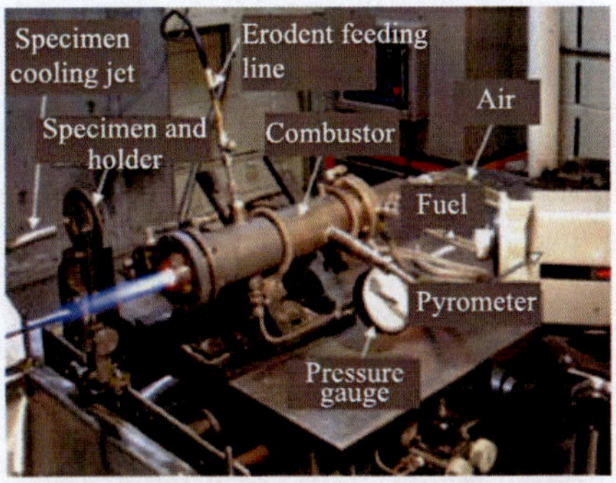

Fig. 14.14 High-velocity combustion-gas simulator with a Mach number of 0.3–1.0 developed by NASA

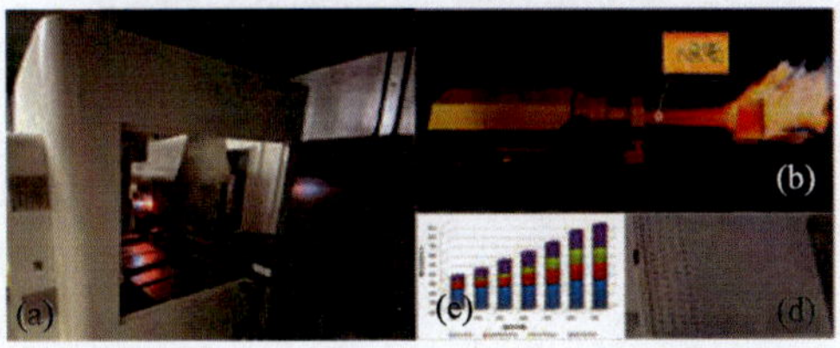

Fig. 14.15 Static model test setup for TBCs: **a** overall setup, **b** supersonic spray gun, **c** AE detection, and **d** spalled blade after the test

service environments [44]. See Chap. 15 for a detailed description of this setup. Here, this setup is briefly described for testing only the thermal insulation effect. This setup consists of a supersonic combustion-gas spray gun capable of generating a high-temperature, high-velocity combustion-gas flow that carries erosive particles and corrosive calcium–magnesium aluminosilicate (CMAS) particles, nondestructive testing systems (e.g., digital image correlation (DIC), acoustic emission (AE), and IR), and a data acquisition and control system with auxiliary modules, as shown in Fig. 14.15. In addition, this setup has the capacity to simulate high-temperature, erosion, and CMAS corrosion service environments in an integrated manner with controllable environmental parameters (e.g., temperature (up to 1700 °C), flame flow rate (up to Mach 1.0), and erosion rate (up to 300 m/s)) as well as with internal cooling and film-cooling conditions found in turbine blades with TBCs. On this basis, we further designed a high-speed rotating rotor system and a turbine model, as well as developed a model test setup for high-speed rotating working blades [45], as shown in Fig. 14.16. In this setup, the working blades of the turbine model are driven by a high-power variable-frequency motor to rotate at a certain high speed. A combustion-gas spray gun is used to apply a high-temperature, high-velocity combustion-gas flow that carries erosive and corrosive particles to the surface of the TBCs on the turbine blades to simulate the real-world service environment in which the TBCs on working blades rotate at a high speed and interact with combustion gases. After being heated to the target temperature by a heater, the high-pressure cooling air is guided into the cooling channels of the working blades through the front air inlet and deflector of the turbine blade to simulate the temperature gradient found in TBCs on real-world

Fig. 14.16 Dynamic model test setup for TBCs: **a** overall setup, **b** turbine model, **c** supersonic spray gun, **d** test process, **e** high-speed rotor system, and **f** control and display system

turbine blades. This setup has the capacity to simulate a combustion-gas environment with a high temperature of 1500 °C, a velocity of Mach 1, an erosion rate of 250 m/s, a rotational speed of 20,000 rpm, and controllable working conditions (temperature of the cooling air: up to 500 °C; flow rate of the cooling air: 500 L/min) for the working blades. These setups provide important tools to researchers in China to facilitate their investigation of the thermal insulation effect of TBCs on aeroengine turbines in working environments (e.g., high-temperature high-velocity combustion gases and high-speed rotation).

14.3.2 Real-Time Temperature Measurement Techniques for Turbine Blades

Real-time temperature measurement is an important basis for analyzing the thermal insulation effect of TBCs on turbine blades. Service environments (e.g., high-temperature, high-pressure, and high-speed rotation) create enormous difficulties for real-time temperature measurement, which necessitates temperature measurement techniques with high accuracy, a long measuring range, an ability to measure temperatures in real time, a fast response speed, a small spatial scale, and high stability. Currently, thin-film thermocouples (TFTCs), phosphor thermometry, and IR thermometry are the main techniques used to measure the temperature of TBCs on turbine blades. The following is an introduction to the progress of these temperature measurement techniques and their applications in TBCs on turbine blades.

(1) **TFTC-based temperature measurement technique**

TFTCs adhere well to curved surfaces, cause little interference with airflow, are resistant to shock and impact, and, therefore, are advantageous over common thermocouples. Figure 14.17 shows the basic structure of TFTCs. In the preparation process, a metallic substrate is coated first with a transition layer, next an electrochemical insulating layer (often made of Al_2O_3), then a temperature measurement functional layer, and finally a protection layer at the top that provides corrosion and shock resistance. The functional layer consists of two different metallic materials connected into a closed loop. A difference between the temperatures of the two metals at the point of connection results in the formation of a corresponding thermoelectric potential. This is the so-called Seebeck effect. Turbine blades operate at high temperatures. As a result, earlier thermocouples made of inexpensive metals (e.g., Cu and nichrome (K type)) were unable to meet the requirements. Lewis Research Center [46] in the United States developed a Pt13% Rh/Pt (R type) film sensor for measuring temperature stresses with an upper measurement limit of 1100 °C and an accuracy of ± 0.3 °C. NASA [47] measured the temperature distribution in a turbine blade at 1250 K with a Pt/Pt-10% Rh (S type) TFTC in the presence of exhaust gas from a combustion chamber. The Rolls-Royce Company [48] in the United Kingdom successfully developed a Pt/Rh TFTC and used it to measure the temperature distribution in the guide vanes of gas turbine engines with a temperature as high as 1200 °C, with an uncertainty of ±2%.

Comparatively, research on the application of TFTCs to aeroengines began relatively late in China. An [49] prepared an integrated structure consisting of a Pt/Rh10-Pt thermocouple and a blade substrate using the vacuum coating technique. However, the measurement process encountered several problems. For example, the TFTC fractured, the film and the filament lead wire became unwelded, and the measurement film and the substrate short-circuited. In addition, 40% of the thermocouple was damaged. In terms of performance, the thermocouple had a measurement error of ± 3% and a cumulative service life of no more than 10 h. To address these problems, Zhang et al. [50] investigated the effects of the thickness of a W:Re TFTC and its substrate thicknesses on its adhesiveness. They found that altering the thickness of the substrate had little effect on the adhesiveness of the TFTC and that suitably increasing the thickness of the TFTC increased its adhesiveness, while further increasing the thickness of the TFTC reduced its Seebeck coefficient.

Professor Zhao's team [51] developed multiple film sensors (including NiCr/NiSi, PtRh/Pt, indium–tin-oxide (ITO)/In_2O_3, and Pt/ITO), of which the PtRh/Pt TFTC is capable of measuring temperatures as high as 1000 °C and exhibits good repeatability and stability even at 1100 °C. These researchers further collaborated with China Gas Turbine Establishment [52]. A Pt/ITO:N TFTC was prepared on a turbine blade and, at a test temperature above 900 °C, was found capable of surviving for more than

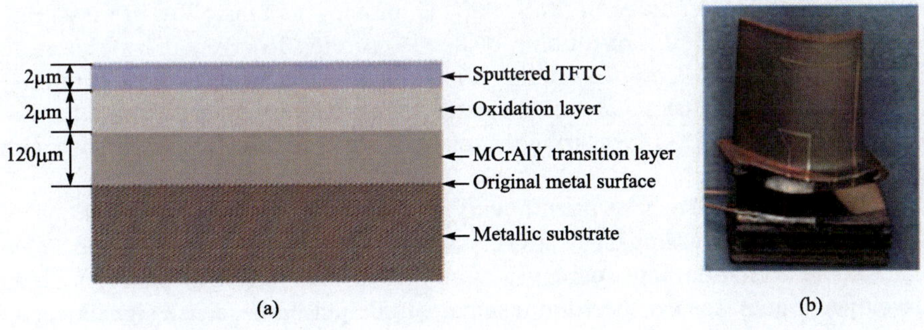

Fig. 14.17 TFTCs for turbine blades

20 h, with a measurement error less than ±1.5%. This TFTC has been successfully employed to measure the surface temperature of turbine engine rotor blades. Overall, TFTCs have been basically developed and used to measure the surface temperature of turbine blades in practice. However, in regard to the investigation of the thermal insulation effect, due to the harsher service environments for turbine blades, future TFTCs face a number of challenges, including accelerated spallation induced by gas thermal shock, erosion, and corrosion; problems related to high-speed rotating blades (i.e., temperature measurement errors and difficulties in attaching lead wires); and the complex match between TBCs and multilayer film-thermocouple systems.

(2) **Phosphor thermometry**

Phosphor thermometry was first developed in the 1980s. Currently, there are three types of phosphor thermometry, namely spectrum, intensity, and lifetime techniques, of which the lifetime technique performs the best in temperature measurement and is most widely used. This technique measures temperature based on photoluminescence. When illuminated by an excitation light, the electrons in a sensitive material jump to high-energy levels. When returning from high-energy levels to the ground state, the electrons generate fluorescence radiation. When an equilibrium state is reached (i.e., the fluorescence emission stabilizes), the fluorescence radiation attenuation time after the disappearance of the excitation light is related to the fluorescence lifetime (i.e., the lifetime in the excited state), which is related to temperature through the following equation:

$$\tau(T) = \frac{1 + e^{-\Delta E/kT}}{R_s + R_T e^{-\Delta E/kT}} \tag{14.36}$$

where Rs, R_T, k, and ΔE are constants and T is temperature. Therefore, the actual surface temperature can be calculated based on the attenuation time of the fluorescence signal. The phosphor lifetime technique is advantageous over conventional measurement techniques in that it is contact-free and has a high spatial resolution. Using this technique, researchers at the Oak Ridge National Laboratory [53] in the United States measured the surface temperature of static and rotating turbine blades in flames with temperatures of 700–1000 °C, proving its feasibility in measuring the temperature of TBCs on turbine blades. According to a report by Yao et al. [54] the Rolls-Royce Company in the United Kingdom developed a 226-nm (Uv) neodymium-doped yttrium–aluminum garnet (Nd:YAG) pulsed laser-based thermometric system capable of measuring the temperature of rotating turbine blades with a higher accuracy.

In recent years, phosphor thermometry has gradually garnered the attention of Chinese researchers. Using Cr^{3+}:YAG crystals as the fluorescent material and a blue LED as the excitation light source, Li et al. [55] of the North University of China developed a fluorescence spectrum technique to output fluorescence signals via an optical fiber, and subsequently used it to measure temperatures of 10–450 °C, with an error of less than ±5 °C. The China Academy of Aerospace Aerodynamics and the

Changchun Institute of Applied Chemistry of the Chinese Academy of Sciences [56] jointly developed a phosphor thermometry system capable of measuring temperatures up to approximately 500 °C. This system has been successfully used to measure the surface temperature and heat flux of a hypersonic-velocity model. Peng and Liu et al. [57–60] of Shanghai Jiaotong University studied a high-temperature measurement technique for gas turbine blades, with which they achieved certain progress in the measurement of temperatures above 1000 °C and, in theory, realized the measurement of the surface and bottom temperatures of TBCs. This is one of the temperature measurement techniques for investigating the thermal insulation effect of TBCs in the future.

(3) **IR radiation thermometry**

IR radiation thermometry is based on Planck's blackbody radiation equation. Specifically, this technique determines the temperature of an object based on the magnitude of the IR radiant energy emitted by it:

$$M(\lambda, T) = \varepsilon(\lambda) \cdot \frac{C_1}{\lambda^5}[e^{\frac{C_2}{\lambda T}} - 1]^{-1} \tag{14.37}$$

where $\varepsilon(\lambda)$ is the emissivity of the test object at temperature T and wavelength λ, C_1 and C_2 are the first and second radiation constants, respectively, λ is the radiation wavelength of the object, and T is the absolute temperature of the object. Due to its advantages such as being noncontact, having a wide measuring range, and yielding a fast response, IR thermometry has been extensively applied to the high-temperature components of aeroengines.

Through a thermometric study conducted with an FLIR T440 IR thermal imager, Skouroliakou et al. [61] noted that emissivity, ambient temperature, and atmospheric humidity are the key factors affecting the relative temperature measurement accuracy. When the IR radiation technique is used to measure the temperature of a turbine blade in its service environment, water vapor and dust affect the measurement accuracy, and an excessively high ambient temperature causes even greater errors. The United Technologies Corporation [62, 63] in the United States successively developed thermometric systems based on two and three bands (i.e., single-, double-, and multicolor pyrometers). Compared to the single-color pyrometer, the double-color pyrometer determines the temperature based on the ratio of IR radiant energy in two bands in adjacent channels and is able to eliminate the adverse effects of the absorption and reflection by dust and water vapor in the ambient environment on measurement accuracy. Li et al. [64] calculated and compared the errors in the temperature of a turbine blade measured by single-, double-, and multicolor pyrometers at wavelengths of 1.2–2.5 μm and simulated the temperature distribution in moving and guide vanes using the computational fluid dynamics (CFD) software. They found similar trends in the errors produced by the three pyrometers from the leading edge to the trailing edge of the blade pressure side—the errors first decreased, then increased, and finally decreased again. The error produced by the double-color pyrometer was smaller than those produced by the other two pyrometers and gradually decreased as the wavelength increased.

For TBCs, the wavelength is a principal factor affecting their emissivity and IR-based thermometric measurements. Manara et al. [65] examined long-wavelength IR thermometry for gas turbines. They found this technique suitable for measuring the temperature of opaque surfaces at near-IR (NIR) and short wavelengths. However, the ceramic material of a TBC is semitransparent in the NIR and short-wavelength regions. As a result, part of the radiation received by the thermometer originates from the alloy substrate, which causes large errors. In the long-wavelength (>10 μm) region, ceramic materials are nontransparent and exhibit high emissivity, making temperature measurements at long wavelengths possible. By combining the principle of radiation thermometry and the advantages of fiber-optic sensors, Hiernaut et al. [66] of a joint research team between the European Commission and the United States developed a submillisecond pyrometer with a measuring range of 727–1327 °C and an accuracy of 1% based on multiwavelength radiation thermometry. As early as the last century, to improve the temperature measurement accuracy, Bird [67] in the United Kingdom researched methods for correcting emissivity and processing signals and developed and manufactured a ROTAMAPII pyrometer capable of measuring temperatures between 550 and 1400 °C with a resolution of ±1 °C.

On this basis, emissivity calibration methods as well as error elimination and compensation algorithms are the focus of Chinese researchers aiming to improve the temperature measurement accuracy for TBCs on turbine blades. Based on an accurate reflection model for discrete irregular surfaces, Gao et al. [68] of the Harbin Institute of Technology calculated the reflected radiation errors in rotor-blade temperature measurements by pyrometers. Wang et al. [69] of the Shanghai Institute of Technical Physics proposed signal collection and background radiation elimination methods for short-wavelength IR radiation thermometry. Researchers at the Guizhou Aeroengine Design Institute and the China Gas Turbine Establishment measured the temperature of gas turbines and aeroengine blades, respectively. We simulated the working conditions of a turbine blade with a TBC under thermal shock and measured the surface temperature field of the coating using an FLIR GF309 system, as shown in Fig. 14.18 [70]. Overall, due to the complexity of the service environments in which TBCs on turbine blades operate, IR radiation thermometry is currently the most important technique for measuring the surface temperature of turbine blades. However, there remains a need to eliminate errors and to improve the measurement accuracy based on specific working conditions when this technique is used in the study of TBCs on turbine blades, particularly those on working blades.

Outside China, IR thermometry is often used in the investigation of the thermal insulation effect of TBCs. IR thermometry can be used in combination with a TFTC to measure the temperature at the interface between a turbine blade and the substrate of its TBC to determine the thermal insulation effect of the TBC on the turbine blade. Developing TFTCs with a high measurement accuracy and high stability as well as improving emissivity correction and eliminating measurement errors in complex environments are currently the key areas of research on the thermal insulation effect of TBCs on turbine blades.

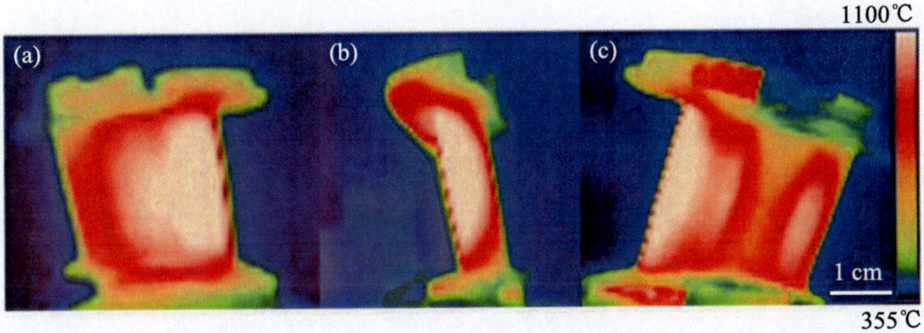

Fig. 14.18 Surface IR temperature contours of the TBC on a turbine blade during the temperature-holding stage: **a** pressure side, **b** leading edge, and **c** suction side

14.3.3 An Experimental Investigation of the Thermal Insulation Effect of a TBC on a Turbine Blade

In an experimental environment simulated based on the similarity principle, Davidson et al. [21] studied the interaction of the overall cooling performance of a TBC and the film cooling using thermometry techniques and further analyzed the influence of the thermal insulation effect of the TBC on the overall cooling efficiency. The results provide a research basis for turbine blade design with optimal cooling efficiency. The following is a description of the process of this study.

(1) Model test parameter design

The overall structure of the model test setup is shown in Fig. 14.12. The main component of the setup is a closed-loop wind-tunnel system driven by a 50-horsepower variable-speed fan. This setup generates a combustion gas with a certain temperature and cooling air with a resistance heater and a high-pressure compressor, respectively. The cooling air is guided into the interior of the blade. To create similar flow conditions for the blade, the test section of the wind tunnel is modified into flow channels capable of accommodating three vanes, as shown in Fig. 14.19. To facilitate the placement of thermocouples in the test section, the simulated C3X vanes in the center are enlarged 12-fold. In addition, the wall flow channel is adjusted to allow the pressure coefficient P in the test section to be similar to that during service.

To allow the Reynolds number Re of the C3X vanes to be similar to that during service, a turbulence generator—a mesh composed of 12 vertical rods 38 mm in diameter spaced at intervals of 85 mm—is placed 0.50 m upstream of the test section. The turbulence generator is capable of generating a fluid with a characteristic length of 3 cm and a turbulence intensity of 20% at the leading edge of the turbine blade. After the turbulence mesh is removed, the turbulence intensity decreases to 5%.

In addition, the high-pressure, low-temperature air in the mainstream wind tunnel is extracted, and its temperature is controlled through heating and liquid N_2 cooling. The air is supplied at a certain flow rate to the turbine blade, forming a closed loop of

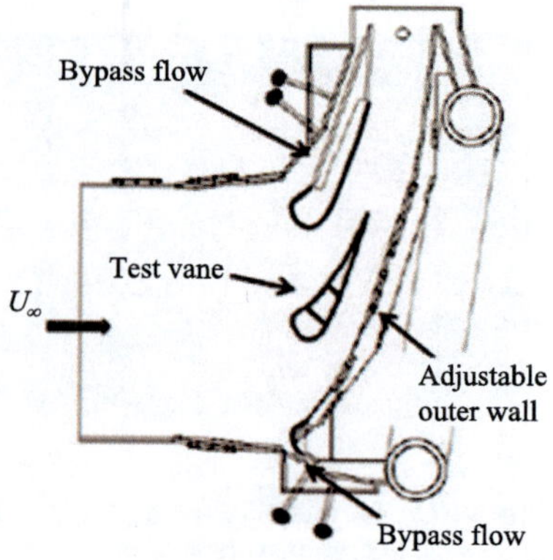

Fig. 14.19 Model test section for the TBC on a turbine blade [21]

cooling air, as shown in Fig. 14.12. To ensure a combustion-gas/cooling-air density ratio of 1.2, the cooling air and N_2 are mixed at a certain ratio and then exported.

(2) Model test specimen and material parameters

Figure 14.20 shows the blade used in the model test. There was a U-shaped internal cooling channel at the front end of the model blade and a radial straight channel at its rear end. In addition, there was a row of 24 cooling holes on the suction side of the test blade at $x/C = 0.23$ along the chord. These holes were angled at 42° to the direction of flow, 4.1 mm in diameter, and spaced at an interval (p/d) of 3.

As mentioned earlier, thermal similarities need to be satisfied in an experimental study of the TBC on a turbine blade. In other words, the Biot number Bi in the experimental environment should be similar to that in the real-world service environment of turbine blades. The turbine vanes used in this study were 12 times as large as the real-world vanes and differed in chord length and TBC thickness from the real-world vanes. Thus, according to Eqs. (14.10) and (14.11), to ensure a similar Bi, the central test vanes were made of epoxy resin with a thermal conductivity k of 1.02 W/m·K. A softwood with a k measured at 0.065 W/(m·K) was used to mimic the TBC material. Table 14.3 compares the geometric and material parameters of the turbine blade and its TBC used in the model test and a turbine blade and its TBC in the real-world

Fig. 14.20 Turbine blade with a TBC used in the model test [21]

Table 14.3 Comparison of the TBC and turbine blade parameters in the model test and the real-world service environment [21]

Parameter	Real turbine	Model turbine	Units
Vane thickness	0.13–0.3	1.27	cm
TBC thickness	0.025	0.23	cm
Vane conductivity	20	1.02	$W/(m \cdot K)$
TBC conductivity	1.5	0.065	$W/(m \cdot K)$
Heat transfer coeff	1500–5000	25–90	$W/(m^2 \cdot K)$
Vane/TBC thickness	5.2	5.52	
Vane/TBC conductivity	13.3	15.7	
Vane Bi	0.3–0.6	0.3–1.1	

service environment. While they differ in size, the simulated and real-world blades have basically similar *Bi*, suggesting that the test results obtained under this set of simulated working conditions are similar to those obtained under real-world service conditions.

(3) Thermometry techniques used in the test

IR thermometry and thermocouples were combined to measure temperatures in this test. Specifically, a ThermaCAM® P20 FLIR system and a P25 IR camera were used for IR temperature measurement. During the test, the surface temperature of the blade in the test section was measured through the NaCl or ZnSe window with a resolution of 1.5 pixels/mm and a field-of-view (FOV) size of 260 pixels × 240 pixels. To improve the measurement accuracy, the Type E thermocouples installed on the model blade were used to calibrate the emissivity for IR thermometry. The size of each measuring point was set to only 1.5 mm × 1.5 mm × 0.1 mm to minimize the disturbance from the thermocouples to the flow field and the interference of the FOV for IR thermometry as well as their influence on the efficiency of the film holes. Figure 14.21 shows the placement of the thermocouples.

(4) Thermal insulation effect analysis

Figure 14.22 shows the variation in the cooling efficiency of the blade in the presence and absence of a TBC with the Reynolds number of the internal cooling air. Here, internal cooling was achieved only through the removal of the heat of the blade via the trailing-edge slots by the cooling air that passed through the interior of the blade. No film holes were involved in the internal cooling process. The *x*-axis shows the location on the blade relative to the chord length (density ratio: 1.2; turbulence intensity: 5%). Both increasing the Reynolds number and adding a TBC can significantly

Fig. 14.21 Placement of thermocouples on the TBC on the turbine blade [21]

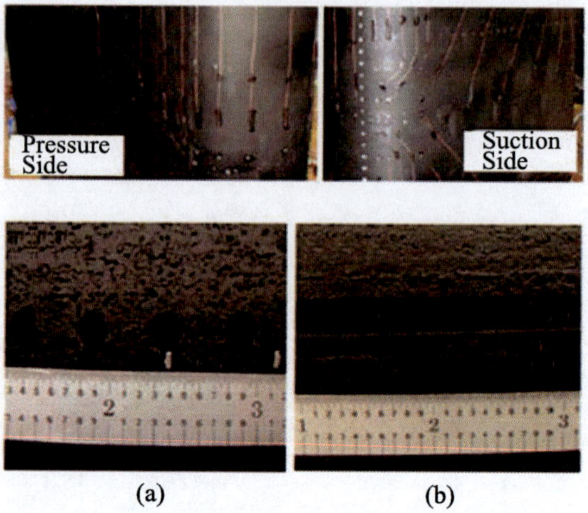

improve the cooling efficiency of the turbine blade. Therefore, increasing the turbulence intensity of the internal cooling air or the thickness of the TBC is an important way of improving the cooling efficiency of a blade.

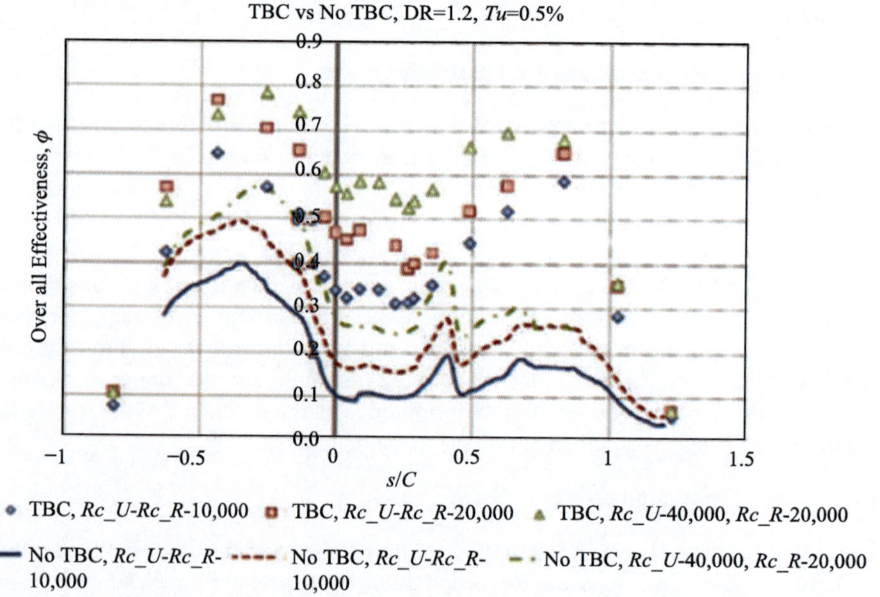

Fig. 14.22 Variation in the surface cooling efficiency (i.e., the overall cooling efficiency ϕ mentioned earlier) of the blade in the presence and absence of a TBC, respectively, with the Reynolds number of the internal cooling air [21] (s represents the location along the chord, C is the chord length, and s/C represents the location relative to the chord length)

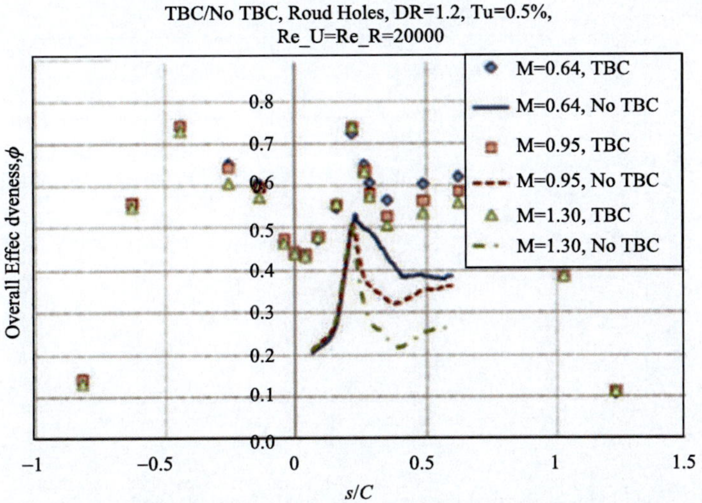

Fig. 14.23 Surface cooling efficiency (i.e., the overall cooling efficiency ϕ mentioned earlier) of the blade with a TBC in the presence and absence of circular film holes, respectively [21] (s represents the location along the chord, C is the chord length, and s/C represents the location relative to the chord length)

Figure 14.23 shows the cooling efficiency at different blowing ratios. We can clearly see that in the absence of a TBC, there is a significant increase in the cooling efficiency as the blowing ratio increased. The addition of a TBC leads to an increase in the overall cooling efficiency of the film on the turbine blade. A comparison shows that in the presence of a TBC, the cooling efficiency of the turbine blade increases at a decreasing rate with an increasing blowing ratio and that the overall cooling efficiency remained the same as that corresponding to a blowing ratio M of 0.64.

Figure 14.24 shows the variation in the thermal insulation efficiency of the TBC with the Reynolds number Re of the internal cooling air. Here, cooling was achieved both by interior cooling, i.e., the cooling air passed through the interior of the blade and exited via the trailing-edge slots to carry away the heat of the blade, and by film hole-enabled cooling, i.e., a cooling-air film was formed on the surface of the blade through the film holes. In Fig. 14.24, the y-axis shows the relative efficiency of the TBC, which has the same meaning as but a different expression than the thermal insulation effect of a TBC mentioned earlier in Eq. (14.8). We can see that the relative efficiency of the TBC increases as the Re of the internal cooling air increases, which is consistent with the earlier finding from the analysis of the thermal insulation effect of the TBC. At $s/C = 0.30$, increasing the Re of the internal cooling air increases the relative efficiency of the TBC by approximately 0.20–0.25.

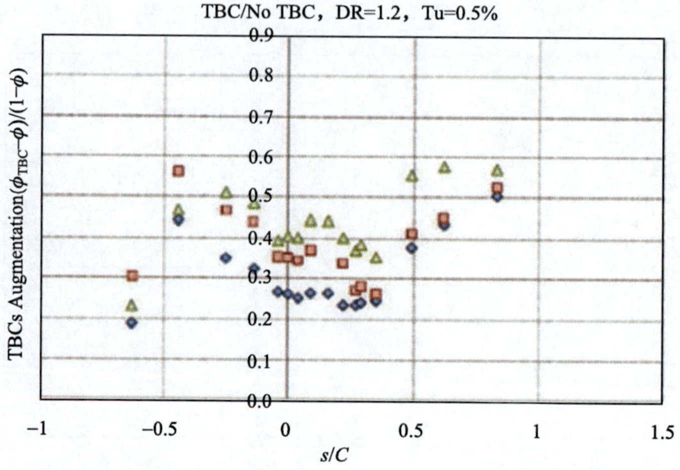

Fig. 14.24 Variation in the thermal insulation efficiency of the TBC with the Reynolds number of the internal cooling air [21]

14.4 Factors Influencing the Thermal Insulation Effect

Extracting the key parameters affecting the thermal insulation effect of TBCs and determining their influences on the thermal insulation effect of TBCs is an inevitable path for designing optimal TBCs for high-performance aeroengine turbine blades. The limitations in high-temperature measurement techniques, particularly the dearth of model test techniques for simulating complex service environments (e.g., high-temperature combustion gases, cooling air, CMAS corrosion, and rotation) as well as the multitude and different patterns of influencing factors (e.g., high-temperature combustion gases, cooling air, film-hole structures, and material parameters), have led to slow progress in the research on the thermal insulation effect of turbine blades with TBCs and its influencing factors. This section gives an overview of the current status of research on the thermal insulation effect of TBCs on turbine blades in two areas, namely film cooling and thermal insulation provided by TBCs. Each area is discussed from three perspectives, namely material parameters, service environments, and cooling structure.

14.4.1 Influential Factors Related to the Material

The primary material parameters affecting the thermal insulation effect of a TBC include the material composition, microstructure, porosity, and thickness of the TBC, with the first three influencing the thermal insulation effect of the TBC by affecting its thermal conductivity. It is easy to understand that reducing the thermal conductivity and increasing the thickness improve the thermal insulation effect of a TBC. Therefore, reducing the thermal conductivity of a TBC, which is a key parameter, through the alteration of the preparation process, microstructure, and material composition to improve its thermal insulation effect is the focus of current research on the thermal insulation effect of TBCs.

Ren et al. [13] prepared two coatings (a yttria-stabilized zirconia (YSZ) coating and a YSZ/Al_2O_3 double-layer coating) and tested their thermal insulation performance using the constant-temperature furnace shown in Fig. 14.25. During the test, a TBC-free substrate specimen and a TBC-coated substrate specimen were each fixed at one end of an Al_2O_3 tube. Three Type K thermocouples were used as sensors and placed inside the furnace to measure the furnace temperature T_0 and to monitor the backside temperature of the coating-free substrate, T_1; the temperature of the single YSZ coating, T_2; and the temperature of the double-layer coating, T_{20}. An Ar flow was supplied at a flow rate of 50 ml/min to cool the substrates. The average heating rate of the resistance furnace was set to 20 °C/min. The average heating rate inside the furnace was 20 °C/min. The recording of T_0, T_1, T_2, and T_{20} started at $T_0 = 450$ °C and stopped at $T_0 = 1150$ °C. The thermal insulation performance of the TBC was determined through the calculation of the difference between T_2 and T_1 (i.e., $\Delta T = T_2 - T_1$). At 1150 °C, the difference between the temperatures at the two ends of the YSZ/Al_2O_3 coating is 6.9% larger than that between the temperatures at the two ends of the YSZ coating, as shown in Fig. 14.26.

Wang et al. [14] simulated the heat transfer behavior of TBCs with different pore spaces and geometries using the finite element method (FEM). The simulation results show the following. The thermal insulation effect of a TBC increases with increasing size, volume fraction of pores, and the number of pore layers in the direction perpendicular to the thickness of the coating. In addition, surface pores do not improve the thermal insulation effect but, instead, facilitate the heat transfer from the external environment to the substrate. In fact, the presence of surface pores reduces the effective thickness of a ceramic coating, thereby reducing its thermal insulation effect. In contrast, the internal pores of a TBC improve its thermal insulation effect, as shown in Fig. 14.27. These studies show that certain quantities of voids and cracks improve the thermal insulation performance of TBCs.

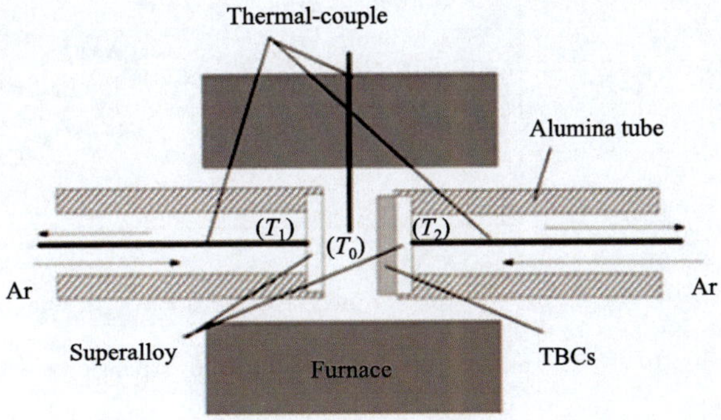

Fig. 14.25 Thermal insulation effects of the single- and double-layer TBCs

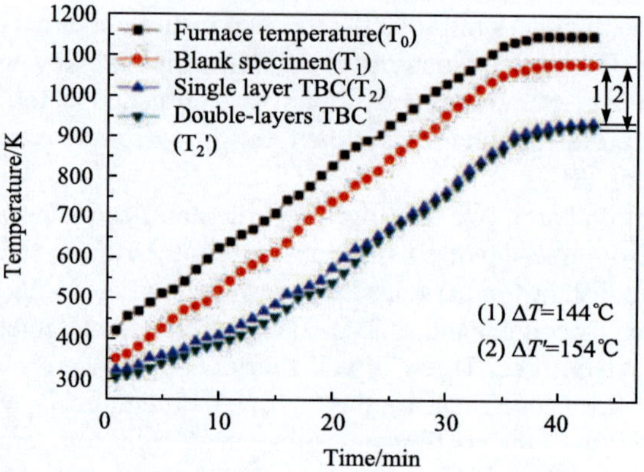

Fig. 14.26 Thermal insulation effects of the single- and double-layer TBCs

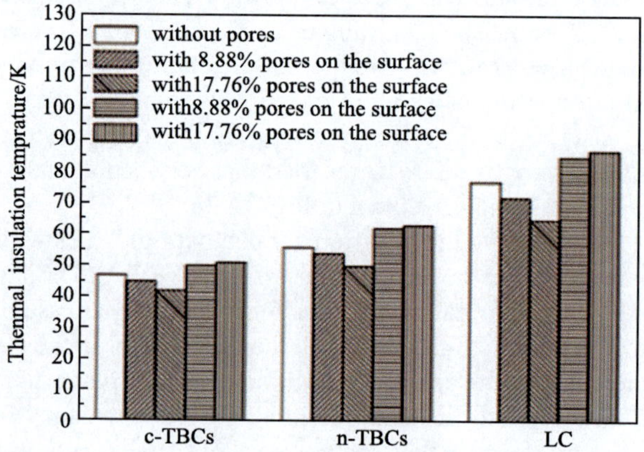

Fig. 14.27 Thermal insulation effects of single- and double-layer TBCs

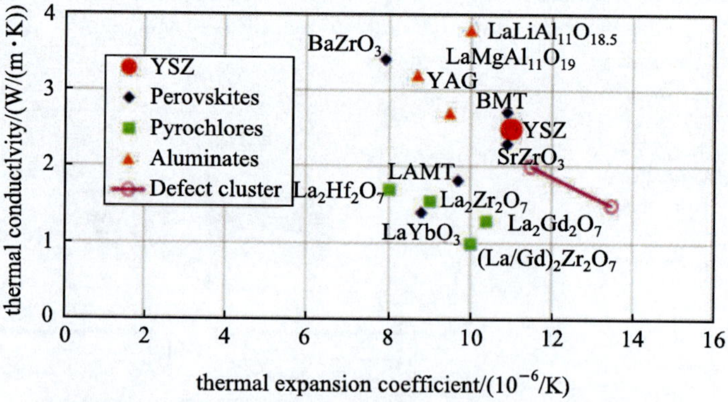

Fig. 14.28 Thermal conductivities and thermal expansion coefficients of new advanced TBCs

For the microstructure of TBCs, Zheng [71] of Tsinghua University established the following relation between the thermal conductivity k_c and porosity φ:

$$k_c = k_m(1 - \phi)^{3/2} \quad (14.38)$$

where k_m is the thermal conductivity of the dense solid phase. Thus, coatings prepared by atmospheric plasma spraying (APS) contain a large number of internal pores, which reduces their density and ensures a low thermal conductivity. As a result, these coatings are often applied to guide vanes serving at higher temperatures. In comparison, due to their more compact structure, physical vapor deposition (PVD) TBCs have a higher thermal conductivity. In addition, heat transfer in a material is based on phonons, the thermal conductivity k_{ph} of which can be expressed as follows:

$$k_{ph} = \frac{1}{3} C_V v_m l \quad (14.39)$$

where C_V is the heat capacity at a constant volume, v_m is the mean velocity at which phonons propagate, and l is the mean free path length of phonons in the material. Hence, reducing the free path length of phonons and increasing phonon scattering have been the main directions of research on the reduction of the thermal conductivity of materials in recent years. In selecting the TBC material with a low thermal conductivity, Clarke [15] found through derivation with the consideration of various factors affecting TBCs that for a material, if $\rho^{1/6}E^{1/2}/(M/m)^{2/3}$ (ρ is the density, E is the elastic modulus, M is the molecular mass, and m is the number of atoms contained in a molecule) takes the minimum value (i.e., if the material has a high molecular mass, a complex crystal structure, nondirectional bonds, and each of its molecules contains a large number of different atoms), it tends to have a low thermal conductivity. In recent years, various methods for enhancing phonon scattering by increasing lattice distortion through the doping of rare-earth zirconates have been widely studied. Xiang et al. [16] found that doping a $La_2Zr_2O_7$ coating with Yb_2O_3 and CeO_2 resulted in defects in its fluorite structure, which, in turn, led to a lower thermal conductivity and a higher coefficient of thermal expansion (CTE) than those of $La_2Zr_2O_7$. Vasen et al. [17] summarized the main relevant materials examined in available studies, as shown in Fig. 14.28. If only the HTC is considered, materials with a pyrochlore structure have the potential to become the most suitable substitutes for YSZ.

14.4.2 Influencing Factors Related to the Service Environment

The service environment is a major factor affecting the thermal insulation effect of a TBC. The main parameters of the service environment include the temperature, velocity, and turbulence intensity of the combustion gas and cooling air. Not only do these parameters affect the temperature field in the TBC on a turbine blade, but their interactions alter the surface temperature field of the turbine blade, thus leading to a long and expensive experimental investigation of the influencing factors. The current studies focus on numerical simulations. Wang and Yang [74] investigated the temperature field of a turbine blade through fluid–solid coupling numerical simulations and found that (1) the thermal insulation effect of the TBC is sensitive to the temperature at the inlet of the turbine and, in the low-temperature stage, increases as the temperature increases; however, at an excessively high temperature, the thermal insulation efficiency of the TBC reaches its limit and remains at approximately 24%, and that (2) the maximum HTC of the surface of the blade is positively correlated with the inlet velocity of the external flow field, with a proportionality coefficient of approximately 8.57, and the thermal insulation effect of the TBC decreases with an increasing velocity of the external flow field. Through numerical simulations, Prapamonthon et al. [72] examined the influence of the cooling efficiency of guide vanes with a TBC on the turbulence intensity of the mainstream. They found that the TBC exerts a stronger thermal insulation effect on the pressure side than the suction side, that the influence of the TBC on the overall cooling efficiency increases as the turbulence intensity increases, and that the effectiveness of the TBC increases with an increasing efficiency of internal cooling achieved by increasing the Reynolds number, as shown in Fig. 14.29. Experimentally, Davidson et al. [21] conducted a low-temperature (400 °C) test with a Biot number equivalent to that in the real-world service environment. In this test, they measured the temperature of a blade using IR thermometry and TFTCs to examine the variation in the thermal insulation performance of the TBC with the turbulence intensity of the internal cooling air. They found that the internal cooling air significantly improved the overall cooling efficiency and that the efficiency of the coating (as shown on the y-axis in Fig. 14.24) increased as the internal cooling efficiency increased. To date, reports on the impact of the service environment on the thermal insulation effect of TBCs have been very limited, and systematic studies are lacking. Comparatively, there are still inadequate experimental data obtained by Chinese researchers from quantitative analyses of the thermal insulation effect. While some tests have been conducted on simple specimens, the obtained data are insufficiently reliable. In particular, data obtained from tests conducted only in high-temperature furnaces are largely not credible.

14.4.3 Influencing Factors Related to the Cooling Structure

The cooling structure alters the thermal insulation effect of the TBC on a turbine blade by changing the film cooling and the flow-field structure to influence the heat flow in the blade and thus is the key to the TBC cooling design for a turbine blade. In high-pressure, high-velocity environments, slight differences in the structural parameters can lead to significant changes in the overall cooling performance of turbine blades. Generally, the structural parameters include those of the internal cooling structure and

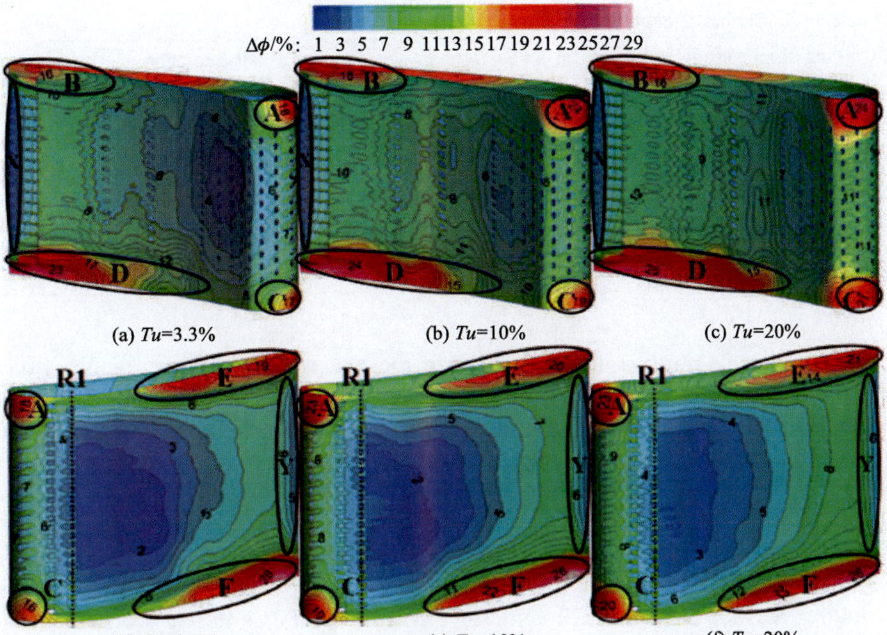

Fig. 14.29 Variation in the thermal insulation effect of a TBC with the turbulence intensity: **a** the pressure side at a turbulence intensity of 3.3%, **b** pressure side at a turbulence intensity of 10%, **c** pressure side at a turbulence intensity of 20%, **d** suction side at a turbulence intensity of 3.3%, **e** suction side at a turbulence intensity of 10%, and **f** suction side at a turbulence intensity of 20%

the film-hole structure as well as the number of film holes. For example, Mensch et al. [73] established a coupled heat transfer model for blade walls. Using the fluid–solid coupling method, they found that a TBC can significantly reduce the temperature of the substrate of a blade and that as the blowing ratio increases, a TBC reduces the heat transfer in the wall, resulting in a higher cooling efficiency. Webb et al. [8] obtained the correlation between heat transfer and wall friction using the wall similarity law in conjunction with an experiment, based on which they investigated the influence of pin fins and ribs on the heat transfer in walls. They found that pin fins and ribs are capable of improving the internal cooling efficiency and lowering the temperature of turbine blades and that the internal cooling efficiency can be optimized based on the rib height and spacing. Bogard and Thole [10] summarized the main factors affecting the cooling performance, as shown in Table 14.4. The blowing ratio, turbulence intensity, and film-hole geometry and angle are important parameters that affect cooling performance. The cooling efficiency decreases as the cooling jet flow becomes separated, and the surface curvature, High free turbulence, and the outlet shape of the holes significantly affect the blowing ratio at the separation of the cooling jet flow and thereby considerably affect the film-cooling performance. While the influences of the parameters of the cooling structure on efficiency have

Table 14.4 Factors affecting film cooling [10]

Cooling/combustion-gas conditions	Film-hole geometry	Blade geometry
Blowing ratio	Film-hole shape	Surface curvature
Turbulence intensity	Film-hole angle	Surface roughness
Cooling-air density ratio Mach number of the combustion gas Rotation	Number and spacing of film holes Film-hole length Film-hole row spacing	Film-hole locations (e.g., leading edge, trailing edge, and blade tip)

been extensively studied, the influences of the structural parameters of film holes on the thermal insulation effect of TBCs have rarely been investigated. Davidson et al. [21] conducted some studies in this area. They found that factors such as film-hole geometry affect the overall cooling efficiency and that TBCs can significantly improve the overall cooling efficiency; however, designing film holes with more complex geometries is of little significance to improving the cooling efficiency, and more consideration should be given to the service life of TBCs while designing film holes. In summary, there are myriad factors that affect TBCs on turbine blades in complex ways. Further experimental investigation of TBCs in real-world environments needs to be conducted, as only reliable data can be used in the design and optimization of TBCs on advanced aeroengine turbine blades.

14.5 Summary and Outlook

This chapter summarizes the thermal insulation effect of TBCs regarding three aspects (i.e., theoretical analysis, experimental testing techniques, and influencing factors) and describes its numerical simulation, experimental research methods, and key influencing parameters, providing a basis for optimizing the preparation process and structural design of TBCs on advanced aeroengine turbine blades. However, further in-depth investigations still need to be conducted in the following areas:

(1) The evaluation models for the interaction between TBCs and film cooling need to be optimized, the correlations between each parameter and the thermal insulation effect need to be established, and numerical simulation methods based on the internal microstructure of coatings and numerical methods for structural optimization need to be developed to provide a theoretical basis for the prediction and optimal design of the thermal insulation effect of TBCs.

(2) Techniques for measuring the surface and interfacial temperatures of TBCs in the service environments of turbine blades need to be developed, and the technical difficulties in the application of IR thermometry and TFTCs to the measurement of the surface and interfacial temperatures of blades rotating at high speeds in high-temperature flames need to be overcome.

(3) Setups for simulating high-speed rotation, thermo–mechano-chemical coupling environments need to be developed, the influences of parameters (e.g., combustion-gas and cooling-air parameters) on the thermal insulation effect of TBCs under high-speed rotation conditions need to be studied, and experimental and numerical simulation results need to be validated against each other to facilitate the optimization and enhancement of the thermal insulation effect of TBCs on moving blades.

References

1. Wang Y, Liu H, Ling X, et al. Effects of pore microstructure on the effective thermal conductivity of thermal barrier coatings. Appl Therm Eng. 2016;102:234–42.
2. Fink PJ, Miller JL, Konitzer DG. Rhenium reduction alloy design using an economically strategic element. JOM. 2010;62(1):55–7.
3. Wahl JB, Harris K. New single crystal superalloys, CMSX®-7 and CMSX-8. Superalloys 2012. Warrendale, PA: TMS, 2012: 179–88.
4. Sato J, Omori T, Oikawa K, et al. Cobalt-base high-temperature alloys. Science. 2006;312:90–1.
5. Lu YZ, Xie G, Wang D, et al. Anisotropy of high temperature creep properties of a Co-base single crystal superalloy. Mater Sci Eng, A. 2018;A720:69–74.
6. Walston WS, O'Hara KS, Ross EW, et al. René N6: third generation single crystal superalloy. Superalloys 1996. Warrenda-Le, PA: TMS;1996. p. 27–34.
7. Pollock TM, Dibbern J, Tsunekane M, et al. New Co-based γ-γ' high-temperature alloys. JOM. 2010;62(1):58–63.
8. Webb RL, Eckert ERG, Goldstein R. Heat transfer and friction in tubes with repeated-rib roughness. Int J Heat Mass Transf. 1971;14(4):601–17.
9. Jubran BA, Hamdan MA, Abdualh RM. Enhanced heat transfer, missing pin, and optimization for cylindrical pin fin arrays. J Heat Transf. 1993;115(3):576–83.
10. Bogard DG, Thole KA. Gas turbine film cooling. J Propul Power. 2006;22(2):249–70.
11. Dai P, Lin F. Recent advances in the study of air-film-cooled gas turbine blades. J Eng Therm Energy Power. 2009;24(01):1–6.
12. Liu ZY, Zhu W, Yang L, et al. Numerical prediction of thermal insulation performance and stresses of thermal barrier coatings coated on a turbine vane. Int J Therm Sci. 2020;158: 106552.
13. Ren C, He Y, Wang D, et al. Fabrication and characteristics of YSZ–YSZ/Al_2O_3 double-layer TBC. Oxid Met. 2011;75(5):325–35.
14. Wang LD, Wang Y, Sun X, et al. Influence of pores on the thermal insulation behavior of thermal barrier coatings prepared by atmospheric plasma spray. Mater Des. 2011;32(1):36–47.
15. Clarke DR. Materials selection guidelines for low thermal conductivity thermal barrier coatings. Surf Coat Technol. 2003;163–164:67–74.
16. Xiang J, Chen S, Huang J, et al. Phase structure and thermophysical properties of co-doped $La_2Zr_2O_7$ ceramics for thermal barrier coatings. Ceram Int. 2012;38(5):3607–12.
17. Vasen R, Jarligo MO, Steinke T, et al. Overview on advanced thermal barrier coatings. Surf Coat Technol. 2010;205(4):938–42.
18. Maikell J, Bogard D, Piggush J, et al. Experimental simulation of a film cooled turbine blade leading edge including thermal barrier coating effects. J Turbomach. 2014;133: 121003.
19. Harrison GF. The Influence of new materials and manufacturing processes on the design of future aero engines. European Propulsion Forum. 1993; ISBN 1857681258.
20. Dees JE, Bogard DG, Ledezma GA, et al. Experimental measurements and computational predictions for an internally cooled simulated turbine vane. J Turbomach. 2012;134(6): 061003.

21. Davidson FT, Dees JE, Bogard DG. An experimental study of thermal barrier coatings and film cooling on an internally cooled simulated turbine vane. Proc ASME Turbo Expo, GT2011: 559–70.
22. Ziaeiasl A, Ramezanlou MT. Thermo-mechanical behavior of gas turbine blade equipped with cooling ducts and protective coating with different thicknesses. Int J Mech Sci. 2019;150:656–64.
23. Zhu W, Wang JW, Yang L, et al. Modeling and simulation of the temperature and stress fields in a 3D turbine blade coated with thermal barrier coatings. Surf Coat Technol. 2017;315:443–53.
24. Wilcox DC. Turbulence modeling for CFD. La Canada, CA: DCW Industries; 1998.
25. Heselhaus A, Vogel D. Numerical simulation of turbine blade cooling with respect to blade heat conduction and inlet temperature profiles. In: 31st joint propulsion conference and exhibit; 1995. p. 3041.
26. Sondak D, Dorney D. Simulation of coupled unsteady fluid dynamics and conduction heat transfer in a turbine stage. In: 35th joint propulsion conference and exhibit, 1999. p. 2521.
27. Eifel M, Caspary V, Honen H, et al. Experimental and numerical analysis of gas turbine blades with different internal cooling geometries. J Turbomach-Trans ASME. 2011;133(1): 011018.
28. Moritz N, Kusterer K, Bohn D, et al. Conjugate calculation of a film-cooled blade for improvement of the leading-edge cooling configuration. Propulsion Power Res. 2013;2(1):1–9.
29. Menter FR. Two-equation eddy-viscosity turbulence models for engineering application. AIAA J. 1994;32(8):1598–605.
30. Yoshiara T, Sasaki D, Nakahashi K. Conjugate heat transfer simulation of cooled turbine blades using unstructured-mesh CFD solver. In: 49th AIAA aerospace sciences meeting including the new horizons forum and aerospace exposition; 2011. p. 498.
31. Dong P. Research on conjugate heat transfer simulation of aero turbine engine air-cooled vane. Harbin Institute of Technology, Harbin, 2009.
32. Du K, Li J. Numerical study on the effects of slot injection configuration and endwall alignment mode on the film cooling performance of vane endwall. Int J Heat Mass Transf. 2016;98:768–77.
33. Zhang K, Li J, Li Z, et al. Effects of simulated swirl purge flow and mid-passage gap leakage on turbine blade platform cooling and suction surface phantom cooling performance. Int J Heat Mass Transf. 2019;129:618–34.
34. Alizadeh M, Izadi A, Fathi A. Sensitivity analysis on turbine blade temperature distribution using conjugate heat transfer simulation. J Turbomach. 2014;136: 011001.
35. Ke ZQ, Wang JH. Numerical investigations of pulsed film cooling on an entire turbine vane. Appl Therm Eng. 2015;87:117–26.
36. Gladden HJ, Livingood JN. Procedure for scaling of experimental turbine vane airfoil temperatures from low to high gas temperatures. Nat Aeronaut Space Administ. 1971; D-6510.
37. Hylton LD, Mihelc MS, Turner ER, et al. Analytical and experimental evaluation of the heat transfer distribution over the surfaces of turbine vanes. NASA Tech Memo. 1983;133(1): 011019.
38. Lynch SP, Sundaram N, Thole KA, et al. Heat transfer for a turbine blade with nonaxisymmetric endwall contouring. J Turbomach. 2011;133(1):01083.
39. Zhang L, Du L, Wang M. An experimental study on film cooling effectiveness over the leading edge of gas turbine blades. Power Eng. 2011;31(11):835–9.
40. Lu Y, Dhungel A, Ekkad SV, et al. Effect of trench width and depth on film cooling from cylindrical holes embedded in trenches. J Turbomach. 2009;131(1): 011003.
41. Li J, Zhu H, Chen D. Experimental study on film cooling effectiveness of the turbine blade with compound-angle fan-shaped holes under wake influence. J Xi'an Jiaotong Univ. 2019;53(09):167–75.
42. Robinson RC. NASA GRC's high pressure burner rig facility and materials test capabilities. NASA Technical Memorandum; 1999. p. 209411.
43. Zhu JD, Miller RA, Kuczmarski MA. Development and life prediction of erosion resistant turbine low conductivity thermal barrier coatings. NASA Technical Memorandum; 2010. p. 215669.

44. Zhou YC, Yang L, Zhong ZC, et al. Type of testing equipment for detecting the failure process of thermal barrier coating in a simulated working environment. U. S. Patent 14, 760, 444, 10 April 2018.
45. Yang L, Zhou Y, Liu Z. An experimental testing system for simulating the working conditions of thermal barrier coatings on turbine blades. CN109682702B, 20 March 2020.
46. Lepicovsky J, Bruckner R, Smith F. Application of thin film thermocouples to localized heat transfer measurements. In: Joint propulsion conference and exhibit; 1995, AIAA-95-2834.
47. Grant HP, Przybyszewski JS, Claing RG. Turbine blade temperature measurements using thin film temperature sensors. Turbine Blades; 1981, NASA, TMX-71844.
48. Lepicovsky J, Bruckner RJ, Smith FA. Application of thin-film thermocouples to localized heat transfer measurements. In: 31st joint propulsion conference and exhibit; 1995. p. 2834.
49. An B. Development and application of thin film temperature sensor. J Propuls Technol. 1992;1:63–7.
50. Zhang Y, Cheng P, Yu K, et al. ITO film prepared by ion beam sputtering and its application in high-temperature thermocouple. Vacuum. 2017;56:31–4.
51. Zhao XH, Wang YR, Chen YZ, et al. Enhanced thermoelectric property and stability of NiCr/NiSi thin film thermocouple on superalloy substrate. Rare Met. 2017;36(6):512–6.
52. Zhao W, Jiang H, Chen Y. Fabrication of Pt/ITO thin-film thermocouple on metal. Measur Control Technol. 2013;32(04):23–5.
53. Tobin KW, Allison SW, Cates MR, et al. High-temperature phosphor thermometry of rotating turbine blades. AIAA J. 2012;28(8):1485–90.
54. Yao Y, Dai J, Huang C. Development for temperature measurement technology in modern aeroengine. Aeronaut Manuf Technol. 2015;12:103–7.
55. Li Y, Wang Y, Wei Y. Research on a method for optical fiber fluorescence spectra temperature measurement. Sens World. 2014;20(11):23–6.
56. Bi Z, Han S, Wu C, et al. Phosphor thermography study in gun tunnel. J Exp Fluid Mech. 2013;27(03):87–92.
57. Cai T, Guo S, Li Y, et al. Ultra-sensitive mechanoluminescent ceramic sensor based on air-plasma-sprayed $SrAl_2O_4:Eu^{2+}$, Dy^{3+} coating. Sens Actuat, A. 2020;315: 112246.
58. Li Y, Peng D, Wang W, et sl. Comparison of PSP and TSP measurement techniques for fast rotating blades. AIAA Aviat. 2018; 3317.
59. Cai T, Peng D, Liu Y, et al. A correction method of thermal radiation errors for high-temperature measurement using thermographic phosphors. J Vis. 2016;19(3):383–92.
60. Peng D, Liu Y. Grid-pattern PSP/TSP system for simultaneous pressure and temperature measurements. Sens Actuat, B Chem. 2016;222:141–50.
61. Skouroliakou AS, Seferis I, Sianoudis I, et al. Infrared thermography imaging: evaluating surface emissivity and skin thermal response to IR heating. e-J Sci Technol. 2014; 3:9–14.
62. Gebhart JR, Kinchen BE, Strange RR. Optical pyrometer and technique for temperature measurement. U. S. Patent 4, 222, 663, 16 September 1980.
63. Gonzalez ES, Oqlukian RL. Triple spectral area pyrometer. U. S. Patent 5, 125, 739, 30 June 1992.
64. Li D, Feng C, Gao S, et al. Effect of pyrometer type and wavelength selection on temperature measurement errors for turbine blades. Infrared Phys Technol. 2018;94:255–62.
65. Manara J, Zipf M, Stark T, et al. Long wavelength infrared radiation thermometry for non-contact temperature measurements in gas turbines. Infrared Phys Technol. 2017;80:120–30.
66. Hernaut JP, Beukers R, Heinz W, et al. Submillisecond six-wavelength pyrometer for high-temperature measurements in the range 2000 to 5000 K. High Temperat High Press. 1986;18:617–25.
67. Bird C, Parrish CJ. Component temperature measuring method. U. S. Patent 7, 003, 425, 21 February 2006.
68. Gao S, Wang LX, Feng C, et al. Analysis and improvement of gas turbine blade temperature measurement error. Meas Sci Technol. 2015;26(10): 105203.
69. Wang Y, Zhu Q, Wang J. Characterization of background radiation in SWIR hyperspectral imager. J Infrared Millimeter Waves. 2011;30(03):279–83.

70. Zhu W, Zhang CX, Yang L, et al. Real-time detection of damage evolution and fracture of EB-PVD thermal barrier coatings under thermal shock: an acoustic emission combined with digital image correlation method. Surf Coat Technol. 2020;399: 126151.
71. Zheng Y. Microstructure observation and thermal conductivity estimation of thermal barrier coating materials. Tsinghua University (Beijing). 2020;1:1–7.
72. Prapamonthon P, Xu H, Yang W, et al. Numerical study of the effects of thermal barrier coating and turbulence intensity on cooling performances of a nozzle guide vane. Energies. 2017;10(3):1–16.
73. Mensch A, Thole KA, Craven BA. Conjugate heat transfer measurements and predictions of a blade endwall with a thermal barrier coating. J Turbomach. 2014;136: 121003.

Chapter 15
Reliability Assessment of TBCs

Regarding the still prominent bottleneck surrounding thermal barrier coating (TBC) spallation, both the material development and application sectors urgently need to understand the time, location, and form of TBC spallation in service environments and to mechanistically determine the key influencing factors and safe service conditions, with the goal of guiding process optimization and safe application. Internationally, studies on the service life assessment of coatings from various perspectives (e.g., strength, damage, and interface crack growth) began as early as the 1980s. For example, a number of service life models have been established, including those developed from a strength perspective by Busso et al. [1] Liu et al. [2] and Wen and Jordan [3] based on residual stresses in thermal growth oxides (TGOs), thermal cycling stresses, and the residual stress threshold, respectively, as well as those developed from the perspective of fracture theory by He et al. [4] and Renusch and Schütze [5] based on interface crack growth and acoustic emission (AE) signals from cracks, respectively. While these models have played a role in advancing the application of TBCs, they have yet to be employed in engineering projects because their predictions differ considerably from real-world engineering situations.

The layers, interfacial properties (e.g., modulus, strength, and fracture toughness), and structural properties (e.g., pores, cracks, and interfacial morphologies) of TBCs are dispersed, and their service conditions (e.g., the temperature and velocity of the combustion gas and corrosive medium) are indeterminate. As a result, the service life of a TBC is inevitably uncertain. Therefore, the service life predicted using single damage parameters unavoidably differs significantly from the actual service life. In view of this problem, we present a reliability assessment method based on failure probability to comprehensively account for the dispersive nature of the properties, structure, and environment of a TBC that can be used to analyze a TBC under various service loading conditions and determine its failure locations and failure probability as well as the relevant influencing factors.

Y. Zhou et al., *Thermal Barrier Coatings: Failure Theory and Evaluation Technology*,
https://doi.org/10.1007/978-981-19-2723-2_15

This chapter focuses on the introduction of theoretical and numerical methods for the reliability assessment of TBCs as well as providing a detailed description of the reliability of TBCs under key conditions such as thermal cycling, interfacial oxidation, and high-temperature erosion.

15.1 Basic Reliability Theory for TBCs

The properties, structure, and environment of a TBC each change in certain patterns within some range. Under these conditions, the failure of TBC (e.g., interfacial oxidation and spallation) is inevitably a probabilistic event. The reliability analysis of a TBC refers to the analysis of the probability that it remains intact under uncertain basic variables (properties, structure, and environment). From another perspective, reliability analysis can be described as the analysis of the occurrence probability of various failure behaviors of the TBC and hence is also referred to as failure probability analysis. The extent to which a basic variable affects the reliability or failure probability is called the sensitivity. This section gives a brief introduction to the basic concepts of reliability, reliability probability, and sensitivity and their mathematical descriptions.

15.1.1 Randomness and Distribution of Property, Structural, and Environmental Parameters

Reliability theory was established and developed based on the uncertainties of structures during design, manufacturing, and use. For TBCs, the uncertainties that affect their reliability are reflected primarily by the randomness of parameters related to their properties, geometry, and service environment in the form of the dispersion of their observed (or measured) values. Therefore, these parameters should be treated as random variables (generally, continuous random variables). The distribution of a continuous random variable can be described using a probability density function (PDF) or a cumulative distribution function (CDF) in conjunction with numerical characteristic parameters (i.e., expectation and variance). Common continuous random variables include those that follow a normal distribution, uniform distribution, or the Weibull distribution.

(1) Normal distribution. This is the most common type of distribution and can be used to approximate the probability distributions of many random variables in practical production and scientific experiments. A random variable is considered to follow or approximately follow a normal distribution if it is influenced by several random factors independent of each other but is not decisively influenced by any single one of these factors. The geometric parameters (e.g., coating

thickness) and some material parameters (e.g., Young's modulus and coefficient of thermal expansion (CTE)) affecting the reliability of a TBC can each be considered to follow a normal distribution.

If a continuous random variable X has the following PDF: [6]

$$f_X(X) = \frac{1}{\sqrt{2\pi}\sigma_X} \exp\left[-\frac{(X-\mu_X)^2}{\sigma_X^2}\right] \quad -\infty < X < +\infty \tag{15.1}$$

then X is said to follow a normal distribution and is denoted by $X \sim N(\mu_X, \sigma_X^2)$, where μ_X and σ_X are the mean and standard deviation, respectively.

The CDF of X is as follows:

$$F_X(X) = \frac{1}{\sqrt{2\pi}\sigma_X} \int_{-\infty}^{X} \exp\left[-\frac{(t-\mu_X)^2}{\sigma_X^2}\right] dt \tag{15.2}$$

When $\mu_X = 0$ and $\sigma_X = 1$, $X \sim N(0, 1)$ is said to follow a standard normal distribution. A random variable X that follows a nonstandard normal distribution can be transformed into a random variable Y that follows a standard normal distribution through the following equation:

$$Y = \frac{X - \mu_X}{\sigma_X} \tag{15.3}$$

Based on the above relation, any variable that follows a nonstandard normal distribution can be transformed into a variable that follows a standard normal distribution.

(2) The Weibull distribution. This was proposed in the 1930s by Swedish engineer Weibull during his study on the strength and fatigue life of structures (e.g., bearings). He explained that these strength and life problems rely on "chain" models, that is, the strength of an entire "chain" depends on the strength of its weakest "link." Assuming that the strengths or lives of different "links" in a chain are mutually independent, as long as the probability distribution of the minimum value of the strength or life of the links is determined, the probability distribution of the chain is as well. On this basis, the Weibull distribution function was developed. Due to its capability to sufficiently reflect the influence of a defect or stress concentration source in a material on its strength or life, the Weibull distribution has been widely applied in reliability engineering [7–9]. Many researchers across the world have used the Weibull parameters to describe the random properties of TBC materials and found that the fracture toughness and interfacial binding energy of TBCs each follow a Weibull distribution [11–15].

The PDF of a random variable X that follows a three-parameter Weibull distribution is as follows:

$$f_X(X) = \begin{cases} \frac{m}{\eta}\left(\frac{X-\gamma}{\eta}\right)^{m-1} \exp\left[-\left(\frac{X-\gamma}{\eta}\right)^m\right] & X \geq \gamma \\ 0 & X < \gamma \end{cases} \tag{15.4}$$

Its CDF is as follows:

$$F_X(X) = P(t \leq X) = \begin{cases} 1 - \exp\left[-\left(\frac{X-\gamma}{\eta}\right)^m\right] & X \geq \gamma \\ 0 & X < \gamma \end{cases} \tag{15.5}$$

where $\gamma \geq 0$ is the location parameter (i.e., the shortest life), $\eta > 0$ is the characteristic life or scale parameter, and $m > 0$ is the shape parameter. m is a key parameter that describes the dispersion of X, with a large m indicating a low dispersion. The curve is close to an exponential distribution when $m = 1$, a Rayleigh distribution when $m = 2$, and a normal distribution when $m = 3 \sim 4$. When $\gamma = 0$, a three-parameter Weibull distribution becomes a two-parameter Weibull distribution, which is primarily used in fatigue testing of materials at high stress levels. For TBCs, the critical stress for interfacial compressive failure and the critical shear strength for shear failure each primarily follow a two-parameter Weibull distribution.

In addition, the service loads (e.g., temperature) to which TBCs are subjected generally each follow a beta distribution, [16] while the size of erosive particles affecting erosion loading follows a Rosin–Rammler distribution [17].

15.1.2 *Definition of Reliability*

The reliability of a TBC refers to the probability that it does not undergo spallation failure under specified conditions (service conditions, structure, and properties) and within a specified time (generalized time, e.g., life and load). Conversely, the probability that a coating spalls is called its failure probability. The boundary between failure and reliability is the often-mentioned critical state. A reliability assessment is also referred to as a limit state assessment. The mathematical expression of the limit state is called the performance function, which can be described using the simplest two-parameter R–S model as follows [15]:

$$Z = R - S = 0 \tag{15.6}$$

where R is the ability of a TBC to withstand loading (i.e., resistance) and S is the generalized load that causes internal forces and deformation in a TBC. When $Z > 0$, the coating is in the safe state; when $Z < 0$, the coating is in the failure state; and when $Z = 0$, the coating is in the critical failure state. Therefore, equation $Z = 0$ is also called the limit state equation.

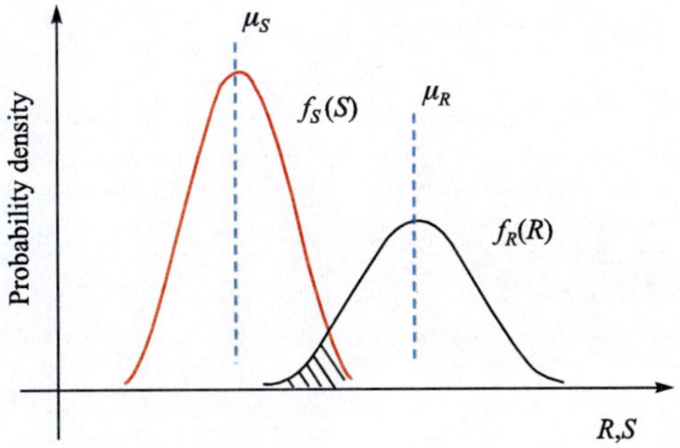

Fig. 15.1 Probability density curves of R and S and illustration of the failure probability

Here, we use Fig. 15.1 to describe the physical meaning of the reliability/failure probability when there is a change in R and/or S. Since R and S are random variables, so is Z. The distributions of R and S are described with PDFs $f_R(R)$ and $f_S(S)$, respectively. μ_R and μ_S are the means of R and S, respectively. The shaded region in the figure indicates $R < S$, i.e., $Z < 0$. Thus, the area of the shaded region can be used to represent the failure probability p_f, and the area of the remaining region is the reliability probability p_r. The PDF of the random variable Z is $f_Z(Z)$.

p_r can be described as the probability that a coating is in a reliable state:

$$p_r = \int_0^{\infty} f_Z(Z)dZ \tag{15.7}$$

The reliability region is the domain of integration for p_r, i.e., $Z > 0$. Correspondingly, p_f is obtained as follows:

$$p_f = \int_{-\infty}^{0} f_Z(Z)dZ \tag{15.8}$$

The failure region is the domain of integration for p_f, i.e., $Z \leq 0$.

Considering that the integration of the probability density over the entire domain of integration is equal to 1, we have

$$p_r + p_f = 1 \tag{15.9}$$

Normally, the resistance and load of a TBC cannot be described with two simple parameters R and S. Suppose that R and S are influenced by basic variables X_i, such as the temperature and velocity of the combustion gas and the velocity and angle of erosion particles, all of which are related to S, and Young's modulus and fracture toughness of the coating, both of which are related to R. Then, the limit state equation can be described as a function of basic variables X_i:

Fig. 15.2 Limit state surface

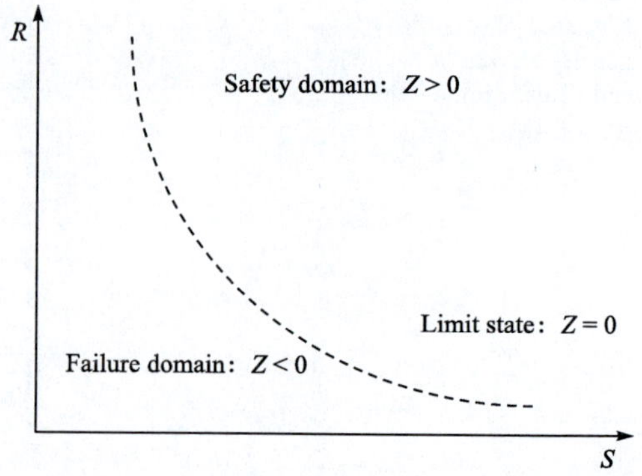

$$Z(X_1, X_2, X_3, \ldots X_n) = 0 \tag{15.10}$$

Then, Z can be viewed as a hypersurface formed with n variables in n-dimensional space, also called a limit state surface, as shown in Fig. 15.2.

p_r can be expressed as follows:

$$p_r = \int_0^{\infty} f_Z(z)dz = \int \cdots \int_{\Omega} f_X(X_1, X_2, X_3, \ldots X_n)dX_1dX_2 \cdots dX_n \tag{15.11}$$

Its domain of integration is the reliability region ($Z > 0$). The random variable $\boldsymbol{X} = (X_1, X_2, \ldots X_n)^T$ is a parameter that affects TBC failure, and $f_X(X_1, X_2, X_3, \ldots X_n)$ is the joint PDF of the random variable $\boldsymbol{X} = (X_1, X_2, \ldots X_n)^T$.

The failure probability or unreliability probability p_f is given as follows:

$$p_f = \int_{-\infty}^{0} f_Z(z)dz = \int \cdots \int_{\Omega} f_X(X_1, X_2, X_3, \ldots X_n)\, dX_1dX_2 \cdots dX_n \tag{15.12}$$

Its domain of integration is the failure region ($Z < 0$).

If $X_1, X_2, \ldots X_n$ are independent of each other, then p_f can be expressed as follows:

$$p_f = \int \cdots \int_{\Omega} f_{X_1}(X_1) f_{X_2}(X_2) \cdots f_{X_n}(X_n)dX_1dX_2 \cdots dX_n \tag{15.13}$$

In the case where the failure criterion (i.e., limit state equation) is simple and contains only a few random variables, after the PDF of each random variable is obtained, p_f can be determined through integration based on Eq. (15.13), or p_r can be determined based on Eq. (15.11).

In the case where a TBC has m failure modes that are independent of each other and each can cause the whole TBC system to fail, the p_f of the whole TBC system is obtained as follows:

$$p_f = 1-\prod_{i=1}^{m}(1-p_{fi}) \tag{15.14}$$

where $p_{fi}(i = 1, 2, ..., m)$ is the failure probability of a single failure mode of the TBC.

15.1.3 Reliability Index and Its Geometric Meaning

Assuming that the random variable Z corresponding to the performance function follows a normal distribution with a mean of μ_Z and a standard deviation of σ_Z, its PDF can be described as follows:

$$f_Z(Z) = \frac{1}{\sqrt{2\pi}\sigma_Z}\exp\left[-\frac{(Z-\mu_Z)^2}{2\sigma_Z^2}\right] \tag{15.15}$$

$Z' = \frac{Z-\mu_Z}{\sigma_Z}$ (with a mean of 0 and a standard deviation of 1) is defined as the standard normal distribution of Z. Then, p_f is derived as follows:

$$p_f = \int_{-\infty}^{0} f_Z(Z)dZ = \int_{-\infty}^{-\frac{\mu_Z}{\sigma_Z}} \frac{1}{\sqrt{2\pi}}\exp\left[-\frac{Z'^2}{2}\right]dZ'$$
$$= \int_{-\infty}^{-\frac{\mu_Z}{\sigma_Z}} \phi(Z')dZ' = \Phi\left(-\frac{\mu_Z}{\sigma_Z}\right) = \Phi(-\beta) \tag{15.16}$$

where $\varphi()$ and $\Phi()$ are the PDF and CDF of the standard normal distribution, respectively, and $\beta = \mu_Z/\sigma_Z$ is referred to as the reliability index of the structure. A large β indicates a small p_f, that is, a high reliability. Once β is determined, p_f can be directly obtained using the CDF of the standard normal distribution. Thus, β solves the difficulty in calculating p_f through numerical integration.

Now, let us explain the physical meaning of β with Eq. (15.6). Suppose that there are only two random variables, R and S, each of which follows a normal distribution. Thus, $\mu_Z = \mu_R - \mu_S$, and $\sigma_Z = \sqrt{\sigma_R^2 + \sigma_S^2}$. Then, β is obtained as follows: [16]

$$\beta = \frac{\mu_Z}{\sigma_Z} = \frac{\mu_R - \mu_S}{\sqrt{\sigma_R^2 + \sigma_S^2}} \tag{15.17}$$

Now, let us transform R and S into random variables following a standard normal distribution by defining

$$R' = \frac{R - \mu_R}{\sigma_R} \tag{15.18}$$

$$S' = \frac{S - \mu_S}{\sigma_S} \tag{15.19}$$

Substitution of Eqs. (15.18) and (15.19) into Eq. (15.6) yields

$$\sigma_R R' - \sigma_S S' + \mu_R - \mu_S = 0 \tag{15.20}$$

Equation (15.20) is the limit state equation in the standard normal space. Simultaneously dividing both sides of Eq. (15.20) by $-\sqrt{\sigma_R^2 + \sigma_S^2}$ gives the normal equation of the limit state equation as follows:

$$-\frac{\sigma_R}{\sqrt{\sigma_R^2 + \sigma_S^2}} R' + \frac{\sigma_S}{\sqrt{\sigma_R^2 + \sigma_S^2}} S' - \frac{\mu_R - \mu_S}{\sqrt{\sigma_R^2 + \sigma_S^2}} = 0$$

Fig. 15.3a shows its geometric relations:

$$\overline{O'P*} = \frac{\mu_R - \mu_S}{\sqrt{\sigma_R^2 + \sigma_S^2}} = \beta$$

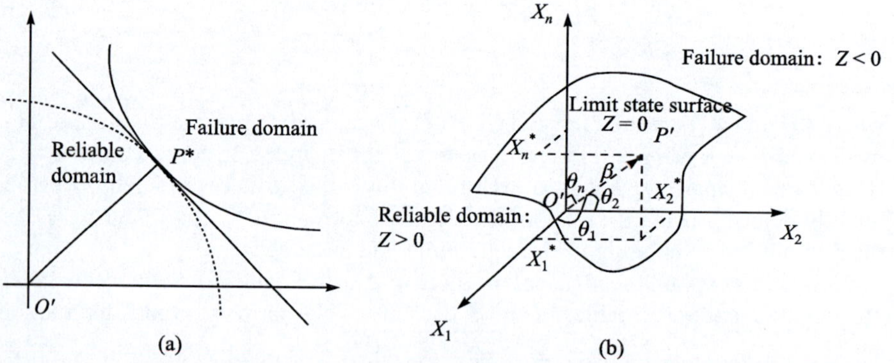

Fig. 15.3 **a** Schematic diagram of the reliability index β. **b** An n-dimensional standard normal space of n variables and the geometric meaning of β

where P^* is the design point, which is the point on the limit state equation closest to the origin O' in the standard normal coordinate system, and the corresponding distance between O' and P^*,$\overline{O'P^*}$, is β. That is, geometrically, β is the shortest distance between the origin O' and the linear failure line in the standard normal coordinate system. For a performance function Z of n variables in an n-dimensional standard normal space, as shown in Fig. 15.3b, the origin represents the mean of Z, and β is the shortest distance between the origin and the limit state surface and can also be understood as the maximum safe distance by which the service state of the TBC can deviate from the mean. Designing the limit state surface (critical failure surface) and its corresponding random variable $\boldsymbol{X}$ using the mean as the origin and β as the safe distance is the key and focus of reliability or safety design.

15.1.4 Reliability Sensitivity

Reliability sensitivity is defined as the extent of the change in p_f caused by the distribution parameters (including μ_{X_i} and σ_{X_i}) of the basic variables and is mathematically described as the partial derivative of p_f with respect to the distribution parameters of the basic variables [15]. The importance of the distribution parameters of the basic variables to β can be ranked according to the dimensionless regularized reliability sensitivity. Based on the relation between p_f and β as well as the relations between β and the distribution parameters of the basic variables, reliability sensitivity can be determined using the rule for finding the derivative of a composite function:

$$\frac{\partial p_f}{\partial \mu_{X_i}} = \frac{\partial p_f}{\partial \beta}\frac{\partial \beta}{\partial \mu_{X_i}} \tag{15.21}$$

$$\frac{\partial p_f}{\partial \sigma_{X_i}} = \frac{\partial p_f}{\partial \beta}\frac{\partial \beta}{\partial \sigma_{X_i}} \tag{15.22}$$

Based on the following equation for calculating p_f:

$$p_f = \Phi(-\beta) = \int_{-\infty}^{-\frac{\mu_Z}{\sigma_Z}} \frac{1}{\sqrt{2\pi}} \exp\left(-\frac{\beta^2}{2}\right) d\beta \tag{15.23}$$

we have

$$\frac{\partial p_f}{\partial \beta} = -\frac{1}{\sqrt{2\pi}} \exp\left(-\frac{\mu_Z^2}{2\sigma_Z^2}\right) \tag{15.24}$$

Further, based on $\beta = \mu_Z/\sigma_Z$ and the relation between the performance function Z and basic variable X_i, the following expression of reliability sensitivity can be determined:

$$\frac{\partial p_f}{\partial \mu_{X_i}} = \frac{\partial p_f}{\partial \beta}\frac{\partial \beta}{\partial \mu_{X_i}} = \frac{1}{\sqrt{2\pi}}\exp\left(-\frac{\mu_Z^2}{2\sigma_Z^2}\right)\frac{\partial\left(\frac{\mu_Z}{\sigma_Z}\right)}{\partial \mu_{X_i}} = -\frac{\left(\frac{\partial Z}{\partial X_i}\right)}{\sqrt{2\pi}\sigma_Z}\exp\left(-\frac{\mu_Z^2}{2\sigma_Z^2}\right) \tag{15.25}$$

$$\begin{aligned}\frac{\partial p_f}{\partial \sigma_{X_i}} &= \frac{\partial p_f}{\partial \beta}\frac{\partial \beta}{\partial \sigma_{X_i}} = -\frac{1}{\sqrt{2\pi}}\exp\left(-\frac{\mu_Z^2}{2\sigma_Z^2}\right)\frac{\partial\left(\frac{\mu_Z}{\sigma_Z}\right)}{\partial \sigma_{X_i}} \\ &= -\frac{\left(\frac{\partial Z}{\partial X_i}\right)^2_{\mu_X}\sigma_{X_i}\mu_Z}{\sqrt{2\pi}\sigma_Z^3}\exp\left(-\frac{\mu_Z^2}{2\sigma_Z^2}\right)\end{aligned} \tag{15.26}$$

The reliability sensitivity corresponding to the μ_{X_i} and σ_{X_i} of each parameter X_i when they take any value (i.e., the local reliability sensitivity) can be determined. In addition, the variation curve, or functional relation, of the local reliability sensitivity as X_i (i.e., its μ_{X_i} and σ_{X_i}) changes continuously (i.e., the global reliability sensitivity) can be determined. For a TBC, reliability sensitivity analysis allows the identification of all the variables that affect its p_f (i.e., the influencing parameters in the performance function), the analysis of its p_f, a further analysis of the local and global reliability sensitivities upon changes in the parameters, and, on this basis, the extraction of the key influencing parameters, which provides a basis for its application and optimal design.

15.2 Reliability Calculation Methods for TBCs

In the above section, two conditions must be met when β is used to calculate p_f, reliability, and reliability sensitivity: (1) each variable that affects the performance function Z of the TBC must follow a normal distribution, and (2) Z must be a linear function of these influencing parameters. However, in reality, these parameter variables do not necessarily follow normal distributions, and the performance function is not necessarily a linear function of each parameter. Therefore, it is impossible to exactly calculate the mean and standard deviation of the performance function as well as β. To address this problem, a series of engineering methods have been developed for calculating reliability. This section introduces several methods for calculating the reliability of TBCs.

15.2.1 Second-Moment Methods

The second-moment methods include the first- and second-order second-moment methods (the FOSM and SOSM methods, respectively), where the latter is developed on the basis of the former. The basic concept of the FOSM method [15, 16, 18] is to transform each variable that does not follow a standard normal distribution into one that does, then to approximate the performance function to a linear function with each random variable following a normal distribution, and further to obtain the approximate value of β.

Based on the criterion of equal CDFs, that is, the PDFs are different, but the CDFs are equal for each variable before and after the transformation, each variable X_i is transformed to a standard normally distributed variable X_i': [19]

$$X_i' = \begin{cases} \dfrac{X_i - \mu_{X_i}}{\sigma_{X_i}} & X_i \text{ is a normally distributed variable} \\ \Phi^{-1}\left[F_{X_i}(X_i)\right] & X_i \; is \text{ not a normally distributed variable} \end{cases} \tag{15.27}$$

where $F_{X_i}()$ is the CDF of X_i, $\Phi()$ is the PDF of a standard normal variable, and $\Phi^{-1}()$ is the inverse function of $\Phi()$.

Correspondingly, the use of standard normalized random variables leads to the transformation of the limit state equation to the form as a function of standard normal variables:

$$Z(X_1, X_2, X_3, \ldots X_n) = G\left(X_1', X_2', X_3', \ldots X_n'\right) = 0 \tag{15.28}$$

An initial trial point $\mathbf{X}'^0$ is selected. Generally, $\mathbf{X}'^0$ can be set to $\left(\mu_{X_1}, \mu_{X_2}, \mu_{X_3}, \ldots \mu_{X_n}\right)$. By performing a Taylor series expansion on the nonlinear performance function G at the initial trial point $\mathbf{X}'^0$ and retaining up to the linear term, we have

$$G(\mathbf{X}') = G(\mathbf{X}'|)_{\mathbf{X}'=\mathbf{X}'^0} + \sum_{i=1}^{n} \left.\frac{\partial G(\mathbf{X}')}{\partial X_i'}\right|_{\mathbf{X}'=\mathbf{X}'^0} \left(X_i' - X_i'^0\right) \tag{15.29}$$

Further, the mean μ_{G_0} and standard deviation σ_{G_0} of the performance function G can be calculated as follows:

$$\mu_{G_0} = G(\mathbf{X}')|_{\mathbf{X}'=\mathbf{X}'^0} - \sum_{i=1}^{n} \left.\frac{\partial G(\mathbf{X}')}{\partial X_i'}\right|_{\mathbf{X}'=\mathbf{X}'^0} X_i'^0 \tag{15.30}$$

$$\sigma_{G_0} = \sqrt{\sum_{i=1}^{n} \left[\left.\frac{\partial G(\mathbf{X}')}{\partial X_i'}\right|_{\mathbf{X}'=\mathbf{X}'^0} \right]^2} \tag{15.31}$$

Correspondingly, β is calculated as follows:

$$\beta_0 = \frac{\mu_{G_0}}{\sigma_{G_0}} \tag{15.32}$$

In the standard normal coordinate system, $G = 0$ represents the hyperplane tangent at the trial point $\mathbf{X}'^0$. Based on the geometric meaning of β (i.e., the shortest distance between the origin O' and the failure surface in the standard normal coordinate system), the value of β (i.e., the minimum value of $\beta_0 = \left(\mathbf{X}'^{\mathrm{T}}\mathbf{X}'\right)^{1/2})$ that meets the constraints can be calculated.

The direction cosine of random variable X_i' is defined as follows:

$$\cos\theta_{X_i'} = -\frac{\frac{\partial G(\mathbf{X}')}{\partial X_i'}}{\sqrt{\sum_{i=1}^{n}\left[\frac{\partial G(\mathbf{X}')}{\partial X_i'}\right]^2}} \tag{15.33}$$

where $\cos\theta_{X_i'}$ represents the relative effect of the ith parameter on the whole standard deviation. A new trial point can be expressed as follows:

$$X_i' = \beta_0 \cos\theta_{X_i'} \tag{15.34}$$

This process is repeated until the errors for two consecutive trial points meet the requirements.

The procedure for using the FOSM method to calculate the final design point X* and β through iteration is summarized as follows:

(1) Establish a performance function or limit state equation based on the failure mechanism of the structure (coating).
(2) Determine the influencing variables X_i ($i = 1, 2, \ldots, n$) in the limit state equation, analyze their distributions, and subject each variable to an equivalent normalization based on Eq. (15.27).
(3) Set an initial trial point $X_i'^0$, generally to the mean of each random variable.
(4) Perform a Taylor series expansion on the performance function near the trial point and retain up to the linear term, and calculate β and $\cos\theta_{X_i'}$ based on Eqs. (15.32) and (15.33), respectively.
(5) Calculate a new trial point X' using Eq. (15.34), and calculate β and $\cos\theta_{X_i'}$ based on X'.
(6) Repeat Steps (4) and (5) until the errors for two consecutive trial points are smaller than the set value $\|X\| < \varepsilon$, and obtain the final trial point and β.

The FOSM method retains only up to the linear term of the Taylor series expansion of the performance function. This approach is suitable for cases where the performance function is not highly nonlinear. However, the TBC involves a large number of variables, and its performance function is often highly nonlinear. As a result, the higher-order terms of the Taylor series cannot be neglected.

In view of this issue, Breitung [20, 21] further proposed an SOSM method that retains up to the quadratic terms of the Taylor series expansion of the performance function:

$$G(\mathbf{X}') = Z(\mathbf{X}'^{*}) + \nabla G(\mathbf{X}'^{*})^{T}(\mathbf{X}' - \mathbf{X}'^{*}) + \frac{1}{2}\nabla^{2} G(\mathbf{X}'^{*})^{T}(\mathbf{X}' - \mathbf{X}'^{*}) \quad (15.35)$$

where

$$\nabla G(\mathbf{X}'^{*}) = \frac{\partial G(\mathbf{X}'^{*})}{\partial \mathbf{X}'}$$

The unit vector and matrix are defined as follows:

$$\boldsymbol{\alpha}_{x^{*}} = -\frac{\nabla G(\mathbf{X}'^{*})}{\|\nabla G(\mathbf{X}'^{*})\|} \quad (15.36)$$

$$\boldsymbol{Q}_{\mathbf{X}^{*}} = -\frac{\nabla^{2} G(\mathbf{X}'^{*})}{\|\nabla G(\mathbf{X}'^{*})\|} \quad (15.37)$$

where

$$\|\nabla G(\mathbf{X}'^{*})\| = \sqrt{\sum_{i=1}^{n}\left[\frac{\partial G(\mathbf{X}'^{*})}{\partial X'_{i}}\right]^{2}}$$

Using the orthonormalization technique, an orthogonal matrix $\boldsymbol{H}$ can be established based on the arrangement of unit vector $\alpha_{x'^{*}}$ in the n-dimensional matrix as follows:

$$\mathbf{H} = \left(\mathbf{H}_1\ \mathbf{H}_2 \cdots \mathbf{H}_n\ \boldsymbol{\alpha}\right) \quad (15.38)$$

Its fundamental system of solutions is as follows:

$$\boldsymbol{H}_1 = \begin{bmatrix} 1 \\ -\frac{\alpha_1}{\alpha_2} \\ 0 \\ \vdots \\ 0 \end{bmatrix}, \boldsymbol{H}_2 = \begin{bmatrix} 0 \\ 1 \\ -\frac{\alpha_2}{\alpha_3} \\ \vdots \\ 0 \end{bmatrix}, \ldots, \boldsymbol{H}_{n-1} = \begin{bmatrix} 0 \\ \vdots \\ 0 \\ 1 \\ -\frac{\alpha_{n-1}}{\alpha_n} \end{bmatrix}$$

The unit orthogonalization of $\boldsymbol{H}$ is carried out to obtain the orthogonal matrix to be solved. Further, the SOSM p_f is obtained as follows: [20]

$$p_f = \frac{\Phi(-\beta)}{\sqrt{\det\left(\mathbf{I} - \beta\left(\mathbf{H}^{T}\mathbf{Q}\mathbf{H}\right)_{n-1}\right)}} \quad (15.39)$$

where β is the reliability index determined using the FOSM method.

However, it remains impossible to solve problems with a high degree of nonlinearity even with a Taylor series expansion up to quadratic terms. Overall, second-moment methods have a low accuracy for problems involving highly nonlinear performance functions.

15.2.2 Monte Carlo Methods

Monte Carlo methods, also known as random sampling methods, [15] probabilistic simulation methods, and statistical experimental methods, are a type of approximate numerical method that calculates structural reliability through random simulations and statistical experiments. The basic idea of Monte Carlo methods is to, based on probability and mathematical statistics theories, transform a multidimensional integration problem involving the calculation of reliability and reliability sensitivity to the form of a mathematical expectation, which is then estimated using the sample mean.

Sampling, i.e., generating samples of random variables that follow various probability distributions, is the basic concept of Monte Carlo methods. The most basic sampling involves the extraction of uniformly distributed random variable samples $(r_1, r_2, \ldots, r_n)$ from the interval of [0, 1]. Each individual sample is called a random number. Random variable samples following other types of distribution are generally realized by transforming them to uniform, independent random numbers in the interval of [0, 1].

Random numbers can be generated using a multitude of methods, such as random number table, physical, and mathematical methods [22, 23]. Currently the most often used, mathematical methods generate random numbers based on mathematical recursive formulas. Because the algorithms are deterministic, random numbers generated in this manner are only approximately random and hence often called pseudorandom numbers. Nevertheless, the resulting random numbers have good randomness properties as long as the recursive formulas and parameters for generating random numbers are chosen appropriately.

Currently, the congruential method is widely used to generate pseudorandom numbers. Its recursion formula is as follows: [15]

$$\begin{cases} x_i == (\lambda x_{i-1} + c)(\text{mod } M) \\ r_i = x_i / M \end{cases} i = 1, 2, \ldots \tag{15.40}$$

where the first expression means that x_i is the remainder obtained from dividing $(\lambda x_{i-1} + c)$ by M, and the multiplier λ, modulus M, increment c, and random source x_0 are all preselected nonnegative integers.

An analysis of Eq. (15.40) shows that because x_i is the remainder of the division with M as the divisor, $0 \leq x_i < M$ must hold. Thus, $0 \leq r_i < 1$, which ensures that r_i is within the internal of [0, 1]. In addition, we know from Eq. (15.40) that there can be at most M different x_i. Therefore, similarly, there can be at most M different r_i. The generated sequences $\{x_i\}$ and $\{r_i\}$ each has a period of L $(L \leq M)$. A cycle occurs after L numbers are generated. Consequently, the congruential method can generate only L random numbers. However, as long as L is sufficiently large, the numbers within one period are likely able to pass the independence and homogeneity tests in mathematical statistics, depending completely on the selection of parameters λ, M, c, and x_0.

In practical applications, the congruential method has two types, namely the mixed multiplicative congruential type ($c > 0$) and the multiplicative congruential type ($c = 0$). For the mixed multiplicative congruential method, it is generally recommended to set M to 2^k, c to an odd number, λ to a number that is exactly divisible by 4, and x_0 to any integer between 0 and M. To improve the applicability of the multiplicative congruential method, a prime modulus multiplicative congruential (PMMC) method has been developed, in which M is the maximum prime smaller than 2^k. The following are two high-performance PMMC generators that have been tested: [23]

$$x_i = 5^5 x_{i-1}\left(\mathrm{mod}\left(2^{35} - 31\right)\right) \tag{15.41}$$

$$x_i = 7^5 x_{i-1}\left(\mathrm{mod}\left(2^{31} - 1\right)\right) \tag{15.42}$$

In fact, the distribution of X_i in the basic variable $\boldsymbol{X} = (X_1, X_2, \ldots X_n)^T$ that affects the reliability is very different and not uniform, but its probability density $f_{X_i}(\mathbf{X})$ is a number within the interval of [0, 1]. Thus, $f_{X_i}(\mathbf{X})$ can correspond to the random numbers generated using Eq. (15.40), that is, let $f_{X_i}(\mathbf{X}) = r_i$. Then, we have [24]

$$\mathbf{X} = f_{X_i}^{-1}(r_i) \quad i = 1, 2, \ldots n \tag{15.43}$$

For each r_i, a corresponding set of values of the basic variables $\boldsymbol{X} = (X_1, X_2, \ldots X_n)^T$ can be obtained. Substitution of this set of values into performance function $Z(X_1, X_2, X_3, \ldots X_n)$ yields a value of Z, which is then compared with 0. If the value is less than 0, the number of failures N_f increases by 1; otherwise, no change is made to N_f. Note that the preset number of cycles cannot exceed the number of generated pseudorandom numbers. Let N be the total number of cycles run by the computer. According to Bernoulli's theorem in the law of large numbers, the frequency of event $Z < 0$ in N independently repeated tests, $\frac{N_f}{N}$, converges to the occurrence probability p_f of the event. Then, p_f is estimated as follows:

$$\widehat{p}_f = \frac{N_f}{N} \tag{15.44}$$

where $\widehat{p}_f$ is the estimated value of p_f. In the calculation process, $Z \leq 0$ must be realized N_f times to ensure the required computational accuracy.

In general, the performance functions in the reliability analysis of TBCs are highly nonlinear, making Monte Carlo sampling methods advantageous. However, in cases where the coating has a very small p_f, to ensure the calculation accuracy, a very large number of calculations (even up to tens or hundreds of thousands of calculations) are needed, resulting in a very large computational burden. Thus, this type of method is suitable for reliability predictions and sensitivity analyses when coatings have a large p_f or are under failure test conditions.

15.2.3 Mean Value Method and Advanced Mean Value Method

In many cases, a reliability analysis can determine only the factors related to a certain failure mode but is unable to provide a specific functional relation. In other words, the specific equation of the failure criterion or the performance function cannot be established through a reliability analysis. A failure criterion that can be used to determine factors but cannot be used to determine a specific equation is often also called an implicit performance function. Evidently, when even the performance function cannot be determined, the FOSM method and Monte Carlo methods are no longer applicable. For these cases, a mean value (MV) method and an advanced mean value (AMV) method have been developed [25].

The basic concept is described as follows [25]. Despite its unknown form, performance function Z is still considered to be a function with basic variables that are continuous everywhere. Thus, a Taylor series expansion can be performed at the MV point.

$$\begin{aligned} Z(\boldsymbol{X}) &= Z(\boldsymbol{\mu}) + \sum_{i=1}^{n} \left(\frac{\partial Z}{\partial X_i}\right) \cdot (X_i - \mu_i) + H(\boldsymbol{X}) \\ &= \left[Z(\boldsymbol{\mu}) - \sum_{i=1}^{n} \left(\frac{\partial Z}{\partial X_i}\right)\mu_i\right] + \sum_{i=1}^{n} \left(\frac{\partial Z}{\partial X_i}\right)X_i + H(\boldsymbol{X}) \\ &= a_0 + \sum_{i=1}^{n} a_i X_i + H(\boldsymbol{X}) \\ &= Z_{MV}(\boldsymbol{X}) + H(\boldsymbol{X}) \end{aligned} \tag{15.45}$$

where μ_i is the mean of each random variable, a_0 is the constant term of the Taylor series expansion, $\sum_{i=1}^{n} a_i X_i$ is the linear term of the Taylor series expansion, Z_{MV} is the sum of the constant and linear terms of the Taylor series expansion and is also a random variable, $H(\boldsymbol{X})$ represents the higher-order terms, and coefficient a_i can be calculated using the numerical differentiation method or the least squares method. By considering the Taylor series expansion up to the linear term while neglecting the higher-order terms (i.e., using the MV method), we have

$$Z_{MV} = Z_{MV}(\boldsymbol{X}) \tag{15.46}$$

If the mean and standard deviation of each random variable $\boldsymbol{X}$ are known, we have

$$\mu_Z \approx a_0 + \sum_{i=1}^{n} a_i \mu_{X_i} \tag{15.47a}$$

$$\sigma_Z^2 \approx \sum_{i=1}^{n} a_i^2 \sigma_{X_i}^2 \tag{15.47b}$$

where μ_Z and σ_Z are the mean and variance of Z, respectively, and μ_{X_i} and σ_{X_i} are the mean and variance of the random variable X_i, respectively. For a continuously distributed random variable with a PDF of $f(x)$, $\mu = \int_{-\infty}^{+\infty} f(x)x\mathrm{d}x$, and $\sigma^2 = \int_{-\infty}^{+\infty} f(x)[x-\mu]^2\mathrm{d}x$. Then, after being transformed to the standard normal space, the distance between the point on function $Z_{MV} = Z_{MV}(\boldsymbol{X})$ that is the closest to the origin (i.e., the most probable point (MPP)), $\boldsymbol{x}$*, and the origin is as follows:

$$L = \left|\frac{\mu_Z}{\sigma_Z}\right| = \left|\frac{a_0 + \sum_{i=1}^{n} a_i \mu_{X_i}}{\sqrt{\sum_{i=1}^{n} \left[a_i \sigma_{X_i}\right]^2}}\right| \tag{15.48}$$

At different functional levels (e.g., $Z = 0$ and 1), there may be different MPPs. When there is a change in the functional level, the moving locus of the MPP is called the MPP locus (MPPL). Based on the geometric meaning of β, as long as the MPP on function Z is determined, the probability for $Z < 0$ is as well. The MV method neglects the higher-order terms and calculates the corresponding probability with the MPP on an approximate function as a substitute for the real point. Evidently, for a highly nonlinear function Z, the MV method produces a large error. Hence, an AMV method has been developed to improve the MV method [15].

The AMV method corrects the error produced by the MV method due to the neglecting of the higher-order terms, as shown in the equation below:

$$Z_{AMV} = Z_{MV}(\boldsymbol{X}) + H_{AMV} \tag{15.49}$$

where H_{AMV} is defined as the difference between the value of Z_{MV} and the value of Z calculated at the MPP $\boldsymbol{x}^*$ of Z_{MV} (i.e., $H_{AMV} = Z(\boldsymbol{x}*) - Z_{MV}(\boldsymbol{x}*)$) and is no longer a random variable but a simple function. The key to the AMV method lies in the use of a simple linear function H_{AMV} to replace the higher-order terms $H(\boldsymbol{X})$, with the goal of reducing the truncation error. Ideally, the truncation error of function H_{AMV}

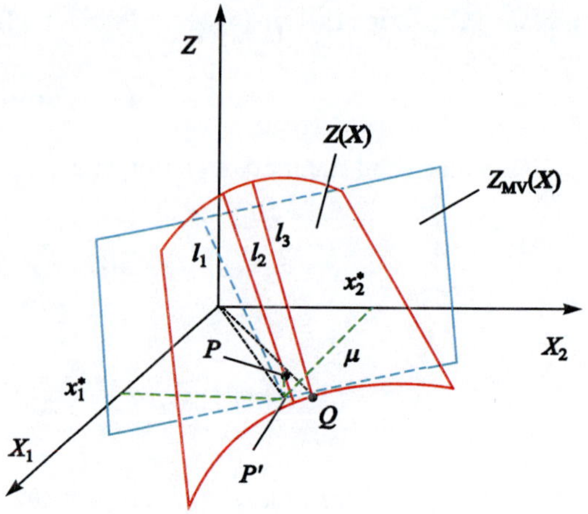

Fig. 15.4 Schematic diagram of the principles of the MV and AMV methods

should be optimized based on the MPP of function Z. The AMV method simplifies the computational procedure by using the locus of the design point for $Z_{\mathrm{MV}}(\boldsymbol{X})$. The truncation error in the approximation is not optimum. However, because the correction point for function Z is often very close to the design point, the AMV method is able to produce reasonable CDF estimates for most engineering problems.

The computational procedure of the AMV method is summarized as follows:

1. Perform a Taylor series expansion on function Z at the MV point to obtain the linear approximation $Z_{\mathrm{MV}}(\boldsymbol{X})$.
2. Calculate the MPP $\boldsymbol{x}^*$ of $Z_{\mathrm{MV}}(\boldsymbol{X})$ (i.e., the linear approximation of function Z).
3. Calculate the value of Z($\boldsymbol{x}^*$), and use H_{AMV} to correct the error.

Figure 15.4 shows a schematic diagram of the computation using the MV and AMV methods for the case in which function Z has only two random variables ($Z_{\mathrm{MV}}(\boldsymbol{X})$ and $Z(\boldsymbol{X})$ represent a flat surface and a curved surface, respectively; $Z(\boldsymbol{X})$ is the real failure surface; $Z_{\mathrm{MV}}(\boldsymbol{X})$ is the approximate failure surface; points P′ and Q are the MPPs on failure surfaces Z_{MV} and Z, respectively; line segment $\overline{PP'}$ represents H_{AMV}; straight lines l_1, l_2, and l_3 are the MMPLs of $Z_{\mathrm{MV}}(\boldsymbol{X})$, $Z_{\mathrm{AMV}}(\boldsymbol{X})$, and Z($\boldsymbol{X}$), respectively).

15.2.4 *Software-Based Numerical Calculation of the Reliability*

For TBCs, complex environments, structures, and failure criteria can make reliability calculations very difficult. Approximate analytical methods similar to the FOSM method are, to a certain extent, suitable for solving problems involving specimens with simple shapes and not highly nonlinear failure criteria. However, in most cases, problems need to be solved using sampling methods similar to Monte Carlo methods and through statistical analysis on computers. More importantly, according to the analysis of TBC failure due to oxidation, erosion, and calcium–magnesium aluminosilicate (CMAS) corrosion in Part I of the book, the analysis and calculation of the loads (e.g., particle velocity and angle, temperature field, stress field, and SIF) in the performance function is extremely difficult and often requires the aid of FE software (e.g., ABAQUS and COMSOL) or programming. Therefore, software for numerically determining the reliability of TBCs must have both powerful computational capacity (e.g., parallel computing) with rich algorithms and interfaces for finite element (FE) software programs such as ABAQUS, ANSYS, and COMSOL.

Reliability analysis software has long been developed worldwide, with commercial programs already available from several countries, including the United States, Germany, France, and Austria. The most important criterion in the evaluation of reliability software is the ability to connect with deterministic software (e.g., FE software), thereby allowing the use of reliability software to truly analyze reliability in general engineering and academic problems. Table 15.1 summarizes several common reliability analysis software applications developed internationally based on whether they are equipped with FE interfaces. In addition, the inherent computational capacity (e.g., parallel computing mode) and the integrated reliability calculation methods are important metrics for reliability software.

NESSUS is a reliability analysis software tool developed by the Southwest Research Institute in the United States for the probabilistic design of the main propulsion systems of NASA spacecraft. The NESSUS reliability software is equipped with connection interfaces for common FE software programs such as ANSYS, ABAQUS, and NASTRAN. In addition, integrated with multiple reliability algorithms (e.g., the

Table 15.1 Several reliability software applications and their interfaces for FE software [15]

Software	Developer	FE software interface
UNIPASS	UNIPASS, United States	NASTRAN
NESSUS	Southwest Research Institute, United States	NASTRAN, ABAQUS, ANSYS
PROFES	Applied Research Associates Inc., United States	NASTRAN, ANSYS
COSSAN	Leopold-Franzens-Universität Innsbruck, Austria	FE-RV
PERMAS-RA/STRUREL	University of Munich, Germany	/
PHIMECA	LaMI/Blaise Pascal University, France	FE Code
CalREL/FERUM/OpenSees	University of California, Berkeley, United States	FEAP
PROBAN	DNV, Norway	FE Code
ANSYS	ANSYS, United States	/

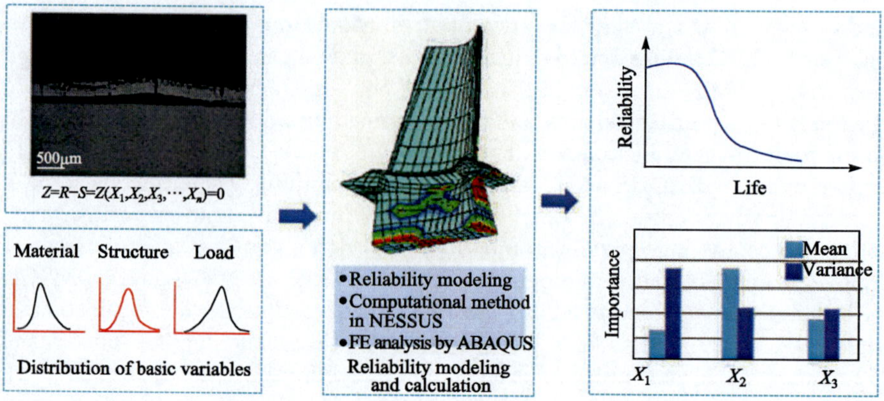

Fig. 15.5 Basic idea for the use of NESSUS to analyze the reliability and reliability sensitivity

FOSM method and Monte Carlo methods introduced in Sect. 15.1), NESSUS offers great convenience for numerically calculating the reliability of complex structures (e.g., TBCs on turbine blades) and in different failure modes.

Figure 15.5 illustrates the basic idea of using NESSUS to analyze reliability and reliability sensitivity.

(1) Establish a failure criterion. Determine the key failure mode for the key service environment (e.g., oxidation, erosion, or CMAS corrosion) based on experimental and theoretical studies and establish its failure criterion, that is, a performance function; the failure criterion can be either explicit or implicit.
(2) Analyze the basic variables (including the properties, structure, and loading) in the performance function and their distributions. Performance variables (e.g., fracture toughness and strength) can be determined through experimental characterization. Structural variables (e.g., those related to pores and intercolumn gaps) can be determined through microscopic observation. Loading variables (e.g., temperature and stress fields) can be determined using analytical or numerical methods.
(3) Establish a numerical model for reliability. This step includes the geometric modeling and meshing of the TBC on the turbine blade and the implementation of the interface between the reliability software and FE data.
(4) Calculate the key damage parameters. Analyze the damage parameters (e.g., stress and strain fields, energy release rate, erosive-particle velocity, and crack parameters) in the key failure mode of the TBC using the finite element method (FEM).
(5) Calculate the reliability. This step includes grouping the samples of the basic variables (e.g., the basic variables determined using the multiplicative congruential method), inputting the results as boundary conditions and material and structural parameters into the FEs to determine whether the performance function is less than 0, and inputting the feedback of the result to NESSUS for statistical analysis.

(6) Analyze the proportion of the failure model, calculate the reliability, and further perform a reliability sensitivity analysis.

15.3 Reliability Prediction for TBCs Under Thermal Cycling Stresses

Thermal cycling is the most common technique for simulating engine service conditions to which TBCs are subjected during takeoff (heating), flight (temperature holding), and landing (cooling). High-temperature oxidation and the thermal mismatch stress between the ceramic layer and transition layer/substrate play key roles in accelerated thermal cycling tests on TBCs. The results obtained from the real-time AE characterization and thermal stress analysis of TBCs under thermal cycling in Chap. 10 show that spallation during the cooling stage is the primary failure mode of TBCs under this type of loading. Thus, with plasma spraying (PS) TBCs subjected to thermal cycling as an example, this section introduces the basic procedure for the reliability assessment of TBCs.

15.3.1 Failure Criterion and Limit State Equation

Figure 15.6 shows the failure mode of a PS TBC under thermal cycling, as reflected by interface crack initiation and growth. The crack formation condition (i.e., failure criteria) can be described as when the interfacial energy release rate reaches the interfacial fracture toughness. Then, the limit state equation is expressed as follows [26]:

$$Z = \Gamma_{TBC} - G = \Gamma_{TBC} - E_{TBC}L(\Delta\alpha\Delta T)^2k(N - N_0) \tag{15.50}$$

where E_{TBC} is Young's modulus of the ceramic layer of the TBC (as shown in Fig. 15.6), L is the half-wavelength of the sinusoidal curve that describes the interfacial roughness, $\Delta\alpha$ is the difference between the CTEs of the coating and substrate,

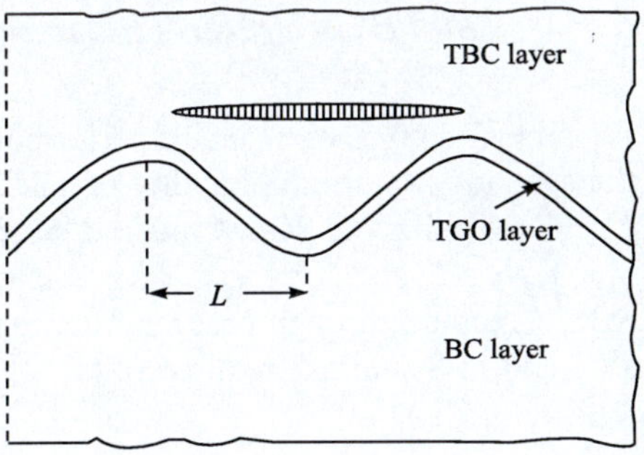

Fig. 15.6 Crack initiation and growth at the interface of a TBC

Table 15.2 Distributions of basic variables [27–35]

Parameter	Mean	Standard deviation	Distribution type
Γ_{TBC}(J/m^2)	50	10	Weibull
E_{TBC}(GPa)	40	8	Weibull
L(μm)	20	4	Normal
$\Delta\alpha$(10^{-6}/°C)	4	0.2	Normal
ΔT(°C)	1000	12	Normal
k	0.009	0.002	Normal
N	400	20	Normal

ΔT is the difference between the highest and lowest temperatures during each thermal cycle, k is a coefficient dependent on the growth of the interfacial TGO layer (i.e., TGO growth strain coefficient), N is the number of thermal cycles at TBC failure, N_0 is the initial number of thermal cycles, and Γ_{TBC} is the fracture toughness of the TBC. When $Z > 0$, the TBC is in a safe and reliable state (i.e., in the safe domain); when $Z < 0$, the TBC is in a failure state (i.e., in the failure domain); and when $Z = 0$, the TBC is in a critical failure state.

15.3.2 Distributions of Basic Variables

Now, let us analyze the distribution of each basic variable in Eq. (15.50), including its mean, standard deviation, and statistical distribution function. Material parameters (e.g., Γ_{int} and E_{TBC}) can be experimentally measured or statistically analyzed based on the available results in the literature. According to previous reports, both Γ_{int} [27–29] and E_{TBC} [30, 31] follow a Weibull distribution. For the sake of convenience, we set N_0 to 0. In addition, we assume that other variables each follow a normal distribution. Table 15.2 summarizes the mean and standard deviation of each variable [32–35].

15.3.3 Prediction of the Spallation Failure Probability of TBCs Under Thermal Cycling, $p_{f,\,tc}$

Equation (15.50) shows that the performance function is highly nonlinear. As a result, the spallation failure probability of TBCs under thermal cycling, $p_{f,\,tc}$, needs to be calculated using the SOSM method. Specifically, the $p_{f,\,tc}$ values at different E_{TBC}, $\Delta\alpha$, k, N, ΔT, and L values were predicted using the computational procedure of the SOSM method through programming in MATLAB. Tables 15.3, 15.4, 15.5, 15.6, 15.7 and 15.8 summarize the results.

Table 15.3 $p_{f,\,tc}$ values predicted for different E_{TBC} values

Parameter (GPa)	Mean	Standard deviation	Distribution type	$p_{f,\,tc}$
E_{TBC}	40	8	Normal	40.64%
E_{TBC}	30	6	Normal	17.71%
E_{TBC}	20	4	Normal	3.23%

Table 15.4 $p_{f,\,tc}$ values predicted for different $\Delta\alpha$ values

Parameter	Mean	Standard deviation	Distribution type	$p_{f,\,tc}$
$\Delta\alpha(10^{-6}/°C)$	5	0.25	Normal	84.74%
$\Delta\alpha(10^{-6}/°C)$	4	0.2	Normal	40.64%
$\Delta\alpha(10^{-6}/°C)$	3	0.15	Normal	5.61%
$\Delta\alpha(10^{-6}/°C)$	2	0.1	Normal	7.27e-4

Table 15.5 $p_{f,\,tc}$ values predicted for different k values

Parameter	Mean	Standard deviation	Distribution type	$p_{f,\,tc}$ (%)
k	0.01	0.002	Normal	51.03
k	0.009	0.002	Normal	40.64
k	0.008	0.002	Normal	30.55
k	0.006	0.002	Normal	13.92
k	0.004	0.001	Normal	1.96

Table 15.6 $p_{f,\,tc}$ values predicted for different N values

Parameter	Mean	Standard deviation	Distribution type	$p_{f,\,tc}$
N	400	20	Normal	40.64%
N	350	17.5	Normal	28.76%
N	300	15	Normal	17.72%
N	200	10	Normal	3.23%
N	150	7.5	Normal	0.72%
N	100	5	Normal	7.27e-4

Table 15.7 $p_{f,\,tc}$ values predicted for different ΔT values

Parameter (°C)	Mean	Standard deviation	Distribution type	$p_{f,\,tc}$
ΔT	1000	12	Normal	40.64%
ΔT	900	10	Normal	22.82%
ΔT	800	10	Normal	9.77%
ΔT	700	10	Normal	2.93%
ΔT	500	10	Normal	8.21e-4

Table 15.8 $p_{f,\,tc}$ values predicted for different L values

Parameter (μm)	Mean	Standard deviation	Distribution type	$p_{f,\,tc}$
L	20	4	Normal	40.64%
L	15	3	Normal	17.72%
L	10	2	Normal	3.23%
L	5	1	Normal	7.27e-4

15.3.4 Reliability Sensitivity Analysis

Tables 15.3, 15.4, 15.5, 15.6, 15.7 and 15.8 show the $p_{f,\,tc}$ values during thermal cycling at various values of variables (e.g., material properties, structure, and loading). The $p_{f,\,tc}$ of the coating increases with increasing N, ΔT, interfacial roughness, and oxidation rate. However, it is not yet possible to quantify the influence of each parameter on $p_{f,\,tc}$. Thus, a further reliability sensitivity analysis needs to be performed.

Based on the definition of reliability sensitivity as well as the computational procedure shown in Eqs. (15.21)–(15.26), the reliability sensitivity of each basic variable can be determined. Here, using Γ_{TBC} as an example, the $p_{f,\,tc}$ of the TBC at various values of the mean and standard deviation of Γ_{TBC} was calculated. Figure 15.7 shows the variation in $p_{f,\,tc}$ with Γ_{TBC}.

From Fig. 15.7, the derivative of $p_{f,\,tc}$ at each mean and standard deviation of Γ_{TBC} can be calculated. Based on the reference mean and standard deviation of Γ_{TBC} in Table 15.2, the reliability sensitivity of the mean and standard deviation of Γ_{TBC} is calculated to be −1.1411 (the minus sign is neglected in Fig. 15.7) and 0.0401, respectively.

Figure 15.8 shows the reliability sensitivity of each of the other parameters calculated using the same method. A comparison shows that the mean of ΔT and the mean of CTE have the most significant influence on coating failure, that is, thermal stress is

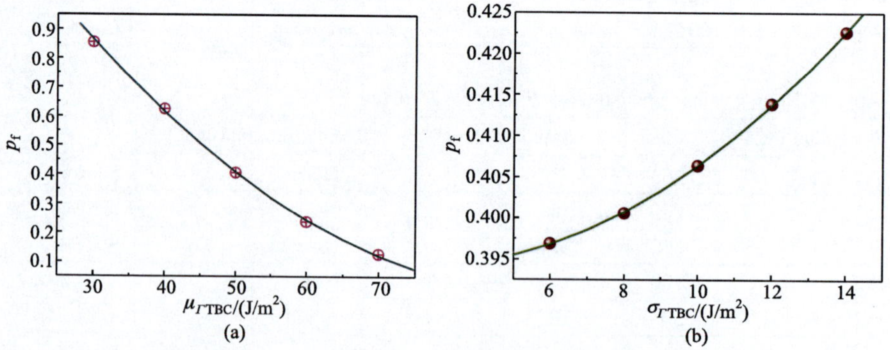

Fig. 15.7 Influences of the **a** mean and **b** standard deviation of Γ_{TBC} on $p_{f,\,tc}$

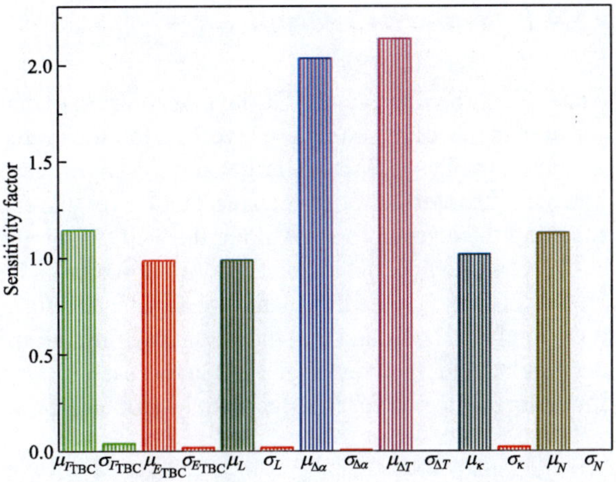

Fig. 15.8 Relative magnitudes of the sensitivity factors of different influencing factors

the most important factor in coating failure. In addition, changes in the mean values of N, Γ_{TBC}, oxidation rate, E_{TBC}, and interfacial roughness each has a large influence on the spallation failure of the TBC under thermal cycling [36, 37]. In comparison, the change in the standard deviation of each parameter has little influence on coating failure.

15.4 Reliability Assessment of TBCs Under Interfacial Oxidation

Interfacial oxidation refers to the formation of a TGO layer from the chemical reaction between the oxygen that diffuses from the ceramic layer and external environment to the interface and the metal elements that diffuse from the transition layer to the interface. The TGO layer increases the thermal mismatch between the layers of a TBC, causing it to spall. Miller [38] and DeMasi et al. [39] established life prediction models for interfacial oxidation conditions based on the weight and thickness of the TGO layer, respectively, with the same concept—a coating spalls when the extent of interfacial oxidation reaches a certain critical level. In addition, interfacial oxidation is considered the most important factor that causes coating spallation. This section analyzes in detail the reliability under interfacial oxidation and its influencing factors from various perspectives (e.g., simple specimens, turbine blade structure, and nonuniform temperature fields in coatings).

15.4.1 Failure Criterion

The CTE of the interfacial TGO layer is lower than that of the ceramic layer and much lower than that of the transition layer and the substrate. Consequently, during cooling, as shown in Fig. 15.9a, very large compressive stresses are formed in the TGO layer and near the interfaces between the TGO layer and the ceramic and transition layers, inducing the coating to spall along the TGO layer and nearby interfaces.

Here, we do not consider the deformation of a TBC system along its thickness; that is, we consider it to be in a plane stress state. Assuming that each layer of the TBC is a linear elastic body and that interface cracking failure occurs when the strain energy accumulated in the interfacial TGO layer reaches that required for interface cracking, the failure criterion for interfacial oxidation can be written as follows: [40, 41]

$$Z = \Gamma_{\text{int}} - G = \Gamma_{\text{int}} - E_{TGO}\, h_{TGO}\, (\Delta\alpha\, \Delta T)^2/(1 - \nu_{TGO}) \tag{15.51}$$

where Γ_{int} is the fracture toughness at the TGO/transition interface, E_{TGO} is Young's modulus of the TGO layer, h_{TGO} is the thickness of the TGO layer, $\Delta\alpha$ is the difference between the CTEs of the TGO layer and substrate, ΔT is the difference between the oxidation temperature and room temperature, and ν_{TGO} is Poisson's ratio of the TGO layer. When $Z > 0$, the TBC is in a safe and reliable state (i.e., in the safe domain); when $Z < 0$, the TBC is in a failure state (i.e., in the failure domain); and when $Z = 0$, the TBC is in a critical failure state.

When the oxidation temperature (e.g., surface temperature) of the TBC on a turbine blade is nonuniform, there is a nonuniform increase in the TGO thickness h_{TGO}. Then, h_{TGO} can be described as follows: [32]

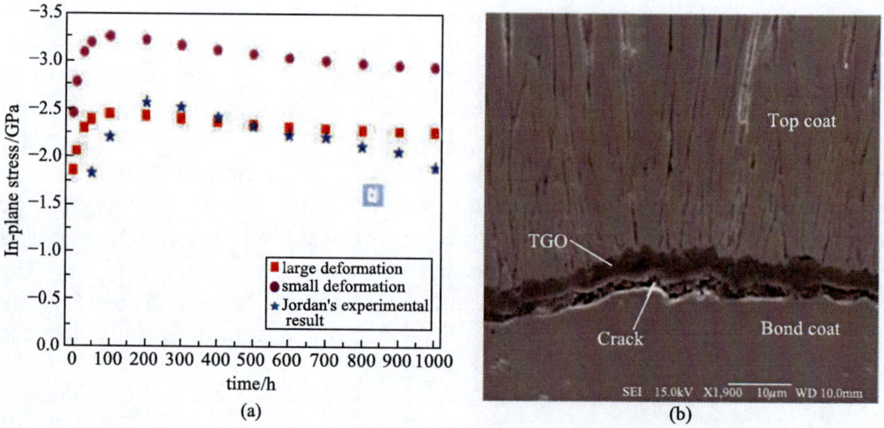

Fig. 15.9 **a** Compressive stresses in the interfacial TGO layer. **b** Coating spallation induced by compressive stresses

$$h_{TGO} = C\exp\left(-\frac{Q}{RT}\right)t^n \quad (15.52)$$

where C, n, and Q are constants obtained by fitting test results (for TBCs, $C = 289\ \text{m/s}^{0.5}$, $Q = 60210$ J/mol, and $n = 0.5$) [32] and R is the gas constant ($R = 8.31$ $\text{J}/(\text{mol}\cdot\text{K})$). Then, the failure criterion in Eq. (15.51) can be written as follows:

$$Z = \Gamma_{\text{int}} - \left[E_{TGO}(\Delta\alpha\Delta T)^2/(1-\nu_{TGO})\right]Ce^{(-Q/(RT))}t^n \quad (15.53)$$

15.4.2 Analysis of the Statistical Characteristics of Parameters Influencing Interfacial Oxidation

The statistical characteristics of the random variables in the failure criterion need to be obtained for reliability analysis under interfacial oxidation. To examine the correctness of the reliability assessment method, the failure probability of TBC due to interfacial oxidation, $p_{f,io}$, is determined using the spallation failure ratio (i.e., the ratio of the spallation failure area to the total area) through a high-temperature oxidation experiment and then compared with theoretical predictions. The oxidation experiment was conducted in a high-temperature furnace in the presence of oxygen. The oxidation temperature was set to 1100 °C. After oxidation, the coating was allowed to cool to approximately 20 °C. In other words, the temperature difference ΔT was 1080 °C (with a standard deviation of 5 °C). An extensive literature analysis shows that the fracture roughness at the TGO-layer/transition-layer interface, Γ_{int}, follows the Weibull distribution [29] and that Poisson's ratio of the TGO layer is set to a constant of 0.23. For the sake of convenience, we assume that each of the other parameters varies approximately near its mean value in the form of a normal distribution. Table 15.9 summarizes the mean and standard deviation of each parameter according to experimental data on E_{TGO}, [32–35, 42, 43] and ΔT as well as those obtained by Jackson [44].

Table 15.9 Statistical characteristics of influencing parameters

Parameter	Mean	Standard deviation	Distribution type
Γ_{int}(J/m^2)	50	5	Weibull
E_{TGO}(GPa)	380	100	Normal
h (μm)	2.85	0.81	Normal
$\Delta\alpha$(10^{-6}/°C)	7.3	0.37	Normal
ΔT(°C)	1080	5	Beta

15.4.3 Reliability and Sensitivity Analysis of TBCs Under Interfacial Oxidation Based on the SOSM Method

Tables 15.10, 15.11, 15.12 and 15.13 summarize the $p_{f,io}$ predicted using the SOSM method for calculating reliability [45]. $p_{f,io}$ increases with increasing E_{TGO}, $\Delta\alpha$, ΔT, and h_{TGO}, consistent with previous findings [36, 37, 46]. Reliability analysis can further provide the variation trend of reliability with each parameter, which is of greater guiding significance to engineering applications.

The importance of the abovementioned parameters affecting interfacial oxidation failure is further determined based on the sensitivity factor. Figure 15.10 shows the reliability sensitivity of each parameter of the TBC under interfacial oxidation obtained using the method introduced in Sect. 15.1.3 in conjunction with the same procedure used in Sect. 15.3.4. The mean values of ΔT and $\Delta\alpha$ have the most significant influence on the oxidation failure of a coating, i.e., thermal stress is an important factor causing high-temperature oxidation failure. In addition, changes in the mean values of Γ_{int}, critical strain energy, E_{TGO}, and h_{TGO} have a significant influence on the high-temperature oxidation failure of a TBC. In comparison, a change in the standard deviation of each parameter has little influence on coating failure.

Table 15.10 $p_{f,io}$ predicted with different E_{TGO} values

Parameter (GPa)	Mean	Standard deviation	Distribution type	$p_{f,io}$ (%)
E_{TGO}	390	102	Normal	49.65
E_{TGO}	380	100	Normal	46.92
E_{TGO}	370	98	Normal	44.16
E_{TGO}	360	94	Normal	41.30
E_{TGO}	350	92	Normal	38.46

Table 15.11 $p_{f,io}$ predicted with different $\Delta\alpha$ values

Parameter	Mean	Standard deviation	Distribution type	$p_{f,io}$
$\Delta\alpha(10^{-6}/°C)$	8	0.4	Normal	66.10%
$\Delta\alpha(10^{-6}/°C)$	7.3	0.37	Normal	46.92%
$\Delta\alpha(10^{-6}/°C)$	6	0.3	Normal	12.92%
$\Delta\alpha(10^{-6}/°C)$	5	0.25	Normal	1.44%
$\Delta\alpha(10^{-6}/°C)$	4	0.2	Normal	2.06e-4

Table 15.12 $p_{f,io}$ predicted with different ΔT values

Parameter (°C)	Mean	Standard deviation	Distribution type	$p_{f,io}$
ΔT	1200	10.8	Normal	68.90%
ΔT	1080	9.61	Normal	46.92%
ΔT	900	8.1	Normal	14.62%
ΔT	800	7.2	Normal	4.26%
ΔT	600	5.4	Normal	2.84e-4

Table 15.13 $p_{f,io}$ predicted with different h_{TGO} values

Parameter (μm)	Mean	Standard deviation	Distribution type	$p_{f,io}$ (%)
h_{TGO}	3	0.86	Normal	52.31
h_{TGO}	2.85	0.81	Normal	46.92
h_{TGO}	2.4	0.69	Normal	29.88
h_{TGO}	2	0.57	Normal	14.32
h_{TGO}	1.5	0.5	Normal	4.2

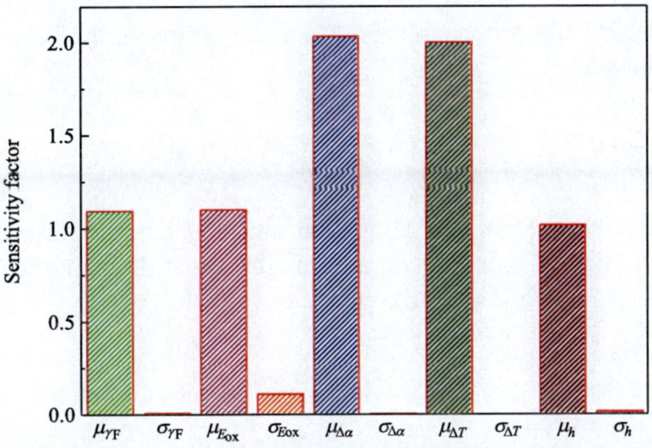

Fig. 15.10 Reliability sensitivity of each parameter influencing the interfacial oxidation failure

15.5 Reliability Assessment of TBCs Against Erosion Failure

15.5.1 Erosion Rate Model and Reliability Analysis Criterion for TBCs on Turbine Blades

Erosion rate E_m, which generally has units of mg/g, is defined as the mass of the exfoliated part of a coating due to the impact of a unit mass of particles. Li [47]

established an E_m model for TBCs under erosion. As shown in Fig. 15.11, assuming that the effect of the particle impact angle on E_m is independent of the effect of any of the other factors, the E_m model can be described as follows:

$$E_m = g(\alpha)E_{90} \tag{15.54}$$

where g(α) represents the effect of the impact angle on Em and E90 is the erosion rate at an impact angle of 90°. Let us assume that after particles vertically impact a coating at a certain velocity and then rebound, a proportion of their kinetic energy (ke) is absorbed by the coating. Then, after Np particles impact a coating, the total kinetic energy of the particles absorbed by the coating is as follows:

$$E_n = \frac{1}{2}k_e N_p m v_0^2 \tag{15.55}$$

where m and v_0 are the mass and velocity of an erosive particle, respectively.

Let h and S_0 be the average thickness and unit area of the microlamellae in the ceramic layer, respectively. In addition, let us assume that all the kinetic energy that N_p impacting particles have lost is used to exfoliate N_c lamellae from the coating and is not dissipated by other mechanisms. Then, according to the law of conservation of energy, we have

$$E_n = 2\lambda\gamma_c N_c S_0 \tag{15.56}$$

where γ_c is the surface energy of a lamellar unit and λ is the interlamellar binding rate ($\Gamma = 2\lambda\gamma_c$, where Γ is the fracture toughness of the ceramic layer) [47]. By combining Eqs. (15.55) and (15.56), we have

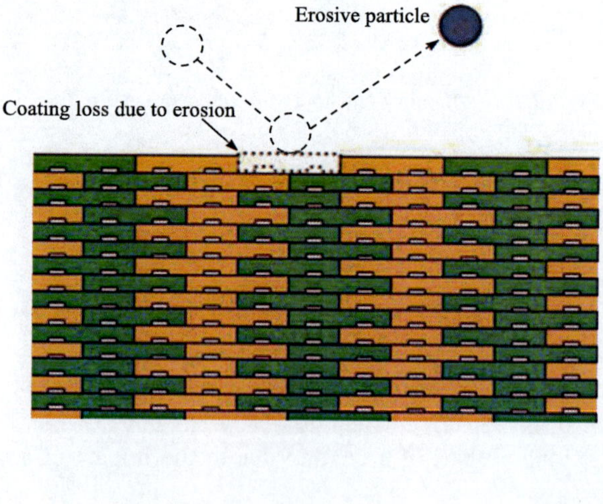

Fig. 15.11 Schematic diagram of the mass loss of a TBC due to erosion

$$\frac{N_p m}{N_c} = \frac{4\lambda\gamma_c S_0}{k_e v_0^2} \tag{15.57}$$

It was experimentally found that the mass lost by a coating due to erosion, W_c, is directly proportional to the total mass of the erosive particles, W_p. W_c and W_p are expressed as follows:

$$W_c = N_c S_0 h \rho_c \tag{15.58}$$

$$W_p = N_p m \tag{15.59}$$

Since $E_{90} = W_c/W_p$, combining Eqs. (15.55)–(15.59) yields

$$E_{90} = \frac{k_e v_0^2 h \rho_c}{2\Gamma} \tag{15.60}$$

The influence of the impact angle on Em can be described by the following function:

$$g(\alpha) = (\sin\alpha)^{n_1}\big(1 + \big(H_c/H_p\big)(1 - \sin\alpha)\big)^{n_2} \tag{15.61}$$

where α is the impact angle, n_1 and n_2 are exponential terms determined by fitting test data, and H_c and H_p are the hardness of the coating and particles, respectively. Figure 15.12 shows the data obtained by Nicholls and Tabakoff et al. [48, 49] from TBC erosion tests in high-temperature wind tunnels and the results obtained from fitting Eq. (15.61) to the data. Specifically, $n_1 = 2.246$ and $n_2 = 0.688$. Then, the E_m model for TBCs can be expressed as follows:

$$E_m = (\sin\alpha)^{2.246}\left(1 + \frac{H_p}{H_c}(1 - \sin\alpha)\right)^{0.688}\frac{k_e v_0^2 h \rho_c}{2\Gamma} \tag{15.62}$$

Assuming that a TBC fails when its thickness in a local area decreases to zero, the local failure criterion for a TBC is expressed as follows:

$$Z = \rho_c H - M_p E_m \tag{15.63}$$

where ρ_c and H are the density and thickness of the ceramic layer, respectively, and M_p is the total mass of the particles impacting the area. Substitution of the E_m model in Eq. (15.62) into Eq. (15.63) yields the following local erosion failure criterion:

$$Z = H - M_p(\sin\alpha)^{2.246}\left(1 + \frac{H_p}{H_c}(1 - \sin\alpha)\right)^{0.688}\frac{k_e v_0^2 h \rho_c}{2\Gamma} \tag{15.64}$$

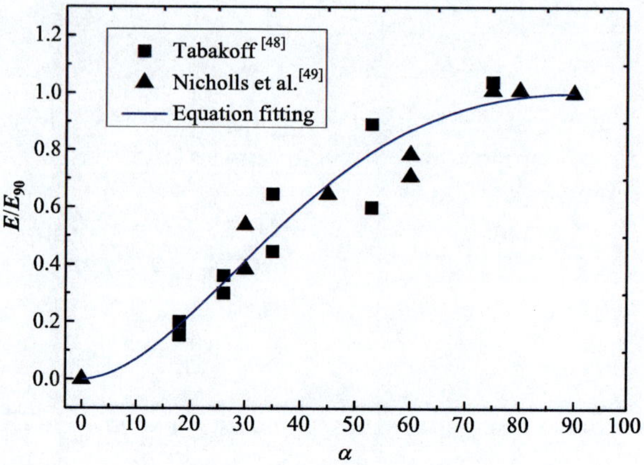

Fig. 15.12 Erosion rates of TBC at different impact angles

When $Z > 0$, the TBC is in a safe and reliable state (i.e., in the safe domain); when $Z < 0$, the TBC is in a failure state (i.e., in the failure domain); and when $Z = 0$, the TBC is in a critical failure state.

15.5.2 Method for Calculating the Erosion Reliability of TBCs on Turbine Blades

Two important conditions need to be met when Eq. (15.64) is used to calculate the reliability of TBCs on turbine blades: (1) each parameter in the equation can be known, and (2) the reliability calculation method can be used to calculate the reliability under each parameter. However, in a real-world engine, erosive particles of various sizes randomly impact the surface of the coating at various velocities and angles with the high-temperature combustion gas. Combustion-gas parameters (including temperature and velocity) all vary with the service state of turbine blades and cannot be directly given. Therefore, the calculation of the erosion reliability of the TBC on a turbine blade should involve two steps, namely the calculation of the erosion process and the reliability, which are discussed one-by-one below:

(1) **Numerical simulation and calculation of the erosion process of the TBC on a turbine blade**

This process was primarily carried out using the computational fluid dynamics software CFX and the FE software ANSYS. First, the Angle/Thickness method in the submodule BladeGen in ANSYS was used to establish a 3D geometric model for the guide and working vanes of a turbine under the action of a combustion gas based on the midline direction angle of the vanes and the variation in the vane thickness along the midline. See elsewhere [50] for the detailed modeling method. Figure 15.13 shows the geometric model of the fluid and solid domains. In the figure, the grid represents the fluid; the hollow parts in the middle represent the solid vanes, i.e., the guide vane (on the left-hand side) and the working vane (on the right-hand side); and the arrows indicate the moving direction of the gas. The metagrid for the fluid domain comprises H-grids, which are eight-node hexahedral elements. The internal channels in the vanes are the key locations for the heat transfer between the combustion gas (fluid domain) and the turbine vanes (solid domain) as well as particle erosion. Therefore, the grid density is much higher in the channels than in other locations.

The 3D fluid field was calculated based on the compressible Reynolds-averaged Navier–Stokes equations in combination with the $k - \omega$ oscillation model while oscillation randomness was neglected. The total temperature and total pressure at the gas inlet were set to 1678 K and 1303 kPa, respectively, while the static pressure at the outlet was set to 571 kPa. In addition, periodic boundary conditions were applied to the guide and rotating vanes [51, 52].

Figure 15.14 shows the distribution of the Mach number at the middle section of the flow field. Generally, a large Mach number indicates a high surface flow rate. Figure 15.14 shows that the Mach number at the inlet on the left-hand side is small (only approximately 0.1). The leading edge of the guide vane blocks the combustion gas, resulting in a zone with a Mach number close to 0 at this location. Toward the outlet, the Mach number first gradually increases and reaches its maximum value (approximately 1.2) on the suction side of the rotating vane (i.e., in the area of the convex side of the vane close to the leading edge) and then gradually decreases to approximately 0.5 at the outlet. In real-world situations, a combustion gas flows out of a combustion chamber at a rate of only approximately 50 m/s, and the maximum design Mach number for the interior of turbine cascade channels is generally 1.2. Thus, this simulation is acceptable.

Figures 15.15 and 15.16 show contour plots of the temperature and pressure distributions at the middle section of the flow field, respectively. Both the temperature

Fig. 15.13 Geometric model of the guide and working vanes of a turbine

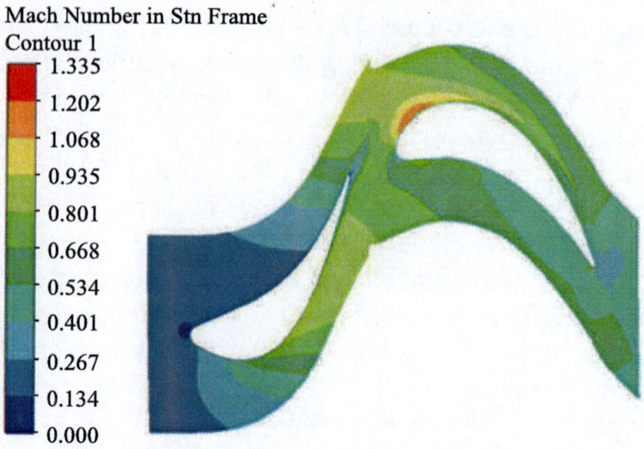

Fig. 15.14 Contour plot of the Mach number at the middle section of the flow field

and pressure are the highest at the inlet and gradually decrease toward the outlet. In addition, their minimum values both appear near the trailing edge of the convex side of the guide vane. Overall, the pressure and temperature on the concave side of each of the guide and rotating vanes are higher than those on the convex side. This is why the concave and convex sides of a vane are also called its pressure and suction sides, respectively.

When moving at a high velocity in the flow channel in a blade, a particle is likely to collide multiple times with the TBC-covered surface of the blade. After each collision, the particle rebounds and subsequently collides with the surface of the blade in other locations. Therefore, it is necessary to determine the rebound velocity and direction of a particle after it collides with the surface of the TBC at different angles. Swar [53] determined the following empirical relations between incident and rebound angles and between incident and rebound velocities for small particles impacting TBC surfaces at different angles based on measurements taken using a high-resolution, high-speed photography technique in high-temperature wind-tunnel tests:

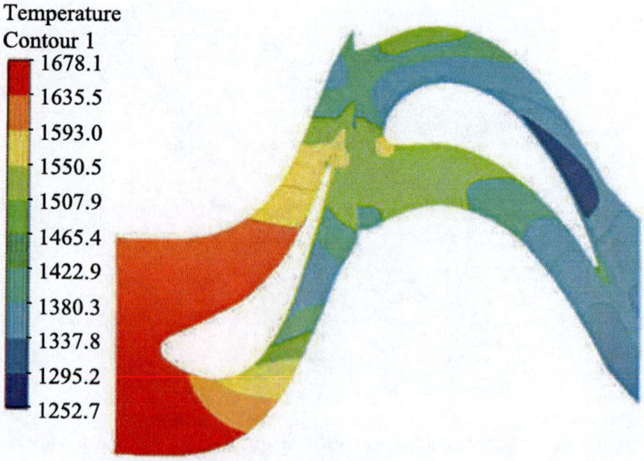

Fig. 15.15 Temperature distribution in the middle section of the flow field

Fig. 15.16 Pressure distribution in the middle section of the flow field

$$e_v = v/v_{re} = 0.963 + 0.3957\alpha - 2.2994\alpha^2 + 1.4276\alpha^3 \tag{15.65a}$$

$$e_\alpha = \alpha/\alpha_{re} = 0.2204 + 3.7169\alpha - 6.6829\alpha^2 + 3.5283\alpha^3 \tag{15.65b}$$

where a and v are the incident angle and velocity, respectively, α_{re} and v_{re} are the rebound angle and velocity, respectively, e_v is the velocity coefficient, and e_α is the angle coefficient. The two equations in Eqs. (15.65a, 15.65b) can be written into a user-custom motion locus program for particles. Further, information (e.g., velocity, angle, and location) of an erosive particle can be obtained through simulation of its motion locus. Here, we make the following assumptions: the aerodynamic force on a particle originates primarily from its sliding velocity relative to the fluid, the initial velocity of a particle is 50% of the gas flow rate at the inlet, the direction of a particle is the same as that of the combustion gas, and the particle flow is continuous at the combustion-gas inlet.

Figure 15.17a, b, c, and d shows the simulated motion loci of particles 2.5, 10, 25, and 60 μm in diameter, respectively. Figure 15.17 shows that the motion loci of smaller particles are more similar to the streamline of the fluid, which is parallel to the outer surface of the turbine vanes. As a result, the intersecting angles between the loci of smaller particles and the surface of the vanes are smaller; i.e., smaller particles impact the surface of the vanes at smaller angles. This is because small particles have relatively low inertia and thus do not tend to detach from the flow of the fluid and impact the surface of the vanes. As shown in Fig. 15.17a, particles 2.5 μm in diameter almost do not collide with the surface of the guide vane and, later, collide only with the concave side of the rotating vane at very small angles and approach the surface of the vane after rebounding. In comparison, larger particles collide with the vanes at larger angles and may impact an adjacent vane after rebounding. In a more serious case, as shown in Fig. 15.17d, particles 60 μm in diameter collide nearly vertically with the leading edge of the convex side of the rotating vane and may move against the flow after rebounding and impact the trailing edge of the convex side of the guide

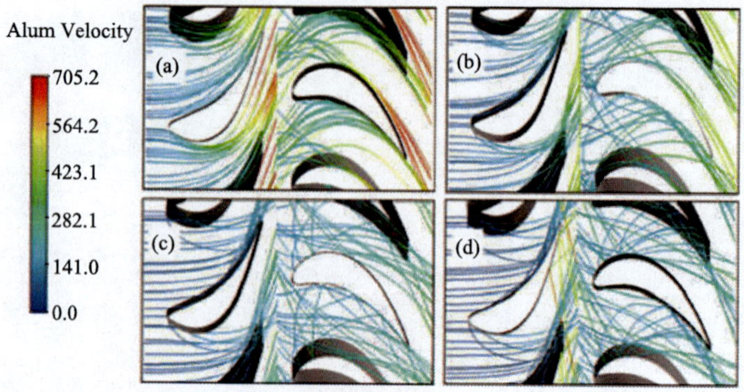

Fig. 15.17 Motion loci of particles with different diameters: **a** 2.5 μm, **b** 10 μm, **c** 25 μm, and **d** 60 μm

vane. As a result, these particles impact the vanes multiple times. The erosion effects of single particles on the TBC on the surface of the vanes are expected to be the most pronounced in this case.

Further, the E_m of the TBC on the vanes can be estimated through simulation by substituting particle parameters (e.g., velocity and size) into Eq. (15.62). Figure 15.18 shows the simulated erosion damage inflicted by particles that have a diameter with a mean of 40 μm and a standard deviation of 18 μm and follow the Rosin–Rammler distribution on the guide and rotating vanes of the engine. Figure 15.18 shows that the area of the convex side of the rotating vane close to the crankcase experiences the most severe erosion, followed by the trailing edge of the concave side of the rotating vane and the area of the trailing edge of the convex side of the guide vane close to the crankcase and, to an even lesser degree, the area of the concave side of the guide vane close to the trailing edge, where erosion damage is uniformly distributed along the height of the vane instead of being concentrated in the area close to the crankcase. This is because the high-speed rotation of the rotating vane around the turbine shaft inevitably imparts an increasing circumferential rotational speed to both the fluid that flows past it and the particles that the fluid carries; correspondingly, there is an increase in the centrifugal force on the particles, and as a result, the particles begin to converge toward the crankcase far from the shaft and impact the surface of the vane close to the crankcase. Similarly, the area of the convex side of the guide vane close to the crankcase sustains severe erosion, which is caused by the impact of large particles that move against the flow after impacting the leading edge of the convex side of the rotating vane and rebounding. Figure 15.19 shows photographs of the working vanes after a whole-turbine erosion test [54]. For the actual turbine, the coating in the front area of the convex side of the rotating vane close to the crankcase similarly sustained the most severe erosion damage, followed by that in the area of the trailing edge of the concave side of the rotating vane close to the crankcase. The simulation results agree well with the test results.

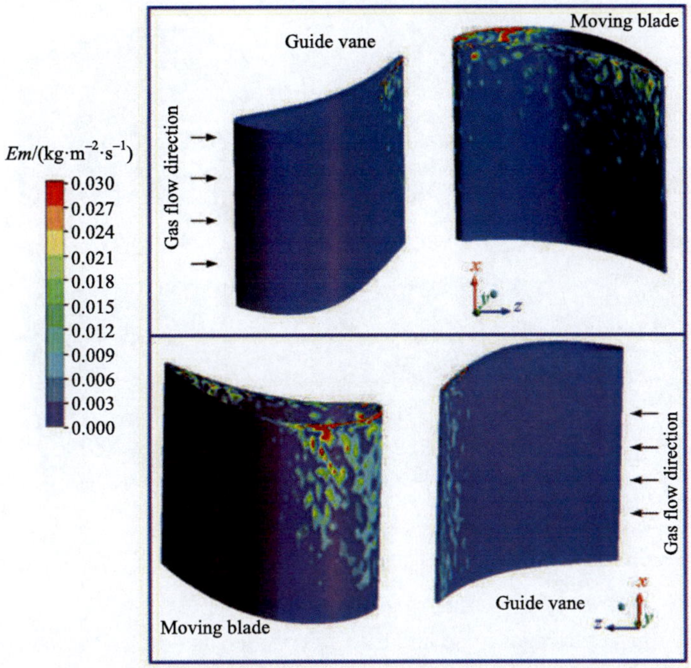

Fig. 15.18 Contour plots of the erosion rate E_m of the TBC on the vane surface

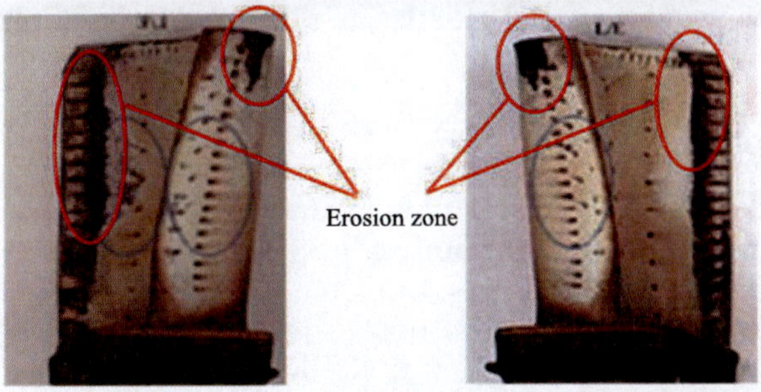

Fig. 15.19 Erosion conditions of the TBC on the surface of the vanes of a whole turbine after an erosion test

To further examine the influence of particle size on the erosion damage sustained by the TBC on the vane surface, erosion of the TBC caused by particles of different sizes was simulated. The results show that the erosion effect of particles on the coating decreases significantly when they are smaller than a certain size. The upper left image in Fig. 15.20 shows the simulated vane eroded by particles 2.5 μm in diameter. The eroded zone exhibits a regular radial pattern that is caused by the multiple impacts of the particles that move toward the trailing edge of the vane and the crankcase under

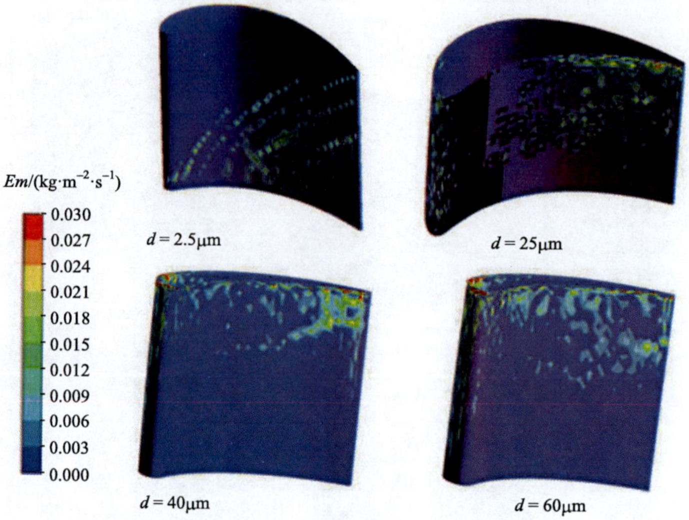

Fig. 15.20 Erosion damage of the TBC on a turbine blade caused by particles of different sizes

the action of the centrifugal force and the drag of the gas flow on the vane surface at small angles. Because these particles impact the vane surface at small angles, their erosion effects are relatively weak. There is no significant difference between the severely eroded zones caused by particles 10, 25, and 60 μm in diameter.

(2) **Numerical simulation and calculation of the erosion reliability of the TBC on a turbine blade**

From Figs. 15.18, 15.19 and 15.20, we know that the distributions of the erosion-induced failure locations and E_m of the TBC on a turbine blade are nonuniform and that local erosion of the TBC on a turbine blade does not necessarily indicate its failure. Therefore, the surface reliability of turbine blades needs to be defined from an overall perspective. Generally, a coating is considered to have lost its protective effect when its spalled area reaches 10% of its total area. Now that the TBC on a turbine blade is considered as a whole, we can transform the erosion reliability under impacts from single particles to the reliability under repeated impacts from multiple erosive particles at different angles, velocities, and locations for analysis while considering the erosion process of a single particle in the first part of the section as a subunit. Because E_m depends on factors such as the inlet pressure, temperature, and outlet pressure of the combustion gas, the rotational speed of the working vane, and the particle diameter, so does the erosion reliability. Then, the failure criterion in Eq. (15.64) can be written in the following implicit form: [55]

$$Z = H - M_{\text{p}} f(P_{\text{in}}, P_{\text{out}}, T_{\text{in}}, R_{\text{PM}}, D)\frac{K_e h}{2\Gamma} \quad (15.66)$$

where H is the thickness of the ceramic layer, M_p is the mass of the erosive particles, P_{in} is the total pressure at the inlet, P_{out} is the static pressure at the outlet, T_{in} is the total temperature at the inlet, R_{PM} is the rotational speed of the vane, K_e is the proportionality coefficient of energy, h is the thickness of each lamella in the ceramic layer, Γ is the fracture toughness of the ceramic layer, and D is the particle diameter. The analysis in the previous section shows that E_m is implicitly related to each of these parameters and thus also to the failure criterion, which is described using function f. However, specific values under these implicit relations can be obtained using the numerical calculation method for the erosion process discussed in the previous section.

As mentioned in Sect. 15.2.3, methods such as the SOSM method and Monte Carlo methods are no longer applicable when only the influencing factors can be determined from the failure criterion but there is a lack of specific equations. In this case, the MV and AMV methods introduced in Sect. 15.2.3 can be used to analyze the reliability. Because a turbine blade has a complex geometry and the erosion failure of its TBC is affected by many influencing parameters, NESSUS is also needed to calculate the reliability through the following main steps: [56]

(1) Analyze the distribution of each basic variable based on the failure criterion in Eq. (15.66).
(2) Establish a geometric model and grid for the TBC on the turbine blade, sample each basic variable, and input the results as the boundary conditions and material and structural parameters into the FE analysis. Determine whether the performance function is less than 0 under this condition using the AMV method, and return the result back to NESSUS for statistical analysis.
(3) Determine the proportion of the failure model, and calculate the reliability.
(4) Analyze the relation between the reliability and each basic variable, i.e., reliability sensitivity.

15.5.3 Statistical Analysis of Parameters Affecting Erosion Failure

Equation (15.66) shows that the erosion failure of a TBC depends primarily on its parameters as well as the particle and environmental parameters. Therefore, the basic variables for reliability analysis also include three types of parameters, namely (1) coating parameters (including the fracture toughness, elastic modulus, and total thickness of the coating as well as the thickness of each lamella), (2) particle parameters (including the diameter or weight of the particles), and (3) combustion-gas parameters (including the temperature and pressure of the combustion gas at the inlet as well as the rotational speed of the working vane, which affects the erosion angle and velocity of the particles as well as the locations at which the particles erode the coating).

Figure 15.21 shows the sampling locations for the total thickness of a coating as well as the results of regression analysis of the thickness measurements based on a normal distribution. Each data point corresponds to a sampled thickness measurement, and its x-axis value is the measured thickness of the coating. All the measurements are sorted in ascending order and denoted by i. Based on mathematical statistics, the quantile of the standard normal distribution corresponding to each sample is as follows:

$$Q_i = \Phi^{-1}\left(\frac{i - 0.5}{N}\right) \tag{15.67}$$

where $\Phi^{-1}(\cdot)$ is the inverse function of the cumulative probability function (CPF) of the standard normal distribution and N is the total number of sampled measurements. The plot in Fig. 15.21 is also called a quantile–quantile (QQ) plot. In a QQ plot, a higher degree of linearity of the data points indicates that the random variable is closer to a normal distribution, the slope of the fitted straight line is the standard deviation of the random variable, and the y-axis value of the point on the fitted straight line corresponding to the x-axis value of 0 is the mean of the random variable. Based on Fig. 15.21, the mean and standard deviation of the thickness of the coating were 268 and 11.46 μm, respectively.

The lamellar thickness of the lamellar structure of the TBC was analyzed using the same regression analysis method based on a normal distribution. Figure 15.22 shows the sampling of the thickness and an analysis of the thickness measurements. The lamellar thickness of the coating also follows a normal distribution with a mean of 5.2 μm and a standard deviation of 0.25 μm.

Based on the experience with the analysis of the brittleness of TBCs as well as their reliability under interfacial oxidation and thermal cycling, we know that their fracture toughness and Young's modulus follow a Weibull distribution. Table 15.14 summarizes the statistical characteristics of T_{in}, P_{in}, P_{out}, R_{PM}, and erosive-particle parameters determined from an extensive literature review.

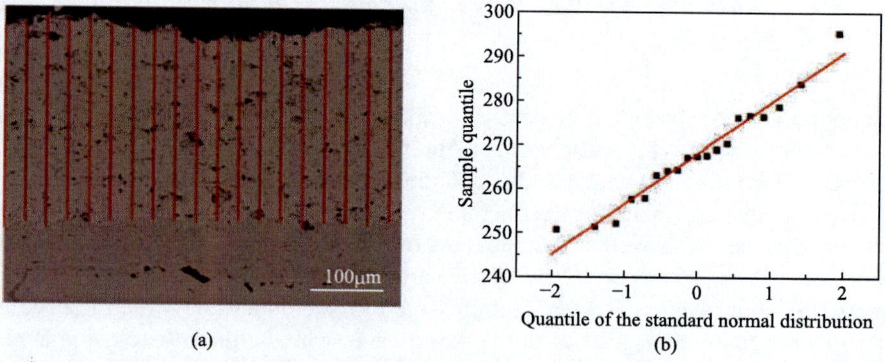

Fig. 15.21 Analysis of the thickness pattern of a ceramic layer: **a** sampling of the thickness; **b** regression analysis based on a normal distribution

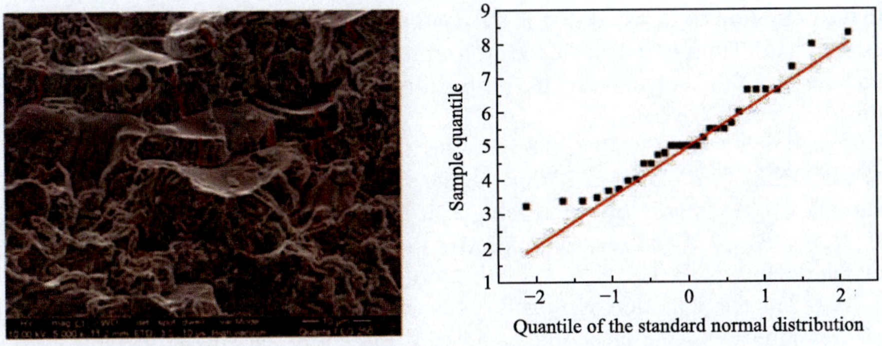

Fig. 15.22 Analysis of the thickness pattern of the lamellae in the ceramic layer: **a** sampling of the thickness; **b** regression analysis based on a normal distribution

Table 15.14 Selected numerical characteristics of relevant random variables in the erosion failure criterion

Name	Unit	Mean	Standard deviation	Distribution type
H	μm	268	11.46	Normal
h	μm	5.2	0.25	Normal
K_e [57]	1	0.026	0.01	Normal
Γ [58]	J/m2	50	10	Weibull
M_p	g	90	5	Normal
P_{in} [13, 51, 59]	kPa	1303	100	Beta
T_{in} [13, 51, 59]	K	1613	39.5	Beta
P_{out} [13, 51, 59]	kPa	578	20	Beta
R_{PM} [13, 51, 59]	r/min	11,710	249.5	Beta
D [60]	D	40	18	Log-normal

15.5.4 *Erosion Failure Probability Prediction and Sensitivity Analysis of TBCs on Turbine Blades*

The erosion process and erosion reliability can be calculated using the method described in Sect. 15.5.2 for each variable in Table 15.14 and each value near the mean value within one standard deviation. The results can be described as the distribution of the erosion failure probability $p_{f,e}$ of the TBC on a turbine blade under a certain parameter, as shown in Fig. 15.23a, which visually displays the locations where the TBC on a turbine blade tends to spall and fail under this set of parameters. In addition, it is possible to calculate the value of the performance function (i.e., failure criterion Z) under each parameter and to describe the variation in the reliability with the performance function, as shown by the CDF in Fig. 15.23b. According to the definition of Z, when $Z \leq 0$, the coating fails; when $Z = 0$, the coating is in a critical state. Thus, based on the location with $Z = 0$ in Fig. 15.23b, we can determine that the $p_{f,e}$ of the coating (i.e., the probability that 10% of the coating spalls) under this condition is 0.67%.

In addition, we can analyze the variation in $p_{f,e}$ with each basic variable, as shown in Fig. 15.24 (see Table 15.14 for the mean μ and standard deviation σ of each variable). $p_{f,e}$ increases with increasing h, M_p, and K_e and decreases with increasing Γ, H, P_{in}, P_{out}, T_{in}, and rotational speed of the blade (ω), while D has little influence on the reliability.

Based on the definition of reliability sensitivity, we can further analyze the extent of the influence of the mean and standard deviation of each basic variable on the reliability. Figure 15.25 shows the results. $p_{f,e}$ is most significantly affected by P_{out} and ω (the rotor speed), followed by H, h, and M_p, while changes in parameters

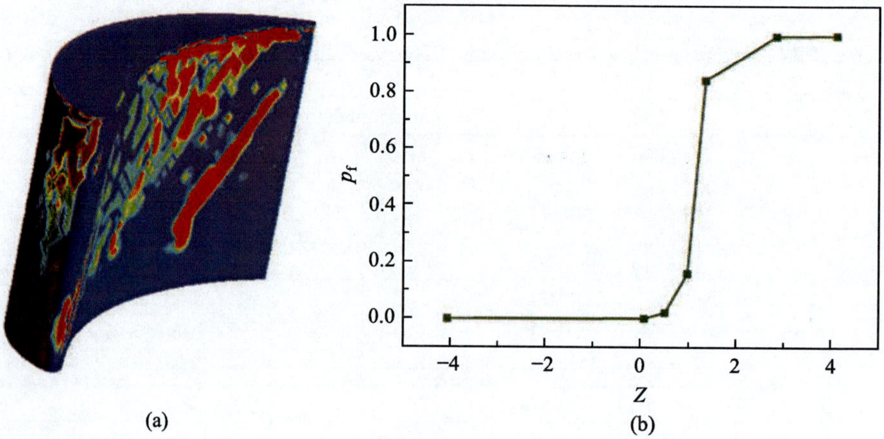

Fig. 15.23 $p_{f,e}$ of a TBC. **a** Distribution of $p_{f,e}$. **b** Cumulative distribution of $p_{f,e}$

Fig. 15.24 Variation in the $p_{f,e}$ of a TBC with basic variables

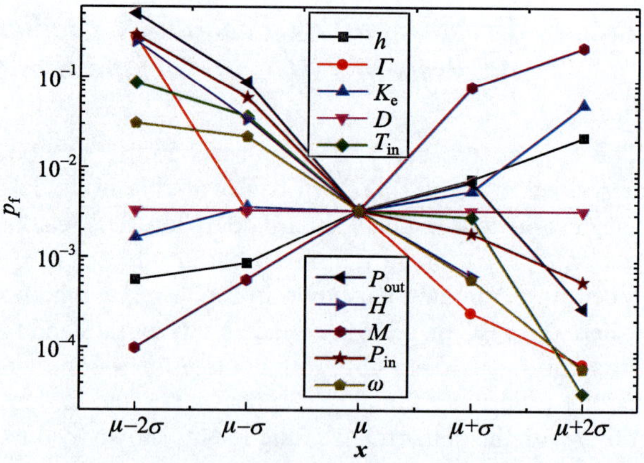

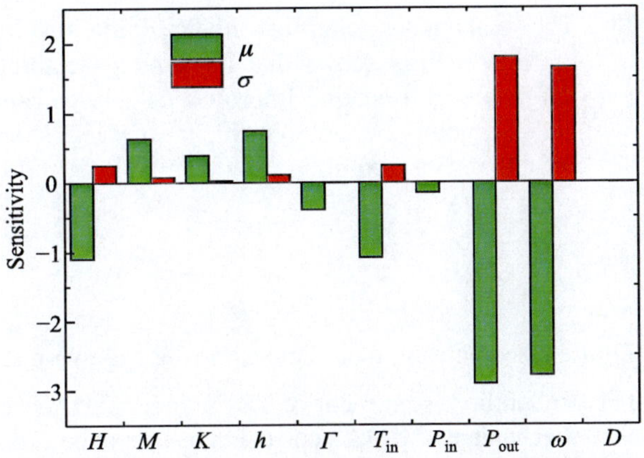

Fig. 15.25 Reliability sensitivity of the mean and standard deviation of each basic variable

such as T_{in} and Γ have little influence on $p_{f,e}$ and D has almost no influence on the spallation of the whole blade surface under erosion. During service, the P_{out} and ω of the turbine of a certain stage of an engine cannot be controlled. Thus, to improve the erosion resistance of a coating or its reliability under erosion conditions, it is possible to appropriately increase the coating thickness or reduce the thickness of each of the coating lamellae prepared by atmospheric plasma spraying (APS) through alteration of the preparation process while ensuring that other properties of the coating are unaffected.

15.6 Summary and Outlook

15.6.1 Summary

This chapter describes an assessment model and calculation methods for the reliability of TBCs in service, as well as analyzes the reliability of TBCs and their key influencing factors in key failure modes (e.g., thermal cycling, interfacial oxidation, and high-temperature erosion). The following is a summary of this chapter:

(1) A coating reliability prediction model that accounts for environmental, structural, and property dispersiveness is established, and the corresponding calculation methods are presented.
(2) The reliability analysis criteria for key TBC failure modes (e.g., thermal cycling, interfacial oxidation, and high-temperature erosion) are established, and key influencing factors are extracted.

(3) The analysis of reliability under oxidation, high-temperature erosion, and thermal cycling shows that increasing the thickness and interfacial fracture toughness of a coating improves its service reliability. In addition, appropriately reducing the lamellar thickness while keeping the total thickness of the coating unchanged improves the reliability of the coating.

15.6.2 Outlook

Future research should be focused on the following areas:

(1) Reliability assessment of TBCs under CMAS corrosion. Because the damage mechanism of CMAS corrosion has yet to be sufficiently studied and the knowledge about key failure modes and their criteria remains inadequate, the reliability of TBCs under high-temperature CMAS corrosion and its key influencing factors require further analysis and investigation.
(2) Reliability assessment of TBCs in multiple failure modes. The real-world service process of TBCs is accompanied with multiple failure modes (e.g., gas thermal shock, interfacial oxidation, CMAS corrosion, and erosion). Currently, reliability assessments are focused primarily on single modes. While concepts for the reliability analysis of multiple failure modes have been made available, further research is required to specifically carry out the prediction and analysis of the reliability under these conditions.
(3) Integrated assessment of reliability and thermal insulation performance. Reliability is the basis and necessary condition for ensuring the safe service of TBCs, while thermal insulation is an inevitable requirement for TBCs. However, the two objectives often compete with each other. For example, increasing porosity improves thermal insulation performance but may reduce reliability. Hence, carrying out integrated assessments of the reliability and thermal insulation performance of TBCs is an important development trend.

References

1. Busso EP, Lin J, Sakurai S. A mechanistic study of oxidation-induced degradation in a plasma-sprayed thermal barrier coating system part II: life prediction model. Acta Mater. 2001;49:1529–36.
2. Liu Y, Persson C, Wigren J. Experimental and numerical life prediction of thermally cycled thermal barrier coatings. J Therm Spray Technol. 2004;13(3):415–24.
3. Wen M, Jordon EH. Remaining life prediction of thermal barrier coatings based on photoluminescence piezospectroscopy measurements. J Eng Gas Turb Power. 2006;128:610–6.
4. He MY, Hutchinson JW, Evans AG. Simulation of stresses and delamination in a plasma-sprayed thermal barrier system upon thermal cycling. Mater Sci Eng A. 2003;345:172–8.

5. Renusch D, Schütze M. Measuring and modeling the TBC damage kinetics by using acoustic emission analysis. Surf Coat Technol. 2007;202:740–4.
6. Mao S, Tang Y, Wang L. Reliability statistics. Beijing: Higher Education Press; 2008.
7. Shen W, Wang FC, Fan QB, Ma Z, Yang XW. Finite element simulation of tensile bond strength of atmospheric plasma spraying thermal barrier coatings. Surf Coat Technol. 2011;205:2964–9.
8. Ren F, Case ED, Timm EJ, Jabobs MD, Schock HJ. Weibull analysis of the biaxial fracture strength of a cast p-type LAST-T thermoelectric material. Philos Mag Lett. 2006;86(10):673–82.
9. Cattell MJ, Clarke RL, Lynch EJ. The transverse strength, reliability and microstructural features of four dental ceramics-Part I. J Dent. 1997;25(5):399–407.
10. Darija M, Dragan J, Mirta B. Nonlinear weighted least squares estimation of a three-parameter Weibull density with a nonparametric start. J Comput Appl Math. 2009;228:304–12.
11. Wang L, Wang Y, Sun XG, He JQ, Pan ZY, Wang CH. Microstructure and indentation mechanical properties of plasma sprayed nano-bimodal and conventional ZrO_2-8wt%Y_2O_3 thermal barrier coatings. Vacuum. 2012;86:1174–85.
12. Lugscheider E, Bobzin K, Barwulf S, Etzkorn A. Mechanical properties of EB-PVD-thermal barrier coatings by nanoindentation. Surf Coat Technol. 2001;138:9–13.
13. Holland FA. A simple method for estimating the parameters of the Beta distribution applied to modeling uncertainty in gas turbine inlet temperature. In: Proceedings of the ASME turbo expo 2002: power for land, sea, and air, F, 2002. American Society of Mechanical Engineers Digital Collection.
14. Macıas-Garcıa A, Cuerda-Correa EM, Dıaz-Dıez M. Application of the Rosin-Rammler and Gates–Gaudin–Schuhmann models to the particle size distribution analysis of agglomerated cork. Mater Charact. 2004;52(2):159–64.
15. Lu Z, Song S, Li H, Yuan X. Analysis of structural reliability and reliability sensitivity. Beijing: Sciences Press; 2009.
16. Gong J. Computational methods for the reliability of engineering structures. Dalian: Dalian University of Technology Press; 2003.
17. Hu M. A study on methods for calculating structural reliability. Master's thesis. Guangzhou: South China University of Technology; 2010.
18. Zhao G, Jin W, Gong J. Theory of structural reliability. Beijing: China Architecture & Building Press; 2000.
19. Rackwitz R, Fiessler B. Structural reliability under combined random load sequences. Comput Struct. 1978;9(5):489–94.
20. Breitung K. Asymptotic approximation for multinormal integrals. J Eng Mech. 1984;110(3):357–66.
21. Breitung K. Asymptotic approximation for probability integrals. Probab Eng Mech. 1989;4(4):187–90.
22. Jin Y, Jiang Y, Li X. Sampling techniques. Beijing: China Renmin University Press; 2002.
23. Xiong G, Xiao T, Zhang Y. Simulation of continuous systems and discrete event systems. Beijing: Tsinghua University Press; 2003.
24. Lohr SL. Sampling: design and analysis. Pacific Grove: Duxbury Press; 1999.
25. Thoft-Cristensen P, Baker MJ. Structural reliability theory and its applications, Springer Science & Business Media; 2012.
26. Tilmann B, Roland H, Trunova O. Damage mechanisms and lifetime behavior of plasma-sprayed thermal barrier coating systems for gas turbines-part II: modeling. Surf Coat Technol. 2008;202:5901–8.
27. Sfar K, Aktaa J, Munz D. Numerical investigation of residual stress fields and crack behavior in TBC systems. Mater Sci Eng, A. 2002;333:351–60.
28. Thurna G, Schneider GA, Bahr HA, Aldinger F. Toughness anisotropy and damage behavior of plasma sprayed ZrO_2 thermal barrier coatings. Surf Coat Technol. 2000;123:147–58.
29. Wang X, Wang XJ, Atkinson A. Interface fracture toughness in thermal barrier coatings by cross-sectional indentation. Acta Mater. 2012;60:6152–63.

30. Mao WG, Chen Q, Dai CY, Yang L, Zhou YC, Lu CS. Effects of piezo-spectroscopic coefficients of 8 wt.% Y_2O_3 stabilized ZrO_2 on residual stress measurement of thermal barrier coatings by Raman spectroscopy. Surf Coat Technol. 2010; 204:3573–77.
31. Guo SQ, Kagawa Y. Effect of thermal exposure on hardness and Young's modulus of EB-PVD yttria-partially-stabilized zirconia thermal barrier coatings. Ceram Int. 2006;32:263–70.
32. Ma K, Schoenung JM. Isothermal oxidation behavior of cryomilled NiCrAlY bond coat: homogeneity and growth rate of TGO. Surf Coat Technol. 2011;205:5178–85.
33. Che C, Wu GQ, Qi HY, Huang Z, Yang XG. Uneven growth of thermally grown oxide and stress distribution in plasma-sprayed thermal barrier coatings. Surf Coat Technol. 2009;203:3088–91.
34. Shen W, Wang FC, Fan QB, Ma Z. Lifetime prediction of plasma-sprayed thermal barrier coating systems. Surf Coat Technol. 2013;217:39–45.
35. Tomimatsu T, Zhu S, Kagawa K. Effect of thermal exposure on stress distribution in TGO layer of EB-PVD TBC. Acta Mater. 2003;51:2397–405.
36. Busso EP, Wright L, Evans HE, McCartney LN, Saunders SR, Osgerby S, Nunn J. A physics-based life prediction methodology for thermal barrier coating systems. Acta Mater. 2007;55:1491–14919.
37. Wu F, Jordan EH, Ma X, Gell M. Thermally grown oxide growth behavior and spallation lives of solution precursor plasma spray thermal barrier coatings. Surf Coat Technol. 2008;202:1628–35.
38. Miller RA. Oxidation-based model for thermal barrier coating life. J Am Ceram Soc. 1984;67(8):517–21.
39. Evans HE. Oxidation failure of TBC systems: an assessment of mechanisms. Surf Coat Technol. 2011;206:1512–21.
40. Demasi JT, Sheffler KD, Ortiz M. Thermal barrier coating life prediction model development phase I final report. NASA Technical Memorandum, 182230; 1989.
41. Guo JW, Yang L, Zhou YC, He LM, Zhu W, Cai CY, Lu C. Reliability assessment on interfacial failure of thermal barrier coatings. Acta Mech Sin. 2016;32(5):915–24.
42. Guo SQ, Kagawa Y. Effect of thermal exposure on hardness and Young's moduli of EB-PVD yttria-partially-stabilized zirconia thermal barrier coating system. Ceram Int. 2006;32:263–70.
43. Zhou CG, Wang N, Wang ZB, Gong SK, Xu HB. Thermal cycling life and thermal diffusivity of a plasma-sprayed nanostructured thermal barrier coating. Scripta Mater. 2004;51:945–8.
44. Jackson RD. The effect of bond coat oxidation on the microstructure and endurance of two thermal barrier coaings. The University of Birmingham, UK; 2010.
45. Yang L, Guo J, Zhu W, Zhou Y, Cai C. A method for assessing the reliability of thermal barrier coatings against interfacial oxidation failure based on the JC algorithm. Chinese National Invention Patent (No. ZL201310142934.4).
46. Martena M, Botto D, Fino P, Sabbadini S, Gola MM, Badini C. Modelling of TBC system failure: stress distribution as a function of TGO thickness and thermal expansion mismatch. Eng Fail Anal. 2006;13:409–26.
47. Li CJ, Yang GJ, Ohmori A. Relationship between particle erosion and lamellar microstructure for plasma-sprayed alumina coatings. Wear. 2006;260:1166–72.
48. Nicholls J, Deakin M, Rickerby D. A comparison between the erosion behaviour of thermal spray and electron beam physical vapour deposition thermal barrier coatings. Wear. 1999;233:352–61.
49. Tabakoff W. High-temperature erosion resistance of coatings for use in turbomachinery. Wear. 1995;186:224–9.
50. Xiao Y. A study on the reliability assessment of thermal barrier coatings on turbine blades and their application. Doctoral Dissertation, Xiangtan University, 2019.
51. Timko L. Energy efficient engine high pressure turbine component test performance report. Aircr Propuls Power. 1984;199(19):237.
52. Zeng J, Wang B, Qing X. Three-dimensional analysis of two-stage high pressure turbine. J Aerosp Power. 2012;27(11):2553–6.
53. Swar R. Particle erosion of gas turbine thermal barrier coating. University of Cincinnati; 2009.

54. Darolia R. Thermal barrier coatings technology: critical review, progress update, remaining challenges and prospects. Int Mater Rev. 2013;58(6):315–48.
55. Xiao YQ, Yang L, Zhou YC, Wei YG, Wang NG. Dominant parameters affecting the reliability of TBCs with turbine blade substrate during the erosion of hot gas stream. Wear. 2017;390–391:166–75.
56. Yang L, Xiao Y, Zhou Y, Zhu W. An erosion rate model for thermal barrier coatings and a method for simulating the erosion conditions of coating-containing turbine blades. Chinese National Invention Patent (No. ZL201610256953.3).
57. Cernuschi F, Lorenzoni L, Capelli S, et al. Solid particle erosion of thermal spray and physical vapour deposition thermal barrier coatings. Wear. 2011; 271(11):2909–18.
58. Thurn G, Schneider GA, Bahr HA, et al. Toughness anisotropy and damage behavior of plasma sprayed ZrO_2 thermal barrier coatings. Surf Coat Technol. 2000;123(2):147–58.
59. Pech K, Downing N. Development of a retracting vane/centrifugal main fuel pump for gas turbine engines. J Fluids Eng. 1976;98(4):619–25.
60. Kirschner M, Wobst T, Rittmeister B, et al. Erosion testing of thermal barrier coatings in a high enthalpy wind tunnel. J Eng Gas Turbines Power. 2015;137(3): 032101.

Chapter 16
Experimental Simulators for the Service Environments of TBCs

Experimental simulation and testing facilities for the key service environments of thermal barrier coatings (TBCs) are essential for Estudying failure mechanisms and reliability. However, extreme environments (e.g., gas thermal shock, erosion, calcium–magnesium aluminosilicate (CMAS) corrosion, and high-speed rotation) cannot be effectively simulated using conventional loading methods such as tensile, compressive, and flexural loading or high-temperature furnace heating. In the 1970s, NASA verified the effectiveness of TBCs through test runs on a J-75 turbine engine and, on this basis, optimized their composition and structural design [1]. However, test runs on turbines are extremely costly and are unable to generate test data on the evolution of damage parameters in the failure process. Therefore, the development of experimental simulators for service environments is a key and topical area of research on TBCs around the world and a core area on which technological embargos are often imposed.

The extreme environments in which TBCs operate can be grouped into thermal, mechanical, and chemical loads as well as their coupling, including the following. (1) Temperature fields. Gas thermal shock, film cooling, and their resulting temperature gradients and extreme cold and hot conditions all cause changes in the temperature field and represent thermal loads. (2) Stress fields. High-speed rotational centrifugal forces and the impact from erosive particles represent mechanical loads. (3) Chemical fields. Interfacial oxidation and CMAS corrosion are chemical reactions and represent chemical loads. (4) Coupling effects of the above three types of fields. For example, the thermal mismatch stresses induced by a temperature field; the growth stresses caused by chemical reactions; and the hot spots, turbulent flow, and wakes resulting from the interaction between rotation (a mechanical load) and the fuel gas (temperature) generate new thermal and mechanical loads. Therefore, experimental simulation of TBC service environments is in fact the simulation and control of the decomposition and coupling of thermal, mechanical, and chemical loads and their coupling. In addition, real-time testing of the failure process of coatings in service

Y. Zhou et al., *Thermal Barrier Coatings: Failure Theory and Evaluation Technology*,
https://doi.org/10.1007/978-981-19-2723-2_16

environments produces direct test data that facilitate an understanding of their failure mechanisms and reliability and is a key and difficult area to be considered in the development of experimental simulators.

Given the above issues, this chapter describes in detail the concepts and progress in the design of simulators for TBC service environments in a progressive manner, from thermal loading (temperature fields) to thermomechanical loading (stress fields), to thermo–mechano-chemical loading (chemical fields), and to thermal–mechano-chemical coupling that accounts for high-speed rotation, in combination with the implementation of the methods for testing the failure process of TBCs in service environments in real time.

16.1 Experimental Simulators for Thermal Loads on TBCs

As the most basic service condition for TBCs, high temperatures can promote various TBC failure modes (e.g., thermal cycling, thermal mismatch stress, creep, and thermal fatigue) and induce interfacial oxidation in TBCs. Applying CMAS to the surface of a coating can also lead to corrosion failure. Therefore, high temperatures are the most basic, simplest service condition simulated in TBC tests. This section is focused on the introduction of high-temperature thermal cycling and high-temperature CMAS corrosion testing facilities for TBCs.

16.1.1 Experimental Simulator for Automatic High-Temperature Thermal Cycling

High-temperature furnaces are the simplest, most common experimental devices used to simulate the high-temperature environments to which TBCs are subjected. The pattern and mechanism of the interfacial oxidation of a TBC can be analyzed by controlling the oxygen content (e.g., by adding oxygen or by exhausting oxygen with inert gases) in a high-temperature furnace. Methods such as heating and holding temperatures in high-temperature furnaces as well as cooling outside furnaces (by natural cooling in air, shock cooling with cold air, and water cooling) can be employed to simulate the alternating hot and cold service environments in which TBCs are subjected, i.e., the thermal load to which TBCs on aircraft turbine blades are subjected during takeoff, flight, and landing. However, the automatic mode of operation of thermal cycling needs to be considered for TBCs because they generally do not spall until after thousands or even tens of thousands of thermal cycles. The process of heating in a furnace, cooling outside the furnace, and counting in a manual manner is extremely time-consuming. Fully automatic operation (heating, temperature holding, and cooling) and counting can significantly improve the experimental efficiency. In

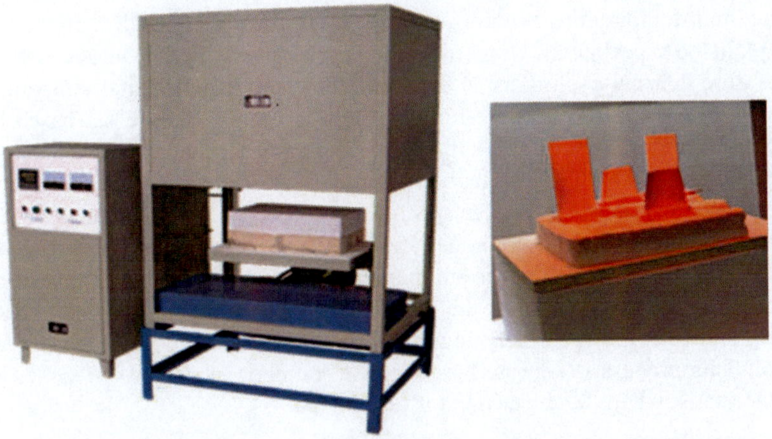

Fig. 16.1 An automatic high-temperature thermal-cycling furnace

addition, long-term stable, reliable operation is an important factor that must be considered in high-temperature experiments on TBCs.

In the following, we introduce the basic components and implementation of an automatic high-temperature thermal-cycling furnace independently developed by our research team. This furnace consists primarily of three major modules, namely a high-temperature heating system, a power system, and a program control system (Fig. 16.1).

(1) High-temperature heating system (i.e., high-temperature furnace body). Similar to the case of a conventional resistance furnace, different heating rods (e.g., silicon molybdenum rods, which are capable of operating for an extended period of time at 1600 °C) can be selected for the furnace body based on the required experimental temperature. The required experimental temperature inside the chamber is ensured by setting the number of silicon molybdenum rods and by conducting temperature measurements using thermocouples.
(2) Automatic closing drive system for the furnace door. Automatic closing of the furnace door can be realized through a mechanical lift. The lifting time, distance, and speed are all controlled through a control system.
(3) Automatic control system. Realized through a programmable logic controller (PLC), this system performs logical operations to open and close the furnace door and executes commands (e.g., sequential control, time keeping, counting, and arithmetic operations) through an internally stored time-control program, as well as controls the processes in which the furnace door is closed and opened through digital input/output (e.g., time and number of cycles). In addition, automatic temperature control inside the chamber is realized through thermocouple-based temperature measurements and heating power control.

This high-temperature furnace can be used to perform simulation experiments with a single temperature as the primary load, such as high-temperature thermal cycling, high-temperature oxidation, and high-temperature CMAS corrosion. Here, we explain the experimental implementation procedure using this furnace with high-temperature thermal cycling as an example. First, the experimental temperature as well as the heating, temperature-holding, and cooling times are set based on the experimental requirements. Then, the high-temperature furnace is turned on to allow the required temperature to be reached inside the furnace, after which the automatic cycling system is started. Next, the specimen is placed on the specimen stage after it lowers. Subsequently, the specimen is automatically brought into the furnace by the drive system. The furnace door is automatically closed, followed by the heating and temperature-holding phases. At the preset time, the furnace door is automatically opened, and then the specimen stage lowers, after which the specimen is allowed to cool. This process is repeated until the set number of cycles is reached or the specimen fails, when the experiment is terminated.

16.1.2 Facilities for Measuring the High-Temperature Contact Angle of CMAS During the CMAS Corrosion Process

As mentioned earlier, high-temperature CMAS corrosion is the primary risk factor causing TBC spallation. The failure process of a TBC under CMAS corrosion is in fact a process in which CMAS becomes adsorbed onto the coating, melts, and then infiltrates into and corrodes the coating, causing the coating to undergo microstructural changes, phase transformation, and deformation and to accumulate damage [2–5]. The less CMAS is adsorbed onto the coating surface, melts, and infiltrates into the interior of the coating, the safer it is. In addition to the control of service conditions such as the CMAS content (the total amount available for adsorption) and temperature (which determines whether CMAS can melt), another important mode of control is to equip the coating itself with the ability to resist CMAS adsorption and infiltration. For example, drawing inspiration from hydrophobic lotus leaves, the surface morphology of a coating can be optimized to reduce the infiltration ability of CMAS, i.e., to increase the contact angle. However, to realize the adsorption resistance of a coating, it is necessary to first quantitatively characterize the contact angle between the coating and CMAS.

(1) Basic principle of contact angle measurement

The contact angle is measured based on the principle of the droplet method [6], as shown in Fig. 16.2. In the figure, R, H, and θ are the radius, height, and contact angle of the droplet, respectively, and points A(x_a, y_a), B(x_b, y_b), and C(x_c, y_c) are the highest point of the droplet and the endpoints of the contact surface of the solid (coating), respectively. The contact angle can be described as the maximum angle between the outline of the droplet and the surface of the solid. When the surface of

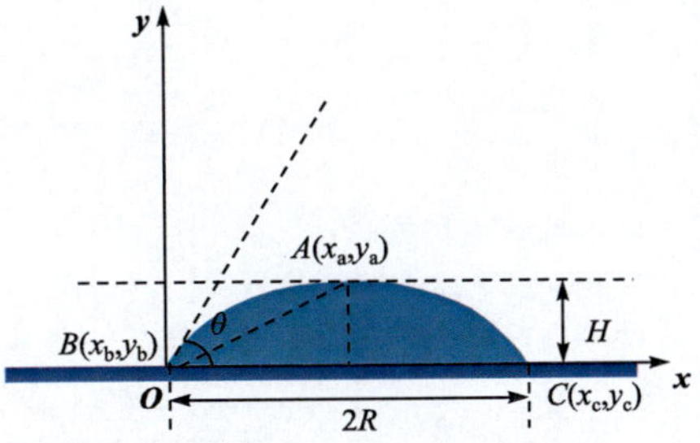

Fig. 16.2 Schematic diagram of the principle of the droplet method for contact angle measurement

the solid is horizontal and the droplet is in an equilibrium state, the left and right endpoints of the contact surface have the same y-axis coordinate, i.e., $y_b = y_c$. Let $2R$ be the diameter of the droplet that spreads on the substrate. Then, the contact angle between the droplet and substrate is expressed as follows:

$$\frac{\theta}{2} = \tan^{-1}\left(\frac{H}{R}\right) \tag{16.1}$$

(2) Design concept and implementation of facilities for measuring the high-temperature contact angle of CMAS

For CMAS that melts on the surface of a TBC and then infiltrates into its interior, three core issues need to be addressed to measure the contact angle of CMAS, θ_{CMAS}. (1) Melting CMAS. As mentioned earlier, CMAS, which is a solid mixture of calcium, magnesium, and aluminum silicates, needs to be melted into droplets at temperatures above 1250 °C. (2) Dripping CMAS. Even when melted, CMAS remains an extremely viscous medium that cannot be dripped with conventional fluid containers. In addition, a turbine blade is not flat. As a result, molten CMAS may drip onto the surface of the coating of a turbine blade at different angles. Therefore, it is necessary to drip CMAS onto the coating surface at different angles. (3) High-temperature imaging. Based on Eq. (16.1), to obtain the contact angle, it is necessary to measure the width and height of the liquid in contact with the coating surface. Therefore, the morphology during the high-temperature dripping process, particularly during the spreading process on the coating surface, needs to be imaged to determine the contact width and height.

However, the available contact angle measurement facilities are primarily designed for room temperature and do not meet the θ_{CMAS} measurement requirements. To solve the above three core issues, we developed a facility for measuring the high-temperature θ_{CMAS} of TBCs [6]. Figure 16.3 shows a schematic diagram describing the principle of this facility. The facility consists of a high-temperature heating system, a CMAS-loading module, a specimen test bed, and an imaging system. A TBC specimen is placed on the specimen test stage. Solid CMAS is fixed above the coating through the CMAS-loading module. Then, the location, height, and angle of the CMAS and coating are adjusted. The CMAS-loading module and

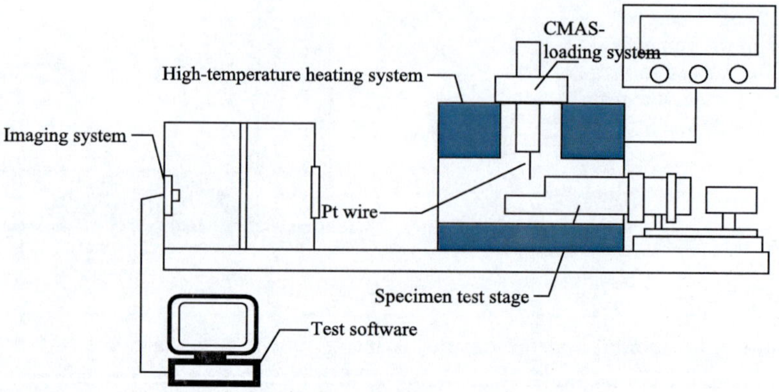

Fig. 16.3 Schematic diagram of a facility for measuring the θ_{CMAS} of TBC

test stage are heated simultaneously through the high-temperature heating system to melt the CMAS and to drip the molten CMAS onto the coating surface along a Pt wire contained in the loading module. The chamber of the high-temperature heating system is equipped with a transparent test window made of quartz glass. A CCD imaging system is placed at the window to facilitate the real-time observation and testing of the wetting of the coating surface by the molten CMAS as well as the recording of the CMAS-spreading process. A program written in MATLAB is used to calculate the width and height of the CMAS after it spreads on the coating surface. Subsequently, θ_{CMAS} is calculated using Eq. (16.1).

In the following, we briefly describe the implementation of the key modules.

High-temperature heating system. This system is composed of a tube-type resistance furnace with a maximum temperature of 1300 °C, and its temperature and heating rate can be precisely controlled to melt CMAS mixtures with different melting points due to different compositions. An opening is prepared at the top of the furnace chamber for the block CMAS-loading module. In addition, an access opening is prepared on the right-hand side of the furnace chamber for the specimen test bed. During the test process, a closed environment can be established inside the furnace chamber through the seal covers of the loading module and test bed. The distance between the CMAS and coating should be minimized to reduce the impact of the gravitational acceleration. The left-hand side of the furnace chamber is equipped with an image observation and acquisition window (made of quartz glass capable of filtering out light noise) to facilitate the observation of changes in CMAS droplets.

Specimen test bed. As shown in Fig. 16.4, the specimen test stage includes a base, specimen-loading stage with a U-shaped groove, and a thermocouple. In real-world service environments, CMAS does not necessarily vertically drip onto coating surfaces. Therefore, a U-shaped specimen stage is designed to allow the coating specimen to be placed at a certain angle relative to the U-shaped groove. This way, the contact angle between the droplet and the surface of the specimen can be adjusted. A Pt/Rh thermocouple is placed underneath the base to measure the temperature.

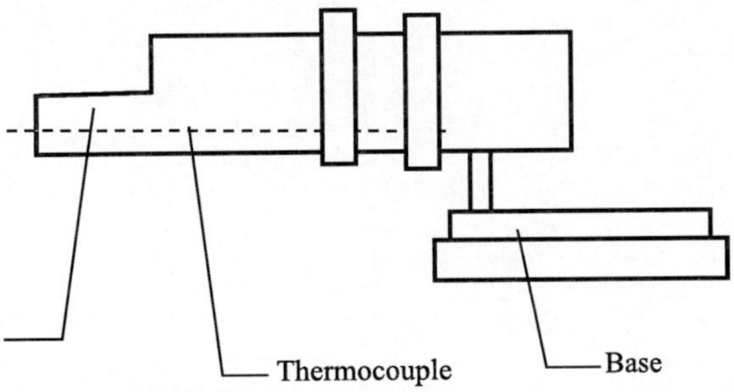

Fig. 16.4 Schematic diagram of the specimen test bed

CMAS-loading module. As shown in Fig. 16.5, this module includes an end cap that fixes the Pt wire used to fix the CMAS, a height-adjustment module, and an alundum tube used to protect the Pt wire. The suspension height of the CMAS is adjusted through a nut in the height-adjustment module to control the height from which the CMAS drips. Because Pt wires are stable and do not react with CMAS at high temperatures, a Pt wire is used to wind solid CMAS. The contact area between the Pt wire and CMAS is reduced to weaken the adhesive force between them and, thereby, to fix the CMAS above the coating. An alundum tube is used to fix and protect the Pt wire, thereby addressing the difficulty in dripping molten CMAS.

Imaging system. This system includes a parallel light source, a camera obscura, a convex lens, an imaging screen, and a CCD camera. Based on the pinhole imaging principle, the specimen is first magnified by the convex lens and then reversely imaged and projected on the imaging screen. The image is subsequently collected with a 5 M CCD camera. Figure 16.6 shows the optical path. A grid is prepared on the imaging screen to facilitate quantitative observation of the volume and morphology of the CMAS droplet.

Fig. 16.5 Schematic diagram of the CMAS-loading module

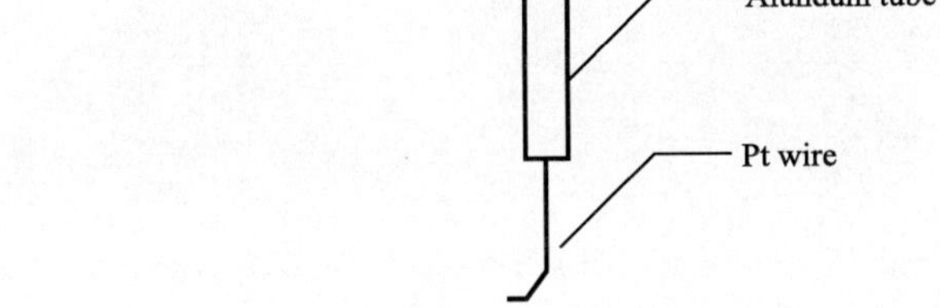

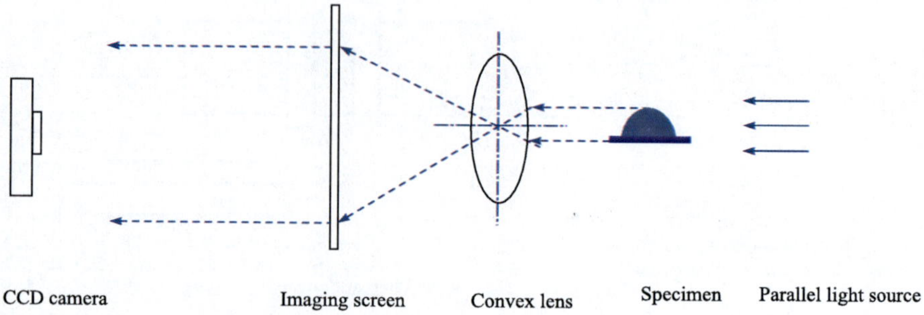

Fig. 16.6 The optical path of the imaging system

Contact angle analysis software. Contact angle calculation based on Eq. (16.1) requires accurate determination of the spreading width and height. If the obtained image has a blurry boundary due to factors such as high-temperature thermal radiation, the outline can be extracted by binarization, Zheng [33] as shown in Fig. 16.7. See Sect. 3.4 of Chap. 3 for the detailed method. After the outline is extracted, the width and height of the spreading area of the CMAS droplet are measured and substituted into Eq. (16.1) to yield a contact angle θ_{CMAS} value. Through the real-time recording of the CMAS-spreading process, a curve depicting the evolution of θ_{CMAS} with time is obtained. On this basis, a stable θ_{CMAS} value is eventually obtained.

(3) Performance of the facility in measuring the θ_{CMAS} of TBCs

Using this facility, we tested the wetting of a TBC by CMAS at a high temperature. CMAS was compressed through static pressure molding into a solid block, which was then wound with a Pt wire protected and fixed by an alundum tube. An atmospheric plasma spraying (APS) TBC was placed on the U-shaped test stage. Subsequently, both the test stage and CMAS were placed in the furnace and then heated to and held at 1250 °C until the CMAS melted and dripped. The imaging system was used to image this process in real time, as shown in Fig. 16.8. In addition, the spreading width and height were measured to determine the evolution of θ_{CMAS} with time [7].

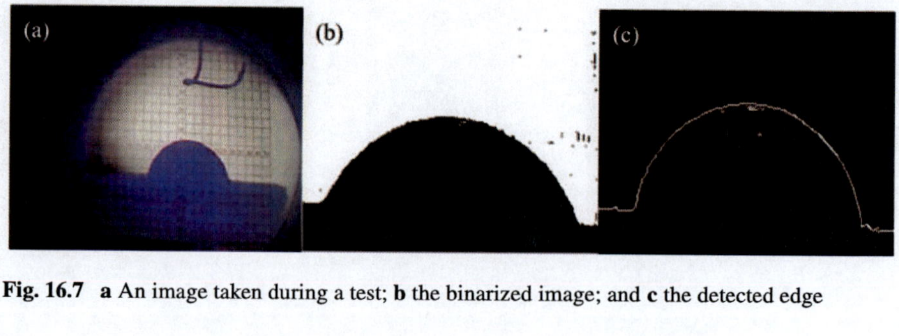

Fig. 16.7 **a** An image taken during a test; **b** the binarized image; and **c** the detected edge

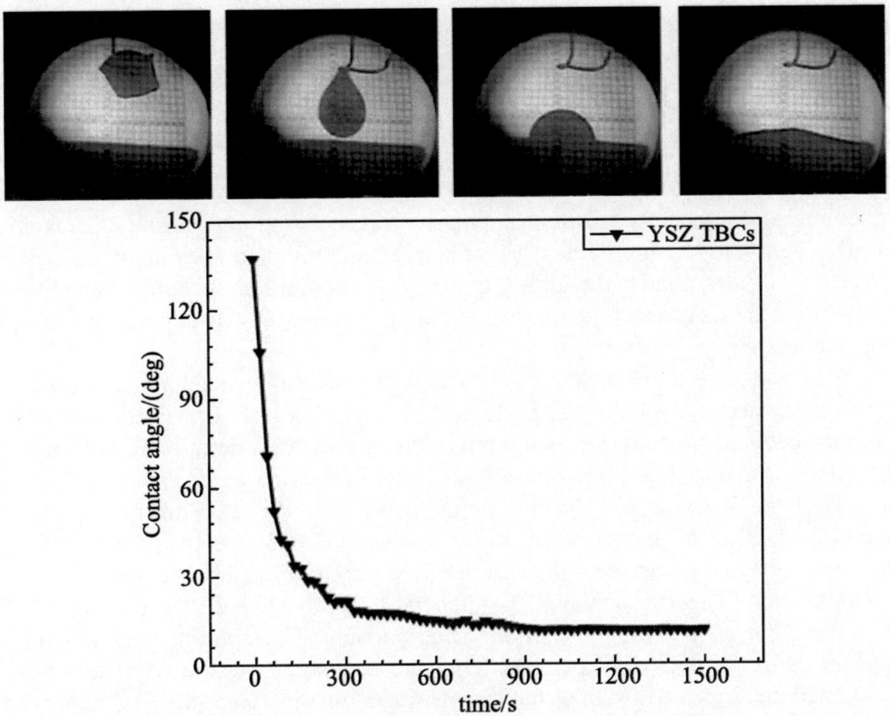

Fig. 16.8 Melting and dripping of CMAS, spreading of the molten CMAS on the coating surface, and evolution of the contact angle θ_{CMAS} [7]

16.2 Combined Thermomechanical Loading Facility for TBCs and Buckling Failure Mechanism of TBCs Under Thermomechanical Loading

During the service process of a TBC, the difference between the coefficients of thermal expansion (CTEs) of the ceramic layer, thermal growth oxide (TGO) layer, and substrate results in the formation of thermal mismatch stresses. In addition, TGO growth, CMAS corrosion, and phase transformation can cause stresses. In particular, high-speed rotational centrifugal forces generate large tensile stresses within the system. Application of mechanical loads (e.g., tensile, compressive, and bending loads) at high temperatures can simulate and accelerate the simulation of thermal mismatch stresses in TBCs and centrifugal forces on TBCs and was an important means for studying TBC failure mechanisms in earlier studies. In addition, through tension, compression, and bending tests at high temperatures, the mechanical properties of coatings and their interfaces can be quantitatively characterized. Thus, combined thermomechanical loading facilities offer an effective means for studying TBCs. This section introduces experimental methods for simultaneously applying high temperatures and mechanical loads as well as important relevant conclusions.

16.2.1 Combined Thermomechanical Loading Testing Facilities

(1) Material testing machine + combustion-gas heating + real-time testing

The basic principle of this mode of loading is described as follows. A material testing machine is used to apply tensile, compressive, and bending loads to a TBC. On this basis, a high-temperature flame is generated through rapid combustion of acetylene and oxygen and applied to the surface of the TBC. Then, a thermocouple is employed to measure the temperature, based on which the fuel gas flow is adjusted to control the temperature.

As shown in Fig. 16.9 [8], a universal material testing machine was used to apply a compressive load (or a tensile load or a bending load as needed) to a TBC. An oxyacetylene flame (generated through the ignition of a mixture of oxygen and acetylene, which was supplied by a mixer pipe connected to an oxygen pipe and an acetylene pipe, at a certain ratio adjusted via a valve) acted on the surface of the coating. Thermocouples were used to measure the surface temperature of the coating, the bottom surface temperature of the substrate, and the temperature at a cross-section (T_3), based on which the temperature field was adjusted and controlled.

During the test process, in addition to thermocouple-based temperature measurements, real-time detection methods (e.g., the acoustic emission (AE) technique discussed in Chap. 11 as well as the high-temperature digital speckle technique and infrared thermography (IRT) introduced in Chap. 13) can be used to quantitatively characterize the TBC failure process in real time.

(2) Material testing machine + high-temperature furnace + real-time testing [9]

There are two primary problems in heating the surface of a coating with a fuel gas: (1) the surface temperature of the coating is nonuniform and (2) even an extremely small

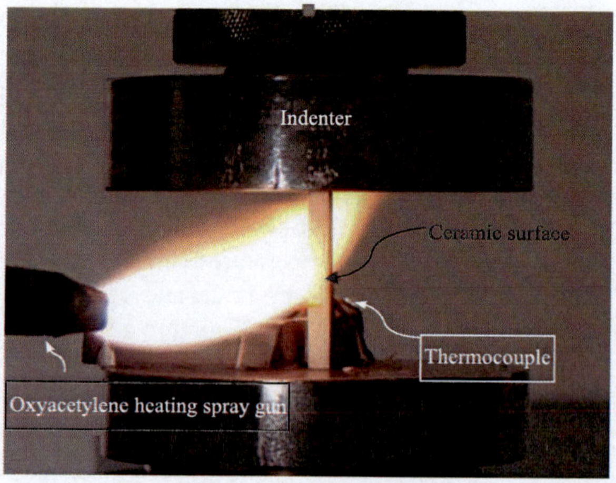

Fig. 16.9 Implementation of thermomechanical loading on a TBC [8]

difference in the fuel gas flow can cause a change in the surface temperature of the coating by several tens of or even over a hundred degrees. More importantly, simply constructed combustion-gas spray guns often have deficiencies such as uncontrollable temperature, a low degree of automation, and poor safety and reliability. To address these issues, a high-temperature furnace can be installed at the clamps of a material testing machine to realize precise loading and temperature control for the specimen. To achieve real-time testing of the TBC failure process, a testing window needs to be designed for the furnace body. For example, a quartz glass window can be used to enable the testing of coating deformation using the high-temperature digital speckle technique. As another example, the waveguide wire technique (see Chap. 11) can be employed to realize real-time AE detection of crack initiation and propagation.

Figure 16.10 shows a schematic diagram of the combined thermomechanical loading with high-temperature furnace heating as the heating mode and high-temperature three-point bending loading as a mechanical loading example [9] as well as real-time testing. The facility includes a universal testing machine (as a loading system), a high-temperature furnace, an ARAMIS optical strain measurement system (digital image correlation (DIC)), a high-temperature pressure bar and clamps, and an observation zone. During the test, the high-temperature furnace is used to heat the TBC specimen coated with high-temperature speckles while, at the same time, a compressive load is applied to the TBC specimen to promote its spallation failure. In addition, the digital speckle technique is used to measure this process in real time. Alternatively, detection techniques (e.g., AE) can be used to detect cracks during the coating failure process in real time.

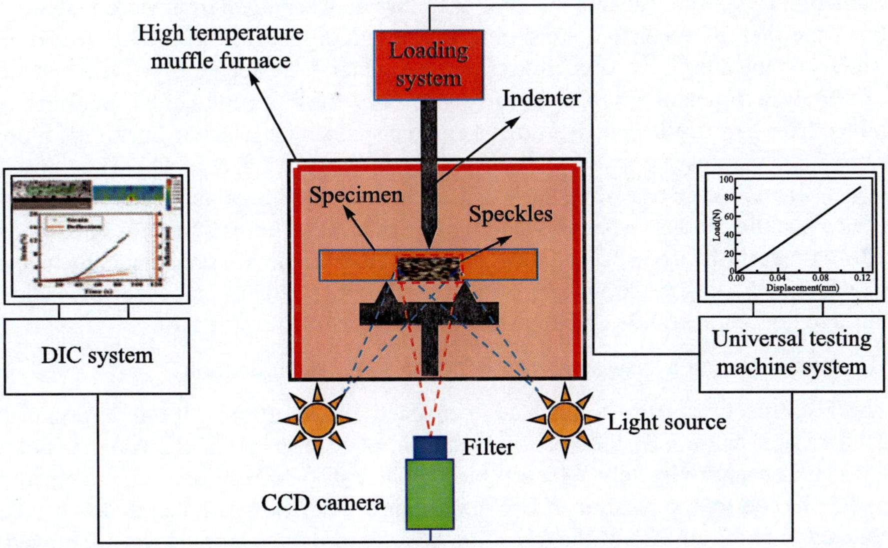

Fig. 16.10 Implementation of the thermomechanical loading on a TBC [9]

When using this method for combined thermomechanical loading, a standard material testing machine can be employed as long as it meets the TBC loading requirements. A high-temperature resistance furnace is typically used as the heating furnace. Due to factors such as high-temperature clamps and temperature drifts, the maximum temperature that can be achieved worldwide is approximately 1500 °C. The heating rate and temperature stability rely on the development level of high-temperature furnaces. The observation window on the high-temperature furnace chamber needs to be designed with quartz or other transparent high-temperature materials. The high-temperature pressure bar and clamps are selected based on the required temperature in the high-temperature furnace, with SiC as a common high-temperature ceramic material. There are no special requirements for the DIC optical strain measurement system and AE system, and the test methods described in Chaps. 11 and 13, respectively, can be followed.

A combined thermomechanical loading facility can be used to analyze the failure mechanism of coatings under combined thermomechanical loading and to perform high-temperature mechanical property characterization tests; these are briefly introduced in the following.

16.2.2 Buckling Failure Modes of TBCs Under Thermomechanical Loading

Thermomechanical buckling refers to a phenomenon in which a TBC becomes unstable and then separates and spalls from the substrate under a compressive stress. Compressive stresses causing this type of failure may originate from various sources (e.g., the thermal mismatch stress during the cooling stage, the CMAS corrosion stress, and the thermal mismatch stress further generated after TGO growth) and are extremely detrimental to TBCs. As early as in 1999, the first author of this book determined from a theoretical analysis that a coating spalls mainly due to buckling failure caused by compressive stresses. Zhou and Hashida [10] Choi et al. [11] Hutchinson et al. [12] and we (in an earlier study) [8] all theoretically found that a coating undergoes a buckling failure when the length of its interfacial defects reaches or exceeds 10–16 times its thickness. Using the PS TBC as an example, we systematically investigated the buckling mode and mechanism of TBCs under thermomechanical loading through high-temperature compression testing [8].

(1) Preparation of a TBC specimen with interfacial delamination

The TBC specimen used in this study consisted of a substrate (40 mm × 5 mm × 5 mm in size) made of SUS304 stainless steel, a 100-μm-thick PS NiCrAlY transition layer, and an approximately 300-μm-thick 8wt% Y_2O_3-ZrO_2 ceramic layer prepared by PS. To control the location of buckling failure, a defect with a length of a can be sprayed onto the surface of the transition layer before the ceramic layer is sprayed. Many methods (e.g., spraying an Al_2O_3 layer [8] and preparing a Ni film [13]) can be employed to prepare the defect. The following are the details of the procedure for preparing a defect-containing TBC with an Al_2O_3 layer as an example. First, the surface of the substrate material to be sprayed is cleaned through processes such as degreasing, thermal treatment, shot blasting, and ultrasonic cleaning. Then, a NiCrAlY BC layer is sprayed onto the surface of the substrate using the PS technique. Subsequently, the surface of the transition layer is roughened, after which both ends of the specimen are tightly covered with soft high-temperature-resistant metal thin slices or high-temperature tape so that only part of the transition layer with a width a is exposed. Afterward, a very thin layer of Al_2O_3 is sprayed onto the surface of the transition layer to represent an interfacial defect of the same thickness. The covering is then removed, and the specimen is sprayed with a ceramic layer. The Al_2O_3 layer is very brittle and thus can be considered to be a very weak bond zone between the ceramic coating and transition layer and susceptible to damage. In fact, this interfacial defect preparation concept has also been applied to the investigation of the interfacial fracture toughness of TBCs as well as other coating and thin-film systems [8, 13].

(2) Coating spallation mode under thermomechanical loading

A TBC specimen with an interfacial defect of length a was subjected to a sustained compressive load of 2000 N using the thermomechanical loading mode shown in Fig. 16.9. At the same time, the spraying gun moved back and forth at a rate of 2 cm/s along the length of the specimen to uniformly heat the ceramic surface. This way, the surface temperature of the coating was controlled at approximately 1000–1500 °C. Based on the buckling criterion for a coating, the ratio of the length of the interface crack to the length of the coating is >10, and the minimum interface crack length required for a 300-μm-thick coating to undergo buckling is 3 mm. The experimental results show that it is very difficult to induce buckling failure in the TBC when the design interfacial delamination length a value is too small ($\leq$3.0 mm), that buckling failure can be observed only when the design a value reaches a certain level ($\geq$5.0 mm), and that the TBC delaminates as a whole when there are a large number of defects at the interface. In the following, we explain these three scenarios in detail.

(a) Buckling failure under thermomechanical loading

Figure 16.11 shows a typical buckling failure mode of the TBC, with $a = 10$ mm (a ratio of crack length to coating thickness $\approx$ 30) and the surface temperatures of the coating and substrate at 1370 °C and 985 °C, respectively. Hutchinson et al. [12] found that for a coating or thin-film system, the critical interfacial delamination length, L_b, is approximately more than 20 times the coating thickness. Figure 16.11 shows that the length of the buckling failure zone is approximately 14.0 mm and is approximately 40 times the coating thickness (300 μm), which is consistent with the theoretical analysis. Figure 16.11b and c shows the morphology of a polished cross-section of the buckled specimen, from which we can more clearly observe the morphology of the specimen at the site of buckling. After spallation, a very thin ceramic coating remains on the surface of the transition layer, suggesting that buckling occurred within the ceramic coating instead of at the interface.

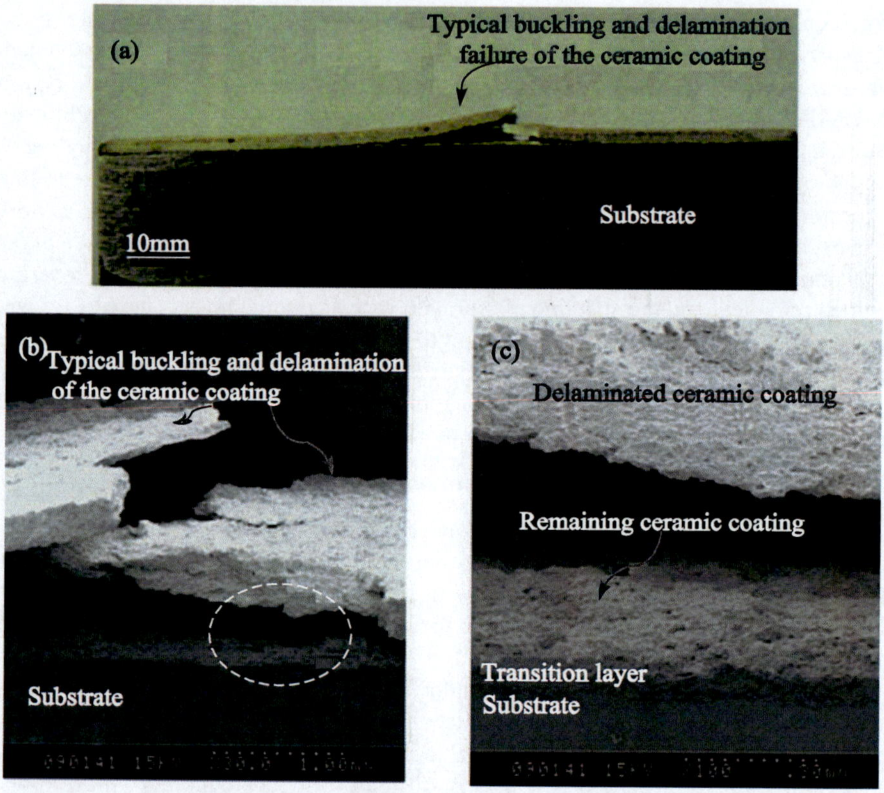

Fig. 16.11 Buckling failure mode of a TBC specimen under combined thermomechanical loading: **a** photograph; **b** SEM image; **c** close-up of the encircled region in (**b**)

A TBC can also undergo buckling failure at low temperatures, for example, during the cooling stage of the thermomechanical loading process. Figure 16.12 shows a photograph of a buckled coating specimen ($a = 8$ mm; the surface temperatures of the coating and the substrate are 277 °C and 214 °C, respectively). This also indicates that when a certain threshold is reached, the compressive stress in a TBC with a pre-existing interface crack during the cooling stage, after oxidation and CMAS corrosion or even thermal cycling alone, may cause the coating to undergo buckling failure [14, 15].

In addition, it is experimentally found that the buckling of the ceramic coating and the propagation of the interface cracks occur very rapidly. As soon as the buckling threshold is reached, the interface cracks rapidly propagate and coalesce, and the ceramic coating quickly delaminates from the substrate. Therefore, it is difficult to observe the postbuckling behavior of the ceramic coating.

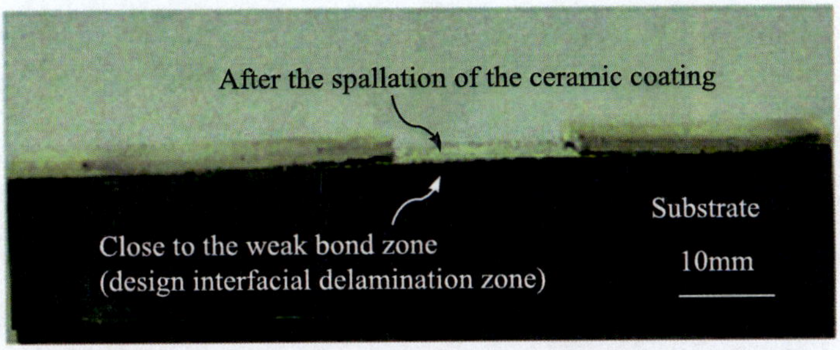

Fig. 16.12 Buckling failure of a TBC specimen at low temperatures

(b) Boundary delamination failure mode

If a TBC contains some defects at the boundary, it may first undergo boundary delamination failure, instead of buckling delamination failure. Boundary delamination refers to delamination that begins from a certain defect at the boundary. The criterion for boundary delamination is that the interfacial energy release rate reaches the interfacial fracture toughness [16]. Figure 16.13a and b shows the morphology of coating specimens that underwent typical boundary delamination failure. For the specimen in Fig. 16.13a, $a = 5$ mm, the surface temperatures of the coating and substrate are 1420 and 1015 °C, respectively, and the length and deflection of boundary delamination are approximately 30 and 3 mm, respectively. For the specimen in Fig. 16.13b, $a = 8$ mm, the surface temperatures of the coating and substrate are 1290 and 992 °C, respectively, and the length and deflection of boundary delamination are approximately 22 and 5 mm, respectively.

The cross-sectional morphology of the TBC specimen that underwent boundary delamination, as shown in Fig. 16.13b, is clearly observed in scanning electron

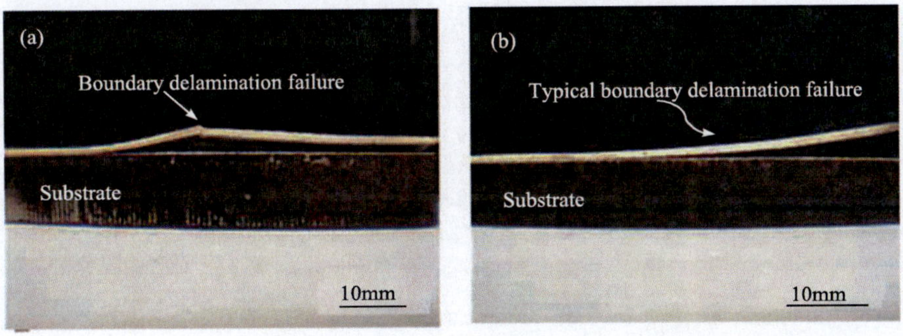

Fig. 16.13 Morphological images of TBC specimens undergoing typical boundary delamination failure

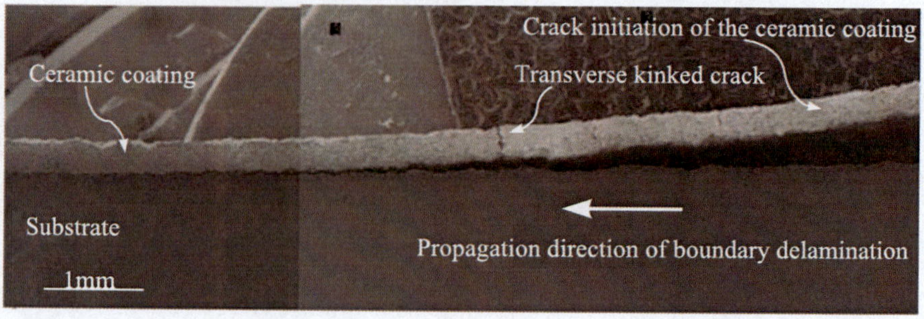

Fig. 16.14 SEM image of a TBC specimen after boundary delamination failure under the combined thermomechanical loading

microscopy (SEM) (Fig. 16.14). Along the propagation direction of boundary delamination, there are several transverse crack branches directly penetrating the ceramic coating from the interface toward the coating surface. Further propagation or expansion of a certain branched transverse crack would cause the ceramic coating to fracture and spall. At the right end, the cracked part of the ceramic layer is approximately 1 mm from the upper surface of the substrate.

To determine the location of boundary delamination failure, two opposite fracture surfaces after spallation were observed using SEM and analyzed using energy dispersive X-ray spectroscopy (EDX). Figure 16.15a shows an SEM image of the surface of the ceramic coating after spallation. It can be seen that a large number of microcracks and a typical irregular layered structure are formed during the spallation process. Figure 16.15b shows the surface morphology of the transition layer exposed after the spallation of the ceramic coating. In addition to large cracks, a very thin ceramic coating remains on the surface of the transition layer, indicating that the fracture occurred within the part of the ceramic coating close to the interface.

(c) Overall buckling failure mode of TBCs

If the design interface defect length is too small (e.g., $\leq$ 3 mm), it is very difficult to induce buckling or boundary delamination failure. However, continuing loading or increasing the temperature gradient may cause overall buckling, as shown in Fig. 16.16. Under this condition, the coating spalls because the applied load exceeds the load-carrying capacity of the substrate, that is, the critical buckling load of the substrate material, so the substrate tends to buckle first, which in turn promotes the delamination between the ceramic coating and the substrate.

(2) Characterization of the high-temperature mechanical properties of TBCs

We can characterize the key mechanical properties parameters (e.g., elastic modulus, fracture toughness, and interfacial fracture toughness) of a TBC at high temperatures by loading it at high temperatures using the high-temperature three-point bending loading and real-time testing method, as shown in Fig. 16.10, while examining its

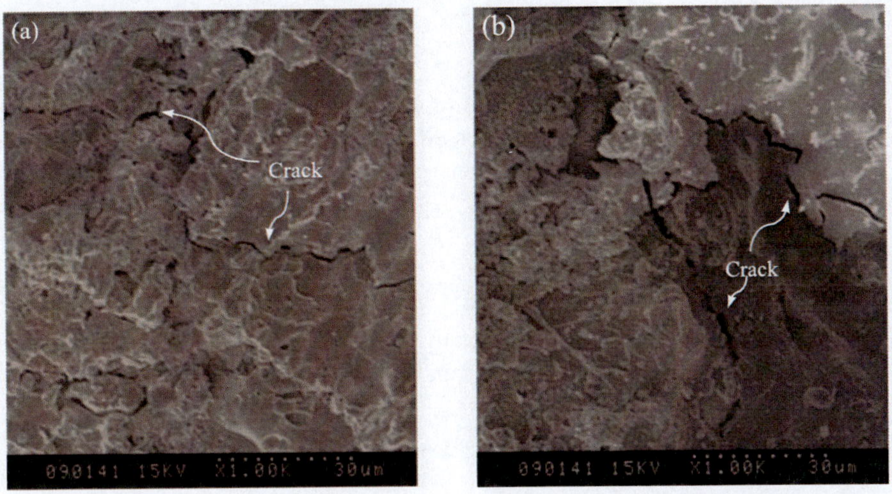

Fig. 16.15 Morphological images of two fracture surfaces after the boundary delamination failure of a TBC specimen: **a** bottom surface of the spalled ceramic coating. **b** Exposed surface of the transition layer

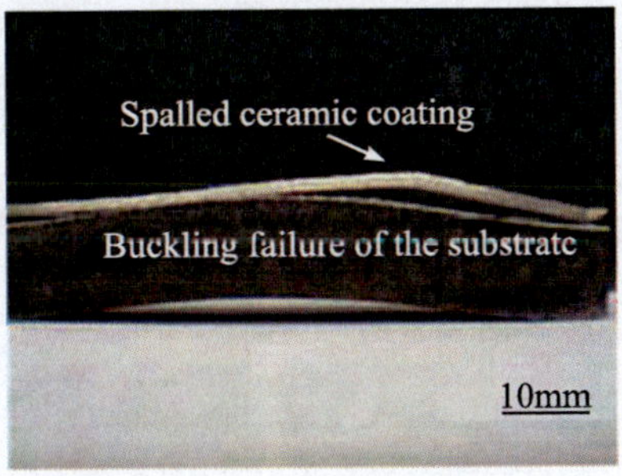

Fig. 16.16 Overall buckling failure of a TBC system under the combined thermomechanical loading

deformation process in real time using the DIC technique [9]. Despite its description in Chap. 9, the test method and results are briefly described in the following with the fracture toughness of a coating as an example.

A 50-mm-long, 6-mm-wide PS TBC specimen was used in the test. The three-point bending frame had a span of 30 mm. The Ni-based high-temperature alloy (GH536) substrate, the NiCoCrAlY BC layer, and the yttria-stabilized zirconia (YSZ) ceramic layer were 2 mm, 80 μm, and 1 mm in thickness, respectively. To allow a single surface crack and a steadily propagating interface crack to be formed in the TBC, it is necessary to prepare a thorough precrack by cutting the surface of the ceramic layer to a certain depth.

The surface fracture toughness K_{TC} of the TBC can be obtained using the following expression:

$$K_{TC} = Y\sigma_{cr}\sqrt{\pi a_0} \tag{16.2}$$

where σ_{cr} is the critical stress for crack propagation, a_0 is the length of the precrack, and Y is a geometric factor. σ_{cr} is expressed as follows:

$$\sigma_{cr} = \frac{E_c y}{\rho} + \sigma_r \tag{16.3}$$

where σ_r is the residual stress in the TBC, E_c is the elastic modulus of the ceramic layer, y is the distance between the crack tip and the neutral axis, and ρ is the radius of curvature of the neural axis. y and ρ can be expressed as follows:

$$y = l + \frac{l_c}{2} \tag{16.4}$$

$$\rho = \frac{[(L/2)^2 + \delta_{cr}^2]}{2\delta_{cr}} \tag{16.5}$$

Figure 16.17 shows the real-time digital speckle test results for the deformation field in the TBC during the three-point bending deformation process at different temperatures. The variation in the strain and deflection at the crack tip with time can be further extracted. As shown in Fig. 16.18, the strain slowly increases with time. At 37 s, the strain at the crack tip increases sharply, indicating crack initiation and propagation. The corresponding critical deflection δ_{cr} is 0.21 mm (Fig. 16.18a). Using the same method, the δ_{cr} values at 400, 600, and 800 °C are found to be 0.12, 0.13, and 0.15 mm, respectively. Based on Eqs. (16.2) and (16.3) in combination with the δ_{cr} values, the K_{TC} of the TBC can be determined. The residual stress is released when the temperature exceeds 400 °C.

Figure 16.19 shows the evolution of the K_{TC} of the TBC with temperature. The K_{TC} of the TBC decreases as the temperature increases. Without the consideration of the residual stress at room temperature, the K_{TC} values of the TBC are 2.54 ± 0.04, 1.28 ± 0.03, 1.23 ± 0.04, and 1.16 ± 0.03 MPa·m$^{1/2}$ at 30, 400, 600, and 800 °C, respectively (see Fig. 16.19a). The trend of the K_{TC} of the TBC with temperature considering the residual stress is similar to that without considering the residual stress. Specifically, with the consideration of residual stress, the K_{TC} values of the TBC are 1.31 ± 0.04, 1.28 ± 0.03, 1.23 ± 0.04, and 1.16 ± 0.03 MPa·m$^{1/2}$ at 30, 400, 600, and 800 °C, respectively (see Fig. 16.19b).

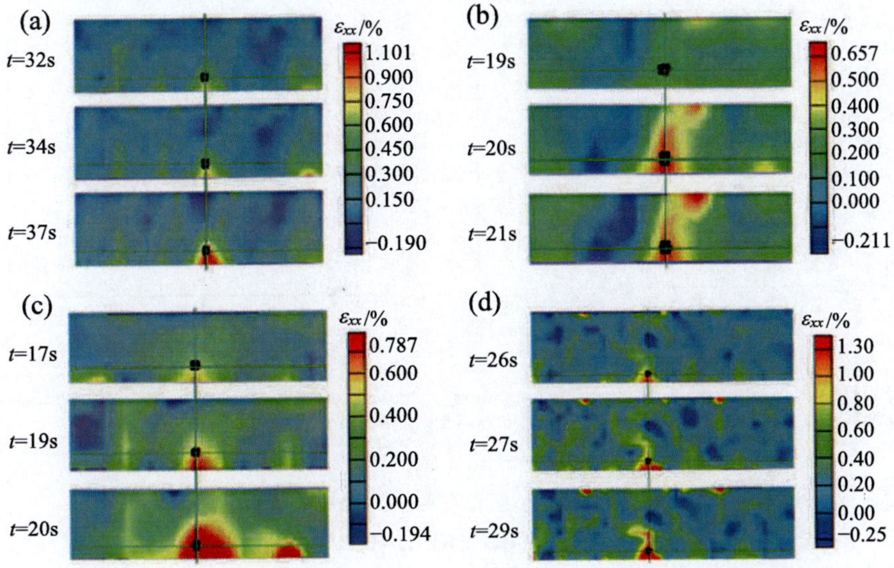

Fig. 16.17 Evolution of the stress field with time under three-point bending loading: **a** 30 °C, **b** 400 °C, **c** 600 °C, and **d** 800 °C

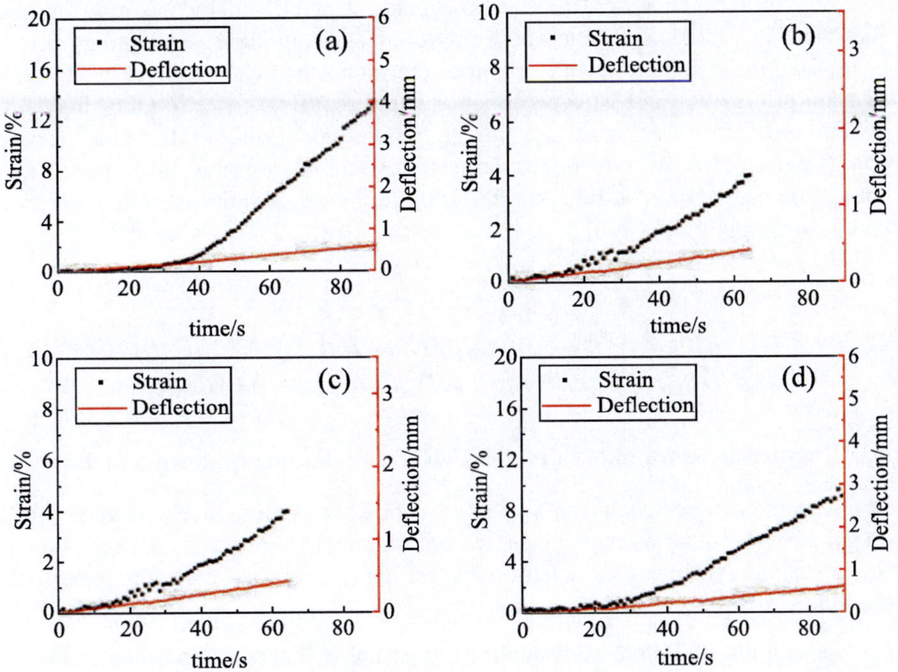

Fig. 16.18 Evolution of the strain and deflection at the crack tip on the surface of the ceramic layer with time: **a** $T = 30$ °C, **b** $T = 400$ °C, **c** $T = 600$ °C, and **d** $T = 800$ °C

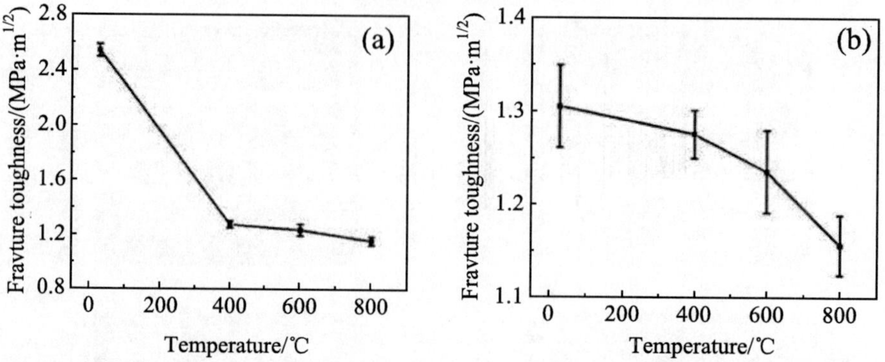

Fig. 16.19 Evolution of the fracture toughness of the coating with temperature: **a** Without the consideration of residual stress; **b** with the consideration of the residual stress

16.3 Static Thermo–Mechano-Chemical Coupling Simulators for TBCs on Turbine Blades

For TBCs on turbine blades, high temperatures and mechanical forces are only the most basic loads. Fuel gases carrying erosive and corrosive particles that act on TBCs on turbine blades, cold air in internal blade channels, and the resulting thermal shock loads are more important. These environmental conditions and loadings cannot be realized through thermal and combined thermomechanical loadings. In addition, relevant information needs to be obtained using real-time detection techniques to determine whether there exist new coating failure modes and mechanisms in these complex environments. Given these issues, this section describes the experimental simulation and testing facilities for TBCs on guide vanes under thermo–mechano-chemical coupling (TMCC).

16.3.1 Overall Design of an Experimental TMCC Simulation and Testing Facility for TBCs on Turbine Blades

(1) Requirements for experimental TMCC simulation and testing facilities

Based on the data requirements for the analysis of the failure mechanism of TBCs on guide vanes (in a stationary state, i.e., not rotating at high speed) in service environments, an experimental simulation and testing facility should be able to perform the following tasks:

(1) Decoupling of TMCC and coupling of thermal, mechanical, and chemical loadings. The high-temperature, high-velocity fuel gas produced in the combustion chamber of an engine as well as the hard erosive particles and corrosive CMAS

medium carried by the fuel gas are all factors to be considered in the simulation of the gas thermal shock on a TBC. In addition, the facility should be able to couple and decouple these three factors.

(2) Simulation of the temperature gradient loading from film cooling. In an actual turbine engine, a cooling airflow with a temperature of 300–500 °C is guided into the internal cooling channels of TBC-covered blades to form a temperature gradient between the coating and substrate, which is essential for ensuring the thermal insulation performance of the TBC. Therefore, the combined action of the fuel gas and cooling air should be simultaneously simulated in the experimental simulation of the environment to which the TBC on a turbine blade is subjected.

(3) Integration of real-time detection techniques for the failure process. Coating spallation is only the final outcome. TBCs have multiple failure modes (e.g., interfacial oxidation, sintering, phase transformation, corrosion, compositional and microstructural evolution, and cracking) under service conditions (e.g., high temperatures, erosion, and CMAS corrosion). In each failure mode, cracks are generated in the coating and its interfaces and subsequently propagate and coalesce, eventually causing coating spallation. Evidently, damage processes (e.g., the evolution of coating and interfacial structures, crack initiation and propagation, and interfacial delamination) are the key to understanding failure mechanisms.

(2) **Design concept and working principle of experimental TMCC simulation and testing facilities**

Figure 16.20 shows the facility design concept and implementation principle formulated based on the above requirements [17]. A high-temperature gas flow is generated through sufficient combustion of a medium (e.g., aviation kerosene, propane, and methane) capable of producing a high heat flux with oxygen and compressed air as an auxiliary oxidizer to simulate a high-temperature, high-velocity fuel gas. A spray gun designed based on the de Laval nozzle is used to impact the TBC specimen with supersonic fuel gas flow. On this basis, internal channels are designed for the spray gun to input CMAS and erosive particles. These particles are delivered into the spray gun via a nitrogen or combustion-supporting gas flow to produce a high-temperature gas flow carrying erosive and corrosive particles, with the goal of simulating high-temperature, erosion, and CMAS corrosion service conditions. A heating device is used to instantaneously heat the compressed air to generate an airflow at a certain temperature. The interior of the turbine blade (or the substrate surface of the specimen) is cooled at abnormal temperatures. Under these conditions, nondestructive detection systems (e.g., AE, digital speckle, IRT, and CCD imaging systems) are used to test the failure process of the TBC in real time.

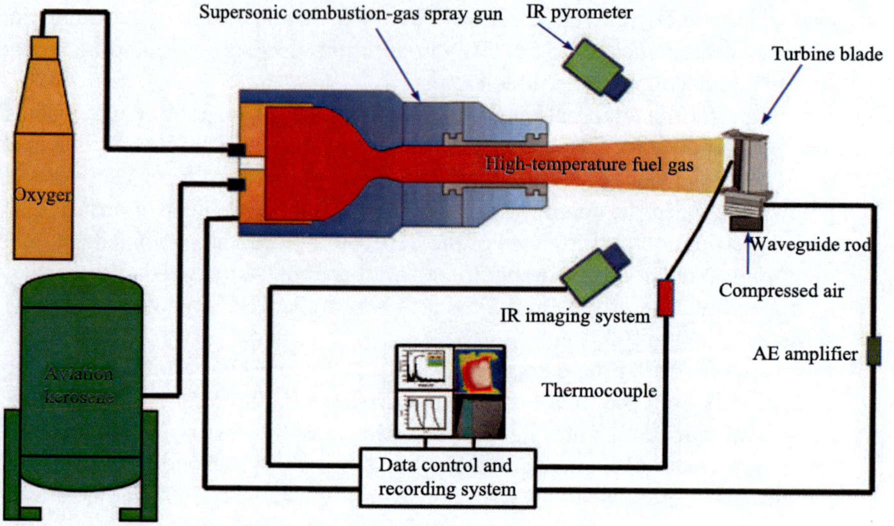

Fig. 16.20 Design concept and working principle of a static simulator for TBCs on turbine blades

(3) Hardware components of an experimental TMCC simulation and testing facility

Figure 16.21 shows the main structure of an experimental TMCC simulation and testing facility for TBCs on turbine blades developed based on the above design concept and working principle. This facility consists mainly of the following components.

A supersonic kerosene spray gun is used to simulate high temperatures, erosion, and CMAS corrosion (hardware). The high-temperature combustion-gas module is composed of an aviation kerosene storage tank, a combustion-supporting gas (oxygen) tank, an air compressor, a pressure regulator valve, a flow valve, a spray gun, and a combustion chamber. Erosive particles and corrosive CMAS particles are delivered into the spray gun via a spiral particle feeder. The spray gun is composed of an air compressor, a precise particle feeder, a servo motor, a flow controller, a pressure regulator valve, and a feeding tube.

Experimental operation platform (hardware). The experimental operation platform is used to operate the hardware modules, including the supersonic kerosene spray gun, the test station for TBC specimens on turbine blades (or other simple-shaped specimens), and nondestructive detection devices and includes an observation and testing window.

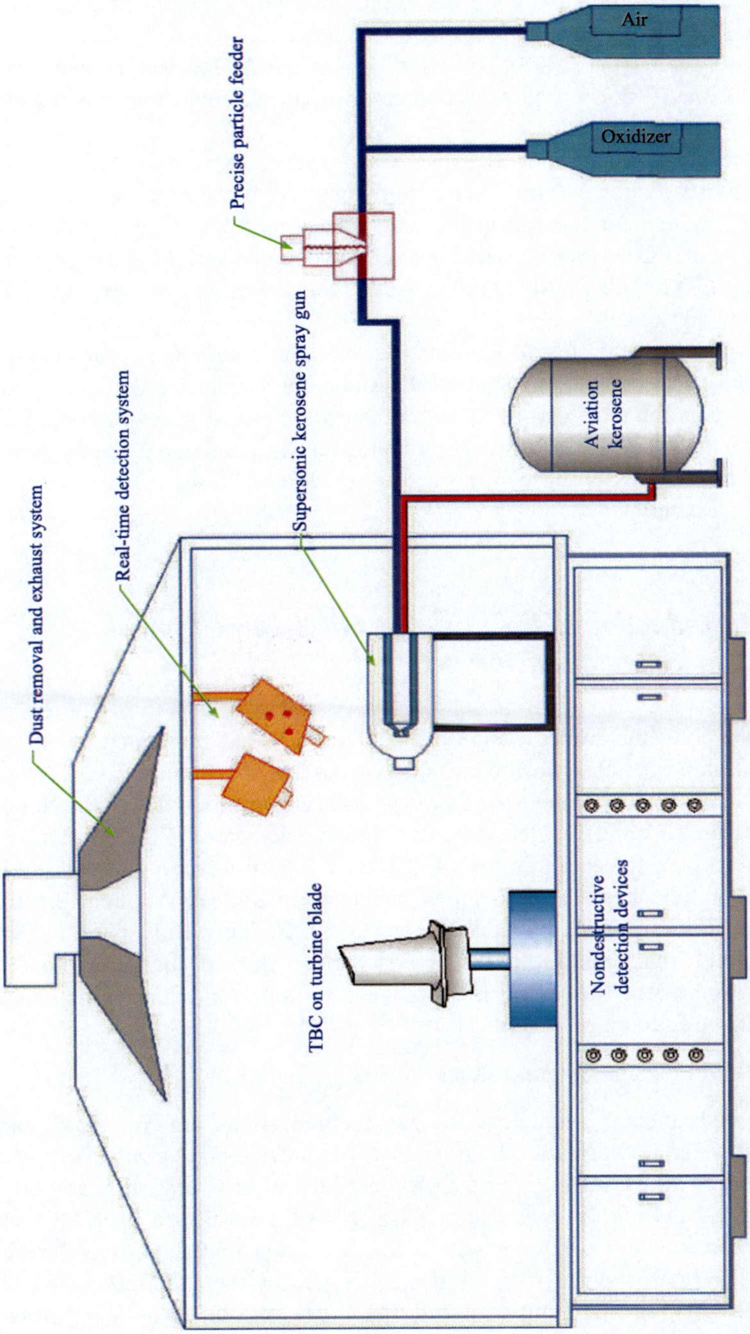

Fig. 16.21 Schematic diagram of the overall structure of a static simulator for TBCs on turbine blades

Cooling systems (hardware). The cooling systems include the water-cooling systems for the spray gun and the experimental platform as well as the air-cooling system for the cooling channels in the turbine blade.

Auxiliary systems (hardware). These systems include the lubrication, overrotation protection, exhaust, noise reduction, and erosive-particle recycling systems of the simulator.

Real-time detection module (hardware and software). This module includes nondestructive detection systems such as an AE system for monitoring crack evolution, an IRT system for measuring the temperature field and detecting interfacial delamination, a DIC system for detecting surface damage and measuring the strain field, and a CCD imaging system and is used to test the failure process of the TBC specimen in real time.

Data acquisition and control systems (hardware and software). These systems include test sensors and automatic control systems for various experimental parameters (e.g., temperature, pressure, flow, and particle velocity); the transmission of the experimental parameter data to the nondestructive detection systems; and the computer control and setting of the various operating statuses of the facility and imaging processing.

16.3.2 Introduction to the Functions of Several Typical Experimental Facilities

Based on the abovementioned design concept and mode of implementation, many aviation research institutes worldwide, such as NASA, the Netherlands Aerospace Center (Dutch: *Nationaal Lucht- en Ruimtevaartlaboratorium* (NLR)), the National Energy Research Center of Germany, the National Research Council of Canada (NRCC) Aerospace Research Center (ARC), the Chinese Academy of Agricultural Mechanization Sciences (CAAMS), the Changchun Institute of Applied Chemistry (CIAC), the Chinese Academy of Sciences (CAS), Beihang University, Xiangtan University, and Shanghai Jiaotong University, have studied simulators for TBC service environments. Several representative simulators and their functions are described in the following.

(1) **Simulators for gas thermal shock**

Currently, experimental simulators for gas thermal shock are the most mature approaches available worldwide to simulate TBC service conditions. Most of the experimental facilities, including the high-pressure fuel gas simulation and testing facility (termed the high-pressure burner rig (HPBR)) developed by NASA [18], the high-speed fuel gas simulator with a Mach number of 0.3–0.7 developed by Zhu et al. [19] and experimental facilities developed by the CIAC, the CAS [20], the CAAMS [21], and Beihang University [22], use various gases (e.g., propane, methane, acetylene, and natural gas) as combustion media.

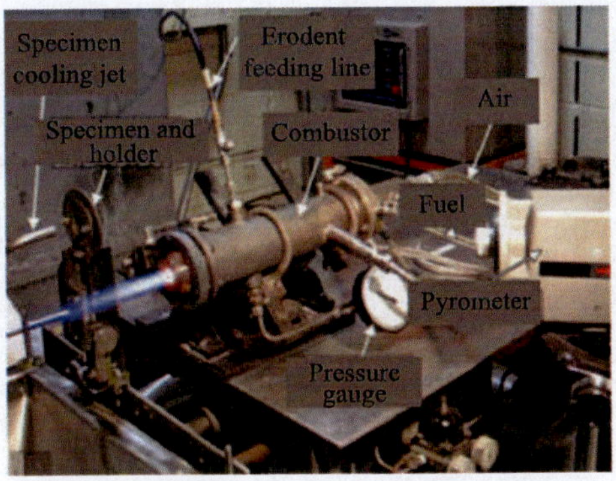

Fig. 16.22 HPBR with a Mach number of 0.3–1[18]

Here, we take the NASA-developed HPBR, with a Mach number range of 0.3–1, as an example, as shown in Fig. 16.22 [18]. With a gas gun for heating, this facility is composed primarily of a combustion chamber, a system for feeding erosive particles and a corrosive gas flow, a cooling system, and a pyrometer. The HPBR is capable of simulating the flow rates and working pressures of fuel gases that impact TBCs in real-world situations and thus can be used to simulate service conditions such as thermal shock, temperature gradients, and thermal fatigue. The morphological changes in a TBC specimen can be observed through the quartz observation window and image acquisition system. In addition, the thermal shock and oxidation resistance of a TBC specimen can be evaluated through the measurement of its weight and the observation of its surface morphology and microstructure after testing.

(2) **Simulators for high-temperature erosion**

An experimental simulation system for high-temperature erosion of TBCs primarily simulates high-temperature conditions, feeds erosive particles to impact the specimen at a controllable speed, and measures damage parameters (e.g., the weight, thickness, and temperature of the specimen). Generally, a high-temperature erosion simulator is developed by adding an erosive-particle feeder to a thermal shock simulator, such as the thermal shock simulator developed at Cranfield University in the United Kingdom, which can reach a temperature of 1400 °C and allows the addition of erosive particles and corrosive salt solutions during testing, and the NASA-developed HPBR, with a Mach number range of 0.3–1 [18]. Here, we take the TBC erosion facility developed at the Juelich Research Center in Germany [23] as an example (Fig. 16.23). This facility jet-sprays a high-speed gas flow carrying erosive particles (Al_2O_3 or other hard particles) onto the surface of a TBC through fuel gas heating to simulate TBC erosion failure. With this facility, the surface temperature of a TBC specimen can be heated up to 1500 °C, while a difference of approximately 400 °C can be achieved between the temperatures of the substrate and the surface of the ceramic layer through the cooling system. Thus, this facility is capable of simulating TBC service environments (e.g., erosion, high-temperature thermal shock, and temperature gradients). In addition, the morphological changes in a specimen can be observed through the quartz observation window and image acquisition system of this facility. Moreover, the erosion failure mode and mechanism of a specimen can be analyzed through the measurement of its weight and thickness as well as the observation of its cracks and microstructure after testing.

(3) **Integrated fuel gas shock, erosion, and CMAS corrosion simulation and real-time testing facilities**

The abovementioned facilities are unable to realize CMAS corrosion loading as an environmental condition and generally use inflammable gases as combustion media, which differ from aviation kerosene in terms of combustion environment. More importantly, it is very difficult to use these facilities to perform real-time testing and quantitative analysis of the coating failure process. To address these issues, we

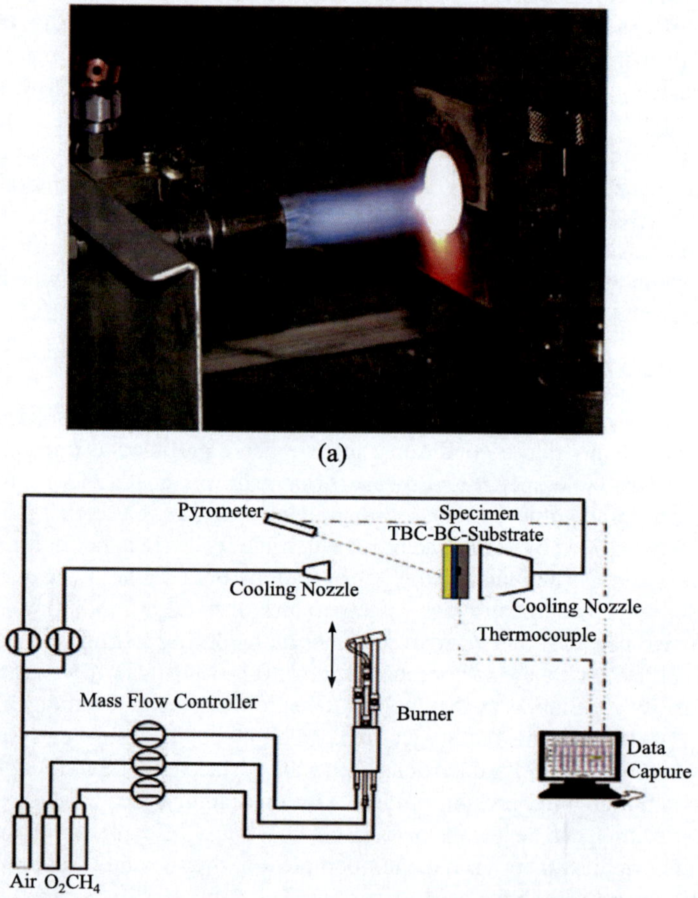

Fig. 16.23 **a** Photograph and **b** schematic diagram of an erosion facility [23]

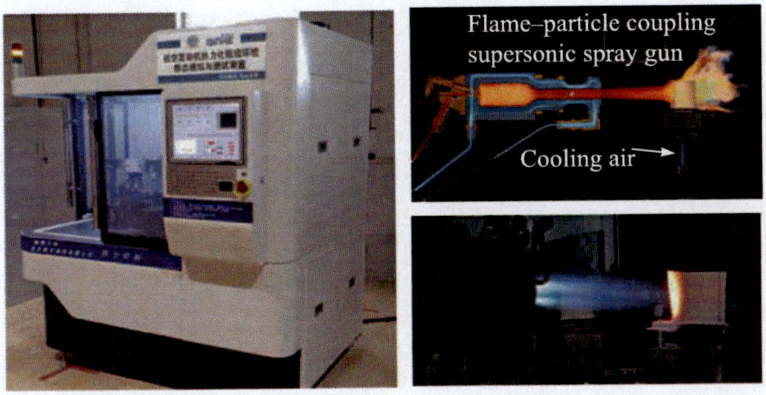

Fig. 16.24 The overall design of a static simulator

developed an experimental service environment simulation and testing facility for TBCs on turbine guide vanes [17, 24–31], as shown in Fig. 16.24. This facility is capable of simulating high-temperature, erosion, and CMAS corrosion service conditions in an integrated manner, producing controllable and adjustable environmental conditions (e.g., high temperatures up to 1700 °C, flame flows with a Mach number up to 2.0, and erosion at a rate up to 300 m/s) and air-cooling the substrate of a TBC specimen. In addition, this facility integrates nondestructive detection systems (e.g., an IRT system, a CCD camera, a digital speckle system, and an AE system) and thus can be used to examine the evolution of the morphology and damage of a specimen. Moreover, this facility enables the simulation of alternating temperature cycling, hard-particle impact, and/or corrosive gas erosion, which are service conditions to which real-world aeroengines are subjected; additionally, the conditions can be highly automatically controlled, and the facility allows the integration of various nondestructive detection devices.

The core components and features of this facility are described as follows:

(1) **Supersonic kerosene spray gun used to simulate high-temperature, erosion, and CMAS corrosion conditions in an integrated manner**. A high-temperature, high-speed gas flow is generated through the reaction between aviation kerosene and combustion-supporting gases (oxygen and compressed air). Then, erosive particles and corrosive CMAS particles are delivered to a reserved tubing inside the spray gun via a particle feeder system and subsequently heated, accelerated, and jet-sprayed, together with the high-temperature flame, onto the blade. The pressure inside the combustion chamber is altered through adjustment of the fuel gas flow to control the flame flow rate. Spray gun nozzles of various sizes are designed and used to adjust the flame size, with the goal of accommodating specimens of various sizes.

(2) **Key experimental parameter testing and control system**. This system measures and automatically controls the key parameters of the experimental facility, including the simulated service temperature as well as the erosive and

CMAS particle velocities. Temperature is a key parameter and includes the temperatures of the flame, the TBC, and the cooling air. The temperature field is primarily measured through a combination of thermocouples, IRT, IR thermal imaging, and thin-film thermocouples. Flow and velocity are the main parameters of erosive particles to be measured and controlled. The flow is set and adjusted through the control of the pressure of the particle-feeding air, and the velocity is calibrated using a particle velocimeter. CMAS particles are controlled through the adjustment of their flow. Parameters (e.g., temperature, particle flow, and the involved gas pressure and flow) are set and controlled through a PLC. The measurement data for these key parameters are used as the input signals for the PLC. The PLC program analyzes the parametric states corresponding to the signals, performs computations, and outputs signals via the output interface of the PLC to control the pressure and flow regulator valves and parameters such as the distance between the spray gun and the turbine blade. In this way, the experimental parameters are set, measured, and controlled.

(3) **Integration techniques for the real-time detection module**. The facility integrates nondestructive detection techniques such as AE, DIC, CCD imaging, complex impedance spectroscopy (CIS), and IR thermal imaging. In terms of hardware, a storage and detection window for each nondestructive detection device is set on the experimental operation platform. Structurally, these windows form a part of the facility. In terms of software, a synchronous control system is developed for the facility to simultaneously activate the detection devices. In regard to signal acquisition and analysis, the failure process is tested and quantitatively characterized using the established methods (see the nondestructive detection sections in Chaps. 10–12 of Part II). The integration of nondestructive detection techniques allows real-time detection of crack propagation, TGO growth, deformation and surface damage, interfacial delamination, and morphological evolution during tests conducted to investigate TBCs on turbine blades under simulated conditions, providing direct basic experimental data for analyzing failure modes and mechanisms.

(4) **Small wind-tunnel facilities.** Simulation tests can be conducted in wind tunnels to investigate the interactions between airflow and TBCs on turbine blades to understand the aerodynamic properties of TBCs. In addition, wind tunnels can be used to simulate TBC service environments. A wind-tunnel test platform with the capability of simulating high-temperature, high-speed gas flow conditions was designed by researchers at Xi'an Jiaotong University [32]. Solid-phase particles can also be added to this wind-tunnel platform. A blade erosion wind-tunnel testing system capable of simulating a gas flow with a maximum temperature of 750 °C and a Mach number of <0.3 was developed at the Shanghai University of Engineering Science [33]. Researchers at the Institute of Mechanics at the CAS developed a small arc-plasma wind tunnel with power as high as 150 kW [34], as shown in Fig. 16.25, which is primarily used to evaluate the ablation resistance of thermal protective materials for hypersonic aircraft engines.

Fig. 16.25 A small arc-plasma wind-tunnel platform

The high-enthalpy erosion wind-tunnel facility developed at the University of Cincinnati in the United States [35] is capable of carrying erosive particles and can be used to conduct erosion simulation tests at room and high temperatures.

16.3.3 Experimental TMCC Simulation and Real-Time Testing Methods

The experimental simulation and testing facility developed by our research team can be used to test TBCs of various structures (e.g., sheet-like, banded, and cylindrical TBCs and TBCs on turbine blades) under gas thermal shock, erosion, and/or CMAS corrosion [25–27]. Figure 16.26a and b shows the testing of a TBC specimen under combined fuel gas and CMAS corrosion conditions and the TBC on a turbine blade under thermal shock, respectively. Various types of loading can be achieved through processes such as thermal shock, thermal cycling, and long-term sustained action. Key damage can be detected and analyzed in real time using nondestructive detection systems [28, 30, 31]. In the following, we give a brief introduction to the experimental methods associated with the TMCC experimental and real-time testing facility as well as the important conclusions obtained from testing on this facility.

(1) **Experimental simulation and testing method**

Here, the TBC on a turbine blade was tested under high-temperature and gas thermal shock conditions as an example. The failure process of the TBC was examined in real time using an ARAMIS noncontact 3D deformation measurement system, an AE nondestructive detection system, and an IR thermal imaging system [36]. The detailed procedure is described as follows. A DB5012 high-temperature adhesive and

(a)

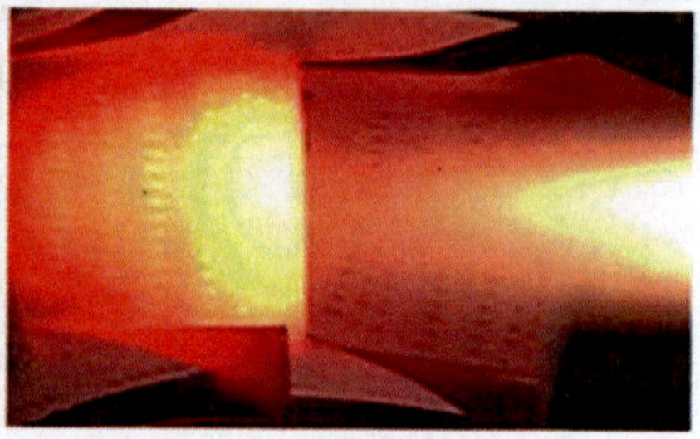

(b)

Fig. 16.26 Tests under thermal shock conditions: **a** specimen under thermal shock; **b** a blade under thermal shock

CoO were sprayed onto the surface of the coating to form high-temperature speckles with high reflective properties. The speckle patterns were subsequently detected in real time during testing using the ARAMIS noncontact 3D deformation measurement system. A spot-welding machine was used to weld one end of a waveguide wire to the free surface of the substrate, and the other end of the waveguide wire was connected to the AE nondestructive detection system. Afterward, the specimen to be tested was fixed with the clamps, and the angle between the specimen and the vertical shaft as well as the distances between the specimen and the probe of the IR imaging system, the CCD camera, and the probe of the digital speckle system were adjusted.

The experimental parameters were subsequently set. For example, the surface temperature of the coating, the temperature-holding time, the cooling time, and the cooling airflow were set to 1100 °C, 30 s, 40 s, and 60 L/min, respectively. Each detection parameter (e.g., the signal frequency acquisition range, sampling rate, and threshold of the AE system, the detection accuracy and acquisition rate of the ARAMIS system, and the number of frames and rate of the CCD camera) was set for each nondestructive detection system.

After the above settings were established, the simulator and its auxiliary systems as well as all the nondestructive detection systems were started to begin the thermal shock test.

(2) **Failure process data measurement**

The specimen investigated in this test was a physical vapor deposition (PVD) TBC for a certain type of turbine blade. Specifically, the TBC specimen consisted of a CMSX-4 substrate, an approximately 80-μm-thick NiCoCrAlY transition layer, and a 150-μm-thick YSZ ceramic layer. The TBC was subjected to the following thermal shock treatment: it was first heated to 1100 °C within 10 s, then held at 1100 °C for 30 s, and subsequently allowed to cool in the air for 40 s. When the crack length reached 10 mm or the spalled area of the TBC reached 10% of its total area, the TBC was considered to have failed and then the test was terminated.

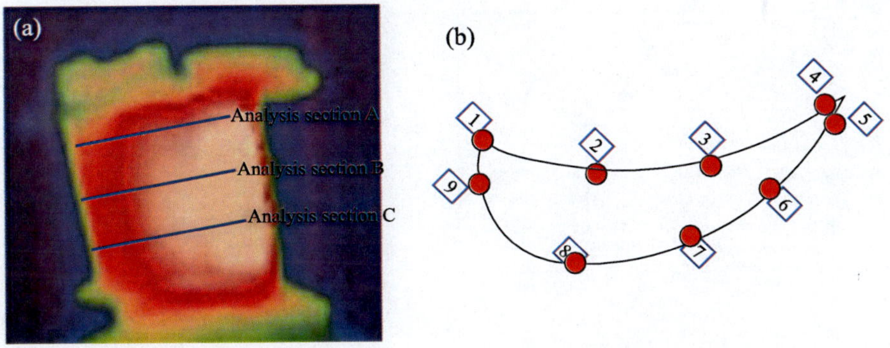

Fig. 16.27 Analysis of the turbine blade: **a** analysis sections and **b** checkpoints

(1) Real-time measurement of the temperature field in the TBC on the turbine blade

The temperature field in the TBC on the turbine blade during the gas thermal shock process was measured in real time through IR thermal imaging. Figure 16.27 shows the specific measurement sites. Three sections were selected from the top to the bottom of the turbine blade as analysis sections (denoted A, B, and C, respectively), with section B being directly impacted by the high-temperature fuel gas. Nine checkpoints were selected on each analysis section. On the pressure side, the checkpoints from the leading edge to the trailing edge of the turbine blade were denoted by 1–4, respectively. On the suction side, checkpoints from the trailing edge to the leading edge of the turbine blade were denoted by 5–9, respectively. The temperature field was measured through IRT. See elsewhere [37, 38] for the detailed method.

Figure 16.28 shows the variation in the temperature in section B with the thermal shock time. At the leading edge (checkpoint 1), the heating rate reached 100 °C/s, and the temperature increases from room temperature to 1100 °C, is maintained at 1100 °C for 30 s, and finally decreases to room temperature after 40 s of cooling. A similar temperature trend can be found at checkpoint 2, where the maximum temperature is approximately 1000 °C. The temperatures at the other two points on the pressure side (checkpoints 3 and 4) increase from room temperature to 900 and 883 °C, respectively. In addition, the heating rates on the suction side (at checkpoints 5–9) are significantly lower than those on the pressure side. The temperature on the suction side increase with time, with maximum temperatures at checkpoints 5–9 of 755, 790, 791, 825, and 896 °C, respectively, which is directly related to the diverting effect of the guide turbine blade on the fuel gas.

The IR imager is able to produce thermal images of only one plane. Thus, the turbine blade was divided into three parts, namely the suction side, the leading edge of the blade, and the suction side. Three images were obtained, as shown in Fig. 16.29. It can be seen that high temperatures are mainly distributed at the leading edge of the blade and the front end of the suction side. A semi-elliptical high-temperature zone

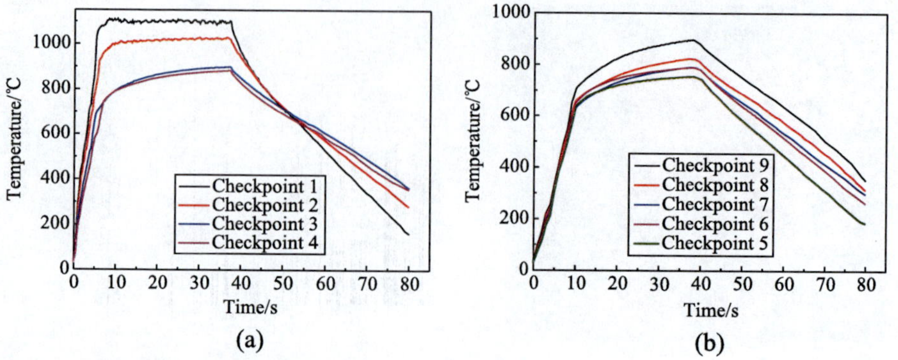

Fig. 16.28 Variation in the temperature in section B during thermal shock cycling: **a** pressure side and **b** suction side

and a rectangular high-temperature zone can be found at the leading edge of the blade and on the pressure side, respectively. The blade diverts the high-temperature fuel gas that impacts its leading edge through its unique geometric structure. The temperature on the pressure side is higher than that on the suction side. Figure 16.30 shows the temperature curves for different locations on the three sections at the same time point. Overall, the temperatures at different checkpoints in the three check sections exhibit basically similar trends. The temperatures at checkpoints 1–4 on each of the three check sections decrease considerably at a high rate, and the temperature at check section B decreases most significantly. The temperature change at each of the three check sections is less significant on the suction side than on the pressure side. The temperatures on the pressure side (at checkpoints 1–4) are significantly higher than those on the suction side (at checkpoints 5–9). The temperature at the site of the impact of the fuel gas (i.e., check section B) is higher than those at other sections (check sections A and C) by approximately 80 °C.

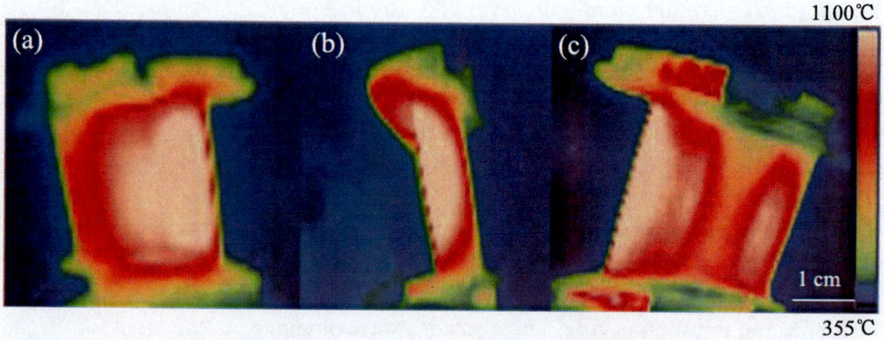

Fig. 16.29 Temperature distribution at each location during the temperature-holding stage of the thermal shock process: **a** pressure side, **b** leading edge of the blade, and **c** suction side

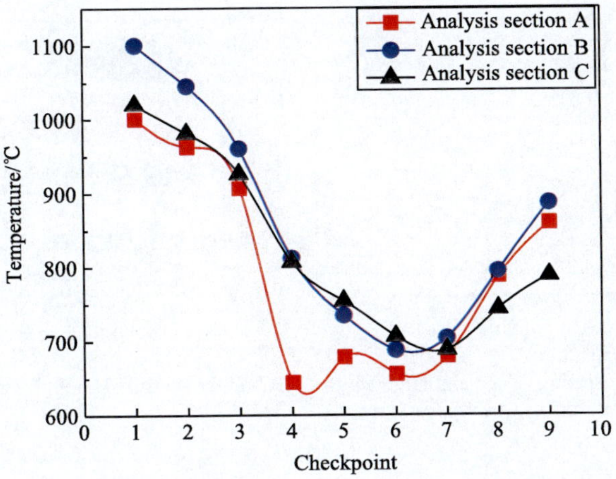

Fig. 16.30 Temperature distribution along the height of the turbine blade during the temperature-holding stage of the thermal shock process

(2) AE signals detected during the crack evolution process

The high-temperature AE detection and signal analysis method described in Chap. 11 was employed to test and analyze the TBC on the turbine blade throughout the thermal shock-induced failure process. While these results are presented in Chap. 11, they are discussed here for different purposes.

First, the AE signals generated from the TBC on the turbine blade under gas thermal shock were analyzed through clustering and wavelet spectrum analysis. Figure 16.31 shows the clustered AE signals. As shown in Fig. 16.31a, the silhouette value is the highest at $k = 5$, suggesting the presence of five damage modes. According to the peak frequencies in Fig. 16.31b, these five damage modes correspond to substrate deformation (90–110 kHz), vertical surface cracks (200–220 kHz), sliding interface cracks (280–325 kHz), and opening interface cracks (400–450 kHz), respectively. The noise signals are in the frequency band of 20–60 kHz.

Throughout the failure process, there are few AE signals during the heating and temperature-holding stages. Most of these signals are generated due to the formation of vertical surface cracks. The signals generated during the cooling stage are due to the formation of interface cracks. Figure 16.32a and b shows the variation in the AE event count and cumulative AE event count for each damage mode with the number of thermal shock cycles, respectively. The number of AE events increases with the number of thermal shock cycles. At 100 cycles, vertical surface cracks and sliding interface cracks are the primary damage modes. At 300 cycles, there is a rapid increase in the number of opening interface cracks. This result suggests that under the thermal mismatch stress, vertical surface cracks are formed in the coating due to the tensile stress during the heating stage and gradually propagate toward the interface, and, upon reaching the interface, these cracks propagate along the interface due to the interfacial tensile stress caused by the buckling of the coating under the

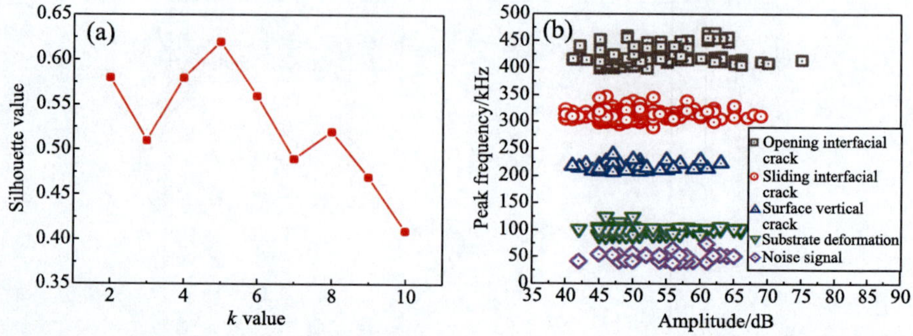

Fig. 16.31 **a** Clustering of the signals from the TBC under thermal shock; **b** amplitude and peak frequency distributions of five types of signals

compressive stress during the cooling stage as well as the interfacial shear stress, gradually causing the coating to spall. This mechanism can be verified based on the macroscopic spallation of the TBC (Fig. 16.33), particularly its microstructural evolution (Fig. 16.34).

(3) Measurement of deformation and surface damage using the digital speckle technique

The variation in the principal surface strains in the TBC on the turbine blade with the number of thermal shock cycles was determined using the real-time digital speckle detection method, as shown in Fig. 16.35. Overall, as the number of thermal shock cycles increases, the strains on the pressure and suction sides both increase gradually, and the surface strain of the coating gradually changes from compressive to tensile strain. Eventually, due to the increase in the tensile strain and the occurrence of stress concentration, the TBC spalls and fails.

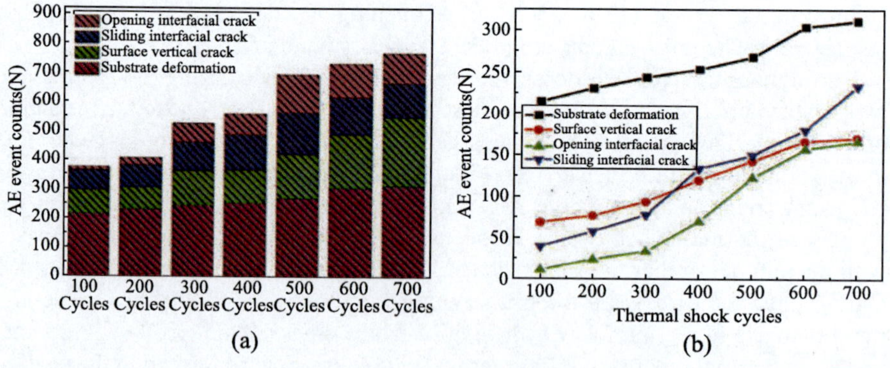

Fig. 16.32 Variation in the number of AE signals generated by each damage mode with the number of thermal shock cycles: **a** AE event count; **b** cumulative AE count

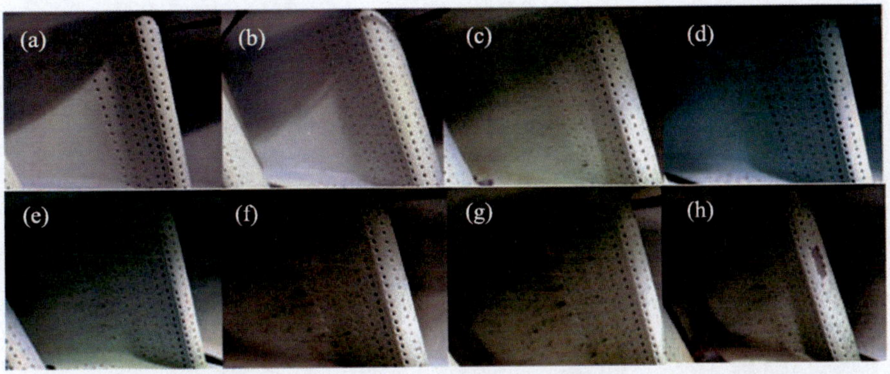

Fig. 16.33 Surface morphology of the TBC at different thermal shock cycles: **a** 0; **b** 100; **c** 200; **d** 300; **e** 400; **f** 500; **g** 600; **h** 710

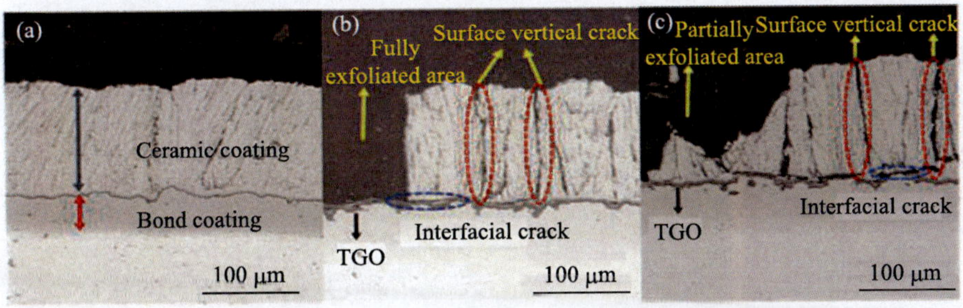

Fig. 16.34 Microstructure of the TBC: **a** nonspalled zone at the leading edge; **b** fully spalled zone; **c** partially spalled zone

One checkpoint was selected from the top, middle, and bottom of the pressure and suction sides, respectively, as shown in Fig. 16.35. The checkpoints on the pressure side are denoted A, B, and C, respectively, while the checkpoints on the suction side are denoted A', B', and C', respectively. Figure 16.36 shows the variation in the principal strain at each checkpoint with the number of cycles. The principal strains are significantly higher in the middle (at checkpoints B and B') than at other locations, primarily due to the high rate of change in the temperature and the presence of a very large temperature gradient in the middle of the TBC. As the number of cycles increases from 20 to 200, the principal strain at checkpoint B increases from –0.18 to 0.68%. In addition, due to the large temperature gradient on the pressure side, the strains on the pressure side are slightly greater than those on the suction side. Generally, compressive stresses can inhibit crack initiation, whereas tensile stresses can promote crack initiation. For a TBC system, as the tensile strain increases to the critical fracture strain, cracks are initiated in the ceramic layer and subsequently propagate, causing delamination in the system.

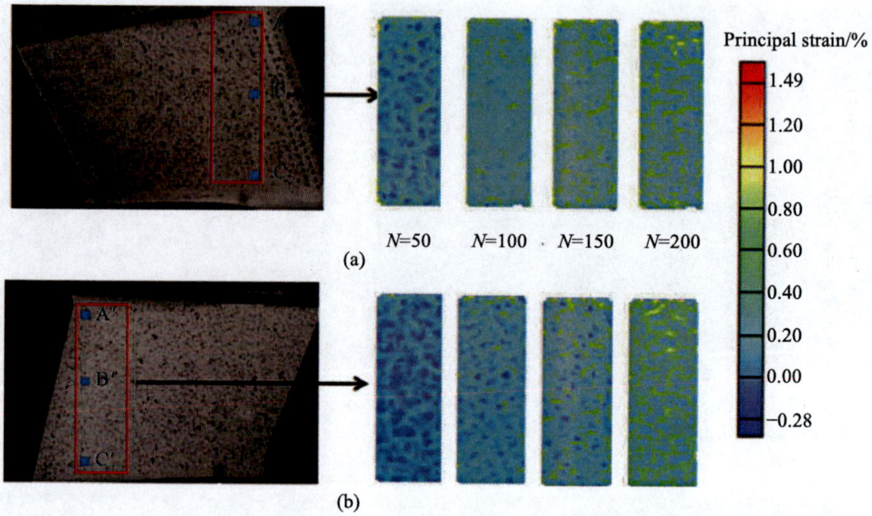

Fig. 16.35 Contour plots showing the variation in the principal surface strains in the TBC on the turbine blade with the number of thermal shock cycles *N*: **a** pressure side and **b** suction side

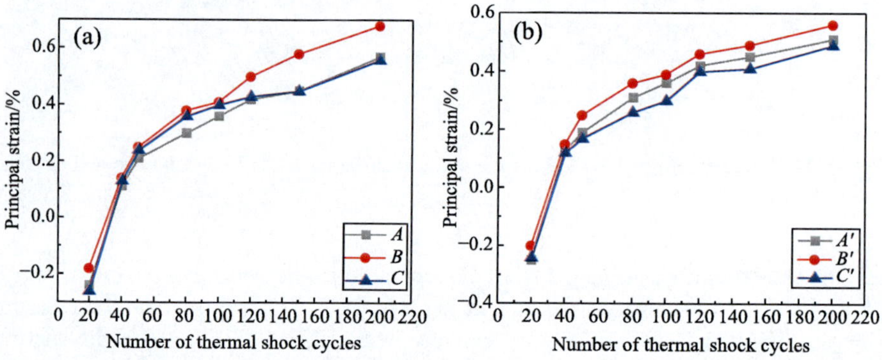

Fig. 16.36 Variation in principal strains at different heights with the number of thermal shock cycles: **a** pressure side and **b** suction side

In addition, this facility can be used to simulate high-temperature erosion [39], high-temperature CMAS corrosion [40], and their combination with gas shock and to conduct real-time testing under these conditions. The details are not repeated here.

16.4 Dynamic Experimental TMCC Simulation and Testing Facilities for TBCs on Turbine Blades

In addition to being exposed to high-temperature, erosion, and CMAS corrosion conditions similar to those of stationary components (e.g., guide vanes), TBCs on working blades are subjected to rotation at speeds as high as 10,000–50,000 rpm. The centrifugal and aerodynamic forces generated due to the high-speed rotation of a blade result in the formation of a complex stress field on the surface of the TBC system. In addition, the complex interactions between a blade in high-speed rotation and a fuel gas (associated with high-temperature, erosion, and CMAS corrosion conditions) lead to results such as the formation of a turbulent flow with a randomly varying speed, pressure, and temperature and the formation of local ultrahigh-temperature zones (i.e., hot spots) [41]. Consequently, the coating sustains damage or even spalls due to local overheating. These complex interactions cannot be produced by simulating fuel gases and stationary components alone. Therefore, developing dynamic TMCC simulation and testing facilities that are capable of simulating high-temperature fuel gas (associated with erosion and high-temperature CMAS corrosion) and high-speed rotation for TBCs on turbine blades can provide an effective tool for understanding TBC failure mechanisms and poses a tremendous challenge in this research field.

16.4.1 Overall Design of Dynamic Experimental TMCC Simulation and Testing Facilities for TBCs on Turbine Blades

(1) Requirements for and difficulties in the development of dynamic experimental TMCC simulators

Simulating gas thermal shock, erosion, and high-temperature CMAS corrosion conditions similar to those simulated for guide vanes is the basic requirement for working blades. In addition, it is necessary to simulate the operating state (i.e., high-speed rotation) of working blades in order to simulate the environmental loads (e.g., centrifugal and aerodynamic forces, hot spots, and turbulence) on TBCs in this state. Moreover, it is necessary to be able to adjust and control these loads to understand the mechanisms by which they induce coating spallation. Based on the development of static simulators, it is relatively easy to simulate gas thermal shock accompanied by erosion and CMAS corrosion, while difficulties lie in the following areas:

(1) Design and dynamic balancing of a high-speed rotor system. As an experimental platform, a rotor system should be, as much as possible, able to generate loads (e.g., centrifugal forces and linear velocities) that meet the needs for studying TBCs on the working blades of various models. However, test models should be, as much as possible, prepared according to specifications, simple, and low cost.

These factors lead to extremely stringent requirements for the rotational speed and its adjustment range as well as for stability, safety, and balance, presenting a tremendous challenge.

(2) Compatibility between the rotor blade model and high-temperature fuel gas module. To simulate the interaction of high-speed rotating blades and fuel gas, the fuel gas module and the rotor blade need to be similar to real-world engines in terms of angle and speed and to be adjustable. This is to ensure that effects (e.g., hot spots, aerodynamic forces, and wakes) similar to those found in TBCs on real-world working blades can be produced. Thus, the hardware and parametric compatibility between the rotor blade model and high-temperature fuel gas module is both a key and a difficult area.

(3) Real-time testing of the TBC failure process. For specimens rotating at high speeds, signal detection and transmission, noise shielding due to high-speed rotation and gas shock near the critical Mach number, and pattern recognition and quantitative analysis of signals are all tremendously difficult.

(2) Design requirements and working principle of experimental TMCC simulation and testing facilities

It is extremely difficult to develop dynamic experimental TMCC simulators. As a consequence, their design concepts, core technologies, and even parameters constitute a core area blockaded by countries across the world.

Figure 16.37 shows a structural design drawing of a dynamic facility developed by our research team [42–47], which comprises the following five components: (1) Supersonic fuel gas spray guns. These spray guns are mainly used to produce a high-temperature, high-velocity fuel gas flow that carries erosive particles and corrosive CMAS particles. (2) A high-speed rotating system. Equipped with a high-power motor, a rotating shaft, and a turbine model, this rotating system is used to rotate the turbine blade at a high speed and to produce an airflow environment similar to that surrounding a real-world engine turbine. (3) An experimental operation platform. This platform includes the rotor system, the turbine model, the supersonic kerosene spray guns, and various auxiliary modules, as well as an observation and detection window. (4) Real-time detection systems. These systems (e.g., an AE system, a digital speckle system, and a CCD high-speed camera) are used to examine the coating failure process. (5) A data acquisition and control system. This system is used to measure, control, and store data on the temperature and pressure of the fuel gas, the particle velocity, the surface temperature of the specimen, and the temperature and pressure of the cooling air. In addition, it is necessary to design a film-cooling system for the interior of the blade; water- or air-cooling systems for the spray guns, the rotating shaft, and the experimental platform; and auxiliary systems (e.g., a vibration control system, a lubrication system, an overload protection system, an exhaust system, and a particle recycling system).

Figure 16.38 shows the working principle of the facility. A high-power motor is used to drive the turbine model, which consists of a guide vane, a working blade, and a nondestructive detection window, to rotate at a high speed via one end of a

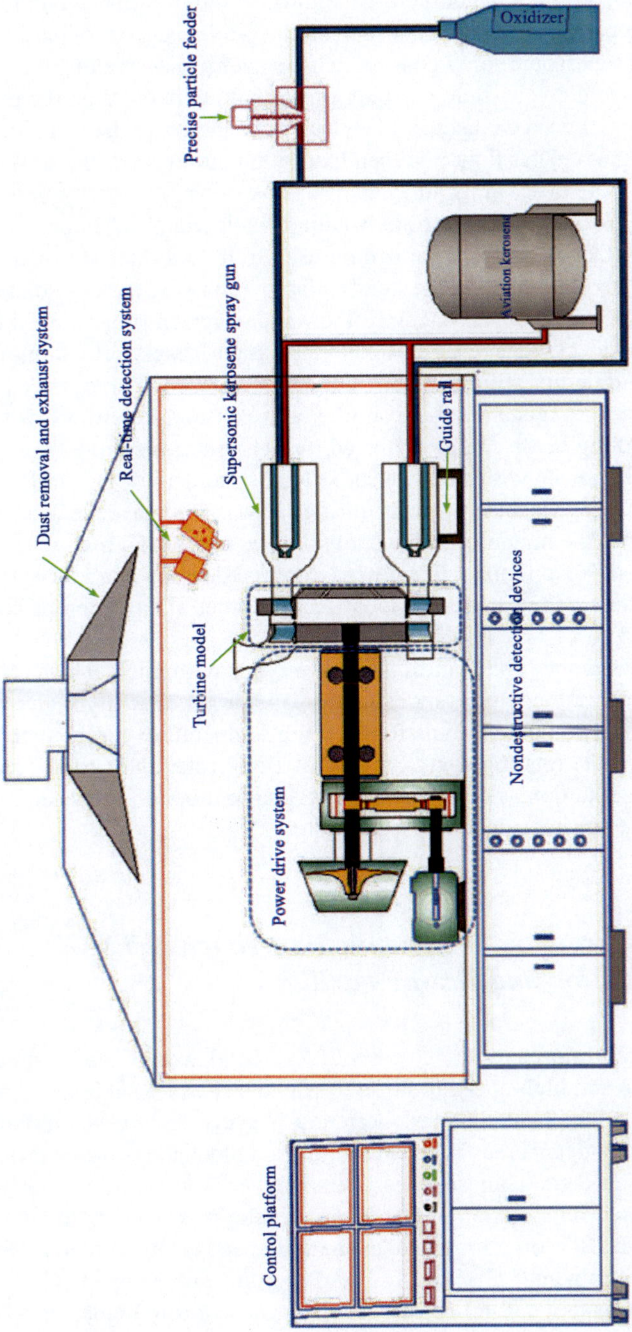

Fig. 16.37 Schematic diagram of the overall structure of the facility

multistage speed-increasing gearbox. The other end of the gearbox drives a load compressor to absorb the load output of the turbine and to produce compressed air to cool high-temperature components such as the guide vanes, moving blades, and spray guns. The supersonic spray guns, which are evenly placed along the periphery, jet-spray a high-velocity, particle-carrying fuel gas flow that acts on the guide vane of the turbine model via a connector. The guide vane increases the velocity, changes the direction of the fuel gas flow, and then loads it on the rotating blade. The cooling air enters the turbine disk via its air inlet and, subsequently, partially flows into the guide vane and partially flows into the rotating blade via a deflector. Under these conditions, the CCD camera system can be used to record the test process and the morphology of the specimen via the quartz glass window on the closed chamber of the turbine. In addition, the AE system, IR thermal imager, and digital speckle system are employed to detect damage signals generated due to crack formation, interfacial delamination, and deformation. Notably, only stationary components (i.e., the TBC on the guide vane) can generally be examined in real time. Real-time testing of the TBC on the working blade can be achieved through a turbine model design with a stationary working blade with a tensile stress applied and a reverse guide vane. The drawback of this design is that a uniform tensile stress is applied to the working blade, which differs from the nonuniform centrifugal force to which a real-world working blade is subjected. Nevertheless, this design allows a detailed analysis of the failure mechanism of the coating under various tensile forces (i.e., different centrifugal forces).

To accurately simulate and control the four major environmental loads (i.e., high-speed rotation, high temperature, shock, and corrosion), the facility is also equipped with water- and air-cooling systems for the high-temperature components, a lubrication system for the rotating shaft, an overrotation protection system, an exhaust system, a noise reduction system, a testing system, a monitoring system, a control system, and a data processing and display system.

16.4.2 Main Progress in Dynamic Experimental TMCC Simulation and Testing Facilities

Thus far, few dynamic facilities have been developed worldwide to simulate the high-temperature and high-speed rotation to which TBCs are subjected, and only a few countries (e.g., the United States, Germany, Canada, and the Netherlands) have made progress in relevant areas but have established blockades on core technologies and parameters. To date, Chinese researchers have achieved minimal advancement in this field. Based on a static simulator, we developed a facility to simulate gas thermal shock on TBCs on turbine blades in high-speed rotation, thereby providing an important experimental platform for studying the failure mechanism, thermal insulation performance, and reliability of TBCs on working blades as well as for examining their simulation.

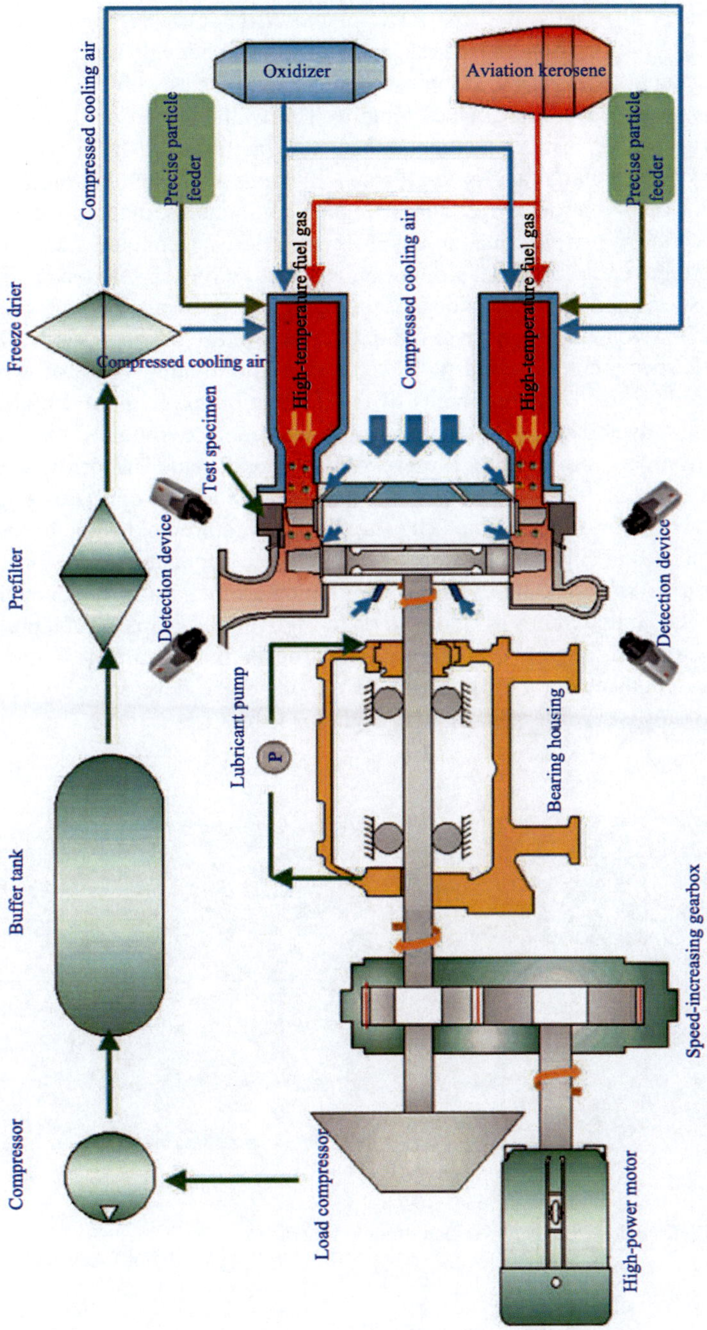

Fig. 16.38 Schematic diagram of the working principle of the facility

(1) Main progress achieved worldwide in dynamic simulators

The NLR (the Netherlands) and the ARC of the NRC (Canada) have developed LCS-4B and LCS-4C high-velocity fuel gas facilities, respectively [48], as shown in Fig. 16.39. Schematic diagrams of their working principles as well as their specific structures have yet to be reported in the literature. The available brief written introduction can help us to understand some functions of these facilities and their experimental simulation processes. Each facility simulates a dynamic working environment of a TBC through a combination of specially designed components, including a rotating device, a mechanical drive device, an airflow circulation system, and a combustion chamber. During the experimental simulation process, jet fuel or kerosene is ignited in the combustion chamber to provide a heat source—a high-velocity, high-pressure gas flow with a Mach number of over 0.8 and a temperature of more than 1300 °C. Multiple TBC specimens are fixed with the clamps, the rotating speed of which is adjusted according to the requirements of the simulation experiment. In addition, cooling air is delivered to the interior of each specimen to simulate the thermal gradient environment surrounding a real-world working blade. A corrosive gas is applied to the exterior of each specimen to simulate the chemical corrosion environment inside a real-world aeroengine. The surface temperature field and morphology are monitored in real time through an IR temperature acquisition system and an image acquisition system. The dynamic simulator developed at the University of Cincinnati in the United States [49], as shown in Fig. 16.40, is capable of simulating a centrifugal force through the high-speed rotation of the specimen. It has been reported that a high-temperature hyperspeed rotating stage developed by Schenck RoTec in Germany is capable of simulating high-speed rotation at 100,000 rpm below 1000 °C [50].

(a)

(b)

Fig. 16.39 LCS-4B and LCS-4C high-velocity fuel gas facilities

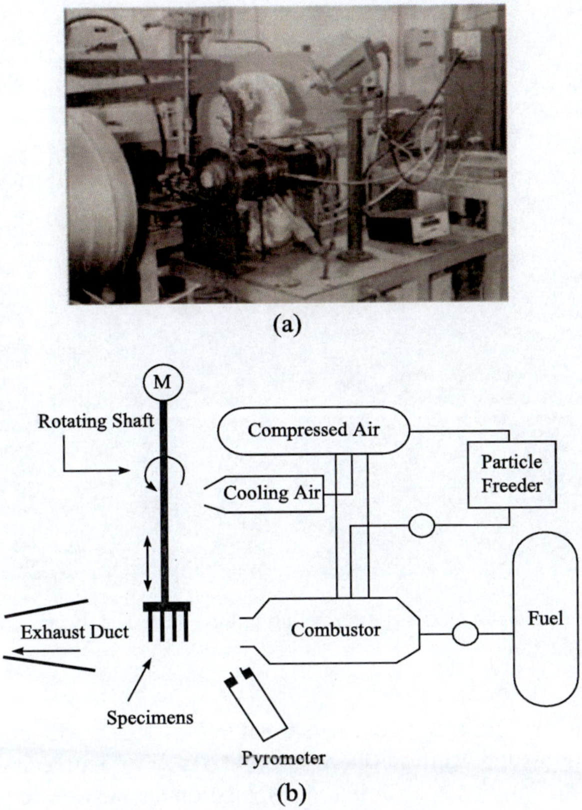

Fig. 16.40 **a** Photograph and **b** schematic diagram of a simulator for the environmental conditions of TBCs under centrifugal forces

(2) Experimental high-temperature, high-speed rotation simulation and testing facility for TBCs on turbine blades

Based on the development of the static simulator, we made a further breakthrough in high-temperature, high-speed rotation design and dynamic balancing techniques and, as a result, developed a high-temperature, high-speed rotating turbine blade and dynamic TMCC simulation and testing facility [42–47], as shown in Fig. 16.41. This facility is capable of producing controllable and adjustable conditions of high-temperature (up to 1500 °C), gas shock (up to a Mach number of 1), erosion (up to 300 m/s), CMAS corrosion (up to 10 mg/min), rotation (up to 20,000 rpm), and film cooling (300–500 °C) and can be used to test TBCs on stationary blades in real time. The operation and display of all experiments in this facility are automated.

The core components of this facility are briefly described in the following:

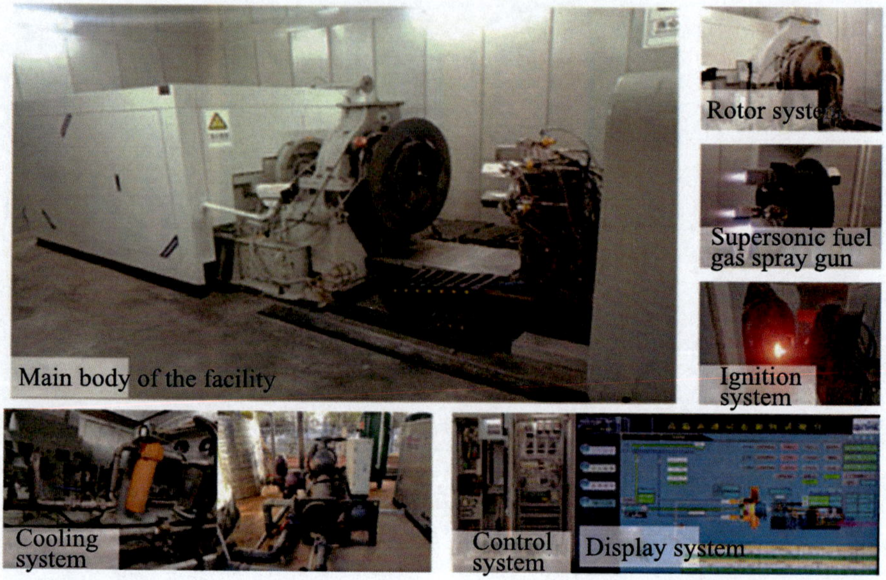

Fig. 16.41 High-temperature, high-speed rotating turbine blade and dynamic TMCC simulation and testing facility

(1) Supersonic fuel gas spray gun. As shown in Fig. 16.42, with compressed air as a combustion-supporting gas, kerosene and oxygen are mixed and ignited using an automatic ignition system in the combustion chamber of the spray gun. The product of combustion expands violently and is transported over a long distance to the nozzle of the spray gun, where it is sprayed, forming a high-velocity flame, which is then jet-sprayed onto the surface of the specimen. The flame velocity can be adjusted through an automatic-induction, self-regulating pressure valve. Erosive and CMAS particles are delivered by compressed air to the particle channel from a particle reservoir via a particle feeder. The particle channel is connected to the spray gun via a connector and a mechanical adjustment system. The particles are accelerated and heated by the high-velocity, high-temperature gas flow inside the spray gun and then sprayed together with the flame out of the spray gun via the nozzle. A water-cooling system is used to cool the fuel gas spray gun to ensure that it operates within the allowable temperature range.
(2) High-speed rotor system. As shown in Fig. 16.43, the rotor system is composed of a high-power motor, a load compressor, a gear system, a bearing housing for power protection, and a turbine model. The high-temperature, high-velocity, particle-carrying fuel gas flow produced by the supersonic fuel gas spray gun is jet-sprayed onto the guide vanes via a fuel gas diversion chamber. After the guide vanes accelerate the fuel gas flow and change its angle, the fuel gas flow acts on the TBC on the high-speed rotating working blades. Under these conditions, cooling air is guided into the guide vanes as well as the turbine disk dovetail via

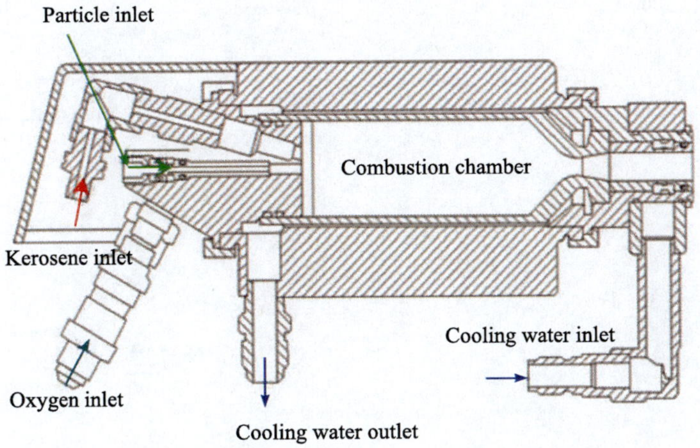

Fig. 16.42 Schematic diagram of the high-temperature fuel gas spray gun

components such as a cooling channel and a deflector to cool the working blade. This way, high-temperature, erosion, CMAS corrosion, and rotation are simulated. The mode of connection and working principle are detailed as follows. The working blade-containing turbine disk is connected to the power shaft of the turbine. The whole-turbine disk and the power shaft of the turbine are driven by a motor to rotate at a high speed. The aerodynamic load output of the high-speed rotation is absorbed by a load compressor to ensure that the high-speed rotating turbine is in a dynamic balance. The load absorbed by the load compressor is used to drive the air compressor to produce clean compressed air via a series of filtering and drying processes. The compressed air is then used to cool the high-temperature components such as the guide vanes and working blades of the turbine model as well as the spray gun. To prevent the vanes and blades or fragments from flying off due to an unstable center of gravity in the rotor system, the rotor system is sealed and equipped with an experimental observation window made of quartz glass. A protection module (i.e., bearing housing) for the core components related to the rotating power is set at the joint of the rotating shaft and the model to prevent the flying debris from damaging the power components and shaft. In addition, the whole rotor system is fixed on the support plate of the experimental platform, which is secured to the ground via anchor rods and a vibration isolator to ensure sufficient stability.

(3) Turbine model. Figure 16.44 shows a schematic diagram of the rotation and structure of two types of models. In the mode of stationary guide vanes and rotating working blades, the turbine model is structurally identical to a real-world turbine disk and comprises components such as a turbine disk, guide vanes, working blades, and a turbine casing. Both the turbine disk and vanes/blades are fabricated using precision casting processes. The overall structure of the turbine model is identical to that of a real-world engine. The turbine

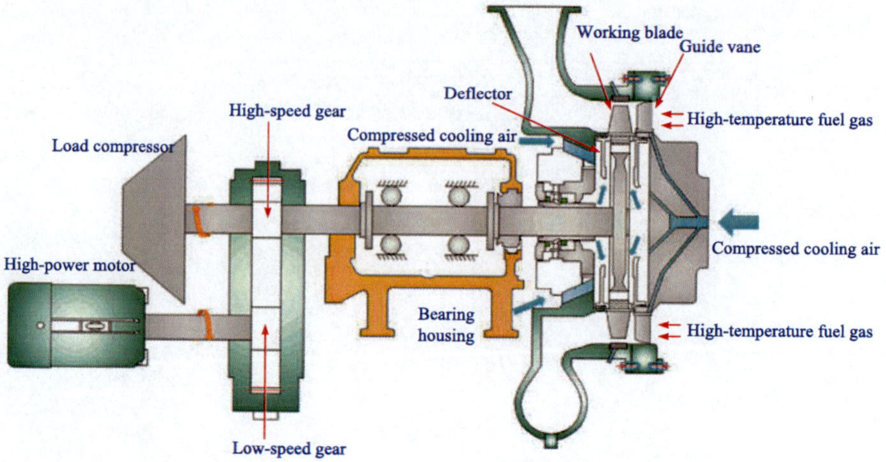

Fig. 16.43 A rotor system for simulating the working conditions of a high-speed rotating turbine blade

model can be scaled down or up, and its specific size can be designed based on the centrifugal force and tangential speed of the vanes/blades as well as the safe operational rotational speed of the rotor system to ensure that its working status is identical to that of a real-world engine. To reduce costs, the number of guide vanes and working blades can be scaled down, while the velocity and angle of the gas flow are ensured to be identical to those in a real-world engine. The vanes and disk can be either cast as a whole or fabricated as connecting pieces as needed (in this case, precise assembling at the joints should be ensured). The fabricated vanes/blades are subsequently coated with a TBC (including a transition layer and a ceramic coating) using PS, PVD, or new coating preparation processes. Then, the preparation of the turbine model is completed. In the mode of stationary working blades and rotating guide vanes, the model is composed of a hollow turbine disk, working blades, guide vanes, and a turbine disk that drives guide vanes to rotate. The hollow turbine disk that houses the working blades is fixed onto the experimental platform and does not come in contact with the rotating shaft. The guide vanes are installed on the rotating shaft via another disk to allow them to rotate at a high speed, generating a wake. High-temperature-resistant alloys (identical to those used to produce the disks and blades/vanes of the actual model) can be selected to fabricate these two types of turbine models. Alternatively, considering the cost and the actual needs, common materials can be used to produce these two types of turbine models.

(4) Measurement and control techniques for key experimental parameters. The key parameters (including the temperature of the simulated service environment and the velocity of the erosive and CMAS particles) of the experimental facility need to be measured and automatically controlled. Temperature is a key parameter and includes the temperatures of the flame, the TBCs on the guide vanes and working

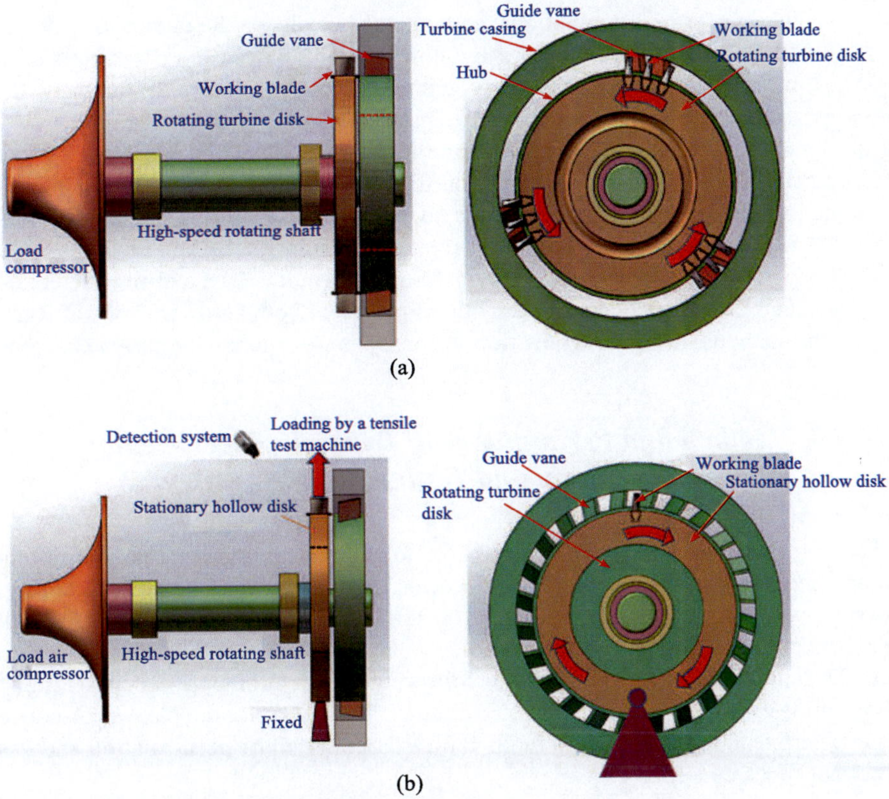

Fig. 16.44 Two rotation modes and turbine models

blades, the cooling air, and the turbine chamber. The temperatures of the flame, the turbine chamber, the cooling airflow, and the stationary TBC components are mainly measured using thermocouples. The temperature of the specimen in the high-speed rotation is difficult to measure and, in practice, is indirectly measured through the measurement of the temperature of the flame using a thermocouple and the periodic measurement of the temperature of the specimen chamber via the flame connector using an IR pyrometer. The flow and velocity are the main erosive-particle parameters to be measured and controlled. The flow is set and adjusted through the pressure of the particle-feeding air, and the particle velocity is measured and calibrated using a particle image velocimetry system. Since CMAS damages the coatings through corrosion instead of impact, flow is the key parameter of CMAS, and hence, CMAS can also be controlled through its flow. Parameters such as the rotational speed, temperature, particle flow, and the involved air pressure and flow are all set and controlled through a PLC, and the measurement data for these key parameters are used as the input signals for the PLC. The PLC program analyzes the parametric states corresponding to the signals, performs computations, and outputs signals via the output interface of the PLC to control the pressure and flow regulator valves and parameters such as the distance between the spray gun and the turbine vanes. In this way, the experimental parameters are set, measured, and controlled.

(5) Integration techniques for the real-time detection module. The deformation field, crack initiation and propagation, interfacial oxidation, and interfacial delamination in a stationary specimen can be examined in real time using the digital speckle technique, the AE characterization technique, CIS, and the IR thermal imaging technique, respectively, via the quartz window. The real-time detection systems are independent of each other in terms of both hardware and software. The methods for signal acquisition and analysis are described in earlier chapters.

16.4.3 Method and Performance of Dynamic Experimental TMCC Simulation and Testing

With the dynamic simulator, it is possible to use the high-speed rotation system alone, the fuel gas (erosion and CMAS corrosion) system alone, or both systems to perform various simulation experiments on TBCs. In addition, it is possible to install only rotating working blades or both guide vanes and working blades on the shaft to conduct simulation experiments. In the following, with a turbine model consisting of both guide vanes and working blades covered with TBCs as an example, we introduce the experimental and testing methods for TBCs under high-speed rotation and gas thermal shock conditions.

(1) **Experimental simulation and testing method**

To perform an experiment on the TBC on turbine blades/vanes under high-speed rotation and gas thermal shock (erosion and CMAS corrosion) conditions, it is necessary to first design a turbine model based on the experimental requirements by preserving the characteristics of the gas flow over the TBC on the actual turbine engine blades/vanes. See elsewhere [45, 46] for the detailed design method. More importantly, it is necessary to ensure a dynamic balance in the rotating system, which requires that the blades/vanes be strictly symmetrical in terms of quantity, size, and weight. In addition, prior to the experiment, a dynamic balance test is required to ensure safety during the experimental process.

Based on the experimental requirements, the key experimental parameters and coating failure are tested. A stationary specimen is tested using the same method as that for the static facility, which is not repeated here. Considering that the interactions between the high-temperature fuel gas and high-speed rotating blades/vanes may result in the formation of hot spots and wakes, the temperature field is measured in real time through IR thermal imaging.

The following parameters were used in the experiment: rotating speed, 19,700 rpm; kerosene flow, 15 L/h; oxygen pressure, 1.55 MPa; oxygen flow, 700

L/min; cooling air pressure, 0.8 MPa; cooling airflow, 400 L/min; and cooling air temperature, 500 °C. In addition, the parameters (e.g., emissivity) of the IR thermal imager were set.

After the above settings were established, the simulator and its auxiliary systems were started to begin the experiment.

(2) **Partial test results**

Figure 16.45 shows partial test results for the TBC on the turbine blades under high-speed rotation and thermal shock conditions. Figure 16.45a shows the high-temperature fuel gas flame generated by a supersonic spray gun. The flame has 3–4 Mach knots, suggesting that a supersonic fuel gas velocity has been reached. Figure 16.45b shows the corresponding rotating speed (19,704 rpm), which is close to the set value. Figure 16.45c and d shows the temperature fields in the cooling air channel and on the surface of the specimen during the thermal shock test. The maximum surface temperature of the specimen reaches 1236 °C, and the temperature in the cooling air channel is approximately 126 °C.

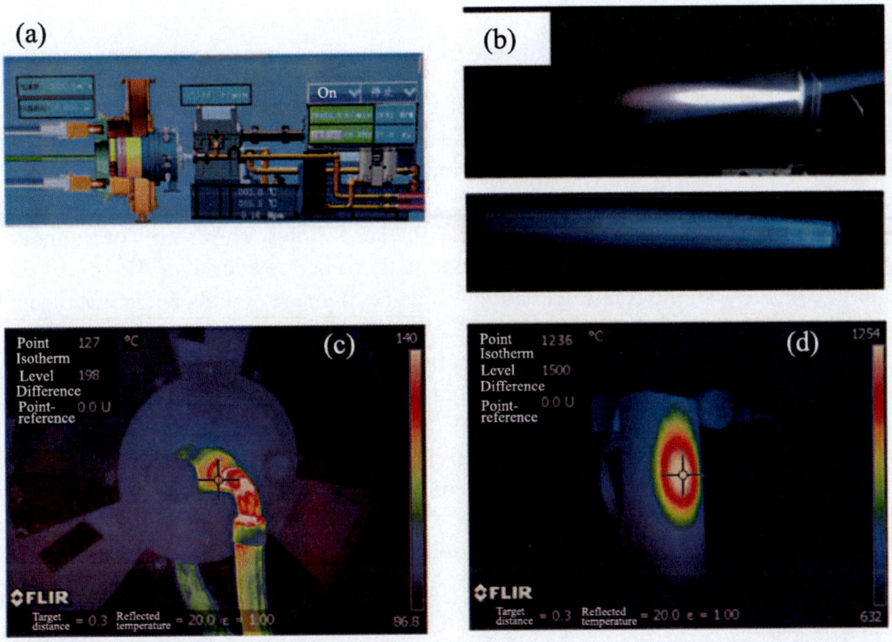

Fig. 16.45 Test results obtained using the experimental high-temperature, high-speed rotation facility: **a** rotating speed, **b** high-temperature fuel gas flame with a Mach cone, **c** IR-measured temperature field in the cooling air channel, and **d** IR-measured temperature field on the blade surface

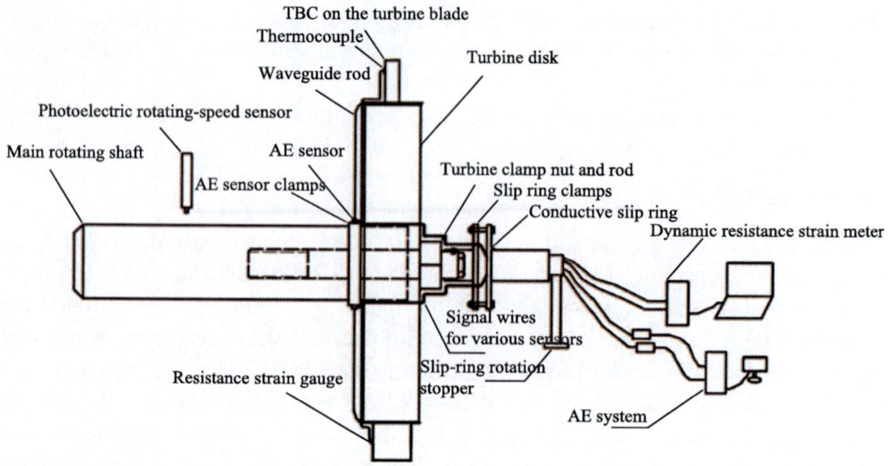

Fig. 16.46 Schematic diagram of the real-time testing of the TBC on a turbine blade under high-speed rotation

(3) Real-time testing method for the failure process under high-speed rotation

With the in-depth accumulation of real-time testing techniques, we investigated methods for testing the failure process of coatings in high-speed rotation in real time in addition to the real-time testing of stationary specimens. We custom-made a set of microminiature multichannel, high-rotating speed, long-life cap-type flange slip rings for the testing. The testing principle is described as follows. The "rotor" of a conductive slip ring is connected to a strain gauge, a thermocouple, and an AE sensor that are in contact with the rotating coating to receive damage signals and convert them into current/differential voltage signals. Through the contact between the "rotor" and "stator" of the conductive slip ring, different types of current/differential signals are transmitted to remotely connected signal acquisition platforms (e.g., a dynamic resistance strain meter and an AE system), thus realizing the nondestructive detection of the signals of strain and temperature fields and crack propagation in the TBC on the working blade rotating at a high speed.

As shown in Fig. 16.46, the conductive slip ring testing system for the failure process of coatings in the high-speed rotation is composed primarily of a high-speed conductive slip ring, a thermocouple, an AE sensor, an AE system, and a dynamic resistance strain meter. During testing, the TBC on the turbine blade, the rotating shaft, the conductive slip ring, and the AE sensor must be coaxial. A waveguide wire used for transmitting AE signals is welded onto the dovetail of the working blade, from which it is extended to the shaft and fixed and coupled with the sensor. The strain gauge and thermocouple are adhered onto the surface of the TBC and are extended via lead wires to the terminal blocks on the shaft. The turbine disk clamp rod is connected with the slip ring clamps. The slip ring clamps drive the conductive slip ring to rotate at a high speed. Various types of rotation signal wires are led via the high-speed slip ring to connect with the corresponding test systems. After the above connection is completed, the damage signal in the layer failure process can be collected.

16.5 Experimental High-Temperature Vibration Simulators for TBCs on Turbine Blades

In addition to adverse conditions such as high-temperature gas shock, film cooling, and high-speed rotation, turbine blades face another type of loading—vibration. Statistics show that vibration-induced failures account for more than 60% of all the aeroengine failures and that over 70% of vibration-induced failures occur in blades. Most of the engines currently in use in China have had blade-fracture failures due to vibration [51], evidently suggesting that vibration-induced blade failure is a severe problem. The extent of the impact of vibration on TBCs remains unclear and receives inadequate attention. In fact, the TBCs on turbine blades are highly susceptible to low- or high-frequency vibration due to factors such as unstable meshing and aerodynamic loading [52]. Therefore, the ability to simulate high-temperature vibration loads is also a requirement for experimental simulators for TBCs.

16.5.1 High-Temperature Vibration Facilities

As its name implies, an experimental high-temperature vibration simulator simulates the combined action of high temperatures and vibration, which can be easily achieved through a high-temperature furnace and vibration load. However, since TBCs under high-temperature vibration conditions have rarely been studied worldwide, there are almost no reports on high-temperature vibration facilities for TBCs. Researchers at the University of California in the United States developed a high-temperature vibration furnace with a maximum temperature of approximately 1200 °C and a maximum vibration frequency of 12,000 Hz. However, this facility can produce a thrust of only 6 N and thus is unable to generate the vibration needed for TBCs on turbine blades.

Given the above issues, we developed a high-temperature vibration facility for TBCs [53], as shown in Fig. 16.47. This facility uses a high-temperature resistance silicon-carbide-rod furnace with a maximum temperature of 1400 °C as the heating source. The interior of the furnace is equipped with a vibration table with a maximum thrust of 6000 N and a frequency up to 5000 Hz. Clamps for banded, cylindrical, and leaf-like TBCs are designed for the vibration loading table to meet the test requirements of various specimens.

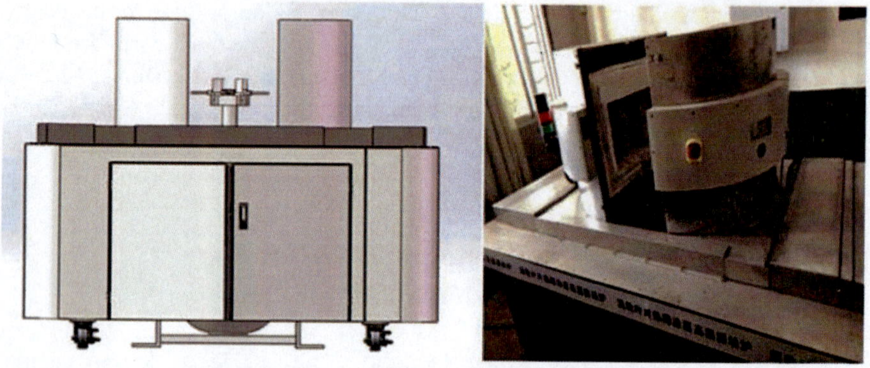

Fig. 16.47 Schematic diagram and photograph of a high-temperature vibration facility for TBCs

16.5.2 Testing of TBCs Under High-Temperature Vibration

With the abovementioned experimental high-temperature vibration facility, we preliminarily studied the failure of a PS TBC at a high temperature of 900 °C and a vibration frequency of 3000 Hz, with the results shown in Fig. 16.48. After 150 h of oxidation at 900 °C in the absence of vibration, no cracks have been formed in the ceramic layer. However, when the other conditions are kept the same, the application of vibration load leads to the formation of a large number of cracks in the ceramic layer. Specifically, it is observed that, under vibration loading, the surface layer spalls significantly, the whole coating loosens, a large number of transverse cracks are formed in the coating, and the coating is extremely prone to spallation.

Under the same test conditions, we analyzed a PVD TBC. Figure 16.49 shows the results. Similar to the PS TBC, vibration loading induces premature spallation of the PVD coating and causes it to loosen and crack.

In short, TBCs are extremely prone to spallation under vibration loading. However, the failure mechanism of coatings under vibration conditions, particularly their control mechanisms, requires further in-depth study and is a direction that demands attention in the TBC research field in the future.

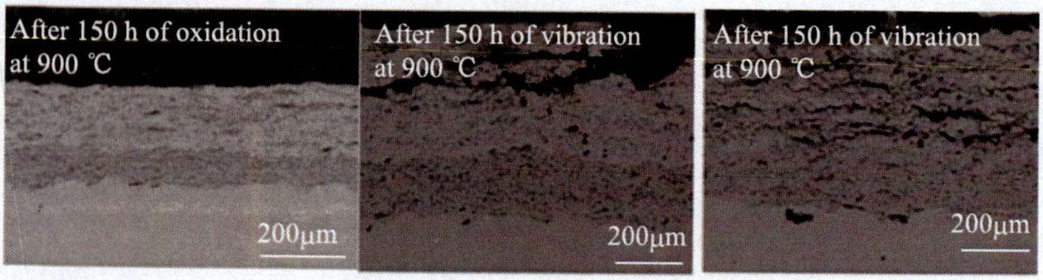

Fig. 16.48 Influence of high-temperature vibration on a PS TBC

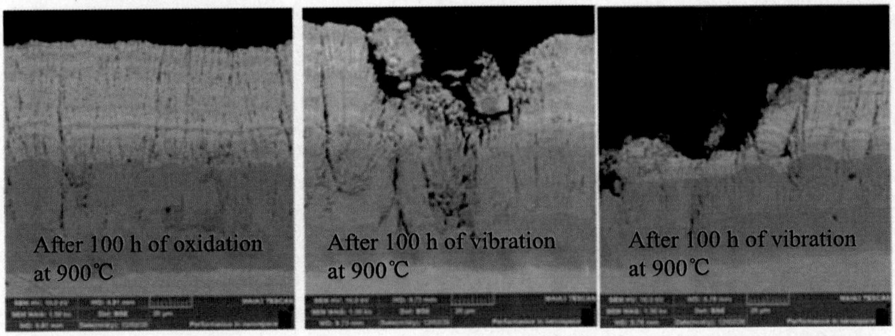

Fig. 16.49 Influence of high-temperature vibration on a PVD TBC

16.6 Summary and Outlook

16.6.1 Summary

This chapter elucidates the main research results from experimental simulation and testing facilities for TBC service environments (including thermal loading, thermomechanical loading, static and dynamic TMCC, and high-temperature vibration) as well as the typical failure modes of TBCs obtained using these facilities. The following provides a summary of this chapter:

(1) Experimental thermal loading facilities, represented by those for simulating automatic thermal cycling and measuring high-temperature contact angles of CMAS, are simple but important platforms for studying the high-temperature oxidation, thermal cycling, and high-temperature CMAS corrosion mechanisms of TBCs.
(2) Based on the experimental simulators that each combine a material testing machine and a high-temperature source (e.g., a high-temperature furnace and fuel gas), the thermomechanical buckling failure mechanism of TBCs is determined and elucidated. These facilities also enable the characterization of the high-temperature mechanical properties of TBCs.
(3) Static TMCC simulators for TBCs on turbine blades, which are extensively developed and used worldwide, make it possible to analyze the failure mechanism and thermal insulation performance of TBCs and to assess their simulations.
(4) The development of high-temperature high-speed rotation simulators for TBCs on turbine blades represents a challenge in the TBC research field, and Chinese researchers have achieved important progress.

16.6.2 Outlook

Future research should be focused on the following areas:

(1) It is necessary to investigate techniques for evaluating the thermal insulation performance of TBCs on turbine blades as well as techniques for coevaluating thermal insulation performance and reliability based on experimental static and dynamic TMCC simulators for TBCs.
(2) Methods for testing the failure process of TBCs under high-temperature and high-speed rotation conditions in real time are necessary. High-speed rotation is a primary load on the TBCs on working blades. The current real-time testing and analysis methods for high-speed rotation conditions require further investigation.
(3) The failure mechanisms of TBCs under high-temperature vibration need to be elucidated. Based on the available preliminary results, high-temperature vibration is another key load causing spallation in TBCs on turbine blades but has yet to receive widespread attention.
(4) The failure mechanism of new TBCs needs to be investigated. Currently, YSZ coatings can operate only at temperatures below 1200 °C and thus are unable to meet the demand for engine development. Studying the failure mechanism of next-generation TBCs using simulators can provide valid data for the development and application of new coatings.

References

1. Zhou Y, Liu Q, Yang L, Wu D, Mao W. Failure mechanisms and life prediction of thermal barrier coatings. Acta Mech Solida Sin. 2010;31(5):504–31.
2. Wu RT, Osawa M, Yokokawa T, Kawagishj K, Harada H. Degradation mechanisms of an advanced jet engine service-retired TBC component. J Solid Mech Mater Eng. 2010;4(2):119–30.
3. Drexler JM, Gledhill AD, Shinoda K, Vasiliev AL, Reddy KM, Sampath S, Padture NP. Jet engine coatings for resisting volcanic ash damage. Adv Mater. 2011;23(21):2419–24.
4. Krause AR, Garces HF, Wivedi GD, Ortiz AL, Sampath S, Padture NP. Calcia-magnesia-alumino-silicate (CMAS)-induced degradation and failure of air plasma sprayed yttria-stabilized zirconia thermal barrier coatings. Acta Mater. 2016;105:355–66.
5. Xu GN, Yang L, Zhou YC, Pi ZP, Zhu W. A chemo-thermo-mechanically constitutive theory for thermal barrier coatings under CMAS infiltration and corrosion. J Mech Phys Solids. 2019;133: 103710.
6. Yang L, Yin B, Zhou Y, Zhu W. An apparatus for testing the wettability of molten CMAS corroding thermal barrier coatings and the corresponding method. China: CN105241791A, 13 January 2016.
7. Yin BB, Liu ZY, Yang L, Zhou YC. Factors influencing the penetration depth of molten volcanic ash in thermal barrier coatings: Theoretical calculation and experimental testing. Res Phys. 2019;13: 102169.

8. Mao W. Interfacial failure analysis of thermal barrier coatings under combined thermomechanical loading. Doctoral dissertation of Xiangtan University;2006.
9. Zhu W, Wu Q, Yang L, Zhou YC. In situ characterization of high temperature elastic modulus and fracture toughness in air plasma sprayed thermal barrier coatings under bending by using digital image correlation. Ceram Int. 2020;46:18526–33.
10. Zhou YC, Hashida T. Thermal fatigue failure induced by delamination in thermal barrier coating. Int J Fatigue. 2002;24:407–17.
11. Choi SR, Hutchinson JW, Evans AG. Delamination of multilayer thermal barrier coatings. Mech Mater. 1999;31(7):431–47.
12. Hutchinson JW, He MY, Evans AG. The influence of imperfections on the nucleation and propagation of buckling driven delaminations. J Mech Phys Solids. 2000;48(4):709–34.
13. Zhu W, Yang L, Guo JW, Zhou YC, Lu C. Determination of interfacial adhesion energies of thermal barrier coatings by compression test combined with a cohesive zone finite element model. Int J Plast. 2015;64:76–87.
14. Zhou YC, Hashida T. Coupled effects of temperature gradient and oxidation on thermal stress in thermal barrier coating system. Int J Solids Struct. 2001;38(24–25):4253–64.
15. Yang L, Yang TT, Zhou YC, Wei YG, Wu RT, Wang NG. Acoustic emission monitoring and damage mode discrimination of APS thermal barrier coatings under high temperature CMAS corrosion. Surf Coat Technol. 2016;304:272–82.
16. Evans AG, Mumm DR, Hutchinson JW. Mechanisms controlling the durability of thermal barrier coatings. Prog Mater Sci. 2001;46(5):505–53.
17. Zhou Y, Yang L, Zhong Z, Cai C. An experimental facility for simulating the service environments of thermal barrier coatings and for detecting their failure in real time. China: CN103091189A, 8 May 2013.
18. Robinson RC. NASA GRC's high pressure burner rig facility and materials test capabilities, NASA Technical Memorandum, 209411;1999.
19. Zhu JD, Miller RA, Kuczmarski MA. Development and life prediction of erosion resistant turbine low conductivity thermal barrier coatings. NASA Technical Memorandum, 215669;2010.
20. Cao X. Automatic thermal cycling meter with controllable temperature for thermal barrier coatings. Opt Fine Mech. 2004;2:20.
21. Wang R, Zhang T, Liu Y, He Q, Wu P. An experimental facility for evaluating the thermal shock life of thermal barrier coatings. China: CN105865961A, 17 August 2016.
22. Gong S, Zhang C, Xu H. A facility for simulating the service environments of thermal barrier coatings and a method for controlling simulated environments. China: CN1699994, 23 November 2005.
23. Vassen R, Kagawa Y, Subramanian R, Zombo P, Zhu D. Testing and evaluation of thermal-barrier coatings. MRS Bull. 2012;37(10):911–6.
24. Zhou Y, Yang L, Chang S, Cai C. A simulation and real-time testing facility for the corrosion of thermal barrier coatings by deposits at high temperatures. China: CN103063563A, 24 April 2013.
25. Yang L, Wang J, Zhou Y, Cai C. A spray gun device for simulating the high-temperature, erosion, and corrosion service environments of thermal barrier coatings. China: CN103091237A, 8 May 2013.
26. Yang L, Zhou C, Zhou Y, Cai C. An integrated dynamic and static service environment testing platform for thermal barrier coatings on turbine blades. China: CN103091238A, 8 May 2013.
27. Yang L, Qi S, Zhou Y, Cai C. A simulation and real-time testing facility for the erosion of thermal barrier coatings on turbine blades. China: CN103063534A, 24 April 2013.
28. Yang L, Wang J, Zhou Y, Cai C. A control and display apparatus for synchronous data acquisition and multi-view display. China: CN103092100A, 8 May 2013.
29. Yang L, Tang M, Zhou Y, Zhou W, Zhu W, Li C. A flame jet spraying apparatus for simulating the service environments of thermal barrier coatings and the corresponding method. China: CN108302531B, 9 August 2019.

30. Yang L, Zhou W, Zhou Y, Tan M, Zhu W, Li C. A working condition simulation and real-time monitoring facility for thermal barrier coatings. China: CN108254275A, 6 July 2018.
31. Zhou YC, Yang L, Zhong ZC, Cai CY. Type of testing equipment for detecting the failure process of thermal barrier coating in a simulated working environment, the United States, 14/760,444, 8 November 2013
32. Wang S, Liu G, Mao J. Modeling experimental system for solid particle erosion on the steam turbine nozzle blades and measuring methods. Proc Chin Soc Electr Eng. 2007;27(11):103–8.
33. Zheng L. Design of turbine blade thermal erosion wind tunnel and experimental study. Shanghai University of Engineering Science;2016.
34. Huang H, Pan W, Fu Z, Wu C. A small arc-heated plasma wind tunnel for thermal protection materials testing;2010.
35. Tabakoff W, Shanov V. Erosion rate testing at high temperature for turbomachinery use. Surf Coat Technol. 1995;76:75–80.
36. Zhu W, Zhang CX, Yang L, Zhou YC, Liu ZY. Real-time detection of damage evolution and fracture of EB-PVD thermal barrier coatings under thermal shock: An acoustic emission combined with digital image correlation method. Surf Coat Technol. 2020;399: 126151.
37. Zhang C. Real-time testing and analysis of the failure process of thermal barrier coatings on turbine blades under thermal shock from high-temperature combustion gases. Master's thesis of Xiangtan University;2020.
38. Yang L, Zhang C, Li C, Zhu W, Zhou Y. A method for detecting the damage sustained by thermal barrier coatings on turbine blades during simulation tests in real time. China: CN109459286B, 31 July 2020.
39. Zhu W, Tan Z, Yang L, Zhou Y. A method for testing high-temperature erosion of thermal barrier coatings. China: CN109900577B, 20 March 2020.
40. Zhu W, Li C, Yang L, Zhou Y, Ge L. A method for testing the high-temperature CMAS corrosion of thermal barrier coatings. China: CN109883938A, 14 June 2019.
41. Povey T, Qureshi I. A hot-streak (combustor) simulator suited to aerodynamic performance measurements. Aerosp Eng. 2008;222(6):705–20.
42. Yang L, Zhou Y, Liu Z, Luo Y, Zhu W. An experimental testing system for simulating the working conditions of thermal barrier coatings on turbine blades. China: CN109682702B, 20 March 2020.
43. Yang L, Shi L, Liu Z, Zhu W, Zhou Y. An apparatus for applying the load equivalent to the service load of thermal barrier coatings on turbine blades and the corresponding method. China: CN109855977B, 3 January 2020.
44. Yang L, Liu Z, Zhu W, Zhou Y. A facility for applying cooling conditions to thermal barrier coatings on turbine blades. China: CN109596371B, 3 January 2020.
45. Zhu W, Luo Y, Yang L, Zhou Y. A turbine model for experimentally simulating the service conditions of thermal barrier coatings. China: CN109339868B, 5 May 2020.
46. Yang L, Luo Y, Zhu W, Zhou Y. A turbine model for experimentally simulating the service conditions of thermal barrier coatings. China: CN109443773B, 7 February 2020.
47. Zhu W, Shi L, Yang L, Zhang C, Zhou Y. An apparatus for applying the load equivalent to the service load of thermal barrier coatings on working blades and the corresponding method. China: CN109580410B, 7 February 2020.
48. Wanhill R, Mom A, Hersbach H. NLR experience with high velocity burner rig testing, 1979–1989. High Temp Technol. 1989;7(4):202–11.
49. Bruce RW. Development of 1232°C(2250 F) erosion and impact tests for thermal barrier coatings. Tribol Trans. 1998;41(4):399–410.
50. High speed balancing and ovespeed test facility. https://schenckotec.com/services/balancing-and-spinning-service/high-speed-balancing.html.
51. Huang Q. Aeroengine design manual. Beijing: Aviation Industry Press;2001.
52. Razaaly N, Persico G, Gori G, Congedo PM. Quantile-based robust optimization of a supersonic nozzle for organic rankine cycle turbines. Appl Math Model. 2020;82:802–24.
53. Yang L, Liu Z, Yan G, Zhu W, Zhou Y. A high-temperature vibration simulator. China: CN109827735B, 6 November 2020.